MW01052924

Discrete Mathematics with Proof

Discrete Mathematics with Proof

Second Edition

Eric Gossett
Bethel University

A JOHN WILEY & SONS, INC., PUBLICATION

Published by John Wiley & Sons, Inc., Hoboken, New Jersey.
Published simultaneously in Canada.

Library of Congress Cataloging-in-Publication Data:

Gossett, Eric.
Discrete mathematics with proof / Eric Gossett. — 2nd ed.
p. cm.
Includes bibliographical references and index.
ISBN 978-0-470-45793-1 (cloth)
1. Mathematics. 2. Computer science—Mathematics. I. Title.
QA39.3.G68 2009
510—dc22 2008055959

Contents

Preface

my wife, Florence Kuofang Gossett,
daughter, Rachel Shinching Holst,
d my son, Nathan Mui Gossett,
son-in-law, Andy Holst,
daughter-in-law, Laura Gossett,
d my grandchildren.

This book has been written for a sophomore-level course in Discrete Mathematics. The material has been directed toward the needs of mathematics and computer science majors, although there is certainly material that is of use for other majors. Students are assumed to have completed a semester of college-level calculus. This assumption is primarily about the level of mathematical maturity of the readers. The material in a calculus course will not often be used in the text.

This textbook has been designed to be suitable for a course that requires students to read the textbook. Many students find this challenging, preferring to just let the instructor tell them "everything they need to know" and using the textbook as a repository of homework exercises and corresponding examples. A typical course in Discrete Mathematics will require much more from the students. Consequently, the textbook needs to support this transition toward greater mathematical maturity. Textbook features that support this goal will be discussed a little later.

This text has been successfully used by requiring students to read a section and submit some simple exercises from that section at the start of a class period, where the material is discussed for the first time. The following class period, the students will submit more difficult exercises. Consequently, extra care has been taken to ensure that students can follow the presentation in the book even before the material is presented in class. While most instructors do not structure their course in this manner, a textbook that has been written to stand on its own will certainly be of value to the students.

This book should work well with a distance education format. However, personal interaction between the student and the instructor (or a knowledgeable teaching assistant) greatly enhances the learning experience.

DISTINGUISHING CHARACTERISTICS OF THIS TEXT

There are currently many textbooks on the market for a course in Discrete Mathematics. Although there is an assumed common core of topics and level, there is still sufficient variation to provide instructors with viable options for choosing a textbook. Here are some of the features that characterize this book.

- There is a heavy emphasis on proof throughout the text (as indicated by the book's title). The formal setting is introduced in Chapter 2 as sets, logic, and Boolean algebras are discussed. Chapter 3 discusses axiomatic mathematics as a system and subsequently focuses on proof techniques. The proof techniques are extensively illustrated in the rest of the text. For example: proof by contradiction in Chapter 4 with The Halting Problem; constructive proofs in Chapter 8 with "a finite projective plane of order n iff $n-1$ mutually orthogonal Latin squares of order n"; complete induction in Chapter 3 with the "optimality (for suitors) of the Deferred Acceptance Algorithm". Combinatorial proof is introduced in Chapter 5 and used in Chapter 8 to establish the necessary conditions for the existence of a balanced incomplete block design.
  Many of the more difficult proofs are accompanied by illustrative examples that can be read in parallel with the proof. For example, Theorem 8.63 on page 504 and Examples 8.50 and 8.51 that appear after the proof. Theorem 3.9 and Example 3.11 (pages 96 and 97) also fit this pattern.

- The text has been written for students to read actively. The text contains more detailed explanations than some competing texts. Homework problems have been designed to reinforce reading. Most cannot be completed by merely finding a clone example to copy and modify. The chapters include Quick Check problems at critical points in the reading. These are problems that should be solved before continuing to read. Detailed solutions are presented at the end of the chapter.
- I have endeavored to eliminate unnecessary page-turns by the reader. As much as possible, related material and the bodies of proofs appear on facing pages. When this is not possible, key figures or expressions are repeated after a page-turn. Examples of this can be seen on pages 122 and 642.
  I have also prevented (in most cases) mathematical expressions from spanning a line break. The reader should not need to scan back-and-forth several times to understand such an expression.
  As an encouragement to the reader to turn back to a previous definition or theorem when it is referenced elsewhere, the definition/theorem number is accompanied by a page number. For example, the proof of Proposition 3.29 on page 102 refers to Definition 3.8 on page 93.
- Technology is introduced when it will enhance understanding. For example: a web page powered by a simple Perl script for testing regular expressions in Chapter 9; a Java Application (and Applet) that allows students to rubber-band graphs to check for planarity in Chapter 10; several applications that explore the inner workings of recursion in Chapter 9; and a Java Application (and Applet) for checking Quine–McCluskey minimizations of boolean expressions in Chapter 12. There are also web links to Applets that animate critical algorithms and structures: Boyer–Moore in Chapter 4; finite automata in Chapter 9; and logic circuits in Chapter 12. Of special note is a collection of animations, *Visualizing Combinatorial Proofs*, designed to help students understand some combinatorial proofs. These can all be found at:

  http://www.mathcs.bethel.edu/~gossett/DiscreteMathWithProof/

- Combinatorics receives a full chapter (Chapter 8) beyond the standard "combinations and permutations" material presented in Chapter 5. The nonstandard topics include Latin squares, finite projective planes, balanced incomplete block designs, coding theory, knapsack problems, Ramsey numbers, partitions, occupancy problems, Stirling numbers, and systems of distinct representatives. The chapter begins with an overview of the major themes that unify the field of combinatorics.
- There are several major examples that present significant algorithms. Examples include: Chapter 1: the Deferred Acceptance Algorithm (the Stable Marriage Problem); Chapter 4: Boyer–Moore algorithm for pattern matching; and Chapter 7: recursive algorithms for Sierpinski curves, persian rugs, and adaptive quadrature.
  Other examples cover problems with a significant history. Examples include: Chapter 7: the Josephus problem; Chapter 8: Kirkman's Schoolgirl Problem; and Chapter 10: the five regular polyhedra and a proof of the Five Color Theorem.
  There are also some important examples from the field of computer science. These include: Chapter 4: The Halting problem; Chapter 9: Shannon's mathematical model of information; Chapter 11: XML; and Chapter 12: Normal Forms in relational databases.
- The Discrete Mathematics course at Bethel University is equally populated with mathematics majors and computer science majors. Consequently, this text was designed to be appropriate for courses for mathematics majors, courses for computer science majors, and courses with bimodal populations. There is sufficient material to design a one-semester course for any of these three options. It is also possible to design a two-semester course that covers the entire book.

An example of using the text with a bimodal group can be found in the chapter on recursion. Chapter 7 starts with an algorithmic approach and then presents recurrence relations (a mathematical approach). Both sets of students see the concept (recursion) in a form that is oriented towards their own major. In addition, they are exposed to an alternative viewpoint.

- Several other topics receive more coverage than is typical. These include: Chapter 4: expressing algorithms, the Halting Problem; Chapter 6: Bayes' Theorem; several topics in Chapter 8; and Chapter 9: regular expressions.
- The text begins with an introductory chapter that provides some explanation and examples of what discrete mathematics is about. This is rare among discrete mathematics textbooks.
- Limitations are discussed where appropriate. For example, the solution of linear homogeneous recurrence relations with constant coefficients that is presented in Chapter 7 requires the factorization of polynomials. There is no general factorization technique for polynomials of degree 6 or higher. The text also contains guidelines to determine whether recursion is an appropriate solution technique for a particular problem.
- An Instructor's Solution Manual with detailed solutions to every problem is available for use by student graders and professors. Appendix H is a free Student's Solution Manual containing detailed solutions to selected exercises.

CHANGES FROM THE FIRST EDITION

The following changes are the major differences between the First and Second editions. However, almost every page contains minor revisions as I have sought to make the exposition as clear and helpful as I am able.

- The numbering of definitions, theorems, propositions, corollaries, and lemmas has been merged. This should help the reader to locate them with less effort. The numbering sequence for examples and Quick Checks have remained as separate sequences.
- Several new sections have been added: 1.3.1; 1.3.6; 3.4 (Applications of Elementary Number Theory); 3.5.5 (Multidimensional Induction); 5.2.2 (Counting Tulips); 6.5 (The Binomial Distribution); and Appendix G (Writing Mathematics).
- The material on elementary number theory has been moved to its own section (3.2). Sections 8.4 (The Knapsack Problem), 8.6 (Ramsey Numbers), and 12.3 (Relational Databases) have all been extensively re-written.
- The material on analyzing claims has been moved from Chapter 2 to Appendix C.
- Just over 30 new examples have been introduced throughout the text.
- Over 250 additional exercises or exercise subitems have been added.

TEXT ORGANIZATION

The chapters in the book are briefly summarized in the following paragraphs.

Chapter 1: Introduction

Chapter 1 provides a working definition of *discrete mathematics* and then offers the reader some brief glimpses at some of the topics that will be covered in the remaining chapters. The chapter also introduces the Stable Marriage Problem and the Deferred Acceptance Algorithm. This material is covered in some detail and appears again in several other chapters.

The exposition of the Stable Marriage Problem introduces a non-trivial algorithm and some proofs. The problem, the algorithm, and the proofs are all fairly intuitive. They prepare the reader for the more detailed expositions of algorithms and proofs that will follow in future chapters. The problem also shows the reader that the material in this course may be different from what they have studied in previous mathematics courses.

Chapter 2: Sets, Logic, and Boolean Algebras

Much of the material in this chapter is not what students tend to rate as most interesting. However, it is foundational to much of what follows. It is even more important than in previous decades because many students are now graduating from high school without ever learning the basics of set theory. Many have never been exposed to either the basic terminology (element, union, intersection) or the standard notation ($\in$, $\cup$, $\cap$).

The basic concepts of propositional and predicate logic are introduced in this chapter. They also serve as a basis for the proof strategies introduced in Chapter 3.

The basic properties of sets and logic are presented in a parallel style to emphasize the similarities. This parallel exposition provides a natural introduction to Boolean algebras. Boolean algebras serve to unify some important aspects of set theory and logic. The early introduction also provides a nontrivial example of an axiomatic system. This example can be recalled when the axiomatic system is formally introduced in Chapter 3.

Chapter 3: Proof

Chapter 3 provides a careful introduction to proof. The chapter starts with a discussion of axiomatic mathematics, which provides the student with information about the context in which proofs exist. It is necessary to have some content in which to give examples of various proof strategies and provide exercises for practice. This is accomplished by introducing much of the standard material from elementary number theory. This introduction also fills in some of the gaps in the student's background knowledge. There is an optional section containing several interesting applications of number theory (for example, pseudorandom number generators and public-key encryption).

The chapter contains a discussion of the major proof strategies and also has a section that provides hints and suggestions for creating proofs. There is also a careful introduction to mathematical induction, including a brief look at multiple induction.

Chapter 4: Algorithms

Chapter 4 is about algorithms. The two major topics are: expressing algorithms and measuring algorithm efficiency. Section 4.1 provides a fairly complete introduction to pseudocode. Courses taught to sophomore computer science majors can either skip this section or else do a quick review. I have found that students who have not yet taken a programming course really need the detailed descriptions found in this section. As a side benefit, my students tell me that this section was very helpful when they enrolled in a programming course after taking discrete math.

Section 4.2 introduces big-$\mathcal{O}$ and big-Θ. My students tend to vote this material as their least favorite in the course. Since this material, and the ability to apply it, is so important in computer science courses (such as data structures), I have expended extra effort to help the students gain a good intuitive understanding of the basic definitions, the reason those definitions are important, and how to apply them to real algorithms.

The chapter ends with two interesting examples. Section 4.3 compares three different algorithms for solving the same problem (finding a substring in a longer text). They provide an interesting example illustrating the practical difference in finding an algorithm with a better big-Θ reference function. The final algorithm (Boyer–Moore) is also worth studying purely for the cleverness of the ideas that are used.

The short section at the end of the chapter examines a problem for which no algorithm can ever exist: the Halting problem. It also provides a very nice example of a proof by contradiction.

Chapter 5: Counting

Chapter 5 presents the standard material about counting. The notions of independent tasks, mutually exclusive tasks, permutation and combinations (with or without repetition) are all present. In addition, the pigeon-hole principle, inclusion–exclusion, and the multinomial counting theorems are presented. The chapter also contains a section that introduces the notion of a combinatorial proof, aided by some computer visualizations.

Chapter 8 expands the counting repertoire with a discussion of occupancy problems.

Chapter 6: Finite Probability Theory

Chapter 6 provides the basic definitions and properties of finite probability. It discusses sample spaces, events, independent and mutually exclusive events, and conditional probability. There is a section that applies many of the counting techniques found in Chapter 5.

More advance topics include expected value, the Binomial distribution, and Bayes' Theorem.

Chapter 7: Recursion

Chapter 7 introduces recursion, first from a computer science perspective (recursive algorithms), and then from a mathematics perspective (recurrence relations). The discussion of recursive algorithms includes numerous nontrivial applications.

Techniques for solving recurrence relations include: back substitution, using the roots of a characteristic equation to solve linear homogeneous recurrence relations with constant coefficients, and the use of generating functions.

There is also a section dedicated to the Master theorems for finding big-Θ reference functions for divide-and-conquer recursions.

As a bonus, the chapter contains a brief discussion of the Josephus problem. The historical origins are explored, and a simplified version of the problem is solved.

Chapter 8: Combinatorics

Chapter 8 presents a brief overview of some common defining characteristics (existence, enumeration, optimization) of the field of combinatorics. It then explores some sample illustrative topics. The topics include: partitions, occupancy problems and Stirling numbers, Latin squares and finite projective planes, balanced incomplete block designs, the knapsack problem, error-correcting codes, and systems of distinct representatives and Ramsey numbers.

Much of the material in this chapter will stretch typical sophomores. The easier sections are 8.1 (partitions, occupancy problems, and Stirling numbers), 8.3 (balanced incomplete block designs), and 8.5 error-correcting codes. However, Section 8.3 does make occasional references to material in Section 8.2.

Chapter 9: Formal Models in Computer Science

Whereas Chapter 8 is oriented towards math majors, this chapter is mainly oriented toward computer science majors. The chapter begins with a mathematically motivated derivation of Shannon's mathematical model of *information*. It also contains his familiar model of communication. The material in this initial section is not needed for subsequent sections and so can be omitted without any break in continuity. I have placed it first in the hope that these models will gain more exposure in discrete mathematics courses.

The chapter continues with discussions of finite-state machines and finite automata. Formal languages are introduced next, with most of the discussion centered on regular grammars. A fairly detailed discussion of regular expressions is presented next. The notation is from the standard Unix/Perl conventions.

Section 9.5 presents Kleene's theorem and the equivalence of finite automata, regular sets, regular expressions, and regular grammars. Nondeterministic finite automata are introduced in an optional section that contains the proof of Kleene's theorem.

The chapter concludes with a brief introduction to the Chomsky hierarchy of grammars, pushdown automata, Turing machines, and the Church–Turing thesis.

Chapter 10: Graphs

Chapter 10 is a fairly lengthy introduction to graph theory. The chapter introduces the basic terminology, numerous examples of graphs and graph families, and the standard material on connectivity and adjacency. Euler circuits and Hamilton cycles are explored, as well as alternative mechanisms for representing graphs in a computer. The notion of graph isomorphism is also discussed. Weighted graphs and Dijkstra's shortest path algorithm are also presented.

The chapter also contains a section that presents four of the most famous theorems in graph theory: Euler's formula, the characterization of regular polyhedra, Kuratowski's theorem, and the four color theorem. A simple Java applet/application is available for click-and-drag exploration of planarity. In addition, the chapter contains a proof of the five color theorem.

Chapter 11: Trees

There is sufficient material about trees to warrant a separate chapter. The chapter starts with the standard definitions (root, leaf, balanced, etc.), and some of the basic counting theorems for trees. The notions of tree traversal and searching and sorting are also presented. Section 11.3 contains three interesting applications of trees: parse trees, Huffman compression, and XML. I typically present just one of the three in any given semester. The chapter ends with spanning trees.

Chapter 12: Functions, Relations, Databases, and Circuits

The material in Chapter 12 is connected by the concepts of functions and relations. The first two sections present the foundational ideas and the final three sections provide nontrivial applications.

The foundational ideas include the definitions of *function* and *relation*, as well as the standard properties that relations might exhibit (reflexive, symmetric, transitive, etc.). There is also a discussion of equivalence relations and a discussion of what it means for a binary operator on a set of equivalence classes to be well-defined.

Much of the material in Sections 12.1 and 12.2 is often presented in the early chapters of other discrete math texts. I have placed it at the end to take advantage of some of the topics presented elsewhere in the book (such as the counting techniques in Chapter 5). I also wanted to keep this material close to the applications in Sections 12.3, 12.4, and 12.5.

The nontrivial applications are: a discussion of normal forms in relational databases, boolean functions and disjunctive normal form, and the design of combinatorial circuits (including the minimization of boolean expressions via the Quine–McCluskey algorithm).

For a one-semester course, I would recommend covering Sections 12.1 and 12.2 and then presenting either Section 12.3 or Sections 12.4 and 12.5.

Appendices

There are several appendices. Appendix A contains a brief review of the standard number systems. Appendix B contains a very brief review of summation notation and Appendix E contains a short introduction to some matrix terminology and arithmetic. Appendix C contains some logic puzzles and is intended as a supplement to Chapter 2. Appendix D contains some background information on the Golden Ratio. Appendix F contains a summary of the Greek alphabet. Appendix G provides some guidelines for writing mathematics in homework exercises and proofs.

Finally, Appendix H will be the most frequently used appendix. It contains detailed solutions to selected Exercises. Consequently, it serves as a free student solution manual. By placing it in the text, students gain in convenience. Of greater importance, instructors can be assured that solutions to problems that are not in Appendix H are not available for purchase by their students. There is no other student solution manual.

SAMPLE COURSES

This text is suitable for courses taught within a math department for mathematics majors, for courses taught within a computer science department for computer science majors, and for courses with bimodal populations. Sample one-semester courses are shown for each of those options. There is also sufficient material in the text to fill a leisurely two-semester course (using one class period for most sections in the text and two class sessions for a few others).

Bimodal Course

The course I teach consists of 45% mathematics and mathematics with secondary licensure majors and 45% computer science majors. The remaining 10% are elementary education majors with a mathematics specialization and physics, engineering, and chemistry majors. The outline below shows the sections I typically cover. You may wish to be a bit less aggressive and delete a few sections.

Ch 1	Ch 2	Ch 3	Ch 4	Ch 5	Ch 7
1.1–1.3	2.1–2.6	3.1–3.3, part of 3.4, 3.5–3.6	4.1–4.4	5.1–5.3	7.1–7.2

Ch 8	Ch 9	Ch 10	Ch 11
8.1–8.3 or 8.1, 8.5	9.2–9.4, summarize 9.5–9.6	10.1–10.5	11.1, 11.2, one part of 11.3 or 11.4

Mathematics Majors

The sections listed here are appropriate for a course populated with mainly mathematics majors and minors. The list needs to be trimmed to fit in a semester.

Ch 1	Ch 2	Ch 3	Ch 4	Ch 5	Ch 6
1.1–1.3	2.1–2.6	3.1–3.6	4.1–4.4	5.1–5.3	6.1–6.6

Ch 7	Ch 8	Ch 9	Ch 10	Ch 11	Ch 12
7.1, 7.2, 7.4, 7.5	8.1–8.6	9.1	10.1–10.6	11.1, 11.4	12.1, 12.2, 12.4

Computer Science Majors

The sections listed here are appropriate for a course populated with mainly computer science majors and minors. The list needs to be trimmed to fit in a semester.

Ch 1	Ch 2	Ch 3	Ch 4	Ch 5	Ch 6
1.1–1.3	2.1, 2.3, 2.4–2.6	3.1–3.5	4.2–4.4	5.1, 5.3	6.1–6.3

Ch 7	Ch 8	Ch 9	Ch 10	Ch 11	Ch 12
7.1–7.3	8.5	9.2–9.6	10.1–10.6	11.1–11.3	12.1, 12.2 and either 12.3 or 12.4, 12.5

ACKNOWLEDGMENTS

Several colleagues at Bethel University made helpful comments on selected chapters of the second edition. They are Nathan Gossett, Benji Shults, and David Wetzell.

In addition, I am grateful for all the students and professors who took the time to point out errors and misprints in the first edition, or who made helpful suggestions.

I wish to thank John Wiley and Sons for choosing to publish this new edition. In particular, Susanne Steitz-Filler, Editor, has been my primary contact and advocate at Wiley. Working with her has been a very positive experience. Christine Punzo, Senior Production Editor, has also been a key contact during the final phase of publication. I appreciate her prompt, helpful replies to my questions. Jacqueline Palmieri, Editorial Program Coordinator, and the Creative Services Department at Wiley took some very rough ideas I had for a front cover and created the attractive final design. I would also like to thank Jeannette Stiefel, the copy editor, for diligently reading the entire book to ensure consistency with accepted writing practices. Kristen McGowan, Assistant Marketing Manager, has contributed by making the book known to potential adopters and readers. My heartfelt thanks to all these people and many others at Wiley whose names I do not know. Without these people, you would not be reading this book.

ERRORS

If you have read this far, you should have realized that there were many people besides the author who contributed significantly to the project. As the author, I am privileged to have my name on the front cover and I also have the duty to take final responsibility for the contents of the book. However, many very competent professionals joined together to convert the original manuscript into something much better: a professionally published textbook. I am deeply grateful for all of these people. Thank you.

I received help and good advice from many people. However, the final content of this text is a direct product of my choices. Consequently, any remaining mistakes are mine. To keep this text as accurate as possible, I am offering a reward of one Sacagawea dollar coin to the first person who informs me of each mistake or misprint in the book. Mistakes can be reported by using a web browser to look at:

http://www.mathcs.bethel.edu/~gossett/DiscreteMathWithProof/

and selecting the *Errors* link.

Eric Gossett
Bethel University
3900 Bethel Drive
St. Paul, MN 55112
gossett@bethel.edu

To the Student

Before plunging into the details of this book, it is helpful to do a brief survey of the entire book. A scan through the table of contents will orient you toward some of the topics (mysterious as they may seem before you study them), and familiarize you with some of the other features of the book. In particular, notice the index. A serious attempt has been made to make the index complete and useful. It will also be helpful to do a quick scan of the appendices.

Also note that every chapter has a section named "Quick Check Solutions", followed by one named "Chapter Review".

The Chapter Review typically contains a brief summary of the main ideas in the chapter and a list of new notation. There will also be a reference to a web site containing lists of the definitions and theorems introduced in the chapter, some sample exam questions (with solutions), and a short collection of projects.

Expectations

My goal in writing this textbook is to equip you, the student, to be able to actually apply the material. This has influenced my attitude toward both exposition and homework exercises. Unlike your high school texts and perhaps your calculus text, this book is not merely a collection of prototype examples to copy and submit as solutions to homework exercises. The subject matter for this course will require you to change your approach to learning mathematics. There are still numerous examples, but they often illustrate concepts rather than procedures. As such, the examples do not often provide simple templates for solving homework exercises. In fact, many of the homework exercises ask you to create proofs or to solve problems that do not have clone cousins among the examples.

The level of precision that will be expected from you will also increase. This precision is essential for mathematics majors, who will eventually encounter courses in algebraic structures and real analysis. It is also essential for computer science majors, who need to develop the ability to create precise and correct algorithms and express them as computer programs.

Reading

Reading a mathematics text requires a different approach than reading other kinds of material. You need to read actively. Have paper and pencil ready. At the end of each sentence or paragraph, ask yourself if you could reproduce the ideas just presented. Do not assume you understand an example until you can reproduce the solution with the book closed. At various places, the reading is interrupted by Quick Check exercises. Quick Check problems provide a chance for you to make sure you understand the current material before reading any further. To receive the greatest benefit, you should write out complete solutions. After (and only after) you have completed your solution, you should compare your solution to the detailed solution at the end of the chapter. Do not proceed with the reading until you understand the textbook solution.

Exercises

Mathematics cannot be learned passively, so it is necessary to provide opportunities for you to practice the concepts presented. However, I do not see much value in extensive sets of drill exercises. In fact, I find that large collections of "clone exercises" tend to be counterproductive. Such exercises do not require you to determine which math techniques to use, so you seldom get a chance to master the first steps in real-world problem solving. When was the last time you were faced with a real problem that came with the instructions "use the quadratic formula to find the roots of the following polynomials ..."? In real problems, you first need to recognize that the quadratic formula is appropriate.

You will also find that homework sets tend to have a number of proof-oriented problems. That is appropriate (and necessary) at this point in your mathematical development. Constructing proofs requires you to move beyond applying the results of theorems; you need to carefully examine the underlying principles that make the theorems true.

You have reached a point in your mathematics career where you need to start determining whether your work is correct without the aid of an answer in the back of the book. Many computational problems can be checked by substituting the solution into the original statement of the problem. It might be helpful to have a classmate examine proofs you have constructed. (But check with your instructor about the guidelines for collaboration.)

You will find it helpful to read the brief suggestions for writing mathematics in Appendix G. The suggestions will help you to submit properly written solutions to homework exercises.

Time

The material in this course is very interesting. However, the material is not trivial. It will take time and effort to master. In fact, it will probably take more time and effort than an average college course. There are some rewards for this extra effort. The material you master and the mathematical maturity level that results are both profitable for future mathematics and computer science courses. You will save yourself some stress if you start the semester by setting a schedule that allows sufficient time for this course. Do not get behind or skip assignments.

Memorize definitions as they are encountered. They are critical to understanding the material. You might start a set of flash cards for definitions, major theorems, and important formulas and algorithms. Record them as they are encountered instead of the night before an exam.

Textbook Errors

Although significant effort has been expended to ensure that this textbook is error-free, it is inevitable that some errors will still persist. In order to keep this text as accurate as possible, I am offering a reward of one Sacagawea dollar coin to the first person who informs me of each mistake or misprint in the book. Mistakes can be reported using the form found at the web address listed below. You may also view a list of previously reported errors:

http://www.mathcs.bethel.edu/~gossett/DiscreteMathWithProof/

CHAPTER

1 Introduction

It is a truth universally acknowledged, that a single man in possession of a good fortune must be in want of a wife.

—Jane Austen — Pride and Prejudice

A typical student just beginning a course in discrete mathematics has no clear idea about the content of the course. This introductory chapter will provide some initial ideas. You will find some of the content of this course to be mathematical ideas that are unfamiliar to you. You will have seen other topics as early as middle school or junior high school.

This chapter contains an introductory section titled What Is Discrete Mathematics? and several interesting problems, one covered in some detail. The others will be covered in more depth in later chapters. The problems represent a sample of the topics covered in discrete mathematics courses; they certainly do not represent all the topics.

1.1 What Is Discrete Mathematics?

The word *discrete* has the following definitions[1] that are relevant to our topic:

1: Constituting a separate entity: individually distinct.
2a: Consisting of distinct or unconnected elements: NONCONTINUOUS.
b: Taking on or having a finite or countably infinite number of values: not mathematically continuous.

Sometimes it is easier to understand a concept by first seeking to comprehend an opposite or a complementary concept. For example, it is easier to grasp the notion of an *irrational number* if we first understand the definition of a *rational number*.

DEFINITION 1.1 ***Rational Number***

A real number, r, is called *rational* if it can be expressed as a ratio of two integers: $r = \frac{p}{q}$, where p and q are integers and $q \neq 0$.

Thus, the rational numbers are just the familiar fractions. In decimal form the fractions are the real numbers that have a finite or a repeating decimal expansion.[2] The irrational numbers are those real numbers that are not rational. They cannot be expressed as a ratio of integers; nor do their decimal expansions ever terminate or form a repeating pattern.

When first trying to understand what discrete mathematics consists of, it may be helpful first to consider what it is not about.[3]

[1] *Webster's Ninth New Collegiate Dictionary*, First Digital Edition.

[2] For example, $\frac{5}{4} = 1.25$ and $\frac{4}{7} = 0.\overline{571428}\overline{571428}$.

[3] My first instinct is to think about what *indiscrete mathematics* would be. I immediately think about mathematics that "tells all". However, the word that fits such an imprudent approach is spelled *indiscreet*, not *indiscrete*. The word *indiscrete* means "not separated into distinct parts."

Mathematically continuous objects cannot be separated into pieces that are not joined together. You are familiar with many continuous mathematical objects: geometric lines and shapes, the graphs of functions used in Calculus [such as $\sin(x)$], and the real numbers. The real numbers are considered continuous because we can never find two real numbers that have a gap between them that does not contain other real numbers. That is, given any two distinct real numbers, r and t, we can always find another number, s, with $r < s < t$. We can then find another number between r and s, and then another real number in the new interval. The process never terminates.

In contrast, discrete mathematical objects consist of separate pieces. For example, the collection of all words on this page is a discrete set. The set of all integers is discrete (although infinite) since it is not possible to find another integer between 3 and 4 (or any other adjacent pair).

A thermometer is a good illustration of the difference between continuous and discrete. A mercury thermometer is a continuous device. It is capable of expressing every temperature in its range (although we will be unable to visually see the difference between 25.67° and 25.6698°). In contrast, a digital thermometer might only show temperatures to the nearest tenth of a degree. It has a discrete set of temperatures it can represent.

A digital thermometer has something in common with computers and calculators: They are only able to concurrently represent a finite (but very large) set of numbers. There will always be numbers that cannot be exactly represented. For example, it is impossible to represent the number one-tenth as a finite binary fraction. The reason is that its binary expansion is infinite, but a computer cannot store an infinite number of bits.[4] This means that computers and calculators are inherently discrete devices.[5]

One additional idea needs to be explored before an initial definition of discrete mathematics can be stated. The definition of *discrete* quoted previously used the phrase "countably infinite". What does that mean?

DEFINITION 1.2 *Countably Infinite*

A set is called *countably infinite* if its elements can be placed in one-to-one correspondence with the positive integers. That is, every element in the set can be labeled by exactly one positive integer and every positive integer is the label for exactly one element of the set.

You may be surprised to learn that the set of rational numbers which includes the integers as a proper subset, is a countably infinite set.[6] However, the set of all real numbers is not countably infinite (although it is certainly infinite, since it contains the rational numbers as a proper subset).[7]

It is now time for a working definition of discrete mathematics.

DEFINITION 1.3 *Discrete Mathematics*

Discrete mathematics is a collection of mathematical topics that examine and use finite or countably infinite mathematical objects.

Although this definition is fairly good, it does not completely capture the true nature of a typical discrete mathematics text or course. In reality, such a course usually

---

[4] In the decimal system, we can never write the number one-third exactly since the decimal expansion contains an infinite number of 3s. One-tenth works the same way in binary.

[5] Therefore you should not be surprised to see many connections with computer science in a discrete mathematics text.

[6] This assertion will be proved at the end of the chapter.

[7] This means that there is more than one size of *infinite*! *Proper subset* is formally defined on page 20.

consists of a number of mostly unrelated topics that often involve finite or countably infinite mathematical objects. However, probably because the topics are already only tenuously related, several other kinds of topics are usually included in a discrete mathematics course. For example, this text contains a chapter on logic and another on strategies for proving theorems. Neither topic fits neatly into the definition given previously. Nevertheless, they seem to work well as a portion of a discrete mathematics course.

One other point bears notice: Continuous mathematical objects *do* appear in a discrete mathematics course. For example, when linear homogeneous recurrence relations are examined in Chapter 7, the real and complex roots of polynomial equations are quite important. Both number systems are continuous. The original problem, however, is certainly motivated by discrete ideas.

1.1.1 A Break from the Past

Many of the topics in this text may cause some students to feel a bit disoriented. Perhaps it is because this course breaks from the kinds of mathematics you have been studying in the past. Most of your previous mathematics has been centered on the continuous real numbers and the continuous geometry in the Euclidean tradition. Even calculus is an extension of previous work that introduces the continuous notion of a limit.

In this course, you will get a chance to expand your mathematical horizons. You will be exposed to mathematics that does not use any numbers, equations, or functions. You will expand your notion of such seemingly simple topics as counting and graphs.

Another difference from most of your previous mathematical learning is the inclusion of some topics that have become prominent as a result of the computer revolution. Topics, such as the study and analysis of algorithms, recursion, and finite-state machines, all fit nicely under the working definition of discrete mathematics.

Those students who have been successful in previous mathematics classes may initially feel unsettled because their old study strategies may not be completely successful with this new material. However, with a few adjustments (in particular the need to *read the text thoroughly*), they should continue to do well.

Students who have struggled in previous mathematics classes should consider this course as a fresh chance to succeed. Read and discuss the text with friends, get help from the instructor as soon as you need it, and promise yourself to never let the homework slide. With proper effort and course management, this could be a class you enjoy.

1.2 The Stable Marriage Problem

In a remote valley there is a village with some rather strange customs. The community is very close knit and the people desire to preserve their traditional customs. One of the customs that helps preserve community is their unique marriage custom. Rather than relying on the modern Western custom of dating, or the older custom of matchmakers or arranged marriages, the village council decides on marriage partners. Their custom differs from traditional arranged marriages because the arrangements are made for an entire group of young people. The village council waits until all young people of the current generation are old enough to marry. They then assign marriage partners.

The young people certainly have preferences about who they would like to marry; the village council clearly must pay attention to their wishes. However, it is unlikely that everyone will get their first (or even second) choice. The village council also has to consider the larger community. If they make assignments that the young people do not like, it is possible that one or more pairs may elope and marry someone other than the partner chosen by the council. In that case, the new couple will have disobeyed the decision of the community leaders. If they stay in the valley, the unity of the community will be broken. If they leave the valley, the future vitality of the community will be diminished.

How should the council make assignments so that no pair can successfully elope while still respecting the preferences of the young people (as much as possible)? Is such an assignment always possible?

Before starting to find a solution, it will be helpful to create a useful definition.

DEFINITION 1.4 ***Stable Assignment***

Given a collection of n men, $m_1, \ldots, m_n$, and n women, $w_1, \ldots, w_n$, we wish to associate every person with a mate. Suppose that each person ranks the people of the opposite gender by preference with no ties. An *assignment* is one of the possible collections of n couples. An assignment is *stable* if there does not exist a man, m_i, and a woman, w_j, who are not partners but m_i prefers w_j to his assigned bride and w_j prefers m_i to her assigned groom.

1.2.1 Seeking a Solution

Where do we start to obtain a solution to this problem? One helpful approach is to look at some small examples to help guide our intuition. One example will be presented here.

EXAMPLE 1.1

A Set of Preferences with Three Stable Assignments

Consider the set of preferences (Table 1.1) for a group of three women (A, B, C) and three men (X, Y, Z). The first table indicates that A would prefer to marry Y, but if that is not possible Z would be her second choice. The second table indicates that Y prefers B, with C as second choice and A as his least desirable mate.

Let R $\longleftrightarrow$ S represent the statement: "R and S have been assigned to be married". There are six distinct marriage assignments. Three of them are stable (Table 1.2); the other three are not (Table 1.3).

TABLE 1.1 Preferences for a Small Stable Marriage Problem

Female preferences

	1	2	3
A	Y	Z	X
B	Z	X	Y
C	X	Y	Z

Male preferences

	1	2	3
X	A	B	C
Y	B	C	A
Z	C	A	B

TABLE 1.2 The Stable Assignments

Female 1st choice

A	$\longleftrightarrow$	Y
B	$\longleftrightarrow$	Z
C	$\longleftrightarrow$	X

All 2nd choice

A	$\longleftrightarrow$	Z
B	$\longleftrightarrow$	X
C	$\longleftrightarrow$	Y

Male 1st choice

A	$\longleftrightarrow$	X
B	$\longleftrightarrow$	Y
C	$\longleftrightarrow$	Z

TABLE 1.3 The Unstable Assignments

A and Z elope

A	$\longleftrightarrow$	X
B	$\longleftrightarrow$	Z
C	$\longleftrightarrow$	Y

C and Y elope

A	$\longleftrightarrow$	Y
B	$\longleftrightarrow$	X
C	$\longleftrightarrow$	Z

B and X elope

A	$\longleftrightarrow$	Z
B	$\longleftrightarrow$	Y
C	$\longleftrightarrow$	X

Note that in the middle stable assignment (all 2nd choice), A would like to elope with Y, but Y prefers C to A. Similarly, none of the others (in that assignment) can find a willing partner for an elopement. Everyone must stay with the assigned partner, so the assignment is stable. The "all 2nd choice" assignment demonstrates that an assignment can be stable even though every person would prefer to marry someone else.

In the first of the unstable assignments, A prefers Z to X and Z prefers A to B. Thus, A and Z can successfully elope, causing the assignment to be unstable. A would really like to elope with Y, but Y likes C better than A, so Y will not cooperate. The six distinct assignments were found by exhaustive enumeration (i.e., listing all possibilities).[8] ■

[8] This approach is not practical if there are too many more candidates for marriage. A better approach is to develop an algorithm that will always lead to one of the stable assignments.

The example shows that multiple stable assignments may exist. Additional examples would demonstrate that the number of stable and unstable assignments need not be the same. We would also discover that sometimes only one stable assignment is possible (with all other assignments being unstable). No matter how many examples we look at, we would not find any that did not have at least one stable assignment. This suggests that we may be able to prove that a stable assignment always must exist. Let us leave that proof until later and concentrate on finding an algorithm that will produce a stable assignment.

Although mathematics contains a great store of formulas to memorize, most of them were arrived at by people in the past who saw the important relationships. They were often guided by intuition and common sense (coupled with an understanding of the mathematical discoveries that occurred previously). Where might we look for intuition?

One good source of ideas would be current courtship practices. In a Western society, a young man typically must ask a young woman if she is willing to marry him. Most young men ask the young women they find more desirable before asking others they prefer less. The young women accept or reject proposals based on their own preferences. At times, they may ask for additional time to consider the proposal. During that additional time, they need to decide whether to settle with someone who is not their top candidate, to hope that a better proposal arrives real soon, or to simply reject the proposal.

The algorithm that will be presented next draws on these ideas. It was developed in the 1960s by Gale and Shapley [39], who were considering a slightly more general version of the problem.[9]

1.2.2 The Deferred Acceptance Algorithm

The algorithm operates in a series of rounds. The members of one gender become the suitors and the members of the other gender are the suitees. Each suitor proposes to his or her highest-ranked suitee. The suitees wait for all proposals during that round, then reject all but the proposal from the highest ranking suitor among their current string of suitors. That suitor is told to wait for an answer (hence the name *the Deferred Acceptance Algorithm*). The suitors who are waiting for an answer may still be rejected in a later round in favor of a better proposal.

At the start of subsequent rounds, the suitors who were rejected in the previous round each propose to their next highest-ranking suitee. Suitees then reject all but the current favorite, telling that suitor to wait until later for an answer. This concludes the current round; the next round is ready to begin. The process ends when every suitee has exactly one proposal pending. The village council then steps in and declares that the current pairings are final.[10]

DEFINITION 1.5 ***Unattached, Viable***

A suitor will be called *unattached* if he or she is not currently waiting for a suitee to respond to a proposal. A suitee is *viable* for a suitor if that suitee has not already rejected a proposal from that suitor.

The algorithm (which, for now, assumes equal numbers of suitees and suitors) can be restated in pseudocode:[11]

[9] The notation [39] is a bibliographic reference.

[10] You should make a mental distinction between the stable marriage (or stable assignment) *problem* and the deferred acceptance *algorithm*. The algorithm is just one procedure for solving the general problem.

[11] Pseudocode is described in Section 4.1. You do not need the details right now.

The Deferred Acceptance Algorithm

round = 1
All suitors are initially considered unattached.
while at least one suitee has no pending proposal, **repeat** the following:
    Each unattached suitor proposes to his or her highest ranked viable suitee.
    Each suitee examines her or his string of suitors, rejecting all except
        the highest ranked suitor. That suitor is told to wait while the
        proposal is considered. (The waiting suitor therefore becomes
        attached for the next round.)
    Add 1 to round.
end while
All suitees accept the proposal of the single suitor waiting for an answer.

Two issues need to be addressed before we can confidently use this algorithm. We must show that the algorithm always terminates, and we must show that it always leads to a stable assignment.

THEOREM 1.6 ***Termination***

The Deferred Acceptance Algorithm always terminates.

Proof: Once a suitee has at least one proposal, that suitee will never have an empty string of suitors (since at each round, the best of the available suitors is kept). At the start of each round, the rejected suitors move to the next suitee on their preference lists. Since there is only a finite number of suitees available, the proposals cannot continue indefinitely.[12] Since no suitee ever keeps more than one suitor, and since the numbers of suitors and suitees are equal, eventually every suitee will have a proposal. □

THEOREM 1.7 ***Stability***

The Deferred Acceptance Algorithm always produces a stable assignment.

Proof 1—Suitor's Perspective: Suppose a suitor, X, is unhappy with the mate, A, assigned by the Deferred Acceptance Algorithm. There may be a suitee, B, that X would prefer. The key relationships in the current assignment are

$$X \longleftrightarrow A \qquad\qquad Y \longleftrightarrow B.$$

However, B will already have rejected X at some round of the algorithm (since X would have proposed to every higher-ranking suitee before proposing to A). The rejection was not made foolishly: B rejected X in favor of a valid proposal from a suitor, Y, whom B prefers to X. Thus, any suitee that X prefers to A will already have a mate that ranks higher than X. Consequently, no suitor will find a higher ranking, willing partner with whom to elope.

Proof 2—Suitee's Perspective: Suppose a suitee, A, is unhappy with the mate, X, assigned by the Deferred Acceptance Algorithm. There may be a suitor, Y, that A likes better than X. However, since suitors propose in the order of their preferences, Y must never have proposed to A (otherwise, A would have rejected X in favor of Y). If Y never proposed to A, then Y must have been accepted by a suitee, B, that Y prefers to A. Thus, A will be unable to find a suitor (ranked higher than X) who is willing to elope. The conclusion is that no suitee will be able to find a higher-ranking suitor who is willing to elope. □

---

[12] We will prove in Example 5.1 that at most $n^2 - 2n + 2$ rounds are possible.

1.2.3 Some Concluding Comments

The Stable Marriage Problem and the Deferred Acceptance Algorithm fit nicely into the notion of discrete mathematics presented earlier. The problem involves a finite collection of people, and the algorithm consists of a finite number of well-defined steps. Intuition and careful thinking were used to develop the algorithm. In addition, the presentation has used some other important elements of mathematics: definitions, theorems, and proofs. An interesting aspect of the problem (so far at least) is that no formulas or equations were used, yet the solution process clearly was of a mathematical nature.

The Stable Marriage Problem has not been exhausted; it will appear again in this text.[13] In the exercises for this chapter, you will be asked to extend the algorithm to solve (or attempt to solve) more general versions of the problem. This feature is also common in mathematical investigations. Once a problem has been solved, it is interesting (and often useful) to see if the solution can be generalized. It is also important to abstract the essential features of the problem and solution so that they can be recognized and applied in other situations that are essentially the same.

For example, the problem that Gale and Shapley originally sought to solve did not involve producing stable marriage assignments. Their original interest was in matching graduating high school seniors with colleges. Their problem involved graduates ranking various colleges by preference. Colleges would have quotas for the number of new freshmen to admit. After reading applications, the colleges would rank the graduates. The problem was then to assure matchings of students and colleges so that a student would not receive a scholarship from one school only to accept an offer from a rival college at a later date.

The Deferred Acceptance Algorithm has actually been used in a third setting: matching medical students who are ready to start an internship or residency with cooperating hospitals. The key features of the problem clearly hold: Two groups to match, with each group having clear preferences toward members of the other group.

1.3 Other Examples

The examples in this section will be discussed in more detail in later chapters. They are presented here to indicate some of the diverse areas that this text will cover. The six subsections provide a very small sample of the topics that lie ahead. It might be helpful to pause at this point and make a quick scan of the table of contents to get a more complete idea of the available topics.

1.3.1 A Simple Counting and Probability Example

The simplest way to connect a dvd player to a television is to use an audio video composite cable. The cable has three wires: a yellow video line and a white and a red audio line for stereo audio (see Figure 1.1). Suppose I buy a cable, but am completely color blind, so I cannot properly match the colored cables to the colored connectors on the dvd player and television.

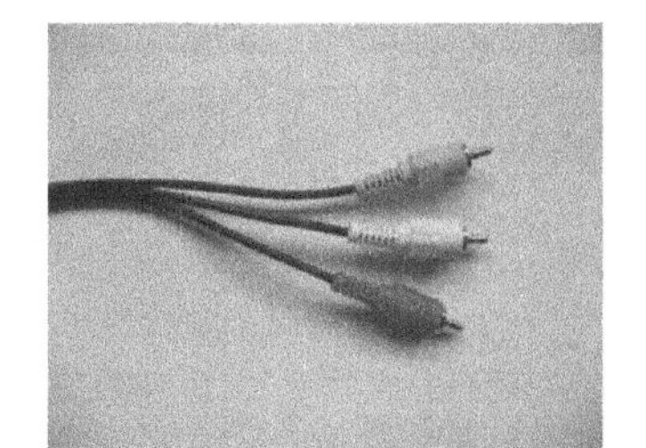

Figure 1.1. A composite cable.

There are two ways I can correctly plug in the cable (since the cable is symmetric on both ends, it does not matter which end is plugged into the television). In how many ways can I incorrectly plug in the cable? What is the probability that I randomly plug it in correctly?

To help think about the question, assume that one end of the cable is labeled A and the other end is labeled B. In how many ways can I plug the A end into the television? I have three choices of wire to plug into the yellow receptacle on the television. Once I have made a choice, there are two wires left that can be plugged into the white receptacle. After I plug one of them in, there is only one choice left for the wire to plug into

[13] See pages 144 and 225 .

the red receptacle.

For example, suppose I plug the red wire into the yellow receptacle. Then I can either plug the yellow wire into the white receptacle and the white wire into the red receptacle, or else I can plug the white wire into the white receptacle and the yellow wire into the red receptacle. So once I make the first choice (red wire into yellow receptacle), there are two ways to plug the rest of the cable into the television. I have three choices for the first wire, and two options for each of those choices. After reading Chapter 5, it will be easy to conclude that there are $3 \cdot 2 \cdot 1 = 6$ ways to plug the cable into the television. Table 1.4 lists all of them (the columns represent the six ways to plug in the cable, with the first being the correct arrangement).

TABLE 1.4 Plugging the Cable into the Television

	Cable Wire					
Yellow Receptacle	Y	Y	R	R	W	W
White Receptacle	W	R	Y	W	Y	R
Red Receptacle	R	W	W	Y	R	Y

Now the B side of the cable needs to be plugged into the dvd player. It should be clear that there are six ways to do that. However, each of those 6 ways can be coupled with each of the 6 ways to plug in the A side of the cable. That results in 36 distinct ways to plug in the cable if the A end is connected to the television. If the B side were to be connected to the television, 36 more possibilities would result.

In all, there are 72 ways to plug in the cable. Only two of those arrangements result in a correct connection. That is, only $\frac{2}{72} \simeq 3\%$ of the time would I randomly connect the cable properly. It would be better to have a friend plug it in for me.

1.3.2 Sierpinski Curves

During the late 1800s, Georg Cantor was attempting to find an infinite set that was larger[14] than the open interval (0, 1) on the real number line [26]. The obvious set to look at was the open unit square, having corners at the points $(0, 0)$, $(0, 1)$, $(1, 1)$, and $(1, 0)$ in the plane. After extensive effort trying to show that the unit square was indeed larger than the unit interval, Cantor started to suspect that they might be the same size! He eventually proved that they were indeed the same size.

Other mathematicians were able to provide a very elegant kind of proof of this counterintuitive result. They constructed functions that map the points on the interval [0, 1] onto points in the closed unit square in such a manner that the function establishes a one-to-one correspondence. However, such functions must be discontinuous [79, Chapter 1]. It is possible to construct continuous functions that map the interval [0, 1] onto the unit square. (There will be points in the unit square that are mapped onto by multiple points in the unit interval.) Such functions are called space-filling curves. Space-filling curves can be more formally defined by requiring the image of the function to have a nonzero area.[15]

One very attractive space-filling curve is the Sierpinski curve, which was defined by Waclaw Sierpinski. The Sierpinski curve is actually the limit of a sequence of curves. The curves will be defined in more detail in Section 7.1.5. For now, it is sufficient to consider the first few members of the sequence.

The curves will be labeled $S_0, S_1, S_2, S_3, \ldots$.

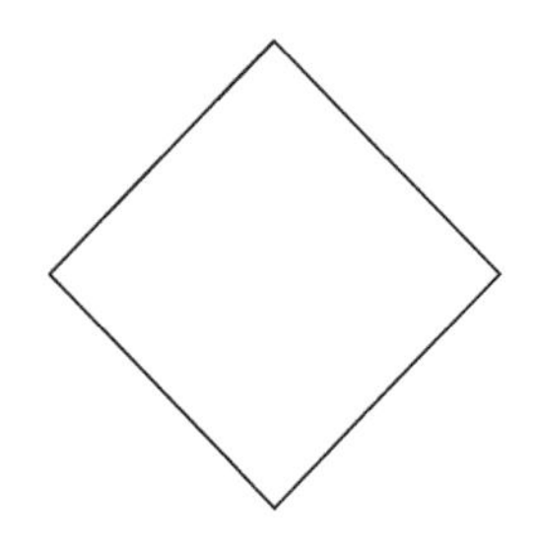

Figure 1.2. S_0.

Consider the curve S_0 in Figure 1.2. We can think of the diamond shape of the curve as depicting a function that successively maps the subintervals [0, 0.25], [0.25, 0.5],

[14] A more precise definition of *larger* and *the same size* can be made using Definition 12.2 on page 732.

[15] See Sagan, Chapter 1 for more details.

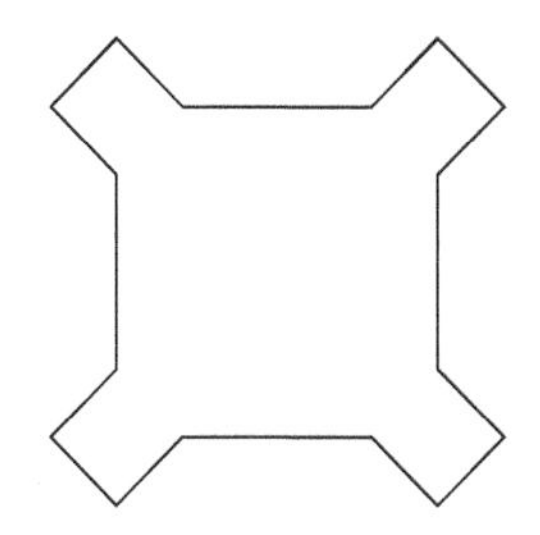

Figure 1.3. S_1.

[0.5, 0.75], and [0.75, 1] onto the points in the plane that lie on the line segments joining (0.5, 1) to (1, 0.5), (1, 0.5) to (0.5, 0), (0.5, 0) to (0, 0.5), and (0, 0.5) to (0.5, 1) in the unit square.

Similarly, we can view S_1 (Figure 1.3) as the image of a function that maps [0, 1], subdivided into 20 equal-length intervals, onto the 20 line segments that constitute the curve S_1. (Notice that the horizontal and vertical line segments in S_1 are double length and hence count as two segments each.)

The curves in the sequence are sets that are the same size as the interval [0, 1] on the real number line. In addition, each curve in the sequence contains more of the points in the unit square than do the previous curves. Also, none of the curves cross themselves. It can be shown that as n approaches infinity, S_n uses more and more of the points in the unit square. The limit, S, of the sequence uses all the points and establishes a mapping from the interval [0, 1] onto the unit square (proving that the interval cannot be smaller than the square). The graph of the curve would appear to be a solid black square in the plane, but the function would match each point in [0, 1] with exactly one point in the square. The curve S_4 in Figure 1.4 demonstrates this limit to some extent.

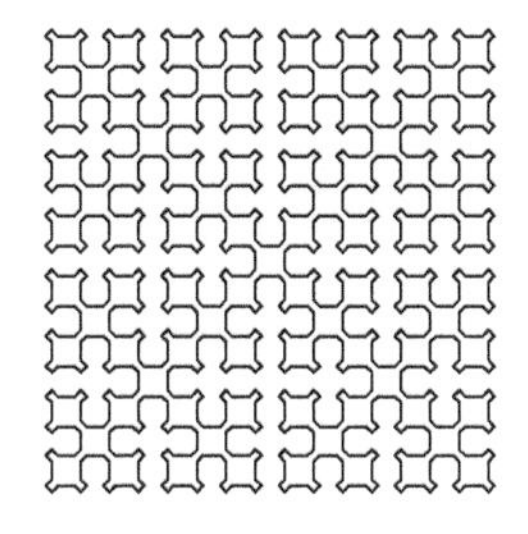

Figure 1.4. S_4.

An alert student might wonder why this topic is in this text, since it clearly is motivated by continuous ideas (rather than discrete ideas). The connection is the mechanism used to draw the curves. That mechanism is a technique called *recursion*. You might consider how to write a program to draw S_4 (or S_6 or, in general, S_n). It does not seem to be an easy task. Using standard methods, it is not. However, using a set of recursive functions, it is quite simple.[16] Recursion is a major topic in typical discrete mathematics courses.

1.3.3 The Bridges of Königsberg

An important branch of mathematics had its origin in a clever puzzle whose solution occurred in 1736 in the town of Königsberg (now named Kaliningrad). The town is built on both sides of the Pregel River and includes an island and a section of land where the river forks. The crude sketch in Figure 1.5 shows the seven bridges that connected the various land masses. The landmasses have been labeled A, B, C, and D.

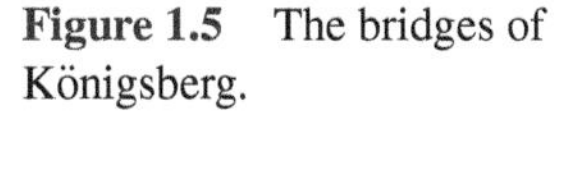

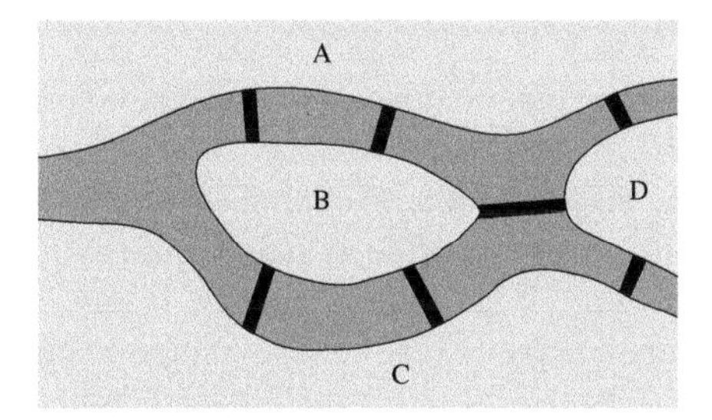

Figure 1.5 The bridges of Königsberg.

A popular puzzle in the town was to produce a tour of the town that crossed each of the bridges exactly once. A really nice solution would provide a tour that began and ended at the same place. Many people tried (and failed) to find a solution. The problem was waiting for Leonhard Euler's masterful solution, which incidentally was the origin of *graph theory*. The details will be presented in Chapter 10.

1.3.4 Kirkman's Schoolgirls

Imagine a schoolmistress with 15 young girls at her boarding school. Each day the schoolmistress lines the girls up in 5 rows of 3 girls each to go for a walk. Is it possible to group the girls so that in the course of 7 walks, each girl will have been in a row

[16]The details appear in Section 7.1.5.

with every other girl exactly once? This problem was proposed in 1850 by the Reverend Thomas P. Kirkman (predating Madeline and Miss Clavel by 90 years).

A solution (there are many) that lists the 5 rows for each of the 7 days is an example of a *balanced incomplete block design*, which is in turn an example of a *combinatorial design*. The techniques for producing combinatorial designs range from guess-and-check through some very sophisticated uses of other mathematical structures (such as vector spaces and finite geometries).

A solution for the Kirkman schoolgirl problem will be presented as Example 8.28 in Section 8.3. A solution to a smaller version of the problem is presented here.

Suppose that there are only 9 schoolgirls and they only walk 4 days each week (perhaps they have a field trip on Wednesdays). It is possible to group them into lines of size 3 so that every girl is with every other girl exactly once per week. If the girls are labeled 1–9, Table 1.5 shows one solution.

TABLE 1.5 An Arrangement of Nine Girls for Weekly Walks

Row	Monday	Tuesday	Thursday	Friday
1	1 2 3	1 4 7	1 5 9	1 6 8
2	4 5 6	2 5 8	2 6 7	2 4 9
3	7 8 9	3 6 9	3 4 8	3 5 7

This solution does not require any sophisticated ideas to generate. The girls can be arbitrarily grouped on Monday. So put 1, 2, and 3 together; then group 4, 5, and 6. This leaves 7, 8, and 9 as the final row. For Tuesday, we must make sure that no girl is with a previous companion. Starting with girl 1, the next available companion is 4. We cannot add either 5 or 6 to the row since girl 4 has already walked with them on Monday. Thus, 1, 4, and 7 become a row. The next available girl is 2. She is matched with 5 and then 8. Girl 3 is matched with 6 and 9. For Thursday, we match girl 1 with girl 5. The third girl cannot be 2, 3, 4, or 7 (since 1 has already been grouped with them), nor can it be 4, 6, 2, or 8 (since 5 has walked with them). That leaves only girl 9. In a similar fashion, 2 and 6 must be grouped with 7. Luckily, the remaining girls (3, 4, and 8) have all been unmatched so far. You should verify that the rows for Friday properly complete the solution.

1.3.5 Finite-State Machines

Researchers in computer science find it helpful to have mathematical models to help study and think about computation and the process of translating a high-level program (perhaps in Java or C) into native machine language. In Section 9.2, we will look at some of the models used. One of the simplest is called a *finite-state machine*. These models can also be used to help programmers think about how to solve a programming task and are used by software engineers to discuss, explore, and communicate critical portions of an algorithm with nonprogrammer clients.

EXAMPLE 1.2

A Simple Finite-State Machine

Suppose we want to write a program to tease a small child named Percival. The program will repeatedly ask for a name. If the name entered is "Percival", the program is to print "Percival, it is your bedtime." If any other name is entered, it will respond with "*the name*, you may stay up late tonight."

A finite-state machine uses a discrete set of *states* to reflect one of a finite number of possibilities. In this case, there are three states: "Percival", "any other name", and "waiting for a name". The finite-state diagram in Figure 1.6 uses bubbles to represent states and labeled arrows to represent the transition between states. The arrows are labeled with an input and an output: (*input*, *output*).

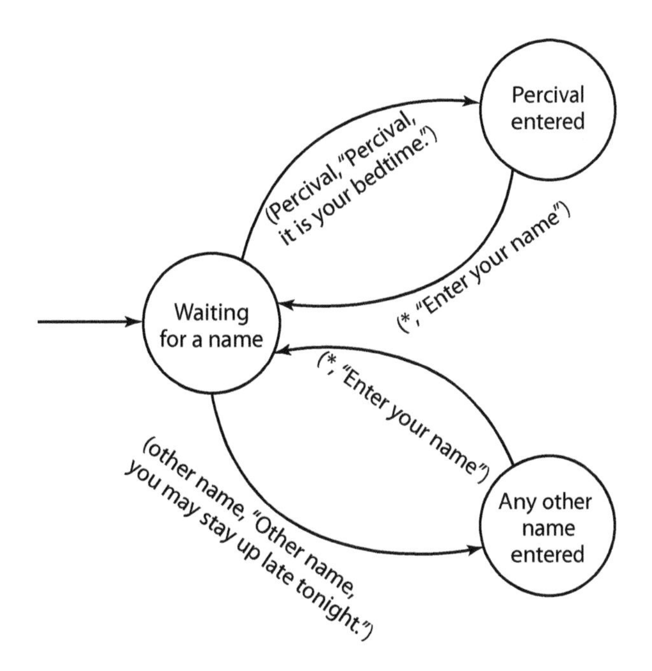

Figure 1.6 A finite-state diagram for the Percival program.

In the finite-state diagram, an input will move from the "waiting for a name" state to either the "Percival" state or the "any other name" state. In each case, an appropriate output is sent to the computer screen. From either state, any input (including no input for at least 30 seconds) causes a new prompt to be printed and moves back to the "waiting for a name" state. On the diagram, the symbol * has been used to indicate "any input". ■

The previous example is trivial, but it does introduce the notion of a finite-state machine without the need for much explaining. The diagram is fairly clear with little need for formal definitions (although formal definitions *will* be presented later). We will be able to model more complicated situations using the simple tool introduced here.

Many other applications to computer science will be discussed. Some examples include (but are not limited to): the analysis of the complexity of algorithms, pattern matching, pseudorandom number generators, encryption, XML, relational databases, and circuit minimization.

Discrete mathematics is a key component in the standard computer science curriculum. In fact, the GRE subject exam in computer science contains more questions from the discrete mathematics course than from almost any other single course in a typical computer science major.

A course in discrete mathematics is also an important component in a typical mathematics major. It is usually where students encounter their first careful examination of logic, axiomatic mathematics, and proof techniques. A more complete look at elementary set theory, elementary number theory, and counting are also presented. Additional topics of interest include (but again are not limited to) Boolean algebras, finite probability theory, recurrence relations, combinatorics, generating functions, and graph theory.

1.3.6 The Set of Rational Numbers Is Countably Infinite

As a final preview example, a proof will be given that the set of rational numbers is countably infinite. This example will illustrate the use of definitions in proofs and also may help you to understand more clearly this counterintuitive result.

The definition of countably infinite (repeated below) is inspired by the buddy system that is often used when swimming at a summer camp. Each swimmer is paired with a partner. If the lifeguard blows a whistle, the partners must find each other and raise their clasped hands. This allows the lifeguard to ensure that no swimmer is without a partner (and so no single swimmer is missing).

In the same way, the definition ensures that every element in the countably infinite set is paired with exactly one integer. This correspondence indicates that the two sets should be considered to be of the same size.

DEFINITION 1.2 *Countably Infinite*
A set is called *countably infinite* if its elements can be placed in one-to-one correspondence with the positive integers. That is, every element in the set can be labeled by exactly one positive integer and every positive integer is the label for exactly one element of the set.

THEOREM 1.8
The set of rational numbers is countably infinite.

Proof: The proof will first demonstrate that the set of positive rationals is countably infinite.

An initial attempt to list the positive rationals in an orderly fashion is seen in Figure 1.7.

1/1 2/1 3/1 4/1 5/1 ...
1/2 2/2 3/2 4/2 5/2 ...
1/3 2/3 3/3 4/3 5/3 ...
1/4 2/4 3/4 4/4 5/4 ...
1/5 2/5 3/5 4/5 5/5 ...
⋮ ⋮ ⋮ ⋮ ⋮

Figure 1.7 A first attempt to list the positive rationals.

The problem with this attempt is that each rational number will appear multiple times. For example, one half will be listed as $\frac{1}{2}, \frac{2}{4}, \frac{3}{6}, \ldots$, which can be corrected by traversing the listing diagonally as shown in Figure 1.8. Whenever a duplicate is encountered, it can be crossed out.

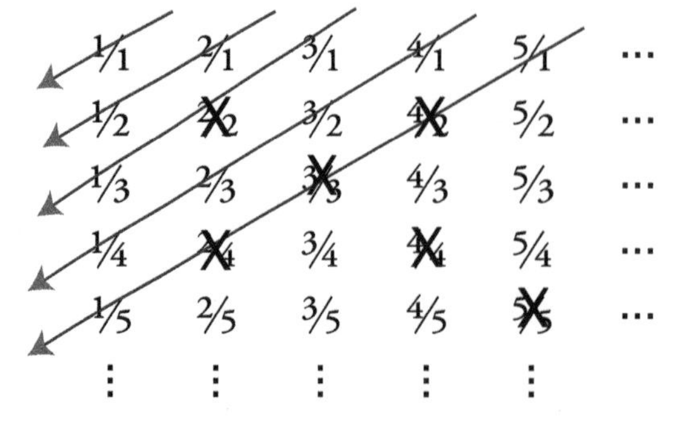

Figure 1.8 Listing the positive rationals.

TABLE 1.6

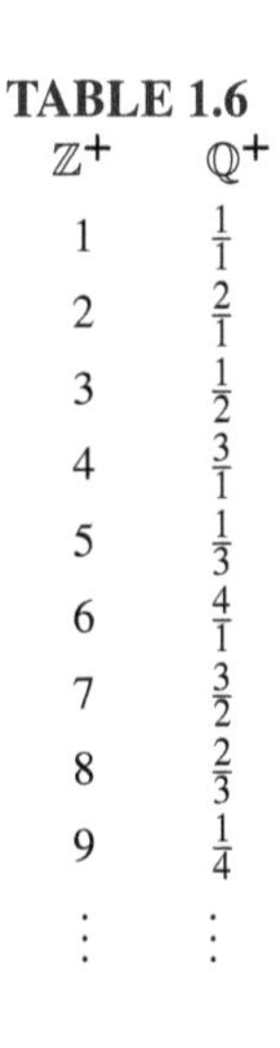

$\mathbb{Z}^+$	$\mathbb{Q}^+$
1	$\frac{1}{1}$
2	$\frac{2}{1}$
3	$\frac{1}{2}$
4	$\frac{3}{1}$
5	$\frac{1}{3}$
6	$\frac{4}{1}$
7	$\frac{3}{2}$
8	$\frac{2}{3}$
9	$\frac{1}{4}$
⋮	⋮

Now start traversing the diagonals in order, assigning the next available positive integer to each remaining entry. As can be seen in Figure 1.9, each integer will eventually be assigned a positive rational number and each positive rational number will have a unique positive integer attached.

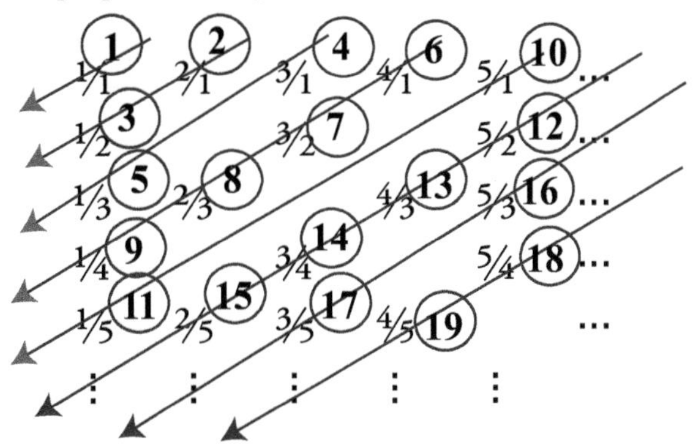

Figure 1.9 Pairing positive integers and positive rationals.

TABLE 1.7

$\mathbb{Z}^+$	$\mathbb{Q}$
1	$\frac{0}{1}$
2	$\frac{1}{1}$
3	$-\frac{1}{1}$
4	$\frac{2}{1}$
5	$-\frac{2}{1}$
6	$\frac{1}{2}$
7	$-\frac{1}{2}$
8	$\frac{3}{1}$
⋮	⋮

Thus, the set of positive rational numbers is countably infinite. The pairing begins as in Table 1.6.

We can form the one-to-one correspondence between the set of positive integers and the set of all rational numbers by adjusting the right-hand column. Insert the number $0 = \frac{0}{1}$ next to 1 and push all other rational numbers down one position. Then, after each positive rational, insert its negation, again pushing all other numbers down one position. The result looks like Table 1.7.

This pairing completes the proof. □

1.4 Exercises

The exercises marked with ★ have detailed solutions in Appendix H. Appendix G contains suggestions to help you submit properly written solutions to homework exercises.

1. Consider the preferences for the following group of four women (C, D, E, F) and four men (L, M, N, O).

Female Preferences

	1	2	3	4
C	O	L	M	N
D	O	M	L	N
E	N	M	O	L
F	O	N	M	L

Male Preferences

	1	2	3	4
L	C	D	E	F
M	D	F	E	C
N	C	E	D	F
O	D	C	E	F

Decide whether the following assignment is stable or unstable. Provide some justification for your answer.

C	⟷	O
D	⟷	M
E	⟷	N
F	⟷	L

2. Find a stable marriage assignment for the following group of four women (A, B C, D) and four men (W, X, Y, Z).

Female Preferences

	1	2	3	4
A	Z	X	Y	W
B	X	Z	W	Y
C	W	Y	Z	X
D	Y	W	X	Z

Male Preferences

	1	2	3	4
W	D	A	B	C
X	B	D	C	A
Y	A	C	D	B
Z	A	C	D	B

3. Suppose that there is a group of four men and four women who are eligible for marriage. Create a set of preferences such that there exists exactly one stable assignment among all the possible assignments that could be made. (Assume that there cannot be ties in the preference ratings.)

4. ★ Suppose that there is a group of two women and only one man who are eligible for marriage. Suppose that the man gets paired up with the woman who rated second on his preference chart. Could such an assignment be stable? Provide adequate justification for your answer. (Assume that there cannot be ties in the preference ratings.)

5. Suppose that there are n young men and m young women (with $n \neq m$) who are eligible for marriage. How can the Deferred Acceptance Algorithm be modified to produce a stable assignment with these conditions? Can you prove that the revised algorithm terminates and always produces a stable assignment?

6. The notion of stable assignments has relevance in other settings. One such setting is the stable roommate problem. In this problem, there are n students who need to be assigned roommates (i.e., two students per room). Each student has ranked the other students in preference order with no ties (that is, a student cannot assign the same preference rank to two potential roommates). Is there an algorithm that will produce a stable assignment of roommates?

To see that no such algorithm is possible, it is sufficient to produce an example where such an assignment is impossible. Your task is to create such an example. [*Hint*: Let $n = 4$ and create a set of preferences so that every possible assignment is unstable.]

7. I plan to make some oatmeal cookies, but I am considering three possible modifications to the recipe: I can cut the sugar in half, I could add some walnuts, or I can add some raisins. I can make all or just some of the changes (or I could decide to make no changes to the recipe). How many different kinds of oatmeal cookies can I make?

8. Provide a different solution for the 9-schoolgirl walking problem in Section 1.3.4. Leave the rows for Monday unchanged. Your solution should not be merely a rearrangement of the days. This means that on at least 1 day of the week, at least one row should contain 3 girls who are never together (as a group of 3) in the previous solution.

9. For each claim, determine whether it is always true or else false in some cases. Then give some justification for your answer. Read carefully.

(a) There is no place for continuous mathematical objects in a discrete mathematics course.

(b) Although not usually the case, some topics in mathematics can be studied (at least in part) without using any equations or functions.

(c) The Deferred Acceptance Algorithm, also called the Stable Marriage Problem, has been used in matching medical students with cooperating hospitals.

(d) ★ John proposed to Mary, but she rejected him in favor of Fred. With the Deferred Acceptance Algorithm, Mary is no longer viable for John.

10. For each claim, determine whether it is always true or else false in some cases. Then give some justification for your answer. Read carefully.

(a) An assignment (in the context of the Stable Marriage Problem) is unstable if there exists a man, m_i, and a woman, w_j, who are not partners but m_i prefers w_j to his assigned bride.

(b) The unit square, having corners at the points $(0, 0)$, $(0, 1)$, $(1, 1)$, and $(1, 0)$ in the plane, is no different in size than the open unit interval $(0, 1)$ on the real number line.

(c) ★ A stable assignment could exist in which each member of one of the groups is paired with the person that he or she ranked lowest in the opposite group. (Assume that there are equal numbers of people in both of the groups and that there are no ties in the preference ratings.)

(d) Suppose that there are two groups of people, and that each person is paired with the individual that he or she ranked lowest in the opposite group. Then this assignment could be stable. (Assume that there are equal numbers of people in both of the groups and that there are no ties in the preference ratings.)

11. Classify each of the following sets/objects as discrete or continuous mathematical objects.

(a) ★ The set of integers greater than 1000.

(b) ★ The set of y-coordinates on the graph of

$$f(x) = 4x^2 + 2x + 5$$

(c) The collection of fans at the baseball game.

(d) The set of angles (in radians, relative to 12 o'clock) attained by the second hand of an analog clock.

(e) The set of real numbers greater than 0 but less than 50.

12. Show that the set of integers, $\mathbb{Z}$, and the set of even integers (a proper subset of $\mathbb{Z}$), are the same size. Your proof should exhibit a function that matches every element of the integers with exactly one even integer, and every even integer with exactly one integer.

13. ★ Suppose that Percival (Example 1.2) notices the inequity of the program. He wants everyone to be given the same message. Modify the finite-state machine to accommodate this change.

14. Suppose that we wish to design a program that repeatedly asks for the professions of people. Any teacher should be told that he or she will receive an apple, while an individual in any other profession should be told that he or she will receive a new car. Thus, if the profession "teacher" is entered, the program will respond with "you will receive an apple". If any other profession is entered, it will respond with "you will receive a new car". Create a finite-state diagram to accommodate this new idea for a computer program.

15. Looking at the town of Königsberg, find 10 different tours that cross any six of the seven bridges exactly once per bridge. List the starting and ending landmass in each case. The bridges have been labeled for the purpose of specifying which bridge is being visited. For this problem, consider a tour and its reverse ordering to be two representations of the same tour.

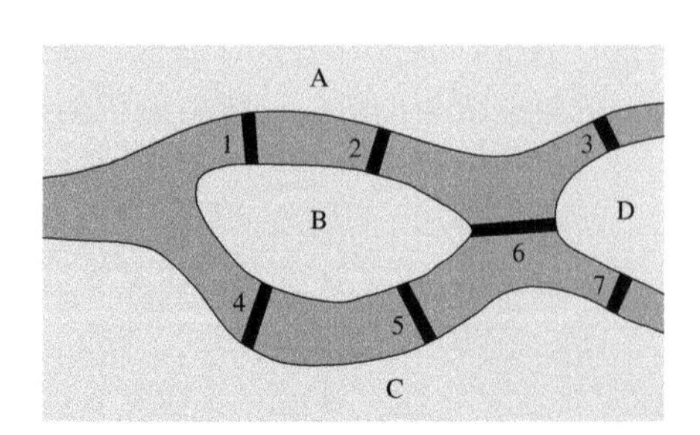

16. Prove that the following sets are countably infinite.

(a) The set of positive even integers.

(b) The set of integers.

17. Locate a definition of *discrete mathematics* in some source other than this book. Write the definition and clearly cite the source.

1.5 CHAPTER REVIEW

1.5.1 Summary

This chapter exists mainly to provide an overview for the rest of the course. Most of the examples in this chapter will be presented in greater detail in future chapters. You may wait until that time to worry about the details.

The exception is the discussion about the Stable Marriage Problem and the Deferred Acceptance Algorithm. That material will be supplemented elsewhere in the book, but the initial discussion presented here will not be repeated.

1.5.2 Notation

The following notation has been introduced in this chapter.

Notation	Page	Brief Description
[39]	5	A bibliographic reference

CHAPTER

2

Sets, Logic, and Boolean Algebras

I think, therefore I am.
—René Descartes (1596 – 1650)

As a young man, René Descartes set himself the task of determining a solid basis for knowledge. He had the advantages of a good education and a broad experience of many cultures and ideas. Inspired by the certainty and self-evidence of the proofs used in geometry, Descartes chose to reason his way carefully to what was true. He eventually picked four rules to guide his search:[1]

1. Never accept anything as true unless I recognize it to be certainly and evidently true.
2. Divide each of the difficulties that I encounter into as many parts as possible and necessary for an easier solution.
3. Think in an orderly fashion, beginning with the things that are simplest and easiest to understand and gradually reaching toward more complex knowledge.
4. In both the process of searching and in reviewing when in difficulty, make enumerations so complete, and reviews so general, that I will be certain nothing is omitted.

In addition to these rules, Descartes chose to continue his study of mathematics, primarily to "accustom my mind to work with truths and not to be satisfied with bad reasoning".

> Since this truth, *I think, therefore I am*, was so firm and assured that all the most extravagant suppositions of the sceptics were unable to shake it, I judged that I could safely accept it as the first principle of the philosophy I was seeking.

Descartes was not the first person to attempt to explicitly and deliberately use reason to arrive at valid conclusions. His attempt has generated criticism. However, he does represent those who find great value in using reason and logic to aid our understanding and keep us from erroneous conclusions.

When most people mention the word *logic*, they are thinking about stringing together a sequence of ideas that lead to a correct conclusion. Such a string of ideas will be called an *argument*.[2] In this sense, *logic* consists of the attempt to make valid conclusions given a collection of assertions and a collection of ideas that relate the assertions. For example, a prosecuting attorney will attempt to develop a sequence of ideas that will compel a jury to conclude that a defendant is guilty. The participants in a political debate will try to use carefully reasoned arguments to support their position on an issue

[1] Descartes's account is found in his *Discourse on the Method of Rightly Conducting the Reason and Seeking Truth in the Sciences*, first published in 1637 [23].

[2] Notice that this is not the same as the common use of the word *argument* to mean a verbal fight.

under debate. In fact, any time someone needs to convince someone else to adopt a course of action, the potential for using logic exists.

There are alternatives to using logic. One alternative is to use an emotional appeal. In this alternative, reason is not an important element. For example, a member of your family may propose fumigating your house to get rid of all the insects that live there. The justification might be that "insects are disgusting!". A careful chain of reasoning about the need for insects to fill an ecological niche may go unheeded.

Suppose you ask a friend for a cup of juice. The friend brings a container filled with liquid. There are two requirements for this to be an acceptable fulfilment of your request. The first is that the container should have the proper form: It should be of a proper size and shape for drinking and it should have no leaks. The second requirement is that it contains juice (not coffee or water). In a similar fashion, logic is concerned with both the *form* or structure of a chain of reasoning and with the *content* (the validity of the individual assertions in the chain). The outer shell or form of the chain of reasoning is called *formal logic*. The content is called *informal logic*. The emphasis is on the *in* in *in*formal to indicate that we are not discussing the *form* of the argument.

Two examples of formal logic are *syllogistic logic* and *symbolic logic*. Both are concerned with the structure of the argument. As can be seen in Appendix C.2.1, it is possible to make errors in how the ideas are connected. It is also possible, as will be examined in Section 2.2.2, to make errors in the content of the argument.

Syllogistic logic was brought to maturity by the Greek philosopher Aristotle, who lived from 384 to 322 B.C. Syllogistic reasoning was the dominant version of formal logic for the next 2000 years in the Western World. A *syllogism* consists of a *major premise*, a *minor premise*, and a *conclusion*. The following example illustrates these components:

All planets in the solar system orbit the sun.	The major premise
Earth is a planet in the solar system.	The minor premise
Earth orbits the sun.	The conclusion

The goal of formal logic is to determine correct principles of reasoning. Using correct principles of reasoning will ensure the validity of the conclusion *assuming the premises are true*. In the preceding example, both premises are true and a valid form of reasoning was used. The conclusion is therefore correct.

Consider another example:

Some planets in the solar system are gas giants.	The major premise
Earth is a planet in the solar system.	The minor premise
Earth is a gas giant.	The conclusion

The conclusion in this case is unfounded, even though the premises are both true. Jupiter, Saturn, Neptune, and Uranus are gas giants; Earth is not. The problem is with the *form* of the syllogism; in this case, an invalid form of reasoning was used. Just because *some* planets are gas giants, we cannot conclude that *all* planets are.

The second variety of formal logic to be examined here is an elementary type of symbolic logic. Symbolic logic started to emerge in the late 1600s. The developers desired to provide mathematical tools for logic that were similar to the manipulations available in algebra. They also wanted the certainty provided by an axiomatic system with proofs, such as was presented in Euclid's geometry.

Symbolic logic introduces letters that are similar to variables in algebra. There are rules that guide correct manipulation of symbolic representations of statements, just as there are rules in algebra that guide the correct manipulation of equations. In its most rigorous forms, symbolic logic provides the mathematical tools to enable chains of reasoning to be verifiably established.

Later in this chapter, informal logic and symbolic logic will be introduced in more detail. Syllogistic logic appears in a modified form in Appendix C.2.1.

Prior to a more detailed look at logic, the very important notion of sets will be introduced. The seemingly unconnected topics of logic and sets will be jointly explored when Boolean algebras are introduced in Section 2.5.

2.1 Sets

It is nearly impossible to study modern mathematics at a nontrivial level without encountering the notion of a set. Sets appear frequently in discrete mathematics. It is therefore useful to review briefly some of what you already know about sets and also to introduce some ideas that you may not be familiar with.

2.1.1 Definitions and Notation

This section will introduce several basic definitions and some essential notation from set theory. There is much more to set theory than is presented here.

The terms that will be discussed next do not have formal definitions. They are considered to be *undefined terms*.[3]

INFORMAL DEFINITION 2.1 ***Set; Element; Member; Universal Set***

Informally, a *set* is a collection of distinct objects, each thought of as a single entity. The set is the aggregate collection. The objects in the set are called the *elements* or *members* of the set. The potential elements are considered to be members of a set U, called the *universal set*.

The notation $x \in A$ is used to indicate that x is an element of the set A. The notation $y \notin A$ indicates that y is not an element of A.

EXAMPLE 2.1

A Set of Integers

The universal set might be the set $\mathbb{Z}$ of all integers.[4] The collection of positive integers can be considered as a set. Call this set P. The number 4 is an element of P. The numbers -5 and 0 are not elements of P. Symbolically, $4 \in P$, but $0 \notin P$. Note that both $\frac{1}{2} \in P$ and $\frac{1}{2} \notin P$ are invalid expressions. This is because $\frac{1}{2}$ is not an element of $\mathbb{Z}$, the universal set for this example. ■

The next two examples introduce some notational conventions for specifying sets.

EXAMPLE 2.2

Small Sets

A common way to represent a set with only a few elements is to list the elements between a pair of braces. Thus, the set containing the letters A through D would be written {A, B, C, D}. Since the focus is on the *collection* of elements, the order in which the elements are listed is unimportant. Thus, the sets {A, B, C, D} and {B, C, A, D} are the same. Since sets consist of *distinct* elements, there is a firm convention that elements are only listed once. It is therefore incorrect to write {A, B, A}. Instead, write this set as {A, B}. ■

---

[3] See page 87 for more about undefined terms.

[4] See Appendix A.2 on page A2 for a definition of the integers.

EXAMPLE 2.3

Large Sets

There are two common ways to list larger sets. In the first, a sufficient number of elements are listed to show the general pattern. The remaining elements are represented by three dots (formally called an *ellipsis*).[5] The set of all positive, even integers could be listed as $\{2, 4, 6, 8, \ldots\}$. The positive, odd integers less than 100 could be specified as $\{1, 3, 5, 7, \ldots, 95, 97, 99\}$.

The other way to specify large sets is to describe the set using *set-builder* notation. The set of positive, even integers is

$$\{n \mid n \in \mathbb{Z}, n > 0, \text{ and } n \text{ is even}\}$$

This is read as "the set of elements n such that n is an integer, n is greater than 0, and n is even". The variable to the left of the vertical bar represents a typical element of the set. The conditions to the right of the vertical bar specify the requirements for set membership.

The set of positive, odd integers less than 100 can be specified as

$$\{n \mid n \in \mathbb{Z}, 0 < n < 100, \text{ and } n \text{ is odd}\}$$

or as

$$\{n \in \mathbb{Z} \mid 0 < n < 100, \text{ and } n \text{ is odd}\}$$ ■

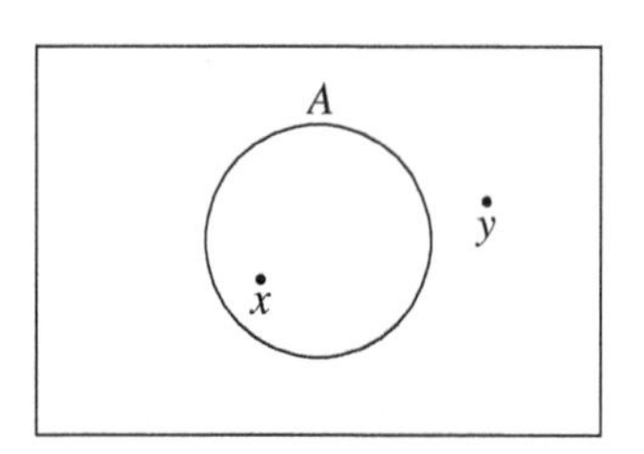

Figure 2.1. A Venn diagram with $x \in A$ but $y \notin A$.

A convenient visual representation of sets is a *Venn diagram*. A Venn diagram consists of a rectangle, representing the universal set, some circles and/or ellipses[6] and some dots. The circles represent sets. The dots represent elements. There is nothing of interest outside the rectangle. A dot is inside a circle whenever the element it represents is a member of the set. Often, the dots are not explicitly shown. Figure 2.1 shows a simple Venn diagram.

DEFINITION 2.2 ***Subset***

A set, B, is a *subset* of a set, A, if every element of B is also an element of A. This is denoted $B \subseteq A$.

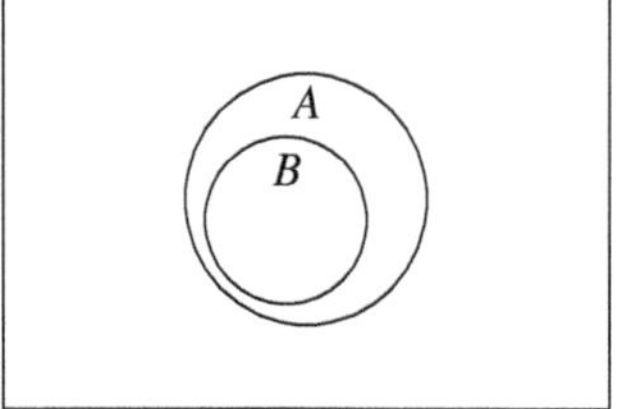

Figure 2.2. $B \subseteq A$.

The Venn diagram in Figure 2.2 illustrates the subset relationship. Notice that every dot that might be placed inside the B circle would also be inside the A circle. It is also possible to imagine the B circle being identical to the A circle. Every dot inside the B circle would still also be a dot inside the A circle. This illustrates the notion that every set is a subset of itself.

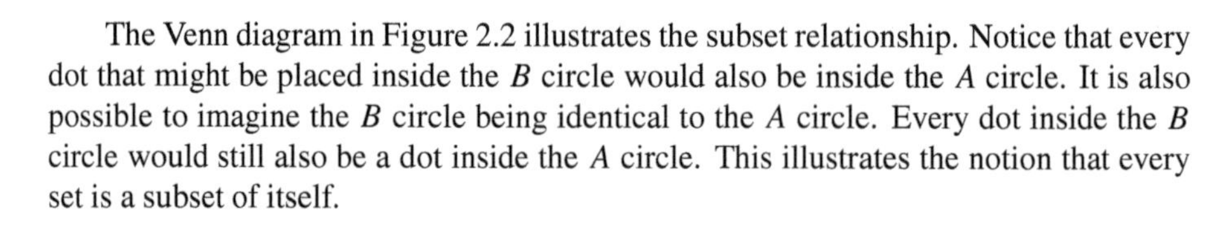

DEFINITION 2.3 ***Proper Subset***

A set, B, is a *proper subset* of the set A, denoted $B \subset A$, if every element of B is an element of A, but there is at least one element of A that is *not* an element of B.

An alternative notation, $B \subsetneq A$, is also used to emphasize the "at least one element of A that is *not* an element of B" part of the definition. In this book, the two notations are considered to mean the same thing.[a]

[a] Some authors use $B \subset A$ to mean "B is a subset of A and it is not important to know whether the inclusion is proper or an equality".

[5] A sentence does not require a terminating period if the final item in the sentence is an ellipsis and another sentence follows. The next (terminal) sentence is correct. *The positive even integers are 2, 4, 6, …* The (terminal) sentence that follows is incorrect. *The positive even integers are 2, 4, 6, ….*

[6] "Circles and/or ellipses" may be replaced by "any convex set". A *convex set* is a closed region such that any two points in the region can be joined by a line segment that remains completely inside the region.

Figure 2.2 shows B as a proper subset of A since there is room to place a dot inside A that would not be inside B. When using Venn diagrams we would typically draw the B circle inside the A circle as in Figure 2.2. However, you need to remind yourself that unless you know that B is a *proper* subset of A, it is possible that the two circles coincide.

EXAMPLE 2.4

A Proper Subset

The set of positive integers P is a proper subset of the integers. This is because every positive integer is an integer ($P \subseteq \mathbb{Z}$), but many integers are not positive integers (for example, 0 and -1). ■

Every set of elements from the universal set U is a subset of U. There is a special set, denoted $\emptyset$, that represents the *empty set*.

DEFINITION 2.4 ***The Empty Set, $\emptyset$***

The *empty set*, denoted by $\emptyset$, is the unique set that does not contain any elements.

The empty set is a subset of every set. This is because the requirement for B to be a subset of A is that every element of B is also an element of A. Since $\emptyset$ has no elements, the requirement can be considered to have been met. Thus $\emptyset \subseteq A$ is always true for every set A.

DEFINITION 2.5 ***Set Equality***

Sets A and B are said to be equal, denoted $A = B$, if and only if $A \subseteq B$ and $B \subseteq A$.

The definition says $A = B$ if and only if every element of A is also in B, and every member of B is also in A. That is, A and B consist of identical collections of elements.

EXAMPLE 2.5

Equal Sets

The sets $A = \{3, 6, 9\}$ and $B = \{n \in \mathbb{Z} \mid 0 < n < 10 \text{ and } n \text{ is evenly divisible by } 3\}$ are equal. ■

It is often necessary to discuss the size of a set.

DEFINITION 2.6 ***Cardinality; Infinite Set***

The *cardinality* of a set is the number of elements in the set. The cardinality of the set, S, is denoted by $|S|$.

A set, S, is said to be *infinite* if for each positive integer, n, there is a proper subset, $S_n \subset S$, with $|S_n| = n$.

Thus, $|\{a, b, c, d, e\}| = 5$, whereas $\mathbb{Z}$ is an infinite set (the requirements of the definition can be met by setting $\mathbb{Z}_n = \{1, 2, 3, \ldots, n\}$). Another common notation for the cardinality of a set S is $o(S)$.

DEFINITION 2.7 *Complement*

The *complement*,[a] $\overline{A}$, of the set A is the set of all elements of the universal set that are not elements of A.[b] Thus, for every x in the universal set, $x \in \overline{A}$ if and only if $x \notin A$.

[a]Note the spelling. The word *complement* carries the idea of "completion". The word *compliment* (with an *i*) means "approval".
[b]Some commonly used alternative notations for the complement of A are A^c and A'.

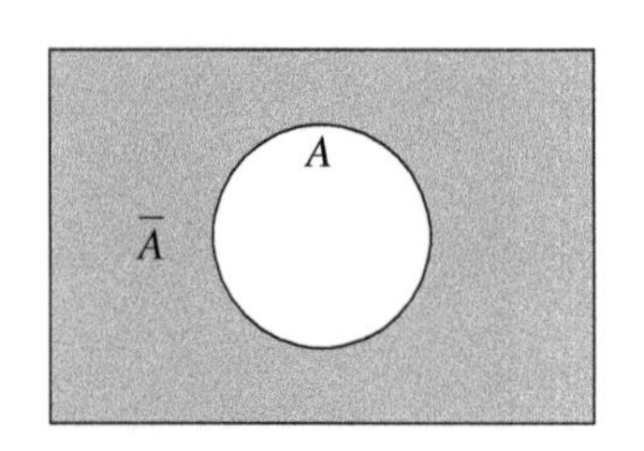

Figure 2.3. A and $\overline{A}$.

The Venn diagram in Figure 2.3 shows this relationship. Notice the convention of using shading rather than a collection of dots. The shaded area is to be considered as the totality of elements in the set that is being distinguished ($\overline{A}$ in this diagram).

EXAMPLE 2.6

A Complement

If the integers, $\mathbb{Z}$, are taken as the universal set, then the complement of the positive integers is the set $\overline{P} = \{0, -1, -2, -3, \ldots\}$. ■

If more than one set is under consideration, additional relationships can be defined. Several will be presented here.

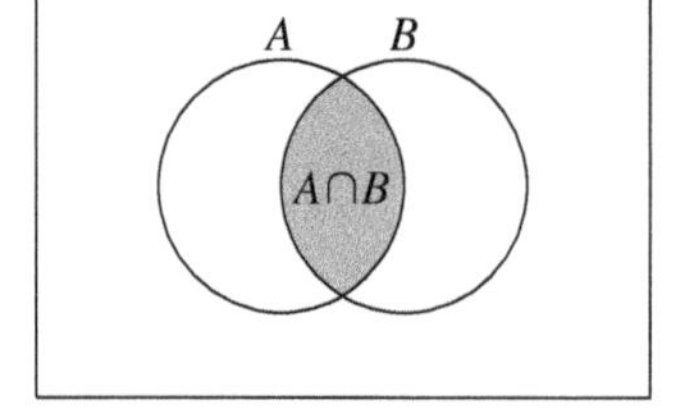

Figure 2.4. $A \cap B$.

DEFINITION 2.8 *Intersection*

The *intersection* of the sets A and B, denoted $A \cap B$, is the set of all elements that are in both A and B. That is, $x \in A \cap B$ if and only if $x \in A$ and $x \in B$.

The Venn diagram in Figure 2.4 illustrates this concept.

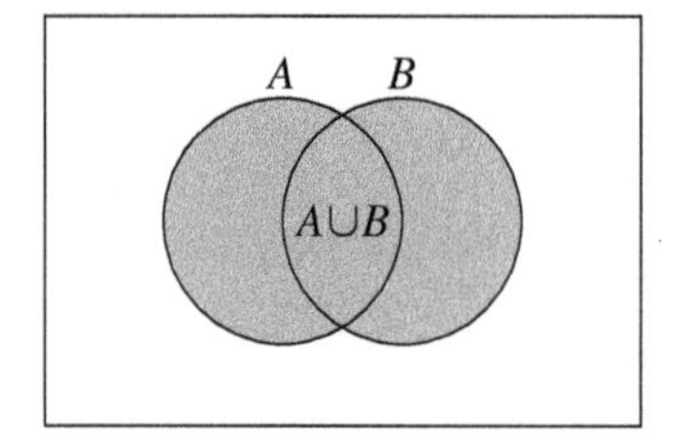

Figure 2.5. $A \cup B$.

DEFINITION 2.9 *Union*

The *union* of the sets A and B, denoted $A \cup B$, is the set of all elements that are either in A or in B (or in both). That is, $x \in A \cup B$ if and only if either $x \in A$ or $x \in B$.

A Venn diagram for a set union is shown in Figure 2.5.

EXAMPLE 2.7

A Union and Intersection

Let E be the even integers: $E = \{\ldots, -6, -4, -2, 0, 2, 4, 6, \ldots\}$ and let P be the set of positive integers. Then $P \cap E = \{2, 4, 6, \ldots\}$ and $P \cup E = \{\ldots, -6, -4, -2, 0, 1, 2, 3, 4, 5, 6, \ldots\}$. ■

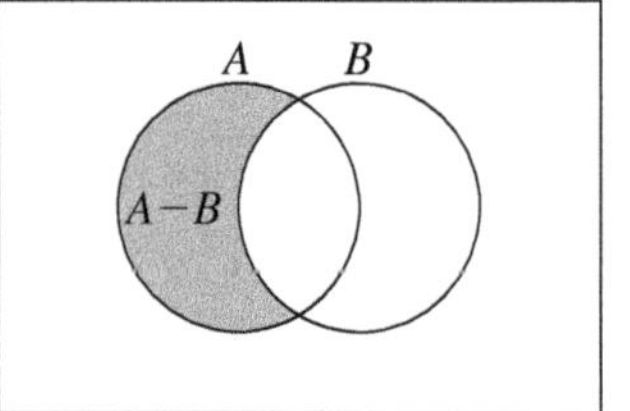

Figure 2.6. $A - B$.

DEFINITION 2.10 *Difference*

The *set difference*, $A - B$, is the set of all elements that are in A but are not in B. That is, $x \in A - B$ if and only if $x \in A$ and $x \notin B$.[a]

[a]Another common notation for set difference is $A \setminus B$.

Figure 2.6 shows a Venn diagram for a set difference.

EXAMPLE 2.8

A Set Difference

Let T be the set of all positive integers that are divisible by 2. Let F be the set of all positive integers that are divisible by 5. Then $T - F = \{2, 4, 6, 8, 12, 14, 16, 18, 22, 24, 26, 28, 32, \ldots\}$. ■

DEFINITION 2.11 ***Disjoint***

The sets A and B are *disjoint* if they have no elements in common. That is, A and B are disjoint if $A \cap B = \emptyset$.

The Venn diagrams in Figures 2.7 and 2.8 show that intersection and union of disjoint sets are still valid concepts.

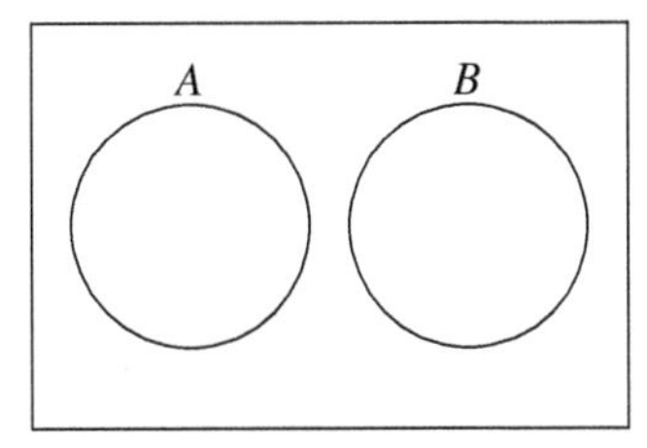

Figure 2.7 The intersection of two disjoint sets is empty.

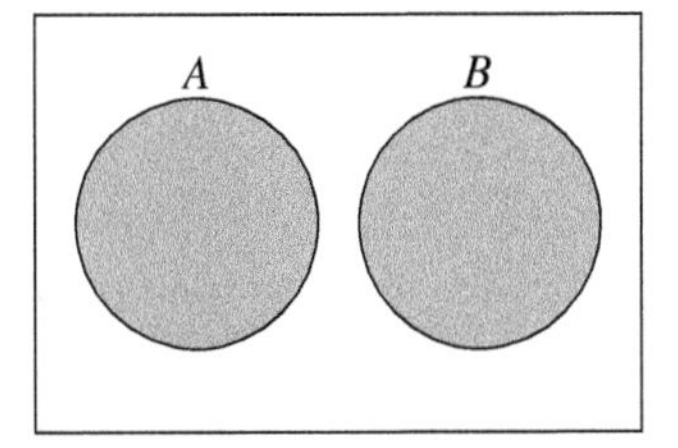

Figure 2.8 The union of these two disjoint sets has been shaded.

EXAMPLE 2.9

A Disjoint Union and Intersection

Let O be the odd integers: $O = \{\ldots, -5, -3, -1, 1, 3, 5, \ldots\}$. Then $E \cap O = \emptyset$ and $E \cup O = \mathbb{Z}$. ■

✔ Quick Check 2.1

Assume the universal set, U, is the set $\{1, 2, 3, 4, 5, 6, 7, 8, 9\}$. Let $D = \{2, 4, 6, 8\}$ and $F = \{1, 2, 3, 4, 5\}$.

1. Is $5 \in D$? Is $5 \in F$?
2. Is $D \subseteq F$?
3. What is $D \cap F$?
4. What is $D \cup F$?
5. What is $\overline{D}$?
6. What are $|D|$ and $|F|$?
7. Find $D - F$ and $F - D$
8. Are D and F disjoint?

Note: Quick check solutions are available at the end of the chapter. However, you will short-circuit the learning process if you look at the answer before you have completed the problems on your own. ☑

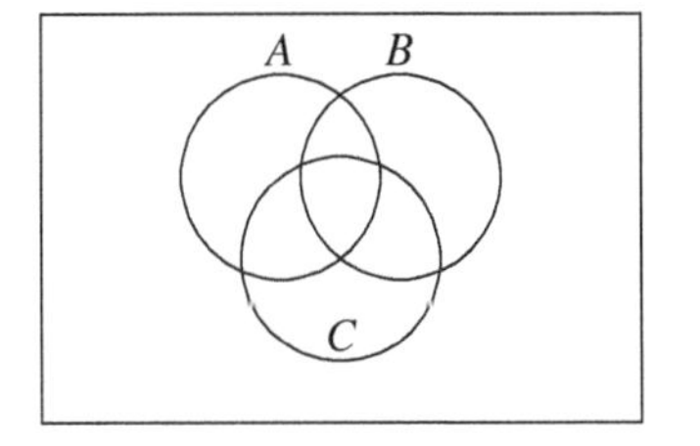

Figure 2.9. A Venn diagram for three sets.

Multiple Sets

The Venn diagram in Figure 2.9 shows the most general relationship between three sets.[7] Imagine the sets representing the three groups of people, "students", "renters", and "waitresses". Some people may be members of all three groups. Some people may be in none of the groups. Others will be in one or two of the groups. The diagram contains regions to represent all possibilities.

It should be apparent that the notions of union and intersection can easily be generalized to include multiple sets. The definitions introduce the concept of an index set. An *index set* serves as a collection of subscripts used to distinguish members from a large collection of other sets. For example, the collection, $\{S_1, S_2, S_3, S_4\}$, of four sets has index set $\{1, 2, 3, 4\}$. More commonly, the index set might be an infinite set such as the set of all positive integers or the set of all real numbers.

Index sets indicate just how deeply the notion of "set" permeates mathematics; these sets are used to describe collections whose members are other sets.

[7]When many more than three sets are being considered, Venn diagrams that show all potential relationships among the sets are difficult to draw.

DEFINITION 2.12 ***Union and Intersection of Multiple Sets***

Let $\{S_i \mid i \in \Upsilon\}$ be a collection of sets with index set Υ. Their *union* is the set

$$\cup_{i \in \Upsilon} S_i = \{s \mid s \in S_j \text{ for some } j \in \Upsilon\}$$

Their *intersection* is the set

$$\cap_{i \in \Upsilon} S_i = \{s \mid s \in S_j \text{ for every } j \in \Upsilon\}$$

Notice that the only difference between the two definitions is the use of the words *some* and *every*.

The Venn diagrams in Figures 2.10 and 2.11 illustrate these definitions.

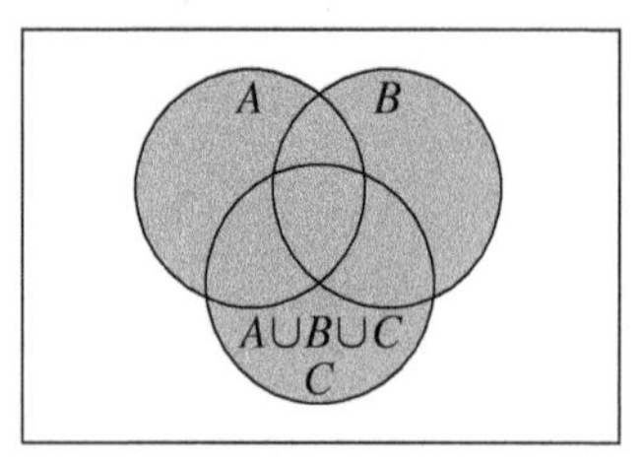

Figure 2.10 The union of three sets.

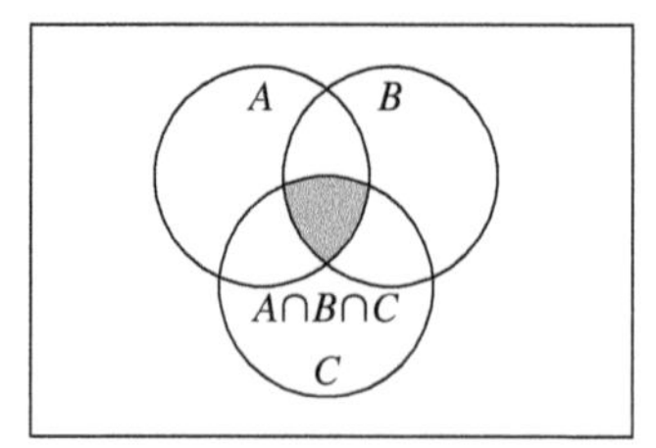

Figure 2.11 The intersection of three sets.

There is an important extension of the concept of disjoint sets.

DEFINITION 2.13 ***Partition***

A collection, $\{S_i \mid i \in \Upsilon\}$, of nonempty subsets of a set S, is said to be a *partition* of S if

- $S = \cup_{i \in \Upsilon} S_i$.
- $S_i \cap S_j = \emptyset$ whenever $i \neq j$.

Notice that every element of S is in exactly one of the subsets, S_i. The set S is a disjoint union of the subsets in the partition.

EXAMPLE 2.10

Some Partitions

The set of even integers and the set of odd integers form a partition of the integers.

For a more interesting example, consider the subsets of the nonnegative integers, W, that are formed by considering each nonnegative integer's remainder when divided by 4. If numbers with the same remainder are placed in the same subset, then there will be four subsets:

$$R_0 = \{0, 4, 8, 12, 16, 20, \dots\}$$
$$R_1 = \{1, 5, 9, 13, 17, 21, \dots\}$$
$$R_2 = \{2, 6, 10, 14, 18, 22, \dots\}$$
$$R_3 = \{3, 7, 11, 15, 19, 23, \dots\}$$

Since $W = R_0 \cup R_1 \cup R_2 \cup R_3$ and $R_i \cap R_j = \emptyset$ whenever $i \neq j$, the collection $\{R_0, R_1, R_2, R_3\}$ forms a partition of W. ■

New Sets from Old

There are additional techniques for creating new sets from existing sets. Two are presented here. In particular, new sets can be created by forming Cartesian products and power sets.

The inspiration for the next definition is the familiar notion of ordered pairs in the real plane. That idea is generalized into ordered triples in 3-space. The Cartesian product, defined next, will generalize this idea to ordered collections (called n-tuples) containing n items.

DEFINITION 2.14 ***Cartesian Product***

Let $\{S_i \mid i = 1, 2, \ldots, n\}$ be a finite collection of two or more sets. The *Cartesian product* of the collection, denoted $S_1 \times S_2 \times S_3 \times \cdots \times S_n$, is the set of all ordered n-tuples with coordinate a_i a member of S_i. That is,

$$S_1 \times S_2 \times S_3 \times \cdots \times S_n = \{(a_1, a_2, \ldots, a_n) \mid a_i \in S_i\}$$

The Cartesian product $\overbrace{S \times S \times \cdots \times S}^{n \text{ times}}$ is often abbreviated as S^n.

Note that there are finitely many sets in the Cartesian product, but each of those sets might be an infinite set. For example, S_1 might be the integers, and S_2 might be the real numbers. If any of the sets, S_i, in the collection is an infinite set, then $S_1 \times S_2 \times S_3 \times \cdots \times S_n$ will also be an infinite set unless $S_i = \emptyset$ for some $i \in \{1, 2, \ldots, n\}$. (Suppose S_1 is infinite. Then there will be an infinite number of distinct elements as first coordinates of the ordered pairs in $S_1 \times S_2 \times S_3 \times \cdots \times S_n$.)

Once the Independent Tasks Principle (page 227) has been understood, it is easy to see that if all of the S_i are finite, then so is $S_1 \times S_2 \times S_3 \times \cdots \times S_n$. In fact, if $|S_i| = k_i$ for $i = 1, 2, \ldots, n$, then $|S_1 \times S_2 \times S_3 \times \cdots \times S_n| = k_1 k_2 \cdots k_n$.

EXAMPLE 2.11 **Some Cartesian Products**

Let $A = \{a, b\}$ and $X = \{x, y\}$. Then $A \times X = \{(a, x), (a, y), (b, x), (b, y)\}$.

If $A = \{a, b, c\}$ and $X = \{x, y\}$, then $A \times X = \{(a, x), (a, y), (b, x), (b, y), (c, x), (c, y)\}$.

If $A = \{a, b\}$ and $X = \{x, y, z\}$, then $A \times X = \{(a, x), (a, y), (a, z), (b, x), (b, y), (b, z)\}$. ■

DEFINITION 2.15 ***Power Set***

Let S be a set. The *power set of S*, denoted $\mathcal{P}(S)$, is the set of all subsets of S (including the empty set and S itself).

If S is finite, it is not difficult to show that $\mathcal{P}(S)$ is also finite. More precisely, Corollary 5.12 on page 249 asserts that if $|S| = n$, then $|\mathcal{P}(S)| = 2^n$. The next example illustrates these ideas.

EXAMPLE 2.12 **A Power Set**

Let $S = \{a, b, c\}$. Then $\mathcal{P}(S) = \{\emptyset, \{a\}, \{b\}, \{c\}, \{a, b\}, \{a, c\}, \{b, c\}, \{a, b, c\}\}$. Notice that $|S| = 3$ and $|\mathcal{P}(S)| = 2^3 = 8$. ■

If S is an infinite set, it is possible (in more advanced courses) to show that $|\mathcal{P}(S)| > |S|$ is always true.[8]

---

[8]Consequently, there is an infinite collection of cardinalities of infinite sets.

✔ Quick Check 2.2

Let $A = \{1, 2, 3\}$, $B = \{2, 3, 4\}$, and $C = \{3, 4, 5\}$.

1. What is $A \cap B \cap C$?
2. What is $A \cup B \cup C$?
3. Is $\{A, B, C\}$ a partition of $\{1, 2, 3, 4, 5\}$?
4. What is $A \times B$?
5. What is $|A \times B \times C|$?
6. What is $\mathcal{P}(A \cap B)$? ☑

More Venn Diagrams

It is possible to use Venn diagrams to aid our intuition about more complex set relationships. However, they may easily mislead us by causing us to focus on just one or two of the possible arrangements of circles and ellipses. For example, consider the claim "$(B \cap C) \subseteq (A \cap C)$ is true only when $B \subseteq A$". It is tempting to investigate this claim by comparing a pair of Venn diagrams. The key idea is to use a collection of parallel lines to shade regions. You might shade $B \cap C$ with lines having positive slope and $A \cap C$ with lines having negative slope. Then determine whether the $B \cap C$ region is entirely within $A \cap C$. The Venn diagrams in Figure 2.12 are natural choices for this investigation.

The diagrams seem to support the claim. Why bother with a formal proof?[9] One good reason is that the claim is false. The Venn diagrams have lead us to an incorrect conclusion. (Very embarrassing!)

Here is an example that shows the claim is false: let $A = \{x, z\}$ $B = \{y, z\}$ $C = \{z\}$. Then $(B \cap C) \subseteq (A \cap C)$ is true, but $B \not\subseteq A$.

The Venn diagram in Figure 2.13 correctly illustrates this example.

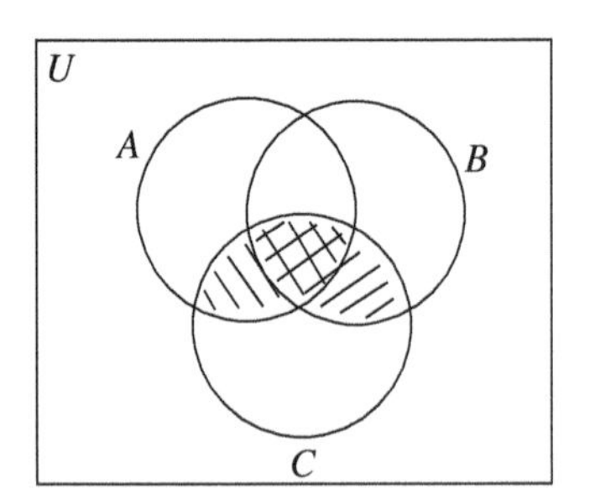

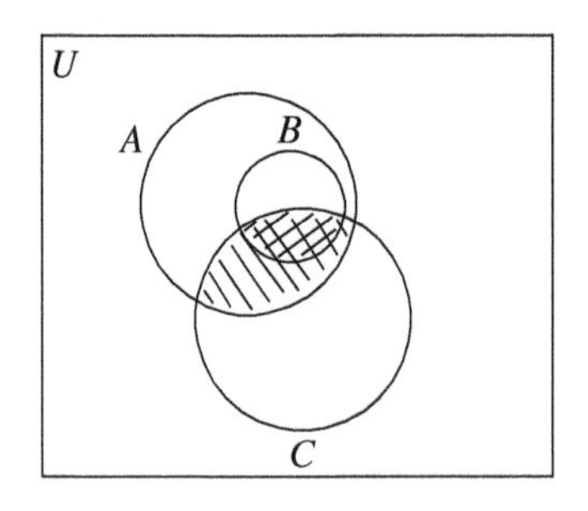

Figure 2.12 Two Venn diagrams to investigate the claim "$(B \cap C) \subseteq (A \cap C)$ is true only when $B \subseteq A$".

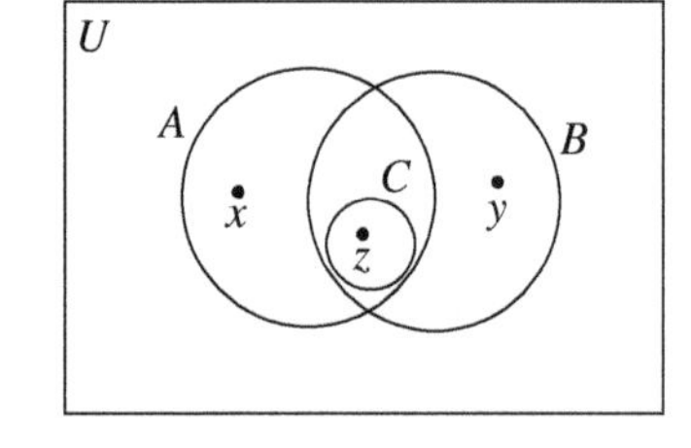

Figure 2.13 A better Venn diagram for the claim "$(B \cap C) \subseteq (A \cap C)$ is true only when $B \subseteq A$".

2.1.2 Exercises

The exercises marked with ★ have detailed solutions in Appendix H. Appendix G contains suggestions to help you submit properly written solutions to homework exercises.

1. Let $A = \{x \mid \cos(x) = 0\}$ and $B = \{x \mid \sin(x) = 1\}$. Prove that $B \subsetneq A$. That is, show that $B \subset A$ but $B \neq A$.

2. Let the universal set be $\{a, b, c, d, e, 1, 2, 3, 4, 5\}$. For each pair of sets, A and B, create a table containing the following information.

Is $d \in A$?	Is $A \subseteq B$?	Is $B \subseteq A$?	Is $A = B$?
Are A and B disjoint?	$\|A\|$	$\|B\|$	$\overline{A}$
$A \cup B$	$A \cap B$	$A - B$	$B - A$

(a) ★ $A = \{a, b, c, 1, 2, 3\}$ and $B = \{c, d, 1, 3\}$.
(b) ★ $A = \{a, c, 1, 2, 3\}$ and $B = \{a, c, d, 1, 2, 3, 4\}$.
(c) $A = \{b, c, e, 4, 2, 1\}$ and $B = \{c, e, b, 1, 2, 4\}$.
(d) $A = \{a, b, d, 1, 4, 5\}$ and $B = \{c, e, 2, 3\}$.
(e) $A = \{a, c, d, 1, 2, 3\}$ and $B = \{c, d, 1, 3\}$.
(f) $A = \{a, b, 1, 2\}$ and $B = \emptyset$.

3. Let P be the set of positive integers and N be the set of negative integers. Is $\{P, N\}$ a partition of $\mathbb{Z}$?

4. Let $X = \{a, c, d, e\}$, $Y = \{b, c, d, g\}$, and $Z = \{a, b, d, f\}$. What are $(X \cap Y) - Z$ and $(X \cup Y) - Z$?

5. Let $U = \{1, 2, 3, 4, 5, 6, 7, 8, 9\}$, $A = \{1, 2, 3, 4, 5\}$ and $B = \{2, 4, 6, 8\}$. What are $\overline{A \cup B}$ and $\overline{A \cap B}$?

[9] Formal proofs will be introduced in Section 2.1.3 on page 29.

6. Let the universal set contain abbreviations of the states in the 13 original colonies: $\{CT, DE, GA, MA, MD, NC, NH, NJ, NY, PA, RI, SC, VA\}$. For each pair of sets, X and Y, create a table containing the following information.

Is $CT \in X$?	Is $X \subseteq Y$?	Is $Y \subseteq X$?	Is $X = Y$?
Are X and Y disjoint?	$\|X\|$	$\|Y\|$	$\overline{X}$
$X \cup Y$	$X \cap Y$	$X - Y$	$Y - X$

(a) $X = \{CT, DE, RI, VA\}$ and $Y = \{CT, VA, MA, SC, NH, RI\}$.

(b) ★ $X = \emptyset$ and $Y = \{NJ, NY, NC, NH\}$.

(c) $X = \{NJ, NY, PA, SC, RI, VA\}$ and $Y = \{NC, GA, DE, MA, CT\}$.

(d) $X = \{MD, MA, SC, CT, NC\}$ and $Y = \{MA, NC, MD, SC\}$.

(e) $X = \{VA, NJ, PA, NY, DE, GA\}$ and $Y = \{NY, NJ, VA, GA, PA, DE\}$.

(f) $X = \{CT, VA, MA, NJ, PA, GA, DE\}$ and $Y = \{PA, VA, NJ, SC\}$.

7. Convert the following sets from set-builder notation to a simple enumeration of the elements, using the "three dots" convention when necessary.

(a) $\{x \in \mathbb{Z} \mid -2 \leq x < 10\}$

(b) $\{x \in \mathbb{Z} \mid x$ is evenly divisible by $5\}$

(c) ★ $\{n \in \mathbb{Z} \mid n = 3k$ for some $k \in \mathbb{Z}\}$

(d) ★ $\{(x, y) \mid x, y \in \mathbb{Z}$ and $x^2 + y^2 \leq 4\}$

(e) {months in the Gregorian calendar | there is an "r" in the month name}

(f) $\{(x, y) \mid x, y \in \mathbb{Z}, 1 \leq x \leq 6,\ 1 \leq y \leq 6$, and $x + y = 8\}$

8. Convert the following into set-builder notation.

(a) $\{-2, -1, 0, 1, 2\}$

(b) ★ $\{\ldots, -6, -4, -2, 0, 2, 4, 6, \ldots\}$

(c) $\{\ldots, (-3, 9), (-2, 4), (-1, 1), (0, 0), (1, 1), (2, 4), (3, 9), \ldots\}$

(d) $\{5, 6, 7, 8, 9, 10, \ldots\}$

(e) {Monday, Tuesday, Wednesday, Thursday, Friday}

9. Convert the following into set-builder notation.

(a) $\{\ldots, -125, -64, -27, -8, -1, 0, 1, 8, 27, 64, 125, \ldots\}$

(b) {a, e, i, o, u}

(c) $\{0, 1, 16, 81, 256\}$

(d) ★ $\{\ldots, -12, -6, 0, 6, 12, 18\}$

(e) $\{\ldots, (-9, -3), (-6, -2), (-3, -1), (3, 1), (6, 2), (9, 3), \ldots\}$

10. Use English sentences to describe the following sets.

(a) ★ $\{\ldots, -21, -15, -9, -3, 3, 9, 15, 21, \ldots\}$

(b) $\{x \mid x \in \mathbb{R}$ and $\cos(x) = 1\}$

(c) $\{\frac{n}{d} \in \mathbb{Q} \mid d = 5k$ for some $k \in \mathbb{Z}, k \neq 0\}$

11. Construct a properly labeled Venn diagram for each of the following triples of subsets from $U = \{a, b, c, d, e, f, g\}$. Place each element in the proper region of the diagram. Then complete the additional parts of the problem.

(a) ★ $A = \{a, b, c\}, B = \{b, c, f, g\}, C = \{b, e, f\}$

i. Shade in the region $(A \cap B) - C$.

ii. List the sets $(A \cap C) - B$ and $A \cup (B \cap C)$.

(b) $A = \{a, b, c, e\}, B = \{c, d, g\}, C = \{b, e, f, g\}$

i. Shade in the region $A \cap B \cap C$.

ii. List the sets $A \cup C$ and $C - (A \cup B)$.

(c) $A = \{a, b, c, d\}, B = \{a, b, c\}, C = \{a, d, e, f\}$

i. Shade in the region $(A - B) \cap C$.

ii. List the sets $\overline{B}$ and $\overline{A \cup C}$.

12. Construct a properly labeled Venn diagram for each of the following triples of subsets from $U = \{w, x, y, z, 1, 2, 3, 4\}$. Place each element in the proper region of the diagram. Then complete the additional parts of the problem.

(a) ★ $A = \{w, x, y, z\}, B = \{1, 2, 3, 4, z\}, C = \{w, x, 1\}$

i. What should be the proper value for $B - C - A$? (Be careful!)

ii. List the sets $\overline{A \cap B}$ and $(A \cap C) \cup B$.

(b) $A = \{w, 1, 3\}, B = \{w, y, 2, 3\}, C = \{y, z, 3\}$

i. Shade in the region $\overline{A} \cap (B - C)$.

ii. List the sets $A \cup B \cup C$ and $B \cap \overline{(C - A)}$.

(c) $A = \{z, 2, 4\}, B = \{w, x, y, 2\}, C = \{x, 1, 2, 4\}$

i. Shade in the region $(A \cup B) - C$.

ii. List the sets $\overline{A} - C$ and $\overline{C} - B$.

13. Let U = {undergraduate college students at a four-year college}, A = {all freshmen}, B = {all female students}, C = {all math majors}. Draw a properly labeled Venn diagram. Then shade in the region $\overline{A} - (B \cup C)$. Describe the shaded region.

> **DEFINITION 2.16** ***Symmetric Difference***
>
> The *symmetric difference* of the sets A and B is denoted $A \triangle B$ and is defined by
>
> $$A \triangle B = (A - B) \cup (B - A)$$

14. Find the symmetric difference for each of the following pairs of sets. (The definition precedes this problem.)

(a) ★ $A = \{1, 2, 4, 5\}, B = \{1, 3, 5\}$.

(b) $A = \{1, 2, 3\}, B = \{1, 2, 3\}$.

(c) $A = \{2, 4, 6\}, B = \{1, 3, 5\}$.

(d) ★ $A = \{1, 2, 3, 4, 5\}, B = \{1, 3, 5\}$.

(e) $A = \{1, 4, 5\}, B = \{1, 2, 3, 5\}$.

15. Find the symmetric difference for each of the following pairs of sets. (The definition precedes Exercise 14.)

(a) $A = \{a, b, c, f, g, h\}, B = \{a, f, g, h\}$.

(b) $A = \emptyset, B = \{c, d, g, h, i\}$.

(c) $A = \{b, c, e, f\}, B = \{a, d, e, f, h, i\}$.

(d) $A = \{a, c, e, g, i\}, B = \{a, c, e, g, h\}$.

16. Form the indicated Cartesian products.

(a) ★ $A \times B$, where $A = \{a, b\}$, $B = \{1, 2, 3, 4\}$.

(b) ★ $A \times B \times C$, where $A = \{a, b\}$, $B = \{1, 2\}$, $C = \{\alpha, \beta, \gamma\}$.

(c) $B \times A$, where $A = \{a, b\}$, $B = \{1, 2, 3, 4\}$.

(d) $C \times A \times B$, where $A = \{a, b\}$, $B = \{1, 2\}$, $C = \{\alpha, \beta, \gamma\}$.

(e) $A \times B \times C$, where $A = \{a, b, c\}$, $B = \{1, 2\}$, $C = \{\alpha, \beta, \gamma\}$.

(f) $A \times B$, where $A = \{(a, x), (a, y), (b, x), (b, y)\}$, $B = \{1, 2\}$.

17. Form the indicated Cartesian products.

(a) $A \times B$, where $A = \{a, b, c\}$, $B = \{w, x, y, z\}$.

(b) $B \times A \times C$, where $A = \{a, b, c\}$, $B = \{x, y, z\}$, $C = \{\alpha, \beta, \gamma\}$.

(c) $C \times A \times B \times D$, where $A = \{a, b\}$, $B = \{1, 2\}$, $C = \{\alpha, \beta\}$, $D = \{3, 4\}$.

18. Form the power set.

(a) ★ $A = \emptyset$.

(b) $A = \{\alpha\}$.

(c) ★ $A = \{\alpha, \beta\}$.

(d) $A = \{\alpha, \beta, \gamma\}$.

(e) $A = \{\alpha, \beta, \gamma, \delta\}$.

19. Use two Venn diagrams to give a visual indication that, in general, $[A \cup (B \cap C)] \neq [(A \cup B) \cap C]$.

20. Use Venn diagrams to give a visual indication for the following relationships.

(a) $A \cup (B \cap C) = (A \cup B) \cap (A \cup C)$.

(b) If $C \subseteq B \subseteq A$, then $A - (B - C) = (A - B) \cup C$.

(c) ★ $(A - B) \cup (A - C) = A - (B \cap C)$.

(d) $(A \cup B) - (A \cap B) = A \triangle B$ (see Definition 2.16).

21. Design a Venn diagram that shows the most general relationships among 4 sets. [*Hints*: Use ellipses instead of circles. Including the area outside the four ellipses, there should be 16 regions.]

22. For each claim, determine whether it is always true or else false in some cases. Then give some justification for your answer.

(a) If $A \subseteq B$, then $A \neq B$.

(b) If $A \not\subseteq B$, then there is at least one element, x, in A such that $x \notin B$.

(c) The empty set is an element of every set.

(d) ★ $|\{\emptyset\}| = 1$ but $|\emptyset| = 0$.

(e) ★ $\overline{\overline{A}} = \overline{(\overline{A})} = A$ (The complement of the complement of A equals A.)

(f) $\overline{U} = \emptyset$.

(g) For all sets, A and B, $(A - B) = (B - A)$.

(h) If A and B are disjoint, and if $C \subseteq A$, then B and C are disjoint.

(i) For all sets, A and B, $A \times B = B \times A$.

23. For each claim, determine whether it is always true or else false in some cases. Then give some justification for your answer.

(a) If A and B are sets, then $A - B \subseteq A$.

(b) $\mathcal{P}(\emptyset) = \mathcal{P}(\{\emptyset\})$.

(c) The empty set has no complement.

(d) ★ If $|A| = |B| = n$ for some positive integer, n, and $A \subseteq B$, then $A = B$.

(e) ★ For any set, A, $\emptyset \times A = A \times \emptyset = \emptyset$.

(f) If A is a set, then $A - \overline{A} = A$.

(g) If $A \subseteq B$ and $B \subseteq C$, then $A \subset C$.

(h) Suppose that A is any set and $B = \{A, \{A\}\}$. Then $A \subseteq B$.

24. Complete the following sentences.

(a) ★ If $(A \cap B) = (A \cup B)$, then ....

(b) If $(A - B) = (B - A)$, then ....

(c) If $A - B = A$, then ....

(d) If $A - B = \emptyset$, then ....

(e) If $|\mathcal{P}(S)| = 32$, then $|S| = ...$

25. Complete the following sentences.

(a) If $A \subseteq B$ and $B \not\subseteq A$, then $A - B = ...$

(b) If $A \times B = B \times A$, then ....

(c) If $\overline{B} \subseteq \overline{A}$, then ....

(d) If $A \triangle B = A$, then .... (See Definition 2.16.)

(e) If $A \subseteq B$ and $C \subseteq D$, then ... $A \cup C$ ... and $A \cap C$ ....

26. List all partitions of each of the following sets.

(a) ★ $S = \{\text{Fall, Spring, Summer, Winter}\}$.

(b) $S = \{1, 3, 5\}$.

(c) $S = \{a, b, c, d\}$.

(d) $S = \{\text{Apple, Banana}\}$.

27. Which of the following collections of subsets form a partition of the set of integers? Provide some justification for your answer.

(a) ★ The set of positive integers and the set of negative integers.

(b) The set of integers less than -200, the set of integers greater than 200, and the set of integers with absolute value not exceeding 200.

(c) The set of integers evenly divisible by 3, the set of integers leaving a remainder of 1 when divided by 3, and the set of integers leaving a remainder of 2 when divided by 3.

(d) The set of integers not evenly divisible by 2, the set of even integers, and the set of integers leaving a remainder of 3 when divided by 6.

(e) The set of odd integers and the set of integers divisible by 4.

2.1.3 Proofs about Sets

In Section 2.1.1 (page 26), Venn diagrams were used to give an intuitive justification for the incorrect claim "$(B \cap C) \subseteq (A \cap C)$ is true only when $B \subseteq A$". This intuitive justification failed because it depended on the manner in which the Venn diagrams were composed. A more thorough justification would require additional diagrams to be produced. For example, several diagrams with $A \subseteq C$, others with $A \subseteq B$, some with $A \cap C = \emptyset$, a few with both $C \subseteq A$ and $C \subseteq B$, and so on. This large collection of possible diagrams is really not the proper way to formally justify the claim.

What is needed is a formal proof. Although a more detailed discussion of proofs and proof strategies will not occur until Chapter 3, it is appropriate to mention here some strategies for formal proofs about sets.

Proofs about Membership

If a set is specified using an exhaustive listing of all the elements, it is trivial to determine whether or not something is an element of the set. For example, a is clearly an element of $\{a, b, c, d\}$, but x is not.

If the set is too large to list exhaustively, it is often still fairly easy to determine set membership. Consider, for example, the set $S = \{n \in \mathbb{Z} \mid n = 6k \text{ for some } k \in \mathbb{Z}\}$. It is easy to see that $828 \in S$ because $828 = 6 \cdot 138$. It is also not difficult to see that $35 \notin S$ because 35 is not a multiple of 6.

A common strategy for showing membership in sets specified using set-builder notation is to verify that a candidate for membership actually satisfies all the requirements (as was done to show $828 \in S$). This strategy is not always easy.

EXAMPLE 2.13 **Irrational Numbers**

Appendix A.3 (page A2) introduces the set of rational numbers, $\mathbb{Q}$, and the set of irrational numbers. These two sets form a partition of the real numbers. Thus, any real number is in exactly one of these two sets.

It is trivial to show that any fraction is in the set of rationals (look at the definition of $\mathbb{Q}$). It is not difficult to show that any repeating decimal (such as $12.\overline{145}$) is a member of $\mathbb{Q}$. There is even a simple algorithm to convert the repeating decimal into an equivalent fraction.[10]

It is also only moderately challenging to show that $\sqrt{2}$ is an irrational number.[11] Proving that π and e are irrational numbers is much harder. ■

Proofs about Subsets and Equality

The most direct strategy for proving that one set, A, is a subset of another set, B, is to simply verify the conditions presented in Definition 2.2. That is, consider a generic[12] element in A and then demonstrate that it is also an element in B. If this can be demonstrated, then clearly every element of A is an element of B and the definition implies that $A \subseteq B$.

A proof that $A = B$ can be achieved by proving that $A \subseteq B$ and also that $B \subseteq A$. This might be done in two separate cases (using the strategy presented in the previous paragraph), or it might be done using a sequence of equivalent statements: $x \in A$ if and only if ... if and only if $x \in B$.

---

[10] For example, $1000 \cdot 12.\overline{145} - 12.\overline{145} = 12133$, So $12.\overline{145} = \frac{12133}{999}$.

[11] See Proposition 3.35 on page 104.

[12] A generic element is one that represents any (every) possible element in the set. The use of generic elements is sometimes called the *choose method.* Additional details can be found on page 106.

EXAMPLE 2.14 **Two Subsets of the Positive Integers**

Let T be the set of all nonnegative powers of 2, and let F be the set of all nonnegative powers of 4. That is, $T = \{2^n \mid n \geq 0\}$ and $F = \{4^k \mid k \geq 0\}$. Then $F \subseteq T$.

Proof: Let $x \in F$ be a typical element. Then we know there is some positive exponent, m, such that $x = 4^m$. But that means $x = 4^m = (2^2)^m = 2^{2m}$, so x is a nonnegative power of 2 and hence $x \in T$. Therefore, every element of F is also an element of T. Definition 2.2 implies that $F \subseteq T$. □

■

This strategy will also work in many situations where the detailed descriptions of the sets are not available. This will be illustrated by proving several propositions[13] of a more general nature.

PROPOSITION 2.17
Let A and B be sets. Then $A \cap B \subseteq A$.

Proof: Let $x \in A \cap B$. Then (using the definition of *set intersection* on page 22) we know that $x \in A$ and $x \in B$. In particular, the claim that $x \in A$ establishes the conclusion of the proposition. That is, any element of the left-hand set, $A \cap B$, must also be an element of the right-hand set, A. □

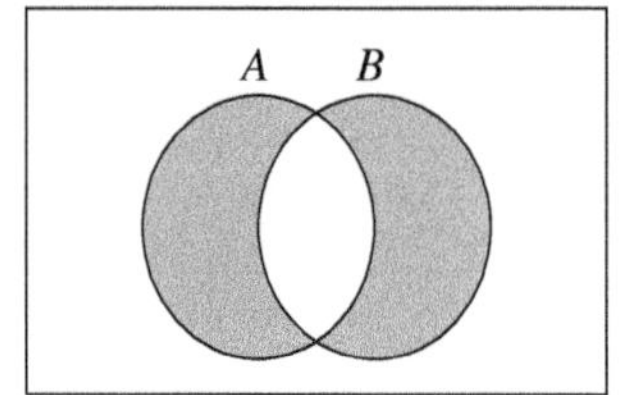

Figure 2.14.
$(A - B) \cup (B - A)$
$= (A \cup B) - (A \cap B)$.

The next proof uses the first strategy for proving set equality.

PROPOSITION 2.18
Let A and B be sets. Then

$$(A - B) \cup (B - A) = (A \cup B) - (A \cap B)$$

The Venn diagram in Figure 2.14 shows that this proposition has strong intuitive justification. The formal proof follows.

Proof of Proposition 2.18:

Phase 1: $(A - B) \cup (B - A) \subseteq (A \cup B) - (A \cap B)$

Suppose $x \in (A - B) \cup (B - A)$. Then either $x \in (A - B)$ or $x \in (B - A)$ (or it is in both sets). Assume for the moment that $x \in (A - B)$ (the other case can be completed in a similar fashion). Then x is in A, but is *not* in B. So $x \notin (A \cap B)$. Since $x \in A$, it is true that $x \in (A \cup B)$. Combining $x \in (A \cup B)$ and $x \notin (A \cap B)$ leads to the assertion that $x \in (A \cup B) - (A \cap B)$. Because every element in $(A - B) \cup (B - A)$ is also in $(A \cup B) - (A \cap B)$, Definition 2.2 on page 20 implies that
$(A - B) \cup (B - A) \subseteq (A \cup B) - (A \cap B)$.

Phase 2: $(A \cup B) - (A \cap B) \subseteq (A - B) \cup (B - A)$

Suppose $y \in (A \cup B) - (A \cap B)$. Then y is *not* in both A and B, but it *is* in one of them. Assume for the moment that it is in B (the other case has a similar proof). Then $y \in B$ but $y \notin A$. So $y \in (B - A)$. This implies that $y \in (A - B) \cup (B - A)$. Consequently, $(A \cup B) - (A \cap B) \subseteq (A - B) \cup (B - A)$.

Conclusion:
Since $(A - B) \cup (B - A) \subseteq (A \cup B) - (A \cap B)$ and also $(A \cup B) - (A \cap B) \subseteq (A - B) \cup (B - A)$, it must be true that $(A - B) \cup (B - A) = (A \cup B) - (A \cap B)$. □

[13]The term *proposition* is similar in meaning to *theorem.* It is a mathematical claim of some generality and interest. See Section 3.1.2 on page 87 for some additional discussion about this terminology.

The next proposition shows how to use a sequence of reversible assertions to prove a set equality.

> **PROPOSITION 2.19** ***A De Morgan Law for Sets***
> Let A and B be sets. Then
> $$\overline{A \cup B} = \overline{A} \cap \overline{B}$$

Proof: The proof uses several of the previous definitions. The phrase "if and only if" will be abbreviated to "iff".

$x \in \overline{A \cup B}$	iff	$x \notin A \cup B$	definition of *complement*
	iff	$(x \notin A)$ and $(x \notin B)$	definition of *union*
	iff	$\left(x \in \overline{A}\right)$ and $\left(x \in \overline{B}\right)$	definition of *complement* (twice)
	iff	$x \in \overline{A} \cap \overline{B}$	definition of *intersection*

Since every element of $\overline{A \cup B}$ is an element of $\overline{A} \cap \overline{B}$, and vice versa, the two sets are equal.[14] □

Here is another useful identity. It shows how to replace *set difference* with *intersection* and *complement*.

> **PROPOSITION 2.20** ***Eliminating Set Difference***
> Let A and B be sets. Then $A - B = A \cap \overline{B}$.

Proof:

$x \in (A - B)$	iff	$(x \in A)$ and $(x \notin B)$	definition of *set difference*
	iff	$(x \in A)$ and $(x \in \overline{B})$	definition of *complement*
	iff	$x \in (A \cap \overline{B})$	definition of *intersection*

□

✔ Quick Check 2.3

1. Suppose that $C \subseteq B$ and also that $B \subseteq A$. Provide a formal proof that $C \subseteq A$.
2. Suppose $B \subseteq A$. Use a sequence of "if and only if" statements to prove $\overline{A - B} = (\overline{A} \cup B)$. ☑

A proof that uses a sequence of assertions can take advantage of other theorems and propositions that have been previously proved.

EXAMPLE 2.15

Using Previous Propositions

It is easy to use Propositions 2.18 and 2.19 to show that

$$\overline{(A \cup B) - (A \cap B)} = \overline{(A - B)} \cap \overline{(B - A)}$$

Proof:

$\overline{(A \cup B) - (A \cap B)}$	$=$	$\overline{(A - B) \cup (B - A)}$	Proposition 2.18
	$=$	$\overline{(A - B)} \cap \overline{(B - A)}$	Proposition 2.19

□

■

Notice that the previous proof did not discuss elements (as was done in the proofs of Propositions 2.19 and 2.20). Proofs that use previously proved properties often may be done without referring to elements. This process is made much easier if you know some fundamental relationships. Table 2.1 summarizes some of them.[15]

---

[14] Step 2, which appeals to the definition of union, actually uses one of the De Morgan's laws from propositional logic (page 53). For now, a Venn diagram will show that the claim is reasonable.

[15] The table is also on page 83 and at http://www.mathcs.bethel.edu/~gossett/DiscreteMathWithProof/ .

TABLE 2.1 Fundamental Set Properties

Idempotence	**Domination**
$A \cup A = A$	$A \cup U = U$
$A \cap A = A$	$A \cap \emptyset = \emptyset$
Associativity	**Identity**
$(A \cup B) \cup C = A \cup (B \cup C)$	$A \cup \emptyset = A$
$(A \cap B) \cap C = A \cap (B \cap C)$	$A \cap U = A$
Commutativity	**De Morgan's Laws**
$A \cup B = B \cup A$	$\overline{A \cup B} = \overline{A} \cap \overline{B}$
$A \cap B = B \cap A$	$\overline{A \cap B} = \overline{A} \cup \overline{B}$
Distributivity ($\cap$ **over** $\cup$)	**Distributivity** ($\cup$ **over** $\cap$)
$A \cap (B \cup C) = (A \cap B) \cup (A \cap C)$	$A \cup (B \cap C) = (A \cup B) \cap (A \cup C)$
$(A \cup B) \cap C = (A \cap C) \cup (B \cap C)$	$(A \cap B) \cup C = (A \cup C) \cap (B \cup C)$
Complement	**Complement (continued)**
$A \cup \overline{A} = U$	$\overline{\emptyset} = U$
$A \cap \overline{A} = \emptyset$	$\overline{U} = \emptyset$
Involution	
$\overline{\overline{A}} = A$	

EXAMPLE 2.16

Using the Fundamental Properties

A previous Quick Check problem asked you to prove that if $B \subseteq A$, then $\overline{A - B} = (\overline{A} \cup B)$. The fundamental properties can be used to prove this is true even if $B \not\subseteq A$. The proof uses a sequence of equalities (rather than a sequence of "if and only if" assertions).

$$\begin{aligned} \overline{A - B} &= \overline{A \cap \overline{B}} && \text{Proposition 2.20} \\ &= \overline{A} \cup \overline{\overline{B}} && \text{De Morgan} \\ &= \overline{A} \cup B && \text{involution} \end{aligned}$$

■

One important idea that some students miss is that the fundamental set properties are valid in both directions. In particular, the distributive properties and De Morgan's laws are often useful in the right-to-left direction. Learn the properties in both directions.

EXAMPLE 2.17

Reversing the Distributive Property

Even with a Venn diagram it takes some effort to see that $(A \cap B) \cup \overline{(A \cup \overline{B})} = B$. Here is a formal proof.

$$\begin{aligned} (A \cap B) \cup \overline{(A \cup \overline{B})} &= (A \cap B) \cup (\overline{A} \cap \overline{\overline{B}}) && \text{De Morgan} \\ &= (A \cap B) \cup (\overline{A} \cap B) && \text{involution} \\ &= (A \cup \overline{A}) \cap B && \text{distributive (in reverse)} \\ &= U \cap B && \text{complement} \\ &= B \cap U && \text{commutativity} \\ &= B && \text{identity} \end{aligned}$$

The second–to–last step used commutativity so that the universal set, U, is on the right in order to exactly match the identity property. Many students have been using the commutativity of the real numbers for so long that they tend to do this mentally, thus forgetting that step in their proofs. Make an effort to become conscious of your use of commutativity (it will also be helpful in several future mathematics classes).[16]

Producing a proof that uses a sequence of equalities requires some effort. It will become easier with experience, but as you are learning you should expect to run into some dead ends. Expect to consume lots of scratch paper during the process. ■

Counterexamples

A counterexample is an example that shows an assertion to be false.

EXAMPLE 2.18

A False Assertion

It is easy to show that the *associativity* and *commutativity* properties cannot be used arbitrarily. Consider, for example, the assertion "$(A \cap B) - C = A \cap (C - B)$". The Venn diagram in Figure 2.15 indicates that this should not always be true.

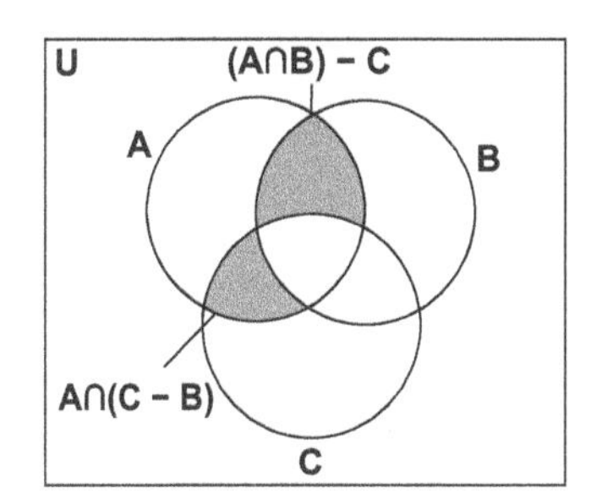

Figure 2.15 The assertion "$(A \cap B) - C = A \cap (C - B)$" is false.

A more formal approach to showing the assertion is false involves creating a specific counterexample. One possibility is to let $A = \{1, 2, 3, 4\}$, $B = \{2, 3, 6, 7\}$, and $C = \{3, 4, 5, 6\}$. Then $(A \cap B) - C = \{2\}$ but $A \cap (C - B) = \{4\}$. The sets are not equal, so the assertion is not always true. ■

2.1.4 Exercises

The exercises marked with ★ have detailed solutions in Appendix H.

1. Provide a formal proof that $A \cup (B - A) = A \cup B$. [*Hint*: Notice that A and $B - A$ are disjoint.]
2. ★ Suppose that $C \subseteq B \subseteq A$. Provide a formal proof that $C \subseteq (A - (B - C))$.
3. Suppose $B \subseteq A$. Provide a formal proof that $(B \cap C) \subseteq (A \cap C)$.
4. Formally prove that for any sets, A and B, the sets $A - B$, $B - A$, and $A \cap B$ form a partition of $A \cup B$. Do you need to make special cases for when either A or B is the empty set?
5. Suppose that $B \subseteq A$ and $C \subseteq A$. Provide a formal proof that $(B \cap C) \subseteq A$.
6. Find counterexamples for each of the following (incorrect) assertions.
   (a) $A - B = B - A$.
   (b) $(A \cap B) \cup C = (A \cap C) \cup B$.
   (c) ★ If $A \subseteq B$ and $A \cap C = \emptyset$, then $B \cap C = \emptyset$.
   (d) Set difference distributes over union: $A - (B \cup C) = (A - B) \cup (A - C)$.
   (e) Associativity of set difference: $(A - B) - C = A - (B - C)$.
   (f) Associativity of mixed union and intersection: $(A \cup B) \cap C = A \cup (B \cap C)$.
   (g) If $A \cap B = A \cap C$, then $B = C$.
7. Find counterexamples for each of the following (incorrect) assertions.
   (a) ★ If $A - C = A - B$, then $C = B$.
   (b) If $A \subseteq B$ and $A \cap C = D$ (with $D \neq \emptyset$), then $B \cap C = D$.
   (c) If $A \subseteq B$, then $\overline{B} \subset \overline{A}$.
   (d) If $A \cup B = A \cup C$, then $B = C$.
   (e) If $|A| = 1$, then $|A \cup B| = |B|$.
   (f) $A \subseteq A \cap B$.
   (g) If $A \subseteq B$, then $A \cup B \neq A \cap B$.
8. What are the most general conditions that make $(A \cap B) - C = A \cap (C - B)$ true? Prove your claim. (This is not a trivial task.)

[16] Of course, you could also prove a series of simple theorems that establish the validity of statements like $U \cap A = A$, but the discipline of explicitly using the commutativity property is probably better at this point in your study of mathematics.

9. What set is $\emptyset - \emptyset$? Use the fundamental set properties and propositions to prove your claim.

10. ★ Formally prove the fundamental set properties. Use sequences of "if and only if" assertions wherever possible.

11. Use the previously proved propositions, the fundamental set properties, and a sequence of equalities to prove the following relationships. Do not skip any steps.

(a) ★ $(A \cap B) - C = A \cap (B - C)$.

(b) ★ $\overline{A} \cap (A \cup B) = B - A$.

(c) $A - (A \cap B) = A \cap \overline{B}$.

(d) $\overline{A \cap (\overline{A} \cup B)} \cup B = U$.

(e) $\overline{(A \cap B) \cup C} = (\overline{A \cup C}) \cup (\overline{B \cup C})$.

12. Use the previously proved propositions, the fundamental set properties, and a sequence of equalities to prove the following relationships. Do not skip any steps.

(a) $\overline{A \cup (B \cap C)} = (\overline{C} \cup \overline{B}) \cap \overline{A}$.

(b) $\overline{A} \cup \overline{B} \cup \overline{C} = \overline{A \cap B \cap C}$.

(c) $(A - C) \cap (C - B) = \emptyset$.

(d) $(A \cap B) \cup (A \cap \overline{B}) = A$.

(e) $(A - C) - (B - C) = (A - B) - C$.

13. Provide formal proofs for the following relationships.

(a) If $C \subseteq A$, then $A - (B - C) = (A - B) \cup C$.

(b) $(A - B) \cup (A - C) = A - (B \cap C)$.

(c) The sets $(A \cap B) - C$ and $(A \cap C) - B$ are disjoint.

14. Provide formal proofs for the following relationships.

(a) ★ $A - \emptyset = A$ and $\emptyset - A = \emptyset$.

(b) ★ If A and B are finite sets with $\mathcal{P}(A) = \mathcal{P}(B)$, then $A = B$.

(c) $A - (A - B) \subseteq B$.

(d) $A - (A - B) = A \cap B$.

15. Use Proposition 2.20 and the fundamental set properties to prove the following set equalities.

(a) ★ $\overline{A} \cup B = \overline{A - B}$.

(b) $(A \cup B) - C = (A - C) \cup (B - C)$.

(c) $(A - B) - C = A - (B \cup C)$.

(d) $A - (B - C) = (A - B) \cup (A \cap C)$.

16. Find a simpler expression for the set $(A - B) - (B - A)$, then provide a formal proof that your expression represents the same set.

17. Provide formal proofs for the following symmetric difference relationships. (The definition of *symmetric difference*, $\triangle$, is on page 27.)

(a) $A \triangle B = B \triangle A$.

(b) $A \triangle A = \emptyset$.

(c) $A \triangle \emptyset = A$.

(d) $(A \cup B) - (A \cap B) = A \triangle B$.

(e) ★ $A \triangle \overline{A} = U$.

(f) $A \cap (B \triangle C) = (A \cap B) \triangle (A \cap C)$ [*Hint*: Work both ends toward a common intermediate expression. Do this as separate sequences of equalities.]

(g) $(A \triangle B) \triangle C = A \triangle (B \triangle C)$ [*Hint*: Work both ends toward a common intermediate expression. Do this as separate sequences of equalities.]

18. Prove the following proposition.

PROPOSITION 2.21 ***Counting a Partition***

Let S be a finite set and let $\{S_1, S_2, \ldots, S_k\}$ be a partition of S. Then

$$|S| = \sum_{i=1}^{k} |S_k|$$

2.2 Logic in Daily Life

Although logic is an essential tool for the mathematician (to prove theorems), it may appear to be of only minor interest to other people. However, this is not the case. We all frequently encounter logical (or seemingly logical) presentations of facts and conclusions. We need to be able to identify which claims do have logical support and which do not.

Many arguments are not easily converted to a form where syllogistic or symbolic logic can be used to examine the validity of the argument form. It is still possible to analyze claims carefully for believability. The subsections that follow present some simple principles to help analyze arguments and some examples of common informal errors in reasoning.

2.2.1 General Guidelines for Analyzing Claims

Some general guidelines for analyzing claims are the following [14]:

Evaluate factual claims. When a claim is made, check to see if there is evidence that it is true. For example, suppose a tour guide tells you that Lake Superior is the largest body of fresh water on our planet. Your friend from the former Soviet Union may wish to challenge the tour guide. Lake Baikal, in Russia, is the deepest lake on Earth. It contains a greater *volume* of water than any other lake. Lake Superior has a larger *surface area* than any other freshwater lake. The Caspian Sea has the largest surface area of any lake if saltwater lakes are included [28]. This illustrates the importance of properly defining terms. In this example, the factual claim was mostly correct. Many other claims are actually false.

Look for flaws in logic or logical inconsistency. This will be addressed in more detail later in this section, and again in Appendix C.2.1. Flaws in logic are usually called *logical fallacies*. *Logical inconsistencies* occur when an argument permits you to reach contradictory conclusions. When you feel that some "amazing offer" is too good to be true, it is likely that you have identified a logical inconsistency. For example, an advertisement for a seminar claims that the seminar can show you how to become a millionaire with no financial investment on your part. Your experience tells you that you need money to invest, or productive work, or some inheritance or prize, or some dishonest activity to make a million dollars. The advertisement and your experience give inconsistent information. Something is wrong.[17]

Question assumptions. This is a very powerful tool. For example, suppose you read that your division-winning basketball team will probably win next year also, since all the starting players will be back. You might consider some assumptions implied by this claim, such as "next year's performance will be like this year's performance". However, the situation will *not* be the same next year. Perhaps one of the starters will get injured. Another might look to the future and spend more time studying than practicing basketball. Perhaps the team will have debilitating internal dissension. Even if they do perform as well next year, why assume that no other team will be even better?

2.2.2 Informal Fallacies

You should know that there are many common ways to make mistakes in the content of arguments. I will briefly discuss some of them. This list does not exhaust the possibilities. You might benefit from describing other mistakes.

Invited inference Suppose you are told "9 out of 10 doctors report that no other over-the-counter pain killer is better than Morphonol". You are subtly invited to infer that Morphonol is *better* than all other over-the-counter pain killers. This is not what the statement actually says. Perhaps a number of doctors were asked, and 90% said "all over-the-counter pain killers are about the same". The other 10% might have said "Morphonol is the least effective".

Appeal to authority It is permissible to have an expert testify, but only if the testimony is related to the expert's area of expertise. Having a famous athlete or a Nobel prizewinning chemist endorse real estate should be no more compelling than having a shoe salesperson make the same endorsement. In either case, the real estate may or may not be a wise investment. The celebrity or authority figure adds no information to the endorsement.

Shifting the focus Some people are very good at changing the topic when their argument is beginning to appear weak. A common way to shift the focus is to call attention to the faults of the other person (rather than weaknesses in their argument). If attention can be focused on the fact that the opponent was once a member of a polka band, perhaps her sound reasoning can be forgotten.

Using rules in an inappropriate context If the sign says, "No swimming", we do not arrest someone for swimming to save a drowning child. There is little danger in making an error in this case. Not all situations are this apparent, however. For

[17] You might also wonder why the seminar leader is being so kind as to tell (for a small admission fee) strangers like yourself how to earn all this money.

example, if you try to induce vomiting in a child who has swallowed drain cleaner because you have heard that is the proper procedure for an accidental poisoning, you may cause even more harm to the child. The "induce vomiting" procedure was never intended for this situation. In the context of poisoning due to ingestion of drain cleaner, other procedures apply.

No one knows, so I must be right This fallacy occurs when we make a definite claim based upon the answer not being known for sure. For example, "It has never been proved that Puppy Yums cause cancer. Therefore, Puppy Yums are safe." The conclusion in this example is unwarranted. Puppy Yums *may* cause cancer; we just haven't proved it yet. It is also possible that Puppy Yums do not cause cancer, but may start a *Salmonella* epidemic.

This fallacy can be used by both sides in an opposing argument. "No one has ever proved there is a God, so God must not exist." "No one has ever proved God does not exist, so God must exist."

Inappropriate generalization In this fallacy, a special case is inappropriately extended to cover other cases. "Taking a few pain killers helped my headache. Taking half a bottle will completely block the pain." The conclusion may be true, but not for the reason given.

Circular reasoning One way to "prove" your claim is to assume it is true, then argue that it must be true by using the assumption. "I don't believe in miracles. Any supposed miracle must have another explanation. I know that miracles don't happen because I have never seen one. I have never seen a miracle, because miracles don't happen." There are less obvious versions of this faulty reasoning. The point to remember is that the *reasoning* is faulty. The *conclusion* may or may not be true.

Confusing the whole and the parts These fallacies are committed by incorrectly attributing a property of the whole to one of the parts, or incorrectly attributing a property of a part to the whole.

Attributing a property of the whole to one of the parts can lead to errors like the following. "Our college is famous for turning out excellent scholars. Joe graduated from our college, so Joe is an excellent scholar." Perhaps Joe had great potential when he was admitted, but spent most of his time attending parties.

Attributing a property of the parts to the whole can also lead to errors. "Every member of our basketball team is an all-American so our team is outstanding." Perhaps every team member is egotistical and wants all the glory. The team cannot work together and has a losing season.

Incorrectly using averages It is important to recognize that average behavior does not fully describe the behavior of the individuals whose average is being considered. "Men are stronger than women; therefore Tom is stronger than Betty." Actually, some women are stronger than most men. Some men are weaker than most women. It is only *on average* that men are stronger than women.

Equivocation This form of reasoning is good for making jokes, but poor for forming valid conclusions. Equivocation occurs when we use the same word in an argument, but with two different meanings. Here is a great joke based upon equivocation:

> Which is better, eternal happiness or a ham sandwich? It would appear that eternal happiness is better, but this is really not so! After all, nothing is better than eternal happiness, and a ham sandwich is certainly better than nothing. Therefore a ham sandwich is better than eternal happiness [87].

This example is popular with children: "How do you get down from an elephant? You don't, you get down from a goose."

Nonsequitur A nonsequitur is a conclusion or statement that does not follow logically from what has been said before. This is a favorite of advertisers and politicians. Imagine the following ad (you supply the visuals). "Ah, home-cooked biscuits, just like grandma used to make. We at Shark Savings and Loan also believe in the old down-home values. (Change of scene.) There has never been a better time to see us for a home-improvement loan. For only half your current monthly house payment and your next unborn child . . .".

As the preceding scenario shows, nonsequiturs are often used to impute virtue by association. The hope is that the good feelings you have toward grandma and apple pie (or whatever) will be unconsciously transferred to the product or person that appears next.

In a less sinister context, unintended nonsequiturs may be an amusing verbal trait for a comic-relief character in your next play or novel.

2.2.3 Everyday Logic versus Symbolic Logic

The manner in which logic is used in daily life does have limitations. Because there is no formal structure it is easier to be ambiguous and imprecise. Systems of formal logic do offer more precision. That precision does not come for free. It takes time to become proficient with a formal system of logic. Many arguments from daily life are quite complex. It may be hard to translate such an argument into a form the formal system can deal with.

Most readers of this text will not need to use symbolic logic on a daily basis. However, I believe it is beneficial to develop a rudimentary familiarity with a few ideas from a system of formal logic. The skills needed to achieve this familiarity will enhance your ability to spot errors in both the form and the content of an argument encountered in daily life. These skills will also help you avoid errors while proving theorems or developing computer programs. The sections that follow provide an introduction to symbolic logic.

✔ Quick Check 2.4

1. Section 2.2 can be grouped into 3 general guidelines and 11 deceptive forms of argument. Under which one of the 3 guidelines do all of the deceptive argument forms most properly fit?
2. Analyze the following claims:
   (a) No manager ever got fired for buying IBM. (An old IBM advertising slogan.)
   (b) Our carrots have no cholesterol!
   (c) This is the best show at the theater. We should see it. ☑

2.2.4 Exercises

The exercises marked with ★ have detailed solutions in Appendix H.

1. Analyze the following claims, using the categories in the text whenever possible. If the general guidelines and informal fallacies don't fit, make up an appropriate category.
   (a) Beautiful Wanda LaWow uses "Beauty Aid". Wouldn't you like to use it too?
   (b) ★ Noted law professor Sir Percival Winthrop says that Puppy Yums is the most nutritious puppy food on the market.
   (c) ★ Crime, drugs, and rebellious teens are rampant. There has never been a time when a police chief like John Rockjaw was needed more.
   (d) ★ Our minivan is sold with seating for five. With the optional seating arrangement, it can carry seven passengers.
   (e) Support our troops in Iraq and Afghanistan (flags waving in the background). A public service announcement brought to you by Sludge Malt Liquor.
   (f) One in five women suffers from hideously disfiguring short eyelashes. Now there is relief from this embarrassing condition. Use Elegant Lashes fake eyelashes.
   (g) I believe all college students should learn some college level mathematics.
   (h) My child is 10 pounds below the average for her age. There must be a medical problem.

(continued on the next page)

(i) My opponent's proposals for foreign policy are worthless. How can we seriously consider ideas from a man who has such a bad haircut?

(j) My grandpa had only a sixth-grade education. My dad never completed high school. They both successfully provided for their families. So I don't need much education either.

(k) My last two dates were creeps. All men are rotten!

(l) Susie sure is pretty. She'll make a great class president.

(m) Poor people are poor because they are lazy. I won't hire any poor people because I don't want lazy workers.

(n) My stock broker made a bundle on the market last year. He's buying a Ford truck. Ford's must be the best.

(o) This amusing example is from [1]: "Daughter returning home at three in the morning: 'Daddy, you said I should be in at a quarter of twelve, and three is a quarter of twelve.' "

(p) My dentist says that regular flossing will help prevent gum disease.

(q) We only use 10% of our brain, so if I can just use another 4% I can become a genius.

(r) Our group of 10 friends celebrated a birthday at a fancy restaurant. I heard you should pay a 15 – 20% tip, so I left a $40 tip on the table.

2. Make up an example for each of the informal fallacies. Do not mimic the examples in the text.

2.3 Propositional Logic

Systems of symbolic logic are foundational structures[18] upon which modern mathematics is built. In this section, an elementary system of symbolic logic called *propositional logic* will be presented. This system is built around statements (propositions).[19] A *statement* is an assertion that may be labeled either *true* or *false*. True will be abbreviated as T and false will be abbreviated F. True and false are formally undefined, but intuitively correspond to common usage. An alternate term for "statement" is *proposition*.

The following are statements:

It is raining in Death Valley.
$3^2 = 9$
All have sinned and fall short of the glory of God.
There is sentient life on at least one other planet.

The following are not statements:

If John is sick.
Assign 7 to y.
The rain in Spain.
Daddy, read me a story.

The first group of examples are statements because they can (at least abstractly) be assigned a value of either T or F. Notice that *we need not know what the proper* T/F *value is* to recognize that a sentence is a statement. The notion of a statement is formally specified in the next definition.

DEFINITION 2.22 ***Statement***
A *statement* is an assertion that may be labeled either *true* or *false*.

As you read this section, you should bear in mind that systems of symbolic logic are mathematical models. As such, they capture some aspects of human thought, but are not complete descriptions. In particular, systems of symbolic logic are designed to eliminate ambiguity.

[18] Two others are the axiomatic method (Section 3.1.2) and set theory (Section 2.1).

[19] Other versions of symbolic logic (such as *predicate logic*) provide mechanisms to work with pieces of statements. (See Section 2.6 for more details.)

2.3.1 Truth Tables

Assume that P and Q are statements.[20] We are often interested in more complex statements that are built from P and Q. Just as compound sentences in English need conjunctions to connect the constituent phrases, so compound statements in logic need something to connect their simpler constituent statements. The objects that build compound statements from simpler statements are called *logic operators*. You may already be familiar with the logic operators AND, OR, and NOT. A sentence that has been constructed by properly connecting simpler statements is still a statement. As such, it is either true or it is false. We usually cannot be sure which it is until we know the T/F values of its constituent statements. One of the easiest methods for determining the T/F value of the compound statement from the T/F values of its constituent statements is to use a *truth table*. A truth table is a two-dimensional table whose entries are the values T and F. Each row represents a combination of T/F values for the constituent statements, together with the resulting T/F value of the compound statement.

Subsections 2.3.2, 2.3.3, and 2.3.5 contain numerous examples of truth tables. Truth tables will actually be used in two ways. In Sections 2.3.2 and 2.3.5, truth tables are used to *define* the action of various logic operators. In sections such as 2.3.3 and Appendix C.1, truth tables will be used as a *computational tool* to determine the T/F values of compound statements.

2.3.2 The Operators NOT, AND, OR, and XOR

TABLE 2.2 NOT

P	$\neg P$
T	F
F	T

TABLE 2.3 AND

P	Q	$P \wedge Q$
T	T	T
T	F	F
F	T	F
F	F	F

TABLE 2.4 OR

P	Q	$P \vee Q$
T	T	T
T	F	T
F	T	T
F	F	F

The operators AND, OR, and XOR combine two statements into a single compound statement. The operator NOT modifies a single statement. The operator NOT is commonly denoted using the symbol $\neg$.[21] Its definition (in truth table form) is in Table 2.2. Thus, if P happens to be true, then $\neg P$ will be false. If P happens to be false, then $\neg P$ will be true.

The operator AND is commonly denoted $\wedge$ and is also known as the *conjunction* operator. Its truth table is presented in Table 2.3. For example, if P is false and Q is true (as represented by the third row), $P \wedge Q$ will be false. Notice that $P \wedge Q$ has the value T only when both of the constituent statements P, Q are true.

The statements P and Q each have specific truth values (even if we do not know them). Thus, only one of the four rows in the truth table for the compound statement $P \wedge Q$ actually applies. The table gives the value for all possible choices of the constituent statements P and Q. For example, if P represents the statement "I have written a math text" and Q represents the statement "I am a millionaire", the second row of the truth table is the row that actually applies to the author of this text. The compound statement "I have written a math text and I am a millionaire" is false, since the author is not a millionaire.

The operator OR is denoted $\vee$ and is also known as the *disjunction* operator. The truth table is presented in Table 2.4.

The truth tables for AND and NOT agree with popular usage of the terms. The table for the *operator* OR may seem somewhat different from popular usage of the *word* "or". Mathematicians include the possibility that both P and Q might be true. In this case $P \vee Q$ is defined to have the T/F value T. Many people neglect to consider the possibility that both P and Q might be true. For example, in common usage we might say "I am either going for a walk, or I will read a book". We understand this to mean an exclusive choice: I will not read while I am walking. The mathematical definition of OR includes the option of doing both (and hence is called an *inclusive* OR). If we wish to exclude the possibility of doing both, we should use the XOR logic operator. The operator XOR (*exclusive OR*) is denoted $\oplus$. The truth table is presented in Table 2.5.

[20]There is nothing special about the letters P and Q. Any other letter can symbolically represent a statement.

[21]Other notations are also common. The most common alternative is $\sim$.

TABLE 2.5 XOR

P	Q	$P \oplus Q$
T	T	F
T	F	T
F	T	T
F	F	F

The XOR operator is especially useful in the design of digital electronics. It is not as frequently used by mathematicians or computer programmers.[22] The XOR operator is therefore not as important for the purposes of this text and will not be used much.[23]

Note that these four tables *define* the logical operators. The definitions used are those that have been found consistent with our intuitive notions of AND, OR, XOR, and NOT.

There is a significant difference in the meanings of AND and OR. It makes a difference whether an apartment rental agreement reads "first and last month rent, *and* a $500 damage deposit are required before you may move in" or reads "first and last month rent, *or* a $500 damage deposit is required before you may move in". The second contract costs you a lot less initially.

2.3.3 Negations of AND, OR, and NOT

TABLE 2.6 $\neg(\neg P)$

P	$\neg(\neg P)$
T	T
F	F

TABLE 2.7 Intermediate Columns for $\neg(\neg P)$

P	$(\neg P)$	$\neg(\neg P)$
T	F	T
F	T	F

The three basic operators can be combined with each other in many ways. The simplest way to combine the basic operators is to take the negation (NOT) of another operator. The negation of NOT is presented in Table 2.6.

Table 2.6 can be derived by adding an intermediate step, $(\neg P)$ and then applying the truth table of NOT to the new statement $(\neg P)$. Recall that the truth table of NOT sets the new truth value to be the opposite of the original statement's truth value. Thus, negating the $(\neg P)$ column causes the truth values to become the values in the third column of Table 2.7.

Intermediate columns can be used to produce the truth tables for the compound negations $\neg(P \wedge Q)$ (Table 2.8) and $\neg(P \vee Q)$ (Table 2.9).

TABLE 2.8 Intermediate Columns for $\neg(P \wedge Q)$

P	Q	$P \wedge Q$	$\neg(P \wedge Q)$
T	T	T	F
T	F	F	T
F	T	F	T
F	F	F	T

TABLE 2.9 Intermediate Columns for $\neg(P \vee Q)$

P	Q	$P \vee Q$	$\neg(P \vee Q)$
T	T	T	F
T	F	T	F
F	T	T	F
F	F	F	T

Do these tables make intuitive sense? Try making up examples to compare the tables with your intuition. For example, you might let P represent the statement "I like my cat" and Q represent the statement "I am a philosopher".

✔ Quick Check 2.5

1. Which of the following are statements?
   (a) As the world turns.
   (b) An apple a day keeps the doctor away.
   (c) All that glitters is not gold.
   (d) Sleep tight and don't let the bedbugs bite.
2. Suppose you are married and wish to open a joint checking account. Which of the following would you prefer to have printed on your checks? Why?
   - John and Jane Doe
   - John or Jane Doe
3. Let P be the statement "I like popcorn". Let Q be the statement "I like jalapeños".
   (a) Write an English sentence for $\neg(P \wedge Q)$.
   (b) Write an English sentence for $\neg(P \vee Q)$.

[22]Most common programming languages do not have a built-in XOR operator, but do have AND, OR, and NOT.

[23]Other useful logic operators for digital electronics are NAND (*not-and*) and NOR (*not-or*). See the Exercises for more information.

(c) Use these sentences to test the intuitive appeal of the truth tables for $\neg(P \wedge Q)$ and $\neg(P \vee Q)$.

4. Translate the English sentences that follow into symbolic notation using statements and logic operators.

(a) I love Betty or Sue.

(b) I love Colin and Tom.

(c) I love Betty or I love Sue.

(d) I love Colin and I don't love Tom. ☑

One fun way to practice working with truth tables and reinforce the definitions of the logic operators is to solve logic puzzles. Appendix C.1 provides that opportunity.

English (and other human languages) can sometimes be ambiguous. For example, a statement such as "we will loan you the money to buy a new car if you qualify for a loan and pay 10% down or trade in a car worth at least $2000" can be understood in two ways. These are illustrated in the next example.

EXAMPLE 2.19

Ambiguous Language

The compound statement $A \wedge B \vee C$ can be interpreted as $(A \wedge B) \vee C$ or as $A \wedge (B \vee C)$. The truth value of the two compound statements is not always the same, as Table 2.10 demonstrates.

TABLE 2.10 The Statements $(A \wedge B) \vee C$ and $A \wedge (B \vee C)$ Are Not Identical

A	B	C	$(A \wedge B) \vee C$	$A \wedge (B \vee C)$
T	T	T	T	T
T	T	F	T	T
T	F	T	T	T
T	F	F	F	F
F	T	T	T	F
F	T	F	F	F
F	F	T	T	F
F	F	F	F	F

If the interpretation of the statement "we will loan you the money to buy a new car if you qualify for a loan and pay 10% down or trade in a car worth at least $2000" is that of column 4, a person who does not qualify for a loan but does have a car to trade in would still receive the loan (independent of any down payment). The interpretation in column 5 is most likely what the lender intended. ■

A more detailed discussion of logic operator precedence can be found in Section 2.3.6 on page 46.

There is a standard ordering for the initial columns in a truth table. If the truth table contains the propositions $P_1, P_2, \ldots, P_n$, then there will be n initial columns that list all 2^n possible combinations of the truth values for these propositions. The standard ordering requires the rightmost such column to contain alternating T and F values, starting with T. Moving toward the left, each new column will have the T's and F's alternating in groups that are twice as large as the groups in the column to its immediate right. Thus, the column on the immediate left of the rightmost such column will alternate pairs of T's and pairs of F's, starting with a pair of T's. The table in Example 2.19 illustrates the general pattern.

2.3.4 Exercises

The exercises marked with ★ have detailed solutions in Appendix H.

1. ★ If there are 4 propositions (A, B, C, and D), how many possible combinations of T/F values are there? Assume that A and B both T, C and D both F is a different combination than A and B both F, C and D both T. [*Hint*: List all the combinations in a systematic manner.]

2. I will be happy if I get a date with either Susie or Penelope. If this does not happen, I will be unhappy. What will make me unhappy?

3. My parents promised to buy me a car for Christmas if I get A's in Calculus and Intro to Soap Operas. I received an A in Calculus and a B in Intro to Soaps. Are my parents obligated to buy me a car?

4. ★ Do the following sentences mean the same thing? Explain.
   - I don't want bread and butter.
   - I don't want bread or butter.

5. Suppose you order a turkey sandwich at a restaurant. The waitress tells you that the sandwich comes with soup or salad. Is the waitress most likely to be using an inclusive OR or an exclusive OR?

6. For each of the following statements, explain the meaning when an inclusive OR is used and when an exclusive OR is meant. Then decide which of the meanings best fits the original context of the sentence.
   (a) ★ The concert will be canceled if not enough tickets are sold or if one of the singers gets sick.
   (b) Before taking the standardized test, you must present a driver's license or a birth certificate.
   (c) When you open a checking account at the bank, you get a new cooler or a $25 gift certificate to a restaurant.
   (d) The complementary hotel breakfast includes three items from list A or four items from list B.

7. In the state of Arizona, a defendant may be found innocent due to insanity if either
   (a) He or she does not understand the consequences of his or her actions
   or
   (b) He or she does not know that his or her actions are wrong.

   I was on a jury for a case where the (guilty) defendant demonstrated some understanding that the act might be wrong. He did not understand the consequences of his actions. Should we have convicted him or found him innocent due to insanity?

8. Describe how the logic operators AND, OR, and NOT are used with an online library card catalog.

9. ★ Suppose the requirements for earning a Bachelor's degree are
   (A) Passing a set of 12 courses in the major.
   (B) Passing a set of 10 courses in general education.
   (C) Completing at least 122 semester credits.
   (D) (i) Passing a proficiency exam in a foreign language
   or
   (ii) successfully completing two courses in a foreign language.
   (E) Having at least a 2.0 GPA overall.
   (F) Having at least a 2.25 GPA in your major.

   (a) Write a symbolic expression using logic operators to represent these requirements.
   (b) Suppose you have completed all the required courses in both the major and in general education. Suppose you also have 123 semester credits, an overall GPA of 2.2, and a GPA in the major of 2.15. Since you had 4 years of high school Spanish, you passed the foreign language proficiency exam. Will you get the degree?

10. Suppose that before Jane goes to the mall with her friends, she has to complete one of the following:
    Option 1: Mowing the lawn (M).
    Option 2: (i) Washing and drying the dishes (D)
    and
    (ii) folding the towels (T).
    Option 3: Dusting the house (H).
    Option 4: Washing the floors (W).
    Option 5: (i) Doing the grocery shopping (S)
    and
    (ii) picking up the dry cleaning (P).

    (a) Write a symbolic expression using logic operators to represent these task descriptions.
    (b) Suppose that Jane refuses to dust, wash floors, and mow the lawn. Jane is also tired of washing and drying the dishes, but she does decide to fold the towels. She also goes grocery shopping but forgets to pick up the dry cleaning on the way home. Should Jane be allowed to go to the mall with her friends?

11. Suppose that the students in Mrs. Clocksworth's sixth-grade classroom are going on a scavenger hunt for the following items:
    (A) A piece of grass that is at least 4 inches in length.
    (B) (i) A 16-ounce water bottle
    or
    (ii) a 32-ounce water bottle.
    (C) (i) A rock weighing one-half of a pound
    or
    (ii) a pile of sand weighing three-quarters of a pound.
    (D) (i) A book with 10 – 20 pictures
    or
    (ii) a textbook with more than 500 pages.
    (E) A purple flower with exactly four petals.

    (a) Write a symbolic expression using logic operators to represent successfully completing all five sections of the scavenger hunt.

(b) Suppose that a student in Mrs. Clocksworth's sixth-grade class will win a prize for successfully completing the scavenger hunt. Brian ends up finding a book with 10 pictures and a textbook with 450 pages in the library. From the school yard, this student gathers a piece of grass 5 inches long, a purple flower with four petals, and a rock weighing exactly one-half of a pound. He does not want to take the time to measure out a pile of sand, so he does not collect any sand at all. He borrows both a 16-ounce water bottle and a 32-ounce water bottle from one of the fifth-grade teachers in the school. Will Brian win a prize?

12. The logic operators NAND and NOR are defined as the negations of AND and OR, respectively. They are used in digital circuit design to replace two circuit components, called *gates*, by a single component. For example, an AND gate followed by a NOT gate can be replaced by a NAND gate. Create the truth tables for NAND and NOR.

13. Each statement is either true or false. Identify which case is correct, and then give some justification for your answer.

(a) It is essential to know a sentence's proper T/F value before the sentence can be recognized as a statement.

(b) The logic operators, AND and OR, combine two statements into a single compound statement.

(c) ★ The statements $\neg(P \vee Q)$ and $(\neg P) \vee (\neg Q)$ have the same truth values for all possible values for P and Q.

(d) ★ The compound proposition $P \vee Q \vee R$ can be interpreted as either $P \vee (Q \vee R)$ or as $(P \vee Q) \vee R$.

(e) No matter what truth values P and Q have, the statements $P \wedge Q$ and $Q \wedge P$ will always have identical truth values.

(f) $P \vee Q$ is true only when both P and Q are true.

(g) The symbol $\vee$ represents "inclusive OR".

14. In English, the phrase "A and/or B" is sometimes used. How could we capture the intended meaning using propositional logic?

15. I promised my wife that tonight we would go out for dinner and also watch a movie and also take a walk. The order of occurrence for these three events is not important. List all the ways I could fail to keep my promise.

16. Is a question a statement?

17. Each statement is either true or false. Identify which case is correct, and then give some justification for your answer.

(a) Popular usage of the word *or* is confined to exclusive OR.

(b) For any statement, P, the expression "P" will be true and the expression "$\neg P$" will be false.

(c) Consider a statement, P, and the statement $\neg(\neg P)$. The truth tables for both statements will be identical, as will their translations into English.

18. An old *Ziggy* cartoon, created by Tom Wilson, placed Ziggy in a restaurant looking at a menu. The waiter is telling him, "You must read the menu carefully sir ... it says bacon, lettuce or tomato sandwich".

(a) Does the cartoonist's use of the word *or* agree with the truth table definition of the logic operator OR?

(b) Write the waiter's sentence symbolically. Clearly identify the meanings of your symbolic statements (P, Q, R, etc.).

(c) Is there any ambiguity in the waiter's sentence? (Of course you need to do more than just write yes or no on your homework paper. If there is any ambiguity, you should describe it. For possible extra credit you could write an essay on how symbolic logic helps eliminate such ambiguity.)

19. This story concerns a caravan going through the Sahara desert. One night they pitched tents. Our three principal characters are A, B, and C. A hated C and decided to murder him by putting poison in the water of his canteen (this would be C's only water supply). Quite independently of this, B also decided to murder C, so (without realizing that C's water was already poisoned) he drilled a tiny hole in C's canteen so that the water would slowly leak out. As a result, several days later C died of thirst. The question is, who was the murderer, A or B?

According to one argument, B was the murderer, since C never did drink the poison put in by A; hence, he would have died even if A had not poisoned the water. According to the opposite argument, A was the murderer, since B's actions had absolutely no effect on the outcome; once A poisoned the water, C was doomed, hence C would have died even if B had not drilled the hole. Which argument is correct?

Reprinted from *What is the name of this book?*, c 1978 by R. M. Smullyan, by permission of Collier Associates, P.O. Box 20149, West Palm Beach, FL 33416 USA.

2.3.5 Implication and the Biconditional

We often encounter statements in the form "If P is true, then Q is also true". This can be abbreviated "If P, then Q", where the "is true" is implicitly included. For example, "If it is raining, then I will carry an umbrella". A statement in this form is called an *implication*. A common notation is $P \rightarrow Q$. An implication asserts the inevitability of Q if P occurs; that is, whenever we observe P, we always find Q as well.[24]

[24]This does not mean that P necessarily *causes* Q. In the implication "If it rains, then the ground gets wet", the wetness *is* caused by the rain. But in the implication "If there are lots of boats in Minnesota, then there are lots of mosquitos in Minnesota", the mosquitos are not caused by the boats. Both are "caused" by the existence of so many lakes. Interpreting the phrase "if ... then ..." in our symbolic logic model is completely defined by the truth table; cause and effect are captured using other mechanisms (such as *modus ponens* on page 54). However, in everyday usage this phrase is often intended as an assertion of explicit cause and effect.

Implications are so important that the parts have been given names. The statement P is called the *hypothesis* and the statement Q is named the *conclusion*.[25] The terms *hypothesis* and *conclusion* indicate the manner in which mathematicians tend to use implications: as an assertion that knowing that P is true guarantees that Q is also true.[26] Many theorems are stated as an implication. The proof of the theorem is what establishes the inevitability of Q given P.

The values in the final column of the truth table depend on the truth values of the hypothesis and conclusion. If both P and Q are true, then the implication is a true assertion. If P is true and Q is false, then the implication is false (Q is not an inevitable event even though P occurs). We therefore have the partial truth table in Table 2.11.

TABLE 2.11 Partial Truth Table for Implication
Preliminary, Incomplete

P	Q	$P \to Q$
T	T	T
T	F	F
F	T	?
F	F	?

What values should we define for the remaining two rows? In one sense, it does not matter. For if P is not true, we have nothing to say about the inevitability of Q. Statement Q may be true or it may be false; we have no extra knowledge from P to help us. Consider again the example "If it is raining, then I will carry an umbrella". If it is *not* raining, I may choose to leave my umbrella at home. However, I may carry it for another reason (perhaps to use as a parasol to keep the sun off my head). The nonoccurrence of P means that the implication can make no contribution to our knowledge.

There is another way to look at the truth values in the final two rows. We would like the values to be consistent with the other truth tables. Thus it is worth considering this table. We shall do so by starting with an example. Suppose your rich friend makes the following promise:

If you get an A in math, then I will buy you a new sports car.

What would it take to make your friend a liar? Certainly, if you get an A in math, but your friend does not buy you a new sports car, the promise would not have been kept. But what if you do not get an A in math? Then your friend is under no obligation to buy you anything. You cannot justly accuse your friend of lying unless you get an A but no sports car. If you do not get an A, you cannot accuse your friend of lying, so you may as well assume your friend told the truth. If this seems reasonable, then because a statement must be either true or false, an implication must be true whenever the hypothesis is false.

The net effect is that the only way you can justly accuse your friend of lying is if the hypothesis (an A in math) is true, but the conclusion (buy a sports car) is false. Thus the negation of $P \to Q$ (for this example) is $P \wedge (\neg Q)$. The truth table for $P \wedge (\neg Q)$ is as follows:

P	Q	$\neg Q$	$P \wedge (\neg Q)$
T	T	F	F
T	F	T	T
F	T	F	F
F	F	T	F

Since $P \wedge (\neg Q)$ should be the negation of $P \to Q$, its negation should be the same as $P \to Q$. Thus the truth table[27] for $P \to Q$ is

P	Q	$P \wedge (\neg Q)$	$\neg(P \wedge (\neg Q))$	$P \to Q$
T	T	F	T	T
T	F	T	F	F
F	T	F	T	T
F	F	F	T	T

which simplifies to the final version of the truth table (Table 2.12).

TABLE 2.12 Implication

P	Q	$P \to Q$
T	T	T
T	F	F
F	T	T
F	F	T

[25]Other writers may use the terms *antecedent* and *consequent*.

[26]Thus in most mathematical uses, we *do* have a cause–effect relationship.

[27]Many texts refer to this definition as *material implication* to emphasize the absence of any *cause–effect* considerations.

TABLE 2.12 Implication

P	Q	$P \rightarrow Q$
T	T	T
T	F	F
F	T	T
F	F	T

A final comment about implications is in order. Implication is an operator, just as NOT, AND, and OR are operators. Implication takes two statements (the hypothesis and conclusion) and creates a new, compound statement whose truth values are defined by the truth table for implication. If we assert that the implication is true, we do not automatically know that either the hypothesis or conclusion is true. In fact, if you look at the truth table, you will see that neither is necessarily true just because the compound statement is true. This observation indicates a major difference between the implication operator and the AND operator. If $P \wedge Q$ is known to be true, then we automatically know that both P and Q must be true.

Therefore, the assertion $P \wedge Q$ is much more specific than the claim $P \rightarrow Q$. In the first case, there is only one possible pair of truth values if the compound statement, $P \wedge Q$, is true. In the second case, there are three possible pairs of truth values for P and Q that are consistent with $P \rightarrow Q$ being true.

TABLE 2.13 Biconditional

P	Q	$P \leftrightarrow Q$
T	T	T
T	F	F
F	T	F
F	F	T

The final logic operator we will examine is the *biconditional*, denoted $P \leftrightarrow Q$. The truth table for the biconditional is shown in Table 2.13.

As can be seen from the truth table, a compound statement formed by the use of the biconditional logic operator is true exactly when the two constituent simpler statements have identical T/F values.

Many mathematical statements involving the biconditional are stated as "P if and only if Q". A biconditional might also be written as "Q is a necessary and sufficient condition for P". The double arrow $\leftrightarrow$ should remind you of two arrows, $\leftarrow$ and $\rightarrow$, joined together. In summary, the biconditional represents two implications, commonly expressed in one of three ways, as shown by the three columns of (Table 2.14).

TABLE 2.14 Common Ways to Express a Biconditional

$P \leftrightarrow Q$	P if and only if Q	Q is a necessary and sufficient condition for P
$Q \rightarrow P$	P if Q	Q is a sufficient condition for P
$P \rightarrow Q$	P only if Q	Q is a necessary condition for P

✔ Quick Check 2.6

1. Show that the compound statement $(P \rightarrow Q) \wedge (Q \rightarrow P)$ has the same truth table as the statement $P \leftrightarrow Q$.
2. Let P be the statement "I am an earthling", and let Q be the statement "I am not a space alien". Translate the statement $P \leftrightarrow Q$ into English. ☑

EXAMPLE 2.20

The Biconditional

Let P be the statement "I am an employed teacher", and let Q be the statement "I have a job". Then $P \rightarrow Q$ is true but $Q \rightarrow P$ is false. (An employed truck driver has a job but is not a teacher.) Using the result in Quick Check 2.6, we see that $P \leftrightarrow Q$ is false. Another way to say this is that the statement "I am a teacher if and only if I have a job" is false.

Let R be the statement "I have students". Then $P \rightarrow R$ is true and $R \rightarrow P$ is true (I can't teach if there are no students). Using the result in Quick Check 2.6 again, we see that $P \leftrightarrow R$ is true. That is, the statement "I am a teacher if and only if I have students" is true. ■

2.3.6 Operator Precedence

You are familiar with the need for an agreement about precedence rules[28] for the standard arithmetic operators $+, -, \cdot, \div$. Thus, $3 + 4 \cdot 5$ has traditionally been given the value 23 rather than the value 35.

TABLE 2.15 Logic Operator Precedence

Higher Precedence
( )
$\neg$
$\wedge, \vee, \oplus$
$\rightarrow, \leftrightarrow$
Lower Precedence

It is similarly convenient to establish a mathematical convention for precedence of logic operators. The common agreement is to evaluate expressions inside parentheses first. Negation is done next (changing the logic value of the logic variable or expression in parentheses to its immediate right). The operators AND, OR, and XOR are applied next, and finally, implication and the biconditional are applied. Table 2.15 summarizes this information. Operators near the top of the table are applied before operators near the bottom.[29]

For example, the expression

$$(A \vee \neg B) \wedge C \rightarrow ((D \vee E) \wedge F) \vee (\neg G \wedge H)$$

is understood as

$$((A \vee (\neg B)) \wedge C) \rightarrow (((D \vee E) \wedge F) \vee ((\neg G) \wedge H))$$

You may use parentheses to change the normal order. For example, $A \wedge (B \rightarrow C)$ causes the implication to be applied before the AND. It is often desirable to add extra parentheses to make an expression easier to read. For example, $A \rightarrow \neg B$ may be easier to read if it is written $A \rightarrow (\neg B)$.

2.3.7 Logical Equivalence

The binary logic operators $\wedge, \vee, \rightarrow$, and $\leftrightarrow$ all take two statements and produce another statement. The unary logic operator $\neg$ changes one statement into another statement. There is another operator that plays a different role: that of a *meta-operator*. The operator $\Leftrightarrow$, which is read "logically equivalent", takes two statements and makes a declaration about whether they are really, at their core, essentially the same statement, or whether they are fundamentally different statements. For example, in Quick Check 2.6, you showed that $(P \rightarrow Q) \wedge (Q \rightarrow P)$ and $P \leftrightarrow Q$ are only cosmetically different; for every pair of T/F values for P and Q, the two compound statements are assigned the same T/F value.

This notion of being essentially the same deserves a name. That definition will be delayed until another useful definition has been made.

DEFINITION 2.23 ***Tautology, Contradiction, Conditional***
A statement is called a *tautology* if every entry in its truth table is T. A statement is called a *contradiction* if every entry in its truth table is F. A statement that is neither a tautology nor a contradiction is called a *conditional* statement.[a]

[a]An alternative name is *contingency*.

The statement $P \vee (\neg P)$ is a tautology (look at its truth table). The statement $P \vee Q$ is a conditional statement because if P and Q are both false, then $P \vee Q$ is also false, but if P and Q are both true, then $P \vee Q$ is true.

[28]You may know this as "order of operations".

[29]Many computer languages carry these precedence rules one step further. The logic operators within a row of the table may be applied in a left-to-right order in the absence of parentheses.

DEFINITION 2.24 ***Logical Equivalence***

Two statements, A and B, are called *logically equivalent* if and only if $A \leftrightarrow B$ is a tautology. Logical equivalence is denoted by the meta-operator $\Leftrightarrow$. The definition of logical equivalence can also be stated as

$$A \Leftrightarrow B \text{ if and only if } A \text{ and } B \text{ have the same truth table}$$

For example,

$$[(P \to Q) \land (Q \to P)] \Leftrightarrow [P \leftrightarrow Q]$$

because $[(P \to Q) \land (Q \to P)] \leftrightarrow [P \leftrightarrow Q]$ is a tautology (think carefully about this sentence).

Appendix C.1.2 contains logic puzzles related to implication, biconditional, and equivalence.

2.3.8 Derived Implications

For every implication, there are three other implications that can be easily derived. As will be seen, the ease of derivation does not necessarily mean that these three derived implications have the same truth value as the original implication. Table 2.16 lists these derived implications.

TABLE 2.16 Derived Implications

Symbolic Form	Name
$P \to Q$	The original implication
$\neg Q \to \neg P$	The contrapositive
$Q \to P$	The converse
$\neg P \to \neg Q$	The inverse

Consider the following three examples.

EXAMPLE 2.21 **Easy Derived Implications**

Let P be the statement "today is Monday". Let Q be the statement "we have math class today". The derived implications are as follows:

Symbolic Form	Name
$P \to Q$	The original implication
If today is Monday, then we have math class today.	
$\neg Q \to \neg P$	The contrapositive
If we don't have math class today, then today is not Monday.	
$Q \to P$	The converse
If we have math class today, then today is Monday.	
$\neg P \to \neg Q$	The inverse
If today is not Monday, then we do not have math class today.	

Suppose that your math class meets on Monday, Wednesday, and Friday. Then under normal conditions, the original implication can be considered a true statement. We would also agree that the contrapositive is a true statement. However, the converse need not be true; we might have math class because it is Wednesday. Similarly, the inverse need not be true (why?). For this example, the biconditional $P \leftrightarrow Q$ is false. ■

EXAMPLE 2.22

Divisibility and Evenness

Consider the statement "If a positive integer n is divisible by 2, then n is even". Let P be the statement "a positive integer, n, is divisible by 2 (no remainder)". Let Q be the statement " n is an even integer". The derived implications are as follows:

Symbolic Form	Name
$P \rightarrow Q$	The original implication
If a positive integer n is divisible by 2, then n is even.	
$\neg Q \rightarrow \neg P$	The contrapositive
If n is not even, then n is not divisible by 2.	
$Q \rightarrow P$	The converse
If n is even, then n is divisible by 2.	
$\neg P \rightarrow \neg Q$	The inverse
If n is not divisible by 2, then n is not even.	

In this example, all four implications are true (because *even* is defined in terms of divisibility by 2). For this example, the biconditional $P \leftrightarrow Q$ is true. ■

EXAMPLE 2.23

More Complex Derived Implications

Let P be the statement "I am a man". Let Q be the statement "I am not a mother". The derived implications should be read carefully.

Symbolic Form	Name	English Translation
$P \rightarrow Q$	implication	If I am a man, then I am not a mother.
$\neg Q \rightarrow \neg P$	contrapositive	If I am a mother, then I am not a man.
$Q \rightarrow P$	converse	If I am not a mother, then I am a man.
$\neg P \rightarrow \neg Q$	inverse	If I am not a man, then I am a mother.

In this example, the original implication and the contrapositive are both true implications, while the converse and inverse need not be (I can be a woman who is not a mother). ■

It can be shown (Exercise 1 in Exercises 2.3.9) that an implication and its contrapositive are logically equivalent statements. Similarly (Exercise 2), the converse and inverse are logically equivalent. Thus the original implication and the contrapositive are either both true, or both false. The converse and the inverse are also either both true or both false. However, as these examples demonstrate, knowing that the original implication is true does not automatically mean that the converse (or inverse) is true. The converse always needs to be investigated separately.

✔ Quick Check 2.7

1. Use truth tables to show that $[\neg(P \wedge Q)] \Leftrightarrow [(\neg P) \vee (\neg Q)]$
2. Write the original implication, the contrapositive, the converse, and the inverse for each of the following statements. Indicate which of the four implications in each set are true.
   (a) If I am an alien, then I am from Mars.
   (b) If I am not an extrovert, then I have no friends.

2.3.9 Exercises

The exercises marked with ★ have detailed solutions in Appendix H.

1. ★ Use truth tables to show that an implication and its contrapositive are logically equivalent statements.
2. Use truth tables to show that the converse and inverse are logically equivalent.
3. Use truth tables to show that the converse and the original implication are *not* logically equivalent.
4. Write and identify the derived implications for the following original implications. You may need to modify the original statement so that it reads more like an implication. If so, write the new form at the beginning of your answer.
   (a) ★ If $n > 1$, then $\sqrt{n} < n$.
   (b) When it rains it pours.
   (c) If V is a vector space, then V has a basis.
   (d) $(P \wedge Q) \to (R \vee S)$.
   (e) If M is a planar map, then M can be colored with at most four colors.
5. Write and identify the derived implications for the following original implications. You may need to modify the original statement so that it reads more like an implication. If so, write the new form at the beginning of your answer.
   (a) ★ On every day that is sunny, we go to the beach.
   (b) It is necessary for us to walk seven miles to arrive at the cave entrance.
   (c) I will attend the banquet only if I am not sick.
   (d) Jill goes to class whenever there will be a quiz.
   (e) Working 40 hours each week is sufficient for me to pay my bills.
6. Complete the following table by filling in the truth values for $P \wedge Q$ and $P \to Q$. The first two columns indicate the statements that P and Q represent.

P	Q	$P \wedge Q$	$P \to Q$
2 is an integer	2 is an even integer		
2 is an integer	2 is less than 1		
2.5 is an integer	2.5 is less than 3		
2.5 is an integer	2.5 is an even integer		

7. Which of the following statements are tautologies?
   (a) ★ $P \to [(\neg P) \to Q]$.
   (b) ★ $(P \wedge Q) \vee Q$.
   (c) $(P \wedge Q) \to Q$.
   (d) $[P \to Q] \leftrightarrow [\neg(P \wedge (\neg Q))]$.
   (e) $[P \to Q] \leftrightarrow [(\neg P) \vee Q]$.
   (f) $([P \to Q] \leftrightarrow [\neg(P \wedge (\neg Q))]) \leftrightarrow [(\neg P) \vee Q]$.
   (g) $(P \vee Q) \to (P \wedge Q)$.
8. Which of the following statements are tautologies?
   (a) $[(\neg P) \leftrightarrow (\neg Q)] \leftrightarrow [Q \leftrightarrow R]$
   (b) $P \to [(\neg Q) \vee R]$
   (c) $[(\neg P) \wedge (P \vee Q)] \to Q$
   (d) $[Q \wedge (P \vee Q)] \to (\neg P)$
9. Determine whether the following pairs represent logically equivalent statements.

(a) ★	$(P \wedge Q)$	$\neg P \vee \neg Q$
(b)	$(P \to Q) \vee P$	$(P \vee \neg Q) \wedge Q$
(c)	$(P \wedge Q) \to P$	$(P \wedge Q) \to Q$
(d)	$\neg(P \wedge Q)$	$(\neg P) \vee (\neg Q)$

10. Is the following statement a tautology? "If the manna in the wilderness was popcorn or my cat is lazy, and also if the manna in the wilderness was popcorn or my cat is not lazy, then the manna in the wilderness was popcorn."
11. $P \vee (\neg P)$ and $\neg[P \wedge (\neg P)]$ are both tautologies. State them in normal English. Do they express the same concept?
12. Which (if any) of the following statements are equivalent to $\neg(P \wedge Q)$?
   (a) $(\neg P) \wedge (\neg Q)$
   (b) $P \vee Q$
   (c) $(\neg P) \wedge Q$
   (d) $(\neg P) \vee (\neg Q)$
13. (a) ★ Produce the truth table for the following proposition.

$$(P \wedge \neg Q) \to (P \vee Q)$$

   (b) ★ Is this a tautology? Explain.
14. (a) Produce the truth table for the following proposition.

$$(P \leftrightarrow Q) \to (P \vee Q)$$

   (b) Is this a tautology? Explain.
15. ★ Let P be the proposition "a man has discovered something he will die for" and let Q be the proposition "he is fit to live". Consider the implication $(\neg P) \to (\neg Q)$: "If a man hasn't discovered something he will die for, then he isn't fit to live" (Martin Luther King, Jr.).
   (a) Write the three derived implications (both symbolically and in English).
   (b) Assume that the original implication is true. Briefly discuss what we know about the truth of the derived implications.
16. Let P be the proposition "you will forgive another" and let Q be the proposition "you break the bridge over which you must pass". Consider the implication $(\neg P) \to Q$: "If you will not forgive another, then you break the bridge over which you must pass" (George Herbert, adapted).
   (a) Write the three derived implications (both symbolically and in English).
   (b) Assume that the original implication is true. Briefly discuss what we know about the truth of the derived implications.
17. Let P be the proposition "I won a prize in the raffle" and let Q be the proposition "I had a winning ticket". Consider the implication $P \to Q$: "If I won a prize in the raffle, then I had a winning ticket".
   (a) Write the three derived implications (both symbolically and in English).
   (b) Assume that the original implication is true. Briefly discuss what we know about the truth of the derived implications.

18. Let P be the proposition "The groundhog sees his shadow" and let Q be the proposition "There will be six more weeks of winter". Consider the implication $P \rightarrow Q$: "If the groundhog sees his shadow, then there will be six more weeks of winter".

(a) Write the three derived implications (both symbolically and in English).

(b) Assume that the original implication is true. Briefly discuss what we know about the truth of the derived implications.

19. Let P be the proposition "I live in the United States of America" and let Q be the proposition "I live in the state of Minnesota". Consider the implication $P \rightarrow Q$: "If I live in the United States of America, then I live in the state of Minnesota".

(a) Write the three derived implications (both symbolically and in English).

(b) Assume that the original implication is false. Briefly discuss what we know about the truth of the derived implications.

20. Are the propositions $\neg(P \wedge \neg Q)$ and $\neg P \vee Q$ logically equivalent? (Be sure to give reasons and show your work!)

21. Each statement is either true or false. Identify which case is correct, and then give some justification for your answer.

(a) ★ Suppose the final column in the truth table for a statement contains an F. Then the statement is not a tautology, but it is a contradiction.

(b) When asserting that an implication is true, we cannot automatically assume that the hypothesis is true, nor can we automatically assume that the conclusion is true.

(c) The logic operator AND has higher precedence than the operator OR.

(d) If both the original implication, $P \rightarrow Q$, and its inverse, $\neg P \rightarrow \neg Q$, are true, then all four derived implications are true.

(e) Two statements are logically equivalent if $P \leftrightarrow Q$ is not a conditional statement.

(f) If the hypothesis of an implication is false, then the implication is true, independent of the truth value of the conclusion.

22. (a) Produce the truth table for the following proposition. Use our standard row ordering. Include all intermediate steps in the table.

$$[P \wedge (P \rightarrow Q)] \rightarrow Q$$

(b) Is this a tautology? Explain.

(c) (Extra credit) Why might I consider this proposition to be significant?

23. Consider the following requirements to vote:

To become a qualified voter, you

Version A: must be at least 18 years old and not have been convicted of a felony.

Version B: must not

- be under 18 years old

or

- have a felony conviction.

(a) Create propositions for the major assertions of the voter qualifications. Then express each version symbolically. (Notice that versions A and B share common pieces.)

(b) Show that version A and version B are logically equivalent. Produce a tight, logically valid proof, not hand waving.

2.4 Logical Equivalence and Rules of Inference

TABLE 2.17 $[\neg(P \wedge Q)] \leftrightarrow [(\neg P) \vee (\neg Q)]$ **Is a Tautology**

P	Q	$[\neg(P \wedge Q)] \leftrightarrow [(\neg P) \vee (\neg Q)]$
T	T	T
T	F	T
F	T	T
F	F	T

Recall that a statement is called a *tautology* if and only if its T/F value is T for all T/F assignments to its component statements. In a truth table, the statement would be represented by the final column and the component statements by the other columns. The statement is a tautology if the final column only has T's. It should be clear that if P and Q are logically equivalent statements, then the statement $(P \leftrightarrow Q)$ is a tautology. For example, consider $\neg(P \wedge Q)$ and $(\neg P) \vee (\neg Q)$. Because these are equivalent statements, $[\neg(P \wedge Q)] \leftrightarrow [(\neg P) \vee (\neg Q)]$ is a tautology. This is summarized in Table 2.17.

Logical equivalences are not the only statements that lead to tautologies. Another example is the statement $P \vee (\neg P)$. With a tautology we can concentrate on the *form* of the statement; we already know it is true for any combination of T/F values of its component statements.

Tautologies provide the "rules" of logic that are used in proofs. If the tautology can be converted to a logical equivalence, it can be used as a substitution rule. If the tautology includes an implication, it is often useful to convert it into a meta-statement called a rule of inference.

DEFINITION 2.25 ***Inference; Rule of Inference***
Let A and B be two statements. Then B may be *inferred* from A, denoted by $A \Rightarrow B$, if $A \rightarrow B$ is a tautology. The symbol, $\Rightarrow$, is the *inference* meta-operator. The meta-statement $A \Rightarrow B$ is called a *rule of inference*.

The key idea is that whenever A has a T in its truth table, so does B. Therefore, if A can be verified as true, then B must be true also.

To utilize these rules, it is reasonable to make the observations contained in the following principles.

The Substitution Principles

Substituting an Equivalent Statement:
If $A \Leftrightarrow B$, and A is a component of a statement, C, then B may be substituted for A without changing the T/F value of C.

Replacing a Logic Variable in a Tautology:
If B is a logic variable in a tautology, C, and A is *any* statement, then A may be substituted for every occurrence of B in C and C will still be a tautology.

Using a Rule of Inference:
If $A \Rightarrow B$, A evaluates to T, and A is a component of a statement, C, then B may be substituted for A without changing the T/F value of C.

Notice that the third substitution principle (using a rule of inference) contains several preconditions. In particular, A *must* evaluate to true before the substitution principle is valid. The next example shows why this is important.

EXAMPLE 2.24 **Misusing a Rule of Inference**

Consider the following collection of statements, where we assume $n \in \mathbb{Z}$ limits the universe of discourse:

A: n is a positive integer
B: n is a nonnegative integer
D: n is a negative integer
C: n is a positive integer or n is a negative integer ($C = A \vee D$).

The implication $A \rightarrow B$ is a tautology so $A \Rightarrow B$. Substituting B for A in C we get

$B \vee D$: n is a nonnegative integer or n is a negative integer,

which is a tautology.

However, C is a conditional statement; it is false when $n = 0$. It is not a coincidence that $n = 0$ makes A false.

The attempted substitution of B for A is therefore invalid. ■

The next example illustrates proper usage of the three substitution principles.

EXAMPLE 2.25

Using the Substitution Principles

Substituting an Equivalent Statement: Let A be the proposition $[\neg(P \wedge Q)]$ and let B be $[(\neg P) \vee (\neg Q)]$ (so $A \Leftrightarrow B$). Let C be the statement $[(\neg P) \vee (\neg Q)] \wedge R$. The first substitution principle asserts that C is logically equivalent to $[[\neg(P \wedge Q)]] \wedge R$. Table 2.18 demonstrates that this is true.

TABLE 2.18 Substituting an Equivalent Statement

P	Q	R	$[(\neg P) \vee (\neg Q)] \wedge R$	$[\neg(P \wedge Q)] \wedge R$
T	T	T	F	F
T	T	F	F	F
T	F	T	T	T
T	F	F	F	F
F	T	T	T	T
F	T	F	F	F
F	F	T	T	T
F	F	F	F	F

Replacing a Logic Variable in a Tautology: Let C be the tautology $B \vee (\neg B)$. Let A be the conditional statement $P \vee Q$. Then replacing B with A leads to another tautology:

$$[P \vee Q] \vee [\neg(P \vee Q)]$$

Table 2.19 provides a confirmation.

TABLE 2.19 Replacing a Logic Variable

B	P	Q	$B \vee (\neg B)$	$[P \vee Q] \vee [\neg(P \vee Q)]$
T	T	T	T	T
T	T	F	T	T
T	F	T	T	T
T	F	F	T	T
F	T	T	T	T
F	T	F	T	T
F	F	T	T	T
F	F	F	T	T

Using a Rule of Inference: Let A be the statement $[\neg P \wedge (P \vee Q)]$ and let B be Q. Example 2.26 on page 54 will show that $A \rightarrow B$ is a tautology. Thus, $A \Rightarrow B$. Let C be the statement $Q \wedge [\neg P \wedge (P \vee Q)]$. The third substitution principle asserts that C can be replaced by $Q \wedge Q$ when $[\neg P \wedge (P \vee Q)]$ is known to be true. The third row of the last two columns of Table 2.20 demonstrates that this is the case. The first row demonstrates that when $[\neg P \wedge (P \vee Q)]$ is false, the substitution is not necessarily valid.

TABLE 2.20 Using a Rule of Inference

P	Q	$P \vee Q$	$\neg P$	$\neg P \wedge (P \vee Q)$	$Q \wedge [\neg P \wedge (P \vee Q)]$	$Q \wedge Q$
T	T	T	F	F	F	T
T	F	T	F	F	F	F
F	T	T	T	T	T	T
F	F	F	T	F	F	F

■

2.4.1 Important Logical Equivalences and Rules of Inference

Some of the more useful logical equivalences and tautologies are presented in Table 2.21.[30] Most are based on intuitively clear ideas (you should discover the idea in each case). They can be formally justified by using truth tables.

TABLE 2.21 Fundamental Logical Equivalences

Idempotence[a]	**Domination**
$(P \vee P) \Leftrightarrow P$	$(P \vee \mathbf{T}) \Leftrightarrow \mathbf{T}$
$(P \wedge P) \Leftrightarrow P$	$(P \wedge \mathbf{F}) \Leftrightarrow \mathbf{F}$
Associativity	**Identity**
$[(P \vee Q) \vee R] \Leftrightarrow [P \vee (Q \vee R)]$	$(P \vee \mathbf{F}) \Leftrightarrow P$
$[(P \wedge Q) \wedge R] \Leftrightarrow [P \wedge (Q \wedge R)]$	$(P \wedge \mathbf{T}) \Leftrightarrow P$
Commutativity	**De Morgan's Laws**
$(P \vee Q) \Leftrightarrow (Q \vee P)$	$[\neg(P \vee Q)] \Leftrightarrow [(\neg P) \wedge (\neg Q)]$
$(P \wedge Q) \Leftrightarrow (Q \wedge P)$	$[\neg(P \wedge Q)] \Leftrightarrow [(\neg P) \vee (\neg Q)]$
Distributivity ($\wedge$ **over** $\vee$)	**Distributivity** ($\vee$ **over** $\wedge$)
$[P \wedge (Q \vee R)] \Leftrightarrow [(P \wedge Q) \vee (P \wedge R)]$	$[P \vee (Q \wedge R)] \Leftrightarrow [(P \vee Q) \wedge (P \vee R)]$
$[(P \vee Q) \wedge R] \Leftrightarrow [(P \wedge R) \vee (Q \wedge R)]$	$[(P \wedge Q) \vee R] \Leftrightarrow [(P \vee R) \wedge (Q \vee R)]$
Law of the Excluded Middle	**Law of Contradiction**
$[P \vee (\neg P)] \Leftrightarrow \mathbf{T}$	$[P \wedge (\neg P)] \Leftrightarrow \mathbf{F}$
Law of Double Negation (Involution)	**Law of Addition**
$\neg(\neg P) \Leftrightarrow P$	$[P \to (P \vee Q)] \Leftrightarrow \mathbf{T}$
Law of Simplification	
$[(P \wedge Q) \to P] \Leftrightarrow \mathbf{T}$	
$[(P \wedge Q) \to Q] \Leftrightarrow \mathbf{T}$	

[a] An idempotent is an algebraic element for which $x^2 = x$.

Properties of implication and the biconditional can be expressed in a second group of important logical equivalences and rules of inference (Table 2.22).

TABLE 2.22 Logical Equivalences and Rules of Inference for Implication and the Biconditional

Implication	**Negation Of An Implication**
$(P \to Q) \Leftrightarrow [\neg(P \wedge (\neg Q))] \Leftrightarrow [(\neg P) \vee Q]$	$[\neg(P \to Q)] \Leftrightarrow [P \wedge (\neg Q)]$
The Biconditional	**Transitivity of Biconditional**
$(P \leftrightarrow Q) \Leftrightarrow [(P \to Q) \wedge (Q \to P)]$	$[(P \leftrightarrow S_1) \wedge (S_1 \leftrightarrow S_2) \wedge \cdots \wedge (S_n \leftrightarrow Q)]$
$(P \leftrightarrow Q) \Rightarrow (P \to Q)$	$\Leftrightarrow (P \leftrightarrow Q)$
$(P \leftrightarrow Q) \Rightarrow (Q \to P)$	

One rule of inference is so important that it has been given the Latin name *modus ponens*, which means "the proposing mode". This inference rule captures the essence of how an implication is used: First show that the hypothesis is true, next show that the implication is true, and finally, assert that the conclusion is true. The final collection of

[30] The table is also on page 83 and at http://www.mathcs.bethel.edu/~gossett/DiscreteMathWithProof/ .

tautologies (Table 2.23) expresses this and other ideas that form the basis for most of the proof strategies covered in Chapter 3.

Once again, you should pay attention to the intuitive content. For example, the second version of proof by contradiction (Table 2.23) can be understood as follows: If we assume that P is true and Q is false, we are then led to the conclusion that P is false (contradicting the assumption that P is true).[31] Therefore, it must be the case that P is true and Q is true, so $P \to Q$ is true.

TABLE 2.23 Logical Equivalences and Rules of Inference Related to Theorems

Modus Ponens (Law of Detachment)	**Law of Hypothetical Syllogism**
$[P \wedge (P \to Q)] \Rightarrow Q$	$[(P \to Q) \wedge (Q \to R)] \Rightarrow (P \to R)$
Contrapositive (Indirect Proof)	**Proof by Contradiction (Reductio Ad Absurdum)**
$[P \to Q] \Leftrightarrow [(\neg Q) \to (\neg P)]$	$[P \to Q] \Leftrightarrow [(P \wedge (\neg Q)) \to (R \wedge (\neg R))]$
	$[P \to Q] \Leftrightarrow [(P \wedge (\neg Q)) \to (\neg P)]$
	$[P \to Q] \Leftrightarrow [(P \wedge (\neg Q)) \to Q]$
Laws of Disjunction	**Proof by Cases**
$[(P \vee Q) \wedge (\neg P)] \Rightarrow Q$	$[(P \to R) \wedge (Q \to R)] \Rightarrow [(P \vee Q) \to R]$
$[(P \vee Q) \wedge (P \wedge (\neg Q))] \Rightarrow P$	$[(P \to Q) \wedge (P \to R)] \Rightarrow [P \to (Q \wedge R)]$

2.4.2 Proving that a Statement is a Tautology

There are now two methods we can use to prove that a statement is a tautology: produce the truth table for the statement and observe the final column (looking for all T's), or use the logical equivalences and rules of inference from the previous tables to create a sequence of equivalent statements, ending with a statement that is obviously true.

EXAMPLE 2.26

Proving $[\neg P \wedge (P \vee Q)] \to Q$ is a Tautology

You should already be proficient at using truth tables to show that $[\neg P \wedge (P \vee Q)] \to Q$ is a tautology. The following proof uses substitution and several of the fundamental logical equivalences (Table 2.21) and one logical equivalence from Table 2.22.

$$
\begin{array}{rcrl}
[\neg P \wedge (P \vee Q)] \to Q & \Leftrightarrow & [(\neg P \wedge P) \vee (\neg P \wedge Q)] \to Q & \text{distributivity} \\
& \Leftrightarrow & [(P \wedge \neg P) \vee (\neg P \wedge Q)] \to Q & \text{commutativity} \\
& \Leftrightarrow & [F \vee (\neg P \wedge Q)] \to Q & \text{law of contradiction} \\
& \Leftrightarrow & [(\neg P \wedge Q) \vee F] \to Q & \text{commutativity} \\
& \Leftrightarrow & (\neg P \wedge Q) \to Q & \text{identity} \\
& \Leftrightarrow & \neg(\neg P \wedge Q) \vee Q & \text{implication} \\
& \Leftrightarrow & [\neg(\neg P) \vee \neg Q] \vee Q & \text{De Morgan} \\
& \Leftrightarrow & (P \vee \neg Q) \vee Q & \text{double negation} \\
& \Leftrightarrow & P \vee (\neg Q \vee Q) & \text{associativity} \\
& \Leftrightarrow & P \vee (Q \vee \neg Q) & \text{commutativity} \\
& \Leftrightarrow & P \vee T & \text{law of the excluded middle} \\
& \Leftrightarrow & T & \text{domination}
\end{array}
$$

[31] This only illustrates one-half of the logical equivalence in that assertion.

The preceding sequence of steps is not the only valid sequence. As you work on proofs of this kind, you will probably find yourself trying more than one option at various points. This will involve a liberal use of scratch paper. Eventually, you will find a sequence that works, and you can transfer it to your official page. ■

Proofs using truth tables are certainly easier to use for problems of this size. However, that technique does not adapt well to more complicated proofs. In particular, when proving theorems in various branches of mathematics, the second technique is what is used.[32]

EXAMPLE 2.27

Tautologies with Biconditionals

Recall that a biconditional is a tautology if and only if its major components are logically equivalent. We can simplify the proof somewhat by starting with one-half of the biconditional and showing that it is logically equivalent to the other half.

Notice the linear nature of the following proof; we start with one-half and derive the second half. This is in contrast to a two-column proof, where we manipulate both columns simultaneously and arrive at the same statement on both sides at the bottom. Two-column proofs can always be turned into linear proofs by starting down the left column, then working back up the right column. Linear proofs are generally considered to be more elegant. Two-column proofs are acceptable on your scratch paper but should be translated for final work.

The following sequence of substitutions shows that

$$[(R \vee P) \rightarrow (R \vee Q)] \leftrightarrow [R \vee (P \rightarrow Q)]$$

is a tautology.

$[(R \vee P) \rightarrow (R \vee Q)]$	$\Leftrightarrow$	$\neg(R \vee P) \vee (R \vee Q)$	implication
	$\Leftrightarrow$	$[(\neg R) \wedge (\neg P)] \vee (R \vee Q)$	De Morgan
	$\Leftrightarrow$	$([(\neg R) \wedge (\neg P)] \vee R) \vee Q$	associativity
	$\Leftrightarrow$	$(R \vee [(\neg R) \wedge (\neg P)]) \vee Q$	commutativity
	$\Leftrightarrow$	$([R \vee (\neg R)] \wedge [R \vee (\neg P)]) \vee Q$	distributivity
	$\Leftrightarrow$	$(\mathbf{T} \wedge [R \vee (\neg P)]) \vee Q$	law of the excluded middle
	$\Leftrightarrow$	$([R \vee (\neg P)] \wedge \mathbf{T}) \vee Q$	commutativity
	$\Leftrightarrow$	$[R \vee (\neg P)] \vee Q$	identity
	$\Leftrightarrow$	$R \vee [(\neg P) \vee Q)]$	associativity
	$\Leftrightarrow$	$R \vee (P \rightarrow Q)$	implication

This tautology can be interpreted as saying that OR distributes over implication: $R \vee (P \rightarrow Q) \Leftrightarrow (R \vee P) \rightarrow (R \vee Q)$. However, Exercise 11 in Exercises 2.4.3 on page 57 shows that a similar assertion does not hold for AND. ■

EXAMPLE 2.28

Verifying Proof by Contradiction

It is possible to prove one of the proof by contradiction logical equivalences using only the fundamental logical equivalences (Table 2.21 on page 53) and the logical equivalences and rules of inference for implication and the biconditional (Table 2.22). The

[32] Actually, as will be seen in Chapter 3, a more informal presentation is usually adopted. However, in essence, the sequence of logical equivalences and rules of inference provides the justification for the more informal presentation.

tautology of interest is

$$[P \to Q] \Leftrightarrow [(P \land (\neg Q)) \to (\neg P)]$$

$$\begin{aligned}
[(P \land (\neg Q)) \to (\neg P)] &\Leftrightarrow [\neg(P \to Q)] \to (\neg P) && \text{negation of an implication} \\
&\Leftrightarrow \neg[\neg(P \to Q)] \lor (\neg P) && \text{implication} \\
&\Leftrightarrow (P \to Q) \lor (\neg P) && \text{double negation} \\
&\Leftrightarrow (\neg P \lor Q) \lor (\neg P) && \text{implication} \\
&\Leftrightarrow (\neg P) \lor ((\neg P) \lor Q) && \text{commutativity} \\
&\Leftrightarrow ((\neg P) \lor (\neg P)) \lor Q && \text{associativity} \\
&\Leftrightarrow (\neg P) \lor Q && \text{idempotence} \\
&\Leftrightarrow P \to Q && \text{implication}
\end{aligned}$$

Here is a slightly different proof:

$$\begin{aligned}
[(P \land (\neg Q)) \to (\neg P)] &\Leftrightarrow [\neg(P \land \neg Q)] \lor (\neg P) && \text{implication} \\
&\Leftrightarrow ((\neg P) \lor \neg(\neg Q)) \lor (\neg P) && \text{De Morgan} \\
&\Leftrightarrow ((\neg P) \lor Q) \lor (\neg P) && \text{double negation} \\
&\Leftrightarrow (\neg P) \lor ((\neg P) \lor Q) && \text{commutativity} \\
&\Leftrightarrow ((\neg P) \lor (\neg P)) \lor Q && \text{associativity} \\
&\Leftrightarrow \neg P \lor Q && \text{idempotence} \\
&\Leftrightarrow P \to Q && \text{implication}
\end{aligned}$$

■

✔ Quick Check 2.8

1. The law of simplification asserts that the implication $(P \land Q) \to P$ is a tautology. It *does not* mean that $P \land Q$ may always be replaced by P. That is, the rule of inference, $(P \land Q) \Rightarrow P$, must be used carefully. Verify the warning in the previous sentence by showing that $P \land Q$ cannot be replaced by P when Q is false.
2. Using the substitution principles and only the fundamental logical equivalences (Table 2.21) and the logical equivalences and rules of inference for implication and the biconditional (Table 2.22), prove that the following are tautologies.
   (a) $[P \land (P \to Q)] \to Q$ (the tautology underlying modus ponens)
   (b) $[P \to Q] \leftrightarrow [(P \land (\neg Q)) \to Q]$ (proof by contradiction) ☑

2.4.3 Exercises

The exercises marked with ★ have detailed solutions in Appendix H.

1. Prove that the statement $[(P \lor Q) \land R] \leftrightarrow [P \lor (Q \land R)]$ is *not* a tautology, even though it looks like an associative law. You may use a truth table if you wish.
2. ★ Write the contrapositive, converse, and inverse of $(P \land Q) \to (R \lor S)$. Use De Morgan's laws to simplify.
3. One version of proof by contradiction (also called *reductio ad absurdum*) is based on $[P \to Q] \Leftrightarrow [(P \land (\neg Q)) \to (\neg P)]$. Write a few sentences describing the intuitive idea behind this tautology. For extra credit, translate the Latin phrase *reductio ad absurdum* into English.
4. Use truth tables to prove the underlying tautologies for each of the fundamental logical equivalences (Table 2.21). Show the intermediate expressions.
   (a) Idempotence
   (b) Domination
   (c) ★ Associativity
   (d) Identity
   (e) Commutativity
   (f) De Morgans's laws
   (g) ★ Distributivity ($\land$ over $\lor$)
   (h) Distributivity ($\lor$ over $\land$)
   (i) Law of the excluded middle
   (j) Law of contradiction
   (k) Law of double negation
   (l) Law of addition
   (m) Law of simplification
5. Use truth tables to prove the following logical equivalences and rules of inference for implication and the biconditional (Table 2.22). Work with the underlying tautologies and show the intermediate expressions.
   (a) ★ Implication
   (b) Negation of an implication
   (c) The biconditional

6. Use truth tables to prove the logical equivalences and rules of inference related to theorems (Table 2.23). Work with the underlying tautologies and show the intermediate expressions.

(a) Modus ponens
(b) Law of hypothetical syllogism
(c) Contrapositive (d) Proof by contradiction
(e) Laws of disjunction (f) Proof by cases

7. Use only the substitution rules and logical equivalences in Tables 2.21 and 2.22 to prove these logical equivalences and rules of inference related to theorems from Table 2.23.

(a) Contrapositive
(b) Laws of disjunction (Work with the underlying tautologies.)

8. Without using truth tables, show that

$$[(A \vee B) \wedge (\neg A)] \wedge [(A \vee B) \wedge (\neg B)]$$

is logically equivalent to the statement F (false). Conclude that the original statement is false for all values of A and B. (Thus, it is a contradiction, as defined in Definition 2.23.)

9. Show (perhaps using a truth table) that

(a) ★ $[(P \to Q) \wedge Q] \to P$ is not a tautology. (The *fallacy of affirming the consequent.*)
(b) $[(P \to Q) \wedge (\neg P)] \to (\neg Q)$ is not a tautology. (The *fallacy of denying the antecedent.*)
(c) But $[(P \to Q) \wedge (\neg Q)] \to (\neg P)$ is a tautology.

10. Prove that each of the following statements is a tautology. Do not use truth tables.

(a) $P \to P$ (b) ★ $P \to [(\neg P) \to Q]$
(c) ★ $[P \to (Q \wedge (\neg Q))] \to (\neg P)$
(d) $[(P \vee Q) \wedge (P \vee (\neg Q))] \to P$
(e) $[P \to (R \to Q)] \leftrightarrow [(P \wedge R) \to Q]$
(f) $[P \to Q] \to [(R \vee P) \to (R \vee Q)]$
(g) $[(P \vee Q) \wedge (P \to R) \wedge (Q \to R)] \to R$
[*Hints*: Start by associating $(P \to R) \wedge (Q \to R)$. Derive the subexpression $[(\neg P \wedge \neg Q) \vee R]$. Look for the law of simplification at the end.]

11. Show that $R \wedge (P \to Q)$ and $(R \wedge P) \to (R \wedge Q)$ are *not* logically equivalent (so AND does not distribute over implication).

12. Prove that each of the following statements is a tautology. Do not use truth tables.

(a) $\neg[P \vee ((\neg P) \wedge Q)] \leftrightarrow [(\neg P) \wedge (\neg Q)]$
(b) $(P \wedge Q) \to (P \to Q)$ (c) $(P \wedge Q) \to (P \vee Q)$
(d) $(\neg P) \to (P \to Q)$
(e) $\neg[P \leftrightarrow Q] \leftrightarrow [(P \wedge (\neg Q)) \vee (Q \wedge (\neg P))]$
(f) ★ $\neg(P \to Q) \to P$

13. Consider the following collection of statements, where we assume $n \in \mathbb{Z}$ limits the universe of discourse:

A: n is a positive integer
B: n is a nonnegative integer
D: $n \geq 1$
C: if n is a positive integer, then $n \geq 1$
$(C = A \to D)$.

As in Example 2.24 (page 51), $A \to B$ is a tautology so $A \Rightarrow B$. Is it valid to use the third substitution principle to replace A by B in C? Discuss your answer in detail.

14. Consider the following demonstration that $(P \wedge Q) \vee (\neg P)$ is a tautology.

$(P \wedge Q) \vee (\neg P)$	$\Leftrightarrow P \vee (\neg P)$	law of simplification
	$\Leftrightarrow T$	law of the excluded middle

However, the following truth table indicates that $(P \wedge Q) \vee (\neg P)$ is *not* a tautology (the second row contains an F in the final column). Resolve the contradiction.

P	Q	$\neg P$	$P \wedge Q$	$(P \wedge Q) \vee (\neg P)$
T	T	F	T	T
T	F	F	F	F
F	T	T	F	T
F	F	T	F	T

15. Consider the following demonstration that $\neg[\neg(P \to Q)] \leftrightarrow [P \leftrightarrow Q]$ is a tautology.

$\neg[\neg(P \to Q)] \leftrightarrow [P \leftrightarrow Q]$	$\Leftrightarrow$	$\neg[\neg(P \to Q)] \leftrightarrow [P \to Q]$	$(P \leftrightarrow Q)$ true implies that $(P \to Q)$ is true
	$\Leftrightarrow$	$(P \to Q) \leftrightarrow (P \to Q)$	double negation
	$\Leftrightarrow$	T	$A \to A$ has been proved in several exercises

However, the following truth table indicates that $\neg[\neg(P \to Q)] \leftrightarrow [P \leftrightarrow Q]$ is *not* a tautology (the third row contains an F in the final column). Resolve the contradiction.

P	Q	$P \leftrightarrow Q$	$P \to Q$	$[\neg(P \to Q)]$	$\neg[\neg(P \to Q)]$	$\neg[\neg(P \to Q)] \leftrightarrow [P \leftrightarrow Q]$
T	T	T	T	F	T	T
T	F	F	F	T	F	T
F	T	F	T	F	T	F
F	F	T	T	F	T	T

16. Suppose that in a homework assignment, Mary is asked to use the law of addition to help prove that $(P \wedge (Q \wedge R)) \rightarrow [((R \wedge P) \wedge Q) \vee Q]$ is a tautology. Mary quickly concludes that since $(P \wedge (Q \wedge R))$ and $((R \wedge P) \wedge Q)$ are different compound statements, the law of addition cannot be used. Additionally, she makes the assumption that $(P \wedge (Q \wedge R)) \rightarrow [((R \wedge P) \wedge Q) \vee Q]$ is not a tautology because she could not use the law specified on her homework assignment.

However, the following truth table indicates that $(P \wedge (Q \wedge R)) \rightarrow [((R \wedge P) \wedge Q) \vee Q]$ *is* a tautology (the final column contains all T's). Resolve the contradiction.

P	Q	R	$Q \wedge R$	$(P \wedge (Q \wedge R))$	$R \wedge P$	$((R \wedge P) \wedge Q)$	$[((R \wedge P) \wedge Q) \vee Q]$	$(P \wedge (Q \wedge R)) \rightarrow [((R \wedge P) \wedge Q) \vee Q]$
T	T	T	T	T	T	T	T	T
T	T	F	F	F	F	F	T	T
T	F	T	F	F	T	F	F	T
T	F	F	F	F	F	F	F	T
F	T	T	T	F	F	F	T	T
F	T	F	F	F	F	F	T	T
F	F	T	F	F	F	F	F	T
F	F	F	F	F	F	F	F	T

17. Consider the following demonstration that $[P \leftrightarrow ((\neg Q) \wedge (\neg R))] \rightarrow (\neg(Q \wedge R) \rightarrow P)$ is a tautology.

$$[P \leftrightarrow ((\neg Q) \wedge (\neg R))] \rightarrow (\neg(Q \wedge R) \rightarrow P)$$

$$\Leftrightarrow [P \leftrightarrow ((\neg Q) \wedge (\neg R))] \rightarrow (((\neg Q) \wedge (\neg R)) \rightarrow P) \quad \text{De Morgan}$$

$$\Leftrightarrow T \quad \text{biconditional}$$

However, the following truth table indicates that $[P \leftrightarrow ((\neg Q) \wedge (\neg R))] \rightarrow (\neg(Q \wedge R) \rightarrow P)$ is *not* a tautology (the sixth and seventh rows contain an **F** in the final column). Resolve the contradiction.

P	Q	R	$((\neg Q) \wedge (\neg R))$	$[P \leftrightarrow ((\neg Q) \wedge (\neg R))]$	$Q \wedge R$	$\neg(Q \wedge R)$	$(\neg(Q \wedge R) \rightarrow P)$	$[P \leftrightarrow ((\neg Q) \wedge (\neg R))] \rightarrow (\neg(Q \wedge R) \rightarrow P)$
T	T	T	F	F	T	F	T	T
T	T	F	F	F	F	T	T	T
T	F	T	F	F	F	T	T	T
T	F	F	T	T	F	T	T	T
F	T	T	F	T	T	F	T	T
F	T	F	F	T	F	T	F	F
F	F	T	F	T	F	T	F	F
F	F	F	T	F	F	T	F	T

2.5 Boolean Algebras

You probably have noticed the similarity between the fundamental set properties on page 32 and several of the tautologies in Section 2.4. For example, both have versions of De Morgan's laws:

$$\overline{(A \cup B)} = \overline{A} \cap \overline{B}$$

$$\neg(P \vee Q) \Leftrightarrow (\neg P) \wedge (\neg Q)$$

This commonality is not an accident. The goal of this section is to develop an abstraction (called a Boolean algebra), which describes many seemingly disparate examples. The two most important examples, for our purposes, are sets and propositions.

The abstraction is the result of work by George Boole, which he completed in the mid-1800s. The definition given here follows the axiomatic definition given by E. V. Huntington in 1904 [53].

The *axioms* are properties that any example of a Boolean algebra must satisfy. A more detailed discussion of axioms will be provided in Section 3.1.2.

DEFINITION 2.26 ***Boolean Algebra***

A *Boolean algebra*, $\mathbb{B}$, consists of an associated set, B, together with three operators and four axioms. The binary operators $+$ and $\cdot$ map elements of $B \times B$ to elements of B. The unary operator *complement*, $\overline{}$, maps elements of B to elements of B.

Parentheses have the highest precedence. The complement operator, $\overline{}$, has the second highest precedence, followed by $\cdot$ and then by $+$, the operator with lowest precedence.

The axioms are as follows:

Identity There exist distinct elements, 0 and 1, in B such that for every $x \in B$

$$x + 0 = x$$
$$x \cdot 1 = x$$

Complement For every $x \in B$, there exists a unique element $\overline{x} \in B$ such that

$$x + \overline{x} = 1$$
$$x \cdot \overline{x} = 0$$

Commutativity For every pair of (not necessarily distinct) elements $x, y \in B$

$$x + y = y + x$$
$$x \cdot y = y \cdot x$$

Distributivity For every three elements $x, y, z \in B$ (not necessarily distinct)

$$x \cdot (y + z) = x \cdot y + x \cdot z$$
$$x + y \cdot z = (x + y) \cdot (x + z)$$

You should note that the operators $+$ and $\cdot$ are not the usual arithmetic operators "plus" and "times"; nor is the complement operator, $\overline{}$, necessarily set complement. The symbols 0 and 1 do not represent numbers on the real line. However, the names *plus*, *times*, *complement*, *zero*, and *one* are commonly used informally when discussing Boolean algebras. Pay attention to the use of the words *distinct* and *unique*.

The statement "$+$ and $\cdot$ map elements of $B \times B$ to elements of B" is just a formal way to say that if $x, y \in B$, then $x + y$ and $x \cdot y$ are also in B.[33]

DEFINITION 2.27 ***Boolean Expression***

Let $\mathbb{B}$ be a Boolean algebra with associated set, B. A *Boolean expression over* $\mathbb{B}$ is any algebraic expression that is composed using elements from B, the operators $+$, $\cdot$, and $\overline{}$, and variables whose possible values are elements of B.

The elements of a Boolean algebra, $\mathbb{B}$, are understood to be the elements of the associated set, B.

[33] Think of $(x, y) \in B \times B$ being mapped to some $z \in B$.

EXAMPLE 2.29 **The Simplest Boolean Algebra**

Here is the simplest Boolean algebra. Let $B = \{0, 1\}$ and define the operators by the following equations and tables.

- $\overline{0} = 1$ and $\overline{1} = 0$
-

+	0	1
0	0	1
1	1	1

·	0	1
0	0	0
1	0	1

Notice the unusual definition for $1 + 1$.

The four axioms are easy to verify.

Identity Using the tables,

- $0 + 0 = 0$ and $1 + 0 = 1$
- $0 \cdot 1 = 0$ and $1 \cdot 1 = 1$

Complement Using the uniquely defined values for complements and the tables,

- $0 + \overline{0} = 0 + 1 = 1$ and $1 + \overline{1} = 1 + 0 = 1$
- $0 \cdot \overline{0} = 0 \cdot 1 = 0$ and $1 \cdot \overline{1} = 1 \cdot 0 = 0$

Commutativity Commutativity is built into the tables (notice the symmetry).

Distributivity An exhaustive verification would require eight cases for each of the distributive axioms. Only one case, $x = 0, y = z = 1$, will be examined for the two distributive axioms. You should try at least one other case on your own.

- $0 \cdot (1 + 1) = 0 \cdot 1 = 0 = 0 + 0 = 0 \cdot 1 + 0 \cdot 1$
- $0 + 1 \cdot 1 = 0 + 1 = 1 = 1 \cdot 1 = (0 + 1) \cdot (0 + 1)$

This completes the verification that B is a Boolean algebra.

Let a, b, c be variables whose potential values are in $B = \{0, 1\}$. Then the following are all valid Boolean expressions over B.

Expression	**Value when $a = 0, b = 1, c = 1$**
$1 \cdot a + (0 + b) \cdot c$	1
$(a \cdot b) + (c \cdot c) \cdot \overline{b}$	0
$\overline{(a + b) \cdot c}$	0

(2.1)

Notice that for any $x \in B$, $x + x = x$ and $x \cdot x = x$. Both are true by the tables used to create this Boolean algebra. This pair of equalities will soon be shown to be valid for *any* Boolean algebra. ■

✔ Quick Check 2.9

1. Use the following values of x, y, z to verify both parts of the distributivity axiom for the Boolean algebra defined in Example 2.29.
   (a) $x = 1, y = 0, z = 1$
   (b) $x = 1, y = 1, z = 1$ ☑

2.5.1 Sets and Propositions as Boolean Algebras

It is not difficult to use both set theory and propositional logic to provide important examples of Boolean algebras. In fact, they will provide an unlimited supply of examples. This section will provide the details.

Sets as Boolean Algebras

Let S be any nonempty set. The following correspondences will set up the interpretation as a Boolean algebra:

- Let $B = \mathcal{P}(S)$. That is, the elements of the Boolean algebra will be all the subsets of S.
- Let 1 be the set S and 0 be $\emptyset$. Observe that for every nonempty set, S, the power set, $\mathcal{P}(S)$, contains these two distinct subsets.
- Let $+$ be $\cup$ and $\cdot$ be $\cap$.
- Let the Boolean complement be set complement.

Note well: In Section 2.1, the *elements* of S were much more in the forefront. For this construction, the elements of S are only of interest because they are used to define the subsets of S. The *subsets* of S are the items of main interest because the subsets of S are the elements of the Boolean algebra.

It is easy to see that the four axioms for a Boolean algebra are satisfied for the assignments listed above. Recall that the elements of $\mathcal{P}(S)$ are the subsets of S. Also, in this context, S has the role of the universal set, U, because no set in $\mathcal{P}(S)$ contains any element that is not in S.

Identity There exist distinct elements, $\emptyset$ and S, in $\mathcal{P}(S)$ such that for all elements $A \in \mathcal{P}(S)$

$$\begin{aligned} A \cup \emptyset &= A \quad \text{set identity properties} \\ A \cap S &= A \end{aligned}$$

Complement For every $A \in \mathcal{P}(S)$, there exists a unique (by the definition of set complement) element $\overline{A} \in \mathcal{P}(S)$ such that

$$\begin{aligned} A \cup \overline{A} &= S \quad \text{set complement properties} \\ A \cap \overline{A} &= \emptyset \end{aligned}$$

Commutativity For every pair of (not necessarily distinct) elements $A, B \in \mathcal{P}(S)$

$$\begin{aligned} A \cup B &= B \cup A \quad \text{set commutativity properties} \\ A \cap B &= B \cap A \end{aligned}$$

Distributivity For every three elements $A, B, C \in \mathcal{P}(S)$ (not necessarily distinct)

$$\begin{aligned} A \cap (B \cup C) &= (A \cap B) \cup (A \cap C) \quad \text{set distributivity properties} \\ A \cup (B \cap C) &= (A \cup B) \cap (A \cup C) \end{aligned}$$

EXAMPLE 2.30 **Another Small Boolean Algebra**

Let $S = \{a, b\}$. Then $B = \mathcal{P}(S) = \{\emptyset, \{a\}, \{b\}, \{a, b\}\}$. As defined previously, let 1 be the set S and 0 be $\emptyset$, let $+$ be $\cup$ and $\cdot$ be $\cap$. Finally, let the Boolean complement be set complement. Then B, together with these operations, is a Boolean algebra with four elements.

Let $C, D, E \subseteq S$ (that is, $C, D, E \in \mathcal{P}(S)$). The following are valid Boolean expressions.

Boolean Notation	Set Notation	Value when $C = \{a\}$, $D = \{a, b\}$, $E = \emptyset$
$(1 + C) \cdot (D + 0)$	$(S \cup C) \cap (D \cup \emptyset)$	$\{a, b\}$
$\overline{C + D \cdot E}$	$\overline{C \cup (D \cap E)}$	$\{b\}$
$C \cdot \overline{E}$	$C \cap \overline{E}$	$\{a\}$

■

Collections of Propositions as Boolean Algebras

What correspondences are appropriate to show that a collection of propositions can be viewed as a Boolean algebra? The easiest part is identifying the corresponding operations. Before reading the next paragraphs, try to determine which logic operators should correspond to $+$, $\cdot$, and $\overline{}$, respectively.

Determining the appropriate elements of B takes a bit more thought. The special identity elements, 0 and 1, will correspond to F and T, respectively. Should B contain additional elements? What role should propositions play? One other issue needs to be addressed: There is no "=" in propositional logic. Which logic operator is the proper replacement?

The full set of correspondences is as follows:

- $B = \{F, T\}$.
- $+$ becomes $\vee$, $\cdot$ becomes $\wedge$, $\overline{}$ becomes $\neg$.
- $=$ is replaced by logical equivalence $\Leftrightarrow$ or by the biconditional $\leftrightarrow$.
- The Boolean variables are propositions.

It is easy to verify that these associations satisfy the four Boolean axioms. Here are the details.

Identity There exist distinct elements, F and T, in B such that for all propositions $P \in B$

$$P \vee F \Leftrightarrow P \quad \text{identity tautologies}$$
$$P \wedge T \Leftrightarrow P$$

Complement For every $P \in B$, there exists a unique (by definition of negation) proposition $(\neg P) \in B$ such that

$$P \vee (\neg P) \Leftrightarrow T \quad \text{law of the excluded middle}$$
$$P \wedge (\neg P) \Leftrightarrow F \quad \text{law of contradiction}$$

Commutativity For every pair of (not necessarily distinct) propositions $P, Q \in B$

$$P \vee Q \Leftrightarrow Q \vee P \quad \text{commutativity laws}$$
$$P \wedge Q \Leftrightarrow Q \wedge P$$

Distributivity For every three propositions $P, Q, R \in B$ (not necessarily distinct)

$$P \wedge (Q \vee R) \Leftrightarrow (P \wedge Q) \vee (P \wedge R) \quad \text{distributive laws}$$
$$P \vee (Q \wedge R) \Leftrightarrow (P \vee Q) \wedge (P \vee R)$$

The next example illustrates the change in notation when propositions are viewed as Boolean expressions.

EXAMPLE 2.31 **Some Propositional Expressions**

Boolean Notation	**Propositional Notation**	**Value when P = T, Q = T, R = F**
$(1 + P) \cdot (Q + 0)$	$(\text{T} \vee P) \wedge (Q \vee \text{F})$	T
$\overline{P + Q \cdot R}$	$\neg(P \vee (Q \wedge R))$	F
$P \cdot \overline{R}$	$P \wedge (\neg R)$	T

■

2.5.2 Proving Additional Boolean Algebra Properties

One of the advantages in having an abstraction like "Boolean algebra" is that we can prove properties of the abstraction and never need to re-prove them for each of the multitude of concrete examples of the abstraction. Thus, if we prove the De Morgan's laws using only the axioms of a Boolean algebra, there would be no need to prove them for sets and propositional logic.[34]

This section will demonstrate the process of proving propositions that hold for all Boolean algebras. Before starting, notice that Boolean algebras have some properties that do not correspond to any of the properties you have used for many years. For example, recall that in the real numbers, whenever $xy = 0$, either $x = 0$ or $y = 0$ (or both). This is often described by saying that there are no *zero divisors* in $\mathbb{R}$.

DEFINITION 2.28 ***Zero Divisor***

Two nonzero elements, x and y, in an algebraic system are called *zero divisors* if $xy = 0$.

EXAMPLE 2.32

Zero Divisors in Boolean Algebras

In a Boolean algebra, the fact that $xy = 0$ does not necessarily mean that $x = 0$ or $y = 0$. For example, consider the Boolean algebra in Example 2.30. Neither the element $F = \{a\}$ nor the element $G = \{b\}$ is the empty set (0). However, $F \cap G = \emptyset$ $(F \cdot G = 0)$. ■

Here is a simple extension to Definition 2.26.

PROPOSITION 2.29 ***The Uniqueness of 0 and 1***

In any Boolean algebra, the distinct elements 0 and 1 are unique.

Proof: To see that 0 is unique, suppose that there is another element, $0'$, that satisfies the identity axiom. That is, for all $x \in B$, $x + 0' = x$. In that case,

$$\begin{aligned} 0' &= 0' + 0 && \text{0 is an identity element for +} \\ &= 0 + 0' && \text{commutativity} \\ &= 0 && \text{0}' \text{ is an identity element for +} \end{aligned}$$

Thus $0' = 0$ and the identity element for + is unique. A similar proof establishes the uniqueness of 1 for the $\cdot$ operator. □

One significant consequence of the definition of Boolean algebras has not yet been mentioned. It is important enough to be given prominence.

The Duality Principle for Boolean Algebras

Let T be a theorem that is valid over a Boolean algebra. Then if all 0s and 1s are exchanged, and if all $+$ and $\cdot$ are exchanged (with a suitable change in parentheses to preserve operator precedence), the result is a theorem that is also valid over the Boolean algebra.

Proof: The proof is an immediate consequence of the symmetry between 0 and 1 and between $+$ and $\cdot$ in the definition of a Boolean algebra. □

[34]There are pedagogical reasons why this strategy has not been followed in this text.

Note well: The duality principle is valid only if the Boolean expression represented by T is always true. The use of parentheses is illustrated by the Absorption property in Table 2.24 (below).

EXAMPLE 2.33

Duality in Action

It is easy to show that $(y \cdot \overline{x}) + (y \cdot x) = y$ for any elements, x and y, in a Boolean algebra. The proof follows.

$$\begin{aligned} (y \cdot \overline{x}) + (y \cdot x) &= y \cdot (\overline{x} + x) && \text{distributivity} \\ &= y \cdot (x + \overline{x}) && \text{commutativity} \\ &= y \cdot 1 && \text{complement} \\ &= y && \text{identity} \end{aligned}$$

A similar proof would show that $(y + \overline{x}) \cdot (y + x) = y$ is always true. However, a simple appeal to the duality principle is enough to establish this. ■

The axioms of Boolean algebras can be used to prove the fundamental properties of Boolean algebras listed in Table 2.24.[35] Notice the duality principle at work in this list. The axioms and fundamental properties can be viewed together on page 84.[36]

TABLE 2.24 Fundamental Boolean Algebra Properties

Idempotence	**Domination**
$x + x = x$	$x + 1 = 1$
$x \cdot x = x$	$x \cdot 0 = 0$
Associativity	**De Morgan's Laws**
$(x + y) + z = x + (y + z)$	$\overline{x + y} = \overline{x} \cdot \overline{y}$
$(x \cdot y) \cdot z = x \cdot (y \cdot z)$	$\overline{x \cdot y} = \overline{x} + \overline{y}$
Involution	**Absorption**
$\overline{\overline{x}} = x$	$x + x \cdot y = x$
	$x \cdot (x + y) = x$

One consequence of associativity is that we can be informal and write $x + y + z$ instead of $(x + y) + z$ or $x + (y + z)$. Since $(x + y) + z = x + (y + z)$, there is no ambiguity in the informal form.

Most of the proofs will be left as exercises. Here is a proof that $\overline{\overline{x}} = \overline{(\overline{x})} = x$.

Proof of the Involution Property:

$$\begin{aligned} \overline{x} + x &= x + \overline{x} && \text{commutativity} \\ &= 1 && \text{complement} \end{aligned}$$

This means that the unique complement of $\overline{x}$ is x (look carefully at the complement axiom). That is, $\overline{\overline{x}} = x$. □

[35] See Exercise 6 on page 67.

[36] The tables are also available in pdf format at http://www.mathcs.bethel.edu/~gossett/DiscreteMathWithProof/ .

✔ Quick Check 2.10

Prove that in any Boolean algebra

1. $\overline{0} = 1$ **2.** $\overline{1} = 0$ ☑

The fundamental Boolean algebra properties can be used to prove additional properties. The style of proof should look familiar.

PROPOSITION 2.30

Let x and y be elements of a Boolean algebra. Then

$$x \cdot y = x \quad \text{iff} \quad x \cdot \overline{y} = 0$$

Proof: Assume that $x \cdot \overline{y} = 0$. Then

$$\begin{aligned} x \cdot y &= (x \cdot y) + 0 && \text{identity} \\ &= (x \cdot y) + (x \cdot \overline{y}) && \text{by assumption} \\ &= x \cdot (y + \overline{y}) && \text{distributivity} \\ &= x \cdot 1 && \text{complement} \\ &= x && \text{identity.} \end{aligned}$$

Now assume $x \cdot y = x$. Then

$$\begin{aligned} x \cdot \overline{y} &= (x \cdot y) \cdot \overline{y} && \text{by assumption} \\ &= x \cdot (y \cdot \overline{y}) && \text{associativity} \\ &= x \cdot 0 && \text{complement} \\ &= 0 && \text{domination.} \end{aligned}$$

□

COROLLARY 2.31

Let x and y be elements of a Boolean algebra. Then

$$x + y = x \quad \text{iff} \quad x + \overline{y} = 1$$

Proof: True by the duality principle. □

The concept of "symmetric difference" is valid for Boolean algebras. The definition coincides, after using Proposition 2.18 (page 30), with the definition for *sets* (page 27).

DEFINITION 2.32 *Symmetric Difference*

Let x and y be elements of a Boolean algebra. Then their *symmetric difference* is denoted $x \,\Delta\, y$ and is defined by

$$x \,\Delta\, y = x \cdot \overline{y} + \overline{x} \cdot y$$

The following proposition is quite useful. It permits manipulations similar to a familiar practice when solving equations over the real numbers. Over the reals, if we know that $x = y$, then it is valid to conclude that $x - y = 0$.

PROPOSITION 2.33

Let x and y be elements of a Boolean algebra. Then

$$x = y \quad \text{iff} \quad x \,\Delta\, y = 0$$

$x = y$ iff $x \triangle y = 0$.

Proof: Suppose that $x = y$. Then

$$
\begin{aligned}
x \triangle y &= x \cdot \overline{y} + \overline{x} \cdot y && \text{definition of } \triangle \\
&= x \cdot \overline{x} + \overline{x} \cdot x && \text{by assumption} \\
&= x \cdot \overline{x} + x \cdot \overline{x} && \text{commutativity} \\
&= 0 + 0 && \text{complement (twice)} \\
&= 0 && \text{identity.}
\end{aligned}
$$

Conversely, assume that $x \triangle y = 0$. Then

$$
\begin{aligned}
x &= x \cdot 1 && \text{identity} \\
&= x \cdot (y + \overline{y}) && \text{complement} \\
&= x \cdot y + x \cdot \overline{y} && \text{distributivity} \\
&= (x \cdot y + x \cdot \overline{y}) + 0 && \text{identity} \\
&= (x \cdot y + x \cdot \overline{y}) + (x \cdot \overline{y} + \overline{x} \cdot y) && \text{by assumption} \\
&= (x \cdot y + (x \cdot \overline{y} + x \cdot \overline{y})) + \overline{x} \cdot y && \text{associativity (twice)} \\
&= (x \cdot y + x \cdot \overline{y}) + \overline{x} \cdot y && \text{idempotence} \\
&= (x \cdot y + x \cdot \overline{y}) + (\overline{x} \cdot y + \overline{x} \cdot y) && \text{idempotence} \\
&= (x \cdot y + (x \cdot \overline{y} + \overline{x} \cdot y)) + \overline{x} \cdot y && \text{associativity (twice)} \\
&= (x \cdot y + 0) + \overline{x} \cdot y && \text{by assumption} \\
&= x \cdot y + \overline{x} \cdot y && \text{identity} \\
&= y \cdot x + y \cdot \overline{x} && \text{commutativity (twice)} \\
&= y \cdot (x + \overline{x}) && \text{distributivity} \\
&= y \cdot 1 && \text{complement} \\
&= y && \text{identity.}
\end{aligned}
$$

□

COROLLARY 2.34

Let x and y be elements of a Boolean algebra. Then

$$x = y \quad \text{iff} \quad (x + \overline{y}) \cdot (\overline{x} + y) = 1$$

Proof: True by the duality principle. (Convince yourself that this is true.) □

✔ Quick Check 2.11

1. Prove the right distributivity properties:

$$(x + y) \cdot z = x \cdot z + y \cdot z \quad \text{and} \quad x \cdot y + z = (x + z) \cdot (y + z)$$

☑

Boolean algebras will appear again in Chapter 12 when the design of circuits is discussed.

2.5.3 Exercises

The exercises marked with ★ have detailed solutions in Appendix H. In the following exercises, assume that $\mathbb{B}$ is a Boolean algebra with associated set, B.

1. Let a and b be elements in a Boolean algebra. Prove that
$$a \cdot b + \overline{a} \cdot b = b$$
and
$$(a + b) \cdot (\overline{a} + b) = b$$
2. ★ Let a and b be elements in a Boolean algebra. Prove that
$$(a + (\overline{a} \cdot (\overline{b} + b))) \cdot b = b$$
and
$$(a \cdot (\overline{a} + (\overline{b} \cdot b))) + b = b$$
3. Let a, b, and c be elements in a Boolean algebra. Prove that
$$(a \cdot b \cdot c) + (b \cdot c) = b \cdot c$$
and
$$(a + b + c) \cdot (b + c) = b + c.$$
4. Let a, b, and c be elements in a Boolean algebra. Prove that
$$\overline{(a + c) \cdot (\overline{b} + c)} = (\overline{a} + b) \cdot \overline{c}$$
and
$$\overline{(a \cdot c) + (\overline{b} \cdot c)} = (\overline{a} \cdot b) + \overline{c}$$
5. ★ Let a and b be elements in a Boolean algebra. Prove that
$$a + b = a + \overline{a} \cdot b$$
and
$$a + b = b + a \cdot \overline{b}$$
6. Prove the remaining parts of the fundamental Boolean algebra properties. Do not use any properties that you have not already proved.
Hints:
   - Idempotence may be easier if you start with the right-hand side.
   - Prove domination and then absorption before proving associativity. Domination: start by using the identity axiom and then the complement axiom.
   - Prove associativity before De Morgan's laws.
   - For De Morgan, show that $x \cdot y$ and $\overline{x} + \overline{y}$ make both parts of the complement axiom true. Then use the uniqueness of complements.
   - Associativity is not trivial. Here is a good strategy: Start with the expression $[(x+y)+z]\cdot[x+(y+z)]$ and use the distributivity axiom to expand the expression. If you start by distributing $[(x+y)+z]$ over $[x+(y+z)]$, you will eventually end up with $x+(y+z)$. On the other hand, if you start by distributing $[x+(y+z)]$ over $[(x+y)+z]$, you will eventually end up with $(x+y)+z$. You will need to use absorption and commutativity many times (as well as some additional distributivity). Since it is possible to derive both $(x+y)+z$ and $x+(y+z)$ from the same initial expression, they must be equal.
7. Prove that if a, b, c, and d are elements of B, then
$$(a + b) + (c + d) = (a + (b + c)) + d$$
and
$$(a \cdot b) \cdot (c \cdot d) = (a \cdot (b \cdot c)) \cdot d$$
8. Let $S = \{a, b, c, d\}$. Prove that $B = \{\emptyset, \{a, b\}, \{c, d\}, S\}$ with $+$ as $\cup$, $\cdot$ as $\cap$, and $\overline{}$ as set complement form a Boolean algebra.
9. ★ Let $B = \{0, 1, 2, \ldots, n\}$. Define $+$ by $x + y = \max\{x, y\}$, $\cdot$ by $x \cdot y = \min\{x, y\}$, and complement by $\overline{x} = n - x$. Does this define a Boolean algebra? Support your answer.
10. Let $S = \{a, b, c\}$ and let $B = \mathcal{P}(S)$. Define $+$ by $x + y = x \cup y$ and define $\cdot$ by $x \cdot y = x \Delta y$, for $x, y \in B$. Define $\overline{}$ as standard set complement. Do these definitions define a Boolean algebra? Support your answer. $x \Delta y$ refers to Definition 2.16 on page 27.
11. Let x and y be elements of a Boolean algebra such that $x + y = 0$. Prove that $x = y = 0$. Then prove that $x \cdot y = 1$ implies that $x = y = 1$.
12. ★ Let x and y be elements of a Boolean algebra.
   (a) Prove $x \cdot (x + y) = x + x \cdot y$.
   (b) What is the dual of this equation?
13. Let x, y, and z be elements of a Boolean algebra. Prove
   (a) $x \cdot (y + x \cdot z) = x \cdot y + x \cdot z$
   (b) $x + y \cdot (x + z) = (x + y) \cdot (x + z)$
14. For each claim, determine whether it is always true or else false in some cases. Then give some justification for your answer.
   (a) In a Boolean algebra, whenever $x \cdot y = 0$, at least one of x or y must also be the zero element.
   (b) In every Boolean algebra, $1 + 1 = 1$.
   (c) The involution property is true by the duality principle.
   (d) ★ If $x \cdot y = 0$, then the duality principle guarantees that $x + y = 1$.
   (e) A one-element Boolean algebra exists for which $0 = 1$.
15. For each claim, determine whether it is always true or else false in some cases. Then give some justification for your answer.
   (a) If an assertion, T, is valid for some of the elements of a Boolean algebra, then if all the 0s and 1s are exchanged, and all $+$ and $\cdot$ are exchanged, the result is an assertion that is valid for the same elements in the Boolean algebra.
   (b) ★ There does not exist a Boolean algebra with exactly three elements.
   (c) Suppose that x, y, z of a Boolean algebra are all elements in $\{0, 1\}$. If $xz = yz$, then $x = y$.
   (d) Suppose that x, y, z of a Boolean algebra are all elements in $\{0, 1\}$. If $x + z = y + z$, then $x = y$.
   (e) In a Boolean algebra, it is never the case that $x = \overline{x}$.

16. It has been shown that the system of propositional logic can be considered as an example of a Boolean algebra. However, the logic operator "implication" was not given a corresponding Boolean algebra operator. How can the proposition $P \to Q$ be expressed in Boolean notation?

17. The following Boolean assertions are not theorems. In each case, find a counterexample. That is, find a Boolean algebra for which the expression is not always valid.

(a) ★ $x + y = x \cdot y$ (b) $x \cdot (y + z) = x \cdot y$

(c) $\overline{(x + y)} = \overline{x} + \overline{y}$ (d) $x \cdot \overline{y} = x + \overline{y}$

(e) $x \cdot y = x$ iff $x = y$

18. The following Boolean assertions are not theorems. In each case, find a counterexample. That is, find a Boolean algebra for which the expression is not always valid.

(a) ★ $x \cdot (y + z) = x + y \cdot z$

(b) $\overline{x} \cdot \overline{y} \cdot \overline{z} = x + y + z$

(c) $\overline{(x \cdot y)} = \overline{x} \cdot \overline{y}$

(d) $(x \cdot y) + z = x \cdot (y + z)$

(e) $x + y + z = x \cdot y \cdot z$

19. Let $x, y \in B$. Prove

$$x + y = x \cdot y \quad \text{iff} \quad x = y.$$

20. Let $x, y, z \in B$. Prove

$$x + y = y \cdot z \quad \text{iff} \quad x \cdot \overline{y} + x \cdot \overline{z} + y \cdot \overline{z} = 0.$$

In the set context, this property essential translates to $x \cup y = y \cap z$ iff $x \subseteq y \subseteq z$. [*Hint*: Use Proposition 2.33.]

21. Let $\mathbb{B}$ be a Boolean algebra. If $x, y, z \in B$, is it always true that $x \cdot y = x \cdot z$ if and only if $y = z$?

22. ★ Suppose that x, a, and b are elements of a Boolean algebra. Show that the equation $x + a = b$ does not always have a unique solution for x.

[*Hint*: Let $S = \{\alpha, \beta, \gamma\}$ and form the Boolean algebra with $B = \mathcal{P}(S)$ in the usual manner. Now let $a = \{\alpha, \beta\}$ and $b = \{\alpha, \beta, \gamma\}$.]

23. Suppose that x and a are elements of a Boolean algebra. Show that the equation $x \cdot a = a$, where $a \neq 0$, does not always have a unique solution for x.

24. Suppose that x, a, and b are elements of a Boolean algebra. Does the equation $\overline{x \cdot a} = b$ always have a solution for x?

25. What is wrong with the following purported proof that $x + y = x$ for all x and y in a Boolean algebra?

$x + y = x + y \cdot y$	idempotence
$= (x + x) \cdot (x + y)$	distributivity (+ over ·)
$= x \cdot (x + y)$	idempotence
$= x$	absorption

2.6 Predicate Logic

Many statements in mathematics involve the notion of quantification. Some examples, whose meaning you need not fully understand, are as follows:

- There are exactly five regular polyhedra.
- All triangles in the Euclidean plane have interior angles that sum to 180°.
- Any planar map can be colored with at most four colors.
- The equation $x^2 + 1 = 0$ has no real roots.
- The equation $3x^4 - 2x^3 + 7x - 5 = 0$ has a complex root.

Notice that the examples involving equations are not statements if the words *has no real roots* and *has a complex root* are removed. Thus "$x^2 + 1 = 0$" is not a statement (x has not been specified), but "$x^2 + 1 = 0$ has a real root" or "all real numbers x satisfy $x^2 + 1 = 0$" are statements.

It is convenient to express statements involving quantifiers using a symbolic decomposition into a *predicate* and a modifying quantifier. The predicate represents the property; the quantifier describes how many objects have the property. A predicate is like a proposition that contains one or more variables. Until the values of the variables are known, it is not possible to assign a truth value to the proposition. For example, let the predicate $P(x)$ be the property[37] "$x \in \mathbb{R}$ and $x^2 + 1 = 0$." If we substitute the value 1 for x, we can determine that the predicate $P(1)$ is false. If we determine that $P(x)$ is false for every choice of x, then the statement "for all $x \in \mathbb{R}$, $P(x)$ is false" is true and the statement "there exists an $x \in \mathbb{R}$ such that $P(x)$ is true" is a false statement.

Notice the important role that context plays. If we change the set to which x belongs

[37] See Appendix A if the symbol $\mathbb{R}$ is unfamiliar.

and set $P(x)$ as the property "$x \in \mathbb{C}$ and $x^2 + 1 = 0$," then there are values of x that make $P(x)$ true (namely, i and $-i$). In this case, "there exists an $x \in \mathbb{C}$ such that $P(x)$ is true" is a true statement.

Because context is important, it is always necessary to define clearly the *universe of discourse* when predicates are introduced. This can be done by writing a sentence describing the set of legal values for the variables in the predicate, or by building that information into the statement. The second approach was taken with "$x \in \mathbb{C}$ and $x^2 + 1 = 0$".

2.6.1 Quantifiers

Quantifiers allow us to specify how many objects have some property. Since this number can be any value, it seems that many quantifiers are needed. Actually, only two are necessary: the *existential* and the *universal* quantifiers presented in the following subsections. Any quantified sentence can be written using these two quantifiers (possibly together with other logic connectives such as NOT or AND). If a finite nonzero number appears (such as 5 in the first of the preceding examples), it may be necessary to use several existential quantifiers and other logic connectives to write the statement symbolically.

The Existential Quantifier

The existential quantifier is abbreviated $\exists$. It represents the concept "there is at least one". Many other phrases are used, for example, "there is", "there exists", "there are", and "for some".

The (false) statement "$x^2 + 1 = 0$ has a real root" can be written symbolically $\exists x \in \mathbb{R},\ P(x)$, where $P(x)$ is the predicate "$x^2 + 1 = 0$".

The (true) statement "$x^2 + 1 = 0$ has no real roots" can be written symbolically $\neg[\exists x \in \mathbb{R},\ P(x)]$.

If x is a member of a finite set, then the statement $\exists x,\ S(x)$ can also be written as $S(x_1) \vee S(x_2) \vee \cdots \vee S(x_n)$.[38] This equivalence enables truth tables to be used with the existential quantifier whenever x has a finite domain.

EXAMPLE 2.34

An Existentially Quantified Predicate

Let $P(x)$ be the predicate $x \leq 0$. Then the statement $\exists x \in \mathbb{R},\ P(x)$ evaluates to true, since $x = -4$ (among many other choices) makes the predicate true. ■

The Universal Quantifier

The universal quantifier is abbreviated $\forall$. It represents the concept "for every element in some set". Other phrases used are "for each", "for all", and "for every".

The statement "all real numbers x satisfy $x^2 + 1 = 0$" can be written symbolically as $\forall x \in \mathbb{R},\ P(x)$.

If $Q(x)$ represents the predicate "the planar map x can be colored with at most four colors", then the statement "any planar map can be colored with at most four colors" is denoted by $\forall x,\ Q(x)$.[39]

If x can have only a finite number of values, then the statement $\forall x,\ S(x)$ can be written as $S(x_1) \wedge S(x_2) \wedge \cdots \wedge S(x_n)$, providing a way to use truth tables with the universal quantifier.

EXAMPLE 2.35

A Universally Quantified Predicate

Let $P(x)$ be the predicate $x \leq 0$. Then the statement $\forall x \in \mathbb{R},\ P(x)$ evaluates to false, since $x = 25$ (among many other choices) makes the predicate false. ■

---

[38] The lack of parentheses is justified by the associativity tautology.

[39] In this example, it is understood that the universe of discourse for x is the set of all planar maps.

Bound Variables and Free Variables

It is possible to write a predicate and then quantify some of the variables and not quantify other variables. The variables whose values have been quantified are said to be *bound*. The other variables are said to be *free*. For example, in the expression $\forall x, P(x, y, z)$, the variable x is bound, but y and z are free.

A predicate for which all variables are bound is a statement. If any variables are free, the expression will not be a statement unless it is possible to assign a T/F value.

EXAMPLE 2.36

Bound and Free

Let $P(x, y, z)$ be the predicate $x \cdot y = z$, where the universe of discourse is the set of real numbers. If we assert $\exists x, [x \cdot y = z]$, there is no way to assign a truth value without quantifying y and z. That is, the expression $\exists x, [x \cdot y = z]$ is a predicate but not a proposition.

The expression $\exists x, \exists y, \exists z, [(x = 2) \wedge (z = 6) \wedge (x \cdot y = z)]$ is a proposition because all three variables are bound. In fact, it has truth value T because $y = 3$ makes the predicate $[(x = 2) \wedge (z = 6) \wedge (x \cdot y = z)]$ true. ■

✔ Quick Check 2.12

1. Translate the following English sentences into "quantified predicate" form. Describe an appropriate universe of discourse. Assign the proper truth value to the first statement.
   (a) Every real number is larger than its own square root.
   (b) There is a student in this class who does not own a car.
2. Translate the following expressions into English.
   (a) $\forall x \in \mathbb{Z}, \ x < 2x$
   (b) Let the universe of discourse be the set of all living trees on planet Earth. Let $P(t)$ be the predicate "tree t is more than 1000 years old." The expression is $\exists t, P(t)$. ☑

Negations of Quantifiers

Consider the predicate $R(x)$: "My friend x has read *Pilgrim's Progress*". What is the negation of the statement $\forall x, R(x)$? Assume I have three friends. The statement is false if at least one of my three friends has not read *Pilgrim's Progress*. Thus the negation of $\forall x, R(x)$ is $\exists x, \neg R(x)$.

What is the negation of $\exists x, R(x)$? If one or more of my friends have read the book, the statement is true. To make the statement false, none of my friends can have read *Pilgrim's Progress*. Thus the negation of $\exists x, R(x)$ is $\forall x, \neg R(x)$.

Negating a Quantified Statement

To negate a universally or existentially quantified statement, interchange the quantifier and negate the predicate.

EXAMPLE 2.37

Life in College

Suppose (perhaps after looking at the grade on a returned exam) Joe has vowed that for the next five days he will finish all homework each day and also get at least seven hours of sleep a night. Suppose Joe has completed the five days. If we let D be the set of five days and let $H(x)$ represent the predicate "Joe finished all homework on day x" and let $S(x)$ represent the predicate "Joe got at least seven hours of sleep on day x", then we can write his self-promise as $\forall x \in D, (H(x) \wedge S(x))$.

What would the negation look like? Using the negation rule just presented and also De Morgan's laws, we have

$$\begin{aligned}\neg[\forall x \in D, (H(x) \wedge S(x))] &\Leftrightarrow \exists x \in D, \neg(H(x) \wedge S(x)) \\ &\Leftrightarrow \exists x \in D, (\neg H(x) \vee \neg S(x))\end{aligned}$$

This can be translated into normal English as "On at least one day, either Joe failed to complete all homework, or Joe failed to get at least seven hours of sleep, or he failed to do both tasks on that day". That is, Joe failed to keep his vow if there was a day during which he did not complete one of the tasks. ■

EXAMPLE 2.38

Translating Isn't Trivial

In Shakespeare's play *The Merchant of Venice*, Portia's father has decided that she must marry the first suitor who correctly chooses which of three small chests (called caskets) contains her portrait. (See Example C.3 on page A15.) The caskets were made of gold, silver, and lead, respectively. One suitor was foolish enough to choose the gold casket. Instead of Portia's portrait, he found a scroll with a message that began "All that glisters is not gold, . . . ".

Using modern terminology, consider the statement

All that glistens is not gold.

There are two natural predicates. Let $G(x)$ indicate that x glistens, and let $Au(x)$ indicate that x is (or is made from) gold,[40] where the universe of discourse, U, is the set of all physical entities.

The English statement appears to be

$$\forall x \in U, G(x) \rightarrow \neg Au(x)$$

A moment's thought will indicate that this is surely not what Shakespeare meant.[41] It asserts that any object that glistens cannot be gold or made of gold. But Shakespeare and his contemporaries certainly considered gold to be something that glistens.

Shakespeare's meaning is better captured by the following assertion:

$$\exists x \in U, G(x) \wedge \neg Au(x)$$

This says that there are some objects that glisten but are nevertheless not gold.

Note that

$$\begin{aligned}\neg [\forall x \in U, G(x) \rightarrow Au(x)] &\Leftrightarrow \neg [\forall x \in U, \neg (G(x) \wedge \neg Au(x))] \\ &\Leftrightarrow [\exists x \in U, \neg (\neg (G(x) \wedge \neg Au(x)))] \\ &\Leftrightarrow \exists x \in U, G(x) \wedge \neg Au(x)\end{aligned}$$

so the negation in the original English statement is more closely connected with the word *all* than with the word *gold*.

A truly modern (nonpoetic) English version of the scroll would start with

Not all that glistens is gold, . . . ■

[40] The symbol for gold on the Periodic Table of Elements is Au.

[41] This example originated in a comment by Arthur Hobbs.

Multiple Quantifiers

Predicates may contain multiple variables, so it is useful to specify a convention for reading predicates with multiple quantifiers. The convention is to read from left to right. The expression can be parenthesized to show the nesting. Thus, $\forall x, \exists y, \forall z, P(x, y, z)$ is parenthesized as $\forall x, (\exists y, (\forall z, P(x, y, z)))$. Reading from left to right, every value for x is considered, and the predicate $\exists y, (\forall z, P(x, y, z))$ is examined for its truth value. To determine this truth value, values of y are examined in order. For each y, the truth of the predicate $\forall z, P(x, y, z)$ is examined. This is accomplished by looking at each z and deciding whether $P(x, y, z)$ is true or false for the current values of x, y, and z. As soon as a y is found for which $\forall z, P(x, y, z)$ is known to be true, we can move to the next x without considering other values for y.

It is common to remove the collection of nested parentheses [so $\forall x, \exists y, \forall z, P(x, y, z)$ is usually preferred over $\forall x, (\exists y, (\forall z, P(x, y, z)))$].

For a more concrete example, let the set C represent all audio CDs and let the set S represent all songs. Finally, let the predicate $P(c, s)$ represent "CD c contains song s." Then the proposition $\forall c \in C, \exists s \in S, P(c, s)$ can be read "every CD contains at least one song." Notice that incorrectly reading this in a right-to-left order changes the meaning (and the truth value): "There is a song that is on every CD."

To investigate the truth of $\forall c \in C, \exists s \in S, P(c, s)$, we would start examining CDs. For each CD, we would then cycle through songs in S, checking to see if a song is on the CD. In the process, every CD would be visited once, but many songs would be examined multiple times (but only once per CD).

EXAMPLE 2.39

Multiple Quantifiers

Let M represent the set of all math classes and let S represent the set of all students at your school. Let the predicate $E(s, m)$ mean that student s is enrolled in class m. Let the predicate $C(s)$ indicate that student s owns a cat.

Then the statement "there is at least one math student who owns a cat" can be represented as

$$\exists s \in S, \exists m \in M, [E(s, m) \wedge C(s)]$$

A computer program that would verify the truth of this statement would contain nested loops, with the outer (slower) loop cycling through the students. That is, for each student, we would examine math classes in succession to see if the student is enrolled in that class and if the student owns a cat. Then we would do the same with the next student, and so on.[42]

The negation of the symbolic statement is

$$\neg(\exists s \in S, \exists m \in M, [E(s, m) \wedge C(s)]) \Leftrightarrow \forall s \in S, \forall m \in M, [\neg E(s, m) \vee \neg C(s)]$$

The previous equivalence was derived in the following manner.

$$\begin{aligned}\neg(\exists s \in S, \exists m \in M, [E(s, m) \wedge C(s)]) &\Leftrightarrow \forall s \in S, \neg(\exists m \in M, [E(s, m) \wedge C(s)]) \\ &\Leftrightarrow \forall s \in S, \forall m \in M, \neg[E(s, m) \wedge C(s)] \\ &\Leftrightarrow \forall s \in S, \forall m \in M, [\neg E(s, m) \vee \neg C(s)]\end{aligned}$$

This can be restated in English as "for each student, if we examine all math classes we see that either the student is not in the math class or else the student does not own a cat (or perhaps both)". This is very awkward. A bit of thought should show that this is essentially the same as "no math student owns a cat" or "all cat owners are not enrolled in any math class".

[42] It might be more efficient to consider $\exists s \in S, [C(s) \wedge (\exists m \in M, E(s, m))]$. As each student is examined, in turn, if the student does not own a cat, the student can be rejected without the need to examine all math classes to see if that student is enrolled.

The first of these alternative expressions can be justified as follows. Recall that $(P \to Q) \Leftrightarrow \neg(P \wedge \neg Q) \Leftrightarrow (\neg P \vee Q)$. If we let $P = E(s, m)$ and $Q = \neg C(s)$, then

$$\forall s \in S, \forall m \in M, [\neg E(s, m) \vee \neg C(s)]$$
$$\Leftrightarrow \forall s \in S, \forall m \in M, [E(s, m) \to \neg C(s)] \quad \blacksquare$$

Notice that the order of multiple quantifiers is important when the quantifiers are not all universal or not all existential. For example, let $P(x, y)$ be the predicate $x + y = 0$, where the universe of discourse is the set of real numbers. The statement $\forall x, \exists y, (x + y = 0)$ is true because every real number, x, has an additive inverse. However, the statement $\exists x, \forall y, (x + y = 0)$ is false because it is not possible to find a real number, x, such that adding it to any other real number always results in 0.

✔ Quick Check 2.13

1. Write the negation of the statement $\forall x \in U, \neg T(x)$.
2. Write the negation of the statement $\exists x \in U, \forall y \in U, [P(x) \to Q(x, y)]$. Simplify as far as possible.
3. Let $P(x, y)$ be the predicate $x \cdot y = y \cdot x$ for real numbers x and y. What are the truth values of $\forall x \in \mathbb{R}, \exists y \in \mathbb{R}, P(x, y)$ and $\exists x \in \mathbb{R}, \forall y \in \mathbb{R}, P(x, y)$? ☑

The following implication is always true:

$$[\exists x, \forall y, P(x, y)] \to [\forall y, \exists x, P(x, y)]$$

Because if there is an x that makes the predicate true for all values of y, it is always possible to use that particular x for the right-hand statement $\forall y, \exists x, P(x, y)$. That is, if x_0 is a value for x that makes $\forall y, P(x_0, y)$ true, then $\forall y, \exists x, P(x, y)]$ must be true for $x = x_0$.

However, the implication $[\forall x, \exists y, P(x, y)] \to [\exists y, \forall x, P(x, y)]$ is not true in general. To see this, let the universe of discourse be the real numbers and let $P(x, y)$ be the predicate $x = 2y$. Observe that choosing $y_0 = \frac{x}{2}$ always makes $\forall x, \exists y, P(x, y)$ true; that is, $\forall x, P(x, \frac{x}{2})$ is true. However, there is no choice of $y = y_0$ that makes $\forall x, x = 2y_0$ true.

The two logical equivalences and one rule of inference in Table 2.25 can be verified using truth tables when x has a finite domain. Other methods need to be used to verify them in general. The meta-operators, $\Leftrightarrow$ and $\Rightarrow$, need to be extended to work with quantified statements. Let A and B be quantified statements. Then $A \Leftrightarrow B$ if and only if A and B have the same truth values. Similarly, $A \Rightarrow B$ if and only if B is true whenever A is true.

TABLE 2.25 Logical Equivalences and a Rule of Inference Involving Quantifiers

Negating an Existential Quantifier	Negating a Universal Quantifier
$\neg[\exists x \in U, P(x)] \Leftrightarrow [\forall x \in U, \neg P(x)]$	$\neg[\forall x \in U, P(x)] \Leftrightarrow [\exists x \in U, \neg P(x)]$
Simplifying Universal to Existential	
$[\forall x \in U, P(x) \wedge (U \neq \emptyset)] \Rightarrow [\exists x \in U, P(x)]$	

EXAMPLE 2.40 **Illustrating the Simplifying Universal to Existential Inference Rule**

If $U = \{a, b, c\}$, the rule of inference $[\forall x \in U, P(x) \wedge (U \neq \emptyset)] \Rightarrow [\exists x \in U, P(x)]$ is easy to illustrate using a truth table. The key step is the realization that $[\forall x \in U, P(x)] \Leftrightarrow [P(a) \wedge P(b) \wedge P(c)]$ and $[\exists x \in U, P(x)] \Leftrightarrow [P(a) \vee P(b) \vee P(c)]$

$P(a)$	$P(b)$	$P(c)$	$P(a) \wedge P(b)$	$(P(a) \wedge P(b)) \wedge P(c)$	$P(a) \vee P(b)$	$(P(a) \vee P(b)) \vee P(c)$
T	T	T	T	T	T	T
T	T	F	T	F	T	T
T	F	T	F	F	T	T
T	F	F	F	F	T	T
F	T	T	F	F	T	T
F	T	F	F	F	T	T
F	F	T	F	F	F	T
F	F	F	F	F	F	F

Now, replace $(P(a) \wedge P(b)) \wedge P(c)$ with $\forall x \in U, P(x)$ and $(P(a) \vee P(b)) \vee P(c)$ with $\exists x \in U, P(x)$.

$\forall x \in U, P(x)$	$\exists x \in U, P(x)$	$[\forall x \in U, P(x)] \rightarrow [\exists x \in U, P(x)]$
T	T	T
F	T	T
F	T	T
F	T	T
F	T	T
F	T	T
F	T	T
F	F	T

You probably recognize the tautology $[\forall x \in U, P(x)] \rightarrow [\exists x \in U, P(x)]$ as a generalization of the law of simplification on page 53. ■

2.6.2 Exercises

The exercises marked with ★ have detailed solutions in Appendix H. You should clearly define any symbolic statements (P, Q, R, etc.) you create for the following exercises.

1. Translate each of the English sentences into quantified predicate form. Clearly define the predicates, variables, and universes of discourse.
   (a) ★ All dogs go to heaven.
   (b) ★ The sum of two integers is an integer.
   (c) At least one student in this class has a pilot's license.
   (d) Every rational number has a multiplicative inverse. Use two variables. (See Appendix A.3 for a refresher on multiplicative inverse.)
   (e) There is a math class on campus in which only A students are enrolled. (There are two universes of discourse. Specify both of them.)
   (f) Every hospital in town has at least one pediatrician on staff. (There are two universes of discourse. Specify both of them.)
   (g) There is a pediatrician who is on staff at every hospital in town. (There are two universes of discourse. Specify both of them.)
2. Translate each of the English sentences into quantified predicate form. Clearly define the predicates, variables, and universes of discourse.
   (a) Every faculty member has access to the Internet.
   (b) ★ Someone in a group of friends has either talked out of turn during a conversation or broken a promise.
   (c) At least one child in this neighborhood watches television every day of the week. (There are two universes of discourse. Specify both of them.)
   (d) All of Jackie's roommates drink coffee in the morning.
   (e) Every woman in attendance at the raffle on Monday won some prize. (There are two universes of discourse. Specify both of them.)
   (f) A student in this school speaks both Chinese and Japanese.
   (g) There is a man who has visited some park in every state of the United States of America. (There are three universes of discourse. Specify all three of them.)

3. Let the universe of discourse be the set of real numbers. Translate each statement into English. Also, determine the proper truth value for the statement.

(a) $\exists x, \exists y, [(x = 2) \wedge (x \cdot y = 4x)]$

(b) ★ $\forall x, \forall y, (x \cdot y = 4x)$ (c) ★ $\forall x, \exists y, (x \cdot y = 4x)$

(d) $\exists x, \forall y, (x \cdot y = 4x)$ (e) $\exists x, \exists y, (x \cdot y = 4x)$

4. Let the universe of discourse be the set of real numbers. Translate each statement into English. Also, determine the proper truth value for the statement.

(a) $\exists x, \forall y, [(x = 3) \wedge (x + y = x - y)]$

(b) $\forall x, \forall y, (x + y = x - y)$ (c) $\forall x, \exists y, (x + y = x - y)$

(d) $\exists x, \forall y, (x + y = x - y)$ (e) $\exists x, \exists y, (x + y = x - y)$

5. Let the universe of discourse be the set of integers. Translate each statement into English. Also, determine the proper truth value for the statement.

(a) ★ $\exists x, \exists y, [(y > 3) \wedge (x^2 + y^2 = 5)]$

(b) $\forall x, \forall y, (x^2 + y^2 = 5)$ (c) $\forall x, \exists y, (x^2 + y^2 = 5)$

(d) $\exists x, \forall y, (x^2 + y^2 = 5)$ (e) $\exists x, \exists y, (x^2 + y^2 = 5)$

6. Let the universe of discourse be the set of integers. Translate each statement into English. Also, determine the proper truth value for the statement.

(a) $\exists x, \forall y, [(x < 0) \wedge (y > 0) \wedge (3x \cdot y = 4x^2 \cdot y)]$

(b) $\forall x, \forall y, (3x \cdot y = 4x^2 \cdot y)$

(c) $\forall x, \exists y, (3x \cdot y = 4x^2 \cdot y)$

(d) $\exists x, \forall y, (3x \cdot y = 4x^2 \cdot y)$

(e) $\exists x, \exists y, (3x \cdot y = 4x^2 \cdot y)$

7. Consider the proposition $\forall x \in U, [P(x) \rightarrow Q(x)]$. Symbolically write the negation of the proposition. You should move $\neg$ as far to the inside as possible.

8. Write the negations of the quantified statements. Move $\neg$ as far inside the predicate as possible.

(a) ★ $\exists x \in U, [Q(x) \vee (\neg P(x))]$

(b) $\exists x \in U, \exists y \in U, [P(x, y) \rightarrow Q(x, y)]$

(c) $\exists x \in \mathbb{R}, \left[x^3 - 3x^2 + 7 = 0\right]$

(d) $\forall \epsilon > 0, \exists \delta > 0, \forall x \in \mathbb{R}, [(|x - x_0| < \delta) \rightarrow (|f(x) - f(x_0)| < \epsilon)]$

9. Write the negations of the quantified statements. Move $\neg$ as far inside the predicate as possible. Assume that x, y, and z are all in the universe U unless otherwise specified.

(a) $\exists x \in U, [(\neg S(x)) \wedge T(x)]$

(b) ★ $\exists x, \forall y, \forall z, [(F(x, y) \wedge G(x, z)) \rightarrow H(y, z)]$

(c) $\exists x \in \mathbb{Z}, \exists y \in \mathbb{Z}, [(x + y = 4) \wedge (x - y = 2)]$

(d) $\forall x \in U, \exists y \in U, [(\neg P(x, y)) \rightarrow Q(x, y)]$

10. Use the statement $\forall s \in S, \forall m \in M, [E(s, m) \rightarrow \neg C(s)]$ from Example 2.39 to justify the interpretation "all cat owners are not enrolled in any math class." [*Hint*: Use the contrapositive tautology from Table 2.23 on page 54.]

11. Use the predicates from Example 2.39 to translate the English statement "there is at least one student in every math class who owns a cat" into symbolic notation. Then negate the symbolic expression. Finally, translate the negation into English.

12. Write the negations of the quantified statements. Move $\neg$ as far inside the predicate as possible.

(a) $\forall x \in \mathbb{R}, [(x \geq 100) \vee (x < 100)]$

(b) $\exists y \in \mathbb{Z}, \left[y \leq y^2\right]$

(c) $\forall x \in A, \exists y \in B, [(\neg M(x, y)) \rightarrow (\neg N(x, y))]$

(d) $\forall x \in \mathbb{R}, \forall y \in \mathbb{R}, \exists z \in \mathbb{R}, \left[z = \frac{(x+y)}{2}\right]$

13. Write the negations of the quantified statements. You may find it helpful to translate the English statement into symbolic notation before negating.

(a) ★ All Cretans are liars. (Attributed to Epimenides, a poet from Crete.[43])

(b) ★ There are no good men.

(c) There is at least one female in every math class.

(d) Someday my prince will come.

(e) Not a creature was stirring, not even a mouse.

(f) There are at least 10 righteous people in Sodom. (Do this one without translating into symbolic notation.)

14. Write the negations of the quantified statements. You may find it helpful to translate the English statement into symbolic notation before negating.

(a) Every family member ordered dessert at the restaurant.

(b) There is a child who does the laundry at least one day of the week.

(c) There was a boy who participated in every game at the party.

(d) There is no bear that can talk.

(e) Some mother wanted to go bowling on Tuesday night.

(f) Each person in the race ran at least 12 miles. (Do this one without translating into symbolic notation.)

15. Suppose I am writing a loop[44] in a computer program. (Imagine, for this problem, that the loop is controlled by our mathematical notation.) Suppose I want to examine a list of names and check if the people in the list are enrolled in my discrete mathematics class. Let the predicate $E(x)$ mean "student x is enrolled in the class." For each goal listed, state when the loop should terminate. That is, should the loop run through the entire list or quit partway through the task? If it should halt partway through the list, what criterion should be used to decide when to quit prematurely? How do you know if the goal has been met?

(a) $\forall x, E(x)$ (b) $\exists x, E(x)$

(c) $\forall x, \neg E(x)$ (d) $\exists x, \neg E(x)$

[43] Was Epimenides telling a lie when he made that statement?

[44] A set of statements that are repeated as a group.

16. Let G represent the set of all game shows on television and let P represent the set of people in your neighborhood. Let the predicate $C(p, g)$ mean that person p has been a contestant on game show g. Let the predicate $D(p)$ indicate that person p is a doctor. Use these predicates to translate the English statement "there is a person in your neighborhood who has been a contestant on a game show but is not a doctor" into symbolic notation. Then negate the symbolic expression. Finally, translate the negation into English.

17. ★

(a) Negate the proposition $\forall s, [(\exists d, [P(s, d)]) \vee Q(s)]$.

(b) Let

$$s = \text{Bethel University student}$$
$$d = \text{Bethel University dorm}$$
$$P(s, d) = \text{student } s \text{ lives in dorm } d$$
$$Q(s) = \text{student } s \text{ rents a book locker.}$$

Write the original proposition and its negation in English.

18. For each claim, determine whether it is always true or else false in some cases. Then give some justification for your answer.

(a) The expression $\forall x \in \mathbb{Z}, (x + y = x \cdot y)$ is a proposition (rather than merely a predicate).

(b) ★ The negation of "every good boy does fine" is "no good boy does fine".

(c) ★ The truth value of $\forall x, \exists y, [x^2 = -y^2]$ depends on the choice for the universe of discourse.

(d) One way to bind a free variable is to assign it a value.

(e) Let S be the set of all odd integers that are divisible by 2. Then $\forall x \in \mathbb{Z}, [(x \in S) \rightarrow (x + x \neq 2x)]$ is a valid assertion.

19. For each claim, determine whether it is always true or else false in some cases. Then give some justification for your answer.

(a) Since $\mathbb{Z} \subseteq \mathbb{R}$, if $\forall x, P(x)$ is true when the universe of discourse is $\mathbb{Z}$, then $\forall x, P(x)$ is true when the universe of discourse is $\mathbb{R}$.

(b) Although the written forms are different, $\exists z, \exists y, \exists x, S(x, y, z)$ and $\exists x, \exists y, \exists z, S(x, y, z)$ will have the same truth values.

(c) If at least one variable is not bound, then a predicate is not a statement.

(d) Let $U = \{x \in \mathbb{R} \mid x = x + 1\}$. Then for all y, z in U, $y + z = z + y$ is a valid assertion.

(e) If for any choice of $x \in U$, a $y \in U$ can be found such that $R(x, y)$ is true, then $\exists y \in U, \forall x \in U, R(x, y)$ is true.

20. ★ Find a counterexample to the claim

$$[\forall x \in X, \exists y \in Y, P(x, y)] \rightarrow [\forall y \in Y, \exists x \in X, P(x, y)]$$

is always true. Clearly specify X and Y.

21. Find a counterexample to the claim

$$[(\exists x \in U, P(x)) \wedge (\exists x \in U, Q(x))] \rightarrow [\exists x \in U, P(x) \wedge Q(x)]$$

is always true. Clearly specify U.

22. Find a counterexample to the claim

$$[\forall x \in U, P(x)] \rightarrow [\exists x \in U, P(x)]$$

is always true.

2.7 Quick Check Solutions

You will learn more if you make an honest attempt to solve the Quick Check problems *before* you look at my solutions. Don't give up too soon! If you do need to look at my answer, make sure you can reproduce the solution without opening your book.

Quick Check 2.1

1. $5 \notin D$ but $5 \in F$
2. $D \not\subseteq F$ because $6 \in D$ but $6 \notin F$
3. $D \cap F = \{2, 4\}$
4. $D \cup F = \{1, 2, 3, 4, 5, 6, 8\}$
5. $\overline{D} = \{1, 3, 5, 7, 9\}$
6. $|D| = 4, |F| = 5$
7. $D - F = \{6, 8\}, F - D = \{1, 3, 5\}$
8. D and F are not disjoint because $D \cap F = \{2, 4\}$.

Quick Check 2.2

1. $A \cap B \cap C = \{3\}$
2. $A \cup B \cup C = \{1, 2, 3, 4, 5\}$

3. $\{A, B, C\}$ is not a partition of $\{1, 2, 3, 4, 5\}$—for example, $A \cap B \neq \emptyset$.
4. $A \times B = \{(1, 2), (1, 3), (1, 4), (2, 2), (2, 3), (2, 4), (3, 2), (3, 3), (3, 4)\}$
5. $|A \times B \times C| = 3 \cdot 3 \cdot 3 = 27$
$$\begin{aligned} A \times B \times C &= \{(1, 2, 3), (1, 2, 4), (1, 2, 5), (1, 3, 3), (1, 3, 4), (1, 3, 5), \\ &\quad (1, 4, 3), (1, 4, 4), (1, 4, 5), (2, 2, 3), \ldots, (3, 4, 5)\} \end{aligned}$$
6. $\mathcal{P}(A \cap B) = \mathcal{P}(\{2, 3\}) = \{\emptyset, \{2\}, \{3\}, \{2, 3\}\}$

Quick Check 2.3

1. Suppose that $C \subseteq B$ and also that $B \subseteq A$. Let $x \in C$. Then since $C \subseteq B$, it is also true that $x \in B$ (using Definition 2.2). But then, since $B \subseteq A$ is also true, $x \in B$ implies $x \in A$.

   Since for every $x \in C$ it is always true that $x \in A$, the definition of subset implies that $C \subseteq A$.
2. Suppose $B \subseteq A$.

$$\begin{array}{llll} x \in \overline{A - B} & \text{iff} & x \notin (A - B) & \text{definition of complement} \\ & \text{iff} & x \notin A \ \text{ or } x \in (A \cap B) & \text{definition of set difference} \\ & \text{iff} & x \in \overline{A} \ \text{ or } x \in (A \cap B) & \text{definition of complement} \\ & \text{iff} & x \in \overline{A} \ \text{ or } x \in B & B \subseteq A \\ & \text{iff} & x \in (\overline{A} \cup B) & \text{definition of union} \end{array}$$

   Since every element of $\overline{A - B}$ is an element of $(\overline{A} \cup B)$, and vice versa, the two sets are equal.

Quick Check 2.4

1. They actually make sense under all three guidelines, but number 2 (look for flaws in logic or logical inconsistency) is the most appropriate.
2. (a) You could consider this an improper generalization. How do they know no manager was fired? (Perhaps a manager at a competing company bought an IBM and was fired.)

   There is an invited inference here. Since no manager has been fired for buying an IBM, an IBM must be the best product. However, it could be that IBM made acceptable computers but not the best. (It is also possible that IBM made the best.)

   (b) No one else's carrots have cholesterol either. This is a variation of a nonsequitur. The appeal is to our (reasonable) desire to eat healthy food. Cholesterol received bad press in the 1980s. It makes a great advertising pitch due to the emotional linkage with health. The implied implication is that the lack of something bad means the presence of something good. Arsenic has no cholesterol either.

   If the claim was "our eggs have an acceptably low amount of cholesterol", there might be some validity to the claim.

   (c) The general guideline "analyze assumptions" might be worth using. The assumption is "some movie should be watched". Under this assumption, the best of the lot is the proper choice. However, it may be the case that all the movies are exploitative trash with no artistic merit. You may be better off playing a game, talking, or reading a book together.

Quick Check 2.5

1. (a) This is neither true nor false. If the word *as* were deleted, it would be a (true) statement. (There is no verb, so it is not even an English sentence.)

(b) This is a statement. We can imagine examples where someone eats an apple every day of their life but still gets ill or injured. The statement is false.

(c) Despite the Shakespearean English, this is a statement. A modern translation is "appearances may be deceiving". (I prefer the original.) This is a tricky sentence to translate into a mathematical statement (see Example 2.38 on page 71).

(d) This is not a statement. It basically means "good night". There is no possible assignment of true or false. (There is a verb, so it is an English sentence, but it is a command, not a claim. That is, the verb is used in an imperative, rather than an indicative, mode.)

2. John or Jane Doe. With this version, only one would need to sign a check. It is inconvenient to always require both signatures.

3. There are several acceptable ways to express these statements. Human languages tend to be ambiguous. Think about how much more precise the mathematical version is.

(a) It is not true that I like both popcorn and jalapeños.

(b) It is not true that I like either popcorn or jalapeños.
(We will see later that an equivalent statement is "I dislike both popcorn and jalapeños".)

(c) Suppose I do like popcorn but I do not like jalapeños. Then it is not the case that I like both, so the statement $\neg(P \wedge Q)$ is true. This agrees with the second row of the truth table. There are seven other combinations to try (four involving $\neg(P \vee Q)$).

4. Let B represent the statement "I love Betty" and let S represent the statement "I love Sue". Similarly, let C represent the statement "I love Colin" and let T represent the statement "I love Tom".

(a) $\mathrm{B} \vee \mathrm{S}$ (b) $\mathrm{C} \wedge \mathrm{T}$

(c) $\mathrm{B} \vee \mathrm{S}$ [Notice that this is the same answer as for part (a).]

(d) $\mathrm{C} \wedge (\neg\mathrm{T})$

Quick Check 2.6

1. The two statements have the same truth table because their truth tables agree for every combination of T/F values for P and Q:

P	Q	$P \to Q$	$Q \to P$	$(P \to Q) \wedge (Q \to P)$
T	T	T	T	T
T	F	F	T	F
F	T	T	F	F
F	F	T	T	T

P	Q	$P \leftrightarrow Q$
T	T	T
T	F	F
F	T	F
F	F	T

This observation can also be written as

$$[(P \to Q) \wedge (Q \to P)] \leftrightarrow [P \leftrightarrow Q]$$

2. I am an earthling if and only if I am not a space alien. (This is not a true statement unless we restrict the collection of beings we are discussing to consist only of earthlings and space aliens.)

Quick Check 2.7

1. Since the two final columns are identical, the two statements are equivalent.

P	Q	$\neg P$	$\neg Q$	$P \wedge Q$	$\neg(P \wedge Q)$	$(\neg P) \vee (\neg Q)$
T	T	F	F	T	F	F
T	F	F	T	F	T	T
F	T	T	F	F	T	T
F	F	T	T	F	T	T

2. (a) Let A represent "I am an alien" and M represent "I am from Mars".

Symbolic Form	**Name**	**English Translation**
$A \to M$	Implication	If I am an alien, then I am from Mars.
$(\neg M) \to (\neg A)$	Contrapositive	If I am not from Mars, then I am not an alien.
$M \to A$	Converse	If I am from Mars, then I am an alien.
$(\neg A) \to (\neg M)$	Inverse	If I am not an alien, then I am not from Mars.

In this problem, the original implication and the contrapositive are false. The converse and the inverse are true statements.

(b) Let E represent "I am an extrovert" and A "I have friends". (I have used the Spanish word *amigo*, A, instead of the English word *friend* so that F maintains its meaning "false.") Notice the effect of the negations.

Symbolic Form	**Name**	**English Translation**
$(\neg E) \to (\neg A)$	Implication	If I am not an extrovert, then I have no friends.
$A \to E$	Contrapositive	If I have friends, then I am an extrovert.
$(\neg A) \to (\neg E)$	Converse	If I have no friends, then I am not an extrovert.
$E \to A$	Inverse	If I am an extrovert, then I have friends.

All four statements are false. Being an extrovert and having friends are independent characteristics; neither determines the other.

Quick Check 2.8

1. When P is true and Q is false, $(P \wedge Q)$ is false, so the truth values differ.

2. (a) One approach is to prove directly that the statement is a tautology.

$[P \wedge (P \to Q)] \to Q$	$\Leftrightarrow [P \wedge (\neg P \vee Q)] \to Q$	implication
	$\Leftrightarrow [(P \wedge (\neg P)) \vee (P \wedge Q)] \to Q$	distributivity
	$\Leftrightarrow [\mathbf{F} \vee (P \wedge Q)] \to Q$	law of contradiction
	$\Leftrightarrow [(P \wedge Q) \vee \mathbf{F}] \to Q$	commutativity
	$\Leftrightarrow (P \wedge Q) \to Q$	identity
	$\Leftrightarrow \mathbf{T}$	law of simplification

(b) One strategy is to show that both sides of the biconditional are logically equivalent.

$$\begin{aligned}
[(P \wedge (\neg Q)) \rightarrow Q] &\Leftrightarrow [\neg(P \wedge \neg Q)] \vee Q && \text{implication} \\
&\Leftrightarrow ((\neg P) \vee \neg(\neg Q)) \vee Q && \text{De Morgan} \\
&\Leftrightarrow ((\neg P) \vee Q) \vee Q && \text{double negation} \\
&\Leftrightarrow (\neg P) \vee (Q \vee Q) && \text{associativity} \\
&\Leftrightarrow (\neg P) \vee Q && \text{idempotence} \\
&\Leftrightarrow P \rightarrow Q && \text{implication}
\end{aligned}$$

Quick Check 2.9

1. (a) $x = 1, y = 0, z = 1$

$$1 \cdot (0 + 1) = 1 \cdot 1 = 1 = 0 + 1 = 1 \cdot 0 + 1 \cdot 1$$
$$1 + 0 \cdot 1 = 1 + 0 = 1 = 1 \cdot 1 = (1 + 0) \cdot (1 + 1)$$

(b) $x = 1, y = 1, z = 1$

$$1 \cdot (1 + 1) = 1 \cdot 1 = 1 = 1 + 1 = 1 \cdot 1 + 1 \cdot 1$$
$$1 + 1 \cdot 1 = 1 + 1 = 1 = 1 \cdot 1 = (1 + 1) \cdot (1 + 1)$$

Quick Check 2.10

1.
$$\begin{aligned}
\overline{0} &= \overline{0} + 0 && \text{identity} \\
&= 0 + \overline{0} && \text{commutativity} \\
&= 1 && \text{complement}
\end{aligned}$$

2. Apply the duality principle to the previous result.

Quick Check 2.11

1.
$$\begin{aligned}
(x + y) \cdot z &= z \cdot (x + y) && \text{commutativity} \\
&= z \cdot x + z \cdot y && \text{distributivity} \\
&= x \cdot z + y \cdot z && \text{commutativity (twice)}
\end{aligned}$$

To prove $x \cdot y + z = (x + z) \cdot (y + z)$, apply the duality principle to the previous result.

Quick Check 2.12

1. (a) The universe of discourse is the set of real numbers. The quantified predicate is $\forall x \in \mathbb{R},\ x > \sqrt{x}$. The statement is false (try $x = 1$ or $x = \frac{1}{4}$).

(b) The universe of discourse is the set of all students in this class. Let s represent students, and let the predicate $C(s)$ represent "student s owns a car". The quantified predicate is $\exists s, \neg C(s)$.

2. (a) Every integer is less than twice itself.

(b) There is a tree on planet Earth that is more than 1000 years old.

Quick Check 2.13

1. $\neg(\forall x \in U, \neg T(x)) \Leftrightarrow \exists x \in U, \neg(\neg T(x)) \Leftrightarrow \exists x \in U, T(x)$

2. Recalling that $\neg(A \rightarrow B) \Leftrightarrow (A \wedge (\neg B))$,

$$\begin{aligned}
&\neg(\exists x \in U, \forall y \in U, [P(x) \rightarrow Q(x, y)]) \\
&\quad \Leftrightarrow \quad \forall x \in U, \exists y \in U, [\neg(P(x) \rightarrow Q(x, y))] \\
&\quad \Leftrightarrow \quad \forall x \in U, \exists y \in U, [P(x) \wedge (\neg Q(x, y))]
\end{aligned}$$

3. Both are true because $\forall x \in \mathbb{R}, \forall y \in \mathbb{R}, (x \cdot y = y \cdot x)$ is true (by the commutativity of real numbers).

2.8 CHAPTER REVIEW

2.8.1 Summary

This chapter may have been a bit of a shock to you. There are few algebraic manipulations, almost no arithmetic, and nothing that resembles the standard content of a trigonometry or calculus textbook. Instead, you have encountered many new definitions and axioms, together with an emphasis on proof (perhaps much like what you encountered in a high school geometry course).

You might be wondering why this material has been chosen as part of this textbook. The answer is quite simple: Sets and logic permeate modern higher mathematics—including discrete mathematics. The rest of this book will assume familiarity with sets. They are particularly important in Chapter 5 (about counting) and Chapter 8 (more counting and also combinatorial designs). Propositions, predicates, and quantifiers are essential for an adequate understanding of proof techniques (Chapter 3).

Boolean algebras are a mathematical formalization that encompasses both sets and propositions. They also serve as a simple example of an axiomatic system—a topic that will be looked at again in Chapter 3. Boolean algebras will be important when binary functions and combinatorial circuits are examined in Chapter 12.

As you review the material in this chapter, you should make a point of knowing the definitions well. That means you should memorize them and think about the precise ideas that are expressed in the definitions. In fact, you should get into the habit of memorizing definitions as they are encountered. Do not wait until the night before an exam. You also need to master the notation that has been introduced (see Section 2.8.2).

In contrast to definitions, the various collections of fundamental properties are best learned as you use them. You probably used the printed tables extensively as you completed homework assignments. The properties you used most frequently are probably already available to you without the need of a table.

There are also a few additional theorems and propositions with which you should be familiar. Ask your instructor to define the word *familiar*. Review the notation, definitions, and theorems before looking at sample exam questions.

This chapter concentrates on finite sets. Very little is said about infinite sets. In some ways, finite sets and infinite sets have very different properties. For example, a finite set never has the same cardinality as any of its proper subsets. However, infinite sets do contain proper subsets having the same cardinality as the full set.

There are two major tasks for your review of sets: Learn the definitions and notation, and start to become comfortable with the manner in which proofs are completed.

The main concerns in the logic sections are to become comfortable with the notation and major concepts (propositions, logic operators, truth tables, predicates, quantifiers, etc.) and to notice the similarity between the style of proof used in the section about sets and the style of proof that uses a sequence of logical equivalences. It is also important that you understand the process of negating a quantified statement. It will be very useful for certain kinds of proof strategies.

The section on Boolean algebras provides a chance to start becoming comfortable working in an abstract mathematical setting. It is of great importance that you understand the role of the axioms as a set of assumptions that *define* Boolean algebras. To assert that some mathematical object is a Boolean algebra, it must be shown that the object satisfies all the axioms. In contrast, once a mathematical object is known to be a Boolean algebra, all the properties and theorems proved about Boolean algebras are known to apply to the current object. This includes the important duality principle.

The sections entitled "Logic in Daily Life" (Section 2.2) and "Analyzing Claims" (Appendix C.2) are not essential to understanding the rest of this book. They have been included because you are likely to encounter (mostly nonmathematical) situations where people discuss "logic" using the terminology those sections introduce.

2.8.2 Notation

The following notation has been introduced in this chapter.

Sets

Notation	Page	Brief Description
$x \in A$	19	element x is a member of (is in) set A
$x \notin A$	19	element x is *not* a member of (is not in) set A
U	19	default notation for the universal set
$\{\ldots\}$	19	the pair of symbols, { and }, enclose the elements of a set
$\{\ldots \mid \ldots\}$	20	set-builder notation: "the set of all ... such that ..."
$\lvert S \rvert$ or $o(S)$	21	the cardinality of S (the number of elements in S)
Venn diagram	20	a visual tool for representing sets and their elements
$B \subseteq A$	20	set B is a subset of set A
$B \subset A$	20	set B is a proper subset of set A
$\emptyset$	21	the empty set
$A = B$	21	sets A and B are equal
$\overline{A}$	22	the complement of set A
$A \cap B$	22	the intersection of sets A and B
$A \cup B$	22	the union of sets A and B
$A - B$	22	the set difference of A and B
Υ	24	default notation for an index set
$\cup_{i \in \Upsilon} S_i$	24	the union of a collection of sets
$\cap_{i \in \Upsilon} S_i$	24	the intersection of a collection of sets
$S_1 \times S_2 \times \cdots \times S_n$	25	the Cartesian product of sets $S_1, \ldots, S_n$
$\mathcal{P}(S)$	25	the power set of S
$A \triangle B$	27	the symmetric difference of sets A and B

Logic

Notation	Page	Brief Description
T	38	True
F	38	False
P, Q, R	39	default variable names for statements
$\neg$	39	the logic operator, NOT
$\wedge$	39	the logic operator, AND
$\vee$	39	the logic operator, OR
$\oplus$	40	the logic operator, XOR
$\rightarrow$	44	the logic operator, implication
$\leftrightarrow$	45	the logic operator, biconditional
$\Leftrightarrow$	47	the logical equivalence meta-operator
$\Rightarrow$	51	the inference meta-operator
$\exists$	69	the existential quantifier
$\forall$	69	the universal quantifier

Boolean Algebras

Notation	Page	Brief Description
$+$	59	the Boolean addition operator
$\cdot$	59	the Boolean multiplication operator
$\overline{}$	59	the Boolean complement operator
$x \triangle y$	65	the symmetric difference of elements x and y in a Boolean algebra

2.8.3 Fundamental Properties

Fundamental Set Properties

Idempotence
$A \cup A = A$
$A \cap A = A$

Domination
$A \cup U = U$
$A \cap \emptyset = \emptyset$

Associativity
$(A \cup B) \cup C = A \cup (B \cup C)$
$(A \cap B) \cap C = A \cap (B \cap C)$

Identity
$A \cup \emptyset = A$
$A \cap U = A$

Commutativity
$A \cup B = B \cup A$
$A \cap B = B \cap A$

De Morgan's Laws
$\overline{A \cup B} = \overline{A} \cap \overline{B}$
$\overline{A \cap B} = \overline{A} \cup \overline{B}$

Distributivity ($\cap$ **over** $\cup$)
$A \cap (B \cup C) = (A \cap B) \cup (A \cap C)$
$(A \cup B) \cap C = (A \cap C) \cup (B \cap C)$

Distributivity ($\cup$ **over** $\cap$)
$A \cup (B \cap C) = (A \cup B) \cap (A \cup C)$
$(A \cap B) \cup C = (A \cup C) \cap (B \cup C)$

Complement
$A \cup \overline{A} = U$ $\quad \overline{\emptyset} = U$
$A \cap \overline{A} = \emptyset$ $\quad \overline{U} = \emptyset$

Involution
$\overline{\overline{A}} = A$

Fundamental Logical Equivalences

Idempotence
$(P \vee P) \Leftrightarrow P$
$(P \wedge P) \Leftrightarrow P$

Domination
$(P \vee \mathbf{T}) \Leftrightarrow \mathbf{T}$
$(P \wedge \mathbf{F}) \Leftrightarrow \mathbf{F}$

Identity
$(P \vee \mathbf{F}) \Leftrightarrow P$
$(P \wedge \mathbf{T}) \Leftrightarrow P$

Associativity
$[(P \vee Q) \vee R] \Leftrightarrow [P \vee (Q \vee R)]$
$[(P \wedge Q) \wedge R] \Leftrightarrow [P \wedge (Q \wedge R)]$

Law of Simplification
$[(P \wedge Q) \rightarrow P] \Leftrightarrow \mathbf{T}$
$[(P \wedge Q) \rightarrow Q] \Leftrightarrow \mathbf{T}$

Commutativity
$(P \vee Q) \Leftrightarrow (Q \vee P)$
$(P \wedge Q) \Leftrightarrow (Q \wedge P)$

De Morgan's Laws
$[\neg(P \vee Q)] \Leftrightarrow [(\neg P) \wedge (\neg Q)]$
$[\neg(P \wedge Q)] \Leftrightarrow [(\neg P) \vee (\neg Q)]$

Distributivity ($\wedge$ **over** $\vee$)
$[P \wedge (Q \vee R)] \Leftrightarrow [(P \wedge Q) \vee (P \wedge R)]$
$[(P \vee Q) \wedge R] \Leftrightarrow [(P \wedge R) \vee (Q \wedge R)]$

Distributivity ($\vee$ **over** $\wedge$)
$[P \vee (Q \wedge R)] \Leftrightarrow [(P \vee Q) \wedge (P \vee R)]$
$[(P \wedge Q) \vee R] \Leftrightarrow [(P \vee R) \wedge (Q \vee R)]$

Law of the Excluded Middle
$[P \vee (\neg P)] \Leftrightarrow \mathbf{T}$

Law of Contradiction
$[P \wedge (\neg P)] \Leftrightarrow \mathbf{F}$

Law of Double Negation (Involution)
$\neg(\neg P) \Leftrightarrow P$

Law of Addition
$[P \rightarrow (P \vee Q)] \Leftrightarrow \mathbf{T}$

DEFINITION 2.26 *Boolean Algebra*

A *Boolean algebra*, $\mathbb{B}$, consists of an associated set, B, together with three operators and four axioms. The binary operators $+$ and $\cdot$ map elements of $B \times B$ to elements of B. The unary operator *complement*, $\overline{\ }$, maps elements of B to elements of B.

Parentheses have the highest precedence. The complement operator, $\overline{\ }$, has the second highest precedence, followed by $\cdot$ and then by $+$, the operator with lowest precedence.

The axioms are as follows:

Identity There exist distinct elements, 0 and 1, in B such that for every $x \in B$

$$x + 0 = x$$
$$x \cdot 1 = x$$

Complement For every $x \in B$, there exists a unique element $\overline{x} \in B$ such that

$$x + \overline{x} = 1$$
$$x \cdot \overline{x} = 0$$

Commutativity For every pair of (not necessarily distinct) elements $x, y \in B$

$$x + y = y + x$$
$$x \cdot y = y \cdot x$$

Distributivity For every three elements $x, y, z \in B$ (not necessarily distinct)

$$x \cdot (y + z) = x \cdot y + x \cdot z$$
$$x + y \cdot z = (x + y) \cdot (x + z)$$

The Duality Principle for Boolean Algebras

Let T be a theorem that is valid over a Boolean algebra. Then if all 0s and 1s are exchanged, and if all $+$ and $\cdot$ are exchanged (with a suitable change in parentheses to preserve operator precedence), the result is a theorem that is also valid over the Boolean algebra.

Fundamental Boolean Algebra Properties

Idempotence
$x + x = x$
$x \cdot x = x$

Domination
$x + 1 = 1$
$x \cdot 0 = 0$

Associativity
$(x + y) + z = x + (y + z)$
$(x \cdot y) \cdot z = x \cdot (y \cdot z)$

De Morgan's Laws
$\overline{x + y} = \overline{x} \cdot \overline{y}$
$\overline{x \cdot y} = \overline{x} + \overline{y}$

Involution
$\overline{\overline{x}} = x$

Absorption
$x + x \cdot y = x$
$x \cdot (x + y) = x$

2.8.4 Additional Review Material

Go to http://www.mathcs.bethel.edu/~gossett/DiscreteMathWithProof/review.xhtml for additional review material, including a sample chapter exam, with solutions.

CHAPTER

3 Proof

Euclid taught me that without assumptions there is no proof.
Therefore, in any argument, examine the assumptions.
Eric Temple Bell (1883 – 1960)

If you were to ask a randomly chosen person what he or she thinks mathematics is, he or she would probably mention arithmetic and numbers. A randomly chosen college student might mention calculus or trigonometry. A mathematician will give a very different answer.

Much of the mathematics you have encountered in the past has not been the kind of mathematics that a typical mathematician would describe as "doing mathematics". What you have usually seen (with the exception of geometry) has been a collection of techniques that may be used to solve various problems. For example, the law of sines and the law of cosines enable you to solve triangles that may not contain a right angle. You have also learned how to solve systems of equations. When you simplify algebraic expressions such as $(3x + 4(x + y) - 7y)$, you probably do not spend time thinking about why the simplification steps are valid (as was done in Example 2.33 on page 64).

Your geometry course was a much better picture of the way mathematicians view their subject. People who use mathematics as a tool tend to focus on the techniques. Mathematicians tend to focus on the structure that underlies the techniques. They are interested in why the techniques work. In addition, mathematicians are curious about more fundamental questions, such as "What are numbers?", "How many numbers are there?", and "Can we classify all objects that share certain properties?".

To summarize, the techniques used to solve problems do not constitute what mathematicians consider to be the soul of mathematics. Mathematicians are more interested in answering questions about the mathematical universe. The answers to these questions, the process of arriving at these answers, and the process of proving that these answers are correct are what mathematicians tend to view as true mathematics.

The goal of this chapter is to introduce you to some standard approaches for producing a formal proof of a mathematical statement. The chapter starts with a section that outlines the context to which proofs most naturally belong. To practice the proof techniques that are eventually presented, some simple mathematical statements need to be available. Some standard results from elementary number theory provide excellent examples of such statements. Consequently, a brief introduction to elementary number theory is provided before moving to the proof techniques.

3.1 Introduction to Mathematical Proof

The normal context for mathematical proofs is the realm of axiomatic mathematics. Before discussing proofs, it is necessary to review the basic outline of axiomatic mathematics.

3.1.1 Mathematics and Proof: The Big Picture

Mathematics is a broad, diverse collection of fields of study. For example, geometry focuses on points and lines, trigonometry on angles. Calculus is concerned with limits and functions, combinatorics with arrangements and counting. These fields can be quite different in content; however, the same system of study is used in all these fields: mathematical proof based on mathematical logic.

Mathematics can be viewed as a system of ideas that logically follow from other ideas and concepts. An informal approach to this development soon becomes unmanageable. Without sufficient confidence in the validity of the earlier concepts, the later concepts are of dubious value. A formal proof is designed to provide confidence in the correctness of the statement being proved.

An Informal Definition of *Proof*

It will be useful to have an informal definition of the term *proof*. (A more precise definition will appear at the end of this section.)

> **DEFINITION 3.1** ***Proof: An Informal Definition***
> A **proof** is a demonstration of the validity of some precise mathematical statement.
> The demonstration should contain sufficient detail to convince the intended audience of its validity.

Often the statement to be proved is of the form "if A is true, then B is also true", where A and B are propositions.

A Very Short History of Proofs

The need for proving mathematical statements was recognized by the ancient Greeks. They were presenting proofs 2500 years ago! Much of what you saw in geometry is a translation of mathematical activity done around 300 B.C.

However, the standard for what constitutes a valid proof has been refined over the centuries. The mathematical *statements* proved by the ancient Greeks are still considered valid (in the proper context), but some of their *proofs* of those statements are no longer considered valid. As successive generations of mathematicians have thought about mathematics, they have found they need to express mathematical ideas with increasing precision. The mental tools needed to explore mathematical ideas at more subtle levels have also needed to become more powerful.

Modern mathematics has become a technical language (with an accompanying cultural philosophy) in much the same way that computer languages have. As such, some training is necessary in order for you to be able to communicate in the language of mathematics. The goal of this chapter is to provide you with the fundamental ideas you will need to begin learning and to start becoming proficient in this language.

A proof is a means by which the truth or validity of most statements is established.[1] Let S_n be the statement whose truth we want to show. A proof typically consists of a sequence of statements $S_1, S_2, \ldots, S_n$, where the truth of S_k can be logically derived from the truth of some or all of $S_1, S_2, \ldots, S_{k-1}$. *Logically derived* means using rules based on tautologies such as those presented in Chapter 2. S_1 is usually an hypothesis of the theorem. $S_2, \ldots, S_{n-1}$ are other hypotheses, undefined terms, axioms, definitions, statements that have been previously proved, or statements that can be logically inferred from any of these.

As was mentioned in the informal definition of *proof* given earlier, the level of detail represented by the sequence of statements $S_1, S_2, \ldots, S_{n-1}$ depends primarily on

[1] The exceptions are undefined terms, axioms, and definitions; they do not need proving. These objects will be discussed in Section 3.1.2.

the intended readers of the proof. A reader who is mathematically advanced may need much less detail than a beginner. In research journals only the bare minimum is typically presented (essentially, an outline of the major steps in the proof is given).

3.1.2 Mathematical Objects Related to Proofs

In the next two subsections, a number of mathematical objects are presented. These are the essential components of the mathematical approach called *the axiomatic method.* The main idea is to start at some (usually small) agreed on set of concepts. The truth of these initial concepts is assumed. Everything else is deduced from the initial concepts using the rules of logic, which are also agreed on ahead of time.

Axioms, Undefined Terms, and Definitions

The axiomatic method is a technique of deduction from prior concepts. In order for such a system to get started, some prior concepts must exist that do not need to be based on still earlier concepts. The physicists have The Big Bang as their beginning concept; the theologians have God. Mathematicians start with undefined terms. *Undefined terms* are usually ideas that have enough intuitive appeal that we may safely use them as a starting place. For example, in geometry, *point*, *line*, and *a point is on a line* are undefined but intuitively clear concepts.

Having undefined terms is not sufficient. In addition, a set of properties that the undefined terms satisfy is needed. Such properties, which are *assumed* to be true, are called *axioms* (from the Greek word meaning "worthy"). Axioms are also called *postulates.*[2] The valid deductions that exist in the system are directly dependent on the set of axioms chosen.

Euclid's Postulates

The most famous set of axioms were those in Euclid's *Elements*. Because of their significant historical impact, his postulates are listed here (using modern terminology).[3]

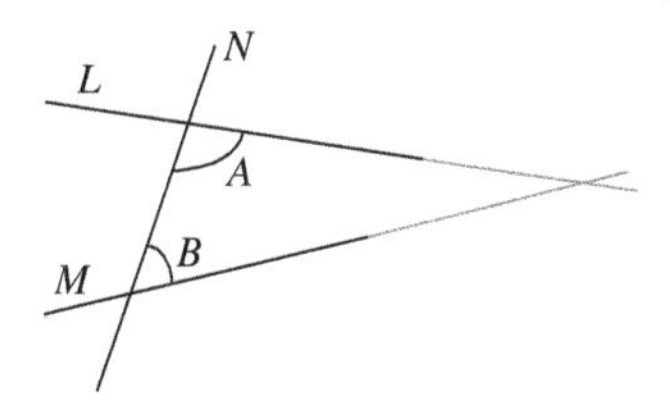

Figure 3.1. Euclid's fifth postulate.

P1 A straight line (segment) can be drawn from any point to any other point.

P2 A line segment can be extended continuously to form a line.

P3 A circle can be created having any center and radius.

P4 All right angles are equal to one another.

P5 Suppose line N crosses lines L and M (Figure 3.1), and the angles A and B (both on the same side of N) add to less than two right angles. Then lines L and M will eventually meet, on the side of N where A and B are.

The fifth axiom seems more awkward than the first four. An alternative (but equivalent) form of the fifth postulate (called Playfair's postulate) is *through a given point can be drawn one and only one line parallel to a given line.* For hundreds of years people tried to show that the fifth postulate was unnecessary. Finally, in the first half of the nineteenth century, Lobachevsky, Bolyai, Gauss, and others showed that the fifth postulate could not be proved from the other four. For example, there exist geometries that satisfy the other postulates but in which *every* pair of distinct lines intersect in exactly one point. There are other geometries in which there exist more than one distinct line through a given point that is parallel to a given line. ■

[2]Some mathematicians (including Euclid) have made a distinction between axioms and postulates. The terms will be used interchangeably in this book.

[3]Euclid actually used some axioms that he did not explicitly state. Around the turn of the last century, the German mathematician Hilbert created a set of axioms that accurately describe Euclid's geometry.

EXAMPLE 3.2 **Axioms for a Boolean Algebra**

The definition of a Boolean algebra on page 59 describes a set of undefined objects, some undefined operators, and four axioms that specify how the objects and operators interact.

It is not necessary to know what the objects are nor what the operators actually do (other than map objects to other objects). The axioms provide all the necessary information to determine the primary shape of a Boolean algebra.

In Chapter 2, this simple definition made it possible to unite two very different mathematical systems: sets and propositional logic. In the first case, the objects of interest were subsets and the operations were union, intersection, and complement. In the second case, the objects were propositions and the operators were OR, AND, and NOT. ■

Not only do we want our axiom systems to be intuitively reasonable,[4] we require at least two additional characteristics. The first is that the axioms be *independent*; that is, none of the axioms should be deducible from the other axioms.[5] The other requirement is that the axioms be *consistent*. Two contradictory valid deductions from the axioms would be unacceptable.

The issue of independence has already made an implicit appearance in the definition of Boolean algebra. In Definition 2.26, only four axioms were listed. Many textbooks include several more—most notably, an associativity axiom. In this textbook, associativity was proved using the other axioms. Thus, associativity is *not* independent of the four axioms listed in the definition.[6]

To be able to make deductions from the undefined terms and axioms, it is necessary to agree on the rules of logic that are to be used. The rules presented in Chapter 2 are the standard ones.

As the axioms lead to additional concepts, it is convenient to give some of them names, especially if the concept is nontrivial. A *definition* is the means for binding a concept, a name for the concept, and a set of associated properties that describe the concept. A definition is a form of aliasing. The concept (name) is equated with the properties. As such, the rules of inference $(C \leftrightarrow P) \Rightarrow (C \rightarrow P)$ and $(C \leftrightarrow P) \Rightarrow (P \rightarrow C)$ enable us to substitute the properties for the concept whenever the concept is known to occur, or to conclude that the concept occurs if the properties are shown to hold.

A definition should be interpreted as "concept if and only if properties". Notice that it works in both directions; a definition is more like an equivalence than like an implication. The *if* means "if the properties hold, then the concept occurs" (concept if properties).

The *only if* means "only those objects that satisfy all the properties are representatives of the concept" (concept only if properties).

Mathematicians seldom write definitions in the form "concept if and only if properties". Instead, they write "if properties then concept" or "concept if properties". For example, in elementary number theory we might make the definition "an integer n is even if and only if there is an integer k such that $n = 2k$". This is typically shortened to "a number, n, is even if it can be written $2k$ for some integer, k". When you encounter *if* used in a definition, you should always mentally substitute the phrase "if and only if".

[4] The previous footnote provides a cautionary warning: Sometimes our intuition can mislead us. The non-Euclidean geometries (geometries that omit postulate 5) are important.

[5] The essence of the Euclidean fifth postulate controversy was a (futile) attempt to show that the fifth axiom was not independent of the others.

[6] One reason other authors include associativity as an axiom is because it is quite tricky to prove. In addition, it is such a fundamental property that they seek to give it prominence by listing it as an axiom.

A Brief Sermon about Definitions

Mathematical definitions are endowed with a level of significance unmatched by definitions in other disciplines. It is not possible to talk with precision about any technical subject without many agreed-on terms. However, in mathematics these technical terms are more numerous and also tend to be much more detailed.

This has ramifications for you as a student. It is tempting to read a mathematical definition and feel you have mastered it if you have only a general notion of what it is about. Succumbing to that temptation will needlessly complicate your life.

Successful mathematics students have learned that definitions need to be memorized in precise detail, as soon as they are encountered. Many theorems or other definitions will be stated using previous definitions. It is not possible to understand them if the prerequisite definitions are still only partially understood. In addition, many proofs require an ability to switch back and forth between concept and properties in a definition.

If you want to make progress in mastering the material in advanced mathematics classes, you will want to be deliberate about learning definitions. You might use a set of flash cards or a definition notebook to list definitions as they are encountered. Review them on a regular basis (rather than the night before an exam).

Propositions, Theorems, Lemmas, and Corollaries

Any statement that is not an axiom or definition needs to be proved. The objects discussed next are all objects that need proving. These statements will be categorized according to their importance and use.

Important statements that have been proved are called *theorems*. Less important (true) statements are called *propositions*.[7] Sometimes the proof of a theorem is quite long. It is often useful to modularize long proofs by introducing subtheorems called *lemmas*. A lemma is usually considered to be a mini-theorem whose main purpose for existing is to help prove part of a more important theorem or proposition. A *corollary* is a statement whose truth is an immediate consequence of some other theorem or proposition.

Deciding which category to place a statement in is somewhat subjective. Observing how other mathematicians make such decisions and understanding the mathematical subject you are working with are the best ways to refine your subjective feelings.

EXAMPLE 3.3 **Theorems and propositions for a Boolean Algebra**

Proposition 2.30 on page 65 is a simple proposition in the Boolean algebra system.

Although no theorems were explicitly named as such, the fundamental Boolean algebra properties on page 64 are prime examples of theorems in the system of Boolean algebra. ■

EXAMPLE 3.4 **Lemmas and Corollaries**

Lemma 3.75 on page 161 will be used to prove Theorem 3.73 (page 161).

Corollary 3.76 on page 161 is an immediate consequence of Theorem 3.73.

Corollary 2.31 on page 65 is an immediate consequence of Propostion 2.30. ■

---

[7]This is not universally the case. Euclid always used the term *proposition*, never the term *theorem*.

EXAMPLE 3.5

A Simple Geometry

A simple example may be helpful.[8] The example introduces a few undefined terms and four axioms. These are not Euclid's axioms; consequently, the theorems that will be proved may seem strange. In fact, the notion of a line is quite different from what you encountered in high school geometry. In this example, a line contains only two points!

Undefined terms Point, line, a point is on a line (and synonymous ideas).

Definitions

Two lines are *parallel* if and only if they have no point in common.

Axioms

1. Any two points determine a line.
2. Any line contains exactly two points.
3. For any point P and any line L not containing P, there is exactly one line through P and parallel to L.
4. There are at least three points.

Theorem W There are at least four points.

Proof: Produce three distinct points and label them A, B, and C (Axiom 4).

Denote the line determined by A and B by AB (Axiom 1).

There is a line through C and parallel to AB (Axiom 3). (In fact, only one such line.)

The line through C and parallel to AB contains a second point; call that point D (Axiom 2).

D is different from C because together they form a line. D cannot be the same point as either A or B since it is on a line that is parallel to AB (definition of *parallel*).

Thus D is a fourth point.

QED [9]

Theorem X There are at least six lines.

Proof: There are at least four points: A, B, C, and D (Theorem W).

The following pairs of points form lines: AB, AC, AD, BC, BD, CD (Axiom 1).

Thus there are at least six lines.

QED

Theorem Y There cannot be more than four points.

Proof: Suppose there is a fifth point E. Then lines CD and DE are parallel to AB by definition of *parallel*. Thus CD and DE are two distinct lines parallel to AB and passing through D. This contradicts Axiom 3. The assumption that E exists must be wrong.[10]

QED

Corollary There are exactly four points.

Proof: Theorems W and Y.

QED ■

Theorems versus Scientific Theories

Mathematics typically utilizes deductive reasoning to logically infer theorems by considering the consequences of prior axioms and theorems. The truth of a theorem *necessarily* follows from the prior information and the rules of logic.

---

[8] This example is from [20, pages 165, 166]. Note the nonstandard use of the terms *line* and *parallel*.

[9] The letters QED represent the end of a proof. (*Quod Erat Demonstrandum* is Latin for "which was to be demonstrated".) The more commonly used modern notation is a small box, □ or ■.

[10] This is a proof by contradiction (see Section 3.3.4).

In contrast, scientists typically utilize inductive reasoning to take a mass of experimental evidence and try to infer general principles that can explain the evidence. If the general principle (initially called an hypothesis) does explain all of the available experimental and observable evidence, if it is able to accurately predict new evidence, and if the principle seems as simple and elegant as possible, then the hypothesis may be elevated to the status of a scientific theory. A scientific theory is a general principle that is accepted as true by a significant majority of the people who are considered competent in the discipline.

A scientific theory's validity is generally attacked by considering the evidence that it is supposed to explain, or by observing new phenomena that the theory does not adequately explain. A mathematical theorem's validity is attacked by finding an error in the proof or by finding faulty assumptions on which some part of the proof depends. Because of inherent human limitations, incorrect scientific theories have been accepted by many intelligent people as true. Similarly, competent mathematicians have been embarrassed by having their "proof" shown to be incorrect after it was published.

Formal proofs should be read (and written) with great care. When writing a proof, it is important to be familiar with the mathematics other investigators have developed. For example, the Greeks tried to devise a method for using a compass and straightedge to trisect *any* angle. In 1837 a mathematician named Pierre Laurent Wantzel proved that such a procedure cannot exist. Yet each year uninformed individuals still spend many hours developing specious techniques for trisecting angles with compass and straightedge.[11]

A Formal Definition of *Proof*

We now have sufficient background to give a formal definition of the term *proof.*

DEFINITION 3.2 ***Mathematical Proof***

A *mathematical proof* of the statement S is a sequence of logically valid statements that connect axioms, definitions, and other already validated statements into a demonstration of the correctness of S. The rules of logic and the axioms are agreed on ahead of time. At a minimum, the axioms should be independent and consistent. The amount of detail presented should be appropriate for the intended audience.

Note: As you read the proofs in this book, notice that proofs are *narratives* that use words to connect mathematical expressions. They are not just a sequence of equations. (See Appendix G for some helpful hints on writing mathematics.)

3.1.3 Exercises

The exercises marked with ★ have detailed solutions in Appendix H.

1. Can we prove the following additional theorem for Example 3.5?[12]

Theorem Z: There are exactly six lines.

2. If we change Axiom 3 in Example 3.5 to "For any point P and any line L, there is a line through P and parallel to L," are theorems W through Z (see previous exercise) still valid?

3. ★ Suppose we change Axiom 4 in Example 3.5 to "There are at least five points". Show that the new set of axioms is inconsistent.

4. Each statement is either true or false. Identify which case is correct, and then give some justification for your answer.

(a) ★ Every definition is either explicitly or implicitly an "if and only if" statement.

(b) Axioms are logical consequences of the primary definitions in a mathematical system.

(c) Suppose the following axioms are used in a revised definition of Boolean algebra: identity, idempotence, complement, commutativity, distributivity. The resulting axiom set is independent.

(d) The amount of detail in a proof should be the same for all audiences.

(e) ★ If your instructor has a Ph.D. in mathematics, then you only need to submit a brief outline of your proof.

(f) Scientists and mathematicians typically use the same kind of reasoning.

(g) When using a definition, we always start with the concept that the definition defines, and then assert (and use) the properties attached to that concept.

[11] See Example 4.32 on page 214 for more details.

[12] An implicit assumption in this kind of question is "if the answer is yes, then you are to supply the proof".

5. ★ In any axiomatic system, some terms must remain undefined. Why is this necessary?

The remaining exercises in this section ask you to prove some simple theorems related to an axiomatic system called a group. *The definition is presented prior to the exercises. That definition includes the notion of a binary operation. A* binary operation *takes two elements from a set and produces a third element in the set. Common examples of binary operations are ordinary addition and ordinary multiplication. Notice that the definition does* not *include a commutativity axiom.*

DEFINITION 3.3 ***Group***

Let G be a nonempty set, together with a binary operation, $\diamond$. G is a *group* if the following four axioms are satisfied.

Closure For all $a, b \in G$, $a \diamond b \in G$.

Associativity For all $a, b, c \in G$, $a \diamond (b \diamond c) = (a \diamond b) \diamond c$.

Identity There is an element $e \in G$, called the *identity*, such that $a \diamond e = e \diamond a = a$ for all $a \in G$.

Inverses For each element $a \in G$, there is an element $b \in G$, called the *inverse* of a, such that $a \diamond b = b \diamond a = e$.

6. ★ Prove that a group has a *unique* identity element. [*Hint*: assume that another element, i, is also an identity. Show that $i = e$.]

7. Prove that the right and left cancelation laws are valid in any group: for all $a, b, c \in G$, if $a \diamond c = b \diamond c$ then $a = b$, and if $c \diamond a = c \diamond b$ then $a = b$.

8. Let G be a group and let $a \in G$. Prove that a has a *unique* inverse in G. [*Hint*: use the previous exercise.]

9. Let G be a group and let $a, b \in G$. Prove: if $(a \diamond b) \diamond (a \diamond b) = (a \diamond a) \diamond (b \diamond b)$, then $a \diamond b = b \diamond a$. (The axioms don't guarantee that elements commute.) Explicitly show all your applications of the associativity axiom.

10. Let G be a group with $a, b \in G$. What is the inverse of the element $a \diamond b$? Verify that your assertion is correct. Explicitly show all your applications of the associativity axiom.

11. Let $S = \{2^k \mid k \in \mathbb{Z}\}$. Show that S forms a group, where the operation $\diamond$ is multiplication of rational numbers.

12. Verify that this multiplication table defines a group.

$\diamond$	e	a	b	c
e	e	a	b	c
a	a	e	c	b
b	b	c	e	a
c	c	b	a	e

3.2 Elementary Number Theory: Fuel for Practice

Number theory is one of the oldest branches of formal mathematics. It is a rich subject that can easily sustain an entire course, either at the undergraduate level (elementary number theory) or at the graduate school level. It is an area of active research with many interesting questions still unanswered. The information in this section is presented with two goals in mind:

- To add to the student's cultural background. These ideas have been part of the assumed knowledge of an educated person for over two thousand years.
- To provide some theorems that are easy to understand and to prove. They will be used for examples and for exercises in the rest of this chapter.

Additional theorems and ideas from elementary number theory will appear in Sections 3.3 and 3.4.

3.2.1 The Integers and Other Number Systems

Elementary number theory is mostly about the natural numbers, the integers, and the rational numbers. These number systems will be defined next. It is strongly suggested that you take a detour to Appendix A for a more comprehensive review of number systems.

DEFINITION 3.4 ***The Natural Numbers***

The set of *natural numbers* is denoted by $\mathbb{N}$, and is defined[a] by

$$\mathbb{N} = \{0, 1, 2, 3, 4, \ldots\}$$

[a]Some authors define the natural numbers to be the set $\{1, 2, 3, \ldots\}$.

The integers extend the natural numbers by including the negations of numbers in $\mathbb{N}$.

DEFINITION 3.5 ***The Integers***

The set of *integers* is denoted by $\mathbb{Z}$, and is defined by

$$\mathbb{Z} = \{\ldots, -4, -3, -2, -1, 0, 1, 2, 3, 4, \ldots\}$$

The set of nonnegative integers is the familiar set of natural numbers. The set of positive integers is the set $\{1, 2, 3, \ldots\}$.

The rational numbers (the familiar collection of fractions) contain (almost) all ratios of integers.

DEFINITION 3.6 ***The Rational Numbers***

The set of *rational numbers* is denoted by $\mathbb{Q}$, and is defined by

$$\mathbb{Q} = \left\{ \frac{p}{q} \mid p \in \mathbb{Z}, q \in \mathbb{Z}, \text{ and } q \neq 0 \right\}$$

The set of real numbers is denoted by $\mathbb{R}$. The set of positive real numbers is denoted by $\mathbb{R}^+$. A precise definition requires a careful examination of a set of axioms, but will not be done here. A brief summary can be found in Appendix A.4 on page A4.

DEFINITION 3.7 ***Irrational Numbers***

A real number that is not rational is called *irrational.*

Unless explicitly told otherwise, the discussion in the remainder of this section will be understood to be about the integers, $\mathbb{Z}$. All quantified statements will therefore implicitly assume that the universal set is $\mathbb{Z}$. One consequence is that fractions should not be used unless they are explicitly called for.

3.2.2 Divisibility

A key idea in number theory is divisibility. Keep in mind that the next definition is about the integers. Rational numbers are *not* part of the discussion.

DEFINITION 3.8 ***Divisible***

The integer a is *divisible* by the nonzero integer b if $a = bc$ for some integer c. We denote this by $b \mid a$ and also say that *b divides a*.

Sometimes the phrase *evenly divisible* is used for emphasis. If a is *not* divisible by b (denoted by $b \nmid a$), then there must be a nonzero remainder when dividing a by b.[13] This can be expressed very precisely in the next theorem (which will be proved at the end of this section). You might recall a less formal version of this theorem from when you first studied division in elementary school and were asked to find a quotient and a remainder.

THEOREM 3.9 ***The Quotient–Remainder Theorem***

Let a and b be integers with $b \neq 0$. Then there exist unique integers q and r such that $a = bq + r$ and $0 \leq r < |b|$.

The next definition is one you have known since childhood. Notice the use of divisibility.

[13] In this context, we are not interested in fractions. We want an integer quotient and remainder.

DEFINITION 3.10 ***Even and Odd***

An integer, n, is *even* if there exists an integer, k, such that $n = 2k$. An integer is *odd* if it is not even.

Notice that the definition implies that 0 is even, since we can write $0 = 2 \cdot 0$. Observe also that if the term *even number* is used, the implied assumption is that the number is an integer. Often the implied assumption is even stronger and the context requires the number to be a nonnegative integer or a positive integer.

One useful idea that will be used many times in the last half of this chapter is the following direct application of the definitions of even and odd.

- If n is an even number, then we can write n as $2k$ for some integer k.
- If n is an odd number, then we can write n as $2k + 1$ for some integer k.

Note well: If there are two numbers under discussion, then we need to write them as $2k$ and $2m$ (if they are both even). It is incorrect to write both as $2k$. This seems obvious here, but in the middle of a proof with more going on to distract you, it is easy to make this mistake. More on this later.

✔ Quick Check 3.1

1. Definition 3.10 provides the justification for the claim "If n is an even number, then we can write n as $2k$ for some integer k". Provide a proof for the claim "If n is an odd number, then we can write n as $2k + 1$ for some integer k". ☑

The next few definitions use the notion of divisibility.

DEFINITION 3.11 ***Greatest Common Divisor***

Let a and b be integers that are not both 0. The *greatest common divisor* (gcd) of a and b is a positive integer d such that

- $d \mid a$ and $d \mid b$.
- If c divides both a and b, then $c \mid d$.

The greatest common divisor of a and b is denoted by $\gcd(a, b)$. An alternative notation is (a, b).

The first of the two conditions on d ensures that d is a *common* divisor. The second condition ensures that it is the largest of all common divisors. Notice that the definition of greatest common divisor is symmetric in a and b: $\gcd(a, b) = \gcd(b, a)$.

EXAMPLE 3.6 **Common Divisors**

The integers 1, 2, 3, and 6 are all common divisors of 12 and 18. You can verify that no larger integer is a common divisor (you only need to check elements in $\{7, \ldots, 12\}$). Since 6 is divisible by 1, 2, and 3, it fulfills the requirements for being $\gcd(12, 18)$. ■

Several methods exist for finding the greatest common divisor of two integers. One of the best algorithms will be introduced in Section 3.4.1. A simpler method (for small numbers) will be explained in Example 3.9.

Perhaps the most important theorem about greatest common divisors is the next theorem. Two proofs will be given, one on page 110 and the other on page 115.

THEOREM 3.12 $\gcd(a, b) = as + bt$

Let a and b be integers such that at least one is not 0. Then there are integers, s and t, such that $\gcd(a, b) = as + bt$.

The next definition is a natural partner to the notion of a greatest common divisor.

DEFINITION 3.13 ***Least Common Multiple***

The *least common multiple* (lcm) of two integers a and b is a nonnegative integer, m, such that

- $a \mid m$ and $b \mid m$.
- If both a and b divide c, then $m \mid c$.

The least common multiple of a and b is denoted by $\text{lcm}(a, b)$. An alternative notation is $[a, b]$.

EXAMPLE 3.7

Common Multiples

The integers 36, 72, and 108 are all common multiples of 12 and 18. You can verify that 36 is the smallest integer that is a common multiple (you need to check numbers in $\{18, \ldots, 35\}$). Thus, $\text{lcm}(12, 18) = 36$. ■

3.2.3 Primes

Perhaps the most important definition in all of number theory is that of a prime number.

DEFINITION 3.14 ***Prime, Composite***

A positive integer p, with $p > 1$, is said to be *prime* if its only positive integer divisors are 1 and p. A positive integer n, $n > 1$, that is not prime is called *composite*. The integer 1 is neither prime nor composite. (In more advanced contexts it is called a *unit*.)

The next theorem is too important to omit from this discussion. The proof will be presented in two pieces. The first part of the proof (existence) is presented in Example 3.40 on page 140. The second part of the proof (uniqueness) is on page 111.

THEOREM 3.15 ***The Fundamental Theorem of Arithmetic***

Every integer n, with $n \geq 2$, can be uniquely written as a product of primes in ascending order.

Note: We consider prime numbers to be products of one prime.

EXAMPLE 3.8

Prime Factorization

The number 12 can be factored into primes as $12 = 2 \cdot 2 \cdot 3$. This is typically summarized as $12 = 2^2 \cdot 3$.

The number 18 can be factored into primes as $18 = 2 \cdot 3 \cdot 3 = 2 \cdot 3^2$. ■

One way to find the greatest common divisor of two integers is to factor each one individually and then, for each common prime factor, keep the minimum of the two exponents.

EXAMPLE 3.9 **Calculating the gcd by Factoring**

Since $12 = 2^2 \cdot 3$ and $18 = 2 \cdot 3^2$, $\gcd(12, 18) = 2 \cdot 3 = 6$.
Since $45 = 3^2 \cdot 5$, $\gcd(12, 45) = 3$ and $\gcd(18, 45) = 3^2 = 9$. ■

DEFINITION 3.16 ***Relatively Prime***
Positive integers a and b are *relatively prime* if $\gcd(a, b) = 1$.

Notice that 4 and 21 are relatively prime, but neither 4 nor 21 is a prime number.

The notion of prime factorization (i.e., expressing a positive integer as a product of primes) enters into the next definition.

DEFINITION 3.17 ***Pythagorean Triple***
The set of integers $\{a, b, c\}$ is called a *Pythagorean triple* if $a^2+b^2 = c^2$. It is called a *primitive Pythagorean triple* if a, b, and c have no common prime factor.

EXAMPLE 3.10 **Pythagorean Triples**

It is not too difficult to generate a few Pythagorean triples:

$$3^2+4^2 = 5^2 \qquad (2\cdot3)^2+(2\cdot4)^2 = (2\cdot5)^2 \qquad 5^2+12^2 = 13^2 \qquad 8^2+15^2 = 17^2$$ ■

3.2.4 The Well-Ordering Principle

One of the axioms of the integers is the well-ordering principle. It is simple but has significant consequences.

AXIOM 3.1 ***The Well-Ordering Principle***
Every nonempty set of natural numbers has a smallest element.

Notice the requirements:

- The set must contain at least one element.
- The set must contain only nonnegative elements.

The second requirement avoids sets like $\{\ldots, -8, -6, -4, -2, 0\}$, which clearly will not have a smallest element.

Since the well-ordering principle is an axiom, we *assume* it is true; we do not attempt to prove it.

The well-ordering principle plays a crucial role in the proof of the previously stated Theorem 3.9.

THEOREM 3.9 ***The Quotient–Remainder Theorem***
Let a and b be integers with $b \neq 0$. Then there exist unique integers q and r such that $a = bq + r$ and $0 \leq r < |b|$.

Proof: The proof will be done in two parts. The first part will show that q and r exist. The second part will then show that q and r are unique.

Existence
The key insight is to focus on r by rearranging the equation $a = bq + r$ as $r = a - bq$. We are looking for an r which is positive and small (perhaps even 0). Thus it makes

sense to define the set[14]

$$S = \{a - bq \mid q \in \mathbb{Z} \text{ and } a - bq \geq 0\}$$

Clearly, any elements in S are nonnegative. Is S nonempty? Notice that we can choose (in many ways) the magnitude and sign of q so that $bq \leq a$. For such a q, $a - bq \geq 0$ and therefore S is not empty.

Since S is a nonempty set of nonnegative integers, the well-ordering principle implies that it has a smallest element. Let that element be $a - bq_0$ and set $r_0 = a - bq_0$. Do q_0 and r_0 meet the requirements of the theorem? Clearly, $a = bq_0 + r_0$ and $0 \leq r_0$ by the way S was constructed. Suppose that $r_0 \geq |b|$. Let $q_1 = q_0 + \frac{|b|}{b}$ (so $q_1 = q_0 \pm 1$ and $bq_1 = bq_0 + |b|$).

We can therefore write $a = bq_0 + r_0 = bq_0 + |b| + (r_0 - |b|) = bq_1 + (r_0 - |b|)$. Thus $a = bq_1 + (r_0 - |b|)$ and $(r_0 - |b|) \geq 0$. Consequently, $(r_0 - |b|)$ must be in S. In addition, $r_0 > (r_0 - |b|)$ because $|b| > 0$. This contradicts the knowledge that r_0 is the smallest element of S. The assumption that led to the contradiction was $r_0 \geq |b|$. We must conclude that $r_0 < |b|$. Thus, q_0 and r_0 satisfy the requirements of the theorem; existence has been demonstrated.

Uniqueness

Suppose that $a = bq_2 + r_2$ with $0 \leq r_2 < |b|$ and also $a = bq_3 + r_3$ with $0 \leq r_3 < |b|$. Then $bq_2 + r_2 = bq_3 + r_3$ and so $b(q_3 - q_2) = r_2 - r_3$. Since $0 \leq r_2 < |b|$ and $0 \leq r_3 < |b|$, we know that $0 \leq |r_2 - r_3| < |b|$.[15] Thus

$$|b| > |r_2 - r_3| = |b| \cdot |q_3 - q_2| \geq 0$$

This can only be true if $q_3 - q_2 = 0$ since q_2 and q_3 are integers. That implies that $q_2 = q_3$ and consequently that $r_2 = r_3$, establishing uniqueness. □

EXAMPLE 3.11

Illustrating Existence in the Previous Proof

Let $a = 13$ and $b = 5$. The set S defined in the proof of Theorem 3.9 is the collection of all nonnegative integers of the form $a - bq$, where q is any integer. The next table helps determine S.

q	...	−3	−2	−1	0	1	2	3	4	...
$a - bq$	...	28	23	18	13	8	3	−2	−7	...

Thus, $S = \{3, 8, 13, 18, 23, 28, 32, \ldots\}$. The number r_0 from the proof is $r_0 = 3 = a - b \cdot 2$. Clearly, $13 = 5 \cdot 2 + 3$, and $0 \leq 3$. Also, $3 < |5|$, so $q = 2$ and $r = 3$ are the values promised by the theorem.

To illustrate the remainder of the existence proof, it is necessary to pretend, for the moment, that $r_0 \geq |b|$. Thus, pretend that $q_0 = 1$ and $r_0 = 8$ (the next smallest value in S). Then $q_1 = q_0 + \frac{|b|}{b} = q_0 + 1 = 1 + 1 = 2$. Consider $r_1 = r_0 - |b|$. We can write $a = bq_1 + (r_0 - |b|) = 5 \cdot 2 + (8 - |5|) = 13$, and $r_1 = r_0 - |b| = 8 - |5| = 3 > 0$. A smaller nonnegative value (r_1) for r has been produced, completing the contradiction.

It would be helpful if you were to repeat this example with $a = 13$ and $b = -5$. In particular, the expression $q_1 = q_0 + \frac{|b|}{b}$ would now become $q_1 = q_0 - 1$ (using -1 as the pretend value for q_0). ■

---

[14] You may wish to look at Example 3.11 as you read this proof.

[15] Since $r_2 - r_3 < |b| - r_3 \leq |b|$ and $r_3 - r_2 < |b| - r_2 \leq |b|$, it follows that $-|b| < r_2 - r_3 < |b|$. But "$|x| < y$ if and only if $-y < x < y$" then implies that $|r_2 - r_3| < |b|$.

3.2.5 Congruence, Factorials, Floor and Ceiling Functions

The next two definitions provide a more sophisticated look at divisibility.

DEFINITION 3.18 *a* mod *m*

Let m be a positive integer. Then $a \bmod m$ is the remainder when a is divided by m. The integer m is called the *modulus*.

EXAMPLE 3.12 **mod**

The Quotient–Remainder theorem can be used to calculate $a \bmod b$. For example, $17 \bmod 5 = 2$ because $17 = 5 \cdot 3 + 2$. ■

DEFINITION 3.19 $a \equiv b \pmod{m}$

We say that a is *congruent to* b, mod m, if m divides $a - b$. This is often expressed as: $a \equiv b \pmod{m}$ if and only if there is an integer k for which $a - b = km$.

Intuitively, a is congruent to b, mod m, if they both have the same remainder when divided by m (and so they both belong to the same "club" with regard to division by m).

Notice that a statement about congruence can be easily converted into a characterization similar to writing even and odd integers in the form $2k$ or $2k + 1$. For example, $n \equiv 1 \pmod{3}$ is equivalent to $n = 3k + 1$ for some k. This representation clearly shows that n has remainder 1 when divided by 3.

EXAMPLE 3.13 **A Simple Congruence**

Since (17 mod 5) and (32 mod 5) are both equal to 2, $17 \equiv 32 \pmod{5}$.

We can also write $17 - 32 = (-3) \cdot 5$ or $32 = 3 \cdot 5 + 17$. ■

✔ Quick Check 3.2

1. Which of the numbers $\{100, 101, 102, 103, 104, 105\}$ are prime?
2. Let $a = 1600$ and $b = 450$.
   (a) Find the q and r (from the Quotient–Remainder theorem) that make $a = bq + r$.
   (b) Factor a and b as products of primes.
   (c) Find $\gcd(a, b)$.
   (d) Find $\operatorname{lcm}(a, b)$.
   (e) Find $a \bmod b$.
   (f) Is $a \equiv b \pmod{11}$? ☑

Note: Whenever we are discussing primes, or divisibility, or the mod operator, the context is understood to be the integers. It is not proper to introduce fractions into a proof in this context. Such a lack of propriety is mainly a matter of style: there is no need to leave the system of integers (entering the larger system of rational numbers) to complete the proof. The following example illustrates this.

EXAMPLE 3.14 **Improper Use of Fractions**

Suppose d is a common divisor for the integers m and n. Prove that d divides $(m + n)$.

An Improper Proof Since d divides m and n, there exist integers j and k such that $j = \frac{m}{d}$ and $k = \frac{n}{d}$. Adding these equations leads to $j + k = \frac{m+n}{d}$. Multiplying both sides by d leads to $m + n = d(j + k)$, which implies that $d \mid (m + n)$. □

The introduction of fractions in the improper proof is entirely unnecessary.

A Proper Proof Since d divides m and n, there exist integers j and k such that $m = dj$ and $n = dk$. Thus $m + n = d(j + k)$, which implies that $d \mid (m + n)$. □

This example seems trivial, but many students find the improper approach tempting. ■

The following two definitions will be used throughout the remainder of this book.

DEFINITION 3.20 $n!$

$n! = n \cdot (n-1) \cdot (n-2) \cdots 3 \cdot 2 \cdot 1$, for positive integers n and is pronounced "n factorial".
Also, $0! = 1$ by definition.

DEFINITION 3.21 ***Floor Function; Ceiling Function***

The *floor* and *ceiling* functions are defined for all real numbers x by

$$\text{floor}(x) = \lfloor x \rfloor = \text{the largest integer in the interval } (x-1, x]$$

and

$$\text{ceiling}(x) = \lceil x \rceil = \text{the largest integer in the interval } [x, x+1)$$

3.2.6 Exercises

The exercises marked with ★ have detailed solutions in Appendix H.

1. Each statement is either true or false. Identify which case is correct, and then give some justification for your answer.
   (a) The number 1 is a prime.
   (b) If a divides b, then there is an integer, c, such that $a = bc$.
   (c) If a divides b, then there is an integer, k, such that $b = ak$.
   (d) The well-ordering principle states that every set of natural numbers contains a smallest element.
   (e) ★ If $x, y \in \mathbb{Z}$ and $x - y$ is divisible by z, then $x \equiv y \pmod{z}$.
2. Each statement is either true or false. Identify which case is correct, and then give some justification for your answer.
   (a) Every positive integer is divisible by at least two distinct positive integers.
   (b) For any nonzero integer, a, the remainder on dividing $5a^2$ by a is always 0.
   (c) The Quotient–Remainder theorem plays a crucial role in the proof of the well-ordering principle.
   (d) ★ If m and n are odd integers, then $m = 2k + 1$ and $n = 2k + 1$ for some integer, k.
3. For each pair of integers, x and y, find the q and r (from the Quotient–Remainder theorem) that make $x = yq + r$. Express the results in a table.

	x	y	q	r
(a) ★	2961	987		
(b)	567	450		
(c)	2388	309		
(d)	1135	39		

4. For each integer, x, factor x as a product of primes. Express the results in a table.

	x	**x (as a product of primes)**
(a) ★	2548	
(b)	4116	
(c)	2366	
(d)	420	

5. For each pair of integers, x and y, find $\gcd(x, y)$. Express the results in a table. [*Hint*: It may be helpful to factor x and y into products of primes first.]

	x	y	**gcd(x, y)**
(a) ★	688	108	
(b)	33	616	
(c)	444	1098	
(d)	224	196	

6. For each pair of integers, x and y, find $\text{lcm}(x, y)$. Express the results in a table. [*Hint*: It may be helpful to factor x and y into products of primes first.]

	x	y	**lcm(x, y)**
(a) ★	999	93	
(b)	207	46	
(c)	34	343	
(d)	1065	104	

7. For each pair of integers, x and y, find $x \bmod y$. Express the results in a table.

	x	y	**x mod y**
(a) ★	57	701	
(b)	1091	786	
(c)	1085	239	
(d)	2002	34	

8. For each pair of integers, x and y, do all the following.
   i. Find the q and r (from the Quotient–Remainder theorem) that make $x = yq + r$.
   ii. Factor x and y as products of primes.
   iii. Find $\gcd(x, y)$.
   iv. Find $\text{lcm}(x, y)$.
   v. Find $x \bmod y$.
   vi. Is $x \equiv y \pmod 5$?

   Express the results in a table.

	x	y	(i)	(ii)	(iii)	(iv)	(v)	(vi)
(a)	684	96						
(b)	1212	895						
(c)	1002	102						
(d)	18	56						

9. List five integers that are congruent to 5 mod 11.
10. Is 4.5 divisible by 1.5? Justify your answer.

11. Consider the integers 32 and 107. By the Quotient–Remainder theorem, there exist unique integers, q and r, such that $32 = 107q + r$. However, note that $32 = 107 \cdot 0 + 32$ and $32 = 107 \cdot (-1) + 139$. Resolve the contradiction.

12. Suppose that a and b are positive integers. How many positive integers not exceeding a are divisible by b?

13. Suppose that you are really excited about your vacation to Florida and have calculated that you are leaving in 224 hours. If it is 5 P.M. right now, what time will it be then? Use material from this section to justify your answer.

14. Prove Proposition 3.22.

> **PROPOSITION 3.22**
> Let $a, b, c, d, m \in \mathbb{Z}$ with $m > 0$. If $a \equiv b \pmod{m}$ and $c \equiv d \pmod{m}$ then
>
> **1.** $a + c \equiv b + d \pmod{m}$.
> **2.** $ac \equiv bd \pmod{m}$.

15. Use Proposition 3.22 to prove Proposition 3.23.

> **PROPOSITION 3.23**
> Let $a, k, m \in \mathbb{Z}$ with $k > 0$ and $m > 1$. Then $a^k \equiv (a \bmod m)^k \pmod{m}$.

16. Prove Proposition 3.24.

> **PROPOSITION 3.24**
> Let $a, b, m \in \mathbb{Z}$ with $m > 0$. Then
>
> **1.** $((a \bmod m) + (b \bmod m)) \bmod m = (a+b) \bmod m$.
> **2.** $((a \bmod m) \cdot (b \bmod m)) \bmod m = ab \bmod m$.

17. Prove Proposition 3.25.

> **PROPOSITION 3.25** ***Composition of mod***
> Let $m > k > 1$ be two integers. Then
>
> $(x \bmod m) \bmod k = x \bmod k$ if and only if $k | m$

18. Prove Proposition 3.26.

> **PROPOSITION 3.26** ***Cancelation with mod***
> Let $a, b, c, m \in \mathbb{Z}$ with $m > 1$. If $\gcd(c, m) = 1$ and $ac \equiv bc \pmod{m}$, then $a \equiv b \pmod{m}$.

19. Proposition 3.26 is in the form "if x and y, then z", where x is $\gcd(c, m) = 1$. Show that if x is changed to $\gcd(c, m) > 1$, then the conclusion, z, may or may not be true. [*Hint*: two well-chosen examples will suffice.]

20. Prove Proposition 3.27.

> **PROPOSITION 3.27**
> Let $a, b, m, n \in \mathbb{Z}$ with $m, n > 1$. If $a \equiv b \pmod{mn}$ then $a \equiv b \pmod{m}$ and $a \equiv b \pmod{n}$.

21. Let $\gamma(x) = \lceil x \rceil - \lfloor x \rfloor$. Show that

$$\gamma(x) = \begin{cases} 0 & x \in \mathbb{Z} \\ 1 & x \notin \mathbb{Z} \end{cases}$$

22. Prove Proposition 3.28.

> **PROPOSITION 3.28**
> For all real numbers x, $\lfloor 2x \rfloor \le \lfloor x \rfloor + \lceil x \rceil \le \lceil 2x \rceil$.

3.3 Proof Strategies

The following subsections present a number of strategies that are used frequently in mathematical proofs. In this chapter, a strategy usually represents the *form* of the proof, independent of the mathematical content of the implication. There are two reasons for learning these strategies.

The first reason is probably obvious: It is useful to have a variety of approaches to try when we are presented with an implication we would like to prove. If one strategy does not seem to work, we can try a different strategy. The various strategies are based on the rules of logic presented in Section 2.4; if the strategy is correctly followed, we can be confident that our proof is valid.

The second reason for learning these strategies is related to reading proofs done by other people. Most published proofs do not state explicitly the strategy that was used in the proof. The reader is assumed to be capable of discovering that from the concisely written proof that is printed. Without some prior knowledge of the strategy used, the proof may be harder to understand.

3.3.1 Trivial Proof

TABLE 3.1 Implication

P	Q	$P \to Q$
T	T	T
T	F	F
F	T	T
F	F	T

The simplest of the proof techniques is based upon the truth table for implication (Table 3.1). An implication is true as long as either the hypothesis is false or the conclusion is true. Suppose we are trying to prove that the implication $P \to Q$ is true. If the conclusion Q is already known to be true, then the truth table indicates that the implication is true. No additional work is needed. Similarly, if the hypothesis P is known to be false, then again, the truth table indicates that the implication is true. In either of these cases, the proof is called a *trivial proof*. The case where P is known to be false is also called a *vacuous proof* because the hypothesis is not true, so there is nothing to prove.

The following implications can be shown to be true using a trivial proof.

- If x is a real number with $x^2 + 1 = 0$, then $x^4 = \pi$.
- Any five-headed human is a genius.
- If $n > 0$ and n is even, then $n > 0$.

3.3.2 Direct Proof

TABLE 3.2 The Law of Hypothetical Syllogism

$[(X \to Y) \wedge (Y \to Z)] \Rightarrow (X \to Z)$

TABLE 3.3 Modus Ponens

$[X \wedge (X \to Y)] \Rightarrow Y$

The most common proof strategy is based on the law of hypothetical syllogism (Table 3.2). The truth of the theorem is demonstrated by a series of intermediate implications that eventually lead to the implication stated in the theorem.

Suppose that the theorem is $P \to Q$. We might find it fairly easy to prove the implication $P \to S_1$. Since we are assuming the truth of P, and we have now established the truth of $P \to S_1$, *modus ponens* (Table 3.3) ensures the truth of S_1. Perhaps it is now possible to prove the implication $S_1 \to S_2$. *Modus ponens* then ensures the truth of S_2 and the law of hypothetical syllogism leads us to the valid implication $P \to S_2$. Continuing in this fashion, we should eventually arrive at $S_n \to Q$. The law of hypothetical syllogism then allows us to conclude $P \to Q$.[16]

This proof strategy is usually called *direct proof* since the written version proceeds directly from P through the S_k's to Q. This name is somewhat misleading because while the proof is evolving in the mind of the mathematician, Q is not ignored. In fact, it is often the case that the mathematician will work backward from Q, trying to ask questions about Q that will lead to statements that look similar to P. While the proof is forming, the mathematician may change several times between moving forward from P and moving backward from Q.[17] It is only in the final, written version of the proof that the progress of the proof appears to be direct.

An example of a direct proof has already appeared in Example 3.5. Restating Theorem W (page 90) as "If the axioms are assumed, then there are at least four points", we see that the proof consists of $P \to S_1 \to S_2 \to S_3 \to S_4 \to S_5 \to Q$. (Actually, the proof is a little more complex than was just indicated.) A more complete description will now be given.

Since the axioms all are assumed true (P), the laws of simplification allow us to assert the truth of axiom 4 (A_4). From A_4 we can produce the points A, B, and C (denote the existence of these three points by S_1). Using S_1 and A_1, we conclude that there is a line AB between A and B (S_2). S_1, S_2, and A_3 imply the existence of a unique line through C and parallel to AB (S_3).[18] S_3 and A_2 assert the existence of a point D (S_4). The final step is the most complex in the proof. In essence, it asserts that D is distinct from A, B, and C. To do this, the results of the previous steps are used as well as appealing to the implication "if two lines are parallel, then they have no points in common" (the "concept $\to$ properties" form of a definition). Since D is distinct from the other points (which are also distinct from each other), there must be at least four points (Q).[19]

---

[16] Often the proof is not as directly linear as this paragraph suggests. For example, S_6 may follow from $(S_2 \wedge S_5 \wedge T_8) \to S_6$, where T_8 is some previously proved theorem. The linear string of implications $P \to S_1 \to S_2 \to \cdots \to S_n \to Q$ is merely suggestive of the general shape of the proof.

[17] The book by Solow ([89]) contains several examples of this forward-backward process.

[18] Notice the use of *modus ponens*.

[19] This expanded description of the proof illustrates a difficult decision: "How much detail is needed?", or "what constitutes a complete proof?". The presentation in the expanded proof does not reach the limit of detail possible, yet it is probably too much already. It is not easy to decide how much detail is needed to ensure mathematical rigor. A related question is, "When is a complete proof necessary?". For example, does a first-year calculus student who plans to major in engineering need to understand the proof of the first fundamental theorem of calculus, or is the ability to apply the theorem sufficient?

You may find it interesting to study the various philosophies developed by mathematicians and philosophers that attempt to provide a firm foundation for doing mathematics. One of these schools of thought is called *logicism*. Logicism attempts to reduce all of mathematics to logic. The most energetic attempt toward

EXAMPLE 3.15 **Divisibility of a Sum**

> **PROPOSITION 3.29**
> Let a, b, and c be integers, with $a \neq 0$. If $a \mid b$ and $a \mid c$, then $a \mid (b + c)$.

Proof: This proof will not only illustrate a direct proof, but will also illustrate a proof that is heavily definition oriented. You may want to review Definition 3.8 on page 93.

Since $a \mid b$, we know that there is an integer k such that $b = ak$ (see page 88: "concept → properties"). Similarly, there is an integer m such that $c = am$.[20]

Therefore, $b + c = ak + am = a(k + m)$. We know that $k + m$ is an integer, so the equation shows that $b + c$ is an integer multiple of a. Therefore, $b + c$ is divisible by a ("properties → concept"). The proof is complete. □ ■

You should pay careful attention to what did *not* appear in the proof of Proposition 3.29. Nowhere in the proof do any fractions appear. The context of the proposition is the set of integers. It is considered poor taste to drag rational numbers into the proof. You should avoid introducing $\frac{b}{a}$ into a proof when you are told that $a \mid b$. Instead, use the definition of *divisibility* and assert the existence of an integer k such that $b = ak$.

✔ Quick Check 3.3

1. Use a direct proof to prove Proposition 3.30. Clearly denote your use of definitions.

> **PROPOSITION 3.30**
> Let a, b, and c be any integers, with $a \neq 0$. If $a \mid b$, then $a \mid (bc)$.

2. Use a direct proof to prove Proposition 3.31.

> **PROPOSITION 3.31**
> Let a, b, and c be integers with $a \neq 0$ and $b \neq 0$. If $a \mid b$ and $b \mid c$, then $a \mid c$.

☑

EXAMPLE 3.16 **Searching for Prime Factors**

One method for determining if a positive integer $n > 1$ is prime is to see if it is divisible by any of the numbers $2, 3, 4, \ldots, (n - 1)$. If not, it must be prime. This strategy can be made more efficient by using the next proposition.

> **PROPOSITION 3.32**
> If n is a positive composite number, then n has at least one prime factor p with $1 < p \leq \sqrt{n}$.

Proof: Since n is composite and positive, there are integers a and b with $1 < a < n$, $1 < b < n$ and $n = ab$. We can assume, without loss of generality,[21] that $a \leq b$. Thus $n = a \cdot b \geq a \cdot a = a^2$, so $a \leq \sqrt{n}$. If a is a prime, we are done. Otherwise, a must

---

this goal was *Principia Mathematica* by Russell and Whitehead, 1910–1913. The book is in three volumes, consisting mostly of symbols. One of the goals of this work was to put arithmetic on a firm foundation.

In 1931 Gödel found a major flaw in the axiomatic method: "Within a rigidly logical system such as Russell and Whitehead had developed for arithmetic, propositions can be formulated that are undecidable or undemonstrable within the axioms of the system. That is, within the system there exist certain clear-cut statements that can be neither proved nor disproved. Hence one cannot, using the usual methods, be certain that the axioms of arithmetic will not lead to contradictions" [13, p. 655].

[20] It is essential that we use different letters for the two integers (k and m); otherwise, we are indirectly asserting that $b = c$.

[21] The phrase "without loss of generality" (often abbreviated "wlog") means that the assumption being made does not introduce anything that would change the validity of the proof. In this case, if $a > b$, we would interchange the respective roles of the letters in the remainder of the proof.

itself have a prime divisor p and $p < a \leq \sqrt{n}$ must be true. Proposition 3.31 implies that p is a prime divisor of n. □

The more efficient algorithm for testing for primes is to see if any of the integers $2, 3, 4, \ldots, \lfloor \sqrt{n} \rfloor$ are divisors. For example, if $n = 13$, then $\sqrt{n} \simeq 3.606$, so we only need to see if 2 or 3 are divisors. Under the original algorithm, we would need to check for divisibility by $2, 3, 4, \ldots, 11, 12$. ■

3.3.3 Indirect Proof: Proving the Contrapositive

The logical equivalence $[P \to Q] \Leftrightarrow [(\neg Q) \to (\neg P)]$ implies that the contrapositive of a valid theorem is automatically true.

COROLLARY 3.33 ***Corollary to Proposition 3.32***

If a positive integer $p > 1$ has no divisor d with $1 < d \leq \sqrt{p}$, then p is prime.

Proof: This is the contrapositive of Proposition 3.32. □

It is often convenient to use the logical equivalence $[P \to Q] \Leftrightarrow [(\neg Q) \to (\neg P)]$ as a proof strategy; we *prove the contrapositive* of the implication we really want. This strategy is often helpful when Q is a statement with an easy to use negation.

As an example, recall the definition of *even* given on page 94, and consider the proposition:

PROPOSITION 3.34

If the integer n is not even, then n^2 is not even.

Proof: The concept "not even" is not as easy to work with as "even" (we have a nicer property to associate with "even"). The contrapositive reads "If n^2 is even, then n is even." We will prove the contrapositive.

Thus we assume that n^2 is even. Using the "concept $\to$ properties" part of the definition of even, we may assert the existence of an integer k such that $n^2 = 2k$. Since we can factor a 2 out of the right-hand side of the previous equation, we must also be able to factor a 2 out of the left-hand side. Thus 2 is a factor of n^2. But then 2 must be a factor of n, so $n = 2j$ for some j.[22] Using the "properties $\to$ concept" form of the definition of even, we conclude that since $n = 2j$, n is even. This proves the contrapositive of Proposition 3.34, and simultaneously proves Proposition 3.34. □

3.3.4 Proof by Contradiction

Sometimes, we do not seem to be able to make progress moving from P to Q using a direct proof or using the contrapositive. *Proof by contradiction* is based on the logical equivalences:

$$
\begin{aligned}
[P \to Q] &\Leftrightarrow [(P \wedge (\neg Q)) \to (R \wedge (\neg R))] \\
[P \to Q] &\Leftrightarrow [(P \wedge (\neg Q)) \to (\neg P)] \\
[P \to Q] &\Leftrightarrow [(P \wedge (\neg Q)) \to Q]
\end{aligned}
$$

We begin by assuming that the hypothesis, P, is true and also assuming that the conclusion, Q, is false. This gives us an additional piece of information (the negation of the conclusion) that we hope will enable us to make progress toward a completed proof. Working forward from $(P \wedge (\neg Q))$, we hope to eventually arrive at a contradiction (such

[22] I have used Proposition 3.39 on page 110.

as Q, $(\neg P)$, or $(R \wedge (\neg R))$, where R can be anything). The logical equivalences above then ensure the truth of $P \to Q$.

TABLE 3.4 Proof by Contradiction Eliminates the Second Row of the Implication Truth Table as a Possibility

P	Q	$P \to Q$
T	T	T
T	**F**	**F**
F	T	T
F	F	T

On an intuitive level, proof by contradiction is based upon the following ideas. We assume that P is true (if P is false there is nothing to prove, the implication is vacuously true). We then assume that Q is false, placing us in row two of Table 3.4. By a series of valid steps we arrive at a contradiction. Something must have gone wrong. The only place possible was when we assumed that Q was false. It follows therefore that Q is actually true. But then since whenever P is true, it necessarily follows that Q is true, the implication $P \to Q$ is true.[23]

The most difficult part of a proof by contradiction is that we do not know ahead of time what contradiction we will eventually arrive at. In fact, if we are trying to prove a "theorem" that is actually false, we will *never* arrive at a contradiction![24]

Recall that Theorem Y in Example 3.5 on page 90 was proved using contradiction and that contradiction played a key role in the existence proof for Theorem 3.9 on page 93. As another example, consider the following proposition.

PROPOSITION 3.35
The number $\sqrt{2}$ is irrational.

Proof: The concept "irrational" is much harder to work with than "rational".

We might consider proving the contrapositive, but what is the contrapositive? In fact, what is the implication? We can restate the implication as "if a number is $\sqrt{2}$, then the number is irrational". The contrapositive is "if a number is rational (not irrational), then it is not the number $\sqrt{2}$". Proving the contrapositive would thus involve showing that something is not $\sqrt{2}$. If you try showing this, you will probably end up using a proof by contradiction, so we will begin by using a proof by contradiction on the original theorem.

Thus we assume that the number we have is $\sqrt{2}$, and that it is a rational number. Therefore, there are integers p and $q \neq 0$ such that $\sqrt{2} = \frac{p}{q}$. We will assume that p and q have no common factors. (If there were any, we could factor them out and cancel, leaving us with a new p and q.) Since q is not 0, we can multiply the equation by q, leaving $\sqrt{2}q = p$. Squaring both sides produces $2q^2 = p^2$. Since the left-hand side has a factor of 2, the right-hand side does also. This means that 2 is a factor of p.[25] Hence, $p = 2r$, for some integer r. The equation $2q^2 = p^2$ can be rewritten as $2q^2 = 4r^2$. Dividing by 2 produces $q^2 = 2r^2$, from which we conclude (in a now familiar way) that 2 is a factor of q. This is the contradiction we need: 2 is a factor of both p and of q, but p and q have no common factors.
The trouble began with the assumption that $\sqrt{2}$ is rational. Hence, $\sqrt{2}$ is irrational. □

[23] We have ruled out the one row in the truth table for implication in which the implication is false.

[24] Direct proof has a safeguard that proof by contradiction does not. If you make a mistake in a direct proof, you will have a very hard time arriving at a proof. If you make a mistake in a proof by contradiction, it may actually be easier to reach a contradiction (due to your mistake), perhaps with disastrous consequences; there may be no legitimate contradiction. Disaster occurs if the theorem you try to prove is actually false. When you make a mistake in a proof by contradiction, you probably will arrive at a contradiction (due entirely to your mistake and not to the content of the theorem). You will assume that you have proved the theorem when, in fact, the theorem is actually false.

A dramatic example of this occurred in connection with Euclid's fifth postulate. In 1733, Giovanni Saccheri tried to prove that the fifth postulate could be proved from the other four. That is, he tried to show there was no such thing as a *non-Euclidean geometry* (a geometry that did not follow all five of Euclid's axioms). He assumed that the other postulates were true and that the fifth postulate was false. He then sought to produce a contradiction. If a contradiction occurred, he could then conclude that it was impossible to have all the postulates except the fifth. Saccheri did arrive at a contradiction (due to a mistake). In the course of his "proof" he had produced many of the major theorems of non-Euclidean geometry. He never realized that he had actually created geometries that satisfied all but the fifth postulate!

[25] I have again used Proposition 3.39 on page 110.

✔ Quick Check 3.4

1. Use a proof by contradiction to provide another proof of Proposition 3.32. ☑

3.3.5 Proof by Cases

Sometimes it is helpful to partition the proof into several disjoint parts whose union is the complete theorem and then prove each part individually.

EXAMPLE 3.17 **A Simple Proof by Cases**

If $n \in \mathbb{Z}$, then $n^3 - n$ is even.

Proof: Consider the two cases n even and n odd. Every integer fits one of these cases so by showing the claim is true for each case, the claim will be shown true for all integers.

n even

If n is even, then there is an integer k such that $n = 2k$. Therefore, $n^3 - n = (2k)^3 - (2k) = 8k^3 - 2k = 2(4k^3 - k) = 2m$, where $m = 4k^3 - k$ is an integer. Thus, $n^3 - n$ is even.

n odd

If n is odd, then there is an integer k such that $n = 2k + 1$. Thus, $n^3 - n = (2k+1)^3 - (2k+1) = 8k^3 + 12k^2 + 4k = 2(4k^3 + 6k^2 + 2k) = 2m$. So $n^3 - n$ is again an even integer. □ ■

Many mathematicians feel that proofs with more than a few cases are less elegant than a proof using some other strategy. An extreme case is the proof of the four color theorem (see Theorem 10.52 on page 645). The first proof, by Kenneth Appel and Wolfgang Haken in 1976, involved the use of over 1000 hours of computer time to examine around 2000 cases (each of which resulted in up to 100,000 subcases). More recent proofs have reduced the number of cases to around 600.

Appel and Haken's proof initiated a controversy in the mathematics community: Should we accept as valid a proof that no human has read unaided by a machine? How do we know that the computer did not make a mistake?

The current majority opinion is that the proof is valid. However, a more elegant proof (without the use of computers) would be preferred.

3.3.6 Implications with Existential Quantifiers

Many implications contain conclusions involving an existential quantifier. For example, "If x and y are two real numbers, then there is a real number r that is between them." The preferred method for proving theorems of this type (called *proof by construction* or a *constructive proof*) is to construct the object that is supposed to exist. This is often done by creating a candidate and showing that it satisfies all the properties of the required object.

PROPOSITION 3.36

If x and y are real numbers with $x < y$, then there exists a real number z with $x < z < y$.

Constructive Proof: Since $x < y$, $2x < x + y$. Dividing by 2 gives $x < \frac{x+y}{2}$. Similarly, since $x < y$, $x + y < 2y$ and so $\frac{x+y}{2} < y$. Thus, let $z = \frac{x+y}{2}$. Then $x < z < y$. □

Sometimes a proof by construction may not be possible. Most mathematicians will accept a proof by contradiction in that case. Initially, assume that the object does not exist, then arrive at a contradiction. Finally, conclude that the object must therefore exist. The actual object is never constructed.[26]

An alternative nonconstructive strategy, introduced by Paul Erdös, is to prove that there is a probability of 1 that the mathematical object exists.

The following proof of Proposition 3.32 (page 102) is nonconstructive. It shows that the theorem is true but does not indicate how the prime factor can be found.

Nonconstructive Proof of Proposition 3.32: Assume the proposition is false. Since n is composite, the fundamental theorem of arithmetic (page 95) implies that n has at least two prime factors. It is therefore possible to write $n = p_1 \cdot p_2 \cdot m$, where $m \geq 1$ may be composite, and p_1 and p_2 are primes. Since the proposition is assumed false, $p_1 > \sqrt{n}$ and $p_2 > \sqrt{n}$. This means that $n = p_1 \cdot p_2 \cdot m > \sqrt{n} \cdot \sqrt{n} \cdot m = n \cdot m \geq n$. That is, $n > n$, a contradiction. The contradiction means the proposition is true. □

3.3.7 Implications with Universal Quantifiers

Many implications contain one or more universal quantifiers. For example, "The interior angles of any triangle sum to 180 degrees" or "For all real numbers $x > 2$, there is a $y < 0$ such that $x = \frac{2y}{y+1}$".

When the universal quantifier appears in the hypothesis of an implication, it is often appropriate to use what is sometimes called the *choose method* as part of the proof strategy. The implication usually reads "for all objects A having properties B, C happens". In the choose method we pick an arbitrary member, x, of the universally quantified set A and show that C happens. Since x was a representative of *any* object in A, we have established that C happens for all elements in A.

We must be careful that in using x, only the properties B are used; we must not use any properties that are true for some elements of A but not for others. Mistakes of this sort are commonly seen in "proofs", such as the first attempt in the next example.

EXAMPLE 3.18

Using the Choose Method

The Proposition For all real numbers $x > 2$, there is a $y < 0$ such that $x = \frac{2y}{y+1}$.

Incorrect Proof Let $x = 4$ (an incorrect use of *choose*). Using the construction method, let $y = -2$. Then $x = 4 = \frac{2(-2)}{-2+1} = \frac{2y}{y+1}$. **Incorrect!** ⍁

The problem with the incorrect proof is that it shows nothing about what to do with any number except $x = 4$.

Preliminary Analysis We will eventually use a constructive proof, but first a preliminary investigation is in order. Suppose (temporarily) that we already have a $y < 0$ for which the proposition is true. We could then solve $x = \frac{2y}{y+1}$ for y as a function of x. Doing so yields $y = -\frac{x}{x-2}$. This seems like the most likely choice to use for y in the constructive proof. The actual proof is presented next.

Correct Proof Let x be a real number greater than 2. We will show that $y = -\frac{x}{x-2}$ satisfies all the necessary conditions. Routine algebra shows that $y = -\frac{x}{x-2}$ satisfies the equation $x = \frac{2y}{y+1}$. It remains to show that $-\frac{x}{x-2} < 0$. But since $x > 2$, the numerator and the denominator are both positive. The negation of the fraction must be negative. □

In the correct proof, no property of x other than $x > 2$ was used. In this sense we "chose" a generic x. ■

[26]One of the schools of mathematical philosophy, called *intuitionism*, considers such an approach unacceptable. Intuitionism admits the existence of a mathematical object only if it can be constructed in a finite number of steps. You may find more on this subject in the references by Boyer or Berlinghoff ([13, 9]). The article by Mandelkern ([68]) is a good, nontechnical introduction to current approaches to constructive mathematics.

EXAMPLE 3.19 **A Subset of the Even Integers**

Let A be the set of all positive integers that are divisible by 4, and let B be the set of all positive integers that are divisible by 2. Then $A \subseteq B$.

The universal quantifier is implicitly contained in the claim "$A \subseteq B$." This is because Definition 2.2 on page 20 contains an assertion of the form $\forall x \in A,\ x \in B$.

Proof: Let $n \in A$. Then (using the definition of *divisible*) we can write $n = 4k$ for some positive integer k. But then $n = 2 \cdot (2k)$ and so n is divisible by 2. This means that $n \in B$.

No special properties of n were used (other than its divisibility by 4), so n really does represent any/every element of A. Thus, using the definition of *subset*, $A \subseteq B$ must be true. □ ■

Up to this point we have always assumed that the implication you are trying to prove actually *is* true. Research mathematicians do not always have the luxury of knowing for sure that an implication is true (until they prove it is—which is one of the main reasons proofs exist). Often students do not have this luxury either.

After spending time unsuccessfully trying to prove a theorem, you might start doubting its truth. What can you do then? You try to prove its negation! If the theorem contains a universal quantifier, the negation is very nice. For example,

> **Proposition** For all objects A having property B, C happens.
>
> **Negation of the Proposition** There is an object A having property B such that C does not happen.[27]

A proof of the negation would seek to find an object x having property B for which C does not hold. The object x so constructed is usually called a *counterexample* to the original (false) proposition. To disprove an implication containing a universal quantifier in the hypothesis, it suffices to find just one counterexample. For example,

> **Proposition** All functions that are defined at a real number c are differentiable at c.
>
> **Counterexample** Let $f(x) = |x - c|$. This function is not differentiable at c. □

Some general observations regarding implications with universal quantifiers:

- It is a big task to prove such an implication is true: You must verify its truth for every possible value in the domain of the quantifier. This often requires you to verify the implication for an infinite number of values. The choose method makes this manageable.
- It is generally easier to show such an implication is false; one single counterexample is all that is necessary.[28] Of course, this assumes that the implication is actually false (or else no counterexample will ever be found).
- Universally quantified implications in a form similar to "$\forall n \in \mathbb{N},\ p(n)$" are often proved using mathematical induction, which will be introduced in Section 3.5. (The key feature here is that the domain for n is the set of natural numbers.)

✔ Quick Check 3.5

1. Prove that every nonzero rational number has a multiplicative inverse. (You might want to review Definition 3.6 on page 93 and the field properties in Appendix A.3.) ☑

[27] Recall the logical equivalence $\neg[\forall x,\ p(x)] \Leftrightarrow [\exists x,\ \neg p(x)]$.

[28] The situation is similar to advocating ideas in public. It is easier to stand on the sidelines and look for flaws (counterexamples) in a speaker's presentation than to be the speaker and carefully construct the ideas.

3.3.8 Proofs Involving the Biconditional and Logical Equivalence

Many theorems are stated using a biconditional: $A \leftrightarrow B$. For example, "A real number is irrational if and only if its decimal expansion is infinite and nonrepeating."

Recalling the biconditional logical equivalence

$$(P \leftrightarrow Q) \Leftrightarrow [(P \rightarrow Q) \wedge (Q \rightarrow P)]$$

it should be clear that we can prove the biconditional $A \leftrightarrow B$ by proving the implications $A \rightarrow B$ and $B \rightarrow A$. In the example at the start of this section, we would prove "if $r \in \mathbb{R}$ is irrational, then its decimal expansion is infinite and nonrepeating," and also prove "if $r \in \mathbb{R}$ has an infinite and nonrepeating decimal expansion, then r is irrational." The desired biconditional would then have been established.

Some theorems are presented as a collection of mutual equivalences:

Theorem The following statements are equivalent:

A
B
C
and so on.

Any proof will consist of proving a series of implications. The critical requirement is that we prove a sufficient number of the implications so that we can "travel" from any statement (A, B, C, etc.) to any other by following the implications. One way to represent this is to produce a diagram containing the statements A, B, C, and so on. An arrow is added for each implication proved; the tail of the arrow leaving the hypothesis, the head of the arrow pointing to the conclusion. If a path exists from any statement to any other statement, then the collection of implications is sufficient to establish the theorem. Two of the possible collections of implications for a theorem with three statements are $\{A \rightarrow B, B \rightarrow C, C \rightarrow A\}$ and $\{A \rightarrow B, B \rightarrow A, A \rightarrow C, C \rightarrow A\}$. The respective diagrams are shown in Figures 3.2 and 3.3.

Figure 3.2. $\{A \rightarrow B, B \rightarrow C, C \rightarrow A\}$.

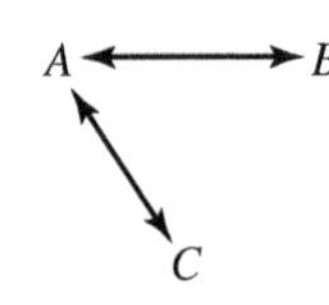

Figure 3.3. $\{A \rightarrow B, B \rightarrow A, A \rightarrow C, C \rightarrow A\}$.

The actual set of implications used (within the constraints mentioned previously) is a matter of convenience. Choose implications that can be proved easily.

The process that has just been outlined can be used to prove the following proposition.

> **PROPOSITION 3.37**
>
> Let a and b be any two distinct real numbers. Then the following are equivalent:
>
> **1.** $a < b$
>
> **2.** $a < \frac{a+b}{2}$
>
> **3.** $\frac{a+b}{2} < b$

Proof: The proposition will be shown as $1 \rightarrow 2 \rightarrow 3 \rightarrow 1$. An alternative strategy will be presented in Exercise 18 in Exercises 3.6.3 (page 162).

1 → 2 If $a < b$, then $\frac{a+b}{2} > \frac{a+a}{2} = a$.

2 → 3 If $a < \frac{a+b}{2}$, then $a + b < \frac{a+b}{2} + b$. Subtracting $\frac{a+b}{2}$ from both sides leads to $\frac{a+b}{2} < b$.

3 → 1 If $\frac{a+b}{2} < b$, then $a + b < 2b$ and so $a < b$. □

A special kind of equivalence is one in which we attempt to characterize some concept. The equivalence is often written as

B is a *necessary and sufficient condition* for A

The statement says that B is necessary for A to even have a chance to occur. That is, without B, A cannot happen. It also claims that the conditions B are sufficient for A to occur. No other conditions are needed; having B guarantees that A occurs. We can write this symbolically as:

$$(\neg B \to \neg A) \wedge (B \to A)$$

3.3.9 Some Important Examples

The three examples presented next provide additional opportunity for you to practice reading proofs. Try to identify the strategies that are used. Each of the theorems is a significant result in elementary number theory.

The Infinitude of the Primes

The next theorem appears in Euclid's *Elements*. The proof is quite clever.

THEOREM 3.38 ***The Infinitude of the Primes***

There is an infinite number of distinct primes.

Proof: Suppose instead that there is only a finite number of primes. Denote them as $\{p_1, p_2, p_3, \ldots, p_n\}$, where n is a positive integer. Consider the positive integer $q = (p_1 p_2 \cdots p_n) + 1$. Since $q \notin \{p_1, p_2, p_3, \ldots, p_n\}$, q must be composite. What are the prime factors of q?

If $p_k \mid q$, then there is an integer m with $q = mp_k$ so $mp_k = (p_1 p_2 \cdots p_n) + 1$. Thus $(p_1 p_2 \cdots p_n) - mp_k = -1$ and so $p_k(p_1 p_2 \cdots p_{k-1} p_{k+1} \cdots p_n - m) = -1$. This implies $p_k | (-1)$, which is impossible.

Since none of the primes p_k divide q and $\{p_1, p_2, p_3, \ldots, p_n\}$ are all the primes that exist, q cannot be composite. It must be the case that q is a new prime. This contradicts the assumption that there is only a finite number of primes and they were all listed in $\{p_1, p_2, p_3, \ldots, p_n\}$.

Therefore, there must be an infinite number of primes. □

Note: The proof led to the conclusion that $(p_1 p_2 \cdots p_n) + 1$ is a prime only because we assumed that there were a finite number of primes. The conclusion is not valid in general. For example, 3, 5, and 7 are all primes but $(3 \cdot 5 \cdot 7) + 1 = 106 = 2 \cdot 53$ is composite.

Characterizing the GCD

The next theorem shows that the greatest common divisor of two integers can be written as a simple expression containing the two integers. This can be a very useful tool. The theorem has already been stated on page 95.

THEOREM 3.12 **$\gcd(a, b) = as + bt$**

Let a and b be integers such that at least one is not 0. Then there are integers, s and t, such that $\gcd(a, b) = as + bt$.

The proof that follows is an elegant proof that uses the well-ordering principle (recall the proof of Theorem 3.9 on page 96). However, it does not indicate how to find s and t. The constructive proof on page 115 provides an efficient algorithm for calculating the gcd (see Example 3.9 on page 96 for another method).

Elegant Nonconstructive Proof of Theorem 3.12:

Let $S = \{ax + by \mid x, y \in \mathbb{Z} \text{ and } ax + by > 0\}$ be the set of all positive integers that can be written in the form $ax + by$ for some choice of x and y.

The well-ordering principle will apply if $S \neq \emptyset$. To show this, let

$$x_0 = \begin{cases} 1 & \text{if } a \geq 0 \\ -1 & \text{if } a < 0 \end{cases} \qquad y_0 = \begin{cases} 1 & \text{if } b \geq 0 \\ -1 & \text{if } b < 0 \end{cases}$$

Clearly $ax_0 + by_0 > 0$ since either $a \neq 0$ or $b \neq 0$. Thus, S is not empty.

The well-ordering principle implies that there is a smallest element, d, in S. Suppose $d = as + bt$. It remains to show (1) d is a common divisor of a and b and (2) d is the *greatest* common divisor of a and b.

(1) If $a = 0$, then clearly $d \mid a$. To see that $d \mid a$ when $a \neq 0$, notice that the Quotient–Remainder theorem asserts the existence of integers q and r with

$$a = dq + r, \quad \text{where } 0 \leq r < |d|$$

However, this implies that $r = a - dq = a - (as + bt)q = a(1 - sq) + b(-tq) = ax_1 + by_1$. Since r is strictly less than d, the smallest element of S, it must be that $r = 0$ (or else r would also be in S, creating a contradiction to the minimality of d). Since $r = 0$, $a = dq$, which implies that $d \mid a$.

A similar argument shows that $d \mid b$.

(2) Suppose c is any other common divisor of a and b. Then there are integers m and n such that $a = cm$ and $b = cn$. But then $d = as + bt = (cm)s + (cn)t = c(ms + nt)$. So $c \mid d$.

The conclusion is that $\gcd(a, b)$ can be expressed in the form $as + bt$. □

The Uniqueness Part of the Fundamental Theorem of Arithmetic

Theorem 3.15 (page 95) claims that every integer $n > 1$ can be written uniquely as a product of primes in ascending order. The existence of such a prime factorization will eventually be proved in Example 3.40 on page 140. It is now time to prove the uniqueness of the factorization.

Notice that $12 = 2 \cdot 2 \cdot 3$ and $12 = 2 \cdot 3 \cdot 2$ are technically different factorizations that we would like to consider as the same. We therefore assume that in any factorization $n = p_1 p_2 \cdots p_k$ into a product of primes, the inequalities $p_1 \leq p_2 \leq p_3 \leq \cdots \leq p_k$ all hold. We may then unambiguously discuss uniqueness.

A preliminary proposition is needed.

PROPOSITION 3.39 ***Prime Divisibility Property***

Let p be a prime. If p divides the product $a_1 a_2 \cdots a_n$, then p divides at least one of the factors a_i.

Proof:

Case 1: $n = 2$ If $p \mid a_1$, we are done. Otherwise, since the only divisors of p are 1 and p and $p \nmid a_1$, we must have $\gcd(p, a_1) = 1$. Theorem 3.12 then implies that integers s and t exist with $1 = ps + a_1 t$. Multiply both sides of this equation by a_2 to yield $a_2 = pa_2 s + (a_1 a_2)t$. Since $p \mid p$ and, by hypothesis, $p \mid (a_1 a_2)$, Propositions 3.29 and 3.30 (page 102) imply that $p \mid (pa_2 s + a_1 a_2 t)$. Thus $p \mid a_2$ and case 1 has been verified.

Case 2: $n > 2$ If $p \mid a_1$, we are done. Otherwise, case 1 implies that $p \mid (a_2 a_3 \cdots a_n)$. Now repeat the process. Eventually, we must find that $p \mid a_i$ for some i with $i \leq n - 2$ or else case 1 will show that p divides either a_{n-1} or a_n.[29] □

[29] This can be made precise using mathematical induction, which is presented in Section 3.5.

THEOREM 3.15 ***The Fundamental Theorem of Arithmetic***

Every integer n, with $n \geq 2$, can be uniquely written as a product of primes in ascending order.

Proof of Uniqueness: Suppose $n \geq 2$ factors into a product of primes in two ways: $p_1 p_2 \cdots p_k = n = q_1 q_2 \cdots q_m$, with $p_1 \leq p_2 \leq \cdots \leq p_k$ and $q_1 \leq q_2 \leq \cdots \leq q_m$. By Proposition 3.39 we know that p_1 must divide one of $q_1, q_2, \ldots, q_m$. By suitably renaming if necessary,[30] we can assume that $p_1 \mid q_1$. But since both p_1 and q_1 are primes, it must be that $p_1 = q_1$. Canceling leads to $p_2 p_3 \cdots p_k = q_2 q_3 \cdots q_m$. Repeat the process: p_2 must divide $q_2 q_3 \cdots q_m$, so (again after renaming if necessary) $p_2 = q_2$. Continue removing identical prime factors.

Eventually, one side will run out of factors. Then either $p_{m+1} p_{m+2} \cdots p_k = 1$ and we conclude that $\{p_{m+1}, p_{m+2}, \ldots, p_k\}$ must be empty, or $q_{k+1} q_{k+2} \cdots q_m = 1$ and we conclude that $\{q_{k+1}, q_{k+2}, \ldots, q_m\}$ is empty. In either case we have concluded that $k = m$ and that the two factorizations are the same. □

3.3.10 Exercises

The exercises marked with ★ have detailed solutions in Appendix H.

1. Prove: If $n \in \mathbb{Z}$, then $n^3 - n$ is even. *Note*: Part (a) is the preferred proof. Parts (b) and (c) are just for practice [and may freely use any intermediate results derived in part (a)].
   (a) ★ Use a direct proof.
   (b) Use an indirect proof that utilizes a vacuous proof. [*Hint*: Factor $n^3 - n$ after writing the contrapositive. Show that it cannot be an odd number.]
   (c) Use a proof by contradiction.
2. Find the error in the proof of the given claim. Additionally, determine whether the claim is true or false. If it is true, provide a correct proof. If it is false, find a counterexample.

   Claim:
   Let a and b be nonzero integers. If $a \mid b$ and $b \mid a$, then $a = b$.

   Proof:
   If $a \mid b$ and $b \mid a$, then $b = ak$ and $a = bm$ for some integers, k and m. By a simple substitution, it is apparent that $b = (bm)k = b(mk)$. Since $b \neq 0$, it is possible to conclude from the last equation that $mk = 1$. Since both m and k are integers, this implies that $m = k = 1$. Thus, $a = b$.

The remaining exercises are grouped by required proof strategy.

Trivial Proofs

3. ★ Prove the following implication. "If $p > 2$ is an even prime, then $p > 2^{100} + 1$." [*Hint*: You may assume that the result of Exercise 14 is true.]
4. Prove the following implication. "If π is an irrational number, then 2 is a prime number."
5. Every rational solution of $x^2 - 2 = 0$ is an integer.
6. Let A and B be sets. Prove: "If $A \cap B = \emptyset$, then $\emptyset \subseteq A$."
7. Prove the implication: If $x \in \mathbb{R}$ and $|x|$ is negative, then x^3 is negative.

Direct Proof

8. Let a and b be positive integers with $a \bmod 3 = 1$ and $b \bmod 3 = 2$. Use a direct proof to show that $(ab) \bmod 3 = 2$.
9. ★ Let $a, b \in \mathbb{Q}$. Use the field axioms (Appendix A.3) and direct proof to prove that $(a + b)^2 = a^2 + 2ab + b^2$.
10. Let a be a real number with $0 < a < 1$. Use a direct proof to show that $a > a^2$.
11. Let $a, b \in \mathbb{R}$ with $a \geq 0$ and $b \geq 0$. Use a direct proof to show that $\frac{a+b}{2} \geq \sqrt{ab}$, with equality if and only if $a = b$.
12. Let a and b be integers. Prove: If $\gcd(a, b) > 1$, then $\gcd(a^2, b^2) > 1$.
13. Let $a, b \in \mathbb{Z}$ and let $d = \gcd(a, b)$. Prove that $\gcd(a/d, b/d) = 1$.

Indirect Proof

14. Use an indirect proof to prove the following assertion. "If $n > 2$ is a prime, then n is odd." [*Hint*: You can write the hypothesis as "$n > 2$ and n is prime".]
15. Let A and B be sets. Use an indirect proof to prove the following assertion. "If $x \in (B - A)$, then $x \notin (A \cap B)$".
16. Let $a > 0$ be a real number. Use an indirect proof to show "If $a < 1$, then $\sqrt{a} > a$".
17. ★ Let $c \in \mathbb{Z}$. Prove: If $c^5 + 7$ is even, then c is odd.
18. Let $a, b, m, n \in \mathbb{Z}$. Prove: If $a \equiv b \pmod{mn}$ then $a \equiv b \pmod{m}$.

---

[30] It is conceivable that $q_1 \leq q_2 \leq \cdots \leq q_j < p_1$ for some $j \geq 1$. The renaming may destroy the assumption that $q_1 \leq q_2$, but the remainder of the proof will effectively show that no renaming is actually needed.

Proof by Contradiction

19. Use a proof by contradiction to prove that an even integer times an odd integer is an even integer.

20. Let A and B be sets. Use a proof by contradiction to show that $(A - B) \cap B = \emptyset$.

21. Let $a, b, c \in \mathbb{Z}$ and $a \neq 0$. Use a proof by contradiction to show that if $a \nmid bc$, then $a \nmid b$.

22. ★ Let $a, b \in \mathbb{R}$ with $a \geq 0$ and $b \geq 0$. Use a proof by contradiction to show that $\frac{a+b}{2} \geq \sqrt{ab}$.

23. Let a, b, and c form a primitive Pythagorean triple with a being odd and $a^2 + b^2 = c^2$. Then

$$a^2 = c^2 - b^2 = (c + b)(c - b).$$

(a) Use a proof by contradiction to show that $c+b$ and $c-b$ have no common prime factors.

(b) Now prove (using any strategy) that $c + b$ and $c - b$ are both squares.

Proof by Cases

24. ★ Use a proof by cases to show that if $n \in \mathbb{Z}$, then $n^2 + n$ is even.

25. Prove: If n is an integer and $3 \mid n^2$, then $3 \mid n$. [*Hint*: Look at remainders mod 3.]

26. Prove that for all real numbers x and y, $|x + y| \leq |x| + |y|$. [*Hint*: Use four cases.]

DEFINITION 3.40 ***max; min***

Let a and b be real numbers. Then

$$\max(a, b) = \begin{cases} a & \text{if } a \geq b \\ b & \text{if } a < b \end{cases}$$

and

$$\min(a, b) = \begin{cases} a & \text{if } a \leq b \\ b & \text{if } a > b \end{cases}.$$

27. Let a and b be real numbers. Use Definition 3.40 to prove that $\max(a, b) + \min(a, b) = a + b$. [*Hint*: Use the cases $a \leq b$ and $a > b$.]

28. Use Definition 3.40 and a proof by cases to show that $\max(\max(x, y), z) = \max(\max(x, z), y)$ with $x, y, z \in \mathbb{R}$.

29. Use Definition 3.40 and a proof by cases to show that if x and y are positive integers and $x \mid y$, then

$$\max(\gcd(x, y), y) = \max(x, \operatorname{lcm}(x, y)).$$

Implication with Existential Quantifiers—Constructive Proof

30. ★ Use a constructive proof to show that there exist two infinite subsets, A and B, of the integers such that $A \cap B = \emptyset$ and $A \cup B = \mathbb{Z}$.

31. Use a constructive proof to show that there must be an infinite number of Pythagorean triples. (Not necessarily primitive Pythagorean triples.)

32. Let $\frac{p}{q} \in \mathbb{Q}$. Prove (constructively) that there exists an integer, n, with $\frac{p}{q} < n$. [*Hint*: use the Quotient–Remainder theorem. Either p or q (or both) might be negative.]

33. Prove that there exist rational numbers, r and s, such that r^s is a positive integer and s^r is a negative integer.

34. Use a constructive proof to show that there exist two distinct positive integers whose sum and difference are both squares.

Implication with Existential Quantifiers—Nonconstructive Proof

35. ★ Use a nonconstructive proof to prove the following assertion. Let n be a positive integer. Then there exists a prime, p, with $n < p$.

36. Let $\frac{p}{q} \in \mathbb{Q}$. Provide a nonconstructive proof that there exists an integer, n, with $\frac{p}{q} < n$.

37. Prove: If A and B are finite sets with $A \subseteq B$ and $|A| \neq |B|$, then there exists an element $x \in B$ such that $x \notin A$.

38. Let $c > 0$ be a real number. Use a nonconstructive proof to show that there exists a real number b such that $x^2 + bx + c$ has real roots.

Implication with Universal Quantifiers

39. Let f and g be differentiable, real-valued functions. Find a counterexample to the assertion $(fg)'(x) = f'(x) \cdot g'(x)$. For extra credit, find an example where the assertion *is* true.

40. ★ Find a counterexample to the "Freshman Theorem": $(a + b)^n = a^n + b^n$, where $n \geq 2$ and a and b are any real numbers.

41. Use the choose method to prove: "If p is a prime with $p > 2$, then $p+1$ is not prime." [*Hint*: Use the result of Exercise 14.]

42. Prove or find a counterexample: Let a be a positive integer and let b, c be integers. If $a \mid (bc)$, then either $a \mid b$ or $a \mid c$ (or both).

43. Exercise 22 on page 68 shows that $x + a = b$ cannot always be solved uniquely for x if x, a, and b are elements of a Boolean algebra. Use the choose method to prove the following. "Let a and b be elements of a Boolean algebra, B. Then $\forall x \in B, [(x + a = b) \rightarrow (x \cdot \bar{a} = b \cdot \bar{a})]$".

Proofs Involving Equivalence

44. Let x and y be integers. Prove that $x - y$ is odd if and only if $x + y$ is odd.

45. Let A and B be sets. Prove that $A \subseteq B$ if and only if $A - B = \emptyset$.

46. ★ Let a be a positive integer. Prove that a is composite if and only if the sum of the positive divisors of a is greater than $a + 1$.

47. Let $n \in \mathbb{Z}$. Prove that the following are equivalent.

- n is even
- $n + 1$ is odd
- n^2 is even
- $(n - 1)(n + 1)$ is odd

48. Let $p \in \mathbb{Z}$, $p > 1$. Prove that the following are equivalent.

- p is a prime
- for all $a, b \in \mathbb{Z}$, $p \mid (ab) \rightarrow [(p \mid a) \vee (p \mid b)]$
- for all $d \in \mathbb{Z}$ with $1 < d < p^2$, $(d \mid p^2) \rightarrow (d = p)$

Incorrect Proofs

49. Suppose that d and c are both common divisors of a and b and suppose that $c \nmid d$. The following argument is an attempt to prove that dc is a common divisor of a and b. You have a twofold task:

(a) Find a counterexample to the assertion.

(b) Find the error in the supposed proof.

Since d is a common divisor of a and b, there are integers, x and y, such that $a = dx$ and $b = dy$. Since c is a common divisor of a and b, but $c \nmid d$, it must be that $c \mid x$ and $c \mid y$. Consequently, there are integers, u and v, such that $x = cu$ and $y = cv$. Thus, $a = dx = d(cu) = (dc)u$ and $b = dy = d(cv) = (dc)v$. Therefore, dc is a common divisor of a and b. ■

50. Let x and a be elements of a Boolean algebra. The following argument is an attempt to prove that if $x \cdot a = a$, then $x = 1$. You have a twofold task:

(a) Find a counterexample to the assertion.

(b) Find the error in the supposed proof.

Since $x \cdot a = a$, one of De Morgan's laws implies that $\overline{x} + \overline{a} = \overline{a}$. Subtracting a from both sides leads to $\overline{x} = \overline{a} - \overline{a} = 0$. Now use De Morgan again to arrive at $\overline{\overline{x}} = \overline{0}$. The involution property and Quick Check 2.10 (page 65) imply that $x = 1$.

3.4 Applications of Elementary Number Theory

This section contains several interesting applications of elementary number theory. These applications will not be used elsewhere in the book.

3.4.1 The Euclidean Algorithm: Calculating gcd(a,b)

The Quotient–Remainder theorem can be used to create an efficient algorithm for finding $\gcd(a, b)$.[31] It may be helpful to look at two examples before formally stating the algorithm.

EXAMPLE 3.20 **Calculating the GCD—a Short Example**

Suppose we wish to calculate $\gcd(12, 45)$. In phase 1 of the algorithm, use the Quotient–Remainder theorem to write $a = bq + r$. Next, the theorem will be applied to the previous divisor (b initially) divided by r. The process will continue until a remainder of zero is produced. The quotient at each step is not used in successive steps of phase 1. It is helpful to distinguish the quotient by placing it in parentheses.

$$45 = 12 \cdot (3) + 9$$
$$12 = 9 \cdot (1) + 3$$
$$9 = 3 \cdot (3) + 0$$

It is easy to show that the last nonzero remainder, 3, is the greatest common divisor of 12 and 45. First, observe that 3 is a common divisor. The justification for this claim follows (showing it is greatest will be done in phase 2).

The final equation, $9 = 3 \cdot (3)$, shows that $3 \mid 9$. Propositions 3.29 and 3.30 (page 102), and the second equation then imply that $3 \mid 12$. Finally, the propositions and the first equation then imply that $3 \mid 45$. Thus 3 is a common divisor of 12 and 45.

In phase 2, start near the bottom of the collection of divisions, solving for the nonzero remainders and substituting and collecting terms. Parenthesize numbers which are not a or b or one of the remainders from phase 1. The goal is to arrive at an equation in the form $3 = 12 \cdot (s) + 45 \cdot (t)$ for some s and t.

$3 = 12 - 9 \cdot (1)$	solve for remainder
$3 = 12 - (45 - 12 \cdot (3)) \cdot (1)$	substitute using previous equation
$3 = 12 \cdot (4) + 45 \cdot (-1)$	simplify

There is now an equation, $3 = 12 \cdot (4) + 45 \cdot (-1)$, that expresses the candidate gcd in the proper form. It only remains to show that 3 is actually the *greatest* common divisor.

Let d be any other common divisor of 12 and 45. Then $12 = kd$ and $45 = md$ for some k and m. Thus, $3 = kd(4) + md(-1) = d(4k - m)$ and so $d \mid 3$. ■

[31] Other algorithms are presented in Example 3.9 on page 96 and Exercise 7 on page 350.

EXAMPLE 3.21 **Calculating the GCD—a Longer Example**

Suppose gcd(245, 90) is sought. The Quotient–Remainder theorem will again produce a collection of equations with quotients and remainders. To describe this process during the subsequent constructive proof of the theorem, it is useful to introduce some notation. Temporarily denote 245 (the larger of the two numbers) as r_{-1} and denote 90 (the smaller number) as r_0.

The calculation will again proceed in two phases. The first phase will find the candidate value for gcd(245, 90) by a sequence of applications of the Quotient–Remainder theorem. The second phase will express that value in the form $245(s) + 90(t)$.

The constructive proof of Theorem 3.12 (presented after this example) will show that $\gcd(245, 90) = 5$, the last nonzero remainder.

Phase 2 will be easier if all the phase-1 equations are rearranged before starting phase 2.

Phase 1:

The Quotient–Remainder Theorem	**The Rearranged Equations**
$\overset{r_{-1}}{245} = \overset{r_0}{90} \cdot (\overset{q_1}{2}) + \overset{r_1}{65}$	$\overset{r_1}{65} = \overset{r_0}{90} \cdot (\overset{-q_1}{-2}) + \overset{r_{-1}}{245}$
$\overset{r_0}{90} = \overset{r_1}{65} \cdot (\overset{q_2}{1}) + \overset{r_2}{25}$	$\overset{r_2}{25} = \overset{r_1}{65} \cdot (\overset{-q_2}{-1}) + \overset{r_0}{90}$
$\overset{r_1}{65} = \overset{r_2}{25} \cdot (\overset{q_3}{2}) + \overset{r_3}{15}$	$\overset{r_3}{15} = \overset{r_2}{25} \cdot (\overset{-q_3}{-2}) + \overset{r_1}{65}$
$\overset{r_2}{25} = \overset{r_3}{15} \cdot (\overset{q_4}{1}) + \overset{r_4}{10}$	$\overset{r_4}{10} = \overset{r_3}{15} \cdot (\overset{-q_4}{-1}) + \overset{r_2}{25}$
$\overset{r_3}{15} = \overset{r_4}{10} \cdot (\overset{q_5}{1}) + \overset{r_5}{5}$	$\overset{r_5}{5} = \overset{r_4}{10} \cdot (\overset{-q_5}{-1}) + \overset{r_3}{15}$
$\overset{r_4}{10} = \overset{r_5}{5} \cdot (\overset{q_6}{2}) + \overset{r_6}{0}$	$\overset{r_6}{0} = \overset{r_5}{5} \cdot (\overset{q_6}{-2}) + \overset{r_4}{10}$

Phase 2:

Starting with the equation $\overset{r_5}{5} = \overset{r_4}{10} \cdot (\overset{-q_5}{-1}) + \overset{r_3}{15}$, which expresses gcd(245, 90) in terms of r_4 and r_3, a sequence of substitute/simplify steps will replace the intermediate remainders and eventually lead to an expression involving the original numbers 245 and 90. The simplification steps will preserve the remainders r_i until they are eliminated by a substitution.

$$
\begin{aligned}
\overset{r_5}{5} &= \overset{r_4}{10} \cdot (\overset{-q_5}{-1}) + \overset{r_3}{15} \\
&= (\overset{(r_3(-q_4)+r_2)}{15(-1) + 25}) \cdot (\overset{(-q_5)}{-1}) + \overset{r_3}{15} && \text{substitute} \\
&= \overset{r_3}{15} \cdot (\overset{t_3}{2}) + \overset{r_2}{25} \cdot (\overset{s_3}{-1}) && \text{simplify} \\
&= (\overset{(r_2(-q_3)+r_1)}{25(-2) + 65}) \cdot (\overset{t_3}{2}) + \overset{r_2}{25} \cdot (\overset{s_3}{-1}) && \text{substitute} \\
&= \overset{r_2}{25} \cdot (\overset{t_2}{-5}) + \overset{r_1}{65} \cdot (\overset{s_2}{2}) && \text{simplify} \\
&= (\overset{(r_1(-q_2)+r_0)}{65(-1) + 90}) \cdot (\overset{t_2}{-5}) + \overset{r_1}{65} \cdot (\overset{s_2}{2}) && \text{substitute} \\
&= \overset{r_1}{65} \cdot (\overset{t_1}{7}) + \overset{r_0}{90} \cdot (\overset{s_1}{-5}) && \text{simplify} \\
&= (\overset{(r_0(-q_1)+r_{-1})}{90(-2) + 245}) \cdot (\overset{t_1}{7}) + \overset{r_0}{90} \cdot (\overset{s_1}{-5}) && \text{substitute} \\
&= \overset{r_0}{90} \cdot (\overset{t_0}{-19}) + \overset{r_{-1}}{245} \cdot (\overset{s_0}{7}) && \text{simplify}
\end{aligned}
$$

Thus, $\gcd(245, 90) = 245 \cdot (7) + 90 \cdot (-19)$. ■

The algorithm used in the previous example is called the *Euclidean Algorithm*.

The Euclidean Algorithm: A Constructive Proof of Theorem 3.12:
Without loss of generality, assume that $a \geq b$ and if $b = 0$ then $a > b$. Temporarily rename a and b as r_{-1} and r_0, respectively. Using the division algorithm multiple times leads to the following sequence of equations. Since the remainder is always nonnegative and is strictly smaller than the divisor, the sequence will certainly be finite, with the final equation having a remainder of 0.

The Quotient–Remainder Theorem	**The Rearranged Equations**
$r_{-1} = r_0 \cdot (q_1) + r_1$	$r_1 = r_0 \cdot (-q_1) + r_{-1}$
$r_0 = r_1 \cdot (q_2) + r_2$	$r_2 = r_1 \cdot (-q_2) + r_0$
$r_1 = r_2 \cdot (q_3) + r_3$	$r_3 = r_2 \cdot (-q_3) + r_1$
$\vdots \quad \vdots \qquad \vdots$	$\vdots \quad \vdots \qquad \vdots$
$r_{n-2} = r_{n-1} \cdot (q_n) + r_n$	$r_n = r_{n-1} \cdot (-q_n) + r_{n-2}$
$r_{n-1} = r_n \cdot (q_{n+1}) + 0$	$0 = r_n \cdot (-q_{n+1}) + r_{n-1}$

The final nonzero remainder, r_n, is a common divisor of a and b. This can be seen by using the sequence of equations on the left, working from the bottom toward the top, and using Propositions 3.29 and 3.30 (page 102). The next paragraph provides the details.

The final equation, $r_{n-1} = r_n \cdot (q_{n+1}) + 0 = r_n \cdot (q_{n+1})$, shows that $r_n \mid r_{n-1}$. The equation $r_{n-2} = r_{n-1} \cdot (q_n) + r_n$ can be used to show that $r_n \mid r_{n-2}$. This assertion is valid because it is now known that $r_n \mid r_n$ and $r_n \mid r_{n-1}$ and so Proposition 3.30 implies that $r_n \mid (r_{n-1} \cdot (q_n))$. Proposition 3.29 then implies that $r_n \mid (r_{n-1} \cdot (q_n) + r_n) = r_{n-2}$. In the same way, once it is known that $r_n \mid r_{k+1}$ and $r_n \mid r_k$, it is possible to conclude that $r_n \mid r_{k-1}$. This process will eventually show that $r_n \mid r_0$ and $r_n \mid r_{-1}$.

Now that it has been verified that r_n is a common divisor of $a = r_{-1}$ and $b = r_0$, it is time to show that integers s and t exist with $r_n = a \cdot (s) + b \cdot (t)$. The sequence of rearranged equations (appearing previously as the sequence on the right) can be used to establish this.

Again working from the bottom toward the top, a sequence of substitutions will replace each remainder (except r_n, the gcd) by an expression involving the previous two remainders. Eventually, r_n will be expressed in terms of r_0 and r_{-1}. The sequence of equations (after simplification) will look like

$$\begin{aligned} r_n &= r_{n-1} \cdot (-q_n) + r_{n-2} \\ &= r_{n-2} \cdot (t_{n-2}) + r_{n-3} \cdot (s_{n-2}) \\ &= r_{n-3} \cdot (t_{n-3}) + r_{n-4} \cdot (s_{n-3}) \\ &\ \vdots \qquad\qquad \vdots \\ &= r_1 \cdot (t_1) + r_0 \cdot (s_1) \\ &= r_0 \cdot (t_0) + r_{-1} \cdot (s_0) \end{aligned}$$

Setting $s = s_0$ and $t = t_0$ and recalling that $a = r_{-1}$ and $b = r_0$ leads to the final expression $r_n = a \cdot (s) + b \cdot (t)$.

It only remains to show that r_n is the *greatest* common divisor. Suppose that c is any other common divisor of a and b. Then $a = ck$ and $b = cm$ for integers k and m. Thus $r_n = a \cdot (s) + b \cdot (t) = cks + cmt = c(ks + mt)$ and so $c \mid r_n$. □

✔ Quick Check 3.6

1. Use the Euclidean Algorithm to calculate gcd(78, 60). 

3.4.2 Hashing

Much thought has been expended towards devising ways to store information on a computer so that the retrieval of specific pieces of information is both easy and fast. One of the simplest ideas is to store each piece of information in a large linear list. If we happen to know that we want the 812th piece of information, we just need to access position 812 in the list. However, we usually do not have a nice numbering scheme for the pieces of information.

Data that is stored on a computer typically comes in two parts: a *key* and a *value*. The key uniquely identifies that piece of data. The value is the information of interest. For example, you probably have a student ID number (a key) that is used by your school to retrieve information about you (such as your address and your gpa).

A commonly used idea is to find a function h, called a *hash function*, that converts keys into list positions. The position numbers are formally called *indices*. The index produced by h determines the spot in the list to store the value part of the information. To make this work properly, it is necessary to use the mod operator.

Suppose enough storage on the computer is allocated to store 10,000 pieces of information. The positions of the list are numbered from 0 to 9,999. Then h is defined as:

$$h(\text{key}) = f(\text{key}) \bmod 10000$$

where f is some expression or algorithm that converts keys to natural numbers. The mod operator ensures that the key will always map to a legal index. It is desirable to choose f so that keys will be fairly evenly dispersed among the list positions.

It is quite difficult to find a function f that does a perfect job of dispersing keys evenly across the list. Usually, there may be two keys such that $h(\text{key}_1) = h(\text{key}_2)$. When this happens, a *collision* is said to have occurred. One simple strategy for handling collisions[32] is to see if position $(h(\text{key}_2) + 1) \bmod 10000$ is available. This linear searching can be continued until an empty spot in the list is found. Of course this makes retrieving the value for key_2 more complex. Details can be found in any *Data Structures* textbook.

EXAMPLE 3.22

Hashing

TABLE 3.5 The English Alphabet

letter	pos	letter	pos
a	1	n	14
b	2	o	15
c	3	p	16
d	4	q	17
e	5	r	18
f	6	s	19
g	7	t	20
h	8	u	21
i	9	v	22
j	10	w	23
k	11	x	24
l	12	y	25
m	13	z	26

Suppose I want to store information about kinds of pets. The key will be the species of the pet (dog, cat, snake, cricket, etc.) and the value will be the animal kingdom of the pet (mammal, reptile, bird, fish, insect, etc.). For this example, suppose my list has size 10 (an unrealistically small size).

I will define a simple hash function (which would not be a good choice for actual use). Suppose key is an n-character word: key = $c_1c_2c_3 \cdots c_{n-1}c_n$. Then

$$h(\text{key}) = \left(\sum_{i=1}^{n} \text{position in alphabet}(c_i)\right) \bmod 10$$

Using Table 3.5, we see that $h(\texttt{cat}) = (3 + 1 + 20) \bmod 10 = 4$.

Now suppose that I have already stored the information for cat (index 4), dog (index 6), cricket (index 9), and snake (index 0) (Table 3.6). When I try to enter goldfish, I find that $h(\texttt{goldfish}) = (7+15+12+4+6+9+19+8) \bmod 10 = 0$. But position 0 is already filled. Using a simple linear search for an open position, I would place goldfish at position 1.

[32]This is usually not the best strategy to use.

TABLE 3.6 Partially Filled Hash Table

index	kingdom	index	kingdom
0	snake: reptile	5	
1		6	dog: mammal
2		7	
3		8	
4	cat: mammal	9	cricket: insect

The only complication that might arise would be if I removed snake from the table. To properly find goldfish I would need to put a special marker in position 0 (perhaps the word deleted). This is because when looking for goldfish, h would indicate it should be in position 0. If that position is empty, then I should conclude that goldfish is not in the list. However, if the deleted message is found, I would start a linear search for goldfish. ■

3.4.3 Pseudorandom Numbers

Computers are often used to simulate real-world events. For example, a store might wish to simulate a typical business day and experiment with how many checkout lines to have open at various times during the day. To do this well, it is necessary to have the simulated customers arrive at the checkouts at random intervals. It is also desirable to have the number of customers in the store at any time also be randomly determined. In such a simulation, the randomness typically follows particular patterns (called *probability distributions*).

The mechanism that enables the computer to work with random events and probability distributions is called a *pseudorandom number generator*. The word *pseudo* appears because, as will be seen shortly, the numbers that are generated only appear to be random. In reality they are created by a deterministic algorithm, typically using prime numbers and the mod operator.

Pseudorandom number generators typically produce a sequence, $\{x_n\}$, of integers. The sequence may consist of several million integers, but it will eventually arrive back at the number, x_0, at which it started and then the sequence will repeat. Each number in the sequence is calculated by using the previous number and some simple arithmetic. A simple example will illustrate the general pattern.

EXAMPLE 3.23 **A Small Pseudorandom Number Generator**

Suppose I want a pseudorandom number generator that produces integers between 0 and 10, inclusive. One way to do this is to define

$$x_{n+1} = (7x_n + 1) \bmod 11$$

If $x_0 = 0$ the sequence will be 0, 1, 8, 2, 4, 7, 6, 10, 5, 3. The sequence will wrap back to 0 after 3 and then repeat. We can visualize this by placing the sequence around a circle in clockwise order, as in Figure 3.4.

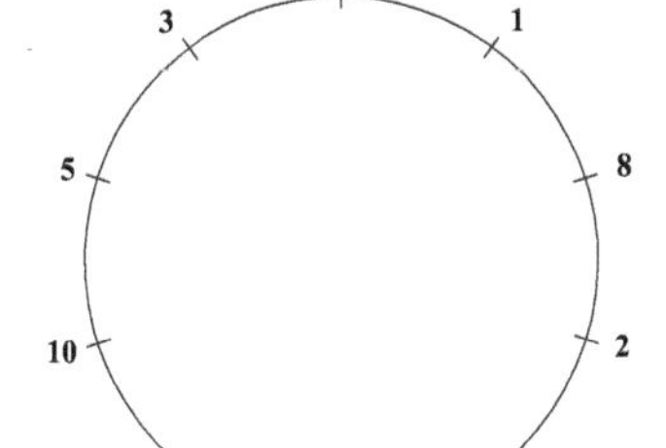

Figure 3.4. A small pseudorandom sequence.

Notice that if we set $x_0 = 4$, the sequence will be 4, 7, 6, 10, 5, 3, 0, 1, 8, 2. That is, we can pick any number on the circle as the starting point and the sequence will then be determined.

The initial position on the circle is often called the *seed* of the pseudorandom number generator.

Observe also that the sequence in this example contains all the integers in the desired range. That is not always the case. If the generator function is changed to $x_{n+1} = (7x_n + 1) \bmod 16$, the sequence becomes 0, 1, 8, 9, and then starts repeating. ■

The sequence in Example 3.23

0, 1, 8, 2, 4, 7, 6, 10, 5, 3

Although a pseudorandom number generator will produce a random collection of integers, it is usually more useful to have a random collection of real numbers in some interval. It is easy to make the conversion. If the pseudorandom number generator is done mod M ($M = 11$ in Example 3.23), then we can produce a sequence of real numbers in the interval [0, 1) by setting $r_n = \frac{x_n}{M}$. In Example 3.23 we would have the sequence (rounded to 4 digits after the decimal) 0, 0.0909, 0.7273, 0.1818, 0.3636, 0.6363, 0.5455, 0.9091, 0.4545, 0.2727.

A sequence of real numbers in [0, 1) are said to be *uniformly distributed* if every number in [0, 1) is equally likely to occur next. Of course, a pseudorandom number generator will not perfectly meet this criterion (even ignoring the fact that a finite collection of integers, converted to fractions, will miss all the irrational and many rational numbers in the interval). However, it is possible to come close enough.

EXAMPLE 3.24 **Testing A Uniform Distribution**

TABLE 3.7 Testing Uniformness.

bucket	count
0	1013
1	1019
2	991
3	1008
4	1022
5	1024
6	997
7	930
8	1009
9	987

The programming language Java provides a good pseudorandom number generator. It also provides a nice function that produces a sequence of real numbers in the interval [0, 1), as previously described. Here is a simple way to test whether this function is approximately uniformly distributed. Generate 10,000 pseudorandom numbers. For each number, determine which of 10 "buckets" it belongs in. The buckets are the subintervals $[.1n, .1n + .1)$, where $n = 0, 1, 2, \ldots, 9$. Keep track of how many pseudorandom numbers land in each bucket. If the distribution is uniform, then the final counts for each bucket should be very close to 1,000 per bucket. The result of one experiment is listed in Table 3.7.

Two refinements to this experiment would be to use more buckets (perhaps 100 instead of 10) and to use more pseudorandom numbers (perhaps 100,000 instead of 10,000). ■

It is not sufficient that the pseudorandom number generator produce a uniformly distributed sequence. There are other kinds of randomness properties that are also desired. For example, if the sequence in Example 3.23 were 0, 1, 2, 3, 4, 5, 6, 7, 8, 9, 10, we would not consider it to be very random (even though it *is* uniformly distributed). Devising pseudorandom number generators that pass a number of these kinds of tests is an active area of research.

EXAMPLE 3.25 **Java's Pseudo-Random Number Generator**

The pseudorandom number generator used by Java is based upon material in *The Art of Computer Programming*, Volume 2 by Donald Knuth [63, Section 3.2.1].

The generator is in the form

$$x_{n+1} = (ax_n + b) \bmod c$$

where $a = 25,214,903,917$, $b = 11$, and $c = 2^{48}$. The generator returns only the 32 high order bits (producing a standard 32-bit integer), but retains the 48-bit version internally to generate the next pseudorandom number. Part of the reason for using the extra 16 bits is to keep the generator from alternating between even and odd numbers. This generator produces all 2^{32} possible 32-bit integers before restarting the sequence. It also does pretty well on the randomness tests. ■

Pseudorandom number generators that follow the model presented in Example 3.25 are called *linear congruential* generators. The theory behind them has been extensively studied. Here is one theorem (from [63]), presented without proof, that provides conditions for which the pseudorandom sequence will include all the potential integers in its range. Such a sequence is said to have *maximum period.* Notice the use of primes and the mod operator.

THEOREM 3.41 ***Creating Maximum Period Linear Congruential Pseudorandom Number Generators***

Let the sequence $\{x_n\}$ be defined by

$$x_{n+1} = (ax_n + b) \bmod c$$

for nonnegative integers a, b, c with $0 < c$, $0 \le a < c$, $0 \le b < c$, and $0 \le x_0 < c$. The sequence has period c if and only if

1. b is relatively prime to c,
2. $a - 1$ is a multiple of p, for every prime p dividing c,
3. $a - 1$ is a multiple of 4 if c is a multiple of 4.

✔ Quick Check 3.7

1. Show that the pseudorandom number generator in Example 3.25 satisfies the conditions of Theorem 3.41.

It is tempting to add your own "enhancements" to a pseudorandom number generator. For example, perhaps at some semi-random intervals, you could restart the sequence at a different part of the circle. That is, pick a new seed every once in a while. Or, you could chain two pseudorandom number generators together. Suppose the generators are f and g. You might pick a seed x_0 and then define $x_{n+1} = f(g(x_n))$.

There is extensive research that indicates that all such attempts actually make the resulting sequence *less* random (in the sense that it will do poorly on many of the tests mentioned earlier).

Researchers have also shown that humans do a very poor job of creating random sequences in their heads (again when evaluated by the randomness tests). People try too hard to avoid any sense of pattern. Real random sequences do exhibit various patterns for short stretches (see [4] for an interesting example).

3.4.4 Linear Congruence and the Chinese Remainder Theorem

You have encountered many linear equations, which can be expressed in the form $ax = b$ with x as the variable. It is also possible to consider linear relationships that involve congruence.[33]

DEFINITION 3.42 ***Linear Congruence***

Let a and b be integers and let m be a positive integer. An expression with variable x of the form $ax \equiv b(\bmod m)$ is called a *linear congruence*. A solution is an integer value for x.

Unlike linear equations, a linear congruence need not have a solution. For example, the linear congruence $5x \equiv 4 \pmod{10}$ has no solution because no multiple of 5 has remainder 4 when divided by 10. Corollary 3.45 (page 120) provides a criterion that insures a solution. In fact, a somewhat unique solution (there will be exactly one non-negative solution that is less than m). The proof of the corollary uses the next definition and theorem.

DEFINITION 3.43 ***The Inverse of a, mod m***

Let a be a nonzero integer. An integer $\tilde{a}$ is said to be an inverse of a, mod m if $a\tilde{a} \equiv 1 \pmod{m}$.

[33] It will be helpful to review Definition 3.19 on page 98.

THEOREM 3.44 ***Inverses mod m***

Let $a, m \in \mathbb{Z}$ with $m > 1$. Then

1. if $\gcd(a, m) > 1$, a has no inverse, mod m.
2. if $\gcd(a, m) = 1$ then a has an inverse mod m.
3. if $\tilde{a}$ and $\hat{a}$ are both inverses for a mod m, then $\tilde{a} \equiv \hat{a} \pmod{m}$.

Proof: The proof for parts 1 and 3 will be left as Exercise 8 on page 130. To prove part 2, assume that a and m are relatively prime (so $a \neq 0$). By Theorem 3.12 there exist integers s and t such that $as + mt = 1$. Thus $as - 1 = m(-t)$ so $as \equiv 1 \pmod{m}$, completing the constructive proof. □

COROLLARY 3.45 ***A Sufficient Condition for a Solution to a Linear Congruence***

Let $a, b, m \in \mathbb{Z}$ with $m > 1$. An expression with variable x of the form $ax \equiv b \pmod{m}$ has an integer solution for x if a and m are relatively prime.

Proof: Theorem 3.44 ensures an inverse, $\tilde{a}$, for a, mod m. Thus, using commutativity and Proposition 3.22 (page 100) twice, $x \equiv a\tilde{a}x \equiv b\tilde{a} \pmod{m}$. □

EXAMPLE 3.26 **What Corollary 3.45 Does not Tell Us**

Theorem 3.44 completely settles the existence of inverses mod m. However, the corollary is not as complete. The corollary is an implication. If we know that a and m are relatively prime, then we know a solution exists. However, if a and m are not relatively prime, the corollary has nothing to say. The converse of the corollary may or may not be true.

For example, the linear congruence $5x \equiv 2 \pmod{10}$ has no solution, but the linear congruence $2x \equiv 2 \pmod{10}$ does, namely $x = 1$. In both examples, $\gcd(a, m) > 1$.

The Chinese Remainder Theorem

You have previously studied systems of linear equations. It is also useful to investigate systems of linear congruences. The following example goes back (at least) to the 4th century, AD [2]. A solution will be presented after the Chinese Remainder Theorem is introduced.[34]

EXAMPLE 3.27 **Sunzi's System of Linear Congruences**

The Chinese mathematician Sunzi recorded the following problem.

> There are certain things whose number is unknown.
> Repeatedly divided by 3, the remainder is 2;
> by 5 the remainder is 3;
> and by 7 the remainder is 2.
> What will be the number?

This translates into the following system of linear congruences.

$$
\begin{aligned}
x &\equiv 2 \pmod{3} \\
x &\equiv 3 \pmod{5} \\
x &\equiv 2 \pmod{7}
\end{aligned}
$$
■

[34] If you are interested in the origin of the theorem's name, do an internet search for "Chinese Remainder Theorem" and "Sun-Tsu". Sun-Tsu's name in pinyin is Sunzi (or Sun Zi). Sunzi (孫子) was a mathematician, not to be confused with the military strategist with the same name.

The following puzzle presents another system of linear congruences.

EXAMPLE 3.28 **Buckets of Water**

I have three water buckets. The first holds 2 liters, the second 5 liters, and the third holds 7 liters of water. I would like to purchase a fourth, much larger bucket. This x-liter bucket should have the property that when I fill the large bucket and then start transferring water to one of the smaller buckets, the amount of leftover water will be in convenient sizes. In particular, if the larger bucket is poured into the 2-liter bucket several times, there will be 1 liter left in the large bucket. There should 3 liters left after repeatedly pouring into the 5-liter bucket. Finally, pouring into the 7-liter bucket will result in 4 liters being left in the large bucket. With this arrangement, I can easily measure the following amounts of water (and many others).

TABLE 3.8 Amounts that Can Measured

Liters	*How*	*Liters*	*How*
1	left after pouring into 2-liter bucket	6	4 liters as above, then add 2 liters
2	2-liter bucket	7	7-liter bucket
3	left after pouring into 5-liter bucket	8	3 liters as above, then add 5 liters
4	left after pouring into 7-liter bucket	9	use 3-liter bucket to fill larger bucket to 9
5	5-liter bucket	10	use 5-liter bucket to fill larger bucket to 10

What size should the large bucket be? The problem translates into the following system of linear congruences. The second congruence in the system indicates that the large bucket equals several copies of the 5-liter bucket, plus an additional 3 liters.

$$\begin{aligned} x &\equiv 1 \pmod{2} \\ x &\equiv 3 \pmod{5} \\ x &\equiv 4 \pmod{7} \end{aligned}$$

■

The solution to the system of linear congruences in Examples 3.27 and 3.28 can be found using the Chinese Remainder theorem.

THEOREM 3.46 ***The Chinese Remainder Theorem***

Let $a_1, a_2, \ldots, a_n, m_1, m_2, \ldots, m_n, x \in \mathbb{Z}$ where $m_1, m_2, \ldots, m_n$ are pairwise relatively prime and all positive. Let $m = m_1 \cdot m_2 \cdots m_n$.
The system of linear congruences

$$\begin{aligned} x &\equiv a_1 \pmod{m_1} \\ x &\equiv a_2 \pmod{m_2} \\ \vdots &\equiv \vdots \qquad \vdots \\ x &\equiv a_n \pmod{m_n} \end{aligned}$$

has a unique solution x, mod m, with $0 \le x < m$. Any other solution, $\hat{x}$, satisfies $\hat{x} \equiv x \pmod{m}$. It is possible to express the solution as

$$x = \left(\sum_{i=1}^{n} a_i M_i \tilde{M_i} \right) \bmod m$$

where $M_i = \frac{m}{m_i}$ and $M_i \cdot \tilde{M_i} \equiv 1 \pmod{m_i}$.

TABLE 3.9 Key Ideas

$m = m_1 \cdot m_2 \cdots m_n$

$x \equiv a_i \pmod{m_i}$

$x = \left(\sum_{i=1}^{n} a_i M_i \tilde{M}_i\right) \bmod m$

$M_i = \frac{m}{m_i}$

$M_i \cdot \tilde{M}_i \equiv 1 \pmod{m_i}$

Proof: Some key ideas from the theorem statement are repeated in Table 3.9.

Existence

The motivation for the expression $x = \left(\sum_{i=1}^{n} a_i M_i \tilde{M}_i\right) \bmod m$ is based upon the following observations.

1. The congruence $a_i M_i \tilde{M}_i \equiv a_i \pmod{m_i}$ is true for all i.
2. $M_k \equiv 0 \pmod{m_i}$ if $k \neq i$ because m_i is a factor of M_k. Thus, $a_k M_k \tilde{M}_k \equiv 0 \pmod{m_i}$ for $k \neq i$.
3. The mod operation ensures that $0 \leq x < m$ is true.

These observations, together with Proposition 3.24 on page 100 ensure that

$$\sum_{i=1}^{n} a_i M_i \tilde{M}_i \equiv a_k \pmod{m_k} \qquad \text{for } k = 1, 2, \ldots, n$$

so $\sum_{i=1}^{n} a_i M_i \tilde{M}_i$ satisfies all the congruences in the system.

Proposition 3.25 implies that

$$\left(\left(\sum_{i=1}^{n} a_i M_i \tilde{M}_i\right) \bmod m\right) \bmod m_k \equiv \left(\sum_{i=1}^{n} a_i M_i \tilde{M}_i\right) \bmod m_k \equiv a_k \pmod{m_k}$$

so $x = \left(\sum_{i=1}^{n} a_i M_i \tilde{M}_i\right) \bmod m$ is the solution we seek.

Uniqueness

Suppose that $\hat{x}$ is another solution to the system of linear congruences. Then for $i = 1, 2, \ldots, n$ integers v_i and w_i exist such that $x = m_i v_i + a_i$ and $\hat{x} = m_i w_i + a_i$. Thus, for each i, $x - \hat{x} = m_i(v_i - w_i)$ and therefore $\hat{x} \equiv x \pmod{m_i}$. Proposition 3.52 (Exercise 9 on page 130) implies that $\hat{x} \equiv x \pmod{m}$. □

EXAMPLE 3.29

Buckets of Water - Part 2

The system of linear congruences in Example 3.28 is

$$x \equiv 1 \pmod{2}$$
$$x \equiv 3 \pmod{5}$$
$$x \equiv 4 \pmod{7}$$

Note that

- $M_1 = \frac{70}{2} = 35$ and $35 \cdot 1 \equiv 1 \pmod{2}$ so $\tilde{M}_1 = 1$
- $M_2 = \frac{70}{5} = 14$ and $14 \cdot 4 \equiv 1 \pmod{5}$ so $\tilde{M}_2 = 4$
- $M_3 = \frac{70}{7} = 10$ and $10 \cdot 5 \equiv 1 \pmod{7}$ so $\tilde{M}_3 = 5$

I used guess-and-check to find the inverses. Theorem 3.12 and the Euclidean Algorithm can be used for harder examples. Theorem 3.46 indicates that

$$x = (1 \cdot 35 \cdot 1 + 3 \cdot 14 \cdot 4 + 4 \cdot 10 \cdot 5) \bmod 70 = 403 \bmod 70 = 53$$

It is easy to verify that 53 is a solution to the system of linear congruences. ■

✔ Quick Check 3.8

1. Solve Sunzi's system.
2. Solve the following system of linear congruences.

$$x \equiv 1 \pmod{4}$$
$$x \equiv 2 \pmod{3}$$
$$x \equiv 4 \pmod{5}$$

3. Does the following system of linear congruences have a solution? If it does, find one.

$$x \equiv 2 \pmod{3}$$
$$x \equiv 5 \pmod{6}$$

☑

EXAMPLE 3.30 **A More General System of Linear Congruences**

Consider the following system of linear congruences.

$$\begin{aligned} 3x &\equiv 1 \pmod{2} \\ 3x &\equiv 4 \pmod{5} \\ 4x &\equiv 2 \pmod{7} \end{aligned}$$

The Chinese Remainder Theorem does not apply directly. However, since $\gcd(3, 2) = 1$, $\gcd(2, 5) = 1$, and $\gcd(4, 7) = 1$, Corollary 3.45 ensures that it is possible to turn each congruence into one with only x on the left.

In this case, since $3 \cdot 1 \equiv 1 \pmod{2}$, the first congruence becomes $x \equiv 1 \pmod{2}$. Similarly, $3 \cdot 2 \equiv 1 \pmod{5}$ so multiplying both sides of the second congruence by 2 and reducing mod 5 yields $x \equiv 3 \pmod{5}$. After simplifying the third congruence, the new, equivalent system is

$$\begin{aligned} x &\equiv 1 \pmod{2} \\ x &\equiv 3 \pmod{5} \\ x &\equiv 4 \pmod{7} \end{aligned}$$

From Example 3.29, we know that 53 is a solution. ■

THEOREM 3.47 ***Decomposing a Linear Congruence***

Let $m > 1$ be an integer with factorization $m = p_1^{e_1} p_2^{e_2} \cdots p_k^{e_k}$ as a product of distinct primes. The linear congruence $ax \equiv b \pmod{m}$ has the same solution set as the system of linear congruences

$$\begin{aligned} ax &\equiv b \pmod{p_1^{e_1}} \\ ax &\equiv b \pmod{p_2^{e_2}} \\ \vdots &\equiv \vdots \qquad \vdots \\ ax &\equiv b \pmod{p_k^{e_k}} \end{aligned}$$

Proof: Let r and s be any integers. Then

$$\begin{gathered} r \equiv s \pmod{m} \\ \text{if and only if} \\ r - s = mt = p_1^{e_1} p_2^{e_2} \cdots p_k^{e_k} t \text{ for some } t \\ \text{if and only if} \\ p_i^{e_i} \mid (r - s) \text{ for } i = 1, 2, \ldots, k \\ \text{if and only if} \\ r - s = p_i^{e_i} u_i \text{ for } i = 1, 2, \ldots, k \\ \text{if and only if} \\ r \equiv s \pmod{p_i^{e_i}} \text{ for } i = 1, 2, \ldots, k \end{gathered}$$

Substituting ax for r and b for s completes the proof. □

EXAMPLE 3.31 **Decomposing a Linear Congruence**

The linear congruence $7x \equiv 13 \pmod{360}$ could be solved by finding an inverse for 7, mod 360. It is also possible to use Theorem 3.47. Notice that $360 = 2^3 \cdot 3^2 \cdot 5$. The solution set is therefore the same as the solution set to

$7x \equiv 13 \pmod{360}$

$$\begin{aligned} 7x &\equiv 13 \pmod{2^3} \\ 7x &\equiv 13 \pmod{3^2} \\ 7x &\equiv 13 \pmod{5} \end{aligned}$$

The inverses for 7 are easier to determine for these three congruences than for the congruence mod 360. In particular, $7 \cdot 7 \equiv 1 \pmod{8}$, $7 \cdot 4 \equiv 1 \pmod{9}$, and $7 \cdot 3 \equiv 1 \pmod{5}$. Notice that $7 \cdot 13 = 91 \equiv 3 \pmod{8}$, $4 \cdot 13 = 52 \equiv 7 \pmod{9}$, and $3 \cdot 13 = 39 \equiv 4 \pmod{5}$. The system is thus equivalent to

$$\begin{aligned} x &\equiv 3 \pmod{2^3} \\ x &\equiv 7 \pmod{3^2} \\ x &\equiv 4 \pmod{5} \end{aligned}$$

The Chinese Remainder Theorem implies that the smallest positive solution is[35]

$$x = (3 \cdot 45 \cdot 5 + 7 \cdot 40 \cdot 7 + 4 \cdot 72 \cdot 3) \bmod 360 = 3499 \bmod 360 = 259$$

It is easy to verify that $x = 259$ solves the original linear congruence. It would not be so easy to find this solution (or any other solution) by trial and error. ■

3.4.5 Fermat's Little Theorem and Fermat's Last Theorem

The next theorem links primes and the mod operator. The word "little" in the name is there to distinguish this theorem from what is considered to be Fermat's big theorem, better known as Fermat's Last Theorem.

THEOREM 3.48 ***Fermat's Little Theorem***

Let $a, p \in \mathbb{Z}$ with p a prime. Then

$$a^p \equiv a \pmod{p}$$

and

$$a^{p-1} \equiv \begin{cases} 1 \pmod{p} & \text{if } \gcd(a, p) = 1 \\ 0 \pmod{p} & \text{if } \gcd(a, p) \neq 1 \end{cases}$$

EXAMPLE 3.32 **Illustrating Theorem 3.48**

Before proving Fermat's Little Theorem, it might be helpful to see a few examples that illustrate the congruences in the theorem.

Let $c = 45 = 9 \cdot 5$ and let $p = 5$. Then c, c^5, and c^4 are all divisible by p, so $c^5 \equiv c \pmod{5}$ and $c^4 \equiv 0 \pmod{5}$ are both true.

Now let $a = 42$ and $p = 5$. Then $42^5 = 130691232$ and 42 are both congruent to 2 mod 5, so $42^5 \equiv 42 \pmod{5}$. Also, $42^4 = 3111696$ which is clearly congruent to 1 mod 5, so $42^{5-1} \equiv 1 \pmod{5}$.

The proof will introduce two sets that are worth considering prior to the proof. Let $A = \{a, 2a, 3a, 4a\} = \{42, 84, 126, 168\}$ and let R be the set consisting of the remainders of the elements of A after division by 5. So $R = \{2, 4, 1, 3\} = \{1, 2, 3, 4\}$. ■

[35] The Euclidean Algorithm can be used to find an inverse for 45, mod 8. Since $\gcd(45, 8) = 1$, the algorithm can be used to find s and t such that $45 \cdot s + 8 \cdot t = 1$. The inverse will be $s = 5$.

Proof of Theorem 3.48 The three congruences in the theorem statement will be verified by examining the two conditions $\gcd(a, p) \neq 1$ and $\gcd(a, p) = 1$.

$\gcd(a, p) \neq 1$

If $\gcd(a, p) \neq 1$ then $p \mid a$ (either $a = 0$ or $\gcd(a, p) = p$). Consequently, a, a^{p-1}, and a^p are all multiples of p. Thus, all three numbers are congruent to 0, mod p. Therefore $a^{p-1} \equiv 0 \pmod{p}$ and $a^p \equiv a \pmod{p}$ when $\gcd(a, p) \neq 1$.

$\gcd(a, p) = 1$

If $\gcd(a, p) = 1$ then the validity of $a^{p-1} \equiv 1 \pmod{p}$, together with Proposition 3.22 (page 100), implies that $a^p \equiv a \pmod{p}$. The proof is thus complete once the validity of "$a^{p-1} \equiv 1 \pmod{p}$ when $\gcd(a, p) = 1$" is established.

To that end, assume that $\gcd(a, p) = 1$. Consider the set A of $p-1$ distinct integers $A = \{a, 2a, 3a, \ldots, (p-1)a\}$ (see Exercise 17 on page 131). The Quotient–Remainder theorem implies that there are integers q_i, r_i with $0 \leq r_i < p$ such that $ia = pq_i + r_i$ for $i = 1, 2, \ldots, p-1$. Let $R = \{r_i \mid i = 1, 2, \ldots, p-1\}$. Even though $|A| = p-1$, it is not necessarily the case that $|R| = p-1$ because there might be integers i and j with $i \neq j$ such that $r_i = r_j$. However, that is not the case.

To verify that the remainders are all distinct, suppose that $r_i = r_j$ were true for some $i, j \in \{1, 2, \ldots, p-1\}$. Notice that $ia \equiv r_i \pmod{p}$ and $ja \equiv r_j \pmod{p}$, so the assumption $r_i = r_j$ implies that $ia \equiv ja \pmod{p}$. This implies that $ia - ja = kp$ for some $k \in \mathbb{Z}$. But then $p|a(i-j)$. Since $\gcd(a, p) = 1$, Proposition 3.39 (page 110) implies that $p \mid (i-j)$ and so $i - j = 0$ must be true ($0 < i < p$ and $0 < j < p$ imply $-p < i - j < p$). The conclusion is that $r_i \neq r_j$ if $i \neq j$. Thus $|R| = p-1$ and hence $R = \{1, 2, \ldots, p-1\}$ ($r_i \neq 0$ because $\gcd(a, p) = 1$).

Proposition 3.22 (page 100) can be applied $p-1$ times to derive the congruence

$$a \cdot (2a) \cdot (3a) \cdots ((p-1)a) \equiv r_1 \cdot r_2 \cdot r_3 \cdots r_{p-1} \pmod{p}$$

After rearranging the remainder terms on the right, this becomes

$$a \cdot (2a) \cdot (3a) \cdots ((p-1)a) \equiv 1 \cdot 2 \cdot 3 \cdots (p-1) \pmod{p}$$

This can also be written as:

$$a^{p-1}(p-1)! \equiv (p-1)! \pmod{p}$$

Notice that $\gcd((p-1)!, p) = 1$, so Proposition 3.26 (page 100) implies that

$$a^{p-1} \equiv 1 \pmod{p}$$

□

The next theorem is quite famous. For centuries it was only a conjecture. It was finally proved by Andrew Wiles in 1994. Wiles built upon the work of many other mathematicians.

The theorem was inspired by Pythagorean triples (page 96). There are infinitely many nonzero integers x, y, z such that $x^2 + y^2 = z^2$. However, no solution in nonzero integers has ever been found for the equation $x^n + y^n = z^n$ for $n > 2$, n an integer. In 1637, Pierre de Fermat made the conjecture that no solutions have been found because there aren't any. In fact, he made a note in the margin of his translated copy of an ancient Greek mathematics book by Diophantus. The note read: *I have a truly marvelous proof of this proposition which this margin is too narrow to contain.*[36]

Given the complexity of the modern proof, Fermat's proof was almost certainly incorrect. The theorem is formally stated next. No proof will be given.

THEOREM 3.49 ***Fermat's Last Theorem***

Let $n \in \mathbb{Z}$ with $n > 2$. Then there are no solutions in nonzero integers x, y, z to the equation $x^n + y^n = z^n$.

---

[36] Fermat actually wrote the note in Latin: *Cuius rei demonstrationem mirabilem sane detexi. Hanc marginis exiguitas non caperet.* [104]

3.4.6 Encryption

The practice of encrypting messages has a long history. The goal is to prevent unauthorized people from being able to correctly decrypt the message. Various techniques have been used to accomplish this goal.

Caesar Cyphers

TABLE 3.10 The Caesar Cypher

letter	pos	letter	pos
A	D	N	Q
B	E	O	R
C	F	P	S
D	G	Q	T
E	H	R	U
F	I	S	V
G	J	T	W
H	K	U	X
I	L	V	Y
J	M	W	Z
K	N	X	A
L	O	Y	B
M	P	Z	C

One of the earliest documented strategies is a substitution cipher. With this strategy, each letter in the alphabet is replaced by another letter. To accomplish this without the need to write down a list of the substitutions, the substitution can be done using a cyclic shift. Each letter can be replaced by the letter n places down the alphabet. The final letters of the alphabet can be replaced by letters that are at the front of the alphabet. The number n can be considered to be an *encoding/decoding key*. Any person who knows that key has the ability to encode and decode messages in this system. Of course, the goal is to keep both the key and the strategy out of the hands of unauthorized persons.

Julius Caesar used just such a strategy. If he had used the English alphabet, his encoding strategy would have been $E_n(x) = (x + n) \bmod 26$. He used a shift of $n = 3$, so letter E would have been encoded as H (see Table 3.10). To decode, just reverse the process: $D_n(x) = (x - n) \bmod 26$.

This strategy may have been effective during Caesar's era, but it is easy to break now. One common strategy for breaking a substitution code is to capture a moderately-long encoded message and measure the frequency of the symbols that appear. These frequencies can be compared to letter frequencies in the host language and some pretty astute guesses can be made about certain letters. Determining the rest of the substitutions then becomes easier. However, another strategy may be even more useful: use a computer to do a brute force "decoding" of the captured message using every possible substitution. It will not take long for a person to scroll through the results and identify the proper decoding.

You can read more about the substitution strategy on Wikipedia [101].

There are two uses of such codes that have survived: secret decoder rings for children, and the Unix rot13 utility. The rot13 utility uses the encryption $E_{13}(x)$, which has the convenient property (in English) that $D_{13}(x) = E_{13}(x)$, so no separate function is needed for decoding.[37] The utility is typically used to hide the solution of puzzles, to hide potential spoilers (perhaps in a movie review), or to scramble the contents of a potentially offensive joke published on usenet (an internet newsgroup system).

Figure 3.5. A three rotor enigma cypher machine used by Luftwaffe in 1944 [57]. Photo by Jszigetvari.

More Elaborate Secret Strategies

If substitution codes are not secure, perhaps a more elaborate system can be developed. If both the strategy and the necessary encoding and decoding keys are unknown to unauthorized people, then messages can be securely transmitted between authorized people.

Perhaps the most famous such scheme was based upon the Enigma machine (Figure 3.5) and was used by the Nazis during World War II. The machine was quite ingenious. However, through a combination of captured machines, messages, and code books, together with some clever code breaking, the opposing side was able to decode many messages [102].

This example illustrates a major flaw in any system that assumes that both the key and the encoding/decoding algorithm are secret: you may think the system is secure and unbreakable but the enemy may have already found a way to break your code.

[37] This feature is the basis for a humorous "research paper" entitled *On the 2ROT13 Encryption Algorithm*.[3]

A Better Strategy: Public Key Encryption

The initially counterintuitive idea behind public key encryption is that only the encryption key is kept secret. Both the decryption key and the encoding–decoding algorithm are publicly advertised. This seems, at first glance, to be completely insecure. There are two ideas that make this strategy the current scheme of choice.

First, if the encoding/decoding algorithm is publicly known, many people will take up the challenge to break the code. Some of these people will be associated with universities. If they break the code, they will publish a paper describing their success.[38] This means that someone using the strategy will know that their code is no longer secure. Therefore, they will not naively continue to use the scheme.

The second idea will take a bit longer to explain. Consider the effect of a strategy that uses different keys, a private key and a public key, for encoding and decoding. What one key accomplishes, the other undoes. Suppose you wish to send a secret message to me. You write the message, apply my public key to the message, and send the resulting encrypted message. I receive the message, apply my private key to the encrypted message, and obtain a copy of the original message. Another person (even you) cannot decrypt the message unless they know my private key.

If I wish to send a message to you, I use your public key to encrypt it. This seems like a good strategy. However, there is still a major problem. One of my enemies could use my public key to send a message, but the message could claim to be from you. Perhaps the message reads: *The business deal we discussed has been approved. Please send $100,000 to the following secret bank account...* The message is not really from you, but I have no way to verify that. A small modification to this strategy can give me a high degree of certainty that a message was really sent by you.

The idea is to have you use my public key to encode the message. Then you use your private key to encrypt a second time before sending the doubly encrypted message to me. After receiving the message that claims to be from you, I use your public key to decrypt, producing a still-encrypted file. I then use my private key to decrypt that file. If the second decryption produces a message I can read, then I have a high degree of certainty that the message was from you and that nobody else has read it. That is because your public key will not properly decrypt a message that was not encoded with your private key. The only way something could go wrong is if one or both of our private keys falls into the hands of unauthorized people.

A more efficient approach is to have you use your private key to encrypt your signature (perhaps your name and some other identifying information. Then append this encrypted information to the end of your message to me. Now encrypt the extended message with my public key and send the result to me. I then use my private key to decrypt. That produces a message with a bit of encrypted stuff at the end. I use your public key to decrypt that final portion and verify that you were the sender.

RSA Encryption

The first really nice public key encryption system was published in 1977. The developers were Ron Rivest, Adi Shamir, and Leonard Adleman, all working at MIT. The algorithm is named by combining the last letters of their names. The algorithm is predicated on the idea that very large prime numbers are fairly easy to produce on a computer, but it is quite difficult to factor a product of two large unknown primes. The details are presented next.

[38]There was a period of time when several governments took the opposite stance. There were laws that declared the exportation of information about the strongest encryption schemes to be illegal. In particular, it was a crime to publish information on how to break the codes. You can find some interesting discussion on the wisdom of such laws by doing an internet search on "export of cryptography" and "Bernstein v. United States".

Preliminary Steps

The first step is to choose two very large primes, p and q, and calculate $n = pq$.[39] It is fairly easy to find large primes (often using probabilistic computer algorithms).

The next step is to calculate[40] $(p\text{-}1)(q\text{-}1)$ and then find an integer e satisfying $1 < e < (p\text{-}1)(q\text{-}1)$ and $\gcd(e, (p\text{-}1)(q\text{-}1)) = 1$. It is best to not choose a small value for e. A popular choice is the prime $e = 2^{16} - 1 = 65537$.

The final step in setting up the algorithm is to find the inverse for e, mod $(p\text{-}1)(q\text{-}1)$. That is, find an integer d such that $de \equiv 1 \pmod{(p\text{-}1)(q\text{-}1)}$. This can be accomplished because $\gcd(e, (p\text{-}1)(q\text{-}1)) = 1$, so by Proposition 3.12 there exist integers c, d such that $de + c(p\text{-}1)(q\text{-}1) = 1$. Values for c and d can be found using the Euclidean Algorithm (page 113) and perhaps some modular arithmetic to make sure that $0 < d < (p\text{-}1)(q\text{-}1)$.

RSA Encoding/Decoding

The public key will be the numbers n and e. The private key will be d, together with the publicly available n.

The original message, M (called the *plaintext*), is converted into numeric form, m, using a collection of blocks of a predetermined number of bits.[41] In addition, some carefully chosen extra bits are typically used to pad the message prior to encryption to foil some potential code breaking attacks [105]. The numeric form of the (padded) message, m, must satisfy[42] $0 < m < n$ and $\gcd(m, n) = 1$. In the unlikely event that $\gcd(m, n) \neq 1$ (so $\gcd(m, n) = p$ or $\gcd(m, n) = q$), some additional padding can be added to m. The encrypted message, c (called the *cyphertext*), is calculated as

$$c = m^e \pmod{n}$$

Theorem 3.50 on page 129 implies that the message can be decrypted as

$$m = c^d \pmod{n}$$

Since d is not publicly available, only someone in possession of that value can correctly decrypt the message. The security of the system depends on d being very difficult for someone to calculate if they only know n, e and perhaps have a number of encrypted messages to experiment with.[43]

EXAMPLE 3.33 **Using RSA**

A simple, but unrealistic, example of RSA encoding might be helpful. Let the "very large" primes be $p = 3$ and $q = 13$. Then $n = 39$ and $(p\text{-}1)(q\text{-}1) = 24$. Choose $e = 5$ (which is a prime). The Euclidean Algorithm can be used to find d. Phase 1 produces:

$$\begin{aligned} 24 &= 5 \cdot (4) + 4 \\ 5 &= 4 \cdot (1) + 1 \\ 4 &= 1 \cdot (4) + 0 \end{aligned}$$

Phase 2 of the algorithm is summarized next.

$$\begin{aligned} 1 &= 5 - 4(1) \\ &= 5 - (24 - 5 \cdot (4)) \cdot (1) \\ &= 24 \cdot (-1) + 5 \cdot (5) \end{aligned}$$

[39] Special software or hardware is used to ensure that large numbers can be represented exactly. The numbers are actually stored in binary. Current estimates are that 2048 to 4096 bits are needed for n to ensure a reasonable degree of security.

[40] Newer versions of the algorithm use $\text{lcm}(p\text{-}1, q\text{-}1)$ instead of $(p\text{-}1)(q\text{-}1)$.

[41] One possible conversion is to just use the bit patterns for the unicode representation of the characters.

[42] The message can always be partitioned into several pieces if $m \geq n$.

[43] A useful code breaking technique is to use the public key to encrypt a number of known messages and then examining the results, hoping to gain some hints as to the value of d.

TABLE 3.11 Letter Positions

letter	pos	letter	pos
a	1	n	14
b	2	o	15
c	3	p	16
d	4	q	17
e	5	r	18
f	6	s	19
g	7	t	20
h	8	u	21
i	9	v	22
j	10	w	23
k	11	x	24
l	12	y	25
m	13	z	26

So 5 is its own inverse, mod 24.

To create m we could break the plaintext into blocks of one letter and use the numeric alphabet positions of the letter (Table 3.11) to convert to a number.

Suppose the plaintext message is cat. We would encrypt in three packages: $m_1 = 03$, $m_2 = 01$, $m_3 = 20$. The three encrypted cyphertexts would be

$$c_1 = m_1^e \bmod n = m_1^5 \bmod 39 = 3^5 \bmod 39 = 243 \bmod 39 = 9$$
$$c_2 = m_1^e \bmod n = m_2^5 \bmod 39 = 1^5 \bmod 39 = 1$$
$$c_3 = m_1^e \bmod n = m_3^5 \bmod 39 = 20^5 \bmod 39 = 3200000 \bmod 39 = 11$$

Upon receiving the three cyphertexts, the decryption algorithm indicates that

$$m_1 = c_1^d \bmod n = c_1^5 \bmod 39 = 9^5 \bmod 39 = 59049 \bmod 39 = 3$$
$$m_2 = c_1^d \bmod n = c_2^5 \bmod 39 = 1^5 \bmod 39 = 1$$
$$m_3 = c_1^d \bmod n = c_3^5 \bmod 39 = 11^5 \bmod 39 = 161051 \bmod 39 = 20$$

We then convert 3 1 20 back to cat. ■

✔ Quick Check 3.9

1. Use the values for p, q, e, and d in Example 3.33 to perform the following tasks.
   (a) Produce the cyphertexts for the plaintext dog.
   (b) Suppose the cyphertexts 25 05 28 are received and form a single plaintext message. What is that message?

THEOREM 3.50 ***RSA Encryption–Decryption***

Let $p, q, e, d, m, c \in \mathbb{Z}$ with p and q prime and let $n = pq$. If
(a) e satisfies $1 < e < (p\text{-}1)(q\text{-}1)$ and $\gcd(e, (p\text{-}1)(q\text{-}1)) = 1$
(b) $ed \equiv 1 \pmod{(p\text{-}1)(q\text{-}1)}$ with $1 < d < (p\text{-}1)(q\text{-}1)$
(c) $m < n$ and $\gcd(m, n) = 1$
(d) $c = m^e \bmod n$
then $m = c^d \bmod n$.

Note: Notice the symmetry of e and d in the theorem statement. (Theorem 3.44 on page 120) ensures that $\gcd(d, (p\text{-}1)(q\text{-}1)) = 1$.) This indicates that starting with m we can produce a cyphertext $c = m^d \bmod n$ and then decrypt as $m = c^e \bmod n$. Thus, either the private or public key can be used for the initial encryption.

Proof: The goal is to start with c and show that $c^d \bmod n = m$, or since $0 < m < n$, to show that $c^d \equiv m \pmod{n}$.

Notice that $c^d \bmod n = (m^e \bmod n)^d \bmod n \equiv m^{ed} \bmod n$. The final step used Proposition 3.23 on page 100.

By assumption, $ed \equiv 1 \pmod{(p\text{-}1)(q\text{-}1)}$, so Proposition 3.27 on page 100 implies that $ed \equiv 1 \pmod{p\text{-}1}$ and $ed \equiv 1 \pmod{q\text{-}1}$. Consequently there exist integers j and k such that

$$ed = j(p\text{-}1) + 1$$
$$ed = k(q\text{-}1) + 1$$

Therefore

$$m^{ed} = m^{j(p\text{-}1)+1} = \left(m^{p\text{-}1}\right)^j \cdot m$$
$$m^{ed} = m^{k(q\text{-}1)+1} = \left(m^{q\text{-}1}\right)^k \cdot m$$

Recall that $\gcd(m, n) = 1$ and that p and q are the only factors of n. It follows that $\gcd(m, p) = 1$ and $\gcd(m, q) = 1$. Fermat's Little Theorem implies that

$$m^{p\text{-}1} \equiv 1 \pmod{p}$$
$$m^{q\text{-}1} \equiv 1 \pmod{q}$$

Thus

$$m^{ed} = \left(m^{p\text{-}1}\right)^j \cdot m \equiv 1^j \cdot m \equiv m \pmod{p}$$
$$m^{ed} = \left(m^{q\text{-}1}\right)^k \cdot m \equiv 1^k \cdot m \equiv m \pmod{q}$$

This establishes the system of linear congruences

$$m^{ed} \equiv m \pmod{p}$$
$$m^{ed} \equiv m \pmod{q}$$

which in turn implies that $p \mid \left(m^{ed} - m\right)$ and $q \mid \left(m^{ed} - m\right)$. But p and q are distinct primes, so their product, n, must also divide $\left(m^{ed} - m\right)$.

Thus, $c^d = m^{ed} \equiv m \pmod{n}$ and the proof is complete. □

DES and AES

The United States government utilized two successors to RSA encryption: DES and AES. DES stands for Data Encryption Standard and AES is an acronym for Advanced Encryption Standard. The DES standard was adopted in 1976 and replaced by AES in 2002. Both systems are more complex than the RSA algorithm. AES is a shared-key algorithm: there is a single key that needs to be known by both sender and receiver. You can find more detail in Wikipedia or in a number of books.

3.4.7 Exercises

The exercises marked with ★ have detailed solutions in Appendix H.

1. Use the Euclidean Algorithm (from the constructive proof of Theorem 3.12 on page 115) to find the greatest common divisors of the following pairs of numbers. Then find s and t so that the gcd can be expressed in the form $as + bt$.
   (a) 24 and 148 (b) ★ 346 and 1056 (c) 63 and 178
   (d) 55 and 77 (e) 165 and 1089 (f) 1820 and 385
2. Use the hash function h from Example 3.22 to populate the table, assuming the information is entered in the following key order (using linear searching in case of collisions):
   hat, sat, rat, cat, mat, bat
3. Use the hash function h from Example 3.22 to populate the table, assuming the information is entered in the following key order (using linear searching in case of collisions). However, modify h to use a table with only eight positions.
   hat, sat, rat, cat, mat, bat
4. Use the pseudorandom number generator in Example 3.23 to generate a sequence of 5 real numbers in $[0, 1)$, using $x_0 = 10$ as the initial seed.
5. (a) The pseudorandom number generator

   $$x_{n+1} = (6x_n + 2) \bmod 16$$

   produces integers in the interval [0,15]. What is the sequence produced if the initial seed is $x_0 = 7$?

   (b) The pseudorandom number generator

   $$x_{n+1} = (6x_n + 2) \bmod 17$$

   produces integers in the interval [0,16]. What is the sequence produced if the initial seed is $x_0 = 7$?

   (c) What feature makes the second generator more useful than the first generator?
6. Design a pseudorandom number generator in the form $x_{n+1} = (ax_n + b) \bmod 100$ that has maximum period. Justify your answer.
7. Prove Proposition 3.51 without using Proposition 3.39.

   LEMMA 3.51 ***Euclid's Lemma***
   Let $a, b, c \in \mathbb{Z}$ with $a > 0$ and $\gcd(a, b) = 1$. If $a \mid (bc)$, then $a \mid c$.
8. Prove parts 1 and 3 of Theorem 3.44.
9. Prove Proposition 3.52.

   PROPOSITION 3.52
   Let $a, b, m_1, m_2, \ldots, m_n \in \mathbb{Z}$. Suppose $m_i > 1$ for each i and $m_1, m_2, \ldots, m_n$ are pairwise relatively prime. If $a \equiv b \pmod{m_i}$ for all i, then $a \equiv b \pmod{m}$, where $m = m_1 \cdot m_2 \cdots m_n$.

10. ★ Solve the following system of linear congruences:

$$x \equiv 4 \pmod{6}$$
$$x \equiv 21 \pmod{35}$$
$$x \equiv 6 \pmod{11}$$

11. Solve the following system of linear congruences:

$$x \equiv 2 \pmod{3}$$
$$x \equiv 5 \pmod{7}$$
$$x \equiv 8 \pmod{10}$$
$$x \equiv 0 \pmod{13}$$

12. Solve the following system of linear congruences. You can use the Euclidean Algorithm to find inverses, or you could use software to list $an \bmod m$ for $n = 1, 2, \ldots, m\text{-}1$ to see which value of n produces a remainder of 1. The Mathematica command `Table[{n,Mod[a n, 49]},{n,1,48}]` accomplishes this, where `a` should be replaced by the number whose inverse, mod 49 is desired.

$$x \equiv 1 \pmod{4}$$
$$x \equiv 3 \pmod{9}$$
$$x \equiv 5 \pmod{25}$$
$$x \equiv 7 \pmod{49}$$

13. A teacher in elementary school wants to place her students in groups. When she tried to form groups of size two, there was one child left over. Groups of size three would produce a final group of size two. Groups of size four left one child without a group, and groups of size five would require one group of size four. Use Theorem 3.46 to determine the most likely number of students in her class. The theorem will not apply directly, but the teacher has produced more information than is needed.

14. The following puzzle can be found in *Brahma's Correct System* by Brahmagupta, published in the early part of the 7th century AD [2].

An old woman goes to market and a horse steps on her basket and crushes the eggs. The rider offers to pay for the damages and asks her how many eggs she had brought. She does not remember the exact number, but when she had taken them out two at a time, there was one egg left. The same happened when she picked them out three, four, five, and six at a time, but when she took them seven at a time they came out even. What is the smallest number of eggs she could have carried?

15. Which of these linear congruences does Corollary 3.45 guarantee to have a solution?

(a) $6x \equiv 5 \pmod{48}$

(b) $5x \equiv 6 \pmod{48}$

(c) $13x \equiv 260 \pmod{1001}$

(d) $42x \equiv 0 \pmod{55}$

16. Use Theorem 3.47 to find the smallest positive solution to the following linear congruences.

(a) ★ $5x \equiv 24 \pmod{48}$

(b) $25x \equiv 304 \pmod{308}$

(c) $31x \equiv 394 \pmod{1001}$

(d) $77x \equiv 1064 \pmod{1080}$

17. Prove that the set $A = \{a, 2a, 3a, \ldots, (p-1)a\}$ from the proof of Fermat's Little Theorem has $p-1$ distinct elements.

18. Let p be a prime. Prove that $2^{p-1} - 1$ is also prime if and only if $p = 3$.

19. Prove the following assertion:
Let $a, p \in \mathbb{Z}$ with p a prime. If $\gcd(a, p) = 1$, then a^{p-2} is an inverse for a, mod p.

20. Use Fermat's Little Theorem, Exercise 19 and the Chinese Remainder theorem to prove that if p and q are distinct primes, then $p^{q-1} + q^{p-1} \equiv 1 \pmod{pq}$.

21. Show that the converse of Fermat's Little Theorem is not true in general. That is, find integers a and b such that $\gcd(a, b) = 1$ and $a^{b-1} \equiv 1 \pmod{b}$, but b is not a prime. [*Hint*: keep b composite but small and start checking potential values for a.]

22. Use the values for p, q, e, and d in Example 3.33 to perform the following tasks.

(a) Produce the cyphertexts for the plaintext quid.

(b) Produce the cyphertexts for the plaintext molt.

(c) Suppose the cyphertexts 32 06 12 10 are received and form a single plaintext message. What is that message?

(d) Suppose the cyphertexts 37 18 03 22 are received and form a single plaintext message. What is that message?

23. Suppose I have established an RSA system for communicating with friends and associates. I have published the public key $n = 91, d = 5$. Assume that I will use Table 3.11 on page 129 to convert from letters to numbers.

(a) ★ How would you encrypt the message work using my public key?

(b) How would you encrypt the message music using my public key?

(c) ★ How would you decrypt the cyphertext sequence 31 33 01 13?

(d) How would you decrypt the cyphertext sequence 75 21 81 52?

(e) What is my private key?

24. Leela has published the public RSA key $n = 391, d = 5$. Assume that she uses Table 3.11 to convert from letters to numbers. (You will need software or a high-end calculator.)

(a) How would you encrypt the message tickle using her public key?

(b) How would you decrypt the cyphertext sequence 299 359 90 174 241 35?

(c) What is her private key?

25. Your enemy has published a public RSA key with $n = 943$ and $d = 403$. What is your enemy's private key?

3.5 Mathematical Induction

Let $P(i)$ be some statement that depends on the integer i. A theorem of the form "$\forall i \in \mathbb{N}$ with $i \geq i_0$, $P(i)$" is usually proved using mathematical induction. A familiar example is the formula

$$P(n): 1+2+3+4+\cdots+n = \sum_{k=1}^{n} k = \frac{n(n+1)}{2} \quad \text{for } n \geq 1$$

Mathematical induction is one of the most commonly used proof techniques in discrete mathematics. As will be seen, an amazing variety of claims can be validated using this technique.

3.5.1 The Principle of Mathematical Induction

An inductive proof is done in two steps. First, we show that the statement $P(1)$ is true. This is often called the *base step*. Then, we prove the implication $[P(i) \rightarrow P(i+1)]$ [read this as "if $P(i)$ is true, then $P(i+1)$ is also true"]. This is called the *inductive step*. These two conditions are enough to conclude that $P(k)$ is true for all positive integers k. The reason is related to making domino chains. The inductive step verifies that each domino will knock the next one down if *it* falls. The base step knocks the first domino over. The inevitable result is that all the dominos fall. This should remind you of *modus ponens* (Chapter 2): $[P \wedge (P \rightarrow Q)] \rightarrow Q$.

THEOREM 3.53 ***Mathematical Induction***

If $\{P(i)\}$ is a set of statements such that

1. $P(1)$ is true and
2. $P(i) \rightarrow P(i+1)$ for $i \geq 1$

then $P(k)$ is true for all positive integers k. This can be stated more succinctly as

$$[P(1) \wedge (\forall i, P(i) \rightarrow P(i+1))] \rightarrow [\forall k, P(k)]$$

Proof: This theorem follows directly from the well-ordering principle. Assume that we have verified that $P(1)$ is true and that for all $i \geq 1$, the truth of $P(i)$ implies that $P(i+1)$ is also true. Let S be the set of all integers $n \geq 1$ such that $P(n)$ is true, and let T be the set of all integers $k > 1$ such that $P(k)$ is false.

Suppose that T is not empty. Then by the well-ordering principle, T must have a smallest element, k_0. Since $k_0 > 1$, we know that $k_0 - 1 \in S$. This means that $P(k_0 - 1)$ is true and since $k_0 - 1 \geq 1$, we also know that $P(k_0 - 1) \rightarrow P(k_0)$. Using *modus ponens*,[44] we conclude that $P(k_0)$ is true, contradicting the assumption that $k_0 \in T$. The trouble started with the assumption that T is not empty. Therefore, $T = \emptyset$ and $S = \{n \in \mathbb{N} \mid n \geq 1\}$; $P(n)$ is true for all positive integers n. □

It should be clear that in step one of Theorem 3.53 there is nothing special about the integer 1. We could have started by showing that $P(0)$ is true and ended by concluding that $P(k)$ is true for all nonnegative integers k. Sometimes we may have a set of statements that is true for $k > 2$. We would then start our inductive proof by using $P(3)$ in step 1. The principle remains the same.

You may recall from a previous course that step 2 is the messy step. We must keep the form of $P(i+1)$ in mind as we transform $P(i)$. In essence, step 2 is a proof. While doing the proof in step 2, you must not lose sight of the major strategy embodied in the theorem of mathematical induction.

[44] $[P(k_0 - 1) \wedge (P(k_0 - 1) \rightarrow P(k_0))] \rightarrow P(k_0)$

The version of mathematical induction described in Theorem 3.53 is also called *finite induction* or *weak induction*. The primary alternative version (complete induction) will be introduced in Section 3.5.2.

Mathematical induction is illustrated in Example 3.34.

EXAMPLE 3.34

An Informal Example

We can use mathematical induction to show that $2+4+6+\cdots+2(n-1)+2n = n(n+1)$ for all positive integers n.

Proof: As an aid to discussion, let $P(i)$ be the statement $2+\cdots+2i = i(i+1)$. When $i = 1$, the statement $P(1)$ becomes $2 = 1(1+1)$, which is obviously true.

Assume that $P(i)$ is true. We must prove the implication $[P(i) \rightarrow P(i+1)]$, that is, $[2+\cdots+2i = i(i+1)] \rightarrow [2+\cdots+2(i+1) = (i+1)(i+2)]$. We may add $2(i+1)$ to both sides of $P(i)$ to obtain

$$2+\cdots+2i+2(i+1) = i(i+1)+2(i+1)$$

The left-hand side of this equation is the same as the left-hand side of $P(i+1)$. If we can show that $i(i+1)+2(i+1) = (i+1)(i+2)$, then the truth of $P(i+1)$ will follow. Simple algebra suffices.

We have shown that $P(1)$ is true and that $P(i) \rightarrow P(i+1)$. We conclude, using mathematical induction, that $P(n)$ is true for all positive integers n. □ ■

While working on the inductive step, you may often encounter an algebraic expression that you hope to show equals another expression. For example, you may wish to show that $(n+1)^2 + \frac{(n+1)(n+2)}{2} = \frac{3n^2+7n+4}{2}$. A common (but unacceptable) approach is to work in a two-column format, transforming each side independently until you reach a line where the two sides are equal. The example just presented might be manipulated in this manner:

Do not do this!

$$\begin{aligned}
(n+1)^2 + \frac{(n+1)(n+2)}{2} &\stackrel{?}{=} \frac{3n^2+7n+4}{2} \\
\frac{2(n+1)^2 + (n+1)(n+2)}{2} &\stackrel{?}{=} \frac{(n+1)(3n+4)}{2} \\
\frac{(n+1)[2(n+1)+(n+2)]}{2} &\stackrel{?}{=} \frac{(n+1)(3n+4)}{2} \\
\frac{(n+1)(3n+4)}{2} &\stackrel{\surd}{=} \frac{(n+1)(3n+4)}{2}
\end{aligned}$$

In essence, we are manipulating both sides of an equation that we are not yet sure is valid. At the bottom we find some statement that we know is true and then wave our hands and say that we could reverse the sequence and have a valid argument. Many students forget to place the question marks above the equal signs, making a claim of equality that cannot be supported at that point in the presentation. It would be better to present the valid sequence. An easy way to achieve this goal is to write the sequence of steps with the question marks over the "=" signs on scratch paper. For the formal presentation you wish to hand in, start at the top left, move down to the bottom, then start back up the right-hand side. The manipulations for the example can be written as

Do this:

$$\begin{aligned}(n+1)^2 + \frac{(n+1)(n+2)}{2} &= \frac{2(n+1)^2 + (n+1)(n+2)}{2} \\ &= \frac{(n+1)[2(n+1) + (n+2)]}{2} \\ &= \frac{(n+1)(3n+4)}{2} \\ &= \frac{3n^2 + 7n + 4}{2}\end{aligned}$$

This approach is illustrated in the proof of Theorem 3.54. Notice the linear style: Start with one side of the claim $P(i+1)$ and work through a sequence of known equalities that terminates with the other side of $P(i+1)$. Notice also the use of the colon. It is incorrect to write $P(i) = 1 + 2 + \cdots + i = \frac{i(i+1)}{2}$. In this example, $P(i)$ is the entire equality, not just the left or the right side. The colon indicates that $P(i)$ *is* the equation that follows.

THEOREM 3.54 ***Sum of the First n Positive Integers***

$$\sum_{j=1}^{n} j = \frac{n(n+1)}{2} \quad \text{for all positive integers, } n\text{, with } n \geq 1$$

Proof: Let $P(i)$ be the claim[45] $\sum_{j=1}^{i} j = \frac{i(i+1)}{2}$.

Base Step

We need to show $P(1)$: $\sum_{j=1}^{1} j = \frac{1(1+1)}{2}$. But clearly, $\sum_{j=1}^{1} j = 1 = \frac{1(1+1)}{2}$.

Inductive Step

We will *assume* that $P(i)$ is true and show that $P(i+1)$ must also be true under this assumption. [The assumption that $P(i)$ is true is called the *inductive hypothesis*.] Because we are assuming that $P(i)$ is true, we can use the equation $\sum_{j=1}^{i} j = \frac{i(i+1)}{2}$ later in the inductive step. [Remember: We only need to show "*if* $P(i)$ is true, then $P(i+1)$ is true".]

It will be helpful to write down what we want to show (so that we will recognize it when it appears later). However, we must be careful not to try and use this equation, because we do not yet know that it is true.

$$P(i+1): \sum_{j=1}^{i+1} j = \frac{(i+1)(i+2)}{2}$$

The sequence of equalities that follows starts with the left-hand side of $P(i+1)$ and ends with the right-hand side. At each step, the equalities are valid. Since the assumed truth of $P(i)$ is used in the sequence, $P(i+1)$ must be true *if* $P(i)$ *is true*. That is all we need to do to establish the inductive step.

$$\sum_{j=1}^{i+1} j = \left(\sum_{j=1}^{i} j\right) + (i+1) \quad \text{isolate the final number in the sum}$$

$$= \frac{i(i+1)}{2} + (i+1) \quad \text{by the inductive hypothesis}$$

$$= \frac{i(i+1)}{2} + \frac{2(i+1)}{2} = \frac{i(i+1) + 2(i+1)}{2} = \frac{(i+1)(i+2)}{2}$$

[45] See Appendix B for a review of summation notation.

Conclusion
We now know that $P(1)$ is true and that $P(i+1)$ is true whenever $P(i)$ is true. The mathematical induction theorem implies that $P(n)$ is true for all $n \geq 1$. □

EXAMPLE 3.35

A Formal Example

This example will follow the preferred method of presentation shown in the proof of Theorem 3.54. However, most of the pedagogical comments will be omitted. This example should show the general pattern to follow when doing an induction proof. As a student handing a proof in for grading, you should be fairly complete and detailed (as opposed to a researcher or writer of an upper-division or graduate level textbook).

Show that $2^n < n!$ for all integers, n, with $n \geq 4$.

Proof: Let $P(i)$ be the inequality when $n = i$. $P(i) : 2^i < i!$.

Base Step
Since $2^4 = 16 < 24 = 4!$, $P(4)$ is true.

Inductive Step
Assume that $P(i)$ is true for some $i \geq 4$. We want to show that $P(i+1) : 2^{i+1} < (i+1)!$ must then also be true.

$$\begin{aligned} 2^{i+1} &= 2 \cdot 2^i \\ &< 2 \cdot i! && \text{by the inductive hypothesis} \\ &< (i+1) \cdot i! && \text{because } i \geq 4 \\ &= (i+1)! \end{aligned}$$

Conclusion
Since $P(4)$ is true and $P(i) \to P(i+1)$ for $i \geq 4$, mathematical induction ensures that $P(n)$ is true for all integers n, with $n \geq 4$. □

✔ Quick Check 3.10

1. Let $P(i) : 1+3+\cdots+(2i-1) = i^2$. Then $P(i+1)$ is the equality
   (a) $1+3+\cdots+(2i-1) = (i+1)^2$.
   (b) $1+3+\cdots+(2i+1) = i^2$.
   (c) $1+3+\cdots+(2i+1) = (i+1)^2$.
   (d) $1+3+\cdots+(2i-1) = i^2$.
2. Use mathematical induction to show that the sum of the first n odd integers is n^2:
   $$1+3+\cdots+(2n-1) = \sum_{i=1}^{n}(2i-1) = n^2$$
   for all integers, n, with $n \geq 1$.
3. You are familiar with the distributive property of the real numbers: $a(b+c) = ab+ac$.[46] Use mathematical induction to prove the *generalized distributive property*:
   $$a\left(\sum_{i=1}^{n} b_i\right) = \sum_{i=1}^{n} ab_i$$
   for $a, b_i \in \mathbb{R}$, and $n \in \mathbb{N}$, $n \geq 2$.

✔

Note: The key step in mathematical induction is the proof that the implication $P(i) \to P(i+1)$ is true. You should think of this implication in fairly generic terms: If P is true at some integer, then it is also true at the next integer. Thus, we can write the

[46] See Appendix A.3 for more details.

implication as $P(k) \rightarrow P(k+1)$ or $P(n) \rightarrow P(n+1)$ or even as $P(i-1) \rightarrow P(i)$. All represent the same implication. To complete the hypotheses of the theorem of mathematical induction, we need a starting value, i_0, where we know $P(i_0)$ is true (most commonly, $i_0 = 1$).

Recall that the implication $P(i) \rightarrow P(i+1)$ may be true, but $P(i)$ may be false. If we cannot find an integer i_0 for which $P(i_0)$ is true, then the conclusion that $P(n)$ is true for all $n \geq i_0$ would be false. Thus, the base step is essential (even though it is often trivial to verify).

EXAMPLE 3.36

Failed Inductions

Suppose we want to show $\forall a \in \mathbb{R}, \forall n \in \mathbb{N}, [((a > 1) \wedge (n \geq 0)) \rightarrow (a^n = 1)]$. This claim can be informally stated as "show $a^n = 1$ for $a > 1$ and $n \geq 0$". We can choose a to be any real number with $a > 1$ and attempt a proof by mathematical induction on n for the assertion: $\forall n \in \mathbb{N}, [(n \geq 0) \rightarrow (a^n = 1)]$. If no special properties (other than $a > 1$) are used, this will establish the original assertion.

We can easily verify the base step ($n = 0$) for this claim since $a^0 = 1$ is true for any $a \in \mathbb{R}$ with $a > 1$ (in fact, any $a \in \mathbb{R}$ with $a \neq 0$).

What we *cannot* verify in this case is the implication $P(0) \rightarrow P(1)$. For example, there is no valid way to use the assertion $2^0 = 1$ to conclude that $2^1 = 1$.

It is also possible to examine a situation where the inductive step *can* be completed for all n, but for which the base step fails. To that end, consider the claim that for all positive integers, n, $\sum_{i=1}^{n} 2i = (n-1)(n+2)$.

Suppose the sum of the first n positive integers is $(n-1)(n+2)$. Then

$$\sum_{i=1}^{n+1} 2i = 2(n+1) + \sum_{i=1}^{n} 2i = 2(n+1) + (n-1)(n+2) = n(n+3)$$

Therefore, if the claim properly calculates the sum of the first n even positive integers, then it also properly calculates the sum of the first $n+1$ positive even integers.

Notice, however, that the claim is *wrong* for every positive integer:

$$\sum_{i=1}^{n} 2i = 2\sum_{i=1}^{n} i = 2 \cdot \frac{n(n+1)}{2} = n(n+1)$$

and $(n-1)(n+2) = n(n+1)$ implies $-2 = 0$. ■

You may be wondering why we should go to all this trouble to prove formally that a mathematical pattern is true for all n greater than some initial value. Would it not be sufficient just to verify that the pattern holds for 5 or 6 values (perhaps even 9 or 10 if you want to be really cautious)? The following example vividly illustrates why this procedure is inadequate.

EXAMPLE 3.37

Euler's Pretty-Good Prime Function

It has been a goal of mathematicians for a very long time to produce an algorithm or a function that produces all prime numbers. A more modest version of this problem is to produce an algorithm or a function that produces an infinite number of primes. It may skip some primes, but it should never produce a value that *is not* prime. Here is a very simple function discovered by Leonhard Euler.

$$ep(n) = n^2 - n + 41$$

The following table shows the first 42 values of the function.

This marvelous function actually does produce a prime for the first 41 values ($n = 0 \ldots 40$). However, $ep(41) = 1681 = 41^2$, so the pattern fails.

n	0	1	2	3	4	5	6	7	8	9	10	11	12	13	14	15
$ep(n)$	41	41	43	47	53	61	71	83	97	113	131	151	173	197	223	251

n	16	17	18	19	20	21	22	23	24	25	26	27	28	29
$ep(n)$	281	313	347	383	421	461	503	547	593	641	691	743	797	853

n	30	31	32	33	34	35	36	37	38	39	40	41
$ep(n)$	911	971	1033	1097	1163	1231	1301	1373	1447	1523	1601	1681

Even a conservative "pattern tester" would probably give up checking the pattern before arriving at the problematic 42nd value. Even if you verify a mathematical pattern for a million different cases, there is still an infinite number of cases left to check. A proof using mathematical induction provides certainty that the pattern always holds, and it can be done in a finite amount of time. ■

Where do the formulas that are typically proved by induction come from? They are usually first found by someone looking for a pattern. Once a sufficient number of examples is available, it is possible to make an educated guess at the correct pattern. Once such a guess has been made, it can be compared to a few new examples. If it continues to hold, it is worth attempting a proof by mathematical induction. If the proof succeeds, the guess has been validated. If not, there are two explanations for the failure. One possibility is that the guess was incorrect. Additional examples may produce a case for which the pattern fails (see Example 3.37). The other possibility is that there is a mistake in the attempted proof.

EXAMPLE 3.38

Guessing a Pattern

I wanted to find a nice formula for the partial sums

$$S_n = 1 - \frac{1}{2} + \frac{1}{2^2} - \frac{1}{2^3} + \cdots \pm \frac{1}{2^n} = \sum_{k=0}^{n} \left(\frac{-1}{2}\right)^k$$

I started by calculating S_n for a few small values of n.

n	1	2	3	4	5
S_n	$\frac{1}{2}$	$\frac{3}{4}$	$\frac{5}{8}$	$\frac{11}{16}$	$\frac{21}{32}$

What patterns did I observe? A bit of thought made it clear (and, after the fact, pretty obvious[47]) that the denominator of S_n is 2^n.

What pattern emerged in the numerators? The sequence of numerators (so far) was 1 3 5 11 21. One thing I noticed (after a while) was that starting with the 5, each new entry equals the previous entry plus twice the entry before that.[48] However, it was not clear[49] how to make use of that information, so I looked for another pattern.

I noticed next that the nth numerator (starting with $n = 2$) seemed to follow the pattern: numerator $= n(n-1) \pm 1$. However, extending the table one more entry showed that this pattern fails.

n	1	2	3	4	5	6
S_n	$\frac{1}{2}$	$\frac{3}{4}$	$\frac{5}{8}$	$\frac{11}{16}$	$\frac{21}{32}$	$\frac{43}{64}$

[47] Just put all the fractions in the sum over a common denominator.

[48] I arrived at this by looking at the differences between successive numerators: $21 - 11 = 10 = 2 \cdot 5$.

[49] It will be clear after looking at Chapter 7.

The pattern seemed elusive. Instinct caused me to look for patterns that might involve powers of 2. However, the numbers do not appear to be related to powers of 2. After a while, I decided to multiply all the numerators by 3: 3 9 15 33 63 129. This was interesting. Compare these values with the powers of 2.

n	1	2	3	4	5	6
$3 \cdot$ numerator	3	9	15	33	63	129
2^{n+1}	4	8	16	32	64	128

A good guess seemed to be that the nth modified numerator is equal to $2^{n+1}+(-1)^n$ (subtract 1 for odd n and add 1 for even n). The real nth numerator would then be $\frac{2}{3}2^n + \frac{1}{3}(-1)^n$. Since the denominator appeared to be 2^n, the value of S_n (if this were the correct pattern) would simplify to

$$S_n = \frac{1}{2^n}\left(\frac{2}{3} \cdot 2^n + \frac{1}{3} \cdot (-1)^n\right) = \frac{2}{3} + \frac{1}{3} \cdot \left(\frac{-1}{2}\right)^n$$

It was time to do one last experiment before proceeding to a proof: See what happens for $n = 0, 7, 8$.

n	0	1	2	3	4	5	6	7	8
S_n	1	$\frac{1}{2}$	$\frac{3}{4}$	$\frac{5}{8}$	$\frac{11}{16}$	$\frac{21}{32}$	$\frac{43}{64}$	$\frac{85}{128}$	$\frac{171}{256}$
$\frac{2}{3} + \frac{1}{3} \cdot \left(\frac{-1}{2}\right)^n$	1	$\frac{1}{2}$	$\frac{3}{4}$	$\frac{5}{8}$	$\frac{11}{16}$	$\frac{21}{32}$	$\frac{43}{64}$	$\frac{85}{128}$	$\frac{171}{256}$

The result indicated that it was time for an attempt at a formal proof using mathematical induction. That is, I predicted

$$\sum_{k=0}^{n} \left(\frac{-1}{2}\right)^k = \frac{2}{3} + \frac{1}{3} \cdot \left(\frac{-1}{2}\right)^n \quad \text{for } n \geq 0$$

The formal proof provides an opportunity for you to practice mathematical induction (see Exercise 3 on page 152). ■

The next example demonstrates mathematical induction in conjunction with products. Some (possibly) new notation needs to be introduced. Recall that

$$\sum_{i=0}^{n} a_i \quad \text{is shorthand for the sum} \quad a_0 + a_1 + \cdots + a_n$$

In a similar fashion, the expression

$$\prod_{i=0}^{n} a_i \quad \text{is used to represent the product} \quad a_0 \cdot a_1 \cdots a_n$$

EXAMPLE 3.39 **Induction with a Product**

Let $n \in \mathbb{N}$. Mathematical induction can be used to show that

$$\prod_{k=1}^{n} \left(1 - \frac{1}{2^k}\right) \geq \frac{1}{4} + \frac{1}{2^{n+1}} \quad \forall n \geq 1$$

Proof: Let $P(n)$ be the claim $\prod_{k=1}^{n}\left(1-\frac{1}{2^k}\right) \geq \frac{1}{4}+\frac{1}{2^{n+1}}$.

Base Step n = 1
$P(1)$ is true because

$$\prod_{k=1}^{1}\left(1-\frac{1}{2^k}\right)=\frac{1}{2}=\frac{1}{4}+\frac{1}{2^2}$$

Inductive Step Suppose

$$\prod_{k=1}^{n}\left(1-\frac{1}{2^k}\right)\geq\frac{1}{4}+\frac{1}{2^{n+1}}$$

is true for some $n \geq 1$. Then

$$\begin{aligned}
\prod_{k=1}^{n+1}\left(1-\frac{1}{2^k}\right) &= \left(1-\frac{1}{2^{n+1}}\right)\cdot\prod_{k=1}^{n}\left(1-\frac{1}{2^k}\right)\\
&\geq \left(1-\frac{1}{2^{n+1}}\right)\cdot\left(\frac{1}{4}+\frac{1}{2^{n+1}}\right) \quad \text{by the inductive hypothesis}\\
&= \frac{1}{4}+\frac{1}{2^{n+1}}-\frac{1}{2^{n+3}}-\frac{1}{2^{2n+2}}\\
&= \frac{1}{4}+\frac{4}{2^{n+3}}-\frac{1}{2^{n+3}}-\frac{1}{2^{2n+2}}\\
&= \frac{1}{4}+\frac{2}{2^{n+3}}+\left(\frac{1}{2^{n+3}}-\frac{1}{2^{2n+2}}\right)\\
&= \frac{1}{4}+\frac{1}{2^{n+2}}+\frac{1}{2^{n+3}}\cdot\left(1-\frac{1}{2^{n-1}}\right)\\
&\geq \frac{1}{4}+\frac{1}{2^{n+2}}
\end{aligned}$$

Hence, $P(n+1)$ is also true.

Conclusion
Since $P(1)$ is true, and whenever $P(n)$ is true, $P(n+1)$ is also true, the theorem of mathematical induction implies that $P(n)$ is true for all $n \geq 1$. □ ■

3.5.2 Complete Induction

Sometimes the inductive step is very hard to prove. Knowing that $P(i)$ is true may not be sufficient to show easily that $P(i+1)$ is true. There is a second (but equivalent) form of mathematical induction that may work.

THEOREM 3.55 ***Complete Induction***

If $\{P(i)\}$ is a set of statements such that

1. $P(1)$ is true and
2. $[P(1) \wedge P(2) \wedge \cdots \wedge P(i)] \to P(i+1)$ for $i \geq 1$

then $P(k)$ is true for all positive integers k. This can be stated more succinctly as

$$[P(1) \wedge (\forall i, [P(1) \wedge P(2) \wedge \cdots \wedge P(i)] \to P(i+1))] \to [\forall k, P(k)]$$

Proof: This theorem also follows directly from the well-ordering principle. The proof used for Theorem 3.53 requires a few minor modifications.

Assume that we have verified that $P(1)$ is true and that for all $i \geq 1$, the truth of all of $\{P(1), P(2), \ldots, P(i)\}$ implies that $P(i+1)$ is also true. Let S be the set of all integers $n \geq 1$ such that $P(n)$ is true, and let T be the set of all integers $k > 1$ such that $P(k)$ is false.

Suppose that T is not empty. Then by the well-ordering principle, T must have a smallest element, k_0. Since $k_0 > 1$, we know that $k_0 - 1 \in S$. This means that $P(1), P(2), \ldots, P(k_0 - 1)$ are all true (k_0 is the smallest integer that is not in S). Since $k_0 - 1 \geq 1$, we also know that $P(1) \wedge P(2) \wedge \cdots \wedge P(k_0 - 1) \rightarrow P(k_0)$. Using *modus ponens*,[50] we conclude that $P(k_0)$ is true, contradicting the assumption that $k_0 \in T$. The trouble started with the assumption that T is not empty. Therefore, $T = \emptyset$ and $S = \{n \in \mathbb{N} \mid n \geq 1\}$; $P(n)$ is true for all positive integers n. □

This should be even easier to use than the theorem of mathematical induction, because we are assuming more before attempting to reach the same conclusion. Complete induction is also called *strong induction*, because it uses a stronger hypothesis.

EXAMPLE 3.40

The Fundamental Theorem of Arithmetic: Part 1 of Proof

The fundamental theorem of arithmetic states that "every integer n, with $n \geq 2$, can be written uniquely as a product of primes in ascending order." Using complete induction, it is not hard to prove the first half:

Every integer n, with $n \geq 2$, can be written as a product of primes.

(The word *uniquely* has been dropped. That part of the theorem was proved on page 111.)

Proof: Let $P(k)$ be the claim that k can be written as a product of primes.

Base Step

The claim is clearly true for $k = 2$ since 2 is itself a prime (the smallest prime) and is therefore a product of one prime. Thus $P(2)$ is true.

Inductive Step

Assume that all integers k for $k = 2, 3, \ldots, n$ can be written as a product of primes. That is, $P(2) \wedge P(3) \wedge \cdots \wedge P(n)$ is true. Consider the integer $n + 1$. Either $n + 1$ is a prime (in which case, $P(n + 1)$ is true and the inductive step has been completed), or else $n + 1$ is composite.

If $n + 1$ is composite, then $n + 1 = a \cdot b$, where $2 \leq a \leq n$ and $2 \leq b \leq n$. By the inductive hypothesis, we can write both a and b as products of primes. Suppose the products are $a = p_1 p_2 \cdots p_s$ and $b = q_1 q_2 \cdots q_t$. Then $n+1 = p_1 p_2 \cdots p_s \cdot q_1 q_2 \cdots q_t$ is also a product of primes.

Conclusion

It has been shown that $P(2)$ is true and also that whenever $P(2) \wedge P(3) \wedge \cdots \wedge P(n)$ is true, then $P(n + 1)$ is also true. The theorem of complete induction implies that $P(k)$ is true for all $k \geq 2$. □

You should spend a few minutes thinking about how you would prove this using only the theorem of mathematical induction. That is, assuming only that n is a product of primes, how would you show that $n + 1$ is also a product of primes. It is not easy. ■

The next example illustrates the need for the base case in an induction proof.

EXAMPLE 3.41

A Failed Induction, Revisited

Suppose we wish to prove that $a^n = 1$ for some $a > 1$ and for all $n \geq 1$ (rather than for $n \geq 0$, as in Example 3.36). This can be formally written as $\exists a \in \mathbb{R}, \forall n \in \mathbb{N}, [((a > 1) \wedge (n \geq 1)) \rightarrow (a^n = 1)]$. Let $P(i) : a^i = 1$ for $a > 1$. Ignoring the base case for the moment, notice that

[50] $[A \wedge (A \rightarrow B)] \rightarrow B$, where $A = P(1) \wedge P(2) \wedge \cdots \wedge P(k_0 - 1)$ and $B = P(k_0)$.

$P(1) \to P(2)$ is true

Proof: Assuming that $P(1)$ is true, $a^2 = a \cdot a = 1 \cdot 1 = 1$, so $P(2)$ is also true.

$(P(1) \wedge P(2) \wedge \cdots \wedge P(i)) \to P(i+1)$ for $i > 1$ is true.

Proof: Assuming that $P(k)$ is true for $k = 1, 2, 3, \ldots, i$

$$a^{i+1} = a^1 \cdot a^i = 1 \cdot 1, \text{ so } P(i+1) \text{ is true}$$

The inductive step has been completed without any problems. (In fact, it was not necessary to break it into two cases.) However, we know that the claim is false! We have failed to show that the base step can be completed. It cannot. The claim *is* false. ∎

✔ **Quick Check 3.11**

1. This is not an Earth-shaking problem, but it is a good first exercise in complete induction.
   Jedediah and Ebenezum have just cooked a large bowl of popcorn. They are sitting in front of their television watching the test pattern. They alternate taking some popcorn out of the bowl. They always take at least one piece of popcorn, but might take as much as a large handful. Prove that eventually the bowl will be empty.

2. Use complete induction to prove that every positive integer can be expressed as a sum of distinct powers of 2. For example, $7 = 2^2 + 2^1 + 2^0$ and $11 = 2^3 + 2^1 + 2^0$. Notice that $4 = 2^1 + 2^1$ is not a sum of distinct powers of 2. ☑

3.5.3 Interesting Mathematical Induction Problems

Mathematical induction is very useful. The following examples amply demonstrate the versatility of this technique. They also demonstrate that the inductive step does not follow any set pattern of algebraic manipulations. (In fact, the chess board and stable marriage examples do not involve *any* algebra!)

Geometric and Arithmetic Progressions

> **DEFINITION 3.56** ***Geometric Progression***
> A sequence is called a *geometric progression* if each term in the sequence (after the first) is a constant multiple of the previous term. Thus, if the terms are $\{a_i\}$ for $i = 0, 1, 2, 3, \ldots$, then $a_{i+1} = ra_i$ for some constant, r.

Notice that we choose the values of a_0 and r. All other elements of the sequence are then unambiguously determined: $a_i = a_0 r^i$. An alternative name for a geometric progression is *geometric sequence*.

> **DEFINITION 3.57** ***Arithmetic Progression***
> A sequence is called an *arithmetic progression* if each term in the sequence (after the first) is obtained by adding a constant to the previous term. Thus, if the terms are $\{a_i\}$ for $i = 0, 1, 2, 3, \ldots$, then $a_{i+1} = a_i + d$ for some constant, d.

Once we choose the values for a_0 and d, all other elements of the sequence are then unambiguously determined: $a_i = a_0 + i \cdot d$. An alternative name for an arithmetic progression is *arithmetic sequence*.

The sum of a geometric progression and the sum of an arithmetic progression are worth knowing. Alternative names for these sums are *geometric series* and *arithmetic series*, respectively. The next theorem is important; you should memorize it.

THEOREM 3.58 ***Partial Sum of a Geometric Progression***

The sum of the first $n+1$ elements of a geometric progression depends upon the value of $r \in \mathbb{R}$.

$$\sum_{i=0}^{n} r^i = \frac{1-r^{n+1}}{1-r} = \frac{r^{n+1}-1}{r-1} \qquad \text{if } r \neq 1$$

$$\sum_{i=0}^{n} 1 = (n+1) \qquad \text{if } r = 1$$

Note: The first formula implicitly assumes that $0^0 = 1$. That is, when $r = 0$, we want $\sum_{i=0}^{n} 0^i = 1 + 0^1 + 0^2 + \cdots + 0^n = 1$.

A problem arises because 0^0 is generally regarded as indeterminate. That is, in different contexts, different values for the expression can be derived. Since $\lim_{x \to 0} x^x = 1$, the assumption $0^0 = 1$ is not without merit. See [106] for additional discussion.

The entire issue could be circumvented by writing the $r \neq 1$ equation as

$$1 + \sum_{i=1}^{n} r^i = \frac{1-r^{n+1}}{1-r} = \frac{r^{n+1}-1}{r-1} \qquad \text{if } r \neq 1$$

The version in the theorem has the weight of tradition behind it.

DEFINITION 3.59 0^0

In the contexts encountered in this book, $0^0 = 1$.

Proof of Theorem 3.58: The second case ($r = 1$) needs no further explanation at this point of your mathematics education.

The case $r \neq 1$ can be proved by mathematical induction. Let $P(n)$ be the equation for the sum of the first $n+1$ terms.

$$P(n)\colon \ \sum_{i=0}^{n} r^i = \frac{1-r^{n+1}}{1-r}$$

Base Step

When $n = 0$, $\sum_{i=0}^{0} r^i = r^0 = 1$ and $\frac{1-r^{0+1}}{1-r} = 1$ since $r \neq 1$.

Inductive Step

Assume that $P(n)$ is true. We want to show that

$$P(n+1)\colon \ \sum_{i=0}^{n+1} r^i = \frac{1-r^{n+2}}{1-r}$$

is also true. This should be familiar by now:

$$\begin{aligned}\sum_{i=0}^{n+1} r^i &= \left(\sum_{i=0}^{n} r^i\right) + r^{n+1} \\ &= \left(\frac{1-r^{n+1}}{1-r}\right) + r^{n+1} \quad \text{by the inductive hypothesis} \\ &= \left(\frac{(1-r^{n+1}) + (1-r)r^{n+1}}{1-r}\right) \\ &= \frac{1-r^{n+2}}{1-r}\end{aligned}$$

Conclusion
Both the base step and the inductive step are valid. Therefore, $P(n)$ is true for all $n \geq 0$. □

Notice that the theorem could also be stated as

$$\sum_{i=0}^{n-1} r^i = \frac{1-r^n}{1-r}$$

COROLLARY 3.60 ***Sum of a Geometric Progression***
If $r \in \mathbb{R}$ and $|r| < 1$, then

$$\sum_{i=0}^{\infty} r^i = \frac{1}{1-r}$$

Proof: Since $|r| < 1$, the partial sum $S_n = \sum_{i=0}^{n} r^i$ can be found using Theorem 3.58. Then take the limit as $n \to \infty$:

$$\lim_{n\to\infty} S_n = \lim_{n\to\infty} \frac{1-r^{n+1}}{1-r} = \frac{1}{1-r}$$

because $\lim_{n\to\infty} r^{n+1} = 0$ when $|r| < 1$. □

The theorem for the partial sum of an arithmetic progression will be left as Exercise 4 on page 152.

Chess Boards

A standard chess board has 64 squares, arranged in an 8-by-8 grid. Since $8 = 2^3$, one possible generalization of a chess board is to have a 2^n-by-2^n grid. If we look only at such generalized chess boards, we can prove the following result.

> Let $n \geq 1$. Suppose we have a 2^n-by-2^n chess board, with one square missing, and a box full of L-shaped tiles. Each tile can cover 3 squares on the chess board. No matter which square on the chess board is missing, we can entirely cover the remaining squares with the tiles.

Figure 3.6. An L-shaped tile.

Proof: The tiles look like the diagram in Figure 3.6.

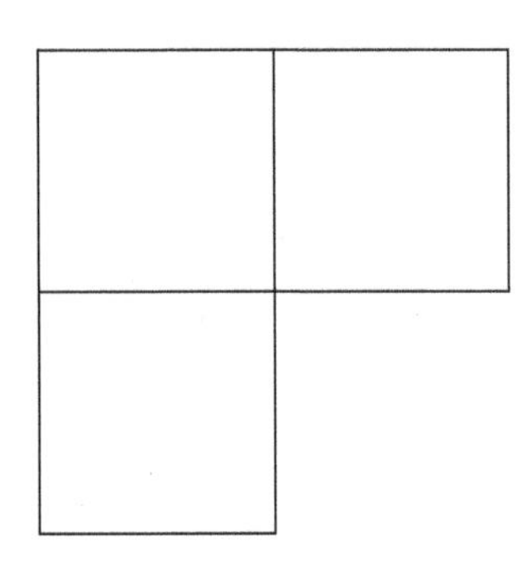

Figure 3.7. The 2^1-by-2^1 chess board.

Base Step
When $n = 1$, the chess board looks like Figure 3.7. No matter which square is removed, we can clearly orient the tile so that the remaining squares are covered.

Inductive Step
Suppose that we can cover the remaining squares on *any* 2^n-by-2^n chess board having one square missing. Now consider a 2^{n+1}-by-2^{n+1} chess board with one square missing.

How do we reduce this to a 2^n-by-2^n chess board (so that we can use the inductive hypothesis)? The chess boards in this problem are easy to divide into four equal-sized parts. Since each side will be divided in half, the two halves will each contain 2^n squares on a side (Figure 3.8).

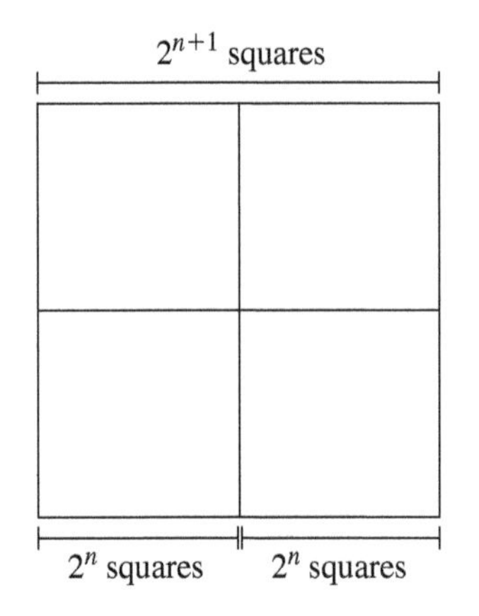

Figure 3.8. The 2^{n+1}-by-2^{n+1} chess board.

Thus the four parts will each be a 2^n-by-2^n chess board. The missing square must be in one of these four quadrants. Assume it is in the top left quadrant (we can rotate the chess board to make this true). Figure 3.9 shows how we proceed. The gray area represents the missing square. (It could be on an edge of the top left quadrant. The proof does not depend on its exact location within that quadrant.)

We start by placing a tile over the center squares in the other three quadrants.

If we consider the squares that are covered by the tile as if they were missing, then each of the four 2^n-by-2^n quadrants of the chess board has a missing square. By the inductive hypothesis, each of these four quadrants can be covered by tiles. Since the tile at the center covers the squares we were pretending were missing, we have a covering of the 2^{n+1}-by-2^{n+1} chess board.

Conclusion
The theorem of mathematical induction now guarantees that we can tile any 2^n-by-2^n chess board with a single missing square, for $n \geq 1$. □

Notice that the proof would fail for a 3-by-3 chess board, since we can't subdivide it into 2-by-2 pieces.

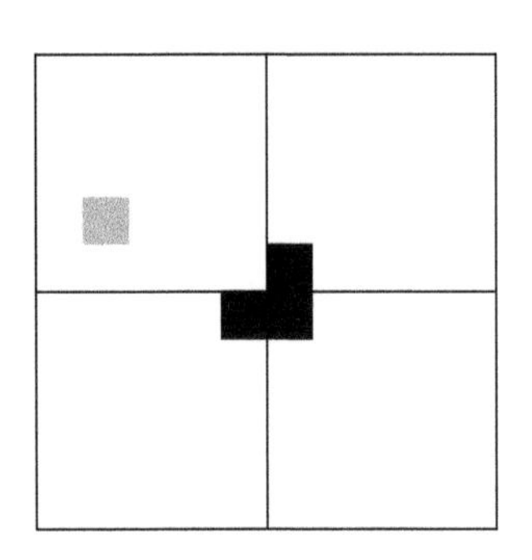

Figure 3.9. Applying the inductive hypothesis.

Optimality of the Deferred Acceptance Algorithm

You will need to review Section 1.2 on page 3 before reading this example.

Recall that for a given set of male and female preferences, there may be multiple stable assignments that can be produced.[51] The Deferred Acceptance Algorithm is one mechanism for choosing one of the (potentially) many stable assignments. We already saw that the algorithm produces a stable assignment. Even more is true.

> **DEFINITION 3.61** ***Optimal***
> A stable assignment is called *optimal for suitors* if every suitor is at least as well off in this assignment as in any other stable assignment.

In Example 1.1 (page 4), the assignment labeled "male 1st choice" is optimal for suitors if the males are the suitors.

The definition says that among all stable assignments, an optimal one places each suitor with a mate that is as high on that suitor's list as the mates (for that suitor) in other stable assignments.

It is not clear at this point that optimality is possible. Perhaps Will has his best mate in stable assignment P, but Zed might have his best mate in stable assignment Q. It might be the case that no one stable assignment is best for *all* suitors. However, the next theorem will show that the Deferred Acceptance Algorithm *always* produces an optimal assignment for suitors.

[51] See Example 1.1 on page 4.

DEFINITION 3.62 ***Possible***

A potential mate is called *possible* for a suitor if there is a stable assignment that pairs them.

In Example 1.1, all mates are possible. This is not true for other sets of preferences.

THEOREM 3.63 ***Optimality of the Deferred Acceptance Algorithm***

The Deferred Acceptance Algorithm produces an assignment that is optimal for every suitor.

Proof Using Complete Induction: The inductive hypothesis will be

> Prior to round k of the Deferred Acceptance Algorithm, no suitor has been rejected by a possible mate.

That is, any potential mate who rejects the suitor must be one that can never exist in a stable assignment with that suitor.[52]

I will use letters near the end of the alphabet to represent suitors, and letters near the front of the alphabet to represent suitees. It might be helpful to draw diagrams as you read the details of the proof. Also, the word *possible* has a technical meaning here, so don't treat it as a normal word.

Base Step:

After the first round, no suitor has been rejected by a possible mate. For suppose that suitor S has just been rejected by potential mate C, in favor of suitor T (who has also proposed to C). Suitor S could never be in any stable assignment with C, because C has just shown a preference for T over S, and T has indicated C as first choice. Thus, if C and S were assigned to be married, C and T would elope. Hence, C is not possible for S.

Since S represents any suitor who was rejected in round 1, it is clear that no suitor has been rejected by a possible mate at this point in the algorithm.

Inductive Step

Using complete induction, assume that in rounds $1 \ldots k-1$, no suitor has been rejected by a possible mate.

We now want to consider a suitor, X, who has just been rejected by A in round k. If A has rejected X, then A must have a proposal from some other suitor, Y, whom A prefers to X.

Think about the suitor Y for a moment. Y has (perhaps) proposed to other potential mates before arriving at A. So at this stage, Y has been rejected by all potential mates that Y prefers above A. By the inductive hypothesis, we know that no person who has rejected Y prior to this round is a possible mate for Y. Consequently, any person Y prefers to A is not a possible mate for Y.

Now consider A and X. Could there be an assignment among the collection of stable assignments in which A and X are paired for marriage? Suppose that such a hypothetical assignment exists. In that hypothetical assignment, Y must inevitably be paired with a possible mate. But all mates Y prefers over A are not possible for Y. Thus Y is paired with a mate, B, that Y would gladly leave for A. In addition, we know from the actual activities of round k of the Deferred Acceptance Algorithm that A prefers Y over X. Hence, A and Y will elope. This means that the hypothetical pairing of A and X is not stable.

The conclusion is that any suitor who is rejected in round k has been rejected by someone who is *not* a possible mate. Thus, the inductive hypothesis continues to be true after round k.

---

[52] Even if some other algorithm or mechanism is used to produce the assignment.

Summary
It has been shown that in round 1, no suitor is rejected by a possible mate. It has also been shown that if prior to round k, no suitor has been rejected by a possible mate, then in round k itself, no suitor is rejected by a possible mate. Using the theorem of complete induction, we conclude that no suitor is ever rejected by a possible mate when the Deferred Acceptance Algorithm is used.

Finally, consider the way the algorithm works. Suitors start with their first choice and work down by order of preference. They are never rejected by a possible mate. Consequently, each suitor must end up paired with the highest ranking possible mate (relative to the suitor's rankings). This proves the theorem. □

3.5.4 The Well-Ordering Principle, Mathematical Induction, and Complete Induction

Although in this book the well-ordering principle was assumed as an axiom and the two theorems on mathematical induction were shown to be direct consequences of that axiom, it is possible to change the roles. It is possible to assume one of the two mathematical induction theorems as an axiom, and then prove the well-ordering principle as a theorem (and also prove the other induction theorem).

THEOREM 3.64 ***The Well-Ordering Principle, Mathematical Induction, and Complete Induction Are Equivalent***

The following are equivalent:

- The well-ordering principle (WOP).
- The theorem of mathematical induction (MI).
- The theorem of complete induction (CI).

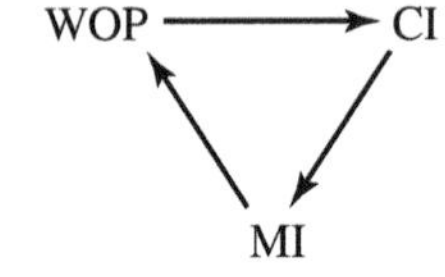

Figure 3.10. Proving Theorem 3.64.

Proof: The implications will be organized as shown in Figure 3.10.

The three implications allow us to move from any one of the principles/theorems to any other, so all three are equivalent.

The implications will be proved as separate theorems. □

THEOREM 3.55 ***WOP → CI***

The well-ordering principle implies the theorem of complete induction.

Proof: The implication WOP → CI was already proved as Theorem 3.55 on page 139. □

THEOREM 3.65 ***CI → MI***

The theorem of complete induction implies the theorem of mathematical induction.

Proof: Assume that Theorem 3.55 is true. Suppose we have verified that the two hypotheses of Theorem 3.53 (page 132) hold for some claim.

That is, we know that $P(1)$ is true and $P(n) \to P(n+1)$ is true for all $n \geq 1$. We want to conclude that $P(n)$ is true for $n \geq 1$.

If we knew that $(P(1) \wedge P(2) \wedge \cdots \wedge P(n)) \to P(n+1)$ is also true for all $n \geq 1$, the theorem of complete induction (which we are assuming is true) would imply $P(n)$ is true for $n \geq 1$ and we would be done. So we need to investigate the implication

$$(P(1) \wedge P(2) \wedge \cdots \wedge P(n)) \to P(n+1)$$

Suppose one or more of $P(1), P(2), \ldots, P(n)$ is false. Then $P(1) \wedge P(2) \wedge \cdots \wedge P(n)$

will also be false, so the implication $(P(1) \wedge P(2) \wedge \cdots \wedge P(n)) \to P(n+1)$ would be true. Finally, consider the case where all of $P(1), P(2), \ldots, P(n)$ are true. Since $P(n) \to P(n+1)$ is true and $P(n)$ is true, we know $P(n+1)$ is true.[53] Thus, since $T \to T$ has the truth value T,

$$(P(1) \wedge P(2) \wedge \cdots \wedge P(n)) \to P(n+1)$$

is true.

In all cases,

$$(P(1) \wedge P(2) \wedge \cdots \wedge P(n)) \to P(n+1)$$

is true and $P(1)$ is true. Complete induction implies that $P(n)$ is true for $n \geq 1$.

We started by assuming that the two hypotheses of mathematical induction were true for P and concluded that $P(n)$ is true for $n \geq 1$. The theorem of mathematical induction is valid. □

The final implication to be proved will complete the circuit in Figure 3.10.

THEOREM 3.66 ***MI → WOP***

The theorem of mathematical induction implies the well-ordering principle.

Proof: A proof by contradiction will be used.

Let S be a nonempty set of nonnegative integers. Assume that the theorem of mathematical induction (Theorem 3.53) is true, but that the well-ordering principle is not true for S. That is, S has no smallest element. Since S has no smallest element, it should be possible to show that S does not contain any numbers in the initial collection of nonnegative integers. Thus, let

$$P(n)\colon\ \{0, 1, 2, \ldots, n\} \cap S = \emptyset$$

The theorem of mathematical induction can be used to show that $P(n)$ is true for all $n \geq 0$.

Base Step

Consider the set $\{0\} \cap S$. Since S has no smallest element, the number 0 cannot be in S. Thus $\{0\} \cap S = \emptyset$.

Inductive Step

Assume that $P(n)$: $\{0, 1, 2, \ldots, n\} \cap S = \emptyset$ is true for some $n \geq 0$. Observe that

$$\begin{aligned}
\{0, 1, 2, \ldots, n, n+1\} \cap S &= (\{0, 1, 2, \ldots, n\} \cup \{n+1\}) \cap S \\
&= (\{0, 1, 2, \ldots, n\} \cap S) \cup (\{n+1\} \cap S) \\
&= \emptyset \cup (\{n+1\} \cap S) \quad \text{by the inductive hypothesis} \\
&= \{n+1\} \cap S
\end{aligned}$$

Therefore, $\{0, 1, 2, \ldots, n, n+1\} \cap S \neq \emptyset$ would imply that $n+1 \in S$. But the inductive hypothesis then leads to the contradiction that $n+1$ is the smallest element of S. Thus, $\{0, 1, 2, \ldots, n, n+1\} \cap S = \emptyset$, completing the inductive step.

Conclusion

Assuming that S has no smallest element but that the theorem of mathematical induction is valid leads to the conclusion that $\{0, 1, 2, \ldots, n\} \cap S = \emptyset$ is true for all $n \geq 0$.

End of the Induction

The induction has established that S does not contain any nonnegative integers as elements. This means that S must be empty, a contradiction. This final contradiction leads back to the assumption that S has no smallest element. We must therefore conclude that S *does* have a smallest element, proving the theorem. □

[53] *Modus ponens*: $[P(n) \wedge (P(n) \to P(n+1))] \to P(n+1)$.

3.5.5 Multidimensional Induction

It is possible to use induction with more than one variable. The next theorem is an example of one such generalization. The proof will be omitted, but the examples that follow should indicate why it is true.

THEOREM 3.67 ***Two-dimensional Mathematical Induction***

If $\{P(i, j)\}$ is a set of statements such that

1. $P(0, j)$ is true for $j \geq 0$ and $P(i, 0)$ is true for $i \geq 0$ and
2. $(P(i\text{-}1, j) \wedge P(i, j\text{-}1)) \to P(i, j)$ for $i \geq 1, j \geq 1$,

then $P(m, n)$ is true for all nonnegative integers m and n. This can also be stated as

$$\begin{aligned}&[(\forall j \geq 0, P(0, j)) \wedge (\forall i \geq 0, P(i, 0))\\&\wedge (\forall i \geq 1, \forall j \geq 1, P(i\text{-}1, j) \wedge P(i, j\text{-}1) \to P(i, j))]\\&\to [\forall m \geq 0, \forall n \geq 0, P(m, n)]\end{aligned}$$

The next example provides a more visual look at induction. The setting is the first quadrant of the plane, with each unit-length square considered to be a pixel. The example starts with a base case that requires three base conditions. The inductive hypothesis assumes that a pixel of unknown color has three specific neighbors whose colors *are* known. The color of the unknown pixel is then shown to be unambiguously determined.

This example will differ from previous inductions because the inductive step will be accomplished by a pre-specified formula, not by a small proof.

EXAMPLE 3.42 **Visualizing Induction**

Consider quadrant 1 of the plane, and assume that each unit-length square is a pixel. Let $p(i, j)$ denote the pixel whose lower-left corner is at the point (i, j) where $i, j \geq 0$ and $i, j \in \mathbb{Z}$.

Choose any 16 colors, and identify the colors using the integers $0, 1, 2, \ldots, 15$. We can choose the colors of $p(0, 0)$, $p(0, 1)$, and $p(1, 0)$ in any manner we wish (from within the set of 16 colors). The other pixels are then determined[54] by using the formulas:[55]

$$i = 0, j \geq 2 \quad p(0, j) = p(0, 1)$$

$$i \geq 2, j = 0 \quad p(i, 0) = p(1, 0)$$

$$i \geq 1, j \geq 1 \quad p(i, j) = \left(\left\lfloor \frac{p(i\text{-}1, j\text{-}1) + p(i, j\text{-}1) + p(i\text{-}1, j)}{3} \right\rfloor + s \cdot p(i\text{-}1, j\text{-}1)\right) \bmod 16$$

where $0 \leq s \leq 15$ is a constant. (Notice that in the division by 3, any fractional part is ignored.)

PROPOSITION 3.68

In the construction just outlined, all pixels will be assigned a unique color.

[54] Do we have free will, or are all our thoughts and actions predetermined in some manner? People have debated this question for centuries. Some have used theological arguments [108] and others have appealed to philosophy, psychology, and human physiology. This example provides an amusing variation on this debate. The initial three pixels have colors determined through the use of free will (within the limitations of the mathematical universe for the construction). All other pixel colors are then predetermined.

[55] The java application at http://www.mathcs.bethel.edu/~gossett/DiscreteMathWithProof/VisualizingInduction/ can be used to see what actually happens with various choices for $p(0, 0)$, $p(0, 1)$, $p(1, 0)$ and s.

Proof:

Base Step

In a typical proof by induction, one or more base cases will be shown to satisfy the claim. In this example, the value for s and the colors for the first three pixels will be assigned by our choice. Thus, by choice (or random selection), $p(0, 0)$, $p(0, 1)$, and $p(1, 0)$ all will have known colors and s is a known integer in $[0, 15]$. In addition, since the color for $p(0, 1)$ and $p(1, 0)$ are known, the first two formulas enable the colors for the pixels along the vertical and horizontal axes to be uniquely specified.

Inductive Step

Assume that for some i, j with $i, j \geq 1$, the colors of $p(i\text{-}1, j\text{-}1)$, $p(i\text{-}1, j)$, and $p(i, j\text{-}1)$ are all known. In a typical proof by induction, we would then look for a means to show that those colors can be used to uniquely determine the color of pixel $p(i, j)$. In this example, that is easy since the formula

$$p(i, j) = \left(\left\lfloor \frac{p(i\text{-}1, j\text{-}1) + p(i, j\text{-}1) + p(i\text{-}1, j)}{3} \right\rfloor + s \cdot p(i\text{-}1, j\text{-}1)\right) \mod 16$$

uniquely determines $p(i, j)$ when $i \geq 1$ and $j \geq 1$.

Conclusion

The three corner pixels $p(0, 0)$, $p(0, 1)$, and $p(1, 0)$ have known colors. These can then be used (by the inductive step) to determine the color of $p(0, j)$ for $j \geq 2$ and $p(i, 0)$ for $i \geq 2$. The third formula then enables the colors to be determined by filling in successive 135° diagonals.[56] For example, $p(1, 1)$ would be the first such diagonal. Then $p(1, 2)$ and $p(2, 1)$ would be the next. The third diagonal would be $p(1, 3)$, $p(2, 2)$, $p(3, 1)$. The inductive step shows that this pattern continues indefinitely, so all pixels will have a unique color. □ ■

EXAMPLE 3.43

Pascal's Triangle

You are probably familiar with Pascal's Triangle from your high school algebra courses. The first six rows of Pascal's triangle are shown in Table 3.12.

TABLE 3.12 Pascal's Triangle

row 0						1					
row 1					1		1				
row 2				1		2		1			
row 3			1		3		3		1		
row 4		1		4		6		4		1	
row 5	1		5		10		10		5		1

Each interior entry is the sum of the two entries that are above it. A way to present this triangle that follows the pattern in Example 3.42 is to define a collection of numbers $t(i, j)$ for $i \geq 0$, $j \geq 0$ and $i, j \in \mathbb{Z}$. The numbers are defined as:

$$t(i, j) = \begin{cases} 1 & i = 0, j = 0 \\ 1 & i = 0, j \geq 1 \\ 1 & i \geq 1, j = 0 \\ t(i\text{-}1, j) + t(i, j\text{-}1) & i \geq 1, j \geq 1 \end{cases}$$

The rows from the original triangle are now diagonals with constant row and column sums (see Table 3.13 on page 150).

[56] These diagonals are characterized by having constant row+column sums.

row 0						1					
row 1					1		1				
row 2				1		2		1			
row 3			1		3		3		1		
row 4		1		4		6		4		1	
row 5	1		5		10		10		5		1

TABLE 3.13 Pascal's Triangle

row 5	1					
row 4	1	5				
row 3	1	4	10			
row 2	1	3	6	10		
row 1	1	2	3	4	5	
row 0	1	1	1	1	1	1
	col 0	col 1	col 2	col 3	col 4	col 5

This is now in a form that conforms to Theorem 3.67. ■

It is possible to express the value of the entry $t(i, j)$ in Pascal's Triangle in a more direct manner.

PROPOSITION 3.69 ***Entries in Pascal's Triangle***

Let $t(i, j)$ be the entry in the ij^{th} position of Pascal's Triangle, for $i \geq 0, j \geq 0$ and $i, j \in \mathbb{Z}$. Then

$$t(i, j) = \frac{(i+j)!}{i! \cdot j!}$$

Proof:

Base Step

There are three cases to consider for the base step. All three cases produce the desired equalities.

$$i = 0, j = 0 \quad \frac{(0+0)!}{0! \cdot 0!} = \frac{1}{1 \cdot 1} = 1 = t(0, 0)$$

$$i = 0, j \geq 1 \quad \frac{(0+j)!}{0! \cdot j!} = \frac{j!}{1 \cdot j!} = 1 = t(0, j)$$

$$i \geq 1, j = 0 \quad \frac{(i+0)!}{i! \cdot 0!} = \frac{i!}{i! \cdot 1} = 1 = t(i, 0)$$

Inductive Step

Suppose that for some $i, j \in \mathbb{Z}$ with $i \geq 1, j \geq 1$ the following two equalities are valid:

$$t(i\text{-}1, j) = \frac{((i-1)+j)!}{(i-1)! \cdot j!}$$

$$t(i, j\text{-}1) = \frac{(i+(j-1))!}{i! \cdot (j-1)!}$$

Then

$$\begin{aligned} t(i, j) &= t(i\text{-}1, j) + t(i, j\text{-}1) \\ &= \frac{((i-1)+j)!}{(i-1)! \cdot j!} + \frac{(i+(j-1))!}{i! \cdot (j-1)!} \quad \text{by the inductive hypothesis} \\ &= \frac{i \cdot ((i-1)+j)!}{i! \cdot j!} + \frac{j \cdot (i+(j-1))!}{i! \cdot j!} \\ &= \frac{(i+j)(i+j-1)!}{i! \cdot j!} \\ &= \frac{(i+j)!}{i! \cdot j!} \end{aligned}$$

which establishes the validity of the formula for $t(i, j)$.

Conclusion
The claim is valid when either i or j is 0 and whenever the claim is true for $t(i\text{-}1, j)$ and $t(i, j\text{-}1)$ it is also true for $t(i, j)$. Theorem 3.67 implies that the claim is valid for all $i, j \in \mathbb{Z}$ with $i \geq 0, j \geq 0$. □

EXAMPLE 3.44

A Generalized Pascal Triangle

The construction in Example 3.43 can be generalized in many ways. The following version simply replaces the 1s on the border with consecutive integers.

The entries will be denoted by $g(i, j)$ and defined by

$$g(i, j) = \begin{cases} 0 & i = 0, j = 0 \\ j & i = 0, j \geq 1 \\ i & i \geq 1, j = 0 \\ g(i\text{-}1, j) + g(i, j\text{-}1) & i \geq 1, j \geq 1 \end{cases}$$

The initial portions of the first seven rows and columns are shown in Table 3.14.

TABLE 3.14 A Generalized Pascal Triangle

	col 0	col 1	col 2	col 3	col 4	col 5	col 6
row 6	6	22	64	162	372	792	1584
row 5	5	16	42	98	210	420	792
row 4	4	11	26	56	112	210	372
row 3	3	7	15	30	56	98	162
row 2	2	4	8	15	26	42	64
row 1	1	2	4	7	11	16	22
row 0	0	1	2	3	4	5	6

Exercise 28 on page 153 provides a closed-form formula for $g(i, j)$ (a formula involving only i and j but no other values $g(r, s)$). Exercise 21 in Exercises 5.2.3 (page 258) provides a more elegant formula (which is algebraically equivalent to the formula in Exercise 28). ■

Multidimensional induction will be used later in the book for the proof of Theorem 8.74. That proof will use an induction on three variables.

3.5.6 Exercises

The exercises marked with ★ have detailed solutions in Appendix H.

1. The following formulas can all be proved by mathematical induction.
   (a) $1^2 + 2^2 + 3^2 + \cdots + n^2 = \frac{n(n+1)(2n+1)}{6}$ for all natural numbers, n with $n \geq 1$.
   (b) ★ $2^m > m$ for all natural numbers, m with $m \geq 1$.
   (c) $a^n < 1$ for all real numbers, a, with $0 \leq a < 1$ and all natural numbers, n with $n \geq 1$ Clearly identify where you have used the assumption $0 \leq a$ and then explain why the proof would fail if $a < 0$.
   (d) Use the hint to prove

$$\sum_{k=1}^{n} \frac{1}{k^2} < 2 - \frac{1}{n} \text{ for all } n \in \mathbb{N} \text{ with } n \geq 2$$

Hint:
$$\frac{n^2+n+1}{n(n+1)^2} = \left(\frac{n^2+n}{n(n+1)^2} + \frac{1}{n(n+1)^2}\right) = \frac{1}{n+1} + \frac{1}{n(n+1)^2} > \frac{1}{n+1}$$

2. *There are many variations to the following legend.* The inventor of an early incarnation of chess presented the game to the Persian Shah. The Shah was so impressed by the game that he gave the inventor permission to name his own reward. The inventor told the Shah that he would ask for one grain of rice for the first square of the game board, two grains of rice for the second square, and four grains for the third square. The number of grains would continue to double until the final (64th) square of the board. How many grains of rice would this require? If each grain of rice weighs 20 mg, how many metric tons of rice would this be? (One milligram is 10^{-9} metric tons.)

3. Prove that

$$\sum_{k=0}^{n}\left(\frac{-1}{2}\right)^k = \frac{2}{3}+\frac{1}{3}\cdot\left(\frac{-1}{2}\right)^n$$

for all natural numbers, n with $n \geq 0$.

4. Formulate a theorem for the sum $\sum_{i=0}^{n}(a+id)$ of the first $n+1$ elements in an arithmetic progression. Indicate how you arrived at the formula, and then use mathematical induction to prove the formula.

5. ★ Find a formula for

$$\frac{1}{2}+\frac{1}{2^2}+\frac{1}{2^3}+\cdots+\frac{1}{2^n} \quad n \geq 1$$

Indicate how you arrived at the formula, and then use mathematical induction to prove the formula.

6. Find a formula without summations or $+\cdots+$ for

$$\frac{1}{a}+\frac{1}{a^2}+\frac{1}{a^3}+\cdots+\frac{1}{a^n}$$

for $n \in \mathbb{N}, n \geq 1$ and $a \in \mathbb{R}, a > 0, a \neq 1$. Indicate how you arrived at the formula, and then use mathematical induction to prove the formula.

7. Find a formula for

$$\sum_{i=1}^{n}\frac{1}{i(i+1)} \quad n \in \mathbb{N}, n \geq 1$$

Indicate how you arrived at the formula, and then use mathematical induction to prove the formula.

8. Find a formula for

$$1\cdot 2+2\cdot 3+\cdots+n(n+1) \quad n \in \mathbb{N}, n \geq 1$$

Indicate how you arrived at the formula, and then use mathematical induction to prove the formula.

9. Find a formula for

$$1\cdot 1!+2\cdot 2!+\cdots+n\cdot n! \quad n \in \mathbb{N}, n \geq 1$$

Indicate how you arrived at the formula, and then use mathematical induction to prove the formula.

10. ★ Let $n \in \mathbb{N}$, with $n \geq 1$. Show that

$$1+(1+2)+(1+2+3)+\cdots+(1+2+\cdots+n) = \sum_{k=1}^{n}\left(\sum_{i=1}^{k} i\right) = \frac{n(n+1)(n+2)}{6}$$

11. Let $A, B_1, B_2, \ldots, B_n$ be sets, with $n \geq 2$. Prove that

$$A\cap(B_1\cup B_2\cup\cdots\cup B_n) = (A\cap B_1)\cup(A\cap B_2)\cup\cdots\cup(A\cap B_n)$$

12. Let $n \in \mathbb{N}$, with $n \geq 1$. Use mathematical induction to show that

$$1+2+3+\cdots+n = \sum_{i=1}^{n} i < \frac{(2n+3)^2}{7}$$

13. ★ Use mathematical induction to show that x^2-1 is divisible by 8 when x is any positive odd integer.

14. Prove that every integer, n, can be written in the form $5\cdot a+7\cdot b$, where $a, b \in \mathbb{Z}$. Use mathematical induction to show this for all integers $n \geq 0$. Then think of another way to validate the claim for all integers $n < 0$.

15. Let $n \in \mathbb{N}$, with $n \geq 0$ and let $x, y \in \mathbb{R}$ with $x \neq -y$. Use mathematical induction to show that $x^{2n} - y^{2n}$ is divisible by $x+y$.

16. Let $n \in \mathbb{N}$, with $n \geq 2$. Show that $n! < n^n$.

17. Prove that

$$1+\frac{1}{\sqrt{2}}+\frac{1}{\sqrt{3}}+\cdots+\frac{1}{\sqrt{n}} = \sum_{i=1}^{n}\frac{1}{\sqrt{i}} > 2(\sqrt{n+1}-1)$$

for $n \in \mathbb{N}$, with $n \geq 1$.

18. Let $x, y_1, y_2, \ldots, y_n$ be elements in a Boolean algebra, with $n \geq 2$. Prove

$$x+(y_1\cdot y_2\cdots y_n) = (x+y_1)\cdot(x+y_2)\cdots(x+y_n)$$

19. Let n be a positive integer. Prove that

$$\frac{1}{2n} \leq \frac{1\cdot 3\cdot 5\cdots(2n-3)\cdot(2n-1)}{2\cdot 4\cdot 6\cdots(2n-2)\cdot(2n)}$$

20. ★ Let $n \in \mathbb{N}$. Prove that

$$\frac{2n+1}{2n+2} \leq \frac{\sqrt{n+1}}{\sqrt{n+2}}$$

[*Hint*: This is a nasty induction problem, but is fairly simple to show algebraically.]

21. Let n be a positive integer. Use Exercise 20 to prove that

$$\frac{1\cdot 3\cdot 5\cdots(2n-3)\cdot(2n-1)}{2\cdot 4\cdot 6\cdots(2n-2)\cdot(2n)} \leq \frac{1}{\sqrt{n+1}}$$

22. Prove

$$\prod_{k=2}^{n}\left(1-\frac{1}{k^2}\right) = \frac{n+1}{2n} \quad \forall n \in \mathbb{N} \text{ with } n \geq 2$$

The results of the next two problems will be used in Chapter 11.

23. ★ Let $h \geq 0$ be an integer. Prove that

$$\sum_{i=0}^{h}(h-i)2^i = 2^{h+1}-h-2$$

24. ★ Let $h \geq 0$ be an integer. Prove that

$$\sum_{i=0}^{h} i2^i = (h-1)2^{h+1}+2$$

25. Use mathematical induction to prove Proposition 3.70.

> **PROPOSITION 3.70**
>
> Let $a, b, k, m \in \mathbb{Z}$ with $m > 1$ and $k \geq 0$. If $a \equiv b \pmod{m}$, then $a^k \equiv b^k \pmod{m}$.

26. What is wrong with the following proof that all horses have the same color?

Proof Let n be the number of horses. When $n = 1$, the statement is clearly true, that is, one horse has the same color, whatever color it is. Assume that any group of n horses has the same color. Now consider a group of $(n+1)$ horses. Taking any n of them, the inductive hypothesis states that they all have the same color, say brown. The only issue is the color of the remaining "uncolored" horse. Consider, therefore, any other group of n of the $(n+1)$ horses that contains the uncolored horse. Again, by the inductive hypothesis, all of the horses in the new group must have the same color. Then, since all of the colored horses in this group are brown, the uncolored horse must also be brown.[57]

27. The sequence of numbers $1, 1, 2, 3, 5, 8, 13, 21, 34, 55, \ldots$ is called the *Fibonacci* sequence. It can be generated by setting $f_0 = 1, f_1 = 1$ and setting $f_n = f_{n-1} + f_{n-2}$ for $n \geq 2$. Notice that all numbers in the sequence are integers.

The following formulas all relate to the definition of the Fibonacci sequence.[58] You will need to remember to use your inductive hypothesis in each case. You will also need to use the definition of the sequence: $f_n = f_{n-1} + f_{n-2}$ for $n \geq 2$.

(a) Prove that $f_0^2+f_1^2+f_2^2+\cdots+f_n^2 = f_n f_{n+1}$ for $n \geq 0$.

(b) ★ Show that $f_0 + f_2 + \cdots + f_{2n} = f_{2n+1}$ for $n \geq 0$.

(c) Prove that $f_{n-1}f_{n+1} - f_n^2 = (-1)^{n+1}$ for $n \geq 1$.

(d) It is an amazing fact that f_n, the nth element of the sequence, can be given by a formula that involves $\sqrt{5}$. The formula is given by

$$f_n = c_1 a^n + c_2 b^n$$

where $n \geq 0$ and

$$c_1 = \frac{1+\sqrt{5}}{2\sqrt{5}} \qquad c_2 = \frac{-(1-\sqrt{5})}{2\sqrt{5}}$$

and

$$a = \frac{1+\sqrt{5}}{2} \qquad b = \frac{1-\sqrt{5}}{2}$$

Note that a and b are the solutions to $x^2 = x + 1$, (the defining equation in Appendix D for the Golden Ratio). Use complete induction to prove that the formula is correct. Verify the formula for both f_0 and f_1 in the base step.

28. Use Theorem 3.67 to prove that $g(i, j)$ in Example 3.44 satisfies:

$$g(i,j) = \begin{cases} 0 & i=0, j=0 \\ j & i=0, j\geq 1 \\ i & i \geq 1, j=0 \\ \frac{(i+j)!}{(i-1)!\cdot(j+1)!} + \frac{(i+j)!}{(i+1)!\cdot(j-1)!} & i\geq 1, j\geq 1 \end{cases}$$

The algebraic simplification in the inductive step is a bit long, but not difficult. The base step must include $g(1, 1)$. The inductive step requires 3 cases: $g(i, 1)$ for $i \geq 2$, $g(1, j)$ for $j \geq 2$ and $g(i, j)$ for $i, j \geq 2$.

29. Consider the generalized Pascal Triangle whose entries, $u(i, j)$, are defined by $u(0, 0) = 1$, $u(i, 0) = 1$ when $i \geq 1$, $u(0, j) = 1$ when $j \geq 1$ and

$u(i, j) = u(i\text{-}1, j) + u(i, j\text{-}1) + 1$ for $i \geq 1, j \geq 1$

Use multiple induction to prove

$$u(i,j) = 2\cdot\frac{(i+j)!}{i!\cdot j!} - 1 \text{ for } i \geq 0, j \geq 0$$

3.6 Creating Proofs: Hints and Suggestions

The earlier parts of this chapter have presented numerous strategies for proofs (for example, direct proof and mathematical induction). Learning those strategies will help you to read intelligently proofs created by other people as well as help you to create proofs on your own. However, knowledge of those strategies is often not all that is needed.

Creating proofs cannot be done by following a set of rules. It requires insight, instinct based on experience, and sometimes creativity and ingenuity. At this point in your mathematical career, the element of "instinct based on experience" is still in a formative stage. Fortunately, this is not a double-bind situation.[59] There are some general suggestions that will help you gain that experience while still succeeding at creating proofs. The goal of this section is to acquaint you with some of these suggestions. Your instructor may have additional suggestions.

3.6.1 A Few Very General Suggestions

There are a number of fairly simple habits and ideas that can enhance your ability to create proofs. They are presented in no particular order.

---

[57] I do not know the original source of this "proof." The version presented here is from [89, p. 58]. There is also a pun associated with this proof (which I mercifully cannot remember at the moment).

[58] Exercise 22 on page 258 contains another formula.

[59] A common double bind is for a new college graduate to find that the job listings in a field all require two to three years of experience. The graduate doesn't qualify for a job due to lack of experience, but cannot gain experience due to lack of a job in the field.

Know the Definitions and Theorems

Most proofs will require knowledge of one or more definitions just to understand the statement that is to be proved. The proof itself will often require the use of other definitions (either in concept-to-property or property-to-concept form). In addition, the proof will often be simplified if you use the results of one or more previously proved theorems.

If you have memorized the definitions (both at an intuitive and at a precise level), you will recognize more easily when they should be inserted into the proof. If you also have memorized the major theorems, you will spend less time flipping through pages of the textbook looking for a random theorem that might be of use.

EXAMPLE 3.45 **The Quotient–Remainder Theorem**

Recall the proof of the Quotient–Remainder theorem on page 96. The proof requires the use of the well-ordering principle (an axiom), some basic notions from set theory (empty, nonempty, set-builder notation), an understanding of the integers (the nature of integers, properties of integer addition and multiplication, the definition of absolute value), properties of absolute value (a theorem), and familiarity with proof by contradiction (existence) and direct proof (uniqueness).

If that material was familiar to you, you probably found the proof easy to read and understand. However, if you had not memorized and/or understood some of the background material, you may have found places where the flow of ideas was a mystery. ■

The main idea in this suggestion is to memorize and understand (deeply and precisely) definitions and theorems. The more you know in this way, the easier it will be to read and create proofs.

Use Lots of Scratch Paper

Colleges and universities produce and recycle lots of paper that has only been used on one side. Grab a stack and use the blank side for scratch paper. You can send it back to the recycling bin after you have used it.

Scratch paper is very helpful for trying out ideas without the need to keep it neat or to erase dead ends. Knowing that mistakes and dead ends can be effortlessly discarded frees you to experiment.

Writing your proof on scratch paper first also enables you to do some editing and revision as you copy the correct proof onto the final paper. The clean copy might then be an improvement over your first correct proof.

Analyze Other People's Proofs

As you read proofs in textbooks or published articles, you will certainly be thinking about the content of the proof. It will also be beneficial to think carefully about the manner in which the proof is presented. You can learn a great amount from a well-presented proof. In particular, you might see ideas that you would not have considered. For example, in the proof of the Quotient–Remainder theorem on page 96, the set $S = \{a - bq \mid q \in \mathbb{Z} \text{ and } a - bq \geq 0\}$ is not one that many students would think to examine. If you imprint in your memory the clever ideas you encounter, the ideas will be available to you in the future.

Unfortunately, published proofs seldom provide you with information about how the proof's creator arrived at the final version. Even if that information is missing, you should realize that the printed version was usually not immediately created in the form you see. The author will typically try a few approaches and settle on the final version after a few detours and dead ends.[60]

[60] Of course, if the proof is fairly simple and the author has been creating proofs for many years, the proof might be written in final form immediately. Experience makes a difference.

Don't Be Afraid to Try Multiple Strategies

If you start using one proof strategy and reach a dead end, try a different strategy. If that also leads nowhere, try another. You might even cycle back to one of the earlier strategies. Perhaps some of the additional thoughts you have during the detour will bear fruit on the return visit.

Many assertions can be proved in more than one way. One of the alternatives might make more sense to you than the others. You just need to find the proper alternative.

EXAMPLE 3.46

Several Approaches

Suppose you need to prove that $(P \vee Q) \wedge R \rightarrow R$ is a tautology. There are several options. You might first try a few truth values for P, Q, and R to see if the statement appears to really be a tautology. You could also think about whether it makes intuitive sense.

Once you decide to proceed with a proof, you might list the alternative approaches available. You could use a truth table or you could use the fundamental logical equivalences. Suppose your instructor has decided that the fundamental logical equivalences and the logical equivalences and rules of inference for implication and the biconditional are the preferred approach. If you do not recognize that the law of simplification immediately completes the proof, you still have two initial options. You might start with right distributivity or you might start by using a substitution using an implication logical equivalence. For such a simple assertion, either approach will quickly lead to a completed proof. ■

The proof of the next assertion has been expanded to show the process of considering more than one of the strategies from Section 3.3.

EXAMPLE 3.47

A Cautionary Tale

Suppose I want to prove the following assertion:

Let $x \in \mathbb{R}$. If $\sqrt{x}$ is irrational, then x is irrational.

I might try to do a direct proof. However, I get stuck pretty quickly. How can I convert "$\sqrt{x}$ is irrational" into something I can manipulate?

Recalling that rational numbers are much easier to write in useful ways, I next consider an indirect proof (since it involves negations of the hypothesis and conclusion of the original implication). Thus, I might try to prove the following:

Let $x \in \mathbb{R}$. If x is rational, then $\sqrt{x}$ is rational.

This looks more promising. I can start by writing $x = \frac{p}{q}$ for integers p and q with $q \neq 0$. Now, what does $\sqrt{x}$ look like? It must be $\sqrt{x} = \frac{\sqrt{p}}{\sqrt{q}}$. Hmm ... I cannot assume anything else about p and q, so I do not know much about $\frac{\sqrt{p}}{\sqrt{q}}$. I am stuck again.

Next, I will try a proof by contradiction. I start by assuming that $\sqrt{x}$ is irrational, but x is rational. I hope to arrive at a contradiction. I again proceed to write $x = \frac{p}{q}$ with p and q integers and $q \neq 0$. Now I need to contradict the irrationality of $\sqrt{x}$. This leads to the same dead end as the previous strategy.

It is time to back off and do something I should have done at the beginning: Look at a few examples to get a feel for the validity of the proposition. I can create a small table of values to try with the proposition. The creative part is to choose values for $\sqrt{x}$ that I know are really irrational. It would also help if it were easy to tell if x itself is irrational. Thinking back to Proposition 3.35 on page 104, my first table entry will be $\sqrt{2}$.

$\sqrt{x}$	x
$\sqrt{2}$	2

Oops! It seems that the assertion is false. No wonder I was getting nowhere with the proof.

The final version of my "proof" is quite simple.

The assertion is false. The numbers $\sqrt{2}$ and 2 provide a counterexample. ■

Incubation

Sometimes you spend a significant amount of time on a problem and do not seem to be making progress. It is often a good idea to move on to something else for a while. You may need some time to free yourself from revisiting the same stale ideas over and over again.

Many people find that after a time of intense work with a problem, a period of incubation is necessary. During that time, their minds can focus on other things. After a full night of sleep, or even a few days of tending to other issues, they will be able to find a new approach that solves the problem.

Incubation is of no practical value if you delay working on your homework until the hour before class, or at 1 A.M. the night before class. There will be no time during which the incubation can occur (remember that incubation needs a previous intense period of concentration on the problem).

Start your math homework on the day it is assigned. That will allow time for incubation, time to meet with classmates to discuss the current text section, and time to attend your math tutoring lab or the instructor's office hours.

3.6.2 Some Specific Tactics

The following suggestions are more narrowly focused on ways to keep making progress as you search for a proof. They may not all apply in any one case, but they do provide a rich set of helpful approaches.

Look for Common Characteristics

Many assertions are stated in ways that give valuable hints about profitable solution strategies. For example, if the assertion contains the term *rational number*, it is likely that the rational number should be expressed in the form $\frac{p}{q}, q \neq 0$, with $p, q \in \mathbb{Z}$. Another example would be an assertion that claims that some set, A, is a subset of another set, B. The presence of the symbol $\subseteq$ suggests that you try choosing a generic element in A and show that it is also in B. In essence, the symbol $\subseteq$ should lead you back to the definition of *subset*[61] as an initial approach to the proof.

Table 3.15 (page 157) lists some more general common characteristics, together with recommended initial strategies.[62]

Forward–Backward

You do not always need to proceed from point A to point B as you construct a proof. It is often useful to start at the beginning and move forward until you get stuck. At that point, you could start at the end and work backward toward the starting point. You might even bounce back and forth a few times. If the two attempts meet in the middle, you can then rework the proof to start at the beginning and proceed all the way to the end.

More specifically, let the assertion be in the form of an implication: $A \to B$. In the forward direction you assume that A is true and use that information to show that B is also true. The motivation is the familiar modus ponens tautology: $[P \wedge (P \to Q)] \to Q$.

In the backward phase you are asking questions about what is necessary for B to be true.

---

[61] See page 20.

[62] The table is also available in pdf format at http://www.mathcs.bethel.edu/~gossett/DiscreteMathWithProof/ .

TABLE 3.15 Some General Proof Strategies

If the assertion ...	Then try ...
claims something is true for all integers $n \geq n_0$	mathematical induction or complete induction
is stated explicitly or implicitly as an implication	direct; indirect; contradiction
contains an existential quantifier	a constructive proof; a nonconstructive proof
contains a universal quantifier	finding a counterexample; the choose method
contains the phrase "if and only if"	to prove the two implications separately; to produce a sequence of equivalent statements linking the two sides of the biconditional
is stated as an equivalence	to look for a complete set of implications that are relatively easy to prove
can be easily split into a collection of independent assertions	proof by cases
is an implication with a true conclusion	trivial proof
is an implication with a false hypothesis	trivial (vacuous) proof
is about membership in a set	direct proof: verify that the element satisfies the set membership requirements
asserts one set is a subset of another	to show that a generic element of the first set is also a member of the second set
asserts the equality of two sets	to show that each set is a subset of the other; to use a sequence of reversible statements with the fundamental set properties and other theorems

EXAMPLE 3.48 **Forward-Backward**

Consider the assertion

$$\text{If } x \in \mathbb{R} \text{ and } 2 \leq x \leq 5, \text{ then } x^2 \leq 7x - 10$$

The proof process might begin in the forward direction. I first try substituting a few values of x into the inequality $x^2 \leq 7x - 10$. I might try $x = 2, 3, 4, 5$. They all make the inequality true. So do the values $x = 2.5$, $\sqrt{3}$, and π. In addition, the inequality is false when $x = 1, 6$, and $\sqrt{2}$. These experiments raise my confidence in the truth of the assertion, but they do not constitute a proof.

As I start thinking about what *will* constitute a valid proof, I enter the backward phase. What do I need to do to show that $x^2 \leq 7x - 10$? I might start by finding an equivalent form of the inequality. A more useful form is $x^2 - 7x + 10 \leq 0$. This is an improvement because the left-hand side is in the familiar form of a degree-2 polynomial. I am quite familiar with several approaches for finding values of x that make $x^2 - 7x + 10$ nonpositive.

It is tempting to grab a graphing calculator and have it graph $x^2 - 7x + 10$. That will add to the confidence level in the validity of the assertion, but it is not a formal proof.

There are at least two other very familiar techniques for deciding which values of x make the polynomial nonpositive. One approach uses calculus.[63] Find the first derivative, locate the critical point(s) and use the first derivative test to show that $f(x) = x^2 - 7x + 10$ is decreasing on the interval $(-\infty, 3.5)$ and increasing on the interval $(3.5, \infty)$. Then factor the polynomial as $x^2 - 7x + 10 = (x - 2)(x - 5)$ to show that $f(x) = 0$ at $x = 2$ and $x = 5$. Combining this information shows that $f(x)$ is nonpositive precisely when $2 \leq x \leq 5$, proving more than the original assertion.

[63] The other approach uses material from algebra to calculate the vertex and verify that the degree-2 polynomial is concave up because the leading coefficient is positive.

The final version of the proof would be attached to the restated assertion.

$$\text{If } x \in \mathbb{R}, \text{ then } x^2 \le 7x - 10 \text{ if and only if } 2 \le x \le 5$$

Proof (outline): Let $f(x) = x^2 - 7x + 10$. Then f is strictly decreasing on the interval $(-\infty, 3.5)$ and strictly increasing on the interval $(3.5, \infty)$. In addition, $f(x) = 0$ when $x = 2$ and $x = 5$ (and at no other real numbers). Since f is a polynomial, it is a continuous function. Therefore, f must be positive on the intervals $(-\infty, 2)$ and $(5, \infty)$. It is nonpositive on $[2, 5]$. □ ■

Try Some Small Examples

Most mathematical assertions that you have encountered have appeared in textbooks. Those that you need to prove are typically found in collections of exercises. This is pedagogically necessary. However, assertions that need proving are not all found in homework sets. In the future, you may encounter problems in your career that are best solved mathematically. In that context, it is not even clear initially what assertion should be proved. You will need to experiment, think about the problem, and, perhaps, write down several competing assertions. You will then spend some more time deciding which assertion is the best.

Even on prestated problem sets, it is often helpful to experiment with a few small examples before starting a general proof. The experiments might give you some needed insight into the more general behavior.

EXAMPLE 3.49 **A Pythagorean Triple Property**

Suppose the positive integers a, b, and c form a primitive Pythagorean triple. Then $a^2 + b^2 = c^2$ (see Definition 3.17 on page 96). In addition, we may assume that $a < b$.[64] It is always true that $c > b$ (and so $c > a$).[65]

It might be interesting to see how the differences between the three integers are related. In particular, is $c - b \ge b - a$, or is $b - a \ge c - b$? A few examples will be helpful. Example 3.10 on page 96 provides three examples of primitive Pythagorean triples.

$$3^2 + 4^2 = 5^2 \qquad 5^2 + 12^2 = 13^2 \qquad 8^2 + 15^2 = 17^2$$

It is clear that $c - b \not\ge b - a$ in the second and third example. However, in all three cases, $b - a \ge c - b$. Finding additional primitive Pythagorean triples is not trivial. Another example is $65^2 + 72^2 = 97^2$. Notice that $b - a \not\ge c - b$.

So neither inequality is always true. Can anything be said about what is necessary to make one or the other occur?

Notice that $c - b \ge b - a$ is equivalent to $\frac{a+c}{2} \ge b$ and $b - a \ge c - b$ is equivalent to $b \ge \frac{a+c}{2}$. Perhaps it will be profitable to investigate three cases: $b = \frac{a+c}{2}$, $b < \frac{a+c}{2}$, and $b > \frac{a+c}{2}$.

Case 1: $\boldsymbol{c - b = b - a \ \left(b = \frac{a+c}{2}\right)}$

If $b = \frac{a+c}{2}$, then $c^2 - a^2 = b^2 = \frac{(a+c)^2}{4}$. Therefore, $3c^2 = 5a^2 + 2ac$, or $0 = 5a^2 + 2ac - 3c^2 = (5a - 3c)(a + c)$.

Since both a and c are greater than 0, it must be that $5a = 3c$. Consequently, $b = \frac{a+c}{2} = \frac{a + \frac{5}{3}a}{2} = \frac{4}{3}a$. That is, $4a = 3b$. Combining $20a = 12c$ and $20a = 15b$ leads to $5b = 4c$.

The three conditions $5a = 3c$, $4a = 3b$, and $5b = 4c$ can be solved simultaneously (as a system of three linear equations). The solutions are all of the form $a = \frac{3}{5}c$ and $b = \frac{4}{5}c$, where c can be anything. However, c is also required to be a positive integer.

[64] If $a = b$ then $2a^2 = c^2$ so $c = 2j$ and then $b = a = 2k$, contradicting primitivity.

[65] All three are at least 1. If $1 \le c \le b$, then $c^2 \le b^2 < a^2 + b^2$, a contradiction.

Thus, $c \in \{5, 10, 15, 20, \ldots\}$. Any solution other than $c = 5$ would make the triple nonprimitive.

Thus, $c - b = b - a$ if and only if $a = 3, b = 4, c = 5$.

Case 2: $\boldsymbol{c - b > b - a}$ $\left(b < \frac{a+c}{2}\right)$

If $b < \frac{a+c}{2}$, then $c^2 - a^2 = b^2 < \frac{(a+c)^2}{4}$. Therefore, $3c^2 < 5a^2 + 2ac$, or $0 < 5a^2 + 2ac - 3c^2 = (5a - 3c)(a + c)$.

Since both a and c are greater than 0, it must be that $5a > 3c$. Consequently, $b < \frac{a+c}{2} < \frac{a+\frac{5}{3}a}{2} = \frac{4}{3}a$. That is, $4a > 3b$.

Thus, if $c - b > b - a$, then $a < b < \frac{4}{3}a$ and $\frac{3}{5}c < a < c$. Also, since $a < b < c$, $\frac{3}{5}c < b < c$.

Case 3: $\boldsymbol{b - a > c - b}$ $\left(b > \frac{a+c}{2}\right)$

Reversing the inequalities in the previous case leads to $5a < 3c$ and $4a < 3b$. Thus $a < \frac{3}{4}b$ and $a < \frac{3}{5}c$. ■

Try Proving a Related But Simpler Theorem

If you are stuck, you might gain valuable insight if you prove a related theorem which is simpler. Consider, for example, the formula for the distance between two points in $\mathbb{R}^3$ (3-space).

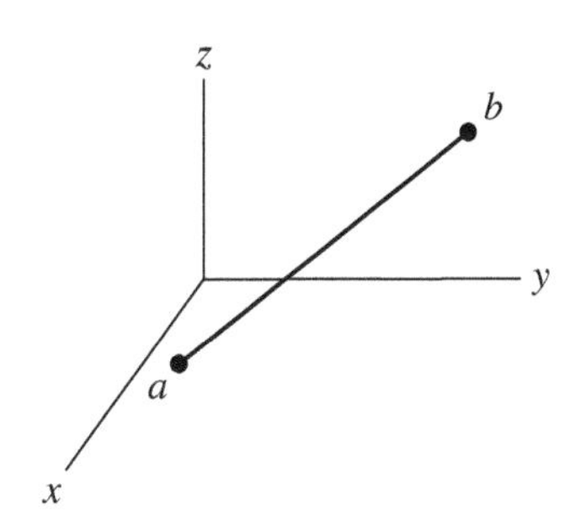

Figure 3.11. The line segment from a to b in $\mathbb{R}^3$.

THEOREM 3.71 ***Distance in*** $\mathbb{R}^3$

Let $a = (x_1, y_1, z_1)$ and $b = (x_2, y_2, z_2)$ be points in $\mathbb{R}^3$. Then the distance from a to b is

$$d(a, b) = \sqrt{(x_1 - x_2)^2 + (y_1 - y_2)^2 + (z_1 - z_2)^2}$$

Figure 3.11 shows a line segment that represents the desired distance.

The theorem will be easier to prove if a 2-space version is examined first.

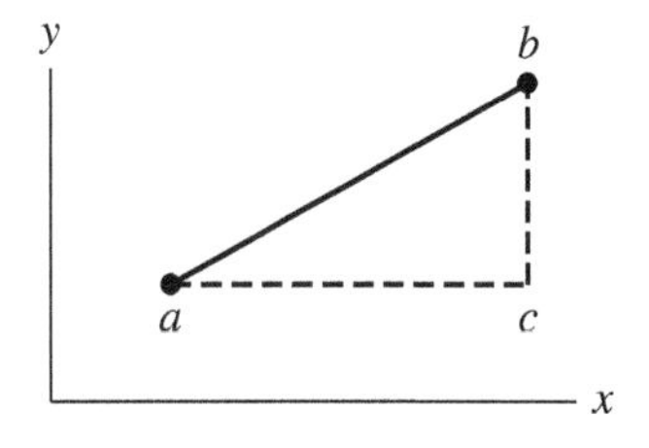

Figure 3.12. The line segment from a to b in $\mathbb{R}^2$.

THEOREM 3.72 ***Distance in*** $\mathbb{R}^2$

Let $a = (x_1, y_1)$ and $b = (x_2, y_2)$ be points in $\mathbb{R}^2$. Then the distance from a to b is

$$d(a, b) = \sqrt{(x_1 - x_2)^2 + (y_1 - y_2)^2}$$

Figure 3.12 shows a line segment that represents the desired distance.

Proof: The proof starts by observing that the point, c, in the graph is (x_2, y_1). It is important to notice that this identification does not depend on the relative placements of the points a and b; the point a could be down and to the right of b, and c would still have coordinates (x_2, y_1). That is, c has the x-coordinate of b and the y-coordinate of a, independent of the locations of a and b.

If a and b are the same point, then the theorem correctly indicates that the distance is 0. Otherwise, the distance calculation breaks naturally into three cases.

Case 1: $\boldsymbol{x_1 = x_2}$

In this case, the length of a vertical line is sought. That is just $|y_2 - y_1|$, which can be written as $\sqrt{(y_1 - y_2)^2}$. Since $(x_1 - x_2)^2 = 0$, the theorem is valid in this case.

Case 2: $\boldsymbol{y_1 = y_2}$

This case seeks the length of a horizontal line. The result is $\sqrt{(x_1 - x_2)^2}$ and the theorem is again valid for this case.

Case 3: $\boldsymbol{x_1 \neq x_2}$ ***and*** $\boldsymbol{y_1 \neq y_2}$

The graph shown previously represents this case. As was done in the first two cases, it is

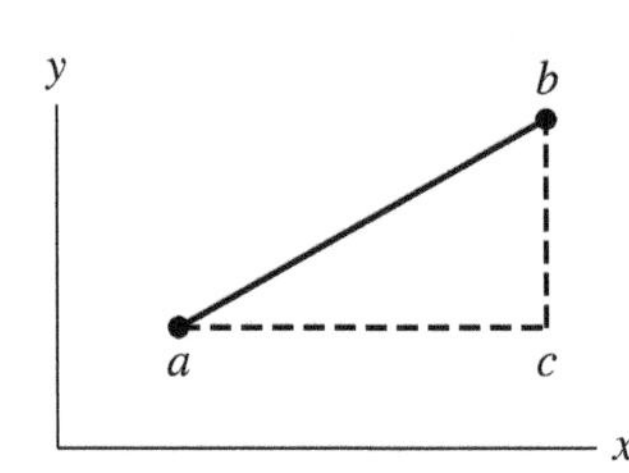

Figure 3.12. The line segment from a to b in $\mathbb{R}^2$.

easy to determine the lengths of the horizontal and vertical dashed lines. Those lengths are $|x_2 - x_1|$ and $|y_2 - y_1|$, respectively. It is again useful to express these lengths as $\sqrt{(x_1 - x_2)^2}$ and $\sqrt{(y_1 - y_2)^2}$.

The length of the solid line can be found using the Pythagorean theorem.

$$d^2 = \left(\sqrt{(x_1 - x_2)^2}\right)^2 + \left(\sqrt{(y_1 - y_2)^2}\right)^2 = (x_1 - x_2)^2 + (y_1 - y_2)^2$$

Consequently,

$$d = \sqrt{(x_1 - x_2)^2 + (y_1 - y_2)^2}$$

□

Both the methodology and the results of Theorem 3.72 can be used to prove Theorem 3.71. The key idea is to try and find the Pythagorean theorem hiding in the graph.

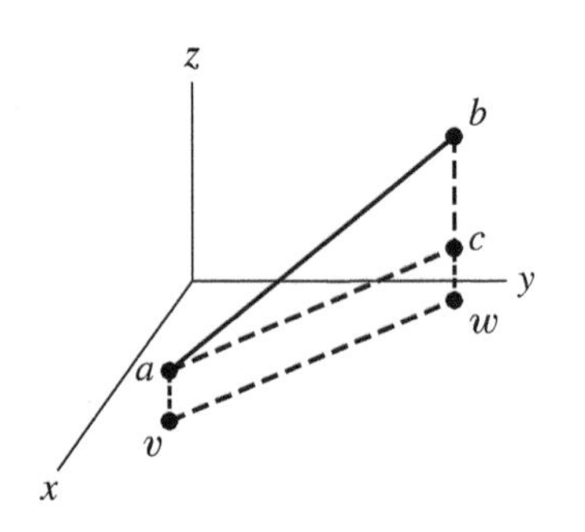

Figure 3.13. Proving Theorem 3.71.

Proof of Theorem 3.71 If the points a and b are the same point, then the theorem correctly produces a distance of 0. Otherwise, produce the point $c = (x_2, y_2, z_1)$ having the same x-coordinate and y-coordinate as b, but the same z-coordinate as a (Figure 3.13). The line segment $\overline{bc}$ will be parallel to the z-axis and the line segment $\overline{ac}$ will be parallel to the xy-plane. If the line segment $\overline{ac}$ is perpendicularly projected onto the xy-plane, the segment $\overline{vw}$ will result. The lengths of $\overline{ac}$ and $\overline{vw}$ will be the same.

The length[66] of vertical line segment $\overline{bc}$ is $|z_2 - z_1| = \sqrt{(z_1 - z_2)^2}$.

On the other hand, Theorem 3.72 implies that the length of $\overline{vw}$ is

$$\sqrt{(x_1 - x_2)^2 + (y_1 - y_2)^2}$$

Thus, the length of $\overline{ac}$ is also $\sqrt{(x_1 - x_2)^2 + (y_1 - y_2)^2}$.

The Pythagorean theorem is valid for the triangle abc. The length of the line segment joining points x and y will be denoted by $d(x, y)$. Thus,

$$\begin{aligned} d(a,b)^2 = d(a,c)^2 + d(b,c)^2 &= \left(\sqrt{(x_1 - x_2)^2 + (y_1 - y_2)^2}\right)^2 + \left(\sqrt{(z_1 - z_2)^2}\right)^2 \\ &= (x_1 - x_2)^2 + (y_1 - y_2)^2 + (z_1 - z_2)^2 \end{aligned}$$

and so

$$d(a,b) = \sqrt{(x_1 - x_2)^2 + (y_1 - y_2)^2 + (z_1 - z_2)^2}$$

□

Prove a More General Theorem

Often it is useful to prove something else instead of (or prior to) proving the desired theorem. Usually the alternative theorem will be a related but simpler theorem (as in the previous suggestion). It is hoped that the proof of the related theorem will give insight into how to prove the original.

Perhaps the most dramatic variation of this idea is to first prove a "harder" theorem! If the "harder" theorem is a generalization of the original, then the original need never be proved. Sometimes the more general theorem is also easier to prove. Consider the following example, from [98] (the assertion has already been proved by contradiction on page 104). We will actually proceed through a series of alternative theorems.

> **PROPOSITION 3.35**
> The number $\sqrt{2}$ is irrational.

Instead of proving the proposition directly, the following more general, claim will be proved first.

[66] It is still important to recognize that this value doesn't depend on the graph. In particular, if $z_1 = z_2$, the line segments $\overline{ab}$ and $\overline{ac}$ are the same line segment.

THEOREM 3.73 $\sqrt{n} \notin (\mathbb{Q} - \mathbb{Z})$

For any positive integer n, $\sqrt{n}$ is either a positive integer, or it is irrational.

Some background work is needed before proving Theorem 3.73. The next definition is of value for this example but need not be memorized.

DEFINITION 3.74 ***Honest***

A rational number $\frac{p}{q}$ is honest if and only if p and q have no common factors and $q > 1$.

LEMMA 3.75

If $\frac{p}{q}$ is an honest fraction, then so is $\left(\frac{p}{q}\right)^2$.

Proof: If p and q have no common factors, then $p \cdot p$ and $q \cdot q$ don't either. Since $\frac{p}{q}$ is honest, $q > 1$. Thus, $q^2 > 1$. Hence, $\left(\frac{p}{q}\right)^2$ is an honest fraction. □

COROLLARY 3.76

If $\left(\frac{p}{q}\right)^2$ is not an honest fraction, then neither is $\frac{p}{q}$.

Proof: This is the contrapositive of the lemma. □

Proof of Theorem 3.73: The real numbers can be partitioned into the disjoint sets of integers, honest fractions, and irrationals, so n (an integer) is *not* an honest fraction. By the corollary to the lemma, if n is not an honest fraction, then neither is $\sqrt{n}$. Since $\sqrt{n}$ is not an honest fraction, it must be an integer or an irrational. □

It is now easy to prove Proposition 3.35.

Proof of Proposition 3.35: By Theorem 3.73, since 2 is an integer, $\sqrt{2}$ is either an integer or it is irrational. We know that it is not an integer, so it must be irrational. □

3.6.3 Exercises

The exercises marked with ★ have detailed solutions in Appendix H.

1. Prove, without using mathematical induction, that $\sum_{k=1}^{2m} k = m(2m+1)$ for all nonnegative integers, m. [*Hint*: Add 1 and $2m$, 2 and $2m-1$, 3 and $2m-2$, etc.]
2. ★ Use the result of Exercise 1 to
   (a) Find the value of $\sum_{k=1}^{2m+1} k$ for $m \geq 0$, where $m \in \mathbb{N}$.
   (b) Provide a noninductive proof that $\sum_{k=1}^{n} k = \frac{n(n+1)}{2}$ for $n \geq 1$, where $n \in \mathbb{N}$.
3. Find the errors in the proof of the given claim. Additionally, determine whether the claim is true or false. If it is true, provide a correct proof. If it is false, find a counterexample.

   Claim:
   Let x be any even natural number. Let y be any odd natural number greater than x. Then $(xy) \bmod (y-x) = xy$.

   Proof:
   Given that x is any even natural number, $x = 2k$ for some natural number, k. Similarly, since y is any odd natural number greater than x, there is a natural number, k, such that $y = 2k+1$. Thus,

$$\begin{aligned}(xy) \bmod (y-x) &= (2k(2k+1)) \bmod ((2k+1)-2k)\\ &= (4k^2+2k) \bmod 1\\ &= (4k^2+2k)\\ &= xy\end{aligned}$$

4. Prove that if n is an even integer and m is an odd integer, then either 4 divides mn or 4 does not divide n.
5. ★ Use an indirect proof to prove the following: "If $2^n - 1$ is prime, then n is prime". [*Hint*: Try some examples. Factoring $2^n - 1$ is easier when n is an even composite integer.]

6. Let a, b, and c be integers with $a \neq 0$. Prove that if $a \mid b$ and $a \mid (b+c)$, then $a \mid c$.
7. Prove that the square of any integer can be written in one of the following forms: $4k$ or $4k+1$.
8. Use a proof by contradiction to show that if $n \in \mathbb{Z}$, then n^2-3 is not divisible by 4. You may use the result of Exercise 7.
9. Prove that the sum of two odd integers is an even integer.
10. Prove that the sum of an odd number of odd integers is odd. You may use the result of Exercise 9.
11. ★ Prove that the sum of two rationals is a rational.
12. Prove that the product of two rationals is a rational.
13. Prove that the sum of a rational and an irrational is an irrational.
14. If x is irrational, then $x + (-x) = 0$ is a rational. Suppose that y is also an irrational with $y \neq -x$. Prove or find a counterexample: $x + y$ is irrational.
15. Prove or find a counterexample: The product of a nonzero rational and an irrational is irrational.
16. Prove or find a counterexample: The product of two irrational numbers is irrational.
17. Let $x \in \mathbb{Z}$. Prove that x is divisible by 3 if and only if $x^2 - 1$ is not divisible by 3.
18. Prove Proposition 3.37 using the strategy $1 \leftrightarrow 2$ and $2 \leftrightarrow 3$.
19. Suppose $n \in \mathbb{Z}$ is not divisible by 3. Prove that
$$n^2 \bmod 3 = 1$$
20. Let $p > 5$ be a prime. Prove that $p+2$ is not a prime.
21. ★ Let $x, y \in \mathbb{R}$. Prove that $|xy| = |x| \cdot |y|$.
22. Use a proof by contradiction to show that the equation $x^3 + 3x + 3 = 0$ has no solutions in $\mathbb{Q}$. [*Hint*: Use a proof by cases inside the proof by contradiction.]
23. Let a and b be real numbers. Prove that
$$\max(a, b) + \min(a, b) = a + b$$
24. Let a and b be real numbers. Prove
    (a) $\max(a, b) = \frac{a+b}{2} + \left|\frac{a-b}{2}\right|$
    (b) $\min(a, b) = \frac{a+b}{2} - \left|\frac{a-b}{2}\right|$
25. Let a and b be positive integers where $a = p_1^{a_1} p_2^{a_2} \cdots p_k^{a_k}$ and $b = p_1^{b_1} p_2^{b_2} \cdots p_k^{b_k}$ with $a_i, b_i \geq 0$. Prove that
    (a) $\gcd(a, b) = p_1^{\min(a_1,b_1)} p_2^{\min(a_2,b_2)} \cdots p_k^{\min(a_k,b_k)}$
    (b) $\operatorname{lcm}(a, b) = p_1^{\max(a_1,b_1)} p_2^{\max(a_2,b_2)} \cdots p_k^{\max(a_k,b_k)}$
26. ★ Let $0 < a \leq b$, where a and b are positive integers. Prove that $\gcd(a, b) = \gcd(b \bmod a, a)$.
27. Show that if a and b are positive integers, then
$$ab = \gcd(a, b) \cdot \operatorname{lcm}(a, b)$$
28. Let $x \in \mathbb{N}$. Prove: If the sum of the digits of x is divisible by 3, then x is also divisible by 3.
29. Suppose that p and $p+2$ are both primes (called *twin primes*). What can you say about $p \bmod 3$?
30. Prove: If n is odd, then $n^2 \equiv 1 \pmod 8$.
31. Let $q \in \mathbb{Z}$. Prove that the following are equivalent.
    - 7 divides q
    - $q^2 \neq 7c+1$, $q^2 \neq 7c+2$, and $q^2 \neq 7c+4$ for any $c \in \mathbb{Z}$
    - 7 divides q^2
32. Let $m, n \in \mathbb{R}$. Prove: $m^2 = n^2 \leftrightarrow ((m = n) \text{ or } (m = -n))$. Do not trivialize this exercise. You need to use the zero product principle (Appendix A.1) to complete this proof.
33. ★ Let S be a set having $n \geq 0$ elements. Prove that S has 2^n subsets (including $\emptyset$ and S itself).
34. Prove that every odd integer can be written as the difference of two squares. (All numbers here are integers.)
35. Let n be a positive integer. Prove the following.
    (a) If $n \equiv 1 \pmod 3$, then $n(n+1) \equiv 2 \pmod 3$. Otherwise, $n(n+1) \equiv 0 \pmod 3$.
    (b) $n(n+1) \not\equiv 1 \pmod 3$.
36. Let a, b, and c form a primitive Pythagorean triple (i.e., $a^2 + b^2 = c^2$ and no prime divides all three). Prove that c is always odd, one of a and b is odd and the other is even.
37. ★ Let a, b, and c form a primitive Pythagorean triple. Show that one of a or b is a multiple of 3. [*Hint*: Use the conclusions of Exercises 35 and 36.]
38. Use a proof by cases to show that if a is an integer, then $a^5 - a$ is divisible by 5.
39. Prove that the product of any four consecutive integers is divisible by 8.
40. Provide a proof for the following claim: Every integer of the form $6^{3k} + 1$ is composite, where k is a positive integer.
41. Let $n \geq 2$ be a positive integer. Prove that n and $n+1$ have no common prime factors.
42. Let x and a be elements in a Boolean algebra. Prove that if $x \cdot a = a$ then $\overline{x} \cdot a = 0$.

3.7 Quick Check Solutions

Quick Check 3.1

1. Let n be odd. Then Definition 3.10 indicates that there does not exist an integer, k, such that $n = 2k$. That is, n is not divisible by 2. The Quotient–Remainder theorem asserts that n can be uniquely expressed in the form $n = 2q + r$, where r is an integer with $0 \leq r < 2$. Thus, $r \in \{0, 1\}$. Since n is not divisible by 2, the only admissible choice is $r = 1$. Thus, $n = 2q + 1$, with q an integer. The letter used to denote the quotient is not important, so this can be stated as $n = 2k + 1$, for some integer, k.

Quick Check 3.2

1. $100 = 2^2 \cdot 5^2$, 101 is prime, $102 = 2 \cdot 3 \cdot 17$, 103 is prime, $104 = 2^3 \cdot 13$, $105 = 3 \cdot 5 \cdot 7$

2. (a) $q = 3$, $r = 250$, $1600 = 3 \cdot 450 + 250$

   (b) $a = 1600 = 2^6 \cdot 5^2$, $b = 450 = 2 \cdot 3^2 \cdot 5^2$

   (c) $\gcd(1600, 450) = 2 \cdot 5^2 = 50$

   (d) To find the least common multiple, take the maximum power of each prime factor in the two factorizations (it is not necessary that a prime appear in both factorizations).

$$\text{lcm}(1600, 450) = 2^6 \cdot 3^2 \cdot 5^2 = 14400$$

   (e) Use the remainder from part (a): $a \bmod b = 250$.

   (f) $1600 = 145 \cdot 11 + 5$, $450 = 40 \cdot 11 + 10$
   They are not congruent mod 11 because the remainders are different. Another way to see this is to notice that $1600 - 450 = 1150$ is not divisible by 11.

Quick Check 3.3

1. **Proof:**
   Since $a \mid b$ we know that there is an integer k such that $b = ak$ (concept $\rightarrow$ properties). Therefore, $bc = (ak)c = a(kc)$. Since kc is an integer this shows that bc is an integer multiple of a, so bc is divisible by a (properties $\rightarrow$ concept).

2. **Proof:**
   Since $a \mid b$, we know that there is an integer k such that $b = ak$ (concept $\rightarrow$ properties). Also, since $b \mid c$, we know that there is an integer m such that $c = bm$ (concept $\rightarrow$ properties again). Therefore, $c = bm = (ak)m = a(km)$. Since km is an integer, this shows that c is divisible by a (properties $\rightarrow$ concept).

Quick Check 3.4

1. **Proof:**
   Since n is positive and composite, there are integers a and b with $1 < a < n$, $1 < b < n$ and $n = ab$. Suppose that $a > \sqrt{n}$ and $b > \sqrt{n}$. Then $n = ab > \sqrt{n} \cdot \sqrt{n} = n$, a contradiction. Therefore, either $a \le \sqrt{n}$ or $b \le \sqrt{n}$ (or both). Without loss of generality, we may assume $a \le \sqrt{n}$. If a is a prime, we are done. Otherwise, a must itself have a prime divisor p and $p < a \le \sqrt{n}$ must be true. Proposition 3.31 implies that p is a prime divisor of n.

Quick Check 3.5

1. **Proof:**
   Let $\frac{p}{q} \in \mathbb{Q}$ be any rational number with $\frac{p}{q} \neq 0$. Since $\frac{p}{q} \neq 0$, we know that $p \neq 0$. Therefore, $\frac{q}{p}$ is also a rational number. We also know that $q \neq 0$ (since $\frac{p}{q}$ is a rational number). Thus $\frac{p}{q} \cdot \frac{q}{p} = \frac{pq}{qp} = 1$. The number $\frac{q}{p}$ is the multiplicative inverse we sought.

   Notice that no special properties of $\frac{p}{q}$ were used except the fact that $p \neq 0$ (which is part of the proposition's hypothesis and must be used).

Quick Check 3.6

1. gcd(78, 60) = 6

Phase 1:

$$\begin{aligned} 78 &= 60 \cdot (1) + 18 \\ 60 &= 18 \cdot (3) + 6 \\ 18 &= 6 \cdot (3) + 0 \end{aligned}$$

Phase 2:

$$\begin{aligned} 6 &= 60 - 18 \cdot (3) && \text{solve for remainder} \\ &= 60 - (78 - 60 \cdot (1)) \cdot (3) && \text{substitute using previous equation} \\ &= 60 \cdot (4) + 78 \cdot (-3) && \text{simplify} \end{aligned}$$

Quick Check 3.7

1. The generator in Example 3.25 meets the requirements in Theorem 3.41.

1. Clearly, $b = 11$ and $c = 2^{48}$ are relatively prime.
2. The only prime that divides c is 2, and $2 \mid (a - 1)$ since $a - 1$ is even.
3. Finally, $4 \mid (a - 1)$ since $\frac{a-1}{4} = 6303725979$.

Quick Check 3.8

1. Theorem 3.46 applies since 2, 3, and 5 are all primes (and hence relatively prime). The theorem indicates that

$$x = (2 \cdot 35 \cdot 2 + 3 \cdot 21 \cdot 1 + 2 \cdot 15 \cdot 1) \bmod 105 = 233 \bmod 105 = 23$$

where the modular inverses are all easy to guess.

2. Theorem 3.46 applies, even though $m_1 = 4$ is not a prime. All that is necessary is that $\gcd(m_i, m_j) = 1$ when $i \neq j$. The theorem indicates that

$$x = (1 \cdot 15 \cdot 3 + 2 \cdot 20 \cdot 2 + 4 \cdot 12 \cdot 3) \bmod 60 = 269 \bmod 60 = 29$$

3. Since 3 and 6 are not relatively prime, Theorem 3.46 does not apply. However, in this case there is still a solution: $x = 5$.

Quick Check 3.9

1. (a) The numeric values of the three messages are 4 15 7.

$$\begin{aligned} c_1 &= 4^5 \bmod 39 = 1024 \bmod 39 = 10 \\ c_2 &= 15^5 \bmod 39 = 759375 \bmod 39 = 6 \\ c_3 &= 7^5 \bmod 39 = 16807 \bmod 39 = 37. \end{aligned}$$

(b)

$$\begin{aligned} m_1 &= 25^5 \bmod 39 = 9765625 \bmod 39 = 25 \\ m_2 &= 5^5 \bmod 39 = 3125 \bmod 39 = 5 \\ m_3 &= 28^5 \bmod 39 = 17210368 \bmod 39 = 19. \end{aligned}$$

The plaintext is yes.

Quick Check 3.10

1. (c) $1 + 3 + \cdots + (2i + 1) = (i + 1)^2$
2. Following the standard pattern, define $P(n)$ as

$$P(n):\ \sum_{i=1}^{n}(2i-1) = n^2$$

Base Step
Since $\sum_{i=1}^{1}(2i-1) = 1$ and $1^2 = 1$, $P(1)$ is true.

Inductive Step
Assume that $P(k)$ is true for some $k \geq 1$. We want to show the $P(k+1)$ must also be true under that assumption. Observe that $P(k+1)$ can be written

$$P(k+1):\ \sum_{i=1}^{k+1}(2i-1) = (k+1)^2$$

The following equalities show that $P(k+1)$ is true if $P(k)$ is true.

$$\begin{aligned}
\sum_{i=1}^{k+1}(2i-1) &= \sum_{i=1}^{k}(2i-1) + (2(k+1)-1) \\
&= \sum_{i=1}^{k}(2i-1) + (2k+1) \\
&= k^2 + (2k+1) && \text{by the inductive hypothesis} \\
&= (k+1)^2 && \text{by a simple factorization}
\end{aligned}$$

Conclusion
Since $P(1)$ is true, and $P(k) \rightarrow P(k+1)$ is a valid implication, we conclude that $P(n)$ is true for $n \geq 1$.

3. Let $P(n)$ be the equation $a\left(\sum_{i=1}^{n} b_i\right) = \sum_{i=1}^{n} ab_i$.

Base Step
When $n = 2$, the statement is the normal distributive property of the real numbers, so it is certainly true.

$$a\left(\sum_{i=1}^{2} b_i\right) = a(b_1 + b_2) = ab_1 + ab_2 = \sum_{i=1}^{2} ab_i$$

Inductive Step
Assume that $P(n)$ is true for some n. Consider $P(n+1)$.

$$\begin{aligned}
a\left(\sum_{i=1}^{n+1} b_i\right) &= a\left(\sum_{i=1}^{n} b_i + b_{n+1}\right) \\
&= a\left(\sum_{i=1}^{n} b_i\right) + ab_{n+1} && \text{using the normal distributive law} \\
&= \sum_{i=1}^{n} ab_i + ab_{n+1} && \text{by the inductive hypothesis} \\
&= \sum_{i=1}^{n+1} ab_i
\end{aligned}$$

Conclusion
The base step ($P(2)$ is true) and the inductive step ($P(n) \to P(n+1)$ is true) are both valid. The theorem of mathematical induction implies that $P(n)$ is true for all $n \geq 2$.

Quick Check 3.11

1. You might have chosen either one of the following base steps (both are acceptable). Let n represent the initial number of pieces of popcorn in the bowl. Let $P(n)$ be the claim "If the bowl starts with n pieces of popcorn, then eventually the bowl will become empty".

Base Step $n = 0$
If the bowl starts empty, then there is nothing to show.

Base Step $n = 1$
If the bowl starts with one piece of popcorn, then when Jedediah takes his first "handful" he will grab that piece (since he must take at least one piece). The bowl will now be empty.

Inductive Step
Assume that when the bowl starts with k pieces of popcorn, for $0 \leq k \leq n$, the bowl will eventually become empty.

Let the next bowl start with $n+1$ pieces of popcorn. Jedediah takes his first handful. He will take at least one piece of popcorn. The bowl will now contain k pieces of popcorn, with $0 \leq k \leq n$. By the inductive hypothesis, this revised bowl will eventually become empty.

Conclusion
The base step [$P(0)$ is true] and the inductive step [$(P(0) \wedge P(1) \wedge \cdots \wedge P(n)) \to P(n+1)$ is true] are both valid. The theorem of complete induction implies that $P(n)$ is true for all $n \geq 0$.

2. This result should not be a surprise; it just states that every positive integer has a base 2 representation.

Base Step $k = 1$
This is easy: $1 = 2^0$.

Inductive Step
Assume that every integer k, where $1 \leq k \leq n$, can be expressed as a sum of powers of 2. Consider the integer $n+1$.

We know that $1 < n+1$, so $n+1 \geq 2^i$ for at least one nonnegative value of i. Let j be the largest exponent for which $n+1 \geq 2^j$. If $n+1 = 2^j$, the inductive step is complete. Otherwise, let $m = (n+1) - 2^j$.

Since $j > 0$, $m < n+1$. Suppose that $m \geq 2^j$. Then

$$n + 1 = m + 2^j \geq 2^j + 2^j = 2^{j+1}$$

which contradicts the maximality of j. Therefore, $m < 2^j$.

By the (complete) inductive hypothesis, $m = 2^{i_1} + 2^{i_2} + \cdots + 2^{i_s}$, where all the powers of 2 are distinct and strictly less than 2^j.

Therefore, $n + 1 = 2^j + 2^{i_1} + 2^{i_2} + \cdots + 2^{i_s}$ is also a sum of distinct powers of 2.

Conclusion
The claim is true for $n = 1$ and whenever it is true for $1 \leq k \leq n$, it is also true for $n+1$. The principle of complete induction asserts that every positive integer is a sum of distinct powers of 2.

3.8 CHAPTER REVIEW

3.8.1 Summary

There are two primary goals for this chapter. The first, and easier, of the goals is to help you become proficient at reading proofs. The more difficult goal is to help you start the process of becoming proficient at producing proofs.

Proofs are an essential part of this course and will be a central feature in most of your future mathematics courses. A proof not only establishes the truth of some assertion, but at its best, the proof will also help you gain understanding as to *why* the assertion is true.

The task of reading a proof designed by another person is not trivial. In most cases, the proof's author will assume that you are familiar with the strategies and techniques that have been presented in this chapter. The proof may contain few overt signposts that inform you about which strategy is being used. You may need to "read between the lines" and find that information on your own.

Producing proofs is harder. You need to discover the key relationships for yourself. Then you need to find an appropriate way to express the proof in a clear, logical manner. You also need to follow accepted conventions in mathematics (use of notation, presentation style, etc.) and design the presentation for the intended audience. When you first begin creating your own proofs, it is not easy to evaluate the quality of the final product. Have you really proved what you intended, or does your "proof" contain some errors? You really need a second opinion. Having your instructor or a more advanced student evaluate your work will provide essential feedback. Do not get discouraged if your first attempts seem pretty dismal. Learning to create proofs is a process. If you stick with it, you will improve over time.

The chapter begins with two kinds of material that provide a foundation for the rest of the chapter. The first section presents an overview of the environment in which proofs naturally belong: axiomatic mathematics. A brief overview of some basic definitions and theorems in elementary number theory is then provided. Elementary number theory provides a simple context in which to practice the proof techniques presented in the rest of the chapter.

The main core of the chapter is the description of a number of proof strategies (direct proof, proof by contradiction, mathematical induction, etc.), followed by some suggestions on how to produce proofs.

There is an additional section (Section 3.4) containing a number of applications of elementary number theory. That section illustrates the use of several of the proof strategies presented in Section 3.3. In addition, several of the applications and theorems are quite interesting and very useful.

One useful way to begin your review process for this chapter is to create a detailed outline of the proof strategies presented in Sections 3.3 and 3.5. Then you may wish to review the list of general proof strategies in Table 3.15 on page 157. Finally, there is no substitute for spending time creating your own proofs.

3.8.2 Notation

Notation	Page	Brief Description
$\mathbb{N}$	92	the set of natural numbers
$\mathbb{Z}$	93	the set of integers
$\mathbb{Q}$	93	the set of rational numbers
$\mathbb{R}$	93	the set of real numbers
$b \mid a$	93	the integer b divides the integer a
$b \nmid a$	93	the integer, b *does not* divide the integer a
gcd	94	greatest common divisor
lcm	95	least common multiple
$a \bmod b$	98	the remainder when the integer, a, is divided by the integer, b
$a \equiv b \pmod{m}$	98	a is congruent to b, mod m
$\sum_{i=0}^{n} a_i$	132	shorthand for the sum $a_0 + a_2 + \cdots + a_{n-1} + a_n$ (see Appendix B)
$\prod_{i=0}^{n} a_i$	138	shorthand for the product $a_0 \cdot a_2 \cdots a_{n-1} \cdot a_n$
$n!$	99	n factorial
$\lfloor x \rfloor$	99	the floor function
$\lceil x \rceil$	99	the ceiling function
f_n	153	the nth Fibonacci number
$\max(a, b)$	112	the maximum (larger) of real numbers, a and b
$\min(a, b)$	112	the minimum (smaller) of real numbers, a and b
$\tilde{a}$	119	the inverse of a, mod m

3.8.3 Additional Review Material

Go to http://www.mathcs.bethel.edu/~gossett/DiscreteMathWithProof/review.xhtml for additional review material, including a sample chapter exam, with solutions.

CHAPTER

4

Algorithms

An algorithm must be seen to be believed.

Donald Knuth (born 1938)

An algorithm is a process for solving a problem. You have been using formal algorithms since you were quite young. For example, in elementary school you learned how to do multidigit addition with paper and pencil. The standard algorithm has you start with the rightmost column and work toward the left, carrying into the next column when the result is greater than 10.

Some algorithms are so complex that it is necessary to devise a notation to describe them clearly and unambiguously. We will look at one such notation in the first section of this chapter. It is often critically important that we have a way to compare the efficiency of two algorithms that solve the same problem. If one algorithm can be completed in 10 minutes, whereas the other will take 10 days, we will prefer the former. Mechanisms for measuring algorithm efficiency will be presented in Section 4.2.

Creating good algorithms is a challenging (and fun) activity. There will be many opportunities to work with algorithms in the remainder of this book. Section 4.3 will present several algorithmic solutions to a common software task: determining whether (and where) a pattern of characters appears in a document. The efficiency of each algorithm will also be determined.

Before proceeding, it will be helpful to establish a formal definition for the term *algorithm*.

DEFINITION 4.1 ***Algorithm***

An *algorithm* is a finite sequence of unambiguous steps for solving a problem or completing a task in a finite amount of time.

The following is *not* an algorithm, even though it consists of a finite sequence of unambiguous steps:

EXAMPLE 4.1

Not an Algorithm

Suppose we would like to calculate the sum $\sum_{n=1}^{\infty} \frac{1}{n^2}$. An obvious procedure uses two variables, s (the sum), and n (the index variable).

Step 1 Set s to 0 and n to 1.

Step 2 Add $\frac{1}{n^2}$ to s (and store the result in s).

Step 3 Add 1 to n (and store the result in n).

Step 4 Go back to step 2 and continue.

This procedure is certainly correct but of no practical value because the process never ends (and is therefore not an algorithm).[1] ■

[1] The first person to calculate this sum was Leonard Euler. The somewhat surprising answer is $\frac{\pi^2}{6}$ [26].

4.1 Expressing Algorithms

There are many ways to express algorithms. For example, in Section 1.2.2, the deferred acceptance algorithm was expressed using a paragraph of normal English prose and then it was expressed (without explanation) using pseudocode. Pseudocode more fully achieves the goal of avoiding ambiguity. However, it does take some explanation before all the conventions become clear. This section will introduce those conventions.

DEFINITION 4.2 *Pseudocode*

Pseudocode is a semiformal language used to describe algorithms. It is more precise than a prose description, but contains less syntactic structure than a compilable computer language. There are no required syntax rules for pseudocode, but there are many useful notational conventions.

The intermediate level of structure in pseudocode makes it ideal for communicating algorithms. The structure helps to eliminate ambiguity and aids in clearly communicating the steps. On the other hand, since pseudocode does not require strict adherence to a formal syntax, we do not need to spend effort making sure that every required semicolon and closing parenthesis is in the proper place.

Even though there is not a rigid syntactic structure to pseudocode, there are some conventions of notation that are helpful. These conventions come in two major areas: flow of control and flow of information. Flow of control is concerned with the order in which steps are completed. Flow of information is concerned with what data are handed to the algorithm, and what data the algorithm ultimately produces.

4.1.1 Flow of Control

In 1966, Böhm and Jacopini published a paper entitled *Flow Diagrams, Turing Machines and Languages with Only Two Formation Rules* [11]. It was not long before the computer science community realized that their work implied that any algorithm can be expressed using what are now called *structured control* constructs: sequence, selection, and repetition. The essential idea of structured control is that subcollections of steps need to have a single entry and a single exit. This was in sharp contrast with the then current practice of using many *goto* statements (step 4 in Example 4.1 on page 169). Pseudocode allows goto's, but they are discouraged. The three major structured control categories are described next.

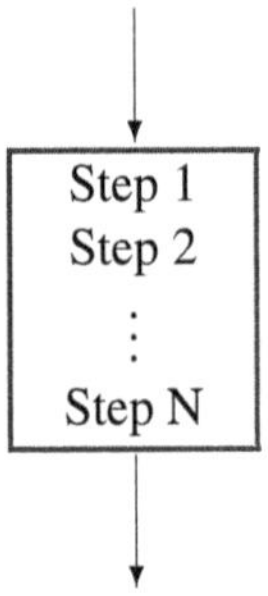

Sequence

The simplest control structure is *sequence*. The convention is to read algorithms from top to bottom, completing each step in order unless directed otherwise. Sequential control is the default. You would most likely perform a purely sequential algorithm without even realizing that you were following this simple convention.

A sequential block of instructions (Figure 4.1) clearly has a single entry (just above the first instruction in the sequence) and a single exit (just after the final instruction in the block).

Step 1
Step 2
⋮
Step N

Figure 4.1. A sequential block.

Selection

Selection control constructs allow the algorithm to take different paths for different initial data. For our purposes, it is useful to consider one-way, two-way, and multiway selection.

One-Way Selection One-way selection allows some steps to be completed conditionally. That is, sometimes the steps will be completed and other times they will be skipped. The pseudocode that achieves this is the *if-then* construct:

```
if condition then
   Step 1
   Step 2
     ⋮
   Step N
```

The indented steps ($1 - N$ here) are called the *body* of the if-then construct. The steps in the body are conditionally executed[2] (depending on the value of *condition*).

Often, the word *then* is omitted in the pseudocode notation. (That convention will be followed in this text.)

```
if condition
   Step 1
   Step 2
     ⋮
   Step N
```

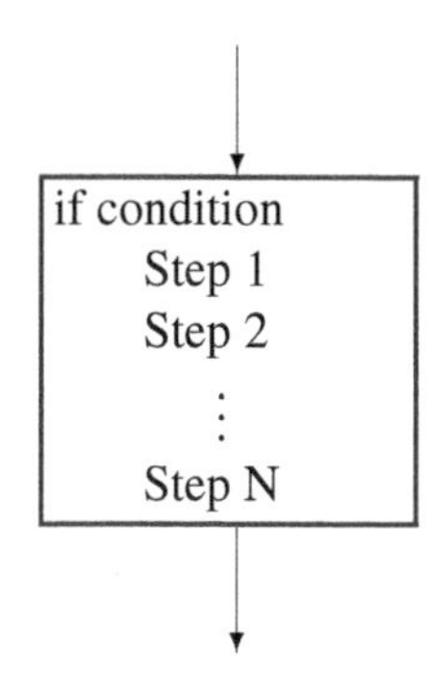

Figure 4.2. An if-then block.

Steps 1–N are completed only if the proposition *condition* evaluates to *true*. The indentation is used to indicate that all N of the steps depend on *condition* being true.

The if-then construct (Figure 4.2) is a single-entry, single-exit structure; the exit is the step immediately following the final step in the if-then (independent of whether the statements in the if-then body are actually executed).[3]

EXAMPLE 4.2

An Algorithm with a Simple If-Then

One simple way to calculate the absolute value of a number is to check its sign. Suppose the number x is a number that can be entered via keyboard and we want to display its absolute value on a computer monitor. The following algorithm will accomplish this. The line numbers are included as a convenience to the reader and are not part of the algorithm.

```
1: read x from the keyboard
2: if x < 0
      set x to −x
3: display x on the monitor
```

Notice the single entry (line 2) and single exit (line 3 even if the "change the sign of x" step is skipped). ■

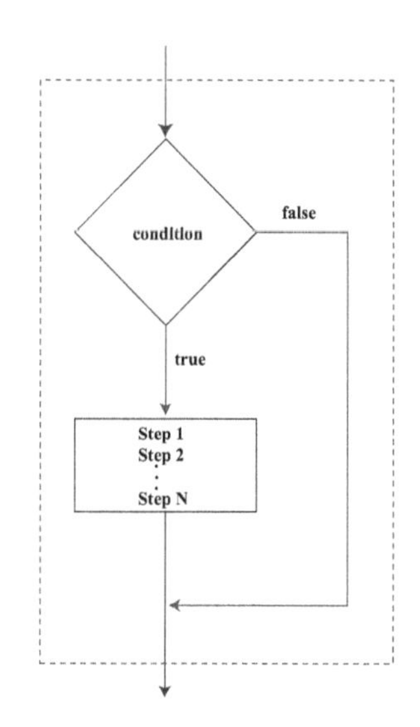

Figure 4.3. Flowchart if-then.

The absolute value algorithm has combined sequence and selection. Think of the if-then selection as a single step. The algorithm is a sequence with three steps (one of which happens to be a selection).

Assignment Operators A useful pseudocode notational convention is to use the compound symbol := to indicate "set the variable on the left to the value on the right." This symbol is therefore called an *assignment operator*. The absolute value algorithm could then be represented as

```
1: read x from the keyboard
2: if x < 0
      x := −x
3: display x on the monitor
```

[2] The term *executed* is another way to say "do what the step indicates".

[3] If you are familiar with flowcharts, which will not be used in this book, the single-entry, single-exit requirement must be imposed by imagining a dotted box around the diamond structure (Figure 4.3).

The computer language C has introduced the symbol = as an alternate notation for the assignment operator. This is a bit unfortunate because we are used to thinking of the symbol = as representing the question "is the left-hand side equal to the right-hand side?". To avoid confusing the assignment operator with the equality operator, C uses the symbol == to indicate the claim that the left-hand side equals the right-hand side.

Because C and its derivatives have essentially won the notational wars, we will use = to represent the assignment operator and the compound symbol == to represent the equality operator. You may still see the := notation in older literature. The final version of the absolute value algorithm is thus

```
1: read x from the keyboard
2: if x < 0
       x = −x
3: display x on the monitor
```

Note: The line $x = -x$ may seem strange. Remember, the assignment operator "sets the variable on the left, x, to the value of the expression on the right, $-x$". If x starts with the value -4, the statement negates that value and places the result, 4, back in x.

Two-Way Selection In a two-way selection, if the proposition *condition* is *true*, the first set of steps are completed; otherwise, the second set of steps are completed. The indentation is again an important part of the notation.

```
if condition then
   Step 1
   Step 2
     ⋮
   Step N
else
   Step N+1
   Step N+2
     ⋮
   Step N+M
```

or

```
if condition
   Step 1
   Step 2
     ⋮
   Step N
else
   Step N+1
   Step N+2
     ⋮
   Step N+M
```

EXAMPLE 4.3

Calculating Paychecks

Suppose all workers at a company are either salaried or hourly workers. A simplified paycheck calculation might look like the following. The symbol * is used to represent multiplication and / represents division.

```
if salaried
  determine yearly_salary
  determine number_of_payperiods
  paycheck = yearly_salary/number_of_payperiods
else
  determine hourly_wage
  determine hours_worked
  determine overtime_hours
  paycheck = hourly_wage*hours_worked
             + 1.5*hourly_wage*overtime_hours
```

■

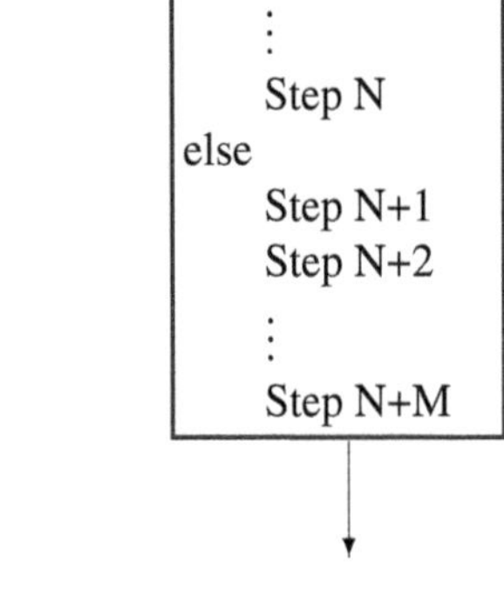

Figure 4.4. An if-else block.

The two-way selection (also called an *if-else* construct) is a single-entry, single-exit control structure (Figure 4.4). It doesn't matter whether the *true* path (then) or the *false* path (else) is chosen; the step after the if-else will always be the one immediately after the structure.

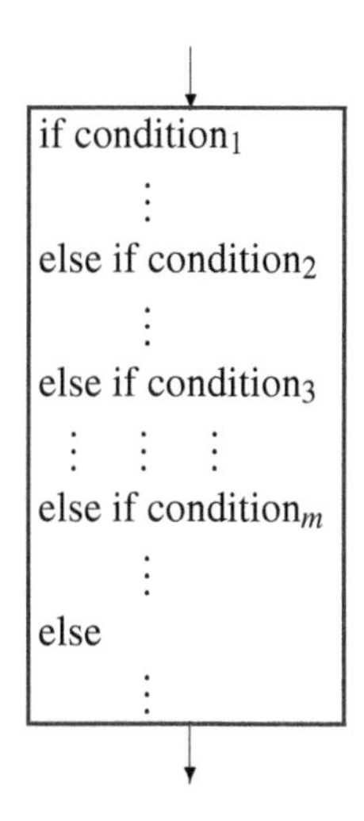

Figure 4.5. A multiway selection block.

Multiway Selection A multiway selection construct provides more than two mutually exclusive options (Figure 4.5). In a multiway selection, we can pick from among many mutually exclusive options. We may also optionally provide a default choice if none of the other options are appropriate. One way to express this in pseudocode is as follows:

```
if condition1
      ⋮
else if condition2
  ⋮   ⋮   ⋮
else if conditionm
      ⋮
else
      ⋮
```

or

```
if condition1
      ⋮
else if condition2
  ⋮   ⋮   ⋮
else if conditionm
      ⋮
```

where the final *else* section may be omitted if there is no default collection of steps.

As usual, there is a single entry and a single exit. Once one of the conditions is true, the steps in that section are executed and then the first step after the multiway structure is the next executed.

EXAMPLE 4.4

Tuition

This simple example illustrates a three-way selection. It will be extended when nesting is discussed. (Notice that scholarships cancel out-of-state tuition.)

```
if student is on a scholarship
  fee = in-state tuition - amount of scholarship
else if student is a state resident
  fee = in-state tuition
else
  fee = out-of-state tuition
```
■

✔ Quick Check 4.1

1. Write an algorithm fragment that prints the sum of two numbers, a and b. Before performing the addition, it should check to see if a equals 0. If it does, a should be converted to 1 before the addition is done.
2. Write an algorithm fragment that determines which of two numbers, a and b, is the smaller. In case of a tie, either one can be chosen. ☑

Repetition

The final control structure category consists of those that repeat a collection of statements. These structures can repeat for either a predetermined number of times or for a variable number of times (determined by a condition).

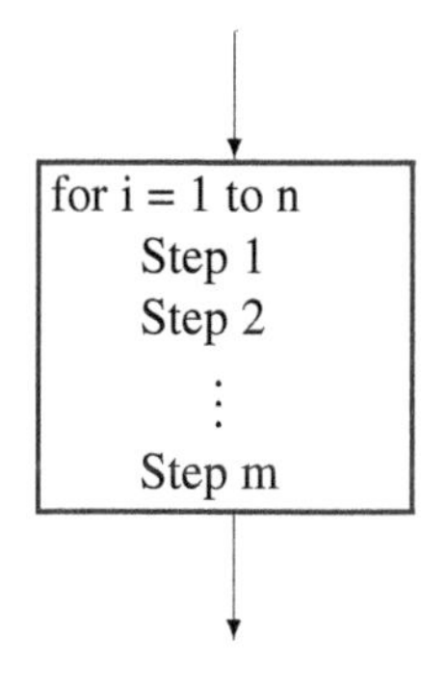

Figure 4.6. A for loop.

Fixed Iteration When the number of times a collection of steps needs to be executed is known (or is stored in a variable), the proper repetition structure is a fixed iteration. One of the many possible pseudocode expressions of this structure is a simple *for* loop (Figure 4.6). The for loop uses an index variable to count how many times the loop has been executed.

Each time through the loop, the index variable (`i` in Figure 4.6) is incremented. The loop is complete when the index variable gets larger than its terminal value (`n` in the pattern). Every time final step `m` is executed, control goes back to the top of the for loop. If `i` is still less than or equal to `n`, the loop body is executed again. Otherwise, the next statement will be the one immediately after step `m`.

The indentation highlights the body of the loop (the statements that are repeated). Note that the body of the loop is using the sequence structure.

EXAMPLE 4.5

Adding

To add the numbers 1–100, the following for loop works well.

```
1:   sum = 0
2:   for i = 1 to 100
2a:    sum = sum + i
3:   display the value of sum
```

Notice the convention used in the assignments; place the value of the right-hand side into the variable on the left. Thus, the statement `sum = sum + i` means, add `i` to the current value of `sum`, then replace the current value of `sum` by the result.

I have chosen to number the steps rather than numbering lines (as is typical) for the purpose of highlighting the single entry, single exit nature of the for loop. As soon as `i` becomes greater than 100, steps 2a and 2b are skipped and step 3 is executed. ■

In the previous example, I implicitly incremented the index variable inside the loop body. Some people use an explicit increment. If you see a for loop and the index variable is not explicitly incremented, the convention is that it is incremented by 1 after the final step in the body (but before returning to the top of the loop to compare with the terminal value).

Indefinite Iteration There are two common control structures that allow a loop to be executed until some condition is met. The more common is the *while* loop, which uses a pretest to determine when to quit. The less common *repeat-until* loop uses a post-test to determine when to exit the loop.

```
while condition                 repeat
   Step 1                          Step 1
   Step 2                          Step 2
     :                               :
   Step N                          Step N
                                until condition
```

The while loop tests the condition before entering the body of the loop. If the condition is false, the loop body is not executed and the first statement after step `N` is the next executed.[4] Otherwise, the loop body is executed and then control goes back to the top, where the condition is once again checked.

The repeat-until loop always executes the loop body at least once. After the loop body has been executed, the condition is checked. If the condition is false, the loop body is repeated again; otherwise the loop is terminated and the first step after the until line is the next to be executed.

Note the difference: A while loop keeps executing as long as the condition is true; a repeat-until loop keeps executing as long as the condition is false.

EXAMPLE 4.6

Count Down

The next two loops accomplish the same objective: Count down to blastoff.

```
i = 10                          i = 10
while i > 0                     repeat
   display i                       display i
   i = i - 1                       i = i - 1
display "Blast Off!!"           until i = 0
                                display "Blast Off!!"
```

■

[4]It is even possible that the condition will be false when the while loop is first encountered. In that case, the loop body will never be executed.

All three of the loop structures are single-entry, single-exit (Figures 4.6 – 4.8).

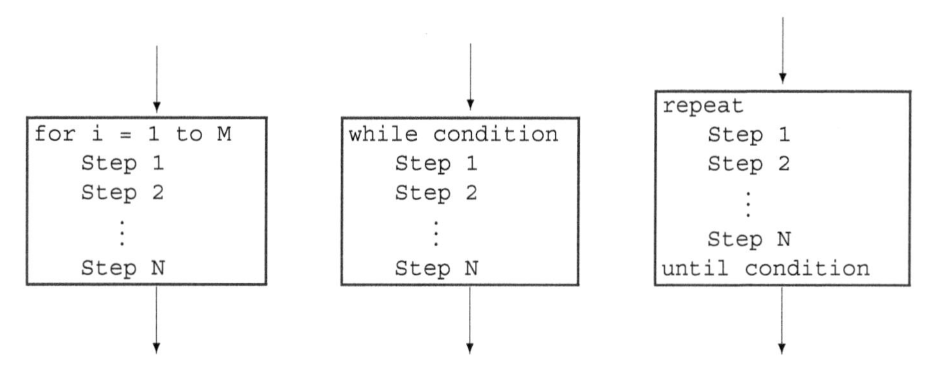

Figure 4.6 A for block. **Figure 4.7** A while block. **Figure 4.8** A repeat block.

There is another, more complex, mechanism that causes a collection of steps to be repeated. This more complex construct is called *recursion* and will be introduced in Chapter 7.

Nesting

Some of the power of the control mechanisms just presented is the ability to combine them. I have already shown examples of combining them by placing one structure after another (sequencing structures as well as steps in an algorithm). It is also possible to combine them using nesting (composition). Any step in one of the patterns can be replaced by a control structure. A second look at the tuition example illustrates this.

EXAMPLE 4.7

Tuition Revisited

This algorithm cycles through all students in the school. It also allows out-of-state students to have scholarships. The algorithm also calculates the total amount of tuition to be collected.[5]

```
total = 0
for student = first_student to last_student
   if student is a state resident
      if student is on a scholarship
         fee = in-state tuition - amount of scholarship
      else
         fee = in-state tuition
   else
      if student is on a scholarship
         fee = out-of-state tuition - amount of scholarship
      else
        fee = out-of-state tuition
   total = total + fee
display total
```

Figure 4.9 on page 176 shows the nesting. The boxes are all single-entry, single-exit constructs. ■

[5]For pedagogical reasons, I have used more nesting than is necessary in this example.

```
total = 0
for student = first_student to last_student
    if student is a state resident
        if student is on a scholarship
            fee = in-state tuition - amount of scholarship
        else
            fee = in-state tuition
    else
        if student is on a scholarship
            fee = out-of-state tuition - amount of scholarship
        else
            fee = out-of-state tuition
    total = total + fee
display total
```

Figure 4.9 Nesting of structured constructs.

A style of computer programming arose after the control structures we have been examining were found to be sufficient for expressing any algorithm. The style was named *structured programming*. Structured programming stipulates that programs (algorithms) be expressed using only single-entry, single-exit control structures from the categories *sequence, selection*, and *repetition*. These control structures may be nested.[6]

Quick Check 4.2

1. Write an algorithm fragment that calculates the sum of the first n odd integers.
2. Write an algorithm fragment that reads a finite sequence of characters and prints the total number of vowels encountered. ☑

4.1.2 Flow of Information

Algorithms need to specify more than just the order in which the steps are carried out. They also need to specify what information will be given at the start (the *input parameters*) and what information will be produced (the *return values*). At times, it is useful to break an algorithm into two or more communicating pieces. It then becomes necessary to identify the pieces and to define the information they will share. Some notation loosely borrowed from computer programming will provide the mechanisms we need.

The pseudocode for an algorithm may optionally give the algorithm a name and define the data values it needs to start and also specify the data values that will be produced. We do this by placing an extra line at the beginning of the algorithm. The general pattern is[7]

return_values **algorithm_name**(input_parameters)

It is often important to specify the *type* of the data elements. This can be done by specifying a data type (such as *integer*, *real* number, *complex* number), followed by the name of the variable that holds the value. The naming of the data types is informal but should be unambiguous. If there are no specific return values, that part can be omitted

algorithm_name(input_parameters)

[6]The control structures have been extended by defining two mechanisms for leaving a loop from the middle of its body. The *break* statement causes the loop to terminate prematurely and control to go to the next step after the loop. The *continue* statement causes the body to terminate prematurely, but control goes back to the top of the loop (so the loop body may be done again). We will not need these extensions.

[7]You do not need to use a bold font for the algorithm name.

or the word *void* can be used to indicate there is no return value (perhaps we just print something).

```
void algorithm_name(input_parameters)
```

EXAMPLE 4.8

Adding a List of Numbers

Suppose we have a list of numbers that we wish to add. The algorithm below will accomplish that task and produce the sum as its return value. The integer n specifies the size of the list.

```
real add_List(integer n, real {a_1, a_2, ..., a_n})
   sum = 0
   for i = 1 to n
      sum = sum + a_i
   return sum
end add_List
```

The first line specifies that the algorithm `add_List` will return a real number. It requires an integer and a collection of real numbers. The line `return sum` instructs the algorithm to terminate and send the current value of `sum` to the program or person that started the algorithm. One or more `return` lines are required unless the algorithm has the `void` return value. The final line, `end` **add_List**, is optional. It helps to visually delimit the algorithm.

Another algorithm can use `add_List` in one of its steps. The next algorithm will display the sums of the first n integers, for values of n from 1 up to 20.

```
for n = 1 to 20
   display add_List(n, {1, 2, ..., n})
```

Since `add_List` returns a real number, we can use the name `add_List` as if it is that number. ■

The previous example illustrates another notational convention we will need to observe. When a named algorithm is *defined*, the types of the parameters and return value are specified. The entire algorithm is also written out. When a named algorithm is *used*, the types are not specified.[8] In addition, only the first line is written. The details of the steps are not written.

EXAMPLE 4.9

Adding Even Numbers

Suppose we wish to add the even numbers from 0 to $2n$. The algorithm `add_Evens` is a simple solution.

```
integer add_Evens(integer n)
   sum = 0
   for i = 1 to n
     sum = sum + 2*i
   return sum
end add_Evens
```

In mathematical notation, `add_Evens(n)` $= \sum_{i=1}^{n} 2i$.

[8]The parameters in the definition are called *formal* parameters. When the algorithm is used, the parameters are called *actual* parameters.

Suppose I want to know for which values of n the sum $\sum_{i=1}^{n} 2i$ is evenly divisible by 3.

The algorithm `threes` provides the answer by using `add_Evens` (which is repeated to the right of `threes`).

```
void threes(integer max)
   for i = 1 to max
      sum = add_Evens(i)
      if sum is divisible by 3
         display i and sum
end threes
```

```
integer add_Evens(integer n)
   sum = 0
   for i = 1 to n
      sum = sum + 2*i
   return sum
end add_Evens
```

■

✔ Quick Check 4.3

1. Write a complete algorithm that takes a sequence of numbers $a_1, a_2, \ldots, a_n$ and returns the alternating sum $a_1 - a_2 + a_3 - a_4 + \cdots \mp a_{n-1} \pm a_n$. (The final term will be added if n is odd and subtracted if n is even.) ☑

Some people prefer to have something more than just indentation to delimit the scope of a selection or repetition statement. One popular notation is to use opening and closing braces for this task. Algorithm `threes` might be represented as follows:

```
void threes(integer max)
   for i = 1 to max {
      sum = add_Evens(i)
      if sum is divisible by 3 {
         display i and sum
      }
   }
end threes
```

The algorithms in this text are generally simple enough that I have chosen to rely solely on the indentation level.

One very useful addition to pseudocode is the ability to add comments. These comments help the human reader understand the code but are not executed when the algorithm is used. Many notations have been used for comments.[9] One simple option is to start a comment with the symbol # and consider everything between the # and the end of the current line to be a comment.[10]

```
void threes(integer max)
  # This could go on forever, so quit at the number max.

  for i = 1 to max
    sum = add_Evens(i)          # add even numbers from 0 to 2*max
    if sum is divisible by 3
      display i and sum         # i meets the requirements
end threes
```

One final note about pseudocode. The `return` statement works in a preemptive fashion: It causes the algorithm to terminate immediately.

---

[9]For example, // comment, /* comment */, { comment }, and (* comment *) have all been used.

[10]The programming language Perl uses this notation.

EXAMPLE 4.10 **Return**

The second `return` in algorithm `preemtive` will never be reached because as soon as `i` reaches 10, the algorithm terminates and sends out the value 10. The for loop never completes. Notice that this algorithm does not require any input data so it has an empty parameter list.

```
integer preemptive()
   for i = 1 to 20
      if i == 10
         return 10
   return 20
end preemptive
```

■

4.1.3 Exercises

The exercises marked with ★ have detailed solutions in Appendix H.

1. ★ Write an algorithm that makes change for a purchase. Assume that the purchase price is between \$0.01 and \$1.00 and that the customer has paid exactly \$1.00. The algorithm should return a set containing the number of quarters, dimes, nickels, and pennies to return (in that order). The return set should be declared as {integer, integer, integer, integer}.
2. Write an algorithm that makes change for a purchase. The algorithm should return a set containing the number of quarters, dimes, nickels, and pennies to return (in that order). The return set should be declared as {integer, integer, integer, integer}. You may assume that the customer does not pay less than the price. Use as many quarters as necessary.
3. Write a complete algorithm that takes as input a calendar year (from the Gregorian calendar) and returns the number of days in February. The leap year rules for the Gregorian calendar are as follows:
   - A leap year occurs in most years that are evenly divisible by 4.
   - A year that is evenly divisible by 100 is usually *not* a leap year.
   - If the year is evenly divisible by 400, then it *is* a leap year.

   For example, 1984 is a leap year; 1983 and 1900 are not. The year 2000 is a leap year.

   Use one multiway selection structure in your algorithm. You are not required to check divisibility in the same order as the rules just listed.
4. Write an algorithm that determines how many positive integers evenly divide the positive integer n. The algorithm should return the number of divisors (not the actual divisors). For example, 4 has three divisors: 1, 2, 4. Call the algorithm `numberOfDivisors`.

   `integer` **numberOfDivisors**`(integer n)`
5. Write an algorithm to determine whether an integer is prime or composite. The algorithm should return a 0 if the integer is prime, a 1 if the integer is composite, and a 2 if the integer is neither prime nor composite and a 3 if the integer is less than 1. You may assume that the algorithm `numberOfDivisors`, defined in Exercise 4, is available.
6. ★ Recall the definition of greatest common divisor (Definition 3.11 on page 94). Write an algorithm that displays the greatest common divisor of two natural numbers, not both zero. If both natural numbers are zero, display an error message saying that this is invalid input.
7. Suppose that Jane does different activities during the various seasons of the year. When it is fall, she goes bike riding, but only if the temperature is at least 50° F. If the temperature is not warm enough in the fall, she reads a book instead. During the winter, Jane always plays in the snow. In both the summer and spring, she goes running unless the temperature exceeds 70° F. If the temperature is over 70° and not over 90° F, Jane walks around the lake. When the temperature gets above 90° F, the only activity she can do is swimming. Write an algorithm that returns the activity that Jane should carry out based on the season and temperature (a real number).
8. A family rolls a standard pair of fair, six-sided dice to help determine which person will do the dishes on Monday night. If the sum of the digits on the pair of dice is odd, then Mom automatically has to do the dishes. A roll of the dice in which the sum of the digits is divisible by 4 means that either Dad or Brother Joe will do the dishes, depending on whether the digits are the same (Dad) or different (Brother Joe). All other rolls of the dice will appoint Sister Sue to do the dishes. Write an algorithm to display the name of the person on dishes duty for Monday night. The algorithm should accept as input two values, which are the the numbers of dots on the top faces of the pair of dice.
9. ★ Write an algorithm that takes a sequence of real numbers $x_1, x_2, \ldots, x_n$ and returns the absolute value of the average of these numbers.
10. Suppose that you are making 40 gingerbread men for a Christmas party. For each cookie, you must put the cutter in the dough, place the cookie on the pan, and then add three raisins for buttons. After this is complete, the 40 cookies must be baked. (Assume that you have a big oven and all of the cookies can bake at the same time.) Write an algorithm to describe the process you must go through for bringing gingerbread men to the party. Input and output values are not necessary in this algorithm.

11. Write an algorithm that takes a sequence of letters $a_1, a_2, \ldots, a_n$ and first displays the list in the order given and then in reverse order. To illustrate, if the input sequence is "g, i, r, l", the algorithm will display "girllrig".

12. Recall that two points are collinear if they lie on the same line. Write an algorithm that takes a series of ordered pairs, representing points on the real plane, and determines whether or not *all* of the points are collinear. It returns true if the points are collinear and false if they are not. For example, if the input is the ordered pairs for four points and only two of the points lie on the same line, the algorithm will return false. Assume that at least two ordered pairs will be used in the algorithm. [*Hint*: Think about the standard formula for slope. You may also assume that there will be no undefined slopes (representing points that are vertically aligned) when using the slope formula.]

13. Jake works at the school cafeteria and is charged with the duty of preparing bag lunches for some students going on a field trip to the zoo. In each bag lunch, he must enclose a turkey sandwich and a granola bar. The beverage and fruit he encloses varies according to student gender: Boys get apple juice and an orange, while girls get orange juice and an apple. Write an algorithm that accepts as input the number of boys and the number of girls going on the zoo field trip and describes the process Jake goes through in preparing the bag lunches for these students.

14. Write an algorithm that accepts as input a positive integer, n, and performs an alternating sequence of multiplications and additions for all the positive integers up through n. For instance, if the integer 6 is entered, the alternating sequence would be $1{\cdot}2{+}3{\cdot}4{+}5{\cdot}6$. Note that the algorithm will only return the final result after performing the multiplications and additions in this sequence. In your algorithm, assume that $n \geq 1$ and remember the general precedence rules for multiplication and addition. If n is odd, the final operation will be an addition (so if $n = 5$, calculate $1 \cdot 2 + 3 \cdot 4 + 5$).

15. ★ Two integers are relatively prime if 1 is their only common factor. Write the algorithm `relativelyPrime`. You may assume that $a \geq 1$ and $b \geq 1$.
`boolean` **relativelyPrime** `(integer a, integer b)`

16. The Euler totient function, $\phi(n)$, determines the number of positive integers that are less than the integer n and are also relatively prime to n (i.e., have no common factors with n other than 1). Write an algorithm to calculate and return $\phi(n)$. For example, $\phi(5) = 4$ since 1, 2, 3, and 4 are relatively prime to 5. Also, $\phi(6) = 2$ (1 and 5 are relatively prime to 6). Assume that the algorithm `relativelyPrime`, defined in Exercise 15, is available.

17. A positive integer, n, is said to be *square free* if its prime factorization contains no repeated factors. That is, $n \neq p^k q$ where $k > 1$, p is prime, and q is a positive integer. Write an algorithm, `squareFree`. You may assume $n \geq 1$.
`boolean` **squareFree** `(integer n)`

18. The *Möbius function*, $\mu(n)$, is defined for positive integers by

$$\mu(n) = \begin{cases} 1 & \text{if } n \text{ is square free with an} \\ & \text{even number of prime factors} \\ -1 & \text{if } n \text{ is square free with an} \\ & \text{odd number of prime factors} \\ 0 & \text{if } n \text{ is not square free} \end{cases}$$

where "square free" is defined in Exercise 17. Write an algorithm `integer` μ `(integer` n`)` that uses the algorithm `squareFree`. The algorithm must effectively determine whether the number of prime factors is even or odd. Note that $\mu(1) = 1$ because 0 is an even number. Treat $n == 1$ as a special case in your algorithm.

4.2 Measuring Algorithm Efficiency

Section 4.1 showed a method for expressing the details of an algorithm. The next task is to establish a mechanism for measuring the efficiency of an algorithm. The standard mechanism creates a measure of relative efficiency: An algorithm is given a rating that declares it to be in the same efficiency category as one of a collection of standard reference functions.

How do functions enter the discussion? When considering the efficiency of an algorithm, there are two resources that are essential: time and space. Space might consist of how many pieces of paper you need to carry out the steps of an algorithm, but more commonly consists of bytes (or megabytes) of computer memory. An algorithm that uses 1 megabyte of memory is preferred (all other things being equal) to one that requires 8 megabytes of memory.[11]

The time efficiency of an algorithm has been the more commonly measured resource. What is typically done is to designate some steps in an algorithm to be the most critical steps. We then count how many times those steps are executed. For most algorithms, the number of times the critical steps are executed depends on the size of the

[11]This is not an exaggerated example; it is possible to take some commercial software and shrink the memory usage by a factor of 8 and still have exactly the same time and functionality performance. This is due to what programmers call *bloatware*—programs that use more memory than is necessary. See the Bloatbusters link at http://www.mathcs.bethel.edu/~gossett/DiscreteMathWithProof/ for some examples.

initial data set. For example, if the algorithm sorts a list of names, the size of the data set will be the number of names to sort. One critical aspect of most sorting algorithms is the need to compare two names to see which comes lexicographically first. For a simple algorithm, the number of times two names are compared may be about n^2 times for a list of n names. As the number of names grows, the time the algorithm requires will grow (approximately) as the square of the number of names; double the list and take about four times as long to complete the sort.

In the sorting example, the function $g(n) = n^2$ serves as a well-known reference function. The actual sorting algorithm might have a time growth of

$$f(n) = 120n^2 - 11n + 450$$

The reference function $g(n)$ is simpler to write, and we know what it looks like without needing to draw the graph. In addition, as n gets larger and larger, it becomes easier to see that the functions have essentially the same shape. Therefore we can think of g as representing the time behavior of the algorithm; not an exact description, but a good approximation.

These ideas will now be formalized. Three definitions will be presented: Big-$\mathcal{O}$, Big-$\boldsymbol{\Omega}$, and Big-$\boldsymbol{\Theta}$, which combines the requirements of the first two definitions.

4.2.1 Big-Θ and Its Cousins

In this section, several definitions will be given. They formalize notions, such as "function f does not grow any faster than function g" or "function f grows just like g for all practical purposes". The definitions will allow for some youthful indiscretion before the comparison is enforced (we will ignore the relative growth rates for "small" values of n). Note that the definitions use a stretched version of g when doing the actual comparisons. Observe the use of n rather than x as the independent variable. The definitions are valid with any real valued independent variable; however, in the context of algorithm analysis, the independent variable represents the size of a data set and is therefore almost always a positive integer. Recall that $\mathbb{R}^+$ denotes the set of positive real numbers.

DEFINITION 4.3 ***Big-$\mathcal{O}$***

The function f is said to be in big-$\mathcal{O}$ of g, pronounced "big oh" and denoted $f \in \mathcal{O}(g)$, if there are positive constants, c and n_0, such that, for all $n \geq n_0$, $|f(n)| \leq c|g(n)|$. Another way to express this is

$f \in \mathcal{O}(g)$ if and only if

$$\exists c \in \mathbb{R}^+, \exists n_0 \in \mathbb{R}^+, \forall n \in \mathbb{R}^+, [(n \geq n_0) \rightarrow (|f(n)| \leq c|g(n)|)].$$

The definition is stating that f is one of the many functions that belong, in a specific sense, to the set of all functions that don't grow faster than g. The constants n_0 and c in the definition are not unique (as Example 4.11 will demonstrate).

EXAMPLE 4.11

Big-$\mathcal{O}$ Basics

It is not hard to show that $f(n) = 120n^2 - 11n + 450$ is in $\mathcal{O}(n^2)$. One approach is to graph $f(n)$ and $cg(n)$ for various values of c. Once we find a choice of c that keeps $cg(n)$ above the graph of $f(n)$, we can try to prove formally that the definition holds. That is, find numbers $n_0 > 0$ and $c > 0$ such that

$$\left|120n^2 - 11n + 450\right| \leq c\left|n^2\right|$$

for all $n \geq n_0$.

The graphs in Figures 4.10 and 4.11 show the stretched reference function $cg(n)$ as a dashed line. They provide different choices for the constants n_0 and c, but both will lead to valid verifications of the definition: There is not just one correct choice of the pair (n_0, c).

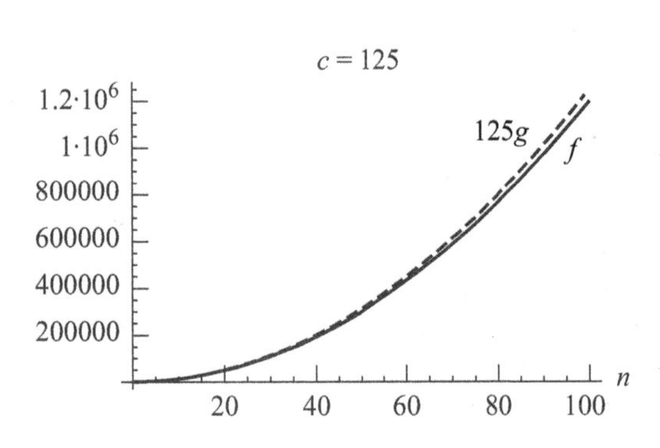

Figure 4.10 Function $f(n)$ is solid, 125 $g(n)$ is dashed.

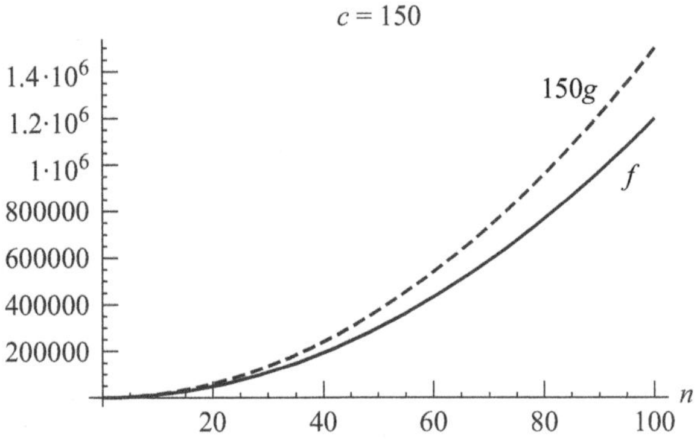

Figure 4.11 Function $f(n)$ is solid, 150 $g(n)$ is dashed.

A possible choice for n_0 in the first graph is 40, since clearly the graph of $125g(n)$ stays above that of $f(n)$ for all values of n greater than 40 (actually, for all values of n greater than about 8.45). The second graph appears to work for $n > 20$ (actually $n > 3.69$).

A zoomed-in view of Figure 4.10 shows why we ignore the relative behavior for small values of n: $f(n)$ is larger for $n \leq 8$ (Figure 4.12).

Figure 4.12 A zoomed-in view of Figure 4.10.

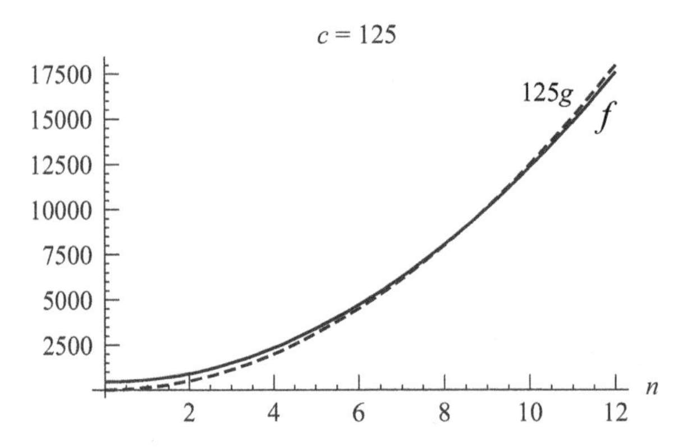

How can the definition be formally verified? Some standard manipulations of inequalities are needed. I initially assume $n \geq 0$ and drop the absolute value symbols. This is permissible because $f(n) > 0$ for $n > 0$ (but see Quick Check 4.4).

$$
\begin{aligned}
120n^2 - 11n + 450 &\leq 120n^2 + 450 && \text{since } -11n \text{ is not positive} \\
&\leq 120n^2 + n && \text{if } n \geq 450 \\
&\leq 120n^2 + n^2 && \text{since } n \leq n^2 \text{ if } n \geq 1 \\
&= 121n^2
\end{aligned}
$$

I have made several assumptions about n: $n \geq 0$, $n \geq 450$, and $n \geq 1$. The most restrictive assumption (which also guarantees that the others hold) is $n \geq 450$. We may therefore use $n_0 = 450$ and $c = 121$ to verify the definition. (I could have worked harder and established $n_0 = 17$ for this c, but there is no compelling reason since big-$\mathcal{O}$ estimates are about what happens for large values of n.) ■

✔ Quick Check 4.4

1. Algebraically verify that $2n^2+5n+7$ is in $\mathcal{O}(n^2)$.
2. Algebraically verify that $101n+73$ is in $\mathcal{O}(n)$.
3. Algebraically verify that $n^2 - 5n - 4$ is in $\mathcal{O}(n^2)$.

EXAMPLE 4.12

Disproving Big-$\mathcal{O}$

How would we show that $f(n) = n^3$ is *not* in $\mathcal{O}(n^2)$?

One useful method is a proof by contradiction. We assume that $n^3 \in \mathcal{O}(n^2)$ and then arrive at a contradiction. We conclude that the assumption $n^3 \in \mathcal{O}(n^2)$ led us astray and so reject the assumption.

On to the proof. Assume that $n^3 \in \mathcal{O}(n^2)$. Then, according to the definition of big-$\mathcal{O}$,

$$\exists c \in \mathbb{R}^+, \exists n_0 \in \mathbb{R}^+, \forall n \in \mathbb{R}^+, \left[(n \geq n_0) \rightarrow \left(|n^3| \leq c|n^2|\right)\right]$$

To find a contradiction, it is necessary to determine what the negation of the claim is. You should carefully verify that

$$\neg\left(\exists c \in \mathbb{R}^+, \exists n_0 \in \mathbb{R}^+, \forall n \in \mathbb{R}^+, \left[(n \geq n_0) \rightarrow \left(|n^3| \leq c|n^2|\right)\right]\right)$$

is logically equivalent to

$$\forall c \in \mathbb{R}^+, \forall n_0 \in \mathbb{R}^+, \exists n \in \mathbb{R}^+, \left[(n \geq n_0) \wedge \left(|n^3| > c|n^2|\right)\right]$$

Given c and n_0, we are looking for an n such that $n \geq n_0$ and $|n^3| > c|n^2|$. Since $n \in \mathbb{R}^+$, $n > 0$. We can drop the absolute value signs since both functions are positive for $n > 0$.

We can then rewrite the inequality as $n^3 - cn^2 > 0$, or $n^2(n - c) > 0$. Clearly, the factor n^2 is not negative, so we must have $n - c > 0$, or $n > c$. No matter what values c and n_0 have, it is always possible to find an n with $n > \max(c, n_0)$. The desired contradiction has been found, so we reject the assumption that $n^3 \in \mathcal{O}(n^2)$. ■

EXAMPLE 4.13

Big-$\mathcal{O}$ Is Deficient

There is a deficiency with the big-$\mathcal{O}$ definition (at least for the purposes of algorithm analysis). To see the deficiency, it is easy to observe graphically (Figure 4.13) that $f(n) = 101n + 73 \in \mathcal{O}(n^2)$.

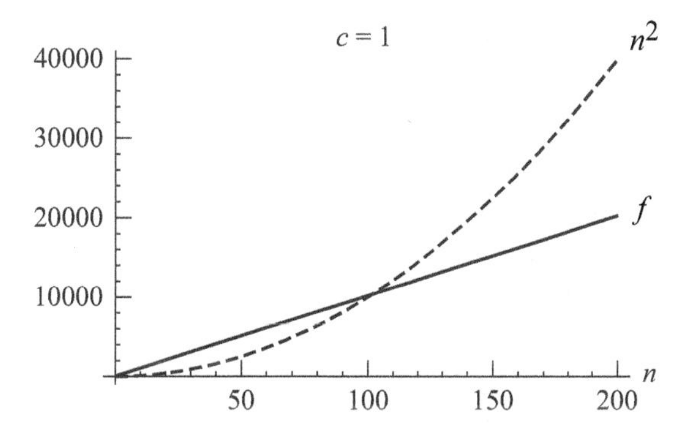

Figure 4.13 Big-$\mathcal{O}$ isn't sufficient.

However, as n gets large, $f(n) = 101n + 73$ does not grow nearly as fast as $g(n) = n^2$. To say that $101n + 73$ grows no faster than n^2 for large n is quite an understatement.

What we really want is a reference function that grows at the same rate as f. (For $f(n) = 101n + 73$, $g(n) = n$ is the best choice.) ■

✔ Quick Check 4.5

1. Algebraically verify that $101n + 73$ is in $\mathcal{O}(n^2)$. ☑

One way to remedy the deficiency in the big-$\mathcal{O}$ definition is to create another definition.[12] The definition we really want is big-Θ, but another useful definition needs to be presented first.

DEFINITION 4.4 ***Big-Ω***
The function f is said to be in big-Ω of g, pronounced "big omega" and denoted $f \in \Omega(g)$, if there are positive constants, c and n_0, such that, for all $n \geq n_0$, $|f(n)| \geq c|g(n)|$. Another way to express this is

$$f \in \Omega(g) \quad \text{if and only if}$$
$$\exists c \in \mathbb{R}^+, \exists n_0 \in \mathbb{R}^+, \forall n \in \mathbb{R}^+, [(n \geq n_0) \rightarrow (|f(n)| \geq c|g(n)|)]$$

So f is in big-$\Omega(g)$ if f eventually stays larger than some positive constant multiple of g.

EXAMPLE 4.14

Big-Ω Basics

It is not hard to show (graphically and algebraically) that $f(n) = 120n^2 - 11n + 450$ is in $\Omega(n^2)$. Figure 4.14 shows that $c = 100$ and $n_0 = 20$ verify the definition ($c = 1$ also works, but the graph is rather boring).

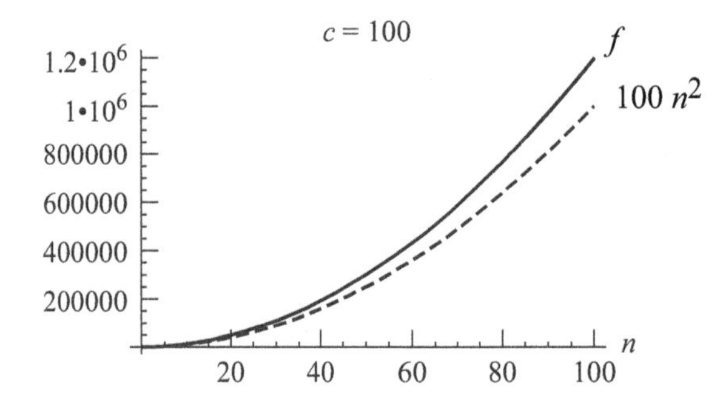

Figure 4.14 Verifying that $f(n) = 120n^2 - 11n + 450$ is in $\Omega(n^2)$.

The algebraic verification is also easy. Assume $n > 0$. Then

$$\begin{aligned} 120n^2 - 11n + 450 &\geq 120n^2 - 11n \\ &\geq 120n^2 - 11n^2 \quad \text{if } n \geq 1 \\ &= 109n^2. \end{aligned}$$

Since $109n^2 > 0$, taking absolute values of both sides will not reverse the inequality. Thus

$$|120n^2 - 11n + 450| \geq |109n^2| = 109|n^2|$$

Let $n_0 = 1$ and $c = 109$ in the big-Ω definition. ■

✔ Quick Check 4.6

1. Use a graphing calculator or a software package, such as *Mathematica* or *Maple*, to find choices for c and n_0 that suggest $\frac{1}{4}\log_2(\frac{n}{2}) \in \Omega(\log_2(n))$. ☑

[12]The big-$\mathcal{O}$ definition has been around too long to revise it.

The next definition is the most useful for algorithm analysis.

DEFINITION 4.5 ***Big-Θ***
The function f is said to be in big-Θ of g, pronounced "big theta" and denoted $f \in \Theta(g)$, if $f \in \mathcal{O}(g) \cap \Omega(g)$.

EXAMPLE 4.15

Big-Θ Basics

Examples 4.11 and 4.14 demonstrated that $f(n) = 120n^2 - 11n + 450$ is in $\mathcal{O}(n^2)$ and also in $\Omega(n^2)$. Therefore, $f \in \Theta(n^2)$. The graph in Figure 4.15 shows that f can be contained between two different multiples of n^2, once n is large enough.

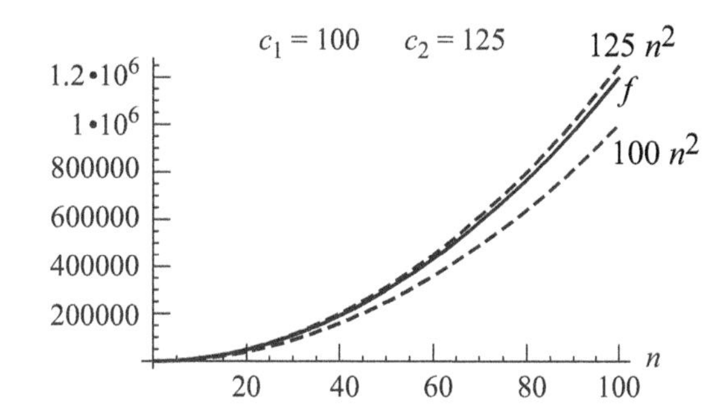

Figure 4.15
$f(n) = 120n^2 - 11n + 450$ is in $\Theta(n^2)$.

By establishing that $f \in \Theta(n^2)$, we know that f grows in essentially the same manner as n^2, a reference function we are quite familiar with. In addition, we have demonstrated that f grows substantially faster than a function that is in $\Theta(n)$, but fundamentally slower than one that is in $\Theta(n^3)$. ■

Social Convention Wins over Correctness

The notion that an algorithm is in $\Theta(g)$, for some reference function $g(n)$, is the one that properly captures the notion "the algorithm's efficiency changes like $g(n)$ as the data size n increases". The big-$\mathcal{O}$ definition leaves a loophole that was described in Example 4.13. Unfortunately, for many years and in many scholarly books and papers, the concept of big-$\mathcal{O}$ has been used in a somewhat loose manner. Many authors make the statement $f \in \mathcal{O}(g)$ but really intend you to recognize that even more is true (namely, $f \in \Theta(g)$). I will try to use the more complete notation whenever it has been verified. Nevertheless, you need to recognize that the phrase "big-$\mathcal{O}$" is used at times as if it were big-Θ.

It is also common practice to say that f *is* $\mathcal{O}(g)$ instead of f *is in* $\mathcal{O}(g)$. It is even common for less precise notation to be used. For example, you may often see claims such as $f = \mathcal{O}(g)$, which might be read out loud as "f is big-$\mathcal{O}$ of g". Another variation would be to write something like "$f(n) = 5n^2 + \mathcal{O}(n)$", meaning that f is a sum of the function $5n^2$ and other terms that are in $\mathcal{O}(n)$.

4.2.2 Practical Big-Θ Tools

For normal evaluation of an algorithm's performance, it is helpful to have some shortcuts that enable us to skip the algebraic details involved in proving the details of the definitions. You have encountered a similar situation in calculus: You first learned (for very good reasons) the definition of the derivative:

$$f'(x) = \lim_{h \to 0} \frac{f(x+h) - f(x)}{h}$$

You then studied some theorems that provided shorter ways to determine f', given f.

For example, if $f(x) = x^n$, then $f'(x) = nx^{n-1}$. You also learned more complex shortcuts such as the product rule and the chain rule. Fortunately, shortcut theorems exist for big-Θ comparisons. They will be examined after a collection of important reference functions are presented.

A Brief Introduction to the World's Favorite Reference Functions

Most algorithms can be classified as being similar (in a big-Θ sense) to one of the functions listed in Table 4.1. The functions are listed with the slower-growing functions at the top and faster-growing functions at the bottom. In the context of algorithm analysis, slow growth is good (as n increases, the algorithm takes very little additional time to complete). The top entry seems too good to be true. In fact, an algorithm in $\Theta(1)$ takes the same amount of time to complete no matter how large the input data set. It is almost certainly an algorithm that ignores the data and is consequently useless.

TABLE 4.1 Standard Reference Functions

Category	Reference Function
Constant	1
Logarithmic	$\log_2(n)$
Linear	n
$n \log n$	$n \log_2(n)$
Quadratic	n^2
Cubic	n^3
Exponential	a^n for $a > 1$

As will be shown shortly, an algorithm with exponential runtime is essentially an impractical algorithm. You may be able to complete a few simple cases, which could be done by hand quicker than you can write a program to implement the algorithm, but it will take the algorithm too long to finish if given any mildly interesting set of data.

The graphs in Figure 4.16 show the standard reference functions. Notice that the vertical axes are not the same scale. The function $n \log_2(n)$ appears on both graphs.

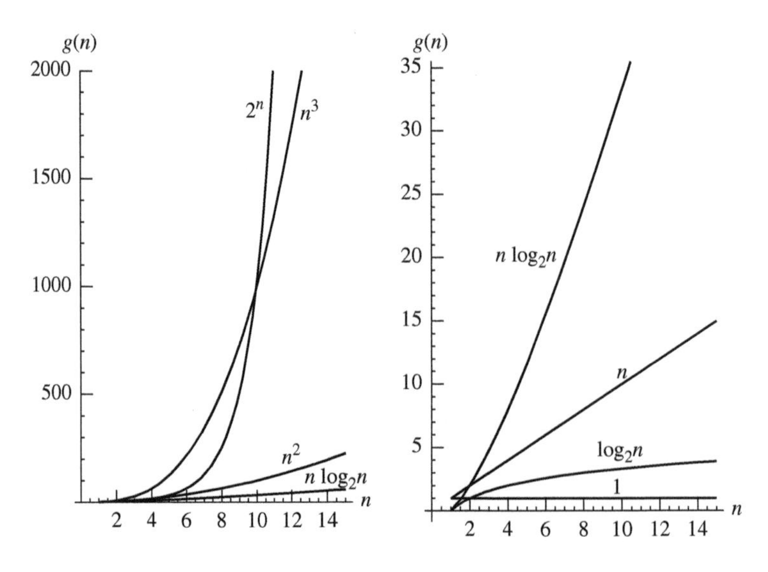

Figure 4.16 Comparing the standard reference functions.

A Brief Review of Logarithmic Functions

You should have previously memorized the definition and properties of logarithmic functions, but just for completeness, they are repeated here. If you have not memorized them, you will profit from doing so now.

DEFINITION 4.6 ***Logarithmic Functions***

The *logarithmic function with base b*, denoted $\log_b(x)$, is defined by the relationship

$$\log_b(x) = y \quad \text{for } x > 0 \quad \text{if and only if} \quad b^y = x, \qquad \text{where } 0 < b \text{ and } b \neq 1$$

The most commonly used logarithmic functions are

common logs $\log_{10}(x)$, usually denoted $\log(x)$ on calculators
natural logs $\log_e(x) = \int_1^x \frac{dt}{t}$, usually denoted $\ln(x)$
base 2 logs $\log_2(x)$, the prime candidate in discrete mathematics

All logarithmic functions share some common properties:

THEOREM 4.7 ***Properties of Logarithmic Functions***

- $\log_b(1) = 0$
- $\log_b(b) = 1$
- $\log_b(x^n) = n \log_b(x)$ for any real number n and $x > 0$
- $\log_b(xy) = \log_b(x) + \log_b(y)$ for $x > 0$ and $y > 0$
- $\log_b(\frac{x}{y}) = \log_b(x) - \log_b(y)$ for $x > 0$ and $y > 0$

Note well: There are no theorems for simplifying $\log_b(x + y)$ or $\log_b(x - y)$.

The final theorem in this brief review is the change of base formula. It shows how you can use the common log or natural log functions built into a calculator to find a base 2 log.[13]

THEOREM 4.8 ***Change of Base Formula***

For all legal bases a and b and all $x > 0$,

$$\log_b(x) = \frac{\log_a(x)}{\log_a(b)}$$

Thus, $\log_2(10) = \frac{\ln(10)}{\ln(2)} \simeq \frac{2.3026}{0.69315} \simeq 3.3219$.

Theorem 4.8 makes it possible to replace the standard reference function, $\log_2(n)$, by any other base logarithm (see Exercise 12 in Exercises 4.2.3). It is common to see natural logs, as well as logs with unspecified base, used as reference functions.

The next two examples will provide two views of the practical implications of the vast differences in growth represented by the reference functions.

EXAMPLE 4.16 **Comparing Reference Functions—Part 1**

Suppose we have a computer that can execute one instruction per nanosecond. That is, it takes 10^{-9} seconds to execute one machine instruction. Thus it can execute 1 billion

[13] If you are using the log functions that come with most computer languages, they will be natural logs, independent of the notation. Computer systems such as *Mathematica* have more flexible log notation. For example, *Mathematica* uses `Log[b,x]` to represent base b logs.

instructions per second.

Table 4.2 compares six hypothetical algorithms. The row labels represent the time complexity of the respective algorithms. For example, the fourth row represents a quadratic algorithm. The columns represent several different sizes of the data set.

Making the simplifying assumption that algorithm i uses $g_i(n)$ machine instructions for a data set of size n, entry (i, j) in the table represents the amount of time it will take algorithm i to run on a data set of size j. For example, $g_4(n) = n^2$. If $n = 10,000$ then $10,000^2 = 100,000,000$ instructions are needed, each requiring 10^{-9} seconds to execute. The total time is therefore 0.1 second.

TABLE 4.2 Time Needed to Process n Items for Six Algorithms (at 10^{-9} seconds per operation).

$g_i(n)$	$n = 10,000$	$n = 100,000$	$n = 1,000,000$	$n = 250,000,000$
$\log_2(n)$	13 nanoseconds	17 nanoseconds	20 nanoseconds	28 nanoseconds
n	0.00001 seconds	0.0001 seconds	0.001 seconds	0.25 seconds
$n \log_2(n)$	0.00013 seconds	0.00166 seconds	0.01993 seconds	6.97434 seconds
n^2	0.1 second	10.0 seconds	16 minutes, 40 seconds	$\simeq$ 1.98 years
n^3	16 minutes, 40 seconds	$\simeq$ 11 days, 14 hours	$\simeq$ 32 years	$\simeq$ 500 million years
2^n	$\simeq 6 \times 10^{2993}$ years	——— Too long to contemplate ———		

To put Table 4.2 in perspective, consider sorting the census data for the United States (population close to 300 million). A standard result in a course on data structures states that the simplest sorting algorithms are in $\Theta(n^2)$. The better algorithms for this task are in $\Theta(n \log_2(n))$, or for some kinds of data, in $\Theta(n)$. Would you prefer to wait 3 years, or a bit under 10 seconds?[14] ■

We can look at the previous example from another viewpoint.

EXAMPLE 4.17

Comparing Reference Functions—Part 2

Suppose that we want to run each of the algorithms in Example 4.16 for 1 minute. What size data set would each be able to process?

Consider the quadratic algorithm. We want to solve the equation below for n.

$$(10^{-9} \text{ seconds/instruction}) \cdot (n^2 \text{ instructions}) = 60 \text{ seconds}$$

It is simple to determine that $n = \sqrt{60{,}000{,}000{,}000} \simeq 244{,}949$.

The equation

$$(10^{-9} \text{ seconds/instruction}) \cdot (n \log_2(n) \text{ instructions}) = 60 \text{ seconds}$$

is harder to solve since it reduces to

$$n \log_2(n) = 60{,}000{,}000{,}000$$

Using a numerical approximation technique such as Newton's method, we find $n \simeq 2$ billion data items. Table 4.3 shows the size data set each of the standard reference

[14]The actual times will be longer in both cases because the assumption "algorithm i uses $g_i(n)$ machine instructions for a data set of size n" is not really true.

functions can complete in one minute.

TABLE 4.3 Number of Items per Minute

$g_i(n)$	data items in 1 minute
$\log_2(n)$	$2^{6 \times 10^{10}}$
n	60 billion
$n \log_2(n)$	$\simeq$ 2 billion
n^2	244,949
n^3	3,915
2^n	36

■

✔ Quick Check 4.7

1. Suppose that you are able to use pencil and paper to process the steps of an algorithm. Compare how long you will take to process 10 items and 100 items using two different algorithms. Algorithm 1 is in $\Theta(n \log_2(n))$ and algorithm 2 is in $\Theta(n^2)$. You may assume that the number of steps is $n \log_2(n)$ or n^2, respectively, and that you take 30 seconds per step. Convert your answers to appropriate time units (minutes or hours).

2. Using the information from the previous question, how many data items can you process in one hour using algorithm 1? How many items in one hour using algorithm 2? ☑

Big-Θ Shortcuts

It is now time to introduce the shortcut theorems that allow us to make practical big-Θ evaluations. The proof of the first theorem requires a few ideas you may have encountered in a calculus class.

DEFINITION 4.9 ***Limit of a Sequence***

Consider the sequence of real numbers $b_0, b_1, b_2, \ldots$ The *limit*, b, as n goes to infinity is denoted by $\lim_{n\to\infty} b_n = b$. It is defined if and only if for every real number $\epsilon > 0$ there is an integer, N, such that

$$|b_n - b| < \epsilon \quad \text{for all } n \text{ with } n \geq N$$

The sequence *converges to* ∞, denoted by $\lim_{n\to\infty} b_n = \infty$, if for every positive integer, M, there is an integer, N, such that

$$|b_n| > M \quad \text{for all } n \text{ with } n \geq N$$

So the sequence has a limit if it stays arbitrarily close to b as n gets larger and larger.

EXAMPLE 4.18 **Simple Sequence Limits**

The sequence $0, \frac{1}{2}, \frac{2}{3}, \frac{3}{4}, \ldots$, defined by $b_n = 1 - \frac{1}{n+1}$, for $n = 0, 1, 2, \ldots$ has the limit 1.

The sequence $0, \frac{1}{2}, \frac{4}{3}, \frac{9}{4}, \ldots$, defined by $b_n = \frac{n^2}{n+1}$, for $n = 0, 1, 2, \ldots$ converges to ∞. ■

The proof of the first shortcut theorem uses the triangle inequality. See Exercise 26 in Exercises 3.3.10 on page 112 for a hint about how to prove this important result.

THEOREM 4.10 ***The Triangle Inequality***

If a and b are real numbers, then $|a + b| \leq |a| + |b|$.

THEOREM 4.11 ***Big-Θ and Polynomials***

Let f be the polynomial function $f(n) = a_k n^k + a_{k-1}n^{k-1} + \cdots + a_1 n + a_0$, where $a_k \neq 0$. Then $f \in \Theta(n^k)$.

Proof: To show $f \in \Theta(n^k)$, it must first be established that $f \in \mathcal{O}(n^k)$. It is easy to generalize the procedure used in Quick Check 4.4 to establish that $2n^2 + 5n + 7$ is in $\mathcal{O}(n^2)$.

Assume that $n > 0$. Then each of the terms n^i is positive so $|a_i n^i| = |a_i| n^i$. Also, if $n \geq 1$, then $n^i \leq n^j$ whenever $i < j$. Therefore (using the triangle inequality),

$$\begin{aligned} |a_k n^k + a_{k-1}n^{k-1} + \cdots + a_1 n + a_0| &\leq |a_k| n^k + |a_{k-1}| n^{k-1} + \cdots + |a_1| n + |a_0| \\ &\leq |a_k| n^k + |a_{k-1}| n^k + \cdots + |a_1| n^k + |a_0| n^k \\ &= \left(\sum_{i=0}^{k} |a_i| \right) n^k \end{aligned}$$

We can take $c = \sum_{i=0}^{k} |a_i|$ and $n_0 = 1$ in Definition 4.3 (page 181).

To complete the proof that $f \in \Theta(n^k)$, it must also be shown that $f \in \Omega(n^k)$. The ideas from Example 4.14 cannot be generalized unless a_k plus the sum of the negative coefficients is a positive number (true in Example 4.14, but not true in general).[15]

It will be necessary to use limits to complete this part of the proof. Since $a_k \neq 0$, the following limit makes sense. Assume that $n > 1$. Then

$$\begin{aligned} \lim_{n\to\infty} \frac{|f(n)|}{\left|\frac{a_k}{2}\right| n^k} &= \lim_{n\to\infty} \left| \frac{a_k n^k + a_{k-1}n^{k-1} + \cdots + a_1 n + a_0}{\frac{a_k}{2} n^k} \right| \\ &= \lim_{n\to\infty} \left| 2 + \sum_{j=0}^{k-1} \frac{2a_j}{a_k} \frac{1}{n^{k-j}} \right| = 2 \end{aligned}$$

since $\frac{1}{n^{k-j}} \to 0$ as $n \to \infty$.

Therefore, using Definition 4.9 (page 189), for every $\epsilon > 0$, there is some integer N (that depends on the choice of ϵ) such that

$$\left| \frac{|f(n)|}{\left|\frac{a_k}{2}\right| n^k} - 2 \right| < \epsilon \quad \text{for all } n \geq N$$

It will be useful to assume that $\epsilon < 1$.[16] In that case, $2 - \epsilon > 1$.

Recalling that $|x| < y$ if and only if $-y < x < y$, we now know that

$$-\epsilon < \frac{|f(n)|}{\left|\frac{a_k}{2}\right| n^k} - 2 < \epsilon$$

Focusing on the left-hand inequality leads to

$$|f(n)| > (2 - \epsilon) \left| \frac{a_k}{2} \right| n^k > \left| \frac{a_k}{2} \right| n^k$$

Therefore, Definition 4.4 (page 184) is satisfied by taking $c = \left|\frac{a_k}{2}\right|$ and $n_0 = N$.

Both requirements in Definition 4.5 have now been verified, so $f \in \Theta(n^k)$. □

[15]Try the techniques used in Example 4.14 for $f_2(n) = -120n^2 - 11n + 450$ or for $f_3(n) = 120n^2 - 120n$. They will fail.

[16]This assumption does not invalidate the proof because there is an N for every ϵ. It would be better to use the notation N_ϵ.

The use of limits in the proof of Theorem 4.11 suggests an alternative approach to showing that $f \in \Theta(g)$. This approach is explored in Exercises 24 – 26 on page 194. The main ideas are simple. If f and g grow at essentially the same rate, then $\lim_{n\to\infty} \frac{f(n)}{g(n)}$ should have a limit that is neither 0 nor infinity (probably related to a reasonable candidate for the constant, c, in Definitions 4.3 and 4.4). On the other hand, if g grows at a faster rate than f [and hence f is trivially in $\mathcal{O}(g)$], then $\lim_{n\to\infty} \frac{f(n)}{g(n)}$ should be 0. Finally, if f grows at a faster rate than g [and is therefore in $\Omega(g)$], then $\frac{f(n)}{g(n)}$ should converge to ∞ as $n \to \infty$.

EXAMPLE 4.19 **Polynomials Are Easy**

Theorem 4.11 makes finding the proper reference function for a polynomial into a trivial task. For example,

$$0.001n^6 + 10{,}000n^5 + 45{,}876n^3 + 85n + 1{,}000{,}000{,}000 \in \Theta(n^6)$$

Notice that the sizes of the coefficients are irrelevant when looking for the reference function. Even though the leading coefficient is small, the degree 6 polynomial function $0.001n^6 + 10{,}000n^5 + 45{,}876n^3 + 85n + 1{,}000{,}000{,}000$ will eventually shoot above *any* polynomial of degree 5 or less. ■

Recall that whenever $f_1(x)$ and $f_2(x)$ are defined, the sum $f_1 + f_2$ is defined by $(f_1 + f_2)(x) = f_1(x) + f_2(x)$ and the product $f_1 \cdot f_2$ by $(f_1 \cdot f_2)(x) = f_1(x) \cdot f_2(x)$.[17]

The next two theorems are similar in spirit to some derivative theorems in calculus: $(f + g)'(x) = f'(x) + g'(x)$ and $(f \cdot g)'(x) = f'(x) \cdot g(x) + f(x) \cdot g'(x)$. However, in the big-$\mathcal{O}$ and big-Θ context, the product theorem is the simpler rule.

THEOREM 4.12 ***Big-Θ and Products***

Suppose that $f_1 \in \Theta(g_1)$ and $f_2 \in \Theta(g_2)$. Then $f_1 \cdot f_2 \in \Theta(g_1 \cdot g_2)$.

Proof: Once again, it is necessary first to show that $f_1 \cdot f_2 \in \mathcal{O}(g_1 \cdot g_2)$ and then show that $f_1 \cdot f_2 \in \Omega(g_1 \cdot g_2)$.

Since $f_1 \in \Theta(g_1)$, it is also true that $f_1 \in \mathcal{O}(g_1)$. Therefore, there must exist constants c_1 and n_1 so that $|f_1(n)| \le c_1|g_1(n)|$ for all $n \ge n_1$. Similarly, there are constants c_2 and n_2 such that $|f_2(n)| \le c_2|g_2(n)|$ for all $n \ge n_2$.

Set $n_0 = \max\{n_1, n_2\}$ and $c = c_1 \cdot c_2$. Then for all $n > n_0$

$$\begin{aligned}
|(f_1 \cdot f_2)(n)| &= |f_1(n) \cdot f_2(n)| \\
&= |f_1(n)| \cdot |f_2(n)| \\
&\le (c_1|g_1(n)|) \cdot (c_2|g_2(n)|) \\
&= (c_1 \cdot c_2)|g_1(n) \cdot g_2(n)| \\
&= c|(g_1 \cdot g_2)(n)|
\end{aligned}$$

establishing that $f_1 \cdot f_2 \in \mathcal{O}(g_1 \cdot g_2)$.

A very similar calculation establishes that $f_1 \cdot f_2 \in \Omega(g_1 \cdot g_2)$. □

DEFINITION 4.13 ***The Pointwise Maximum Function***

Suppose $g_1(x)$ and $g_2(x)$ are defined on some domain. Then the function $\max\{g_1, g_2\}$ is defined as the pointwise maximum of g_1 and g_2:

$$\max\{g_1, g_2\}(x) = \max\{g_1(x), g_2(x)\}$$

The definition says that at every x, $\max\{g_1, g_2\}(x)$ is assigned the value of whichever original function is larger at that x.

[17] Read the first statement as "the new function $f_1 + f_2$ sends the number x to $f_1(x) + f_2(x)$".

THEOREM 4.14 ***Big-Θ and Sums***

Suppose that $f_1 \in \Theta(g_1)$ and $f_2 \in \Theta(g_2)$. Assume also that there is a positive integer, n_0, such that for all $n > n_0$, $f_1(n) > 0$, $f_2(n) > 0$, $g_1(n) > 0$ and $g_2(n) > 0$. Then $(f_1 + f_2) \in \Theta(\max\{g_1, g_2\})$.

Proof of Theorem 4.14: The proof begins just like the proof of Theorem 4.12.

Since $f_1 \in \Theta(g_1)$, it is also true that $f_1 \in \mathcal{O}(g_1)$. Therefore, there must exist constants c_1 and n_1 so that $|f_1(n)| \le c_1|g_1(n)|$ for all $n \ge n_1$. Similarly, there are constants c_2 and n_2 such that $|f_2(n)| \le c_2|g_2(n)|$ for all $n \ge n_2$.

Set $n_3 = \max\{n_1, n_2\}$ and $c_3 = c_1 + c_2$. Set $g(n) = \max\{g_1(n), g_2(n)\}$. Then for all $n > n_3$

$$\begin{aligned}
|(f_1 + f_2)(n)| &= |f_1(n) + f_2(n)| \\
&\le |f_1(n)| + |f_2(n)| \\
&\le (c_1|g_1(n)|) + (c_2|g_2(n)|) \\
&\le (c_1|g(n)|) + (c_2|g(n)|) \\
&= (c_1 + c_2)|g(n)| \\
&= c_3|g(n)| \\
&= c_3 \cdot |\max\{g_1, g_2\}(n)|
\end{aligned}$$

This shows that $(f_1+f_2) \in \mathcal{O}(\max\{g_1, g_2\})$. To show that $(f_1+f_2) \in \Omega(\max\{g_1, g_2\})$, it will be necessary to assume that $f_1(n) > 0$ and $f_2(n) > 0$ for all $n > n_0$.

Since $f_1 \in \Theta(g_1)$, it is also true that $f_1 \in \Omega(g_1)$. Therefore, there must exist constants c_4 and n_4 so that $|f_1(n)| \ge c_4|g_1(n)|$ for all $n \ge n_4$. Similarly, there are constants c_5 and n_5 such that $|f_2(n)| \ge c_5|g_2(n)|$ for all $n \ge n_5$.

Set $n_6 = \max\{n_0, n_4, n_5\}$, $c_6 = \min\{c_4, c_5\}$, and $g(n) = \max\{g_1(n), g_2(n)\}$. Then for all $n \ge n_6$

$$\begin{aligned}
|(f_1 + f_2)(n)| &= |f_1(n) + f_2(n)| \\
&= f_1(n) + f_2(n) && \text{since the functions are positive for } n > n_0 \\
&= |f_1(n)| + |f_2(n)| \\
&\ge (c_4|g_1(n)|) + (c_5|g_2(n)|) \\
&\ge c_6|g_1(n)| + c_6|g_2(n)| \\
&\ge c_6|g(n)| && \text{since the sum of two positives is larger than either} \\
&= c_6 \cdot |\max\{g_1, g_2\}(n)|
\end{aligned}$$

□

To see why it was necessary to make the assumption that f_1 and f_2 were positive for all $n > n_0$, consider the functions $f_1(n) = 3n^2$ and $f_2(n) = -3n^2$. Both are easily seen to be in $\Theta(n^2)$, so $g_1(n) = n^2$ and $g_2(n) = n^2$. Thus $\max\{g_1, g_2\}(n) = n^2$. However, $(f_1 + f_2)(n) = 0$ for all n. This will clearly *not* be $\ge cn^2$ for a *positive* constant c. Hence, $(f_1 + f_2) \notin \Omega(n^2)$.

You may also be wondering why the function $\max\{g_1, g_2\}(n)$ is used as the reference function instead of the more natural $(g_1 + g_2)(n)$. It *is* true that $(f_1 + f_2) \in \Theta(g_1 + g_2)$ under the assumptions of the theorem. Notice, however, that $\max\{g_1, g_2\}(n) \le (g_1 + g_2)(n)$. Thus, when finding a big-$\mathcal{O}$ estimate for $(f_1 + f_2)(n)$, we have $|(f_1 + f_2)(n)| \le c|\max\{g_1, g_2\}(n)| \le c|(g_1 + g_2)(n)|$ for all $n \ge n_0$. The reference function that stays closer to $f_1 + f_2$ is preferred.

The Big-Θ shortcut theorems are summarized on page 224.

EXAMPLE 4.20 **A New Reference Function**

Suppose an algorithm's time performance is $3n(7\log_2(n) + 11n^2) + 12n\log_2(n)$. What is a good reference function (in a big-Θ sense)?

By using Theorem 4.14, a good reference function for the sum $7\log_2(n) + 11n^2$ is n^2 (notice that both terms are positive for $n > 1$).

The term $3n(n^2)$ has n^3 as its reference function, using Theorem 4.12. (It is easy to see that $3n \in \Theta(n)$ and $11n^2 \in \Theta(n^2)$; take the product of the two reference functions n and n^2.)

We now have the function $n^3 + 12n\log_2(n)$. Using the ranking of the common reference functions (Section 4.2.2), it is clear that n^3 is the larger. Thus, $3n(7\log_2(n) + 11n^2) + 12n\log_2(n) \in \Theta(n^3)$. ■

A Social Convention When choosing a reference function, you should never use any coefficients (other than 1). Thus, n^3 is an acceptable reference function, but $5n^3$ is not. [We can easily show that $5n^3 \in \Theta(n^3)$, so the 5 is unnecessary.]

Also, using Theorems 4.11 and 4.14, it is not acceptable to leave lower ranking terms in the reference function. For example, the reference function $n\log_2(n) + 5n + 9$ can be replaced by $n\log_2(n)$ since for large enough n, $n\log_2(n) > 5n + 9$.

✔ Quick Check 4.8

1. Find good reference functions for the following:
   (a) $6n + 7n(\log_2(n) + 9)$
   (b) $(2n + \log_2(n))(3\log_2(n) + 8)$
   (c) $\log_2(n^2 + n)$ [*Hint*: Use properties of logs.] ☑

4.2.3 Exercises

The exercises marked with ★ have detailed solutions in Appendix H.

1. Use a graphing calculator or a software package to graphically experiment with the big-$\mathcal{O}$ definition (Definition 4.3). In each case, either find values for c and n_0 that show $f_i \in \mathcal{O}(n^2)$, or else give some justification that $|f_i(n)| > cn^2$ will always occur for every $c > 0$, for large enough values of n [and so $f_i \notin \mathcal{O}(n^2)$].
   (a) $f_1(n) = 6n + 9$
   (b) $f_2(n) = n^2 + 1000$
   (c) ★ $f_3(n) = 3n\log_2(n)$
   (d) ★ $f_4(n) = \frac{n^3}{8}$
   (e) $f_5(n) = 2^n$
2. Use a graphing calculator or a software package to graphically experiment with the big-Ω definition (Definition 4.4). In each case, either find values for c and n_0 that show $f_i \in \Omega(n^2)$, or else give some justification that $|f_i(n)| < cn^2$ will always occur for every $c > 0$, for large enough values of n [and so $f_i \notin \Omega(n^2)$].
   (a) $f_1(n) = 6n + 9$
   (b) $f_2(n) = n^2 + 1000$
   (c) $f_3(n) = 3n\log_2(n)$
   (d) ★ $f_4(n) = \frac{n^3}{8}$
   (e) $f_5(n) = 2^n$
3. Recall that in Definition 4.5, n is any real number (although we usually consider it as an integer). Define $f(x) = \lfloor x \rfloor \cdot \lceil x \rceil$. Show graphically that $f(x) \in \Theta(x^2)$. Write the values of c_1, c_2, and n_0 on the graph.
4. ★ For each function in Exercise 1, show algebraically that it is in $\mathcal{O}(n^2)$, or else show algebraically that it is *not* in $\mathcal{O}(n^2)$. [For part (e) you may want to use limits and L'Hôpital's rule.]
5. ★ For each function in Exercise 2, show algebraically that it is in $\Omega(n^2)$, or else show algebraically that it is *not* in $\Omega(n^2)$.
6. Which of the functions in Exercise 1 are in $\Theta(n^2)$? Justify your answer.
7. Algebraically verify that $f(n) = 120n^2 - 11n + 450$ from Example 4.14 is in $\Omega(n^2)$ with $c = 1$.
8. For each pair of functions, show algebraically that $f \in \Theta(g)$ or $f \notin \Theta(g)$. Note that to prove $f \in \Theta(g)$, it must be shown that $f \in \mathcal{O}(g)$ *and* $f \in \Omega(g)$. Show your work in investigating both of these requirements.
   (a) $f(n) = 12n^5 + 3n^4 - n - 7$; $g(n) = n^5$
   (b) ★ $f(n) = \frac{n^2+4}{n+4}$; $g(n) = n$
   (c) ★ $f(n) = n!$; $g(n) = n^n$
   (Recall that $\lim_{n\to\infty} \frac{n!}{n^n} = 0$.)
   (d) $f(n) = \frac{5n^4+2n^2+3}{n^2+1}$; $g(n) = n^2$
   (e) $f(n) = 3^n + n^6$; $g(n) = n^6$
9. ★ Prove that $\lfloor x \rfloor \in \Theta(x)$.
10. Let $x, y, z \in \mathbb{R}$ with $x, y, z > 0$ and $y \neq 1$. Prove that
$$x^{\log_y(z)} = z^{\log_y(x)}$$
[*Hint*: $x = y^{\log_y(x)}$]
11. Prove that $\lfloor \log_2(x) \rfloor \in \Theta(\log_2(x))$.
12. Show that $\log_b(n) \in \Theta(\log_2(n))$ for any $b > 0$ with $b \neq 1$. Generalize this result.

13. Suppose someone proves that an algorithm is in $\Theta(1)$.
    (a) Write what this means in terms of Definitions 4.3 and 4.4.
    (b) Explain intuitively what this means in terms of the sizes of data sets.
14. Prove Theorem 4.8 on page 187.
15. Extend the table in Example 4.16 (page 187) to include a column for $n = 300{,}000{,}000$. You may omit the 2^n row.
16. Suppose the algorithms in Example 4.17 (page 188) are moved to a computer that requires a microsecond per instruction (10^{-6} seconds). How many data items can each algorithm complete in 30 seconds?
17. Suppose three algorithms are run on a machine that can execute one instruction every microsecond (10^{-6} seconds). They require, respectively, $n\log_2(n)$, $n\log_2(n) + 5n$, and n^2 operations for a data set of size n. Compare the time the algorithms require for data sets of size
    (a) 100
    (b) 100,000
    (c) 100,000,000.

    Comment on any patterns you see in your table of results.
18. Prove the big-Ω part of Theorem 4.12 on page 191.
19. Use the big-Θ theorems to find good reference functions for the following:
    (a) $f_1(n) = 3n^2 + 5n(2n + 7)$
    (b) $f_2(n) = \frac{n(n+1)}{2}$
    (c) ★ $f_3(n) = 121(\log_2(n) + n)(n + 3n\log_2(n)) + 6n^2$
    (d) $f_4(n) = 3^n + 3n^6 + 5\log_2(n)$
    (e) $f_5(n) = \frac{n(n-1)}{4}(6n^2 + \log_2(n))$
20. Use the big-Θ theorems to find good reference functions for the following:
    (a) ★ $f_1(n) = (2n^2 + 7)\log_2(n) + 2^n(n^3 + 4)$
    (b) $f_2(n) = (n^2 + 5 \cdot 2^n)\log_2(n^7 \cdot 2^n) + 2^{(n+2)}$
    (c) $f_3(n) = (n + 2^{(4n)})(n3^n + \log_2(n) + n^2)$
    (d) $f_4(n) = (n^2\log_2(n))^2 + n(\log_2(n) + n^2)$
    (e) $f_5(n) = 5n^3 + (\frac{n}{2} + n^3)(4n + \frac{3}{2}n^2)$
21. Show that $\log_2(n!) \in \mathcal{O}(n\log_2(n))$.
22. ★ Which function grows faster, 2^n or $n!$? Justify your answer algebraically.
23. Let $f_1(x) = -\lfloor x \rfloor$ and $f_2(x) = \lceil x \rceil$.
    (a) Show that $\max(f_1, f_2)(x) = f_2(x)$ for $x \geq 0$.
    (b) Show that $(f_1 + f_2) \notin \Omega(\lceil x \rceil)$, that is, that $(f_1 + f_2) \notin \Omega(\max\{f_1, f_2\})$.
24. Prove Theorem 4.15 and Propositions 4.16 and 4.17.

THEOREM 4.15 $f \in \Theta(g)$

Let f and g be real-valued functions for which $g(n) \neq 0$ for $n \geq n_0$, for some integer, $n_0 \geq 0$. If there is a real number, $r \neq 0$, such that

$$\lim_{n\to\infty} \left|\frac{f(n)}{g(n)}\right| = r$$

then $f \in \Theta(g)$.

PROPOSITION 4.16 $f \in \mathcal{O}(g)$

Let f and g be real-valued functions for which $g(n) \neq 0$ for $n \geq n_0$, for some integer, $n_0 \geq 0$. If

$$\lim_{n\to\infty} \left|\frac{f(n)}{g(n)}\right| = 0$$

then $f \in \mathcal{O}(g)$.

PROPOSITION 4.17 $f \in \Omega(g)$

Let f and g be real-valued functions for which $g(n) \neq 0$ for $n \geq n_0$, for some integer, $n_0 \geq 0$. If

$$\lim_{n\to\infty} \left|\frac{f(n)}{g(n)}\right| = \infty$$

then $f \in \Omega(g)$.

25. The theorem and propositions in Exercise 24 are not "if and only if" assertions. Simple counterexamples show that their converses are false. Find counterexamples for each of the following claims. [*Hint*: Think about the paragraph that follows the proof of Theorem 4.11, on page 191.]
    (a) ★ Let f and g be real-valued functions for which $g(n) \neq 0$ for $n \geq n_0$, for some integer, $n_0 \geq 0$. If $f \in \mathcal{O}(g)$, then
    $$\lim_{n\to\infty} \left|\frac{f(n)}{g(n)}\right| = 0$$
    (b) Let f and g be real-valued functions for which $g(n) \neq 0$ for $n \geq n_0$, for some integer, $n_0 \geq 0$. If $f \in \mathcal{O}(g)$, then
    $$\lim_{n\to\infty} \left|\frac{f(n)}{g(n)}\right| = r$$
    for some real number, $r \neq 0$.
    (c) Let f and g be real-valued functions for which $g(n) \neq 0$ for $n \geq n_0$, for some integer, $n_0 \geq 0$. If $f \in \Omega(g)$, then
    $$\lim_{n\to\infty} \left|\frac{f(n)}{g(n)}\right| = \infty$$
    (d) Let f and g be real-valued functions for which $g(n) \neq 0$ for $n \geq n_0$, for some integer, $n_0 \geq 0$. If $f \in \Theta(g)$, then there is a real number, $r \neq 0$, such that
    $$\lim_{n\to\infty} \left|\frac{f(n)}{g(n)}\right| = r$$
    [*Hint*: f and g can stay close together in a big-Θ sense without the quotient $\frac{f(n)}{g(n)}$ ever converging. One of the functions might dance around the other forever.]
26. Use Theorem 4.15 in Exercise 24 to provide an alternative proof for Theorem 4.11 on page 190.
27. ★ Use Theorem 4.15 in Exercise 24 to verify your solutions to Exercise 19 (you may need to use L'Hôpital's rule).

4.2.4 Big-Θ in Action: Searching a List

As stated earlier in this chapter, the main point of big-Θ is to provide a mechanism for comparing the efficiency of competing algorithms. In this section, two algorithms for searching a list will be compared. Some additional ideas will also be presented: best case, worst case, and average case behavior.

The problem will first be described, then the two algorithms explained and analyzed. You should review Theorem 3.54 on page 134. This result is used often enough in the analysis of algorithms that you should memorize it.

The second algorithm requires the following additional theorem.

THEOREM 4.18 ***Logarithms Are Order Preserving***

If $0 < x < y$, then $\log_b(x) < \log_b(y)$

Searching a List

Suppose that a list, $\{a_0, a_1, a_2, \ldots, a_{n-1}\}$, of n items has been produced.[18] An item with the value x exists, and we need to determine whether x is in the list. (Perhaps x represents a student and the list is a school's database of student information.)

Sequential Search

The simplest solution is to start at the beginning of the list, and compare each item, in turn, to x. As soon as x is found, the algorithm could return the position in the list where x resides. If x is not found, the algorithm could return a "not found message".[19] This algorithm is called a *sequential search*[20] and can be expressed as

```
1: integer sequentialSearch(x, {a0, a1, a2, ..., an-1})
2:     for i = 0 to n - 1
3:         if x == ai
4:             return i              # x == ai so exit and return i
5:     return "not found"           # x did not match any of the as
6: end sequentialSearch
```

The critical step in the algorithm is at line 3. The amount of time the algorithm needs is strongly related to the number of times this step is executed. Since the comparison is inside a loop, the number of times it executes is related to the size of the list.

Best Case If x happens to be in the first position of the list ($x = a_0$), then line 3 is executed only one time. The best case behavior is therefore in $\Theta(1)$.

Worst Case The worst case occurs in two ways: Either x is found in the final position ($x = a_{n-1}$), or else x is not in the list. In either case, line 3 will be executed n times. Thus, the worst case behavior is in $\Theta(n)$.

Average Case Average case behavior is usually harder to analyze than best case or worst case. For the search problem, it is made more troublesome by the need to worry about items that are not found. What proportion of all searches result in "not found"? That is something that we do not know without knowing more about the list and its intended use. The analysis that will be done here will just consider average behavior for successful searches. In addition, it will be assumed that every item in the list is equally

[18] Most popular programming languages start lists at subscript 0.

[19] In a traditional programming language like C, we could return -1 to represent "not found". In a language such as Java, we could throw an exception.

[20] Also called *linear search*.

likely to be the target of the search.

Intuitively, we would expect that on average, about half the list should be searched before x is found. More formally, since items are equally likely to be the target of the search, we can just average the cost of finding each item. To find that $x = a_i$ requires line 3 to be executed $i+1$ times. We want the sum of these costs, divided by the number of costs that are being averaged. This is

$$\frac{1+2+3+\cdots+n}{n} = \frac{1}{n}\sum_{i=1}^{n} i = \frac{1}{n}\cdot\frac{n(n+1)}{2} = \frac{n+1}{2} \in \Theta(n)$$

The average case behavior for a successful search is in $\Theta(n)$.

Binary Search

The more sophisticated approach is to use a binary search. To understand the motivation for this algorithm, think about our list as a phone book. The sequential search would start at the first page and look at names in alphabetical order until the name you wanted was found. This is a poor strategy for looking up a name in a large phone directory. A better approach is to open the directory to the middle page and determine whether the name you are looking for is in the first half or the second half of the phone book. Suppose the name is in the first half. You could then open to the page that is one-quarter of the way through the directory and determine whether the name comes before or after (or on) that page. Each step of the way, you are rejecting lots of names without ever looking at them.[21]

A *binary search* assumes that the list is in some kind of lexicographical order (perhaps alphabetic or numerical order). The concept of successively dividing the list in half is a simple one. However, properly expressing the algorithm is not trivial: There are many subtle errors that can occur. There are several correct variants (depending on when we determine whether a match has occurred). The one given here uses a multiway selection. Notice the use of the floor function, $\lfloor x \rfloor$, in line 5 to round the quotient down to an integer.

```
 1: integer binarySearch(x, {a_0, a_1, a_2, ..., a_{n-1}})
 2:     low = 0                 # index of left edge of the list's active portion
 3:     high = n - 1            # index of right edge of the list's active portion
 4:     while low ≤ high
 5:         mid = ⌊(low + high)/2⌋  # index of middle of active portion (rounded down)
 6:         if x > a_mid
 7:             low = mid + 1     # ignore the left half next iteration
 8:         else if x < a_mid
 9:             high = mid - 1    # ignore the right half next iteration
10:         else
11:             return mid        # found x at position mid
12:     return "not found"        # x is not in the list
13: end binarySearch
```

You should spend some time thinking carefully about this algorithm. You might try it by hand for a small list and look for a few entries in the list. Notice the difference between the index *position*, `mid`, and the number at that position in the list, a_{mid}.

The critical steps are the comparisons at lines 4, 6, and 8.

Best Case The best case would be if x was at the middle position. Three comparisons are made, so the best case is again in $\Theta(1)$.

[21] You might think about an even better strategy: Look at the first letter in the name, then estimate the distance from the front of the directory that the letter will be listed. This is called an *interpolation search*.

Worst Case Finding x in the last spot examined or not finding x at all both require almost the same number of comparisons. To get a feel for the amount of work required, first consider a list with 2^k items. Suppose that x is not in the list. How many times can we chop the list in half before we end up with a sublist having only one element?

The first division produces a list with $\frac{2^k}{2} = 2^{k-1}$ elements. The next division produces a list with 2^{k-2} elements. We stop when the list has $2^0 = 1$ element. Each division results in a decrease of 1 in the exponent on 2. Therefore, we can subdivide k times. This is the maximum number of times through the loop. Each time through the loop causes comparisons at line 4 and 6, and possibly another at line 8. The worst case would always use three. In addition, once `low` becomes greater than `high`, line 4 will be checked one last time when x is not present. So there are $3k + 1$ comparisons in the worst case for an unsuccessful search and $3k$ for a successful search.

How is k related to the size of the list? For this list (size $n = 2^k$) it is easy to see that $k = \log_2(n)$. What if the list's size is not a power of 2? The first subdivision produces a list of size $\lceil \frac{n}{2} \rceil$.[22] The next produces a list of size $\left\lceil \frac{\lceil \frac{n}{2} \rceil}{2} \right\rceil \simeq \frac{n}{2^2}$. The next division produces a list of approximately $\frac{n}{2^3}$ items. We can continue as long as $\frac{n}{2^k} \geq 1$. Thus, $2^k \leq n$ so $k \leq \log_2(n)$. Since k must be an integer (representing the number of times the list is bisected), $k = \lfloor \log_2(n) \rfloor$. In the worst case, there would be $3\lfloor \log_2(n) \rfloor + 1$ comparisons.

Thus, the worst case behavior is in $\Theta(\log_2(n))$.[23]

Average Case Once again there is the difficulty of determining what proportion of the time x will not be in the list. Even if we restrict the analysis to searches for items that are in the list, this analysis will not be easy. Assume that every item in the list is equally likely to be the target of the search.

We already know that the average case behavior must be between the best case and worst case. Is the average closer to the best case or to the worst case? One way to build some intuition is to do some experiments. One useful experiment is to fill a list with the numbers $0, 1, 2, \ldots, n-1$ and look for each in turn, counting the number of times lines 4, 6, and 8 are executed in each case. The average can then be calculated. Table 4.4 contains the results for some representative choices for n.

TABLE 4.4 Average Number of Comparisons in a List of Size n

n	1000	2000	3000	4000	5000	6000	7000	8000	9000	10000
average	22.912	25.399	26.890	27.892	28.691	29.385	29.897	30.388	30.787	31.188

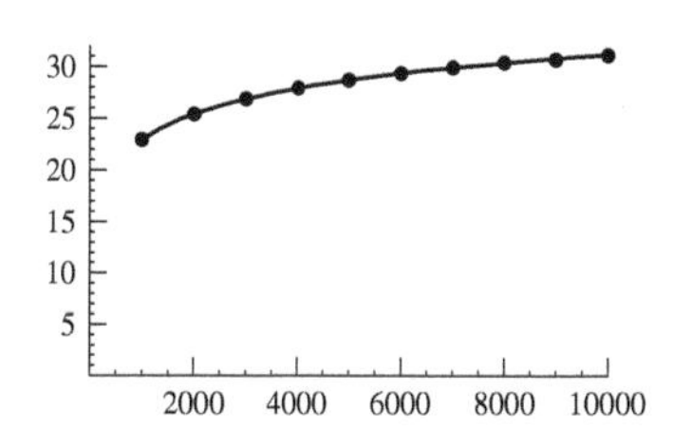

Figure 4.17. The average number of comparisons appears to be logarithmic.

A quick plot (Figure 4.17) shows that the data seems to be logarithmic. A least-squares fit to the data produces the function $f(n) = 2.489 \log_2(n) - 1.883$. The graph shows the data from Table 4.4 and the least squares function. Clearly, the function fits the data very well.

The experiment indicates that the average case is almost certainly in $\Theta(\log_2(n))$. We could try some additional investigations before attempting a formal proof of this result. One possibility is to look at lists of size $n = 2^k$ for various values of k. With this choice of n, every subdivision results in equal sized sublists on either side of the new midpoint, avoiding the messy detail of different sized sublists.[24] After each bisection of a sublist, the algorithm will always execute lines 4 and 6. Line 8 will be skipped some times. Assume (for the purpose of simplification) that, on average, each step takes 2.5

[22] For a list of size 7, the sublists have sizes 3 and 4. We would take the list of size 4 for a worst case search.

[23] Proving that $3\lfloor \log_2(n) \rfloor + 1 \in \Theta(\log_2(n))$ is left to the reader.

[24] For example, a list of size 6 has 2 items to the left of the midpoint but 3 items to the right.

comparisons. Then with a bit of work[25] it can be shown that for a list of size $n = 2^k$

$$\text{average number of comparisons} \simeq \frac{1}{2^k - 1} \cdot \sum_{j=0}^{k-1} \left(2^j\right) \cdot \left(\frac{5}{2}(j+1)\right)$$
$$\to 2.5 \log_2(n) - 2 \quad \text{as } k \to \infty$$

This is quite close to the function that resulted from a least squares fit to the experimental averages. We have enough evidence to justify trying to prove that the average case is in $\Theta(\log_2(n))$. Unfortunately, a formal proof (Corollary 11.27 on page 689) requires some ideas that will be presented in Chapter 11, so the formal proof will be deferred.

Summary

How do the sequential search and binary search algorithms compare? Table 4.5 indicates that the binary search is to be preferred if the list is already in order.

TABLE 4.5 Sequential Search versus Binary Search

Algorithm	Best Case	Average Case	Worst Case
Sequential Search	$\Theta(1)$	$\Theta(n)$	$\Theta(n)$
Binary Search	$\Theta(1)$	$\Theta(\log_2(n))$	$\Theta(\log_2(n))$

If the list is unordered, it must first be sorted in order for the binary search to work. The work needed to do this sorting is not included in the previous big-Θ estimates. The work needed to sort the list and then search for just a few items may actually be greater than just doing a few sequential searches on the list. However, as the number of searches increases, especially for large values of n, it is much more efficient to spend the extra effort to sort the list once and then use a binary search algorithm.

A Reality Check

So far, the analysis has focused entirely on how many times the algorithms make various comparisons. Is this a realistic measure? One way to validate the analysis experimentally is to program both algorithms and measure the total time needed to execute (rather than counting comparisons).

Table 4.6 summarizes the results for lists of size 1000 – 10,000, in increments of 1000. Each item in the list was the target of the search. Because timing is difficult to measure in such small units (milliseconds), each test was run 50 times and the total timings were averaged for each n. Times on a different computer will certainly be different, but their relative sizes should not change much.

TABLE 4.6 Timing Comparisons for Sequential and Binary Searches

n	1000	2000	3000	4000	5000	6000	7000	8000	9000	10000
Sequential (millisecs)	13	48	118	215	300	440	620	828	1012	1240
Binary (millisecs)	3	4	7	12	14	18	21	24	27	33

A least-squares fit for the sequential timings produces the function

$$s(n) = 0.0000122n^2 + 0.00309n - 4.217$$

The binary data has a least squares fit of $b(n) = 0.000242n \log_2(n) - 0.515$. Figure 4.18 on page 199 shows the graphs.

Recall that the program measured the time to find all n items in the list, so the table shows the measures for n times the average search time. Thus, the sequential search has an average search time that approaches $0.0000122n + 0.00309$ milliseconds as n

[25] The details have been suppressed here. Just take the result on faith—or derive it yourself.

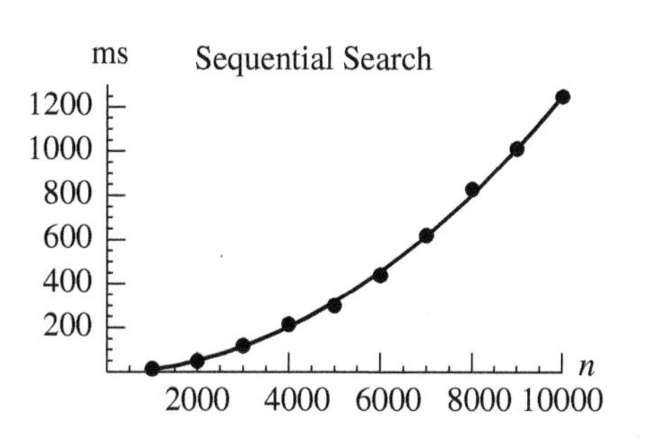

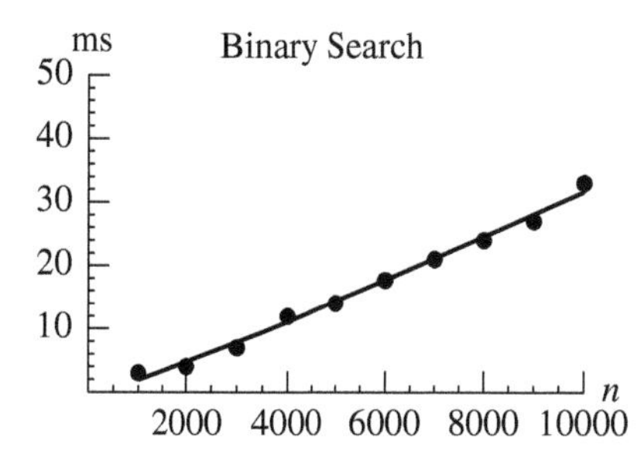

Figure 4.18 Visual timing comparisons.

gets large. This is clearly in $\Theta(n)$. Similarly, the binary search had an average search time that approaches $0.000242\log_2(n)$ milliseconds as n gets large. This is clearly in $\Theta(\log_2(n))$, as expected.

The experiment shows that measuring the number of comparisons is an accurate indicator of the actual time the algorithm requires to execute. The experiment also allows us to give a very crude estimate of the average time needed to search for a number in the list if the list were really huge.[26] If $n = 300{,}000{,}000$, the sequential search would take approximately $s(300{,}000{,}000)$ milliseconds (about 35 years) to search for all the items in the list. The binary search would take approximately $b(300{,}000{,}000)$ milliseconds (about 30 minutes).

EXAMPLE 4.21

Counting Critical Operations

The following fragment of pseudocode represents part of an algorithm. The critical operations are the comparison on line 3 and the swap on line 4. It is easier to focus on line 3, since it will be executed every time. What is a good big-Θ estimate for the number of comparisons?

```
1: for i = 1 to n
2:     for j = i to n
3:         if a_i < a_j
4:             swap a_i and a_j
```

Notice that the loop at line 1 (with index i) will be executed n times. The inner loop (line 2, with index j) will also be executed between 1 and n times. Each time line 3 is executed we need to add 1 to the comparison count. An easy way to keep track of this is to observe that the inner loop will add $c_i = \sum_{j=i}^{n} 1$ to the comparison count for each value of the index i. The outer loop will therefore add $\sum_{i=1}^{n} c_i = \sum_{i=1}^{n}\left(\sum_{j=i}^{n} 1\right)$ comparisons to the total. The total number of comparisons will be

$$\begin{aligned}\sum_{i=1}^{n}\left(\sum_{j=i}^{n} 1\right) &= \sum_{i=1}^{n}(n-i+1)\\ &= \sum_{i=1}^{n} n - \sum_{i=1}^{n} i + \sum_{i=1}^{n} 1\\ &= n^2 - \frac{n(n+1)}{2} + n\\ &= \frac{n(n+1)}{2}\end{aligned}$$

times.[27] Thus, this nested loop is in $\Theta(n^2)$. ■

[26] If the list cannot be stored entirely in the computer's memory, different algorithms (usually focused on minimizing the number of accesses to external storage) must be used and so these estimates may not be valid.

[27] Notice the use of the "fence post" rule in the evaluation $\sum_{j=i}^{n} 1 = n - i + 1$: To count the number of fence posts, count the number of fence sections and add one because there are two outer posts.

✔ Quick Check 4.9

1. Find a good big-Θ estimate for the number of assignment statements in this algorithm fragment. Provide adequate justification for your answer.

```
1: primeCount = 0
2: compositeCount = 0
3: for i = 2 to n
4:   if i is prime
5:      primeCount =
        primeCount + 1
6:   else
7:     compositeCount =
       compositeCount + 1
```

2. Find a good big-Θ estimate for the number of comparison statements in this algorithm fragment. Find a separate big-Θ estimate for the number of assignment statements. Provide adequate justification for your answers.

```
1: powerCount = 0
2: for i = 1 to n
3:   if i is a power of 2
4:      powerCount =
        powerCount + 1
```

✔

4.2.5 Exercises

The exercises marked with ★ have detailed solutions in Appendix H.

1. Use Theorem 3.54 to calculate the following. Assume k, m, and n are all positive integers.

   (a) $\sum_{i=1}^{1000} i$ (b) $\sum_{i=0}^{n} i$

   (c) ★ $\sum_{i=k}^{m} i$ (d) $\sum_{i=-k}^{m} i$

2. Find a formula for

$$\sum_{i=1}^{n} i^2$$

   You may look in books (including this one) or on the internet, but it would be even better if you experiment and guess the formula yourself. You need not provide a formal proof of correctness.

3. ★ Analyze the average case behavior of the sequential search algorithm if half the time x is not in the list. Assume that items in the list are equally likely to be the target of the search.

4. Suppose line 7 of the binary search algorithm is changed to

```
7:          low = mid
```

   Will the algorithm still work correctly? If not, give an example of where it will fail. If so, explain why it is still correct and determine whether it will still be as efficient.

5. ★ Suppose line 5 of the binary search algorithm is changed to

```
5:          mid = ⌈(low + high)/2⌉
```

   Will the algorithm still work correctly? If not, give an example of where it will fail. If so, explain why it is still correct and determine whether it will still be as efficient.

6. Suppose that your friend knows a different version of the binary search algorithm. He claims that his version is accurate and is more efficient than the variant provided in this textbook. Your friend's version is as follows:

```
1: integer newBinarySearch(x,
     {a_0, a_1, a_2, ..., a_{n-1}})
2:   low = 0        # index of left edge
                      of the list's
                      active portion
3:   high = n - 1  # index of right
                      edge of the list's
                      active portion
4:   while low < high
5:      mid = ⌊(low + high)/2⌋  # index of
            middle of active portion
            (rounded down)
6:      if x > a_mid
7:         low = mid + 1  # ignore
                  the left half next
                  iteration
8:      else
9:         high = mid - 1 # ignore
                  the right half next
                  iteration
10:  if x = a_low
11:     return low    # found x at
                         position low
12:  else
13:     return "not found"   # x is
                      not in the list
14: end newBinarySearch
```

   Does your friend's algorithm actually work correctly? If not, give an example of where it will fail. If so, explain why it is correct and determine whether it is more or less efficient than the version in this textbook.

7. Count (exactly) the number of critical operations in the following pseudocode fragments. Then, find good big-Θ reference functions for the number of critical operations.

(a) ★ The critical operation is the sum in line 5. (For the big-Θ part, replace 20 by n.)

```
1: i = 1
2: sum = 0
3: while i ≤ 20
4:     for j = 1 to i
5:         sum = sum + a_ij
6:     i = i + 1
```

(b) The critical operation is the multiplication in line 5. For the big-Θ part, assume $p = m = n$. You might recognize this as the pseudocode for a matrix multiplication.

```
1: for i = 1 to n
2:     for j = 1 to m
3:         c_ij = 0
4:         for k = 1 to p
5:             c_ij = c_ij + a_ik · b_kj
```

(c) The critical operations are the sums at lines 6, 8, and 9. The fragment is adding various entries in a matrix.

```
1: allEntriesSum = 0
2: lowerTriangleSum = 0
3: diagonalSum = 0
4: for i = 1 to n
5:    for j = 1 to n
6:       allEntriesSum
         = allEntriesSum + a_ij
7:    for k = 1 to i
8:       lowerTriangleSum
         = lowerTriangleSum + a_ik
9:    diagonalSum = diagonalSum+a_ii
```

(d) The critical operations are at lines 7, 8, and 9. Use the result of Exercise 2. The algorithm calculates

$$(1!)(1!+2!)(1!+2!+3!)\cdots(1!+2!+\cdots+n!)$$
$$=\prod_{i=1}^{n}(1!+2!+\cdots+i!)$$
$$=\prod_{i=1}^{n}\left(\sum_{j=1}^{i} j!\right)$$

```
1: product = 1
2: for i = 1 to n
3:    sum = 0
4:    for j = 1 to i
5:       factorial = 1
6:       for k = 1 to j
7:         factorial = k*factorial
8:       sum = sum + factorial
9:    product = product*sum
```

8. Count (exactly) the number of critical operations in the following pseudocode fragments. Then find good big-Θ reference functions for the number of critical operations.

(a) ★ The critical operations are at lines 6 and 7. For the big-Θ part, replace 30 by n.

```
1: i = 0
2: count = 0
3: sumterms = 0
4: while i < n
5:     for j = 1 to 30
6:         sumterms = sumterms + j
7:         count = count + 1
8:     i = i + 1
```

(b) The critical operations are at lines 3 and 5. Assume that addition and division are equally expensive. For the big-Θ part, assume $w = y$.

```
1: for i = 0 to w
2:   for j = 0 to i
3:      total = total + purchase_ij
4: for k = 1 to y
5:    sum = sum + (total ÷ k)
```

(c) The critical operations are at lines 6, 7, and 9. Assume that addition and multiplication are equally expensive. For the big-Θ part, assume $m = p$.

```
1: lowersum = 0
2: uppersum = 0
3: total = 0
4: for i = 1 to m
5:   for j = 1 to p
6:     lowersum = lowersum + x_ij
7:     uppersum = uppersum + y_ij
8:     for k = 1 to j
9:         total = total
           + k · (lowersum
                + uppersum)
```

(d) The critical operations are at lines 4, 5, 7, and 8. For the big-Θ part, assume $a = h = c = d$.

```
1: for h = 1 to a
2:     for i = 1 to h
3:        for j = 1 to c
4:           term_hi = term_hi + term_ij
5:           sum1 = sum1 + term_hi
6:        for k = 1 to d
7:           term_hk = term_hk + term_ik
8:           sum2 = sum2 + term_hk
```

4.3 Pattern Matching

The example in this section illustrates many of the ideas presented earlier in this chapter. The main point is

> **Observation**
>
> One of the best ways to improve performance is to find a better algorithm.

The problem to be solved in this section is to find the first occurrence of a substring in a longer document. You might consider what a word processor needs to do to search a document for a word or sentence.

The document will be called the *text* and denoted as a string of n characters: $T_1T_2T_3 \cdots T_n$. The substring of m characters that needs to be found (if it is present) is called the *pattern* and will be denoted $p_1p_2p_3 \cdots p_m$.[28] The algorithm will return the index in the text where the first occurrence of the pattern begins, or the number -1 if the pattern is not present.

Choosing a proper algorithm depends somewhat on the kind of text and pattern you are expecting to encounter. A search for the word *soldier* in the text of Shakespeare's *Macbeth* may require different optimizations than a search for the pattern "AAGTCT" in a text representing a strand of DNA. Searching for the code of a computer virus inside a binary computer file may require other kinds of optimization.

Before worrying too much about such differences, it is usually helpful to have *some* correct solution. Once that is available, we can start looking for better solutions.[29]

4.3.1 The Obvious Algorithm

The simplest solution requires the characters of the pattern to be matched with successive characters in the text. Once a mismatch is found, the pattern is shifted by one character and the matching started over from the beginning of the pattern. When the pattern character is the same as the text character, we say that a *hit* has occurred. When the characters are different, we say that a *miss* has occurred.

Suppose the text is "mathematics" and the pattern is "math". If the pattern is lined up under the second "m" in the text, hits occur for the first three letters in the pattern, but a miss occurs at the "i-h" pairing.

```
m a t h e m a t i c s
          m a t h
```

The algorithm can be expressed in pseudocode as

```
 1: integer obvious(string T, string p)
 2:     n = the length of T # i represents the text position where the pattern
 3:     m = the length of p # currently begins; j is the position in the pattern
 4:     i = 1
 5:     while i ≤ n − m + 1  # pattern will extend past text if i > n − m + 1
 6:         j = 1
 7:         while (j ≤ m) and (T[i+j−1] == p[j])  # still in pattern and another hit
 8:             j = j + 1
 9:         if j == m + 1       # has the entire pattern been matched?
10:             return i        # yes, exit and return the position of the match
11:         i = i + 1           # no, shift the pattern and try again
12:     return −1               # pattern not present
13: end obvious
```

Notice that characters in the text may be examined multiple times.

[28] Notice the break with previous convention—the strings in this section start with subscript 1, not with 0.

[29] An extremely fast but incorrect solution is never better than a very slow but correct solution.

EXAMPLE 4.22

The Obvious Algorithm

Suppose we want to find the first occurrence of the pattern "mama" in the text "mammals amaze mama".[30] The table shows the pattern as it slides along the text. The character in the pattern where a miss occurs is in a bold font. The final columns show the numbers of hits and misses.

Text and Pattern																		Hits	Misses
m	a	m	m	a	l	s	␣	a	m	a	z	e	␣	m	a	m	a		
m	a	m	**a**															3	1
	m	a	m	a														0	1
		m	**a**	m	a													1	1
			m	a	**m**	a												2	1
				m	a	m	a											0	1
					m	a	m	a										0	1
						m	a	m	a									0	1
							m	a	m	a								0	1
								m	a	m	a							0	1
									m	a	**m**	a						2	1
										m	a	m	a					0	1
											m	a	m	a				0	1
												m	a	m	a			0	1
													m	a	m	a		0	1
														m	a	m	a	4	0
																		–	–
																	Total	12	14

■

Notice that if the pattern ever sticks out beyond the final character of the text, there can be no match (line 5 in the pseudocode). We can effectively cause the pattern to slide along the text by judicious use of the subscripts. If the pattern starts at character i in the text, then character p_j in the pattern is under character T_{i+j-1} in the text.

EXAMPLE 4.23

Worst Case for the Obvious Algorithm

Consider looking for the pattern "aaab" in the text "aaaaaaab".

Text and Pattern								Hits	Misses
a	a	a	a	a	a	a	b		
a	a	a	**b**					3	1
	a	a	a	**b**				3	1
		a	a	a	**b**			3	1
			a	a	a	**b**		3	1
				a	a	a	b	4	0
								–	–
Total								16	4

Characters that are not among either the first $m - 1$ characters or the final $m - 1$ characters of the text are examined m times. Those at the front and back of the text (other than the first and last) are also examined multiple times. ■

[30] The symbol "␣" represents a space.

The conclusion is that the worst case performance of the obvious algorithm is in $\Theta((n-m)m)$. This is because all but the $2(m-1)$ characters at the ends of the text are examined m times apiece.

However, in a typical word processor search, the behavior is more like that in Example 4.22, where most of the time a miss happens at the first character of the pattern. If you consider finding the pattern "bb" in the text "aaaaaaaabb", you can easily see that the best case performance of the obvious algorithm (for a match at the end of the text) is in $\Theta(n)$ because the misses occur at the first character, thus avoiding any multiple examinations of text characters.

The average performance depends on the nature of the text and pattern, but will certainly be between the best and worst case performance.

4.3.2 KMP: Knuth–Morris–Pratt

One of the first improvements on the obvious algorithm was presented by Knuth, Morris, and Pratt (KMP). It avoids the need to examine characters in the text multiple times by first doing some preprocessing of the pattern (which needs to be counted in the big-Θ estimate). Since it needs to examine each of the n text characters at least once (and in the best case only once), and since it needs to first examine each of the m pattern characters, the KMP algorithm is in $\Theta(n+m)$ in the best case.

The key idea is that when a miss occurs somewhere after the first character in the pattern, we already know something about the earlier characters in the pattern. We may already know (implicitly) that we need to shift more than one character to the right. For example, consider the first miss in Example 4.22.

m	a	m	m	a	l	s	␣	a	m	a	z	e	␣	m	a	m	a
m	a	m	**a**														

Consider the situation from the perspective of the pattern. It knows that its first three characters match the text. It also knows that the fourth character does not match. It is unaware of what the rest of the text contains. We can denote these ideas by using a "?" to represent a character whose value is unknown, and "[x]" to indicate that the character is *not* "x":[31]

m	a	m	[a]	?	?	?	?	?	?	?	?	?	?	?	?	?	?
m	a	m	**a**														

Using only information inherent in the pattern, we can conclude that the pattern can be shifted three characters to the right. This is because

- If the pattern were to shift by 1 character, the initial "m" would be aligned with the "a" in the second position of the pattern (recall that the second character of the pattern had a hit with the text).
- If the pattern were to shift by 2 characters, the "a" in position 2 of the pattern would be aligned with the character in the text that we know is *not* an "a".

Denoting characters we no longer care about by a "#", we can jump to

#	#	#	[a]	?	?	?	?	?	?	?	?	?	?	?	?	?	?
			m	a	m	a											

The pattern has shifted three characters to the right and is ready for more processing.

The KMP algorithm begins by building a table with one column per pattern character. The table indicates how many positions the pattern may shift right if there is a miss at that character of the pattern (attempting to match from left to right). For the example

[31] This notation is from [91].

we have been examining, the table is

pattern position	1	2	3	4
pattern	m	a	m	a
shift	1	1	3	3

If a miss occurs at the first character, we know nothing about the next text character and so can only shift by 1. If a hit occurs on the first pattern character and a miss at the second, we only know that the text does not have an “a” at the second position, so a shift of 1 is all that is possible. However, if the first two characters in the pattern are hits and the third is a miss, we can shift by 3. This is because a shift of one will align an “m” in the pattern with an “a” in the text. A shift of 2 would line an “m” in the pattern up with a character in the text which is known to not be an “m”. The final shift value has already been explained.

There is one more clever observation: **After a shift, there is no need to reexamine any text characters to the left of the miss.** This is because the shift guarantees that either the newly shifted pattern characters will still be hits, or else the pattern will have shifted past the text character. This should often give the KMP algorithm an advantage over the obvious algorithm.

Quick Check 4.10

1. Show the details of the KMP search for the pattern “mama” in the text “mammals amaze mama”. Ignore the optimization in the previous paragraph. ☑

EXAMPLE 4.24

KMP Shift Table

Consider Example 4.23 again. We can construct the KMP shift table as follows:

If a miss occurs on the first character, we can only shift one position. If the miss occurs on the second character, we have

a	[a]	?	?
a	**a**	a	b

A shift of 1 character will line the first “a” in the pattern up with something that is not an “a”, so we need to shift by 2.

If the miss occurs on the third character

a	a	[a]	?
a	a	**a**	b

We cannot shift by 1 or 2 (the second or first “a” in the pattern would line up with something which is not an “a”), so a shift of 3 is possible.

Finally, if the miss happens at the “b”

a	a	a	[b]
a	a	a	**b**

we can only shift by 1.

The shift table is therefore

pattern position	1	2	3	4
pattern	a	a	a	b
shift	1	2	3	1

pattern	a	a	a	b
shift	1	2	3	1

The search for "aaab" in "aaaaaaab" using KMP produces

Text and Pattern								Hits	Misses	
a	a	a	a	a	a	a	b			
a	a	a	**b**					3	1	
	a	a	a	**b**				1	1	← Start at the third "a"
		a	a	a	**b**			1	1	since we already know
			a	a	a	**b**		1	1	the first two will match.
				a	a	a	b	2	0	
								—	—	
Total								8	4	

This is better than the obvious algorithm.

If we were to search for "aaab" in "aacaaaab", we would see a small improvement:

Text and Pattern								Hits	Misses
a	a	c	a	a	a	a	b		
a	a	**a**	b					2	1
			a	a	a	**b**		3	1
				a	a	a	b	2	0
								—	—
Total								7	2

We must add 4 to the total hits and misses to measure the effort to construct the shift table. The totals for the examples are really 16 and 13, respectively. These two examples are not too far from the theoretical best case performance estimate of 8 + 4 = 12 comparisons. ■

The best case behavior of KMP can be shown to be in $\Theta(n + m)$ since in the best case, characters in the text are examined only once.[32] However, in practice, KMP doesn't do much better than the obvious algorithm for most types of text and pattern. This is because most misses occur on or near the first character in the pattern.

KMP will not be fully developed here because there is a better alternative.

4.3.3 BM: Boyer–Moore

The Boyer–Moore algorithm uses two shift tables. In addition, it uses the clever approach of checking the pattern against the text in a right-to-left direction. These innovations enable the algorithm to sometimes shift the pattern past unexamined characters in the text. By not examining every character in the text, the Boyer–Moore algorithm can exhibit dramatic performance increases over the obvious and KMP algorithms. This is true even though it requires more preprocessing than KMP. Boyer–Moore is in $\Theta(n+m)$ in the worst case, whereas KMP is in $\Theta(n+m)$ in the best case. In the best case, Boyer–Moore could be in $\Theta(\frac{n}{m})$ if the match is at the end of the text, all misses occur at the final character in the pattern, and the maximum shift of m is used at every miss.

EXAMPLE 4.25

Boyer–Moore Head-to-Head with KMP and Obvious

Dieter Bühler at the University of Tübingen in Germany has created a very nice animation of the pattern matching algorithms presented in this section. The animation can be

[32]Quick Check 4.10 almost achieves this performance.

reached from a link at

http://www.mathcs.bethel.edu/~gossett/DiscreteMathWithProof/

I compared the performance of the three algorithms by searching for the pattern "soldier" in the text of Shakespeare's play *Macbeth*. The results are summarized in Table 4.7. The second column represents the average number of times each character in the text was compared to a character in the pattern. The third column represents the total number of characters processed (including during shift table creation).

TABLE 4.7 Searching for "soldier" in Macbeth

Algorithm	Exam/Char	Chars
Obvious	1.05	4356
KMP	1.05	4348
BM	0.18	758

■

The *Last* Table

The simpler of the two Boyer–Moore tables is named the *last table*. It handles a miss by focusing on the character *in the text* at which the miss occurred. Since we now know that character, we can shift the pattern right far enough to place that same character under the text character (assuming the text character occurs somewhere in the pattern). If the pattern does not contain the text character, we can shift the entire pattern past the place where the miss occurred.

The last table needs one entry for every distinct character in the text. What is generally done is to create a table with one entry for every possible character that the software system can represent.[33]

EXAMPLE 4.26

A Simple DNA Search

Suppose we want to search for the pattern "AACT" in a DNA text string. How do we build the last table? For this example, it is reasonable to restrict the table to the four possible characters in a DNA string: "A", "C", "G", and "T". We can initially think about the table as it is being used. Imagine that a miss has occurred at the right edge of the pattern (the first to be checked in Boyer–Moore).

```
?   ?   ?   [T]
A   A   C    T
```

If the text contains a "C" at the miss position, we can only shift the pattern right by one place. If the miss happened to be an "A" we could shift right by two places, placing the rightmost "A" under the "A" in the text. However, if the miss occurs at a "G", we can shift the pattern right by four places (since "G" does not occur in the pattern).

We need a table that records, for each distinct character in the text, where the final copy of that character is in the pattern. We can then do some simple arithmetic to determine how far to shift the pattern to line that character up with the known text character. For example, consider the search for "mama".

```
m   a   m   m   ...
m   a   m   a
```

[33]Generally 128 or 256 if ASCII or ANSI characters are used. If Unicode is used, the table will need to be truncated to just the characters in the text or to some sufficiently small subset of Unicode. (Unicode can represent $2^{16} = 65,536$ distinct characters, including Chinese, Hebrew, Arabic, etc.)

```
m  a  m  m  ...
m  a  m  a
```

We want to shift the "m" in position 3 of the pattern under the known "m" in position 4 of the text. The last table informs the algorithm that the final "m" in the pattern is at position 3. The algorithm also knows that the miss occurred at character 4 of the pattern. The required shift is thus the miss position, minus the table value of the text character (i.e., shift one place to the right).

The critical observation is that the suggested shift depends not only on the character in the text that was not matched, but also the position of the character in the pattern where the miss occurs. ■

The last table is very easy to create. The value under text character x is the position of the rightmost occurrence of x in the pattern, or 0 if x does not appear in the pattern.

EXAMPLE 4.27 **The Last Table for the Simple DNA Search**

The last table for a search for the pattern "AACT" in a DNA string is

A	C	G	T
2	3	0	4

■

The algorithm for building the last table is simple. Assume, as usual, that the pattern, p, has m characters. Denote the table by `L` and index the table by the characters in the text alphabet, in lexicographical order. Denote the value for character x by `L[x]`.[34]

```
1: table lastTable(string p)
2:     for each character c in the text alphabet
3:         L[c] = 0        # the default value for characters not in the pattern
4:     for i = 1 to m
5:         L[p_i] = i      # this character's right-most position will be kept
```

Notice that table entries corresponding to characters in the pattern will be first set to 0, then replaced by the positions of that character in the pattern. The final value of `L[c]` will be the position of the final occurrence of `c` in the pattern (or 0 if `c` is not in the pattern). You should verify the table in Example 4.27 using this algorithm.

✔ Quick Check 4.11

1. Assume that the text alphabet consists of the 8 characters "a"–"h". Build the last table for the following patterns.
   (a) "gabbed"
   (b) "bcadgafb" ☑

The following pseudocode fragment indicates how the Boyer–Moore algorithm uses the last table. The text, T, has n characters and the pattern, p, has m characters. The $n - m + 1$ at line 2 causes the loop to terminate as soon as the pattern moves beyond the end of the text. The first character of the pattern (`j = 1`) is aligned with character `i` in the text. As `j` increases, the text position follows (`i+j-1`). Notice that the value of `j - L[`T_{i+j-1}`]` on line 9 could be negative. That is why we make sure that `i` is incremented by at least 1 (line 9).

```
1: i = 1
2: while i ≤ n−m+1           # pattern will extend past text if i > n−m+1
3:     j = m                 # start at the right of the pattern
4:     while (j ≥ 1) and (p_j == T_{i+j−1})  # still in pattern and a hit
5:         j = j−1
6:     if j == 0             # has every character in the pattern been matched?
7:         return i          # yes, exit and return position of match
8:     else                  # no, shift (by at least 1) and try again
9:         i = i + max{1, j - L[T_{i+j−1}]} # shift only enough to align with the miss
10: return −1                # pattern not found
```

[34]This is standard array notation in programming languages, but with the less common convention of using characters as indices.

EXAMPLE 4.28

The Last Table in Action

The Boyer–Moore algorithm using only the last table will be used to find the pattern "AACT" in the text "ACGTAACTG". Example 4.27 calculated the last table to be

A	C	G	T
2	3	0	4.

Using the preceding code fragment and recalling that Boyer–Moore compares the pattern to the text from right to left (in the pattern),

A	C	G	T	A	A	C	T	G	the text
A	A	**C**	T						The miss is at the G in the text and j = 3, shift max{1,3-L[G]} = 3
			A	A	C	**T**			Miss at C and j = 4, shift max{1,4-L[C]}=1
				A	A	C	T		Match, return i = 5

■

Quick Check 4.12

1. Use the Boyer–Moore algorithm with only the last table to find the pattern "mama" in the text "mammals amaze mama". Assume that the text alphabet consists of the seven characters "a", "e", "l", "m", "s", "z", and "␣" (space).

The Boyer–Moore Shift Table

The Boyer–Moore *shift table* works in a manner similar to the KMP shift table, except that it is designed to be used with a right-to-left comparison of pattern and text. The table has one entry for every character in the pattern. The table value indicates the number of places to shift if a miss occurs at the pattern position corresponding to the table index. For this table, we view the situation from the pattern's perspective, so when a miss occurs, all we know is what character is *not* present in the text.

EXAMPLE 4.29

The Shift Table for the Simple DNA Search

The shift table for a search for the pattern "AACT" in a DNA string can be calculated by looking at a miss for each character in the pattern. We start at the right and work toward the left.

If a miss occurs at the fourth character, we have the following situation:

?	?	?	[T]
A	A	C	**T**

We can only shift one place to the right, since the text character could be a "C".

If a miss occurs at the third position, the following is known:

?	?	[C]	T
A	A	**C**	T

We can shift by four positions because we know that neither "C" nor "A" matches the "T" in a shift of less than 4.[35]

Similar observations indicate that the shift values for misses under the first and second characters in the pattern will also result in a shift of 4. The shift table is thus

A	A	C	T
4	4	4	1

■

[35] We know about the "T" because the "T" in the pattern was a hit. We are not really using information about the text when building the shift table. In fact, the shift table is constructed before looking at the text.

✔ Quick Check 4.13

1. Build the Boyer–Moore shift table for the pattern “mama”. ☑

What are the key ideas used for building the shift table? We want the smallest shift such that we

Don’t Repeat the Miss We do not shift the same letter in the pattern to the spot where the miss just occurred.[36]

Do Repeat the Hits We need every character shifted to the right of the miss to match the previous character known to reside in that spot.[37]

For example, suppose a miss occurs in the second position of “mama”.

```
?   [a]   m   a
m    a    m   a
```

Then we need to make sure we don’t shift another “a” under the “[a]” and also need to match the “ma” in positions 3 and 4 (if possible):

```
#   #   m   a
        m   a   m   a
```

Assume again that the pattern, p, has m characters. The variable s will represent the amount by which to shift. The table will be denoted D.[38]

```
 1: table shiftTable(string p)
 2:    # j represents the position in the pattern where a miss occurs
 3:    for j = 1 to m        # for each position in the pattern
 4:       s = 1              # minimum shift is 1
 5:       while s < m        # maximum shift is m
 6:          # make sure that a shift of s won't place the
 7:          # same pattern character in position j
 8:          if (s < j) and (p_{j-s} == p_j)
 9:             s = s+1      # same character, pick a larger shift
10:          else            # a different character
11:             # now make sure the characters that shift to the right of the miss
12:             # match the previous pattern characters in those positions
13:             k = max{j+1,s+1}   # stay in pattern and check to right of the miss
14:             while (k ≤ m) and (p_{k-s} == p_k) # keep checking until a mismatch
15:                k = k + 1
16:             if k == m+1         # did all pattern characters match?
17:                D[j] = s         # yes, s is the smallest viable shift
18:                s = m+1          # break out of the "while s < m" loop
19:             else
20:                s = s+1          # no, try a bigger shift
21:          end if
22:       end while
23:       if s == m
24:          D[j] = m          # need to shift completely past the miss
25:    end for
26: end shiftTable
```

[36]Lines 8 and 9 in algorithm shiftTable ensure this.

[37]Lines 13–15 check this in algorithm shiftTable.

[38]Mainly to correspond to the table in Dieter Bühler’s animation. The algorithm shiftTable differs from his approach. This algorithm is simpler and occasionally produces a more efficient shift table.

The test that ensures a different character is shifted to the miss position (*don't repeat the miss*) need not be done if we will shift the pattern beyond that position. That is, if the shift amount `s` is greater than or equal to the miss position `j`, then do not try to compare the characters p_{j-s} and p_j (there is no character p_{j-s} in the pattern because $j - s < 1$).[39]

Similarly, the check that characters to the right of the miss should match the previous occupants (*do repeat the hits*) need not be done for positions in the text that the pattern shifts beyond (hence the second element in line 13 in the pseudocode). The algorithm `shiftTable` is a straightforward implementation of these observations.

The following pseudocode fragment shows how the shift table would be used.

```
1: i = 1 # start at the first character of the text
2: while i ≤ n−m+1  # pattern will extend past text if i > n−m+1
3:     j = m            # start at the right of the pattern
4:     while (j ≥ 1) and (p_j == T_{i+j−1}) # still in pattern and a hit
5:         j = j − 1
6:     if j == 0   # has every character in the pattern been matched?
7:         return i            # yes, exit and return position of match
8:     else                    # no, shift and try again
9:         i = i + D[j]
10: return −1                # pattern not found
```

EXAMPLE 4.30

The Shift Table in Action

The Boyer–Moore algorithm using only the shift table will be used to find the pattern "AACT" in the text "ACGTAACTG". Example 4.29 calculated the shift table to be

p_j	A	A	C	T
D[j]	4	4	4	1

Using the preceding code fragment, and once again recalling that Boyer–Moore compares the pattern to the text from right to left (in the pattern),

A	C	G	T	A	A	C	T	G	the text
A	A	**C**	T						The miss is at $j = 3$, shift 4
				A	A	C	T		Match, return $i = 5$ ■

✔ Quick Check 4.14

1. Use the Boyer–Moore algorithm with only the shift table to find the pattern "mama" in the text "mammals amaze mama". ☑

The Full Algorithm

You may have noticed that when building the two tables, we ignored some of the available information. When building the last table, we ignored the characters of the text that were hits. This information is neglected in order to keep the algorithms for constructing and using the table as simple as possible. When building the shift table, we ignore the actual value of the text character at a miss. This is done so that the shift table can be constructed completely using only the pattern. It would be much more complicated to build a shift table that also used information about which character is in the text at that position. The extra work is not worth the trouble because the last table focuses on exactly that information in a very simple manner.

[39] See line 8 of `shiftTable`.

Table 4.8 summarizes the contrasting features of the last and shift tables.

TABLE 4.8 Characteristics of the Last and Shift Tables

Table	Size	Indexed by	Goal(s)
last	The number of characters in the text alphabet	Text *character* at the miss	Shift the rightmost matching pattern character to beneath the miss in text
shift	The number of characters in the pattern (m)	The *position* in the pattern at the miss	(A) Don't repeat the miss. (B) Do repeat the hits.

The full Boyer–Moore algorithm can now be presented. The only new feature is choosing the maximum shift from the two tables (line 10). Notice that the presence of $D[j]$ means we no longer need to worry about negative values for $j - L[T_{i+j-1}]$.

```
 1: integer BoyerMoore(string T, string p)
 2:     L = lastTable(p)
 3:     D = shiftTable(p)
 4:     i = 1    # start at the first character of the text
 5:     while i ≤ n − m + 1  # pattern will extend past text if i > n − m + 1
 6:        j = m             # start at the right of the pattern
 7:        while (j ≥ 1) and (T_{i+j−1} == p_j)  # still in pattern and a hit
 8:           j = j − 1
 9:        if j == 0          # has every character in the pattern been matched?
 8:           return i        # yes, exit and return position of match
10:        i = i + max{D[j], j - L[T_{i+j−1}]} # no, shift and try again
11:     return −1             # pattern not found
12: end BoyerMoore
```

EXAMPLE 4.31

The Full Boyer–Moore Algorithm in Action

The full Boyer–Moore algorithm can be used to search for "mama" in the text "mammals amaze mama". Recall that the two tables are

Last Table

a	e	l	m	s	z	␣
4	0	0	3	0	0	0

Shift Table

m	a	m	a
2	2	4	1.

The shifts are as follows. The codes L, S, and B indicate which of the tables won the privilege of determining the amount to shift in the following line (B for both).

Text and Pattern																		Hit	Miss	Win	$j - L[T_{i+j-1}]$
m	a	m	m	a	l	s	␣	a	m	a	z	e	␣	m	a	m	a				
m	a	m	**a**															0	1	B	$4-3$
	m	**a**	m	a														2	1	S	$2-3$
			m	a	m	**a**												0	1	L	$4-0$
							m	a	m	a								3	1	S	$1-0$
									m	a	m	**a**						0	1	L	$4-0$
													m	a	m	**a**		0	1	B	$4-3$
														m	a	m	a	4	0		
																		—	—		
																	Total	9	6		

4.3.4 Exercises

The exercises marked with ★ have detailed solutions in Appendix H.

1. ★ Use Dieter Bühler's animation to compare the total processing (including table construction) for a search for the pattern "When shall we three" in the text of Shakespeare's *Macbeth.* Also indicate the average number of comparisons per character and the position at which the pattern first starts in the text. List all five algorithms using the names in this text (not Dieter's names). Note that the animation looks for all matches, so it continues to search after the pattern is found.
2. Use Dieter Bühler's pattern matching animation to compare the total processing (including table construction) for a search for the pattern "GGGACTGAAAAC" in the text of the "human chromosome 12 DNA". Also indicate the average number of comparisons per character and the position at which the pattern first starts in the text. List all five algorithms using the names in this text (not Dieter's names). Note that the animation looks for all matches, so it continues to search after the pattern is found.
3. Generalize Example 4.23 so that the text consists of $n-1$ a's, followed by one b. Let the pattern be $m-1$ a's, followed by one b. Produce an exact count (in terms of n and m) of the number of comparisons required by the obvious algorithm.
4. Build the KMP shift table for the following patterns.
   (a) "abba" (b) ★ "mimmmi" (c) "minimize"
5. Build the KMP shift table for the following patterns.
   (a) "aardvark" (b) "envelope" (c) "millimeter"
6. Build the Boyer–Moore last table for the following pattern/alphabet pairs.
   (a) "abba", {a, b} (b) ★ "mimimi", {i, m}
   (c) "minimize", {e, i, m, n, z}
7. Build the Boyer–Moore last table for the following pattern/alphabet pairs.
   (a) "connecticut", {c, e, i, n, o, t, u}
   (b) "mississippi", {i, m, p, s}
   (c) "hawaii", {a, h, i, w}
8. Build the Boyer–Moore last table for the following pattern/alphabet pairs.
   (a) "giggling", {g, i, l, n}
   (b) "sustenance", {a, c, e, n, s, t, u}
   (c) "coffee", {c, e, f, o}
9. Build the Boyer–Moore shift table for the following patterns.
   (a) "abba" (b) ★ "mimimi" (c) "minimize"
10. Build the Boyer–Moore shift table for the following patterns.
    (a) "banana" (b) "envelope" (c) "stress"
11. Build the Boyer–Moore shift table for the following patterns.
    (a) "sheepish" (b) "yummm" (c) "swissmiss"
12. ★ Search for the pattern "fum" in the text "femfufofum." Create a table like that used in Example 4.22 and count hits and misses. Graph paper would be helpful.
    (a) Use the obvious algorithm.
    (b) Use the KMP algorithm.
    (c) Use Boyer–Moore with only the last table. Show $j - L[T_{i+j-1}]$.
    (d) Use Boyer–Moore with only the shift table.
    (e) Use the full Boyer–Moore algorithm. Indicate which table determines each shift. Show $j - L[T_{i+j-1}]$.
13. Search for the pattern "AGCCT" in the text

    "AGCTTAGCCAGCCT"

    Create a table like that used in Example 4.22 and count hits and misses. Graph paper would be helpful.
    (a) Use the obvious algorithm.
    (b) Use the KMP algorithm.
    (c) Use Boyer–Moore with only the last table. Show $j - L[T_{i+j-1}]$.
    (d) Use Boyer–Moore with only the shift table.
    (e) Use the full Boyer–Moore algorithm. Indicate which table determines each shift. Show $j - L[T_{i+j-1}]$.
14. Search for the pattern "pie" in the text "pickled peppers". Create a table like that used in Example 4.31 and count hits and misses. Graph paper would be helpful.
    (a) Use the obvious algorithm.
    (b) Use the KMP algorithm.
    (c) Use Boyer–Moore with only the last table. Show $j - L[T_{i+j-1}]$.
    (d) Use Boyer–Moore with only the shift table.
    (e) Use the full Boyer–Moore algorithm. Indicate which table determines each shift. Show $j - L[T_{i+j-1}]$.
15. Search for the pattern "ABCAD" in the text

    "ABCEDADAEFABABCADE"

    Create a table like that used in Example 4.22 and count hits and misses. Graph paper would be helpful.
    (a) Use the obvious algorithm.
    (b) Use the KMP algorithm.
    (c) Use Boyer–Moore with only the last table. Show $j - L[T_{i+j-1}]$.
    (d) Use Boyer–Moore with only the shift table.
    (e) Use the full Boyer–Moore algorithm. Indicate which table determines each shift. Show $j - L[T_{i+j-1}]$.
16. Search for the pattern "eve" in the text "elevator eleven". Create a table like that used in Example 4.22 and count hits and misses. Graph paper would be helpful.
    (a) Use the obvious algorithm.
    (b) Use the KMP algorithm.
    (c) Use Boyer–Moore with only the last table. Show $j - L[T_{i+j-1}]$.
    (d) Use Boyer–Moore with only the shift table.
    (e) Use the full Boyer–Moore algorithm. Indicate which table determines each shift. Show $j - L[T_{i+j-1}]$.

17. Search for the pattern "ZXYX" in the text

"ZWXZXYZXZWYXYXW"

Create a table like that used in Example 4.22 and count hits and misses. Graph paper would be helpful.

(a) Use the obvious algorithm.
(b) Use the KMP algorithm.
(c) Use Boyer–Moore with only the last table. Show $j - L[T_{i+j-1}]$.
(d) Use Boyer–Moore with only the shift table.
(e) Use the full Boyer–Moore algorithm. Indicate which table determines each shift. Show $j - L[T_{i+j-1}]$.

18. Search for the pattern "see" in the text "sally sells seashells." Create a table like that used in Example 4.22 and count hits and misses. Graph paper would be helpful.

(a) Use the obvious algorithm.
(b) Use the KMP algorithm.
(c) Use Boyer–Moore with only the last table. Show $j - L[T_{i+j-1}]$.
(d) Use Boyer–Moore with only the shift table.
(e) Use the full Boyer–Moore algorithm. Indicate which table determines each shift. Show $j - L[T_{i+j-1}]$.

4.4 The Halting Problem

4.4.1 Setting the Stage

Some students find reading and creating proofs to be difficult. They are tempted to skip them when reading a math textbook (or when doing homework). One justification they might make to themselves is "I am not going to get a job doing proofs so I can just understand the statement of the theorem and use it. Someone else has already checked that it is correct." This is not a beneficial attitude.

Proofs are not poison; at their best, they provide insight and enlightenment. The theorem statement is generally very useful but the statement does not tell *why* it is true. That is one of the purposes of the proof.

The goal of this section is to present a very famous theorem that claims something (the halting problem) is impossible. Knowing that something is impossible is a great advantage. If we believe the theorem's proof, we are not tempted to waste time in a doomed effort to achieve the impossible task.

EXAMPLE 4.32 **Trisectors**

There is a famous problem left by the ancient Greek mathematicians: Given an arbitrary angle, trisect it using only a compass and straightedge.[40] In 1837 Pierre Laurent Wantzel proved that this is impossible. If we do not insist on using a compass and straightedge, the problem is not difficult.[41] So there is no solution (using the specified tools) and no practical need for a solution. Yet there are people (called *trisectors*) who every year send math departments long "proofs" of trisections.[42] For most trisectors, no amount of explanation will convince them that they are trying to do the impossible and that there is no use for such a construction even if it could be done. They can be shown the errors in their purported proof, but they will just go home and try to invent a better proof.

There is an important distinction between impossible (trisection using compass and straightedge) and difficult (the four-color theorem – Theorem 10.52 on page 645). We should not waste time attempting the impossible; we should not give up prematurely on the merely difficult. ■

The proof of the halting theorem that will be presented will use a contradiction that is similar to a famous paradox.

[40] A straightedge is like a ruler without any tic marks.
[41] The ancient Greek mathematician Archimedes had a solution by other means.
[42] For a sympathetic, insightful examination of trisectors and their motivations, read [25].

Quick Check 4.15

1. A famous paradox in set theory (due to Bertrand Russell) can be stated in the following form: *In a certain village, the barber shaves every man who does not shave himself. Does the barber (a man) shave himself?* This problem has no trick answers (such as "the barber lives in another village").

The halting problem is properly described in terms of Turing machines, which will not be formally defined until Chapter 9. However, you may think of a Turing machine as a computer program. A computer program ideally takes some data as input, runs for a while, then creates some output and halts.

An interesting observation is that the computer program itself can be considered as data. For example, there are programs, called *pretty printers*, that take the text of another computer program (perhaps written in Java) as input and produce a nicely formatted version of the other program as output. The obvious thing to do once you have written a pretty printer is to run it on its own source text!

4.4.2 The Halting Problem

The halting problem will be described here in terms of computer programs. The ideas in Chapter 9 can be used to describe it more formally in terms of Turing machines. Those ideas are not needed to understand the theorem or its proof.

DEFINITION 4.19 ***The Halting Problem***

The *halting problem* seeks a computer program, H, that accepts as input another computer program, P, together with some input data, d, and then outputs true if P eventually halts when run on d but outputs false if P runs forever when d is used as input.

The program H can be viewed as a function defined by

$$H(P, d) = \begin{cases} \text{true} & \text{if } P(d) \text{ halts} \\ \text{false} & \text{if } P(d) \text{ runs forever} \end{cases}.$$

Notice that the halting problem *does not* seek an algorithm that determines whether a particular program P_0 halts on specific input data d_0. It seeks an algorithm that works for *all* programs and *all* legal input data.

In 1936, Alan Turing proved that the halting problem seeks something that is impossible.[43]

THEOREM 4.20 ***The Halting Theorem***

There will never be an algorithm that solves the halting problem.

More than one proof is available. Turing originally used a proof that adapted Cantor's famous diagonalization proof that the set of real numbers is uncountably infinite.[44]

The alternative proof given here is quite nice.

---

[43] Alan Turing was at the center of the successful attempt to crack the German *Enigma machine* used for secret codes during World War II.

[44] Recall the definition of *discrete mathematics* in Chapter 1. Students typically encounter Cantor's proof in a course in *real analysis*. The result says that the infinite set of real numbers is qualitatively larger than the infinite sets of integers and rational numbers, which are qualitatively the same size.

Proof: The proof begins with a critical assumption: Suppose, for the moment, that a solution to the halting problem exists. That is, there is an algorithm $H(P, d)$, as defined previously, that correctly determines whether any program P eventually terminates when run on any data set d.

The proof continues by cleverly introducing two simple algorithms that use H. The first, called `willHalt`, determines whether an algorithm (program) halts when run on itself. If $H(P, P)$ is true, `willHalt` returns true, otherwise it returns false.

```
1: boolean willHalt(program P)
2:     return H(P, P)
3: end willHalt
```

The next algorithm will enter an infinite loop (thus never halting) if it is run on a program P that halts when given itself as data. If P does not halt when run on itself, `barber` returns a message and halts.

```
1: boolean barber(program P)
2:     if willHalt(P)
3:         start an infinite loop
4:     else
5:         return "I quit!"
6: end barber
```

The *really* clever step in the proof is to run `barber` on itself. Consider the alternatives.

- If `barber` is a program that will halt when run on itself, then `barber(barber)` will enter an infinite loop at line 3. So barber will *not* halt when run on itself! This is a contradiction.
- If `barber` is a program that will *not* halt when run on itself, then `barber(barber)` will return the message "I quit!" and halt. This is also a contradiction!

The algorithm `barber` creates an insurmountable problem. The source of this difficulty traces back to the assumption that algorithm $H(P, d)$ actually exists. The conclusion is that algorithm $H(P, d)$ *cannot exist* (not now, not in 1000 years, not ever). □

The theorem shows that we cannot write a universal algorithm to determine when programs will halt. (We can, however, often do a good job of determining whether particular programs halt on a typical range of data sets.)

One of the major implications of the halting theorem is that there are tasks that no computer program can ever solve. Not even if we create vastly more powerful computers and write magnificently brilliant programs. In fact, the halting problem has become a standard that allows us to determine that many other tasks are not possible. What is typically done is to show that *if* we could solve Problem X, *then* we could use the algorithm that solves Problem X to produce a solution to the halting problem. However, since the halting problem cannot be solved, Problem X must also be unsolvable.

For a second opinion on this entire section, see the brief article *Halting Problem Is Solvable*, by Bala Rajagopalan. It can be found in the rec.humor.funny archives or reached from a link at

http://www.mathcs.bethel.edu/~gossett/DiscreteMathWithProof/

4.5 Quick Check Solutions

Quick Check 4.1

1. One simple solution is

```
read a and b
if a == 0
  a = 1
print a+b
```

2. Notice that we do not need to explicitly check for equality:

```
if a < b
   print a
else
   print b
```

Can you spot the error in the following incorrect algorithm?

```
1: if a < b
2:      smaller = a
3: smaller = b
4: print smaller
```

The problem is that line 3 is always executed. If `a` is less than `b`, line 2 correctly assigns `a` to `smaller`. However, line 3 immediately changes the value of `smaller` to `b`. Line 4 then prints the wrong result.

Quick Check 4.2

1. One solution is

```
count = 0
sum = 0
i = 0
while count < n
   if i is odd
      sum = sum + i
      count = count + 1
   i = i + 1
print sum
```

A better solution is

```
sum = 0
for i = 1 to n
   sum = sum + (2i − 1)
print sum
```

2. An indefinite iteration is appropriate here

```
count = 0
while more characters exist
   if next character is a vowel
      count = count + 1
print count
```

A more detailed version can be written

```
count = 0
while more characters exist
   read the next character into c
   if c ∈ {a, e, i, o, u}
      count = count + 1
print count
```

I have assumed, for this example, that there are only five vowels in English.

Quick Check 4.3

1. There are other solutions that are correct. Notice the convention of writing

```
if even
  do something
```

rather than

```
if even == true
  do something
```

The convention is used since even already has the value true or the value false.

```
real alternatingSum(integer n, real {a1, a2, ..., an})
   even = true
   altsum = a1
   for i = 2 to n
      if even
         altsum = altsum - ai
         even = false
      else
         altsum = altsum + ai
         even = true
   return altsum
end alternatingSum
```

Quick Check 4.4

1. $$\begin{aligned} 2n^2 + 5n + 7 &\le 2n^2 + 5n^2 + 7 && \text{for } n \ge 1, \text{ since } n \le n^2 \text{ if } n \ge 1 \\ &\le 2n^2 + 5n^2 + 7n^2 && \text{if } n \ge 1 \\ &= 14n^2 \end{aligned}$$
Since $2n^2 + 5n + 7 \le 14n^2$, it would be nice to conclude that $|2n^2 + 5n + 7| \le 14|n^2|$. Note, however, that $-5 < 2$ but $|-5| \not< 2$. We need to make sure that $0 < 2n^2 + 5n + 7$ before taking absolute values. Fortunately, the function $2n^2 + 5n + 7$ is concave up (since it is quadratic with positive leading coefficient), so for large enough n, the function *will* be greater than zero. In fact, it is easy to see graphically that this function is positive for all n. Thus $|2n^2 + 5n + 7| \le 14|n^2|$ for $n \ge 1$ (i.e., $n_0 = 1$ and $c = 14$).

2. $$\begin{aligned} 101n + 73 &\le 101n + 73n && \text{if } n \ge 1 \\ &= 174n \end{aligned}$$
Let $n_0 = 1$ and $c = 174$. Then $0 < 101n + 73$ and $|101n + 73| \le 174|n|$.
Another approach is to notice that $101n + 73 = 102n$ when $n = 73$. Thus, the definition can be verified with $n_0 = 73$ and $c = 102$: $|101n + 73| \le 102|n|$ for $n \ge 73$.

3. The term $-5n$ requires some extra caution. It is important that the absolute values in the definition not be dropped. (In Example 4.11, the absolute values were dropped because the function, $120n^2 - 11n + 450$ is positive for all $n \ge 0$—the smallest value of n being considered.) The function, $n^2 - 5n - 4$ does not have this property. For example, when $n = 1$, $n^2 - 5n - 4 = -8 < 1 = n^2$, but $|n^2 - 5n - 4| = 8 \not\le 1 = |n^2|$.
Assume that $n \ge 0$. Then
$$\begin{aligned} |n^2 - 5n - 4| &\le |n^2 - 5 \cdot 5 - 4| && \text{if } n > 5 \\ &= |n^2 - 29| \\ &\le n^2 \end{aligned}$$

Using $n_0 = 6$ and $c = 1$ in Definition 4.2 completes the proof.

An alternative approach is to notice that

$$n^2 - 5n - 4 = \left(n - \frac{5+\sqrt{41}}{2}\right)\left(n + \frac{5+\sqrt{41}}{2}\right)$$

so $n^2 - 5n - 4 > 0$ if $n \geq 6$. Thus, assuming that $n \geq 6$, the absolute values can be temporarily dropped. Then $0 < n^2 - 5n - 4 < n^2 - 5n < n^2$. Consequently, $|n^2 - 5n - 4| < 1|n^2|$, and so $n^2 - 5n - 4 \in \mathcal{O}(n^2)$ (using $n_0 = 6$ and $c = 1$).

Quick Check 4.5

1.

$$\begin{aligned} 101n + 73 &\leq 101n + 73n \quad \text{if } n \geq 1 \\ &= 174n \\ &\leq 174n^2 \quad\quad\quad \text{if } n \geq 1 \end{aligned}$$

Let $n_0 = 1$ and $c = 174$. Since $0 < 101n + 73$, $|101n + 73| \leq 174|n^2|$ for $n \geq 1$.

Quick Check 4.6

1. The following graph suggests that $c = \frac{1}{8}$ and $n_0 = 5$ should work.

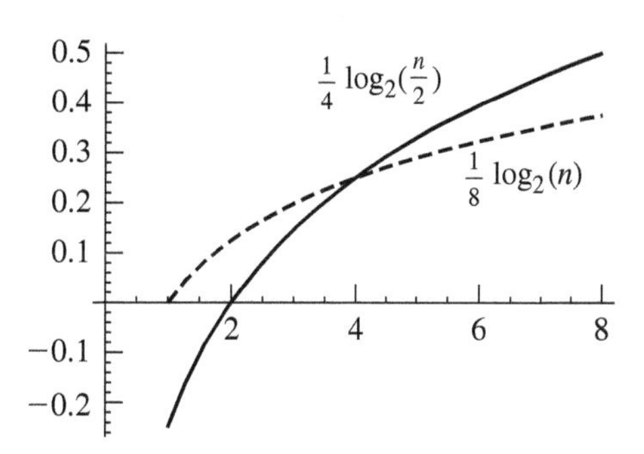

Quick Check 4.7

1. The table represents the approximate times.

Θ	$n = 10$	$n = 100$
$n \log_2(n)$	16.6 minutes	5.5 hours
n^2	50.0 minutes	83.3 hours

2. There are 3600 seconds in 1 hour, so we need to solve the equations

$$\begin{aligned} 30n \log_2(n) &= 3600 \\ 30n^2 &= 3600 \end{aligned}$$

Using your graphing calculator, or Newton's method, or a software function such as *Mathematica*'s NSolve function, the results are

algorithm 1 25–26 items

algorithm 2 almost 11 items

Quick Check 4.8

1. There are alternate paths you might take [for example, in part (a), you might first expand $7n(\log_2(n) + 9)$ as $7n\log_2(n) + 63n$].

   (a) Using Theorem 4.14 (after checking that both functions are positive), the factor $\log_2(n) + 9$ can be replaced by $\log_2(n)$ because $9 \in \Theta(1)$. The new term $(7n)(\log_2(n))$ can be replaced by $n\log_2(n)$ using Theorem 4.12 [and observing that $7n \in \Theta(n)$]. This leaves the function $6n + n\log_2(n)$. Observing that $n\log_2(n)$ grows faster than n, Theorem 4.14 implies that $6n + n\log_2(n) \in \Theta(n\log_2(n))$. The final result is that $6n + 7n(\log_2(n) + 9) \in \Theta(n\log_2(n))$.

   (b) The simpler approach is to notice that $2n + \log_2(n) \in \Theta(n)$ (look at the famous reference functions) and that $3\log_2(n) + 8 \in \Theta(\log_2(n))$. Thus, $(2n + \log_2(n))(3\log_2(n) + 8) \in \Theta(n\log_2(n))$. You could also start by expanding the original product.

   It is not difficult to show that

$$6n\log_2(n) \le (2n + \log_2(n))(3\log_2(n) + 8) \le 33n\log_2(n) \quad \text{for } n \ge 2$$

   so you could graphically verify this result if it seems counterintuitive.

   (c) This requires the properties of log functions. First, observe that

$$\log_2(n^2 + n) = \log_2(n(n + 1)) = \log_2(n) + \log_2(n + 1)$$

   Next, notice that

$$\begin{aligned}\log_2(n) \le \log_2(n + 1) &\le \log_2(n + n)\\ &= \log_2(2) + \log_2(n)\\ &= 1 + \log_2(n) \quad \text{for } n \ge 1\end{aligned}$$

   and

$$1 + \log_2(n) \le \log_2(n) + \log_2(n) = 2\log_2(n) \quad \text{for } n \ge 2$$

   so

$$\log_2(n) \le \log_2(n + 1) \le 2\log_2(n) \quad \text{for } n \ge 2$$

   Thus $\log_2(n + 1) \in \Theta(\log_2(n))$. Using Theorem 4.14, it should now be clear that $\log_2(n^2 + n) \in \Theta(\log_2(n))$.

Quick Check 4.9

1. There are two assignments before the loop. The loop is executed $n - 1$ times. It doesn't matter which branch of the selection statement we choose; there will always be an assignment. Thus there are $n + 1$ assignments, which is in $\Theta(n)$.

2. The comparisons appear easy to count: one comparison for each time through the loop. There is one technical issue: The for loop actually does a hidden comparison on each iteration to make sure that $i \le n$. It makes one more comparison when it determines that $i > n$ and the loop should be terminated. Thus, there are really $n + (n + 1) = 2n + 1$ comparisons. Thus, the number of comparisons is in $\Theta(n)$.

   The number of assignments is a bit more complicated (but not by much). We want to count each of the numbers $2^0, 2^1, 2^2, \ldots, 2^k$, where $2^k \le n$ and $2^{k+1} > n$. You should be comfortable with this by now: We want $k \le \log_2(n)$, where k must be an integer. Thus $k = \lfloor \log_2(n) \rfloor$. The number of assignments is in $\Theta(\log_2(n))$.

Quick Check 4.10

1. Text and Pattern

m	a	m	m	a	l	s	␣	a	m	a	z	e	␣	m	a	m	a	Hits	Misses
m	a	m	**a**															3	1
			m	a	**m**	a												2	1
						m	a	m	a									0	1
							m	a	m	a								0	1
								m	a	m	a							0	1
									m	a	**m**	a						2	1
												m	a	m	a			0	1
													m	a	m	a		0	1
														m	a	m	a	4	0
																		–	–
																	Total	11	8

Quick Check 4.11

1. Set all the table values to 0, and then read the pattern from left to right. At each character, copy its position over the previous value in the table.

(a)

a	b	c	d	e	f	g	h
2	4	0	6	5	0	1	0

(b)

a	b	c	d	e	f	g	h
6	8	2	4	0	7	5	0

Quick Check 4.12

1. The last table is

a	e	l	m	s	z	␣
4	0	0	3	0	0	0

Text and Pattern

m	a	m	m	a	l	s	␣	a	m	a	z	e	␣	m	a	m	a	Hits	Misses
m	a	m	**a**															0	1
	m	**a**	m	a														2	1
		m	a	m	**a**													0	1
						m	a	m	**a**									0	1
							m	a	m	a								3	1
								m	a	m	**a**							0	1
												m	**a**	m	a			2	1
														m	a	m	a	4	0
																		–	–
																	Total	11	7

Quick Check 4.13

1.

m	a	m	a
2	2	4	1

Quick Check 4.14

1. We know from Quick Check 4.13 that the shift table is

m	a	m	a
2	2	4	1

The search produces

Text and Pattern																		Hits	Misses
m	a	m	m	a	l	s	␣	a	m	a	z	e	␣	m	a	m	a		
m	a	m	**a**															0	1
	m	**a**	m	a														2	1
			m	a	m	**a**												0	1
				m	a	m	**a**											0	1
					m	a	**m**	a										1	1
									m	a	m	**a**						0	1
										m	a	m	**a**					0	1
											m	a	m	**a**				0	1
												m	**a**	m	a			2	1
														m	a	m	a	4	0
																		—	—
																	Total	9	9

Quick Check 4.15

1. Consider the problem with either answer:
   - If the barber shaves himself, then he is in the group of people that the barber does not shave—so he does *not* shave himself!
   - If the barber does not shave himself, then he is in the group of people who are shaved by the barber—so he *does* shave himself!

   Both options lead to a contradiction. Something is terribly wrong!

   This paradox (and others) caused mathematicians to reconsider their theories about sets. Here is a related paradox:

   Let A be the set containing all sets that do not contain themselves. Is $A \in A$?

   To see that a set can contain itself, let T be the set of all sets that contain more than two elements. The sets $D =$ the set of all dogs, $C =$ the set of all cats, and $B =$ the set of all birds are all members of T, so $T \in T$.

   Consider the new problem with either answer:
   - If $A \in A$, then A fails to satsify the condition for membership in A, so $A \notin A$.
   - If $A \notin A$, then A satisfies the condition for membership in A so $A \in A$.

   One way to avoid the paradox is to exclude the notion of a set containing itself from set theory.

4.6 CHAPTER REVIEW

4.6.1 Summary

With the advent of computers, algorithms have received significant attention. In that context, two issues are of interest in this chapter: (1) how can we express an algorithm clearly and unambiguously? and (2) How should we compare the relative efficiency of two algorithms that accomplish the same task?

This chapter starts with a presentation of the basic structuring mechanisms used by modern programming languages. The mechanisms are expressed in pseudocode but can easily be translated into your favorite computer language. Pseudocode has been designed to capture the mechanisms that specify the actions of the algorithm without being overly concerned with syntactic issues (such as the placement of semicolons or the use of parentheses).

A major theorem in computer science is that any algorithm can be expressed using only structured control constructs. These constructs can be partitioned into three single-entry, single-exit categories: sequence, selection, and repetition. Constructs in these categories can be nested within other constructs. By using only structured constructs to specify an algorithm, the algorithm becomes easier to comprehend, easier to prove correct, and easier to modify at a later date.

One simple approach to comparing the efficiency of two algorithms is to run both algorithms on many data sets and compare the times needed to finish. This requires that both algorithms must be translated to a computer program and also requires the sample data sets to be created.

An alternative approach uses a mathematical analysis of the algorithms to categorize each as belonging to one of several groups of algorithms whose efficiencies are similar. The main idea is that two algorithms are in the same group if, for very large data sets, both algorithms take about the same time to complete. This notion is made precise by designating some key operations within the two algorithms as the critical operations. A function representing the number of critical operations used to process a data set of size n is produced for each algorithm. The functions are then compared to a collection of standard reference functions. If f is the function that represents the number of critical operations for an algorithm and g is another function (typically a simple reference function), we say that $f \in \Theta(g)$ if there exist positive constants, n_0, c_l, c_u, such that $c_l|g(n)| \leq |f(n)| \leq c_u|g(n)|$ for all $n > n_0$.

Category	**Reference Function**
Constant	1
Logarithmic	$\log_2(n)$
Linear	n
$n \log n$	$n \log_2(n)$
Quadratic	n^2
Cubic	n^3
Exponential	a^n for $a > 1$

Section 4.2 examines this comparison approach in some detail. The basic definitions are introduced and several theorems are introduced that provide alternative ways to determine an appropriate reference function. The section also discusses the process of starting with an algorithm, expressed in pseudocode, and producing a function that counts the number of critical operations. The table in the margin contains many of the common reference functions, moving from desirable to undesirable as the list moves from top to bottom.

Section 4.3 considers an important problem: determining whether a string appears as a substring of some larger text. Several algorithms are presented. An important conclusion is that a better algorithm can make a dramatic difference in efficiency.

The chapter ends with a significant result in theoretical computer science: There will never be an algorithm that solves the halting problem. This theorem shows that there are limits to what can be achieved with computers. The proof provides an interesting example of a proof by contradiction.

4.6.2 Notation

Notation	Page	Brief Description
= or :=	171	the pseudocode assignment operator
==	172	the pseudocode equality operator (assertion)
*	172	the pseudocode multiplication operator
/	172	the pseudocode division operator
#	178	the pseudocode start-of-comment symbol
$f \in \mathcal{O}(g)$	181	f is in big-$\mathcal{O}$ of g
$f \in \Omega(g)$	184	f is in big-Ω of g
$f \in \Theta(g)$	185	f is in big-Θ of g
␣	203	a visible symbol for the space character
[x]	204	represents a character which is *not* x

4.6.3 Big-Θ Shortcuts

Theorem 4.11 Big-Θ and Polynomials Let f be the polynomial function $f(n) = a_k n^k + a_{k-1} n^{k-1} + \cdots + a_1 n + a_0$, where $a_k \neq 0$. Then $f \in \Theta(n^k)$.

Theorem 4.12 Big-Θ and Products Suppose that $f_1 \in \Theta(g_1)$ and $f_2 \in \Theta(g_2)$. Then $f_1 \cdot f_2 \in \Theta(g_1 \cdot g_2)$.

Theorem 4.14 Big-Θ and Sums Suppose that $f_1 \in \Theta(g_1)$ and $f_2 \in \Theta(g_2)$. Assume also that there is a positive integer, n_0, such that for all $n > n_0$, $f_1(n) > 0$, $f_2(n) > 0$, $g_1(n) > 0$ and $g_2(n) > 0$. Then $(f_1 + f_2) \in \Theta(\max\{g_1, g_2\})$.

Theorem 4.15 $f \in \Theta(g)$ Let f and g be real-valued functions for which $g(n) \neq 0$ for $n \geq n_0$, for some integer, $n_0 \geq 0$. If there is a real number, $r \neq 0$, such that

$$\lim_{n\to\infty} \left| \frac{f(n)}{g(n)} \right| = r$$

then $f \in \Theta(g)$.

Proposition 4.16 $f \in \mathcal{O}(g)$ Let f and g be real-valued functions for which $g(n) \neq 0$ for $n \geq n_0$, for some integer, $n_0 \geq 0$. If

$$\lim_{n\to\infty} \left| \frac{f(n)}{g(n)} \right| = 0$$

then $f \in \mathcal{O}(g)$.

Proposition 4.17 $f \in \Omega(g)$ Let f and g be real-valued functions for which $g(n) \neq 0$ for $n \geq n_0$, for some integer, $n_0 \geq 0$. If

$$\lim_{n\to\infty} \left| \frac{f(n)}{g(n)} \right| = \infty$$

then $f \in \Omega(g)$.

4.6.4 Additional Review Material

Go to http://www.mathcs.bethel.edu/~gossett/DiscreteMathWithProof/review.xhtml for additional review material, including a sample chapter exam, with solutions.

CHAPTER

5

Counting

Man is fond of counting his troubles, but he does not count his joys.
If he counted them up as he ought to, he would see
that every lot has enough happiness provided for it.

Fyodor Dostoyevsky (1821 – 1881)

A familiar nursery rhyme[1] reads

As I was going to St. Ives
I met a man with seven wives;
Every wife had seven sacks;
Every sack had seven cats;
Every cat had seven kits.
Kits, cats, sacks, and wives,
How many were going to St. Ives?

Of course, you learned at a young age that you could weasel out of the hard question by noticing that "I" was the only one going to St. Ives. Your carefree younger days are in the past. It is time to deal with the implied problem.

✔ Quick Check 5.1

- How many kittens, cats, sacks, and wives were met on the road to St. Ives? Give a subtotal for each of the four categories, as well as a grand total. ☑

Counting problems are ubiquitous. Some are easy; some are extremely difficult. Most of the material in this chapter explores the techniques for counting and applications of those techniques. One more introductory example illustrates the diversity of counting problems.

EXAMPLE 5.1

The Maximum Number of Rounds in the Deferred Acceptance Algorithm

Recall the Stable Marriage Problem, introduced in Section 1.2. The Deferred Acceptance Algorithm was developed by Gale and Shapely to solve this problem.

Gale and Shapely claim that the Deferred Acceptance Algorithm will end after at most $n^2 - 2n + 2$ stages. Why is this true?

The algorithm ends on the round during which the last unclaimed suitee gets a proposal. This is because once a suitee receives a first proposal, she or he will never have an empty string of suitors. Therefore, once the final suitee receives a proposal, all other suitees must have exactly one suitor (there are equal numbers of males and females).

[1] Variations on this problem have been around for a long, long time. An ancient (circa 1650 B.C.) Egyptian collection of math problems, called the Rhind Papyrus, contains one version. See [34, p. 55] for more details.

How long can this be avoided? There is an initial round when each suitor proposes to his or her first choice. There are then n suitors, each of whom can propose to at most $n-2$ additional suitees and still leave the final suitee unclaimed. Finally, the last unclaimed suitee must be proposed to. This gives a total of at most

$$1 + n(n-2) + 1 = n^2 - 2n + 2$$

rounds.

Exercise 1 in Exercises 5.1.6 asks you to create a set of preferences that results in this upper bound actually occurring. ■

5.1 Permutations and Combinations

Much of the material in this section relates to counting the number of ways it is possible to arrange a subset of elements selected from a finite set. There are two important pairs of concepts you should look for: order, and repetition.[2] The counting formulas that will eventually be examined fit neatly into the grid shown in Table 5.1.

TABLE 5.1 Order and repetition

	With Order	**Without Order**
Without Repetition	Permutations	Combinations
With Repetition	Permutations with repetition	Combinations with repetition

DEFINITION 5.1 ***Permutation; Combination***
Permutations and combinations arise when a subset is to be chosen from a set. A *permutation* is a collection of elements for which an ordering of the chosen elements has been imposed. A *combination* is another name for a subset; order is unimportant.

Permutations and combinations will be examined in detail after some prerequisite ideas have been presented.

5.1.1 Two Basic Counting Principles

There are two important principles for counting when special conditions prevail. The first principle applies when there is a sequence of independent tasks or choices. For example, if I wish to choose one of five books to read and one of three snacks to eat while I read, the two choices are independent. The snack I choose is not influenced by my choice of book.

The second principle applies when there is a sequence of mutually exclusive tasks or choices. For example, if I wish to go to one of three restaurants or one of four delicatessens, the choices are mutually exclusive. I either go to a restaurant or a delicatessen.

These concepts are formalized in the next definition.

DEFINITION 5.2 ***Independent Tasks; Mutually Exclusive Tasks***
The tasks in a collection or sequence of tasks are said to be *independent* if the outcome of any task is not influenced by the outcomes of the other tasks in the collection or sequence. The tasks in a collection are said to be *mutually exclusive* if completing any one of the tasks excludes the completion of the other tasks.

These special conditions will now be examined in more detail.

[2] An alternative term for "repetition" is *replacement*.

The Independent Tasks Principle

If a project can be decomposed into two independent tasks with n_1 ways to accomplish the first task and n_2 ways to accomplish the second task, then the project can be completed in $n_1 \cdot n_2$ ways.

Note: There are other names for this principle. Two of the more common ones are the "multiplication rule" and the "product rule". These names focus on what you should do once you determine that the principle applies. The name used in this book focuses on what you should verify before you start multiplying.

This principle can be visualized by thinking of a large matrix or table with n_1 rows and n_2 columns. The rows are labeled by the possible ways to accomplish the first task. The columns are labeled by the possible ways to accomplish the second task. At the intersection of row i and column j, we place the project option: accomplish task one by method i and task two by method j. Clearly, there are $n_1 \cdot n_2$ entries in the table, they are all distinct, and no other options are available.

Another visualization is to think of a classroom having n_1 rows and n_2 columns of desks. The total number of desks is $n_1 \cdot n_2$, which is also the number of ways a student may choose a desk at which to sit.

EXAMPLE 5.2

Books and Snacks

If I wish to choose one of five books to read and one of three snacks to eat while I read, the two choices are independent. The total number of book–snack combinations is thus $5 \cdot 3 = 15$. This can also be shown by exhaustively listing all the possible combinations. ■

The principle can be easily extended to any finite number of tasks. For example, if I am faced with a series of three choices (choose a book, choose a snack, and choose a room to read and eat in), the total number of distinct possibilities is the product of the number of ways to make each choice.

EXAMPLE 5.3

Books, Snacks, and Rooms

Suppose I wish to choose one of five books to read, one of three snacks to eat while I read, and one of four rooms to read and eat in. The total number of book/snack/room combinations is $5 \cdot 3 \cdot 4 = 60$. This can be visualized as a three-dimensional grid with four horizontal levels (corresponding to the rooms). At each level is a table with five rows (books) and three columns (snacks). The arrangement is similar to a three-dimensional tic-tac-toe board. If I place a marker on the second row, first column of the table at level 3, I have effectively chosen book 2, snack 1, and room 3. ■

A more interesting use of the extended version of the Independent Tasks Principle is the proof of the following proposition (from [64]), which uses ideas derived from the Fundamental Theorem of Arithmetic (see page 95).

Suppose the positive integer a has prime factorization $a = p_1^{e_1} \cdot p_2^{e_2} \cdots p_k^{e_k}$. It is not difficult to show that any divisor b of a must have a prime factorization of the form $b = p_1^{d_1} \cdot p_2^{d_2} \cdots p_k^{d_k}$, where $d_i \leq e_i$, for $i = 1, 2, \ldots, k$. The proof that follows uses this observation.

PROPOSITION 5.3 ***Counting Divisors***

Let a be a positive integer with prime factorization $a = p_1^{e_1} \cdot p_2^{e_2} \cdots p_k^{e_k}$. Then the number, $\nu(a)$, of positive divisors of a (including 1 and a) is

$$\nu(a) = (e_1 + 1) \cdot (e_2 + 1) \cdots (e_k + 1)$$

$a = p_1^{e_1} \cdot p_2^{e_2} \cdots p_k^{e_k}$

Proof: Any divisor b of a can be written $b = p_1^{d_1} \cdot p_2^{d_2} \cdots p_k^{d_k}$, where $d_i \leq e_i$, for $i = 1, 2, \ldots, k$. The choice of value for d_i is independent of the choice for the value of d_j if $i \neq j$. There are $e_i + 1$ choices for d_i, since $d_i \in \{0, 1, 2, \ldots, e_i\}$. Thus, there are $(e_1 + 1) \cdot (e_2 + 1) \cdots (e_k + 1)$ distinct divisors of a. □

EXAMPLE 5.4

The Principle "Fails"

Suppose I wish to choose one of two restaurants. At the restaurant, I will choose one item from the menu. If restaurant A has 15 items on the menu and restaurant B has 25 items, I cannot use the Independent Tasks Principle to count the number of possible meals. If I use $n_1 = 2$, what value should I use for n_2? Clearly, the two choices (restaurant, menu item) are *not* independent. The choice of restaurant determines how many menu items I will have to choose from. ■

The Mutually Exclusive Tasks Principle

If a project can be decomposed into two mutually exclusive tasks with n_1 ways to accomplish the first task and n_2 ways to accomplish the second task, then the project can be completed in $n_1 + n_2$ ways.

Note: The more commonly used names for this principle are: the "addition rule" and "the sum rule". These names focus on what you should do once you determine that the principle applies. The name used in this book focuses on what you should verify before you start adding.

This can be visualized by observing that I can do *either* task 1 (in one of n_1 ways) *or* I can do task 2 (in one of n_2 ways), but I cannot do both. I can therefore write all the possibilities for task 1 on one piece of paper and all the possibilities for task 2 on another piece of paper. If I spread both pieces of paper before me, I can choose exactly one of the possibilities I see. I thus add the number of options.

EXAMPLE 5.5

The Second Principle to the Rescue

Suppose I wish to choose one of two restaurants. At the restaurant I will choose one item from the menu. I can use the Mutually Exclusive Tasks Principle if I think of the two mutually exclusive tasks as "choose an item from restaurant A's menu" and "choose an item from restaurant B's menu". If restaurant A has 15 items on the menu and restaurant B has 25 items, then there are $15 + 25 = 40$ possible restaurant/meal options. ■

EXAMPLE 5.6

Puzzles, Books, and Homework

I have three puzzles, two science fiction novels, and four textbooks in my room. I need to decide whether to assemble a puzzle, read a (recreational) book, or do the homework for one of my four classes. In all, I have $3 + 2 + 4 = 9$ ways to spend the evening. ■

These two basic counting principles do not cover all possible situations, as the next example illustrates.

EXAMPLE 5.7

Both Principles "Fail"

I have three microwave dinners in my refrigerator: a fish dinner, a frozen pizza, and some fried rice. I also have a carton of milk, a container of juice, and a can of root beer. The root beer is only acceptable with the pizza, and milk doesn't go well with pizza. I never drink juice with fried rice. How many acceptable food/drink pairs are there? Principle 1 would predict $3 \cdot 3 = 9$, whereas principle 2 would predict $3 + 3 = 6$. The five viable alternatives are (fish, milk), (fish, juice), (pizza, root beer), (pizza, juice), and (fried rice, milk).

Neither basic principle was correct. This is because the choice of food and the choice of drink are not independent (picking pizza excludes milk). Neither are they mutually exclusive (picking fish does not exclude picking juice). ■

✔ Quick Check 5.2

1. Exhaustively list all pairs of choices from a list of four types of bread, $\{B_1, B_2, B_3, B_4\}$, and a list of four types of meat, $\{M_1, M_2, M_3, M_4\}$.
2. I am in the market for a new refrigerator. I will choose between a side-by-side design or a top-freezer design. There are four acceptable side-by-side models and five acceptable top-freezer models. In how many ways can I select a new refrigerator?
3. A hybrid car is available in nine different exterior colors and three different interior colors. How many distinct color combinations are available?
4. The hybrid car in the previous exercise also is available with either cloth or leather upholstery. Cloth upholstery is available in all three interior colors, but leather upholstery is only available in two colors. How many exterior–interior choices are there? ☑

5.1.2 Permutations

Suppose we start with a set containing n elements. We wish to arrange r of them (with $r \leq n$) in order. That is, order is important, but there is only one copy of each element (no repetition). In how many distinct ways can this be done? The answer will be denoted[3] $P(n, r)$.

EXAMPLE 5.8 **$P(3, 2)$**

Let the three elements be the letters "a," "b," and "c". We wish to arrange two of these letters in order. The possibilities are ab, ac, ba, bc, ca, and cb. Thus $P(3, 2) = 6$.

Notice that "ab" is considered distinct from ba (order is important) and "aa," "bb," and "cc" were not viable choices (no repetition). ■

EXAMPLE 5.9 **A Simple Reading Schedule**

Suppose you need to decide in which order to read three unrelated books. If there is no obvious reason to read any one book before another, we can read the books (conveniently labeled "a," "b," and "c") in the following orders: abc, acb, bac, bca, cab, and cba. Thus $P(3, 3) = 6$. ■

EXAMPLE 5.10 **French Horn Duets**

A band director has four students who play the French horn. She wants to have a French horn duet at the spring band concert. If all four of the students are at about the same level of competence, she might as well assign the two parts (primo and secondo) to two of them randomly. In how many ways can she do this?

Label the students "a," "b," "c," and "d." Let the pair "cb" mean c plays primo and b plays secondo. The parts can be assigned in the following twelve ways: ab, ac, ad, ba, bc, bd, ca, cb, cd, da, db, and dc. Therefore, $P(4, 2) = 12$. ■

The previous examples were easy to solve by exhaustively listing all possibilities. This is not always realistically possible. We need a more sophisticated approach. The

[3] Other common notations for $P(n, r)$ are ${}_nP_r$ and P_r^n.

previous examples do not lead to an obvious pattern, so a more theoretical approach is needed. Fortunately, the Independent Tasks Principle is all that is needed.

Assume we wish to count the number of ways to place in order r items from a set of n distinct items. There are n choices for the first item. After that item has been selected, there are $(n-1)$ items remaining from which the second item can be chosen. Clearly, the choice of first item has no influence on which of the *remaining* $(n-1)$ items is chosen (independence). Once the second item has been selected, there are still $(n-2)$ items remaining from which to choose (independently) the third item. The pattern continues until r items have been chosen. The Independent Tasks Principle (in its expanded version) implies that the total number of ways to make these choices is the product of the number of ways to make the individual choices.

Permutations

Let $0 \le r \le n$. The number of ways to arrange r objects from a set of n objects, in order, but without repetition is

$$\begin{aligned} P(n,0) &= 1 \\ P(n,r) &= n \cdot (n-1) \cdot (n-2) \cdots (n-r+1) = \frac{n!}{(n-r)!} \quad \text{for } r \ge 1 \end{aligned}$$

Note that $P(n,n) = n!$ (since $0! = 1$). That is, the number of ways to arrange n objects in order is $n!$. You may be wondering why there is a "+1" in the final factor, $n-r+1$. The product can be written as $(n-0)(n-1)(n-2)\cdots(n-(r-2))(n-(r-1))$. There needs to be r factors, but the first one subtracts 0 from n, so the final factor should subtract $r-1$ from n.

EXAMPLE 5.11 **Organizing a Ballot**

You are in charge of placing the names of five candidates on a ballot. You have been instructed to place the names in random order. In how many ways can this be done? Using the formula for $P(n,n)$, there are $5! = 120$ ways. ■

EXAMPLE 5.12 **Class Presentations**

An instructor has divided the class into seven groups. She wishes to have three of the groups make their presentations today. In how many ways can she arrange the three presentations? The order in which the presentations are made is important (ask the group members!) so $P(7,3)$ is appropriate.

$$P(7,3) = \frac{7!}{(7-3)!} = \frac{7!}{4!} = 7 \cdot 6 \cdot 5 = 210$$ ■

Notice the cancellation that occurred in the previous example:

$$\frac{7!}{4!} = \frac{7 \cdot 6 \cdot 5 \cdot \not{4} \cdot \not{3} \cdot \not{2} \cdot \not{1}}{\not{4} \cdot \not{3} \cdot \not{2} \cdot \not{1}} = 7 \cdot 6 \cdot 5$$

EXAMPLE 5.13 **Distinct Birthdays**

A group of n people can have *distinct* birthdays in

$$P(365,n) = 365 \cdot 364 \cdots (365-n+1)$$

ways (ignoring leap years). ■

5.1.3 Permutations with Repetition

If the objects that we are arranging come in unlimited (or sufficiently large) quantities, then it is possible to increase the number of arrangements. For example, if the letters "a," "b," and "c" are to be arranged in two-letter sequences, the possibilities are either {ab, ac, ba, bc, ca, cb} or {aa, ab, ac, ba, bb, bc, ca, cb, cc}, depending on whether repetitions are excluded or permitted.

Permutation with Repetition

Let $0 \leq r$ and $1 \leq n$. The number of ways to arrange r objects in order, chosen from a set of n distinct objects if objects may be repeated is

$$n^r$$

The Independent Tasks Principle can be used to prove the previous formula. There are r positions to be filled, each having n possible values. The choices are independent, so there are

$$\underbrace{n \cdot n \cdot n \cdots n}_{r \text{ times}} = n^r$$

ways to arrange the objects.

EXAMPLE 5.14 **Three-Letter Codes**

There are $26^3 = 17{,}576$ ways to designate an inventory item by a code consisting of three lower case letters. There are $10^3 = 1000$ ways to designate an item using three digits. ■

EXAMPLE 5.15 **Birthdays**

A group of n people can have birthdays in 365^n ways (ignoring leap years). ■

✔ Quick Check 5.3

1. Calculate $P(9, 4)$.
2. The owner of a small business has decided to allow the middle managers (Jane, Tom, and Pin) to each have one week of vacation during the month of July. Only one manager is allowed on vacation at a time. Assuming that July has four weeks, in how many distinct ways can the managers sign up for vacations?
3. In how many ways can the genders of the children be arranged in a family with eight children? (Assume that order is important.) ☑

5.1.4 Combinations

Permutations are arrangements where order is important. Often the order of the objects is not important. What is of interest is the *subset* of objects that are chosen. For example, if a mother of five children chooses two of them to pull weeds, the children are concerned with which two are chosen, not what order they are chosen in. Arrangements where order is not important are called *combinations*, and denoted[4] $C(n, r)$. No matter which notation is used, it is usually pronounced "n choose r". (Some students find it helpful to equate the words *combination* and *committee*.)

[4]Other common notations for $C(n, r)$ include the *binomial coefficient* notation $\binom{n}{r}$ and the notations C_r^n and ${}_nC_r$.

EXAMPLE 5.16 **Pulling Weeds**

Suppose the five children who are potential weed pullers are named Annabelle, Bartholomew, Candice, Dorothea, and Ernie, better known to their close associates as A, B, C, D, and E. The two lucky workers can be chosen in 10 ways: AB, AC, AD, AE, BC, BD, BE, CD, CE, or DE. Observe that AB and BA are the same pair of "happy" children. ■

EXAMPLE 5.17 **Scrubbing Floors**

The three children who were not chosen in the previous example have no opportunity to feel smug, because mom has decided that she needs three willing floor scrubbers. It should be clear after a moment of thought that there must be 10 ways to choose three floor scrubbers from a set of five children. (Once the weed pullers are chosen, there are exactly three children left. The weed pullers can be chosen in 10 ways [see Example 5.16].)

It is possible to list all 10 floor scrubbing combinations: ABC, ABD, ABE, ACD, ACE, ADE, BCD, BCE, BDE, and CDE. Observe that ABC, ACB, BAC, BCA, CAB, and CBA are all the same set of floor scrubbers. ■

In the previous example, there were $2! = 2$ ways to list each set of two weed pullers, and $3! = 6$ ways to list each set of three floor scrubbers. This is true in general: There are $r!$ ways to list a set of r objects if order is important. This insight leads to the next counting formula. To count the number of ways to pick a subset of r elements (without repetition) from a set of n objects, count the number of permutations, then divide by the number of times each distinct subset of size r has been counted. For example, there are $P(5, 2) = 20$ ways to list two children in order (permutations): AB, BA, AC, CA, AD, DA, AE, EA, BC, CB, BD, DB, BE, EB, CD, DC, CE, EC, DE, and ED. Since each subset of two children appears twice, the number of combinations is $\frac{20}{2} = 10$.

Combinations

Let $0 \le r \le n$. The number of ways to choose a subset of r objects from a set of n objects without repetition is

$$C(n, r) = \frac{n!}{r! \cdot (n - r)!}$$

If $0 \le n < r$, then $C(n, r) = 0$.

Notice that $C(n, r) = \frac{P(n,r)}{r!}$. (See Exercise 43 on page 237.)

EXAMPLE 5.18 **Choosing a Nominating Committee**

Your club has 35 members. Elections are approaching, and it is necessary to form a nominating committee consisting of 4 members. There are $C(35, 4) = \frac{35!}{4! \cdot 31!} = 52,360$ ways to form this committee. ■

A standard deck of 52 playing cards consists of four suits (clubs ♣, diamonds ♢, hearts ♡, and spades ♠). Diamonds and hearts are red cards, clubs and spades are black cards. Each suit has 13 face values (ace, 2, 3, 4, 5, 6, 7, 8, 9, 10, jack, queen, king). The jack, queen, and king in each suit have "pictures" with faces on them. These 12 cards (3 per suit) are commonly called face cards. Note carefully the distinction between "face value" and "face card".

EXAMPLE 5.19

Crazy Eights

Crazy eights is a popular children's card game. Each player is dealt 7 cards from a standard 52-card deck. How many different "hands" of 7 cards are there? Since the order in which the cards are dealt does not matter and no card appears more than once, this is a combination problem. The solution is thus

$$C(52, 7) = \frac{52!}{7! \cdot 45!} = 133{,}784{,}560$$

■

Most calculators cannot compute $n!$ if $n > 69$ because $69! \simeq 1.71122 \times 10^{98}$, but $70! \simeq 1.19786 \times 10^{100}$. Most calculators only have room for a two-digit exponent on 10. Solving the previous example by first calculating 52! and then dividing the result by 45! and then 7! may not give a completely accurate result. One of my calculators gives (1.3378×10^8) as the solution. The calculator has displayed the answer with the last four digits rounded (in this case because the calculator only displays eight digits).

There is a method that will work even if your calculator has no $\boxed{x!}$ key: First cancel the 45! (symbolically), then cancel common factors, and then use the calculator. The first step produces

$$C(52, 7) = \frac{52!}{7! \cdot 45!} = \frac{52 \cdot 51 \cdot 50 \cdot 49 \cdot 48 \cdot 47 \cdot 46}{7 \cdot 6 \cdot 5 \cdot 4 \cdot 3 \cdot 2 \cdot 1}$$

It is then possible to use the 7 in the denominator to reduce the 49 in the numerator to 7; the 6, 4, and 2 in the denominator to cancel the 48 in the numerator; and the 5 in the denominator to reduce the 50 in the numerator to a 10. The 3 in the denominator will reduce the 51 in the numerator to a 17. This leaves the product $52 \cdot 17 \cdot 10 \cdot 7 \cdot 1 \cdot 47 \cdot 46$.

EXAMPLE 5.20

More Crazy Eights

How many different seven-card crazy eights hands are there that contain exactly one 8?

There are four ways to choose an 8, with $52 - 4 = 48$ cards remaining from which to choose the other six cards for the hand. Consequently, there are

$$4 \cdot C(48, 6) = 4 \cdot \frac{48!}{6! \cdot 42!} = \frac{4 \cdot 48 \cdot 47 \cdot 46 \cdot 45 \cdot 44 \cdot 43}{6 \cdot 5 \cdot 4 \cdot 3 \cdot 2 \cdot 1} = 49{,}086{,}048$$

distinct crazy eights hands containing exactly one 8.

How many different seven-card crazy eights hands are there that contain exactly one 8 and exactly two face cards?

There are four choices for the eight. There are 12 cards from which to choose the two face cards. This can be done in $C(12, 2)$ ways. Finally, there are $52 - 12 - 4 = 36$ cards left that are neither face cards nor 8s from which to choose the remaining four cards for the hand. There are consequently (using the Independent Tasks Principle)

$$\begin{aligned} 4 \cdot C(12, 2) \cdot C(36, 4) &= 4 \cdot \frac{12!}{2! \cdot 10!} \cdot \frac{36!}{4! \cdot 32!} \\ &= 4 \cdot (6 \cdot 11) \cdot (3 \cdot 35 \cdot 17 \cdot 33) \\ &= 15{,}550{,}920 \end{aligned}$$

crazy eights hands containing exactly one 8 and exactly two face cards. ■

EXAMPLE 5.21

Crop Rotation

A farmer has divided his land into nine plots. He typically leaves two of the plots fallow. In how many ways can he choose the two fallow fields?

Since the order in which the two fields are chosen is unimportant, and since a plot of land cannot count twice, the proper number is

$$C(9,2) = \frac{9!}{2! \cdot 7!} = \frac{9 \cdot 8}{2} = 36$$ ■

EXAMPLE 5.22

Choosing a Pair

Given n people, a single pair can be chosen in $C(n,2) = \frac{n \cdot (n-1)}{2} \in \Theta(n^2)$ ways. In Figure 5.1, the gray points represent the value of n, and the black points represent the value of $C(n,2)$. Notice that as the number of people (n) increases, the number of pairs of people increases much faster.

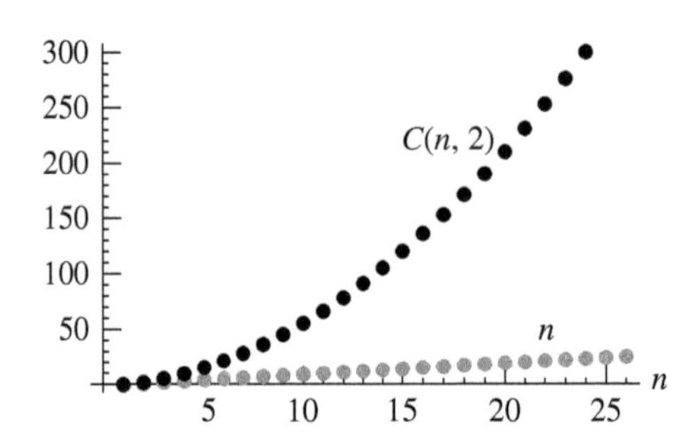

Figure 5.1 Rate of growth for $C(n, 2)$. ■

✔ Quick Check 5.4

1. Compute $C(8,6)$.

2. A catering service offers 10 dinner choices. The campus Gourmet Club plans to have four catered dinner meetings during the school year. Members do not wish to repeat a menu selection during the year.
   (a) How many distinct collections of four dinners can members pick? (The order in which the meals are served is unimportant to the club members.)
   (b) Suppose members also care about the order in which the meals are scheduled. How many distinct schedules are there?

3. In the discussion after Example 5.19 I mentioned that my eight-digit calculator rounded the nine digit number 133,784,560 to 1.3378×10^8. I can recapture the full solution by entering the key sequence ÷ 1 0 = and manually appending a "0" to the displayed result. Why does this work? Will it always work? ☑

5.1.5 Combinations with Repetition

The final counting formula relates to combinations with repetition. A situation that would require this principle is the following.

EXAMPLE 5.23

Planning to Study

Meg has three classes (algebra, biology, chemistry) that need some attention in the next two days. Tonight she has scheduled a study session during which she can work on two classes, or spend the entire session on one class. Each class has sufficient work to fill the entire session. In how many ways can she schedule the evening if all she cares about is which subjects are to be worked on, not the order in which they are tackled?

Observe that repetition is allowed; it is possible to work on the same class for the entire session. Also notice that the order in which the work is scheduled is not important. This problem can be solved by exhaustively listing the possible arrangements: aa, ab, ac, bb, bc, cc. There are six possible ways. ■

Combinations with Repetition

Let $0 \leq r$ and $1 \leq n$. The number of subsets of r elements from a set containing n distinct elements, with repetition permitted, is

$$C(n+r-1, r) = \frac{(n+r-1)!}{r! \cdot (n-1)!}$$

You need not memorize the expression $\frac{(n+r-1)!}{r! \cdot (n-1)!}$. It is sufficient to memorize $C(n+r-1, r)$ and then use the expansion for $C(n, r)$. An alternative is to memorize the process presented in Example 5.24.

Another way to express the notion of repetition is to consider n to represent the number of categories of distinct objects. Each category will contain many identical copies of its object. Whether you consider a single object that can be chosen and then returned so that it can be chosen again (*with replacement*) or you consider multiple identical copies (*with repetition*), the net effect will be the same.

The derivation of the combination with repetition formula is more sophisticated than the derivations of the other formulas. It is easiest to understand if the task is cleverly visualized.[5] An example of the visualization will be presented before the derivation is formalized.

EXAMPLE 5.24

Candy Markers

You are to choose three pieces of candy from a jar containing 12 (or more) pieces of candy. There are four flavors of candy. There are at least three pieces of each flavor in the jar. In how many ways can you choose the candy?

The set of flavors is the important feature. You do not care about the order in which the three pieces are chosen. If a flavor has been chosen, there are sufficient pieces left to choose another of the same flavor, so repetition is allowed.

Assume for the moment that each piece of candy costs a dime. If the four flavors are represented by four empty boxes, the three pieces can be chosen by placing three dimes in the boxes. Thus, if two dimes are placed in the third box and one dime is placed in the fourth box, two pieces of flavor 3 (licorice) and one piece of flavor 4 (cinnamon) have been chosen (Figure 5.2).

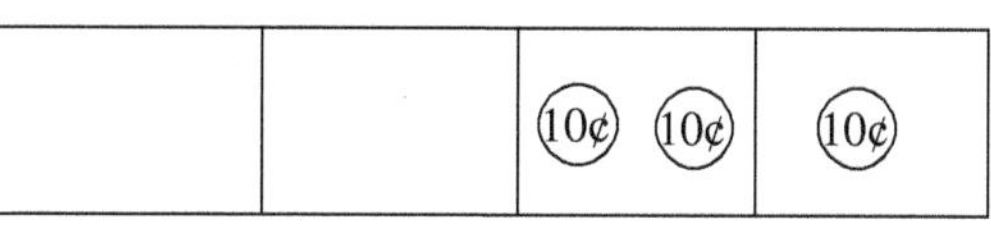

Figure 5.2 Choosing candy.

This can be diagrammed using "|" to represent an interior wall[6] of a box, and "○" to represent a dime. The arrangement just mentioned would be shown as

| | ○ ○ | ○

whereas the selection of two butterscotch and one licorice would be represented by

| ○ ○ | ○ |

[5] The visualization presented next follows the presentation in [76].

[6] Convince yourself that the two exterior walls can be omitted from the diagram.

The number of distinct patterns of |'s and ◯'s represents the number of distinct ways to choose the candy. The patterns are uniquely determined by which positions contain the ◯'s. There are $4 - 1$ interior walls (and hence, three |'s) and three dimes (◯'s). Thus, there are $4-1+3$ symbols to arrange. There are $C(4+3-1, 3) = C(6, 3) = 20$ ways to pick the three positions for the ◯'s (and hence 20 ways to choose the candy).

You should list the 20 distinct ways to choose the candy. ■

Derivation of the Combinations with Repetition Counting Formula: The n distinct elements can be represented by a line of empty, adjacent boxes. Each of the n elements corresponds to a unique box. The r elements can be chosen by placing r markers in the boxes. The number of markers in a box represents the quantity of the corresponding element that has been chosen.

The process of placing markers in the boxes can be carried out in a systematic manner. Start at the first box. At each stage a decision is made: either place a marker in the current box or else move to the next box. Continue until all the markers have been placed in a box. The result can be summarized in a linear diagram by using the symbol "|" to represent a vertical wall of a box and the symbol "◯" to represent a marker. The decision to place a marker results in a ◯ being written to the diagram; the decision to move to the next box results in a | being written to the diagram. The two exterior walls are not important because they are never involved in a decision: the process starts in the first box (so the leftmost wall has already been crossed) and the process ends before the rightmost wall is reached.

The $n - 1$ interior walls and the r markers completely determine which set of r elements will be chosen. The number of distinct combinations is the same as the number of visually distinct linear arrangements of |'s and ◯'s. There are $n - 1$ interior walls, and r markers, so there are $(n - 1) + r = n + r - 1$ positions to fill. It is necessary to choose a subset of r positions in which to place the markers. The actual walls and markers are distinct items, so there is no repetition. The order in which the markers are placed is unimportant (since they are visually indistinguishable). What is important is the subset of positions they occupy. Thus, the original problem has been converted into finding a subset of size r from a set of size $n+r-1$. The combinations formula applies; there are $C(n + r - 1, r)$ ways to place the markers.[7] □

EXAMPLE 5.25 **Selecting Fruit**

Four family members have just completed lunch and are ready to choose their after-lunch fruit. There are bananas, apples, pears, kiwi, apricots, and oranges in the house. In how many ways can a selection of four pieces of fruit be chosen?

Since it has been implied that there is sufficient fruit of each variety for every family member to have his or her first choice, and only the selection of varieties (not which person eats what fruit) is of interest, this is a combination with repetition. The solution is thus

$$C(6 + 4 - 1, 4) = C(9, 4) = \frac{9!}{4! \cdot 5!} = 126$$

■

[7] Notice the problem-solving strategy: A problem with an unknown solution has been transformed into a problem that has already been solved.

✔ Quick Check 5.5

1. Calculate the number of ways to choose 2 items, with repetition, from a set of 4 items.
   (a) Exhaustively list all possibilities.
   (b) Use the combinations with repetition formula.
2. An English teacher has designed extra-credit projects for his students. They may
   - Write a poem.
   - Write a short story.
   - Read a book and complete a book report.
   - Write a one-act play.
   - Write a letter to the editor.
   - Write an article for the school newspaper.

   Multiple projects from a category are acceptable. For example, someone may turn in two poems.
   How many different sets of three projects are possible? ☑

Table 5.2 summarizes the formulas for permutations and combinations. They are organized according to whether order is important and repetition is permitted. This table and several other counting formulas can also be found on page 272. The tables can be found in pdf format at http://www.mathcs.bethel.edu/~gossett/DiscreteMathWithProof/ .

TABLE 5.2 Arranging r elements from a set containing n distinct elements

	With Order	Without Order
Without Repetition	$P(n, r) = \dfrac{n!}{(n-r)!}$	$C(n, r) = \dfrac{n!}{r! \cdot (n-r)!}$
With Repetition	n^r	$C(n+r-1, r) = \dfrac{(n+r-1)!}{r! \cdot (n-1)!}$

5.1.6 Exercises

The exercises marked with ★ have detailed solutions in Appendix H.

1. (a) Design a set of marriage preferences so that a group of four men and four women actually require 10 rounds for the Deferred Acceptance Algorithm.
   (b) Can you determine a general pattern of preferences so that a group of n men and n women require the maximum $n^2 - 2n + 2$ rounds for the Deferred Acceptance Algorithm?
   (c) Produce a big-Θ estimate for the worst-case behavior of the Deferred Acceptance Algorithm, where proposals are the critical operation.
2. ★ This evening, I can either read one of four books, watch one of three videos, or talk on the phone with one of three friends. In how many ways can I spend the evening?
3. ★ Tomorrow night I plan to listen to one of 12 music CDs, and then eat one of four frozen dinners. After dinner I will put together one of three picture puzzles. How long can I maintain such an active schedule? (Actually, in how many ways can I spend the evening?)
4. A boy needs to eat breakfast and lunch, practice piano, mow the lawn, and read a book today. In how many ways can he arrange these activities, assuming he does not care in which order the meals occur?
5. How many tours consisting of a visit (in the order specified) to a park, a museum, a mall, and a restaurant are possible if we can select from five parks, two museums, six malls, and eight restaurants?
6. ★ Suppose that either a doctor or a dentist is chosen to speak about his or her profession to a seventh-grade classroom. How many choices are there for this speaker if there are 70 doctors and 23 dentists available?
7. Mom requires Bobby to eat 10 (not necessarily distinct) servings of grain products each day. Suppose that the grains available to Bobby are cereal, bread, rice, pasta, and crackers. For how many days can Bobby make a different selection of his grain servings, assuming that he obeys Mom's rule?
8. I have five textbooks on the shelf above my desk. In how many ways can I place these books in a line?
9. How many distinct, five-digit zip codes are possible?
10. Assuming that there will never be more than 500 million people in the United States before the year 2050 A.D., how many digits are necessary to provide every person with a personal zip code? (Notice that businesses and institutions are being neglected.)

11. ★ Suppose that for a conference, individuals are given a badge with five features: a color, a shape, an uppercase alphabet letter, an animal, and a one-digit number. What is the largest number of distinct badges, assuming that there are 7 colors, 8 shapes, and 4 animals available?
12. What is the major distinction between permutations and combinations?
13. Suppose that in an experiment you are asked to arrange a penny, a cracker, a washer, and a pencil in a row. In how many ways can the arrangements be made?
14. An individual won a raffle and can choose a prize from one of four lists. The first list contains 19 possible prizes, while the second, third, and fourth lists include 14, 21, and 18 items, respectively. How many prizes does the raffle winner have to choose from?
15. ★ Next week, I intend to visit my old home town. I wish to visit five friends but have time to see only three of them. In how many ways can I schedule the visits?
16. A special school club called the Lions is formed by including either the teacher or one of the three top students from each of 43 classrooms. How many ways are there to form the Lions? Assume that both a teacher and a student from a single classroom are not chosen to be members of the Lions.
17. A local pizza store offers a choice of seven toppings. How many distinct three-topping pizzas do they offer if no topping can be repeated?
18. ★ Recall that a bit is either a "0" or a "1". If 7 bits are used to encode characters, how many different characters can be encoded? [*Hint*: Exhaustive enumeration (0000000, 0000001, 0000010, 0000011, ...) is too tedious for this problem.]
19. ★ A small country (9 million people) is installing a new, state-of-the-art telephone system. How many digits are necessary to allow sufficient phone numbers? Discuss the assumptions you have made, as well as the justification for your answer.
20. Suppose that you are taking a bus trip from your home in Florida to California. There are seven different bus services offering trips from Florida to Arizona, four offering trips from Florida to Texas, eight offering trips from Arizona to California, and seven offering trips from Texas to California. How many possibilities are there for a bus ride from Florida to California, via either Arizona or Texas?
21. I will be sharing an apartment with two friends. The apartment has one large and one small bedroom. Two of us will share the large bedroom and the other will sleep in the smaller room. In how many distinguishable ways can we assign rooms?
22. ★ My young niece will soon turn 11. Suppose I want to send her a dollar for every year old she will be. I have a large stack of crisp new dollar bills and also a large pile of shiny new gold Sacagawea[8] dollar coins. In how many distinct ways can I use these two kinds of dollars to give $11? Solve this using combinations with repetition and then by a more direct (and simpler) method.
23. The day after nine students complete an exam, the teacher passes out their grades. The possible grades are A, B, C, D, and F. In how many ways can the grades be assigned to the students?
24. In how many ways can you purchase three movies from a store that sells 57 distinct movies in unlimited quantities if
    (a) There are no restrictions on your purchase of the three movies
    (b) You must purchase a copy of the owner's home-video footage of her new grandchild.
25. A recording artist plans to place six songs on her new CD.
    (a) In how many ways can she order these songs?
    (b) In how many ways can she order these songs if she has already chosen the first and last selections?
26. A farmer has divided his land into nine plots. He typically plants one plot of soybeans, one of corn, one of alfalfa, and one of potatoes. The rest of the plots are left fallow. How many distinct patterns of crop/plot pairings are there?
27. In how many distinguishable ways can I form a committee of ten people if the committee must contain at least one faculty member, at least one administrator, and at least two students? Assume that there are more than six faculty, more than six administrators, and more than seven students. Also, assume that a committee with three administrators, four students and three faculty is different from a committee with four administrators, three students, and three faculty, but we do not care which four administrators, etc. are chosen.
28. A piano has 88 keys; 52 are white and 36 are black.
    (a) You have 10 fingers. Assuming that your fingers were made of rubber and it was possible to reach any combination of ten keys simultaneously, how many 10-key combinations are there?
    (b) A standard octave contains 12 keys, representing distinct musical notes. Triad chords require three keys to be played simultaneously. Ignoring how agreeable the sound will be, in how many ways can a triad be played within a single octave?
29. A pair of identical six-sided dice are tossed.
    (a) How many distinct rolls are possible? (Note that the dice are identical.)
    (b) Suppose the pair of dice is rolled three times. How many *sequences* of rolls are possible?
    (c) Suppose the pair of dice is rolled three times. How many *sets* of rolls are possible?
    (d) How many distinct rolls are possible if one die is white and the other is red?

[8] For more information, see the Sacagawea link at http://www.mathcs.bethel.edu/~gossett/DiscreteMathWithProof/ .

30. ★ The following excerpt is from *The Mythical Man-Month* by Frederick Brooks [37, p. 78]:

> If there are n workers on a project, there are $\frac{n^2-n}{2}$ interfaces across which there may be communication, and there are potentially almost 2^n teams within which coordination must occur. The purpose of organization is to reduce the amount of communication and coordination necessary; hence organization is a radical attack on the communication problems treated above.

(a) Provide a mathematical justification for the claim: "If there are n workers on a project, there are $\frac{n^2-n}{2}$ interfaces across which there may be communication."
(b) Provide a mathematical justification for the claim: "If there are n workers on a project, there are potentially almost 2^n teams within which coordination must occur."

31. In 1992 the Mattel toy company introduced a talking Teen Barbie doll. The company had compiled a collection of 270 possible expressions for Barbie to "speak". To make the dolls appear more individually distinct, Mattel randomly chose 4 of the 270 possibilities for any particular doll.
(a) How many distinct dolls could they manufacture?
(b) Suppose one of the 270 statements is considered to be undesirable. How many distinct dolls contain the undesirable statement among their 4 exclamations?

32. A local pizza store offers a choice of seven toppings and three sizes (small, medium, and large). How many distinct three-topping pizzas do they offer?

33. A fletcher is making some arrows. Each arrow has three feathers on the tail end, spaced at 120° intervals around the shaft. The feathers come in four colors: red, blue, yellow, and green. How many visually distinct patterns are there if a color can be used in more than one position on an arrow? Assume that rotating the arrow in any direction (including end-to-end) does not result in a different pattern, but changing colors does. Indicate which counting principle or formula most directly applies to this problem.

34. ★
(a) How many ways are there to situate four people in identical chairs at a circular table?
(b) Generalize part (a) to the case when there are n people.

35. A standard "Trivial Pursuit" game contains two boxes of 500 question cards each. Consider one of these boxes. Each question is in one of six categories (such as "Literature" and "Movies"). Each card contains one question from each category. When a card is used, only one question is asked. The card is then placed at the end of the box.

Suppose a game is played that requires exactly 500 questions. In how many ways can the sequence of 500 questions be chosen? (Your calculator will hate this question!)

36. ★ How many different seven-card crazy eights hands are there that contain no 5s but have four cards of the same kind?

37. How many different seven-card crazy eights hands are there that contain exactly two 2s, exactly three 3s, and exactly two cards with a third common face value?

38. How many different eight-card hands are there that contain exactly two suits, with four cards from each suit?

39. How many different eight-card hands are there that contain only face cards?

40. Mrs. Candy has a large box of lollipops, chocolate bars, and caramels. She wants to give each of the nine children in her neighborhood three pieces of candy. Taking into account that each type of candy is available in a quantity greater than 30, in how many ways can Mrs. Candy distribute the candy?

41. How many odd four-digit numbers can be created from the following digits: 1, 5, 6, 7, 8? Assume that each digit can only be used once.

42. ★ What value should $C(n, 0)$ have (intuitively)? What value should $C(n, n)$ have, (again, intuitively)? How do your answers match with the combinations formula on page 232?

43. Explain (intuitively) why $P(n, r) = P(r, r) \cdot C(n, r)$.

44. The game of dominoes uses a set of small rectangular tiles. Each tile is called a *domino*. A domino contains two equal-sized regions that contain zero or more dots, in fixed patterns. (A region with no dots is called a "blank.") The domino set has a maximum number of dots (usually 6, 9, or 12). Every pattern (number of dots) appears exactly one time with every other pattern on a domino. Dominoes with both regions containing the same pattern are called "doubles". The following diagram shows two dominoes: the 1–5 domino and a double 4.

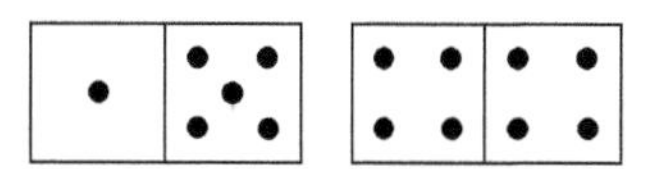

(a) Create a formula $D(n)$ that computes the number of dominoes in a set whose highest tile is a double n. For example, $D(0) = 1$, since there is only one domino (double blanks). Also, $D(1) = 3$ since there will be three dominos: double blanks, blank 1, and double 1s.
(b) A set of dominoes whose highest tile is a double 12 will contain $D(12)$ tiles. Derive this number in as many distinct ways as possible. That is, use different counting techniques to derive this number.

5.1.7 More Complex Counting Problems

It might be helpful to reflect for a moment about the rates at which the numbers in the counting formulas grow. It should be clear that there will be more arrangements when order is important than when order is irrelevant. Thus, for a common n and r, there should be more permutations than combinations. It should also be clear that if repetition is permitted the number of arrangements should increase. These observations can be seen in Table 5.3, where the numbers of permutations and combinations of six objects are compared. Notice the explosive growth in row 2. Permutations with repetition allow many more arrangements than any of the other options.

TABLE 5.3 Permutations and combinations with $n = 6$

		r = 0	1	2	3	4	5	6
Permutations	$P(6, r)$	1	6	30	120	360	720	720
Permutations with repetition	6^r	1	6	36	216	1296	7776	46656
Combinations	$C(6, r)$	1	6	15	20	15	6	1
Combinations with repetition	$C(6+r-1, r)$	1	6	21	56	126	252	462

The examples in this section will typically (but not always) require more than one of the counting principles or formulas to be used. It will be helpful to consider the four criteria used to develop the principles: independence, mutual exclusion, order, and repetition. If the presence or absence of these are noted for each problem, the proper counting principle can be easily identified. It is generally easiest to apply one of the permutation or combination principles whenever possible. Only if these do not apply should one of the two basic principles[9] be tried. Section 8.1 contains some additional, more sophisticated material related to counting.

Several examples and problems will be easier to describe using the next few definitions.

DEFINITION 5.4 ***Digit***

A *digit* is any one of the characters 0, 1, 2, 3, 4, 5, 6, 7, 8, or 9.

DEFINITION 5.5 ***Alphanumeric***

A character is said to be *alphanumeric* if it is either an uppercase or lowercase letter or it is a digit.

A character is said to be *alphanumeric-upper* if it is either an uppercase letter or it is a digit.

EXAMPLE 5.26 **License Plates**

A state wishes to produce license plates that consist of three uppercase letters, a space, and three digits. If no three-letter sequences are excluded, how many license plates are possible?

Order is clearly important. Repetition is permitted. However, there is a letter sequence and a digit sequence. How should these be combined? The choice of letter sequence has no influence on which digit sequence is chosen. The Independent Tasks Principle thus applies.

There are $26^3 = 17{,}576$ letter sequences and $10^3 = 1000$ digit sequences. There are thus $17{,}576 \cdot 1000 = 17{,}576{,}000$ possible license plates. ■

EXAMPLE 5.27 **Alphanumerics**

The sequence of characters "aB3d5" is an alphanumeric sequence. The sequence "3DFG67" is an alphanumeric-upper sequence. The sequence "a34,205b" is not alphanumeric because it contains a comma. ■

[9] The Independent Tasks Principle and the Mutually Exclusive Tasks Principle.

EXAMPLE 5.28

More Inventory Codes

The owner of a small store has decided to use an inventory code that distinguishes between taxable and nontaxable items. Nontaxable items will have a code that starts with the letter "N", followed by any combination of four alphanumeric-upper characters. Taxable items will start with any upper case letter *except* "N", again followed by any combination of four alphanumeric-upper characters. How many distinct inventory codes are there?

The key feature is that the first letter of the code produces two mutually exclusive sets of items (taxable and nontaxable). The Mutually Exclusive Tasks Principle thus applies; add the number of nontaxable codes to the number of taxable codes. There are 36 alphanumeric-upper characters, so there are $36^4 = 1{,}679{,}616$ codes for nontaxable items. There are 25 choices for the first letter of a code for a taxable item (independent of the rest of the code), so there are $25 \cdot 36^4 = 41{,}990{,}400$ codes for taxable items. The sum is 43,670,016. (Can you produce a somewhat simpler solution to this problem?) ■

EXAMPLE 5.29

Attending the Cinema

Four friends are planning a trip to a cinema complex. The complex is showing six different movies (all starting at the same time). In how many ways can the friends view the movies if each views only one?

I will assume that the people each care which movie they view, so order is important. More than one friend can view a movie, so repetition is permitted. This is a permutation with repetition problem. There are $6^4 = 1296$ possible viewing arrangements. ■

EXAMPLE 5.30

Scheduling Presentations

An instructor has divided her math class into eight groups. Each group must make a presentation to the class. She can schedule three presentations per day but will have only two groups on the final day, so that she can serve cookies. In how many ways can this be done?

The simplest way to solve this is to think of scheduling a sequence of eight events. There are $P(8, 8) = 8! = 40{,}320$ ways to do this.

A more complex approach selects a set of three groups for the first day. This can be done in $P(8, 3) = 336$ ways. She can then choose a set of three groups for the second day in $P(5, 3) = 60$ ways. There are now only two groups left for the third day. These can be arranged in two ways. The 3 days must now be properly combined. The days are not mutually exclusive; all three days must be scheduled. Are they independent? The choice of groups for day one does influence the available choices for day 2. This influence has been eliminated by reducing the pool from which to choose on day 2 down to five groups. With this modification, the choices by day are independent. The total number of three day arrangements is thus $336 \cdot 60 \cdot 2 = 40{,}320$. ■

EXAMPLE 5.31

At Most One Eight

In the card game crazy eights, it is advantageous to have eights in your hand. If seven cards are dealt from a standard deck of 52 cards, how many hands have at most one 8?

It will be helpful to consider two mutually exclusive possibilities: no 8s and exactly one 8. If exactly one 8 is in the hand, there are four choices for the 8, and 48 cards from which to choose the other 6 cards. There are $C(48, 6) = 12{,}271{,}512$ ways to choose the six noneights, so there are $4 \cdot 12{,}271{,}512 = 49{,}086{,}048$ hands with exactly one 8. There are $C(48, 7) = 73{,}629{,}072$ ways to deal a hand with no 8s. There are thus $49{,}086{,}048 + 73{,}629{,}072 = 122{,}715{,}120$ ways to deal a hand with at most one 8. ■

✔ Quick Check 5.6

1. Plot the values of $C(4, r)$ for $0 \leq r \leq 4$.
2. Plot the values of $C(5, r)$ for $0 \leq r \leq 5$.
3. Produce an alternate solution to Example 5.28.
4. A high school student needs to listen to the election results on the day of the election, and then read the next-day coverage in a newspaper. There are five television channels and four radio stations that are carrying election coverage. The student has two newspapers to choose from. Assuming the student is loyal to one election-day source for listening, in how many ways can a listening–reading pair be chosen?
5. Suppose the student in the previous problem is instead required either to compare the coverage of two television stations or compare the coverage of two radio stations. The student must then report on the coverage in one newspaper. In how many ways can this be done? ☑

Some apparently difficult counting problems become fairly easy once you adopt the proper viewpoint. The next theorem illustrates this situation.

THEOREM 5.6 ***Ordered Triplets***

Let i, j, k, and n be positive integers. The number of ordered triplets (i, j, k) with $1 \leq i \leq j \leq k \leq n$ is

$$C(n+2, 3) = \frac{n(n+1)(n+2)}{6}$$

Proof: This seems like a problem that requires a triple nested summation:

$$\sum_{k=1}^{n} \sum_{j=1}^{k} \sum_{i=1}^{j} 1$$

This particular triple summation is not really difficult to calculate. However, a more elegant approach to the counting problem uses a visualization involving $n+2$ labeled balls in a container.

The first n balls are labeled $1, 2, \ldots, n$. The final two balls are labeled S and L. Consider the possible sets of three balls that can be pulled out of the container. In most cases, you would pick three balls with distinct numbers as labels. Placing the balls in sorted order represents a triple with strict inequalities $i < j < k$.

If one of the balls is the one labeled S, assign it the value of the *smaller* numeric label on either of the other two balls. Similarly, if the ball labeled L is chosen, assign it the value of the *larger* numeric label among the other two balls. If both the specially labeled balls are chosen, assign them the numeric label on the other ball, resulting in a triple with three identical numbers. In these cases, two or more of the balls will have the same number. Sorting these labels will produce another distinct triple, this time involving one or two $\leq$ signs in the ordering $i \leq j \leq k$.

The key insight is that every possible ordered triple will be produced by exactly one of the three-element subsets of these $n+2$ balls. There must therefore be $C(n+2, 3) = \frac{n(n+1)(n+2)}{6}$ possible ordered triples (i, j, k) with $1 \leq i \leq j \leq k \leq n$. □

A very simple proof that uses neither summations nor labeled balls is possible. Exercise 26 on page 247 asks you to provide an alternative proof.

The Multinomial Counting Theorem

Recall that another common notation for $C(n, r)$ uses the binomial coefficient notation: $\binom{n}{r}$. That notation is more convenient for the material that will be presented next.

Suppose a new car dealer has just received 12 new cars, all of the same model and with identical options. The only difference is the color; five of the cars are red, four are blue, and three are yellow. The dealer wants to park them in a single row, facing the main street. In how many visually distinguishable ways can the cars be arranged?

This is fairly easy to count: There are $\binom{12}{5}$ ways to pick positions for the red cars (we cannot visually tell one red car from another red car, so only the positions in the line-up matter). There are then $\binom{7}{4}$ ways to pick positions for the blue cars (from among the remaining spots). Finally, there are $\binom{3}{3}$ ways to choose positions for the yellow cars. Since the set of available positions has been reduced at each stage, the tasks of placing the red, the blue, and the yellow cars are independent. The Independent Tasks Principle implies that there are

$$\binom{12}{5} \cdot \binom{7}{4} \cdot \binom{3}{3} = 27{,}720$$

ways to line up the cars in a visually distinguishable manner.

Notice that

$$\binom{12}{5} \cdot \binom{7}{4} \cdot \binom{3}{3} = \frac{12!}{5! \cdot 7!} \cdot \frac{7!}{4! \cdot 3!} \cdot \frac{3!}{3! \cdot 0!} = \frac{12!}{5! \cdot 4! \cdot 3!}$$

This example easily generalizes to the task of arranging n items in a visually distinguishable manner.

THEOREM 5.7 ***The Multinomial Counting Theorem–Version 1***

Suppose there exists a set of n items containing n_1 *identical* items of type 1, n_2 *identical* items of type 2, ..., and n_k *identical* items of type k, where $n = n_1 + n_2 + \cdots + n_k$. The number of visually distinguishable ways to arrange the n items in a row is

$$\text{multinomial}(n_1, n_2, \ldots, n_k) = \frac{n!}{n_1! \cdot n_2! \cdots n_k!}$$

The proof is (in essence) the same as the counting process in the new car example.

The multinomial counting theorem can also be used to solve another counting problem.

Suppose we start with n distinct items and want to place some of them in groups. If n_1 items are placed into group 1, n_2 items are placed into group 2, ..., and n_j items are placed into group j, then the task can be accomplished in

$$\text{multinomial}(n_1, n_2, \ldots, n_j, n - \sum_{i=1}^{j} n_i) = \frac{n!}{n_1! \cdot n_2! \cdots n_j! \cdot (n - \sum_{i=1}^{j} n_i)!}$$

ways. This is true because there are $\binom{n}{n_1}$ ways to choose items for the first group, $\binom{n-n_1}{n_2}$ ways to choose items for the second group, and so on. Finally, there are $\binom{n-n_1-n_2-\cdots-n_{j-1}}{n_j}$ ways to choose items for group j. This leaves $\binom{n-\sum_{i=1}^{j} n_i}{n-\sum_{i=1}^{j} n_i} = 1$ way to ignore the remaining items. The groups have been formed in an independent manner, so the total number of ways to accomplish the distribution is

$$\binom{n}{n_1}\cdot\binom{n-n_1}{n_2}\cdots\binom{n-\sum_{i=1}^{j-1}n_i}{n_j}\cdot\binom{n-\sum_{i=1}^{j}n_i}{n-\sum_{i=1}^{j}n_i}$$
$$=\frac{n!}{n_1!\cdot(n-n_1)!}\cdot\frac{(n-n_1)!}{n_2!\cdot(n-n_1-n_2)!}\cdots\frac{(n-\sum_{i=1}^{j-1}n_i)!}{n_j!\cdot(n-\sum_{i=1}^{j}n_i)!}\cdot\frac{(n-\sum_{i=1}^{j}n_i)!}{(n-\sum_{i=1}^{j}n_i)!}$$
$$=\frac{n!}{n_1!\cdot n_2!\cdots n_j!\cdot(n-\sum_{i=1}^{j}n_i)!}$$
$$=\text{multinomial}(n_1,n_2,\ldots,n_j,n-\sum_{i=1}^{j}n_i)$$

EXAMPLE 5.32 **Crazy Eights Hands**

Suppose that four children wish to play a game of crazy eights. They use a standard 52-card deck and each child is dealt seven cards. In how many ways can the cards be dealt? Notice that the children care which of the four hands they receive.

This problem has 52 distinguishable items (the cards), a subset of which is being placed into four distinct groups of size 7. The multinomial counting theorem can be used to count the number of ways to do this:

$$\text{multinomial}(7,7,7,7,52-28)=\frac{52!}{7!\cdot7!\cdot7!\cdot7!\cdot24!}$$
$$=410{,}001{,}479{,}718{,}205{,}060{,}857{,}600$$ ■

If we count the items that are left over as an additional group, the examples just given can be restated in a simple manner.

THEOREM 5.8 ***The Multinomial Counting Theorem—Version 2***

Suppose there exists a set of n *distinguishable* items that is to be partitioned into k *distinguishable* subsets. Subset 1 will have n_1 items, subset 2 will contain n_2 items, ..., and subset k will contain n_k items, where $n = n_1 + n_2 + \cdots + n_k$. The number of ways to accomplish this partitioning is

$$\text{multinomial}(n_1,n_2,\ldots,n_k)=\frac{n!}{n_1!\cdot n_2!\cdots n_k!}$$

EXAMPLE 5.33 **Spring Cleaning**

A family with four children is getting ready for spring cleaning. The parents have decided that they need one child to mop floors, one to vacuum, and two to clean windows. In how many ways can the work crews be formed?

Assume that the children are Abigail, Bertram, Carlos, and Daphne. It is easy (in this case) to list all possibilities.

Mop	A	A	A	B	B	B	C	C	C	D	D	D
Vacuum	B	C	D	A	C	D	A	B	D	A	B	C
Windows	C,D	B,D	B,C	C,D	A,D	A,C	B,D	A,D	A,B	B,C	A,C	A,B

There are 12 ways to form the work crews. The multinomial counting theorem implies that there should be multinomial$(1, 1, 2) = \frac{4!}{1!\cdot1!\cdot2!} = 12$ possible work crews.

Notice that both the elements of the set to be partitioned (the children) and the subsets (the work crews) are distinguishable. ■

EXAMPLE 5.34

Robots

The parents have rented four identical robots to do some outside work. The available tasks are: pull weeds and paint the outer walls of the house. In how many ways can the robot work crews be formed if each task requires two robots?

We can again list all the options (each R represents a robot).

Pull Weeds	RR
Paint House	RR

There is only one way to do this: assign two robots to each task.

Theorem 5.8 seems to imply that there should be multinomial$(2, 2) = \frac{4!}{2! \cdot 2!} = 6$ ways. Why the disagreement? It is because the four robots (items to be partitioned) are *not* distinguishable, so the theorem does not apply. Theorem 5.8 is assuming four distinct robots: R_1, R_2, R_3, R_4. The work crews would then have more options.

Pull Weeds	R_1R_2	R_1R_3	R_1R_4	R_2R_3	R_2R_4	R_3R_4
Paint House	R_3R_4	R_2R_4	R_2R_3	R_1R_4	R_1R_3	R_1R_2

To properly use the theorem, you need to make sure all the hypotheses are fulfilled. ■

EXAMPLE 5.35

Hard Work Pays Off

As a reward for all the hard work during spring cleaning, the parents have purchased three identical copies of a comic book. The comic books need to be distributed to the children to read but two children will need to share. In how many ways can the comic books be dispersed?

It is easy to list all possibilities—just indicate which two children must share. The choices are AB, AC, AD, BC, BD, CD. There are only six ways to distribute the comic books.

Notice that the multinomial counting theorem does not directly apply because the subsets (comic book reading groups) are not all distinguishable. In particular, the two groups of one child each are not distinguishable (one child per group with identical comic books). The group of two children *can* be distinguished from the other two.

The theorem counts the grouping AB twice because it assumes the three comic books are distinguishable. For example, it might think of them as having colors red, blue, and green, with the two-person reading group always getting green. The theorem counts the grouping AB (green) twice—once with C reading the red comic book and once with C reading the blue one. The multinomial counting theorem can be used if we remove the overcounts for the two groups having one child. The result is

$$\frac{\text{multinomial}(1, 1, 2)}{2!} = \frac{12}{2} = 6$$ ■

✔ Quick Check 5.7

1. An elementary school teacher wants to group her class of 29 students into seven groups for a project. She wants each group to have four students. However, one group, the weasels, will need to have five students. The groups will be given animal names (pre-chosen by the teacher). In how many ways can she create the groups? Notice that the children may prefer to be lions rather than skunks.

(continued on the next page)

2. The elementary teacher has decided that giving the groups names has created a large amount of whining and complaining. For the next group project, no names or other distinguishing designations will be used for the groups. Assuming that there will still be seven groups (with one group of five), in how many ways can the groups be formed?
3. How likely is it that a typical elementary school teacher would take time to count how many sets of teams are possible? ☑

5.1.8 Exercises

The exercises marked with ★ have detailed solutions in Appendix H.

1. ★ A boy needs to eat breakfast and lunch, practice piano, mow the lawn, and read a book today. In how many ways can he arrange these activities if breakfast must occur before lunch, and at least one other activity must separate the meals?
2. How many distinct license plates are there with[10]
   (a) four uppercase letters, followed by a space and two digits?
   (b) ★ either six uppercase or lowercase letters or two digits followed by four uppercase letters?
   (c) a digit, followed by two uppercase or lowercase letters, followed by three alphanumeric-upper characters?
   (d) eight alphanumeric characters?
3. There are four common coins in the United States: pennies (1 cent), nickels (5 cents), dimes (10 cents), and quarters (25 cents).
   (a) In how many ways can I form 29 cents?
   (b) How many distinguishable ways can I have three of these coins? Solve this in two ways: by using the appropriate counting principles and formulas and then by listing all the distinguishable collections of three coins.
4. ★ In how many ways can five tomato plants, two rhubarb plants, and seven raspberry bushes be arranged along a garden strip if there does not need to be any distinction among the plants/bushes of the same type?
5. James and a friend played 14 games of tic-tac-toe one night. In how many ways can James end the night with eight wins, two ties, and four losses?
6. In how many ways can you choose four countries to visit in four consecutive months from a list of nine possible countries, including France and Germany, if
   (a) France must be visited?
   (b) France or Germany must be visited but not both?
   (c) France and Germany must be visited but neither country can be visited during the first of the consecutive months?
7. A recording artist plans to place six songs on her new CD. In how many ways can she order the songs if "Heart Like a Donut" must be placed immediately before "I Don't 'Do Lunch' "?
8. How many distinct inventory codes are there if the codes contain exactly three symbols and if
   (a) the symbols can be any combination of alphanumeric-upper characters?
   (b) the symbols can be any combination of alphanumeric-upper characters, but the first symbol must be a letter?
9. ★ A piano has 88 keys; 52 are white and 36 are black. In a standard octave, there are seven white and five black keys. How many chords consisting of three white and two black keys are there in a standard octave?
10. A local pizza store offers a choice of seven toppings and three sizes (small, medium, and large). How many distinct pizzas do they offer if you may use any number of toppings (0–7) but toppings cannot be repeated?
11. ★ Determine the number of "words" (not necessarily words found in the English dictionary) that can be formed using each of the eight letters in the word *lollipop* exactly once.
12. Although repeatedly warned not to play with food at the table, Karen spends time organizing her M&M's. How many ways can she arrange five red, three green, one blue, and three brown M&M's in a line on the table?
13. Suppose that you must choose a password at your work that is five to seven characters long. How many possible passwords are there if
   (a) each password can be any combination of alphanumeric characters?
   (b) each password must contain at least one digit? (The remaining characters are still alphanumeric.)
14. Plot the values of $C(8, r)$, for r between 0 and 8. Why do the points form a symmetric shape?
15. Consider the set $S = \{a, b, c, d\}$.
   (a) Use the multinomial counting theorem to count the number of ways there are to partition S into subsets of sizes 1, 1, and 2, where the subsets are labeled X, Y, and Z, respectively. Then list all the partitions.
   (b) How many ways are there to partition S into unlabeled subsets of sizes 1, 1, and 2? List them.
16. Consider the set $S = \{a, b, c, d, e\}$.
   (a) Use the multinomial counting theorem to count the number of ways there are to partition S into subsets of sizes 1, 1, 1, and 2, where the subsets are labeled W, X, Y, and Z, respectively.
   (b) How many ways are there to partition S into unlabeled subsets of sizes 1, 1, 1, and 2? List them.

[10]Assume that all letter sequences are permissible.

17. ★ Suppose that seven students in Mrs. Corner's fifth grade class and three chaperons are going on a field trip to the swimming pool in three vans that hold two, three, and four passengers, respectively. How many possible ways are there to transport the seven students to the swimming pool using all of the vans? Assume that the three chaperons are the drivers of the vans but that it is not a concern in this problem which chaperon drives which van. A van may be left at the school. [*Hint*: How should you account for the empty seats?]

18. DNA consists of the four nucleotides A (Adenine), C (Cytosine), G (Guanine), and T (Thymine). The DNA strands AAG and AGA are considered to be distinct. Count the number of distinct DNA strands that consist of

(a) 5A, 3C, 1G

(b) 3A, 2C, 4G, 2T

(c) 5A, 5C, 5G, 5T

19. ★ A foreman has a team of 15 workers. He needs 5 of them to unload a truck full of bricks, 3 to start mixing cement, and 4 to clean up yesterday's mess. In how many ways can he create the work groups?

20. How many different five-card hands are there with five distinct face values on the cards?

21. How many different five-card hands are there that contain a straight? A straight occurs when the five cards have five consecutive face values. One example of a straight is 8-9-10-jack-queen. Note that the ace can be used as either the lowest card or the highest card in a straight.

22. How many different eight-card hands are there with no more than three black cards?

23. How many different eight-card hands are there that contain exactly two suits?

24. Suppose that Coach Martin is trying to decide who should play on the church softball team out of a group of 15 women and 21 men. There must be exactly 12 players on the team. In how many ways can Coach Martin create a softball team if

(a) there must be more women than men on the team?

(b) the number of men multiplied by the number of women must be odd?

25. Suppose that the numbers $n_1, n_2, \ldots, n_k$ in Theorem 5.7 are all equal to 1. Show that multinomial$(1, 1, \ldots, 1) = P(k, k)$. Is this something we should expect?

26. Provide an alternate proof for Theorem 5.6 that does not use labeled balls nor summations.

27. What are some differences between combinations with repetition and the multinomial counting theorem (version 1)?

28. What is the difference between versions 1 and 2 of the multinomial counting theorem?

29. Twelve children have been invited to Hitomi's birthday party. Her parents have bought three kinds of party favors for the guests. There are three pencil boards, four colored pencil sets, and five sets of colored marking pens. In how many ways can the party favors be distributed to the guests?

30. Ana and Matsuri are planning a concert for their friends. They are organizing the songs they will sing. There will be four popular songs, five Japanese folk songs, and three British folk songs. Suppose they intend to present the music in three blocks with each block consisting of all the songs of the same genre (so all three British folk songs will be sung one after another).

(a) In how many ways can they arrange the concert?

(b) In how many ways could they arrange the songs if they did not group them in genre blocks?

(c) What percentage of the total number of arrangements, (b), is the collection of blocked arrangements, (a)?

31. You are the leader of a school committee and need to find people to run the ticket booth for one hour after school on each of the seven days before an upcoming school carnival. The booth only requires one worker per day. There are 15 members on your committee that are candidates for working at the ticket booth, and they are not limited to working any specified amount of time. There is a requirement, however, that a person must work the entire hour. A committee member earns points for each hour they work at the ticket booth. If all you care about is how many points to assign to each committee member, and not which days they work, in how many ways can you assign the seven ticket booth work sessions if

(a) ★ There is a persnickety committee member, Laura, who refuses to work at the ticket booth if any of {John, Sue, Richard, Emily} work at the ticket booth?

(b) It was declared that Ben has not been "pulling his weight" on the committee and must work at least four of the days?

32. Six government employees, including the President and the Vice President, are getting their picture taken for a newspaper column. In how many ways can the employees be lined up in a row for the picture if

(a) The President must stand next to the Vice President?

(b) The President and the Vice President cannot be situated on either end?

(c) There must be at least two employees between the President and the Vice President? Make a table to help solve this part.

33. Five male college students are each bringing a date to the orchestra concert. They bought tickets for 10 consecutive seats in row 17. In how many ways can all the students be seated at the concert if

(a) Each male must be seated next to his date?

(b) There must be an alternation of males and females in the row (but dates need not be next to each other)?

(c) No two people on a date can sit next to each other? (Assume there are only three couples for this part. The case with $n = 5$ appears in Exercise 30 on page 266.) [*Hint*: Label the members of the three couples with x, y, and z. Do not distinguish by gender, and do not assign a particular couple to each letter yet. How many patterns of two x's, two y's and two z's are permissible if two of the same letter cannot be adjacent? Now assign couples and genders.]

5.2 Combinatorial Proofs

5.2.1 Introduction to Combinatorial Proofs

There are often several ways to prove a mathematical claim. The next proposition illustrates two very different approaches to the same claim.[11]

PROPOSITION 5.9

Let n and r be nonnegative integers with $r \le n$. Then $\binom{n}{r} = \binom{n}{n-r}$.

Algebraic Proof:

$$\binom{n}{r} = \frac{n!}{r!(n-r)!} = \frac{n!}{(n-r)!r!} = \frac{n!}{(n-r)![n-(n-r)]!} = \binom{n}{n-r} \qquad \square$$

Combinatorial Proof: The number of ways to choose a subset of size r from a set of size n is $\binom{n}{r}$. Each subset of size r automatically determines a subset of size $n - r$ consisting of the elements *not* chosen (and vice versa). The number of ways to choose a subset of size $n - r$ must therefore be the same as the number of ways to choose a subset of size r. $\square$

Both proofs are correct. However, the combinatorial proof is generally considered to be more elegant. The strategy in a combinatorial proof is to count the same thing in two ways, and then assert that the two answers must be the same. For Proposition 5.9, the key observation is that selecting a subset of size r automatically forms a companion subset of size $n - r$ (and vice versa). Thus, counting the distinct subsets of size r will produce the same total as will counting the distinct subsets of size $n - r$.

A combinatorial proof is a method for establishing an identity by counting something in two distinct ways. Combinatorial proofs usually avoid the need for extensive algebraic manipulations. In addition, they often provide insight that is not apparent if an inductive proof or an algebraic proof is used to establish the identity.

THEOREM 5.10 ***Pascal's Theorem***

Let n and r be positive integers with $r \le n$. Then

$$\binom{n+1}{r} = \binom{n}{r-1} + \binom{n}{r}$$

Proof: We already know that there are $\binom{n+1}{r}$ subsets consisting of r elements from a set of size $n + 1$.

Consider these r-element subsets in a different manner. Start by designating one of the original $n + 1$ elements in a special way (perhaps by painting it blue). Then any subset of size r will either contain the blue element or it will not. That is, the r-element subsets can be partitioned into two disjoint groups: those that contain the blue element and those that don't contain the blue element. The Mutually Exclusive Tasks Principle (page 228) implies that the number of subsets of size r is the sum of the number of subsets in each of these disjoint collections.

If a subset of size r contains the blue element, we can choose the remaining elements in $\binom{n}{r-1}$ ways.

If a subset does not contain the blue element, there are only n remaining elements from which to choose the r elements of the subset. This can be done in $\binom{n}{r}$ ways. $\square$

A visualization of this proof can be found at http://www.mathcs.bethel.edu/~gossett/vcp/ .

[11]Many of the claims in this section involve combinations. These theorems are traditionally stated using the binomial coefficient notation.

THEOREM 5.11

Let n be a nonnegative integer. Then

$$\sum_{r=0}^{n} \binom{n}{r} = 2^n$$

Proof: Let S be any set with n elements. The number of subsets of S will be counted in two ways.

Observe that subsets of S come in many sizes. Some have only one element, some have several, and one subset is empty. There are $\binom{n}{r}$ subsets of size r, for $0 \leq r \leq n$. Since a subset cannot have two different sizes, these groupings are mutually exclusive. The Mutually Exclusive Tasks Principle implies

$$|\mathcal{P}(S)| = \sum_{r=0}^{n} \binom{n}{r}$$

The second approach uses an "in-crowd" visualization. Place all the elements of S in a line. A subset can be formed by having some of the elements step forward, while the rest remain to the back. The ones that have stepped forward can be thought of as the "winners"; the others are "losers". The winners form a subset of S. In how many ways can we form a subset of winners? This is really a sequence of independent tasks: Deciding whether the second element is a winner or a loser is not influenced by what was decided about the first element. There are n decisions to be made, each with two possible outcomes so, according to the Independent Tasks Principle on page 227, there are 2^n ways to form a subset of S.

Because we are counting the same collection of subsets, the two answers must be the same. Hence, $\sum_{r=0}^{n} \binom{n}{r} = 2^n$. □

There is a useful corollary that follows from the proof of Theorem 5.11 (but not directly from the theorem itself).

COROLLARY 5.12 $|\mathcal{P}(S)|$

Let S be a finite set with $n \geq 0$ elements. Then $|\mathcal{P}(S)| = 2^n$.

The next theorem has a very simple combinatorial proof. It would be much more difficult to prove using algebraic methods.

THEOREM 5.13 ***Vandermonde's Theorem***

Let n, m, and r be nonnegative integers, with $r \leq n$ and $r \leq m$. Then

$$\binom{n+m}{r} = \sum_{i=0}^{r} \binom{n}{i} \cdot \binom{m}{r-i}$$

Proof: Consider a set S consisting of $n + m$ items. There are $\binom{n+m}{r}$ ways to form a subset of size r.

Now paint n of the items red and paint the other m items blue. A subset of size r will have some red items and some blue items.

Suppose the subset contains i red items. Then it must contain $r-i$ blue items. There are $\binom{n}{i}$ ways to choose i red items from the n possibilities and $\binom{m}{r-i}$ ways to choose the blue items in the set. Since these selections are independent, there are $\binom{n}{i} \cdot \binom{m}{r-i}$ ways to choose a set of size r having i red items. Because $r \leq n$ and $r \leq m$, i can be any number between 0 and r. Since subsets having different numbers of red items cannot occur at the same time, the Mutually Exclusive Tasks Principle applies. Consequently there are $\sum_{i=0}^{r} \binom{n}{i} \cdot \binom{m}{r-i}$ possible subsets of size r.

Equating the two counts finishes the proof. □

A visualization of this proof can be found at http://www.mathcs.bethel.edu/~gossett/vcp/. You should also look at the module about Oresme's Sequence.

✔ Quick Check 5.8

1. Prove that

$$r \cdot \binom{n+1}{r} = (n+1) \cdot \binom{n}{r-1}.$$

(a) Use an algebraic proof.

(b) Use a combinatorial proof. [*Hint*: Consider a group of $n+1$ people who need to select a committee of size r from among themselves and also must appoint a chairperson for the committee (from among the r committee members).] ☑

The binomial coefficient notation, $\binom{n}{r}$, that has been used in this section was named for the coefficients in the well-known binomial theorem.

THEOREM 5.14 ***The Binomial Theorem***

Let n be a nonnegative integer. Then

$$(x+y)^n = \sum_{r=0}^{n} \binom{n}{r} x^{n-r} y^r$$

Proof 1: A store is having a grand opening. The first n people who arrive at the store will receive either a free book or a free DVD. There are x different books and y different DVDs (each with at least n copies on hand). We can count the number of ways these gifts can be distributed in two ways. Notice that the n people are distinguishable.

Observe first that receiving a book and receiving a DVD are mutually exclusive, so each person has $x + y$ possible gifts. Since there are n copies of each gift on hand, the choice that one person makes is independent of another person's choice. There are therefore $(x + y)^n$ possible ways to distribute gifts.

For the second enumeration, suppose that r people choose to receive a DVD. Then $n - r$ people will receive a book. There are $\binom{n}{r}$ ways to select the r people who receive a DVD, which also will determine which people receive a book. Each of the r people who receive a DVD can choose one of y DVDs, for a total of y^r ways to make those selections. The people who receive books can make their choices in x^{n-r} ways. Since all three tasks (deciding which r people pick a DVD, choosing DVDs, and choosing books) are independent, the Independent Tasks Principle implies that there are $\binom{n}{r}x^{n-r}y^r$ ways that exactly r people can receive a DVD. But $0 \le r \le n$ and different choices of r are mutually exclusive. The Mutually Exclusive Tasks Principle implies that the total number of ways to distribute gifts is therefore

$$\sum_{r=0}^{n} \binom{n}{r} x^{n-r} y^r$$

Equating the two counts completes the proof. □

A visualization of this proof can be found at http://www.mathcs.bethel.edu/~gossett/vcp/.

Proof 2: Consider the product $(x+y)^n = \underbrace{(x+y)(x+y)(x+y)\cdots(x+y)}_{n \text{ times}}$.

Each term in the expanded product contains either an x or a y from each of the n factors. Thus, the sum of the exponents on x and y in each term must always be n. How many times does $x^{n-r}y^r$ occur? We can count the occurrences by numbering the factors $(x+y)$ from 1 to n. The term $x^{n-r}y^r$ will result once for every distinct subset of size $n-r$ from the set $\{1, 2, 3, \ldots, n-1, n\}$. The coefficient of $x^{n-r}y^r$ must therefore be $\binom{n}{n-r} = \binom{n}{r}$. □

EXAMPLE 5.36 **A Binomial Expansion**

Proof 2 can be illustrated by considering the product

$$(x_a + y_a)(x_b + y_b)(x_c + y_c) = x_a x_b x_c + x_a x_b y_c + x_a y_b x_c + x_a y_b y_c + y_a x_b x_c + y_a x_b y_c + y_a y_b x_c + y_a y_b y_c$$

Dropping the subscripts and using the commutative and associative laws of the real (or complex) numbers results in the sum

$$x^3 + x^2y + xyx + xy^2 + yx^2 + yxy + y^2x + y^3 = x^3 + 3x^2y + 3xy^2 + y^3$$ ■

EXAMPLE 5.37 **Another Proof of Theorem 5.11**

The binomial theorem can be used to provide yet another proof of Theorem 5.11 on page 249. We already know that for a set S of size n

$$|\mathcal{P}(S)| = \sum_{i=0}^{n} \binom{n}{i}$$

By using the binomial theorem, we know that

$$2^n = (1+1)^n = \sum_{i=0}^{n} \binom{n}{i} 1^{n-i} 1^i = \sum_{i=0}^{n} \binom{n}{i} = |\mathcal{P}(S)|$$ ■

The binomial theorem has a straightforward generalization that uses the multinomial coefficients introduced previously.

THEOREM 5.15 ***The Multinomial Theorem***

Let n be a nonnegative integer. Then

$$(x_1 + x_2 + \cdots + x_k)^n = \sum_{\substack{0 \le n_1, n_2, \ldots, n_k \le n \\ n_1 + n_2 + \cdots + n_k = n}} \text{multinomial}(n_1, n_2, \ldots, n_k)\, x_1^{n_1} x_2^{n_2} \cdots x_k^{n_k}$$

The proof is essentially the same as the proof of the binomial theorem. In this case, version 1 of the multinomial counting theorem is used to calculate the coefficients.

5.2.2 Counting Tulips: Three Combinatorial Proofs

This section contains an extended example that illustrates the use of combinatorial proofs.

A gardener has $r \ge 1$ red tulips and $b \ge 1$ blue tulips, each in its own pot. She plans to plant them in a line along the edge of her driveway. In how many visually distinguishable ways can she arrange them?

The answer to this question certainly depends upon the numbers of each color that must be planted. The solution is summarized in Theorem 5.16 on page 252.

The meaning of *visually distinguishable* in this context can be illustrated by the following examples. Even though red tulips 1 and 2 in Figure 5.3 have changed places in the left-hand diagram, the two rows are considered visually indistinguishable because it is the patterns of colors that are important. In the right-hand example, moving the blue tulip from the right end to the middle causes the two rows to be visually distinguishable.

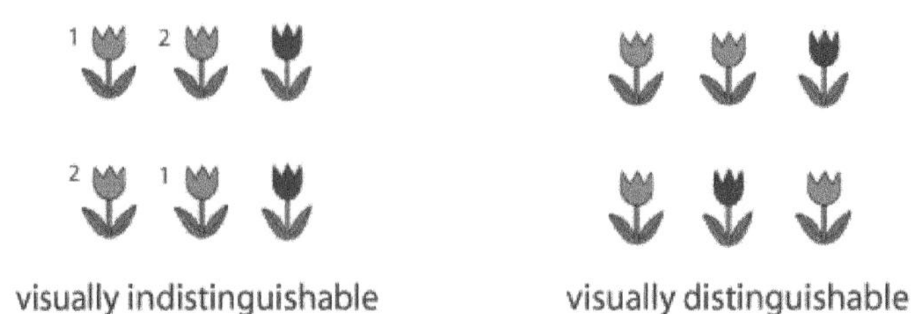

Figure 5.3 Visually distinguishable.

THEOREM 5.16 ***Counting Tulip Patterns***

A gardener has $r \geq 1$ red tulips and $b \geq 1$ blue tulips, each in its own pot. She plans to plant them in a line along the edge of her driveway. The number of visually distinguishable ways she can arrange them is summarized in the following table.

Number of Red Tulips Planted	Number of Blue Tulips Planted	Number of Visually Distinguishable Patterns
r	b	$\binom{b+r}{r}$
r	$0 \leq \text{number planted} \leq b$	$\binom{b+r+1}{r+1}$
$0 \leq \text{number planted} \leq r$	$0 \leq \text{number planted} \leq b$	$\binom{b+r+2}{r+1} - 1$

As an added bonus, the proof of the theorem will yield some very nice combinatorial identities. Three combinatorial proofs will be used to address the three cases mentioned in the theorem. The theorem will be illustrated and then proved.

EXAMPLE 5.38 **Illustrating Theorem 5.16**

Let $r = 2$ and $b = 1$. The number of visually distinguishable patterns in each of the three cases is easy to calculate and to list. Notice that each case contains all patterns in the previous case, plus additional choices.

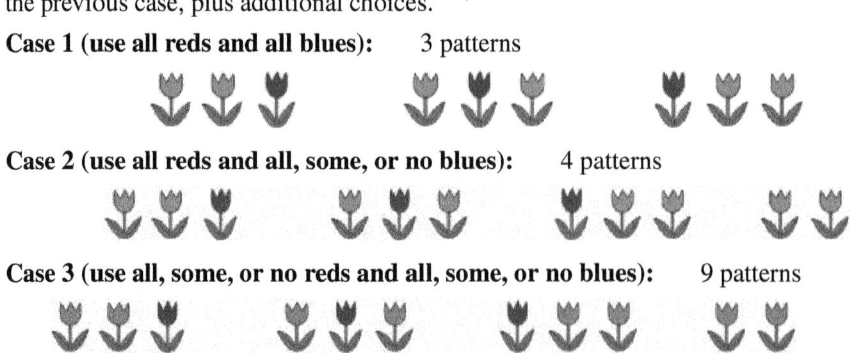

Case 1 (use all reds and all blues): 3 patterns

Case 2 (use all reds and all, some, or no blues): 4 patterns

Case 3 (use all, some, or no reds and all, some, or no blues): 9 patterns

$\varnothing$ ■

All Tulips Must Be Planted

Suppose the gardener has decided to plant all $r + b$ tulips. In how many visually distinguishable ways can she arrange them? Variations on the following pair of enumerations have been around for a while.

The first enumeration

For this case, the first enumeration is more complicated than the second enumeration. For the other two cases, the second enumerations will be more challenging.

The gardener can group patterns by the number, i, of blue tulips planted on the left before the first red tulip is encountered. Different values of i will result in mutually exclusive sets of patterns. If there are i blue tulips and then a red tulip, then there are $(r - 1) + (b - i)$ tulips that still must be arranged, with $b - i$ of them being blue. There are therefore $\binom{(r-1)+(b-i)}{b-i}$ patterns with exactly i blue tulips before the first red is encountered. Thus (using the Mutually Exclusive Tasks Principle) there are

$$\sum_{i=0}^{b} \binom{(r-1)+(b-i)}{b-i} = \sum_{k=0}^{b} \binom{(r-1)+k}{k} \text{ distinct patterns}$$

The left-hand sum is $\binom{(r-1)+b}{b} + \binom{(r-1)+(b-1)}{b-1} + \cdots + \binom{(r-1)+1}{1} + \binom{(r-1)}{0}$ whereas the right-hand sum uses is: $\binom{(r-1)}{0} + \binom{(r-1)+1}{1} + \cdots + \binom{(r-1)+(b-1)}{b-1} + \binom{(r-1)+b}{b}$.

The second enumeration

Since there are $b + r$ tulips, and the pattern is determined by which of the $b + r$ positions contain red tulips, there are $\binom{b+r}{r}$ distinct color patterns.

Combining the enumerations

Since the two enumerations are both counting the number of visually distinguishable patterns of red and blue tulips, the results must be equal, establishing Theorem 5.17.

THEOREM 5.17 ***Plant all tulips***

For all positive integers b and r,

$$\sum_{k=0}^{b} \binom{r-1+k}{k} = \binom{b+r}{r}$$

Theorem 5.17 is often presented in the following form.

COROLLARY 5.18

For all nonnegative integers b and r,

$$\sum_{k=0}^{b} \binom{r+k}{k} = \binom{b+r+1}{b}$$

Proof:

Replace $r - 1$ by s in the theorem, then rename s as r. Now observe that Proposition 5.9 (on page 248) implies $\binom{b+r+1}{r+1} = \binom{b+r+1}{b}$.

Finally, $\sum_{k=0}^{0} \binom{r+k}{k} = \binom{r}{0} = 1 = \binom{0+r+1}{0}$ so the identity holds when $b = 0$. □

Theorem 5.19 on page 255 is a simple consequence of Corollary 5.18, using the identity $\binom{r+k}{k} = \binom{r+k}{r}$. However, it can also be proved directly using another combinatorial proof. A visualization of the Theorem 5.17/Corollary 5.18 proof can be found at http://www.mathcs.bethel.edu/~gossett/vcp/ .

All Of The Red Tulips Must Be Planted

Suppose now that the gardener intends to use all r of the red tulips, but the number of blue tulips can be any number from 0 to b, inclusive.

The first enumeration

The gardener must first decide how many blue tulips to use. Suppose she decides to use a combined total of i tulips. Then there will be $i - r$ blue tulips. She can mark i evenly-spaced positions along the driveway at which to plant the tulips. There are $\binom{i}{r}$ ways to place the r red tulips among the i positions. Each such set of positions leads to a different visually distinguishable pattern.

Since the patterns with j blue tulips and the patterns with $m \neq j$ blue tulips are mutually exclusive and visually distinguishable, the total number of visually distinguishable patterns is

$$\sum_{i=r}^{b+r} \binom{i}{r}$$

The summation starts with $i = r$ because the minimum planting contains all of the red tulips. A simple expansion shows that

$$\sum_{i=r}^{b+r} \binom{i}{r} = \sum_{k=0}^{b} \binom{r+k}{r}$$

The second enumeration

The gardener can also use the following procedure to choose the pattern. First, she places all the pots containing the red tulips along the edge of the driveway, leaving room between and outside the pots for her assistant to place pots of blue tulips.

She then generates a string of $b + r + 1$ letters. The string will contain b copies of the letter B and $r + 1$ copies of the letter M. The string will be a prescription for placing a subset of blue tulips in the spaces that have been reserved.

The assistant starts by standing by the leftmost gap (that is, to the left of all the red tulips). The gardener starts reading the string. She hands one blue tulip to the assistant for each letter B that occurs before an M is encountered. The assistant places those pots in the current gap. Letters are crossed out once they are processed.

When an M is encountered, the assistant *moves* right to the next gap. The gardener then starts handing a blue tulip to the assistant for each letter B contained in the unprocessed portion of the string, stopping when a letter M is encountered. As soon as an M appears, the assistant moves to the next gap. If multiple Ms are adjacent, there will be one or more gaps with no blue tulips.

This process continues until all of the Ms have been processed. The remaining letter Bs (and the remaining pots of blue tulips) will be ignored.

At this point, the gardener can plant the tulips in evenly spaced holes along the driveway.

The key point is that every string of b Bs and $r + 1$ Ms results in a visually distinguishable pattern of tulips and every visually distinguishable pattern corresponds to a unique string. (See Figure 5.4 for two simple examples.) The $r + 1$ Ms allow the assistant to move past each of the $r + 1$ positions that are between or outside the r red tulips. How many strings of letters are possible? Once the positions of the $r+1$ Ms have been chosen, the string is uniquely determined. There are thus $\binom{b+r+1}{r+1}$ distinct patterns.

The pattern BBMMBM corresponds to

The pattern MBBMMB corresponds to

Figure 5.4. Two B-M patterns.

Combining the enumerations

Since the two enumerations are counting the number of visually distinguishable patterns of red and blue tulips, the results must be equal, completing the proof of Theorem 5.19, found on page 255.

THEOREM 5.19 ***Plant all of the red tulips and*** $0 \le$ ***# blue tulips*** $\le b$

For all positive integers n and r,

$$\sum_{k=0}^{b} \binom{r+k}{r} = \binom{b+r+1}{r+1}$$

The following corollary does not involve red and blue tulips, but is a useful form that can be used in other contexts.

COROLLARY 5.20

For all positive integers n and r,

$$\sum_{i=0}^{n} \binom{i}{r} = \binom{n+1}{r+1}$$

Proof:

$$\binom{b+r+1}{r+1} = \sum_{k=0}^{b} \binom{r+k}{r} = \sum_{i=r}^{b+r} \binom{i}{r} = \sum_{i=0}^{b+r} \binom{i}{r}$$

The second equality can be seen to be true by expanding the two middle summations. The final equality is valid because of the convention that $\binom{i}{r} = 0$ whenever $i < r$. Now let $n = b + r$.

□

The Lower Bound for Both Red and Blue Tulips Is Zero

Suppose, finally, that the number of red tulips actually planted can be any number from 0 to r and the number of blue tulips actually planted can be any number from 0 to b. In how many visually distinguishable ways can the gardener arrange the tulips?

The first enumeration

The gardener may first decide how many red tulips and how many blue tulips to use. Suppose she decides to use j red tulips and k blue tulips. Then there will be $j+k$ tulips to plant. Different choices for the numbers of red and blue tulips will lead to visually distinguishable patterns. For a specific choice of j and k, the number of visually distinct patterns will be $\binom{j+k}{k}$ because the positions of the blue tulips will determine the visual pattern.

The total number of visually distinguishable patterns is therefore

$$\sum_{j=0}^{r} \sum_{k=0}^{b} \binom{j+k}{k}$$

The second enumeration

The gardener can use the following procedure for determining the pattern. She starts with $b+1$ letter Bs and $r+1$ letter Rs. These $b+r+2$ letters are randomly arranged in a string. Once the string has been formed, she discards the first R and all Bs to the left of that R. If any Bs remain, she then discards the final B and all Rs to the right of that B. Otherwise, she discards all the remaining Rs (leaving an empty string). The remaining string is a prescription for the pattern of tulips to plant (the empty string indicates that no tulips will be planted). (See Figure 5.5 for several simple examples.)

The pattern RRBRBBR corresponds to

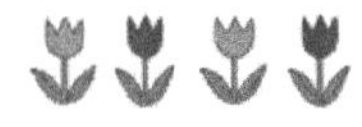

The pattern BRRBRRB corresponds to

The pattern RBBRRRB corresponds to

The pattern BBRBRRR corresponds to

Figure 5.5. Several B-R patterns.

The procedure (with one exception) produces distinct prescriptions. If that were not true, then there must be two distinct strings, X and Y, that form the same prescription after the discards. These two strings must either have a different number of Bs to the left of the first R or a different number of Rs to the right of the final B (or both). Suppose that string X has more Bs to the left of the first R than does string Y. Since both strings have $b + 1$ Bs, string Y (with one exception) must have more Bs than string X does between the first R and the final B. But that means that the resulting prescriptions will be visually distinct, contradicting the assumption that the two strings produced the same prescription.

The previously mentioned exception occurs when string X consists of $b + 1$ Bs followed by $r + 1$ Rs and string Y consists of b Bs, followed by RB and then r Rs. Both lead to the prescription that indicates that no tulips should be planted. See Figure 5.6.

BBBBBBRRRRRR

BBBBBRBRRRRR

Figure 5.6. The Exception.

A similar argument can be used when the two strings have a different number of Rs after the final B.

The procedure will produce every possible pattern. For each possible pattern, write a sequence of Bs and Rs that matches the visual pattern. Denote that string by P. Suppose there are j red tulips and k blue tulips in the pattern. The string consisting of $b - k$ Bs, followed by an R, followed by the string P, followed by a B and then $r - j$ Rs will be a string of length $b + r + 2$ that produces the desired prescription.

There are $\binom{b+r+2}{r+1}$ distinct strings and only two of them produce the same pattern, so there must be

$$\binom{b+r+2}{r+1} - 1 \text{ visually distinct ways to plant the tulips.}$$

Combining the enumerations

The two enumerations can be combined to establish Theorem 5.21.

THEOREM 5.21 $0 \le$ ***# red tulips*** $\le r$ ***and*** $0 \le$ ***# blue tulips*** $\le b$

For all positive integers b and r,

$$\sum_{j=0}^{r}\sum_{k=0}^{b}\binom{j+k}{k} = \binom{b+r+2}{r+1} - 1$$

The inner summation in Theorem 5.21 can be expanded to produce another combinatorial identity.

COROLLARY 5.22

For all positive integers b and r,

$$\sum_{j=0}^{r}\binom{b+j+1}{j+1} = \binom{b+r+2}{r+1} - 1$$

Proof:

$$\binom{b+r+2}{r+1} - 1 = \sum_{j=0}^{r}\sum_{k=0}^{b}\binom{j+k}{k} = \sum_{j=0}^{r}\sum_{k=0}^{b}\binom{j+k}{j} = \sum_{j=0}^{r}\binom{b+j+1}{j+1}$$

The final equality follows from Theorem 5.19. □

Extracting the right-hand sides of Theorems 5.17, 5.19, and 5.21 proves Theorem 5.16.

5.2.3 Exercises

The exercises marked with ★ have detailed solutions in Appendix H.

1. Prove that $\sum_{r=0}^{n}(-1)^r\binom{n}{r} = 0$ if n is a positive integer.
2. ★ Let $n \in \mathbb{N}$. Which sum is larger? (Provide adequate justification for your answer.)

$$\sum_{i=0}^{n}\binom{2n+1}{i} \quad \text{or} \quad \sum_{j=n+1}^{2n+1}\binom{2n+1}{j}$$

3. Prove that if n is a positive integer with $n \geq 2$,

$$\binom{2n}{2} = 2\cdot\binom{n}{2} + n^2$$

   (a) Use an algebraic proof.
   (b) Use a combinatorial proof.
4. ★ What is the coefficient of the term involving x^4y^2 in the expanded form of $(x+y)^6$?
5. What is the coefficient of the term involving $w^4x^3y^2$ in the expanded form of $(w+x+y)^9$?
6. Use mathematical induction to prove the binomial theorem.
7. How many distinct solutions are there to the equation

$$w + x + y + z = 16$$

   where w, x, y, z are nonnegative integers? [*Hint*: Think of distributing 16 identical objects to four people.]
8. Let $n, r \in \mathbb{N}$. Prove

$$\sum_{k=0}^{n}\binom{k}{r} = \binom{n+1}{r+1}$$

   [*Hint*: Prove the assertion directly for $n < r$, and then use mathematical induction on n to prove the assertion when $0 \leq r \leq n$.]
9. Let $n \in \mathbb{N}$.
   (a) ★ Prove that $n^2 = 2\binom{n}{2} + \binom{n}{1}$. Treat the cases $n < 2$ and $n \geq 2$ separately.
   (b) Use part (a) and Exercise 8 to determine $\sum_{k=1}^{n}k^2$.
10. Let $a, b, c \in \mathbb{N}$ with $0 \leq c \leq b \leq a$. Prove that

$$\binom{a}{b}\binom{b}{c} = \binom{a}{b-c}\binom{a-b+c}{c}$$

   (a) Use an algebraic proof.
   (b) Use a combinatorial proof.
11. Let n be a nonnegative integer. Prove that

$$\sum_{k=0}^{n}\binom{n}{k}2^k = 3^n$$

12. Let n be any natural number. Prove that

$$\sum_{k=0}^{n}\binom{n}{k}^2 = \binom{2n}{n}$$

   [*Hint*: $\binom{n}{k}^2 = \binom{n}{k}\cdot\binom{n}{k}$. Now replace the second $\binom{n}{k}$ with an equivalent expression.]
13. Let n be a nonnegative integer. Without using calculus, prove that

$$n(1+y)^{n-1} = \sum_{r=1}^{n}\binom{n}{r}ry^{r-1}$$

   [*Hint*: Start by substituting $n-1$ for n in the binomial theorem. Then multiply both sides of the equation by n.]
14. Use mathematical induction to prove Corollary 5.18 (page 253). [*Hint*: let r be constant and induct on b.]
15. ★ A group of n roommates share a house and car. They need to buy groceries for the week. One or more of them need to do the shopping, but only one will be the driver. Use a combinatorial proof to show that they can form a shopping committee with driver in

$$\sum_{k=1}^{n}k\binom{n}{k} = n2^{n-1}$$

   ways. (The combinatorial proof should be used to establish the validity of the identity.)
16. Let $n \in \mathbb{N}$. Use a combinatorial proof to show that

$$\sum_{k=1}^{n}k\binom{n}{k}^2 = n\binom{2n-1}{n-1}$$

   [*Hint*: $\binom{n}{k}^2 = \binom{n}{k}\cdot\binom{n}{k}$. Now replace the second $\binom{n}{k}$ with an equivalent expression.]
17. Let $k, n \in \mathbb{N}$. Use a combinatorial proof to show that

$$\sum_{j=0}^{k}\binom{k}{j}\binom{n+1}{k-j} = \sum_{j=0}^{k}\binom{k+1}{j}\binom{n}{k-j}$$

   [*Hint*: Both sides are equal to $\binom{n+k+1}{k}$.]
18. Let n and r be integers with $0 \leq r \leq n$. Prove that

$$\sum_{k=0}^{n-r}\binom{r+k}{r} = \binom{n+1}{r+1}$$

19. Let n be a nonnegative integer. Prove that

$$\sum_{r=0}^{n}\binom{2n+1}{r} = 2^{2n}$$

20. Let S be a finite set. Let N_e be the number of subsets of S which have an even number of elements, and let N_o be the number of subsets of S that have an odd number of elements. Prove that $N_e = N_o$. [*Hint*: Consider two cases: $|S| = 2k$ and $|S| = 2k+1$, for $k \in \mathbb{N}$. In each case, show that $N_e - N_o = 0$.]

21. Use Theorem 3.67 (page 148) and Theorem 5.10 (page 248) to prove that $g(i, j)$ in Example 3.44 (page 151) satisfies, for all $i, j \in \mathbb{Z}$ with $i \geq 0, j \geq 0$:

$$\begin{aligned} g(0,0) &= 0 \\ g(i,0) &= i \text{ if } i \geq 1 \\ g(0,j) &= j \text{ if } j \geq 1 \\ g(i,j) &= \binom{i+j}{i-1} + \binom{i+j}{j-1} \end{aligned}$$

The base step must include $g(1, 1)$. The inductive step requires 3 cases: $g(i, 1)$ for $i \geq 2$, $g(1, j)$ for $j \geq 2$ and $g(i, j)$ for $i, j \geq 2$.

22. The Fibonacci sequence is defined in Exercise 27 on page 153 and also in Example 7.15 on page 351. Use induction and Theorem 5.10 to prove for all $n \geq 0$,

$$f_n = \sum_{k=0}^{n} \binom{n-k}{k}$$

[*Hints*: (1) Keep f_n on the left-hand side of the equations. (2) During the inductive step, use two cases: n even, and n odd. (3) Recall that $\binom{a}{b} = 0$ if $b < a$.]

23. Show that for $n \geq 0$

$$\sum_{k=1}^{n} k^2 \binom{n}{k} = \binom{n}{2} 2^{n-1}$$

24. Use a combinatorial proof to show that for $n \geq 1$

$$\sum_{r=0}^{n} \binom{2n}{2r} = 2^{2n-1}$$

25. Recall that $\binom{n}{r} = 0$ whenever $n < r$. Use a combinatorial proof to show that if $n > 0$, then

$$\sum_{r \geq 0} \binom{n}{2r} = \sum_{r \geq 0} \binom{n}{2r+1}$$

[*Hint*: Form all possible committees from a set of n people. Find a one-to-one correspondence between sets of even sizes and sets of odd sizes. A special person may be useful.] This identity is from [8].

5.3 Pigeon-Hole Principle; Inclusion–Exclusion

5.3.1 The Pigeon-Hole Principle

Figure 5.7. Nutches in Nitches.

One of my favorite Dr. Seuss books is *On Beyond Zebra* [80], which introduces the part of the alphabet that comes after Z. One of the new letters is described next.[12]

> And **NUH** is the letter I use to spell Nutches
> Who live in small caves, known as Nitches, for hutches.
> These Nutches have troubles, the biggest of which is
> The fact there are many more Nutches than Nitches.
> Each Nutch in a Nitch knows that some other Nutch
> Would like to move into his Nitch very much.
> So each Nutch in a Nitch has to watch that small Nitch
> Or Nutches who haven't got Nitches will snitch.

This short poem illustrates a very simple yet powerful principle named the *pigeon-hole principle*. The principle states that if there are n pigeon-holes and more than n pigeons, some pigeon hole must contain more than one pigeon.[13]

EXAMPLE 5.39 **Student IDs**

Suppose a school has decided to assign every student a unique five-digit ID number. How many students can attend the school before it becomes necessary to reuse an ID number?

There are $10^5 = 100{,}000$ distinct ID numbers, so the 100,001st student will need to be assigned a previously used number. ■

[12]From *On Beyond Zebra!* by Dr. Seuss, copyright TM & copyright (c) by Dr. Seuss Enterprises, L.P. 1955, renewed 1983. Used by permission of Random House Children's Books, a division of Random House, Inc.

[13]The phrase *pigeon hole* is a popularized name for the *Dirichlet drawer principle*. Instead of pigeons and pigeon holes, think about placing items into the set of open sorting compartments in an old-fashioned roll-top desk.

The principle becomes a little more interesting in its full form.

The Pigeon-Hole Principle

If n objects are distributed into k boxes, then at least one box must contain at least $\lceil \frac{n}{k} \rceil$ objects.

This principle is not hard to intuitively visualize. Suppose that $n = qk + r$, with $0 \leq r < k$. Then $\frac{n}{k} = q + \frac{r}{k}$, where $\frac{r}{k} < 1$. Think of placing the n objects into the boxes in an evenly distributed manner.

After distributing the first qk objects, each box will contain q items. If $r = 0$, then $\lceil \frac{n}{k} \rceil = \frac{n}{k} = q$. There are no more objects to distribute, so every box contains $\frac{n}{k} = q$ objects. If $r > 0$, there will be r more objects to distribute so r of the boxes will contain $q + 1 = \lceil \frac{n}{k} \rceil$ objects and $k - r$ boxes will contain q objects.

The Pigeon-Hole Principle asserts even more. It says that even if you put the objects into the boxes in some random or some extremely clever fashion, there will always be a box which contains at least $\lceil \frac{n}{k} \rceil$ objects. You will be asked to provide a formal proof in Exercise 29 on page 266.

EXAMPLE 5.40 **Host Families**

A choir director is taking a 57-member choir on tour. The director has located 16 families who are willing to take choir members for the night at the first tour stop. Assuming the choir director will stay at a fancy hotel, will it be necessary for some host family to find space for five choir members?

The pigeon-hole principle indicates that there will be a host family that must find room for at least $\left\lceil \frac{57}{16} \right\rceil = 4$ choir members. With proper planning, no family will need to accept five choir members. ■

The next two propositions are from the collaboration of the Hungarian mathematicians P. Erdös and G. Szekeres.

DEFINITION 5.23 ***Monotone Sequence***

A sequence $a_1, a_2, \ldots, a_k$ is called *monotone* if either

$$a_1 \leq a_2 \leq a_3 \leq \cdots \leq a_k$$

or

$$a_1 \geq a_2 \geq a_3 \geq \cdots \geq a_k$$

The sequence is *strictly monotone* if all the inequalities are strict (for example, $<$ rather than $\leq$).

The next proposition (from [30]) uses distinct numbers to keep the discussion and proof simple. If the numbers are not distinct, the proposition will remain valid if the word *strictly* is dropped.

PROPOSITION 5.24 ***Sequences***

Let $x_1, x_2, \ldots, x_{n^2+1}$ be a sequence of $n^2 + 1$ distinct numbers. Then there is a strictly monotone subsequence of at least $n + 1$ of the numbers.

EXAMPLE 5.41 **Illustrating Proposition 5.24**

The sequence 4, 7, 2, 1, 3, 8, 5, 9, 0, 6 contains $3^2 + 1 = 10$ numbers. Proposition 5.24 implies that it must contain either a subsequence of 4 or more elements that is increasing,

4, 7, 2, 1, 3, 8, 5, 9, 0, 6

or a subsequence of 4 or more that is decreasing (or both). In fact, the subsequence 1, 3, 5, 6 is increasing and the subsequence 4, 2, 1, 0 is decreasing.

One systematic way to approach this is to create a set of 10 ordered pairs, corresponding to the 10 numbers in the sequence. The first entry in ordered pair j will denote the number of elements in the longest increasing sequence starting with x_j. The second entry will denote the number of elements in the longest decreasing sequence starting with x_j. For the sequence in this example, the ordered pairs are

$$(4, 4), (3, 4), (4, 3), (4, 2), (3, 2), (2, 3), (2, 2), (1, 2), (2, 1), (1, 1)$$

Notice the occurrences of the number 4. ■

Proof of Proposition 5.24: The proof follows the approach used in the example. Create the set of ordered pairs $\{(i_j, d_j)\}$ for $j = 1, 2, \ldots, n^2 + 1$, where i_j is the length of the longest increasing subsequence starting at x_j and d_j is the length of the longest decreasing subsequence starting at x_j.

Suppose that all of the ordered pairs contain only values less than $n+1$ (and greater than 0). Observe first that the pigeon-hole principle implies that the ordered pairs cannot all be distinct. This is because there are only $n \cdot n = n^2$ distinct ordered pairs with entries in $\{1, 2, \ldots, n\}$, but there are $n^2 + 1$ ordered pairs in this collection.

Suppose that $(i_j, d_j) = (i_k, d_k)$ with $j < k$. There are two cases to investigate.

$\boldsymbol{x_j < x_k}$ There must be an increasing sequence starting with x_j and having x_k as second element that contains $i_k + 1 = i_j + 1$ elements, contradicting the maximality of i_j. (Take the sequence of length i_k starting with x_k and prepend x_j.)

$\boldsymbol{x_j > x_k}$ There must be a decreasing sequence starting with x_j and having x_k as second element that contains $d_j + 1$ elements, contradicting the maximality of d_j.

Both cases lead to a contradiction. The assumption that created this situation was the assumption that $1 \le i_j \le n$ and $1 \le d_j \le n$ for all $j \in \{1, 2, \ldots, n^2 + 1\}$. At least one of the i_j or one of the d_j must be larger, proving the proposition. □

PROPOSITION 5.25 ***Divisibility***

Let n be a positive integer and let S be a subset of $\{1, 2, 3, \ldots, 2n - 1, 2n\}$ that contains $n + 1$ elements. Then S contains elements, a and b, such that a divides b.

EXAMPLE 5.42 **Illustrating Proposition 5.25**

Let $n = 4$ and let $S = \{3, 4, 5, 6, 7\}$. Then $a = 3$ and $b = 6$ would satisfy the conclusion of the proposition.

Notice that we can write each of these numbers in the form $2^k \cdot q_j$, where $0 \le k$ and q_j is a product of odd primes. The choices for a and b have the form $3 = 2^0 \cdot 3$ and $6 = 2^1 \cdot 3$. ■

Proof of Proposition 5.25: Write the $n + 1$ elements in S as $2^{k_j} \cdot q_j$ for $j = 1, 2, \ldots, n + 1$ and $0 < q_j < 2n$. The key observation is that there are only n odd numbers in $\{1, 2, \ldots, 2n\}$.

The set $\{q_1, q_2, \ldots, q_{n+1}\}$ contains more elements than there are odd integers in the set of candidate values. This means that at least two of them are the same number. Suppose that $q_v = q_w$. Then S contains two numbers of the form $2^{k_v} \cdot q_v$ and $2^{k_w} \cdot q_v$. One of these numbers divides the other (depending on which of k_v and k_w is larger). □

The pigeon-hole principle will appear again in Section 8.6.2.

5.3.2 Inclusion–Exclusion

Suppose a math teacher looks at her class list and finds that there are 18 computer science majors, 12 mathematics majors, and 3 students who are double majoring in mathematics and computer science. How many students are in the class? One approach is to add the numbers of students in each major, then subtract one copy of the double counts: $18 + 12 - 3 = 27$ students. This is an example of a counting procedure called *inclusion–exclusion.*

THEOREM 5.26 ***Simple Inclusion–Exclusion***

Let S be a finite set with $S = A \cup B$. Then the number of elements in S is

$$|S| = |A| + |B| - |A \cap B|$$

EXAMPLE 5.43 **Counting Cards**

How many cards are there in a standard 52-card deck that are either face cards or Hearts? There are 12 face cards and 13 Hearts. Since 3 of the Hearts are face cards, there are $12 + 13 - 3 = 22$ cards in the collection. ■

Inclusion-exclusion gets more interesting when more than two subsets are involved.

EXAMPLE 5.44 **Music Genres**

A group of friends are comparing their preferences in music. There are five who like country music, four who like classical, and seven who like rock music. There are two people who like classical and rock, two people who like country and classical, and one person who likes country and rock. Only one person likes all three genres. How many friends are in the group?

Following the strategy used previously, we can add the number who like each genre, then subtract double counts: $5 + 4 + 7 - 2 - 2 - 1$. However, there is an error in this expression. Consider the person who likes all three genres. He has been counted once for each of the three categories, then subtracted once for liking classical and rock, once for liking country and classical, and once for liking country and rock. That person has not been counted. We need to add 1 to the total, correctly calculating that there are $5 + 4 + 7 - 2 - 2 - 1 + 1 = 12$ friends.

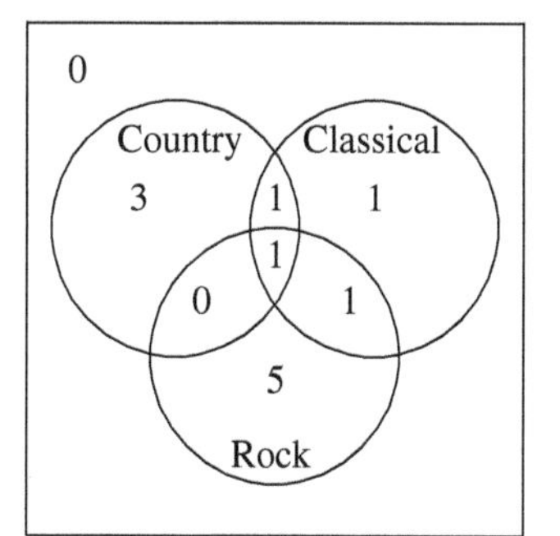

Figure 5.8. Music preferences.

What has been done is to add the number in each single genre, then subtract the common elements from each pair of genres, and finally add the number who are in all genres.

An alternative approach is to use a Venn diagram (Figure 5.8). In this approach, we start with the person who is in all three genres and work outward. For example, since two people like both classical and rock, the region representing classical and rock (but not country) will have one person (thus making the total for classical and rock = 2). The region outside the three circles represents people who do not like any of the three genres. ■

Warning: The Venn diagram approach is limited: If there are more than three or four categories, it is difficult to represent all possible intersections on a diagram using only circles. Replacing circles with ellipses will make it possible to add a few more categories correctly.

THEOREM 5.27 ***Inclusion–Exclusion***

Let $A_1, A_2, \ldots, A_n$ be finite sets. Then

$$\begin{aligned}|A_1 \cup A_2 \cup \cdots \cup A_n| = &\sum_{i=1}^{n} |A_i| - \sum_{1 \le i < j \le n} |A_i \cap A_j| \\ &+ \sum_{1 \le i < j < k \le n} |A_i \cap A_j \cap A_k| \\ &- \cdots \pm |A_1 \cap A_2 \cap \cdots \cap A_n|\end{aligned}$$

Proof: The proof will show that every element on the left-hand side of the equation is counted exactly once by the right-hand side. To that end, consider an element that appears in exactly r of the subsets A_i, for $1 \le i \le n$.

On the right-hand side, the element will be counted $r = \binom{r}{1}$ times in the sum $\sum_{i=0}^{n} |A_i|$ since it appears in exactly r of the subsets $A_1, A_2, \ldots, A_n$. It will appear $\binom{r}{2}$ times in the sum $\sum_{1 \le i < j \le n} |A_i \cap A_j|$ because the r sets it is in will be paired that many times. The right-hand side subtracts this total.

In general, the element will be counted $\binom{r}{s}$ times in the sum

$$\sum_{1 \le i_1 < i_2 < \cdots < i_s \le n} |A_{i_1} \cap \cdots \cap A_{i_s}|$$

for $s \le r$. This total will be subtracted if s is even, and added if s is odd.

The right-hand side therefore counts the element

$$\binom{r}{1} - \binom{r}{2} + \binom{r}{3} - \cdots + (-1)^{s+1}\binom{r}{s} + \cdots + (-1)^{r+1}\binom{r}{r}$$

times.

Exercise 1 on page 257 implies that

$$\binom{r}{1} - \binom{r}{2} + \binom{r}{3} + \cdots + (-1)^{r+1}\binom{r}{r} = \binom{r}{0} = 1$$

This means that the right-hand side counts every element exactly once. □

COROLLARY 5.28 ***Inclusion–Exclusion with a Universal Set***

Let $S = A_1 \cup \cdots \cup A_n$ be a finite set and let $S \subseteq U$. The number of elements in the complement, $\overline{S}$, is

$$\begin{aligned}|\overline{S}| = |U| - &\sum_{i=1}^{n} |A_i| + \sum_{1 \le i < j \le n} |A_i \cap A_j| - \sum_{1 \le i < j < k \le n} |A_i \cap A_j \cap A_k| \\ &+ \cdots \mp |A_1 \cap A_2 \cap \cdots \cap A_n|\end{aligned}$$

Proof: The sets S and $\overline{S}$ are disjoint, so $S \cap \overline{S} = \emptyset$. Theorem 5.26 implies that $|U| = |S \cup \overline{S}| = |S| + |\overline{S}|$. Thus, $|\overline{S}| = |U| - |S|$. Since $|S| = |A_1 \cup A_2 \cup \cdots \cup A_n|$, the expression for $|A_1 \cup A_2 \cup \cdots \cup A_n|$ in Theorem 5.27 can be used to replace $|S|$ in the equation $|\overline{S}| = |U| - |S|$. □

✔ Quick Check 5.9

1. A second set of friends are comparing their preferences in music. There are four who like country music, six who like rock, and three who like classical. They also discovered that two of them like both classical and rock and one likes both classical and country. No one likes both country and rock. Finally, two people do not like any of the genres listed. How many friends are in this group? ✔

EXAMPLE 5.45 **Divisibility**

How many of the integers in $\{1, 2, 3, \ldots, 1000\}$ are divisible by at least one of the primes 2, 3, or 5?

There are $\left\lfloor \frac{1000}{2} \right\rfloor = 500$ numbers that are divisible by 2. There are $\left\lfloor \frac{1000}{3} \right\rfloor = 333$ that are divisible by 3, and $\left\lfloor \frac{1000}{5} \right\rfloor = 200$ that are divisible by 5.

Numbers that are divisible by both 2 and 3 are also divisible by 6. There are $\left\lfloor \frac{1000}{6} \right\rfloor = 166$ such numbers. There are $\left\lfloor \frac{1000}{10} \right\rfloor = 100$ numbers less than or equal to 1000 that are divisible by both 2 and 5. There are $\left\lfloor \frac{1000}{15} \right\rfloor = 66$ divisible by 3 and 5. Finally, there are $\left\lfloor \frac{1000}{30} \right\rfloor = 33$ numbers in $\{1, 2, 3, \ldots, 1000\}$ that are divisible by 2, 3 and 5.

The inclusion–exclusion theorem implies that there are $(500+333+200) - (166 + 100 + 66) + 33 = 734$ numbers in $\{1, 2, 3, \ldots, 1000\}$ that are divisible by at least one of the primes 2, 3, or 5. There are only 266 numbers in that set that are not divisible by any of the three primes. ■

The final result in this section presents a famous function from the field of elementary number theory.

DEFINITION 5.29 ***The Euler Totient Function***

Let n be a positive integer. The *Euler totient function*, denoted $\phi(n)$, is the number of integers in $[1, n]$ that are relatively prime to n.

For example, $\phi(12) = 4$, since 1, 5, 7, 11 are relatively prime to 12.

Exercise 16 on page 180 asked for an algorithm to calculate $\phi(n)$. That algorithm requires a method to determine if two integers are relatively prime. The following theorem provides a better method for calculating $\phi(n)$.

THEOREM 5.30 $\boldsymbol{\phi(n)}$

Let n be a positive integer whose distinct prime divisors are $p_1, p_2, \ldots, p_r$. Then

$$\phi(n) = n\left(1 - \frac{1}{p_1}\right)\left(1 - \frac{1}{p_2}\right)\cdots\left(1 - \frac{1}{p_r}\right)$$

Proof: If $n = 1$, there are no prime divisors, so the expression reduces to $\phi(1) = 1$, which agrees with the definition.

The strategy for finding $\phi(n)$ will be to determine the number, c, of integers in $[1, n]$ that are *not* relatively prime to n and then subtract that total from n. Inclusion-exclusion will be used to calculate c.

Let A_i be the set of integers in $[1, n]$ that are divisible by p_i, for $i = 1, 2, \ldots, r$. Then $c = |A_1 \cup A_2 \cup \cdots \cup A_r|$. The inclusion–exclusion theorem implies that

$$\begin{aligned} c &= \sum_{i=1}^{r} |A_i| - \sum_{1 \le i < j \le r} |A_i \cap A_j| + \sum_{1 \le i < j < k \le r} |A_i \cap A_j \cap A_k| \\ &\quad - \cdots \pm |A_1 \cap A_2 \cap \cdots \cap A_r| \\ &= \sum_{i=1}^{r} \left\lfloor \frac{n}{p_i} \right\rfloor - \sum_{1 \le i < j \le r} \left\lfloor \frac{n}{p_i \cdot p_j} \right\rfloor + \sum_{1 \le i < j < k \le r} \left\lfloor \frac{n}{p_i \cdot p_j \cdot p_k} \right\rfloor \\ &\quad - \cdots \pm \left\lfloor \frac{n}{p_1 \cdot p_2 \cdots p_r} \right\rfloor \\ &= \sum_{i=1}^{r} \frac{n}{p_i} - \sum_{1 \le i < j \le r} \frac{n}{p_i \cdot p_j} + \sum_{1 \le i < j < k \le r} \frac{n}{p_i \cdot p_j \cdot p_k} - \cdots \pm \frac{n}{p_1 \cdot p_2 \cdots p_r} \end{aligned}$$

The floor function, $\lfloor\ \rfloor$, can be dropped because n is divisible by all the p_i. The value we seek is

$$\begin{aligned} \phi(n) = n - c &= n \left(1 - \sum_{i=1}^{r} \frac{1}{p_i} + \sum_{1 \le i < j \le r} \frac{1}{p_i \cdot p_j} \right. \\ &\quad \left. - \sum_{1 \le i < j < k \le r} \frac{1}{p_i \cdot p_j \cdot p_k} + \cdots \mp \frac{1}{p_1 \cdot p_2 \cdots p_r} \right) \\ &= n \left(1 - \frac{1}{p_1}\right)\left(1 - \frac{1}{p_2}\right) \cdots \left(1 - \frac{1}{p_r}\right) \end{aligned}$$

(The final step may be easier to understand in the reverse direction.) □

Since $12 = 2^2 \cdot 3$, Theorem 5.30 asserts that $\phi(12) = 12\left(1 - \frac{1}{2}\right)\left(1 - \frac{1}{3}\right) = 4$, which agrees with the earlier calculation of $\phi(12)$.

5.3.3 Exercises

The exercises marked with ★ have detailed solutions in Appendix H.

1. Many people are paid every other week. Show that there will always be a month in which these people receive three paychecks.
2. (a) How many people need to be gathered to guarantee that at least two have birthdays in the same month?
   (b) ★ How many people need to be gathered to guarantee that at least four have birthdays in the same month?
3. Prove that in any set of three (not necessarily distinct) integers, there will always be two whose sum is even.
4. Suppose that 11 students have contributed money to buy a special present for their teacher. The total amount of money collected was \$29.54. Was it necessary for some student to give at least \$2.70?
5. ★ Suppose there are 15 married couples at a party. All 30 names are placed in a hat and names are picked at random to form a game committee. How many names need to be picked to guarantee that the game committee contains at least one married couple?
6. Consider a room that contains six people. Every pair of people are either friends or are enemies. Prove that there is a subgroup of at least three of the people who are either mutual friends or are mutual enemies. That is, show that there are three people all of whom are friends or there are three people all of whom are enemies. [*Hint*: Pick one special person and consider the friends and enemies of that person.]
7. Suppose the letters A – J are arranged randomly in a string. Prove that after removing up to six well-chosen letters from the string, the remaining letters will either be in alphabetical order or else will be in reverse alphabetical order.
8. Consider the set $\{1, 2, \ldots, 9, 10\}$. Show that if any six of the numbers (all distinct) are chosen, then two of them will add up to 11.
9. Create a sequence of 10 distinct numbers for which there are no increasing subsequences of four or more numbers and no decreasing subsequences of five or more numbers.
10. A group of $n > 1$ people gathered for a party. Everyone shook hands with 0 or more other people. Prove there were at least two people who shook the same number of hands.

11. ★ Prove that in any set of $n + 1$ integers from the set $\{1, 2, 3, 4, \ldots, 2n\}$, two of the numbers must differ by 1.

12. Prove that in every set of 1001 real numbers from the half-open interval $(0, 1]$, at least two of the numbers will differ by less than 0.001.

13. Prove that the decimal expansion of a rational number is a repeating decimal. (A repeating decimal is one for which the decimal expansion eventually starts repeating. For example,

$$\begin{aligned}\frac{123}{42} &= 2.9285714285714285714285714285 71\ldots \\ &= 2.9\overline{285714}\end{aligned}$$

is a repeating decimal. See Appendix A.4 for a bit more discussion about this topic.)

14. Prove that if $k = k_1 + k_2 + \cdots + k_n - n + 1$ objects are placed into n boxes, then for some i with $1 \leq i \leq n$, box i will contain at least k_i objects.

15. Suppose that a party was held for baseball players. However, only players in the following positions were allowed to attend: pitcher, catcher, right field, and left field. Three hundred sixty-four players ended up coming to this party. Additionally, awards were given out to people on the night of the gathering (at most one award per person).

Consider trying to figure out the number of awards that must have been presented to ensure that at least three catchers received awards.

Here is an incorrect solution: *Using the pigeon-hole principle, it is necessary to find the smallest n such that* $\lceil \frac{n}{4} \rceil = 3$, *since there were four positions, and three catchers were to receive awards. Thus,* $n = 9$ *and so 9 players must been given awards to ensure that at least three of them were catchers.*

(a) Find the error in the incorrect solution.

(b) Answer the original question: How many of the 364 players must have received awards to ensure that at least three catchers received awards? [*Hint*: What additional information is needed?]

16. Use simple inclusion–exclusion to count the number of integers between 1 and 100 that include the digit "2".

17. ★ A company wanted to hire some newly graduated college students. They decided to send one of their employees from the personnel department to a job fair. That employee's role was to screen the applicants. Students who passed the screen were invited for interviews. The screening process evaluated three things:

- Was the student's resumé free of grammatical and spelling errors?
- Did the student wear appropriate clothing to the job fair?
- Was the student willing to work for a salary in the range specified by the company?

The employee had contact with 65 students at the job fair. Only 30 of them had mechanically acceptable resumés. There were 44 students who were willing to consider a job at the salary range offered. Clothing was a problem: 37 students *did not* meet the company's dress standards. Some of the students met two of the three standards: 18 met the resumé and dress requirements, 15 met the resumé and salary standards, and 16 met the dress and salary requirements. On the other hand, 5 students failed all three screens.

(a) How many students were offered interviews?

(b) How many students passed the resumé screen but failed the other two?

18. A group of 12 people have gathered for dinner. There are nine U.S. citizens, seven adults, and four who read science fiction. There are five U.S. citizens who are adults and three U.S. citizens who read science fiction. Only three of the adults read science fiction and only two of these adult science fiction readers are U.S. citizens. How many of the children are neither U.S. citizens nor readers of science fiction?

19. A seventh-grade math teacher decides that she would like to treat her two classes by bringing in ice cream. She does not want to have to worry about bringing in ice cream toppings, so she chooses three extremely flavorful types of ice cream as candidates: strawberry-blackberry cheesecake, chocolate caramel brownie, and toffee fudge swirl. She conducts a survey of her 79 students by asking them which of the three candidate ice creams they like. Thirty-eight students claim to enjoy strawberry-blackberry cheesecake. Only 20 students report to disliking chocolate caramel brownie ice cream, while only 29 say that they do not enjoy toffee fudge swirl. Additionally, 26 students say that the flavors of strawberry-blackberry cheesecake and chocolate caramel brownie ice cream are good, while 17 students are fond of strawberry-blackberry cheesecake and toffee fudge swirl. Interestingly, 5 students claim that ice cream is their favorite food and believe that all three candidate flavors sound wonderful. A few students, 2 to be exact, say that they dislike all three of the teacher's ice cream options.

(a) How many of the seventh-grade students like both chocolate caramel brownie and toffee fudge swirl ice cream?

(b) How many of the seventh-grade students who like at least one of the three candidate flavors would be disappointed if the teacher brought in toffee fudge swirl ice cream?

20. An employer decides to give his employees gift bags as a token of appreciation for all their hard work. He chooses to place three types of items in the bags: a gift certificate to a restaurant, movie tickets, and a T-shirt featuring the company name. The employer does not necessarily wish to give the same gift bag to every employee. The problem the employer faces is that after filling and delivering the gift bags rather quickly, he begins to question whether or not he delivered some empty bags (i.e., he may have forgotten to put at least one item in each bag). Of his 101 employees, 46 ended up with a gift certificate, 50 did not end up with movie tickets, and 53 ended up with a T-shirt. Some lucky employees received two of the possible gift bag items: 20 received a gift certificate and a T-shirt, 25 received movie tickets and a T-shirt, and 26 received movie tickets and a gift certificate. Fourteen extremely fortunate employees received all three gift bag items. Did any employees receive solely an empty gift bag or did every employee get at least one item?

21. Consider the set, S, with $S = A_1 \cup A_2 \cup A_3 \cup A_4 \cup A_5$. Use the inclusion–exclusion theorem to generate a formula for the number of elements in the union of the five subsets of S, given that no three of these sets have any elements in common (i.e., the intersection of any three or more sets is empty). Do not include any unnecessary terms in the formula.

22. ★ Use inclusion–exclusion to calculate the number of bit strings of length 7 that do not contain a sequence of five consecutive 1s.

23. Use inclusion–exclusion to calculate the number of bit strings of length 9 that either begin with two 0s, have eight consecutive 0s, or end with a 1 bit.

24. How many integers in $[1, 1000]$ are divisible by at least one of the primes 2, 3, 5, or 7?

25. Example 5.45 and the proof of Theorem 5.30 both used the expression $\lfloor \frac{n}{r} \rfloor$ to represent the number of integers in $[1, n]$ that are divisible by r. Prove that this is correct.

26. Calculate $\phi(n)$ for each n. In each case, verify your answer by listing all the numbers in the interval $[1, n]$ that are relatively prime to n.
(a) 16 (b) 19 (c) 28 (d) 30

27. Show that if n is odd, then $\phi(2n) = \phi(n)$.

28. A bridge hand consists of 13 cards from a standard 52-card deck. How many bridge hands contain a void suit? (A hand contains a void—or missing—suit if none of the cards in the hand are from that suit. Thus, a hand with no hearts contains a void suit.)

29. Use a proof by contradiction to prove the pigeon-hole principle. Do not assume that objects are evenly distributed among the boxes.

30. Recall Problem 33 in Exercises 5.1.8 on page 247. That problem concerned five male college students who bring dates to an orchestra concert. The men bought tickets for 10 consecutive seats in row 17. In how many ways can all the college students be seated at the concert if no two people on a date can sit next to each other? [*Hint*: Let A_i be the set of all seating arrangements in which the members of couple i sit next to each other.]

5.4 Quick Check Solutions

Quick Check 5.1

1. The numbers are

wives	7
sacks	$7^2 = 49$
cats	$7 \cdot 49 = 7^3 = 343$
kittens	$7 \cdot 343 = 7^4 = 2401$
Total	2800

Quick Check 5.2

1. The following table lists all possibilities in a clear manner.

	Bread			
Meat	B_1M_1	B_2M_1	B_3M_1	B_4M_1
	B_1M_2	B_2M_2	B_3M_2	B_4M_2
	B_1M_3	B_2M_3	B_3M_3	B_4M_3
	B_1M_4	B_2M_4	B_3M_4	B_4M_4

2. Using the Mutually Exclusive Tasks Principle, there are $4 + 5 = 9$ ways.

3. The Independent Tasks Principle implies that there are $9 \cdot 3 = 27$ distinct combinations.

4. Since the number of available colors depends on the choice of upholstery, the choices are not independent. Since choosing an exterior color does not exclude the choice of an interior color, the choices are not mutually exclusive. Neither of the basic counting principles applies.

   A careful analysis of the problem does provide a way to use the principles. There are three choices: exterior color, interior color, and upholstery type. If the upholstery type is selected first, we can use the Mutually Exclusive Tasks Principle: The car will either have cloth or leather upholstery, but not both. Thus, if we can count the number of ways, C, to have cloth upholstery and the number of ways,

L, to have leather upholstery, the total number of choices will be $C + L$. The Independent Tasks Principle implies that $C = 9 \cdot 3 = 27$ and $L = 9 \cdot 2 = 18$. The number of choices is therefore $27 + 18 = 45$.

The process can be visualized using the following tree diagram. The diagram represents a set of mutually exclusive decisions as a branch (or fork) in the tree. Independent choices are represented by successive arrows along a path. The number of ways to make a choice is recorded below each arrow. The numbers along a path are multiplied (the Independent Tasks Principle). Finally, the numbers at the end of distinct branches are added (the Mutually Exclusive Tasks Principle).

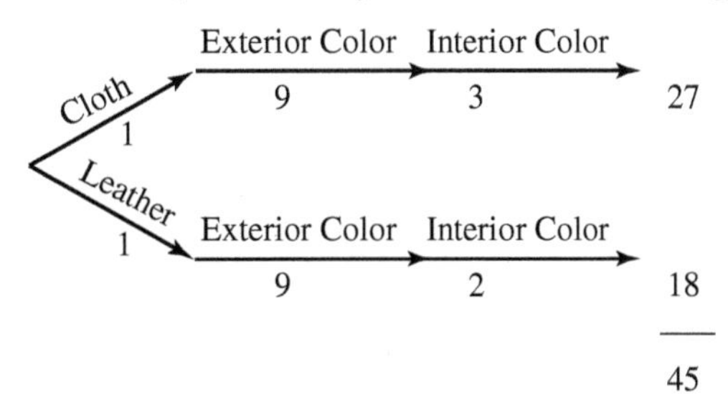

Quick Check 5.3

1. $P(9, 4) = 9!/(9 - 4)! = 9!/5! = 3024$

2. One way to solve this is to write the names of the three managers in a row. Then write the vacation weeks below the names. For example,

Jane	Tom	Pin
2	4	1

would represent Jane being on vacation during week 2, Tom during week 4, and Pin during week 1. Clearly, it matters to the managers which order the numbers are written in, and repetition is not allowed. The problem can be solved by counting the number of ways to arrange three out of four numbers in order ($n = 4$, $r = 3$). This is $P(4, 3) = 4!/(4 - 3)! = 4!/1! = 24$.

3. This is a permutation with repetition. (A girl as oldest child does not exclude another girl being a member of the family.) There are $n = 2$ genders and $r = 8$ children, so $2^8 = 256$ distinct arrangements.

Quick Check 5.4

1. $C(8, 6) = \dfrac{8!}{6! \cdot (8 - 6)!} = \dfrac{8 \cdot 7 \cdot \cancel{6} \cdot \cancel{5} \cdot \cancel{4} \cdot \cancel{3} \cdot \cancel{2} \cdot \cancel{1}}{(\cancel{6} \cdot \cancel{5} \cdot \cancel{4} \cdot \cancel{3} \cdot \cancel{2} \cdot \cancel{1}) \cdot 2 \cdot 1} = 4 \cdot 7 = 28$

2. (a) $C(10, 4) = \dfrac{10!}{4! \cdot 6!} = 210$

   (b) $P(10, 4) = \dfrac{10!}{6!} = 5040$

3. The number 133,784,560 needs nine digits to display. The calculator stores the full number internally but needs to convert to scientific notation to display. When I divide by 10, the internal number becomes 13,378,456, which requires only eight digits to display.

   This is not a trick that works in general. For example,

$$15! = 1{,}307{,}674{,}368{,}000$$

which the calculator displays as 1.3077×10^{12}. If I divide by 100,000, the calculator displays 13,076,744, which is incorrect even if I manually append five zeros. The error occurs because, for this number, the calculator is unable to store the exact number internally.

Quick Check 5.5

1. Let the four objects be "a", "b", "c", and "d".

   (a) aa, ab, ac, ad, bb, bc, bd, cc, cd, dd

   (b) $C(4+2-1, 2) = C(5, 2) = \dfrac{5!}{2! \cdot 3!} = 10$

2. There are $n = 6$ project categories and $r = 3$ projects. Since repetition of categories is permissible, the solution is

$$C(6+3-1, 3) = C(8, 3) = \frac{8!}{3! \cdot 5!} = 56$$

Quick Check 5.6

1. 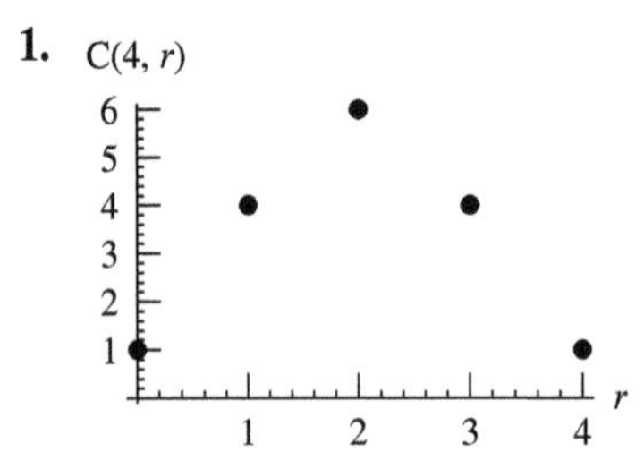

2. 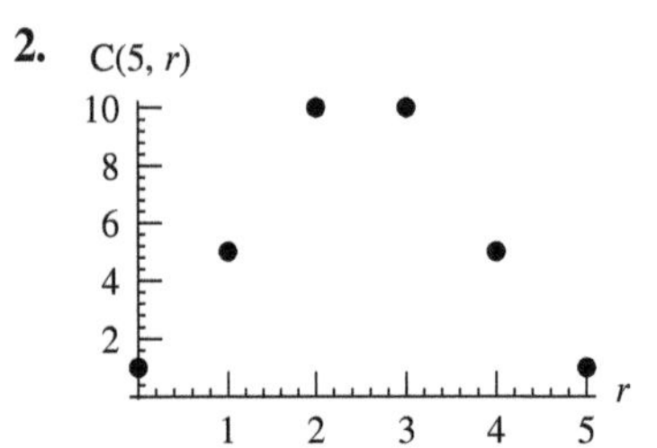

3. There are 26 choices for the first letter and then 36^4 choices for the remaining characters. Thus there are $26 \cdot 36^4 = 43,670,016$ codes.

   An alternate packaging of the original solution is to use a tree diagram:

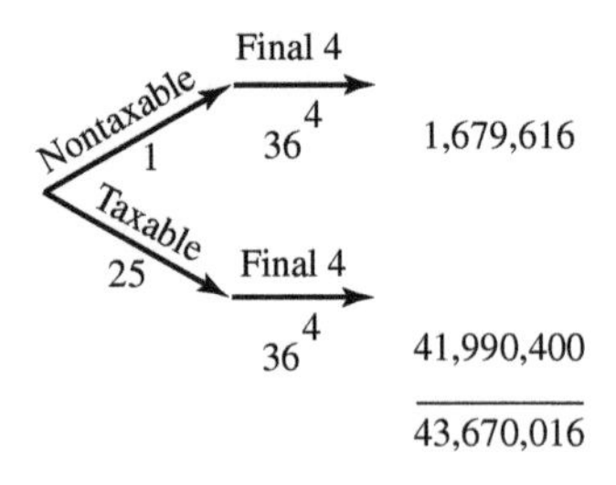

4. There are nine sources for the election-day listening and two sources for the next day reading. Listening and reading are independent, so there are 18 pairs.

5. There are $C(5, 2)$ ways to pick a pair of television stations and $C(4, 2)$ ways to pick a pair of radio stations. The choice of television versus radio leads to mutually exclusive alternatives. The choice of newspaper is independent of the first choice. Combining the two basic counting principles (Section 5.1.1) leads to either of the following two expressions. (Describe in words the difference in strategies.)

$$(C(5, 2) + C(4, 2)) \cdot 2 = 32$$
$$C(5, 2) \cdot 2 + C(4, 2) \cdot 2 = 32$$

   A tree diagram could also be used. The second expression would produce a tree similar to the previous diagrams in this set of solutions.

Quick Check 5.7

1. Because the groups are distinguishable, the multinomial counting theorem (version 2) applies. The groups can be formed in

$$\text{multinomial}(4, 4, 4, 4, 4, 4, 5) = 385{,}558{,}359{,}634{,}498{,}050{,}000$$

ways.

2. The multinomial counting theorem assumes that the groups are distinguishable. For this problem, once the teams have been chosen, they are not additionally distinguished by being given identifying labels. The multinomial theorem counts each collection of six four-member teams 6! times. (The same six groups can be lined up in 6! ways without changing the collection of groups.) The number of groups for this problem is thus a mere

$$\frac{\text{multinomial}(4, 4, 4, 4, 4, 4, 5)}{6!} = 535{,}497{,}721{,}714{,}580{,}625$$

3. Highly unlikely. The purpose of problems like these is to help you master the concepts. There is no claim that these particular problems are practical applications of the multinomial counting theorem.

Quick Check 5.8

1. (a) Algebraic proof 1:

$$\begin{aligned} r \cdot \binom{n+1}{r} &= \frac{r \cdot (n+1)!}{r! \cdot (n+1-r)!} = \frac{(n+1)!}{(r-1)! \cdot (n+1-r)!} \\ &= (n+1) \cdot \frac{n!}{(r-1)! \cdot (n+1-r)!} \\ &= (n+1) \cdot \frac{n!}{(r-1)! \cdot (n-r+1)!} \\ &= (n+1)\binom{n}{r-1} \end{aligned}$$

Algebraic proof 2:

$$\frac{\binom{n+1}{r}}{\binom{n}{r-1}} = \frac{(n+1)!}{r! \cdot (n+1-r)!} \cdot \frac{(r-1)! \cdot (n-r+1)!}{n!} = \frac{n+1}{r}$$

Now cross-multiply.

(b) Two tasks need to be accomplished: (1) Choose a committee, and (2) appoint one of the committee members to be chair. These tasks can be done in either order.

committee first The committee members can be chosen in $\binom{n+1}{r}$ ways. There are then r committee members that are available for the position of chairperson. The two tasks can therefore be accomplished in $\binom{n+1}{r} \cdot r$ ways.

chairperson first The chairperson can be any one of the $n + 1$ people in the group. Once the chairperson has been chosen, there are n people left from which to choose the remaining $r - 1$ committee members. Consequently, there are $(n + 1) \cdot \binom{n}{r-1}$ ways to accomplish the two tasks.

Quick Check 5.9

1. Let W be the set of people who like country (country and western) music, C be the set of people who like classical music, and R be the set of people who like rock music. Finally, let N be the set of people who do not like any of the three genres of music. Notice that any intersection involving N is empty. In addition, since no person likes both country and rock, the intersection $W \cap C \cap R$ must also be empty. The inclusion–exclusion theorem gives the value

$$\begin{aligned}|W \cup C \cup R \cup N| &= |W| + |C| + |R| + |N| \\ &\quad - |W \cap C| - |W \cap R| - |W \cap N| - |C \cap R| \\ &\quad - |C \cap N| - |R \cap N| + |W \cap C \cap R| + |W \cap C \cap N| \\ &\quad + |W \cap R \cap N| + |C \cap R \cap N| - |W \cap C \cap R \cap N| \\ &= (4 + 3 + 6 + 2) - (1 + 0 + 0 + 2 + 0 + 0) \\ &\quad + (0 + 0 + 0 + 0) - 0 \\ &= 12\end{aligned}$$

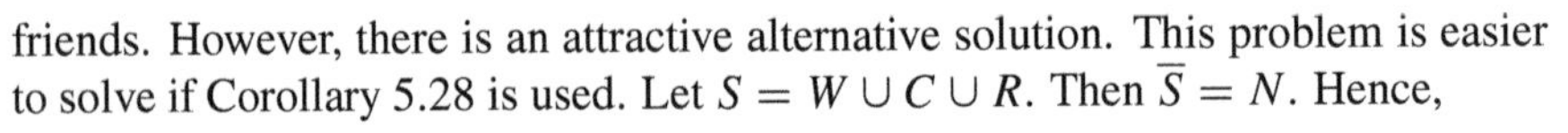

friends. However, there is an attractive alternative solution. This problem is easier to solve if Corollary 5.28 is used. Let $S = W \cup C \cup R$. Then $\overline{S} = N$. Hence,

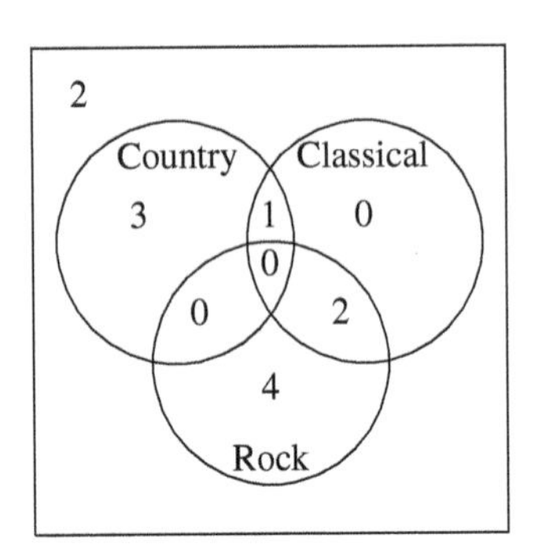

$$\begin{aligned}2 &= |N| \\ &= |U| - |W| - |C| - |R| + |W \cap C| + |W \cap R| + |C \cap R| - |W \cap C \cap R| \\ &= |U| - (4 + 3 + 6) + (1 + 0 + 2) - 0 \\ &= |U| - 10\end{aligned}$$

Consequently, $|U| = 12$.

A Venn diagram can also express the information.

5.5 CHAPTER REVIEW

5.5.1 Summary

This chapter introduces the basic collection of counting techniques. (Additional techniques can be found in Section 8.1.) Two basic counting principles are presented: the Independent Tasks Principle (multiplication) and the Mutually Exclusive Tasks Principle (addition). The Pigeon-Hole Principle is also presented.

Several counting formulas are then derived from the basic principles. The two permutation formulas count the number of ways r items can be chosen, in order, from a set of size n. The two combination formulas count the number of ways that a subset of size r can be chosen from a set of size n. (Recall that subsets are never ordered.) There are two formulas in each case because the counts will differ depending upon whether items can be chosen more than once (repetition) or only once. The following table summarizes these four formulas.

Arranging r elements from a set containing n distinct elements

	With Order	**Without Order**
Without Repetition	$P(n, r) = \dfrac{n!}{(n-r)!}$	$C(n, r) = \dfrac{n!}{r! \cdot (n-r)!}$
With Repetition	n^r	$C(n + r - 1, r) = \dfrac{(n + r - 1)!}{r! \cdot (n-1)!}$

Two additional counting formulas are also introduced: the multinomial formula and the inclusion–exclusion formula. Three interpretations of the multinomial formula are presented. It can be used to calculate the coefficients in a multivariable polynomial expansion (Theorem 5.15 on page 251). The multinomial formula can also be used to count visually distinguishable patterns for collections of objects with several identifiable types of objects, each with several copies available (Theorem 5.7 on page 243). Finally, the multinomial formula can be used to count the number of ways to partition a distinguishable set into distinguishable subsets with predetermined size constraints (Theorem 5.8 on page 244). The inclusion–exclusion formula counts the number of distinct items in a union of possibly nondisjoint sets.

Keep in mind that some problems do not fit neatly into just one of these categories. You may need to combine more than one counting technique to derive the correct count.

This chapter also includes a discussion of the combinatorial proof technique. That technique is appropriate in a chapter devoted to counting because it works by counting something in two different ways and then equating the two answers. Many useful theorems were introduced in that section. For example, Pascal's theorem

$$\binom{n+1}{r} = \binom{n}{r-1} + \binom{n}{r}$$

and the Binomial theorem

$$(x+y)^n = \sum_{r=0}^{n} \binom{n}{r} x^{n-r} y^r$$

Students have differing strategies for mastering all these formulas. One strategy is to memorize every formula, together with a few clues to help decide when each formula might be appropriate. Another strategy is to memorize the two basic counting principles and also the derivation of all the other formulas. If this is understood, then each new problem can be solved by applying the appropriate principle (and hence essentially re-deriving the other formulas each time). This second approach is actually not as difficult as it may first seem. Many people find it is actually easier to apply correctly than the "memorize everything" approach. It also is possible to employ an approach that blends these two extremes.

5.5.2 Notation

Notation	Page	Brief Description
$n!$	99	(footnote 3.20) n factorial
$P(n, r)$	229	the number of permutations of size r from a set of size n (no repetition)
$C(n, r)$	231	the number of combinations of size r from a set of size n (no repetition)
$\binom{n}{r}$	231	the binomial coefficient notation for $C(n, r)$
$\text{multinomial}(n_1, n_2, \ldots, n_k)$	243	the multinomial counting formula (see Theorems 5.7 and 5.8 for interpretations)
$\phi(n)$	263	The Euler totient function

5.5.3 Some Counting Formulas

The Independent Tasks Principle

If a project can be decomposed into two independent tasks with n_1 ways to accomplish the first task and n_2 ways to accomplish the second task, then the project can be completed in $n_1 \cdot n_2$ ways.

The Mutually Exclusive Tasks Principle

If a project can be decomposed into two mutually exclusive tasks with n_1 ways to accomplish the first task and n_2 ways to accomplish the second task, then the project can be completed in $n_1 + n_2$ ways.

TABLE 5.2 Arranging *r* elements from a set containing *n* distinct elements

	With Order	**Without Order**
Without Repetition	$P(n, r) = \dfrac{n!}{(n-r)!}$	$C(n, r) = \dfrac{n!}{r! \cdot (n-r)!}$
With Repetition	n^r	$C(n+r-1, r) = \dfrac{(n+r-1)!}{r! \cdot (n-1)!}$

THEOREM 5.7 ***The Multinomial Counting Theorem—Version 1***

Suppose there exists a set of n items containing n_1 *identical* items of type 1, n_2 *identical* items of type 2, ..., and n_k *identical* items of type k, where $n = n_1 + n_2 + \cdots + n_k$. The number of visually distinguishable ways to arrange the n items in a row is

$$\text{multinomial}(n_1, n_2, \ldots, n_k) = \frac{n!}{n_1! \cdot n_2! \cdots n_k!}$$

THEOREM 5.8 ***The Multinomial Counting Theorem—Version 2***

Suppose there exists a set of n *distinguishable* items that is to be partitioned into k *distinguishable* subsets. Subset 1 will have n_1 items, subset 2 will contain n_2 items, ..., and subset k will contain n_k items, where $n = n_1 + n_2 + \cdots + n_k$. The number of ways to accomplish this partitioning is

$$\text{multinomial}(n_1, n_2, \ldots, n_k) = \frac{n!}{n_1! \cdot n_2! \cdots n_k!}$$

5.5.4 Additional Review Material

Go to http://www.mathcs.bethel.edu/~gossett/DiscreteMathWithProof/review.xhtml for additional review material, including a sample chapter exam, with solutions.

CHAPTER

6

Finite Probability Theory

The probability that we may fail in the struggle ought not to deter us from the support of a cause we believe to be just.

Abraham Lincoln (1809 – 1865)

One of the features of the universe that we often observe is an apparent randomness behind many events. For example, the amount of time between phone calls at a receptionist's phone appears to be random. The number that appears when we roll a pair of fair dice also appears to be random. Whether you believe the universe is a collection of random events or is a completely determined series of events, each individual experiences many events that appear to be random. Mathematics is capable of modeling both *deterministic* and *random* processes (or combinations of the two). The mathematical discipline of probability theory provides models for describing random events.

We are all familiar with predictions that reflect uncertainty. These include weather forecasts, the odds for winning a sweepstakes, the probable number of deaths on the highway next week, and the relative likelihoods of getting lung cancer if you are a heavy smoker or a nonsmoker.

A more subtle example is the results of an opinion poll. A number of people are surveyed. The poll's results are assumed to be a good approximation to the opinions of the whole country. However, there is always a (hopefully very small) chance that the results are completely wrong. For example, the poll takers might accidentally survey the only 10,000 people in the country who *do not* want popcorn to become the national snack. The poll would then reflect an (incorrect) overwhelming consensus against popcorn.

To better utilize models of uncertainty, the differences in human personality can also be considered. Some people are risk takers; others are risk avoiders. Most are somewhere in between. Because of this difference, two people may decide on opposite courses of action, based on the same estimates of an event's likelihood. For example, one may choose to invest in a scheme with a 40% chance of a high return on investment but a 60% chance of a moderate loss. Another person may feel these odds are not good enough to warrant the investment.

Even when we study randomness and uncertainty, patterns emerge. There is a different characteristic to the random distribution of people's heights and the distribution of time lag between phone calls to the reference desk of a library. Mathematicians model these differing patterns by using *probability distributions*.

By using these distributions, it is often possible to make very accurate predictions of the collective behavior of a large number of random individual components. For example, individual gas molecules may behave in highly unpredictable ways, but the collection of many gas molecules in a bottle will have a very predictable aggregate behavior. Thus we can accurately predict the rate of gas expansion given a known temperature change, even though we may not be able to predict the motion of any individual molecule.

The purpose of this chapter is to familiarize you with some of the elementary definitions and concepts from discrete probability theory. You may enroll in a course in probability and statistics to learn about continuous probability distributions.

6.1 The Language of Probabilities

The initial purpose of this section is to introduce a number of basic definitions. The definitions relate to some activity that will produce a random result. For example, the activity might be to flip a coin, or to wake up and observe the temperature. Such an activity is often called a *random experiment*. The first three definitions are highly interrelated. They describe the objects in our model.

6.1.1 Sample Spaces, Outcomes, and Events

DEFINITION 6.1 ***Sample Space, Outcome***
A *sample space* is the set of all *outcomes*. The outcomes are an exhaustive collection of the possible results of some random experiment. Sample spaces are often denoted by S.

EXAMPLE 6.1

Flipping Coins

The sample space for the experiment of flipping a coin contains two outcomes: heads and tails. We can denote the sample space by {H, T}. The sample space for flipping a coin twice (or flipping two coins simultaneously) is {HH, HT, TH, TT}. Thus, in this experiment, there are four outcomes.

Many people find tree diagrams to be helpful tools for constructing sample spaces. Figure 6.1 shows the steps needed to construct the sample space for flipping a coin twice. On the first flip, either a head or a tail will occur. The two lines on the left show this step. The next flip will also result in either a head or a tail. The upper right lines show the possible results of the second flip, given that the first flip is a head. The lower right lines show the possible results of the second flip, given that the first flip is a tail. The four possible outcomes are shown along the right edge.

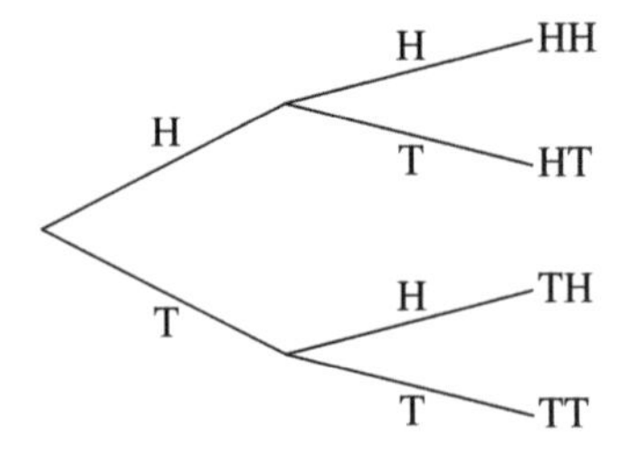

Figure 6.1. Flipping a coin twice.

In the experiment of flipping a coin twice, you might have guessed that the sample space should be {HH, HT, TT}, since a head and a tail seems to be the same as a tail and a head. However, for computing the likelihood of obtaining a head and a tail from two flips, the four-element sample space utilizes simpler computations (using the techniques presented shortly). ■

EXAMPLE 6.2

Rolling a Die

The sample space for the experiment of rolling a standard six-sided die consists of six outcomes: $\{1, 2, 3, 4, 5, 6\}$. ■

EXAMPLE 6.3

Having Children

The sample space for a family with three children can be defined as the eight-element set {BBB, BBG, BGB, BGG, GBB, GBG, GGB, GGG}. The tree diagram in Figure 6.2 illustrates the construction of this sample space. Notice that I have focused on the gender of the children. This sample space is designed to answer questions, such as How likely is it for the children in a family of three children, randomly chosen from among all families with three children, to consist of two girls and one boy?

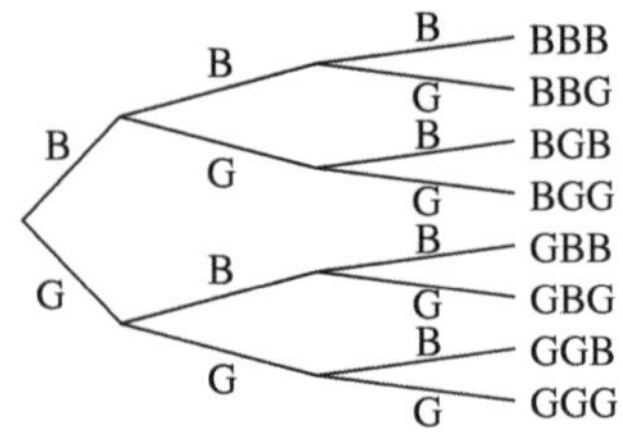

As will be seen later, it will be easier to compute correctly the probability that a family consists of two girls and one boy if we keep the outcomes GGB, GBG, and BGG distinct (in essence, distinguishing the children by birth order). ■

Figure 6.2. A family with three children.

It should be clear that the sample space (and the outcomes it contains) is a set that we are free to choose. Our choice will influence the effectiveness of the model we are constructing.

✔ Quick Check 6.1

1. List the sample space (and hence, all the possible outcomes) of the random experiment "pick a straw". Assume that there are three straws with differing lengths. One person displays the tops of the straws, hiding the lower portions so that all appear to be of equal length. A second person chooses one of the straws.
2. In the game paper, rock, scissors, two opponents each randomly choose one of the listed objects by using one hand to imitate the chosen object. Scissors can cut paper, so in a scissors–paper match scissors is the winning choice. A rock can damage a pair of scissors, so in a scissors–rock match the rock wins. Since paper can wrap a rock, the paper wins in a paper–rock match. If both opponents choose the same object, there is no winner. Each match is performed by counting to 3 together, then displaying the choices simultaneously on "3".
   (a) Design a useful notation for the sample space that consists of a single person's choices. Is there any significant difference between this sample space and the space for "pick a straw" in problem 1?
   (b) Create a representation of the sample space for a single match in paper, rock, scissors. The outcomes are the possible (two-person) results of a single match. ✔

The next definition extends the usefulness of the outcome concept.

DEFINITION 6.2 ***Event***
An *event* is a set of outcomes.

Notice that an event is a subset of the sample space. Recall that the sample space is the set of *all* outcomes. An event can contain any number of outcomes, including just one or none.

When examining the result of a random experiment, we say that an event has occurred if the resulting outcome is in the event. An outcome in the event is called a *favorable outcome* (with respect to that event). Notice that performing the experiment produces a single outcome. All events containing that outcome are said to have occurred.

EXAMPLE 6.4 **Flipping Coins Again**

There are several natural events we can associate with the random experiment of flipping a coin twice. One such event is the event "both the same," which is represented by the set of outcomes {HH, TT}. Another natural event is "both heads," represented by {HH}.[1] ■

EXAMPLE 6.5 **Childhood Events**

Some events built from the sample space of three child families are as follows:

- Two girls, one boy {GGB, GBG, BGG}.
- All boys {BBB}.
- At least one of each gender {BBG, BGB, GBB, GGB, GBG, BGG}.
- All the same gender {BBB, GGG}.

[1] Notice the technical distinction between the *outcome* HH and the *event* {HH}.

Suppose, for each family in the world having three children, we place a card with their names and address in a large drum. A giant shakes the drum and chooses one card. If the lucky family has the pattern BGG, then the events "two girls, one boy" and "at least one of each gender" both occurred. The events "all boys" and "all the same gender" did not occur. ■

It is customary to denote events by an uppercase letter. Thus, the event "roll an even number" associated with the experiment of rolling a die might be denoted E. In most situations where probabilities are of interest, it is the probabilities of events that are desired rather than probabilities of outcomes.

✔ Quick Check 6.2

1. List the event NW: "no winner" in the game paper, rock, scissors.
2. List the event O_1: "opponent 1 wins" in the game paper, rock, scissors. ☑

The next two definitions are related to events.

DEFINITION 6.3 ***Complement of an Event***
The *complement of an event* is an event consisting of all outcomes in the sample space, S, which are not part of the original event. The complement of an event, E, is denoted $\overline{E}$.

A simple example: The complement of the event "at least one of each gender" is "all the same gender".

DEFINITION 6.4 ***Mutually Exclusive Events***
Two events are said to be *mutually exclusive events* if they have no outcomes in common. Another way of defining this is to call two events mutually exclusive if they cannot both occur simultaneously.

Definition 6.4 can be illustrated with a Venn diagram (Figure 6.3).

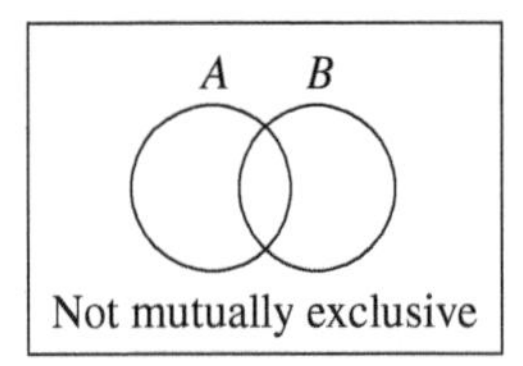

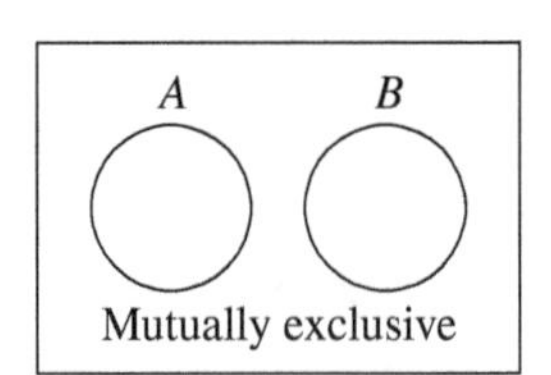

Figure 6.3 Mutually exclusive events.

Notice that two events that each contain a single (distinct) outcome are mutually exclusive.

EXAMPLE 6.6

Picking a Card

Consider the experiment in which you thoroughly shuffle a standard deck of cards,[2] then choose a card at random. The events H: "a heart" and S: "a spade" are mutually exclusive. The events H and F: "a face card" are not mutually exclusive since they contain the jack, queen, and king of hearts in common. ■

[2] See page 232 for a description of a standard deck of cards.

✔ Quick Check 6.3

1. What is the complement, $\overline{O_1}$, of the event O_1: "opponent 1 wins" in the game paper, rock, scissors?
2. List the event PC: "paper was chosen by at least one opponent."
3. Are the events O_1 and NW: "no winner" mutually exclusive?
4. Are the events PC and O_1 mutually exclusive? ☑

6.1.2 Probabilities of Events

With each outcome or event that can be derived from a sample space, we wish to associate a number that measures the likelihood of that outcome or event occurring. The notation is established in the next definition. The mechanisms for determining the value of the number will be presented later.

DEFINITION 6.5 ***The Probabilities of Outcomes and Events***

We denote the probability of an outcome, θ, by $\mathbf{P}(\theta)$, and the probability of an event, A, by $\mathbf{P}(A)$.

The *probability of an event*, A, is a number between 0 and 1, inclusive, which reflects the likelihood that event A occurs. That is,

$$0 \leq \mathbf{P}(A) \leq 1$$

In addition, it is required that the sum of the probabilities of all outcomes in the sample space add to 1. We can express this as

$$\sum_{\theta \in S} \mathbf{P}(\theta) = 1$$

For a discrete sample space, if $\mathbf{P}(A) = 0$, we say A is *impossible*. If $\mathbf{P}(A) = 1$, we say A is *certain*.

For consistency, we require $\mathbf{P}(\{\theta\}) = \mathbf{P}(\theta)$. Thus, $\mathbf{P}(\{H\})$ (the probability of the *event* "heads") is the same as $\mathbf{P}(H)$ (the probability of the *outcome* "heads").

If we flip a fair coin, we expect $\mathbf{P}(H) = \mathbf{P}(T) = \frac{1}{2}$. Even if the coin is not fair, we still require $\mathbf{P}(H) + \mathbf{P}(T) = 1$.

One assumption that makes it easier to compute probabilities is the following:

DEFINITION 6.6 ***Equally Likely Outcomes***

A sample space has *equally likely outcomes* if every outcome has the same probability.

EXAMPLE 6.7 **Probabilities with a Fair Die**

If we roll a fair die, then each outcome is equally likely (in fact, that is exactly what we mean by "fair die"). Thus $\mathbf{P}(1) = \mathbf{P}(2) = \mathbf{P}(3) = \mathbf{P}(4) = \mathbf{P}(5) = \mathbf{P}(6) = \frac{1}{6}$. All six probabilities must be equal, and their sum must equal 1. ■

EXAMPLE 6.8 **Outcomes that Are Not Equally Likely**

The three child family example is not modeled most effectively by assuming equally likely outcomes. There are actually slightly more boys than girls conceived. Thus the outcome BBB should be more likely than the outcome GGG.[3] ■

[3] I am assuming (unrealistically) that all children live to adulthood. In actual fact, the infant mortality rate is higher for boys. This also influences the probabilities of family composition.

Methods for Determining Probabilities

There are three common methods for specifying how probabilities should be determined: theoretical, empirical, and subjective. Deciding which method is most appropriate is largely determined by the sample space and by how well we understand the mechanisms that govern the random experiment of interest.

Theoretical Theoretical methods assume a high degree of knowledge about the mechanisms that govern the random experiment. The determination of probabilities is simplest when the sample space contains equally likely outcomes.

- ***Equally likely outcomes*** If the sample space has n outcomes, the probability of an outcome is $1/n$, for all outcomes. We determine the probability of an event by the proportion of favorable outcomes in the sample space. This can be expressed as

$$\mathbf{P}(E) = \frac{\text{the number of outcomes in } E}{\text{the number of outcomes in the sample space}}$$

- ***Unequally likely outcomes*** The probability of each outcome must be determined by some theoretical consideration that will be determined by the context.[4] In this case the probability of an event is the sum of the probabilities of the outcomes in that event.[5] This can be expressed as

$$\mathbf{P}(E) = \sum_{\theta \in E} \mathbf{P}(\theta)$$

Empirical There are many situations for which we have no theoretical basis for determining the probabilities of the outcomes in the relevant sample space. For example, the probability of the gender of a child cannot be easily determined just prior to conception. It is useful in these situations to determine the probabilities empirically. In the gender example, we could measure the relative frequencies of male and female babies born during a 5-year period at 100 different hospitals. Those relative frequencies will not be exact probabilities, but they may be close enough for most purposes.

Subjective In some situations, we have neither a theoretical basis nor past records from which to determine the probabilities of outcomes. In those situations, the best that can be done initially is to make educated guesses. Future records might provide a basis for how well we guessed.

EXAMPLE 6.9 **Flipping Coins**

A theoretical approach to flipping a coin is to assume that heads and tails are equally likely. In that case, $\mathbf{P}(H) = 1/2 = \mathbf{P}(T)$.

If the coin is not fair, we could flip it 1000 times, recording the number of heads and tails. If 592 heads occurred, then we could empirically estimate that $\mathbf{P}(H) = 592/1000 = 0.592$. ■

EXAMPLE 6.10 **Random Tracks**

Some CD players can be set to randomly play selections (in an equally likely manner). Suppose the current disk has 12 tracks (selections). Suppose also that you like 8 of the selections but you dislike 2 of the selections. You are indifferent about the remaining 2 selections.

---

[4]See Example 6.13 on page 279 for an illustration.

[5]The previous case (equally likely outcomes) also fits this definition, since for any equally likely outcome, $\mathbf{P}(\theta) = 1/n$.

What is the probability that the first randomly selected track contains a selection you like? The total number of outcomes is 12 and the number of favorable outcomes is 8, so the probability is $8/12 = 2/3$.

What is the probability that the first randomly selected track contains a selection you dislike? The total number of outcomes is 12 and the number of favorable outcomes is 2, so the probability is $2/12 = 1/6$.[6] ■

EXAMPLE 6.11

A New Product

Suppose you are head of product development at a medium-sized corporation. Your development team has conceived of a radically new product. It is not clear whether there is a sufficient market for this product. You need to decide the probability of high, moderate, or low future sales. The best you can do initially is compare the success of previous radical new ideas your company (or the competition) has tried, then make a guess (a subjective probability).

If your initial estimate assigns a large probability to "low future sales", you will most likely discontinue development of the product. However, if you feel the combined probabilities of moderate and high sales are large, it would be worth the expense to do some market research (using techniques of inferential statistics) to refine your subjective estimate. ■

EXAMPLE 6.12

A Candy Jar—Equally Likely Outcomes

I have a candy jar that I keep stocked with four flavors of candy. The jar contains three licorice (my favorite), four cinnamon, four butterscotch, and five watermelon flavored candies.

Suppose I want to know the probabilities of choosing various flavors. The initial step is to determine the proper sample space. Should the sample space contain 16 outcomes, or should it contain 4? If it contains 16 outcomes, then I can assume equally likely outcomes. If it contains 4, then the probabilities will differ. For this example, I will use a 16-outcome sample space, with equally likely outcomes. To simplify the discussion, I will number the pieces of candy of each flavor. The sample space is $\{L_1, L_2, L_3, C_1, C_2, C_3, C_4, B_1, B_2, B_3, B_4, W_1, W_2, W_3, W_4, W_5\}$

Since each outcome is equally likely, there is a probability of 1/16 that any particular piece will be randomly chosen. The probability of choosing a licorice candy is 3/16. The probability of choosing a watermelon is 5/16.

I happen to like all the flavors except watermelon. The probability of the event A: "an acceptable flavor" is 11/16, since there are 11 pieces of candy in the event A. ■

EXAMPLE 6.13

A Candy Jar—Unequally Likely Outcomes

The previous example can also be considered using a different sample space. The sample space consists of four unequally likely outcomes: $\{L, C, B, W\}$. The proper assignment of probabilities is $\mathbf{P}(L) = 3/16$, $\mathbf{P}(C) = 1/4$, $\mathbf{P}(B) = 1/4$, and $\mathbf{P}(W) = 5/16$.

The probability of an acceptable flavor is computed as

$$\mathbf{P}(A) = \mathbf{P}(L) + \mathbf{P}(C) + \mathbf{P}(B) = 3/16 + 1/4 + 1/4 = 11/16$$ ■

[6] You may feel the phrase "favorable outcomes" is inappropriate for this situation. An alternative is to mention that the event "a song you dislike" contains two outcomes. Speaking about favorable outcomes is traditional but may be somewhat confusing at times. Speaking about the number of outcomes in an event is more precise and always appropriate.

✔ Quick Check 6.4

In each problem, identify the method for determining the requested probabilities. Then solve the problem. The choices for method are: theoretical—equally likely outcomes; theoretical—unequally likely outcomes; empirical; and subjective.

1. In the random experiment "pick a straw" with three straws, what is $\mathbf{P}(M)$? What assumptions did you make?
2. If the choices of object in paper, rock, scissors are equally likely, what are the probabilities of the following events?
   (a) NW: "no winner"
   (b) PC: "paper was chosen by at least one opponent"
   (c) O_1: "opponent 1 wins"
3. Mr. Van Winkle overslept 23 of the previous 100 work days. What is the probability that he will oversleep tomorrow (a work day)?
4. What is the probability that a randomly chosen member of your class has all four biological grandparents currently living?
5. A class contains 18 freshmen, 12 sophomores, 4 juniors, and 2 seniors. The random experiment is to choose one student. Let the sample space be {Fr, So, Ju, Se}.
   (a) Compute the probability of each outcome.
   (b) Compute the probability of the event UD: "upper division student". (An upper division student is a junior or senior.) ☑

Probability Models

Now you have encountered all the pieces needed to describe the mathematical model for discrete probability.

Discrete Probability Models

A discrete probability model consists of three primary components:

1. A sample space chosen to reflect the possible outcomes.
2. The probabilities assigned to the outcomes.
3. The theoretical requirements that probabilities of outcomes satisfy

$$0 \leq \mathbf{P}(\theta) \leq 1 \quad \text{and} \quad \sum_{\theta \in S} \mathbf{P}(\theta) = 1$$

The first two components are chosen by the model builder. The theoretical requirements have been imposed by the mathematicians who developed the field of probability.

A consequence of the theoretical requirements is that $0 \leq \mathbf{P}(E) \leq 1$ for any event E. This follows because an event is a subset of the sample space and the probability of the event is the sum of the probabilities of the outcomes in the event. Thus

$$0 \leq \sum_{\theta \in E} \mathbf{P}(\theta) \leq \sum_{\theta \in S} \mathbf{P}(\theta) = 1$$

In at least one case, a model served as the standard, and reality was changed to more closely approximate the model. During the late 1960s, the United States government held a draft lottery. Young men of military age were drafted into the military services based on a random ranking by birthday. However, after the lottery was held, it became apparent that some months tended to have a higher proportion of low draft numbers than would be expected if all birthdays were equally likely to be chosen at each stage of the lottery. The actual experiment was not close enough to what the model predicted. The next year's lottery was implemented differently, ensuring that the fairness built into the mathematical probability model was more closely achieved.

The next definition contains terms that you are likely to be familiar with already. The definition serves to remove some potential ambiguity from future examples and homework exercises.

DEFINITION 6.7 ***Vowels; Consonants***

The English alphabet has 26 letters. The five[a] letters, {a, e, i, o, u}, are called *vowels*. The other 21 letters are called *consonants*.

[a]For the examples and exercises in this textbook, the letter y will always be a consonant.

EXAMPLE 6.14 **Choosing Letters**

One experiment consists of randomly choosing a letter from the English alphabet. Two events of interest are the events V: "a vowel was chosen" and C: "a consonant was chosen". If every letter is equally likely to be chosen, then $\mathbf{P}(V)=\frac{5}{26}$ and $\mathbf{P}(C)=\frac{21}{26}$. ■

6.1.3 Exercises

The exercises marked with ★ have detailed solutions in Appendix H.

1. Describe suitable sample spaces for the following random experiments.

(a) Rolling a standard pair of fair, six-sided dice.

(b) ★ Drawing a single straw, if there are two long straws and one short straw.

(c) ★ Flipping three coins.

(d) Calling a local time–temperature phone number and recording the temperature.

(e) Recording the number of games your favorite athletic team wins next season.

(f) Counting the number of dates you go on next semester (or next quarter).

(g) Counting the number of letters (including junk mail) in your mailbox tomorrow.

(h) Observing whether your odometer registers an even or odd number of miles the next time you park your car.

2. Assign a suitable probability to each outcome in the sample spaces for the following random experiments. (Refer to Exercise 1). Describe which method (Section 6.1.2) you used (or which method you would use, given additional information).

(a) Rolling a standard pair of fair, six-sided dice.

(b) ★ Drawing straws, where there are two long and one short straw.

i. For a sample space with three outcomes.

ii. For a sample space with two outcomes.

(c) ★ Flipping three coins.

(d) Calling a local time–temperature phone number and obtaining the temperature.

(e) Recording the number of games your favorite athletic team wins next season.

(f) Counting the number of dates you go on next semester (or next quarter).

(g) Counting the number of letters (including junk mail) in your mailbox tomorrow.

(h) Observing whether your odometer registers an even or odd number of miles the next time you park your car.

3. For each sample space in Exercise 1, list several events that would be of interest. For example, in (a), the event D: "doubles" is $\{(1,1),(2,2),(3,3),(4,4),(5,5),(6,6)\}$.

4. For each of the following random experiments, complete the following tasks.

i. Describe a suitable sample space.

ii. Assign a suitable probability to each outcome in the sample space. Describe which method (theoretical, empirical, subjective) you used (or would use, given additional information).

iii. List several events that would be of interest corresponding to the sample space.

(a) Recording the number of students in attendance on a particular day at a school with 400 students.

(b) Rolling a pair of "unfair", six-sided dice, where it is twice as likely to have the sum of the digits be greater than 6 than for the sum to be less than or equal to 6.

(c) Observing whether or not Dad lets you stay up past bedtime for two nights in a row.

(d) Selecting a lowercase alphabet letter and then choosing a single digit.

(e) Recording whether or not some child in your neighborhood has visited Disney World.

(f) Tracking how many times you eat out at a restaurant next month.

5. Explain the differences and similarities between theoretical, empirical, and subjective probabilities.

6. Consider the random experiment of picking a card from a standard 52-card deck. Describe the complement of each of the following events.

(a) A face card. (b) A red card.

(c) ★ A heart or a face card.

(d) A heart and a face card.

(e) A club or a red face card.

(f) A spade and a heart.

7. Consider the random experiment of selecting an integer from the set of all integers. Describe the complement of each of the following events:

(a) The set of positive integers.

(b) ★ The set of irrational integers.

(c) The set of integers that are divisible by 4.

(d) $\{n \mid n = 2k + 1 \text{ for some } k \in \mathbb{Z}\} \cup \{m \mid m \in \mathbb{Z}, m > 0\}$.

(e) $\{x \in \mathbb{Z} \mid x^2 = 1\}$. (f) $\{x \in \mathbb{Z} \mid x \text{ is odd or } |x| > 4.\}$

8. Which of the following event pairs are mutually exclusive?

- Picking a card.

  (a) ★ (a ten, a diamond)

  (b) ★ (a red card, a black card)

  (c) (a face card, an ace) (d) (a king, a face card)

- Rolling a die.

  (e) (an even number, a number less than 4)

  (f) (an even number, a 5)

  (g) (a number less than 4, a number greater than 3)

  (h) $(\{2, 3, 5\}, \{1, 3, 6\})$

- Choosing Student Officers by Lottery.
  Assume that the student body president, vice president, treasurer, and secretary are to be chosen by lottery.

  (i) (all female, all seniors)

  (j) (president is female, treasurer is male)

  (k) (all seniors, all juniors)

  (l) (all commuters, president is male)

9. Which of the following event pairs are mutually exclusive?

- Picking a card.

  (a) (a jack, a number card)

  (b) (a spade, a black face card)

  (c) (a red five, a red queen)

  (d) (a black king, a club)

- Throwing coins into a pond.
  Assume that three coins are taken from a coin purse containing four pennies, four nickels, and four dimes and are thrown into the pond.

  (e) (all three are nickels, the number of dimes even)

  (f) (no dimes, at most one nickel)

  (g) (at least two different types of coins, exactly one nickel)

  (h) (two pennies, one of each type of coin)

- Choosing pizza toppings.
  Assume that two pizza toppings are chosen from the following list, with repeats permitted: {extra cheese, green peppers, pepperoni, pineapple, olives, sausage, tomatoes}.

  (i) (a vegetable included, two meats)

  (j) (pepperoni and tomatoes included, pineapple or sausage included)

  (k) (meatless, pineapple included)

  (l) (double the extra cheese, no olives)

10. Compute the probabilities of the following events. The random experiment is rolling a pair of fair, six-sided dice.

(a) **P**(the sum is 12) (b) ★ **P**(the sum is 8)

(c) **P**(the sum is less than 7)

11. Compute the probabilities of the following events. The random experiment is rolling a single six-sided die, where any digit, x, with $\sqrt{x}$ an integer, is twice as likely to occur as a digit that does not fit this criterion.

(a) ★ **P**(subtracting 5 from the number produces a negative number)

(b) **P**(the number is even)

(c) **P**(the number is divisible by both 3 and 4)

(d) **P**(the number is less than 3)

12. Compute the probabilities of the following events. The random experiment is flipping a coin, then rolling a fair, six-sided die, and finally flipping a second coin.

(a) **P**(there are no heads)

(b) ★ **P**(the number is 2 and at least one head)

(c) **P**(the number is less than 3 and the same face on both coins)

13. Consider the random experiment of driving toward a street light. The experiment is to determine whether the light will be green, yellow, or red at the time you are 100 feet away from the light. Assume the light stays green for 2 minutes, is yellow for 30 seconds, then stays red for 2.5 minutes.

(a) Describe a suitable sample space.

(b) Assign a reasonable probability to each outcome in the sample space.

(c) Compute the probabilities of the following events. Which method (Section 6.1.2) did you use to determine the event probabilities?

  i. The light is not red. ii. The light is not yellow.

(d) Describe why your sample space (with probabilities) is only a model. [*Hint*: What happens to the model if a bulb burns out or the light stops working?]

14. Consider the random experiment of choosing two students out of {Ben, Jennifer, Kelly, Mark, Peter} to come up to the chalkboard to complete a homework problem. Assume that a pair of students that includes at least one of the two females is three times as likely as a pair with no females.
    (a) Describe a way for the teacher to use a sample space with equally likely outcomes to carry out this experiment in a random manner.
    (b) Describe a suitable sample space having unequally likely outcomes.
    (c) Assign a reasonable probability to each outcome in the unequally likely sample space.
    (d) Compute the probabilities of the following events. Which method (Section 6.1.2) did you use to determine the event probabilities?
        i. Kelly is chosen to come up to the board.
        ii. Peter is not chosen to come up to the board.
15. Consider the random experiment of calculating the number of errors per hour you will make while typing a school document. (An example of an error is typing "teh" instead of "the".) The experiment is to determine whether the number of errors per hour on the next day you type will be exactly 0, 4, 8, 12, or 16. Assume that the previous day, you made exactly 0 errors in 1 of the hours, exactly 4 errors in each of 5 different hours, exactly 8 errors in each of 7 different hours, exactly 10 errors in each of 8 different hours, and exactly 16 errors in each of 3 different hours. (This was a long day of typing!)
    (a) Describe a suitable sample space.
    (b) Assign a reasonable probability to each outcome in the sample space.
    (c) Compute the probabilities of the following events. Which method (theoretical, empirical, subjective) did you use to determine the event probabilities?
        i. You make over 8 typing errors per hour.
        ii. You never achieve typing perfection. (Actually, you never make 0 errors per hour.)
    (d) Describe why your sample space (with probabilities) is only a model.
16. Define (in your own words) the term *probability*.
17. Suppose a random experiment generates numbers in the set $S = \{0, 1, 2, 3, \ldots, 2n\}$. Each even number greater than 0 has probability $\frac{1}{2n}$ of occurring and each odd number has probability $\frac{1}{2n+1}$ of occurring. What is $\mathbf{P}(0)$?
18. Suppose a random experiment generates numbers in the set $Z^+ = \{1, 2, 3, \ldots\}$.
    (a) Create a suitable sample space.
    (b) Show that the outcomes cannot be equally likely.
    (c) Produce acceptable probabilities for the outcomes.
19. Let E be an event in a discrete probability model. Prove that

$$\mathbf{P}(E) + \mathbf{P}(\overline{E}) = 1$$

6.2 Conditional Probabilities and Independent Events

Sometimes we may perform a random experiment but obtain only partial information about the result. It may then be possible to use the partial information to revise the estimate of an event's probability. For example, if a friend flips a nickel and a penny, the probability that both coins land with the head side up is 1/4. If my friend flips the coins and I learn that the nickel landed with the head side up, the probability that both are heads should be revised to 1/2 (the probability that the penny is a head). The reason for the change is because once I know the nickel is a head, I will revise the sample space. The original sample space might be represented as $\{N_H P_H, N_H P_T, N_T P_H, N_T P_T\}$, signifying two heads, a head on the nickel and a tail on the penny, a tail on the nickel and a head on the penny, and two tails. Once I know that the nickel is a head, the sample space reduces to $\{N_H P_H, N_H P_T\}$. In the revised sample space, one of two outcomes has a head for the penny. Thus, the probability of two heads is 1/2.

6.2.1 Definitions

DEFINITION 6.8 ***Conditional Probability***
The probability that an event A has occurred, given the knowledge that event B has occurred, is called the *conditional probability of A, given B*, and denoted by $\mathbf{P}(A|B)$.

EXAMPLE 6.15

Picking Cards

The probability of choosing a queen is $4/52 = 1/13$. If I choose a card (without looking at it) and am told that the card is a face card, I would then revise the probability to $4/12 = 1/3$. That is, $\mathbf{P}(Q \mid F) = 1/3$. Once I know that a face card was chosen, the likelihood that it is a queen is greater.[7] ■

It is helpful to notice that the additional information causes us to consider a subset of the sample space. In the preceding example, the knowledge that a face card has been chosen enables us to conclude that only 12 of the original 52 outcomes are now viable results. The probability is thus 4/12 that the chosen card is a queen.

EXAMPLE 6.16

Picking Queens

The probability that I have chosen a queen, given that a heart has been picked, has the value $\mathbf{P}(Q \mid H) = 1/13$ since there is only one queen among the thirteen hearts. Notice that in this case $\mathbf{P}(Q \mid H) = \mathbf{P}(Q)$. The new information did not cause the probability to change. In essence, the proportion of queens in the suit of hearts is identical to the proportion of queens in the whole deck. ■

Conditional probabilities can also be used in a more sophisticated manner to help refine probabilities in the presence of additional information that is not a direct result of the random experiment. For example, we might revise an empirical or subjective estimate of a new product's marketability based on the results of a market survey. We would then consider conditional probabilities such as **P**(high sales | 75% of those surveyed like the product). This will be explored more fully in Section 6.6.

EXAMPLE 6.17

A Card Scam

Suppose you are offered the opportunity to engage in a "friendly wager". You are given the opportunity to guess the color on the opposite side of a card. The card can be any one of a set of three. One card is red on both sides, one is black on both sides, and the third is red on one side and black on the other. Your opponent chooses a card and places it face down on the table. One side is showing.

Suppose the side that is showing is red. Then you know it can't be the card with black on both sides. It must either be the card with red/black or the card with red/red. Suppose you lose \$10 if the hidden color is red but win \$10 if the hidden color is black. Is this a fair game? Explain why it is fair or why it is not fair. To encourage you to think about the problem yourself, I have postponed presenting the solution until later in this section. ■

[7] You may wish to skip this footnote on your first reading. A few subtle points need clarification in Example 6.15. The probability of a queen *before* I choose a card is 1/13. After I have chosen a card, the probability of a queen is either 0 or 1. I either chose a queen (in which case the card is a queen with probability 1), or I did not choose a queen (in which case there is a probability of 0 that the card is a queen). This will always be the case: An event has some initial probability before the experiment is conducted. After the experiment has been conducted, the event has occurred with probability 1 or 0 (it either occurred or it did not). The confusion occurs if the experiment has been conducted but you do not know the result. In that case, as far as you can tell, the initial probability is still valid. However, it may be the case that you obtain partial information about the result of the experiment. With partial information, you still may not know whether an event of interest occurred, but you may be able to revise your estimate of the event's likelihood in such a way that the new information is adequately accounted for.

In this sense, the number assigned to the conditional probability $\mathbf{P}(A \mid B)$ is a representation of the likelihood of the event A *assuming only the knowledge* that B has occurred. Someone else may know which outcome, θ, actually occurred. For them, the probability $\mathbf{P}(A)$ will be either 1 or 0, depending on whether θ is in A or is not in A. The person who only knows that B has occurred must use $\mathbf{P}(A \mid B)$ as the best estimate of the likelihood of A.

The notion of independence was introduced in Chapter 5 when the Independent Tasks Principle was presented. The notion that one task can occur without any influence from a second task has an analog in probability theory.

DEFINITION 6.9 ***Independent Events***

Event A is *independent* of event B if $\mathbf{P}(A \mid B) = \mathbf{P}(A)$. In such a case, the events, A and B, are said to be *independent events.*

Exercise 27 on page 296 asks you to prove that if $\mathbf{P}(A \mid B) = \mathbf{P}(A)$, then $\mathbf{P}(B \mid A) = \mathbf{P}(B)$ is also true. That is, if A is independent of B, then B is also independent of A. This justifies calling the pair "independent events".

EXAMPLE 6.18

Independent and Not-So-Independent Queens

We have seen that the events Q and F are not independent; knowing that a face card has been chosen causes us to revise the probability that that card is a queen. However, the events Q and H are independent; our estimated probability that the chosen card is a queen does not change if we learn that the card is a heart. ■

EXAMPLE 6.19

Random Books

Suppose you are hurrying through an airport to catch an intercontinental plane. You suddenly remember you have brought nothing to read. You rush into a gift shop, grab two paperback books off the rack at random and buy them. Assume that books come in three varieties (in equal proportions): "poor," "fair," and "good." The sample space can be represented as {GG, GF, GP, FG, FF, FP, PG, PF, PP}.

The event $2ndF$: "the second book is fair" has probability $3/9 = 1/3$ (just count the number of outcomes with an "F" as the second letter). The event $1stG$: "the first book is good" also has probability $3/9 = 1/3$. Notice that $\mathbf{P}(2ndF \mid 1stG) = 1/3$. This is because the knowledge that the first book is "good" has reduced the sample space to {GG, GF, GP}. Only one of the three viable outcomes has the second book "fair." Since $\mathbf{P}(2ndF \mid 1stG) = \mathbf{P}(2ndF)$, the events $2ndF$ and $1stG$ are independent.

The event NG: "neither book is good" has probability 4/9. In this case, $\mathbf{P}(2ndF \mid NG) = 2/4 = 1/2$. (The viable outcomes in the sample space are {FF, FP, PF, PP}.) Since $\mathbf{P}(2ndF \mid NG) \neq \mathbf{P}(2ndF)$, the events $2ndF$ and NG are not independent. If you know that neither book is "good," the second book is more likely to be "fair." ■

EXAMPLE 6.20

Heart Disease

According to the U.S. National Center for Health Statistics [36], the empirical probability that an American who died in 2005 died of heart disease was 0.266. If the randomly chosen person was male, the probability was 0.267. A few other age-adjusted death rates are listed next.

- $\mathbf{P}$(2005 death was by heart disease) = 0.266
- $\mathbf{P}$(2005 death was by heart disease | white) = 0.269
- $\mathbf{P}$(2005 death was by heart disease | black or African American) = 0.253

Compare these to some empirical probabilities from 1980.

- $\mathbf{P}$(1980 death was by heart disease) = 0.382
- $\mathbf{P}$(1980 death was by heart disease | white female) = 0.396
- $\mathbf{P}$(1980 death was by heart disease|black or African American female) = 0.341 ■

EXAMPLE 6.21 **The Gambler's Fallacy**

There is a common misunderstanding that leads to unwise behavior. If we are conducting a sequence of identical random experiments, it is incorrect to assume that the experiments have a "memory".

For example, the experiment might be to flip a coin. Even if the coin is truly fair (heads and tails are equally likely), it *is* possible to flip 10 heads in a row.[8] It is incorrect to assume that the eleventh flip is more likely to be a tail since the heads have "used up their quota." On the next flip, the probability of another head is still 1/2.

More formally, the sample space can be considered to be all distinct sequences of eleven H's and T's (there are $2^{11} = 2048$ of them). If you were to list them all, you would notice that exactly half (1024) of them have an H as the final letter. There are two outcomes with H's in the first 10 positions. One has an H as the eleventh letter and the other has a T as the final letter. In the language of conditional probabilities, $\mathbf{P}(\text{the eleventh letter is an H}) = 1/2 = \mathbf{P}(\text{the eleventh letter is an H}|\text{the first ten are H's})$. Similar observations indicate that the result of the eleventh flip and the result of any sequence for the first ten flips are independent events.

The gambler's fallacy is the strong (but incorrect) conviction that "my luck has been bad for so long, it *must* change to good luck soon". It is in the realm of possibility to have very long streaks of bad luck. (It is also in the realm of possibility to have a long streak of "bad luck" because someone has "stacked the deck" against you!) As long as the event of interest is independent of the outcomes for the previous experiments, there will be a constant probability that the event will occur in the next experiment. ■

SOLUTION 6.22 **A Card Scam**

The card scam in Example 6.17 on page 284 is *not* a fair game! To see this, you must consider the correct sample space. The original sample space (before you see one side of the chosen card) has six outcomes (the hidden side), which I have designated as $\{R_1, R_2, B_1, B_2, T_r, T_b\}$. The two outcomes designated R_1 and R_2 are the two sides of the card for which both sides are red. The two outcomes designated B_1 and B_2 are the two sides of the card with two black sides. The outcomes designated T_r and T_b are the red and black sides (respectively) of the card with one side of each color. Clearly, if the card is chosen randomly, there is a probability of 3/6 that the hidden side has either color.

However, we know that the side that is not hidden is red. The revised sample space for the hidden side is therefore $\{R_1, R_2, T_b\}$. Notice that only one of the three possible outcomes is black. Your probability of winning is thus 1/3, not 1/2! The reason many people predict a probability of 1/2 for black is that they fail to consider that the card with two red sides can be placed with either side down. This means that there are really two ways to have the hidden color be red, not one way.

Using conditional probability notation, $\mathbf{P}(\text{black} \mid \text{red showing}) = 1/3$. ■

✔ Quick Check 6.5

1. Suppose you are a spectator at a game of paper, rock, scissors. The opponents are visible through a soundproof window. You are standing in a position such that you can only see which object opponent 2 chooses. The opponents individually choose each object equally often.

   (a) What is the probability that opponent 1 chooses paper? (Denote this event as P-?.)

   (b) Suppose you see that opponent 2 has chosen rock. Denote this event by ?-R. Your best estimate for the probability that opponent 1 wins is $\mathbf{P}(\text{P-R} \mid \text{?-R})$. What is this probability?

(continued on the next page)

[8] The probability for this occurring is approximately .00098, or close to 1 chance in 1000.

(c) Are the events P-R and ?-R independent?

(d) Are the events P-R and ?-R mutually exclusive?

2. Assume again that paper, scissors, and rock have equally likely probabilities.

(a) What is $\mathbf{P}(O_1 \mid PC)$? (See Quick Check 6.3 on page 277.)

(b) List the reduced sample space used to compute the previous conditional probability.

3. When rolling a single fair die, are the events E: "an even number" and L: "a number less than 6" independent? ☑

6.2.2 Computing Probabilities

Often, the sample space is too large or the event of interest too complicated to enable easy direct use of the theoretical techniques for determining probabilities. If the event can be understood as being constructed from other, simpler events using some of the logic operators introduced in Chapter 2, then there are formulas to compute the probabilities using the probabilities of the simpler events.[9] This is a common technique in mathematics: Break a complex problem into simpler components, solve the simpler problems, then put the pieces back together.

In the probability principles that follow, we can understand the events in terms of logic operators or in terms of set theory. This is because an event is a set of outcomes. For example, if A and B are events, the event $A \wedge B$ is the logic operator notation for the event in which both the event A and the event B occur simultaneously. In set theory notation, this is the same as $A \cap B$ (the intersection of the *sets* A and B). Similarly, the event $A \vee B$ is the logic operator notation for the event in which one (or both) of the events occur. In set theory notation, this is written $A \cup B$ (the union of the *sets* A and B). The set theory version of the event $\neg A$ has already been introduced: the complement of A, denoted $\overline{A}$.

EXAMPLE 6.23

Building Events

I have already introduced the events H: "a heart" and F: "a face card". The event H has 13 outcomes and the event F has 12 outcomes.

The event $H \cup F$ contains 22 outcomes: the 13 hearts together with the 9 face cards that are not hearts.

The event $H \cap F$ contains 3 outcomes: the jack, queen, and king of hearts. ■

Consider the experiment of picking a card. The event $H \vee F$ is the event "heart or face card." There are 13 hearts and 12 face cards. However, 3 of the face cards are also hearts (and are therefore counted once each for being a heart and once each for being a face card). Eliminating the double counts, the event actually contains $13 + 12 - 3 = 22$ outcomes. The probability is therefore $22/52 = 11/26$. The calculation of this probability can be expressed as:

$$\mathbf{P}(H \cup F) = \mathbf{P}(H) + \mathbf{P}(F) - \mathbf{P}(H \cap F) = 13/52 + 12/52 - 3/52 = 11/26$$

It might be helpful to visualize the sample space as a table with 4 rows and 13 columns, corresponding to the 4 suits and 13 cards in a suit. If the H row is highlighted in yellow and the three columns corresponding to face cards are highlighted in blue, the event $H \cup F$ consists of all cards that are highlighted in some color. (There are actually three colors because of the overlap.)

[9] The formulas will be presented shortly. All seven will be summarized on page 292.

This need to avoid double-counting any overlaps leads to the first probability principle:[10]

1. $\mathbf{P}(A \cup B) = \mathbf{P}(A) + \mathbf{P}(B) - \mathbf{P}(A \cap B)$.

It is easy to see[11] that whenever events A and B are mutually exclusive,

2. $\mathbf{P}(A \cap B) = 0$.

The first two formulas can be combined to give the third probability principle:

3. $\mathbf{P}(A \cup B) = \mathbf{P}(A) + \mathbf{P}(B)$, if A and B are mutually exclusive events.

EXAMPLE 6.24 **Hearts and Spades**

Example 6.6 presented the mutually exclusive events H: "a heart" and S: "a spade". In this case, there are no outcomes (cards) that these events have in common. The term $\mathbf{P}(H \cap S)$ in the first probability principle is thus equal to 0 (an impossible event). The third probability principle is thus a special case of the first:

$$\mathbf{P}(H \cup S) = \mathbf{P}(H) + \mathbf{P}(S)$$

■

Quick Check 6.6

1. Use one of the probability principles to compute the probability that either opponent 1 or opponent 2 wins in paper, rock, scissors. Check your answer by examining the full sample space.
2. Which of the probability principles presented so far is appropriate to compute the probability that both opponents win (simultaneously)?
3. What is the probability of picking either a jack or a spade? (Show your work.) ☑

The fourth probability principle can be introduced by considering the probability of choosing a random card that is simultaneously a heart and a face card. You have already concluded that there are three cards that meet this requirement: the jack, queen, and king of hearts. The probability must therefore be 3/52.

Consider an indirect strategy for computing $\mathbf{P}(H \cap F)$. Assume for the moment that the deck of cards has been grouped into four stacks. Each stack of cards consists of all cards in a suit. The choice of a random card is done by a two-step process. First, choose a suit (a stack). Second, pick a random card in the suit. The probability of choosing the suit of hearts in step one is $\mathbf{P}(H) = 1/4$. The probability of choosing a face card in step 2 is $\mathbf{P}(F \mid H) = 3/13$. How should these two numbers be combined to correctly compute the probability $\mathbf{P}(H \cap F)$? Before giving the answer, consider the problem in the next paragraph.

Suppose you have 100 coins. One-fourth of the coins are dimes. How many of the coins are dimes? In elementary school you learned that the proper quantity is computed as $\frac{1}{4} \cdot 100 = 25$. To use a proportion (expressed as a fraction), you multiply. Suppose also that 1/5 of the dimes were minted in 1980. There must be $\frac{1}{5} \cdot 25 = 5$ dimes that were minted in 1980. The proportion of the 100 coins that are dimes minted in 1980 is $\frac{1}{4} \cdot \frac{1}{5} = \frac{1}{20}$. Successive proportions can be combined by multiplying.

In the random card choice, the probability is calculated by multiplying the probabilities at each step:

$$\mathbf{P}(H \cap F) = \mathbf{P}(H) \cdot \mathbf{P}(F \mid H) = \frac{1}{4} \cdot \frac{3}{13} = \frac{3}{52}$$

[10] This formula should remind you of inclusion–exclusion (Theorem 5.26 on page 261).

[11] Just review the definition of *mutually exclusive*.

which agrees with our intuitive calculation.

This two-step process is often helpful when the intuitive approach is hard to use. It is therefore convenient to summarize it as probability principle 4:

4. $\mathbf{P}(A \cap B) = \mathbf{P}(A) \cdot \mathbf{P}(B \mid A) = \mathbf{P}(B) \cdot \mathbf{P}(A \mid B)$

The formula is given in two forms to show that it doesn't matter which conditional probability is known. The important point is to match the proper unconditional probability. Visualizing the two-step process might help: First choose A, then B given A (first choose B, then A given B).

Probability principle 4 can be rearranged (assuming $\mathbf{P}(A) \neq 0$ and $\mathbf{P}(B) \neq 0$) as

$$\mathbf{P}(A \mid B) = \frac{\mathbf{P}(A \cap B)}{\mathbf{P}(B)}$$

and

$$\mathbf{P}(B \mid A) = \frac{\mathbf{P}(A \cap B)}{\mathbf{P}(A)}$$

which are often used as the definition of conditional probability.

EXAMPLE 6.25

Poor and White in America

Around 1990, about 78.1% of the population in the United States was classified as "white" [65]. Also, about 7.8% of the white population was low income. What is the (empirical) probability that a randomly chosen person in America at that time was a low-income white?

Probability principle 4 gives

$$\mathbf{P}(L \cap W) = \mathbf{P}(W) \cdot \mathbf{P}(L \mid W) = 0.781 \cdot 0.078 \simeq 0.061,$$

or about 6%. ■

Recall that events A and B are independent if $\mathbf{P}(B|A) = \mathbf{P}(B)$. Probability principle 5 is thus a special case of formula 4:

5. $\mathbf{P}(A \cap B) = \mathbf{P}(A) \cdot \mathbf{P}(B)$, if A and B are independent events.

EXAMPLE 6.26

The Queen of Hearts

In Example 6.16, it was shown that $\mathbf{P}(Q \mid H) = \mathbf{P}(Q)$. Thus the probability of a queen and a heart is (according to probability principle 5):

$$\mathbf{P}(Q \cap H) = \mathbf{P}(Q) \cdot \mathbf{P}(H) = \frac{1}{13} \cdot \frac{1}{4} = \frac{1}{52}$$

This is in agreement with our knowledge that there is only one queen of hearts. ■

EXAMPLE 6.27

Coin Flipping

We already know (for a fair coin) that $\mathbf{P}(H) = 1/2$. Since the first and second flips of a coin are independent, $\mathbf{P}(HH) = \mathbf{P}(H) \cdot \mathbf{P}(H) = \frac{1}{2} \cdot \frac{1}{2} = \frac{1}{4}$, which is in agreement with Example 6.1 on page 274. ■

The next probability principle captures the idea that an event and its complement must have probabilities that add to 1: either the event **A** must occur, or the event $\overline{A}$ must occur (and both cannot occur simultaneously).

6. $\mathbf{P}(\overline{A}) = 1 - \mathbf{P}(A)$

EXAMPLE 6.28

Heartless

The complement of the event H: "heart" is $\overline{H}$: "not a heart". There are 39 cards that are not a heart. Thus $\mathbf{P}(\overline{H}) = 39/52 = 3/4$. Probability principle 6 could also be used to compute this probability:

$$\mathbf{P}(\overline{H}) = 1 - \mathbf{P}(H) = 1 - \frac{1}{4} = \frac{3}{4}$$ ■

✔ Quick Check 6.7

1. Around 1990, about 10.3% of the people in the United States were classified as "black". Of the black population, 27.8% were considered low income.
   (a) Compute $\mathbf{P}(B \cap L)$, the probability that a randomly chosen American was a low-income black.
   (b) How many times larger was the low-income white population than was the low-income black population? (See Example 6.25.)
2. Consider yet again the game of paper, rock, scissors. Use two methods to compute the probability that opponent 1 chooses paper and opponent 2 chooses scissors. The first method should directly use the sample space. The second method should use either probability principle 4 or 5. Explain which principle you chose and why it is appropriate.
3. Use probability principle 6 to compute $\mathbf{P}(\overline{O_1})$, the probability opponent 1 does not win in paper–rock–scissor.
4. Show that probability principle 5 is a special case of probability principle 4. [*Hint*: I have already provided all the pieces. You just need to assemble the pieces.] ☑

The next example provides a nontrivial use of principle 6. The result is counterintuitive but correct.

EXAMPLE 6.29

Probability Principle 6: Birthday Probabilities

Suppose there is a room with n people. How likely is it that at least two of the people share a common month and day for their birthdays? (Notice that the *year* of birth is of no interest.) Since a year has 365 days (ignoring leap years), it seems unlikely unless n is quite large.

The requirement "at least two" simplifies the problem conceptually, but makes the probability computation much harder. For example, we need to consider the probability that *exactly* two people share a common birthday on January 1, that *exactly* three people share a common birthday on January 1, ..., that *exactly* two people share a common birthday on January 2, that *exactly* three people share a common birthday on January 2, and so on. Then we need to add all these probabilities. The task seems harder than it is worth.

Fortunately, there is a better way: Use complementary events. The complement of the event C: "at least two people share a common birthday" is $\overline{C}$: "no two people share a common birthday." It is easier to compute $\mathbf{P}(\overline{C})$.

What needs to be done is to count the number of ways that n people can have distinct birthdays. Then divide by the number of ways that n people can have birthdays (distinct or not distinct). In Example 5.13 on page 230, it was shown that n people can have *distinct* birthdays in

$$365 \cdot 364 \cdots (365 - n + 1)$$

ways. They can have birthdays in

$$365^n$$

ways (Example 5.15 on page 231). Thus

$$\mathbf{P}(\overline{C}) = \frac{(365 \cdot 364 \cdots (365 - n + 1))}{365^n}$$

and by probability principle 6,

$$\mathbf{P}(C) = 1 - \frac{(365 \cdot 364 \cdots (365 - n + 1))}{365^n}$$

For example, if $n = 4$,

$$\mathbf{P}(\overline{C}) = \frac{(365 \cdot 364 \cdot 363 \cdot 362)}{365^4} = \frac{47831784}{48627125} \simeq 0.984$$

Thus $\mathbf{P}(C) \simeq 0.016$, which is not very likely.

Table 6.1 lists, for a number of values of n, the probabilities that at least two people share a common birthday (ignoring leap years).

TABLE 6.1 Birthday Probabilities

n	P(C)	n	P(C)	n	P(C)
1	0.0000	10	0.1169	35	0.8144
2	0.0027	15	0.2529	40	0.8912
3	0.0082	20	0.4114	45	0.9410
4	0.0164	21	0.4437	50	0.9704
5	0.0271	22	0.4757	51	0.9744
6	0.0405	23	0.5073	52	0.9780
7	0.0562	24	0.5383	53	0.9811
8	0.0743	25	0.5687	54	0.9839
9	0.0946	30	0.7063	55	0.9863

The results may seem surprising. With only 23 people, the probability is greater than one half! If there are more than 53 people, the probability is more than 98%! The reason the probabilities are so high is that *every* pair of people are potential matches. As the number of people increases, the number of pairs increases much faster.[12]

Observe carefully that this example *does not* imply that in a room of 53 people, there is a 98% probability that at least one other person shares *your* birthday. The probabilities are for *some* collection of two or more people. We cannot specify any of the people ahead of time.

Notice also that even in a group of 53 people, there is still a 2% chance that there will be no common birthdays. If you were to pass a calendar around your classroom on which everyone was to indicate their birthday, and there were no common birthdays, you would not have proved the formula for $\mathbf{P}(C)$ to be incorrect. ∎

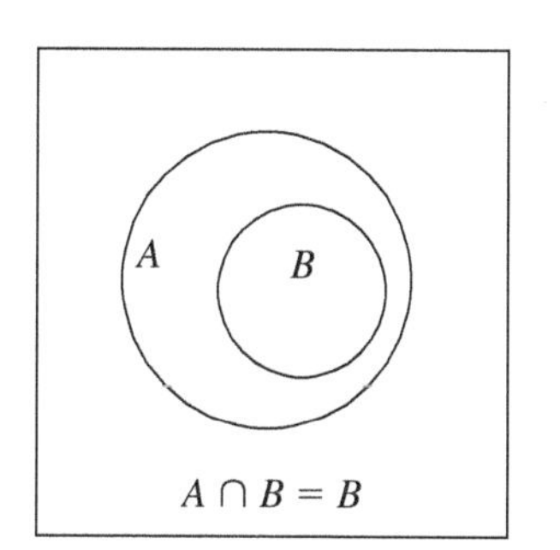

Figure 6.4. $A \cap B = B$.

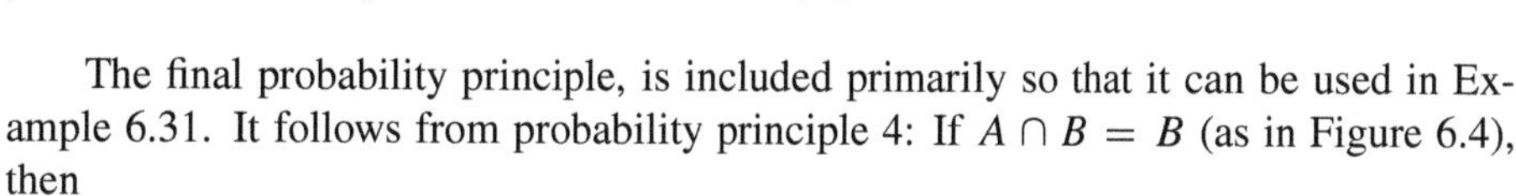

The final probability principle, is included primarily so that it can be used in Example 6.31. It follows from probability principle 4: If $A \cap B = B$ (as in Figure 6.4), then

$$\textbf{7. } \mathbf{P}(B \mid A) = \frac{\mathbf{P}(A \cap B)}{\mathbf{P}(A)} = \frac{\mathbf{P}(B)}{\mathbf{P}(A)}.$$

The first equality is a rearrangement of formula 4. The second equality follows since $A \cap B = B$.

EXAMPLE 6.30

Probability Principle 7

If HF represents the event "a heart face card", then $\mathbf{P}(HF \mid H) = 3/13$, since there are 3 face cards in the 13 card suit of hearts. This can also be computed as $\mathbf{P}(HF)/\mathbf{P}(H) = \frac{3/52}{13/52} = 3/13$. ∎

[12]Review Example 5.22 on page 234.

Probability Principles

In the formulas that follow, assume that A and B are events.

1. $\mathbf{P}(A \cup B) = \mathbf{P}(A) + \mathbf{P}(B) - \mathbf{P}(A \cap B)$.
2. $\mathbf{P}(A \cap B) = 0$, if A and B are mutually exclusive events.
3. $\mathbf{P}(A \cup B) = \mathbf{P}(A) + \mathbf{P}(B)$, if A and B are mutually exclusive events.
4. $\mathbf{P}(A \cap B) = \mathbf{P}(A) \cdot \mathbf{P}(B \mid A) = \mathbf{P}(B) \cdot \mathbf{P}(A \mid B)$.
5. $\mathbf{P}(A \cap B) = \mathbf{P}(A) \cdot \mathbf{P}(B)$, if A and B are independent events.
6. $\mathbf{P}(\overline{A}) = 1 - \mathbf{P}(A)$.
7. $\mathbf{P}(B \mid A) = \frac{\mathbf{P}(B)}{\mathbf{P}(A)}$, if $A \cap B = B$.

The final example in this section serves as a review for many of the concepts and probability principles that were presented. This might be an excellent example to read and discuss with a group. Your goal should be to understand the mathematics I have presented. Once the group has worked through the example and all members of the group understand it, you should celebrate by going out together for some ice cream.

EXAMPLE 6.31 **Probability in the Courtroom**

In 1968, the California Supreme Court reversed the verdict of a trial in which probability was used by the prosecution. The defendants were charged with purse snatching (with a loss of \$35–40). The prosecutor had called a mathematics instructor as an expert witness. The instructor explained how probability principle 5 works with independent events (using the roll of two dice as an example). The prosecutor then asked for estimates of probabilities for a few events related to the case. The instructor was unable (or unwilling) to make such estimates. The prosecutor then suggested some probabilities (listed in Table 6.2), telling the jury they should substitute their own estimates.

The victim and one witness testified that the guilty parties were a white woman with blond hair in a ponytail (who actually stole the purse) and a black male with a beard and mustache, who had driven her away in a yellow car with a whitish top. The police later found a married couple who fit these characteristics, had no known source of income, but who had paid a \$35 traffic fine shortly after the date of the crime.

The prosecutor suggested the (subjective) probabilities listed in Table 6.2.

TABLE 6.2 The Prosecutor's Subjective Probabilities

Characteristic	Individual Probability
Partly yellow automobile	1/10
Man with a mustache	1/4
Woman with a ponytail	1/10
Woman with blond hair	1/3
Black man with a beard	1/10
Interracial couple in car	1/1000

The prosecutor then used formula 5 to conclude that the probability of all these occurring together was about 1 chance in 12 million. He further claimed that his subjective estimates were conservative, so the actual probability could be even less likely, perhaps one in a billion.

The jury, after over 8 hours of deliberation and five ballots, voted to convict. The verdict was appealed.

The California Supreme Court mentioned two major objections to the prosecutor's use of probability.[13] First, the probability estimates were subjective, with no evidence introduced to substantiate those estimates. Even though the members of the jury were told these were subjective estimates, the court felt that the jury was unduly impressed by the suggested numbers.

The second objection related to the unsubstantiated claim that the characteristics mentioned were independent (a necessary assumption for principle 5 to be valid). The court mentioned, in particular, that having a beard and having a mustache were unlikely to be independent events.

The California Supreme Court justices then performed some probability calculations of their own, to show that the defendants were not the only likely couple to fit the description of the witnesses. Their calculations (with notational and expository changes) are presented next [92].

Let C be the event that a random couple (in the crime area) "exhibits all the characteristics the prosecutor assigned probabilities to". Let $\mathbf{P}(C)$ represent the probability of such a couple existing. The prosecutor estimated that $\mathbf{P}(C) = 1/12{,}000,000$. The Supreme Court justices computed the conditional probability that another such couple would exist, given the knowledge that one such couple does exist. Denote this conditional probability as $\mathbf{P}(\text{more than one couple has } C \mid \text{at least one couple has } C)$. The calculations below demonstrate how the justices arrived at their estimate of this conditional probability.

The probability that a randomly chosen couple will *not* exhibit all the characteristics C is $1 - \mathbf{P}(C)$. The justices made the assumption that different couples having C are independent events and then calculated (using probability principle 5)

$$\mathbf{P}(C \text{ does not occur in any of } n \text{ couples}) = (1 - \mathbf{P}(C))^n$$

This implies (using probability principle 6) that

$$\mathbf{P}(\text{at least one couple has } C) = 1 - (1 - \mathbf{P}(C))^n$$

Suppose now that a particular couple out of n possible couples is chosen at random. The probability that that couple, and no other, has C is, again by principle 5, $\mathbf{P}(C) \cdot (1 - \mathbf{P}(C))^{n-1}$. There are n mutually exclusive couples, each of which might be the only couple with C. The probability that C occurs in *exactly one couple* is thus (using probability principle 3)

$$\mathbf{P}(\text{exactly one couple has } C) = n \cdot \mathbf{P}(C) \cdot (1 - \mathbf{P}(C))^{n-1}$$

By subtracting the probability that C occurs in exactly one couple from the probability that C occurs in at least one couple, they obtained the probability that C occurs in more than one couple. That is,

$$\begin{aligned} &\mathbf{P}(\text{more than one couple has } C) \\ &= \mathbf{P}(\text{at least one couple has } C) - \mathbf{P}(\text{exactly one couple has } C) \\ &= [1 - (1 - \mathbf{P}(C))^n] - [n \cdot \mathbf{P}(C) \cdot (1 - \mathbf{P}(C))^{n-1}] \end{aligned}$$

Finally, since the event ("more than one couple has C" **and** "at least one couple has C") = "more than one couple has C," probability principle 7 implies

$$\begin{aligned} &\mathbf{P}(\text{more than one couple has } C \mid \text{at least one couple has } C) \\ &= \frac{\mathbf{P}(\text{more than one couple has } C)}{\mathbf{P}(\text{at least one couple has } C)} \\ &= \frac{[1 - (1 - \mathbf{P}(C))^n] - [n \cdot \mathbf{P}(C) \cdot (1 - \mathbf{P}(C))^{n-1}]}{1 - (1 - \mathbf{P}(C))^n} \end{aligned}$$

[13] A minor objection was that introducing probability placed a mathematically unsophisticated defense attorney at a disadvantage!

This last probability can be used to decide how likely it is that the defendants were the only possible couple matching the characteristics C. The justices noted that as n increases, the probability of another couple with C existing (once we know that such a couple can exist) increases. When n is close to 12 million,

$$\mathbf{P}(\text{more than one couple has } C \mid \text{at least one couple has } C) \simeq 0.418.$$

Even for a smaller pool of couples in the area, the probability is fairly high that another couple exists having the characteristics described by the witnesses. The graph in Figure 6.5 graphs the conditional probability

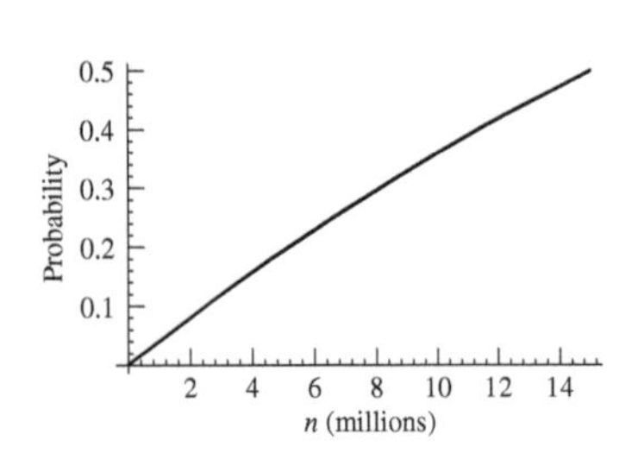

Figure 6.5. $\mathbf{P}$(more than one couple has C | at least one couple has C) as a function of n.

$$\mathbf{P}(\text{more than one couple has } C \mid \text{at least one couple has } C)$$

as a function of n, assuming that $\mathbf{P}(C) = 1/12{,}000{,}000$.

In case someone objected to the use of $\mathbf{P}(C) = 1/12{,}000{,}000$, they also noted that if $\mathbf{P}(C) = 1/n$, then as n gets bigger, the conditional probability approaches 41.8% for n couples. More precisely, if $\mathbf{P}(C) = 1/n$, then (using some calculus)

$$\mathbf{P}(\text{more than one couple has } C \mid \text{ at least one couple has } C) \to \frac{e-2}{e-1} \simeq 0.418$$

This approximation is close, even for n as small as 10.

The California Supreme Court justices concluded as follows:[14]

> Hence, even if we should accept the prosecution's figures without question, we would derive a probability of over 40% that the couple observed by the witnesses could be "duplicated" by at least one other equally distinctive interracial couple in the area, including a black man with a beard and mustache, driving a partly yellow car in the company of a blonde woman with a ponytail. Thus the prosecution's computations, far from establishing beyond a reasonable doubt that the defendants were the couple described by the prosecution's witnesses, imply a very substantial likelihood that the area contained more than one such couple, and that a couple other than the defendants was the one observed at the scene of the robbery. ■

✔ Quick Check 6.8

1. Suppose there is a room with 12 people. What is the probability that at least two of them share a common birthday (ignoring leap years)?
2. Assume in Example 6.31 that $n = 10$ and $\mathbf{P}(C) = 1/n$. Calculate $\mathbf{P}$(more than one couple has C | at least one couple has C). Then describe what these probabilities mean. ☑

6.2.3 Exercises

The exercises marked with ★ have detailed solutions in Appendix H.

1. Define (in your own words) the term *conditional probability*.

2. Compute the following conditional probabilities. The random experiment is picking a card from a standard 52 card deck.
   (a) $\mathbf{P}$(king | face card) (b) ★ $\mathbf{P}$(4 | face card)
   (c) ★ $\mathbf{P}$(4 | a red card) (d) $\mathbf{P}$(face card | a black card)

3. Compute the following conditional probabilities. The random experiment is flipping a fair coin three times.
   (a) ★ $\mathbf{P}$(the third is a tail | the second is a tail)
   (b) $\mathbf{P}$(at least one head | at least two are the same)
   (c) $\mathbf{P}$(they are all the same | the first is not a head)
   (d) $\mathbf{P}$(at least two are tails | at least one is a head)

[14] I have taken the liberty to change a few words and suppress the last name of the defendants.

4. According to adherents.com [52], around August 2007 about 85% of the world's people considered themselves religious. The largest groups and their approximate proportions are listed next. I have modified the proportions slightly so that they add to 100%.

Christianity 32%	*Islam* 20%
Nonreligious 15%	*Hindu* 13%
Primal-indigenous 5%	*Chinese traditional* 5%
Buddhism 5%	*Other* 5%

Calculate the following (mostly) conditional empirical probabilities for a person being a member of a religious grouping. (An adherent of Islam is a *Muslim*.)

(a) **P**(religious)
(b) **P**(Christian | religious)
(c) **P**(Muslim | religious)
(d) **P**(Hindu | religious)
(e) **P**(Buddhist | religious)
(f) **P**(Christian or Muslim | religious)
(g) **P**(Hindu or Buddhist | religious)

5. ★ The random experiment is choosing two letters from the English alphabet. Use an appropriate probability principle to determine **P**(at least one vowel is chosen). Indicate which principle you used and show the details.

6. ★ The random experiment is choosing a month for your vacation during this year and then a month for the next year. Use an appropriate probability principle to determine **P**(no trip in August). Indicate which principle you used and show the details. Assume that all months are equally likely to be chosen.

7. The random experiment is rolling a pair of fair, six-sided dice. Use an appropriate probability principle to determine **P**(one of the dice is a 4). Indicate which principle you used and show the details.

8. Compute the following conditional probabilities. The random experiment is rolling a pair of fair, six-sided dice.

(a) ★ **P**(an even sum | one die is a 3)
(b) **P**(the sum is 12 | the sum is even)
(c) **P**(the sum is 8 | one die is a 2)

9. Compute the following conditional probabilities. The random experiment is rolling a pair of six-sided dice in which an even digit on the die is twice as likely to occur as an odd digit.

(a) ★ Calculate $\mathbf{P}(1, 3) = \mathbf{P}(3, 1)$.
(b) Calculate $\mathbf{P}(2, 4) = \mathbf{P}(4, 2)$.
(c) Calculate $\mathbf{P}(3, 4) = \mathbf{P}(4, 3)$.
(d) **P**(the product is even | the sum is greater than 9)
(e) **P**(no 3s | the product is divisible by 5)
(f) **P**(the sum is divisible by 4 | doubles)

10. Explain the distinction between *mutually exclusive events* and *independent events*.

11. Determine whether the following event pairs are independent:

- Picking a card.
  (a) ★ (a ten, a diamond)
  (b) ★ (a red card, a black card)
  (c) (a face card, an ace)
  (d) (a king, a face card)
- Rolling a pair of fair dice.
  (e) (the first die shows a 1, the second die shows a 3)
  (f) (a sum of 8, one die is a 2)
  (g) (an even sum, at least one die is odd)
  (h) (the first die is even, the second die is odd)
- Flipping Three Fair Coins.
  (i) (the second is a head, the third is a tail)
  (j) (the second is a head, none are tails)
  (k) (at least one is a head, at least one is a tail)
  (l) (at least one is a head, at least two are tails)
  (m) (at least one is a head, at least two are the same)
  (n) (at least two are heads, they are all the same)

12. Determine whether the following event pairs are independent:

- Picking an integer, x, such that $-50 \leq x \leq 50$ and $x \neq 0$.
  (a) (an even integer, a negative integer)
  (b) (an integer y such that $|y| > 10$, a natural number)
  (c) (a positive or negative single digit integer, an integer y such that y^2 is less than 2)
  (d) (an integer divisible by 9, an integer divisible by 6)
  (e) (an integer divisible by 2, an integer divisible by 3)
- Choosing two consecutive days of the week.
  (f) (Monday is chosen, Sunday is not chosen)
  (g) (Saturday or Monday is chosen, Tuesday, Friday, Wednesday, or Sunday is chosen)
  (h) (Wednesday and Thursday are chosen, Thursday is chosen)
  (i) (none of Monday, Wednesday, Friday, and Sunday are chosen, Tuesday is not chosen)
- Picking flavors for a double scoop ice cream cone. Assume that two ice cream flavors are chosen from the following list, with repeats permitted: {bubble gum, chocolate, vanilla}. It does not matter in which order the scoops are placed on the cone.
  (j) (at least one scoop of vanilla, no double chocolate or chocolate with vanilla)
  (k) (two different flavors, a scoop of chocolate or bubble gum)
  (l) (bubble gum without vanilla, two scoops of the same flavor)
  (m) (a scoop of vanilla and a scoop of chocolate, a scoop of vanilla)

13. Give, if possible, examples of the following:

(a) ★ Two events that are mutually exclusive but not independent.

(b) Two events that are mutually exclusive and also independent.

(c) Two events that are neither mutually exclusive nor independent.

(d) Two events that are not mutually exclusive but are independent.

Use the formulas on page 292 to calculate the probabilities in exercises 14–23. In each case, indicate which principle you used, and show the intermediate calculations.

14. Compute the probability of rolling a pair of fair dice and obtaining distinct numbers on the two dice.

15. Consider the random experiment of choosing eight distinct letters of the English alphabet, where exactly two of the letters chosen are vowels and six of the letters are consonants. What is the probability that the letter "a" is among the eight letters and that these same eight letters can be arranged in a row to form the word *children*?

16. ★ Compute the probability of picking a spade or a three (using a standard 52-card deck).

17. Consider the candy jar in Example 6.12 on page 279. Compute the probability that I first choose a watermelon and then choose a cinnamon. Assume that I do not replace any candy once it has been chosen. [*Hint*: What happens to the sample space after the watermelon candy has been chosen?]

18. Compute the probability of picking a diamond with an even number printed on it (i.e., a 2, 4, 6, 8, or 10) from a standard 52-card deck. Use one of the probability principles. Describe which principle you used and why it was appropriate. Check the reasonableness of your answer by directly computing the probability. Describe this second solution.

19. ★ Consider the random experiment of rolling two fair dice. What is the probability that the sum is 10 and both dice show an even number? Solve this three ways [two in part (b)]:

(a) Using a six by six table, representing the 36 possible rolls of the dice, count the possibilities directly.

(b) Use probability principle 4 (both versions).

20. Compute the probability that a family having three children, randomly chosen from all families having three children, has at least one child of each gender and contains at least two girls. Use one of the probability principles. Describe which principle you used and why it was appropriate. What assumptions have you made?

21. The random experiment is flipping a fair coin three times. Compute the probability of obtaining at least one head.

22. The random experiment is flipping a fair coin three times. Compute the probability of a head on the first flip or a tail on at least one of the last two flips.

23. The random experiment is choosing an uppercase letter from the English alphabet.

(a) What is the probability of choosing either a consonant in the first half of the alphabet or else a vowel?

(b) What is the probability of choosing a vowel that is within two letters of the letter L? (The letters R, S, U, and V are within two letters of T.)

24. Consider a random experiment of flipping an unfair coin three times. Assume that $\mathbf{P}(H) = 0.8$ and $\mathbf{P}(T) = 0.2$.

(a) ★ List the sample space.

(b) ★ Decide whether the successive flips are independent. That is, should the probability of a head on the second flip be revised if we know that the first flip produced a tail? Should knowledge of the first two flips cause us to revise the probabilities of a head on the third flip?

(c) ★ Use your answer to part (b), an appropriate probability principle from Section 6.2.2, and a proper choice of method from Section 6.1.2 to compute the probability of each outcome.

(d) Compute the probabilities of the following events:

i. No heads.
ii. At least one head.
iii. All tails.
iv. A tail on the first flip.
v. Identical results on the first and third flips.

25. According to [95], there were approximately 1050 boys born in the United States during 1987 for every 1000 girls born. Thus the probability that a newly born child is a boy is approximately 0.512 (so the probability the child is a girl is 0.488). Assume that the gender of previous children in the family has no influence on the gender of the next child conceived (i.e., assume independence).

(a) Compute the probabilities for all eight outcomes in Example 6.3 on page 274.

(b) Compute the probabilities of the following events:

i. Two girls, one boy.
ii. All boys.
iii. At least one girl. (Is there an "easy way" to do this part?)
iv. At least one of each gender.
v. All the same gender.

26. Suppose there is a room with 28 people. What is the probability that at least two of them share a common birthday (ignoring leap years)?

27. Assume that $\mathbf{P}(A \mid B) = \mathbf{P}(A)$. Prove $\mathbf{P}(B \mid A) = \mathbf{P}(B)$.

28. Show that

$$\lim_{n\to\infty} \frac{\left(1-\left(1-\frac{1}{n}\right)^n\right) - n\cdot\left(\frac{1}{n}\right)\cdot\left(1-\frac{1}{n}\right)^{n-1}}{1-\left(1-\frac{1}{n}\right)^n} = \frac{e-2}{e-1}$$

29. Suppose a reporter read the California Supreme Court opinion in Example 6.31. The reporter summarized the verdict by writing

> The California Supreme Court overturned the verdict because there was not enough evidence to support the prosecution's probability estimate.

Write a letter to the editor showing the inadequacies of the reporter's simplistic summary.

30. According to [35], it is possible to categorize the (nonelderly) people without health insurance during 1989 in the United States by family income level. The following table shows the data.

Family Income Level	Number of People (in Millions)
Less than \$5,000	4.7
\$5,000–9,999	5.0
\$10,000–14,999	5.6
\$15,000–19,999	4.6
\$20,000–29,999	5.9
\$30,000–39,999	3.2
\$40,000–49,999	1.9
\$50,000 or more	3.3

Consider the random experiment of choosing to give health insurance to one of these people, not showing favoritism in any direction with regard to family income level. The events of interest are membership in one of the eight income brackets in the table.

(a) Compute the probabilities for all eight events in the relevant sample space for this random experiment. In other words, determine the probability that the randomly chosen person without health insurance fits into any one of the given income levels.

(b) Suppose that the previous random experiment just involved giving health insurance to a person for one year. Consider the random experiment of choosing two of these people to receive health insurance during two different years. It is permissible for the same person to be chosen twice. Additionally, assume that the person chosen for one year has no influence on the person chosen for the next year (i.e., assume independence). The focus in this experiment is on the income level of the person chosen for each of the two years. Compute the probabilities of the following events:

i. Two people chosen from the same income group.

ii. Neither person has an income below \$5,000.

iii. The person chosen in the first year has an income of at least \$50,000.

iv. Exactly one person from the \$40,000–49,999 income bracket is chosen or exactly one person from the \$15,000–19,999 income bracket is chosen.

v. Both of the people chosen have an income that is at least \$30,000.

31. According to the *Statistics New Zealand* "snapshot" of work, education, and income data collected in the 2001 Census, 6 in 10 people received income from wages and salaries. Consider the random experiment of choosing three people in New Zealand. Let W represent that a person received income from wages and salaries, and N represent that a person received no income from wages and salaries. Assume that whether or not one person received income from wages and salaries has no correlation with the income situation of the other two people (i.e., assume independence). The three people will be chosen in succession. There are eight events of interest: {WWW, WWN, WNW, WNN, NWW, NWN, NNW, NNN}.

(a) Compute the probabilities for all eight events of interest for this random experiment.

(b) Compute the probabilities of the following events:

i. No one received income from wages and salaries or everyone did.

ii. At most one person received income from wages and salaries.

iii. Exactly one person did not receive income from wages and salaries.

iv. Someone did not receive income from wages and salaries.

v. Exactly one person received income from wages and salaries or at least two people did not receive income from wages and salaries.

6.3 Counting and Probability

This section will eventually examine probabilities related to large sample spaces. The technique to be used will first be illustrated with a small sample space.

EXAMPLE 6.32 **Probabilities and Candy**

Example 5.24 on page 235 introduced a candy jar with four flavors of candy. What is the probability of picking at least one cinnamon candy if there are three pieces of each flavor?

This task will be simplified if the complementary event is examined. Denote the complementary event by NC. There are $C(3+3-1,3) = C(5,3) = 10$ ways to choose a set of three pieces of non-cinnamon candy. Since there are $C(4+3-1,3) = 20$ ways to choose a collection of three pieces of candy with no restrictions, the probability is $1 - \mathbf{P}(NC) = 1 - \frac{10}{20} = .5$ that at least one cinnamon was picked.

What is the probability of picking three pieces of candy in which there is either exactly one cinnamon or there is exactly one licorice, (cinnamon and licorice may both be present)?

Let OC be the event "exactly one cinnamon" and OL be the event "exactly one licorice". We want the probability of the event $OC \cup OL$. That probability is

$$\mathbf{P}(OC \cup OL) = \mathbf{P}(OC) + \mathbf{P}(OL) - \mathbf{P}(OC \cap OL)$$

The jar contains identical pieces of candy, so $|OC|$ can be calculated by picking one cinnamon, removing the other pieces of cinnamon from the jar, and then counting the number of ways to choose two more pieces of candy. There are $C(3+2-1, 2) = C(4, 2) = 6$ ways to do this. Similarly, there are 6 ways to pick exactly one licorice. There are two ways to pick one cinnamon, one licorice, and a third piece that is neither cinnamon nor licorice.

Thus, $\mathbf{P}(OC \cup OL) = \mathbf{P}(OC) + \mathbf{P}(OL) - \mathbf{P}(OC \cap OL) = \frac{6}{20} + \frac{6}{20} - \frac{2}{20} = 0.5$ (you might find it helpful to list the 10 favorable events). ■

The second example considers the probability of a card hand.

EXAMPLE 6.33

Crazy Eights Hands

Young crazy eights players[15] might wonder how likely it is that a hand they have just been dealt will contain at most one 8. We have the theoretical basis to compute this probability. It is just the number of seven-card hands that contain at most one 8 divided by the total number of seven-card hands. The numerator has already been computed in Example 5.31 on page 241. There are $C(52, 7) = 133{,}784{,}560$ seven-card hands. Thus $\mathbf{P}(\text{at most one } 8) = \frac{122{,}715{,}120}{133{,}784{,}560} \simeq 0.917$, or 91.7%. Hands with two or more 8s occur only about 8.3% of the time.

Example 5.20 on page 233 calculated the number of crazy eights hands with exactly one 8 and exactly 2 face cards to be 15,550,920. The probability of such a hand occurring at random (assuming every hand is equally likely to occur) is $\frac{15{,}550{,}920}{133{,}784{,}560} \simeq 0.116$ (approximately 12%). ■

In the board game Risk$^{\circledT}$, the opponents each try to conquer the entire world. The mechanism used is to roll fair dice. The attacker can use from one to three dice (depending on the number of armies the attacker has). The defender can use one or two dice (again depending on the number of defending armies). The opponents roll all the dice. They each line their dice up in descending numeric order. They then compare their (respective) highest values until the defender runs out of dice for that roll. For example, suppose they roll A6, A3, A4, D4, and D5. (That is, the attacker rolls a 6, 3, and 4; the defender rolls a 4 and 5.) They would compare the two pairs (A6, D5) and (A4, D4). The pairs are scored independently. The player with the higher number in a pair wins that match. If there is a tie, the defender wins. In the example just mentioned, each opponent would lose one army.

A smart attacker will compute how likely it is to win. I will compute this probability in two special cases.

EXAMPLE 6.34

Risk Dice: One against One

If each player has one die, the sample space is the set of 36 pairs: (attacker's number, defender's number). It is possible to list all the pairs, then determine how many have the first component larger than the second. This seems too tedious. An alternate approach is to write a short computer program to do the search for you. I will take a more mathematical approach.

We need to count how many ways the attacker can win. We will then divide by 36 (the number of possible outcomes) to compute the probability. There are 6 values the

[15] Example 5.19 on page 233.

attacker can roll. For each of these numbers, I need to count how many numbers the defender can roll that are less than the attacker's number (Table 6.3).

There are 15 ways the attacker can win. (Which counting principle or counting formula have I used?) Thus **P**(attacker wins) $= \frac{15}{36} \simeq 0.417$. This is not a favorable situation for the attacker. ■

EXAMPLE 6.35

Risk Dice: Two against One

If the attacker has two dice and the defender has only one, the sample space can be visualized as a table with 36 rows, labeled by the possible pairs of numbers the attacker can roll, and 6 columns, labeled by the 6 rolls the defender can roll. In each of the 216 cells, an A or D could be written, depending on who is the winner of that match. The number of A's could then be counted. This is much too tedious. A computer program would also work nicely in this case. I will again use a mathematical approach.

If the defender rolls a 6, the attacker loses. There are 36 possible ways this can happen (the attacker rolls any number from 1 to 6 on each of two dice, for a total of $6^2 = 36$). If the defender rolls a 5, the defender will win if any number from 1 to 5 is rolled by the attacker. There are $5^2 = 25$ such rolls possible.

If the defender rolls a 4, the attacker will lose if only numbers less than or equal to 4 are rolled. There are $4^2 = 16$ ways this can happen. The pattern continues and is recorded in Table 6.4.

There are thus $36 + 25 + 16 + 9 + 4 + 1 = 91$ rolls on which the defender will win. Thus **P**(defender wins) $= \frac{91}{216} \simeq 0.421$. Consequently **P**(attacker wins) $= 1 -$ **P**(defender wins) $\simeq 0.579$. This is favorable for the attacker. ■

TABLE 6.3 Number of Rolls Smaller than Attacker's

Attacker's roll	Number of smaller rolls
6	5
5	4
4	3
3	2
2	1
1	0

TABLE 6.4 Number of Rolls Defender Wins

Defender's roll	Number of rolls defender wins
6	36
5	25
4	16
3	9
2	4
1	1

✔ Quick Check 6.9

1. If a five-card hand is dealt using a standard 52-card deck, what is the probability of being dealt a hand with all four aces?
2. What is the probability of being dealt a five-card hand that contains three clubs and two diamonds?
3. What is the probability that rolling a pair of dice will result in the sum being even? ☑

6.3.1 Exercises

The exercises marked with ★ have detailed solutions in Appendix H.

1. There are four common coins in the United States: pennies (1 cent), nickels (5 cents), dimes (10 cents), and quarters (25 cents).
   (a) In how many ways can I form 29 cents?
   (b) How many distinguishable ways can I have a set of three of these four kinds of coins? Assume I have many of each kind of coin. Solve this in two ways: by using the appropriate counting principles and formulas and then by listing all the distinguishable collections of three coins.
   (c) If I randomly select three coins from a jar containing 12 coins (three coins of each of the four types), what is the probability that the coins I choose add to at least 29 cents?
2. Consider the random experiment of tossing one fair coin three times in succession and then rolling a single fair, six-sided die. Jake and his parents are carrying out this experiment to help them decide where to go on vacation this summer.
   (a) ★ How many distinct outcomes are there in the sample space for the random experiment in this exercise?
   (b) ★ Suppose that Jake gets to choose the vacation spot if the number of heads multiplied by the digit on the die is greater than 10. What is the probability that Jake's parents get to choose the vacation spot?
   (c) Now suppose that Jake gets to choose the vacation spot if there are at least two tails, but the digit on the die cannot be even. What is the probability that Jake gets to pick the vacation spot?
   (d) Finally, suppose that Jake's parents get to choose the vacation spot if no tails appear or the digit on the die is divisible by 3. What is the probability that Jake's parents do not get to pick the vacation spot?
3. I will be sharing an apartment with two friends. The apartment has one large and one small bedroom. Two of us will share the large bedroom and the other will sleep in the smaller room.
   (a) In how many distinguishable ways can we assign rooms?
   (b) What is the probability that I get the single room, if we choose randomly?

4. Consider the random experiment of tossing eight fair coins simultaneously.
   (a) How many possible outcomes are there in this experiment?
   (b) What is the probability that at least five of the coins are heads?
   (c) ★ What is the probability that the number of heads and the number of tails differ by at most 2?
   (d) What is the probability that not exactly two coins are tails?
5. Suppose that a movie theater places seven distinct positive integers not exceeding 60 on each movie ticket in no particular order. On the day you buy a ticket, the movie theater randomly selects seven distinct positive integers not exceeding 60 and awards a free movie to any person with a ticket containing all seven of these numbers. You receive a ticket containing these seven numbers on it: 7, 15, 16, 27, 44, 45, 49.
   (a) In how many ways can the movie theater select the seven distinct positive integers not exceeding 60?
   (b) What is the probability that you will win a free movie?
   (c) What is the probability that you will have at most one of the numbers correct?
6. ★ To win a prize at the school fair, you must choose exactly 6 out of 7 distinct winning integers, where all numbers are between 10 and 99, inclusive. The order in which you select these integers does not matter, but each participant must choose 7 numbers. Note that people who choose all seven of the winning integers do not qualify for a prize.
   (a) In how many ways can seven numbers be chosen?
   (b) What is the probability that a randomly chosen set of seven numbers will match exactly six of the winning numbers?
7. Suppose that at a raffle, you buy a ticket that contains three distinct integers greater than 150 but less than 200. Assume that the order in which these integers appear on the ticket does not matter. At the time of the raffle, the people running the raffle randomly choose 8 of these viable integers. A person can win a raffle prize by having the 3 numbers on his or her raffle ticket be among the 8 chosen by the people running the raffle. The raffle numbers on your ticket are the following: 151, 174, 199.
   (a) How many distinct ways can the eight numbers be chosen by the people running the raffle?
   (b) What is the probability that you will win a prize?
8. Mom has seven jobs that she needs completed. The jobs are doing the laundry, mowing the lawn, vacuuming, washing the windows, folding the towels, cleaning the garage, and running some errands. She will randomly assign you three of these jobs and will also assign a day on which each job must be completed. The jobs will be completed on Monday, Tuesday, and Wednesday. Assume that Mom will not ask you to do any job more than once in these three days.
   (a) In how many ways can Mom assign the jobs for you to complete?
   (b) What is the probability that you will have to fold the towels?
9. ★ You are at a school carnival and are going to spin a colorful wheel six successive times. The equally distributed colors on this wheel are red, orange, yellow, and green. Having the wheel pointer land in the red section means that you win a cookie, while having the wheel pointer land in the orange, yellow, or green sections indicates winning the following prizes, respectively: a stuffed animal, a book, a piece of paper that says "better luck next time".
   (a) How many spin combinations are possible, assuming that you spin the wheel six different times?
   (b) What is the probability that you will win at least two stuffed animals (i.e., the wheel pointer lands in the orange section at least twice)?
   (c) What is the probability that you will win exactly one cookie and exactly one book (i.e., the wheel pointer lands in the red section exactly once and in the yellow section exactly once)?
10. In 1992 the Mattel toy company introduced a talking Teen Barbie doll. The company had compiled a collection of 270 possible expressions for Barbie to "speak". To make the dolls appear more individually distinct, they randomly chose 4 of the 270 possibilities to program into any particular doll.
    (a) How many distinct dolls could they manufacture?
    (b) Suppose one of the 270 statements is considered to be undesirable. How many distinct dolls contain the undesirable statement among their four exclamations?
    (c) What is the probability of getting a doll that speaks the undesirable phrase?
11. Compute the probabilities for the following crazy eights hands.
    (a) ★ At least three eights. (b) All the same suit.
    (c) All clubs. (d) All red cards.
    (e) Four eights and three diamonds other than the eight.
12. Compute the probabilities for the following crazy eights hands.
    (a) Exactly one ace.
    (b) Exactly one ace and exactly two cards with a face value between 5 and 7, inclusive.
13. Compute the probabilities for the following crazy eights hands.
    (a) No 5s but four cards of the same kind.
    (b) Exactly two face cards and exactly two red cards that are not face cards.
    (c) Exactly two face cards and exactly two red cards.

14. Calculate the following probabilities for a 5-card hand using a normal 52-card deck.

(a) A royal flush. A royal flush is the 10, jack, king, queen, and ace within a single suit.

(b) A straight flush. A straight flush is simply a straight (see Exercise 16) that occurs within a single suit.

15. Calculate the following six probabilities for a 5-card hand using a normal 52-card deck.

(a) All five cards are diamonds.

(b) ★ All five cards are the same suit.

(c) Containing two queens and no other duplicate face values (for example, two of the queens, together with a three, a seven, and a king).

(d) Containing three queens and no other duplicate face values.

(e) Containing three cards with the same face value and two cards that do not repeat previous face values.

(f) Containing three cards with face value A and two cards with face value B (for example, 3 queens and 2 fours, or 3 tens and 2 fives).

16. Calculate the following probabilities for a 5-card hand using a normal 52-card deck.

(a) Containing five distinct face values on the cards.

(b) Containing a straight. A straight occurs when the five cards have five consecutive face values. One example of a straight is 8-9-10-jack-queen. Note that the ace can be used as either the lowest card or the highest card in a straight (before a 2 or after a king).

17. Calculate the following probabilities for an 8-card hand using a normal 52-card deck.

(a) Only face cards.

(b) No more than three black cards.

18. Calculate the following probabilities for an 8-card hand using a normal 52-card deck.

(a) Containing exactly two suits, with four cards from each suit

(b) Containing exactly two suits, with no specification about how the cards are distributed between those suits

19. Calculate the following probabilities for an 8-card hand using a normal 52-card deck.

(a) ★ Exactly two 2s, exactly three 3s, and exactly three cards with a third common face value.

(b) Four pairs. In other words, two of each of four distinct face values.

20. Consider the following two-person games involving a pair of fair, six-sided dice. Use appropriate counting principles and probability principles to derive the solutions.

(a) ★ Player A rolls a pair of dice and x and y are assigned the respective values showing on the top faces of the dice. Player A wins if either $x-3>0$ or $y-3>0$; otherwise, player B wins. Which player is more likely to win?

(b) Player A rolls a pair of dice. Player A wins if the sum of the digits on the dice is 8 or if both dice show odd digits; otherwise, player B wins. Which player is more likely to win?

(c) Player A repeatedly rolls a pair of dice until one of the following two events occurs. If a 3 appears on at least one of the dice, player A wins. If the sum of the digits on the two dice is 8, player B wins. Which player is more likely to win?

(d) Player A repeatedly rolls a pair of dice until one of the following two events occurs. If doubles occur, player A wins, If the sum of the digits on the two dice is either 3 or 9, player B wins? Which player is more likely to win?

21. Consider the random experiment of rolling three fair, six-sided dice. Compute the probability of the following events.

(a) At least one die is a 1 or a 6.

(b) At least two digits are even.

(c) Doubles (i.e., at least two of the dice show the same value).

(d) At most one of the digits on the dice is divisible by 3.

22. It has been determined that 7 out of a group of 11 teachers will be required to work at an after school homework help session. However, two of the teachers, Mrs. Henderson and Mr. Lay, also have an after school commitment with coaching the swim team, and so at most one of these two people can work at the homework help session.

(a) In how many ways can the seven teachers be chosen to work at the after school homework help session?

(b) What is the probability that Mrs. Henderson will work at the after school homework help session?

23. Coach Benson is going to choose 9 out of 16 people to fill the nine distinct positions on his baseball team. If Joe is assigned a position, it must be catcher. Additionally, if Joe is chosen for the team, then his friend Jeff must also be assigned a position.

(a) In how many ways can Coach Benson fill the positions for his baseball team?

(b) What is the probability that Jeff is assigned a position on the baseball team?

24. There are 17 men and twelve women signed up to go on a ski trip. Unfortunately, there is only transportation available for 18 people to go. There is an additional requirement that at least 4 women attend the ski trip.

(a) In how many ways can the people be chosen to attend the ski trip, assuming 18 people will be chosen?

(b) What is the probability that at least 12 men are chosen to attend the ski trip?

25. Suppose that you have decided that you will buy at least one, but no more than four items in a department store that sells 24 distinct product types, including T-shirts and jeans. You know that if you buy any T-shirt(s), you will also buy at least one pair of jeans. Assume that items within a product type are indistinguishable.

(a) ★ How many acceptable distinct purchases of the items can you make at the department store, assuming that you may buy more than one item of any particular product type?

(b) What is the probability that you will buy exactly one T-shirt?

26. Calculate the remaining probabilities that the attacker wins in a game of Risk. (This is not an easy task.)

6.4 Expected Value

It is time to extend the probability model that has been presented. The new concepts are the *value of an outcome* and *random variables*. One of the benefits of introducing these concepts will be a mathematical tool for analyzing sweepstakes and lotteries.

DEFINITION 6.10 ***Value of an Outcome***
To each outcome in a sample space, a real number may be associated. This number is called the *value of the outcome*. The value is a measure of the usefulness or desirability of the outcome.

The values of outcomes can be assigned by whatever criteria you wish. Often they are monetary values. They can also be assigned subjectively, as in the second example.

EXAMPLE 6.36 **Free Tickets**
A radio station is giving away free tickets to a concert. There are three levels of seating quality: \$20 seats, \$15 seats, and \$8 seats. If the station is giving away tickets at every quality level, the numbers 20, 15, 8, and 0 represent the value of being chosen (or not chosen) as the recipient of a ticket. ■

EXAMPLE 6.37 **Candy Values**
Recall the candy jar in Example 6.12. I have given each flavor a rating from 1 to 10, with 10 being most favorable. The ratings are shown in Table 6.5.

TABLE 6.5 Candy Ratings

Flavor	Value
Licorice	10
Cinnamon	8
Butterscotch	7
Watermelon	3

■

Notice that the probability of an outcome and the outcome's value are distinct concepts. If subjective values are used, the value of an outcome might change from person to person. For example, you might rate the four candy flavors differently.

EXAMPLE 6.38 **Flipping Coins**
For the random experiment of flipping a coin, a variable X can be assigned one of the values 1 or 0, depending on whether the result of the flip was a head or a tail. There is no reason (at the moment) to prefer 1 and 0 to any other pair of distinct numbers. The choice of value is a feature of the model that we tailor to the problem at hand.

For example, suppose the random experiment of flipping a coin is linked to a game in which I win 1 point if a head appears, but my opponent wins a point if a tail appears. I can think of my opponent winning a point as being equivalent to my losing a point (and the opponent not keeping a score). It thus makes sense to assign the value 1 to X if a head appears, and a value of -1 if a tail appears. If the game is played many times, I can keep a running sum of the values X attains to see who is winning. ■

DEFINITION 6.11 ***Random Variable***

A variable whose numeric value is assigned as the result of a random experiment is called a *random variable*. Random variables are traditionally denoted by uppercase letters, such as X and Y. The values of the random variable X are denoted by x or by x_i if a series of trials occurs.

EXAMPLE 6.39 **Rolling a Die**

If the random experiment is rolling a die, the random variable X can be assigned one of the values {1, 2, 3, 4, 5, 6}, again depending on the result of the roll. ■

EXAMPLE 6.40 **Picking a Piece of Candy**

If the random experiment is to reach into the candy jar and pull one piece out, the random variable X can be assigned the value of the flavor that is picked. Thus, if a cinnamon is chosen, X will have the value 8. ■

It is often useful to examine the average behavior of a random experiment. For example, if I were to roll a die many times, what number (on average) would appear? In other words, what is the average value of the random variable X, associated with this random experiment? One way to decide is to carry out an actual sequence of rolls, record the values of X after each roll, and then compute the average of these numbers. It is better to consider this theoretically.

Suppose for the moment that the die is rolled 600 times. If the die is fair, then about 1/6 of the time X will be 1, about 1/6 of the time X will be 2, and so on. The theoretical average is thus[16]

$$\begin{aligned}
&\frac{\overbrace{1+\cdots+1}^{100\text{ times}}+\overbrace{2+\cdots+2}^{100\text{ times}}+\cdots+\overbrace{5+\cdots+5}^{100\text{ times}}+\overbrace{6+\cdots+6}^{100\text{ times}}}{600}\\
&= \frac{1\cdot 100+2\cdot 100+3\cdot 100+4\cdot 100+5\cdot 100+6\cdot 100}{600}\\
&= 1\cdot\frac{100}{600}+2\cdot\frac{100}{600}+3\cdot\frac{100}{600}+4\cdot\frac{100}{600}+5\cdot\frac{100}{600}+6\cdot\frac{100}{600}\\
&= 1\cdot\frac{1}{6}+2\cdot\frac{1}{6}+3\cdot\frac{1}{6}+4\cdot\frac{1}{6}+5\cdot\frac{1}{6}+6\cdot\frac{1}{6}\\
&= 1\cdot\mathbf{P}(1)+2\cdot\mathbf{P}(2)+3\cdot\mathbf{P}(3)+4\cdot\mathbf{P}(4)+5\cdot\mathbf{P}(5)+6\cdot\mathbf{P}(6)=3.5
\end{aligned}$$

Notice that the theoretical average value (3.5) is an impossible value. You can never roll a 3.5! What this number signifies is that after many rolls, the average value of X should be near 3.5. If you were to actually roll a die 600 times, you are not likely to roll *exactly* 100 of each number. You *are* likely to roll close to 100 of each number.

EXAMPLE 6.41 **Average Candy Value**

If every day I randomly pick a piece of candy from the jar in Examples 6.12 and 6.37, what should my average value be? It will be helpful to have the probabilities and values of each flavor (Table 6.6 on page 304).

[16] I have taken the liberty of sorting the values X assumes. This will not affect the result of the computation but makes the principles involved clearer.

TABLE 6.6 Candy Values and Probabilities

Flavor	Probability	Value
Licorice	3/16	10
Cinnamon	1/4	8
Butterscotch	1/4	7
Watermelon	5/16	3

If I were to perform this random experiment every day for 160 days (each time replacing the flavor I had just picked), I would expect an average value of

$$\frac{\overbrace{10+10+\cdots+10}^{30 \text{ times}}+\overbrace{8+8+\cdots+8}^{40 \text{ times}}+\overbrace{7+7+\cdots+7}^{40 \text{ times}}+\overbrace{3+3+\cdots+3}^{50 \text{ times}}}{160}$$

$$= \frac{10\cdot 30+8\cdot 40+7\cdot 40+3\cdot 50}{160}$$

$$= 10\cdot\frac{3}{16}+8\cdot\frac{1}{4}+7\cdot\frac{1}{4}+3\cdot\frac{5}{16}$$

$$= 10\cdot \mathbf{P}(10)+8\cdot \mathbf{P}(8)+7\cdot \mathbf{P}(7)+3\cdot \mathbf{P}(3) = 6.5625$$

Notice again that this is a theoretical average value. If I perform the experiment many times, I expect the average value to be near $6\frac{9}{16}$. ■

In both examples, the average value of the (respective) random variable X could be expressed as a sum of the values of X times the probabilities X attains those values. This computation is of sufficient importance to warrant a definition.

DEFINITION 6.12 ***Expected Value***

If X is a random variable, the *expected value* of X, denoted $\mathbf{E}(X)$, is defined to be

$$\mathbf{E}(X) = \sum x\cdot \mathbf{P}(x)$$

where the sum is understood to be over all possible values, x, of X.[a]

[a] Another way to understand this sum is to add the product $x\cdot \mathbf{P}(\theta)$ for each outcome θ in the sample space, since X is assigned one of its values for each outcome.

EXAMPLE 6.42 **Expected Candy Value**

As was demonstrated previously, the expected value for the candy jar is

$$\sum x\cdot \mathbf{P}(x) = 10\cdot \mathbf{P}(10)+8\cdot \mathbf{P}(8)+7\cdot \mathbf{P}(7)+3\cdot \mathbf{P}(3)$$

$$= 10\cdot\frac{3}{16}+8\cdot\frac{1}{4}+7\cdot\frac{1}{4}+3\cdot\frac{5}{16} = 6.5625$$ ■

✔ Quick Check 6.10

1. Compute the expected value for the random variable X.

x	$\mathbf{P}(x)$
2	0.30
4	0.40
5	0.10
6	0.10
10	0.05
20	0.05

2. Compute the expected value for the random variable X.

x	$\mathbf{P}(x)$
100	0.25
−50	0.20
0	0.10
75	0.15
−25	0.30

☑

Expected values can be effectively used as one piece of information to analyze lotteries and sweepstakes. The idea is to define a random variable whose value corresponds to your winnings (or losses). The expected value of this random variable represents what you would win or lose if you were to play the game many times. Alternately, the expected value represents the combined average loss (or winnings) of all people who participate in the lottery.

It will be easier to discuss lotteries and sweepstakes using the following definitions.

DEFINITION 6.13 ***Odds***

The *odds* of an event can be expressed either as a ratio of success to failure of occurrence, or as a ratio of success to total occurrences. That is, the odds of an event are expressed as either $S : F$ or as $S : T$. The corresponding probabilities of success are $\mathbf{P}(S) = \frac{S}{S+F}$ or $\mathbf{P}(S) = \frac{S}{T}$, respectively. Note that $T = S + F$.

If a phrase, such as "odds are 3 to 2 in favor", the odds are being expressed as $S : F = 3 : 2$. If a phrase such as "odds are 5 to 2 against", the odds are in the form $F : S = 5 : 2$. This is the traditional manner for expressing odds. However, most lotteries and sweepstakes are using the $S : T$ form. It is your responsibility to determine which version is being used in any particular context. The example that follows demonstrates one possible way to check.

EXAMPLE 6.43

A Simple Lottery

The Minnesota State Lottery sponsored the lottery shown in Figure 6.6. The first task,

Figure 6.6 Lakes and Loons lottery.

Prizes And Odds For LAKES AND LOONS

(Based on 25,200,000 tickets sold)

If you get	You win	Approx. odds*	Approx. number of winners**
3 Bobbers	Free Ticket	1:8.33	3,024,000
3 Nets	$2	1:11.11	2,268,000
3 Cabins	$5	1:50	504,000
3 Stars	$10	1:125	201,600
3 Fish	$20	1:500	50,400
3 Boats	$50	1:1,200	21,000
3 Trees	$100	1:4,800	5,250
3 Hats	$500	1:12,000	2,100
3 Loons	$5,000	1:240,000	105

* The average overall odds of winning a prize are approximately 1:4.15. The average odds of winning a cash prize are 1:8.26.

** The number of winners may vary based on sales, distribution and number of prizes claimed.

before the expected value can be computed by means of the definition, is to decide which version of odds is being used. Consider the third row. The ratio $1 : 50$ is given, with 504,000 prizes available. Does this mean 1 chance in 50 of winning or 1 chance to win and 50 chances to lose? We can compare both versions to the total expected ticket sales:

$S : T$ Assume the ratio is in the form $S : T$. Then the probability of winning is $\frac{S}{T} = \frac{1}{50}$. The number of winners for this prize is thus $\frac{1}{50} \cdot 25{,}200{,}000 = 504{,}000$, which agrees with the chart.

$S : F$ Assume the ratio is in the form $S : F$. Then the probability of winning is

$$\frac{S}{S+F} = \frac{1}{1+50} = \frac{1}{51}$$

The number of winners for this prize is thus $\frac{1}{51} \cdot 25{,}200{,}000 \simeq 494{,}118$, which does not agree with the chart.

Therefore, the $S : T$ form for expressing odds is being used, not the $S : F$ form.

It is also necessary to interpret the prize in line one and to add the missing prize. Each ticket costs $1, so I will assume a free ticket is worth $1.[17] The missing prize is the

[17] You will not be allowed to cash the free, unscratched ticket in for a dollar. I am making a simplifying assumption here: You win a dollar, which is immediately spent on another ticket.

one for which you win nothing. If we add the entries in the final column of the previous table and subtract the result from 25,200,000, we can compute the approximate number of losers (people who win nothing). In this case, there are approximately 19,123,545 losers. With this settled, we can produce Table 6.7, which summarizes the values and probabilities for the random variable X associated with this lottery. The final column contains the product of the values and their respective (approximate) probabilities. You should compute the values of a few rows yourself, to make sure you understand how I did it.

TABLE 6.7 Expected value for Lakes and Loons

x	$\mathbf{P}(x)$	$x \cdot \mathbf{P}(x)$
0	0.758870833	0.00000
1	0.120048019	0.12000
2	0.090009000	0.18000
5	0.020000000	0.10000
10	0.008000000	0.08000
20	0.002000000	0.04000
50	0.000833333	0.04167
100	0.000208333	0.02083
500	0.000083333	0.04167
5000	0.000004167	0.02083
$\sum x \cdot \mathbf{P}(x) \simeq$		0.6450

The (approximate) expected value is the sum of the final column. This is 0.645, or 64.5 cents. After subtracting the dollar to buy the ticket, we observe that, on average, you lose 35.5 cents every time you buy a lottery ticket. In other words, if all the winnings and losses for all participants were totaled, the average gain (including the cost of the ticket) is -35.5 cents. In this case, it is more accurate to speak of an "average loss" rather than an "average gain". At 25,200,000 tickets, this represents a consumer loss of about \$8,946,000 (a number that will now be computed).

There are 105 lucky first prize tickets that pay (after the ticket cost is subtracted) a total of \$524,895, contributing to a total cash payout of \$13,230,000 for all monetary prizes. The total (pre-payout) income is computed as follows:

Total number of tickets	25,200,000
Number of free tickets	−3,024,000
Total income	\$22,176,000

Since (total consumer loss) = (total income) − (cash payout), the total consumer loss is

$$\$22,176,000 - \$13,230,000 = \$8,946,000$$

Of course, no one will ever lose 35.5 cents on a ticket. They will either lose a dollar, win a free ticket, or win some multiple of a dollar. However, there will be 19,123,545 times someone immediately loses a dollar. ■

Using the following definition, it should be clear that the Lakes and Loons lottery was not a fair game since the game cost \$1 to play but $\mathbf{E}(X) = \$0.645$.

DEFINITION 6.14 ***Fair Game***

A game of chance having an associated random variable X is called a *fair game* if $\mathbf{E}(X)$ = the cost of playing the game.

EXAMPLE 6.44

The Reader's Digest Sweepstakes

The Reader's Digest sweepstakes has been around for many years. The prizes and estimated[18] $S : T$ odds for one sweepstakes are listed in Table 6.8 (with bonus prizes omitted).

[18] The actual odds depend on how many entries the publisher receives. The estimates are based on the number of contestants in previous sweepstakes.

TABLE 6.8 A Reader's Digest Sweepstakes

x (Prize)	Number of Prizes	Odds	P(x)	$x \cdot$ P(x)
\$5,000,000	1	1 : 201,000,000	4.97512×10^{-9}	0.024876
\$150,000	1	1 : 201,000,000	4.97512×10^{-9}	0.000746
\$100,000	1	1 : 201,000,000	4.97512×10^{-9}	0.000498
\$25,000	2	1 : 100,500,000	9.95025×10^{-9}	0.000249
\$10,000	4	1 : 50,250,000	1.99005×10^{-8}	0.000199
\$5,000	8	1 : 25,125,000	3.98010×10^{-8}	0.000199
\$200	25	1 : 8,040,000	1.24378×10^{-7}	0.000025
\$125	200	1 : 1,005,000	9.95025×10^{-7}	0.000124
\$89	53,259	1 : 3,774	2.64971×10^{-4}	0.023582
\$0	200,946,499	1 : 1.00027	0.99973	0.000000
			$\sum x \cdot \mathrm{P}(x) \simeq$	0.050498

The odds for losing are 200,946,499 : 201,000,000, which is approximately 1 : 1.00027. Notice that the probability of losing is over 99.9%! This explains why most people reading this book may not know anyone who has won a prize in this sweepstakes. At the time this sweepstakes was run, it cost 29 cents (for a stamp) to enter. The expected value is a bit over 5 cents, so the average loss was approximately 24 cents. The actual losses were 29 cents; 53,501 lucky people actually gained at least \$89.[19] The expected number of entries (201,000,000) indicates that many people do not mind paying for a stamp in return for a (small) chance to win a much larger prize. ■

✔ Quick Check 6.11

1. A state lottery has posted the following $S : F$ odds. There are seven prizes.

Prize	Odds
\$500	1 : 99
\$200	2 : 98
\$100	4 : 96

   (a) Compute the expected value.
   (b) It costs \$20 to buy a ticket. Is this a fair game?
   (c) What is the probability of losing?
   (d) How much profit does the state expect to make?

2. The 1992 Reader's Digest sweepstakes lists the following prizes and odds. Assuming there are 199,500,000 entries, compute the expected value and compare it to the sweepstakes in Example 6.44. Extra credit: There is an inconsistency in the published figures. What is it?

x (Prize)	Number of Prizes	Odds
\$5,000,000	1	1 : 199,500,000
\$100,000	1	1 : 199,500,000
\$25,000	3	1 : 66,500,000
\$10,000	5	1 : 39,900,000
\$5,000	10	1 : 19,950,000
\$2,500	50	1 : 3,990,000
\$120	250	1 : 665,000
\$109	53,395	1 : 3,736
\$0	?	1 : ?

☑

[19]The \$89 prize was a watch, so the winners of that prize did not actually win money.

6.4.1 Exercises

The exercises marked with ★ have detailed solutions in Appendix H.

1. ★ What is signified by the ratios 1 : 4.15 and 1 : 8.26 in Example 6.43 on page 305? How can these probabilities be computed directly from the prize table?
2. Assign values to the following prizes.
   (a) Two tickets to the Super Bowl.
   (b) Two tickets to Hawaii.
   (c) A college degree.
   (d) A close friend. (e) Good health.
   How did you arrive at these numbers?
3. The proponents of state lotteries emphasize the following reasons for having a lottery.
   (a) A lottery provides entertainment.
   (b) The lottery generates revenue for the state, thus lowering taxes. (Most states predesignate where their share of the profits will be spent: for example, education, environmental protection, or road repair.)
   (c) Legal gambling should make illegal gambling less desirable.
   The opponents of state lotteries emphasize these points:
   (a) The state will need to spend more money for social programs (such as treatment for chronic gamblers).
   (b) The lottery will not generate as much revenue as promised. (A substantial portion of the profit is used to administer the lottery itself.)
   (c) The money will not be used as promised. For example, if the money is designated for education, a corresponding amount of revenue from other sources will be diverted away from education. In other cases, much of the money may be used for other purposes.
   Which arguments do you find most convincing? (Write several paragraphs to support your answer.)
4. Design a spreadsheet model for analyzing lotteries and sweepstakes. The model should contain the following features:
   (a) The data to be entered will be
      i. The cost to participate.
      ii. The values of the various prizes (excluding losing).
      iii. The odds for winning the various prizes (again excluding losing).
   (b) The values produced will include
      i. The probability of winning each prize.
      ii. The probability of losing (displayed as a separate summary value).
      iii. The expected value of the game.
      iv. The average gain or loss.
   (c) You may assume that there are at most 25 prizes.
   (d) You may also assume that odds are computed in the form $S : T$ but are expressed in the reduced form $1 : \frac{T}{S}$ (that is, with a numerator of 1).
5. Compute the expected values for the random variables listed below.

(a) ★

X	P(X)
2	0.35
5	0.40
8	0.25

(b)

Y	P(Y)
−10	0.125
−5	0.25
0	0.25
5	0.25
10	0.125

(c)

X	P(X)
0.524	0.138
0.913	0.322
1.127	0.045
2.418	0.285
3.212	0.160
−8.2	0.050

6. You are at the fair and are about to spin a colorful wheel in attempt to win a prize. If the spinner lands in the green sector of the wheel, you win a cookie, worth \$1. Similarly, having the spinner land in the yellow, red, and blue sectors gives you the following prizes, respectively: a toy car, worth \$3; a stuffed animal, worth \$5; a \$10 bill. The central angles in the colored sectors on the wheel have been measured and are recorded here: green: 180 degrees; yellow: 130 degrees; red: 35 degrees; blue: 15 degrees. It costs you \$2.75 to play this game once (i.e., spin the wheel once). What is the expected value of the game? Is it a fair game?
7. ★ A local club is holding a raffle for a used car valued at \$500. They are selling 2000 tickets for \$2 per ticket. What is the expected value of this raffle? What is the probability of losing? What is the average gain or loss (for the ticket purchasers)?
8. It costs \$2.50 to buy one of four sealed envelopes. Two of the envelopes each contain one dollar. The third envelope contains three dollars, and the fourth contains five dollars. Assuming that each envelope is equally likely to be picked, what is the expected value of the game? Is it a fair game?
9. Suppose that you are playing a card game at one of the school carnival booths. The cost of this game is \$0.90. To play, you draw two cards from a standard 52-card deck. If you obtain two face cards, you win \$8. However, if you get two aces, you owe \$0.50 (in addition to the cost of the game). Drawing an ace and a card with a number from 2 to 10, inclusive, on it will allow you to win \$2.50, while choosing two cards with a number from 2 to 10, inclusive, on each of them will give you \$0.50. In all other cases, nothing happens. What is the expected value of the game? Is it a fair game?

10. There are three types of problems on your final exam. Each of the 55 T/F questions are worth 2 points, while each of the 65 multiple choice questions are worth 3 points and each of the 27 fill-in-the-blank problems are worth 4 points. The probability that class member Jana will answer a T/F question correctly is 0.94, while the probability that she will answer a multiple choice question correctly is 0.89. Jana is twice as likely to get a T/F question correct as a fill-in-the blank problem. Assuming that there is no partial credit for answers given, what is Jana's expected score on her final?

11. Another way to deal with the "free ticket" prize in the Lakes & Loons lottery (page 305) is to assign it the value of $\mathbf{E}(X)$ instead of the value \$1.

(a) Write an equation that expresses this relationship. You may use the table in Example 6.43 to save time.

(b) Solve the equation for $\mathbf{E}(X)$.

(c) Is the new value for $\mathbf{E}(X)$ larger or smaller than the old value? Does this make sense?

12. A loaded die has the following probabilities:

1	.1	4	.125
2	.125	5	.125
3	.125	6	.4

What is the expected value of a roll? What is the expected value of a roll with a fair die?

13. The Reader's Digest has recently started a new form of sweepstakes having only one prize, with no entry fee (since you may enter via the internet instead of via the postal service). One such 2008 sweepstakes, "Volunteer Vacation Giveaway (#424)", had a prize valued at \$17,855 and 1 : 75,000,00 odds of winning.

(a) Calculate the expected value of this sweepstakes.

(b) Is this a fair game?

14. Consider the random experiment of rolling a pair of six-sided dice. Let X be the random variable whose value is the sum of the digits on the dice.

(a) Find the expected value of X, assuming fair dice.

(b) Suppose that any sum greater than 9 is three times as likely to occur as is any sum less than or equal to 9. Compute the expected value of X.

(c) Calculate the expected value of X, assuming that each die is biased so that a 1 is twice as likely as any other digit to come up.

15. ★ Suppose that a seamstress estimates that next year she will make 5,000 shirts of a particular style to sell. Because of the variation in production costs and in the price that she can sell her shirts, her profit per shirt may vary (see the probabilities given in the following table).

Profit Per Shirt	\$ −2	\$0	\$1	\$2	\$5	\$7
Probability	.30	.23	.19	.10	.11	.07

Estimate the profit on the 5,000 shirts.

16. There are 23 students in a classroom. Let S be the random variable whose value is a randomly chosen test score selected from the student scores for the past math test. The following table shows the distribution.

Test Score (%)	**# of Students**
70	6
75	1
82	7
86	2
94	3
99	4

(a) If all students are equally likely to be chosen, what is the expected value of S?

(b) If students scoring 99% are 3 times as likely to be chosen as students scoring 94%, students scoring 94% are 4 times as likely to be chosen as are students scoring in the 80s, students scoring in the 80s are twice as likely to be selected as students scoring 75%, and students scoring 75% and 70% are equally likely to be chosen, what is the expected value of S?

17. ★ Let D be the random variable whose values are 28, 30, or 31, depending on whether a randomly chosen (nonleap year) month contains 28, 30, or 31 days.

(a) If months are equally likely to be chosen, what is the expected value of D?

(b) If months that start with the letter "J" are twice as likely to be chosen as other months, what is the expected value of D?

18. Consider the following definition of gambling [45].

> A gamble is a reallocation of wealth, on the basis of deliberate risk, involving gain to one party and loss to another, usually without the introduction of productive work on either side. The determining process always involves an element of chance and may be only chance.

(a) In a gambling situation (as defined previously), what does expected value quantify?

(b) What is the significance of the logic operator *and* in the phrase "gain to one party *and* loss to another"?

19. Describe any differences between sweepstakes and lotteries. (Does the previous exercise offer any insight?)

20. A tire manufacturer has introduced a new 50,000-mile tire. Their testing indicates that about 5% of the tires will wear out before 45,000 miles. About 15% will wear out between 45,000 and 50,000 miles, while 50% will wear out between 50,000 and 55,000 miles. About 25% will wear out between 55,000 and 60,000 miles. The remaining 5% will wear out between 60,000 and 65,000 miles. What is the approximate expected tire life (in miles)? What assumptions have you made?

21. A lottery advertises three levels of prize money. Your lottery card has five numbers on it. You win if three or more of the numbers match. The prizes and odds are listed.

Match	Prize	Odds (S : T)
5 of 5	\$100,000	1 : 575,757
4 of 5	\$250	1 : 3,386.8
3 of 5	\$10	1 : 102.6

(a) Compute the expected value of this game.
(b) Compute the probability of losing.
(c) If tickets cost \$1, compute the average loss.

22. ★ A lottery charges \$5 per entry. In the lottery, the participants choose six digits (in order, with repetition). Only one person is allowed to choose any particular six-digit number. The lottery officials will award \$5,000,000 to the person who selects the winning combination of numbers. Find the expected value and probability of losing. Is this a fair game? State the assumptions you made to solve this problem.

23. The "Daily 3" lottery consists of four games. The participants pay \$1, choose a game, and then choose (in most cases) a three-digit number. The winning three-digit number is picked in the evening. The winning number may contain repeated digits (for example, 533 is a possible winning number). The games, odds, and payouts are listed.

6-way box: You must pick three distinct digits. You win if a permutation of your three digits matches the winning number. Notice that if 533 is the winning number, no one can win this game.

Straight: You win if your number matches the winning number exactly. You may repeat digits.

3-way box: Two of your digits must be the same (and the other different from them). You win if a permutation of your number matches the winning number.

Front pair: You choose the front two numbers (leaving the third number unspecified). If your two front numbers match those positions in the winning number exactly, you win. You may repeat digits.

Game	Odds (S : T)	Payout
6-way box	1 : 166.67	\$80
Straight	1 : 1000	\$500
3-way box	1 : 333.33	\$160
Front pair	1 : 100	\$50

For each game,

(a) Show how the odds were computed. (This requires material from Chapter 5.) What assumptions have you made?
(b) Compute the expected value.
(c) Compute the average loss.
(d) Compute the net profit (for the state and administrators) if one million people play the game.

6.5 The Binomial Distribution

The introduction to this chapter discussed the notion that randomness can exhibit patterns. Those patterns are formalized using mathematical models called *probability distributions*. This section introduces the most important finite probability distribution: the *binomial distribution*.

The binomial distribution describes the behavior of a random event that has two possible outcomes. The outcomes are traditionally labeled as *success* and *failure*. For example, if a coin is flipped, we might denote the outcome "heads" as a success and "tails" as a failure. In this model, the probability of a success is constant and is denote by p, where $0 \leq p \leq 1$. Consequently, the probability of a failure is $1 - p$.

The binomial distribution describes a situation where a fixed-probability, two-outcome random experiment is independently repeated n times. For example, a coin is flipped five times. The binomial distribution describes the probability that exactly r of the n experiments result in a success, for $r = 0, 1, \ldots, n$.

DEFINITION 6.15 ***Binomial Distribution***

Let R be a random experiment with two outcomes. Denote the outcomes as *success* and *failure* and assume that success has a fixed probability, p, of occurring. Assume also that R may be repeated, with each new trial being independent of previous trials.

Suppose R is repeated n times and that the random variable X records the number of successes. The set of possible values for X, together with their probabilities, is called the *binomial distribution* with parameters n and p. The random variable X is a *binomially distributed random variable* with parameters n and p.

Let $\mathbf{P}(X{=}r)$ denote the probability that there are r successes in n trials, where the values of n and p are implicitly known.

EXAMPLE 6.45 **Three Out of Five**

Suppose you play five games of paper, scissor, rock with a friend, with success being a win by you. What is the probability that you win the first, third, and fifth games?

Quick Check 6.4 on page 280 established your probability of winning to be $p = \frac{1}{3}$. Since the repeated trails are independent, the required probability is $\frac{1}{3} \cdot \frac{2}{3} \cdot \frac{1}{3} \cdot \frac{2}{3} \cdot \frac{1}{3} = \frac{4}{243}$.

What is the probability that you win the first three games, then lose the last two? The trials are still independent, so the probability is $\frac{1}{3} \cdot \frac{1}{3} \cdot \frac{1}{3} \cdot \frac{2}{3} \cdot \frac{2}{3} = \frac{4}{243}$.

It should be easy to convince yourself that the probability will remain the same no matter which three of the five games you win.

There are $\binom{5}{3} = 10$ ways to choose the three games you win. These 10 choices are mutually exclusive, so the probability that $X = 3$ is $\sum_{i=1}^{10} \frac{4}{243}$; that is, $\mathbf{P}(X=3) = \frac{40}{243}$.

✔ Quick Check 6.12

1. Let X be a binomially distributed random variable with $n = 4$. Suppose $p = 0$. Use your intuition to create a table listing the probabilities for the potential values of X.

Theorem 6.16, which is stated next, contains the factors p^r and $(1-p)^{n-r}$. These expressions can be problematic if $p = r = 0$ or $p = 1$ and $r = n$. What value should we assign to 0^0? The choice that makes the theorem agree with our intuition is to define $0^0 = 1$. This agrees with the choice made in the note after the statement of Theorem 3.58 on page 142 and presented as Definition 3.59.

THEOREM 6.16 ***Binomial Distribution Probabilities***

Let X be a binomially distributed random variable with parameters n and p. Then

$$\mathbf{P}(X=r) = \binom{n}{r} p^r (1-p)^{n-r}$$

Proof: If $p = 0$, then $\mathbf{P}(X=0) = \binom{n}{0} 0^0 1^n = 1$ and $\mathbf{P}(X=r) = \binom{n}{r} 0^r 1^{n-r} = 0$ for $r > 0$. If $p = 1$, $\mathbf{P}(X=r) = \binom{n}{r} 1^r 0^{n-r} = 0$ for $r < n$ and $\mathbf{P}(X=n) = \binom{n}{n} 1^n 0^0 = 1$.

If $0 < p < 1$. There are $\binom{n}{r}$ ways to choose which r of the n trials will be successes. These options are mutually exclusive so their probabilities should be added. In each case, there are r trials that are a success and $n-r$ trials that are failures. Since successive trials are independent and multiplication is commutative, the probability of exactly r successes is $p^r(1-p)^{n-r}$. Adding this number $\binom{n}{r}$ times completes the proof. □

Theorem 6.16 provides an exact probability for $\mathbf{P}(X=r)$. However, this is not what is commonly used in practice. The reason is that practical use might involve calculating probabilities like $\mathbf{P}(X \geq 500)$ where $n = 1000$. Calculating $\mathbf{P}(X=r)$ for $r = 500, 501, \ldots, 1000$ would be tedious. If $n \leq 20$ and p is one of a few common values (such as $p \in \{0.1, 0.2, 0.25, \text{ etc.}\}$) you could use a table of pre-calculated values. If both $np \geq 10$ and $n(1-p) \geq 10$, the preferred choice for calculating probabilities like $\mathbf{P}(X \geq 500)$ is to approximate the binomial distribution by using a normal distribution. Details can be found in any probability and statistics textbook.

Suppose we make a table of values for X and their respective probabilities. For instance, with $n = 5$ and $p = \frac{1}{3}$ the result would be Table 6.9.

Notice that the probabilities add to 1. This will always be true.

TABLE 6.9 Binomial Distribution with $n = 5$ and $p = \frac{1}{3}$

r	$\mathbf{P}(X=r)$
0	$\frac{32}{243}$
1	$\frac{80}{243}$
2	$\frac{80}{243}$
3	$\frac{40}{243}$
4	$\frac{10}{243}$
5	$\frac{1}{243}$

THEOREM 6.17

Let X be a binomially distributed random variable with parameters n and p. Then

$$\sum_{r=0}^{n} \mathbf{P}(X{=}r) = 1$$

Proof: The proof follows from Theorem 6.16 and a direct application of Theorem 5.14 on page 250.

$$\begin{aligned}\sum_{r=0}^{n} \mathbf{P}(X{=}r) &= \sum_{r=0}^{n} \binom{n}{r} p^r (1-p)^{n-r} \\ &= \sum_{r=0}^{n} \binom{n}{r} (1-p)^{n-r} p^r \\ &= ((1-p)+p)^n \\ &= 1\end{aligned}$$

□

It is helpful to look at graphs of various binomial distributions. For example, the distribution in Table 6.9 can be visualized by the bar chart in Figure 6.7.

TABLE 6.9 Binomial Distribution with $n = 5$ and $p = \frac{1}{3}$

r	$\mathbf{P}(X{=}r)$
0	$\frac{32}{243}$
1	$\frac{80}{243}$
2	$\frac{80}{243}$
3	$\frac{40}{243}$
4	$\frac{10}{243}$
5	$\frac{1}{243}$

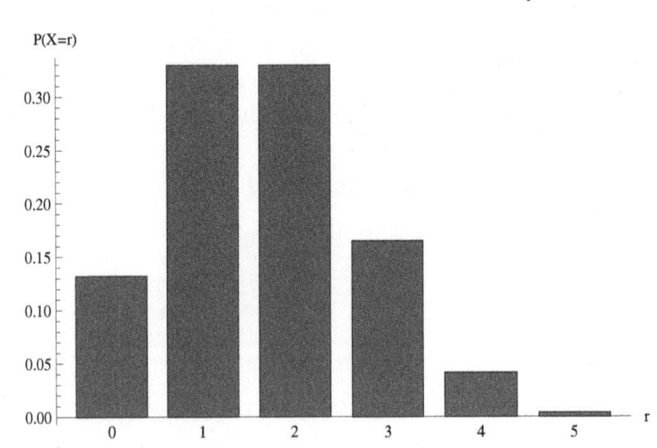

Figure 6.7 Binomial Distribution with $n = 5$ and $p = \frac{1}{3}$.

The binomial distribution with parameters $n = 100$ and $r = 0.2$ is graphed in Figure 6.8. Notice the very low probability for values of X outside the interval [6,35].

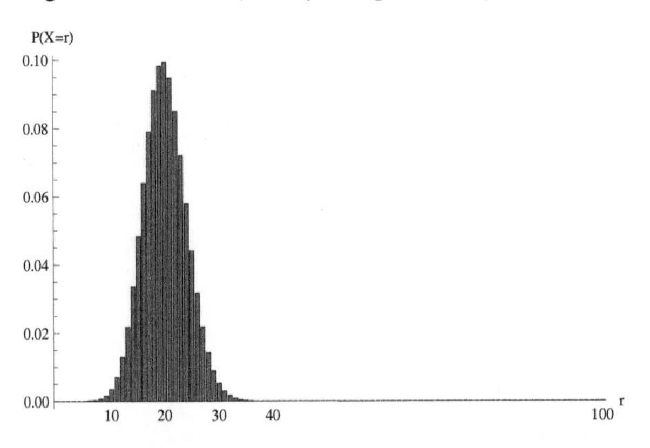

Figure 6.8 Binomial Distribution with $n = 100$ and $p = 0.2$.

✔ Quick Check 6.13

1. Create a table of probabilities for a binomially distributed random variable with parameters $n = 4$ and $p = 0.4$.
2. What is the probability of having $X \leq 2$ in the previous distribution?

A portion of the graph in Figure 6.8 has been reproduced as Figure 6.9.

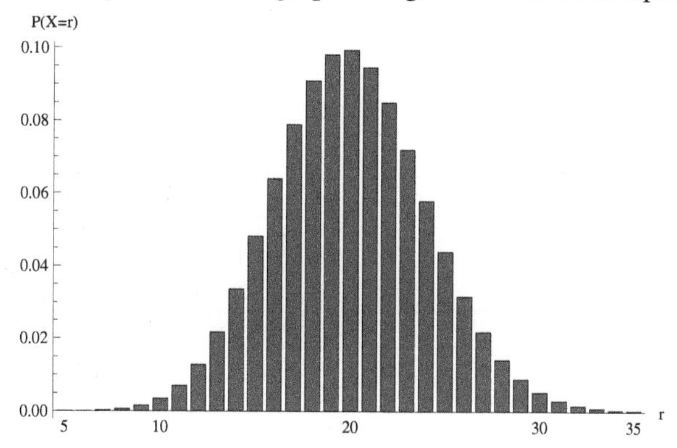

Figure 6.9 Binomial Distribution with $n = 100$ and $p = 0.2$.

Notice that $20 = 100 \cdot 0.2 = n \cdot p$. This is not an accident, as the next example and theorem will show. It should also make intuitive sense; if about 20% of the 100 trials should be a success, then about 20 successes should occur on average.

EXAMPLE 6.46

Expected Value

The expected value for the binomially distributed random variable in Table 6.9 (repeated here) is easy to calculate (recall that $n = 5$ and $p = \frac{1}{3}$).

r	$\mathbf{P}(X=r)$
0	$\frac{32}{243}$
1	$\frac{80}{243}$
2	$\frac{80}{243}$
3	$\frac{40}{243}$
4	$\frac{10}{243}$
5	$\frac{1}{243}$

$$
\begin{aligned}
\mathbf{E}(X) &= \sum_{r=0}^{n} r \cdot \mathbf{P}(X = r) \\
&= 0 \cdot \frac{32}{243} + 1 \cdot \frac{80}{243} + 2 \cdot \frac{80}{243} + 3 \cdot \frac{40}{243} + 4 \cdot \frac{10}{243} + 5 \cdot \frac{1}{243} \\
&= \frac{80}{243} + \frac{160}{243} + \frac{120}{243} + \frac{40}{243} + \frac{5}{243} \\
&= \frac{405}{243} \\
&= \frac{5}{3} \\
&= n \cdot p
\end{aligned}
$$

■

The previous examples indicate that np is a reasonable guess for $\mathbf{E}(X)$, when X is a binomially distributed random variable with parameters n and p. This is indeed the case, as the next theorem shows.

THEOREM 6.18 ***Expected Value of a Binomially Distributed Random Variable***

Let X be a binomially distributed random variable with parameters n and p. Then

$$\mathbf{E}(X) = np$$

Proof: The proof is an algebraic reordering to use Theorem 5.14 on page 250.

$$\begin{aligned}
\mathbf{E}(X) &= \sum_{r=0}^{n} r \cdot \mathbf{P}(X = r) \\
&= \sum_{r=0}^{n} r \binom{n}{r} p^r (1-p)^{n-r} \\
&= \sum_{r=1}^{n} r \binom{n}{r} p^r (1-p)^{n-r} \\
&= \sum_{k=0}^{n-1} (k+1) \binom{n}{k+1} p^{k+1} (1-p)^{n-k-1} \\
&= p \cdot \sum_{k=0}^{n-1} (k+1) \binom{n}{k+1} p^{k} (1-p)^{n-k-1} \\
&= p \cdot \sum_{k=0}^{n-1} \frac{(k+1)n!}{(k+1)!(n-k-1)!} p^{k} (1-p)^{n-k-1} \\
&= p \cdot \sum_{k=0}^{n-1} \frac{n \cdot (n-1)!}{k!(n-1-k)!} p^{k} (1-p)^{n-1-k} \\
&= p \cdot n \cdot \sum_{k=0}^{n-1} \frac{(n-1)!}{k!(n-1-k)!} p^{k} (1-p)^{n-1-k} \\
&= p \cdot n \cdot \sum_{k=0}^{n-1} \binom{n-1}{k} p^{k} (1-p)^{n-1-k} \\
&= p \cdot n \cdot \sum_{k=0}^{n-1} \binom{n-1}{k} (1-p)^{n-1-k} p^{k} \\
&= p \cdot n \cdot ((1-p)+p)^{n-1} \\
&= n \cdot p
\end{aligned}$$

□

✔ Quick Check 6.14

Prize	Odds
\$500	1 : 99
\$200	2 : 98
\$100	4 : 96

1. The lottery in Quick Check 6.11, Exercise 1 (page 307) has probability 0.93 of losing, so a probability of success of $p = 0.07$. Suppose you buy one lottery ticket each day for 30 days. What is the probability that you win some prize at least once?
2. What is the expected number of wins for the lottery in the previous exercise?
3. Discuss the expected value found in the previous exercise when compared with the expected value found in part (a) of Quick Check 6.11, Exercise 1.

☑

6.5.1 Exercises

The exercises marked with ★ have detailed solutions in Appendix H.

1. Let X be a binomially distributed random variable with parameters $n = 6$ and $p = 0.2$.
   (a) ★ Produce a table listing the probabilities for the potential values of X. Do not approximate.
   (b) Verify that these probabilities add to 1.
   (c) Produce a bar chart showing the probabilities.

2. Explain the equality that links lines 2 and 3 in the proof of Theorem 6.18.

3. Let X be a binomially distributed random variable with parameters $n = 6$ and $p = 0.5$.
   (a) Produce a table listing the probabilities for the potential values of X.
   (b) Verify that these probabilities add to 1.
   (c) Produce a bar chart showing the probabilities.

4. Let X be a binomially distributed random variable with parameters $n = 6$ and $p = 0.7$.
   (a) Produce a table listing the probabilities for the potential values of X.
   (b) Verify that these probabilities add to 1.
   (c) Produce a bar chart showing the probabilities.

5. Let X be a binomially distributed random variable with $n = 5$.
   (a) Let $p = 0.2$ and produce a table listing the probabilities for the potential values of X.
   (b) Let $p = 0.8$ and produce a table listing the probabilities for the potential values of X.
   (c) Make some observations about the relationship between the two tables.

6. Let X be a binomially distributed random variable with parameters $n = 5$ and $p = 0.5$.
   (a) ★ Calculate $\mathbf{P}(X \leq 2)$.
   (b) Calculate $\mathbf{P}(2 \leq X \leq 4)$.

7. Let X be a binomially distributed random variable with parameters $n = 6$ and $p = 0.2$.
   (a) Calculate $\mathbf{P}(X \geq 4)$.
   (b) Calculate $\mathbf{P}(2 \leq X \leq 4)$.

8. A blood center has found that 34% of previous blood donors have type A+ blood. Today there will be 10 donors at the center. What is the probability that the number, X, of donors with type A+ blood today will satisfy $3 \leq X \leq 5$? Assume that X is binomially distributed.

9. Suppose you flip a fair coin 100 times. This exercise requires a graphing calculator or mathematical software.
   (a) What is the probability that exactly 50 of the flips will be heads?
   (b) What is the probability that the number of heads will be between 45 and 55, inclusive?
   (c) What is the probability that there will be at most 5 heads?

10. Prove Proposition 6.19.

> **PROPOSITION 6.19**
> Let X be a binomially distributed random variable with parameters n and p and Y a binomially distributed random variable with parameters n and $1 - p$. Then $\mathbf{P}(X{=}r) = \mathbf{P}(Y{=}n{-}r)$.

The remaining exercises introduce another discrete probability distribution - the geometric distribution.

> **DEFINITION 6.20** ***Geometric Distribution***
> Let R be a random experiment with two outcomes. Denote the outcomes as *success* and *failure* and assume that success has a fixed probability, p, of occurring. Assume also that R may be repeated, with each new trial being independent of previous trials.
> Let X be a random variable that records the number of times R is repeated until the first success occurs. Then $X \in \{1, 2, 3, \ldots\}$. The set of possible values for X, together with their probabilities, is called the *geometric distribution* with parameter p. The random variable X is a *geometrically distributed random variable* with parameter p.
> Let $\mathbf{P}(X = r)$ denote the probability that r trials are required to obtain the first success, where the value of p is implicitly known.

11. Prove Proposition 6.21.

> **PROPOSITION 6.21**
> Let X be a geometrically distributed random variable with parameter p. Then $\mathbf{P}(X{=}r) = (1 - p)^{r-1} p$.

12. Let X be a geometrically distributed random variable with $p = 0.1$.
   (a) Create a table showing the probabilities that $X = r$, for $1 \leq r \leq 10$.
   (b) Create a bar chart for the table in part (a).

13. Let X be a geometrically distributed random variable with $p = 0.25$.
   (a) Create a table showing the probabilities that $X = r$, for $1 \leq r \leq 10$.
   (b) Create a bar chart for the table in part (a).

14. Let X be a geometrically distributed random variable with $p = 0.5$.
   (a) Create a table showing the probabilities that $X = r$, for $1 \leq r \leq 10$.
   (b) Create a bar chart for the table in part (a).

15. Let X be a geometrically distributed random variable with $p = 0.8$.

(a) Create a table showing the probabilities that $X = r$, for $1 \le r \le 10$.

(b) Create a bar chart for the table in part (a).

16. Prove Proposition 6.22.

> **PROPOSITION 6.22**
>
> Let X be a geometrically distributed random variable with parameter $p > 0$. Then
>
> $$\sum_{r=1}^{\infty} \mathbf{P}(X{=}r) = 1$$

17. ★ Prove Lemma 6.23. [*Hints*: (1) Use the ratio test from calculus to show convergence. (2) Let $z = (1 - p)$. Then Corollary 3.60 on page 143 implies $\sum_{k=0}^{\infty} z^k = \frac{1}{1-z}$. Now consider $\left(\sum_{k=0}^{\infty} z^k\right)\left(\sum_{k=0}^{\infty} z^k\right)$.]

> **LEMMA 6.23**
>
> Let $0 < p < 1$. Then
>
> $$\sum_{k=0}^{\infty} k(1-p)^k = \frac{1-p}{p^2}$$

18. Use Lemma 6.23 and Corollary 3.60 (on page 143) to prove Proposition 6.24.

> **PROPOSITION 6.24**
>
> Let X be a geometrically distributed random variable with parameter $p > 0$. Then
>
> $$\mathbf{E}(X) = \frac{1}{p}$$

6.6 Bayes's Theorem

Conditional probabilities enable us to revise probability estimates. The knowledge that one event has occurred often causes us to revise our estimate of the probability of another event's occurrence.[20] Sometimes we know the numeric value of a conditional probability in one direction [perhaps $\mathbf{P}(A \mid B)$], but actually need the numeric probability in the other order [$\mathbf{P}(B \mid A)$].

EXAMPLE 6.47 **Diagnosing Tuberculosis**

A convenient test for tuberculosis is the intermediate-strength purified protein derivative (PPD) Mantoux skin test. This test is less expensive than a chest X-ray, but is less reliable. The test sometimes predicts a patient has tuberculosis when in fact he or she does not. It also occasionally predicts someone is healthy when in fact that person does have tuberculosis.

Let T be the event "the patient has tuberculosis". Then $\overline{T}$ is the event "the patient does not have tuberculosis". Let W be the event "Warning: the PPD test predicts the patient has tuberculosis", and $\overline{W}$ be the complementary event "the PPD test predicts the patient does not have tuberculosis".

Clinical studies have been done with two groups of people. The first group contains people known (by other means) to have the disease. The second group contains people who are known (again by other means) to be free of tuberculosis. The results of these clinical studies produced the following approximate, empirical conditional probabilities [78]:

$$\mathbf{P}(W \mid T) = 0.775 \qquad \mathbf{P}(W \mid \overline{T}) = 0.15$$

and consequently (using probability principle 6)

$$\mathbf{P}(\overline{W} \mid T) = 0.225 \qquad \mathbf{P}(\overline{W} \mid \overline{T}) = 0.85.$$

This adds credence to the test but does not answer the questions patients care about the most: What are $\mathbf{P}(T \mid W)$ and $\mathbf{P}(T \mid \overline{W})$? The solution will be presented after some additional theoretical development. ■

[20] See Example 6.15 on page 284 for a quick review.

3. $\mathbf{P}(A \cup B) = \mathbf{P}(A) + \mathbf{P}(B)$, if A and B are mutually exclusive events

4. $\mathbf{P}(A \cap B) = \mathbf{P}(A) \cdot \mathbf{P}(B|A) = \mathbf{P}(B) \cdot \mathbf{P}(A \mid B)$

Probability principles 3 and 4 from Section 6.2.2 (page 292) can be used to reorder conditional probabilities in a useful manner. Rearranging formula 4 gives

$$\mathbf{P}(B \mid A) = \frac{\mathbf{P}(A \cap B)}{\mathbf{P}(A)}$$

Using the other part of the same formula produces the next refinement:

$$\mathbf{P}(B \mid A) = \frac{\mathbf{P}(B) \cdot \mathbf{P}(A \mid B)}{\mathbf{P}(A)}$$

Formula 3 implies

$$\mathbf{P}(A) = \mathbf{P}(A \cap B) + \mathbf{P}(A \cap \overline{B})$$

If formula 4 is once again applied to each of the previous summands,

$$\mathbf{P}(A) = \mathbf{P}(B) \cdot \mathbf{P}(A \mid B) + \mathbf{P}(\overline{B}) \cdot \mathbf{P}(A \mid \overline{B})$$

Putting these equations together leads to Bayes's theorem.

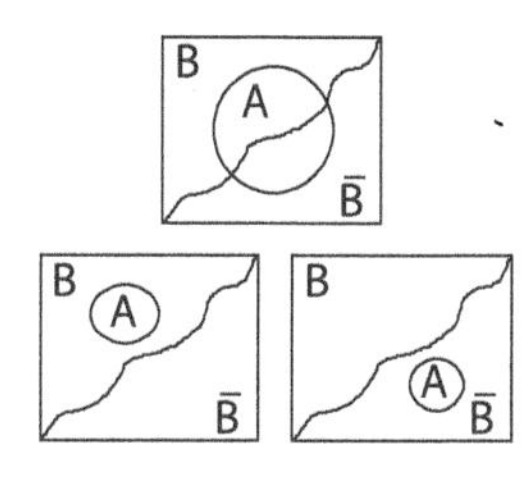

Figure 6.10. Events in Bayes's theorem.

THEOREM 6.25 ***Bayes's Theorem***

Let A and B be events. Then

$$\mathbf{P}(B|A) = \frac{\mathbf{P}(B) \cdot \mathbf{P}(A \mid B)}{\mathbf{P}(A)} = \frac{\mathbf{P}(B) \cdot \mathbf{P}(A \mid B)}{\mathbf{P}(B) \cdot \mathbf{P}(A \mid B) + \mathbf{P}(\overline{B}) \cdot \mathbf{P}(A \mid \overline{B})}$$

B and $\overline{B}$ are mutually exclusive and their union is the entire sample space so the events in Bayes's theorem can be visualized as one of the diagrams in Figure 6.10.[21]

SOLUTION 6.48

Diagnosing Tuberculosis

To apply Bayes's theorem, we need to know $\mathbf{P}(T)$. According to [78], $\mathbf{P}(T) \simeq 0.002$. Bayes's theorem thus implies

$$\mathbf{P}(T \mid W) = \frac{\mathbf{P}(T) \cdot \mathbf{P}(W \mid T)}{\mathbf{P}(T) \cdot \mathbf{P}(W \mid T) + \mathbf{P}(\overline{T}) \cdot \mathbf{P}(W \mid \overline{T})}$$

$$= \frac{0.002 \cdot 0.775}{0.002 \cdot 0.775 + 0.998 \cdot 0.15} \simeq 0.0102.$$

Similarly,

$$\mathbf{P}(T \mid \overline{W}) = \frac{\mathbf{P}(T) \cdot \mathbf{P}(\overline{W} \mid T)}{\mathbf{P}(T) \cdot \mathbf{P}(\overline{W} \mid T) + \mathbf{P}(\overline{T}) \cdot \mathbf{P}(\overline{W} \mid \overline{T})}$$

$$= \frac{0.002 \cdot 0.225}{0.002 \cdot 0.225 + 0.998 \cdot 0.85} \simeq 0.0005.$$

This test does an excellent job of catching people who actually have the disease since only 0.05% are missed ($\mathbf{P}(T \mid \overline{W})$). However, about 99% of those given a warning do not have the disease ($1 - \mathbf{P}(T \mid W)$). These people require additional tests, generally an X-ray, to verify that they are indeed free of tuberculosis. About 15% of the people without tuberculosis will have a false warning ($\mathbf{P}(W \mid \overline{T})$). ■

✔ Quick Check 6.15

1. Calculate $\mathbf{P}(B \mid A)$ if $\mathbf{P}(A) = 0.6$, $\mathbf{P}(B) = 0.4$, and $\mathbf{P}(A \mid B) = 0.8$.
2. Calculate $\mathbf{P}(D \mid J)$ if $\mathbf{P}(J \mid \overline{D}) = 0.40$, $\mathbf{P}(D) = 0.75$, and $\mathbf{P}(J \mid D) = 0.20$. ☑

[21] The diagrams assume that neither A nor B is the entire sample space.

There is a more general version of Bayes's theorem. It assumes that the sample space, S, can be expressed as a union of mutually exclusive events, $B_1, B_2, \ldots, B_n$. In that case, probability principle 3 implies

$$\mathbf{P}(A \cap S) = \mathbf{P}(A \cap B_1) + \mathbf{P}(A \cap B_2) + \cdots + \mathbf{P}(A \cap B_n)$$

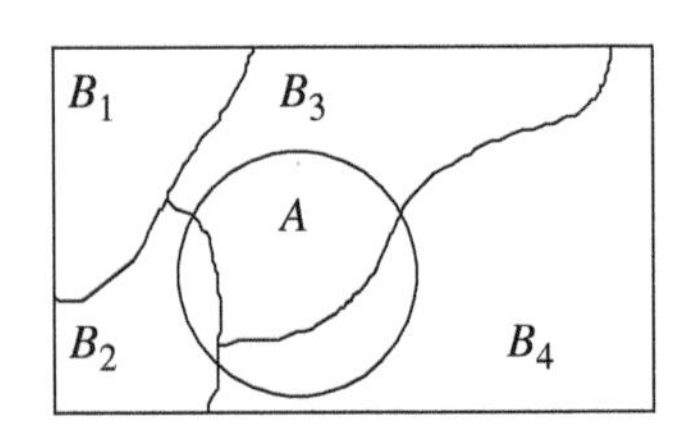

Figure 6.11. Bayes's theorem with $n = 4$.

THEOREM 6.26 ***Generalized Bayes's Theorem***

Suppose the events, $B_1, B_2, \ldots, B_n$, are mutually exclusive and their union is the entire sample space. Then, for $1 \le i \le n$

$$\mathbf{P}(B_i \mid A) = \frac{\mathbf{P}(B_i) \cdot \mathbf{P}(A \mid B_i)}{\mathbf{P}(B_1) \cdot \mathbf{P}(A \mid B_1) + \mathbf{P}(B_2) \cdot \mathbf{P}(A \mid B_2) + \cdots + \mathbf{P}(B_n) \cdot \mathbf{P}(A \mid B_n)}$$

Figure 6.11 shows one of many possible ways that the events might be related, assuming $n = 4$.

EXAMPLE 6.49 **Monty Hall**

A problem whose solution has generated many heated discussions is the "Monty Hall" problem. The problem involves a television game show in which a contestant is presented with three doors. Behind one door is an expensive prize (perhaps a 4-year scholarship to a famous college). Behind the other two doors are inexpensive consolation prizes (perhaps coffee cups bearing the show's logo).

The host knows which door conceals the expensive prize; the contestant does not. The contestant chooses a door, which remains closed. The host opens one of the other doors, displaying one of the consolation prizes. The host decides which door to open in one of two ways:

- If the contestant has chosen the door concealing the expensive prize, the host uses a coin flip to randomly choose one of the remaining doors.[22]
- If the contestant has chosen a door concealing a consolation prize, the host chooses the door that conceals the other consolation prize.

At this time, the contestant is offered the opportunity to switch doors.

Let S_i be the event "the scholarship is behind door i". Let C_j be the event "the host reveals a cup behind door j". We assume that $\mathbf{P}(S_i) = 1/3$ for $i = 1, 2$, and 3.

Suppose the contestant has chosen door 1 and the host has opened door 3, revealing a coffee cup. Is the contestant better off (on average) changing doors or sticking with the initial choice?

Clearly, $\mathbf{P}(C_3|S_2) = 1$ and $\mathbf{P}(C_3|S_3) = 0$, since the host must open a door containing a cup. Also, $\mathbf{P}(C_3 \mid S_1) = 1/2$, since the host flips a coin in this case. We are interested in $\mathbf{P}(S_2 \mid C_3)$. The generalized Bayes's theorem provides the value:

$$\mathbf{P}(S_2 \mid C_3) = \frac{\mathbf{P}(S_2) \cdot \mathbf{P}(C_3 \mid S_2)}{\mathbf{P}(S_1) \cdot \mathbf{P}(C_3 \mid S_1) + \mathbf{P}(S_2) \cdot \mathbf{P}(C_3 \mid S_2) + \mathbf{P}(S_3) \cdot \mathbf{P}(C_3 \mid S_3)}$$

$$= \frac{\frac{1}{3} \cdot 1}{\frac{1}{3} \cdot \frac{1}{2} + \frac{1}{3} \cdot 1 + \frac{1}{3} \cdot 0} = \frac{2}{3}$$

The contestant should change doors.

Notice that from the host's perspective, $\mathbf{P}(S_2)$ is either 1 or 0, depending on whether the scholarship is or is not behind door 2. The contestant however, has only partial knowledge. From her perspective, the scholarship has a probability of 2/3 of being behind door 2.[23] ■

[22]This has been added here to eliminate some of the controversy. If there is any ambiguity about how the host selects a door, the problem will not have a clear solution.

[23]See the extended footnote in Example 6.15 on page 284 for a similar discussion.

Quick Check 6.16

1. Compute $\mathbf{P}(S_1 \mid C_3)$ in Example 6.49.
2. Suppose $\mathbf{P}(A \mid B_1) = 1/4$, $\mathbf{P}(A \mid B_2) = 1/8$, and $\mathbf{P}(A \mid B_3) = 5/8$. If $\mathbf{P}(B_1) = 1/3$, $\mathbf{P}(B_2) = 1/2$, and $\mathbf{P}(B_3) = 1/6$, what are
   (a) $\mathbf{P}(B_1 \mid A)$
   (b) $\mathbf{P}(B_2 \mid A)$
   (c) $\mathbf{P}(B_3 \mid A)$

6.6.1 Exercises

The exercises marked with ★ have detailed solutions in Appendix H.

1. ★ Compute these conditional probabilities from Example 6.47 on page 316.
   (a) $\mathbf{P}(\overline{T} \mid W)$ (b) $\mathbf{P}(\overline{T} \mid \overline{W})$
2. You have invented a lie detector machine and have presented it to the court system for use. Suspects are asked whether or not they have committed a particular crime, and the machine can detect whether or not they are lying. Your invention has the property that 89% of guilty suspects in the court of law are properly judged. However, innocent suspects are incorrectly judged 1.75% of the time.
   (a) Suppose that a suspect is randomly selected from a group of suspects in which it is known that 10% of the people have committed a crime. The lie detector machine indicates that this person is guilty. What is the probability that this person is actually innocent?
   (b) Suppose now that a suspect is randomly selected from a group of suspects in which it is known that 17% of the people have committed a crime. The lie detector machine indicates that this person is innocent. What is the probability that this person is actually guilty?
3. The 1988 Information Please Almanac [55, page 799] offers information concerning the number of female and male arrests for serious crimes in the United States in 1986, categorized by sex and age. According to this source, the total number of arrests for serious crimes in the United States in 1986 was 2,167,071. Of the people arrested, 1,709,919 were male and 457,152 were female. There were 516,494 arrested males under the age of 18, while 124,911 of the arrested females were under 18.

   Let M represent the event "male", F represent the event "female", and $U18$ represent the event "under 18".

   Calculate the following conditional probabilities.
   (a) ★ $\mathbf{P}(M \mid U18)$ (b) $\mathbf{P}(F \mid U18)$
   (c) $\mathbf{P}(M \mid \overline{U18})$ (d) $\mathbf{P}(F \mid \overline{U18})$
4. Suppose that the weather can either be Hot, Mild, or Cold. It is either Sunny or Rainy. Approximately 50% of the days in a year are Mild, with 25% of the days Hot. It rains on about 20% of the days. The following probabilities are also known: $\mathbf{P}(S \mid M) = 0.9$, $\mathbf{P}(S \mid H) = 0.8$, $\mathbf{P}(S \mid C) = 0.6$, and $\mathbf{P}(M \mid R) = 0.25$.

   Compute $\mathbf{P}(M \mid S)$, $\mathbf{P}(H \mid S)$, $\mathbf{P}(C \mid S)$, and $\mathbf{P}(R \mid M)$.
5. The 1988 *Information Please Almanac* [55, page 615] indicates the following empirical probabilities related to the years of education (8 or less, 9–11, 12, 13–15, or 16 or more) of the people of voting age at the time of the 1984 presidential election in the United States and whether or not these people reported that they voted.

   $\mathbf{P}$(8 years or less) $= 0.121$
   $\mathbf{P}$(9–11 years) $= 0.130$
   $\mathbf{P}$(12 years) $= 0.399$
   $\mathbf{P}$(13–15 years) $= 0.182$
   $\mathbf{P}$(16 years or more) $= 0.168$
   $\mathbf{P}$(Did not vote | 8 years or less) $= 0.571$
   $\mathbf{P}$(Did not vote | 9–11 years) $= 0.556$
   $\mathbf{P}$(Did not vote | 12 years) $= 0.413$
   $\mathbf{P}$(Did not vote | 13–15 years) $= 0.325$
   $\mathbf{P}$(Did not vote | 16 years or more) $= 0.209$

   Compute the conditional probabilities listed.
   (a) $\mathbf{P}$(8 years or less | Voted)
   (b) $\mathbf{P}$(9–11 years | Voted)
   (c) $\mathbf{P}$(12 years | Voted)
   (d) $\mathbf{P}$(13–15 years | Voted)
   (e) $\mathbf{P}$(16 years or more | Voted)
6. A symphony orchestra schedules its musical offerings in the following proportions: 30% *B*aroque, 40% *C*lassical, 20% *R*omantic, and 10% *M*odern. The resident *D*irector almost always directs the more modern music, with *G*uest conductors more frequently directing older music. More specifically, for the past few years the approximate probabilities have been $\mathbf{P}(G \mid B) = 0.4$, $\mathbf{P}(G \mid C) = 0.25$, $\mathbf{P}(G \mid R) = 0.2$, $\mathbf{P}(G \mid M) = 0.1$. Compute
   (a) $\mathbf{P}(B \mid G)$ (b) ★ $\mathbf{P}(C \mid G)$ (c) ★ $\mathbf{P}(R \mid D)$
   (d) $\mathbf{P}(M \mid D)$ (e) $\mathbf{P}(G)$ (f) $\mathbf{P}(D)$
7. Design a spreadsheet model for Bayes's theorem.
8. Design a spreadsheet model for the generalized Bayes's theorem.
9. In a small kitchen appliance assembly plant, there are three types of products: *E*lectric Can Openers make up 27% of the production, while *M*icrowaves and *T*oasters make up 55% and 18% of the production, respectively. Not all of the products assembled at this plant work correctly. In fact, it is unfortunate that 1 in 10 *E*lectric Can Openers, 1 in 5 *M*icrowaves, and 25% of *T*oasters are *D*efective. Compute the following probabilities and then answer the questions.
   (a) $\mathbf{P}(M \mid D)$ (b) $\mathbf{P}(T \mid D)$
   (c) $\mathbf{P}(E \mid D)$ (d) $\mathbf{P}(D)$
   (e) Should the plant management be concerned?
   (f) Suppose that a product assembled at this plant is randomly selected. What is the meaning of the probabilities in parts (a)–(c)?

10. The boss of a large company has decided to award his employees with an all-expense-paid trip to Florida. He has 225 employees, each of which will be assigned accommodations at one of the following hotels: Gator Inn, Everglades Inn, Beachside Suites, Sea Breeze Resort. More specifically, 47 employees will be assigned rooms at the Gator Inn, while 59, 79, and 40 employees will be assigned rooms at the Everglades Inn, Beachside Suites, and Sea Breeze Resort, respectively. It is known that the showers do not have hot water in 3% of the rooms at the Gator Inn, in 1.5% of the rooms at the Beachside Suites, in 5% of the rooms at Sea Breeze Resort, and in 2.5% of the rooms at the Everglades Inn. What is the probability that

(a) An employee will be assigned a room with a shower that has hot water?

(b) An individual who has been assigned a room with a shower that has no hot water is staying at Sea Breeze Resort?

(c) An individual who has been assigned a room with a shower that has hot water is staying at the Everglades Inn?

(d) An individual who has been assigned a room with a shower that has no hot water is staying at the Beachside Suites?

(e) An individual who has been assigned a room with a shower that has hot water is staying at the Gator Inn?

11. ★ Both children (that is, people under the age of 18) and adults attended a music concert, with 33% of the people being children. As an added bonus for attending the concert, each guest got to choose exactly one of the following gifts: a ticket to the next concert, a poster autographed by the band members, a video, a T-shirt. Among the adults, 65% chose the ticket, while 21% chose the T-shirt, 10% the video, and 4% the poster. 32% of the children chose the ticket, while 31% chose the video, 9% the poster, and 28% the T-shirt. What is the probability that a randomly selected person from the concert who

(a) Chose the video is an adult?

(b) Chose the ticket is a child?

(c) Chose the poster is an adult?

(d) Chose the T-shirt is a child?

12. Suppose that five workers at a bakery are charged with the duty of stamping the expiration date on the wrapper for each loaf of bread to be sold. Janelle, who is given 22% of the loaves to stamp, fails to stamp the expiration date once in every 100 loaves; Amy, who is given 14% of the loaves to stamp, fails to stamp the expiration date twice in every 99 loaves; Mary, who is given 18% of the loaves to stamp, fails to stamp the expiration date three times in every 88 loaves; Sam, who is given 27% of the loaves to stamp, stamps the expiration date on 45 loaves out of every 50; and Pamela stamps the expiration date on 8 loaves out of every 10.

(a) Suppose that a customer received his or her loaf of bread and complained that it was missing an expiration date stamp. What is the probability that this was a loaf given to Pamela to stamp (i.e., Pamela forgot to place the date on the loaf)?

(b) Suppose that it is uncertain whether or not Amy really fails to stamp the expiration date twice in every 99 loaves. However, it has been determined that if a loaf of bread has no expiration date, the probability that it was Amy who failed to do this is 0.23. Was the original conclusion that Amy fails to stamp the expiration date twice in every 99 loaves correct? If not, give a new estimate for Amy's rate of stamping the expiration date.

13. The 1988 *Information Please Almanac* [55, page 66] offers data on the full-time status of the United States civilian labor force in 1986. The civilian labor force can be divided into three distinct groups of people: *M*ales, 20 years and older, *F*emales, 20 years and older, Persons 16–19 years old. Males, 20 years and older, made up approximately 58.44% of the labor force in 1986, while Females, 20 years and older, made up approximately 38.34% of the labor force. The following conditional probabilities are implied in the text:

$$\mathbf{P}(\text{Unemployed} \,|\, \text{Male, 20+}) = 0.061$$
$$\mathbf{P}(\text{Employed} \,|\, \text{Female, 20+}) = 0.934$$
$$\mathbf{P}(\text{Unemployed} \,|\, \text{Person, 16–19}) = 0.234$$

Compute the following probabilities.

(a) $\mathbf{P}(\text{Male, 20+} \,|\, \text{Employed})$

(b) $\mathbf{P}(\text{Female, 20+} \,|\, \text{Unemployed})$

(c) $\mathbf{P}(\text{Person, 16-19} \,|\, \text{Employed})$

(d) $\mathbf{P}(\text{Person, 16-19} \,|\, \text{Unmployed})$

(e) $\mathbf{P}(\text{Employed})$

(f) $\mathbf{P}(\text{Unemployed})$

14. Consider the following variations of the "Monty Hall" problem (Example 6.49 on page 318). In each case, assume you have chosen door number 1 and the host has opened door number 3. What is $\mathbf{P}(S_2 \,|\, C_3)$?

(a) There are two consolation prizes: an autographed photo of the host, and a coffee cup. Door 3 contained a cup.

(b) Both consolation prizes are cups (as in Example 6.49). There is no coin flip; if the contestant chooses door 1, the host opens door 3 unless the scholarship is behind door 3, in which case door 2 is opened.

15. With the advent of AIDS, it has become essential that donated blood be screened. A screening test developed in the mid-1980s was given the acronym ELISA [42], which is simpler than "enzyme linked immuno sorbent assay". This test correctly produces a warning with a probability of about 0.977 when the donated blood contains AIDS antibodies. The test incorrectly produces a warning with probability near 0.074 when the donated blood does not contain AIDS antibodies.

Suppose one blood sample in ten thousand actually contains AIDS antibodies. (Near the end of the 1980s, about 1 person in ten thousand in the United States was known to have AIDS. You may wish to use a more up-to-date estimate.)

Restate the previous sentences using the notation of conditional probabilities. Then answer the questions.

(a) If ELISA produces a warning, what is the probability that a donor has AIDS antibodies in his or her blood?

(b) If ELISA produces a warning, what is the probability that a donor does not have AIDS antibodies in his or her blood?

(c) If ELISA does not produce a warning, what is the probability that a donor has AIDS antibodies in his or her blood?

(d) If ELISA does not produce a warning, what is the probability that a donor does not have AIDS antibodies in his or her blood?

(e) If a donor is notified that a blood screen produced a warning, should the donor panic? Explain your answer.

16. At a 6th, 7th, and 8th grade middle school of 600 students, each student writes a story and is rated on a scale of *P*oor, *A*verage, and *E*xcellent. There are twice as many 6th graders as there are 7th graders. There are 24 more 7th grade students than 8th grade students and 23% of the 6th graders rate *P*oor, while 44% rate *A*verage. It is also known that 59% of the 7th grade students rate *A*verage, and 29% rate *E*xcellent. In the 8th grade, 11% of the people rate *P*oor, and 11% of the people rate *E*xcellent. Calculate the following conditional probabilities:

(a) $\mathbf{P}(6 \mid E)$

(b) $\mathbf{P}(6 \mid P)$

(c) $\mathbf{P}(7 \mid A)$

(d) $\mathbf{P}(7 \mid E)$

(e) $\mathbf{P}(8 \mid A)$

(f) $\mathbf{P}(8 \mid P)$

17. A company owns three identical fast food restaurants (i.e., they have the same name) at different locations. The fraction of employees who quit, reported by each restaurant, and the causes are shown in the following table. (For example, the fraction $\frac{1}{7}$ means that approximately 1 in 7 people from Restaurant *A* quit due to *L*ow Pay.) Assume that the entries in the table list the #1 reason why people quit, as people can obviously terminate employment due to multiple factors.

	Restaurant *A*	Restaurant *B*	Restaurant *C*
*L*ow Pay	$\frac{1}{7}$	$\frac{1}{5}$	$\frac{1}{12}$
*P*oor Management	$\frac{3}{14}$	$\frac{1}{4}$	$\frac{1}{4}$
*T*oo Busy	$\frac{2}{7}$	$\frac{1}{5}$	$\frac{1}{6}$
*I*rritable Customers	$\frac{5}{14}$	$\frac{7}{20}$	$\frac{1}{2}$

Suppose that out of all of the employees who quit last month, half came from Restaurant *A*, while $\frac{3}{10}$ and $\frac{1}{5}$ of the employees who quit came from Restaurant *B* and Restaurant *C*, respectively. Consider an employee who quit last month.

(a) If it was discovered that this employee quit because of *I*rritable Customers, what is the probability that he or she came from Restaurant *C*?

(b) If it was discovered that this employee quit because of *L*ow Pay, what is the probability that he or she came from Restaurant *A*?

(c) If it was discovered that this employee quit because of being *T*oo Busy, what is the probability that he or she came from Restaurant *B*?

(d) If it was discovered that this employee quit because of *P*oor Management, what is the probability that he or she came from Restaurant *B*?

(e) If it was discovered that this employee quit because of *L*ow Pay, what is the probability that he or she did not come from Restaurant *C*?

(f) If it was discovered that this employee quit because of *I*rritable Customers, what is the probability that he or she did not come from Restaurant *A*?

18. There are nine players on a particular baseball team, each of which has a batting average (see the following table).

Player	Batting Average
Johnson	0.402
Sawtell	0.382
Ito	0.200
Teller	0.310
Carlson	0.457
Anderson	0.210
Patters	0.315
Reeds	0.278
Brookson	0.341

(a) Suppose that a player on this team was selected to go up to bat, and he got a hit. Assume that the first four players in the preceding table were twice as likely to be selected as the other players. What is the probability that

i. It was Johnson at bat?

ii. It was Carlson at bat?

(b) Suppose that a player on this team was selected to go up to bat, and he did *not* get a hit. Assume now that the last four players in the preceding table were twice as likely to be selected as the other players. What is the probability that it was Ito at bat?

19. ★ The *Homeless in America* volume in the Information Series on Current Topics [65] indicates the following empirical probabilities related to a family in 1989 America being considered "low income".

$$\begin{aligned} \mathbf{P}(\text{White}) &= 0.781 \\ \mathbf{P}(\text{Black}) &= 0.103 \\ \mathbf{P}(\text{Other}) &= 0.116 \\ \mathbf{P}(\text{Low income} \mid \text{White}) &= 0.078 \\ \mathbf{P}(\text{Low income} \mid \text{Black}) &= 0.278 \\ \mathbf{P}(\text{Low income} \mid \text{Other}) &= 0.192 \end{aligned}$$

Compute the following conditional probabilities, which indicate the ethnic mix of the low-income families.

(a) $\mathbf{P}(\text{White} \mid \text{Low income})$

(b) $\mathbf{P}(\text{Black} \mid \text{Low income})$

(c) $\mathbf{P}(\text{Other} \mid \text{Low income})$

(d) Is this the ratio you were expecting?

20. The 1992 *World Almanac* [50, page 943] provides some empirical probabilities that a 25- to 34-year-old adult American was living with his or her parents in 1990. Let *MLP* represent "male, living with parents" and *FLP* represent "female, living with parents". Let *SM* stand for "single male" and *SF* represent "single female". The empirical probabilities are

$$\mathbf{P}(MLP) = 0.15 \qquad \mathbf{P}(MLP \mid SM) = 0.32$$
$$\mathbf{P}(FLP) = 0.095 \qquad \mathbf{P}(FLP \mid SF) = 0.20$$

The almanac does not list the probabilities that an adult in that age bracket is single or married. This omission provides you with the opportunity to complete the following activities.

(a) Solve for x:

i. $\mathbf{P}(SM \mid MLP) = x \cdot \mathbf{P}(SM)$

ii. $\mathbf{P}(SF \mid FLP) = x \cdot \mathbf{P}(SF)$.

[*Hint*: Use the simple form of Bayes's theorem.]

(b) State in words what these two equations mean.

(c) State whether more or less than half of the men and women in this age group were single. [*Hint*: Use your answer to part (a).]

6.7 Quick Check Solutions

Quick Check 6.1

1. One possibility is $T = \{\text{L, M, S}\}$, where T is the name of the sample space, L represents "long", M represents "medium", and S represents "short". The outcomes are the three possible choices of straw.

2. (a) A simple choice is $G = \{\text{P, S, R}\}$. There is one outcome for each possible choice. There is no significant difference; both have three distinct outcomes.

(b) A natural attempt is to designate the sample space as:

$$\{\text{P-P, P-R, P-S, R-R, R-S, S-S}\}$$

This is not the easiest to use sample space because it does not adequately represent which opponent wins. A better sample space is

$$\{\text{P-P, P-R, P-S, R-P, R-R, R-S, S-P, S-R, S-S}\}$$

We can agree that the first letter represents opponent 1 and the second letter represents opponent 2. Thus, the outcome P-S represents a win for opponent 2.

Quick Check 6.2

1. $NW = \{\text{P-P, R-R, S-S}\}$

2. $O_1 = \{\text{P-R, R-S, S-P}\}$

Quick Check 6.3

1. $\overline{O_1} = \{\text{P-P, P-S, R-P, R-R, S-R, S-S}\}$

2. $PC = \{\text{P-P, P-R, P-S, R-P, S-P}\}$

3. Yes, the intersection $\{\text{P-R, R-S, S-P}\} \cap \{\text{P-P, R-R, S-S}\}$ is empty. It is impossible simultaneously to have opponent 1 win and also have no winner.

4. No, the intersection is nonempty:

$$\{\text{P-P, P-R, P-S, R-P, S-P}\} \cap \{\text{P-R, R-S, S-P}\} = \{\text{P-R, S-P}\}$$

Therefore, it is possible simultaneously to have opponent 1 win and to have at least one of the opponents choose paper.

Quick Check 6.4

1. Theoretical—equally likely outcomes. Assuming that each outcome is equally likely, $\mathbf{P}(M) = 1/3$.
2. Theoretical—equally likely outcomes.
   (a) $\mathbf{P}(NW) = 3/9 = 1/3$. There are three ways to have no winner and 9 possible outcomes.
   (b) $\mathbf{P}(PC) = 5/9$
   (c) $\mathbf{P}(O_1) = 3/9 = 1/3$
3. Empirical. His current ratio is oversleeping 23/100 of the time. The best estimate (assuming he does not have a sudden transformation of character and habits) is that there is a probability of about 0.23 that he will oversleep tomorrow.
4. Subjective. Your answer to this will be influenced by the ages of your classmates and how much you know about their families. (There is one situation where the subjective probability becomes precise: If you are the only student and all four of *your* grandparents are currently living, the probability is 1.) The next time the class gathers, you can conduct a survey. You will then have a theoretical determination of the probability. It will not be empirical because once the survey is taken, you have complete knowledge of the ratio: (number with all 4 living)/(number in class). An empirical probability is an *estimate* of a current probability that is based on past performance.
5. Theoretical—unequally likely outcomes.
   (a) $\mathbf{P}(\text{Fr}) = 18/36 = 1/2$. $\mathbf{P}(\text{So}) = 12/36 = 1/3$. $\mathbf{P}(\text{Ju}) = 4/36 = 1/9$. $\mathbf{P}(\text{Se}) = 2/36 = 1/18$.
   (b) $\mathbf{P}(\text{UD}) = \mathbf{P}(\text{Ju}) + \mathbf{P}(\text{Se}) = 1/9 + 1/18 = 1/6$. I have used the definition

$$\mathbf{P}(E) = \sum_{\theta \in E} \mathbf{P}(\theta).$$

Quick Check 6.5

1. (a) The probability is $3/9 = 1/3$. This part does not use conditional probability.
   (b) $\mathbf{P}(\text{P-R} \mid \text{?-R}) = 1/3$. The revised sample space is

$$\{\text{P-R, R-R, S-R}\}.$$

   (c) If we didn't know that opponent 2 had displayed rock, the probability estimate would be $3/9 = 1/3$. Therefore, the events "opponent 2 displays rock" and "opponent 1 wins" are independent.
   (d) These events are not mutually exclusive since both contain the outcome P-R.
2. (a) $\mathbf{P}(O_1 \mid PC) = 2/5$. Opponent 1 wins on P-R or S-P (among the five outcomes in PC).
   (b) $PC = \{\text{P-P, P-R, P-S, R-P, S-P}\}$
3. $\mathbf{P}(E) = 3/6 = 1/2$. $\mathbf{P}(E \mid L) = 2/5$. The events are not independent. (Knowing that a 6 was not rolled decreases the possibility of an even number.)

Quick Check 6.6

1. Probability principle 3 is appropriate since these are mutually exclusive events. $\mathbf{P}(O_1 \cup O_2) = \mathbf{P}(O_1) + \mathbf{P}(O_2) = 1/3 + 1/3 = 2/3$.
2. Probability principle 2.
3. $\mathbf{P}(J \cup S) = \mathbf{P}(J) + \mathbf{P}(S) - \mathbf{P}(J \cap S) = 4/52 + 13/52 - 1/52 = 4/13$.

Quick Check 6.7

1. (a) $\mathbf{P}(B \cap L) = \mathbf{P}(B) \cdot \mathbf{P}(L \mid B) = 0.103 \cdot 0.278 \simeq 0.029$

   (b) There were about $\frac{0.061}{0.029} \simeq 2.1$ times as many low-income whites.

2. (a) P-S is the only outcome in the sample space that is in the event. The probability is thus 1/9.

   (b) Using probability principle 5, $\mathbf{P}(\text{P-?} \cap \text{?-S}) = \mathbf{P}(\text{P-?}) \cdot \mathbf{P}(\text{?-S}) = \frac{1}{3} \cdot \frac{1}{3} = \frac{1}{9}$. Principle 5 is more appropriate because the choices of opponent 1 are independent from the choices of opponent 2. This can be seen by comparing $\mathbf{P}(\text{P-?})$ and $\mathbf{P}(\text{P-?} \mid \text{?-S})$. (Principle 4 is not incorrect since it is always valid. Principle 5 is simpler and is thus preferred when it is valid.)

3. $\mathbf{P}(\overline{O_1}) = 1 - \mathbf{P}(O_1) = 1 - \frac{1}{3} = \frac{2}{3}$. (There are three ways for opponent 1 to lose and three ways for a tie. Thus there are six ways out of nine for opponent 1 to not win.)

4. Using the second form of formula 4,

$$\begin{aligned} \mathbf{P}(A \cap B) &= \mathbf{P}(B) \cdot \mathbf{P}(A \mid B) \\ &= \mathbf{P}(B) \cdot \mathbf{P}(A) \quad (\text{since } \mathbf{P}(A \mid B) = \mathbf{P}(A)) \\ &= \mathbf{P}(A) \cdot \mathbf{P}(B) \end{aligned}$$

Quick Check 6.8

1. $\mathbf{P}(C) = 1 - \frac{365 \cdot 364 \cdot 363 \cdots 355 \cdot 354}{365^{12}} = 1 - \frac{365}{365} \cdot \frac{364}{365} \cdot \frac{363}{365} \cdots \frac{355}{365} \cdot \frac{354}{365} \simeq 0.1670$

   The change at the second equal sign was unnecessary for $n = 12$ but was done to show how to keep your calculator from producing a number too large to store (which will happen with large values of n).

2. $\mathbf{P}(\text{more than one couple has } C \mid \text{at least one couple has } C)$

$$= \frac{\left[1-\left(1-\frac{1}{10}\right)^{10}\right]-\left[10 \cdot \frac{1}{10} \cdot \left(1-\frac{1}{10}\right)^{9}\right]}{1-\left(1-\frac{1}{10}\right)^{10}} \simeq 0.405$$

   If the population is 10 people and there is a 1 in 10 chance that a couple with particular characteristics can be found, then if we actually find such a couple, there is about a 40% chance we will find at least one other couple who have those characteristics.

Quick Check 6.9

1. There are 48 ways to choose the card that is not an Ace. There are $C(52, 5) = 2,598,960$ ways to deal a five-card hand. Thus

$$\mathbf{P}(4 \text{ Aces}) = \frac{48}{2{,}598{,}960} \simeq 0.000018.$$

   A .0018% chance indicates a very rare event.

2. There are $C(13, 3)$ ways to choose the clubs and $C(13, 2)$ ways to choose the diamonds. The choices are not mutually exclusive. However, since the problem has predetermined that these two suits will occur, the choices of clubs and of diamonds are independent. Thus there are $C(13, 3) \cdot C(13, 2) = 22,308$ ways to be dealt the required kind of hand. The probability is thus

$$\mathbf{P}(3 \text{ clubs and } 4 \text{ diamonds}) = \frac{22{,}308}{2{,}598{,}960} \simeq 0.0086,$$

   which is still unlikely.

3. Recall that the sum of two numbers is even if both are even or both are odd. Since the two dice are independent, there must be $3 \cdot 3 = 9$ ways to roll two even numbers and $3 \cdot 3 = 9$ ways to roll two odd numbers. (I have used the Independent Tasks Principle.) Rolling two even numbers and rolling two odd numbers are mutually exclusive events, so the Mutually Exclusive Tasks Principle implies that there are $9 + 9 = 18$ pairs that have an even sum. The probability is thus

$$\mathbf{P}(\text{even sum}) = \frac{18}{36} = 0.5,$$

which probably agrees with your intuition.

Quick Check 6.10

1. $\mathbf{E}(X) = 2 \cdot 0.3 + 4 \cdot 0.4 + 5 \cdot 0.1 + 6 \cdot 0.1 + 10 \cdot 0.05 + 20 \cdot 0.05 = 4.8$
2. Using an alternative way to organize the calculation:

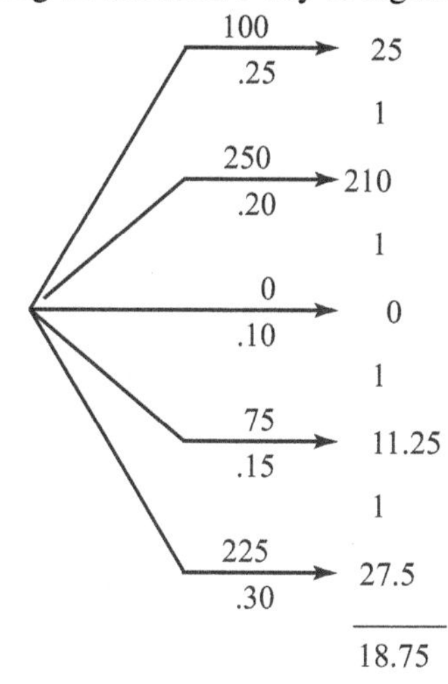

Quick Check 6.11

1. (a) The first step is to convert odds to probabilities. The table in part (a) reflects this conversion. The random variable X will be assigned the value of the prize won. Notice the inclusion of the most common prize: losing.

x	$\mathbf{P}(x)$	$x \cdot \mathbf{P}(x)$
500	0.01	5
200	0.02	4
100	0.04	4
0	0.93	0

$\sum x \cdot \mathbf{P}(x) \simeq 13$

Thus $\mathbf{E}(X) = \$13$.

(b) This is not a fair game. The state expects to make (on average) \$7 per ticket.

(c) $\mathbf{P}(\text{losing}) = 0.93 = 1 - (\mathbf{P}(500) + \mathbf{P}(200) + \mathbf{P}(100))$.

(d) There are only seven prizes, so the odds imply that 100 tickets are to be sold. The state will make $\$7 \cdot 100 = \$700$. A quick way to check this answer is to notice that the state takes in $\$20 \cdot 100 = \$2000$ and pays out \$1300, for a profit of $\$2000 - \$1300 = \$700$.

2. There are 53,715 winners, so $199{,}500{,}000 - 53{,}715 = 199{,}446{,}285$ losers. The odds are expressed in the $S : T$ form. This can be seen from the \$25,000 prize. There should be three such prizes. The product $\frac{1}{66{,}500{,}000} \cdot 199{,}500{,}000 = 3$ matches the assumption $S : T$. (Continued on the next page.)

The inconsistency is the \$120 entry. With the stated odds, there should be three hundred \$120 prizes. From the symmetry of the table of odds, I assumed that the ratio 1 : 665,000 is correct and the number 250 is incorrect. (The sweepstakes actually awarded 300 prizes; the number 250 was incorrect.) Another minor discrepancy: The odds for the \$109 prize are really 1 : 3,736.3, but this does not make any significant difference in the expected value. The rounded number was legitimate to publish.

The odds for losing are 199,446,285 : 199,500,000. This can be converted to 1 : 1.00027 by dividing both sides by 199,446,285.

The probabilities can be calculated from the $S : T$ odds as S/T.

x (Prize)	Number of Prizes	Odds	$\mathbf{P}(x)$	$x \cdot \mathbf{P}(x)$
\$5,000,000	1	1 : 199,500,000	5.01253×10^{-9}	0.02506
\$100,000	1	1 : 199,500,000	5.01253×10^{-9}	0.00050
\$25,000	3	1 : 66,500,000	1.50376×10^{-8}	0.00038
\$10,000	5	1 : 39,900,000	2.50627×10^{-8}	0.00025
\$5,000	10	1 : 19,950,000	5.01253×10^{-8}	0.00025
\$2,500	50	1 : 3,990,000	2.50627×10^{-7}	0.00063
\$120	300	1 : 665,000	1.50376×10^{-6}	0.00018
\$109	53,395	1 : 3,736	2.67666×10^{-4}	0.02918
\$0	199,446,285	1 : 1.00027	0.99973	0.00000
			$\sum x \cdot \mathbf{P}(x) \simeq$	0.05643

The expected value is about half a cent higher than in Example 6.44.

Quick Check 6.12

1. If $p = 0$, then we do not expect any successes, so we want $\mathbf{P}(X=0) = 1$ and $\mathbf{P}(X=r) = 0$ for $r > 0$.

Quick Check 6.13

1. The resulting table is

r	$\mathbf{P}(X=r)$
0	$\binom{4}{0}.4^0.6^4 = 0.1296$
1	$\binom{4}{1}.4^1.6^3 = 0.3456$
2	$\binom{4}{2}.4^2.6^2 = 0.3456$
3	$\binom{4}{3}.4^3.6^1 = 0.1536$
4	$\binom{4}{4}.4^4.6^0 = 0.0256$

2. $\mathbf{P}(X \leq 2) = \mathbf{P}(X=0) + \mathbf{P}(X=1) + \mathbf{P}(X=2) = 0.1296 + 0.3456 + 0.3456 = 0.8208.$

Quick Check 6.14

1. The probability of at least one win is 1 minus the probability of no wins. $\mathbf{P}(X=0) = \binom{30}{0} \cdot 0.07^0 \cdot 0.93^{30} \simeq 0.113367$, so the probability of at least one win is approximately 88.7%.

2. The expected number of wins is $np = 30 \cdot 0.07 = 2.1$.

3. With an expected 2.1 wins there will be an expected 27.9 loses, at \$20 per loss. The tickets will cost $30 \cdot 20 = \$600$. If one of the expected wins is a \$100 prize, your total expected winnings will, in the best case, be just barely above the cost of the tickets. It is more likely that you will not get the \$500 prize, so your expected winnings will be less than the ticket cost.

   Quick Check 6.11, Exercise 1 indicates that you expect to win \$13 per ticket on average. For 30 tickets, this comes out at \$390.

   Of course, with such a small number of tickets, you could do worse than the expected number of wins and never win at all.

Quick Check 6.15

1. $\mathbf{P}(B \mid A) = \frac{\mathbf{P}(B)\cdot\mathbf{P}(A|B)}{\mathbf{P}(A)} = \frac{0.4\cdot 0.8}{0.6} \simeq 0.53$
2. The expanded form of Bayes's theorem is appropriate here ($\mathbf{P}(J)$ was not given). It is necessary to compute $\mathbf{P}(\overline{D}) = 1 - \mathbf{P}(D) = 0.25$.

$$\mathbf{P}(D \mid J) = \frac{\mathbf{P}(D)\cdot\mathbf{P}(J \mid D)}{\mathbf{P}(D)\cdot\mathbf{P}(J \mid D) + \mathbf{P}(\overline{D})\cdot\mathbf{P}(J \mid \overline{D})}$$

$$= \frac{0.75\cdot 0.20}{0.75\cdot 0.20 + 0.25\cdot 0.40} = 0.60$$

Quick Check 6.16

1. Since $\mathbf{P}(S_3 \mid C_3) = 0$ and

$$\mathbf{P}(S_1 \mid C_3) + \mathbf{P}(S_2 \mid C_3) + \mathbf{P}(S_3 \mid C_3) = 1$$

$\mathbf{P}(S_1 \mid C_3) = 1/3$.

2. (a) $$\mathbf{P}(B_1 \mid A) = \frac{\mathbf{P}(B_1)\cdot\mathbf{P}(A \mid B_1)}{\mathbf{P}(B_1)\cdot\mathbf{P}(A \mid B_1) + \mathbf{P}(B_2)\cdot\mathbf{P}(A \mid B_2) + \mathbf{P}(B_3)\cdot\mathbf{P}(A \mid B_3)}$$

$$= \frac{\frac{1}{3}\cdot\frac{1}{4}}{\frac{1}{3}\cdot\frac{1}{4} + \frac{1}{2}\cdot\frac{1}{8} + \frac{1}{6}\cdot\frac{5}{8}} = \frac{1}{3}$$

(b) $$\mathbf{P}(B_2 \mid A) = \frac{\mathbf{P}(B_1)\cdot\mathbf{P}(A \mid B_1)}{\mathbf{P}(B_1)\cdot\mathbf{P}(A \mid B_1) + \mathbf{P}(B_2)\cdot\mathbf{P}(A \mid B_2) + \mathbf{P}(B_3)\cdot\mathbf{P}(A \mid B_3)}$$

$$= \frac{\frac{1}{2}\cdot\frac{1}{8}}{\frac{1}{3}\cdot\frac{1}{4} + \frac{1}{2}\cdot\frac{1}{8} + \frac{1}{6}\cdot\frac{5}{8}} = \frac{1}{4}$$

(c) $$\mathbf{P}(B_3 \mid A) = \frac{\mathbf{P}(B_1)\cdot\mathbf{P}(A \mid B_1)}{\mathbf{P}(B_1)\cdot\mathbf{P}(A \mid B_1) + \mathbf{P}(B_2)\cdot\mathbf{P}(A \mid B_2) + \mathbf{P}(B_3)\cdot\mathbf{P}(A \mid B_3)}$$

$$= \frac{\frac{1}{6}\cdot\frac{5}{8}}{\frac{1}{3}\cdot\frac{1}{4} + \frac{1}{2}\cdot\frac{1}{8} + \frac{1}{6}\cdot\frac{5}{8}} = \frac{5}{12}$$

Notice that the three conditional probabilities add to 1, which is expected since we are assuming that the sample space is a union of the mutually exclusive events B_i.

6.8 CHAPTER REVIEW

6.8.1 Summary

This chapter introduces the basic concepts of finite probability theory. An advanced course in probability or probability and statistics will extend these ideas to continuous (hence infinite) sample spaces. Most of the concepts will remain unchanged, but the mathematical details will change. In particular, integration will become a central tool in continuous probability theory.

Section 6.1 begins with a number of foundational concepts (such as sample space, event, independence). The standard model for probability is also introduced. The key notion is that the probabilities of the outcomes in a sample space should add to 1. Several methods (theoretical, empirical, subjective) for determining the probabilities of events are then discussed. The most important are the theoretical methods. In that context, the notion of equally likely outcomes is important.

Probability theory becomes more useful with the notion of conditional probabilities (Section 6.2). A conditional probability allows us to revise our estimate of an event's likelihood if we gain additional information. Section 6.2 continues by presenting several formulas that summarize some fundamental relationships between the probabilities of two events.

The chapter concludes with four short, but very interesting, sections. Section 6.3 shows how the material in Chapter 5 can be used to calculate theoretical probabilities. Section 6.4 introduces the important notions of random variables and expected value. The basic idea is to create a variable whose value is determined by the outcome of a random experiment. The expected value of a random variable, X, captures the notion of "the average value of X". Section 6.5 introduces the most important finite probability distribution: the binomial distribution. A probability distribution describes a particular pattern that the results of a random experiment might follow if repeated many times. Section 6.6 introduces Bayes's theorem. This theorem shows how to turn a collection of conditional probabilities into a different set of conditional probabilities (which are perhaps of greater interest). Practical applications include use with medical tests that attempt to diagnose diseases such as tuberculosis.

The material in this chapter will make much more sense if you thoroughly understand the definitions. Seek to gain an intuitive understanding of the material. If you can accomplish these tasks, the rest of the chapter will be much easier; conceptual understanding is more important than computational details. It is very useful to be aware that the definitions of outcomes, sample spaces, and events are expressed in the language of sets. This means that notions such as "the complement of an event" are really not new ideas.

The formulas and theorems are motivated by simple ideas. If you understand those ideas, the formulas are easy to memorize (since they are just mathematical shorthand for ideas you have already mastered).

Perhaps the most complex part of the chapter is the discussion of Bayes's theorem. For the generalized Bayes's theorem, it is important to note that the events, $B_1, B_2, \ldots, B_n$, form a partition of the sample space. To use the theorem, it is necessary to know all but one of the probabilities in the formula.

6.8.2 Notation

Notation	Page	Brief Description
S	274	a sample space
θ	277	an outcome
$\overline{E}$	276	the complement of event, E, ($\overline{E}$ is also an event)
$\mathbf{P}(\theta)$	277	the probability of outcome θ
$\mathbf{P}(E)$	277	the probability of event E
$\mathbf{P}(A \mid B)$	283	the conditional probability of event A, given that event B has occurred
$\mathbf{E}(X)$	304	the expected value of random variable X
$\mathbf{P}(X{=}r)$	310	the probability of r success in n binomial trials

6.8.3 Additional Review Material

Go to http://www.mathcs.bethel.edu/~gossett/DiscreteMathWithProof/review.xhtml for additional review material, including a sample chapter exam, with solutions.

CHAPTER

7

Recursion

If you already know what recursion is, just remember the answer. Otherwise, find someone who is standing closer to Douglas Hofstadter than you are; then ask him or her what recursion is.

Andrew Plotkin (born 1970)

One of the simplest optical tricks is to use two mirrors to produce a seemingly infinite chain of images. Each image is a smaller version of its containing image. The images shrink and shift a bit at each stage.

Imagine using a similar idea to solve problems. The key idea is to consider a solution process that uses solutions to one or more smaller versions of the same problem to create the solution to the original problem.

A simple example may help illustrate the idea.

EXAMPLE 7.1

Pascal's Triangle

Pascal's triangle is often introduced in high school algebra classes when discussing the expansion of $(x + y)^n$. The coefficient of the term $x^{n-k}y^k$ is the $(k + 1)$th entry in the $(n + 1)$th row of the triangle. Denote the current row by r and the current column by c. Then $r \geq 0$, $c \geq 0$, and $c \leq r$. Column 0 is always the position of the first entry in the row. Thus, columns are determined relative to the current row (rather than relative to the entire table). The first few rows are shown in Table 7.1.

TABLE 7.1 Standard Visualization of Pascal's Triangle

Row 0					1				
Row 1				1		1			
Row 2			1		2		1		
Row 3		1		3		3		1	
Row 4	1		4		6		4		1

The word *column* becomes more intuitive if an alternative visualization of the triangle[1] is used (Table 7.2).

TABLE 7.2 Alternative Visualization of Pascal's Triangle

Row 0	1				
Row 1	1	1			
Row 2	1	2	1		
Row 3	1	3	3	1	
Row 4	1	4	6	4	1

Notice that the entries in both column 0 and in the final column of each row are always 1. The other entries always equal the sum of the two entries in the previous row

[1] See Example 3.43 on page 149.

```
0:         1
1:       1   1
2:     1   2   1
3:   1   3   3   1
4: 1   4   6   4   1
```

```
0:  1
1:  1 1
2:  1 2 1
3:  1 3 3 1
4:  1 4 6 4 1
```

that are the nearest neighbors in the first visualization. More precisely, let the entry in row r and column c be denoted $t(r, c)$. Then[2]

$$t(r, c) = \begin{cases} 1 & \text{if } c = 0 \text{ or if } c = r \\ t(r\text{-}1, c\text{-}1) + t(r\text{-}1, c) & \text{otherwise} \end{cases}$$

Suppose the value of $t(12, 9)$ was needed. One approach would be to build rows 0 through 12 of the triangle. An alternative approach uses the algorithm **PascalTriangle**.[3] This algorithm directly calculates the answer if c places the entry on the left or right border of the triangle. Otherwise, it uses the definition

$$t(r, c) = t(r\text{-}1, c\text{-}1) + t(r\text{-}1, c)$$

to "pass the buck."

```
1: integer PascalTriangle (integer r, integer c)
2:    if (c == 0) or (c == r)
3:       return 1
4:    else
5:       return PascalTriangle(r−1,c−1) + PascalTriangle(r−1,c)
6: end PascalTriangle
```

PascalTriangle will work as long as

1. it is able to get correct values for `PascalTriangle(r−1,c−1)` and `PascalTriangle(r−1,c)`,
2. the process of "passing the buck" eventually stops.

The second condition is met because at line 5, one or both of r and c gets smaller (with $r \geq c$). These are nonnegative integers, so eventually c must reach 0 or r must reach c.

The first condition is less obviously met. It holds because the algorithm mimics (with a change of variable names) Pascal's theorem on page 248.

Pascal's Theorem
Let n and r be positive integers with $r \leq n$. Then

$$\binom{n+1}{r} = \binom{n}{r-1} + \binom{n}{r}$$

$$t(r, c) = \begin{cases} 1 & \text{if } c = 0 \text{ or if } c = r \\ t(r\text{-}1, c\text{-}1) + t(r\text{-}1, c) & \text{otherwise} \end{cases}$$

■

To explore the nature of the two requirements listed at the end of the previous example, another example will be introduced.

EXAMPLE 7.2

Subset Sums

Let W be a nonempty set of positive integers, and let s be a positive integer. How can we identify all subsets of W whose sum is s?

It is helpful, initially, to examine a particular case. For example, let $s = 6$ and $W = \{1, 2, 3, 4, 6\}$. The subsets of W whose sum is 6 are $\{1, 2, 3\}$, $\{2, 4\}$, and $\{6\}$.

One approach to devising an algorithm to find these subsets is to start with the entire set, W, and determine if it produces the desired sum. If not, throw out one element, and try again. Then add that element back in and discard a different element. Keep going until all elements have been discarded. Now start discarding collections of two elements, then collections of three, and so on.

The reduction to a smaller set suggests an alternative: use the "passing the buck" strategy from the previous example. Here is a preliminary algorithm:[4]

---

[2]Note the similarity to Theorem 5.10 on page 248.
[3]See Chapter 4 for a review of the algorithmic notation used in this text.
[4]See page 22 for a review of the set difference operator.

```
1: void SubsetSum1(set of positive integers W, integer s)
2:     t = sum of integers in W
3:     if t == s
4:       print W
5:     else
6:       for each x ∈ W
7:         V = W − {x}
8:         SubsetSum1(V,s)
9: end SubsetSum1
```

The algorithm properly passes the buck because the loop in lines 6–8 checks each subset that has one less element. The invocation of **SubsetSum1** in line 8 will determine if any smaller subsets will have sum s. However, this algorithm is not ready for use yet. It does not know when to stop. Before calculating a sum, it should check to see if W is empty or if $s \leq 0$. If $W = \emptyset$, the algorithm could just return and do nothing. Here is a complete algorithm:

```
1: void SubsetSum2(set of positive integers W, integer s)
2:     if (W == Ø) or (s ≤ 0)
3:       return
4:     t = sum of integers in W
5:     if t == s
6:       print W
7:       return
8:     else
9:       for each x ∈ W
10:        V = W − {x}
11:        SubsetSum2(V,s)
12: end SubsetSum2
```

This solution is not good enough; important information is being neglected. In particular, if $t \leq s$, no smaller subset can possibly have sum s (since every element is greater than 0). Here is a more efficient solution.

```
1: void SubsetSum3(set of positive integers W, integer s)
2:     if (W == Ø) or (s ≤ 0)
3:       return
4:     t = sum of integers in W
5:     if t < s
6:       return
7:     if t == s
8:       print W
9:       return
10:    else   # t > s, so look at smaller subsets
11:      for each x ∈ W
12:        V = W − {x}
13:        SubsetSum3(V,s)
14: end SubsetSum3
```

Lines 2–6 determine whether there is any point in continuing. Lines 7–9 handle a subset that meets the goal. Finally, lines 11–13 start examining smaller subsets for a match. The solution meets both previously established criteria for a solution: (1) **SubsetSum3**(V, s) will find any matches involving elements in the subset V, and (2) the algorithm will always stop since line 13 examines only proper subsets of W.

This algorithm still has the undesirable trait of identifying some solutions multiple times. For example, **SubsetSum3**($\{1, 2, 3, 4, 6\}$,6) will identify the solution $\{6\}$ multiple times since each subset containing 6 will eventually be reduced to the set $\{6\}$. ■

7.1 Recursive Algorithms

DEFINITION 7.1 ***Recursion***

Recursion is a process of expressing the solution to a problem in terms of a simpler version of the same problem. A *recursive algorithm* is an algorithm that invokes itself during execution.

Another simple example will illustrate these definitions.

EXAMPLE 7.3

Calculating n!

Since $n! = n \cdot (n-1) \cdot (n-2) \cdots 2 \cdot 1$, it should be clear that $n! = n \cdot (n-1)!$.[5]

A recursive procedure to calculate $n!$ can be expressed as

```
1: integer factorial(integer n)
2:   if n == 1
3:     return 1
4:   else
5:     return n·factorial(n − 1)
6: end factorial
```

```
1: integer factorial(3)
2:   if 3 == 1
3:     return 1
4:   else
5:     return 3·factorial(2)
6: end factorial
```

Figure 7.1. Page 1

```
1: integer factorial(2)
2:   if 2 == 1
3:     return 1
4:   else
5:     return 2·factorial(1)
6: end factorial
```

Figure 7.2. Page 2

```
1: integer factorial(1)
2:   if 1 == 1
3:     return 1
4:   else
5:     return 1·factorial(1)
6: end factorial
```

Figure 7.3. Page 3

Notice that the first step checks for values of n for which the solution is trivially known. The final statement calculates the answer by using the assumption that `factorial(n − 1)` is known.

Suppose $n = 3$ for the initial invocation. Because n is not equal to 1, the algorithm will drop to line 5 and attempt to multiply 3 times `factorial(2)`. However, `factorial(2)` is at present unknown. The algorithm must set the current multiplication aside and calculate `factorial(2)`. It does this by starting an entirely new version of the same process.

We can think of the initial invocation `factorial(3)` (Figure 7.1) as being done on page 1 of a stack of blank sheets of paper. The algorithm starts a new page (page 2, Figure 7.2) to calculate `factorial(2)`. At line 2 of the page 2 algorithm, it is determined that 2 is not equal to 1, so line 5 (on page 2) attempts to multiply 2 and `factorial(1)`. But `factorial(1)` is unknown. This means that page 2 is set aside and page 3 (Figure 7.3) is used to calculate `factorial(1)`.

On page 3, line 2 indicates that n *is* equal to 1. Thus, line 3 of page 3 returns the value 1 as the answer. Where does it return that value to? To line 5 of page 2. Now the page 2 algorithm can calculate the product 2 times `factorial(1)` to be $2 \cdot 1 = 2$. Line 5 on page 2 returns the answer 2 to line 5 of page 1. Finally, the product on line 5 of page 1 can be calculated as $3 \cdot 2 = 6$. Line 5 of page 1 sends the number 6 to whatever program originally invoked the algorithm. ■

Recursion is a very powerful technique, but many students find it difficult to understand on first acquaintance. It is generally helpful to look at numerous examples before attempting to write your own recursive algorithm.

Before looking at yet another example, I will give some general guidelines for creating a recursive algorithm. A detailed example will then be discussed. Next, some criteria will be given to help determine which problems are *unsuitable* candidates for a recursive solution. Finally, this section will end with detailed examinations of several examples that *ought* to be solved using recursive algorithms.

[5]Notice that ! has higher precedence than $\cdot$. Thus, $(n-1)!$ is calculated first, then multiplied by n.

7.1.1 General Guidelines for Creating Recursive Algorithms

Creating a recursive algorithm has some similarities with a proof by induction. Two correspondences are especially noteworthy:

- The base case in induction and the trivially solved smaller case in a recursive algorithm.
- The inductive step and the use of a smaller version of the algorithm.

The following steps are helpful guides for creating a recursive algorithm:

Step 1 Identify how to reduce the problem into smaller versions of itself.
Step 2 Identify one or more instances of the problem that can be directly solved.
Step 3 Determine how the solution can be obtained by combining the solutions to one or more smaller versions.
Step 4 Verify that the invocations in step 3 are within bounds.
Step 5 Assemble the algorithm.

These steps will now be examined in greater detail.

Identify how to reduce the problem into smaller versions of itself.

This is not possible for every problem. It is not natural for many other problems. For example, if I want to calculate 6% sales tax on a purchase costing $100, there is no natural way to create smaller versions of the problem. Nevertheless, recursion *is* possible (and appropriate) for many problems.

In the previous examples, the repetitive additions and multiplications of numbers with a common increment were indications that smaller versions of the problem could be easily identified. For example, $(n-1)!$ is a smaller version of $n!$.

Identify one or more instances of the problem that can be directly solved.

It is essential that *some* base case (or cases) can be solved directly. Otherwise, the recursive invocation of smaller versions of the problem will never cease.

Determine how the solution can be obtained by combining the solutions to one or more smaller versions.

This is the step that initially seems troublesome to many students. You need to exercise faith at this step.[6] You *assume* that the recursive invocations of the smaller instances work correctly. Your task is then to assemble those right answers into the correct answer for the original problem.

Consider line 5 of Example 7.3. It assumes that `factorial(`$n-1$`)` correctly returns the value of $(n-1)!$. With that assumption, it is correct to claim that $n! = n\cdot$ `factorial(`$n-1$`)`.

How can you build confidence that your faith is not misplaced? The key is to convince yourself that your assembled solution is correct *whenever* the recursive invocations work correctly. The final vindication of your faith is provided by the next step.

Verify that the invocations in step 3 are within bounds.

Make sure that the smaller versions in step 3 are (a) *really* smaller than the original and (b) not smaller than one of the instances found in step 2. If the recursive invocations seek solutions to bigger versions of the problem (or of the same size problem), you will never reach a base case. The algorithm will never terminate. If a recursive invocation tries to solve a problem that is smaller than every base case, the same problem will occur.

[6]Faith in the sense of a reasoned confidence in a reliable object of faith, not as an unfounded wish that something good will happen.

Once you know that each new invocation works on a smaller problem and that eventually a base case will be reached, it is inevitable that the algorithm must terminate after a finite number of steps.[7]

Assemble the algorithm.

Make sure that the first thing done in the algorithm is to check for the instances found in step 2. Otherwise, you will recurse past the base cases and never terminate.

7.1.2 A Detailed Example

Any nonnegative integer can be represented as a binary (base 2) number. For example, the number five, denoted "5" in decimal notation (base 10), is written as "101" in binary.[8]

Is it possible to determine the number of 1s in the binary representation of the decimal integer n?

One possible algorithm is to convert the decimal number to binary and then count the 1s. Is there another way? Of course! I will now develop a recursive algorithm to solve this problem.

Step 1

The key insight is that every odd number has a 1 in the rightmost position[9] in the binary representation, whereas every even number has a zero in that position.

If the number is even, it can be divided by 2 and the number of 1s will not change. For example, the binary representation of 6 is 110. Dividing by 2 produces 3, which has binary representation 11. Dividing an odd integer by 2 produces a remainder. Ignore the remainder. The number of ones in the quotient will be one less than the number of 1s in the dividend. For example, the binary representation for 13 is 1101.[10] Dividing by 2, the quotient is 6 (the remainder is discarded), with binary representation 110.

It is fairly easy to see that we can obtain the quotient of a binary division by 2 simply by eliminating the rightmost bit (effectively shifting all bits right by one position).

It is now possible to define the basic reduction:

- If n is even, the number of 1s in the binary representation of n is the same as the number of 1s in the binary representation of $\frac{n}{2}$.
- If n is odd, the number of 1s in the binary representation of n is one more than the number of 1s in the binary representation of $\left\lfloor \frac{n}{2} \right\rfloor$.[11]

Step 2

The base cases are fairly easy. Since the reduction in step 1 distinguished between even and odd values for n, perhaps it is necessary to have both an even and an odd base case.

- The number of 1s in the binary representation of 0 is 0.
- The number of 1s in the binary representation of 1 is 1.

After a bit of thought, it should be apparent that the second base case is not really necessary: We can use the reduction rule to determine that the odd number 1 has [1 + (the number of 1s in the binary representation of $\left\lfloor \frac{1}{2} \right\rfloor$)] = [1 + (the number of 1s in the binary representation of 0)] 1s in its representation.

Therefore, our base case is

- The number of 1s in the binary representation of 0 is 0.

---

[7] I am assuming that the problem size can be represented by a nonnegative integer.

[8] This represents $1 \times 2^2 + 0 \times 2^1 + 1 \times 2^0$.

[9] The 2^0 or 1s place.

[10] $2^3 + 2^2 + 2^0$

[11] We lose a 1 when we shift the rightmost bit off the representation to get the quotient.

Step 3

Most of the work for this step has already been done in step 1.

```
if n is even
  return (the number of 1s in n/2)
else
  return 1 + (the number of 1s in ⌊n/2⌋)
```

Step 4

Both recursive invocations involve $\frac{n}{2}$, which is smaller than n and never less than 0 (the base case).

Step 5

```
1: integer numOnesInBin(integer n)
2:   if n == 0
3:     return 0
4:   else if n is even
5:     return numOnesInBin(n/2)
6:   else
7:     return 1 + numOnesInBin(⌊n/2⌋)
8: end numOnesInBin
```

The algorithm is complete and ready to use. It might be helpful to work through a few examples using the RecursionDemo Java application.[12] This program allows you to step through the algorithm and actually see the various recursive calls to the algorithm. The screen capture in Figure 7.4 shows the algorithm part way through determining the number of 1s in the decimal number 10. The original invocation has determined that 10 is even[13] and has recursively invoked the algorithm on $n = 5$. The first copy of the algorithm is waiting for the answer from copy 2. The second copy of the algorithm has determined that 5 is odd and has invoked a third copy of the algorithm to calculate the number of 1s in 2.[14]

The third copy of the algorithm has completed its task and is about to return its answer to copy 2. There were two additional recursive invocations that have already terminated at this point in the process.

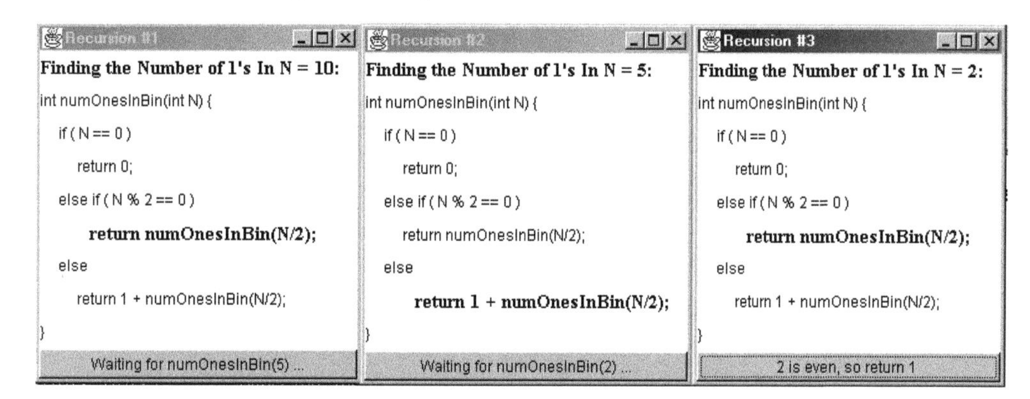

Figure 7.4 RecursionDemo in action.

You should attempt a sufficient number of examples so that you are able to work comfortably through an example by hand (perhaps using several sheets of paper).

---

[12] Available at http://www.mathcs.bethel.edu/~gossett/DiscreteMathWithProof/ .

[13] Because the remainder upon division by 2 is 0. (Java uses the % operator to denote "take the remainder.")

[14] Java automatically rounds down for integer division. Thus, there is no need for the floor function: $N/2$ is 2 when $N = 5$.

7.1.3 When Should Recursion Be Avoided?

Recursion is not always an appropriate solution strategy, even if we can easily think of a way to solve the problem using recursion. This is because recursion may be much less efficient than other strategies.

EXAMPLE 7.4

Adding the First *n* Integers

Suppose you want to add the integers $1, 2, 3, \ldots, n$ but have irresponsibly forgotten the formula $\sum_{k=1}^{n} k = \frac{n(n+1)}{2}$. Adding the numbers one at a time is too boring.

A simple observation leads to a new procedure for calculating the sum: $\sum_{k=1}^{n} k = n + \sum_{k=1}^{n-1} k$.

That is, we can find the sum of the first n integers by adding n to the sum of the first $n-1$ integers!

While this is not a profound observation, it *is* correct. However, it seems of little use. How are we to find the sum of the first $n-1$ integers? The same way: $\sum_{k=1}^{n-1} k = (n-1) + \sum_{k=1}^{n-2} k$. Similarly, $\sum_{k=1}^{n-2} k = (n-2) + \sum_{k=1}^{n-3} k$. This apparently leads nowhere.

The observation that rescues the procedure is that the sum of the first 1 integers is 1. Therefore, the sum of the first two integers is $2 + \sum_{k=1}^{1} k = 2 + 1 = 3$. Now we can calculate the sum of the first three integers: $3 + \sum_{k=1}^{2} k = 3 + 3 = 6$. The sum of the first 4 integers is therefore $4 + 6 = 10$. We can continue until we reach n. ■

The previous example may seem quite ridiculous. To avoid adding the numbers 1 through n, I have done a lot of work. Eventually I still add the numbers 1 through n. There does not seem to be much practicality to this *new! improved!* method.

The algorithm is even more impractical than is readily apparent. When an algorithm invokes a copy of itself, the computer must store all the information necessary to restart the invoking copy at a later time. It must then set aside additional memory for the new copy. If many copies are waiting for results, the computer will have used a noticeable amount of memory. Moreover, the process of creating new copies and sending the results back to invoking copies will take extra time.

There are two easily identified situations where a recursive algorithm should usually be modified: *tail-end recursion* and *redundant recursion*.

```
1: integer factorial(n)
2:   if n == 1
3:     return 1
4:   else
5:     return n·factorial(n-1)
6: end factorial
```

Tail-End Recursion

An algorithm uses tail-end recursion if the only recursive invocation it makes occurs at the last line of the algorithm.[15] An example would be the recursive calculation of $n!$ in Example 7.3. The only recursive invocation, `factorial(n − 1)`, occurs at line 5.

Algorithms that use tail-end recursion can be easily changed into algorithms that use a simple loop. No suspended copies of a recursive algorithm are required.

EXAMPLE 7.5

Efficiently Calculating *n*!

```
1: integer nfact(integer n)
2:   nf = n
3:   i = n
4:   while i ≠ 1
5:     i = i − 1
6:     nf = nf * i
7:   return nf
8: end nfact
```

The multiplication and recursive invocation at line 5 of Example 7.3 has become a simple multiplication at line 6. Lines 4 and 5 accomplish what the recursion previously achieved. The base case check in the recursive form of the algorithm has now become the loop termination test in line 4. ■

[15] That is, a *single* recursive invocation occurs as the last step in the algorithm.

Redundant Recursion

Redundant recursion occurs when an algorithm directly or indirectly invokes multiple instances of the same smaller version. Recall the final comment in Example 7.2. The algorithm **SubsetSum3** generates many of the subsets of W multiple times during the recursion.

A simpler example will provide another illustration of redundant recursion.

EXAMPLE 7.6

Calculating 2^n the Hard Way

For $n \geq 1$,

$$2^n = 2 \cdot 2^{n-1} = 2^{n-1} + 2^{n-1}$$

Since $2^0 = 1$, a recursive algorithm for calculating 2^n is

```
1: integer tn(integer n)
2:    if n == 0
3:       return 1
4:    else
5:       return tn(n − 1) + tn(n − 1)
6: end tn
```

You may have (incorrectly) thought that this is a tail-end recursion. The situation is actually much worse. Notice that the algorithm calculates 2^{n-1} two different times. In addition, when the algorithm invokes the left-hand copy of tn($n-1$) it will calculate 2^{n-2} two times (via invocations of tn($n-2$)). When the algorithm invokes the right-hand copy of tn($n-1$), it will again calculate 2^{n-2} twice.

Thus, 2^{n-1} will be calculated two times and 2^{n-2} will be calculated four times. You should convince yourself that if $n > 2$ the algorithm will calculate 2^{n-3} eight times. This is clearly inefficient. A simple loop would be a better approach.

```
1: integer twoExpn(integer n)
2:    tn = 1
3:    i = 1
4:    while i ≤ n
5:       tn = 2 * tn
6:       i = i + 1
7:    return tn
8: end twoExpn
```

■

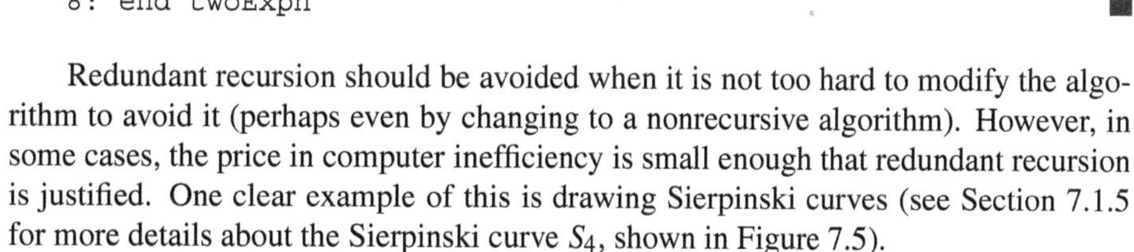

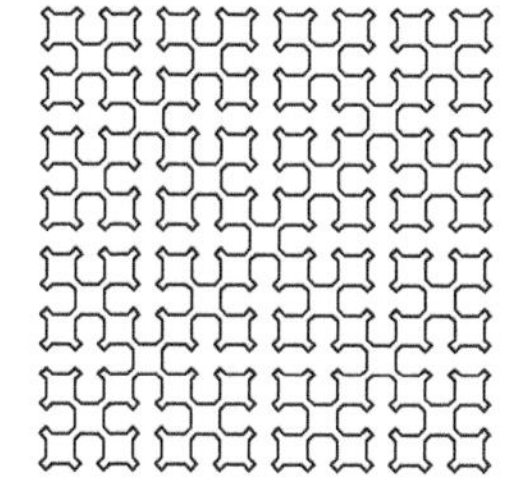

Figure 7.5. S_4.

Redundant recursion should be avoided when it is not too hard to modify the algorithm to avoid it (perhaps even by changing to a nonrecursive algorithm). However, in some cases, the price in computer inefficiency is small enough that redundant recursion is justified. One clear example of this is drawing Sierpinski curves (see Section 7.1.5 for more details about the Sierpinski curve S_4, shown in Figure 7.5).

Recall that the final algorithm in Example 7.2 contained a redundant recursion. It is possible to modify that algorithm to eliminate most of the redundancy, while still using recursion.

EXAMPLE 7.7

New, Improved Subset Sums

```
11: for each x ∈ W
12:    V = W − {x}
13:    SubsetSum3(V,s)
```

The redundancy in algorithm **SubsetSum3** on page 331 results because the distinct subsets generated by the loop in lines 11–13 contain many common elements. As each one is reduced, there will be many ways to start with slightly different larger sets and reduce to the same smaller set.

An insight that leads to a better solution is found in the combinatorial proof of Pascal's theorem on page 248. In that proof, the subsets of a set were divided into those that contain a special element and those subsets that do not contain the special element. This idea can be used to replace the loop in lines 11–13 with a pair of recursive invocations. First, choose any one element $x \in W$. The first recursive invocation will examine all subsets of W that contain x. The second recursive invocation will examine all subsets of W that do not contain x.

Planning the details and order of operations in the new algorithm is not trivial. One important observation is that the section that examines subsets which are required to contain x must then choose another mandatory element before it does another pair of recursive invocations. This suggests that there will be a set, M, of mandatory elements that must be passed off to the subsequent recursions.

The improved algorithm breaks the set whose sum is sought into two disjoint subsets: W, the set of elements that can be removed, and M, the set of mandatory elements. Thus, the sum of elements in $M \cup W$ is compared to the desired goal, s.

```
1: void SubsetSum(set of positive integers M, set of positive integers W, integer s)
2:    if M ∪ W == ∅  # no sum possible
3:      return
4:    t = sum of integers in M ∪ W
5:    if t < s    #  no point in examining any subsets
6:      return
7:    if t == s  # this subset has the desired sum
8:      print M ∪ W
9:      return
10:   else  # t > s, so look at smaller subsets
11:     if W == ∅   # no smaller subsets possible without deleting required elements
12:       return
13:     else   # now for the recursion
14:       choose any x ∈ W    # x is the new mandatory element
15:       SubsetSum(M,W − {x},s)    # examine subsets not containing x
16:       SubsetSum(M ∪ {x},W − {x},s) # examine subsets containing x
17: end SubsetSum
```

The algorithm will terminate because the recursive invocations at lines 15 and 16 each involve $W - \{x\}$, which is smaller than W. Eventually, the second set will be the empty set and line 11 will prevent any additional recursions.

This algorithm will not print any duplicate solutions. It does still have a bit of redundancy. This is because an invocation is done at line 16 using the same set (union) as was examined at lines 2–9. However, at line 16 it is already known that $M \cup W$ is not a solution, so during the next invocation that set will be subdivided and never revisited. ■

✔ Quick Check 7.1

1. Suppose that line 14 in **SubsetSum** always chooses the largest element in W. Create a diagram showing the sequence of recursive invocations when the initial invocation is **SubsetSum**(∅,{4,3,1},4). An invocation can be specified by listing the sets M and W as they are encountered at line 2. Draw a box around all invocations that lead to the sum 4. For uniformity, show the invocation at line 15 to the left and the invocation at line 16 to the right. The diagram will then begin as shown in Figure 7.6.

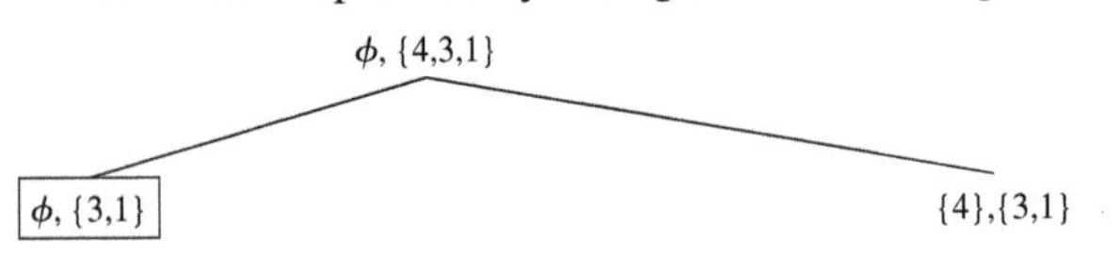

Figure 7.6 Initial invocations for **SubsetSum**(∅, {4, 3, 1}, 4).

Using Recursion to Develop Nonrecursive Algorithms

For some problems, it may be easier to develop a recursive algorithm as a first solution. It is then possible to examine the recursive algorithm and make improvements. In particular, if the algorithm uses tail-end recursion, it is usually easy to translate into a nonrecursive algorithm.

A major thread in Section 7.2 is a formalization of this process in a more mathematical setting.

7.1.4 Persian Rugs

A visually appealing use of recursion (Figure 7.7) was presented in the article titled "'Persian' Recursion", by Anne M. Burns, in the June 1997 issue of *Mathematics Magazine* (vol. 70, No. 3), published by the Mathematical Association of America.

Figure 7.7 A Persian rug.

The image is created by coloring the outer border with an initial color (represented by an integer in the range 0–15). The algorithm then chooses a new color by using a color function on the color values of the four corners (initially all the same). The new color is then used to paint the lines that form the horizontal and vertical centers of the image. The image has thus been partitioned into four subsquares. The algorithm then recursively colors the subsquares using the same procedure.

The color function is

$$\text{new color} = ((\text{top left} + \text{top right} + \text{bottom left} + \text{bottom right})/\,4 + \text{shift}) \bmod 16$$

A pseudocode description of the algorithm is shown on the next page.

Although there are multiple recursive invocations, this is *not* an example of redundant recursion because each invocation colors a different region of the rug. Since there are multiple recursive invocations (rather than a single invocation), I would also not consider this to be tail-end recursion.

By changing the initial color and the value of the "shift" constant, this algorithm can create 256 distinct "Persian rugs".[16]

[16]To see the rugs in color, go to http://www.mathcs.bethel.edu/~gossett/DiscreteMathWithProof/ for a Java applet that generates these Persian rugs.

```
 1: colorRug(integer(0-15) initial_color, integer(0-15) shift)
 2:     color the border using initial_color
 3:     colorSquare(the_rug, shift)
 4:  end colorRug

 6: colorSquare(region current_region, integer(0-15) shift)
 7:   if region has more than 4 pixels
 8:     new_color = ((top_left_color + top_right_color + bottom_left_color +
                      bottom_right_color)/4 + shift) mod 16
 9:     use new_color to color the horizontal and vertical center lines
            of current_region
10:     colorSquare(top_left_subregion, shift)
11:     colorSquare(top_right_subregion, shift)
12:     colorSquare(bottom_left_subregion, shift)
13:     colorSquare(bottom_right_subregion, shift)
14:   else
15:     do nothing  # all pixels in this region have been colored already
16: end colorSquare
```

EXAMPLE 7.8

A Small Persian Rug

The pseudocode algorithm buries some technical details. Rather than fully develop the algorithm, I will instead work through a small example. The example will start with a 9-by-9 grid, representing the rug. Each square in the grid will represent one pixel on a computer screen. I will write the color number in each square as the algorithm progresses.

Let the initial color be 6 and let the shift constant be 12. Then line 2 of the algorithm produces the following grid:

6	6	6	6	6	6	6	6	6
6								6
6								6
6								6
6								6
6								6
6								6
6								6
6	6	6	6	6	6	6	6	6

Line 3 contains the first invocation of the recursive portion of the algorithm. The entire grid is sent to the procedure `colorSquare`.

Line 8 calculates the value of `new_color`. In this case, it is

$$((6+6+6+6)/4+12) \bmod 16 = 2$$

At line 9, the previously uncolored squares in row 5 and in column 5 are colored using color 2.

6	6	6	6	6	6	6	6	6
6				2				6
6				2				6
6				2				6
6	2	2	2	2	2	2	2	6
6				2				6
6				2				6
6				2				6
6	6	6	6	6	6	6	6	6

The `top_left` region consists of the 25 squares in the square whose corners are at (row,column) positions $(1, 1)$, $(1, 5)$, $(5, 1)$, and $(5, 5)$. The `top_right` region is bounded by $(1, 5)$, $(1, 9)$, $(5, 5)$, and $(5, 9)$ (notice the overlap). Similarly, `bottom_left` is bounded by $(5, 1)$, $(5, 5)$, $(9, 1)$, and $(9, 5)$ and `bottom_right` is bounded by $(5, 5)$, $(5, 9)$, $(9, 5)$, and $(9, 9)$.

Line 10 initiates the recursive invocation `colorSquare(top_left,12)`. This effectively starts the process of coloring the subgrid

6	6	6	6	6
6				2
6				2
6				2
6	2	2	2	2

At line 8, the value of `new_color` is found to be $((6{+}6{+}6{+}2)/4{+}12) \bmod 16 = 1$. Line 9 colors the grid squares at $(2, 3)$, $(3, 3)$, $(4, 3)$, $(3, 2)$, and $(3, 4)$ with color 1.

6	6	6	6	6
6		1		2
6	1	1	1	2
6		1		2
6	2	2	2	2

The new value of `top_left` is now bounded by $(1, 1)$, $(1, 3)$, $(3, 1)$, and $(3, 3)$. Line 10 starts a new recursive invocation that will eventually color the square at $(2, 2)$ with color $((6 + 6 + 6 + 1)/4 + 12) \bmod 16 = (4 + 12) \bmod 16 = 0$.[17]

The recursive invocation at line 11 will eventually color the square at $(2, 4)$ with color 15. After lines 12 and 13 have completed, the grid will look like

6	6	6	6	6	6	6	6	6
6	0	1	15	2				6
6	1	1	1	2				6
6	15	1	13	2				6
6	2	2	2	2	2	2	2	6
6				2				6
6				2				6
6				2				6
6	6	6	6	6	6	6	6	6

The initial recursion is now ready to complete line 11, using the `top_right` region bounded by $(1, 5)$, $(1, 9)$, $(5, 5)$, and $(5, 9)$. Lines 12 and 13 will follow. You should verify that the final result is

6	6	6	6	6	6	6	6	6
6	0	1	15	2	15	1	0	6
6	1	1	1	2	1	1	1	6
6	15	1	13	2	13	1	15	6
6	2	2	2	2	2	2	2	6
6	15	1	13	2	13	1	15	6
6	1	1	1	2	1	1	1	6
6	0	1	15	2	15	1	0	6
6	6	6	6	6	6	6	6	6

■

[17] Using integer arithmetic, any remainders are discarded during division. Thus $\frac{19}{4} = 4$.

7.1.5 Drawing Sierpinski Curves

During the late 1800s, Georg Cantor was attempting to find an infinite set that was larger than the open interval (0, 1) on the real line [26]. The obvious set to look at was the open unit square, having corners at the points (0, 0), (0, 1), (1, 1), and (1, 0) in the plane. After extensive effort trying to show that the unit square was indeed larger than the unit interval, Cantor started to suspect that they might be the same size! He eventually proved that they were indeed the same size.

During the early 1890s, Peano and Hilbert were able to provide very elegant constructive proofs of this counterintuitive result. They constructed functions that continuously map the points on the interval [0, 1] onto all the points in the closed unit square.[18] Such functions are called space-filling curves. Space-filling curves can be more formally defined by requiring the image of the function to have a nonzero area. See [79, Chapter 1] for more details.

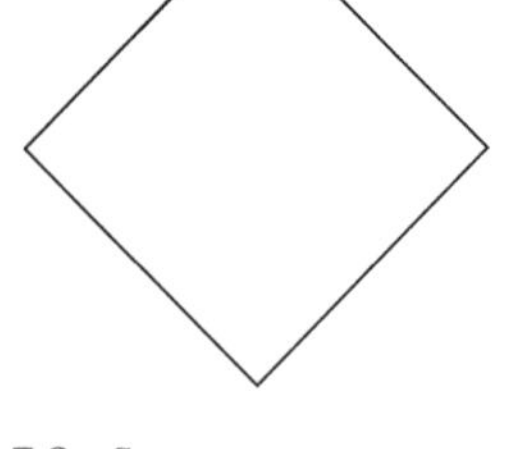

Figure 7.8. S_0.

One very attractive space-filling curve is the Sierpinski curve. The curve, defined by Waclaw Sierpinski, is actually the limit of a sequence of curves. The curves will be defined in more detail in the next section. For now, it is sufficient to consider the first few members of the sequence.[19]

The curves will be labeled $S_0, S_1, S_2, S_3, S_4, S_5, \ldots$.

Consider the curve S_0 in Figure 7.8.

We can think of the diamond shape of the curve as depicting a function that successively maps the subintervals [0, 0.25], [0.25, 0.5], [0.5, 0.75], and [0.75, 1] onto the points in the plane that lie on the line segments joining (0.5, 1) to (1, 0.5), (1, 0.5) to (0.5, 0), (0.5, 0) to (0, 0.5), and (0, 0.5) to (0.5, 1) in the unit square.

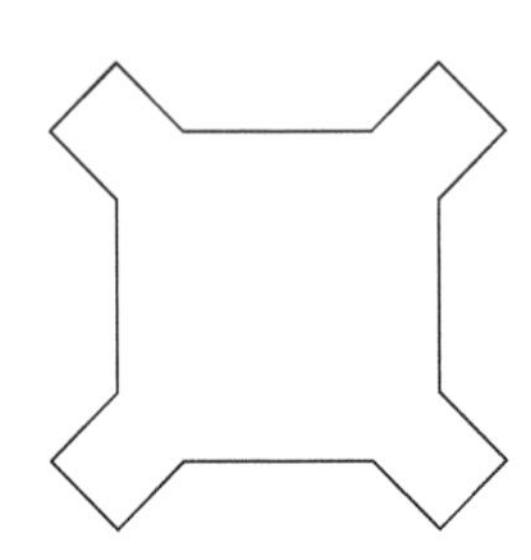

Figure 7.9. S_1.

Similarly, we can view S_1 (Figure 7.9) as the image of a function that maps [0, 1], subdivided into 20 equal length intervals, onto the 20 line segments that constitute the curve S_1. (Notice that the horizontal and vertical line segments in S_1 are double length and hence count as two segments each.)

The curves in the sequence are sets that are the same size as the interval [0, 1] on the real number line.

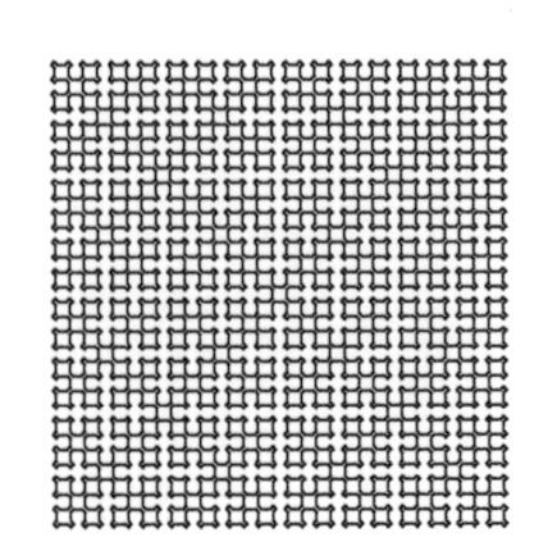

Figure 7.10. S_5.

As can be seen from S_5 (Figure 7.10), each curve in the sequence contains a visually greater set of the points in the unit square than do the previous curves. Also, none of the curves cross themselves. It can be shown that as n approaches infinity, S_n uses more and more of the points in the unit square. The limit, S, of the sequence uses all the points and establishes a correspondence between the interval [0, 1] and the unit square. The graph of the curve would appear to be a solid black square in the plane, but the function would match each point in [0, 1] with exactly one point in the square.

Constructing the Curves

The sequence S_n, whose limit is the Sierpinski curve, can be simply and elegantly defined in terms of four mutually recursive sequences of accessory functions. The accessory sequences will be labeled A_n, B_n, C_n, and D_n.

The accessory sequences are mutually recursive because, for example, the definition of A_n depends on A_{n-1}, B_{n-1}, and D_{n-1}. The basic recursion to be presented is from [109].

Schematic diagrams for each sequence of curves are presented in Table 7.3.

[18]There will be points in the unit square that are mapped onto by multiple points in the unit interval. Any mapping that is both onto and one-to-one must be discontinuous [79, Chapter 1].

[19]You may prefer to work through an interactive version of this material. It can be found at http://www.mathcs.bethel.edu/~gossett/DiscreteMathWithProof/ .

TABLE 7.3 Schematic diagrams for constructing S_n

The Schematic Diagrams

For $n \geq 0$,

$$S_n : A_n \searrow B_n \swarrow C_n \nwarrow D_n \nearrow$$

Notice that in the base case ($n = 0$), the only nonempty contributions are from the line segments represented by the arrows.

A_0 is empty. For $n \geq 1$,

$$A_n : A_{n\text{-}1} \searrow B_{n\text{-}1} \rightarrow\rightarrow D_{n\text{-}1} \nearrow A_{n\text{-}1}$$

B_0 is empty. For $n \geq 1$,

$$B_n : B_{n\text{-}1} \swarrow C_{n\text{-}1} \downarrow\downarrow A_{n\text{-}1} \searrow B_{n\text{-}1}$$

C_0 is empty. For $n \geq 1$,

$$C_n : C_{n\text{-}1} \nwarrow D_{n\text{-}1} \leftarrow\leftarrow B_{n\text{-}1} \swarrow C_{n\text{-}1}$$

D_0 is empty. For $n \geq 1$,

$$D_n : D_{n\text{-}1} \nearrow A_{n\text{-}1} \uparrow\uparrow C_{n\text{-}1} \nwarrow D_{n\text{-}1}$$

Each curve in Table 7.3 is made using line segments of a standard length. The length of the line segments is determined by the value of n for the final curve shown. As n gets larger, the length decreases so that the entire figure will fit inside the unit square. For example, compare the difference in line segment lengths for S_0 and S_2 in Figure 7.11.

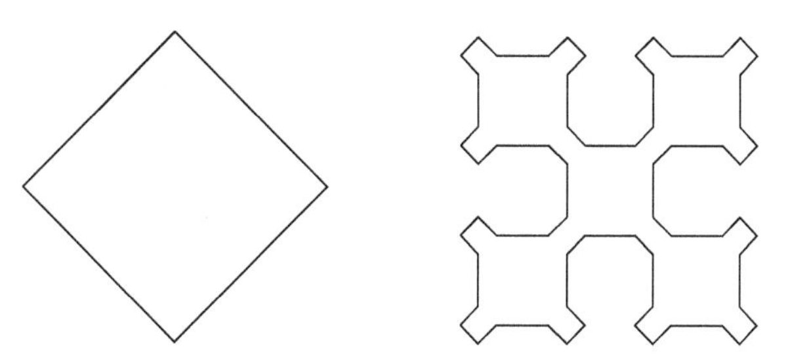

Figure 7.11 Comparing the difference in line segment lengths for S_0 and S_2.

The schematic diagrams contain small arrows. Each arrow represents a line segment (having the standard length) in the direction of the arrow. Some diagrams have two adjacent arrows, indicating the need to draw two line segments in the same direction, creating the double length line segments mentioned earlier.

The schematics also contain symbols for various members of the sequences of curves S_n, A_n, B_n, C_n, and D_n. This indicates the need to (recursively) draw that curve. The curves A_0, B_0, C_0, and D_0 are all defined to be empty (do nothing).

EXAMPLE 7.9

A Detailed Look at S_1

According to the schematic, S_1 is formed by drawing A_1 and then a line segment going down and to the right, and then drawing B_1 followed by a line segment going down and to the left. Next comes a copy of C_1 and a line segment going up and to the left followed by D_1 and a final line segment going up and to the right.

A_1 is formed by drawing A_0 (do nothing) then a line segment going down and to the right, followed by B_0 (again, do nothing) followed by two line segments going to the right. Next draw D_0 (nothing) followed by a line segment going up and to the right and then a final copy of A_0. When completed, A_1 looks like Figure 7.12.

The subfigures B_1, C_1, and D_1, respectively, are formed in a similar fashion (Figures 7.13 – 7.15).

$A_n : A_{n\text{-}1} \searrow B_{n\text{-}1} \rightarrow\rightarrow D_{n\text{-}1} \nearrow A_{n\text{-}1}$

$B_n : B_{n\text{-}1} \swarrow C_{n\text{-}1} \downarrow\downarrow A_{n\text{-}1} \searrow B_{n\text{-}1}$

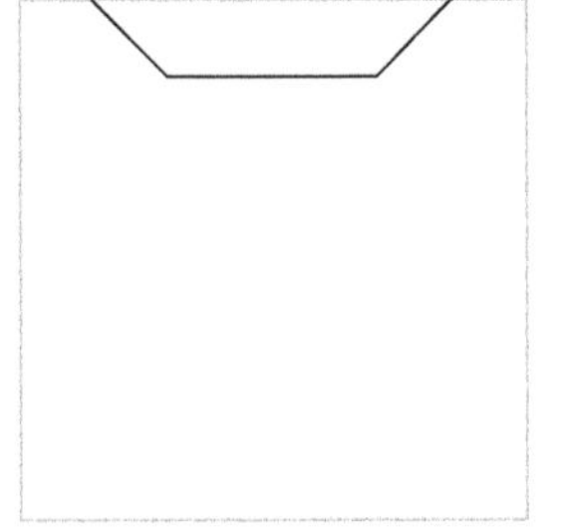

Figure 7.12 A_1.

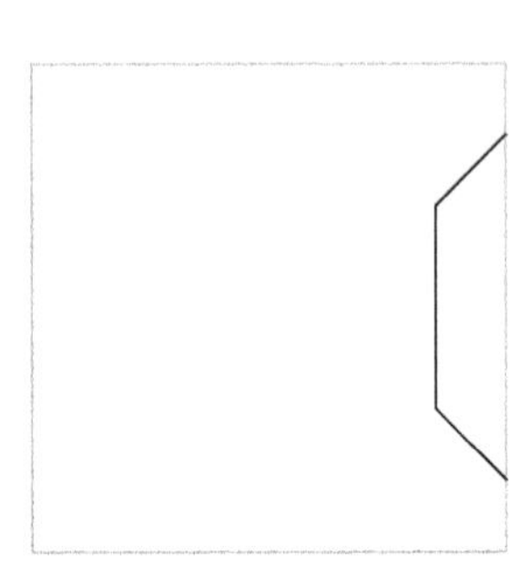

Figure 7.13 B_1.

$C_n : C_{n\text{-}1} \nwarrow D_{n\text{-}1} \leftarrow\leftarrow B_{n\text{-}1} \swarrow C_{n\text{-}1}$

$D_n : D_{n\text{-}1} \nearrow A_{n\text{-}1} \uparrow\uparrow C_{n\text{-}1} \nwarrow D_{n\text{-}1}$

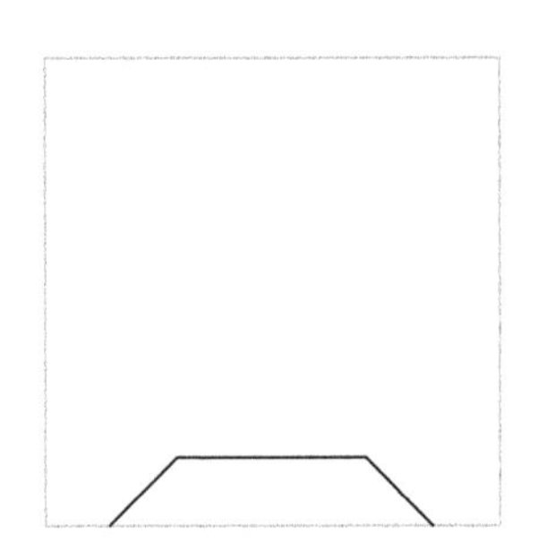

Figure 7.14 C_1.

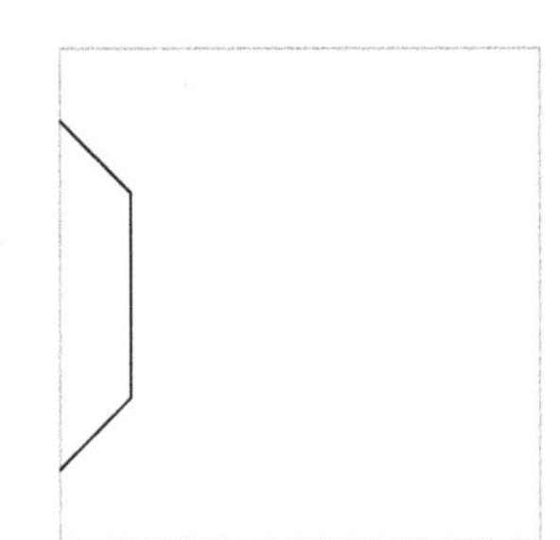

Figure 7.15 D_1.

When these components are assembled in the manner outlined by the schematic, the curve S_1 results (Figure 7.16).

$S_n : A_n \searrow B_n \swarrow C_n \nwarrow D_n \nearrow$

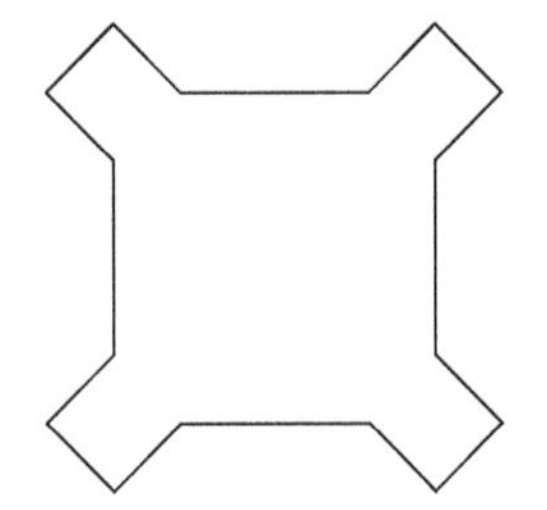

Figure 7.16 S_1.

■

✔ Quick Check 7.2

1. Use the schematic diagrams to construct A_2 from A_1, B_1, C_1, and D_1.

☑

EXAMPLE 7.10 **A Semidetailed Look at S_2**

S_2 is composed from A_2, B_2, C_2, and D_2, which are in turn assembled from A_1, B_1, C_1, and D_1. In all cases, the length of the standard line segment depends on the value of n in the final figure (S_2 in this case). The curves in Figures 7.17 – 7.19, and 7.20 are A_2, B_2, C_2, and D_2, respectively. Make sure you know how each of these were constructed from A_1, B_1, C_1, and D_1.

The curve S_2 is assembled using the schematic (Figure 7.21). ■

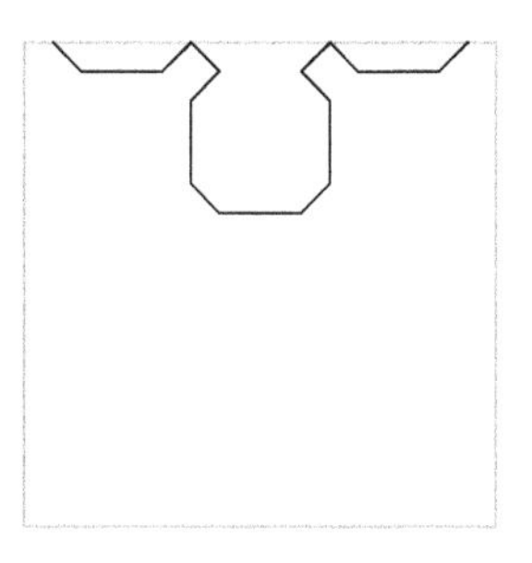

Figure 7.17 A_2.

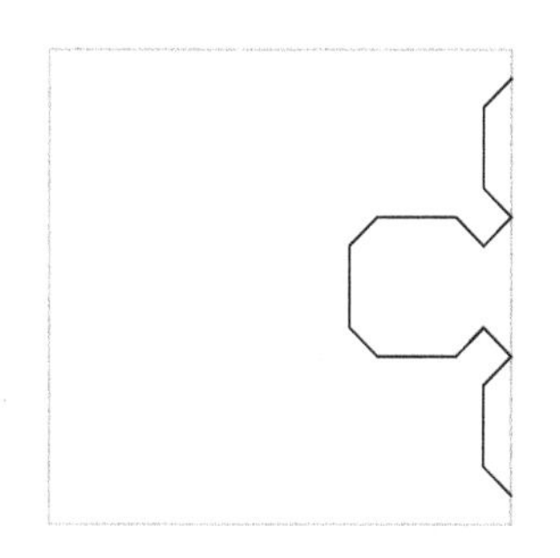

Figure 7.18 B_2.

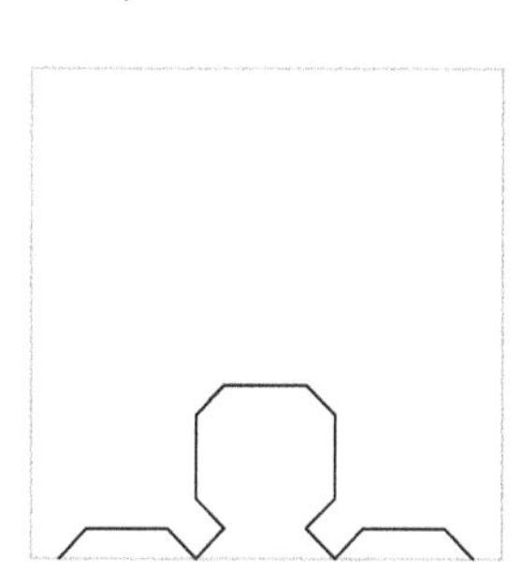

Figure 7.19 C_2.

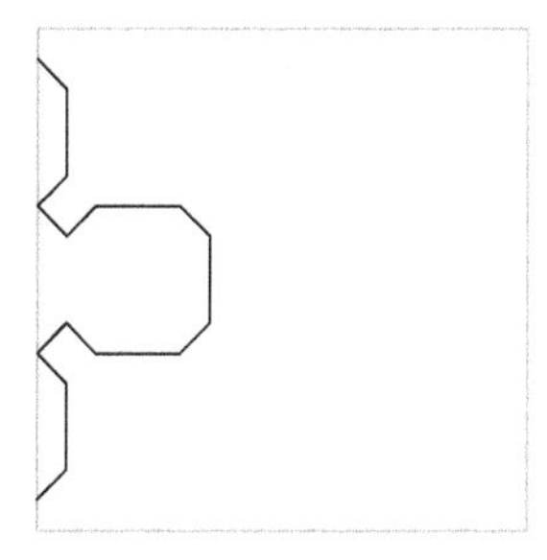

Figure 7.20 D_2.

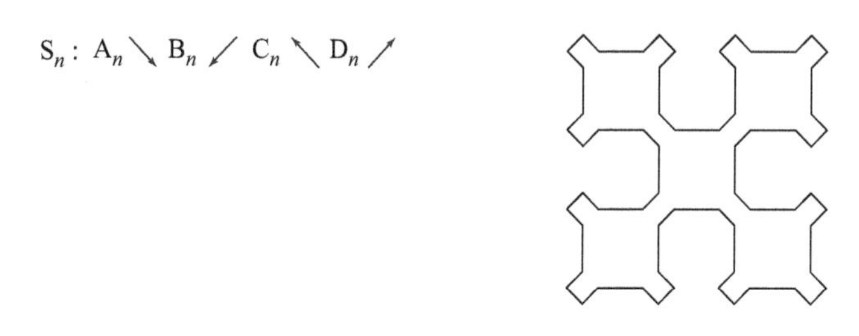

Figure 7.21 S_2.

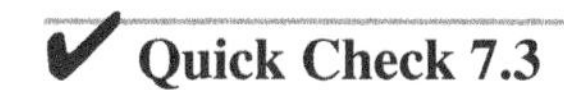

Quick Check 7.3

1. Construct B_3.
2. Make educated guesses about A_3, C_3, and D_3. Then construct S_3. ☑

If you want a more detailed look at how the curves in this section were generated on a computer, look at the *Mathematica* notebook *Sierpinski.nb*.[20]

7.1.6 Adaptive Quadrature

This section contains a very nice use of recursion. However, it does assume that the reader is familiar with some ideas from calculus. There will be an extended detour necessary before the recursive algorithm can be presented. To avoid detracting from the main point of this example (the recursive algorithm at the end), I will keep the background detour as brief as possible. If this material were presented in a numerical analysis course, it would need to be developed in more detail.

Some functions are so intricate that they do not have nice antiderivatives.[21] If we wish to integrate such a function, we will need to use a numerical approximation. A common approach is to approximate the function using a polynomial, integrate the polynomial, and then assume the integral of the polynomial is a good approximation to the integral of the original function.

---

[20] Available as "Sierpinski Curves" at http://www.mathcs.bethel.edu/~gossett/DiscreteMathWithProof/ .

[21] One such function is $f(x) = e^{-x^2}$.

EXAMPLE 7.11

Approximating with a Quadratic Polynomial

Suppose we want to approximate $\int_{-\pi}^{\pi} e^{-x^2}\,dx$. The quadratic polynomial that passes through the two end points and the midpoint is $p(x) = \left(\frac{e^{-\pi^2}-1}{\pi^2}\right)x^2 + 1$.

A brief look at Figure 7.22 shows that the polynomial (the dashed curve) does a poor job of approximating the original function.

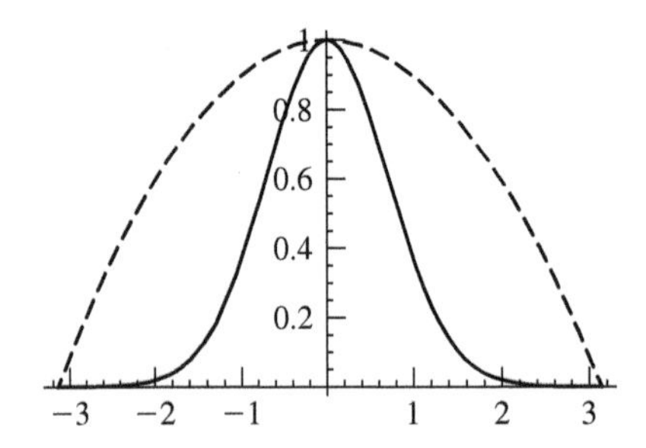

Figure 7.22 Approximating e^{-x^2} (solid) with a quadratic polynomial (dashed).

This approximation can be improved (and will be soon). ■

If we approximate the function $f(x)$ using the quadratic polynomial, $p(x)$, that interpolates the midpoint and two endpoints of the interval $[a, b]$, the integral $\int_a^b f(x)\,dx$ can be approximated by $\int_a^b p(x)\,dx$.

You may have formed the Lagrange interpolating polynomial and performed the integration in one of your calculus classes. The result is known as Simpson's rule:

Simpson's Rule

$$\int_a^b f(x)\,dx \simeq \left(f(a) + 4f\left(\frac{a+b}{2}\right) + f(b)\right)\left(\frac{b-a}{6}\right)$$

One improvement in the numerical approximation is to break the interval $[a, b]$ into a number of subintervals and use Simpson's rule on each subinterval. This technique is known as *composite integration.* If we use more subintervals, we would expect a better approximation. This is usually true.

EXAMPLE 7.12

Composite Integration

If we use two subintervals in the previous example ($[-\pi, 0]$ and $[0, \pi]$), it is clear (Figure 7.23) that the approximation will be much better.

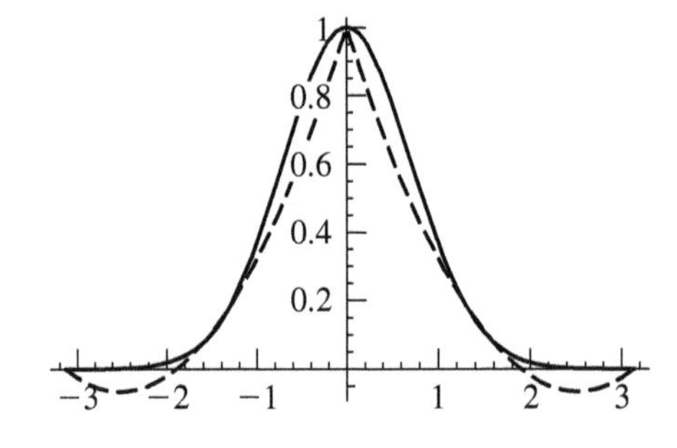

Figure 7.23 Approximating e^{-x^2} (solid) with a composite quadratic polynomial (dashed).

In fact, $\int_{-\pi}^{\pi} e^{-x^2}\,dx \simeq 1.77244$. The single interval Simpson's rule approximates this as 4.1889. Using the two-interval composite integration (with Simpson's rule), the approximation is 1.40248. ■

The big question is, How many subintervals should we use if we want the approximation to be accurate to a predetermined level?

A theorem that is proved in numerical analysis states the following:

THEOREM 7.2 ***Error in Simpson's Rule***

Suppose the interval $[a, b]$ is subdivided into $\frac{N}{2}$ subintervals, where $N = 2m$ for some positive integer m. Also, assume that $f^{(4)}(x)$ is continuous on $[a, b]$. Then a composite Simpson's rule will have an error that is at most

$$\frac{(b-a)^5}{90 \cdot 2 \cdot N^4} \cdot \max_{x\in[a,b]} |f^{(4)}(x)|$$

If we can calculate $\max_{x\in[a,b]} |f^{(4)}(x)|$, then it is not hard to find N so that the error is smaller than 0.00001 (or any other value we choose). However, estimating $\max_{x\in[a,b]} |f^{(4)}(x)|$ may not be trivial. Is there another way to guarantee a certain accuracy? Certainly![22]

A technique known as *adaptive quadrature* can be used to approximate an integral with a predetermined degree of accuracy. The basic idea is to use Simpson's rule to approximate the integral. Then a two-interval composite Simpson's rule is used to get a second opinion. If the two approximations are close enough, the algorithm assumes it is done and returns the composite estimate. Otherwise, it recursively calculates the integral on the two subintervals $[a, \frac{a+b}{2}]$ and $[\frac{a+b}{2}, b]$.

Some notation is needed before formalizing the preceding intuitive ideas.

Let $S(\alpha, \beta) = \left(f(\alpha) + 4f(\frac{\alpha+\beta}{2}) + f(\beta)\right)\left(\frac{\beta-\alpha}{6}\right)$. Thus $S(\alpha, \beta)$ is the result of a Simpson's rule on the interval $[\alpha, \beta]$.

Let $E(\alpha, \beta)$ represent the error in Simpson's rule on $[\alpha, \beta]$. Since Simpson's rule uses a single interval, $N = 2$. Consequently, $E(\alpha, \beta) \le \frac{(\beta-\alpha)^5}{90 \cdot 2^5} \cdot |f^{(4)}(\zeta_{\alpha\beta})|$, where $\zeta_{\alpha\beta}$ is a point where $\max_{x\in[\alpha,\beta]} |f^{(4)}(x)|$ occurs. Then

$$\int_\alpha^\beta f(x)\,dx = S(\alpha, \beta) - E(\alpha, \beta) \tag{7.1}$$

and (letting $\nu = \frac{\alpha+\beta}{2}$)

$$\int_\alpha^\beta f(x)\,dx = (S(\alpha, \nu) - E(\alpha, \nu)) + (S(\nu, \beta) - E(\nu, \beta)) \tag{7.2}$$

Equating the right-hand sides of (7.1) and (7.2) produces

$$S(\alpha, \beta) - S(\alpha, \nu) - S(\nu, \beta) = E(\alpha, \beta) - E(\alpha, \nu) - E(\nu, \beta) \tag{7.3}$$

If α is very close to β, it seems reasonable that $f^{(4)}(\zeta_{\alpha\beta}) \simeq f^{(4)}(\zeta_{\alpha\nu}) \simeq f^{(4)}(\zeta_{\nu\beta})$. Denote this approximately equal value by f_4.

Setting $\gamma = \frac{\beta-\alpha}{4}$ and doing some simple algebra leads to

$$\begin{aligned} |E(\alpha, \beta) - E(\alpha, \nu) - E(\nu, \beta)| &\le |E(\alpha, \beta)| + |E(\alpha, \nu)| + |E(\nu, \beta)| \\ &\simeq \left(\frac{(2\gamma)^5}{90} + \frac{\gamma^5}{90} + \frac{\gamma^5}{90}\right) \cdot |f_4| \\ &\simeq \frac{\gamma^5}{3} \cdot |f_4| \end{aligned} \tag{7.4}$$

[22] You knew that recursion needed to reappear sometime.

In a similar fashion, equation (7.2) leads to

$$\left| \int_{\alpha}^{\beta} f(x)\,dx - S(\alpha, \nu) - S(\nu, \beta) \right| \simeq \frac{\gamma^5}{45} \cdot |f_4| \tag{7.5}$$

Combining equations (7.5), (7.3), and (7.4) produces

$$\left| \int_{\alpha}^{\beta} f(x)\,dx - S(\alpha, \nu) - S(\nu, \beta) \right| \simeq \frac{1}{15} |S(\alpha, \beta) - S(\alpha, \nu) - S(\nu, \beta)| \tag{7.6}$$

Thus, the two-interval estimate $S(\alpha, \nu) + S(\nu, \beta)$ is within τ of $\int_{\alpha}^{\beta} f(x)\,dx$ as long as

$$|S(\alpha, \beta) - S(\alpha, \nu) - S(\nu, \beta)| < 15\tau \tag{7.7}$$

Inequality (7.7) provides the criterion needed for the base step of the recursion. If we want to be conservative, we could choose a number smaller than the "safety constant" 15 produced by the derivation (recall the assumption that $f^{(4)}(\zeta_{\alpha\beta}) \simeq f^{(4)}(\zeta_{\alpha\nu}) \simeq f^{(4)}(\zeta_{\nu\beta})$). A typical value would be in the range 5–15. A smaller constant will cause the algorithm to do more work before accepting a solution, but will be less likely to produce an inaccurate value.

```
 1: real Simpson(function f, real α, real β)
 2:    return (f(α) + 4 f((α+β)/2) + f(β)) · ((β−α)/6)
 3: end Simpson

 5: real AQ(function f, real α, real β, real τ)
 6:    real whole, left, right, ν, safety
 7:    safety = 10
 8:    ν = (α + β)/2
 9:
10:    whole = Simpson(f,α,β)
11:    left  = Simpson(f,α,ν)
12:    right = Simpson(f,ν,β)
13:
14:    if |whole - left - right| < safety · τ
15:      return left + right
16:    else
17:      return AQ(f,α,ν,τ/2) + AQ(f,ν,β,τ/2)
18: end AQ
```

Notice at line 17 that the recursive invocations use $\frac{\tau}{2}$ so that the combined errors do not exceed τ.

The screen grab in Figure 7.24 is from a Java application that shows the algorithm in action.[23] As each subinterval is partitioned, the program draws a line to indicate the new intervals. When a subinterval's integral has been found, the region is colored (using alternate colors to help distinguish the regions). Notice that in regions where the function fluctuates rapidly, smaller intervals are used whereas in regions where the function is calm, much larger intervals can be used. The function used is

$$f(x) = \begin{cases} 1 & |x| < .00001 \\ \dfrac{\sin\left(\frac{\pi|x|}{r}\right)}{\left(\frac{\pi|x|}{r}\right)} & .00001 \le |x| \le r \\ \sin\left(\frac{1}{|x|}\right) & r < |x| \end{cases}$$

where r is implicitly defined by the conditions $\sin(\frac{1}{r}) = 0$ and $0.05 < r < 0.06$.

[23] Available at http://www.mathcs.bethel.edu/~gossett/DiscreteMathWithProof/ .

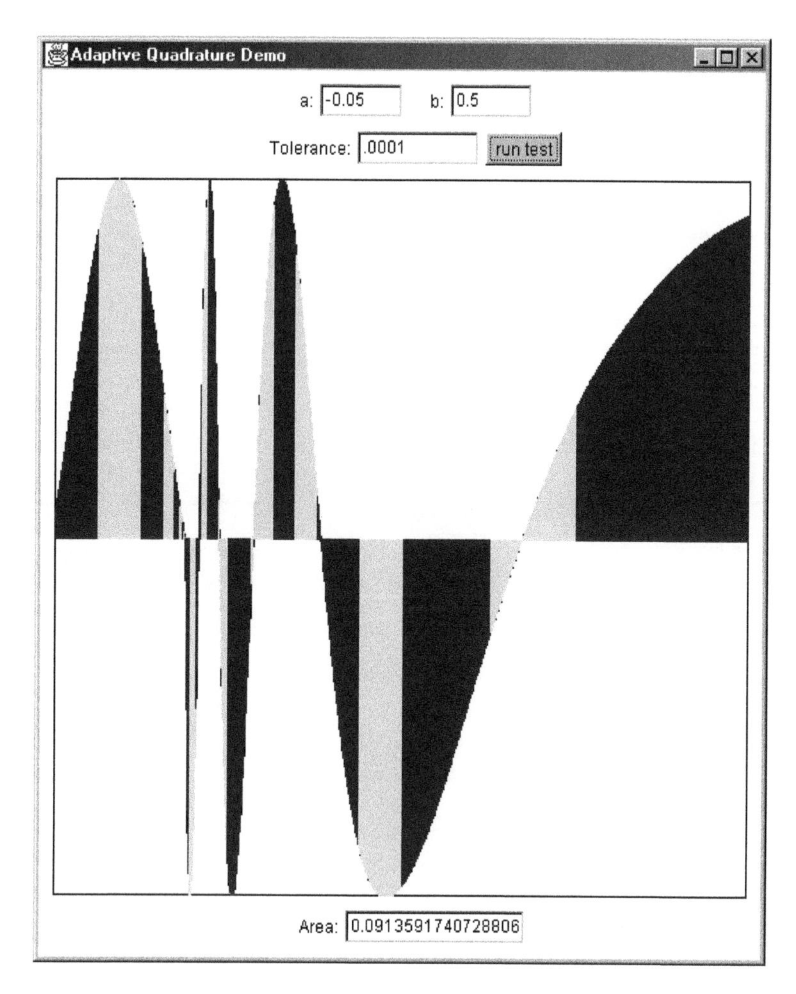

Figure 7.24 Adaptive quadrature in action.

✔ Quick Check 7.4

Use adaptive quadrature to approximate $\int_0^4 \sqrt{4-x}\,dx$ to an accuracy of $\tau = .01$ (using a safety constant of 10). Indicate the intervals on which the recursions terminated. ☑

7.1.7 Exercises

The exercises marked with ★ have detailed solutions in Appendix H.

1. ★ Recall Example 7.4.

(a) Use pseudocode to write a recursive algorithm for calculating the sum of the first n positive integers. You may assume that n will never be less than 1.

(b) Does this algorithm use tail-end recursion?

(c) Use pseudocode to write a nonrecursive algorithm for calculating the sum of the first n integers.

2. (a) Use pseudocode to write a recursive algorithm for calculating a^n, where n is a natural number.

(b) Does this algorithm use tail-end recursion?

(c) Use pseudocode to write a nonrecursive algorithm for calculating a^n, where n is a natural number.

3. (a) Use pseudocode to write a recursive algorithm for calculating $\sqrt{\sqrt{\ldots\sqrt{a}\ldots}}$, where there will be n nested square roots. You may assume that a will be positive.

(b) Does this algorithm use tail-end recursion?

(c) Use pseudocode to write a nonrecursive algorithm for calculating $\sqrt{\sqrt{\ldots\sqrt{a}\ldots}}$, where $a > 0$ and there will be n nested square roots.

4. The algorithms in Examples 7.3 and 7.5 for calculating $n!$ are incomplete. They do not handle the case $0! = 1$. Modify both versions of the algorithm to correctly work if $n = 0$.

5. ★ Repeat Quick Check 7.1 (page 338), but in line 14 of **SubsetSum**, always choose the smallest element of W.

(a) Draw a diagram showing the recursive invocations of **SubsetSum**($\emptyset$,{1, 3, 4}, 4).

(b) Which strategy (largest element or smallest element) seems better? Explain your answer.

6. Write a recursive algorithm to calculate $n \cdot a$, where n is a positive integer and a is a real number.

7. ★ The greatest common divisor of integers a and b (see Definition 3.11) can be recursively calculated (see Exercise 26 on page 162) using the identities

$$\gcd(a, b) = \gcd(b, a)$$
$$\gcd(a, b) = \gcd(b \bmod a, a), \quad \text{where } 0 < a \leq b$$

(a) Determine a sufficient set of base conditions for a recursive algorithm that uses these identities.

(b) Write a recursive algorithm for calculating the gcd of two nonnegative integers.

8. Write a recursive algorithm to find the largest odd number that divides n, where $n \geq 1$.

9. Write a self-contained recursive algorithm for finding the sum of the prime factors of a positive integer. For example, the sum of the prime factors of 12 is $2 + 2 + 3 = 7$.

10. Assume you have already written the following algorithms.

- boolean **isPrime**(integer n) — returns true if n is prime
- integer **oneFactor**(integer n) — returns a factor a of n with $1 < a < n$, where n is known to be composite.

Assume that n is greater than 0. Write a recursive algorithm to print the prime factorization of n. (*Print* means the return value will be `void`.)

11. Write a recursive algorithm for performing a binary search (see Section 4.2.4). Return -1 to indicate "the sought for item is not present."

12. Define the function f by

$$f(n) = \begin{cases} \frac{n}{2} & \text{if } n \text{ is even} \\ 3n + 1 & \text{if } n \text{ is odd} \end{cases} \qquad \text{for integers } n \geq 1$$

Let n_0 be any positive integer and define $n_i = f(n_{i-1})$. The *Collatz conjecture* asserts that $n_i = 1$ for some $i < \infty$.

(a) Write a recursive algorithm that prints the sequence of values generated by f (including the initial n). It should terminate once the value 1 has been printed. (*Print* means the return value will be `void`.)

(b) (Extra Credit) What is the length of the longest sequence that can be generated starting at $n \leq 1000$? What is the initial value of n?

13. Find and fix the error in the following recursive algorithm for finding the product of integers, a and b.

```
1: integer product(integer a, integer b)
2:   if b == 1
3:     return a
4:   return a + product(a, b − 1)
5: end product
```

14. Find and fix the error in the following recursive algorithm for finding a^n, for $a \in \mathbb{R}$ and $n \in \mathbb{N}$.

```
1: real power(real a, natural number n)
2:   if n == 0
3:     return 1
4:   return power(a, n − 1)
5: end power
```

15. ★ Write a recursive algorithm to calculate a^{2^n}. [*Hint*: $a^{2^{n+1}} = (a^{2^n})^2$. Assume that n is a natural number.]

16. Write a recursive algorithm that will reverse the order of the characters in a nonempty string and return the new string.

17. Define $p(n, k)$ to be the number of ways to write the integer n as a sum of exactly k positive integers. Theorem 8.5 on page 422 indicates that $p(n, k) = p(n-1, k-1) + p(n-k, k)$ and that $p(n, k) = 0$ for $k > n$, and also $p(n, 0) = 0$, and $p(n, n) = 1$. Write a recursive algorithm for finding $p(n, k)$.

18. Use a 9-by-9 grid to make a Persian rug starting with color 9 for the border and shift constant 14.

19. Use a 9-by-9 grid to make a Persian rug starting with color 12 for the border and shift constant 7.

20. Copy the 5 schematic diagrams for the Sierpinski curves on a piece of paper. Close your textbook, and then, using only the diagrams, draw S_0, S_1, and S_2.

21. ★ Use adaptive quadrature to approximate $\int_1^{10} \frac{1}{x}\, dx$ to an accuracy of $\tau = 0.01$ (using a safety constant of 10). Indicate the intervals on which the recursions terminated.

7.2 Recurrence Relations

The previous section introduced the notion of recursive algorithms. In this section, a more mathematically oriented variation will be introduced. The multiline algorithm will be replaced by a function. From recursively defined functions, it is easy to move to recursively defined sequences. This introduces what will be called a *recurrence relation*.

DEFINITION 7.3 ***Recursively Defined Functions***

A *recursively defined function* is a function, f, for which

- $f(k)$ is given for one or more specific values of k.
- Elsewhere, $f(n)$ is defined in terms of one or more values $f(k)$ with $k < n$.

Unless otherwise specified, the domain will be assumed to be the set of nonnegative integers.

EXAMPLE 7.13

A Recursive Definition for n!

The factorial function can be defined by

- $f(0) = 1$
- $f(n) = n \cdot f(n-1)$ for $n > 0$ ■

EXAMPLE 7.14

Arithmetic Progression

Definition 3.56 stated that an arithmetic progression is a sequence of numbers of the form $a, a+d, a+2d, \ldots, a+nd, \ldots$. Such a sequence can be easily defined recursively:

- $A(0) = a$
- $A(n) = A(n-1) + d$ for $n > 0$ ■

A common practice for denoting the value of a function defined on the nonnegative integers is to use subscripts rather than parenthesized arguments. The previous example would then be written as

- $a_0 = a$
- $a_n = a_{n-1} + d$ for $n > 0$

It should be clear that a recursively defined function, f, always generates a sequence:

$$\{f_0, f_1, f_2, \ldots, f_n, \ldots\}$$

Sometimes it is possible to move in the other direction.

> **DEFINITION 7.4 *Recurrence Relation***
>
> Let $\{a_n \mid n = 0, 1, 2, \ldots\}$ be a sequence. A *recurrence relation* for $\{a_n\}$ is a formula that expresses a_n in terms of some subset of $\{a_0, a_1, \ldots, a_{n-1}\}$.
>
> Given a recurrence relation and one or more base conditions, the sequence it generates is called the *solution of the recurrence relation.*

The only significant difference between Definition 7.3 and Definition 7.4 is that one is about functions and the other is about sequences. The focus in this section will be on sequences (recurrence relations). Section 7.3 will use the recursive function notation.

EXAMPLE 7.15

The Fibonacci Sequence

Perhaps the most famous recurrence relation arises from the Fibonacci sequence.[24] The recurrence relation is

- $f_0 = 1$
- $f_1 = 1$
- $f_n = f_{n-1} + f_{n-2}$ for $n > 1$.

Each Fibonacci number (after the first two) is the sum of the two previous Fibonacci numbers. The sequence generated is thus

$$1, 1, 2, 3, 5, 8, 13, \ldots$$ ■

Leonardo Fibonacci (Leonardo, the son of Bonaccio) lived around 1175–1250 A.D. He was born in medieval Pisa. As a boy, he traveled to many parts of the Mediterranean region with his father. During this period, he lived for a while in North Africa. He encountered Eastern and Arabic mathematics during this time and became convinced

[24]Some fascinating properties of the Fibonacci sequence were introduced on page 153.

that current European mathematical practices were inferior. In 1202 he published an important book, titled *Liber abaci* (the "book of the abacus"). The book discussed arithmetic and elementary algebra. In the book, Fibonacci presented the problem that was the origin of the Fibonacci sequence.

EXAMPLE 7.16

Fibonacci's Rabbits

A newborn male and a newborn female rabbit are placed inside a fenced area for breeding. Suppose[25] it takes 1 month for a newborn rabbit to reach breeding age. Suppose also that in 1 month, a mature pair can produce one male and one female baby rabbit. How many *pairs* of rabbits will there be in one year?

Table 7.4 indicates the population change inside the fence (assuming no rabbits die).

TABLE 7.4 Rabbit Population

Month	0	1	2	3	4	5	6	7	...
Baby Pairs	1	0	1	1	2	3	5	8	...
Mature Pairs	0	1	1	2	3	5	8	13	...
Total Pairs	1	1	2	3	5	8	13	21	...

Note that each mature pair will reproduce themselves during the next month, so once the initial month has been completed, the top row will look like the second row, shifted right by one position. The entries in the third row are just the sum of the entries for that month in the first two rows. Each entry (after the first month) in the second row will be the sum of the numbers of mature rabbits and baby rabbits in the previous month. Thus, the second row looks like the third row, shifted right by one position. ■

The final introductory example describes a famous puzzle.

EXAMPLE 7.17

The Tower of Hanoi Puzzle

In 1883, the French mathematician Édouard Lucas created an interesting puzzle, called the Tower of Hanoi.

The "game board" (Figure 7.25) consists of eight disks of uniformly decreasing size, together with a board containing three pegs or dowels, labeled A, B, and C, each able to hold all eight disks.

Figure 7.25 The Tower of Hanoi puzzle.

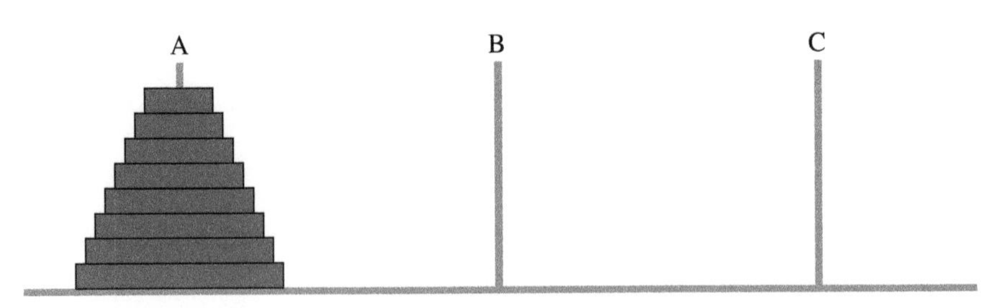

The goal is to move the disks from peg A to peg C. They must be moved one disk at a time (from one peg to another), and no disk may be placed on top of a smaller disk.[26]

Lucas spiced up the problem by inventing a legend about a fancier version of the puzzle. In the legend, some Brahman monks were diligently carrying out a task, set by God at the creation of the world. The task was to solve a Tower of Hanoi puzzle consisting of 64 golden disks, fitting onto three diamond needles. The monks worked

[25] Please do not quote these assumptions on a biology exam!

[26] You can easily find numerous Web sites that contain Java applets that animate the puzzle. A good place to start is the "Tower of Hanoi Puzzle" link at http://www.mathcs.bethel.edu/~gossett/DiscreteMathWithProof/.

at the task day and night (in shifts, of course). Once they completed the task, the world would end. A recurrence relation creeps in if we ask the question:

> What is the minimum number of moves, H_n, it will take to solve a Tower of Hanoi puzzle with n disks?

A good place to begin is with a few small examples. The case $n = 0$ is a bit of a problem (should it be $H_0 = 0$, or $H_0 = 1$?), so leave it for now. Clearly, $H_1 = 1$, since a single move is required. For $n = 2$, move the smaller disk to peg B, then move the larger disk to peg C, and finally move the smaller disk to peg C. Thus, $H_2 = 3$. Notice that the smaller disk required two moves, the larger disk only one move.

For $n = 3$, you will eventually decide that the most efficient solution is to (perhaps counterintuitively) move the smallest disk to peg C. Then, move the middle disk to peg B, followed by a move of the smallest disk to peg B (on top of the middle disk). This has used three moves and involved two disks (recall H_2). We are now free to move the large disk to peg 3. Finally, the two smaller disks need to be moved onto peg C. But this is just a two-disk stack again, requiring three moves (starting by moving the smaller disk to peg A).

You should notice a pattern: To move n disks from peg A to peg C, first move the top $n - 1$ disks to peg B (H_{n-1} moves), then move the bottom disk to peg C (1 move), followed by moving the top $n - 1$ disks from peg B to peg C (another H_{n-1}) moves.

Thus, $H_1 = 1$ and

$$H_n \le 2H_{n-1} + 1 \quad \text{for } n \ge 2$$

To see that

$$H_n \ge 2H_{n-1} + 1 \quad \text{for } n \ge 2$$

recall that a larger disk can never be placed on a smaller disk. To move the largest disk, the other $n - 1$ disks must first be moved. The largest disk must be moved to an empty peg (so that it does not cover smaller disks). Therefore all the smaller disks must first be moved to a single peg, requiring at least H_{n-1} moves before the largest disk can be moved. To reach the final configuration, the $n - 1$ smaller disks must then be moved on top of the larger disk, again requiring at least H_{n-1} moves.

The recurrence relation implies that we should define $H_0 = 0$ and set

$$H_n = 2H_{n-1} + 1 \quad \text{for } n \ge 1$$

A closed-form formula will be derived in Exercise 7 on page 375. ■

7.2.1 Solving Recurrence Relations

In many cases, a recurrence relation arises during the process of solving some problem. Example 7.17 has already illustrated this. Another illustration will be seen shortly in Example 7.19. Once we have a recurrence relation, how do we solve it? In particular, how do we find a_n for any given value of n?

One obvious solution is to start at the beginning and keep substituting into the recurrence relation until the appropriate value of n has been generated. However, if n is very large, this may be time consuming. Is there an alternative? In particular, is it possible to produce a closed-form formula for a_n (a formula that does not have any other a_k on the right-hand side)?

The answer is yes. A closed-form formula can often be found through a process called *back substitution*. The process is illustrated in the next example.

EXAMPLE 7.18

Solving a Recurrence Relation

Let a recurrence relation be defined as

- $a_0 = 2$
- $a_n = 3a_{n-1} + 2 \quad \text{for } n > 0$

The solution is $\{2, 8, 26, 80, \ldots\}$.

The relation $a_n = 3a_{n-1} + 2$ is valid no matter what value of n (greater than 0) is used. Thus, $a_{n-1} = 3a_{n-2} + 2$. Substituting into the first relation produces $a_n = 3(3a_{n-2} + 2) + 2$. This can be carried all the way back to $a_1 = 3a_0 + 2 = 3 \cdot 2 + 2 = 8$.

$$
\begin{aligned}
a_n &= 3a_{n-1} + 2 \\
&= 3(\mathbf{3a_{n-2} + 2}) + 2 \quad \text{substitute for } a_{n-1} \\
&= 3^2 a_{n-2} + (3 \cdot 2 + 2) \quad \text{simplify} \\
&= 3^2(\mathbf{3a_{n-3} + 2}) + (3 \cdot 2 + 2) \quad \text{substitute for } a_{n-2} \\
&= 3^3 a_{n-3} + (3^2 \cdot 2 + 3 \cdot 2 + 2) \quad \text{simplify} \\
&\vdots \qquad\qquad \vdots \\
&= 3^k a_{n-k} + 2 \cdot (3^{k-1} + 3^{k-2} + \cdots + 3 + 1) \\
&\vdots \qquad\qquad \vdots \\
&= 3^n a_0 + 2 \cdot (3^{n-1} + 3^{n-2} + \cdots + 3 + 1) \\
&= 2 \cdot (3^n + 3^{n-1} + 3^{n-2} + \cdots + 3 + 1) \\
&= 2\frac{3^{n+1} - 1}{3 - 1} = 3^{n+1} - 1 \quad \text{for } n \geq 0
\end{aligned}
$$

A few comments are in order. First, notice the alternation between substitute and simplify. Note especially that during a simplify step I do not simplify very far. The goal is to isolate the term(s) with a_k. The rest of the line should be left in expanded form so that it is easy to recognize the pattern that develops. Once the pattern becomes clear, I can write the general step as $3^k a_{n-k} + 2 \cdot (3^{k-1} + 3^{k-2} + \cdots + 3 + 1)$. The process will terminate when $k = n$. The final step requires the recognition of a geometric series.[27] A closed-form expression for the nth element of the sequence is therefore $a_n = 3^{n+1} - 1$. It is easy to see that this formula produces the sequence $\{2, 8, 26, 80, \ldots\}$. ■

Example 7.19 assumes that you have read Section 7.1.5.

EXAMPLE 7.19

The Number of Line Segments in a Sierpinski Curve

We can use a recurrence relation to count the number of line segments in the curve S_n, for $n > 0$.[28]

The schematic diagram $S_n : A_n \searrow B_n \swarrow C_n \nwarrow D_n \nearrow$ indicates that S_n should have as many line segments as there are (combined) in the subcurves A_n, B_n, C_n, and D_n, plus four more line segments for the arrow segments. It will not take long to convince yourself that the subcurves each have the same number of line segments.

[27] The formula for the sum of a geometric series is something you should already have memorized.

[28] This number is useful when determining how long each line segment should be in order for the curve to be drawn inside the fixed-sized square.

Let the number of line segments in A_n be given by a_n. If we let s_n represent the number of line segments in the curve S_n, then

- $s_0 = 4$
- $s_n = 4a_n + 4 \quad$ for $n > 0$
  A_0 is empty. The diagram for A_n,

indicates that

- $a_0 = 0$
- $a_n = 4a_{n-1} + 4 \quad$ for $n > 0$

It is now straightforward to solve the recurrence relation for $\{a_n\}$.

$$\begin{aligned}
a_n &= 4a_{n-1} + 4 \\
&= 4(\mathbf{4a_{n-2} + 4}) + 4 \quad \text{substitute} \\
&= 4^2 a_{n-2} + (4^2 + 4) \quad \text{simplify} \\
&= 4^2(\mathbf{4a_{n-3} + 4}) + (4^2 + 4) \quad \text{substitute} \\
&= 4^3 a_{n-3} + (4^3 + 4^2 + 4) \quad \text{simplify} \\
&\vdots \qquad\qquad \vdots \\
&= 4^k a_{n-k} + (4^k + 4^{k-1} + \cdots + 4^2 + 4) \\
&\vdots \qquad\qquad \vdots \\
&= 4^n a_0 + (4^n + 4^{n-1} + \cdots + 4^2 + 4) \\
&= 4^n + 4^{n-1} + \cdots + 4^2 + 4 \\
&= 4 \cdot (4^{n-1} + 4^{n-2} + \cdots + 4 + 1) \\
&= 4\frac{4^n - 1}{4 - 1} = \frac{4^{n+1} - 4}{3} \quad \text{for } n \geq 0
\end{aligned}$$

TABLE 7.5 The Number of Line Segments in S_n, for $0 \leq n \leq 5$

n	s_n
0	4
1	20
2	84
3	340
4	1364
5	5460

Thus, $s_n = 4a_n + 4 = \frac{4^{n+2} - 4}{3}$, $n \geq 0$. The first few values are listed in Table 7.5. ■

✔ Quick Check 7.5

1. Use back substitution to find a closed-form formula for the recurrence relation $\{b_n\}$, where $b_0 = 0$ and $b_n = 1 - b_{n-1}$ for $n > 0$. ☑

EXAMPLE 7.20

A Harder Recurrence Relation

Consider the recurrence relation defined by

- $a_0 = 0$
- $a_n = na_{n-1} + 1 \quad$ for $n > 0$

Before starting the back substitution, notice that n is now more than merely a subscript. The relationship between n the number and n the subscript needs to be handled carefully. In particular, the recurrence relation implies that $a_1 = 1a_0 + 1 = 1$ and $a_2 = 2a_1 + 1 = 3$. Notice also that $a_{n-1} = (n-1)a_{n-2} + 1$.[29]

The back substitution produces the following equalities.

[29]**Not** $a_{n-1} = na_{n-2} + 1$.

$$
\begin{aligned}
a_n &= n \cdot a_{n-1} + 1 \\
&= n(\mathbf{(n-1) \cdot a_{n-2} + 1}) + 1 \quad \text{substitute} \\
&= n(n-1) \cdot a_{n-2} + (n+1) \quad \text{simplify} \\
&= n(n-1) \cdot (\mathbf{(n-2) \cdot a_{n-3} + 1}) + (n+1) \quad \text{substitute} \\
&= n(n-1)(n-2) \cdot a_{n-3} + (n(n-1) + n + 1) \quad \text{simplify} \\
&\vdots \qquad\qquad \vdots \\
&= \frac{n!}{(n-k)!} \cdot a_{n-k} + \left(\frac{n!}{(n-(k-1))!} + \frac{n!}{(n-(k-2))!} + \cdots + \frac{n!}{(n-1)!} + \frac{n!}{n!} \right) \\
&\vdots \qquad\qquad \vdots \\
&= \frac{n!}{0!} \cdot a_0 + \left(\frac{n!}{1!} + \frac{n!}{2!} + \cdots + \frac{n!}{(n-1)!} + \frac{n!}{n!} \right) \\
&= \sum_{k=1}^{n} \frac{n!}{k!} \quad \text{for } n \geq 1
\end{aligned}
$$

recall that $a_0 = 0$

It is possible (but not trivial) to go a step further. Symbolic mathematics software packages such as *Mathematica* and *Maple* are able to convert this to an expression involving the Euler gamma functions:

$$
a_n = n! \left(\frac{(n+1) \cdot \Gamma(n+1, 1) \cdot e}{\Gamma(n+2)} - 1 \right)
$$

where

$$
\Gamma(z) = \int_0^\infty t^{z-1} e^{-t} \, dt
$$

$$
\Gamma(z, 1) = \int_1^\infty t^{z-1} e^{-t} \, dt
$$

It can be shown that $\frac{(n+1) \cdot \Gamma(n+1,1)}{\Gamma(n+2)} \to 1$ as $n \to \infty$, so a_n asymptotically approaches $n!(e-1)$. ■

The previous example indicates that recurrence relations may not always have simple solutions. The next example indicates that the back substitution method for solving recurrence relations has some limitations.

EXAMPLE 7.21

Solving the Fibonacci Sequence

Suppose we want a closed-form formula for f_n, the nth Fibonacci number. Starting the back substitution,

$$
\begin{aligned}
f_n &= f_{n-1} + f_{n-2} \\
&= (\mathbf{f_{n-2} + f_{n-3}}) + f_{n-2} \quad \text{substitute} \\
&= 2f_{n-2} + f_{n-3} \quad \text{simplify} \\
&= 2(\mathbf{f_{n-3} + f_{n-4}}) + f_{n-3} \quad \text{substitute} \\
&= 3f_{n-3} + 2f_{n-4} \quad \text{simplify} \\
&= 3(\mathbf{f_{n-4} + f_{n-5}}) + 2f_{n-4} \quad \text{substitute} \\
&= 5f_{n-4} + 3f_{n-5} \quad \text{simplify}
\end{aligned}
$$

How do we identify a general pattern? Notice that the coefficients after the simplification are: (2, 1), (3, 2), (5, 3). These seem familiar. They look like pairs of numbers from the Fibonacci sequence! If the next pair is (8, 5), this guess is most likely correct.

$$= 5(\mathbf{f_{n-5} + f_{n-6}}) + 3f_{n-5} \quad \text{substitute}$$
$$= 8f_{n-5} + 5f_{n-6} \quad \text{simplify}$$

The conjecture seems correct. What is the general pattern? Perhaps the following table will help.

n	0	1	2	3	4	5
f_n	1	1	2	3	5	8

It seems that the coefficients appear in the pairs (f_k, f_{k-1}). The generic step is therefore

$$= f_k f_{n-k} + f_{k-1} f_{n-(k+1)}$$

This will terminate when $k = n - 1$:

$$= f_{n-1} f_{n-(n-1)} + f_{(n-1)-1} f_{n-(n-1+1)}$$
$$= f_{n-1} f_1 + f_{n-2} f_0$$
$$= f_{n-1} + f_{n-2}$$

We have arrived back where we started! This attempt to find a closed-form formula for f_n has failed. ■

Fortunately, there are other techniques for solving recurrence relations. A closed-form formula for f_n will be derived in Example 7.26 on page 363.

7.2.2 Linear Homogeneous Recurrence Relations with Constant Coefficients

Recurrence relations with some special properties can be solved by a technique that is more powerful than back substitution. The required special properties are defined next.

DEFINITION 7.5 ***Homogeneous; Constant Coefficients; Linear***

A recurrence relation for the sequence $\{a_n\}$ is a *linear recurrence relation with constant coefficients* if it is in the form

$$a_n = c_1 a_{n-1} + c_2 a_{n-2} + \cdots + c_k a_{n-k} + f(n)$$

for some constants $c_1, c_2, \ldots, c_k$.

The recurrence relation is a *linear homogeneous recurrence relation with constant coefficients* if $f(n) = 0$ for all n.

Some examples should make these ideas clear. Base values (such as a_0) for the sequence $\{a_n\}$ will not be given, since the definitions do not depend on these values.

EXAMPLE 7.22 **Illustrating Definition 7.5**

The following recurrence relations illustrate the terminology introduced in Definition 7.5:

Recurrence Relation	Homogeneous	Linear	Constant Coefficients
$a_n = 5a_{n-1} + 7$	no: $+7$	yes	yes
$a_n = (n^2 + 4)a_{n-1} + 8a_{n-2}$	yes	yes	no: $n^2 + 4$
$a_n = 5a_{n-1} + 3n$	no: $+3n$	yes	yes: $3n$ is not a coefficient of some a_j
$a_n = 3a_{n-1} \cdot a_{n-2} + 4a_{n-3}$	yes	no: $a_{n-1} \cdot a_{n-2}$	yes
$a_n = 4a_{n-1}^2 + 5a_{n-2}$	yes	no: a_{n-1}^2	yes
$a_n = \sin(a_{n-1}) + n$	no: $+n$	no: $\sin(a_{n-1})$	yes

■

Recurrence relations that have all three of the properties in Definition 7.5 are the main topic of this section. The name for such recurrence relations is long, but their form is simple.

DEFINITION 7.6 ***Linear Homogeneous Recurrence Relations with Constant Coefficients of Degree k***

A *linear homogeneous recurrence relation with constant coefficients of degree k* is a recurrence relation that can be written in the form

$$a_n = c_1 a_{n-1} + c_2 a_{n-2} + \cdots + c_k a_{n-k}$$

for some k with $1 \leq k$ and $c_k \neq 0$.

The constant k is called the *degree* of the recurrence relation. The constants, c_j, are called the *coefficients* of the recurrence relations.

EXAMPLE 7.23 **Some Linear Homogeneous Recurrence Relations with Constant Coefficients**

The following recurrence relations are all linear homogeneous recurrence relations with constant coefficients.

Recurrence Relation	Degree
$a_n = 3a_{n-1} + 4a_{n-2}$	2
$a_n = 5a_{n-1}$	1
$a_n = 3a_{n-1} + 4a_{n-3}$	3
$a_n = -3a_{n-2} + 6a_{n-3} - 9a_{n-5}$	5

Notice the convention that the factors a_i are written in descending order. Follow this convention! It will keep you from some errors later in this section. ■

Observe that $a_n = 0$ is always a solution to any linear homogeneous recurrence relation with constant coefficients (usually called the *trivial solution*).

Suppose, for the moment, that the recurrence relation is

$$a_n = c_1 a_{n-1} + c_2 a_{n-2} + \cdots + c_k a_{n-k}$$

and that we were fortunate enough to find a real (or complex) number $r \neq 0$ such that

$$a_n = r^n \quad \text{for all } n > k$$

In that case, we could substitute the explicit values into the recurrence relation and produce

$$\begin{aligned} r^n &= c_1 r^{n-1} + c_2 r^{n-2} + \cdots + c_k r^{n-k} \\ &= r^{n-k}\left(c_1 r^{k-1} + c_2 r^{k-2} + \cdots + c_{k-1} r + c_k\right) \end{aligned}$$

Thus,

$$r^k - c_1 r^{k-1} - c_2 r^{k-2} - \cdots - c_{k-1} r - c_k = 0$$

This leads to Definition 7.7:

DEFINITION 7.7 ***The Characteristic Equation***

The *characteristic equation* of the recurrence relation

$$a_n = c_1 a_{n-1} + c_2 a_{n-2} + \cdots + c_k a_{n-k}$$

is

$$x^k - c_1 x^{k-1} - c_2 x^{k-2} - \cdots - c_{k-1} x - c_k = 0$$

Note that if the final term[30] in the recurrence is $c_k a_{n-k}$ with $c_k \neq 0$, then the characteristic equation will have degree k. The equation is formed by moving every term in the recursion to the left and then replacing a_{n-j} by x^{k-j}, for $0 \leq j \leq k$.

The phrase "characteristic equation" is used in linear algebra when discussing eigenvalues of a square matrix and also in differential equations. Although in each case the context is distinct, nevertheless there is a common feature: The roots of the characteristic equation are the key to the solutions of the various problems.

✔ Quick Check 7.6

1. For each recurrence relation, determine which of the special properties apply (homogeneous, constant coefficients, linear).

(a) $a_n = \cos(n) \cdot a_{n-1} + 3a_{n-2}$

(b) $b_n = 3b_{n-2} + 5b_{n-4}$

(c) $a_n = (a_{n-1})^2 + 4a_{n-2} + 9$

(d) $b_n = 2^{b_{n-1}} + 8b_{n-2}$

2. For each of these linear homogeneous recurrence relations with constant coefficients, determine the characteristic equation.

(a) $a_n = 5a_{n-1} - 9a_{n-2}$

(b) $b_n = b_{n-1} - 7b_{n-3}$

(c) $a_n = 4a_{n-2} + 3a_{n-1}$ Be careful! Something is wrong here. ☑

THEOREM 7.8

Suppose the sequence $\{a_n\}$ is generated by the recurrence relation

$$a_n = c_1 a_{n-1} + c_2 a_{n-2} + \cdots + c_k a_{n-k}$$

If $a_n = \theta r^n$ also generates this sequence, then r is a root of the characteristic equation. Conversely, if r is a root of the characteristic equation, then any expression of the form θr^n generates a sequence that is a solution to the recurrence relation.

The proof is an easy extension of the ideas that motivated Definition 7.7 and will be left as an exercise.

Where did the idea for setting a_n equal to something of the form θr^n come from in the first place? One reason is an experimentally derived one: Suppose you solve (perhaps using back substitution) a number of linear homogeneous recurrence relations with constant coefficients and look for a pattern. The first place to start would be linear homogeneous recurrence relations with constant coefficients of the form

$$a_n = c_1 a_{n-1}$$

It is easy to see, using back substitution, that the solution is $a_n = a_0 c_1^n$. The solution is essentially a power of the coefficient. The next attempt would be to solve linear homogeneous recurrence relations with constant coefficients of the form

$$a_n = c_1 a_{n-1} + c_2 a_{n-2}$$

It is not as easy to see a pattern in the form of the solutions to this recursion. After spending some time getting nowhere, you might eventually try to use the result from the degree one case. Since the solution was expressed in terms of powers of some number, you might try this for the degree 2 case. This leads naturally to Theorem 7.8.

We now know (by Theorem 7.8) that any solution in the form θr^n must have r as a root of the characteristic equation, so there will be only a small number of such solutions. For a linear homogeneous recurrence relation with constant coefficients of degree $k > 1$, the characteristic equation has more than one root.[31] Each of these roots generates a solution to the recurrence relation. What is the most general form of a solution? The next theorem will begin to answer that question.

[30] Assuming subscripts are arranged in decreasing order.

[31] Counting repeated roots.

THEOREM 7.9

Suppose the characteristic equation of the degree k recurrence relation

$$a_n = c_1 a_{n-1} + c_2 a_{n-2} + \cdots + c_k a_{n-k}$$

has k *distinct* roots, $r_1, r_2, \ldots, r_k$. Then for any choice of constants, $\theta_1, \theta_2, \ldots, \theta_k$, the closed-form expression

$$a_n = \theta_1 r_1^n + \theta_2 r_2^n + \cdots + \theta_k r_k^n$$

generates a solution to the recurrence relation.

In addition, if the k initial values, $a_0, a_1, \ldots, a_{k-1}$, are specified, it is always possible to find unique values, $\theta_1, \theta_2, \ldots, \theta_k$, so that the recurrence relation generates the solution that matches those initial values.

Proof: The proof of the first claim is not difficult but does carry some algebraic baggage. I will restrict the demonstration here to the case $k = 2$. The more general proof does not require any additional ideas. What needs to be done then is to show that $a_n = \theta_1 r_1^n + \theta_2 r_2^n$ actually satisfies the recurrence relation. Substituting into the relation produces

$$\begin{aligned} \theta_1 r_1^n + \theta_2 r_2^n &= c_1\left(\theta_1 r_1^{n-1} + \theta_2 r_2^{n-1}\right) + c_2\left(\theta_1 r_1^{n-2} + \theta_2 r_2^{n-2}\right) \\ &= \left(c_1\theta_1 r_1^{n-1} + c_2\theta_1 r_1^{n-2}\right) + \left(c_1\theta_2 r_2^{n-1} + c_2\theta_2 r_2^{n-2}\right) \end{aligned}$$

The equation can be rearranged to produce

$$\theta_1 r_1^{n-2}\left(r_1^2 - c_1 r_1 - c_2\right) + \theta_2 r_2^{n-2}\left(r_2^2 - c_1 r_2 - c_2\right) = 0$$

The expression on the left is equal to 0 because r_1 and r_2 are both roots of the characteristic equation: $x^2 - c_1 x - c_2 = 0$.

Now a proof for the second claim. The constants $\theta_1, \theta_2, \ldots, \theta_k$ need to be chosen so that the first k terms in the sequence match $a_0, a_1, \ldots, a_{k-1}$. That is, so that

$$\begin{aligned} \theta_1 r_1^0 + \theta_2 r_2^0 + \cdots \theta_k r_k^0 &= a_0 \\ \theta_1 r_1^1 + \theta_2 r_2^1 + \cdots \theta_k r_k^1 &= a_1 \\ \theta_1 r_1^2 + \theta_2 r_2^2 + \cdots \theta_k r_k^2 &= a_2 \\ \vdots \quad & \quad \vdots \\ \theta_1 r_1^{k-1} + \theta_2 r_2^{k-1} + \cdots \theta_k r_k^{k-1} &= a_{k-1} \end{aligned}$$

This forms a system of k linear equations in the unknowns $\theta_1, \theta_2, \ldots, \theta_k$ (the values $a_0, a_1, \ldots, a_{k-1}$, and $r_1, r_2, \ldots, r_k$ are all known).

A famous theorem in linear algebra (Vandermonde's matrix theorem) implies that whenever the coefficient matrix[32] of a system of linear equations has the following form (for *distinct* values $v_1, v_2, \ldots, v_k$), the system has a unique solution.

$$\begin{bmatrix} 1 & 1 & \cdots & 1 \\ v_1 & v_2 & \cdots & v_k \\ v_1^2 & v_2^2 & \cdots & v_k^2 \\ \vdots & \vdots & \vdots & \vdots \\ v_1^{k-1} & v_2^{k-1} & \cdots & v_k^{k-1} \end{bmatrix}$$

Thus, $\theta_1, \theta_2, \ldots, \theta_k$ are uniquely determined (and are not hard to calculate in practice). □

[32]The *coefficient matrix* for the linear system of interest here is created by extracting the θs and a_ks (the *coefficients* of the system). The extracted coefficients are listed in rows that correspond to the equations in the system. See Definition E.3 (on page A30) in Appendix E.

It is time for a few examples before developing additional theoretical insights.

EXAMPLE 7.24

An Easy Linear Homogeneous Recurrence Relation with Constant Coefficients

Consider the recurrence relation specified by

- $a_0 = -2$ and $a_1 = 3$
- $a_n = a_{n-1} + 6a_{n-2} \quad$ for $n \geq 2$

The characteristic equation is

$$x^2 - x - 6 = 0$$

having roots[33] $r_1 = 3$ and $r_2 = -2$.

Thus, the general solution is of the form

$$a_n = \theta_1 3^n + \theta_2(-2)^n$$

The final step is to determine values for θ_1 and θ_2 so that the solution matches a_0 and a_1. In order for that to be true, the closed-form expression for a_n must produce the predetermined values for a_0 and a_1:

$$\theta_1 3^0 + \theta_2(-2)^0 = -2$$
$$\theta_1 3^1 + \theta_2(-2)^1 = 3$$

This simplifies to

$$\theta_1 + \theta_2 = -2$$
$$3\theta_1 - 2\theta_2 = 3$$

This system is small enough to be solved by substitution (Gaussian elimination[34] is usually better).

$$\theta_1 = -\theta_2 - 2$$

So

$$3(-\theta_2 - 2) - 2\theta_2 = 3$$
$$-5\theta_2 = 9$$
$$\theta_2 = -\frac{9}{5}$$

Therefore, $\theta_1 = -\theta_2 - 2 = -\frac{1}{5}$.

The solution we seek is

$$a_n = -\frac{1}{5} \cdot 3^n - \frac{9}{5} \cdot (-2)^n \quad \text{for } n \geq 0$$

Warning! It is critical that you use *exact* values for the roots and for the θs. Using decimal approximations produced by a calculator will not lead to a correct formula. More on this soon.

It is very easy to do some simple checking for validity. Form a table that contains the first few values generated by the recurrence relation and the first few values generated

[33]It does not matter which root is labeled r_1 and which is labeled r_2, as long as we are consistent throughout the problem.

[34]Gaussian elimination is a row reduction technique taught in linear algebra courses.

by the closed-form formula. They should match. Table 7.6 shows the validity check for this example.

TABLE 7.6 Validity Checking

n	Recurrence	Closed-Form
0	−2	−2
1	3	3
2	−9	−9
3	9	9
4	−45	−45

■

The previous example illustrates the general approach for solving linear homogeneous recurrence relations with constant coefficients having characteristic equations with distinct roots:

The General Procedure for Solving Linear Homogeneous Recurrence Relations with Constant Coefficients Having Distinct Roots

Step 1 Form the characteristic equation.

Step 2 Find the roots of the characteristic equation.

Step 3 If the roots are distinct, express the general solution in the form

$$a_n = \theta_1 r_1^n + \theta_2 r_2^n + \cdots + \theta_k r_k^n \quad \text{for } n \geq k$$

Step 4 Form the system of linear equations to determine the θs:

$$\begin{aligned} \theta_1 r_1^0 + \theta_2 r_2^0 + \cdots + \theta_k r_k^0 &= a_0 \\ \theta_1 r_1^1 + \theta_2 r_2^1 + \cdots + \theta_k r_k^1 &= a_1 \\ \theta_1 r_1^2 + \theta_2 r_2^2 + \cdots + \theta_k r_k^2 &= a_2 \\ \vdots \qquad & \quad \vdots \\ \theta_1 r_1^{k-1} + \theta_2 r_2^{k-1} + \cdots + \theta_k r_k^{k-1} &= a_{k-1} \end{aligned}$$

Step 5 Solve the linear system and substitute the solution values for the θs into the general solution from step 3.

EXAMPLE 7.25

Erroneous Rounding

Consider the linear homogeneous recurrence relation with constant coefficients defined by

- $a_0 = 1$ and $a_1 = -1$
- $a_n = 2a_{n-2} \quad \text{for } n \geq 2$

The characteristic equation, $x^2 - 2 = 0$, has roots $r_1 = \sqrt{2}$ and $r_2 = -\sqrt{2}$. Suppose we use a calculator and write these as $r_1 = 1.414$ and $r_2 = -1.414$.[35]

The linear system would then be

$$\begin{aligned} \theta_1 + \theta_2 &= 1 \\ 1.414\theta_1 - 1.414\theta_2 &= -1 \end{aligned}$$

Using a calculator to solve this system leads to $\theta_1 = 0.146$ and $\theta_2 = 0.854$. The solution is therefore

$$a_n = 0.146 \cdot 1.414^n + 0.854 \cdot (-1.414)^n$$

[35] Not an uncommon action; many students routinely round all calculator output to two or three digits. This idea is not a good, especially with intermediate values during a long calculation.

Consider Table 7.7, which compares the values generated by the recurrence relation and the values generated by the closed-form formula.

TABLE 7.7 Validity Check when Rounding

n	Recurrence	Formula
0	1	1.00000
1	−1	−1.00111
2	2	1.99940
3	−2	−2.00162
4	4	3.99758
5	−4	−4.00203
⋮	⋮	⋮
20	1,024	1,020.91
21	−1,024	−1,022.05

The formula starts poorly and deteriorates as n increases. The correct solution is

$$a_n = \left(\frac{2-\sqrt{2}}{4}\right)\left(\sqrt{2}\right)^n + \left(\frac{2+\sqrt{2}}{4}\right)\left(-\sqrt{2}\right)^n \quad \text{for } n \geq 0$$ ■

Recall that the method of back substitution was not very helpful for finding a closed-form formula for the Fibonacci numbers. We now have a technique that can find such a formula.

EXAMPLE 7.26 **The Fibonacci Sequence Revisited**

The characteristic equation for the Fibonacci sequence is $x^2 - x - 1 = 0$, since the recurrence relation is $f_n = f_{n-1} + f_{n-2}$ for $n \geq 2$. There is no simple way to factor the left-hand side of the equation, so the quadratic formula is necessary.

$$x = \frac{1 \pm \sqrt{5}}{2}$$

The roots are $r_1 = \frac{1+\sqrt{5}}{2}$ and $r_2 = \frac{1-\sqrt{5}}{2}$. The general solution is of the form

$$f_n = \theta_1 \left(\frac{1+\sqrt{5}}{2}\right)^n + \theta_2 \left(\frac{1-\sqrt{5}}{2}\right)^n$$

The system of linear equations that determine the θs is

$$\theta_1 + \theta_2 = 1$$

$$\left(\frac{1+\sqrt{5}}{2}\right)\theta_1 + \left(\frac{1-\sqrt{5}}{2}\right)\theta_2 = 1$$

This looks messy, but the substitution is not really difficult: Since $\theta_2 = 1 - \theta_1$,

$$\left(\frac{1+\sqrt{5}}{2}\right)\theta_1 + \left(\frac{1-\sqrt{5}}{2}\right)(1-\theta_1) = 1$$

$$\left(\frac{1+\sqrt{5}}{2} - \frac{1-\sqrt{5}}{2}\right)\theta_1 + \frac{1-\sqrt{5}}{2} = 1$$

$$\sqrt{5}\,\theta_1 = \frac{1+\sqrt{5}}{2}$$

$$\theta_1 = \frac{5+\sqrt{5}}{10}$$

Thus $\theta_2 = 1 - \theta_1 = \frac{5-\sqrt{5}}{10}$.

The Fibonacci sequence is generated by

$$f_n = \left(\frac{5+\sqrt{5}}{10}\right)\left(\frac{1+\sqrt{5}}{2}\right)^n + \left(\frac{5-\sqrt{5}}{10}\right)\left(\frac{1-\sqrt{5}}{2}\right)^n \quad \text{for } n \geq 0$$

This result may strike you as a bit odd; how can the Fibonacci sequence (a sequence of *integers*) be generated by a complicated expression involving $\sqrt{5}$?[36] However, if you program this formula into a graphing calculator or a program like *Mathematica* or *Maple*, you will find that it does indeed generate the Fibonacci sequence. ■

✔ Quick Check 7.7

1. Find a closed-form formula for a_n, if
   - $a_0 = 1, a_1 = 2$, and $a_2 = 3$
   - $a_n = 4a_{n-1} + a_{n-2} - 4a_{n-3}$ for $n \geq 3$

2. Find a closed-form formula for a_n, if
   - $a_0 = 3$
   - $a_n = 2a_{n-1} + 5 \quad$ for $n \geq 1$

EXAMPLE 7.27

A Second Look at a Previous Pattern

Recall the attempt in Example 3.38 (page 137) to find a simple formula for the partial sum

$$S_n = 1 - \frac{1}{2} + \frac{1}{2^2} - \frac{1}{2^3} + \cdots \pm \frac{1}{2^n} = \sum_{k=0}^{n}\left(-\frac{1}{2}\right)^n$$

Some of the initial partial sums are listed in Table 7.8.

TABLE 7.8 The Initial Partial Sums

n	0	1	2	3	4	5	6	7	8
S_n	1	$\frac{1}{2}$	$\frac{3}{4}$	$\frac{5}{8}$	$\frac{11}{16}$	$\frac{21}{32}$	$\frac{43}{64}$	$\frac{85}{128}$	$\frac{171}{256}$

One of the observations (which was not pursued in that example) concerned the differences of successive numerators. Let a_n = the nth numerator. Some initial differences are listed in Table 7.9.

TABLE 7.9 Some Initial Differences of Successive Numerators

n	0	1	2	3	4	5	6	7	8
S_n	1	$\frac{1}{2}$	$\frac{3}{4}$	$\frac{5}{8}$	$\frac{11}{16}$	$\frac{21}{32}$	$\frac{43}{64}$	$\frac{85}{128}$	$\frac{171}{256}$
a_n	1	1	3	5	11	21	43	85	171
$a_n - a_{n-1}$		0	2	2	6	10	22	42	86

---

[36] You may also have noticed the unexpected appearance of the golden ratio: $\Phi = \frac{1+\sqrt{5}}{2}$. See Appendix D for more about the golden ratio.

It is easy to notice that $a_n - a_{n-1} = 2a_{n-2}$ for $n \geq 2$. Thus, there is a linear homogeneous recurrence relation with constant coefficients that the numerators seem to follow: $a_n = a_{n-1} + 2a_{n-2}$ for $n \geq 2$.

The characteristic equation is $x^2 - x - 2 = 0$, which has roots $x = 2, -1$. The general solution is therefore

$$a_n = \theta_1 2^n + \theta_2(-1)^n$$

The θs can be found using the initial values $a_0 = a_1 = 1$.

$$\begin{aligned} \theta_1 + \theta_2 &= 1 \\ 2\theta_1 - \theta_2 &= 1 \end{aligned}$$

Thus $\theta_1 = \frac{2}{3}$ and $\theta_2 = \frac{1}{3}$.

The numerator of S_n should be $\frac{2}{3}2^n + \frac{1}{3}(-1)^n$, which agrees with the pattern guessed in Example 3.38.

The recurrence relation was built from a partial table of values for S_n, so it is still only a (good) guess. The proof by mathematical induction at the end of Example 3.38 is still needed. ■

An alternative approach for finding the roots of the characteristic equation in exercise 1 of Quick Check 7.7 is to use the following theorem, which will not help if all the roots are irrational or complex.

THEOREM 7.10 ***Rational Roots Theorem***

Suppose the polynomial $c_n x^n + c_{n-1}x^{n-1} + \cdots + c_1 x + c_0$ has integer coefficients where $c_n \neq 0$ and $c_0 \neq 0$. Then any rational (or integer) zero of the polynomial must be of the form $\pm\frac{p}{q}$, where p evenly divides c_0 and q evenly divides c_n.

EXAMPLE 7.28

Finding Rational Roots

The polynomial equation $3x^3 - 4x^2 - 6x + 8 = 0$ has possible rational roots

$$\pm 1, \pm 2, \pm 4, \pm 8, \pm\frac{1}{3}, \pm\frac{2}{3}, \pm\frac{4}{3}, \pm\frac{8}{3}$$

Brute force substitution into the equation verifies that

$$3\left(\frac{4}{3}\right)^3 - 4\left(\frac{4}{3}\right)^2 - 6\left(\frac{4}{3}\right) + 8 = 0$$

None of the other 15 possibilities work. Since $\frac{4}{3}$ is a root, $(x - \frac{4}{3})$ is a factor. It is also valid to instead assert that $(3x - 4)$ is a factor. Dividing the polynomial $3x^3 - 4x^2 - 6x + 8 = 0$ by $3x - 4$ produces the quotient $x^2 - 2$. This can be factored using the quadratic formula. The full factorization yields $(3x - 4)(x - \sqrt{2})(x + \sqrt{2}) = 0$. ■

✔ Quick Check 7.8

1. Use Theorem 7.10 to list the potential rational roots for $6x^2 + 13x - 5$.
2. Factor $6x^2 + 13x - 5$.

☑

7.2.3 Repeated Roots

EXAMPLE 7.29 **Two Linear Homogeneous Recurrence Relations with Constant Coefficients Whose Characteristic Equations Have Repeated Roots**

Consider the linear homogeneous recurrence relation with constant coefficients defined by

- $a_0 = 2$ and $a_1 = 1$
- $a_n = 2a_{n-1} - a_{n-2}$ for $n \geq 2$

The characteristic equation is $x^2 - 2x + 1 = 0$. This factors as $(x-1)^2 = 0$, so the roots are $r_1 = r_2 = 1$.

Blindly following established procedure would mean the general solution is of the form $a_n = \theta_1 1^n + \theta_2 1^n$. That is, $a_n =$ a constant, for all n. However, the recurrence relation indicates that the sequence starts as $\{2, 1, 0, -2, -4, -6, \ldots\}$! Clearly, blindly following the previous technique was a mistake.

Suppose, instead, that the recurrence relation is defined by

- $a_0 = 2$ and $a_1 = 1$
- $a_n = 4a_{n-1} - 4a_{n-2}$ for $n \geq 2$

In this case, the characteristic equation is $x^2 - 4x + 4 = (x-2)^2 = 0$, having roots $r_1 = r_2 = 2$. The general solution (assuming past procedures apply) is $a_n = \theta_1 2^n + \theta_2 2^n = (\theta_1 + \theta_2)2^n$. What does the system of linear equations look like?

$$\theta_1 + \theta_2 = 2$$
$$2(\theta_1 + \theta_2) = 1$$

Substituting the first equation into the second,

$$2 \cdot 2 = 1$$

Certainly something is wrong again! ■

The problems in the previous example arose because the characteristic equations have repeated roots. Some new technique is needed to handle this situation.

One source of the problem is that there are not enough unknowns in the system of linear equations. Notice that if the substitution $\omega = \theta_1 + \theta_2$ is made for the second illustration in Example 7.29, the system of equations becomes two equations in one unknown (an overdetermined system).

$$\omega = 2$$
$$2\omega = 1$$

What is needed is some way to have θ_1 and θ_2 be multiplied by *different* expressions. It seems reasonable to keep the 2^n in each case. It should also be clear (after a moment's reflection) that an expression of the form $c2^n$, for some constant c, will not improve matters. Somehow, n needs to be involved. Before proceeding to the solution, some background material is necessary.

✔ Quick Check 7.9

1. Let $p(x) = x^4 - 6x^3 + 12x^2 - 10x + 3$.
   (a) Factor $p(x)$ (perhaps using Theorem 7.10). There is a repeated zero, r. What is its multiplicity?
   (b) Find $p'(x)$ and then factor it. Is r still a zero? If so, what is its multiplicity as a zero of $p'(x)$?
   (c) Find $p''(x)$ and then factor it. Is r still a zero? If so, what is its multiplicity as a zero of $p''(x)$?

(d) Find $p^{(3)}(x)$ and then factor it. Is r still a zero? If so, what is its multiplicity as a zero of $p^{(3)}(x)$?

(e) Make a hypothesis about the relationship between a repeated zero of a polynomial and the derivatives of the polynomial. ☑

The explorations in Quick Check 7.9 lead to the next theorem.

THEOREM 7.11 ***Derivatives and Repeated Roots***

Let $p(x)$ be a polynomial with a zero, r, having multiplicity $\nu > 1$. Then r is also a zero of the derivative $p^{(j)}(x)$ for $j = 1, 2, \ldots, \nu - 1$.

Proof: Since r has multiplicity ν, $p(x)$ can be written as $p(x) = (x - r)^\nu q_0(x)$, for some polynomial $q_0(x)$ for which r is *not* a zero. Consider the derivatives

$$\begin{aligned}
p(x) &= (x-r)^\nu q_0(x) \\
p'(x) &= \nu(x-r)^{\nu-1} q_0(x) + (x-r)^\nu q_0'(x) \\
&= (x-r)^{\nu-1}\left(\nu q_0(x) + (x-r) q_0'(x)\right) \\
&= (x-r)^{\nu-1} q_1(x) \\
p''(x) &= (\nu-1)(x-r)^{\nu-2} q_1(x) + (x-r)^{\nu-1} q_1'(x) \\
&= (x-r)^{\nu-2} q_2(x) \\
\vdots \quad &\vdots \qquad \vdots \\
p^{(j)}(x) &= (x-r)^{\nu-j} q_j(x) \\
\vdots \quad &\vdots \qquad \vdots \\
p^{(\nu-1)}(x) &= (x-r) q_{\nu-1}(x)
\end{aligned}$$

Thus, r is a zero for each of the first $\nu - 1$ derivatives. □

The next theorem settles the issue of how to handle repeated roots.

THEOREM 7.12 ***Linear Homogeneous Recurrence Relations with Constant Coefficients Whose Characteristic Equations Have Repeated Roots***

Suppose the characteristic equation of the recurrence relation

$$a_n = c_1 a_{n-1} + c_2 a_{n-2} + \cdots + c_k a_{n-k}$$

has a root, r, of multiplicity ν. Then for any choice of constants, $\alpha_0, \alpha_1, \ldots, \alpha_{\nu-1}$, the closed-form expression

$$a_n = \left(\alpha_0 + \alpha_1 n + \alpha_2 n^2 + \cdots + \alpha_{\nu-1} n^{\nu-1}\right) r^n$$

generates a solution to the recurrence relation.

Notation: Notice the change in notation. When there are k distinct roots, $r_1, r_2, \ldots, r_k$, the coefficients are given corresponding subscripts: $\theta_1, \theta_2, \ldots, \theta_k$. When there is a single root, r, of multiplicity ν, the coefficients are given subscripts that match the corresponding exponent on n: $\alpha_0, \alpha_1, \ldots, \alpha_{\nu-1}$. The Greek letter for the coefficients has also changed. This change in variable name will be extended in Theorem 7.13.

A simple example will be given before looking at the proof of this theorem.

EXAMPLE 7.30 **Repeated Roots Revisited**

Recall the following recurrence relation.

- $a_0 = 2$ and $a_1 = 1$
- $a_n = 4a_{n-1} - 4a_{n-2} \quad$ for $n \geq 2$

The characteristic equation is $x^2 - 4x + 4 = (x - 2)^2 = 0$, having root $r = 2$ of multiplicity $\nu = 2$. A general solution (assuming Theorem 7.12 is correct) is $a_n = (\alpha_0 + \alpha_1 n)\, 2^n$. It is now possible to form a system of linear equations (having a unique solution) to find values for the αs that will match the base values of the recurrence relation:

$$(\alpha_0 + \alpha_1 \cdot 0)2^0 = 2$$
$$(\alpha_0 + \alpha_1 \cdot 1)2^1 = 1$$

This can be expressed as

$$\alpha_0 = 2$$
$$2\alpha_0 + 2\alpha_1 = 1$$

The values of the αs are $\alpha_0 = 2$ and $\alpha_1 = -\frac{3}{2}$. A solution[37] is $a_n = \left(2 - \frac{3}{2}n\right) 2^n$, for $n \geq 0$. ■

Proof of Theorem 7.12: The characteristic equation is

$$x^k - c_1 x^{k-1} - \cdots - c_{k-1}x - c_k = 0$$

where r is a root of multiplicity ν. Define

$$\begin{aligned} p_0(x) &= x^{n-k}\left(x^k - c_1 x^{k-1} - \cdots - c_{k-1}x - c_k\right) \\ &= x^n - c_1 x^{n-1} - \cdots - c_{k-1}x^{n-k+1} - c_k x^{n-k} \end{aligned}$$

Using the same ideas[38] as in the proof of Theorem 7.11, it can be shown that r is a zero of $p_0^{(j)}(x)$, for $0 \leq j \leq \nu - 1$. In particular, r is a zero of multiplicity $\nu - 1$ for the polynomial $p_0'(x)$, and hence a zero of multiplicity $\nu - 1$ for the polynomial $p_1(x) = x \cdot p_0'(x)$.[39] But then r is a zero of multiplicity $\nu - 2$ for the polynomial $p_2(x) = x \cdot p_1'(x)$. This can be continued until arriving at the polynomial $p_{\nu-1}(x) = x \cdot p_{\nu-2}'(x)$, where r is a root of multiplicity 1.

Thus, r is a zero of each of the ν polynomials $p_0(x), p_1(x), \ldots, p_{\nu-1}(x)$. The form of the polynomials $p_j(x)$ is important.

$$\begin{aligned} p_0(x) &= x^n - c_1 x^{n-1} - \cdots - c_k x^{n-k} \\ p_1(x) &= x \cdot \left(n x^{n-1} - c_1(n-1)x^{n-2} - \cdots - c_k(n-k)x^{n-k-1}\right) \\ &= n x^n - c_1(n-1)x^{n-1} - \cdots - c_k(n-k)x^{n-k} \end{aligned}$$

[37] Actually, *the* solution, but this has not been proved yet.

[38] Just consider x^{n-k} as part of $q(x)$. (Note that $r \neq 0$ since $c_k \neq 0$.)

[39] Recall that $r \neq 0$, so multiplying by x does not change the multiplicity.

$$
\begin{aligned}
p_2(x) &= x \cdot \left(n^2 x^{n-1} - c_1(n-1)^2 x^{n-2} - \cdots - c_k(n-k)^2 x^{n-k-1}\right) \\
&= n^2 x^n - c_1(n-1)^2 x^{n-1} - \cdots - c_k(n-k)^2 x^{n-k} \\
&\vdots \\
p_{\nu-1} &= n^{\nu-1} x^n - c_1(n-1)^{\nu-1} x^{n-1} - \cdots - c_k(n-k)^{\nu-1} x^{n-k}
\end{aligned}
$$

To keep the algebraic details from becoming too messy, but still keep sufficient detail to indicate the general case, I will restrict the rest of the proof to the case $\nu = 3$.

Substituting $a_n = \left(\alpha_0 + \alpha_1 n + \alpha_2 n^2\right) r^n$ into the recurrence relation, $a_n = c_1 a_{n-1} + c_2 a_{n-2} + \cdots + c_k a_{n-k}$, leads to

$$
\begin{aligned}
(\alpha_0 + \alpha_1 n + \alpha_2 n^2) r^n = c_1 &\left(\alpha_0 + \alpha_1(n-1) + \alpha_2(n-1)^2\right) r^{n-1} \\
+ c_2 &\left(\alpha_0 + \alpha_1(n-2) + \alpha_2(n-2)^2\right) r^{n-2} \\
+ \cdots + c_k &\left(\alpha_0 + \alpha_1(n-k) + \alpha_2(n-k)^2\right) r^{n-k}
\end{aligned}
$$

Moving all nonzero terms to the same side of the equation and grouping by powers of n yields

$$
\begin{aligned}
&\alpha_0 \left(r^n - c_1 r^{n-1} - \cdots - c_k r^{n-k}\right) \\
&+ \alpha_1 \left(n r^n - c_1(n-1) r^{n-1} - \cdots - c_k(n-k) r^{n-k}\right) \\
&+ \alpha_2 \left(n^2 r^n - c_1(n-1)^2 r^{n-1} - \cdots - c_k(n-k)^2 r^{n-k}\right) \\
&= 0
\end{aligned}
$$

This is the same as

$$\alpha_0 p_0(r) + \alpha_1 p_1(r) + \alpha_2 p_2(r) = 0$$

which is true since r is a zero of each of the polynomials $p_j(x)$.[40] □

EXAMPLE 7.31

A Root of Multiplicity Three

A linear homogeneous recurrence relation with constant coefficients having a root $r = 2$ with multiplicity 3 will have characteristic equation

$$(x-2)^3 = x^3 - 6x^2 + 12x - 8 = 0$$

The recurrence relation will therefore be

$$a_n = 6a_{n-1} - 12a_{n-2} + 8a_{n-3} \quad \text{for } n \geq 3$$

Suppose the base values are $a_0 = 1$, $a_1 = 0$, and $a_2 = 4$.

The general solution is of the form

$$a_n = \left(\alpha_0 + \alpha_1 n + \alpha_2 n^2\right) 2^n$$

The αs are determined by the system of linear equations

$$
\begin{aligned}
\alpha_0 + 0 + 0 &= 1 \\
2\alpha_0 + 2\alpha_1 + 2\alpha_2 &= 0 \\
4\alpha_0 + 8\alpha_1 + 16\alpha_2 &= 4
\end{aligned}
$$

The first two equations imply that $\alpha_2 = -1 - \alpha_1$. Substituting into the third equation leads to $\alpha_1 = -2$, so $\alpha_2 = 1$.

The solution is

$$a_n = \left(1 - 2n + n^2\right) 2^n \quad \text{for } n \geq 0$$ ■

[40] Dropping the assumption that $\nu = 3$ leads to the equation $\alpha_0 p_0(r) + \alpha_1 p_1(r) + \cdots + \alpha_{\nu-1} p_{\nu-1}(r) = 0$.

✔ Quick Check 7.10

1. Find a closed-form formula for the nth term of the recurrence relation
   - $a_0 = -2, a_1 = -6$
   - $a_n = -6a_{n-1} - 9a_{n-2}$ for $n \geq 2$ ☑

The General Case

It is now possible to state a theorem about linear homogeneous recurrence relations with constant coefficients having repeated roots.

THEOREM 7.13 ***Solving Linear Homogeneous Recurrence Relations with Constant Coefficients***

Suppose the characteristic equation of the recurrence relation

$$a_n = c_1 a_{n-1} + c_2 a_{n-2} + \cdots + c_k a_{n-k}$$

has j distinct roots, $r_1, r_2, \ldots, r_j$, having respective multiplicities $\nu_1, \nu_2, \ldots, \nu_j$, with $\nu_1 + \nu_2 + \cdots + \nu_j = k$. Then for any choice of constants, $\alpha_0, \alpha_1, \ldots, \alpha_{\nu_1 - 1}$, $\beta_0, \beta_1, \ldots, \beta_{\nu_2 - 1}, \ldots$, the closed-form expression

$$\begin{aligned} a_n = &\left(\alpha_0 + \alpha_1 n + \alpha_2 n^2 + \cdots + \alpha_{\nu_1 - 1} n^{\nu_1 - 1}\right) r_1^n \\ &+ \left(\beta_0 + \beta_1 n + \beta_2 n^2 + \cdots + \beta_{\nu_2 - 1} n^{\nu_2 - 1}\right) r_2^n \\ &+ \cdots + \left(\omega_0 + \omega_1 n + \omega_2 n^2 + \cdots + \omega_{\nu_j - 1} n^{\nu_j - 1}\right) r_j^n \end{aligned}$$

generates a solution to the recurrence relation.

In addition, if base values $a_0, a_1, \ldots, a_{k-1}$ are specified, then unique values can be found for $\alpha_0, \alpha_1, \ldots, \alpha_{\nu_1 - 1}, \beta_0, \beta_1, \ldots, \beta_{\nu_2 - 1}, \ldots$ so that the closed-form formula matches the sequence generated by the recurrence relation. Thus, the closed-form formula is the only solution; any other solution must be algebraically equivalent.

The proof of this theorem is not conceptually difficult, but the notational aspects are a bit messy. A more precise statement of this theorem would use doubly subscripted coefficients in the expression for the general form. The first subscript would link the coefficient to the root, r_i. The second subscript would link the coefficient to the exponent on n. The proof will not be given. Several examples should serve to indicate how the theorem can be used.

EXAMPLE 7.32

A Simple Example of the General Case

Let $a_0 = 12$, $a_1 = 18$, and $a_2 = 24$ and

$$a_n = 12a_{n-2} + 16a_{n-3} \quad \text{for } n \geq 3$$

The characteristic equation is

$$x^3 - 12x - 16 = (x - 4)(x + 2)^2 = 0$$

The roots are $r_1 = 4$ with multiplicity $\nu_1 = 1$, and $r_2 = -2$ with multiplicity $\nu_2 = 2$. The general solution is of the form

$$a_n = \alpha_0 4^n + (\beta_0 + \beta_1 n)(-2)^n$$

The system of linear equations that determines the unknown coefficients is

$$\begin{aligned} \alpha_0 + \beta_0 + 0 \cdot \beta_1 &= 12 \\ 4\alpha_0 - 2\beta_0 - 2\beta_1 &= 18 \\ 16\alpha_0 + 4\beta_0 + 8\beta_1 &= 24 \end{aligned}$$

One way that this system can be solved is to add 4 times the second equation to the third equation, producing $32\alpha_0 - 4\beta_0 = 96$. Then substitute $\beta_0 = 12 - \alpha_0$ to find that $\alpha_0 = 4$. Thus, $\beta_0 = 8$, and the second of the original equations then implies that $\beta_1 = -9$. The solution is therefore

$$a_n = 4^{n+1} + (8 - 9n)(-2)^n \quad \text{for } n \geq 0$$

■

EXAMPLE 7.33

Two Repeated Roots

Consider the recurrence relation defined by

- $a_0 = 1, a_1 = 1, a_2 = 0, a_3 = 2$
- $a_n = 8a_{n-2} - 16a_{n-4}$

The characteristic equation is $x^4 - 8x^2 + 16$. This polynomial is easily factored:

$$x^4 - 8x^2 + 16 = (x^2 - 4)^2 = (x - 2)^2(x + 2)^2$$

Thus, $r_1 = 2$ with multiplicity 2 and $r_2 = -2$ with multiplicity 2. The general form is thus

$$a_n = (\alpha_0 + \alpha_1 n)\, 2^n + (\beta_0 + \beta_1 n)(-2)^n$$

The system of linear equations that will determine the αs and βs is shown next.

$$\begin{aligned}
\alpha_0 + \beta_0 &= 1 \\
(\alpha_0 + \alpha_1 \cdot 1)\, 2^1 + (\beta_0 + \beta_1 \cdot 1)(-2)^1 &= 1 \\
(\alpha_0 + \alpha_1 \cdot 2)\, 2^2 + (\beta_0 + \beta_1 \cdot 2)(-2)^2 &= 0 \\
(\alpha_0 + \alpha_1 \cdot 3)\, 2^3 + (\beta_0 + \beta_1 \cdot 3)(-2)^3 &= 2
\end{aligned}$$

This simplifies to

$$\begin{aligned}
\alpha_0 + \beta_0 &= 1 \\
2\alpha_0 + 2\alpha_1 - 2\beta_0 - 2\beta_1 &= 1 \\
4\alpha_0 + 8\alpha_1 + 4\beta_0 + 8\beta_1 &= 0 \\
8\alpha_0 + 24\alpha_1 - 8\beta_0 - 24\beta_1 &= 2
\end{aligned}$$

The solution of this system of linear equations is

$$\alpha_0 = \frac{13}{16} \quad \alpha_1 = -\frac{5}{16} \quad \beta_0 = \frac{3}{16} \quad \beta_1 = -\frac{3}{16}$$

The solution for the recurrence relation is thus

$$a_n = \left(\frac{13}{16} - \frac{5}{16}n\right) 2^n + \left(\frac{3}{16} - \frac{3}{16}n\right)(-2)^n \quad \text{for } n \geq 0$$

A small table of values for both the original recurrence relation and the closed-form expression indicate that no arithmetic errors have occurred.

n	Recurrence Relation	Closed-Form
0	1	1
1	1	1
2	0	0
3	2	2
4	−16	−16
5	0	0
6	−128	−128
7	−32	−32
8	−768	−768

■

The General Procedure for Solving Linear Homogeneous Recurrence Relations with Constant Coefficients

Step 1 Form the characteristic equation.

Step 2 Find the distinct roots of the characteristic equation, $r_1, r_2, \ldots, r_j$, having respective multiplicities $\nu_1, \nu_2, \ldots, \nu_j$, with $\nu_1 + \nu_2 + \cdots + \nu_j = k$.

Step 3 Express the general solution as a sum of j terms in the form

$$\left(\delta_0 + \delta_1 n + \delta n^2 + \cdots + \delta_{\nu_i - 1} n^{\nu_i - 1}\right) r_i^n$$

Step 4 Use the result of step 3 to form a system of linear equations to determine the unknown coefficients.

Step 5 Solve the linear system and substitute the solution values for the unknown coefficients into the general solution from step 3.

EXAMPLE 7.34

Linear Homogeneous Recurrence Relations with Constant Coefficients Having Complex Roots

You should recall from your previous mathematics courses that some polynomials have complex numbers[41] as zeros. Theorem 7.13 is still valid. This example involves a linear homogeneous recurrence relation with constant coefficients whose characteristic equation has complex roots.

Let $a_0 = 3$, $a_1 = 3$, $a_2 = 11$, and $a_3 = 34$.
For $n \geq 4$, let $a_n = 3a_{n-1} - a_{n-2} - 4a_{n-4}$.

The characteristic equation is

$$x^4 - 3x^3 + x^2 + 4 = 0$$

Theorem 7.10 (page 365) can be used to find a root. The choices are $\pm 1, \pm 2, \pm 4$. One root is $r = 2$. Dividing $x^4 - 3x^3 + x^2 + 4$ by $x - 2$ produces the quotient $x^3 - x^2 - x - 2$. So

$$x^4 - 3x^3 + x^2 + 4 = (x-2)(x^3 - x^2 - x - 2)$$

Using Theorem 7.10 again, it is easy to see that 2 is a root of $x^3 - x^2 - x - 2$. After dividing $x^3 - x^2 - x - 2$ by $x - 2$, we can write

$$x^4 - 3x^3 + x^2 + 4 = (x-2)^2(x^2 + x + 1)$$

The rational roots theorem provides no help factoring $x^2 + x + 1$. The quadratic formula produces the zeros $x = \frac{-1 \pm \sqrt{-3}}{2} = \frac{-1 \pm \sqrt{3}i}{2}$.

The roots are therefore $r_1 = 2$ with multiplicity $\nu_1 = 2$, $r_2 = \frac{-1+\sqrt{3}i}{2}$ with multiplicity $\nu_2 = 1$, and $r_3 = \frac{-1-\sqrt{3}i}{2}$ with multiplicity $\nu_3 = 1$.

The general form for the solution is

$$a_n = (\alpha_0 + \alpha_1 n)\, 2^n + \beta_0 \left(\frac{-1+\sqrt{3}i}{2}\right)^n + \gamma_0 \left(\frac{-1-\sqrt{3}i}{2}\right)^n$$

[41] Numbers of the form $a + bi$, where $i^2 = -1$. See Appendix A for a brief review of complex numbers.

The system of linear equations is

$$\alpha_0 + 0 \cdot \alpha_1 + \beta_0 + \gamma_0 = 3$$

$$2\alpha_0 + 2\alpha_1 + \left(\frac{-1+\sqrt{3}i}{2}\right)\beta_0 + \left(\frac{-1-\sqrt{3}i}{2}\right)\gamma_0 = 3$$

$$4\alpha_0 + 8\alpha_1 + \left(\frac{-1+\sqrt{3}i}{2}\right)^2\beta_0 + \left(\frac{-1-\sqrt{3}i}{2}\right)^2\gamma_0 = 11$$

$$8\alpha_0 + 24\alpha_1 + \left(\frac{-1+\sqrt{3}i}{2}\right)^3\beta_0 + \left(\frac{-1-\sqrt{3}i}{2}\right)^3\gamma_0 = 34$$

This simplifies to

$$\alpha_0 + \beta_0 + \gamma_0 = 3$$

$$2\alpha_0 + 2\alpha_1 + \left(\frac{-1+\sqrt{3}i}{2}\right)\beta_0 + \left(\frac{-1-\sqrt{3}i}{2}\right)\gamma_0 = 3$$

$$4\alpha_0 + 8\alpha_1 + \left(\frac{-1-\sqrt{3}i}{2}\right)\beta_0 + \left(\frac{-1+\sqrt{3}i}{2}\right)\gamma_0 = 11$$

$$8\alpha_0 + 24\alpha_1 + \beta_0 + \gamma_0 = 34$$

Solving this system by hand is not fun.[42] I used *Mathematica* to determine that $\alpha_0 = \alpha_1 = \beta_0 = \gamma_0 = 1$. The final solution is thus

$$a_n = (1+n)\,2^n + \left(\frac{-1+\sqrt{3}i}{2}\right)^n + \left(\frac{-1-\sqrt{3}i}{2}\right)^n \quad \text{for } n \geq 0$$

The solution sequence is $\{3, 3, 11, 34, 79, 191, 450, 1023, 2303, \ldots\}$. ■

7.2.4 The Sordid Truth

The techniques presented in this section for solving linear homogeneous recurrence relations with constant coefficients depend on two assumptions:

- We must be able to find all the roots of the characteristic equation in exact form.[43]
- We must be able to solve exactly the system of linear equations that determines the αs.

The second assumption becomes a bit of an issue if there are more than 3 or 4 αs. Solving linear systems with more than three or four equations is quite tedious, especially if rounding off is not allowed. However, computer software exists that can handle this task for quite a large number of equations (much more than you will ever need).

The real issue is the first assumption. There is more involved than just being able to factor properly. Many polynomials have no rational roots (that is, all their roots are either irrational or complex). Can computer software help us here? Yes, but only to a limited extent. A brief detour is needed to make this more explicit.

You are familiar with the famous quadratic formula.[44]

---

[42] But it can be done without a totally unreasonable amount of work. You might start by adding the two middle equations.

[43] That is, if the root is $\sqrt{2}$, we may not use an approximation such as 1.41421356. Review Example 7.25.

[44] *Infamous* in the minds of some high school students.

Quadratic Formula

The equation

$$ax^2 + bx + c = 0$$

where a, b, and c are real or complex numbers, has the solutions

$$x = \frac{-b \pm \sqrt{b^2 - 4ac}}{2a}$$

Notice that this formula expresses the roots in terms of the coefficients of the polynomial.

The *content*[45] of the quadratic formula was known in ancient times. During the 1500s A.D., it became a challenge to find such a formula (involving algebraic manipulations of the coefficients) for a general cubic equation. The bizarre story behind the eventual discovery of such a formula can be found in [26]. If the result were to be expressed using modern notation, it would look like the following:

Let $ax^3 + bx^2 + cx + d = 0$ be a polynomial equation with real or complex coefficients. Then the simplest of the three roots looks like the multiline expression in Figure 7.26.

$$-\frac{b}{3\,a} + \frac{\sqrt[3]{-2\,b^3 + 9\,a\,c\,b - 27\,a^2\,d + \sqrt{4\,(3\,a\,c - b^2)^3 + (-2\,b^3 + 9\,a\,c\,b - 27\,a^2\,d)^2}}}{3\,\sqrt[3]{2}\,a}$$

$$-\frac{\sqrt[3]{2}\,(3\,a\,c - b^2)}{3\,a\,\sqrt[3]{-2\,b^3 + 9\,a\,c\,b - 27\,a^2\,d + \sqrt{4\,(3\,a\,c - b^2)^3 + (-2\,b^3 + 9\,a\,c\,b - 27\,a^2\,d)^2}}}$$

Figure 7.26 The simplest root of $ax^3 + bx^2 + cx + d = 0$.

A few years later, a "formula" for the general degree 4 equation was found (actually, it was more like an algorithm prescribing the algebraic manipulations of the coefficients).[46] Progress then came to a halt: No one could find a formula for the general degree 5 polynomial.

During the 1800s two young mathematicians independently proved that no such formula exists for polynomials of degree 5 or higher. The mathematicians were the Norwegian Neils Abel (died in 1829 of tuberculosis at age 26) and the Frenchman Évariste Galois (shot in a duel in 1832 at age 20).

Since no such formula exists[47] for degree 5 or higher polynomial equations, there is no guarantee that we can exactly solve the characteristic equation for recurrence relations of degree 5 or higher. This means that the technique of Theorem 7.13 cannot be applied for every potential recurrence relation.

The quadratic formula is truly amazing: It is simple, and it always produces all the roots.

More could be done with the technique presented in Sections 7.2.2 and 7.2.3. In particular, it can be extended to cover linear *nonhomogeneous* recurrence relations with constant coefficients. Such recurrence relations look like

$$a_n = c_1 a_{n-1} + c_2 a_{n-2} + \cdots + c_k a_{n-k} + g(n)$$

[45]But not the modern formulaic expression.

[46]It produces a formula that is even nastier than the degree 3 case.

[47]For degree 5 or higher polynomial equations, there *is* a technique that does not express the roots as algebraic expressions in the coefficients. The technique uses much more sophisticated ideas. See the Chapter 7 links at http://www.mathcs.bethel.edu/~gossett/DiscreteMathWithProof/ for more details.

for some function, $g(n)$, which does not involve any of the a_js.

Additional work can also be done to simplify the solution when the characteristic equation has complex roots.

Neither of these additions will be pursued here. Instead, a more general approach that uses generating functions will be examined in Section 7.4.

7.2.5 Exercises

The exercises marked with ★ have detailed solutions in Appendix H.

1. In Example 7.16 on page 352, how many rabbits will there be after one year?

2. ★ Write a recursive algorithm for f_n, the nth Fibonacci number. Is this an efficient algorithm?

3. Solve the following recurrence relations (find closed-form formulas).

(a) $a_0 = 5$, and $a_n = 2a_{n-1} - 3$ for $n \ge 1$

(b) ★ $a_0 = 1$, and $a_n = na_{n-1}$ for $n \ge 1$

(c) ★ $a_0 = -1, a_1 = 3$, and $a_n = 2a_{n-2}$ for $n \ge 2$

(d) $a_0 = -3$, and $a_n = a_{n-1} + n$ for $n \ge 1$

4. Solve the following recurrence relations (find closed-form formulas).

(a) $a_0 = 7$, and $a_n = 8a_{n-1}$ for $n \ge 1$

(b) $a_0 = 0$, and $a_n = 4a_{n-1} + 5$ for $n \ge 1$

(c) ★ $a_0 = 0$, and $a_n = 5a_{n-1} + n$ for $n \ge 1$
Full simplification is possible but tricky.

(d) $a_0 = 1$, and $a_n = \frac{a_{n-1}}{n}$ for $n \ge 1$

5. Solve the following recurrence relations (find closed-form formulas).

(a) $a_0 = 4$, and $a_n = 3na_{n-1}$ for $n \ge 1$

(b) $a_0 = 3$, and $a_n = a_{n-1}^2$ for $n \ge 1$

(c) $a_0 = 3$, and $a_n = 3 + 2a_{n-1}$ for $n \ge 1$

(d) $a_0 = 1$, and $a_n = \frac{6a_{n-1}}{n^2}$ for $n \ge 1$

6. Solve the following recurrence relations (find closed-form formulas).

(a) $a_0 = 0$, and $a_n = \frac{8a_{n-1}}{n!}$ for $n \ge 1$

(b) $a_0 = 1$, and $a_n = \frac{8a_{n-1}}{n!}$ for $n \ge 1$

(c) $a_0 = 7$, and $a_n = (n+1)a_{n-1}$ for $n \ge 1$

(d) $a_0 = 2$, and $a_n = a_{n-1} + \frac{n}{6}$ for $n \ge 1$

(e) $a_0 = 5, a_1 = 4$, and $a_n = \frac{1}{3}a_{n-2}$ for $n \ge 2$

7. The Tower of Hanoi (Example 7.17 on page 352)

(a) Find a closed-form formula for H_n, the minimum number of moves to solve the Tower of Hanoi problem with n disks.

(b) Suppose $n = 64$ and that the Brahman monks can move one disk per second. How long will it take to solve the puzzle?

8. Let f_n be the nth Fibonacci number. Prove that

$$f_n = f_k f_{n-k} + f_{k-1} f_{n-k-1} \quad \text{for } k \in \{1, 2, \ldots, n-1\}$$

9. Prove Theorem 7.8 (on page 359).

10. Prove that

$$\lim_{n\to\infty} \frac{f_n}{f_{n-1}} = \frac{1+\sqrt{5}}{2} = \Phi, \quad \text{the golden ratio.}$$

11. Find closed-form formulas for the following linear homogeneous recurrence relations with constant coefficients. Do not round off or use calculator approximations; use exact arithmetic!

(a) ★ $a_0 = 2, a_1 = -2$, and $a_n = -2a_{n-1} + 15a_{n-2}$ for $n \ge 2$

(b) $a_0 = 3$, and $a_n = 7a_{n-1}$ for $n \ge 1$ (Use the technique for solving linear homogeneous recurrence relation with constant coefficients, *not* back substitution.)

(c) $a_0 = 1, a_1 = 1$, and $a_n = 2a_{n-2}$ for $n \ge 2$

(d) $a_0 = -1, a_1 = 0, a_2 = 1$,
and $a_n = 2a_{n-1} + 5a_{n-2} - 6a_{n-3}$ for $n \ge 3$

12. Find closed-form formulas for the following linear homogeneous recurrence relations with constant coefficients. Do not round off or use calculator approximations; use exact arithmetic!

(a) $a_0 = 1, a_1 = 4$, and $a_n = 5a_{n-1} - 6a_{n-2}$ for $n \ge 2$

(b) $a_0 = 2, a_1 = 3$, and $a_n = 8a_{n-1} - 16a_{n-2}$ for $n \ge 2$

(c) ★ $a_0 = 3, a_1 = 4, a_2 = 6$,
and $a_n = 6a_{n-1} - 11a_{n-2} + 6a_{n-3}$ for $n \ge 3$

(d) $a_0 = 0, a_1 = 1, a_2 = 2$, and
$a_n = a_{n-1} + 9a_{n-2} - 9a_{n-3}$ for $n \ge 3$

13. Find closed-form formulas for the following linear homogeneous recurrence relations with constant coefficients. Do not round off or use calculator approximations; use exact arithmetic!

(a) $a_0 = \frac{1}{3}$, and $a_n = \frac{3}{4}a_{n-1}$ for $n \ge 1$ (Use the technique for solving linear homogeneous recurrence relation with constant coefficients, *not* back substitution.)

(b) $a_0 = 3, a_1 = 7$, and $a_n = 6a_{n-1} + 3a_{n-2}$ for $n \ge 2$

(c) $a_0 = 2, a_1 = 5$, and $a_n = -\frac{1}{25}a_{n-2}$ for $n \ge 2$

(d) $a_0 = -4, a_1 = -3, a_2 = 0$, and
$a_n = -3a_{n-1} - 3a_{n-2} - a_{n-3}$ for $n \ge 3$

14. A CD rack contains n slots in which CD cases can be inserted. The slots can hold a single CD, or a double CD case can be inserted into two adjacent slots. Suppose that the pattern of interest is the visually distinguishable arrangements of single and double CD cases in the rack. That is, all single CD cases look the same as each other and all double CD cases look the same as each other. Create, and solve, a recurrence relation that counts the number of visually distinguishable case arrangements there are for a rack with n slots.

15. Suppose that it is possible to climb a set of stairs by taking arbitrary combinations of either one or two stairs at a time. For example, someone could get to stair 3 by taking 3 single steps, or by moving to stair 1 in a single step and then to stair 3 using a double step. It is also possible to move to stair 2 in a double step and then to stair 3 with a single step.

(a) Create a recurrence relation for counting the number of distinct ways to climb n stairs. Include the base conditions for the recurrence relation. [*Hint*: Try a few small values for n by exhaustively listing all possibilities.]

(b) Find a closed-form formula for the number of distinct ways to climb n stairs.

16. In how many ways can a $2 \times n$ rectangular checker board be tiled using combinations of 2×1 and 2×2 tiles?

17. ★ Bit strings of length n.

(a) Create a recurrence relation for counting the number of distinct bit strings of length n that do not contain three consecutive 1s. Include the base conditions for the recurrence relation.

(b) Explain why the process for finding a closed-form formula for the recurrence relation in part (a) may be problematic with current methods. (A closed-form formula will be the goal of a future exercise.)

18. Find closed-form formulas for the following linear homogeneous recurrence relations with constant coefficients whose characteristic equations have repeated roots. Use exact arithmetic!

(a) $a_0 = 5, a_1 = -3$, and $a_n = a_{n-1} - \frac{1}{4}a_{n-2}$ for $n \geq 2$

(b) ★ $a_0 = 1, a_1 = 2, a_2 = 1$, and $a_n = 8a_{n-1} - 21a_{n-2} + 18a_{n-3}$ for $n \geq 3$

(c) $a_0 = 0, a_1 = 0, a_2 = 0, a_3 = 125$, and $a_n = -2a_{n-1} + 11a_{n-2} + 12a_{n-3} - 36a_{n-4}$ for $n \geq 4$

(d) $a_0 = 2, a_1 = 2, a_2 = 4, a_3 = 8$, and $a_n = -a_{n-1} + 9a_{n-2} - 11a_{n-3} + 4a_{n-4}$ for $n \geq 4$

19. Find closed-form formulas for the following linear homogeneous recurrence relations with constant coefficients whose characteristic equations have repeated roots. Use exact arithmetic!

(a) $a_0 = 1, a_1 = 3, a_2 = 8$, and $a_n = 8a_{n-1} - 20a_{n-2} + 16a_{n-3}$ for $n \geq 3$

(b) $a_0 = 0, a_1 = 1, a_2 = -1, a_3 = 2$, and $a_n = 4a_{n-1} + 26a_{n-2} - 60a_{n-3} - 225a_{n-4}$ for $n \geq 4$

(c) $a_0 = 1, a_1 = 1, a_2 = 2, a_3 = 2$, and $a_n = -6a_{n-1} - 12a_{n-2} - 10a_{n-3} - 3a_{n-4}$ for $n \geq 4$

(d) $a_0 = 2, a_1 = 4, a_2 = 6, a_3 = 6, a_4 = 20$, and $a_n = 7a_{n-1} - 9a_{n-2} - 23a_{n-3} + 50a_{n-4} - 24a_{n-5}$ for $n \geq 5$ (Use a matrix-capable calculator or computer software to solve the linear system.)

20. Find closed-form formulas for the following linear homogeneous recurrence relations with constant coefficients whose characteristic equations have repeated roots. Use exact arithmetic!

(a) $a_0 = 1, a_1 = 5, a_2 = 9, a_3 = 12$, and $a_n = 10a_{n-2} - 25a_{n-4}$ for $n \geq 4$

(b) $a_0 = 6, a_1 = 0, a_2 = 8, a_3 = 10, a_4 = 40$, and $a_n = 6a_{n-1} - 11a_{n-2} + 2a_{n-3} + 12a_{n-4} - 8a_{n-5}$ for $n \geq 5$ (Use a matrix-capable calculator or computer software to solve the linear system.)

(c) $a_0 = 2, a_1 = -2$, and $a_n = -a_{n-2}$ for $n \geq 2$

(d) $a_0 = 1, a_1 = 2, a_2 = 3$, and $a_n = -a_{n-1} - 4a_{n-2} + 6a_{n-3}$ for $n \geq 3$

21. Suppose you are given an unlimited supply of red, blue, and green cards.

(a) Write a recurrence relation that counts the number, s_n, of distinguishable ways to form a stack consisting of n colored cards.

(b) Solve the recurrence relation.

(c) Show that

$$s_n = \sum_{i=0}^{n}\sum_{j=0}^{n-i} \text{multinomial}(i, j, n-i-j)$$

(d) Prove that

$$\sum_{i=0}^{n}\sum_{j=0}^{n-i} \frac{n!}{i!j!(n-i-j)!} = 3^n$$

22. Suppose you are given an unlimited supply of red, blue, and green cards. Use recurrence relations to determine the number, a_n, of ways to form a stack consisting of n colored cards in which a green card is never directly on top of another green card.

One of the following strategies may appeal to you. A computer algebra system such as *Mathematica* or *Maple* will be useful.

Strategy 1: There are three possible colors for the top card in a stack of height n that does not have adjacent green cards. Now determine how each of these three cases can be built from smaller stacks that do not have adjacent green cards.

Strategy 2: Work the problem by listing the possibilities for the first few values of n. Then let y_n be the number of stacks of height n which have a nongreen card on top and let x_n be the number of stacks of height n that have a green card on top. Express a_n in terms of y_n and x_n. Then write a recursion for y_n in terms of y_is and x_js. Next, express x_n in terms of y_is. Finally, find a recurrence relation that only contains y_is. Solve that recurrence relation and use its solution to find a closed-form expression for a_n.

23. A board has one row of n square cells. There is a collection of square tiles that can each cover one cell. There are both red square tiles and blue square tiles. There is also a collection of dominoes that can each cover two cells. Dominoes are either yellow, green, or orange. Let a_n represent the number of visually distinct ways to cover the board, assuming that there are distinct left and right ends to the board. Set $a_0 = 1$.

(a) Write a complete recurrence relation for a_n.

(b) Find a closed-form formula for a_n.

7.3 Big-Θ and Recursive Algorithms: The Master Theorem

Section 4.2 introduced the big-Θ mechanism for ranking algorithm efficiency and Section 7.1 introduced recursive algorithms. The goal of this section is to apply big-Θ analysis to a class of recursive algorithms called *divide-and-conquer* algorithms. The name *divide and conquer* has been used for this solution technique because the original problem is divided into smaller problems, each of which is solved, and then the individual solutions are used to solve (conquer) the original problem. A few examples will indicate the primary direction.

EXAMPLE 7.35 **Recursive Binary Search**

The binary search algorithm (page 196) can be rewritten as a recursive algorithm. A return value of "not found" will percolate back up through all pending recursive invocations.[48]

The base condition examines a list of size 1. If the one item is equal to x, then the item's position in the sublist will be returned. Otherwise, the item has not been found.

```
 1: integer recBinarySearch(x, {a_0, a_1, a_2, ..., a_{n-1}})
 2:    if n == 1
 3:       if a_0 == x
 4:          return 0
 5:       else
 6:          return "not found"
 7:
 8:    if x < a_⌊n/2⌋  # is x in the first half of the list?
 9:       return recBinarySearch(x, {a_0, ..., a_{⌊n/2⌋-1}}) # try the 1st half
10:    else
11:       return ⌊n/2⌋ + recBinarySearch(x, {a_⌊n/2⌋, ..., a_{n-1}}) # try the 2nd half
12: end recBinarySearch
```

Notice that inside a recursive invocation of the algorithm (from line 11), the list will be renumbered starting at index 0. So if the original list has 4 elements and the recursion is invoked as

```
11: return 2 + recBinarySearch(x, {a_2, a_3}) # try the 2nd half
```

the recursive next-page version of the algorithm will assume it is working with a list denoted as $\{a_0, a_1\}$ and not as $\{a_2, a_3\}$.

Assume, for the duration of this example, that $n = 2^k$, for some $k \in \mathbb{N}$. Then each of the divisions into two sublists will result in equal-sized sublists.

Let $f(n)$ represent the number of critical operations (in the worst case) to search a list of size n for x. There are two comparisons (lines 2 and 3) when the list has only one element, so $f(1) = 2$. If $n > 1$, then there will still be two comparisons (lines 2 and 8), and one recursive invocation of `recBinarySearch` on a list of size $\frac{n}{2}$ (since $n = 2^k$). Thus

- $f(n) = f(\frac{n}{2}) + 2$
- $f(1) = 2$

This linear nonhomogeneous recurrence relation seems like a good candidate for back substitution. Notice that when a list of size $\frac{n}{2}$ is divided into two equal-sized sub-

---

[48] An additional parameter or some other mechanism (such as throwing an exception) will be needed to keep track of this in an algorithm that is implemented using a typical computer language, such as Java.

lists, each sublist will consist of $\frac{n}{2^2}$ elements. Only one of those lists will be examined.

$$\begin{aligned}
f(n) &= f\left(\frac{n}{2}\right) + 2 \\
&= \left[f\left(\frac{n}{2^2}\right) + 2\right] + 2 \quad \text{substitute} \\
&= f\left(\frac{n}{2^2}\right) + 2 + 2 \quad \text{simplify} \\
&= \left[f\left(\frac{n}{2^3}\right) + 2\right] + (2 + 2) \quad \text{substitute} \\
&= f\left(\frac{n}{2^3}\right) + (2 + 2 + 2) \quad \text{simplify} \\
&\vdots \qquad \vdots \\
&= f(1) + \overbrace{2 + \cdots + 2 + 2}^{k \text{ times}} \\
&= 2(k + 1) \\
&= 2(\log_2(n) + 1) \\
&= 2\log_2(n) + 2
\end{aligned}$$

We can test this formula for a few small values of n. Recalling that $n = 2^k$, the following table can be derived.

n	$2\log_2(n) + 2$
1	2
2	4
4	6
8	8
16	10

When $n = 2$, there will be two comparisons, and then one recursive invocation with two comparisons (for a list of size 1), for a total of four comparisons. This agrees with the table. When $n = 4$, there will be two comparisons, and then a recursive invocation of `recBinarySearch` on a list of size 2, requiring four comparisons, for a total of six comparisons. This also agrees with the table. You should verify that the final values in the table are also correct.

The net result is that when $n = 2^k$, `recBinarySearch` $\in \Theta(\log_2(n))$, in agreement with the prior (worst case) result for the nonrecursive `binarySearch` algorithm (in Section 4.2.4). ■

The next example also involves an algorithm that divides the data list into two equal parts. However, in this example, both sublists need to be processed recursively.

EXAMPLE 7.36

Simple Merge Sort

Consider the following recursive algorithm for sorting a list of items. First, split the list into two equal (or nearly equal) length sublists. Recursively sort each list, and then merge the two sublists into a single, sorted list. The merging can proceed by starting at the front of each of the two sorted sublists and comparing the current front item in each list. Move the smaller of the two items to a new list. Keep moving the smaller of the two front items until one of the lists is empty. Then append all remaining items in the nonempty list onto the end of the new list.

For example, to sort the data set {q, s, p, w, z, r, y, x}, first split it into two sublists of size 4. The sublists are {q, s, p, w} and {z, r, y, x}. The sublists, after sorting, are {p, q, s, w} and {r, x, y, z}. They can be merged by starting at the left of each list. The elements p and r are compared first and p is moved (or copied) to a new list. Next, the new front elements (or first uncopied elements) are compared—q and r in this case. The element q is copied to the new list. Then s and r are compared and r is copied to the new list. The lists now look like: {s, w}, {x, y, z}, and {p, q, r}. The elements s and x are compared and s is moved; then w and x are compared and w is moved. The lists now look like: {}, {x, y, z}, and {p, q, r, s, w}. Finally, the elements x, y, and z are moved to the new list, producing the sorted list: {p, q, r, s, w, x, y, z}.

The algorithm mergeSort utilizes this strategy.

```
 1: sorted list mergeSort({a_0, a_1, ..., a_{n-1}})
 2:    if n == 1
 3:      return {a_0} # a list of length 1 is already sorted
 4:
 5:    mergeSort({a_0, ..., a_{⌊n/2⌋-1}})      # sort the left half
 6:    mergeSort({a_{⌊n/2⌋}, ..., a_{n-1}})    # sort the right half
 7:
 8:    # merge the two sorted sublists
 9:
10:    i = 0           # index for left half
11:    j = ⌊n/2⌋       # index for the right half
12:    k = 0           # index for the new list
13:
14:    # compare the front elements
15:
16:    while i ≤ ⌊n/2⌋ - 1 and j ≤ n - 1
17:       if a_i ≤ a_j
18:         b_k = a_i
19:         i = i + 1
20:       else
21:         b_k = a_j
22:         j = j + 1
23:       k = k + 1
24:
25:    # copy everything remaining in the left list
26:
27:    while i ≤ ⌊n/2⌋ - 1
28:       b_k = a_i
29:       i = i + 1
30:       k = k + 1
31:
32:    # copy everything remaining in the right list
33:
34:    while j ≤ n - 1
35:       b_k = a_j
36:       j = j + 1
37:       k = k + 1
38:
39:    return {b_0, b_1, ..., b_{n-1}}
40: end mergeSort
```

Assume, for the rest of this example, that $n = 2^k$ for some $k \in \mathbb{N}$. The algorithm works by first sorting two sublists of length $\frac{n}{2}$ and then merging the sublists. The merge requires $n + 1$ comparisons (some while neither list has been exhausted, others while the remaining elements in the uncompleted list are copied, and one more to determine that the other list has been completed). The merge also requires n data items to be copied. There will be an additional comparison at the beginning of each invocation of the algorithm.

Suppose, for now, that comparisons are much faster than the data copies, so they can be ignored. Then the function f, which counts the number of data copies, is

- $f(n) = 2f\left(\frac{n}{2}\right) + n$
- $f(1) = 0$

This recurrence relation is also easy to solve using back substitution.

$$\begin{aligned}
f(n) &= 2f\left(\frac{n}{2}\right) + n \\
&= 2\left[2f\left(\frac{n}{2^2}\right) + \frac{n}{2}\right] + n \quad \text{substitute} \\
&= 2^2 f\left(\frac{n}{2^2}\right) + 2n \quad \text{simplify} \\
&= 2^2\left[2f\left(\frac{n}{2^3}\right) + \frac{n}{2^2}\right] + 2n \quad \text{substitute} \\
&= 2^3 f\left(\frac{n}{2^3}\right) + 3n \quad \text{simplify} \\
&\ \ \vdots \qquad \vdots \\
&= 2^k f(1) + kn \\
&= kn \\
&= n\log_2(n)
\end{aligned}$$

This sorting algorithm is therefore in $\Theta(n\log_2(n))$. You might wish to verify this for $n = 1, 2, 4,$ and 8. Notice that best case, worst case, and average case are all the same: This algorithm uses the same number of data copies for every data set with n elements.[49] ■

✔ Quick Check 7.11

1. Show the complete details for a merge sort of the set {h, d, a, c, g, f, b, e}. ☑

One more example will be presented before looking at some theorems. The notation will look nastier but is not conceptually harder than the two previous examples.

EXAMPLE 7.37

Persian Rugs

Section 7.1.4 introduces a recursive algorithm for drawing Persian rug designs. The key step is the algorithm `colorSquare`. In that algorithm, a check for the base condition is made, then a new color is calculated, the central "plus sign" is painted, and then the remaining unpainted pixels are divided into four equal-sized groups and recursively painted.

The actual code that implements this algorithm assumes that there are $(2^k + 1)^2$ pixels in the entire rug.

The outer border (containing $2(2^k+1)+2(2^k-1) = 2^{k+2}$ pixels) is painted before the recursion starts, so the recursive algorithm starts with a grid containing $n = (2^k-1)^2$ pixels.

Assume, for this example, that painting pixels is the critical operation (so the comparisons and color calculations can be ignored). An efficient way to paint the pixels in the central "plus sign" is to paint the horizontal line and then paint the vertical line. The

[49] This has only been proved here when $n = 2^k$.

central pixel will be painted twice, but this is faster (and simpler) than adding tests to see if the next pixel to paint is the central pixel.

The plus sign will therefore require $2(2^k - 1) = 2\sqrt{n}$ pixels to be painted, but only $2(2^k - 1) - 1 = 2^{k+1} - 3$ pixels can be removed from further consideration. This leaves

$$(2^k - 1)^2 - 2^{k+1} + 3 = 2^{2k} - 2^{k+2} + 4 = 4\left(2^{2k-2} - 2^k + 1\right) = 4\left(2^{k-1} - 1\right)^2$$

pixels to color via the four recursive invocations. Each invocation will therefore paint a grid with $\left(2^{k-1} - 1\right)^2 = \frac{(\sqrt{n}-1)^2}{4}$ pixels.

The function f, which counts the number of pixels that are painted after the border has been painted, is therefore

$$f(n) = 4f\left(\frac{(\sqrt{n}-1)^2}{4}\right) + 2\sqrt{n} \quad \text{and} \quad f(1) = 1$$

The back substitution for this recurrence relation is a bit messy. It might be worth making a simplifying assumption and then seeing how the simpler (but incorrect) recursion behaves. If the new recurrence relation does not deviate too much from the correct recurrence relation, perhaps they will have the same big-Θ reference functions.

The new (but incorrect) recurrence relation is

$$g(n) = 4g\left(\tfrac{n}{4}\right) + 2\sqrt{n} \quad \text{and} \quad g(1) = 1$$

The justification is that $(\sqrt{n} - 1)^2 \simeq n$ when n is large. When n is small, the approximation is not the best. For example, when $k = 3$, $n = (2^k - 1)^2 = 49$, but $(\sqrt{n} - 1)^2 = 36$. Ignore this discrepancy for now and assume that $n = (2^k)^2 = 4^k$. Notice that when $n = 4$, there would be no central plus sign to paint. It makes sense for this approximation to make the base condition $g(4) = 4$.

The following table shows that the two recurrence relations have similar orders of magnitude as k increases.

$n_f = (2^k - 1)^2$

$n_g = 4^k$

k	$\left(2^k - 1\right)^2$	$f\left(\left(2^k - 1\right)^2\right)$	4^k	$g\left(4^k\right)$
4	225	246	256	480
8	65025	70486	65536	130560
12	16769025	18167126	16777216	33546240
16	4294836225	4652750166	4294967296	8589803520

Back substitution on g can be used to approximate the closed-form formula for f.

$$\begin{aligned}
g(n) &= 4g\left(\frac{n}{4}\right) + 2\sqrt{n} \\
&= 4\left[4g\left(\frac{n}{4^2}\right) + 2\sqrt{\frac{n}{4}}\right] + 2\sqrt{n} && \text{substitute} \\
&= 4^2 g\left(\frac{n}{4^2}\right) + \left(2^2\sqrt{n} + 2\sqrt{n}\right) && \text{simplify} \\
&= 4^2\left[4g\left(\frac{n}{4^3}\right) + 2\sqrt{\frac{n}{4^2}}\right] + \left(2^2\sqrt{n} + 2\sqrt{n}\right) && \text{substitute} \\
&= 4^3 g\left(\frac{n}{4^3}\right) + \left(2^3\sqrt{n} + 2^2\sqrt{n} + 2\sqrt{n}\right) && \text{simplify} \\
&= 4^3\left[4f\left(\frac{n}{4^4}\right) + 2\sqrt{\frac{n}{4^3}}\right] + \left(2^3\sqrt{n} + 2^2\sqrt{n} + 2\sqrt{n}\right) && \text{substitute} \\
&= 4^4 g\left(\frac{n}{4^4}\right) + \left(2^4\sqrt{n} + 2^3\sqrt{n} + 2^2\sqrt{n} + 2\sqrt{n}\right) && \text{simplify} \\
&\ \ \vdots \qquad\qquad\qquad \vdots
\end{aligned}$$

Continuing the back substitution eventually leads to

$$\begin{aligned}
g(n) &= 4^{k-1} g(4) + 2\sqrt{n} \sum_{i=0}^{k-2} 2^i \\
&= 4^k + 2\sqrt{n} \cdot \frac{2^{k-1} - 1}{2 - 1} \\
&= 4^k + \sqrt{n} \cdot \left(2^k - 2\right) \\
&= n + \sqrt{n} \cdot (\sqrt{n} - 2) \\
&= 2n - 2\sqrt{n}
\end{aligned}$$

The net result is that the number of pixel paintings is in $\Theta(n)$ (if the approximation $(\sqrt{n} - 1)^2 \simeq n$ is legitimate). This should make intuitive sense: Except for a few double paintings with the plus signs, every pixel is painted only once. Thus, $\Theta(n)$ behavior should be expected.

Exercise 7 on page 389 asks you to perform the back substitution with the correct recurrence relation. ■

The examples indicate that back substitution might be useful for analyzing the worst case behavior of recursive algorithms that split the data into a number of equal-sized subsets. However, those examples made some assumptions about n. It is time to assume less. It is also time to start demonstrating that the behavior in these examples fits into some general patterns.

The examples in this section have demonstrated a method for analyzing the computational complexity of recursive divide-and-conquer algorithms. The complexity of many recursive algorithms can be found by noticing that the associated recurrence relation fits a pattern that has been captured in an established theorem.

There are a number of similar theorems that have been developed that all tend to be named the master theorem. The first appeared in [54] in 1980. Two simple versions of the theorem will be proved here. A few reasonable extensions will then be briefly mentioned.

The next lemma will be used many times in this section.

LEMMA 7.14

Let $x, y, z \in \mathbb{R}$ with $x, y, z > 0$ and $y \neq 1$. Then

$$x^{\log_y(z)} = z^{\log_y(x)}$$

Proof: Since $x = y^{\log_y(x)}$,

$$\begin{aligned}
x^{\log_y(z)} &= \left(y^{\log_y(x)}\right)^{\log_y(z)} \\
&= y^{\log_y(x) \cdot \log_y(z)} \\
&= y^{\log_y(z) \cdot \log_y(x)} \\
&= \left(y^{\log_y(z)}\right)^{\log_y(x)} \\
&= z^{\log_y(x)}
\end{aligned}$$

□

DEFINITION 7.15 ***Nondecreasing Function***

A real-valued function is said to be *nondecreasing* on an interval if for all x, y in the interval with $x \leq y$, $f(x) \leq f(y)$.

The first theorem is patterned after the recursion, $f(n) = f\left(\frac{n}{2}\right) + 2$, that arose in Example 7.35. Three of the coefficients have been turned into more general constants. Notice the strong assumption that the problem data set has been subdivided into a subsets of size $\frac{n}{b}$. The recurrence relation also asserts that there are c extra critical operations during each invocation of the algorithm.

THEOREM 7.16 ***The Master Theorem—Version 1***

Let $a, b \in \mathbb{N}$ and $c, d \in \mathbb{R}$, with $a \geq 1$, $b > 1$, $c > 0$, and $d \geq 0$. Let $f(n)$ be a nondecreasing function on the interval $(0, \infty)$, where

- $f(n) = af\left(\frac{n}{b}\right) + c$
- $f(1) = d$
- $f(n) = 0$ if $n < 1$

then $$f \in \begin{cases} \Theta\left(n^{\log_b(a)}\right) & \text{if } a > 1 \\ \Theta\left(\log_b(n)\right) & \text{if } a = 1 \end{cases}$$

Three comments are in order before proceeding to the proof. First, the assumption that f is nondecreasing is not too limiting. We expect algorithms to take longer (or use more memory, or perform more calculations) as n increases. The nondecreasing assumption is used instead of "strictly increasing" because some algorithms might take the same amount of time for 19 items of data as they do for 18 items.

The second comment is that the base of the logarithms in the big-Θ reference functions is not important as long as you are consistent (see Exercise 12 on page 193).

The third comment is that the assertion $f(n) = 0$ if $n < 1$ is necessary so that the recursion has a complete set of base cases. This condition is important whenever n is not a power of b.

The proof will use results from Exercises 2–4 on page 389. The proof uses cases and many algebraic manipulations, but each step is fairly simple.

Proof of Theorem 7.16: The theorem will be proved using two cases.

Case 1: $n = b^k$

Suppose there is a nonnegative integer, k, such that $n = b^k$. Using back substitution, it is possible (Exercise 2) to show that

$$f(n) = da^k + c\sum_{i=0}^{k-1} a^i$$

If $a = 1$, then $a^k = 1$ and $\sum_{i=0}^{k-1} a^i = k$; otherwise $\sum_{i=0}^{k-1} a^i = \frac{a^k-1}{a-1}$. Thus,

$$f(n) = \begin{cases} da^k + c\frac{a^k-1}{a-1} & \text{if } a > 1 \\ d + ck & \text{if } a = 1 \end{cases}$$

The expression $da^k + c\frac{a^k-1}{a-1}$ can be written as $\frac{c+d(a-1)}{a-1}a^k - \frac{c}{a-1}$. Let $\alpha = \frac{c+d(a-1)}{a-1}$ and $\beta = \frac{c}{a-1}$. Note that $\alpha > 0$ and $\beta > 0$.

The expression for $f(n)$ when $a > 1$ can be written as $\alpha a^k - \beta$.

Since $n = b^k$ (so $k = \log_b(n)$), $\alpha a^k - \beta = \alpha a^{\log_b(n)} - \beta$. Lemma 7.14 implies that this can be written as $\alpha n^{\log_b(a)} - \beta$.

If $a = 1$, then $d + ck = d + c\log_b(n)$. These observations can be combined as

$$f(n) = \begin{cases} \alpha n^{\log_b(a)} - \beta & \text{if } a > 1 \\ d + c\log_b(n) & \text{if } a = 1 \end{cases}$$

Consequently,

$$f(n) \in \begin{cases} \Theta\left(n^{\log_b(a)}\right) & \text{if } a > 1 \\ \Theta\left(\log_b(n)\right) & \text{if } a = 1 \end{cases}$$

Case 2: $n \neq b^k$

Suppose that n is not a power of b. Then there must exist a nonnegative integer, k, such that $b^k < n < b^{k+1}$. One immediate observation is that $k < \log_b(n) < k+1$. The proof will proceed by examining two subcases: $a = 1$ and $a > 1$.

$\alpha = \frac{c+d(a-1)}{a-1}$

$\beta = \frac{c}{a-1}$

The definitions of α and β from case 1 will be retained. Other results from the analysis in case 1 will also be used. Both subcases use the assumption that f is nondecreasing.

$a = 1$ Let $m = b^k$. Then $k = \log_b(m)$ and $m < n < bm$.

If $n = b^k$ (case 1)

$$f(n) = \begin{cases} \alpha n^{\log_b(a)} - \beta & a > 1 \\ d + c\log_b(n) & a = 1 \end{cases}$$

$$\begin{aligned} f(n) \le f(bm) &= f(b^{k+1}) \\ &= d + c(k+1) \\ &= (d+c) + ck \\ &= (d+c) + c\log_b(m) \\ &< (d+c) + c\log_b(n) \end{aligned}$$

Exercise 3 implies that $f \in \mathcal{O}\left(\log_b(n)\right)$.

Let $M = b^{k+1}$. Then $k = \log_b(M) - 1$ and $\frac{M}{b} < n < M$.

$$\begin{aligned} f(n) \ge f\left(\frac{M}{b}\right) &= f(b^k) \\ &= d + ck \\ &= d + c(\log_b(M) - 1) \\ &= (d-c) + c\log_b(M) \\ &> (d-c) + c\log_b(n) \end{aligned}$$

Exercise 4 implies that $f \in \Omega\left(\log_b(n)\right)$. Combining the two assertions, it is valid to conclude that $f \in \Theta\left(\log_b(n)\right)$.

$a > 1$ Let $m = b^k$. Then $m < n < bm$ and

$$\left(b^{k+1}\right)^{\log_b(a)} = b^{\log_b(a^{k+1})} = a^{k+1}$$

$$\begin{aligned} f(n) \le f(bm) &= f(b^{k+1}) \\ &= \alpha a^{k+1} - \beta \\ &= (\alpha a)a^k - \beta \\ &= (\alpha a)a^{\log_b(m)} - \beta \\ &= (\alpha a)m^{\log_b(a)} - \beta \\ &< (\alpha a)n^{\log_b(a)} - \beta \end{aligned}$$

Exercise 3 implies that $f \in \mathcal{O}\left(n^{\log_b(a)}\right)$.

Let $M = b^{k+1}$. Then $\frac{M}{b} < n < M$ and

$$\begin{aligned} f(n) \ge f\left(\frac{M}{b}\right) &= f(b^k) \\ &= \alpha a^k - \beta \\ &= \left(\frac{\alpha}{a}\right) a^{\log_b(M)} - \beta \\ &= \left(\frac{\alpha}{a}\right) M^{\log_b(a)} - \beta \\ &> \left(\frac{\alpha}{a}\right) n^{\log_b(a)} - \beta \end{aligned}$$

Exercise 4 implies that $f \in \Omega\left(n^{\log_b(a)}\right)$. Combining the two assertions, it is valid to conclude that $f \in \Theta\left(n^{\log_b(a)}\right)$. □

✔ Quick Check 7.12

1. Let $f(n) = 9f\left(\frac{n}{3}\right) + 5$ for $n > 1$, $f(1) = 2$, and $f(n) = 0$ for $n < 1$. Find a good reference function, $g(n)$, for f. ☑

Theorem 7.16 does well for Example 7.35, but does not cover the recursions in Examples 7.36 and 7.37. All three exhibit the form $f(n) = af(\text{a reduced } n) + s(n)$. The three versions for $s(n)$ in these examples are, respectively, $s(n) = 2$, $s(n) = n$, and $s(n) = 2n^{.5}$. The next version of the master theorem is general enough to cover all three of these versions of $s(n)$, for suitable values of n. It is appropriate for recurrence relations that subdivide a problem of size n into a subproblems, each of size b, with cn^v extra work.

THEOREM 7.17 ***The Master Theorem—Version 2***

Let $a, b \in \mathbb{N}$ and $c, d, v \in \mathbb{R}$, with $a \geq 1, b > 1, c > 0, d \geq 0$ and $v \geq 0$. Let $f(n)$ be a nondecreasing function on the interval $(0, \infty)$, where

- $f(n) = af\left(\frac{n}{b}\right) + cn^v$
- $f(1) = d$
- $f(n) = 0$ if $n < 1$

Then

$$f \in \begin{cases} \Theta\left(n^{\log_b(a)}\right) & \text{if } a > b^v \quad (\log_b(a) > v) \\ \Theta\left(n^{\log_b(a)} \cdot \log_b(n)\right) & \text{if } a = b^v \quad (\log_b(a) = v) \\ \Theta\left(n^v\right) & \text{if } a < b^v \quad (\log_b(a) < v) \end{cases}$$

Notice that when $v = 0$, this theorem gives the same reference functions as Theorem 7.16.

The three cases have a simple intuitive explanation. The critical question is, Which term is more significant, $af\left(\frac{n}{b}\right)$ or cn^v? As will be seen in the proof, this answer is determined by the relative sizes of a and b^v.

Proof of Theorem 7.17: The theorem will be proved using the same two cases as were used to prove Theorem 7.16. The outline of the proof is similar to the outline of the proof of Theorem 7.16. However, the algebraic details are messier and the proof is a bit longer. Nevertheless, each step is reasonably simple, so the proof is not difficult to read.

Case 1: $n = b^k$

Suppose there is a nonnegative integer, k, such that $n = b^k$. Using back substitution, it is possible (Exercise 6) to show that

$$f(n) = da^k + cn^v \sum_{i=0}^{k-1} \left(\frac{a}{b^v}\right)^i$$

Recall that $k = \log_b(n)$ for this case.

If $a = b^v$, then $\frac{a}{b^v} = 1$ and $\sum_{i=0}^{k-1}\left(\frac{a}{b^v}\right)^i = k$; otherwise $\sum_{i=0}^{k-1}\left(\frac{a}{b^v}\right)^i = \frac{\left(\frac{a}{b^v}\right)^k - 1}{\frac{a}{b^v} - 1}$. But $(b^v)^k = b^{vk} = b^{v\log_b(n)} = b^{\log_b(n^v)} = n^v$. Thus,

$$cn^v \frac{\left(\frac{a}{b^v}\right)^k - 1}{\frac{a}{b^v} - 1} = \left(\frac{c}{\frac{a}{b^v} - 1}\right)\left(a^k - n^v\right)$$

Thus

$$f(n) = \begin{cases} da^k + \left(\frac{c}{\frac{a}{b^v} - 1}\right)\left(a^k - n^v\right) & a \neq b^v \\ da^k + cn^v k & a = b^v \end{cases}$$

$a \geq 1, b > 1, c > 0,$
$d \geq 0$ and $v \geq 0$

When $a \neq b^v$, $f(n) = da^k + \left(\frac{c}{\frac{a}{b^v}-1}\right)(a^k - n^v)$.

Set $\alpha = \left(d + \frac{c}{\frac{a}{b^v}-1}\right)$ and $\beta = \left(\frac{c}{\frac{a}{b^v}-1}\right)$. Note that $\alpha > 0$ and $\beta > 0$ when $a > b^v$. Also $\beta < 0$ when $a < b^v$. The previous observations show that $f(n)$ can be expressed as

$$f(n) = \begin{cases} \alpha a^k - \beta n^v & \text{if } a \neq b^v \\ da^k + cn^v k & \text{if } a = b^v \end{cases}$$

Since $a^k = n^{\log_b(a)}$ (using Lemma 7.14 on page 382 and the assumption $n = b^k$), this can also be written as

$$f(n) = \begin{cases} \alpha n^{\log_b(a)} - \beta n^v & \text{if } a \neq b^v \\ dn^{\log_b(a)} + cn^v \log_b(n) & \text{if } a = b^v \end{cases}$$

The three cases in the theorem statement will now be examined.

$\boldsymbol{a = b^v}$ When $a = b^v$, $v = \log_b(a)$. Thus,

$$f(n) = dn^{\log_b(a)} + cn^{\log_b(a)} \log_b(n)$$

Consequently,

$$f(n) \in \Theta\left(n^{\log_b(a)} \log_b(n)\right)$$

$\boldsymbol{a > b^v}$ In this case, $\alpha > 0$ and $\beta > 0$, and since $\log_b(a) > \log_b(b^v) = v$,

$$f(n) = \alpha n^{\log_b(a)} - \beta n^v \in \Theta(n^{\log_b(a)})$$

$\boldsymbol{a < b^v}$ In this case, $\beta < 0$ and $\log_b(a) < v$, so

$$f(n) = \alpha n^{\log_b(a)} - \beta n^v = \alpha n^{\log_b(a)} + |\beta| n^v \in \Theta(n^v)$$

Case 2: $n \neq b^k$

Suppose that n is not a power of b. Then there must exist a nonnegative integer, k, such that $b^k < n < b^{k+1}$. One immediate observation is that $k < \log_b(n) < k+1$. The proof will proceed by examining two subcases: $a = b^v$ and $a \neq b^v$.

The definitions of α and β from case 1 will be retained. Other results from the analysis in case 1 will also be used. Both subcases use the assumption that f is nondecreasing.

$\boldsymbol{a = b^v}$ Note that $v = \log_b(a)$ in this subcase. Also, by Case 1, $f(b^k) = da^k + cn^v k$.

Let $m = b^k$. Then since $m < n < bm$, $\log_b(m) < \log_b(n)$ and also $m^{\log_b(a)} < n^{\log_b(a)}$. Consequently,

$$\begin{aligned} f(n) \leq f(bm) &= f(b^{k+1}) \\ &= da^{k+1} + c(bm)^v(k+1) \\ &= daa^{\log_b(m)} + cb^v m^v(\log_b(m)+1) \\ &= daa^{\log_b(m)} + cb^v m^{\log_b(a)}(\log_b(m)+1) \\ &= daa^{\log_b(m)} + cam^{\log_b(a)}(\log_b(m)+1) \\ &= dam^{\log_b(a)} + cam^{\log_b(a)}(\log_b(m)+1) \\ &= (da+ca)m^{\log_b(a)} + cam^{\log_b(a)} \log_b(m) \\ &< (da+ca)n^{\log_b(a)} + can^{\log_b(a)} \log_b(n) \end{aligned}$$

Exercise 3 implies that $f \in \mathcal{O}\left(n^{\log_b(a)} \log_b(n)\right)$.

Now let $M = b^{k+1}$. Then since $\frac{M}{b} < n < M$, $\log_b(n) < \log_b(M)$ and also $n^{\log_b(a)} < M^{\log_b(a)}$. Let $M - n = \delta$. Then

$$\begin{aligned}
f(n) \geq f\left(\frac{M}{b}\right) &= f(b^k) \\
&= da^k + c\left(\frac{M}{b}\right)^v k \\
&= da^{\log_b(M)-1} + c\left(\frac{M}{b}\right)^{\log_b(a)} \left(\log_b(M) - 1\right) \\
&= \left(\frac{d}{a}\right) a^{\log_b(M)} + \left(\frac{c}{a}\right) M^{\log_b(a)} \left(\log_b(M) - 1\right) \\
&= \left(\frac{d}{a}\right) M^{\log_b(a)} + \left(\frac{c}{a}\right) M^{\log_b(a)} \left(\log_b(M) - 1\right) \\
&= \left(\frac{d-c}{a}\right) (n+\delta)^{\log_b(a)} + \left(\frac{c}{a}\right) M^{\log_b(a)} \log_b(M) \\
&> \left(\frac{d-c}{a}\right) (n+\delta)^{\log_b(a)} + \left(\frac{c}{a}\right) n^{\log_b(a)} \log_b(n)
\end{aligned}$$

$$\left(\frac{1}{b}\right)^{\log_b(a)} = a^{\log_b(b^{-1})} = a^{-1}$$

The factor $(n + \delta)^{\log_b(a)}$ is dominated[50] by $n^{\log_b(a)}$, which does not grow as fast as $n^{\log_b(a)} \log_b(n)$. Exercise 4 implies that $f \in \Omega\left(n^{\log_b(a)} \log_b(n)\right)$. Combining the two assertions, $f \in \mathcal{O}\left(n^{\log_b(a)} \log_b(n)\right)$ and $f \in \Omega\left(n^{\log_b(a)} \log_b(n)\right)$, it is valid to conclude that $f \in \Theta\left(n^{\log_b(a)} \log_b(n)\right)$.

$\boldsymbol{a \neq b^v}$ Let $m = b^k$. Then since $\left(b^{k+1}\right)^{\log_b(a)} = b^{\log_b(a^{k+1})} = a^{k+1}$,

$$\begin{aligned}
f(n) < f(bm) = f(b^{k+1}) &= \alpha a^{k+1} - \beta(bm)^v \\
&= \alpha a a^{\log_b(m)} - \beta b^v m^v \\
&= \alpha a m^{\log_b(a)} - \beta b^v m^v
\end{aligned}$$

If $n = b^k$

$$f(n) = \begin{cases} \alpha n^{\log_b(a)} - \beta n^v & \text{if } a \neq b^v \\ d n^{\log_b(a)} + c n^v \log_b(n) & \text{if } a = b^v \end{cases}$$

If $\boldsymbol{a > b^v}$, then $\alpha > 0$ and $\beta > 0$ and $\log_b(a) > v$, so

$$f(n) = \alpha a m^{\log_b(a)} - \beta b^v m^v < a n^{\log_b(a)}$$

Exercise 3 implies that $f \in \mathcal{O}(n^{\log_b(a)})$.

If $\boldsymbol{a < b^v}$, then $\beta < 0$ and $\log_b(a) < v$. Let $n - m = \gamma$. Then

$$\begin{aligned}
f(n) &= \alpha a m^{\log_b(a)} + |\beta| b^v m^v \\
&= \alpha a (n-\gamma)^{\log_b(a)} + |\beta| b^v m^v \\
&< \alpha a (n-\gamma)^{\log_b(a)} + |\beta| b^v n^v
\end{aligned}$$

Since $(n - \gamma)^{\log_b(a)} \in \mathcal{O}\left(n^{\log_b(a)}\right)$, Exercise 3 implies that $f \in \mathcal{O}(n^v)$.

Now let $M = b^{k+1}$. Then since $\left(b^k\right)^{\log_b(a)} = a^k$,

$$\begin{aligned}
f(n) \geq f\left(\frac{M}{b}\right) = f(b^k) &= \alpha a^k - \beta\left(\frac{M}{b}\right)^v \\
&= \left(\frac{\alpha}{a}\right) a^{\log_b(M)} - \left(\frac{\beta}{b^v}\right) M^v \\
&= \left(\frac{\alpha}{a}\right) M^{\log_b(a)} - \left(\frac{\beta}{b^v}\right) M^v
\end{aligned}$$

[50] This can be made precise by using Newton's binomial theorem on page 397.

$f(n) \geq \left(\frac{\alpha}{a}\right) M^{\log_b(a)} - \left(\frac{\beta}{b^v}\right) M^v$

If $a > b^v$, then $\alpha > 0$ and $\beta > 0$ and $\log_b(a) > v$. Let $M - n = \delta$. Then

$$\begin{aligned} f(n) &\geq \left(\frac{\alpha}{a}\right) M^{\log_b(a)} - \left(\frac{\beta}{b^v}\right) M^v \\ &= \left(\frac{\alpha}{a}\right) M^{\log_b(a)} - \left(\frac{\beta}{b^v}\right) (n+\delta)^v \\ &> \left(\frac{\alpha}{a}\right) n^{\log_b(a)} - \left(\frac{\beta}{b^v}\right) (n+\delta)^v \end{aligned}$$

The factor $(n+\delta)^v$ is in $\Omega(n^v)$. Since $\log_b(a) > v$, $\left(\frac{\alpha}{a}\right) n^{\log_b(a)} - \left(\frac{\beta}{b^v}\right)(n+\delta)^v$ is in $\Omega(n^{\log_b(a)})$. Therefore, Exercise 4 implies that $f \in \Omega(n^{\log_b(a)})$.

If $a < b^v$, then $\beta < 0$ and $\log_b(a) < v$. Thus,

$$\begin{aligned} f(n) &= \left(\frac{\alpha}{a}\right) M^{\log_b(a)} - \left(\frac{\beta}{b^v}\right) M^v \\ &= \left(\frac{\alpha}{a}\right) M^{\log_b(a)} + \left(\frac{|\beta|}{b^v}\right) M^v \\ &= \left(\frac{\alpha}{a}\right) (n+\delta)^{\log_b(a)} + \left(\frac{|\beta|}{b^v}\right) M^v \\ &> \left(\frac{\alpha}{a}\right) (n+\delta)^{\log_b(a)} + \left(\frac{|\beta|}{b^v}\right) n^v \end{aligned}$$

Since $\left(\frac{\alpha}{a}\right)(n+\delta)^{\log_b(a)} \in \Omega\left(n^{\log_b(a)}\right)$, $f \in \Omega(n^v)$.

Thus, when $a > b^v$, $f \in \mathcal{O}(n^{\log_b(a)})$ and $f \in \Omega(n^{\log_b(a)})$. Consequently, $f \in \Theta(n^{\log_b(a)})$. Also, when $a < b^v$, $f \in \mathcal{O}(n^v)$ and $f \in \Omega(n^v)$. Therefore, $f \in \Theta(n^v)$. ☐

✔ Quick Check 7.13

In each case, use Theorem 7.17 to find a good big-Θ reference function.

1. For the complexity function, $f(n) = 2f\left(\frac{n}{2}\right) + n$, $f(1) = 0$ for `mergeSort`.
2. For the approximate complexity function, $f(n) = 4f\left(\frac{n}{4}\right) + 2n^{\frac{1}{2}}$, $f(1) = 1$, for the Persian rugs algorithm. ☑

There are many recursive algorithms for which the complexity function does not fit the description in Theorem 7.17. A few simple extensions seem desirable. One possible extension is to change the term cn^v to a more general term, $s(n)$. Another extension is to be more realistic about the reduction, $\frac{n}{b}$. The `recBinarySearch` algorithm can handle lists with lengths that are not powers of 2. When the list is split, the sublists may have lengths $\lfloor \frac{n}{2} \rfloor$ and $\lceil \frac{n}{2} \rceil$ (see Exercise 1). The recursion of interest would involve either

$$f(n) = f\left(\left\lfloor \frac{n}{2} \right\rfloor\right) + 2$$

or

$$f(n) = f\left(\left\lceil \frac{n}{2} \right\rceil\right) + 2$$

An algorithm like `mergeSort` might involve a recursion that looks like

$$f(n) = f\left(\left\lfloor \frac{n}{2} \right\rfloor\right) + f\left(\left\lceil \frac{n}{2} \right\rceil\right) + n$$

Another reasonable extension is to broaden the base condition. Instead of $f(1) = d$, it might become $f(n) = d$ for $1 \leq n \leq n_0$.

7.3.1 Exercises

The exercises marked with ★ have detailed solutions in Appendix H.

1. Let $n \in \mathbb{N}$. Prove that

$$\left\lfloor \frac{n}{2} \right\rfloor + \left\lceil \frac{n}{2} \right\rceil = n$$

2. Let f be a function that satisfies the hypotheses of Theorem 7.16 (page 383). Use back substitution to show that

$$f(n) = da^k + c\sum_{i=0}^{k-1} a^i$$

whenever $n = b^k$ for some $k \in \mathbb{N}$.

3. ★ Let f, g, and h be real-valued functions. If $f(n) \le h(n)$, for all $n \ge n_0$ and if $h \in \mathcal{O}(g)$, prove $f \in \mathcal{O}(g)$.

4. Let f, g, and h be real-valued functions. If $f(n) \ge h(n)$, for all $n \ge n_0$ and if $h \in \Omega(g)$, prove $f \in \Omega(g)$.

5. ★ Let x, y, z be positive real numbers with $y > 1$, y and z constant. Prove
$x^{\lfloor \log_y(z) \rfloor} \in \mathcal{O}\left(x^{\log_y(z)}\right)$ and $x^{\lfloor \log_y(z) \rfloor} \in \Omega\left(x^{\log_y(z)-1}\right)$.

6. Let f be a function that satisfies the hypotheses of Theorem 7.17 (page 385). Use back substitution to show that

$$f(n) = da^k + cn^v \sum_{i=0}^{k-1} \left(\frac{a}{b^v}\right)^i$$

whenever $n = b^k$ for some $k \in \mathbb{N}$.

7. Assuming that $n = \left(2^k - 1\right)^2$, use back substitution to solve the recurrence relation (from Example 7.37 on page 380)

$$f(1) = 1 \text{ and } f(n) = 4f\left(\frac{(\sqrt{n}-1)^2}{4}\right) + 2\sqrt{n}$$

Then produce a good big-Θ reference function, $g(n)$. [*Hint:* Do not neglect the assumption that $n = (2^k - 1)^2$.]

8. Find good big-Θ reference functions for the following recurrence relations.
   (a) ★ $f(1) = 5$ and $f(n) = 3f\left(\frac{n}{2}\right) + 4n^2$
   (b) $f(1) = 2$ and $f(n) = 2f\left(\frac{n}{2}\right) + 2$
   (c) $f(1) = 0$ and $f(n) = 2f\left(\frac{n}{2}\right) + n\sqrt{n}$
   (d) $f(1) = 1$ and $f(n) = 4f\left(\frac{n}{2}\right) + 2n$

9. Find good big-Θ reference functions for the following recurrence relations:
   (a) $f(1) = 6$ and $f(n) = 8f\left(\frac{n}{2^3}\right) + \frac{1}{3}$
   (b) ★ $f(1) = 4$ and $f(n) = 3f\left(\frac{n}{2}\right) + \frac{4}{3}n$
   (c) $f(1) = 2$ and $f(n) = f\left(\frac{n}{5}\right) + 2\sqrt{n}$
   (d) $f(1) = \frac{3}{5}$ and $f(n) = 9f\left(\frac{n}{3}\right) + 3n^2$

10. Find good big-Θ reference functions for the following recurrence relations:
   (a) $f(1) = \frac{2}{3}$ and $f(n) = 2f\left(\frac{n}{10}\right) + (4\sqrt{n})^{\frac{3}{2}}$
   (b) $f(1) = 7$ and $f(n) = 5f\left(\frac{n}{4}\right) + \frac{1}{4}$
   (c) $f(1) = \frac{1}{6}$ and $f(n) = 7f\left(\frac{n}{2}\right) + \frac{15n^2}{4}$
   (d) $f(1) = 5$ and $f(n) = 3f\left(\frac{n}{7}\right) + 14n^4$

11. Prove Proposition 7.18.

> **PROPOSITION 7.18**
>
> Let $n \in \mathbb{N}$ and $r \in \mathbb{R}$ with $r > 0$ and $r \neq 1$. Then
>
> $$\sum_{i=0}^{n} ir^i = \frac{nr^{n+1}}{r-1} - \frac{r(r^n - 1)}{(r-1)^2}$$

12. Prove the following version of the master theorem. [*Hint:* You will need Proposition 7.18.]

> **THEOREM 7.19** ***The Master Theorem—Version 3***
>
> Let $a, b \in \mathbb{N}$ and $c, d \in \mathbb{R}$, with $a \ge 1$, $b > 1$, $c > 0$, and $d \ge 0$. Let $f(n)$ be defined on the interval $[1, \infty)$ by
>
> - $f(n) = af\left(\frac{n}{b}\right) + c\log_b(n)$
> - $f(1) = d$
>
> If $n = b^k$, for some $k \in \mathbb{N}$, then
>
> $$f \in \begin{cases} \Theta\left(n^{\log_b(a)} \cdot \log_b(n)\right) & \text{if } a > 1 \\ \Theta\left([\log_b(n)]^2\right) & \text{if } a = 1 \end{cases}$$

13. Use Theorem 7.19 to find good big-Θ reference functions for the following recurrence relations:
   (a) ★ $f(1) = 1$ and $f(n) = f\left(\frac{n}{2}\right) + 2\log_2(n)$
   (b) $f(1) = 0$ and $f(n) = 4f\left(\frac{n}{4}\right) + 3\log_4(n)$
   (c) $f(1) = 6$ and $f(n) = 2f\left(\frac{n}{2}\right) + \log_2(n)$

14. Use Theorem 7.19 to find good big-Θ reference functions for the following recurrence relations:
   (a) $f(1) = \frac{2}{3}$ and $f(n) = 3f\left(\frac{n}{2}\right) + \frac{1}{2}\log_2(n)$
   (b) $f(1) = 1$ and $f(n) = 5f\left(\frac{n}{5}\right) + 4\log_5(n)$
   (c) $f(1) = 11$ and $f(n) = 2f\left(\frac{n}{2^2}\right) + \frac{1}{4}\log_4(n)$

15. Consider the following algorithm, which returns a maximum value in a list of integers. The algorithm assumes that the list size, n, is a power of 2.

```
 1: integer recMax({a0, a1, ..., an-1})
 2:     if n == 1
 3:         return a0
 4:
 5:     x = recMax({a0, ..., a⌊n/2⌋-1})
 6:     y = recMax({a⌊n/2⌋, ..., an-1})
 7:
 8:     if x ≥ y
 9:         return x
10:     else
13:         return y
14: end recMax
```

   (a) What is the recurrence relation that counts the number of comparisons for this algorithm? (The critical steps are at lines 2 and 8.)
   (b) What is a good big-Θ reference function for algorithm `recMax`?

16. ★ Consider the following algorithm, which performs a crude shuffle of a list of integers. The algorithm divides the list into three equal sublists, recursively shuffles each list, and then concatenates the shuffled sublists in a rotated order. The algorithm assumes that the list size, n, is a power of 3.

```
1: list of integers
      shuffle ({a_0, a_1, ..., a_{n-1}})
2:   if n == 1
3:      return {a_0}
4:
5:   m = n/3   # assumes that n = 3^k
6:
7:   b = shuffle({a_0, ..., a_{m-1}})
8:   c = shuffle({a_m, ..., a_{2m-1}})
9:   d = shuffle({a_{2m}, ..., a_{3m-1}})
10:
11:  e = {c_0, ..., c_{m-1}, d_0, ..., d_{m-1}, b_0, ..., b_{m-1}}
12:
13:  return e
14: end shuffle
```

(a) What is the recurrence relation that counts the number of data copies for this algorithm? (The critical step is at line 11.)

(b) What is a good big-Θ reference function for algorithm `shuffle`?

17. Consider the following algorithm, which returns the median value in a sorted list of integers. The algorithm assumes that the list size, n, is a power of 3. It keeps grabbing the middle third of the list until only one item remains.

```
1: integer recMedian ({a_0, a_1, ..., a_{n-1}})
2:   if n == 1
3:      return a_0
4:
5:   m = n/3   # assumes that n = 3^k
6:
7:   for i = 0 to m - 1
8:      b_i = a_{i+m}   # copy the
                         middle third
9:
10:  return recMedian({b_0, ..., b_{m-1}})
11: end recMedian
```

(a) What is the recurrence relation that counts the number of data copies for this algorithm? (The critical step is at line 8.)

(b) What is a good big-Θ reference function for algorithm `recMedian`?

18. The following algorithm takes an unsorted list of positive integers, along with two integers, x and y. It returns the largest number, z, in the list such that either $z^x = y$ or $z^y = x$ is true. It returns 0 if no such z exists. The algorithm assumes that the list size, n, is a power of 2 with $n \geq 1$.

```
1: integer xyMax(x,y,{a_0, a_1, ..., a_{n-1}})
2:   if n == 1
3:      if (a_0^x == y) or (a_0^y == x)
4:         return a_0
5:      else
6:         return 0
7:
8:   # process the left half
9:
10:  m_1 = xyMax(x,y,{a_0, ..., a_{⌊n/2⌋-1}})
11:
12:  # process the right half
13:
14:  m_2 = xyMax(x,y,{a_{⌊n/2⌋}, ..., a_{n-1}})
15:
16:  # find the largest
17:
18:  max = m_1
19:  if m_2 > max
20:     max = m_2
21:
22:  return max
23: end xyMax
```

(a) What is the recurrence relation that counts the number of comparisons for this algorithm? (The critical steps are at lines 2, 3, and 19.)

(b) What is a good big-Θ reference function for algorithm `xyMax`?

19. Consider the following algorithm, which squares each entry in a list of integers. The algorithm assumes that the list size, n, is at least 1.

```
1: list of integers
      squareList ({a_0, a_1, ..., a_{n-1}})
2:   if n == 1
3:      return {a_0 · a_0}
4:
5:   for i = 0 to n - 1
6:      {b_i} = squareList({a_i})
7:
8:   # extract each b_i from the
9:   # one-element returned lists
10:
11:  return ({b_0, ..., b_{n-1}})
12: end squareList
```

(a) What is the recurrence relation that counts the number of comparisons and multiplications for this algorithm? (The critical steps are at lines 2 and 3.)

(b) What is a good big-Θ reference function for algorithm `squareList`? (You may assume that $n > 1$ for this part.) No theorems are needed.

20. The following algorithm calculates, for $n = 4^k$ where $k \in \mathbb{N}$ and $a_k \in \mathbb{R}$,

$$\prod_{k=0}^{n-1} a_k^2.$$

```
1: real sp({a_0, a_1, ..., a_{n-1}})
2:     if n == 1
3:         return a_0 · a_0
4:
5:     return sp({a_0, ..., a_{n/4 - 1}})
               · sp({a_{n/4}, ..., a_{2n/4 - 1}})
               · sp({a_{2n/4}, ..., a_{3n/4 - 1}})
               · sp({a_{3n/4}, ..., a_{n-1}})
6: end squareProduct
```

(a) Use Theorem 7.16 to determine the multiplication complexity of this algorithm. Show the details.

(b) Determine the exact number of multiplications by (i) a direct count and (ii) by using the formula derived in the proof of Theorem 7.16.

21. Suppose you are given a square on a computer screen where the number of pixels along each side is a power of 2. If you wish to paint the square gray, you could alternate between black and white pixels to form a "checkerboard" pattern. In computer screen graphics, the origin is at the top left corner of the screen. The x-axis is positive toward the right and the y-axis is positive moving downward.

The following algorithm will draw a checkerboard pattern in a square with upper-left corner at `(topLeftX, topLeftY)`, and having sides of length `sideLength` pixels.

(a) What is the recurrence relation that counts the number of comparisons and pixel paintings for this algorithm? (The critical steps are at lines 2–6.)

(b) What is a good big-Θ reference function for algorithm `paintGray`?

```
1: void paintGray(integer topLeftX,
                  integer topLeftY,
                  integer sideLength)
2:     if sideLength == 2
3:         paintPixel(topLeftX,
                topLeftY, WHITE)
4:         paintPixel(topLeftX + 1,
                topLeftY, BLACK)
5:         paintPixel(topLeftX,
                topLeftY + 1, BLACK)
6:         paintPixel(topLeftX + 1,
                topLeftY + 1, WHITE)
7:         return
8:
9:     paintGray(topLeftX, topLeftY,
                sideLength/2)
10:    paintGray(topLeftX +
                 sideLength/2,
                 topLeftY,
                 sideLength/2)
11:    paintGray(topLeftX,
                 topLeftY +
                 sideLength/2,
                 sideLength/2)
12:    paintGray(topLeftX +
                 sideLength/2,
                 topLeftY +
                 sideLength/2,
13:              sideLength/2)
14:    return
15: end paintGray
```

7.4 Generating Functions

Sometimes a simple change of viewpoint enables us to make a creative breakthrough. This section introduces one such viewpoint change.[51]

The change in viewpoint is to convert a sequence $a_0, a_1, a_2, \ldots$ into a formal power series $a_0 + a_1 z + a_2 z^2 + \cdots = \sum_{k=0}^{\infty} a_k z^k$. The term formal is used to indicate that issues of convergence are not considered relevant. The symbol z is a formal place holder, not a variable for which numbers might be substituted. However, wishing to keep all the advantages but without the responsibility of verifying convergence, formal power series will be manipulated using typical operations such as adding, multiplying, taking derivatives, and integrating.[52]

The purposes of this section are to

- Inform the reader that back substitution and Theorem 7.13 are not the only techniques for solving recurrence relations (nor are they the most powerful).
- Provide just enough detail for the reader to know what generating functions are and a bit about how they are used.

[51] For a thorough presentation, see [43].

[52] It is possible to justify this irresponsibility rigorously.

DEFINITION 7.20 ***Generating Function***

Let $a_0, a_1, a_2, \ldots$ be a sequence of real or complex numbers. The *generating function*, $G(z)$, for the sequence is the formal power series

$$G(z) = a_0 + a_1 z + a_2 z^2 + \cdots = \sum_{k=0}^{\infty} a_k z^k.$$

If the sequence is finite, $a_0, a_1, a_2, \ldots, a_n$, then

$$G(z) = a_0 + a_1 z + a_2 z^2 + \cdots + a_n z^n \text{ is a polynomial.}$$

Define $a_k = 0$ if $k < 0$.

EXAMPLE 7.38 **An Important Generating Function**

The sequence 1, 1, 1, ... of all 1s has the generating function

$$G(z) = 1 + z + z^2 + \cdots = \sum_{k=0}^{\infty} z^k = \frac{1}{1-z}$$

To justify the final equality, notice that

$$\begin{aligned}(1-z)(1+z+z^2+\cdots) = 1 &+ z + z^2 + z^3 + \cdots \\ &- z - z^2 - z^3 - \cdots\end{aligned}$$

■

Much of the power of generating functions results from the dual perspectives they provide: We can consider the generating function as a power series with the terms of the sequence as coefficients, or we can view it as a function in z (such as $\frac{1}{1-z}$). In many problems, the values of the coefficients a_k are unknown. However, it may be possible to manipulate the generating function so that an expression such as $G(z) = \frac{1}{1-z}$ is derived. In that case, we can transform the function into a power series (such as $1+z+z^2+\cdots$) and then read the coefficients, thus solving the original problem.

Before looking at a simple example that illustrates this process, note that a sequence can be shifted right one place by multiplying the generating function by z. That is, to change $\{a_0, a_1, a_2, \ldots\}$ into $\{0, a_0, a_1, a_2 \ldots\}$, simply form $zG(z)$. This is true since

$$zG(z) = z\sum_{k=0}^{\infty} a_k z^k = \sum_{k=0}^{\infty} a_k z^{k+1} = \sum_{j=1}^{\infty} a_{j-1} z^j = 0 + a_0 z + a_1 z^2 + \cdots$$

This shifting process can be easily generalized.

PROPOSITION 7.21 ***Shifting Generating Functions***

Let $G(z) = \sum_{k=0}^{\infty} a_k z^k$ be the generating function for the sequence $\{a_0, a_1, a_2, \ldots\}$. Then

$$z^m G(z) = \sum_{k=0}^{\infty} a_k z^{k+m} = \sum_{k=m}^{\infty} a_{k-m} z^k$$

Proof:

$$\sum_{k=m}^{\infty} a_{k-m} z^k = z^m \sum_{k=m}^{\infty} a_{k-m} z^{k-m} = z^m \sum_{j=0}^{\infty} a_j z^j = z^m G(z)$$

□

Proposition 7.21 can be interpreted as saying that $z^m G(z)$ is the generating function for the sequence $\{\overbrace{0, 0, \ldots, 0}^{m \text{ times}}, a_0, a_1, a_2, \ldots\}$.

EXAMPLE 7.39 **The Tower of Hanoi Revisited**

Recall the Tower of Hanoi puzzle (Example 7.17 on page 352). The recurrence relation that defines the solution is

- $H_0 = 0$, $H_1 = 1$
- $H_n = 2H_{n-1} + 1$ for $n \geq 1$.

Let

$$H(z) = \sum_{n=0}^{\infty} H_n z^n$$

be the generating function for the sequence $\{H_n\}$.

Multiply both sides of the recurrence relation by z^n and form the summation over all positive values of n for which the coefficients, H_n, are defined in all three summations:

$$\sum_{n=1}^{\infty} H_n z^n = 2\sum_{n=1}^{\infty} H_{n-1} z^n + \sum_{n=1}^{\infty} z^n$$

Using Proposition 7.21 and Example 7.38, this can be written as

$$(H(z) - H_0) = 2zH(z) + \left(\frac{1}{1-z} - 1\right)$$

Thus,

$$H(z)(1-2z) = \frac{1}{1-z} - 1$$

so

$$H(z) = \frac{1}{(1-z)(1-2z)} - \frac{1}{1-2z}$$

If the coefficients of the power series representation of the right-hand side could be found, H_n would be determined. Some additional background work is needed before this can be done successfully. ■

To do any serious work with generating functions, it is necessary to have completed the power series expansions of a number of common functions (such as $\frac{1}{1-2z}$). It is then a simple matter of finding the appropriate entry in a table of such expansions. The table can be produced using results from the study of power series expansions. Results can also be proved (once the coefficients have been produced in some manner) by a verification similar to that used with

$$\frac{1}{1-z} = \sum_{k=0}^{\infty} z^k$$

Table 7.10 on page 394 is a small table of useful expansions. (The table is repeated on page 416.[53]) Issues of convergence are ignored.[54]

One additional useful fact will enable the Tower of Hanoi problem to be completed.

THEOREM 7.22 ***Multiplying Generating Functions***

Let $F(z) = \sum_{k=0}^{\infty} f_k z^k$ and $G(z) = \sum_{k=0}^{\infty} g_k z^k$ be two generating functions. The generating function that is the product of F and G is

$$P(z) = F(z) \cdot G(z) = \sum_{k=0}^{\infty} \left(\sum_{j=0}^{k} f_j g_{k-j} \right) z^k$$

Proof: The proof is a routine algebraic manipulation (Exercise 1 on page 401). □

[53]The table is also available in pdf format at http://www.mathcs.bethel.edu/~gossett/DiscreteMathWithProof/.

[54]So are the proofs in some cases.

TABLE 7.10 A Small Table of Useful Generating Functions

Some Useful Generating Functions		
$G(z)$	**Summation Notation**	**Expanded Notation**
$\frac{1}{1-z}$	$\sum_{k=0}^{\infty} z^k$	$1 + z + z^2 + z^3 + \cdots$
$\frac{1}{1+z}$	$\sum_{k=0}^{\infty} (-1)^k z^k$	$1 - z + z^2 - z^3 + \cdots$
$\frac{1}{1-z^m}$	$\sum_{k=0}^{\infty} z^{mk}$	$1 + z^m + z^{2m} + z^{3m} + \cdots$
$\frac{1}{1-cz}$	$\sum_{k=0}^{\infty} c^k z^k$	$1 + cz + c^2z^2 + c^3z^3 + \cdots$
$\frac{1}{(1-z)^m}$	$\sum_{k=0}^{\infty} \binom{m+k-1}{k} z^k$	$1 + mz + \binom{m+1}{2} z^2 + \binom{m+2}{3} z^3 + \cdots$
$\frac{z}{(1-z)^2}$	$\sum_{k=0}^{\infty} k z^k$	$0 + z + 2z^2 + 3z^3 + \cdots$
$(1+z)^c$	$\sum_{k=0}^{\infty} \binom{c}{k} z^k$	$1 + cz + \binom{c}{2} z^2 + \binom{c}{3} z^3 + \cdots$
e^z	$\sum_{k=0}^{\infty} \frac{1}{k!} z^k$	$1 + z + \frac{z^2}{2!} + \frac{z^3}{3!} + \cdots$

EXAMPLE 7.40 **Two Simple Generating Functions**

The equation $\frac{1}{1-z} = \sum_{k=0}^{\infty} z^k$, together with Theorem 7.22, can be used to find the coefficients for $\frac{1}{(1-z)^2}$.

$F(z) \cdot G(z) = \sum_{k=0}^{\infty} \left(\sum_{j=0}^{k} f_j g_{k-j} \right) z^k$

$$\frac{1}{(1-z)^2} = \frac{1}{1-z} \cdot \frac{1}{1-z} = \left(\sum_{k=0}^{\infty} z^k \right) \cdot \left(\sum_{k=0}^{\infty} z^k \right)$$
$$= \sum_{k=0}^{\infty} \left(\sum_{j=0}^{k} 1 \cdot 1 \right) z^k$$
$$= \sum_{k=0}^{\infty} (k+1) z^k$$

Similarly, Table 7.10 and Theorem 7.22 imply

$$\frac{1}{(1-z)(1-2z)} = \frac{1}{1-z} \cdot \frac{1}{1-2z} = \left(\sum_{k=0}^{\infty} z^k \right) \cdot \left(\sum_{k=0}^{\infty} 2^k z^k \right)$$
$$= \sum_{k=0}^{\infty} \left(\sum_{j=0}^{k} 1 \cdot 2^{k-j} \right) z^k$$
$$= \sum_{k=0}^{\infty} \left(\sum_{j=0}^{k} 2^j \right) z^k = \sum_{k=0}^{\infty} \left(\frac{2^{k+1} - 1}{2 - 1} \right) z^k$$

$$= \sum_{k=0}^{\infty} \left(2^{k+1} - 1\right) z^k$$

Note the use of the identity $\sum_{j=0}^{k} 2^{k-j} = \sum_{j=0}^{k} 2^j$, which can be justified by noticing that the right-hand sum is just the left-hand sum in reverse order. ■

It is time to finish the Tower of Hanoi problem.

EXAMPLE 7.41

Tower of Hanoi Continued

Previous progress had derived

$$H(z) = \frac{1}{(1-z)(1-2z)} - \frac{1}{1-2z}$$

Using an identity in Table 7.10 and a result from Example 7.40, this can be expanded.

$$\begin{aligned} H(z) &= \frac{1}{(1-z)(1-2z)} - \frac{1}{1-2z} \\ &= \left(\sum_{k=0}^{\infty} \left(2^{k+1} - 1\right) z^k\right) - \left(\sum_{k=0}^{\infty} 2^k z^k\right) \\ &= \sum_{k=0}^{\infty} \left(2^{k+1} - 1 - 2^k\right) z^k \\ &= \sum_{k=0}^{\infty} \left(2^k - 1\right) z^k \end{aligned}$$

The conclusion is that $H_n = 2^n - 1$ (since H_n is the coefficient of z^n).

Instead of using Theorem 7.22, the coefficients of $H(z)$ could be found using a partial fraction decomposition. Writing

$$\frac{1}{(1-z)(1-2z)} - \frac{1}{1-2z}$$

as

$$\frac{A}{1-z} + \frac{B}{1-2z} - \frac{1}{1-2z}$$

and solving for A and B produces (with the aid of Table 7.10 after A and B are found.)

$$\begin{aligned} H(z) &= \frac{-1}{1-z} + \frac{2}{1-2z} - \frac{1}{1-2z} \\ &= \frac{1}{1-2z} - \frac{1}{1-z} \\ &= \sum_{k=0}^{\infty} 2^k z^k - \sum_{k=0}^{\infty} z^k \\ &= \sum_{k=0}^{\infty} \left(2^k - 1\right) z^k \end{aligned}$$

Therefore, the coefficient of H_n is $2^n - 1$. ■

The derivative of a generating function will be treated as a formal manipulation, rather than the operation defined in calculus. However, the definition is motivated by the standard manipulations derived in calculus.

DEFINITION 7.23 *The Derivative of a Generating Function*

Let $A(z) = \sum_{k=0}^{\infty} a_k z^k$ be a generating function. Then its *derivative* is denoted by $A'(z)$ and is defined by

$$A'(z) = \sum_{k=1}^{\infty} k a_k z^{k-1}$$

EXAMPLE 7.42

Using Derivatives with Generating Functions

A derivation for Table 7.10 row 6 is easy. The first part of Example 7.40 implies that

$$\frac{z}{(1-z)^2} = z\sum_{k=0}^{\infty}(k+1)z^k = \sum_{k=0}^{\infty}(k+1)z^{k+1} = \sum_{j=1}^{\infty} jz^j = \sum_{j=0}^{\infty} jz^j.$$

The previous equality can be proved another way. For that purpose, suppose $H(z) = \sum_{k=0}^{\infty} kz^k$ and $G(z) = \sum_{k=0}^{\infty} z^k = \frac{1}{1-z}$. Definition 7.23 implies that $G'(z) = \sum_{k=1}^{\infty} kz^{k-1}$, and so $zG'(z) = \sum_{k=1}^{\infty} kz^k = \sum_{k=0}^{\infty} kz^k = H(z)$.

On the other hand, $\frac{d}{dz}\left(\frac{1}{1-z}\right) = \frac{1}{(1-z)^2}$, so $G'(z) = \frac{1}{(1-z)^2}$. Thus, $H(z) = zG'(z) = \frac{z}{(1-z)^2}$ and consequently, $\sum_{k=0}^{\infty} kz^k = \frac{z}{(1-z)^2}$. ■

The next example shows how derivatives of generating functions might arise in a natural manner.

EXAMPLE 7.43

A Recurrence Relation with a Generating Function Derivative

Consider the recurrence relation defined by $a_n = (n-1)a_{n-1} + 1$, where $a_0 = 1$.

Let $A(z)$ represent the generating function for this recurrence relation. Using the common practice of multiplying both sides of the recurrence relation equation by z^n and summing, the following identity arises. (Since a_{-1} is undefined, the summation starts with $n = 1$.)

$$\sum_{n=1}^{\infty} a_n z^n = \sum_{n=1}^{\infty}(n-1)a_{n-1}z^n + \sum_{n=1}^{\infty} z^n$$

Thus (dropping the term with $n - 1 = 0$),

$$A(z) - a_0 = z^2\sum_{n=2}^{\infty}(n-1)a_{n-1}z^{n-2} + \left(\frac{1}{1-z} - 1\right)$$

By using the identity $a_0 = 1$ and a change of summation index, this can be written as

$$A(z) = z^2\sum_{k=1}^{\infty} k a_k z^{k-1} + \frac{1}{1-z}$$

Consequently, Definition 7.23 implies that $A(z)$ must satisfy the differential equation

$$A(z) = z^2 A'(z) + \frac{1}{1-z} \quad \text{with } A(0) = 1$$

Solving this differential equation (and converting the solution to a power series) is not trivial (see [43] for an approach using hypergeometric series). The result, which involves the permutation $P(n-1, r)$, is equivalent to (via Exercise 4 on page 401)

$$A(z) = 1 + \sum_{n=1}^{\infty}\left[\sum_{r=0}^{n-1} P(n-1, r)\right] z^n$$

so

$$a_n = \sum_{r=0}^{n-1} P(n-1, r) \quad \text{for } n \geq 1$$

This problem and its solution should remind you of Example 7.20 on page 355. ■

One additional theorem is very useful in this context. The theorem is a generalization, first created by Isaac Newton, of the binomial theorem and was used by him to great effect. The modern notation differs from Newton's original notation.[55] Before the theorem is presented, a definition is required.

Recall that the normal binomial coefficient $\binom{n}{r}$ contains a product of r integers in the numerator (assuming $r > 0$):

$$\binom{n}{r} = \frac{n!}{r!(n-r)!} = \frac{n(n-1)(n-2)\cdots(n-(r-1))}{r!}$$
$$= \frac{n(n-1)(n-2)\cdots(n-r+1)}{r!}$$

and $\binom{n}{0} = \frac{n!}{0!(n-0)!} = 1$.

We can extend this by keeping the form, but allowing n to be any real number.

DEFINITION 7.24 ***Generalized Binomial Coefficients***

Let u be any real number and let r be a nonnegative integer. Then the *generalized binomial coefficient*, u choose r, is defined by

$$\binom{u}{r} = \begin{cases} \frac{u(u-1)(u-2)\cdots(u-r+1)}{r!} & \text{if } r > 0 \\ 1 & \text{if } r = 0 \end{cases}$$

Note that if $u = 0$ and $r > 0$, then $\binom{0}{r} = 0$. In addition, if u is a negative integer, for example $u = -n$ for n a positive integer, and $r > 0$, then

$$\binom{-n}{r} = \frac{(-n)(-n-1)(-n-2)\cdots(-n-r+1)}{r!}$$
$$= \frac{(-1)^r(n)(n+1)(n+2)\cdots(n+r-1)}{r!}$$
$$= (-1)^r\frac{(n+r-1)!}{r!(n-1)!}$$
$$= (-1)^r\binom{n+r-1}{r} \quad \text{(an ordinary binomial coefficient).}$$

THEOREM 7.25 ***Newton's Binomial Theorem***

Let w and z be real or complex numbers with $|\frac{z}{w}| < 1$. Then for any real number, u,

$$(w+z)^u = \sum_{k=0}^{\infty}\binom{u}{k}w^{u-k}z^k$$

When $w = 1$, this reduces to

$$(1+z)^u = \sum_{k=0}^{\infty}\binom{u}{k}z^k \quad \text{for } |z| < 1$$

The proof uses ideas from the theory of infinite series typically covered in second semester calculus.

Newton's theorem can be used to prove some of the identities in the collection of useful identities. For this purpose, we will consider this theorem to be a formal

[55]See [26] for details of the original notation and examples of Newton's use of the theorem in calculus and other areas.

manipulation and not consider issues of convergence. (Therefore, the requirements that $|\frac{z}{w}| < 1$ and $|z| < 1$ will be ignored.)

EXAMPLE 7.44 **Using Newton's Binomial Theorem**

For any real or complex number c, Table 7.10 row 4 can be derived using Theorem 7.25:

$(w+z)^u = \sum_{k=0}^{\infty} \binom{u}{k} w^{u-k} z^k$

$\binom{-n}{r} = (-1)^r \binom{n+r-1}{r}$

$$\frac{1}{1-cz} = (1+(-cz))^{-1} = \sum_{k=0}^{\infty} \binom{-1}{k} (-cz)^k$$
$$= \sum_{k=0}^{\infty} (-1)^k \binom{k}{k} (-1)^k c^k z^k$$
$$= \sum_{k=0}^{\infty} c^k z^k$$

■

EXAMPLE 7.45 **Another Useful Identity**

For positive integers, m, Table 7.10 row 5 is derived by

$$\frac{1}{(1-z)^m} = (1+(-z))^{-m} = \sum_{k=0}^{\infty} \binom{-m}{k} (-z)^k$$
$$= \sum_{k=0}^{\infty} (-1)^k \binom{m+k-1}{k} (-1)^k z^k$$
$$= \sum_{k=0}^{\infty} \binom{m+k-1}{k} z^k$$

■

EXAMPLE 7.46 **Another Recurrence Relation via Generating Functions**

Consider the recurrence relation

- $a_0 = -3$
- $a_n = a_{n-1} + n \quad$ for $n \geq 1$

Let $A(z) = \sum_{k=0}^{\infty} a_k z^k$ be the generating function for the sequence generated by the recurrence relation. The same initial step will convert the recurrence relation into an equation involving generating functions: Multiply both sides by z^n and sum over all appropriate values of n. In this case, the $n-1$ on the right produces a valid subscript as long as $n \geq 1$. Thus

$$\sum_{n=1}^{\infty} a_n z^n = \sum_{n=1}^{\infty} a_{n-1} z^n + \sum_{n=1}^{\infty} n z^n$$
$$= z \sum_{n=1}^{\infty} a_{n-1} z^{n-1} + \sum_{n=1}^{\infty} n z^n = z \sum_{n=0}^{\infty} a_n z^n + \sum_{n=0}^{\infty} n z^n$$

The final step results from a change of variable in the left-hand sum and because $0z^0$ can be added to the right-hand sum without changing its value.

Note that $A(z) = a_0 + \sum_{n=1}^{\infty} a_n z^n$. Using row 6 of Table 7.10 and converting to expressions involving $A(z)$ leads to

$\frac{z}{(1-z)^2} = \sum_{k=0}^{\infty} k z^k$

$$A(z) - a_0 = zA(z) + \frac{z}{(1-z)^2}$$

Solving for $A(z)$ produces

$$A(z) = \frac{a_0}{1-z} + \frac{z}{(1-z)^3}$$

Using the identities in rows 1 and 5 of Table 7.10,

$\frac{1}{1-z} = \sum_{k=0}^{\infty} z^k$

$\frac{1}{(1-z)^m} = \sum_{k=0}^{\infty} \binom{m+k-1}{k} z^k$

$$\begin{aligned} A(z) &= \sum_{n=0}^{\infty} a_0 z^n + z \sum_{n=0}^{\infty} \binom{n+2}{n} z^n \\ &= \sum_{n=0}^{\infty} a_0 z^n + \sum_{n=0}^{\infty} \frac{(n+2)(n+1)}{2} z^{n+1} \\ &= \sum_{n=0}^{\infty} a_0 z^n + \sum_{k=1}^{\infty} \frac{(k+1)(k)}{2} z^k \\ &= a_0 + \sum_{n=1}^{\infty} \left(a_0 + \frac{n(n+1)}{2} \right) z^n \\ &= \sum_{n=0}^{\infty} \left(a_0 + \frac{n(n+1)}{2} \right) z^n \end{aligned}$$

Since $a_0 = -3$, this simplifies to

$$A(z) = \sum_{n=0}^{\infty} \frac{(n+3)(n-2)}{2} z^n$$

This implies that

$$a_n = \frac{(n+3)(n-2)}{2} \quad \text{for } n \geq 0$$

Substituting into the recurrence relation $a_n = a_{n-1} + n$ leads to

$$\frac{(n+3)(n-2)}{2} = \frac{(n+2)(n-3)}{2} + n$$

which is a valid equation. ■

✔ Quick Check 7.14

1. Use Newton's binomial theorem to expand $\frac{1}{1+2z}$ and $\frac{1}{1-3z}$ as power series.
2. Use Theorem 7.22 to find the power series expansion of $\left(\frac{1}{1+2z} \cdot \frac{1}{1-3z}\right)$. Simplify as far as possible.
3. Use generating functions to find a closed-form formula for a_n if
   - $a_0 = 1, a_1 = 2$
   - $a_n = a_{n-1} + 6a_{n-2}$ for $n \geq 2$ ☑

Generating Functions and Counting

Generating functions can be used to solve many counting problems. The next example illustrates one such use.

One new idea is all that is needed. Observe that

$$(1-z)(1+z+z^2+\cdots+z^n) = 1 - z^{n+1}$$

Thus

$$1+z+z^2+\cdots+z^n = \frac{1-z^{n+1}}{1-z}$$

We can think of the polynomial $1+z+z^2+\cdots+z^n$ as the generating function for the sequence, $1, 1, 1, \ldots, 1, 0, 0, \ldots$, with $n+1$ initial 1s and 0s for all other elements of the sequence.

EXAMPLE 7.47

Solving an Integer equation

A homework exercise in Chapter 5 (Exercise 7 on page 257) asked for the number of distinct solutions to the equation $x_1 + x_2 + x_3 + x_4 = 16$, where x_1, x_2, x_3, x_4 are nonnegative integers. This can be solved using a generating function approach.

Consider a solution to the equation. Let the solution be $k_1 + k_2 + k_3 + k_4 = 16$. Since this equation is true, it is also true that $z^{k_1} \cdot z^{k_2} \cdot z^{k_3} \cdot z^{k_4} = z^{16}$. Now consider the product

$$\left(1 + z + z^2 + z^3 + \cdots + z^{16}\right)^4$$

The coefficient of z^{16} is obtained as a sum of products of the form $z^{k_1} \cdot z^{k_2} \cdot z^{k_3} \cdot z^{k_4}$, where $k_1 + k_2 + k_3 + k_4 = 16$. In fact, every such product contributes 1 to the sum which determines the coefficient of z^{16}. So there is a one-to-one correspondence between solutions to the equation $x_1 + x_2 + x_3 + x_4 = 16$ and terms of the form $z^{k_1} \cdot z^{k_2} \cdot z^{k_3} \cdot z^{k_4}$ with $k_1 + k_2 + k_3 + k_4 = 16$ in the expansion of $\left(1 + z + z^2 + z^3 + \cdots + z^{16}\right)^4$.

Consequently, the number of solutions to the equation is the coefficient of z^{16} in the product $\left(1 + z + z^2 + z^3 + \cdots + z^{16}\right)^4$.

With the help of *Mathematica*, the product expands to

$$\begin{aligned} &1 + 4z + 10z^2 + 20z^3 + 35z^4 + 56z^5 + 84z^6 \\ &+ 120z^7 + 165z^8 + 220z^9 + 286z^{10} + 364z^{11} + 455z^{12} \\ &+ 560z^{13} + 680z^{14} + 816z^{15} + 969z^{16} + \text{other terms (up to } z^{64}) \end{aligned}$$

The number of solutions is therefore 969. As a bonus, we also now know that there are 816 solutions to the equation $x_1 + x_2 + x_3 + x_4 = 15$, and 680 solutions to the equation $x_1 + x_2 + x_3 + x_4 = 14$, and so on. This expansion would *not* tell us the number of solutions to the equation $x_1 + x_2 + x_3 + x_4 = 17$ (Why not?). ■

The previous example can be solved more easily if the problem is first generalized.

EXAMPLE 7.48

Example 7.47 Revisited

Let a_n represent the number of nonnegative integer solutions to the equation $x_1 + x_2 + x_3 + x_4 = n$, for $0 \le n$. Let $A(z) = \sum_{k=0}^{\infty} a_k z^k$. Example 7.47 has already established that a_n is the coefficient of z^n in the expansion of $\left(1 + z + z^2 + \cdots\right)^4$.

Thus (using Table 7.10),

$$A(z) = \left(\frac{1}{1-z}\right)^4 = \frac{1}{(1-z)^4} = \sum_{n=0}^{\infty} \binom{4+n-1}{n} z^n = \sum_{n=0}^{\infty} \binom{n+3}{n} z^n$$

In particular, when $n = 16$, $a_n = \binom{19}{16} = 969$. ■

The final example in this section introduces the notion of counting with inequality constraints.

EXAMPLE 7.49

Counting with Inequality Constraints

I have 12 Sacagawea dollars that I wish to distribute to my three nieces. Each niece should get at least two coins. Since Erin is much younger than Grace or May and does not have a job, she should receive at least four coins. In how many ways can I distribute the coins?

There are some natural upper limits on the numbers of coins each niece can receive. Erin can receive at most eight coins if at least four coins must remain for the other two nieces. Similarly, Grace and May can each receive at most six coins so that at least four coins remain for Erin and two coins remain for the other older niece.

Generating functions can be introduced by creating a polynomial in z for each niece. For Erin, the natural choice is the expression $z^4 + z^5 + z^6 + z^7 + z^8$, since she can receive from four to eight coins. Grace and May will share the same expression, namely, $z^2 + z^3 + z^4 + z^5 + z^6$.

The solution to the problem is the coefficient of z^{12} in

$$\left(z^4 + z^5 + z^6 + z^7 + z^8\right)\left(z^2 + z^3 + z^4 + z^5 + z^6\right)\left(z^2 + z^3 + z^4 + z^5 + z^6\right)$$

The product simplifies to

$$z^8 + 3z^9 + 6z^{10} + 10z^{11} + 15z^{12} + 18z^{13} + 19z^{14} + 18z^{15} + 15z^{16} + 10z^{17} + 6z^{18} + 3z^{19} + z^{20}$$

Thus, there are 15 ways to distribute the coins (list them). ■

7.4.1 Exercises

The exercises marked with ★ have detailed solutions in Appendix H.

1. Prove Theorem 7.22 (page 393).
2. Use Theorem 7.22 (page 393) to find the coefficients of the following generating functions. Simplify as far as possible.
   (a) ★ $\left(\sum_{k=0}^{\infty} z^k\right)\left(\sum_{k=0}^{\infty} kz^k\right)$
   (b) $\left(\sum_{k=0}^{\infty} 2^k z^k\right)\left(\sum_{k=0}^{\infty} 5^k z^k\right)$
   (c) $\left(\sum_{k=0}^{\infty} kz^k\right)\left(\sum_{k=0}^{\infty} kz^k\right)$ [*Hint*: Use the result of Exercise 1(a) on page 151.]
3. Use Theorem 7.22 (page 393) to find the coefficients of the following generating functions. Simplify as far as possible.
   (a) $\left(\sum_{k=0}^{\infty} 3^k z^k\right)\left(\sum_{k=0}^{\infty} 3^k z^k\right)$
   (b) $\left(\sum_{k=0}^{\infty} (k+1)z^k\right)\left(\sum_{k=0}^{\infty} 2z^k\right)$
   (c) $\left(\sum_{k=0}^{\infty} kz^k\right)\left(\sum_{k=0}^{\infty} 5^k z^k\right)$ [*Hint*: Use the result of Exercise 4(c) on page 375.]
4. ★ Use back substitution to solve the recurrence relation $a_n = (n-1)a_{n-1} + 1$, where $a_1 = 1$.
5. Use Newton's binomial theorem to expand the following into power series:
   (a) $\frac{1}{1-z^m}$ (b) ★ $(1+3z)^{-2}$ (c) $\sqrt[3]{1+z}$
6. Use Newton's binomial theorem to expand the following into power series:
   (a) $\left(1+\frac{1}{2}z\right)^{-5}$ (b) $\frac{1}{(1-2z)^4}$ (c) $\left(\frac{1}{1-4z^2}\right)^3$
7. Use generating functions to provide an alternative derivation of the solution for the recurrence relation $a_n = 4a_{n-1} + 4$ in Example 7.19 on page 354.
8. Show that $\sum_{j=0}^{k}(k-j)3^j = \frac{3^{k+1}-3-2k}{4}$. [*Hint*: First show that $\sum_{j=0}^{k}(k-j)3^j = \sum_{m=0}^{k-1}\left(\sum_{i=0}^{m} 3^i\right)$.]
9. Use generating functions to solve the following recurrence relations:
   (a) $a_0 = 1$, and $a_n = 3a_{n-1}$ for $n \ge 1$
   (b) ★ $a_0 = 1$, and $a_n = 3a_{n-1} + 7$ for $n \ge 1$
   (c) $a_0 = 1$, and $a_n = 3a_{n-1} + n$ for $n \ge 1$ [*Hint*: Use Exercise 8.]
   (d) $a_0 = 2$, $a_1 = -2$, and $a_n = -2a_{n-1} + 15a_{n-2}$ for $n \ge 2$
10. Use generating functions to solve the following recurrence relations:
    (a) $a_0 = 1, a_1 = 8$, and $a_n = 7a_{n-1} - 12a_{n-2}$ for $n \ge 2$
    (b) $a_0 = 5$, and $a_n = \frac{1}{4}a_{n-1}$ for $n \ge 1$
    (c) $a_0 = 3$, and $a_n = 5a_{n-1} + 1$ for $n \ge 1$
    (d) ★ $a_0 = 3, a_1 = -12$, and $a_n = -5a_{n-1} + 36a_{n-2}$ for $n \ge 2$
11. Use generating functions to solve the following recurrence relations.
    (a) $a_0 = 8$, and $a_n = 24a_{n-1} - 144$ for $n \ge 1$
    (b) $a_0 = -1, a_1 = -2$, and $a_n = a_{n-1} + 20a_{n-2}$ for $n \ge 2$
    (c) $a_0 = 1$, and $a_n = 6a_{n-1} - 5$ for $n \ge 1$
    (d) $a_0 = 4, a_1 = 20$, and $a_n = 4a_{n-1} - 4a_{n-2}$ for $n \ge 2$
12. ★ If an unlimited supply of indistinguishable pennies, indistinguishable nickels, indistinguishable dimes, and indistinguishable quarters is available, how many distinct arrangements of coins can be formed whose sum is 38 cents? Use generating functions in your solution strategy. You may want to use a computer algebra system for this problem.
13. You are about to buy an item in the vending machine that costs \$0.95. You have 7 pennies, 12 nickels, and 8 dimes, where all the coins of the same type are indistinguishable. Determine the number of ways to insert the coins into the vending machine to make the purchase of exactly \$0.95, assuming that the order in which the coins are inserted does not matter. Use generating functions in your solution strategy. You may want to use a computer algebra system for this problem.
14. Find a generating function for the sequence $-1, 1, -1, 1, -1, 1, \ldots$.
15. Find a generating function for the sequence, $1, 2^3, 3^3, 4^3, \ldots$, of cubes. [*Hint*: Use derivatives.]

16. Mary made three dozen identical homemade chocolate chip cookies and is going to distribute them to four families in her neighborhood. Each family must receive at least six cookies. The Landers family cannot receive more than seven cookies because the mother does not want her family eating too many sweets. Mary also knows that there are many children in the Johnson family, so she wants to give them an ample supply of cookies. There are seven people in this family, and Mary wants to make sure that each family member gets at least two cookies. In how many ways can Mary distribute the chocolate chip cookies to the four neighborhood families? Use generating functions in your solution strategy. You may want to use a computer algebra system for this problem.

17. Exercise 17 on page 376 asked you to create a recurrence relation for counting the number of distinct bit strings of length n that do not contain three consecutive 1s. A closed-form solution was not produced in that exercise. It is now possible to make some progress. Find the generating function for this recurrence relation (expressed as a ratio of two polynomials in z).

18. Use partial fractions decompositions to find generating functions for the following expressions.

(a) ★ $\frac{3z}{1-z-6z^2}$ (b) $\frac{5}{1-z-12z^2} - \frac{8}{4z-1}$

(c) $\frac{7+7z}{2(1-\frac{3}{2}z-z^2)}$ (d) $\frac{6z}{1-\frac{36}{5}z+\frac{7}{5}z^2}$

19. Jang Geum needs to cook a meal for the king. She needs a total of 10 dishes. She has the ingredients for 6 meat dishes, 4 fish dishes, and 9 vegetable dishes. Use generating functions to determine the total number of ways she can prepare the meal if

(a) all that matters are the numbers of each type of dish. (2 meat, 3 fish, 4 vegetable vs 3 meat, 2 fish, 4 vegetable)

(b) individual recipes matter. (There are many ways to choose two meat dishes, but no duplicates are allowed.)

20. Generating functions and Newton's binomial theorem can be used to derive the combinations with repetition counting formula on page 235. The proof will examine all legal values for n and r simultaneously. A clever idea is to let the power series

$$\frac{1}{1-z} = 1 + z + z^2 + z^3 + \cdots$$

represent each of the n objects in the set of distinguishable objects. The term z^k will represent choosing k copies of the item. Since there are n items, the expression

$$C(z) = \frac{1}{(1-z)^n} = \left(1 + z + z^2 + z^3 + \cdots\right)^n$$

is the key to the result.

(a) Complete the discussion started previously by showing that the number of ways to choose r items with repetition from a set of n distinct items is the coefficient of z^r in $C(z)$.

(b) Use Newton's binomial theorem to calculate the coefficient of z^r in $C(z)$.

7.5 The Josephus Problem

Many of the problems that mathematicians and computer scientists dearly love have been around for a long time. One such problem is known as *the Josephus problem*, named after the first-century Jewish historian Flavius Josephus. Josephus did not invent the problem. Instead, an event from his life served as the inspiration for the problem statement.

Many current books refer to *Mathematical Recreations and Essays* by W. W. Rouse Ball [7, originally published in 1892] for the problem statement:

> Another of these antique problems consists in placing men around a circle so that if every m^{th} man is killed, the remainder shall be certain specified individuals. Such problems can be easily solved empirically.
>
> Hegesippus[a] says that Josephus saved his life by such a device. According to his account, after the Romans had captured Jotapat, Josephus and forty other Jews took refuge in a cave. Josephus, much to his disgust, found that all except himself and one other man were resolved to kill themselves, so as not to fall into the hands of their conquerors. Fearing to show his opposition too openly he consented, but declared that the operation must be carried out in an orderly way, and suggested that they should arrange themselves round a circle and that every third person should be killed until all but one man was left, who must then commit suicide. It is alleged that he placed himself and the other man in the 31st and 16st place respectively.
>
> [a] *De Bello Judaico*, bk III, chaps. 16–18

The problem (which will be addressed eventually) is quite interesting. However, the story, as quoted above, is not completely accurate. In fact, Hegesippus never existed, and there is no evidence that Josephus and his allies ever sat in a circle and killed every third person. The original event can be found in Josephus's book *The Jewish War*.[56] The Hegesippus that Ball cites was a fourth-century translation of Josephus. Some anonymous translator got the author's name wrong.[57]

The story, as related by Josephus,[58] is as follows:

Josephus was a general for the Jews in a war against the Romans, who were led by Vespacian. Josephus and his troops were surrounded in the city of Jotapata. Eventually the city fell, but Vespacian ordered his troops to capture Josephus (rather than kill him).

Before the city fell, Josephus and 40 others managed to hide in a cave. On the third day after the city fell, the Romans found out about the cave. Vespacian sent two men to offer Josephus safe passage if he would surrender. At first he refused, but eventually started to change his mind. His companions were not pleased when they saw he was starting to consider surrender. They told him he should kill himself instead of surrender, or, if he was not brave enough, they would take the matter into their own hands. Josephus then launched into an articulate speech about why suicide is morally wrong. His speech did not convince his allies. In fact, they were on the verge of killing him and then killing themselves. The story concludes[59] (with Josephus speaking in the third person):

> But in this predicament, his resourcefulness did not forsake him. Trusting in God's protection, he hazarded his life on one last throw, saying: "As we are resolved to die, come, let us draw lots and decide the order in which we are to kill each other in turn. Whoever draws the first lot shall die by the hand of him who comes next; luck will thus take its course down the whole line. In this way we shall be spared taking our lives in our own hands. For it would be unfair when the rest were gone if one man should change his mind and escape." This proposal inspired assurance; his advice was taken, and he drew lots with the rest. Each man in turn offered his throat for the next man to cut, in the belief that his general would immediately share his fate; they thought death together with Josephus sweeter than life. He, however—should we say by fortune or by divine providence—was left with one other man; and, anxious neither to be condemned by the lot, nor, if he were left as the last, to stain his hand with the blood of a fellow countryman, he persuaded this man also, under a pact, to remain alive.

I do not know when or where the mathematical version (with the circle and every third person) originated.

Other Versions of the Problem

A version of the problem, existing in published form at least as early as the 1500s or early 1600s, involves a ship with 15 Turks and 15 Christians. A storm has arisen and in an attempt to save some passengers, everyone agrees that half the passengers need to be thrown into the sea. The passengers are placed into a circle, and every ninth man is tossed overboard. The problem is to find an arrangement so that all members of your favorite religio-ethnic group survive and all members of the other group become fish food.

An Asian variant involves a man with two wives, each of whom is the mother of 15 children. The first wife has died and the man is getting old. The surviving wife convinces him that the estate is too small to divide among 30 children. In fact, it should go to just one child. The wife convinces him to arrange the children in a circle and eliminate (but not kill for a change!) every tenth child. The final child will inherit everything.

---

[56] A good translation into English is [56].

[57] Dr. Laurence Creider provided extensive help researching Hegesippus.

[58] With only one other surviving witness to verify the details.

[59] From Chapter 8 of *The Jewish War* [56].

The second wife arranges the children and the process begins. In an interesting twist, the first 14 to be eliminated are all children of the first wife. The father becomes alarmed, especially after he notices that the only remaining child from the first wife will be eliminated next. He suggests that they should start over, beginning with the sole remaining child of the first wife and travel around the circle in the opposite direction. The second wife cannot object without giving herself away, but she figures that the odds are 15 to 1 in her favor. The end result is that the child of the first wife is the final child, defeating the second wife's evil strategy. Your task, of course is to place the children around the circle to match the story.

Solving the Josephus Problem

The original problem (with every third person being eliminated) is a bit more involved than is appropriate for the level of this text (but not by much). The interested reader should consult [43] for the solution.

Instead, imagine that n people are placed in a circle, and every second person is eliminated.[60] The value we want is the position (start counting at 1) of the final person. Call this position j_n.

A good place to begin is with a few small examples. Table 7.11 shows the order in which people are eliminated and the value of j_n, for several small n. You should draw a few of the circles and verify the numbers.

TABLE 7.11 Order of Elimination with n People

n	Elimination Sequence	j_n
1	—	1
2	2	1
3	2 1	3
4	2 4 3	1
5	2 4 1 5	3
6	2 4 6 3 1	5
7	2 4 6 1 5 3	7
8	2 4 6 8 3 7 5	1
9	2 4 6 8 1 5 9 7	3

What can we observe from these examples? One trend (at least so far) is that j_n seems to be odd. Another trend seems to be that even numbered positions are eliminated first, in order. These two observations are actually related: Since all even positions will be eliminated first (according to the "every second person" rule), the final position will always be an odd number.

It takes just a little bit of creativity (or else a few years worth of mathematical maturity and experience) to make the following observations:

> Since approximately half the people (those in even-numbered positions) are eliminated immediately, it may be profitable to write n in a form that involves the number 2. If n is even, we use up exactly half the people in this first phase, while if n is odd, there will still be one extra person left before wrapping back to the beginning.[61] Because even and odd are apparently significant characteristics, it may be useful to write n as either $n = 2k$ for even n, or $n = 2k + 1$ for odd n.

[60] The solution presented here is also from [43].

[61] For example, when $n = 7$, positions 2, 4, and 6 are eliminated in phase 1, but 7 still remains before getting back to 1.

Consider the case where $n = 2k$ is even. After phase 1, only the odd numbered positions are left. There will be k such numbers and the next available position will be position 1. The problem has effectively been reduced to a problem of size k. There is one pesky detail: a problem of size k has the positions numbered as $1, 2, 3, \ldots, k$, but a problem of size $n = 2k$ has the remaining positions numbered $1, 3, 5, \ldots, 2k - 1$. It is easy to see how the two sequences relate: the old sequence can be grouped in pairs (odd, even). We keep only the first member of each pair. Look at the following table as i goes from 1 to k:[62]

Original sequence	1	2	3	4	5	$\cdots$	$(2i-1)$	$(2i)$	$\cdots$	$(2k-1)$	$(2k)$
Relabeled sequence	1	–	2	–	3	$\cdots$	i	–	$\cdots$	k	–

Suppose we already knew the final position number, j_k, for a circle of size k. Then a circle of size $n = 2k$ would end up in the same place, assuming we could suitably relabel the original odd positions after phase 1 eliminates the even positions. It should be clear from the table that relabeled position i corresponds to original position $2i - 1$.

This leads to a clever strategy: Start with a circle of size $n = 2k$. After the even positions have been eliminated, relabel the positions as $1, 2, 3, \ldots, k$. The final position in this relabeled circle will be j_k. This corresponds to position $2j_k - 1$ in the original circle.

We now have the recursive relations $j_1 = 1$, and $j_{2k} = 2j_k - 1$. What we need is a similar recursive reduction when n is odd.

If $n = 2k + 1$ is odd, phase 1 leaves only the odd positions. There are now $k + 1$ positions, so the reduced problem looks like a circle of size $k+1 = \frac{n+1}{2}$. The relabeling is also a bit more complicated, since the next person is not in the original position 1, but in original position $2k + 1$.[63]

Original	1	2	3	4	5	$\cdots$	$(2i-2)$	$(2i-1)$	$(2i)$	$\cdots$	$(2k-1)$	$(2k)$	$(2k+1)$
Relabeled	2	–	3	–	4	$\cdots$	–	$(i+1)$	–	$\cdots$	$(k+1)$	–	1

The correspondence is a bit messy. Here is a revised idea: Do not end phase 1 until the original position 1 is eliminated (that person will always be the next to go). If we relabel after this point, the table becomes

Original	1	2	3	4	5	$\cdots$	$(2i)$	$(2i+1)$	$(2i+2)$	$\cdots$	$(2k-1)$	$(2k)$	$(2k+1)$
Relabeled	–	–	1	–	2	$\cdots$	–	i	–	$\cdots$	$(k-1)$	–	k

That looks much better! In fact, after the revised phase 1, there will be a circle of size k. The final person will be in relabeled position j_k, corresponding to original position $2j_k + 1$. This leads to the recursive relation $j_{2k+1} = 2j_k + 1$.

The recursive reduction formulas are

- $j_1 = 1$
- $j_{2n} = 2j_n - 1$
- $j_{2n+1} = 2j_n + 1$

Can these recurrence relations be turned into a closed-form formula? If so, by what technique? Notice that they are not homogeneous, so the linear homogeneous recurrence relation with constant coefficients technique is out. If you try to do some back substitution, the need for two distinct relations (even vs odd) will quickly lead to something that is messy and quite awkward. You need to keep track of how many 2s are in the original n to keep this sorted out. We can try this a bit just to see what happens.

[62]Look at the case $n = 8$ if you want something more concrete.

[63]Look at the case $n = 9$ for a concrete example.

Let $n = 2^m q$, where q is odd and $m \geq 1$.

$$\begin{aligned} j_{(2^m q)} &= 2 j_{(2^{m-1} q)} - 1 \quad \text{substitute} \\ &= 2\left(2 j_{(2^{m-2} q)} - 1\right) - 1 \quad \text{substitute} \\ &= 2^2 j_{(2^{m-2} q)} - (2 + 1) \quad \text{simplify} \\ &\vdots \\ &= 2^m j_q - \sum_{i=0}^{m-1} 2^i \\ &= 2^m j_q + 1 - 2^m \end{aligned}$$

At this point, we know that q is odd, so there is an r with $q = 2r + 1$. Then $j_q = 2 j_r + 1$. But what do we do about r? Is it even or odd?

We have reached an apparent dead end, but the experience may still provide some insight later on.

So linear homogeneous recurrence relations with constant coefficients techniques do not work, back substitution seems to fail, and after some messing around, it seems that generating functions may also be difficult to apply. What can be done? One observation is that $j_{2k+1} - j_{2k} = 2$ in all cases. That is, for each odd n, subtracting j_{n-1} from j_n always equals 2. There is no similar constant difference if n is even and the same subtraction is done.

Perhaps a larger table of small cases will help, especially now that the recurrence relations help to reduce the work. For example, $j_{10} = 2 j_5 - 1 = 2 \cdot 3 - 1 = 5$. (Note the duplication for $n = 16$ in the tables.)

n	1	2	3	4	5	6	7	8	9	10	11	12	13	14	15	16
j_n	1	1	3	1	3	5	7	1	3	5	7	9	11	13	15	1

n	16	17	18	19	20	21	22	23	24	25	26	27	28	29	30	31	32
j_n	1	3	5	7	9	11	13	15	17	19	21	23	25	27	29	31	1

Notice the pattern in the second row. In particular, notice that the pattern changes whenever $n = 2^m$. This should not be a big surprise if you consider what was learned in the attempt to use back substitution. The pattern seems to start at 1 when $n = 2^m$, then build by 2 until $n = 2^{m+1}$, where it returns to 1. A bit of thought and experimentation will lead to a simple formula, once the proper characterization of n is found.

The useful way to write n is $n = 2^m + i$, where $0 \leq i < 2^m$. For example,

n	4	5	6	7	8
$2^m + i$	$2^2 + 0$	$2^2 + 1$	$2^2 + 2$	$2^2 + 3$	$2^3 + 0$

THEOREM 7.26 ***The Modified Josephus Problem***

Suppose n people are seated around a circle, numbered from 1 to n. Start counting with the first person and eliminate every second person. Continue until only one person is left. Denote the final position by j_n. Let $n = 2^m + i$ with $0 \leq i < 2^m$. Then

$$j_n = j_{(2^m + i)} = 2i + 1$$

Proof: The theorem can be proved using complete induction.

Base Step $n = 1 = 2^0 + 0$

Since $i = 0$, the theorem predicts $j_1 = 2 \cdot 0 + 1 = 1$, which is correct.

Inductive Step

Assume that the theorem is true for all positive integers less than n.

Suppose first that n is even. Then $n = 2^m + i = 2^m + 2k = 2(2^{m-1} + k)$, for $0 \leq k < 2^{m-1}$. The recursive reduction formulas (page 405) imply that

$$j_n = j_{(2(2^{m-1}+k))} = 2j_{(2^{m-1}+k)} - 1$$

By the inductive hypothesis, $j_{(2^{m-1}+k)} = 2k + 1$. Thus

$$j_n = 2j_{(2^{m-1}+k)} - 1 = 2\,(2k+1) - 1 = 2(2k) + 1 = 2i + 1$$

Now suppose that n is odd. Then $n = 2^m + i = 2^m + 2k + 1 = 2(2^{m-1} + k) + 1$, where $0 \leq k \leq 2^{m-1} - 1$. Using the other recursive reduction formula, and then the inductive hypothesis

$$j_n = j_{(2(2^{m-1}+k)+1)} = 2j_{(2^{m-1}+k)} + 1 = 2\,(2k+1) + 1 = 2i + 1$$

The induction is finished: The theorem is true for $n = 1$, and whenever the theorem is true for all positive integers less than n, it is also true for n. □

7.5.1 Exercises

The exercises marked with ★ have detailed solutions in Appendix H.

1. Consider the problem variant with 30 people on a boat. Designate the two groups the Gs and the Bs (the "good guys" and the "bad guys"). Where should the Gs and Bs be placed around the circle so that all 15 Gs are left after 15 people are thrown overboard?
2. Consider the man with 30 children. How were the children placed around the circle (and which direction was used initially)? Designate the children by F and S ("child of First wife" and "child of Second wife", respectively).
3. You are probably familiar with the story of Scheherazade from the story *1001 Arabian Nights*. In the story, the king, displeased with women in general, decided to spend each night with a new bride and then have her executed the next day (thus ensuring her faithfulness). Many women died. Eventually Scheherazade volunteered to marry the king. On her wedding night she told the king such an interesting story that he delayed her execution by one day. The next night she told another story that was so good, the king delayed her execution again. This continued for 1001 nights, until finally the king realized he should stay married to her and keep her as queen. (He was a slow learner.)

   Suppose instead that the king has 1001 concubines. Each will draw a number and he will select every other concubine in circular order. The last concubine left alive will become queen. What number should Scheherazade pick so that she survives and becomes queen?
4. Let p_n represent the position that Josephus's partner should be in so that he is the second-to-last to be selected for execution (in a circle with n people, and every second person executed).
   (a) ★ Find or guess a formula for p_n.
   (b) Prove that your formula is correct. [*Hint*: Do the same recurrence relations hold?]

7.6 Quick Check Solutions

Quick Check 7.1

1. The complete diagram contains two invocations that lead to the sum 4. Notice how the W (the set on the right at each point of the diagram) gets smaller at each step.

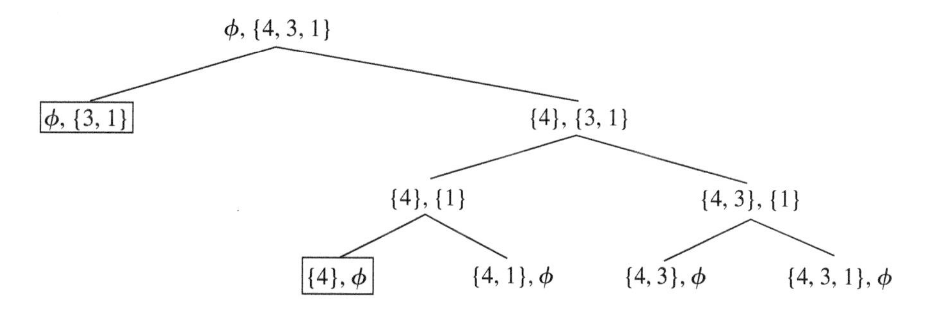

Quick Check 7.2

1. A_2 is created from A_1, B_1, and D_1 using the schematic

$$A_n : A_{n-1} \searrow B_{n-1} \rightarrow\rightarrow D_{n-1} \nearrow A_{n-1}$$

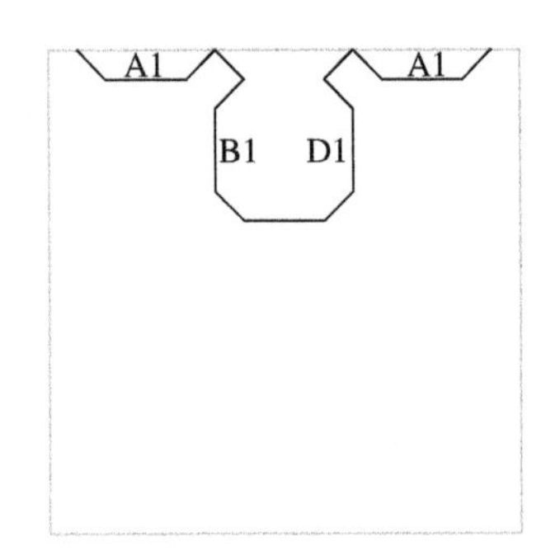

Quick Check 7.3

1. B_3 is created from A_2, B_2, and C_2 using the schematic

$$B_n : B_{n-1} \swarrow C_{n-1} \downarrow\downarrow A_{n-1} \searrow B_{n-1}$$

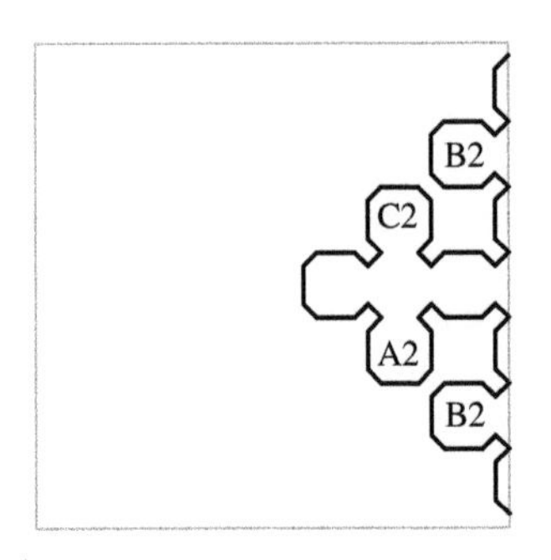

2. A_3, B_3, C_3, and D_3 are

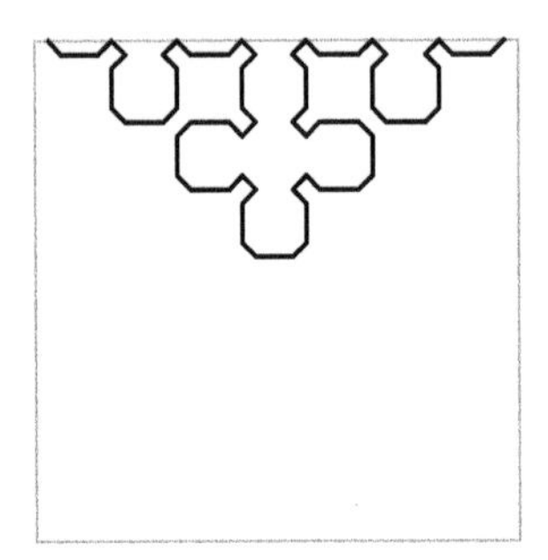

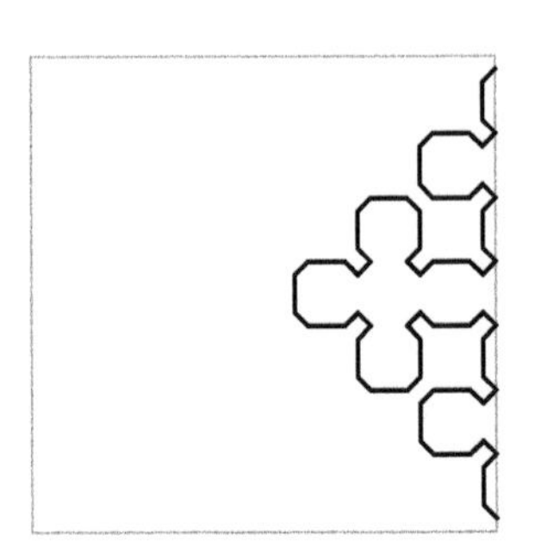

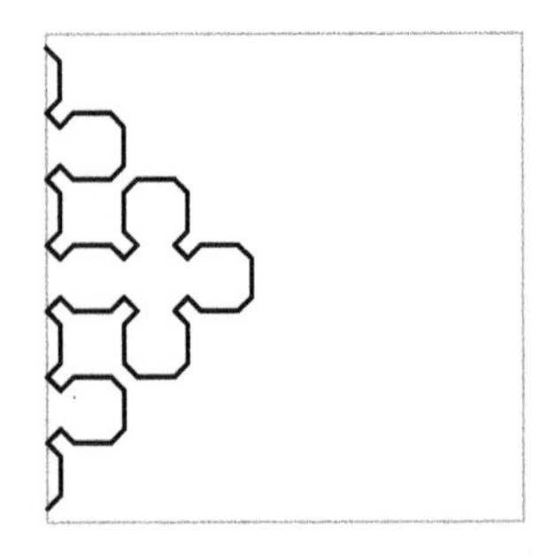

So S_3 is

S_n : $A_n \searrow B_n \swarrow C_n \nwarrow D_n \nearrow$

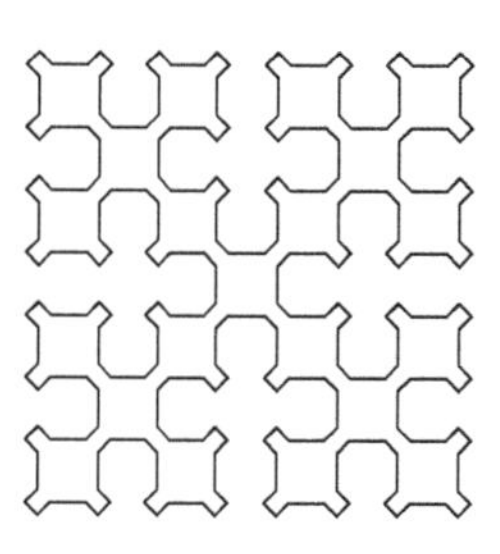

Quick Check 7.4

The graph of the function is

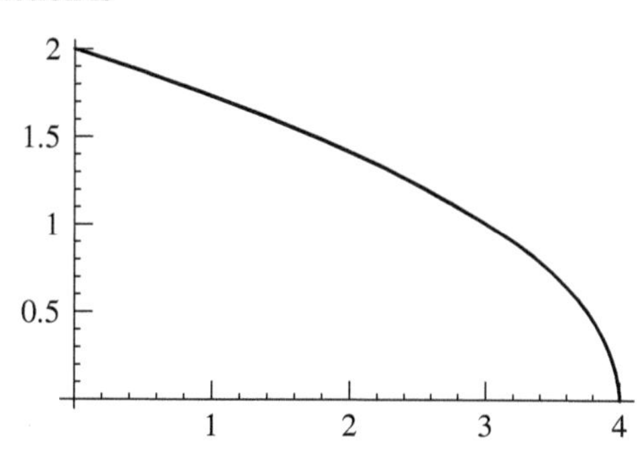

It may help to program Simpson's rule into a graphing calculator or a symbolic system such as *Mathematica* or *Maple*, or else using a traditional programing language such as C or Java.

Page 1: [0, 4] $\tau = 0.01$, $S(f, 0, 4) \simeq 5.10457$, $S(f, 0, 2) \simeq 3.44747$, and $S(f, 2, 4) \simeq 1.80474$, so $|\text{whole} - \text{left} - \text{right}| = 0.147641 > 10 \cdot 0.01$.

Page 2: [0, 2] (Page 3: [2, 4] is pending.)

$\tau = 0.005$, $S(f, 0, 2) \simeq 3.44747$, $S(f, 0, 1) \simeq 1.86923$, and $S(f, 1, 2) \simeq 1.57847$, so $|\text{whole} - \text{left} - \text{right}| = 0.000225281 < 10 \cdot 0.005$. The algorithm returns $1.86923 + 1.57847 = 3.4477$ as the value of $\int_0^2 \sqrt{4 - x}\, dx$.

Page 3: [2, 4] $\tau = 0.005$, $S(f, 2, 4) \simeq 1.80474$, $S(f, 2, 3) \simeq 1.21887$, and $S(f, 3, 4) \simeq 0.638071$, so $|\text{whole} - \text{left} - \text{right}| = 0.0521988 > 10 \cdot 0.005$.

Page 4: [2, 3] (Page 5: [3, 4] is pending.)

$\tau = 0.0025$, $S(f, 2, 3) \simeq 1.21887$, $S(f, 2, 2.5) \simeq 0.660872$, and $S(f, 2.5, 3) \simeq 0.558073$, so $|\text{whole} - \text{left} - \text{right}| = 0.0000796489 < 10 \cdot 0.0025$. The algorithm returns $0.660872 + 0.558073 = 1.218945$ as the value of $\int_2^3 \sqrt{4 - x}\, dx$.

Page 5: [3, 4] $\tau = 0.0025$, $S(f, 3, 4) \simeq 0.638071$, $S(f, 3, 3.5) \simeq 0.430934$, and $S(f, 3.5, 4) \simeq 0.225592$, so $|\text{whole} - \text{left} - \text{right}| = 0.0184551 < 10 \cdot .0025$. The algorithm returns $0.430934 + 0.225592 = 0.656526$ as the value of $\int_3^4 \sqrt{4 - x}\, dx$.

Back to Page 3 Page 3 returns $\int_2^3 \sqrt{4 - x}\, dx + \int_3^4 \sqrt{4 - x}\, dx = 1.218945 + .656526 = 1.87547$.

Back to Page 1 The final result is $3.4477 + 1.87547 = 5.32317$, using the intervals [0, 2], [2, 3], and [3, 4].

The actual value is $5\frac{1}{3}$, so the error is approximately 0.0101633. Even using the more conservative safety factor of 10 (rather than 15) the algorithm did not quite deliver as promised. Changing the safety constant to 5 produces a result that is in error by approximately 0.00129. It uses the intervals [0, 2], [2, 3], [3, 3.5], [3.5, 3.75], and [3.75, 4].

Quick Check 7.5

1. The back substitution is simple. In this example, however, it is better to simplify at each step.

$$\begin{aligned}
b_n &= 1 - b_{n-1} \\
&= 1 - (\mathbf{1 - b_{n-2}}) \quad \text{substitute} \\
&= b_{n-2} \quad \text{simplify} \\
&= \mathbf{1 - b_{n-3}} \quad \text{substitute (no simplification needed)} \\
&= 1 - (\mathbf{1 - b_{n-4}}) \quad \text{substitute} \\
&= b_{n-4} \quad \text{simplify} \\
&\vdots \qquad\qquad \vdots \\
&= \begin{cases} 1 - b_{n-k} & \text{if } k \text{ is odd} \\ b_{n-k} & \text{if } k \text{ is even} \end{cases} \\
&\vdots \qquad\qquad \vdots \\
&= \begin{cases} 1 - b_0 & \text{if } n \text{ is odd} \\ b_0 & \text{if } n \text{ is even} \end{cases} \\
&= \begin{cases} 1 & \text{if } n \text{ is odd, } n \geq 1 \\ 0 & \text{if } n \text{ is even, } n \geq 0 \end{cases}
\end{aligned}$$

Quick Check 7.6

1. (a) Homogeneous and linear; does *not* have constant coefficients [the coefficient $\cos(n)$ depends on n].
   (b) Linear homogeneous recurrence relation with constant coefficients
   (c) This *does* have constant coefficients but is *not* homogeneous (+9) and is *not* linear ($(a_{n-1})^2$).
   (d) This *is* homogeneous and *does* have constant coefficients but is *not* linear ($2^{b_{n-1}}$).
2. (a) $x^2 - 5x + 9 = 0$
   (b) $x^3 - x^2 + 7 = 0$ Notice that the missing term b_{n-2} results in $0x$ in the characteristic equation.
   (c) $x^2 - 3x - 4 = 0$ Note that the subscripts were not in the standard order. They needed to be converted to $a_n = 3a_{n-1} + 4a_{n-2}$ before forming the characteristic equation.

Quick Check 7.7

1. The characteristic equation is $x^3 - 4x^2 - x + 4 = 0$. This can be factored. One approach is to group terms: $(x^3 - x) + (-4x^2 + 4) = 0$. This simplifies to $x(x^2 - 1) - 4(x^2 - 1) = 0$. Combining common terms leads to $(x - 4)(x^2 - 1) = 0$, which factors as $(x - 4)(x - 1)(x + 1) = 0$.

   Thus, the roots are distinct: $r_1 = 4$, $r_2 = 1$, and $r_3 = -1$.

   The general solution is

$$a_n = \theta_1 4^n + \theta_2 1^n + \theta_3 (-1)^n = \theta_1 4^n + \theta_2 + \theta_3 (-1)^n$$

The system of linear equations that determines the values of the θs is

$$\begin{aligned}
\theta_1 + \theta_2 + \theta_3 &= 1 \\
4\theta_1 + \theta_2 - \theta_3 &= 2 \\
16\theta_1 + \theta_2 + \theta_3 &= 3
\end{aligned}$$

Rather than begin a messy set of substitutions, and without assuming you have seen Gaussian elimination, it is still possible to solve this system without much effort. As an initial step, I will add the second equation to the first and third equations (thus eliminating θ_3). This produces the reduced system

$$5\theta_1 + 2\theta_2 = 3$$
$$20\theta_1 + 2\theta_2 = 5$$

Subtracting the first equation from the second (in this reduced system) eliminates θ_2: $15\theta_1 = 2$. Thus $\theta_1 = \frac{2}{15}$.

Substituting this value into the equation $5\theta_1 + 2\theta_2 = 3$ and solving for θ_2 produces $\theta_2 = \frac{7}{6}$.

Finally, substituting the values for θ_1 and θ_2 into $\theta_1 + \theta_2 + \theta_3 = 1$ leads to $\theta_3 = \frac{-3}{10}$.

The solution is generated by

$$a_n = \frac{2}{15}4^n + \frac{7}{6} - \frac{3}{10}(-1)^n \quad \text{for } n \geq 0$$

This formula should certainly be tested to see that it generates the same values as the first five or six iterations of the original recurrence relation.

2. It is tempting to claim that the characteristic equation is $x^2 - 2x - 5 = 0$. However, this recurrence relation is *not* homogeneous. The solution technique under current consideration is *not valid* for this recurrence relation. A back substitution is appropriate.

$$\begin{aligned}
a_n &= 2a_{n-1} + 5 \\
&= 2(\mathbf{2a_{n-2} + 5}) + 5 \quad \text{substitute} \\
&= 2^2 a_{n-2} + (2 \cdot 5 + 5) \quad \text{simplify} \\
&= 2^2(\mathbf{2a_{n-3} + 5}) + (2 \cdot 5 + 5) \quad \text{substitute} \\
&= 2^3 a_{n-3} + (2^2 \cdot 5 + 2 \cdot 5 + 5) \quad \text{simplify} \\
&\ \ \vdots \qquad\qquad \vdots \\
&= 2^k a_{n-k} + 5 \cdot \sum_{j=0}^{k-1} 2^j \\
&\ \ \vdots \qquad\qquad \vdots \\
&= 2^n a_0 + 5 \cdot \sum_{j=0}^{n-1} 2^j \\
&= 3 \cdot 2^n + 5(2^n - 1) \\
&= 2^{n+3} - 5 \quad \text{for } n \geq 0
\end{aligned}$$

Quick Check 7.8

1. The potential rational roots are

$$\pm 1, \pm 5, \pm\frac{1}{2}, \pm\frac{5}{2}, \pm\frac{1}{3}, \pm\frac{5}{3}, \pm\frac{1}{6}, \pm\frac{5}{6}$$

2. The actual roots are $-\frac{5}{2}$ and $\frac{1}{3}$. The linear polynomials $x - \frac{1}{3}$ and $x + \frac{5}{2}$ have the same roots, but their product does not yield the original polynomial. The required factorization is

$$6x^2 + 13x - 5 = (3x - 1)(2x + 5)$$

Quick Check 7.9

1. (a) $p(x) = (x-1)^3(x-3)$, the zero $r = 1$ has multiplicity $\nu = 3$.

(b) $p'(x) = 4x^3 - 18x^2 + 24x - 10 = 2(x-1)^2(2x-5)$ so $r = 1$ has multiplicity $\nu = 2$. An alternative approach is to use the product rule:
$p'(x) = 3(x-1)^2(x-3) + (x-1)^3 = (x-1)^2((3x-9) + (x-1)) = (x-1)^2(4x-10)$.

(c) $p''(x) = 12x^2 - 36x + 24 = 12(x^2 - 3x + 2) = 12(x-1)(x-2)$. $r = 1$ has multiplicity $\nu = 1$.

(d) $p^{(3)}(x) = 24x - 36 = 12(2x-3)$, so $r = 1$ is no longer a zero.

(e) It appears that the multiplicity of the zero decreases by 1 with each new derivative.

Quick Check 7.10

1. The characteristic equation is $x^2 + 6x + 9 = (x+3)^2 = 0$, having root $r = -3$ of multiplicity $\nu = 2$. The general solution has the form $a_n = (\alpha_0 + \alpha_1 n)(-3)^n$. The values of the αs are determined by

$$\begin{aligned} \alpha_0 &= -2 \\ -3\alpha_0 - 3\alpha_1 &= -6 \end{aligned}$$

The solution is $a_n = (-2 + 4n)(-3)^n$ for $n \geq 0$.

Quick Check 7.11

1. The first step is to split the list in half and recursively sort each half. Then the sorted sublists are merged.

Sort {h, d, a, c} Split the list into two lists of length 2: {h, d} and {a, c}.

Sort {h, d} Split {h, d} into two lists of size 1. The recursive invocations just return the single-item lists.
Now merge the lists, forming {d, h}.

Sort {a, c} Split {a, c} into two lists of size 1. The recursive invocations just return the single-item lists.
Now merge the lists, forming {a, c}.

Merge Now merge the two sorted lists, {d, h} and {a, c}, producing {a, c, d, h}.

Sort {g, f, b, e} Split the list into two lists of length 2: {g, f} and {b, e}.

Sort {g, f} Split {g, f} into two lists of size 1. The recursive invocations just return the single-item lists.
Now merge the lists, forming {f, g}.

Sort {b, e} Split {b, e} into two lists of size 1. The recursive invocations just return the single-item lists.
Now merge the lists, forming {b, e}.

Merge Now merge the two sorted lists, {f, g} and {b, e}, producing {b, e, f, g}.

Merge Now merge the sorted sublists: {a, c, d, h} and {b, e, f, g}. The merged list is {a, b, c, d, e, f, g, h}.

Quick Check 7.12

1. Using Theorem 7.16 with $a > 1$,

$$g(n) = n^{\log_3(9)} = n^2$$

Quick Check 7.13

1. For this algorithm, $a = b = 2$, $c = v = 1$, and $d = 0$. Since $a = b^v$, and $\log_2(2) = 1$, `mergeSort` is in $\Theta(n \log_2(n))$.

2. For this recurrence relation, $a = b = 4$, $c = 2$, $d = 1$, and $v = \frac{1}{2}$. Since $a = 4 > 2 = \sqrt{4} = b^v$, and since $\log_4(4) = 1$, $f \in \Theta(n)$.

Quick Check 7.14

1. The expansion for $\frac{1}{1+2z}$ is

$$\frac{1}{1+2z} = (1+2z)^{-1} = \sum_{k=0}^{\infty} \binom{-1}{k} (2z)^k = \sum_{k=0}^{\infty} (-1)^k \binom{k}{k} 2^k z^k = \sum_{k=0}^{\infty} (-2)^k z^k$$

For $\frac{1}{1-3z}$, the expansion is

$$\frac{1}{1-3z} = (1+(-3z))^{-1} = \sum_{k=0}^{\infty} \binom{-1}{k} (-3z)^k = \sum_{k=0}^{\infty} (-1)^k \binom{k}{k} (-3)^k z^k = \sum_{k=0}^{\infty} 3^k z^k$$

2. Make the identification $f(z) = \frac{1}{1+2z}$ and $g(z) = \frac{1}{1-3z}$. Then $f_j = (-2)^j$ and $g_{k-j} = 3^{k-j}$ (Table 7.10 row 4). Therefore, the coefficient of z^k in the product is

$$\begin{aligned}
\sum_{j=0}^{k} (-2)^j 3^{k-j} &= 3^k \sum_{j=0}^{k} \left(-\frac{2}{3}\right)^j \\
&= 3^k \cdot \frac{\left(-\frac{2}{3}\right)^{k+1} - 1}{\left(-\frac{2}{3}\right) - 1} \\
&= \frac{3^{k+1}}{5} \cdot \left(1 - \left(-\frac{2}{3}\right)^{k+1}\right) \\
&= \frac{1}{5}\left(3^{k+1} - (-2)^{k+1}\right)
\end{aligned}$$

Therefore,

$$\frac{1}{1+2z} \cdot \frac{1}{1-3z} = \sum_{k=0}^{\infty} \frac{1}{5}\left(3^{k+1} - (-2)^{k+1}\right) z^k$$

3. Multiply both sides of the recurrence relation by z^n and sum from $n = 2$ (the smallest value for which all the subscripts are valid).

$$\sum_{n=2}^{\infty} a_n z^n = \sum_{n=2}^{\infty} a_{n-1} z^n + 6 \sum_{n=2}^{\infty} a_{n-2} z^n$$

If the generating function for the sequence is $A(z) = \sum_{n=0}^{\infty} a_n z^n$, then the previous expression can be rewritten as

$$A(z) - a_0 - a_1 z = (zA(z) - a_0 z) + 6z^2 A(z)$$

This is true because

$$\sum_{n=2}^{\infty} a_{n-1} z^n = z \sum_{n=2}^{\infty} a_{n-1} z^{n-1} = z \sum_{j=1}^{\infty} a_j z^j = z\left(A(z) - a_0\right)$$

and

$$\sum_{n=2}^{\infty} a_{n-2} z^n = z^2 \sum_{n=2}^{\infty} a_{n-2} z^{n-2} = z^2 \sum_{j=0}^{\infty} a_j z^j = z^2 A(z)$$

Rearranging the terms produces

$$A(z)(1 - z - 6z^2) = a_0 + (a_1 - a_0)z = (1 + z)$$

so [using part (2)]

$$\begin{aligned}
A(z) &= (1+z)\left(\frac{1}{1-z-6z^2}\right) \\
&= (1+z)\left(\frac{1}{(1+2z)(1-3z)}\right) \\
&= (1+z)\left(\sum_{k=0}^{\infty}\frac{1}{5}\left(3^{k+1}-(-2)^{k+1}\right)z^k\right)
\end{aligned}$$

The coefficient of z^n in the preceding product is

$$\begin{aligned}
a_n &= \frac{1}{5}\left(3^{n+1}-(-2)^{n+1}\right) + \frac{1}{5}\left(3^n-(-2)^n\right) \\
&= \frac{1}{5}\left(3^n(3+1)-(-2)^n(-2+1)\right) \\
&= \frac{4}{5}3^n + \frac{1}{5}(-2)^n
\end{aligned}$$

7.7 CHAPTER REVIEW

7.7.1 Summary

This chapter introduces recursion. Recursion is an ingenious problem solving strategy that expresses the solution to a problem in terms of smaller versions of the same problem.

The chapter presents recursion in an algorithmic context (Section 7.1) and also in a functional context (Sections 7.2 – 7.4).

In the algorithmic context, the key ideas to review are the suggested five steps for creating recursive algorithms and the material that indicates when recursion may be inappropriate.

Creating a Recursive Algorithm

Step 1 Identify how to reduce the problem into smaller versions of itself.

Step 2 Identify one or more instances of the problem that can be directly solved.

Step 3 Determine how the solution can be obtained by combining the solutions to one or more smaller versions.

Step 4 Verify that the invocations in step 3 are within bounds.

Step 5 Assemble the algorithm.

When Recursion May Be Inappropriate

Tail-end recursion An algorithm uses tail-end recursion if the only recursive invocation it makes occurs at the last line of the algorithm.

Redundant recursion Redundant recursion occurs when an algorithm directly or indirectly invokes multiple instances of the same smaller version.

Several examples are presented that demonstrate the power of recursion. The algorithms for drawing Persian rugs and Sierpinski curves and the adaptive quadrature algorithm all use multiple recursions.

Section 7.2 introduces recurrence relations. Recurrence relations express the notion of recursion using functions and/or sequences. The chapter presents three general methods for solving recurrence relations: back substitution (Section 7.2.1), a two-phased process for linear homogeneous recurrence relations with constant coefficients (Section 7.2.2), and generating functions (Section 7.4).

Back substitution is the simplest of the three methods but can become algebraically unmanageable if the recurrence relation is too complex.

The second method requires the recurrence relation to be in a special form: linear, homogeneous, with constant coefficients. The solution for this special form of recurrence relation can be expressed in terms of the roots of the characteristic equation. The solution is initially presented for the case where the characteristic equation has distinct roots and then generalized for the case where the characteristic equation has repeated roots. This method has a practical limitation: It requires the exact roots of a polynomial equation. It is possible to extend this technique so that the homogeneous requirement can be dropped. That extension is not presented in this book.

The use of generating function is the most powerful of the three methods. That method is also the most complex of the three. The presentation in this chapter is only an introduction to generating functions, but it contains sufficient material to be practically useful.

Back substitution is used in Section 7.3 to prove several versions of the master theorem. This theorem describes appropriate big-Θ reference functions for a subclass of recursive algorithms. The algorithms for which the master theorem is appropriate are those that divide the original problem into two or more subproblems, each having the same size.

The chapter concludes with an application of recurrence relations to solve a modified version of an ancient math puzzle called the Josephus problem.

The solution technique for linear homogeneous recurrence relations with constant coefficients is summarized in the following five steps.

The General Procedure for Solving Linear Homogeneous Recurrence Relations with Constant Coefficients

Step 1 Form the characteristic equation.

Step 2 Find the roots of the characteristic equation, $r_1, r_2, \ldots, r_j$ having respective multiplicities $\nu_1, \nu_2, \ldots, \nu_j$, with $\nu_1 + \nu_2 + \cdots + \nu_j = k$.

Step 3 Express the general solution as a sum of j terms in the form

$$\left(\delta_0 + \delta_1 n + \delta n^2 + \cdots + \delta_{\nu_i - 1} n^{\nu_i - 1}\right) r_i^n$$

Step 4 Use the result of step 3 to form a system of linear equations to determine the unknown coefficients.

Step 5 Solve the linear system and substitute the solution values for the unknown coefficients into the general solution from step 3.

7.7.2 Notation

Notation	Page	Brief Description
$S_n : A_n \searrow B_n \swarrow C_n \nwarrow D_n \nearrow$	343	The schematic diagram for the nth Sierpinski curve
f_n	351	The nth Fibonacci number
$p^{(j)}(x)$	367	The jth derivative of the function, $p(x)$
Φ	364 (footnote 36)	The golden ratio
$\sum_{k=0}^{\infty} a_k z^k$	392	A generating function
$\binom{u}{r}$	397	A generalized binomial coefficient
j_n	404	The position of the surviving person in a Josephus problem where every second person is eliminated

7.7.3 Generating Function Table

Some Useful Genrating Functions

$G(z)$	Summation Notation	Expanded Notation
$\frac{1}{1-z}$	$\sum_{k=0}^{\infty} z^k$	$1 + z + z^2 + z^3 + \cdots$
$\frac{1}{1+z}$	$\sum_{k=0}^{\infty} (-1)^k z^k$	$1 - z + z^2 - z^3 + \cdots$
$\frac{1}{1-z^m}$	$\sum_{k=0}^{\infty} z^{mk}$	$1 + z^m + z^{2m} + z^{3m} + \cdots$
$\frac{1}{1-cz}$	$\sum_{k=0}^{\infty} c^k z^k$	$1 + cz + c^2z^2 + c^3z^3 + \cdots$
$\frac{1}{(1-z)^m}$	$\sum_{k=0}^{\infty} \binom{m+k-1}{k} z^k$	$1 + mz + \binom{m+1}{2} z^2 + \binom{m+2}{3} z^3 + \cdots$
$\frac{z}{(1-z)^2}$	$\sum_{k=0}^{\infty} kz^k$	$0 + z + 2z^2 + 3z^3 + \cdots$
$(1+z)^c$	$\sum_{k=0}^{\infty} \binom{c}{k} z^k$	$1 + cz + \binom{c}{2} z^2 + \binom{c}{3} z^3 + \cdots$
e^z	$\sum_{k=0}^{\infty} \frac{1}{k!} z^k$	$1 + z + \frac{z^2}{2!} + \frac{z^3}{3!} + \cdots$

7.7.4 Additional Review Material

Go to http://www.mathcs.bethel.edu/~gossett/DiscreteMathWithProof/review.xhtml for additional review material, including a sample chapter exam, with solutions.

CHAPTER

8

Combinatorics

Suffice it to say that combinatorics, a subject that exploded in the twentieth century, has roots running back into the mathematical past. And at least a few of these can be traced to Leonard Euler, who seems never to have met a math problem he didn't like.

William Dunham – Euler: The Master of Us All

Combinatorics is a branch of mathematics that examines many seemingly unrelated ideas. These ideas are collected for several reasons. One reason is that the ideas are much more tightly connected than a casual look would indicate. It is often the case that one combinatorial object will be used to create another. Examples of this will be seen when Latin squares, finite projective planes, balanced incomplete block designs, and error-correcting codes are discussed in this chapter.

A second reason for grouping these seemingly different topics under a common mathematical subdiscipline is the commonality of methods used in their study. Different branches of mathematics tend to have somewhat distinctive collections of mathematical tools and strategies. Combinatorics is characterized by the frequent use of mathematical tools, such as counting, induction, constructive proofs, generating functions, and theorems from linear algebra, algebraic systems, and number theory.

The third reason these topics are given a collective identity is that they tend to be concerned with a fairly small collection of broadly related questions. In particular, combinatorial topics can be categorized as problems concerned with one or more of the following three broad questions:

Existence Many combinatorial topics relate to finding a special configuration of elements from some set. The existence problem seeks to determine if such a configuration is possible. If it is possible, the existence problem seeks ways to construct the configuration.

A very simple example is a search for the existence of magic squares. A *magic square* is an n by n matrix[1] whose entries are the complete set of positive integers $\{1, 2, \ldots, n^2\}$, arranged in such a way that the sum of every row, the sum of every column, and the sum of the two diagonals are all the same number.

The following example of a 3-by-3 magic square illustrates this definition.[2]

$$\begin{matrix} 6 & 1 & 8 \\ 7 & 5 & 3 \\ 2 & 9 & 4 \end{matrix}$$

Enumeration Sometimes it is not too difficult to construct the desired configuration. The interesting question then may be *How many distinct configurations are possible?* This is an enumeration problem.

One simple example might be the determination of the number of distinct 3-by-3 magic squares. Section 8.1 will explore additional enumeration problems.

[1] See Appendix E for a brief introduction to matrices.

[2] The Chapter 8 section at http://www.mathcs.bethel.edu/~gossett/DiscreteMathWithProof/ contains a link to a Web site that has much more information about magic squares.

Optimization Sometimes the configuration of interest has additional properties that enable us to distinguish among acceptable alternatives. It may be that among all configurations of a certain type, some are more useful than others. The problem is then to find an optimal configuration.

The Stable Marriage Problem (Section 1.2) is an example of this type of problem, as is the knapsack problem in Section 8.4.

The previous discussion used the word *configuration* many times. Although combinatorics includes more ideas than merely that of arranging objects into special configurations, exploring configurations is a major thread in the discipline. A more formal terminology to describe such arrangements is to call the configuration a *combinatorial design*. This concept is so important that Herbert Ryser [72, p. 2] uses it as the basis of his informal definition of *combinatorics*.[3]

> Combinatorial mathematics cuts across the many subdivisions of mathematics, and this makes a formal definition difficult. But by and large it is concerned with the study of the arrangement of elements into sets. The elements are usually finite in number, and the arrangement is restricted by certain boundary conditions imposed by the particular problem under investigation.

The boundary conditions for magic squares problems are the requirements that every number in the set $\{1, 2, \ldots, n^2\}$ appears in the magic square and that the row, column, and diagonal sums be equal. The boundary conditions for the Stable Marriage Problem include the requirement that an assignment be stable and also the assumption that there be no ties in the individual preferences.

The first sentence in Ryser's informal definition refers to the connections between combinatorics and other branches of mathematics. As has already been mentioned, those connections are quite strong. This provides a rich assortment of mathematical ideas and tools that may be applied to solving problems in combinatorics and also permits combinatorial ideas to be used for solving problems in other branches of mathematics. However, it creates a problem when discussing combinatorics in a text aimed at lower division undergraduates. The exposition of many of the topics quickly incorporates mathematical ideas that are typically encountered in the junior and senior years of college. This limits how deeply many topics can be explored here. However, there is still a sufficiently rich assortment of accessible combinatorial ideas so that this chapter can provide a nontrivial introduction to combinatorics. Furthermore, the material on graph theory in Chapter 10 is often considered a part of combinatorics.

✔ Quick Check 8.1

1. How many distinct ways can the numbers $1, 2, \ldots, 9$ be arranged in a 3-by-3 matrix?
2. Prove that the common sum in an n by n magic square must be $\frac{n(n^2+1)}{2}$. [*Hint*: Use a combinatorial proof—count the entries in the magic square two different ways.]
3. List all subsets of three integers from $\{1, 2, 3, \ldots, 9\}$ whose sum is 15. Conclude that 1, 3, 7, and 9 cannot appear in the center or in one of the corners, that 2, 4, 6, and 8 must be in one of the corners, and that 5 must be in the center of any 3-by-3 magic square.
4. Using the results of the previous problem, how many potential 3-by-3 magic squares are possible?
5. Find all possible 3-by-3 magic squares. ☑

[3]His informal definition goes on to describe existence and enumeration problems. Those ideas have already been discussed here and so the quotation has been truncated.

8.1 Partitions, Occupancy Problems, Stirling Numbers

This section introduces some interesting problems in enumerative combinatorics. They are collectively gathered into a category of counting problems called *occupancy problems*. Many of the background ideas needed to explore these problems were introduced in Chapter 5. Two additional ideas need to be presented here: partitions of an integer and Stirling numbers of the second kind. For completeness, Stirling numbers of the first kind will also be discussed at the end of the section.

8.1.1 Partitions of a Positive Integer

Leonhard Euler was the first mathematician to make significant progress in answering a seemingly simple counting problem. The problem starts with a positive integer, n, and asks how many ways n can be written as a sum of positive integers, where the order of the summands does not matter.

EXAMPLE 8.1 **Expressing 6 as a Sum of Positive Integers**

The number 6 is small enough to exhaustively list all possible ways to express it as a sum of positive integers (with the order of the summands unimportant). The list (Table 8.1) shows that there are eleven ways to do this.

TABLE 8.1 All Ways to Write 6 as a Sum of Positive Integers (Order of Summands Unimportant)

$6 = 6$	$6 = 5 + 1$	$6 = 4 + 2$
$6 = 4 + 1 + 1$	$6 = 3 + 3$	$6 = 3 + 2 + 1$
$6 = 3 + 1 + 1 + 1$	$6 = 2 + 2 + 2$	$6 = 2 + 2 + 1 + 1$
$6 = 2 + 1 + 1 + 1 + 1$	$6 = 1 + 1 + 1 + 1 + 1 + 1$	

Notice that there is only one way (each) to write 6 as a sum of exactly one term, as a sum of exactly five terms, or as a sum of exactly six terms. There are three ways to write 6 as a sum of two terms ($5 + 1$, $3 + 3$, and $4 + 2$) and three ways to write 6 as a sum of three terms. Finally, there are two ways to write it as a sum of four terms. ■

Example 8.1 introduces ideas that will be easier to discuss if some formal definitions and new notation are introduced. The notation introduced in the next definition is one common choice for this context, but there is no standard notation.

DEFINITION 8.1 ***Partition of an Integer***

A *partition of an integer*, n, is a representation of n as a sum of positive integers, where the order of the summands is not important.

The number of partitions of n is denoted by $p(n)$. The number of partitions of n that contain exactly k summands is denoted by $p(n, k)$.

Notes:

1. The letter p is lowercase: $p(n, k)$ is not the permutation $P(n, r)$.
2. This definition of *partition* is quite different from the set-oriented definition introduced on page 24. The context will usually make clear which definition is intended.

EXAMPLE 8.2 **The Number of Partitions of 6**

The terminology in Definition 8.1 can be applied to Example 8.1. Thus, $p(6) = 11$ and

$$p(6,6) = 1 \qquad p(6,5) = 1 \qquad p(6,4) = 2$$
$$p(6,3) = 3 \qquad p(6,2) = 3 \qquad p(6,1) = 1$$

■

Although it is quite easy to understand what $p(n)$ represents, it is not as easy to calculate $p(n)$ when n gets larger than one or two digits. Some additional relationships need to be presented before any real progress can be made.

PROPOSITION 8.2 ***p(n) and p(n, k)***

Let n be a positive integer. Then

$$p(n) = \sum_{k=1}^{n} p(n, k)$$

Proof: Let S_k be the set of all partitions of n that contain exactly k summands. Then $|S_k| = p(n, k)$. In addition, the sets $\{S_1, S_2, \ldots, S_n\}$ are disjoint and their union is the set, S, of all partitions of n. Proposition 2.21 on page 34 completes the proof. □

A closed-form expression for $p(n)$ is found in Theorem 8.3. This theorem was first recognized by Euler. It uses ideas from Section 7.4, but can be understood without reading that section.

THEOREM 8.3 ***A Generating Function for p(n)***

Let n be a positive integer. Then $p(n)$ is the coefficient of z^n in the generating function

$$\prod_{m=1}^{\infty}\left(\sum_{i=0}^{\infty} z^{im}\right).$$

That is,

$$\sum_{n=1}^{\infty} p(n)z^n = \prod_{m=1}^{\infty}\left(1 + z^m + z^{2m} + z^{3m} + z^{4m} + \cdots\right)$$

EXAMPLE 8.3

Determining $p(3)$

Theorem 8.3 asserts that $p(3)$ is the coefficient of z^3 in the power series $\prod_{m=1}^{\infty}\left(\sum_{i=0}^{\infty} z^{mi}\right)$. The initial values of the product

$$\left(1 + z + z^2 + z^3 + z^4 + z^5 + \cdots\right)\left(1 + z^2 + z^4 + z^6 + \cdots\right)$$
$$\left(1 + z^3 + z^6 + z^9 + \cdots\right)\cdots$$

can be calculated by noticing that any term with a power of z that is larger than 3 cannot contribute to the coefficient of z^3. The partial calculation below does not show all the factors that equal 1 (for example, $1 \cdot 1 \cdot 1 \cdot z^4 \cdot 1 \cdots$ is written as just z^4).

$$\begin{aligned}\left(1 + z + z^2 + z^3 + \cdots\right)\left(1 + z^2 + x^4 + \cdots\right)\left(1 + z^3 + z^6 + \cdots\right)\cdots \\ = 1 + z + \left(z^2 + z^2\right) + \left(z \cdot z^2 + z^3 + z^3\right) + \cdots \\ = 1 + z + 2z^2 + 3z^3 + \cdots\end{aligned}$$

Since the only partitions of 3 are $3 = 3$, $3 = 2 + 1$, and $3 = 1 + 1 + 1$, the assertion in the theorem is correct in this case.

It will be helpful to look at the previous algebraic expansion in more detail. For that purpose, the exponents will be left as products.

$$\left(1+z^{1\cdot 1}+z^{2\cdot 1}+z^{3\cdot 1}+\cdots\right)\left(1+z^{1\cdot 2}+z^{2\cdot 2}+\cdots\right)\left(1+z^{1\cdot 3}+z^{2\cdot 3}+\cdots\right)\cdots$$
$$=1+z^{1\cdot 1}+\left(z^{2\cdot 1}+z^{1\cdot 2}\right)+\left(z^{1\cdot 1}\cdot z^{1\cdot 2}+z^{3\cdot 1}+z^{1\cdot 3}\right)+\cdots$$

Now, let the right-hand factor in an exponent represent the distinct integers in the partition and let the left-hand factor in the exponents represent the number of copies of the right-hand factor that appear in the partition. Thus, the factor $z^{3\cdot 4}$ would correspond to three 4s and the term $z^{2\cdot 3}\cdot z^{3\cdot 5}$ would correspond to the partition $21=3+3+5+5+5$.

Terms	Corresponding Partitions		
$z^{1\cdot 1}$	$1=1$		
$z^{2\cdot 1}+z^{1\cdot 2}$	$2=1+1$	$2=2$	
$z^{1\cdot 1}\cdot z^{1\cdot 2}+z^{3\cdot 1}+z^{1\cdot 3}$	$3=1+2$	$3=1+1+1$	$3=3$

The association between partitions of n and terms in the expansion of the product in Theorem 8.3 is the central idea in the proof of the theorem. ■

Proof of Theorem 8.3 For the duration of this proof, let

$$S_m=\left(1+z^m+z^{2m}+z^{3m}+\cdots\right) \quad \text{and} \quad P=\prod_{m=1}^{\infty} S_m$$

Consider the coefficient, c_n, of z^n in P, for $n \geq 1$. That coefficient is obtained by adding a finite set of terms of the form $z^{i_1 m_1}\cdot z^{i_2 m_2}\cdots z^{i_k m_k}$ for some positive integer, k, with $n=i_1m_1+i_2m_2+\cdots+i_km_k$. In addition, $m_r \neq m_s$ if $r \neq s$, since the m_r's must come from different factors, S_r, in the original product, P.

It is possible to convert each of these terms to a partition of n. The value of m_r can be used as the integer in the sum on the right-hand side of the partition and the value of i_r will be the repeat factor. The equation $n=i_1m_1+i_2m_2+\cdots+i_km_k$ guarantees that this is a partition of n. A different choice of k and of the i's and m's will result in a different partition of n. Since every such term in the expansion corresponds to a different partition, it is clear that $c_n \geq p(n)$.

It is also possible to start with a partition of n and find a corresponding term in the expansion of P. Suppose the partition can be expressed as $n=\sum_{m=1}^{n} a_m m$, where a_m is the number of copies of the integer m that are present in the partition. Some of the a_m's may be zero. Let k be the number of nonzero a_m's in the sum. Label the values of m having a nonzero a_m as $m_1, m_2, \ldots, m_k$ and rename their associated a_m's as $i_1, i_2, \ldots, i_k$. The partition thus corresponds to a term $z^{i_1 m_1}\cdot z^{i_2 m_2}\cdots z^{i_k m_k}$. Each m_j is distinct, so the factor $z^{i_j m_j}$ is actually present in a distinct factor, S_j, in P. Thus, the term $z^{i_1 m_1}\cdot z^{i_2 m_2}\cdots z^{i_k m_k}$ must occur as part of the expansion. This implies that $p(n) \geq c_n$.

The conclusion is that $c_n = p(n)$ and the theorem has been proved. □

The infinite sums and products in Theorem 8.3 are not necessary if all that is needed is the coefficient for a particular value of n. Corollary 8.4 establishes the appropriate modification (which was implicitly used in Example 8.3).

COROLLARY 8.4 ***Calculating $p(n)$***

Let n be a positive integer. Then $p(n)$ is the coefficient of z^n in the polynomial

$$\prod_{m=1}^{n}\left(\sum_{i=0}^{\lfloor \frac{n}{m} \rfloor} z^{im}\right)$$

$p(n)$ is the coefficient of z^n in the polynomial $\prod_{m=1}^{n}\left(\sum_{i=0}^{\lfloor \frac{n}{m} \rfloor} z^{im}\right)$

Proof: The coefficient of z^n is a sum involving terms of the form $z^{i_1 m_1} \cdot z^{i_2 m_2} \ldots z^{i_k m_k}$ for some positive integer, k, with $n = i_1 m_1 + i_2 m_2 + \cdots + i_k m_k$. If $im > n$, then z^{im} cannot be a factor in such a term. Thus, $m \le n$ must certainly hold. In addition, in order that $im \le n$, it is necessary that $i \le \frac{n}{m}$. Since i is an integer, $i \le \lfloor \frac{n}{m} \rfloor$ must also be true. □

EXAMPLE 8.4 **Calculating $p(4)$**

The value of $p(4)$ is the coefficient of z^4 in the expression $\prod_{m=1}^{4}\left(\sum_{i=0}^{\lfloor \frac{4}{m} \rfloor} z^{im}\right)$.

$$\begin{aligned}\prod_{m=1}^{4}\left(\sum_{i=0}^{\lfloor \frac{4}{m} \rfloor} z^{im}\right) &= \left(1+z+z^2+z^3+z^4\right)\left(1+z^2+z^4\right)\left(1+z^3\right)\left(1+z^4\right) \\ &= 1+z+2z^2+3z^3+5z^4+5z^5+6z^6+7z^7 \\ &\quad +7z^8+6z^9+5z^{10}+5z^{11}+3z^{12}+2z^{13}+z^{14}+z^{15}\end{aligned}$$

Note well: The coefficients of z, z^2, z^3,and z^4 in the expansion above are the correct values for $p(1)$, $p(2)$, $p(3)$, and $p(4)$, respectively. However, the remaining coefficients are not the values of $p(n)$ for $n = 5, 6, \ldots, 15$. It was not really necessary to keep track of terms with exponents greater than 4, so when expanding the product we could simply write $= 1+z+2z^2+3z^3+5z^4+\cdots$. ■

It is now time to consider the numbers, $p(n,k)$. A nice combinatorial proof establishes the useful recurrence relation in the next theorem.

THEOREM 8.5 ***A Recurrence Relation for $p(n, k)$***

Let n and k be integers and let $0 < k \le n$. Then

$$p(n,k) = p(n\text{-}1, k\text{-}1) + p(n\text{-}k, k)$$

where $p(n,k) = 0$ for $k > n$ and $p(n,0) = 0$. In addition, $p(n,n) = 1$.

Proof: The boundary conditions will be established first. Since $n > 0$, there is no way that n can be expressed as a sum with no summands. Thus, $p(n,0) = 0$. Also, any sum with $k > n$ positive integer summands will have a sum that is greater than n. Thus, $p(n,k) = 0$ if $k > n$ and the boundary conditions have been established. Finally, the only way to express n as a sum with n summands is to add n 1s. Thus, $p(n,n) = 1$.

Now consider the set, S, of all partitions of n that contain exactly k summands. Define subsets, A and B, of S by

$$A = \{s \in S \mid s \text{ contains at least one 1 as a summand}\}$$
$$B = \{s \in S \mid s \text{ contains no 1's as summands}\}$$

Clearly, $S = A \cup B$ and $A \cap B = \emptyset$, so $\{A, B\}$ is a (set) partition of S. Thus, $|S| = |A| + |B|$.

For each partition in A, remove one of the 1s from the sum. The resulting partition will be a partition of $n-1$ having exactly $k-1$ summands. Also, any partition of $n-1$ having exactly $k-1$ summands can be transformed into a partition of n with k summands by adding an additional 1 to the sum. Therefore, $|A| = p(n\text{-}1, k\text{-}1)$.

For each partition in B, subtract 1 from each summand. The result will still be a partition since every summand will be at least 1. The new partition will have a sum of $n-k$. Conversely, every partition of $n-k$ having exactly k summands (for $k \le n$) can be transformed into a partition of n having k summands by adding 1 to each summand. Therefore, $|B| = p(n\text{-}k, k)$. □

✔ Quick Check 8.2

1. Show the $p(n, 1) = 1$ for $n \geq 1$.
2. Use Theorem 8.5 and Proposition 8.2 (page 420) to fill in the blank spots in Table 8.2.

TABLE 8.2 The Values of $p(n)$ and $p(n, k)$ for $n, k \leq 6$

	$p(n,k)$						$p(n)$
$n \setminus k$	1	2	3	4	5	6	
1		—	—	—	—	—	
2			—	—	—	—	
3				—	—	—	
4					—	—	
5						—	
6							

✔

8.1.2 Occupancy Problems

The value of $p(n, k)$ has other interpretations. One alternative is explored in Exercise 4 on page 433. Here is another interpretation. Suppose we have n identical red balls and k identical buckets. In how many ways can the balls be placed into the buckets if every bucket must receive at least one ball? The answer turns out to be $p(n, k)$. This problem is an example of an *occupancy problem.* This class of counting problems is the subject of this section.[4] Familiarity with the material in Chapter 5 is assumed.

DEFINITION 8.6 *Occupancy Problems*
Occupancy problems are concerned with placing *objects* into *containers*.
Occupancy problems are categorized by whether the objects and containers are distinguishable or indistinguishable and by whether or not containers may be empty.

Historically, objects have been balls and the containers have usually been urns or cells. The more general notions of objects and containers will be used here.

EXAMPLE 8.5

A Small Occupancy Problem—Part 1

Suppose there are six objects and three containers. There are eight possible occupancy problems, depending on whether objects and containers are distinguishable and whether or not containers can be empty. The cases will be abbreviated using the notation O—object, C—container, D—distinguishable, I—indistinguishable, ∅—containers may be empty, ¬∅—containers may not be empty.

OD CD ∅ Suppose the objects are six driver's licenses (for six different people) and the containers are three bins labeled "\$50 fine", "\$100 fine," and "no fine". A court clerk randomly places licenses into the bins and then the judge issues the fines to the hapless drivers. There are $3^6 = 729$ ways to do this since each license has three possible locations and the assignments are independent.

OD CD ¬∅ Suppose the objects are slips of paper with the names of six contestants in a piano contest and the containers are clipboards labeled "outstanding", "commendable", and "participant". The judges confer after each contestant has performed and then attach the name slip to the appropriate clipboard. The sponsoring organization has decided that

[4]The organization that will be used was inspired by the presentation and arrangement in [74].

every clipboard must contain at least one name (judging is relative to all participants, not to some absolute standard).

One tempting counting strategy is to first make sure that the three clipboards each have one name, and then distribute the remaining names randomly. There are $P(6, 3) = 120$ ways to choose a distinct name for each clipboard. The remaining three names can be distributed independently in $3^3 = 27$ ways for a total of $120 \cdot 27 = 3240$ ways. However, this cannot be correct because there would be only 729 ways to attach names to clipboards if we drop the restriction that every clipboard needs to have at least one name.

The error occurs because this strategy has many arrangements that are counted more than once. For example, placing names a, b, c, d on the outstanding clipboard, e on the commendable clipboard, and f on the participant clipboard can be done in four ways (depending on which of a, b, c, and d is chosen in the first round).

This occupancy problem is nontrivial. The strategy that will eventually be used is to first determine in how many ways the six names can be placed onto unlabeled clipboards, and then multiply by 3! (the number of ways to label the three clipboards). There are 90 ways to complete the first phase,[5] so there are 540 ways to distribute the contestant's names onto the clipboards.

OI CD ∅ Suppose there are six oranges and three Christmas stockings hanging by the fireplace. The stockings are labeled "Tom", "Mary", and "Spot". In how many ways can the oranges be placed into the stockings if stockings can be empty?

Instead of placing the oranges directly into the stockings, a marker can first be used to label each orange with the name of the stocking into which it will be placed. There are three distinct labels, and each label can be used up to six times. Therefore the task is to select a set of six labels, from a set of three distinct possibilities. This is thus a combination with repetition problem. The combinations with repetition formula on page 235 can be used with $n = 3$ and $r = 6$.

There are $C(3 + 6 - 1, 6) = 28$ ways to fill the stockings.

OI CD ¬∅ Suppose Bob, Sue, and Bob Junior also have labeled Christmas stockings and six oranges. However, they have decided that each person will get at least one orange in his or her stocking. In how many ways can the oranges be placed into the stockings?

The best strategy is to first place an orange into each stocking. There are then $n = 3$ labels to write on $r = 3$ oranges. Using the strategy from the previous case, there are $C(3 + 3 - 1, 3) = C(5, 3) = C(5, 2) = 10$ ways to fill the stockings.

The remaining four cases will be illustrated in Example 8.6 on page 426. ■

Example 8.5 introduced the case of distributing n distinguishable objects into k distinguishable containers, where every container must receive at least one object. The solution to that case was presented without justification. Before that justification can be given, some new notation needs to be introduced.

DEFINITION 8.7 ***Stirling Numbers of the Second Kind***

The number of ways to distribute n distinguishable objects into k indistinguishable containers with every container receiving at least one object is denoted $S(n, k)$. The numbers, $S(n, k)$, are called the *Stirling numbers of the second kind.*

The Stirling numbers are named after James Stirling, an English mathematician who lived in the 1700s and was a contemporary of Isaac Newton. Section 8.1.3 will more fully develop the Stirling numbers of the second kind, as well as introducing the Stirling numbers of the first kind.

[5] Justification for this claim will be presented later in Example 8.7 on page 428.

A complete set of solutions to the eight categories of occupancy problems is presented in the next theorem and also on page 532.[6]

THEOREM 8.8 ***Occupancy Problems***

Table 8.3 lists solutions to all eight categories of occupancy problems, where n represents the number of objects and k represents the number of containers.

TABLE 8.3 The Number of Ways to Place n Objects into k Containers

		Containers: Distinguishable		Containers: Indistinguishable	
Objects	Distinguishable	$\emptyset$:	k^n	$\emptyset$:	$\sum_{i=1}^{k} S(n,i)$
		$\neg\emptyset$:	$k!S(n,k)$	$\neg\emptyset$:	$S(n,k)$
	Indistinguishable	$\emptyset$:	$\binom{k+n-1}{n}$	$\emptyset$:	$\sum_{i=1}^{k} p(n,i)$
		$\neg\emptyset$:	$\binom{n-1}{k-1}$	$\neg\emptyset$:	$p(n,k)$

$\emptyset$: containers may be empty
$\neg\emptyset$: containers must contain at least one object

Proof: Each of the eight cases will be examined separately. The cases will be abbreviated using the notation O—object, C—container, D—distinguishable, I—indistinguishable, $\emptyset$—containers may be empty, $\neg\emptyset$—containers may not be empty. The cases will be examined starting in the upper-right square of the table and moving in a counterclockwise direction.

OD CI $\neg\emptyset$ The number of ways to distribute n distinguishable objects into k indistinguishable containers with every container receiving at least one object is $S(n, k)$ (true by definition). The proof need not show how to calculate $S(n, k)$.

OD CI $\emptyset$ This case differs from the previous case because some of the containers may be empty. In fact, the objects can all be placed into just one container, or just two of the containers, or any subcollection of the containers (including the full set of containers). There are $S(n, i)$ ways to distribute the objects into exactly i containers. Since the containers are indistinguishable and the subcases of distributions into exactly i or exactly j containers are mutually exclusive if $i \neq j$, a simple sum will complete the count.

OD CD $\emptyset$ Each of the objects can be placed into any container, with no restrictions. There are k choices for the first object, k choices for the second object, ..., k choices for the nth object. Thus, there are k^n ways to distribute the objects.

OD CD $\neg\emptyset$ Erase the labels (or other distinguishing characteristics) from the containers. There are $S(n, k)$ ways to distribute the objects into these unlabeled containers. There are now $k!$ ways to place the labels back onto the containers. These two phases are independent, so there are $k!S(n, k)$ ways to distribute the objects in this case.

OI CD $\emptyset$ The containers are labeled, but the objects are indistinguishable. Place the objects into the containers, and then, for each object, write the label of its container on the object. There will now be a collection of n written labels, from a set of k possible labels. This is a combination with repetition problem. There are $C(k + n - 1, n) = \binom{k+n-1}{n}$ ways to do this.[7]

[6] The table is also available in pdf format at http://www.mathcs.bethel.edu/~gossett/DiscreteMathWithProof/ .

[7] The n in the combinations with repetition formula on page 235 is the number, k, of labels and r is the number, n, of objects.

OI CD ¬∅ Since objects are indistinguishable, it is possible to first place one object into each container. There are now $n - k$ indistinguishable objects to place into the k containers, with no additional restraints. The previous case implies that there are $\binom{k+(n-k)-1}{n-k} = \binom{n-1}{n-k} = \binom{n-1}{(n-1)-(n-k)} = \binom{n-1}{k-1}$ ways to do this. (Proposition 5.9 on page 248 was used in the middle step.)

OI CI ¬∅ There is a one-to-one correspondence between distributions of n indistinguishable objects into k indistinguishable containers and partitions of the integer n with exactly k summands. To see this, first consider a distribution of n indistinguishable objects into k indistinguishable containers. Count the number of objects in each container. This will result in k numbers whose sum is n. That is, a partition of n with exactly k summands (recall that the order of the summands is unimportant). Now consider a partition of n with exactly k summands. The individual summands can be used to determine the number of objects placed into corresponding containers. There is no implied order to the summands, so the corresponding containers are indistinguishable.

OI CI ∅ This case differs from the previous case because some of the containers may be empty. In fact, the objects can be all be placed into just one container, or just two of the containers, or any subcollection of the containers (including all of them). There are $p(n, i)$ ways to distribute the indistinguishable objects into exactly i containers. Since the containers are indistinguishable and the subcases with different choices for i are mutually exclusive, a simple sum will complete the count. □

Table 8.3 contains solutions for all eight cases of the occupancy problem. The only missing piece is a practical method for finding values for $S(n, k)$. That deficiency will be fully redressed in Section 8.1.3 on page 427. The next Quick Check will provide a small collection of values.

✔ Quick Check 8.3

1. Use Definition 8.7 on page 424 and exhaustive enumeration to calculate $S(n, k)$, for $1 \le n \le 4$ and $1 \le k \le n$. ☑

EXAMPLE 8.6

A Small Occupancy Problem—Part 2

The four cases that were omitted in Example 8.5 will be explored here.

OD CI ¬∅ Six co-workers have all arrived at the airport and need to get to the local convention center. Their company has arranged for three taxis to meet them at the airport. It does not make sense for one of the prepaid taxis to drive to the convention center without a passenger. In how many ways can the co-workers ride to the convention center? Notice that they probably care about with whom they share a taxi. (Some people have more in common to discuss during the ride.)

The number of arrangements is $S(6, 3)$, which was previously claimed to be 90.

OD CI ∅ A hostess has invited some friends over for tea and crumpets. Altogether, there are six people. The hostess has arranged seating in three areas: the living room, the dining room, and the patio. Each area is large enough to hold all six people. If the hostess and her guests consider the location unimportant, but the grouping of friends to be significant, in how many ways can the people be arranged? In this problem, the people are the objects and the locations are the containers.

Since a location can be empty, there are $S(6, 1) + S(6, 2) + S(6, 3) = 122$ ways for the people to be distributed.[8]

OI CI ¬∅ Josephine has designed an interactive art project. The project consists of three large bowls, placed at the vertices of an equilateral triangle on the floor. The triangle has been inscribed in a circle, and people can walk freely around the circle.

[8]The values of $S(6, 1)$ and $S(6, 2)$ can be found using techniques introduced in Section 8.1.3.

Josephine has provided a set of six identical wax oranges. Art patrons can complete the installation by placing the wax oranges into the bowls. Josephine has left instructions to specify that every bowl should have at least one orange (in order for the project to have "balance"). How many distinct arrangements are possible?

In this problem, both the bowls and the oranges are indistinguishable. Since every bowl must contain at least one orange, there will be $p(6, 3) = 3$ different arrangements: $\{4, 1, 1\}$, $\{3, 2, 1\}$, $\{2, 2, 2\}$.

OI CI ∅ Even though Josephine has specified that at least one orange should be placed in each bowl, there is nothing to stop people from violating that request. In how many ways can the oranges be arranged if bowls can be empty?

There are $p(6, 1) + p(6, 2) + p(6, 3) = 1 + 3 + 3 = 7$ possible arrangements: $\{6, 0, 0\}$, $[\{5, 1, 0\}, \{4, 2, 0\}, \{3, 3, 0\}]$, $[\{4, 1, 1\}, \{3, 2, 1\}, \{2, 2, 2\}]$. ■

	$S(n,k)$			
$n \setminus k$	1	2	3	4
1	1	–	–	–
2	1	1	–	–
3	1	3	1	–
4	1	7	6	1

	$p(n,k)$					
$n \setminus k$	1	2	3	4	5	6
1	1	–	–	–	–	–
2	1	1	–	–	–	–
3	1	1	1	–	–	–
4	1	2	1	1	–	–
5	1	2	2	1	1	–
6	1	3	3	2	1	1

✔ Quick Check 8.4

Use Theorem 8.8 to answer the following questions:

1. I have four CDs that I wish to give away, and three friends who are potential recipients. In how many ways can I give away the CDs if
   (a) The CDs are all different, and each friend should receive at least one CD.
   (b) The CDs are all different, and I feel no obligation to give every friend at least one CD.
2. I have four new Sacagawea $1 coins that I plan to distribute to my three charming nieces. In how many ways can I do this?
3. A farmer has four sons and three identical potato fields. Each field takes eight person-days to harvest. In how many ways can the sons be assigned fields to harvest on the first day of the harvest? ☑

8.1.3 Stirling Numbers

Some of the basic properties of the Stirling numbers of the first and second kinds will be examined in this section. As a convenience, Definition 8.7 is repeated here.

DEFINITION 8.7 ***Stirling Numbers of the Second Kind***
The number of ways to distribute n distinguishable objects into k indistinguishable containers with every container receiving at least one object is denoted $S(n, k)$. The numbers, $S(n, k)$, are called the *Stirling numbers of the second kind.*

An alternative view of $S(n, k)$ is given in the next proposition.

PROPOSITION 8.9 ***Set Partitions and $S(n, k)$***
Let A be a set with n elements. The number of ways to partition A into a collection of exactly k nonempty subsets is $S(n, k)$.

Proof: Think of the elements of A as objects, and the subsets in the partition as containers. Since every subset is nonempty, this is an occupancy problem with distinguishable objects, indistinguishable containers (the sets in the partition have no implied order), and in which each container must receive at least one object. Theorem 8.8 completes the proof. □

The recurrence relation in the next theorem provides a useful mechanism for calculating $S(n, k)$.

THEOREM 8.10 ***A Recurrence Relation for $S(n, k)$***

Let n and k be integers and let $0 < k \leq n$. Then

$$S(n, k) = S(n\text{-}1, k\text{-}1) + kS(n\text{-}1, k)$$

where $S(n, k) = 0$ for $k > n$ and $S(n, 1) = 1$. In addition, $S(n, n) = 1$ and $S(n, 0) = 0$ for $n > 0$.

Proof: The boundary conditions will be established first. The only way $n > 0$ objects can be distributed into a single container is to place each object into that container and hence $S(n, 1) = 1$. Also, it is impossible to distribute n objects into more than n containers and still require each container to receive at least one object. Thus, $S(n, k) = 0$ if $k > n$. The only way to distribute n objects into n containers if each container must receive at least one object is to place exactly one object in each container. Thus, $S(n, n) = 1$. If $n > 0$, then it is impossible to distribute the n objects into no containers, so $S(n, 0) = 0$ when $n > 0$. The boundary conditions have been verified. A combinatorial proof will be used to establish the recurrence relation.

Consider a collection of n distinguishable objects. Since they are distinguishable, choose one of the objects and paint it red (or some other unique color). Let D be the set of all distributions of these objects into exactly k containers. The elements of D can be partitioned into two subsets:[9]

$$A = \{d \in D \mid \text{the red object is the only object in its container}\}$$
$$B = \{d \in D \mid \text{the red object is not the only object in its container}\}$$

Then $D = A \cup B$ and $A \cap B = \emptyset$. Hence, $|D| = |A| + |B|$.

The distributions in A are in one-to-one correspondence with distributions of $n - 1$ distinguishable objects into $k - 1$ indistinguishable containers. The correspondence can be achieved by removing (or adding) the red object and its enclosing container. Thus, $|A| = S(n\text{-}1, k\text{-}1)$.

Let C be the set of all distributions of $n - 1$ distinguishable (but nonred) objects into k indistinguishable containers. Then $|C| = S(n\text{-}1, k)$. The relationship between distributions in B and distributions in C is more complex. If we start with a distribution in C, then adding the red object to any one of the k containers will create a distribution in B. Because the objects are distinguishable, each of these k possible distributions will be distinct. Conversely, if $d \in B$, then removing the red object will certainly produce a distribution, $x \in C$. But d is not the only distribution in B that will produce x. Since each of the containers in d contains at least one object, moving the red object to any other container will still result in a distribution in B. There are k such distributions, each resulting in x when the red object is removed. Thus, there is a k-to-1 correspondence between B and C. Consequently, $|B| = kS(n\text{-}1, k)$. □

EXAMPLE 8.7 **Calculating $S(6, 3)$**

$n \backslash k$	1	2	3	4
1	1	–	–	–
2	1	1	–	–
3	1	3	1	–
4	1	7	6	1

$S(n, k)$

It is now possible to calculate $S(6, 3)$. The calculation will use Theorem 8.10.

$$\begin{aligned} S(6, 3) &= S(5, 2) + 3 \cdot S(5, 3) \\ &= [S(4, 1) + 2 \cdot S(4, 2)] + 3 \cdot [S(4, 2) + 3 \cdot S(4, 3)] \\ &= [1 + 2 \cdot 7] + 3 \cdot [7 + 3 \cdot 6] = 90 \end{aligned}$$ ■

[9]Exercise 22 on page 434 asks you to illustrate this for $n = 4$ and $k = 2$. It might be helpful to do that problem as you read the proof.

Stirling Numbers of the First Kind

The remainder of this section will introduce the Stirling numbers of the first kind and discuss their connection to the Stirling numbers of the second kind. The results are not about counting, so the rest of this section is not directly related to enumerative combinatorics.

To set the context, consider a function, $f(x)$, that represents some process that can be measured. The function is unknown, but it is possible to approximate the function using the measured values. The simplest approximation technique is to construct an interpolating polynomial, $p(x)$, which matches the measured values of f. Suppose, for example, that the process has been measured at 0, 1, 2, 3, and 4 seconds. Table 8.4 shows the measured values.

TABLE 8.4 The Measured Values of Some Unknown Function, f

x	$f(x)$
0	a_0
1	a_1
2	a_2
3	a_3
4	a_4

One fairly simple way to construct the interpolating polynomial is to write it as

$$\begin{aligned} p(x) &= a_0 + (a_1 - a_0)x + \frac{a_2 - 2a_1 + a_0}{2}x(x-1) \\ &+ \frac{a_3 - 3a_2 + 3a_1 - a_0}{6}x(x-1)(x-2) \\ &+ \frac{a_4 - 4a_3 + 6a_2 - 4a_1 + a_0}{24}x(x-1)(x-2)(x-3) \end{aligned} \tag{8.1}$$

It should be easy to convince yourself that p and f have the same values when $x \in \{0, 1, 2, 3, 4\}$. The coefficients were derived by successively substituting[10] the values of x into the prototype polynomial

$$p(x) = b_0 + b_1 x + b_2 x(x-1) + b_3 x(x-1)(x-2) + b_4 x(x-1)(x-2)(x-3)$$

Thus, b_0 can be found by substituting 0 for x. Then b_1 can be found by substituting 1 for x, giving the equation $a_1 = a_0 + b_1 \cdot 1$. Continuing in this fashion will eventually produce all the coefficients.

It is possible to convert this polynomial into the standard form as a sum of powers of x. The result will be a polynomial of degree 4:

$$\begin{aligned} p(x) &= a_0 + \frac{(-25\,a_0 + 48\,a_1 - 36\,a_2 + 16\,a_3 - 3\,a_4)}{12}x \\ &+ \frac{(35\,a_0 - 104\,a_1 + 114\,a_2 - 56\,a_3 + 11\,a_4)}{24}x^2 \\ &+ \frac{(-5\,a_0 + 18\,a_1 - 24\,a_2 + 14\,a_3 - 3\,a_4)}{12}x^3 \\ &+ \frac{(a_0 - 4\,a_1 + 6\,a_2 - 4\,a_3 + a_4)}{24}x^4 \end{aligned} \tag{8.2}$$

The main idea needed here is that any polynomial of degree 4 or less can be expressed as a sum (with appropriate coefficients) of elements of $B_1 = \{1, x, x^2, x^3, x^4\}$ or as a sum (with different coefficients) of elements in $B_2 = \{1, x, x(x-1), x(x-1)(x-2), x(x-1)(x-2)(x-3)\}$. A sum of this type is called a *linear combination*.

DEFINITION 8.11 ***Linear Combination***

Let $e_1, e_2, \ldots, e_k$ be expressions. A *linear combination* of $\{e_1, e_2, \ldots, e_k\}$ is an expression of the form

$$c_1 e_1 + c_2 e_2 + \cdots + c_k e_k$$

where $\{c_1, c_2, \ldots, c_k\}$ is a set of constants.

[10]A more general method for calculating the coefficients, using *divided differences*, can be found in most numerical methods textbooks.

The polynomial transformation presented just before Definition 8.11 is an example of transforming a linear combination of elements in B_2 into a linear combination of elements in B_1. It is also always possible to transform a linear combination of elements in B_1 into an equivalent linear combination of elements in B_2.[11]

There is an alternative to brute force polynomial multiplication that will transform equation (8.1) into equation (8.2). The idea is to first find expressions that represent each of the polynomials, $1, x, x(x-1), x(x-1)(x-2), x(x-1)(x-2)(x-3)$, as a sum of the polynomials $1, x, x^2, x^3, x^4$. Table 8.5 shows the result (using notation from Definition 8.12).

TABLE 8.5 Transforming $x(x-1)(x-2)\cdots(x-k)$ into a Linear Combination of $x, x^2, \ldots, x^{k+1}$

$(x)_n$	Linear Combination
x	x
$x(x-1)$	$-x+x^2$
$x(x-1)(x-2)$	$2x-3x^2+x^3$
$x(x-1)(x-2)(x-3)$	$-6x+11x^2-6x^3+x^4$

The linear combination

$$c_0 + c_1x + c_2x(x-1) + c_3x(x-1)(x-2) + c_4x(x-1)(x-2)(x-4)$$

can then be written as

$$c_0 + c_1x + c_2\left(-x+x^2\right) + c_3\left(2x-3x^2+x^3\right) + c_4\left(-6x+11x^2-6x^3+x^4\right)$$

and this simpler expression can then be expanded.

The coefficients for the reverse transformation can also be calculated. That is, each of the polynomials in $1, x, x^2, x^3, x^4$ can be written as a linear combination of the polynomials, $1, x, x(x-1), x(x-1)(x-2), x(x-1)(x-2)(x-3)$. Table 8.6 shows the results.

TABLE 8.6 Transforming x^j into a Linear Combination of Expressions of the Form $x(x-1)(x-2)\cdots(x-k)$

x^n	Linear Combination
x	x
x^2	$x+x(x-1)$
x^3	$x+3x(x-1)+x(x-1)(x-2)$
x^4	$x+7x(x-1)+6x(x-1)(x-2)+x(x-1)(x-2)(x-3)$

TABLE 8.7 The Values of $S(n, k)$ for $n, k \leq 4$

	$S(n,k)$			
$n \backslash k$	1	2	3	4
1	1	–	–	–
2	1	1	–	–
3	1	3	1	–
4	1	7	6	1

This long detour has finally come back to Stirling numbers. If the coefficients in the right-hand side of Table 8.6 are examined, they bear a striking resemblance to the numbers in Table 8.7. To demonstrate that this is not a coincidence, some additional notation is needed.[12]

DEFINITION 8.12 *The Falling Factorial, $(x)_n$*
The *falling factorial* is denoted $(x)_n$ and is defined by $(x)_0 = 1$ and

$$(x)_n = \prod_{i=0}^{n-1}(x-i) = x(x-1)(x-2)\cdots(x-n+1) \qquad \text{for } n \geq 1$$

[11] If you have had a course in linear algebra, you will recognize B_1 and B_2 as alternative bases for the set of all polynomials over $\mathbb{R}$ having degree 4 or less. The sums are just linear combinations of the basis elements. The transformations being discussed are the standard change of basis transformations.

[12] The notation in Definition 8.12 is the most commonly used in combinatorics. However, the same notation means something else (but very similar) in other contexts. One suggested alternative notation for the falling factorial is $x^{\underline{n}}$.

The Stirling numbers of the second kind are the coefficients in the transformation of x^n to a linear combination of falling factorials.

THEOREM 8.13 ***x^n as a Linear Combination of Falling Factorials***

Let n be a positive integer. Then

$$x^n = \sum_{k=1}^{n} S(n,k) \cdot (x)_k$$

Proof: The proof is by mathematical induction. If $n = 1$, then $x^1 = x = (x)_1$, so the theorem is true for the base step.

Suppose that $x^{n-1} = \sum_{k=1}^{n-1} S(n\text{-}1,k) \cdot (x)_k$ for some $n \geq 1$. Then

$$\begin{aligned}
x^n &= x \cdot x^{n-1} \\
&= x \cdot \sum_{k=1}^{n-1} S(n\text{-}1,k) \cdot (x)_k && \text{inductive hypothesis} \\
&= \sum_{k=1}^{n-1} S(n\text{-}1,k) \cdot (x-k) \cdot (x)_k && \text{add and subtract the same expression (see the next line)} \\
&\quad + \sum_{k=1}^{n-1} S(n\text{-}1,k) \cdot k \cdot (x)_k \\
&= \sum_{k=1}^{n-1} S(n\text{-}1,k) \cdot (x)_{k+1} && \text{Definition 8.12 and commutativity} \\
&\quad + \sum_{k=1}^{n-1} k \cdot S(n\text{-}1,k) \cdot (x)_k \\
&= \sum_{j=2}^{n} S(n\text{-}1,j\text{-}1) \cdot (x)_j && \text{change of index and Theorem 8.10} \\
&\quad + \sum_{k=1}^{n-1} [S(n,k) - S(n\text{-}1,k\text{-}1)] \cdot (x)_k \\
&= \sum_{k=2}^{n} S(n\text{-}1,k\text{-}1) \cdot (x)_k && \text{change of index and distributivity and summation properties} \\
&\quad + \sum_{k=1}^{n-1} S(n,k) \cdot (x)_k \\
&\quad - \sum_{k=1}^{n-1} S(n\text{-}1,k\text{-}1) \cdot (x)_k \\
&= \sum_{k=1}^{n-1} S(n,k) \cdot (x)_k && \text{commutativity and summation properties} \\
&\quad + [S(n\text{-}1,n-1) \cdot (x)_n - S(n\text{-}1,0) \cdot (x)_1] \\
&= \sum_{k=1}^{n} S(n,k) \cdot (x)_k && S(n\text{-}1,n\text{-}1) = S(n,n) = 1 \text{ and } S(n\text{-}1,0) = 0
\end{aligned}$$

This completes the induction. □

EXAMPLE 8.8 **Illustrating the Proof of Theorem 8.13**

It may be helpful to use a concrete example to illustrate the inductive step in the proof of Theorem 8.13.[13] For this purpose, let $n = 3$ and recall that $S(2, 1) = S(2, 2) = 1$, $S(3, 1) = S(3, 3) = 1$, and $S(3, 2) = 3$.

The inductive hypothesis assumes that $x^2 = \sum_{k=1}^{2} S(2, k) \cdot (x)_k = 1 \cdot x + 1 \cdot x(x-1)$. Then

$$\begin{aligned}
x^3 &= x \cdot [1 \cdot x + 1 \cdot x(x-1)] \\
&= [1 \cdot (x-1) \cdot x + 1 \cdot (x-2) \cdot x(x-1)] + [1 \cdot 1 \cdot x + 1 \cdot 2 \cdot x(x-1)] \\
&= [1 \cdot x(x-1) + 1 \cdot x(x-1)(x-2)] + [1 \cdot 1 \cdot x + 2 \cdot 1 \cdot x(x-1)] \\
&= [1 \cdot x(x-1) + 1 \cdot x(x-1)(x-2)] + [(1-0)x + (3-1)x(x-1)] \\
&= [1 \cdot x(x-1) + 1 \cdot x(x-1)(x-2)] \\
&\quad + [1 \cdot x + 3 \cdot x(x-1)] - [0 \cdot x + 1 \cdot x(x-1)] \\
&= [1 \cdot x + 3 \cdot x(x-1)] + 1 \cdot x(x-1)(x-2) - 0 \cdot x \\
&= 1 \cdot x + 3 \cdot x(x-1) + 1 \cdot x(x-1)(x-2) \\
&= S(3,1) \cdot (x)_1 + S(3,2) \cdot (x)_2 + S(3,3) \cdot (x)_3
\end{aligned}$$

■

Theorem 8.13 characterizes the coefficients that express x^n as a linear combination of falling factorials. The corresponding result for expressing $(x)_n$ as a linear combination of powers of x is provided by the next definition.

DEFINITION 8.14 ***Stirling Numbers of the First Kind***

The coefficients of the expansion of $(x)_n$ as a linear combination of powers of x are the *Stirling numbers of the first kind* and are denoted by $s(n, k)$. That is,

$$(x)_n = \sum_{k=0}^{n} s(n, k) \cdot x^k$$

The notation for the Stirling numbers of the first kind is very similar to the notation for the Stirling numbers of the second kind. They differ only in the case of the s. The Stirling numbers of the first kind have no direct significance in counting problems: some of them are negative integers.

The previously mentioned mechanism for finding the numbers, $s(n, k)$, is not very convenient. It involves expanding the algebraic expression $(x)_n$. A more suitable method is presented in the next theorem.

THEOREM 8.15 ***A Recurrence Relation for $s(n, k)$***

Let n and k be positive integers. Then

$$s(n, k) = s(n\text{-}1, k\text{-}1) - (n\text{-}1) \cdot s(n\text{-}1, k)$$

where $s(0, 0) = 1$, $s(n, 0) = 0$, and $s(n, k) = 0$ if $n < k$. In addition, $s(n, n) = 1$.

Proof: The base conditions are straightforward to verify. First, $(x)_0 = 1 = s(0, 0) \cdot x^0 = s(0, 0)$. Also, for $n > 0$, x is always a factor of $(x)_n$, so $s(n, 0) = 0$ must hold (or else a term with no x's would appear in the sum in Definition 8.14). If $k > n$ and $s(n, k) \neq 0$, then a nonzero term with a power of x greater than n would appear in the expansion of the degree-n polynomial $(x)_n$. This is impossible, so $s(n, k) = 0$ whenever $k > n$.

[13] A pdf file that combines the proof of Theorem 8.13 and this example can be found at http://www.mathcs.bethel.edu/~gossett/DiscreteMathWithProof/ .

The polynomial $(x)_n = x(x-1)(x-2)\cdots(x-n+1)$ has a coefficient of 1 on the x^n term after it is expanded, so $s(n,n) = 1$.

Now assume that $n > 0$ and notice that

$$(x)_n = \sum_{k=0}^{n} s(n,k)x^k$$

and also

$$(x)_n = (x - (n\text{-}1))(x)_{n-1} = \sum_{k=0}^{n-1} s(n\text{-}1,k)x^{k+1} - \sum_{k=0}^{n-1}(n\text{-}1)s(n\text{-}1,k)x^k$$

Thus

$$\begin{aligned}
\sum_{k=0}^{n} s(n,k)x^k &= \sum_{k=0}^{n-1} s(n\text{-}1,k)x^{k+1} - \sum_{k=0}^{n-1}(n\text{-}1)s(n\text{-}1,k)x^k \\
&= \sum_{j=1}^{n} s(n\text{-}1,j\text{-}1)x^j - \sum_{k=0}^{n-1}(n\text{-}1)s(n\text{-}1,k)x^k \\
&= \left[\sum_{k=1}^{n} s(n\text{-}1,k\text{-}1)x^k - \sum_{k=1}^{n-1}(n\text{-}1)s(n\text{-}1,k)x^k\right] \\
&\quad - (n\text{-}1)s(n\text{-}1,0)x^0 \\
&= \sum_{k=1}^{n} s(n\text{-}1,k\text{-}1)x^k - \sum_{k=1}^{n-1}(n\text{-}1)s(n\text{-}1,k)x^k \\
&= \sum_{k=1}^{n} s(n\text{-}1,k\text{-}1)x^k - \sum_{k=1}^{n}(n\text{-}1)s(n\text{-}1,k)x^k \quad \text{since } s(n\text{-}1,n) = 0 \\
&= \sum_{k=1}^{n} \left[s(n\text{-}1,k\text{-}1) - (n\text{-}1)s(n\text{-}1,k)\right] x^k
\end{aligned}$$

The recurrence relation is established by comparing the coefficients of x^k for $k > 0$ on both sides of the equation. □

8.1.4 Exercises

The exercises marked with ★ have detailed solutions in Appendix H.

1. Use Corollary 8.4 (page 421) to find $p(5)$. Show your work, but do not do more than is necessary.

2. Use Theorem 8.5 (page 422) and Table 8.53 (page 519) to calculate the following values of $p(n,k)$.

(a) $p(7,4)$ (b) ★ $p(8,5)$
(c) $p(8,3)$ (d) $p(10,4)$

3. List all partitions of n for the following values of n. Subdivide the partitions into groups with identical numbers of summands. List partitions with the same number of summands in a column.

(a) ★ $n = 5$ (b) $n = 6$
(c) $n = 7$ (d) $n = 8$

4. The goal of this problem is to show that $p(n,k)$ can also be interpreted as the number of partitions of n for which the largest summand is exactly k.

(a) ★ Show that $p(n,k)$ satisfies the recurrence relation

$$p(n,k) = \sum_{i=0}^{k} p(n-k,i) \quad 0 < k \le n$$

with base conditions $p(n,1) = 1$ and $p(n,k) = 0$ if $k > n$.

(b) Define $q(n,k)$ to be the number of partitions of n for which the largest summand is exactly k. Show that

$$q(n,k) = \sum_{i=0}^{k} q(n-k,i) \quad 0 < k \le n$$

with base conditions $q(n,1) = 1$ and $q(n,k) = 0$ if $k > n$.

(c) Conclude that $p(n,k)$ also represents the number of partitions of n for which the largest summand is exactly k.

5. Prove that $p(n) \leq \frac{1}{2}(p(n-1) + p(n+1))$.
Hints:

i. Write the inequality as
$p(n+1) - p(n) \geq p(n) - p(n-1)$.

ii. Interpret $p(n+1) - p(n)$ as the number of partitions of $n+1$ that do not contain a 1 as a summand. (You need to give adequate justification for this assertion.)

iii. Create a similar interpretation for $p(n) - p(n-1)$.

iv. Now complete the proof.

6. Use generating functions (Section 7.4) and Theorem 8.5 to prove that $p(n, 2) = \lfloor \frac{n}{2} \rfloor$ for $n \geq 2$. You may assume that $p(0, k) = 0$ for all $k > 0$.

7. For each of the occupancy problem cases below, make a complete listing of all valid distributions when $n = 4$ and $k = 3$. Label the objects a, b, c, d when they are distinct and label the containers X, Y, Z when they are distinct. Use *'s to indicate indistinguishable objects. Note that **OD CD ∅** (with 81 distributions) has been omitted.

(a) **OD CI ¬∅** (b) **OD CI ∅**
(c) ★ **OD CD ¬∅** (d) **OI CD ∅**
(e) **OI CD ¬∅** (f) **OI CI ¬∅**
(g) **OI CI ∅**

8. I have a bowl of identical candy bars. Four young trick-or-treaters have just knocked on my door and are holding their bags out for some candy. If I grab six candy bars from my bowl, in how many ways can I distribute all six of them? Give two answers: (a) I will try to be fair (b) I need not be fair.

9. ★ Jody has 12 identical boxes of Girl Scout cookies. She plans to sell them all to relatives. If she has six candidate relatives, in how many ways can she unload the cookies? You may assume that some relatives may decline her offer, but that all the boxes will be sold.

10. A technology company is moving to a new building with an excess of space. The company has 11 programmers on staff. If the new building contains two identical office suites, each large enough to hold up to 10 programmers, in how many ways can the programmers be assigned to suites?

11. A library has four identical display cases that are used to promote new acquisitions. This month, the librarians wish to promote nine books. They do not want any empty display cases. In how many ways can the books be displayed?

12. I have seven new quarters that I wish to place into my two front pockets. I really don't care which pocket(s) the coins end up in. In how many ways can I pocket the quarters?

13. An English professor has chosen five novels for a course in contemporary literature. Each student must choose to write an analysis of at least one of the novels. Extra credit will be given for every additional analysis. There are six students enrolled in the class. Give a numeric count of the number of ways the students can submit reports.

14. Mei Ling received three identical laundry bags as high school graduation presents. She has brought all three to college. If she has 20 pieces of laundry, in how many ways can she stuff the bags with all 20 pieces of dirty laundry?

15. Fifteen people are going to attend a picnic. Shirley is packing one ham and cheese sandwich for each of these people. She will carry the sandwiches in all three of her red and white checkered picnic baskets. In how many ways can Shirley bring the sandwiches to the picnic?

16. ★ Jeff just went shopping and bought his parents an anniversary present consisting of eight blue towels. The cashier told him that she had four white bags available for carrying the towels home. In how many ways can Jeff bring his towels home, assuming that Jeff does not have to use all the bags?

17. Suppose that you are going to design six blue T-shirts. You have nine pictures that can be ironed onto the T-shirts: a lion, a cat, a flower, a heart, a moon, a star, a person, a palm tree, and a dolphin. In how many ways can the T-shirts be created, assuming there are no boundaries concerning the number of pictures to be placed on each T-shirt?

18. Suppose that you are the boss of a small company. You have three distinct jobs that need to be completed next week by people on your staff: stuffing envelopes, copying files, and typing. There are seven employees who will each be assigned one of these jobs. At least one employee must be assigned to each task. In how many ways can you assign the jobs?

19. The people in charge of a raffle are going to give out four prize bags to the raffle winners. They want to make sure that each prize bag contains at least one of the 11 gift certificates for a department store. In how many ways can the prize bags be prepared, assuming that all the gift certificates are of the same monetary value?

20. Meg decides to throw her 10-year-old daughter Sue a birthday party and allows her to invite 15 friends to the house. Meg purchases 25 red balloons for the children, since red is Sue's favorite color. In how many ways can the balloons be distributed to the children at the birthday party (including Sue), assuming that no child should go home without a balloon?

21. Solve Exercise 44(b) on page 239 as an occupancy problem.

22. Illustrate (by listing the distributions) the combinatorial proof of Theorem 8.10 (page 428) when $n = 4$ and $k = 2$.

23. Extend Table 8.54 on page 499 so that it includes values for $n = 5$ and $n = 6$.

24. ★ Express $S(9, 4)$ in terms of $S(6, k)$, for $0 < k \leq 6$.

25. (From the 2000 British Mathematical Olympiad.) Find the number of ways in which the seven dwarves could form into four unnamed teams, where there is no order to the teams nor to the dwarves within each team.

26. Let $T(n, k)$ be the number of ways to partition a set of n distinct elements into exactly k nonempty, ordered subsets.[14]

(a) Prove that $T(n, k) = k!S(n, k)$.

(b) Use inclusion–exclusion to prove that

$$T(n, k) = \sum_{i=0}^{k} (-1)^i \binom{k}{i} (k - i)^n$$

for $0 < k \leq n$. [*Hint*: There are k^n ways to create the partitions if empty subsets are allowed. This count contains illegal partitions, so subtract the number of partitions with at least one empty subset. Now adjust for over-subtracting in the previous step, etc.] Finally,

(c) Conclude that

$$S(n, k) = \frac{1}{k!} \sum_{i=0}^{k} (-1)^i \binom{k}{i} (k - i)^n$$

for $0 < k \leq n$.

27. Create a table (similar to Table 8.7 on page 430) of values for $s(n, k)$ with $1 \leq k \leq n \leq 6$.

28. What is the numeric coefficient of the term $x(x-1)(x-2)(x-3)(x-4)$ when x^6 is written as a linear combination of falling factorials?

29. What is the numeric coefficient of the term $x(x-1)(x-2)(x-3)$ when x^7 is written as a linear combination of falling factorials?

30. ★ What is the numeric coefficient of x^3 when $x(x-1)(x-2)(x-3)(x-4)(x-5)$ is written as a linear combination of powers of x?

31. What is the numeric coefficient of x^5 when $x(x-1)(x-2)(x-3)(x-4)$ is written as a linear combination of powers of x?

32. Express the polynomial $3x^4 + 2x^3 - 6x^2 + 11x - 5$ as a linear combination of falling factorials.

33. Express the polynomial $2x^5 - 4x^4 + x^3 + 7x^2 - 9x - 9$ as a linear combination of falling factorials.

8.2 Latin Squares, Finite Projective Planes

This section introduces two combinatorial designs: Latin squares and finite projective planes. It concludes by discussing the connection between them.

8.2.1 Latin Squares

Latin squares are similar to magic squares because both are n-by-n matrices with positive integer entries. However, magic squares contain one copy of each of the numbers $1, 2, \ldots, n^2$, whereas Latin squares contain n copies each of the numbers $1, 2, \ldots, n$. The major defining characteristic (or boundary condition) is described in the formal definition.

DEFINITION 8.16 ***Latin Square***

A *Latin square of order n* is an n-by-n matrix for which every entry is a number in $\{1, 2, \ldots, n\}$. Every number in $\{1, 2, \ldots, n\}$ must appear at least once in every row and at least once in every column.

A Latin square of order n is *standardized* if the elements in the first row are written in increasing numeric order moving left to right, and the elements in the first column are written in increasing numeric order moving top to bottom.

It is easy to prove that every number in $\{1, 2, \ldots, n\}$ must appear *at most* once in every row and in every column (see Exercise 1 on page 457). Combining this observation with the requirement in the definition, it should be clear that every row and every column is a permutation of the set $\{1, 2, \ldots, n\}$, as defined next.

DEFINITION 8.17 ***Permutation of a Set***

A *permutation of the set* $\{a_1, a_2, \ldots, a_n\}$ is an ordered arrangement of the elements $a_1, a_2, \ldots, a_n$.

There are $n!$ distinct orderings (permutations) of the n elements $\{a_1, a_2, \ldots, a_n\}$ (see Section 5.1.2).

[14]This problem is from [15].

EXAMPLE 8.9

Some Small Latin Squares

There is only one Latin square of order 1. There are two distinct Latin squares of order 2. The one on the left is standardized.

$$\begin{matrix} 1 & 2 \\ 2 & 1 \end{matrix} \qquad\qquad \begin{matrix} 2 & 1 \\ 1 & 2 \end{matrix}$$

There are several Latin squares of order 3. One of them (the only standardized Latin square of order 3) is shown next.

$$\begin{matrix} 1 & 2 & 3 \\ 2 & 3 & 1 \\ 3 & 1 & 2 \end{matrix}$$

■

✔ Quick Check 8.5

1. List all distinct Latin squares of order 3.
2. Produce all standardized Latin squares of order 4. ☑

The number, $L(n)$, of distinct Latin squares grows rapidly with n. In fact, Theorem 8.65 on page 508 shows that there are at least $n! \cdot (n-1)! \cdot (n-2)! \cdots 3! \cdot 2! \cdot 1$ distinct Latin squares of order n. Table 8.8 shows some values for this lower bound, as well as the actual values for $L(n)$. We have already verified that the lower bound is exact for $n = 1, 2$, and 3.

TABLE 8.8 Lower Bounds and Actual Values for the Number of Distinct Latin Squares with $n \leq 6$

n	1	2	3	4	5	6
$L(n) \geq$	1	2	12	288	34560	24883200
$L(n)$	1	2	12	576	161280	812851200

Latin Squares and the Design of Experiments

Latin squares can be used to help researchers design experiments that minimize bias in the results. In particular, if the experiment has two factors that might influence the result, a Latin square can be of use.

EXAMPLE 8.10

A New Comic Strip

Suppose that a newspaper editor has decided to replace one of the comic strips in the daily paper. He has found three potential replacement strips but would like to learn a bit about subscriber preferences before making a decision.

He has decided to run each candidate replacement comic strip for a week and then see which is most popular. It has occurred to him that the order in which the strips appear might be a significant factor in how subscribers react. He would therefore like to divide the subscribers into subgroups with each possible ordering of the three candidate strips assigned to a subgroup. This requires $3! = 6$ subgroups.

The editor has realized that there is a practical problem with this scheme. The problem is that the newspaper would need to print six editions of each paper, but currently there are only three editions: metro, northern suburbs, and southern suburbs. Creating the extra editions is expensive.

The editor has found an acceptable alternative that still controls for the "order of appearance" effect. Suppose the candidate comic strips are denoted as A, B, and C, with editions denoted M, N, and S. The following Latin square uses comic strips as row labels and editions as column labels. The entry in the ith row and jth column represents the week in which comic strip i will appear in edition j. In this arrangement, each strip will appear in some edition first, in another edition second, and in the other edition last. Every edition will run each strip for a week.

	M	N	S
A	1	2	3
B	2	3	1
C	3	1	2

This scheme is not as exhaustively thorough as the original scheme (for example, no subscribers will read the strips in the order ACB). However, it is far better than arbitrarily picking only one ordering to run in all editions. ■

The previous method of experimental design can be used in many situations, as the next example illustrates.

EXAMPLE 8.11

An Agricultural Experiment

An agricultural researcher is trying to breed a new variety of popcorn. She has three university-owned farms in the area that each have a small section of field she may use. She wants to study the effects of three different fertilizer treatment regimes on each of three new varieties of seed. She knows that the three fields have different soil characteristics (among other differences), so the choice of field is significant when interpreting the results.

In addition, she is assuming that the variety of seed and the fertilizer regimes are also significant. She does not have the funding to run separate experiments to test seeds and fertilizers, so she needs to study both factors simultaneously. An exhaustive comparison would have each seed variety matched with each fertilizer regime in each field. This would require each section of field to be subdivided into nine plots. Her allocation of field space is not large enough for this to be practical; she only has space for three plots per field.

She has decided to use a Latin square design to minimize bias in the results. In particular, she wants to ensure that every variety of seed is matched with every fertilizer regime, that every variety of seed is planted in a plot of each field, and that every fertilizer regime is used in some plot of every field.

The rows in the Latin square represent the seed varieties: A, B, and C. The columns represent the fertilizer regimes: L, M, and N. The numeric entries in the design represent the fields.

	L	M	N
A	1	3	2
B	2	1	3
C	3	2	1

■

The next example is more complex.

EXAMPLE 8.12

Washing Machines

An industrial chemist is studying the effectiveness of a number of options for washing clothes. He wishes to compare three brands of washing machine, three brands of detergent, as well as three water temperatures. He also wishes to factor in the effects of water hardness. He will use three levels of hardness. He would like to match these characteristics as fairly as possible, but does not have the time or resources to examine every conceivable combination (there are $3^4 = 81$ combinations).

He has decided that the washing machine brand (A, B, and C) and detergent brand (X, Y, and Z) should be represented as the rows and columns, respectively, of a design. He has ranked the water temperatures as 1, 2, and 3 and also ranked water hardness as 1, 2 and 3. He is searching for a design with ordered pairs, (t, h) as entries. An entry of (t, h) in row i and column j means that temperature t and hardness h will be used together in a brand i washing machine with detergent j.

His first attempt at a design is shown next. He has realized that the set of temperatures should form a Latin square and the set of water hardness ratings should also form a Latin square. These Latin squares will ensure that each temperature (water hardness) will be used with each brand of washer and each brand of detergent.

	X	Y	Z
A	$(1, 2)$	$(2, 1)$	$(3, 3)$
B	$(2, 1)$	$(3, 3)$	$(1, 2)$
C	$(3, 3)$	$(1, 2)$	$(2, 1)$

The problem with this design is that temperature 1 and hardness 1 are never tested together, nor is temperature 1 and hardness 3. The researcher revised the design and produced the following arrangement. This time, each temperature is matched with each hardness. In fact, the design ensures that the two elements in every pair of categories (washer, detergent, temperature, hardness) share a common test. Thus, brands A and Y appear together in a test, brand B and hardness 3 share a common test, temperature 1 and hardness 2 occur together, and so on.

	X	Y	Z
A	$(1, 1)$	$(2, 3)$	$(3, 2)$
B	$(2, 2)$	$(3, 1)$	$(1, 3)$
C	$(3, 3)$	$(1, 2)$	$(2, 1)$

■

Orthogonal Latin Squares

Example 8.12 introduced one of the most important aspects of Latin squares: orthogonality. The next definition provides the formal context. The definition uses notational conventions from Definition E.2 on page A29.

DEFINITION 8.18 ***Orthogonal Latin Squares***

Let $L^1 = (a_{ij})$ and $L^2 = (b_{ij})$ be two Latin squares of order n. L^1 and L^2 are said to be *orthogonal* if the set of ordered pairs $\{(a_{ij}, b_{ij}) \mid i = 1, 2, \ldots, n$ and $j = 1, 2, \ldots, n\}$ contains n^2 distinct ordered pairs. That is, $(a_{ij}, b_{ij}) \neq (a_{rs}, b_{rs})$ unless $i = r$ and $j = s$.

A collection of k Latin squares of order n is said to be *mutually orthogonal* if every pair in the collection is orthogonal.

EXAMPLE 8.13 **Example 8.12 Revisited**

Example 8.12 already has established that the following pair of Latin squares is orthogonal. The collection of ordered pairs is also shown.

1	2	3		1	3	2	$(1, 1)$	$(2, 3)$	$(3, 2)$
2	3	1		2	1	3	$(2, 2)$	$(3, 1)$	$(1, 3)$
3	1	2		3	2	1	$(3, 3)$	$(1, 2)$	$(2, 1)$

Notice that all nine ordered pairs are distinct; there are no repetitions. Compare this with the pair of Latin squares that the chemist initially tried. The set of ordered pairs, $\{(1, 2), (2, 1), (3, 3)\}$, contains only three elements.

1	2	3		2	1	3	$(1, 2)$	$(2, 1)$	$(3, 3)$
2	3	1		1	3	2	$(2, 1)$	$(3, 3)$	$(1, 2)$
3	1	2		3	2	1	$(3, 3)$	$(1, 2)$	$(2, 1)$

■

EXAMPLE 8.14

Mutually Orthogonal Latin Squares

The following collection is a set of three mutually orthogonal Latin squares of order 4. You should take the time to verify that each pair is orthogonal.

$$L^1 = \begin{matrix} 1 & 2 & 3 & 4 \\ 2 & 1 & 4 & 3 \\ 3 & 4 & 1 & 2 \\ 4 & 3 & 2 & 1 \end{matrix} \qquad L^2 = \begin{matrix} 1 & 2 & 3 & 4 \\ 4 & 3 & 2 & 1 \\ 2 & 1 & 4 & 3 \\ 3 & 4 & 1 & 2 \end{matrix} \qquad L^3 = \begin{matrix} 1 & 2 & 3 & 4 \\ 3 & 4 & 1 & 2 \\ 4 & 3 & 2 & 1 \\ 2 & 1 & 4 & 3 \end{matrix}$$

■

A brief examination of Example 8.9 on page 436 will convince you that there are no pairs of orthogonal Latin squares of order 2. It *is* possible to find a pair of orthogonal Latin squares of order 3 (Example 8.13) and also possible to find a set of three mutually orthogonal Latin squares of order 4 (Example 8.14). An exhaustive search through the collection of all 12 Latin squares of order 3 (see the Quick Check 8.5 solution on page 521) will prove that there is no collection of three or more mutually orthogonal Latin squares of order 3. Let $m(n)$ represent the maximum number of mutually orthogonal Latin squares of order n. Table 8.9 shows what has been established so far in this discussion.

TABLE 8.9 Partial Results for $m(n)$

n	2	3	4
$m(n)$	1	2	at least 3

Exhaustively verifying (or disproving) that $m(4) = 3$ seems like too much work. Perhaps a better approach is to attempt a conjecture based on the (admittedly meager) results so far. The obvious guesses are that $m(n) = n - 1$ or $m(n) \geq n - 1$ or $m(n) \leq n - 1$.

The following example will provide the key insight needed to identify and prove the proper conjecture.

EXAMPLE 8.15

Transforming Sets of Orthogonal Latin Squares

Consider the pair of orthogonal Latin squares from a previous example.

$$L^1 = \begin{matrix} 1 & 2 & 3 \\ 2 & 3 & 1 \\ 3 & 1 & 2 \end{matrix} \qquad L^2 = \begin{matrix} 1 & 3 & 2 \\ 2 & 1 & 3 \\ 3 & 2 & 1 \end{matrix}$$

It is possible to modify L^2 so that its first row is the same as the first row of L^1 and so that the new pair is still orthogonal. The transformation is very easy.

Notice that the entry in the first row, second column of L^2 is a 3 but a 2 is desired. Simply change every 3 to a 2 and vice versa. The result is shown next.

$$L^1 = \begin{matrix} 1 & 2 & 3 \\ 2 & 3 & 1 \\ 3 & 1 & 2 \end{matrix} \qquad L^{2'} = \begin{matrix} 1 & 2 & 3 \\ 3 & 1 & 2 \\ 2 & 3 & 1 \end{matrix}$$

Because 2s and 3s have been uniformly exchanged, $L^{2'}$ is still a Latin square (every number appears exactly once in each row and column). Also, the exchange has not changed the property that the set of ordered pairs from corresponding positions still contains all nine possible ordered pairs. This is true since, for example, the old pair $(1, 2)$ becomes $(1, 3)$ after the transformation, but the old $(1, 3)$ simultaneously becomes $(1, 2)$.

In a similar fashion, consider the following three mutually orthogonal Latin squares of order 4.

$$L^1 = \begin{matrix} 3 & 2 & 1 & 4 \\ 2 & 3 & 4 & 1 \\ 1 & 4 & 3 & 2 \\ 4 & 1 & 2 & 3 \end{matrix} \qquad L^2 = \begin{matrix} 4 & 1 & 2 & 3 \\ 3 & 2 & 1 & 4 \\ 1 & 4 & 3 & 2 \\ 2 & 3 & 4 & 1 \end{matrix} \qquad L^3 = \begin{matrix} 2 & 4 & 3 & 1 \\ 3 & 1 & 2 & 4 \\ 1 & 3 & 4 & 2 \\ 4 & 2 & 1 & 3 \end{matrix}$$

These can be transformed into a new set of mutually orthogonal Latin squares in which the first rows are all "1 2 3 4". This can be done in three steps (for this example).

In step 1, interchange 1 and 3 in L^1, interchange 1 and 4 in L^2, and interchange 1 and 2 in L^3. The resulting Latin squares are shown.

$$L^{1'} = \begin{matrix} 1 & 2 & 3 & 4 \\ 2 & 1 & 4 & 3 \\ 3 & 4 & 1 & 2 \\ 4 & 3 & 2 & 1 \end{matrix} \qquad L^{2'} = \begin{matrix} 1 & 4 & 2 & 3 \\ 3 & 2 & 4 & 1 \\ 4 & 1 & 3 & 2 \\ 2 & 3 & 1 & 4 \end{matrix} \qquad L^{3'} = \begin{matrix} 1 & 4 & 3 & 2 \\ 3 & 2 & 1 & 4 \\ 2 & 3 & 4 & 1 \\ 4 & 1 & 2 & 3 \end{matrix}$$

In step 2, interchange 2 and 4 in both $L^{2'}$ and $L^{3'}$.

$$L^{1'} = \begin{matrix} 1 & 2 & 3 & 4 \\ 2 & 1 & 4 & 3 \\ 3 & 4 & 1 & 2 \\ 4 & 3 & 2 & 1 \end{matrix} \qquad L^{2''} = \begin{matrix} 1 & 2 & 4 & 3 \\ 3 & 4 & 2 & 1 \\ 2 & 1 & 3 & 4 \\ 4 & 3 & 1 & 2 \end{matrix} \qquad L^{3''} = \begin{matrix} 1 & 2 & 3 & 4 \\ 3 & 4 & 1 & 2 \\ 4 & 3 & 2 & 1 \\ 2 & 1 & 4 & 3 \end{matrix}$$

In the final step, interchange 3 and 4 in $L^{2''}$.

$$L^{1'} = \begin{matrix} 1 & 2 & 3 & 4 \\ 2 & 1 & 4 & 3 \\ 3 & 4 & 1 & 2 \\ 4 & 3 & 2 & 1 \end{matrix} \qquad L^{2'''} = \begin{matrix} 1 & 2 & 3 & 4 \\ 4 & 3 & 2 & 1 \\ 2 & 1 & 4 & 3 \\ 3 & 4 & 1 & 2 \end{matrix} \qquad L^{3''} = \begin{matrix} 1 & 2 & 3 & 4 \\ 3 & 4 & 1 & 2 \\ 4 & 3 & 2 & 1 \\ 2 & 1 & 4 & 3 \end{matrix}$$

It is now possible to make a clever observation about $L^{1'}$, $L^{2'''}$, and $L^{3''}$. Suppose we want to find a fourth Latin square of order 4, L^4, to add to this mutually orthogonal collection. The same kinds of transformations can be used to ensure that the first row of L^4 is "1 2 3 4" before it is added to the collection.

What number can be in the second row, first column of L^4? It can't be a 1, or else L^4 would not be a Latin square. If it is a 2, then L^4 and $L^{1'}$ would not be orthogonal since there would be at least two copies of the ordered pair (2, 2). There are similar problems with a 3 or 4 in that position of L^4. The conclusion is that no such Latin square can exist. There are at most 3 mutually orthogonal Latin squares of order 4. ■

Theorem 8.21 on page 441 will present the major result about sets of mutually orthogonal Latin squares: There are at most $n - 1$ mutually orthogonal Latin squares of order n. Two lemmas will be used in the proof of the theorem.

LEMMA 8.19 ***Interchanging Numbers Preserves Latin Squares***

Let L be a Latin square of order n and let $i, j \in \{1, 2, \ldots, n\}$. If every copy of i in L is changed to a j, and simultaneously, every j in L is changed to an i, then the resulting n-by-n matrix is still a Latin square.

Proof: Since L is a Latin square, every number in $\{1, 2, \ldots, n\}$ appears exactly once in each row and in each column. After the interchange, there will still be exactly one i and one j in each row and column (but now in different positions). Every other number, $k \notin \{i, j\}$, will still be in the same positions in the matrix, so will also appear exactly once in each row and in each column. Consequently, every number in $\{1, 2, \ldots, n\}$ still appears exactly once in each row and in each column. Therefore the new matrix will be a Latin square. □

LEMMA 8.20 ***Interchanging Numbers Preserves Orthogonality***

Let L^1 and L^2 be orthogonal Latin squares of order n and let $i, j \in \{1, 2, \ldots, n\}$. Form an n-by-n matrix $L^{1'}$ by changing every copy of i in L^1 to a j, and simultaneously, changing every j in L^1 to an i. All other entries in L^1 are copied unchanged into $L^{1'}$. Then $L^{1'}$ and L^2 are also orthogonal Latin squares.

Proof: Lemma 8.19 ensures that $L^{1'}$ is a Latin square. Let M be the matrix of ordered pairs from L^1 and L^2 and let M' be the matrix of ordered pairs from $L^{1'}$ and L^2.

Since L^1 and L^2 are orthogonal, the ordered pairs (i, k) and (j, k) (for every choice of k) both appear somewhere in M. After interchanging i and j, the ordered pair (i, k) will appear in M' at the position that (j, k) is at in M. Similarly, the ordered pair (j, k) will appear in M' at the position occupied by (i, k) in M. That is, the ordered pairs (i, k) and (j, k) will swap places, for every choice of k. Ordered pairs that do not have either i or j as first element do not move.

The interchange has moved some ordered pairs to other positions, but no ordered pairs have disappeared and no new ordered pairs have been created. Therefore, $L^{1'}$ and L^2 are orthogonal. □

EXAMPLE 8.16 **Illustrating the Proof of Lemma 8.20**

Consider the following Latin squares. The matrix of ordered pairs is also shown.

$$L^1 = \begin{matrix} 1 & 2 & 3 \\ 2 & 3 & 1 \\ 3 & 1 & 2 \end{matrix} \qquad L^2 = \begin{matrix} 1 & 3 & 2 \\ 2 & 1 & 3 \\ 3 & 2 & 1 \end{matrix} \qquad \begin{matrix} (1,1) & (2,3) & (3,2) \\ (2,2) & (3,1) & (1,3) \\ (3,3) & (1,2) & (2,1) \end{matrix}$$

Suppose the 1s and 2s are exchanged in L^1, creating $L^{1'}$. In the matrix of ordered pairs, the elements $(a^1_{11}, a^2_{11}) = (1, 1)$ and $(a^1_{33}, a^2_{33}) = (2, 1)$ have traded positions. Also, $(1, 2)$ and $(2, 2)$ have swapped places and $(1, 3)$ and $(2, 3)$ have traded places.

$$L^{1'} = \begin{matrix} 2 & 1 & 3 \\ 1 & 3 & 2 \\ 3 & 2 & 1 \end{matrix} \qquad L^2 = \begin{matrix} 1 & 3 & 2 \\ 2 & 1 & 3 \\ 3 & 2 & 1 \end{matrix} \qquad \begin{matrix} (2,1) & (1,3) & (3,2) \\ (1,2) & (3,1) & (2,3) \\ (3,3) & (2,2) & (1,1) \end{matrix}$$

■

THEOREM 8.21 ***Maximal Sets of Mutually Orthogonal Latin Squares***

Let $\{L^1, L^2, \ldots, L^k\}$ be a collection of mutually orthogonal Latin squares of order $n > 1$. Then $k \le n - 1$.

Proof: Use sequences of interchanges to transform each of the Latin squares into a Latin square whose first row is "$123 \cdots n$". Lemmas 8.19 and 8.20 ensure that the new collection, $\{L^{1'}, L^{2'}, \ldots, L^{k'}\}$, is still a set of mutually orthogonal Latin squares.

How many distinct choices are there for the element in the second row, first column? That element cannot be a 1 or the resulting matrix would not be a Latin square (two 1s in column 1). It can, however, be any of the numbers $2, 3, \ldots, n$. Can both $L^{i'}$ and $L^{j'}$ contain the same value in row two, column 1? If so, the matrix of ordered pairs for $L^{i'}$ and $L^{j'}$ would contain an ordered pair (m, m) in row two, column 1 and also in row 1, column m. This contradicts the orthogonality of $L^{i'}$ and $L^{j'}$. Consequently, no two Latin squares in $\{L^{1'}, L^{2'}, \ldots, L^{k'}\}$ have the same element in row 2, column 1.

There are only $n - 1$ possible choices for the number in row 2, column 1 for the Latin squares in $\{L^{1'}, L^{2'}, \ldots, L^{k'}\}$. Consequently, $k \le n - 1$. □

The maximum number, $m(n)$, of mutually orthogonal Latin squares of order n is still an open question. Table 8.10 (extracted from [19]) shows what is known for $2 \le n \le 12$. Note that it is known that there cannot be $n - 1 = 9$ mutually orthogonal Latin squares of order 10, but the maximum is not known exactly. It is not known whether there can be 11 mutually orthogonal Latin squares of order 12. However, the exact maximum is known for many (but not all) values of n greater than 12. In particular, if n is a prime power, then $n - 1$ mutually orthogonal Latin squares do exist [17].

TABLE 8.10 The Maximum Number, $m(n)$, of Mutually Orthogonal Latin Squares of Order n

n	2	3	4	5	6	7	8	9	10	11	12
$m(n)$	1	2	3	4	1	6	7	8	$2 \le m(10) < 9$	10	$5 \le m(12) \le 11$

As will be seen in Section 8.2.2, the lack of even a pair of orthogonal Latin squares of order 6 is quite significant. The following problem is directly related to $m(6) = 1$.

EXAMPLE 8.17 **Euler's 36-Officer Problem**

Leonhard Euler once posed a question about arranging 36 military officers in a 6-by-6 arrangement with some extra conditions. The officers are characterized by two qualities: their rank (lieutenant, colonel, etc.) and their regiment. He stated the problem in this manner:

> A very curious question, which has exercised for some time the ingenuity of many people, has involved me in the following studies, which seem to open a new field of analysis, in particular the study of combinations. The question revolves around arranging 36 officers to be drawn from 6 different ranks and also from 6 different regiments so that they are ranged in a square so that in each line (both horizontal and vertical) there are 6 officers of different ranks and different regiments.

Euler conjectured that it is impossible to construct such an arrangement (with the additional condition that no rank/regiment pair appears more than once). The value $m(6) = 1$ in the previous table shows that his instinct was correct.

Euler also conjectured that $m(n) = 1$ for any n with $n \equiv 2 \pmod 4$. The value $2 \leq m(10)$ shows that this conjecture was incorrect (a rarity for Euler). ■

8.2.2 Finite Projective Planes

You are familiar with the geometry of the plane and the undefined terms *point* and *line*. Euclid's axioms for the plane were presented in Example 3.1 on page 87. A less useful geometry was introduced in Example 3.5 on page 90.

This section introduces a much more useful collection of non-Euclidean geometric axioms. These axioms will define the notion of a *finite projective plane*.

DEFINITION 8.22 ***Finite Projective Plane***

A *finite projective plane* consists of a finite set, $\mathcal{P}$, of *points* and a finite set, $\mathcal{L}$, of *lines*. Lines are finite sets of points. If $L = \{p_1, p_2, \ldots, p_k\}$, then point p_i is said to be *on* line L and L is said to *contain* the point p_i, for $i = 1, 2, \ldots, k$.

The following axioms characterize finite projective planes.

FPP1 Any two distinct points are on one and only one common line.

FPP2 Any two distinct lines contain one and only one common point.

FPP3 There exist four distinct points, no three of which are on a common line.

Axiom FPP2 implies that there is no such thing as "parallel lines" in a finite projective plane.

Notice the symmetry in Axioms FPP1 and FPP2. It is possible to consider a point to be the set of lines that contain that point. This sense of "mirror image interchangeability" between points and lines is called *duality*.[15]

The Duality Principle for Finite Projective Planes

Let T be a theorem that is valid for a finite projective plane. If the terms *point* and *line* and also *on* and *contains* (or their equivalents) are interchanged, the result is also a theorem that is valid for the finite projective plane.

[15] This is similar to the duality principle for Boolean algebras (page 63).

The proof of the duality principle for finite projective planes, depends on the ability to replace axiom FPP3 by its dual. The following proposition proves half of what is needed to verify the equivalence between FPP3 and its dual, FPP3′. The other half is Exercise 17 on page 458. The interchangeability of FPP3 and FPP3′ means that both could be included as part of the definition (adding some redundancy). The four axioms would then contain two fully dual pairs of axioms, ensuring the validity of the duality principle for finite projective planes.

PROPOSITION 8.23 ***FPP3′***

In any finite projective plane there exist four distinct lines, no three of which contain a common point.

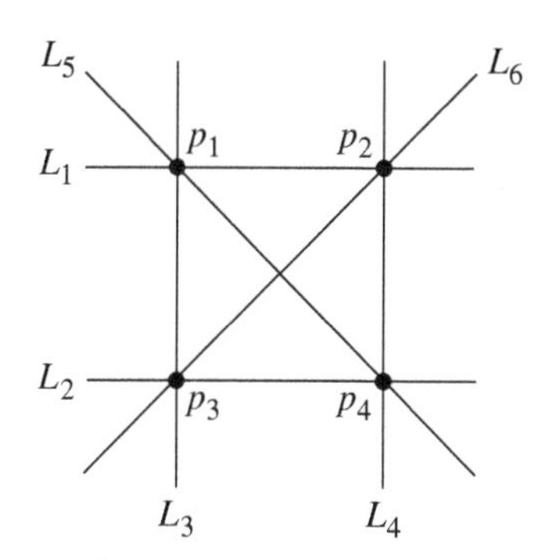

Figure 8.1. A known sub-configuration in $\mathcal{F}$.

Proof: Let $\mathcal{F}$ be a finite projective plane. By FPP3, there exist four distinct points, p_1, p_2, p_3, p_4, such that no three are on a common line. This, together with FPP1 and FPP2, implies that there must be six distinct lines, $L_1, L_2, L_3, L_4, L_5, L_6$, such that every pair of distinct points in $\{p_1, p_2, p_3, p_4\}$ is associated with a distinct line in $\{L_1, L_2, L_3, L_4, L_5, L_6\}$. Without loss of generality,[16] a portion of $\mathcal{F}$ can be depicted using Figure 8.1.[17]

Suppose that the proposition is false. Then in every set of four lines, at least three contain a common point.[18] This is true in the configuration in the diagram for 12 of the 15 possible subsets of four lines. The subsets that need more thought are $\{L_1, L_2, L_3, L_4\}$, $\{L_1, L_2, L_5, L_6\}$, and $\{L_3, L_4, L_5, L_6\}$. Any one of these three subsets will allow a contradiction to be generated.

Consider the lines $\{L_1, L_2, L_3, L_4\}$. Under the assumption that FPP3′ is false, there must be a collection of three of them that contain a common point. This leads to four very similar cases, depending on which three lines are guaranteed to contain a common point.

- Suppose L_1, L_2, L_3 contain a common point, p. Then $p \neq p_3$ because p_3 is not on L_1. However, both p and p_3 are on L_2 and also on L_3. This violates FPP1. (Alternatively, L_2 and L_3 are distinct lines that both contain two distinct points, p and p_3, violating FPP2.)

- Suppose L_1, L_2, L_4 contain a common point, p. Then $p \neq p_4$ because p_4 is not on L_1. However, both p and p_4 are on L_2 and also on L_4. This violates FPP1.

- Suppose L_1, L_3, L_4 contain a common point, p. Then $p \neq p_1$ because p_1 is not on L_4. However, both p and p_1 are on L_1 and also on L_3. This violates FPP1.

- Suppose L_2, L_3, L_4 contain a common point, p. Then $p \neq p_3$ because p_3 is not on L_4. However, both p and p_3 are on L_2 and also on L_3. This violates FPP1.

In every case, a contradiction arises. The assumption that led to the contradiction was the assumption that the proposition is false. Therefore, the proposition must be true. □

[16] It is always possible to rename some points and lines so that this configuration will occur in $\mathcal{F}$.

[17] The diagram in Figure 8.1 shows only a portion of the finite projective plane. The situation is like looking at a very complex map with a magnifying glass. The portion of the map that is under the magnifying glass is clear and easy to read; the words on the rest of the map are too small to read. However, the portion that is readable is all that is necessary to provide the needed information. In a similar fashion, the little bit of the finite projective plane that is illustrated in the diagram is all that is necessary to arrive at the desired conclusion.

[18] Review Section 2.6.1 on page 69 if this negation is not clear.

EXAMPLE 8.18

The Fano Plane

The smallest finite projective plane is known as the Fano plane. It contains seven points and seven lines. Every line contains three points. Figure 8.2 is the standard visual representation for the Fano plane. It shows the plane with and without line labels.

The points are a, b, c, d, e, f, and g. The lines are listed at the top of the figure. Line L_4 is conventionally depicted as a circle for aesthetic reasons.

FPP1 Any two distinct points are on one and only one common line.

FPP2 Any two distinct lines contain one and only one common point.

FPP3 There exist four distinct points, no three of which are on a common line.

FPP3′ There exist four distinct lines, no three of which contain a common point.

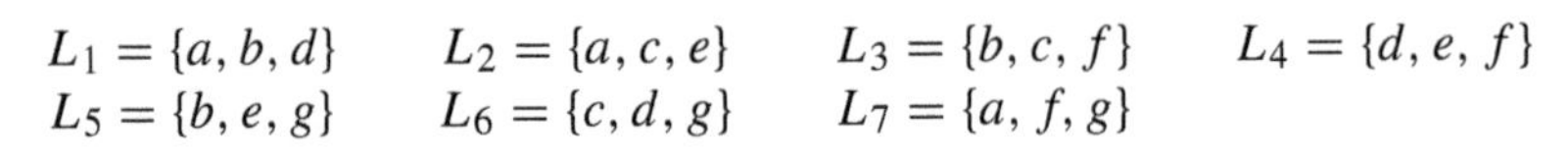

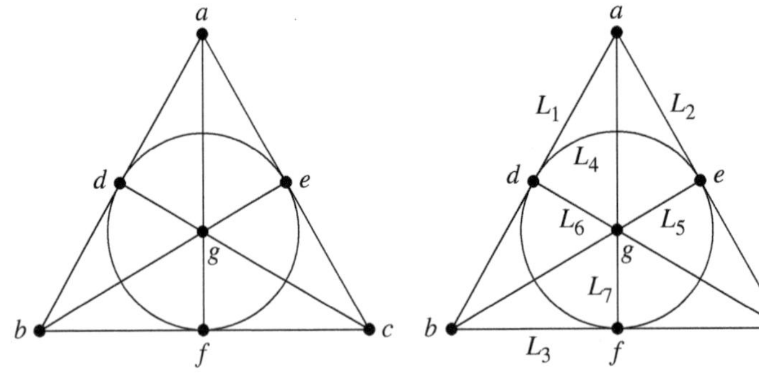

Figure 8.2 The Fano plane.

You should verify that FPP1 and FPP2 hold for this example. Notice that the points a, b, c, and g satisfy FPP3: No three of the four are on a common line. The lines $L_1 L_2$, L_3, and L_4 satisfy the requirements of FPP3′: No three of them contain a common point. ■

Some simple, but useful facts about finite projective planes are captured in the following lemmas.

LEMMA 8.24
Every line, L, in a finite projective plane must contain at least three distinct points.

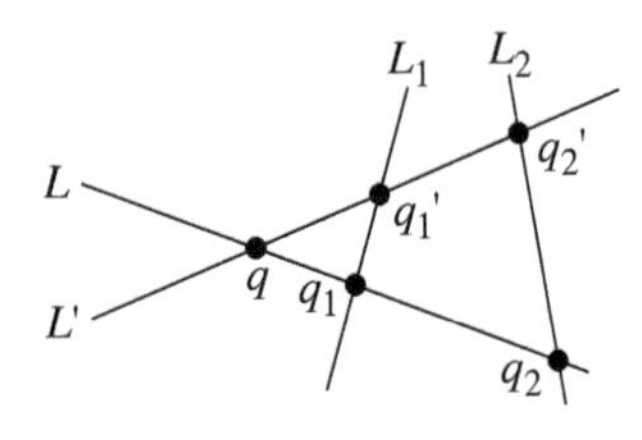

Figure 8.3. L must contain at least three distinct points.

Proof: FPP3′ asserts the existence of four lines, L_1, L_2, L_3, L_4, no three of which contain a common point. At least three of these lines are distinct from L. Assume, with suitable renaming if necessary, that $L \notin \{L_1, L_2, L_3\}$. Let the common point on both L and L_i be denoted q_i, for $i = 1, 2, 3$. If the points q_i are all distinct, the claim is true. Otherwise, there must still be two distinct q_i (since L_1, L_2, and L_3 contain no common point). Assume that $q_1 \neq q_2$ (see Figure 8.3). There must be at least one other point, q_1' on L_1 and at least one other point, q_2' on L_2. They must be on a common line, L', with $L' \neq L$. L and L' must contain a common point, q (FPP2). If $q = q_1$, then L_1 and L' would both contain q_1 and q_1', contradicting FPP1. Similarly, $q \neq q_2$. Therefore, L must contain the three distinct points q_1, q_2, and q. □

LEMMA 8.25
Let $\mathcal{F}$ be a finite projective plane.
(a) Let L be a line in $\mathcal{F}$ that contains exactly $n + 1$ points. If p is any point that is not on L, then p must be on exactly $n + 1$ distinct lines.
(a) Let p be a point in $\mathcal{F}$ that is on exactly $n + 1$ distinct lines. If L is any line that does not contain p, then L contains exactly $n + 1$ distinct points.

Proof: The two parts of the lemma are dual statements, so only one needs to be proved. Nevertheless, it is instructive to prove both separately.

(a) Since p is not on L, the points, $p_1, p_2, \ldots, p_n, p_{n+1}$, on L must each be on a common line with p (by FPP1). The line in common for p_i and p cannot be the same as the line in common for p_j and p $(i \neq j)$. If it was the same, that line would contain both p_i and p_j and so would L, violating FPP2. There must therefore be at least $n+1$ lines that contain p. Denote the line that contains both p and p_i as L_i, for $i = 1, 2, \ldots, n+1$.

Suppose there is another line, M, which contains p. Since $M \neq L$, it must share a common point, p_j, with L (by FPP2). The distinct lines M and L_j both contain p and p_j, violating FPP2. This contradiction means that the only lines that contain p are $L_1, L_2, \ldots, L_{n+1}$, completing the proof.

(b) This is Exercise 18 on page 458. □

The next theorem provides insight into the structure of finite projective planes.

THEOREM 8.26 ***Properties of Finite Projective Planes***

Let $\mathcal{F}$ be a finite projective plane and let $n \geq 2$ be an integer. The following are equivalent:

1. There exists a line in $\mathcal{F}$ that contains exactly $n+1$ points.
2. There exists a point in $\mathcal{F}$ that is on exactly $n+1$ lines.
3. Every line in $\mathcal{F}$ contains exactly $n+1$ points.
4. Every point in $\mathcal{F}$ is on exactly $n+1$ lines.
5. There are exactly n^2+n+1 distinct points in $\mathcal{F}$.
6. There are exactly n^2+n+1 distinct lines in $\mathcal{F}$.

Proof: If $\mathcal{F}$ is a finite projective plane, the three axioms guarantee (details pending) the existence of at least seven distinct points, $p_1, p_2, p_3, p_4, x, y, z$ and at least six distinct lines, $\{L_1, L_2, L_3, L_4, L_5, L_6\}$. Suitable renaming results in Figure 8.4, which provides a visual representation of a portion of $\mathcal{F}$. The points p_1, p_2, p_3, p_4 are those asserted by FPP3. The lines are consequences of FPP1 and FPP2 (as in Proposition 8.23). The points x, y, z must exist since the lines L_1 and L_2 must contain a common point (FPP2), as must the lines L_3 and L_4 and also L_5 and L_6. Notice that several of the lines have been displayed with bends in the middle to show the intersections at points y and z.

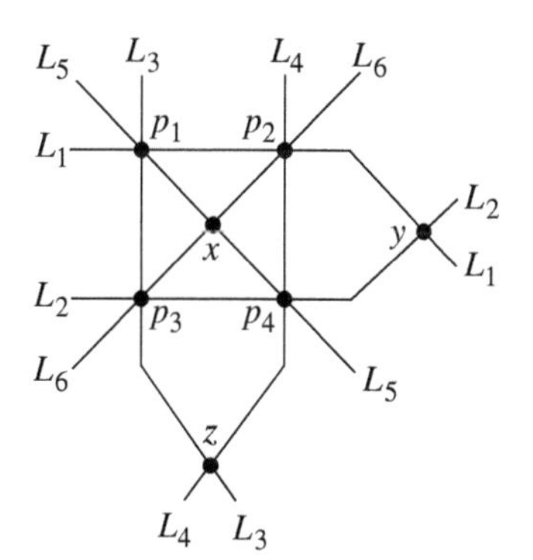

Figure 8.4. A guaranteed subconfiguration in $\mathcal{F}$.

The seven points and six lines cannot have any incidences with each other not shown in Figure 8.4. That is, no pair of these lines will share an additional common point (or FPP2 would be violated). Also, no pair of the seven points can be on an additional common line (or FPP1 would be violated). Note that the lines may contain additional points and other points and lines may exist in $\mathcal{F}$.

The proof will now be presented. The duality principle implies that statements 1 and 2 are logically equivalent. Statements 3 and 4 are also logically equivalent, as are 5 and 6.

The proof will be completed by proving that 1 implies 2, which in turn implies 1, and then proving that 1 and 2 imply 3 and 4. It is then demonstrated that 3 and 4 imply 5 and 6. It should be obvious that 3 implies 1 and 4 implies 2. The mutual equivalence is completed by showing that either 5 or 6 implies 1.

1 → 2 Let L be the line that contains exactly $n+1$ points. Lemma 8.25(a) completes the proof of statement 2 if there is a point that is not on L. Such a point must exist by FPP3.

2 → 1 The proof is a direct consequence of Lemma 8.25(b), assuming that some line exists which does not contain the point. Such a line must exist by FPP3′.

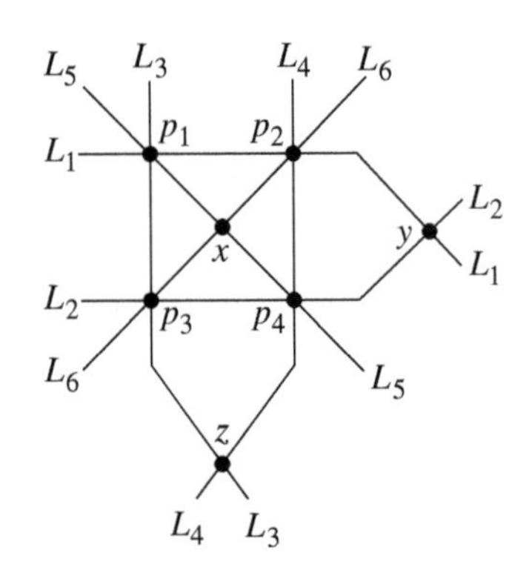

1 and 2 → 3 and 4 This is the most complex part of the proof.

Let L be a line that contains exactly $n + 1$ points. FPP3 guarantees that at least two of the four points p_1, p_2, p_3, p_4 (from the previous diagram) must not be on L. Assume (with suitable renaming if necessary) that $p_1 \notin L$ and $p_2 \notin L$.

Lemma 8.25(a) implies that p_1 is on exactly $n + 1$ lines and that p_2 is on exactly $n + 1$ lines. But then Lemma 8.25(b) implies that any line that doesn't contain p_1 must contain exactly $n + 1$ points. Similarly, any line that doesn't contain p_2 must contain exactly $n + 1$ points. The line that contains both p_1 and p_2 (denoted L_1 in the diagram) is the only line that fails to avoid at least one of the points p_1 and p_2. Consequently, it is the only line for which the number of contained points is in doubt. All others have now been shown to have $n + 1$ points.

The line L_2 in the diagram is distinct from L_1 and so is already known to contain exactly $n + 1$ points. The point x in the diagram is not on L_2, so Lemma 8.25(a) asserts that x is on exactly $n + 1$ lines. Since $x \notin L_1$, Lemma 8.25(b) asserts that L_1 must contain exactly $n + 1$ points. This completes the proof of statement 3.

Now let p be any point. Then FPP3′ guarantees the existence of some line, L, that does not contain p. Since every line is now known to contain exactly $n + 1$ points, Lemma 8.25(a) implies that p is on exactly $n + 1$ lines. This completes the proof of statement 4.

FPP1 Any two distinct points are on one and only one common line.

FPP2 Any two distinct lines contain one and only one common point.

FPP3 There exist four distinct points, no three of which are on a common line.

FPP3′ There exist four distinct lines, no three of which contain a common point.

1 There exists a line in $\mathcal{F}$ that contains exactly $n+1$ points.
2 There exists a point in $\mathcal{F}$ that is on exactly $n+1$ lines.
3 Every line in $\mathcal{F}$ contains exactly $n+1$ points.
4 Every point in $\mathcal{F}$ is on exactly $n+1$ lines.
5 There are exactly n^2+n+1 distinct points in $\mathcal{F}$.
6 There are exactly n^2+n+1 distinct lines in $\mathcal{F}$.

3 and 4 → 5 and 6 A combinatorial proof will be given. Suppose that there are k distinct points in $\mathcal{F}$. Since each pair of points determines a line, there must be $\binom{k}{2}$ such pairs. However, this collection of pairs of points overcounts the lines. In fact, each line has been counted once for each pair of points it contains. There are $n + 1$ points on each line, so each line has been counted $\binom{n+1}{2}$ times. The number of lines is thus

$$\frac{\binom{k}{2}}{\binom{n+1}{2}} = \frac{k \cdot (k-1)}{(n+1) \cdot n}$$

By the duality principle, if there are k points, then there must also be k lines. Hence,

$$k = \frac{k \cdot (k-1)}{(n+1) \cdot n}$$

and consequently, there are $k = n^2 + n + 1$ lines. This proves statement 6. Statement 5 follows from duality.

(5 or 6) → 1 Suppose that either statement 5 or statement 6 is true. Consider any line, $L \in \mathcal{F}$. It must contain at least 2 points. Suppose L contains exactly $m + 1$ points, for some $m \geq 2$. Even if $m \neq n$, it is possible to prove (using the previous parts of this proof) that there are exactly $m^2 + m + 1$ points and exactly $m^2 + m + 1$ lines in $\mathcal{F}$. (Starting with statement 1, all the other statements, including 5 and 6, follow using m instead of n.) It must therefore be the case that $m^2 + m + 1 = n^2 + n + 1$ (since the assumption is that one of 5 or 6 is true). Both m and n are positive integers. Since the function $f(x) = x^2 + x + 1$ is strictly increasing for $x > -\frac{1}{2}$, it must be the case that $m = n$. Thus, L has exactly $n + 1$ points, completing the proof of statement 1. □

The previous theorem leads naturally to the following definition.

DEFINITION 8.27 ***Finite Projective Plane Of Order n***

A finite projective plane, $\mathcal{F}$, is said to have (or be of) *order n* if every line in $\mathcal{F}$ contains $n + 1$ points.

It might seem strange that a finite projective plane with $n + 1$ points on a line is defined to be of order n. One reason will be presented in Corollary 8.31 (on page 457), where finite projective planes of order n are linked with Latin squares of order n.

✔ Quick Check 8.6

1. Prove directly that every point in a finite projective plane must be on at least three distinct lines.
2. Assume statements 3 and 4 in Theorem 8.26 and prove statement 5 directly. (The original proof asserted the duality with statement 6.) ☑

The lines in a finite projective plane are tightly interconnected, as the next proposition demonstrates. The proof is Exercise 20 on page 458.

> **PROPOSITION 8.28**
> Let $\mathcal{F}$ be a finite projective plane of order n. Let p be any point of $\mathcal{F}$, and let the $n+1$ distinct lines that contain p be $L_1, L_2, \ldots, L_{n+1}$. If L is any line in $\mathcal{F}$ with $L \notin \{L_1, L_2, \ldots, L_{n+1}\}$, then each of the $n+1$ points on L is on a distinct line in the set $\{L_1, L_2, \ldots, L_{n+1}\}$.

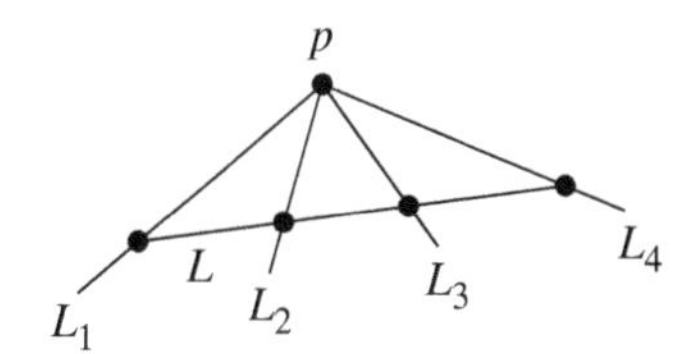

Figure 8.5. A visualization of Proposition 8.28 for $n = 3$.

Note that $p \notin L$. There is a "visual" interpretation of the proposition. Imagine the lines $L_1, L_2, \ldots, L_{n+1}$ fanning out from the common origin, p. Then every other line, L, crosses each of these lines. The $n+1$ intersections are the points on L. Figure 8.5 illustrates this with $n = 3$.

8.2.3 Finite Projective Planes and Latin Squares

This section introduces an example of a common occurrence in combinatorics: using one combinatorial design as the raw material to create another combinatorial design. The main topic in this section is the demonstration of how to create a finite projective plane from a set of mutually orthogonal Latin squares, and then how to create a set of mutually orthogonal Latin squares from a finite projective plane.

EXAMPLE 8.19 **A Finite Projective Plane from Mutually Orthogonal Latin Squares**

It is possible to create a finite projective plane of order 3 by starting with a pair of mutually orthogonal Latin squares of order 3. For this example, let the orthogonal Latin squares be as defined next.

$$L^1 = \begin{array}{ccc} 1 & 2 & 3 \\ 2 & 3 & 1 \\ 3 & 1 & 2 \end{array} \qquad L^2 = \begin{array}{ccc} 1 & 3 & 2 \\ 2 & 1 & 3 \\ 3 & 2 & 1 \end{array}$$

The construction of the finite projective plane will proceed in several steps. The first step will create a 4-by-9 matrix. All but the first two rows will be created from the rows of the orthogonal Latin squares. All numbers in the matrix will be from the set $\{1, 2, 3\}$, with each number repeated three times per row.

$$\begin{array}{ccccccccc} 1 & 1 & 1 & 2 & 2 & 2 & 3 & 3 & 3 \\ 1 & 2 & 3 & 1 & 2 & 3 & 1 & 2 & 3 \\ 1 & 2 & 3 & 2 & 3 & 1 & 3 & 1 & 2 \\ 1 & 3 & 2 & 2 & 1 & 3 & 3 & 2 & 1 \end{array}$$

Notice that row 3 was created by taking each row of L^1, in order. Row 4 was similarly created from L^2. Also, observe how the columns can be naturally grouped into clusters of three columns each.

The next step requires the columns to be labeled with the numbers 1 through 9.

1	2	3	4	5	6	7	8	9
1	1	1	2	2	2	3	3	3
1	2	3	1	2	3	1	2	3
1	2	3	2	3	1	3	1	2
1	3	2	2	1	3	3	2	1

1 2 3	4 5 6	7 8 9
1 1 1	2 2 2	3 3 3
1 2 3	1 2 3	1 2 3
1 2 3	2 3 1	3 1 2
1 3 2	2 1 3	3 2 1

Note that elements from the kth column of L^j appear in the kth positions of the column clusters. For example, column 2 of L^1 contains the numbers 2, 3, and 1. These appear in row 3 at positions 2, 5, and 8, respectively. Positions 2, 5, and 8 are the second positions in their respective clusters of columns.

The labeled matrix will now be used to start creating the lines of the finite projective plane. Each of the four rows of the matrix will contribute three partial lines. This will be done as follows: Define $L'_{r,i}$ to be the set of column labels, j, such that there is an i in column j of row r. For example, $L'_{4,2} = \{3, 4, 8\}$ since the number 2 appears in columns 3, 4, and 8 of row 4. The result is shown in the next table.

r	i = 1	2	3
1	$\{1, 2, 3\}$	$\{4, 5, 6\}$	$\{7, 8, 9\}$
2	$\{1, 4, 7\}$	$\{2, 5, 8\}$	$\{3, 6, 9\}$
3	$\{1, 6, 8\}$	$\{2, 4, 9\}$	$\{3, 5, 7\}$
4	$\{1, 5, 9\}$	$\{3, 4, 8\}$	$\{2, 6, 7\}$

The final step borrows an idea from the use of perspective in a painting. To show perspective, painters use a "vanishing point" or a point at infinity. The majority of lines in the painting will tend toward the vanishing point. This is similar to looking down a long, very straight train track. It appears as if the two tracks converge and meet in the far distance. The connection with this example is the need to have every pair of lines meet at a point. This construction introduces several points at infinity where various lines will meet.

There are 12 partial lines, when $13 = 3^2 + 3 + 1$ are required, and each partial line has one less point than it should. In addition, if numbers 1 through 9 are to be points in the finite projective plane, then four additional points are needed. Since they have been inspired by points at infinity, they will be denoted ∞_1, ∞_2, ∞_3, and ∞_4.

The points in the projective plane will be

$$\{1, 2, 3, 4, 5, 6, 7, 8, 9, \infty_1, \infty_2, \infty_3, \infty_4\}$$

The lines will be

$$\{L_{r,i} \mid r = 1, 2, 3, 4 \text{ and } i = 1, 2, 3\} \cup \{L_\infty\}$$

where

$$L_\infty = \{\infty_1, \infty_2, \infty_3, \infty_4\} \quad \text{and} \quad L_{r,i} = L'_{r,i} \cup \{\infty_r\}$$

For example, $L_{4,2} = L'_{4,2} \cup \{\infty_4\} = \{3, 4, 8\} \cup \{\infty_4\} = \{3, 4, 8, \infty_4\}$.

Table 8.11 shows all 13 lines.

TABLE 8.11 The 13 Derived Lines

r	i = 1	2	3
1	$\{1, 2, 3, \infty_1\}$	$\{4, 5, 6, \infty_1\}$	$\{7, 8, 9, \infty_1\}$
2	$\{1, 4, 7, \infty_2\}$	$\{2, 5, 8, \infty_2\}$	$\{3, 6, 9, \infty_2\}$
3	$\{1, 6, 8, \infty_3\}$	$\{2, 4, 9, \infty_3\}$	$\{3, 5, 7, \infty_3\}$
4	$\{1, 5, 9, \infty_4\}$	$\{3, 4, 8, \infty_4\}$	$\{2, 6, 7, \infty_4\}$
		$\{\infty_1, \infty_2, \infty_3, \infty_4\}$	

It is not easy to create a two-dimensional visual representation for this finite projective plane. In Figure 8.6, several of the lines have been drawn as dashed curves. The table of 13 lines has been repeated to help interpret the diagram. The key observations, which are easy to see from the table, are as follows:

- Every line contains four points.
- Every point is on four lines.
- Every pair of lines contains one point in common.
- Every pair of points is on one common line.

r	i = 1	2	3
1	$\{1, 2, 3, \infty_1\}$	$\{4, 5, 6, \infty_1\}$	$\{7, 8, 9, \infty_1\}$
2	$\{1, 4, 7, \infty_2\}$	$\{2, 5, 8, \infty_2\}$	$\{3, 6, 9, \infty_2\}$
3	$\{1, 6, 8, \infty_3\}$	$\{2, 4, 9, \infty_3\}$	$\{3, 5, 7, \infty_3\}$
4	$\{1, 5, 9, \infty_4\}$	$\{3, 4, 8, \infty_4\}$	$\{2, 6, 7, \infty_4\}$
		$\{\infty_1, \infty_2, \infty_3, \infty_4\}$	

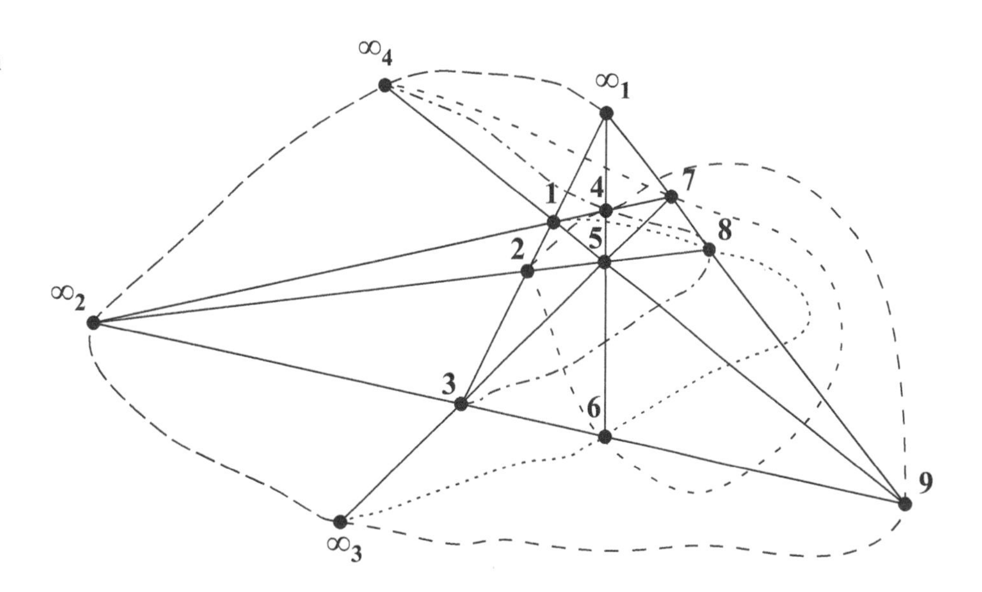

Figure 8.6 Constructing a finite projective plane from Latin squares.

■

The construction used in the previous example started with two mutually orthogonal Latin squares of order 3. The result was a finite projective plane of order 3 (having $n + 1 = 4$ points per line and 4 lines per point and containing $3^2 + 3 + 1 = 13$ points and 13 lines).

This construction can be easily generalized: start with a set of $n-1$ mutually orthogonal Latin squares of order n and create a finite projective plane of order n. Figure 8.7 and Theorem 8.29 (both on page 450) contain this generalization and a proof that it is valid.

Construction 8.1 Constructing a Finite Projective Plane of Order $n > 1$ from $n - 1$ Mutually Orthogonal Latin Squares of Order n

Let the $n - 1$ mutually orthogonal Latin squares be $L^1, L^2, \ldots, L^{n-1}$.

Step 1 Create the $n + 1$ by n^2 matrix, $A = (a_{rc})$, as follows:

$$
\begin{array}{cccccccccccc}
1 & 1 & \cdots & 1 & 2 & 2 & \cdots & 2 & \cdots & n & n & \cdots & n \\
1 & 2 & \cdots & n & 1 & 2 & \cdots & n & \cdots & 1 & 2 & \cdots & n \\
\multicolumn{4}{c}{\text{row 1 of } L^1} & \multicolumn{4}{c}{\text{row 2 of } L^1} & \cdots & \multicolumn{4}{c}{\text{row } n \text{ of } L^1} \\
\multicolumn{4}{c}{\text{row 1 of } L^2} & \multicolumn{4}{c}{\text{row 2 of } L^2} & \cdots & \multicolumn{4}{c}{\text{row } n \text{ of } L^2} \\
\multicolumn{4}{c}{\vdots} & \multicolumn{4}{c}{\vdots} & \vdots & \multicolumn{4}{c}{\vdots} \\
\multicolumn{4}{c}{\text{row 1 of } L^{n-1}} & \multicolumn{4}{c}{\text{row 2 of } L^{n-1}} & \cdots & \multicolumn{4}{c}{\text{row } n \text{ of } L^{n-1}}
\end{array}
$$

Notice that the columns of A are grouped into n natural clusters, with n columns per cluster.

Step 2 Label the columns of A with the numbers 1 through n^2.

1	2	$\cdots$	n	$n+1$	$n+2$	$\cdots$	$2n$	$\cdots$	$(n-1)n+1$	$(n-1)n+2$	$\cdots$	n^2
1	1	$\cdots$	1	2	2	$\cdots$	2	$\cdots$	n	n	$\cdots$	n
1	2	$\cdots$	n	1	2	$\cdots$	n	$\cdots$	1	2	$\cdots$	n
Row 1 of L^1				Row 2 of L^1				$\cdots$	Row n of L^1			
Row 1 of L^2				Row 2 of L^2				$\cdots$	Row n of L^2			
$\vdots$				$\vdots$				$\vdots$	$\vdots$			
Row 1 of L^{n-1}				Row 2 of L^{n-1}				$\cdots$	Row n of L^{n-1}			

Step 3 Create the $(n + 1)n = n^2 + n$ partial lines, $L'_{r,i}$, where $L'_{r,i}$ contains the number c if and only if $a_{rc} = i$.

Step 4 The finite projective plane, $\mathcal{F}$, will now be specified. The points in $\mathcal{F}$ are the $n^2 + n + 1$ elements in the set $\{1, 2, 3, \ldots, n^2, \infty_1, \infty_2, \ldots, \infty_{n+1}\}$. The $n^2 + n + 1$ lines of $\mathcal{F}$ are the lines $\{L_{r,i} \mid r \in \{1, 2, \ldots, n + 1\}$ and $i \in \{1, 2, \ldots, n^2\}\} \cup \{L_\infty\}$, where $L_\infty = \{\infty_1, \infty_2, \ldots, \infty_{n+1}\}$ and $L_{r,i} = L'_{r,i} \cup \{\infty_r\}$.

Figure 8.7 Constructing a finite projective plane of order $n > 1$ from $n - 1$ mutually orthogonal Latin squares of order n.

✔ Quick Check 8.7

1. Use Construction 8.1 (Figure 8.7) to create a finite projective plane of order 2. Include a visual diagram of the plane. ☑

THEOREM 8.29 ***Finite Projective Planes from Mutually Orthogonal Latin Squares***

If a set of $n - 1$ mutually orthogonal Latin squares of order n exists, then a finite projective plane of order n also exists.

Proof: The proof shows that Construction 8.1 produces a finite projective plane of order n. The proof is not elegant, but has the advantage of using only elementary ideas. More elegant proofs exist, but these proofs use some advanced mathematical tools that have not been introduced in this text.

The Rows of A Are Orthogonal As a preliminary observation, notice that every pair of rows in A (the matrix in Construction 8.1) is orthogonal, in the sense that every one of the n^2 possible vertical pairs in the list $\binom{1}{1}, \binom{1}{2}, \ldots, \binom{1}{n}, \binom{2}{1}, \ldots, \binom{2}{n}, \ldots, \binom{n}{1}, \ldots, \binom{n}{n}$ appears exactly once if one row is placed directly above the other.

This claim can been justified by considering a few cases.

Rows 1 and 2 Rows 1 and 2 have the orthogonality property by construction. The vertical pairs are even arranged in the natural ordering.

Rows 1 or 2 and row j, with $j > 2$ Row 1 has the orthogonality property with row j because every cluster of n columns has the same number in row 1, but each number in $\{1, 2, \ldots, n\}$ in the corresponding columns of row j. Each first entry of the vertical pairs appears in a distinct cluster of row 1. The vertical pairs will have $\binom{1}{1}, \binom{1}{2}, \ldots, \binom{1}{n}$, in some order from the first cluster, $\binom{2}{1}, \ldots, \binom{2}{n}$ in some order from the second cluster, an so on. All n^2 vertical pairs will be present.

Notice that row 2 contains one copy of each of the numbers $1, 2, \ldots, n$ in every cluster of n columns. Since row j in A is defined by the rows of a Latin square, every cluster of n columns in that row will be a row of a Latin square. Thus, every cluster of n columns in row j will also contain exactly one copy of each of the numbers $1, 2, \ldots, n$.

Row 2 also has the orthogonality property with row j. This is true because the source for row j is the Latin square, L^{j-2}, and any Latin square contains every number in $\{1, 2, \ldots, n\}$ exactly once in each of its columns. The element in the kth column of row 2 is always a k and the kth column of L^{j-2} is distributed among the kth positions of each column cluster (relative to the start of the cluster), so the second elements in each pair will be distinct.[19]

Rows j and k, with $j, k > 2$ The orthogonality of these rows is a direct result of L^{j-2} and L^{k-2} being members of a set of mutually orthogonal Latin squares. The numbers are arranged in one row of n^2 elements instead of n rows of n elements each, but the set of pairs is the same in either arrangement when checking orthogonality.

Now that the orthogonality of the rows has been verified, the proof may proceed. Notice that Construction 8.1 ensures that the lines each contain $n + 1$ points.

Every point is on exactly $n+1$ lines because the points in $\{1, 2, \ldots, n^2\}$ each appear in exactly one of the lines produced from a row of A, and A has $n+1$ rows.[20] The points $\{\infty_1, \infty_2, \ldots, \infty_{n+1}\}$ are each in L_∞. In addition, ∞_j appears in each of the n lines $L_{j,c}$, for $c = 1, 2, \ldots, n$, but in no other lines.

The construction creates $n^2 + n + 1$ points and $n^2 + n + 1$ lines. If the construction has produced a finite projective plane, then it must be of order n. It remains to verify that the points and lines in $\mathcal{F}$ actually constitute a finite projective plane. This will be done by directly verifying the three axioms.

FPP1 The proof that every pair of points is on exactly one common line will be completed by looking at three cases.

∞_i and ∞_j The only common line for these two points is L_∞.

k and ∞_j The only common line for these two points is $L_{j,i}$, where $a_{jk} = i$.

---

[19] For example, the second positions of the clusters in row j are the elements in the second column of L^{j-2}. These positions, when matched with row 2 of A will produce (probably in a different order) the vertical pairs $\binom{2}{1}, \binom{2}{2}, \ldots, \binom{2}{n}$.

[20] For example, row j of A contains some number, i, in the kth column. The number k will consequently be on line $L_{j,i}$ and no other line produced by row j.

k and j with $k \neq j$ For points k and j to be on at least one common line, the matrix A must contain some row with the same number, i, in both column k and column j.

	$\cdots$	k	$\cdots$	j	$\cdots$
Row r	$\cdots$	i	$\cdots$	i	$\cdots$

If $k \equiv j \pmod{n}$, then this will occur in row 2 of A. If k and j are in the same cluster of columns, it will occur in row 1. Suppose that neither of these cases apply. Then the positions in the original Latin squares, corresponding to columns k and j in A, cannot be in the same row or the same column of the Latin squares. Lemma 8.32 on page 459 implies that there is some row of A where columns k and j have the same value. Therefore, points k and j must be on at least one common line.

Now suppose that points k and j are on two common lines. Then there must be a pair, L'_{r_1,c_1} and L'_{r_2,c_2} of partial lines with $\{k, j\} \subseteq (L'_{r_1,c_1} \cap L'_{r_2,c_2})$.

Suppose $r_1 \neq r_2$. Then the matrix A looks like this:

	$\cdots$	k	$\cdots$	j	$\cdots$
Row r_1	$\cdots$	c_1	$\cdots$	c_1	$\cdots$
	$\vdots$	$\vdots$	$\vdots$	$\vdots$	$\vdots$
Row r_2	$\cdots$	c_2	$\cdots$	c_2	$\cdots$

which violates the orthogonality property established at the beginning of this proof. Therefore $r_1 = r_2$.

Since L_{r,c_1} and L_{r,c_2} are disjoint if $c_1 \neq c_2$, it must be that $c_1 = c_2$, so the assumed pair of lines common to k and j is really a single line.

FPP2 Cases will be used (once again) to show that every pair of lines contain exactly one common point.

L_∞ and $L_{r,c}$ These two lines contain only the point ∞_r in common.

L_{r,c_1} and L_{r,c_2} with $c_1 \neq c_2$ These two lines contain only ∞_r in common, since L'_{r,c_1} and L'_{r,c_2} are disjoint.

L_{r_1,c_1} and L_{r_2,c_2} with $r_1 \neq r_2$ Suppose $c_1 \neq c_2$ and L_{r_1,c_1} and L_{r_2,c_2} contain the points k and j in common (they cannot contain a common point of the form ∞_i). Then the following table must represent part of A, violating the orthogonality property.

	$\cdots$	k	$\cdots$	j	$\cdots$
Row r_1	$\cdots$	c_1	$\cdots$	c_1	$\cdots$
	$\vdots$	$\vdots$	$\vdots$	$\vdots$	$\vdots$
Row r_2	$\cdots$	c_2	$\cdots$	c_2	$\cdots$

If instead, $c_1 = c_2$, then A must contain the following configuration, again violating the orthogonality property.

	$\cdots$	k	$\cdots$	j	$\cdots$
Row r_1	$\cdots$	c_1	$\cdots$	c_1	$\cdots$
	$\vdots$	$\vdots$	$\vdots$	$\vdots$	$\vdots$
Row r_2	$\cdots$	c_1	$\cdots$	c_1	$\cdots$

Therefore, L_{r_1,c_1} and L_{r_2,c_2} can have only one point in common. Since $r_1 \neq r_2$, that point cannot be in the form ∞_i. It remains to show that L_{r_1,c_1} and L_{r_2,c_2} are not disjoint.

By the orthogonality property, the vertical pair $\binom{c_1}{c_2}$ must appear in some column of A when rows r_1 and r_2 are compared. Suppose this occurs in column k.

Then A must look like the following.

	⋯	k	⋯
Row r_1	⋯	c_1	⋯
	⋮	⋮	⋮
Row r_2	⋯	c_2	⋯

The lines L_{r_1,c_1} and L_{r_2,c_2} both contain the point k, showing that they are not disjoint.

FPP3 Since $n > 1$, there are at least three rows and at least four columns in A. Consider the lines constructed in steps 3 and 4. Points 1 and 2 are both in/on $L_{1,1}$ and points $n+1$ and $n+2$ are both in/on $L_{1,2}$. Also, points 1 and $n+1$ are both on $L_{2,1}$ and points 2 and $n+2$ are both on $L_{2,2}$.

Since by FPP1, no two points can be on more than one common line, it is not difficult to verify that $\{1, 2, n+1, n+2\}$ is a set of 4 points in $\mathcal{F}$ for which no three are on a common line. □

The reverse construction is also possible.

EXAMPLE 8.20

Orthogonal Latin Squares from a Finite Projective Plane

It is possible to create a pair of orthogonal Latin squares of order 3 from the finite projective plane of order 3 shown in Figure 8.8. The first step is to create a table (Table 8.12)

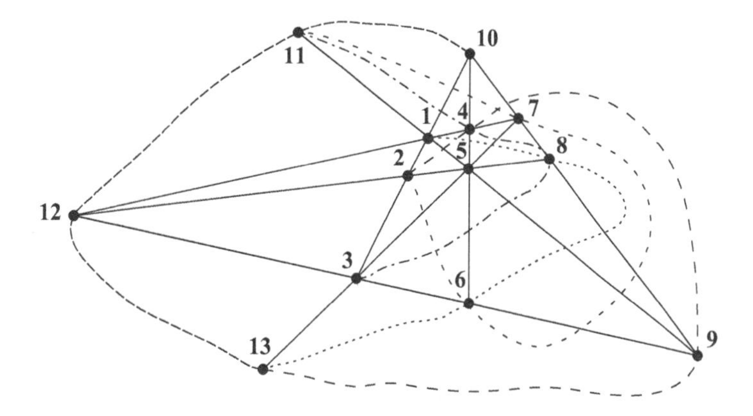

Figure 8.8 Constructing orthogonal Latin squares from a finite projective plane.

with the $n^2 + n + 1 = 13$ lines arranged in five rows. The bottom row will contain only one line, arbitrarily chosen here to be the line $L_{13} = \{10, 11, 12, 13\}$. The other four rows are arranged by the presence of one of the points on the special line in the last row.

TABLE 8.12 Step 1

10	{1,2,3,10}	{4,5,6,10}	{7,8,9,10}
11	{1,5,9,11}	{2,6,7,11}	{3,4,8,11}
12	{1,4,7,12}	{2,5,8,12}	{3,6,9,12}
13	{1,6,8,13}	{2,4,9,13}	{3,5,7,13}
		{10,11,12,13}	

TABLE 8.13 Step 2a (T)

1	2	3
$\{1, 2, 3\}$	$\{4, 5, 6\}$	$\{7, 8, 9\}$
$\{1, 5, 9\}$	$\{2, 6, 7\}$	$\{3, 4, 8\}$
$\{1, 4, 7\}$	$\{2, 5, 8\}$	$\{3, 6, 9\}$
$\{1, 6, 8\}$	$\{2, 4, 9\}$	$\{3, 5, 7\}$

In the next step, the line L_{13} and all points on it will be removed. This will leave $n^2 + n = 12$ partial lines and $n^2 = 9$ points, with the partial lines arranged in an $n + 1 = 4$ by $n = 3$ table, T, with the columns labeled by the numbers 1 through n (with $n = 3$ in this example). Notice (Table 8.13) that the partial lines in each row have been sorted by smallest elements. Thus, $\{2, 5, 8\}$ appears before $\{3, 6, 9\}$ in row three because 2 is less than 3.

The reduced table (Table 8.13) corresponds to the reduced geometric construct in Figure 8.9 on page 454.

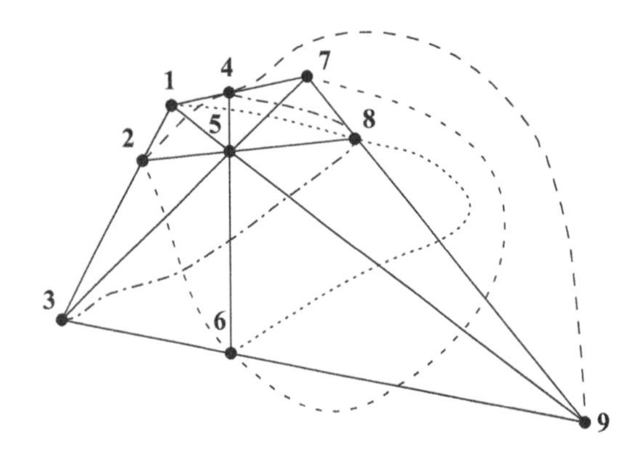

Figure 8.9 The reduced geometric construct.

1	2	3
{1, 2, 3}	{4, 5, 6}	{7, 8, 9}
{1, 5, 9}	{2, 6, 7}	{3, 4, 8}
{1, 4, 7}	{2, 5, 8}	{3, 6, 9}
{1, 6, 8}	{2, 4, 9}	{3, 5, 7}

Notice that the partial lines in row 1 list the points in their natural order: $1, 2, 3, \ldots, n^2$. If this were not so, it would be easy to go back to the original finite projective plane and rename the points so that the points in row 1 would be in natural order. The construction also requires the second row to be in a particular pattern. In this case, the desired pattern is $\{1, 4, 7\}, \{2, 5, 8\}, \{3, 6, 9\}$. This also can be achieved by a more selective renaming of points. The key idea is to swap the names of pairs of points with the condition that a point can only exchange with points where it is on a common line in row 1. Thus, 4 and 5 can exchange names without modifying the line $\{4, 5, 6\}$ in row 1. The desired pattern is achieved by the following circular renaming of the original points: $4 \to 6 \to 5 \to 4,\ 7 \to 8 \to 9 \to 7$. The renaming changes the set $\{4, 5, 6\}$ to $\{6, 4, 5\}$, which is the same set as $\{4, 5, 6\}$. The convention is to write the set in sorted order. The result is shown in Table 8.14.

TABLE 8.14 Step 2b (standardized T)

1	2	3
{1, 2, 3}	{4, 5, 6}	{7, 8, 9}
{1, 4, 7}	{2, 5, 8}	{3, 6, 9}
{1, 6, 8}	{2, 4, 9}	{3, 5, 7}
{1, 5, 9}	{2, 6, 7}	{3, 4, 8}

The third step will transform T into an $n+1 = 4$ by $n^2 = 9$ matrix, A (Table 8.15). The columns of A will be labeled by the points (1 through 9) and the rows will be derived from the rows in T. The entry in row i, column j of A will be the label of the column of T for which the point j appears in a line at row i of T. Thus, row 3 of A will have a 2 in columns 2, 4, and 9, since the line $\{2, 4, 9\}$ appears in the second column of the third row of T.

TABLE 8.15 Step 3 (A)

1	2	3	4	5	6	7	8	9
1	1	1	2	2	2	3	3	3
1	2	3	1	2	3	1	2	3
1	2	3	2	3	1	3	1	2
1	2	3	3	1	2	2	3	1

The columns of T have been separated into three clusters of three columns each. The bottom $n - 1$ rows will form the Latin squares. Each of these rows can be copied cluster-by-cluster into n by n matrices (Table 8.16).

TABLE 8.16 Step 4

$$L^1 = \begin{matrix} 1 & 2 & 3 \\ 2 & 3 & 1 \\ 3 & 1 & 2 \end{matrix} \qquad L^2 = \begin{matrix} 1 & 2 & 3 \\ 3 & 1 & 2 \\ 2 & 3 & 1 \end{matrix}$$

It is easy to verify that L^1 and L^2 are a pair of orthogonal Latin squares of order 3. ■

The steps in Example 8.20 can be generalized. Construction 8.2 (Figure 8.10) describes how to start with a finite projective plane of order n and construct a set of $n-1$ mutually orthogonal Latin squares.

Construction 8.2 Constructing a Collection of $n-1$ Mutually Orthogonal Latin Squares of Order $n>1$ from a Finite Projective Plane of Order n

Let $\mathcal{F}$ be a finite projective plane of order n, consisting of the n^2+n+1 points $1, 2, \ldots, n^2+n, n^2+n+1$, and the n^2+n+1 lines $L_1, L_2, \ldots, L_{n^2+n}, L_{n^2+n+1}$. By suitably renaming points if necessary, assume that $L_{n^2+n+1} = \{n^2+1, n^2+2, \ldots, n^2+n, n^2+n+1\}$.

Step 1 Create a table that contains $n+2$ rows. The final row will consist only of L_{n^2+n+1}. Row i, for $i = 1, 2, \ldots n+1$, will consist of the n lines other than L_{n^2+n+1} that contain the point n^2+i.

Step 2 Create a table, T, having $n+1$ rows and n columns by deleting L_{n^2+n+1} and all the points it contains from the previous table. If necessary, rename points or swap rows so that the first row is in numeric order; that is, it consists of the partial lines

$$\{1, 2, \ldots, n\},\ \{n+1, n+2, \ldots, 2n\},\ \ldots,\ \{(n-1)n+1, (n-1)n+2, \ldots, n^2\}$$

Now rename points again so that the second line consists of the pattern

$$\{1, n+1, 2n+1, \ldots, (n-1)n+1\},\ \{2, n+2, 2n+2, \ldots, (n-1)n+2\},\ \ldots,\ \{n, 2n, \ldots, n^2\}$$

This must be done by renaming a point using only points it is on a common partial line with in row 1. Label the columns of T with the numbers 1 through n.

Step 3 Create an $n+1$ by n^2 matrix, $A = (a_{ij})$, from T. The columns of A will be labeled with the numbers 1 through n^2. Set $a_{ij} = k$ if the partial line in row i, column k of T contains the point j. Divide the columns of A into n clusters of n columns each.

Step 4 A will contain the pattern "$111\cdots1\ 22\cdots2\cdots n\ n\cdots n$" in row 1 and the pattern "$12\cdots n\ 12\cdots n\cdots 12\cdots n$" in row 2. Remove those two rows, forming the $n-1$ by n^2 matrix, A'. The $n-1$ rows of A' will be used to form the Latin squares, with the ith row forming the Latin square L^i. More specifically, the jth row of L^i will be the jth cluster in row i of A'.

Figure 8.10 Constructing a collection of $n-1$ mutually orthogonal Latin squares of order $n>1$ from a finite projective plane of order n.

✔ Quick Check 8.8

1. Use Construction 8.2 to create a Latin square of order 2, starting with the Fano plane.

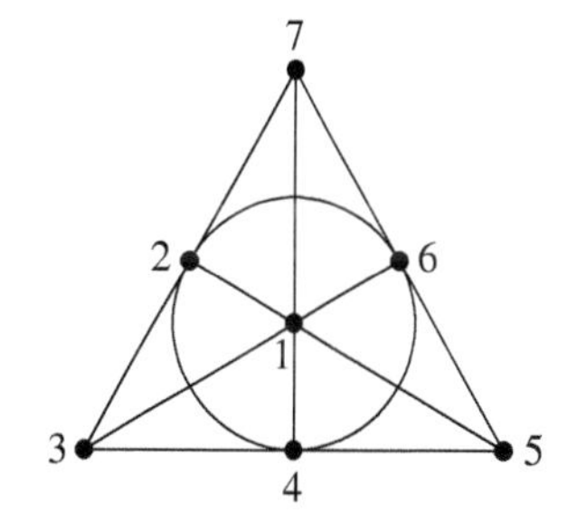

☑

THEOREM 8.30 ***Mutually Orthogonal Latin Squares from Finite Projective Planes***

If a finite projective plane of order n exists, then a set of $n-1$ mutually orthogonal Latin squares of order n also exists.

Proof: The proof will be accomplished by showing that Construction 8.2 always works. Most of the proof will be concerned with the resulting matrices in step 4. However, a few claims need to be verified for the earlier steps.

Step 1 claims it is possible to construct a table with $n+2$ rows, where all but the final row contain n lines, each with a single point common to all lines in that row. This is possible because the points in L_{n^2+n+1} cannot be contained in any other common line (FPP1). In addition, each of the points $n^2+1, \ldots, n^2+n+1$ is on exactly $n+1$ lines.

Step 2 results in a table, T, with $n+1$ rows, each containing n partial lines of n points each. Neither renaming changes any essential features of the original finite projective plane, $\mathcal{F}$. The second renaming is done by exchanging names in such a way that row 1 remains unchanged (recall that the order in which elements are listed inside a set is not important). The second renaming is always possible because in row 2 every point, x, must appear with points that are not on a common line with x in row 1 (or else FPP1 would be violated). This can only happen if the partial lines in row 2 each contain exactly one point from each partial line in row 1. It is therefore possible to rename elements within partial lines in row 1 of T so that the partial lines in row 2 have the desired pattern.

The matrix, A, in step 3 will contain n clusters per row because T contains n partial lines. Each cluster will contain n columns because each partial line in T contains n points. Notice that row i of A must contain each of the numbers 1 through n exactly n times apiece. This is because each column label in T is assigned to the n points in the partial line residing in that column of row i in T. In addition, each point in the set $\{1, 2, \ldots, n^2\}$ appears exactly once per row of T. (Suppose some point, m, were in two partial lines of row i of T. Add n^2+i back to the partial lines, re-creating two of the lines in $\mathcal{F}$. These two lines in $\mathcal{F}$ would contain the pair of common points m and n^2+i, contradicting FPP1 and FPP2.)

Step 2 assures that the pattern "$111\cdots1\ 22\cdots2\cdots nn\cdots n$" will appear in row 1 of A and the pattern "$12\cdots n\ 12\cdots n\cdots 12\cdots n$" will appear in row 2.

It now must be shown that each of rows 3 through $n+1$ forms a Latin square and that those Latin squares are mutually orthogonal.

The rows of A' form Latin squares Consider the transition from T to A and focus on any row, $i > 1$ of A. Each cluster of row i in A contains each of the numbers 1 through n. If this were not so, then some cluster of row i would contain a duplicate value, j. This means that there is a partial line in row i, column j of T that contains two points, k and m, that are both in the same cluster of A. The points k and m will also be on a common partial line in row 1 of T since they are in the same cluster. This contradicts FPP1 and FPP2.

Since it is now clear that each cluster in row i of A' contains every number in $\{1, 2, \ldots, n\}$, it is clear that every row in L^i contains each of those numbers exactly once.

Suppose some column, c, of L^i contains a duplicate value, j. Then there must be a partial line in row i of T that contains a pair of points, k and m. The common column means that $k \equiv m \pmod{n}$ (since k and m are in the same relative positions from the start of their respective clusters). This, in turn, implies that k and m must be on a common partial line in row 2 of T, contradicting FPP1 and FPP2.

Therefore, every column of L^i must also contain every number in $\{1, 2, \ldots, n\}$. Thus, L^i is a Latin square of order n, for $i = 1, 2, \ldots, n-1$.

The Latin squares are mutually orthogonal Suppose that L^i and L^j are not orthogonal. That is, some ordered pair appears in two places when the matrix of ordered pairs from L^i and L^j is created. Then there will be distinct columns, k and m in A', such that the following configuration must occur (it is possible that $a = b$).

$$\begin{array}{r|ccccc} & \cdots & k & \cdots & m & \cdots \\ \hline \text{Row } i & \cdots & a & \cdots & a & \cdots \\ & \vdots & \vdots & \vdots & \vdots & \vdots \\ \text{Row } j & \cdots & b & \cdots & b & \cdots \end{array}$$

This indicates that both k and m are on a common partial line in column a of T and that k and m are also on a common partial line in column b of T. Since $i \neq j$, the two partial lines are distinct. But these partial lines come from distinct lines in $\mathcal{F}$, so they cannot contain two common points. This contradiction means that L^i and L^j must be orthogonal. □

COROLLARY 8.31 ***Finite Projective Plane If and Only If Mutually Orthogonal Latin Squares***

A finite projective plane of order n exists if and only if there is a set of $n-1$ mutually orthogonal Latin squares of order n.

Proof: Theorems 8.29 and 8.30. □

8.2.4 Exercises

The exercises marked with ★ have detailed solutions in Appendix H.

1. Prove that every number in $\{1, 2, \ldots, n\}$ must appear *at most* once in every row and in every column of an n by n Latin square.

2. Produce a standardized Latin square of order 5.

3. Determine if the following pairs of Latin squares are orthogonal by constructing the matrices of ordered pairs.

(a)
$$\begin{matrix} 2 & 1 & 3 \\ 3 & 2 & 1 \\ 1 & 3 & 2 \end{matrix} \qquad \begin{matrix} 2 & 3 & 1 \\ 1 & 2 & 3 \\ 3 & 1 & 2 \end{matrix}$$

(b) ★
$$\begin{matrix} 1 & 2 & 3 \\ 2 & 3 & 1 \\ 3 & 1 & 2 \end{matrix} \qquad \begin{matrix} 3 & 1 & 2 \\ 2 & 3 & 1 \\ 1 & 2 & 3 \end{matrix}$$

(c)
$$\begin{matrix} 3 & 2 & 1 \\ 2 & 1 & 3 \\ 1 & 3 & 2 \end{matrix} \qquad \begin{matrix} 3 & 1 & 2 \\ 2 & 3 & 1 \\ 1 & 2 & 3 \end{matrix}$$

(d)
$$\begin{matrix} 1 & 2 & 3 & 4 \\ 2 & 1 & 4 & 3 \\ 3 & 4 & 1 & 2 \\ 4 & 3 & 2 & 1 \end{matrix} \qquad \begin{matrix} 1 & 3 & 4 & 2 \\ 4 & 2 & 1 & 3 \\ 2 & 4 & 3 & 1 \\ 3 & 1 & 2 & 4 \end{matrix}$$

(e)
$$\begin{matrix} 1 & 2 & 3 & 4 \\ 2 & 4 & 1 & 3 \\ 3 & 1 & 4 & 2 \\ 4 & 3 & 2 & 1 \end{matrix} \qquad \begin{matrix} 1 & 2 & 3 & 4 \\ 2 & 3 & 4 & 1 \\ 3 & 4 & 1 & 2 \\ 4 & 1 & 2 & 3 \end{matrix}$$

4. Fill in the missing elements to produce an orthogonal pair of Latin squares of order 5.

$$\begin{matrix} 1 & 2 & 3 & 4 & 5 \\ 2 & * & * & * & * \\ 3 & * & * & * & * \\ 4 & * & * & * & * \\ 5 & * & * & * & * \end{matrix} \qquad \begin{matrix} 1 & * & * & * & * \\ * & 2 & * & * & * \\ * & * & 3 & * & * \\ * & * & * & 4 & * \\ * & * & * & * & 5 \end{matrix}$$

5. A Latin square is called *self-orthogonal* if it is orthogonal to its transpose. (See page A29 for the definition of *transpose*.) Determine which of the following Latin squares are self-orthogonal.

(a)
$$\begin{matrix} 2 & 1 & 3 \\ 3 & 2 & 1 \\ 1 & 3 & 2 \end{matrix}$$

(b) ★
$$\begin{matrix} 1 & 2 & 3 \\ 2 & 3 & 1 \\ 3 & 1 & 2 \end{matrix}$$

(c)
$$\begin{matrix} 1 & 3 & 4 & 2 \\ 4 & 2 & 1 & 3 \\ 2 & 4 & 3 & 1 \\ 3 & 1 & 2 & 4 \end{matrix}$$

(d)
$$\begin{matrix} 1 & 2 & 3 & 4 \\ 4 & 3 & 2 & 1 \\ 2 & 1 & 4 & 3 \\ 3 & 4 & 1 & 2 \end{matrix}$$

6. Prove that there are no self-orthogonal Latin squares of order 3. (The definition of *self-orthogonal* is presented in Exercise 5.)

7. ★ Let L be a Latin square of order n and assume that $i, j \in \{1, 2, \ldots, n\}$. Prove that if rows i and j are swapped and then columns i and j are swapped, the new matrix, L', is also a Latin square. [*Hint*: Try working with a few 3-by-3 and 4-by-4 Latin squares to gain insight.]

8. Suppose two orthogonal Latin squares of order n both have the same last row. Is it possible for them to have identical entries in an additional common position that is not in the last row? Explain your answer.

9. A nursing supervisor in a small nursing home needs to schedule the four duty nurses (Wendy, Xing, Yolanda, and Zoe) for the next four weeks. There are four 6-hour shifts (morning, afternoon, evening, and graveyard). She wants to be fair, so she intends to schedule every nurse to work every shift for a 1-week period. Design a fair duty roster for the 4 weeks. The roster should be expressed in terms of the problem, not in terms of any mathematical tools that you use to solve the problem.

10. ★ A square dance teacher at a small school has decided to assign dance partners so that the children are not forced to choose. The class consists of 12 boys and 12 girls. The square dance lessons are 3 days per week for 4 weeks. Describe an equitable design for assigning dance partners (you do not need to produce the design). How do you know such a design is possible?

11. According to the web site *Western Square Dance Dancer Terminology* (current Web address available at www.mathcs.bethel.edu/~gossett/DiscreteMathWithProof/), a standard western square dance has four couples arranged in a square. (This information is also available in [82].) The couple whose backs are to the "caller" is designated couple 1. Couples 2, 3, and 4 are numbered going counterclockwise from couple 1. Suppose the teacher in Exercise 10 wants every child to have equal numbers of class periods in each of the four positions, while still matching boys and girls equitably. Describe a design for assigning dance partners that meets the enhanced objectives (you do not need to produce the design). How do you know such a design is possible?

12. A soap manufacturer wishes to develop a new detergent that causes less environmental damage than the old detergent. The new detergent should also work well at various kinds of cleaning. The researchers at the company have four candidate detergents, code named A, B, C, and D. They wish to test them with four kinds of cleaning tasks: normal, grease stains, food stains, and mud stains. They have also decided to use four different brands, W, X, Y, and Z, of washing machine to see if the machine is a significant factor.

Management has decided that doing tests with all 64 combinations is too expensive. Design a fair testing arrangement that exposes each candidate detergent to each washer brand and each type of cleaning task. Present your results using the terminology of the detergent company.

13. Suppose the research team in Exercise 12 also wishes to consider the effect of water hardness. They will use four different levels of hardness, denoted H1, H2, H3, and H4. However, management will only allow 16 tests to be run.

Design a fair testing arrangement that exposes each candidate detergent to each washer brand and each type of cleaning task, as well as each water hardness level. Present your results using the terminology of the detergent company.

14. Solve the "25 officers problem": Arrange 25 officers into a 5-by-5 matrix such that every row and every column contains exactly one officer from each of 5 ranks and from each of 5 regiments. Every rank/regiment pairing should appear exactly one time. [*Hint*: You might try solving the "9 officers problem" just to get a feel for the organizational aspects of the problem.]

15. Each of the following statements is either true (always) or false (at least sometimes). Determine which option applies for each statement and provide adequate explanation for your choice.

(a) In a Latin square of order n, the two diagonals will each contain every number in $\{1, 2, \ldots, n\}$.

(b) It is possible to find a pair of orthogonal Latin squares of order n for all $n \geq 3$.

(c) Any set of mutually orthogonal Latin squares of order $n > 1$ will contain fewer than n matrices.

(d) ★ The number of distinct Latin squares of order $n > 1$ is always greater than the maximum number of mutually orthogonal Latin squares of order n.

16. Theorem 8.65 asserts that the number of distinct Latin squares of order n is at least $n! \cdot (n-1)! \cdot (n-2)! \cdots 3! \cdot 2! \cdot 1$.

(a) Produce (with proof) a lower bound for the number of *standardized* Latin squares of order n, for $n \geq 2$.

(b) Calculate the actual number of standardized Latin squares of order n, for $1 \leq n \leq 6$.

17. Prove that FPP1, FPP2, and FPP3$'$ imply FPP3.

18. ★ Prove part (b) of Lemma 8.25 on page 444. Do not appeal to duality.

19. A configuration similar to that shown in the diagram from the proof of Theorem 8.26 (Figure 8.4 on page 445) must be present in any finite projective plane. In particular, it must be present in the Fano plane. Show this is true by a suitable renaming of the points in Example 8.18 on page 444.

20. Prove Proposition 8.28 on page 447.

21. ★ Let L_1 and L_2 be two distinct lines in a finite projective plane. Prove that there is a point that is on neither L_1 nor L_2.

22. Let p_1 and p_2 be two distinct points in a finite projective plane. Prove that there is a line that contains neither p_1 nor p_2.

23. A finite projective plane of order 1 cannot exist. Why not?

24. Start with a copy of the Fano plane and eliminate lines until the following property is true.

> For every line, L, there exists another line, L', such that L and L' have no common points.

How many lines do you need to eliminate before this property holds? Generalize your conclusion to a statement that is valid for all finite projective planes.

25. A finite projective plane of order 3.

(a) How many points and lines are in a finite projective plane of order 3? How many points are on each line? How many lines contain each point?

(b) Draw a diagram of a finite projective plane of order 3. Do not worry too much about aesthetics. Start with the following partial diagram.

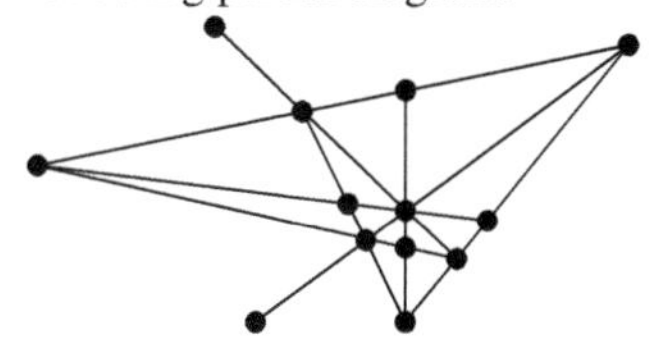

26. Each of the following statements is either true (always) or false (at least sometimes). Determine which option applies for each statement and provide adequate explanation for your choice.

(a) A square (with the four vertices considered to be points) is a geometric figure in which neighboring points are on exactly one common line and intersecting lines contain exactly one common point. There are also four distinct points, no three of which are on a common line. A square therefore satisfies all the axioms for a finite projective plane.

(b) In a finite projective plane, every set of four lines has the property that no three of the lines contain a common point.

(c) Let p_1, p_2, and p_3 be three points in a finite projective plane, $\mathcal{F}$, of order $n \geq 2$. Then there is a line in $\mathcal{F}$ that contains all three points.

(d) ★ The duality principle for finite projective planes states that if $\mathcal{F}$ is a finite projective plane, then it is possible to create another projective plane by changing points in $\mathcal{F}$ into lines and lines in $\mathcal{F}$ into points.

(e) ★ According to Theorem 8.26 (page 445), there exist finite projective planes with 7, 13, 21, 31, 43, and 57 points and lines, respectively.

27. Let p_1, p_2, and p_3 be three points in a finite projective plane, $\mathcal{F}$, that are not on a common line. Prove that they must be the vertices in a triangle within $\mathcal{F}$.

28. Use Construction 8.1 (page 450) to create a finite projective plane of order 4. You do not need to draw a visual diagram, just list all the lines. Start with the pair of orthogonal Latin squares listed and fill in the third Latin square to produce a set of three mutually orthogonal Latin squares.

$$\begin{array}{cccc} 1 & 2 & 3 & 4 \\ 2 & 1 & 4 & 3 \\ 3 & 4 & 1 & 2 \\ 4 & 3 & 2 & 1 \end{array} \qquad \begin{array}{cccc} 1 & 3 & 4 & 2 \\ 4 & 2 & 1 & 3 \\ 2 & 4 & 3 & 1 \\ 3 & 1 & 2 & 4 \end{array} \qquad \begin{array}{cccc} 1 & 4 & 2 & 3 \\ & & & \\ & & & \\ & & & \end{array}$$

29. Use Construction 8.2 (page 455) to create a set of three mutually orthogonal Latin squares of order 4. The lines of a finite projective plane of order 4 are listed next.

$\{3, 8, 9, 15, 19\}$ $\{5, 6, 7, 8, 17\}$
$\{1, 3, 5, 12, 18\}$ $\{1, 7, 13, 15, 20\}$
$\{6, 9, 12, 13, 21\}$ $\{2, 6, 10, 15, 18\}$
$\{1, 8, 10, 14, 21\}$ $\{4, 8, 13, 16, 18\}$
$\{1, 6, 11, 16, 19\}$ $\{2, 5, 13, 14, 19\}$
$\{1, 2, 4, 9, 17\}$ $\{4, 7, 10, 12, 19\}$
$\{12, 14, 15, 16, 17\}$ $\{3, 4, 6, 14, 20\}$
$\{2, 8, 11, 12, 20\}$ $\{4, 5, 11, 15, 21\}$
$\{7, 9, 11, 14, 18\}$ $\{3, 10, 11, 13, 17\}$
$\{5, 9, 10, 16, 20\}$ $\{2, 3, 7, 16, 21\}$
$\{17, 18, 19, 20, 21\}$

30. Finish the verification of FPP3 in the proof of Theorem 8.29 on page 450.

31. Look at page 450. In the proof of Theorem 8.29, it was shown that the rows of the matrix, A, in Construction 8.1 are orthogonal. Verify this directly for the construction in Quick Check 8.7 by exhaustively listing all the vertical pairs for each pair of rows in A.

32. Follow the proof outline, (a)–(d), to prove Lemma 8.32. Example 8.14 on page 439 may be helpful.

> **LEMMA 8.32**
> Let $\{L^1, L^2, \ldots, L^{n-1}\}$ be a set of mutually orthogonal Latin squares of order n. Let $r_1 \neq r_2$ and $c_1 \neq c_2$. Then for some $j \in \{1, 2, \ldots, n-1\}$, $L^j_{r_1c_1} = L^j_{r_2c_2}$.

(a) Explain why it is possible to use sequences of interchanges to transform each of the Latin squares into a Latin square whose first row is "$123 \cdots n$" and still end up with a collection of mutually orthogonal Latin squares. Then note that the interchanges can be reversed, so the transformation process does not change the validity of what follows. Denote the new Latins squares as $L^{1'}, L^{2'}, \ldots, L^{n-1'}$.

(b) Suppose $r_1 \neq r_2$, $c_1 \neq c_2$ and $i \neq j$. Prove: If $L^{i'}_{r_1c_1} = L^{j'}_{r_1c_1}$, then $L^{i'}_{r_2c_2} \neq L^{j'}_{r_2c_2}$.

(c) Suppose $r_1 \neq 1$ and $r_2 \neq 1$. Form the $n-1$ ordered pairs $(L^{l'}_{r_1c_1}, L^{l'}_{r_2c_2})$, for $l = 1, 2, \ldots, n-1$. Denote the list of ordered pairs as $(x_1, y_1), (x_2, y_2), \ldots, (x_{n-1}, y_{n-1})$ (just to keep the notation simple). The previous part of the proof implies that $(x_i, y_i) \neq (x_j, y_j)$ if $i \neq j$. Show that if $k \neq m$, then $x_k \neq x_m$ and $y_k \neq y_m$.

(d) Complete the proof by considering two cases.

i. $r_1 = 1$ or $r_2 = 1$: Suppose $r_2 = 1$ (the other option is handled in a similar manner). Show that $y_i = c_2$ for all i. Notice that $x_j \neq c_1$ for all j. Conclude that for some value of j, $x_j = y_j = c_2$.

ii. $r_1 \neq 1$ and $r_2 \neq 1$: Show that $x_j \neq c_1$ and $y_j \neq c_2$. Conclude that if $y_i \neq x_i$ for all i, then it is impossible to create $n-1$ distinct ordered pairs (x_i, y_i). Consequently, $y_i = x_i$ for some i.

33. Does a finite projective plane of order 6 exist? If so construct one, if not explain why not.

8.3 Balanced Incomplete Block Designs

Recall the fertilizer experiment in Example 8.11 on page 437. You might have found the example a bit contrived: there were three fields, three plots per field, three seed varieties, and three fertilizer regimes. What if the various components of the experiment do not all come in convenient groups of size three? The answer is that sometimes there is an alternative to Latin square designs. The alternative is called a balanced incomplete block design.

Balanced incomplete block designs were briefly introduced in Section 1.3.4 (page 9). That section introduced a schoolmistress with nine schoolgirls in her boarding school. The solution to that puzzle is reviewed in the next example.

EXAMPLE 8.21 **Nine Schoolgirls**

A schoolmistress desires to have the nine girls at her boarding school go for a walk on four days each week. The girls will walk in three rows with three students in each row. The schoolmistress also wants each girl to walk in a common row with every other girl exactly once per week. The solution that was presented in Section 1.3.4 is listed in Table 8.17.

TABLE 8.17 Arranging Nine Schoolgirls for Weekly Walks

	Monday	Tuesday	Thursday	Friday
Row 1	1 2 3	1 4 7	1 5 9	1 6 8
Row 2	4 5 6	2 5 8	2 6 7	2 4 9
Row 3	7 8 9	3 6 9	3 4 8	3 5 7

Table 8.17 provides a convenient example to introduce some terminology that will be used throughout this section. Notice that there are 12 three-girl rows in the table. These rows are examples of *blocks* (the main topic is "balanced incomplete *block* designs"). Each block (row) contains three girls. In the general setting, the items contained in the blocks are called *varieties* (recall the "three seed *varieties*" in the fertilizer experiment). Each block contains three varieties, and each variety is in four blocks (one block per day). Finally, every pair of varieties appears in exactly one common block. ■

EXAMPLE 8.22 **Popcorn Revisited**

Suppose the researcher in Example 8.11 has four fields (each with three plots). Two of the fields are near the city (and adjacent to each other). The other two are in a rural area (and have similar soil conditions). The researcher wishes to compare four seed varieties and only two fertilizer regimes. Is it possible to create an experimental design that allows each seed variety to be tested in a common field with every other variety and also for each seed variety to have both fertilizer regimes and both city and rural fields?

A Latin square will not work with these parameters. However, it is not hard to produce an experimental design that meets the objectives. The four fields will be the blocks, and the seeds will be four varieties. The fields are denoted C_1, C_2, R_1, and R_2 (C for "city" and R for "rural"). Fields C_1 and R_1 will use the first fertilizer regime; C_2 and R_2 will use the second regime. The varieties will be denoted 1, 2, 3, and 4. Each block will contain three varieties (since there are three plots per field). Every variety will appear twice with every other variety, but each variety will be planted in only three blocks. The design is listed next.

C_1	C_2	R_1	R_2
1	1	1	2
2	3	2	3
3	4	4	4

You should check the various claims made prior to the listing of the design. For example, varieties 2 and 3 appear together in blocks C_1 and R_2. Also, variety 2 will appear in both city and rural fields and also with both fertilizer regimes.

Notice that a block (field) does not contain every variety of seed. ■

These examples motivate the formal definition.

DEFINITION 8.33 ***Balanced Incomplete Block Design***

A *balanced incomplete block design*, abbreviated *BIBD*, is a combinatorial design consisting of a finite collection of finite sets (called *blocks*), each consisting of a finite number of elements (called *varieties*). The boundary conditions a BIBD must satisfy are expressed in terms of five parameters, commonly expressed as the 5-tuple of positive integers, (v, b, r, k, λ).

The parameter v represents the number of distinct varieties; the parameter b represents the number of blocks. Every variety is required to be in exactly r blocks, and every block must contain exactly k varieties. Finally, every pair of distinct varieties must appear in exactly λ common blocks.

A combinatorial design which meets these conditions is often referred to as a (v, b, r, k, λ)-design.

A (v, b, r, k, λ)-design with $k = v$ and $r = b$ is called *trivial*.

Notes:

1. All five parameters are positive, so $v \geq 1, b \geq 1, r \geq 1, k \geq 1$, and $\lambda \geq 1$.

2. A trivial BIBD consists of b identical blocks, each containing every variety.

It is convenient to express balanced incomplete block designs using a matrix. The next definition provides the details.

DEFINITION 8.34 ***The Incidence Matrix of a BIBD***

Let D be a (v, b, r, k, λ)-design with varieties $\{u_1, u_2, \ldots, u_v\}$ and blocks $\{B_1, B_2, \ldots, B_b\}$.

The *incidence matrix* of D is the v by b matrix, M, where

$$m_{ij} = \begin{cases} 1 & \text{if } v_i \in B_j \\ 0 & \text{otherwise} \end{cases}$$

The term *incomplete* refers to the fact that not every possible block is present in the design. Since there are v varieties and each block contains k varieties, there are potentially $\binom{v}{k}$ blocks. A typical BIBD has fewer blocks.[21] The term *balanced* refers to the uniform size of blocks, the uniform number of blocks each variety appears in, and the uniform way that pairs of varieties appear in blocks.

[21] In Example 8.21 there are only 12 of the $\binom{9}{3} = 84$ potential blocks in the design. In Example 8.22, all $4 = \binom{4}{3}$ of the possible blocks are used in the design, so this design is actually complete.

✔ Quick Check 8.9

1. Determine the values of the parameters (v, b, r, k, λ) for the BIBD in Example 8.21.
2. Determine the values of the parameters (v, b, r, k, λ) for the BIBD in Example 8.22.
3. Produce the incidence matrix for the agricultural BIBD in Example 8.22.
4. Produce the incidence matrix for the schoolgirl BIBD in Example 8.21.

It should seem likely that the parameters, (v, b, r, k, λ), cannot be chosen arbitrarily. The following theorem expresses the most fundamental relationships that these parameters must satisfy.[22]

THEOREM 8.35 ***The Parameters of a BIBD***

Let D be a balanced incomplete block design with parameters (v, b, r, k, λ). Then

$$bk = vr$$

and

$$r(k-1) = \lambda(v-1)$$

Proof: Let M be the incidence matrix for the (v, b, r, k, λ)-design. Both equations will be proved using combinatorial proofs that count 1s in M.

The first equation is verified by counting all the 1s in M two different ways.

Since there are b blocks, each containing k varieties, each of the b columns of M will contain k 1s, for a total of bk 1s. On the other hand, each of the v varieties is in r blocks, so each of the v rows of M contains r 1s, for a total of vr 1s. Therefore, $bk = vr$.

The second equation is verified by counting the 1s in a submatrix of M. Start by choosing any variety, u. Delete the row of M that corresponds to u and delete every column that corresponds to a block that does not contain u. Now count the 1s in the matrix, M_u, that remains.

Since u is in r blocks, there will be r columns in M_u. Each of those columns will contain $k-1$ 1s (since the 1 in u's row has been removed). On the other hand, u is in λ common blocks with each of the $v-1$ other varieties. So each of those varieties contributes λ 1s to M_u. Consequently, $r(k-1) = \lambda(v-1)$. □

Theorem 8.35 presents a pair of *necessary* conditions for the existence of a (v, b, r, k, λ)-design. These conditions are not *sufficient* conditions. That is, in order for a (v, b, r, k, λ)-design to exist, the two equations in the theorem must be true. However, even if the two equations are true, there may be no BIBD with the given parameters. For example, the parameters $(43, 43, 7, 7, 1)$ satisfy the two equations ($43 \cdot 7 = 43 \cdot 7$ and $7 \cdot 6 = 1 \cdot 42$), but no BIBD with parameters $(43, 43, 7, 7, 1)$ exists. (The nonexistence is a consequence of Theorem 8.38 on page 463.)

DEFINITION 8.36 ***Symmetric; Resolvable***

A balanced incomplete block design is *symmetric* if $v = b$ and $r = k$. Symmetric balanced incomplete block designs are often referred to as (v, k, λ)-designs.

A balanced incomplete block design is *resolvable* if the blocks can be grouped into disjoint collections (of equal numbers of blocks) so that every variety appears exactly once in each group of blocks.

[22] There are other conditions. Some of them will be presented in this text.

The BIBD in Example 8.22 is a symmetric BIBD because $v = b = 4$ and $r = k = 3$. The BIBD in Example 8.21 is resolvable. The blocks are grouped by day of the week. If this design were not resolvable, the design would not have solved the original puzzle of enabling all nine girls to walk in three lines each day.

✔ Quick Check 8.10

1. The original schoolgirl puzzle, proposed by Reverend Thomas Kirkman (see page 9), requires 7 walks with 15 girls arranged in 5 rows of 3 girls each.
   (a) A solution to his puzzle requires a resolvable BIBD. What are the parameters, (v, b, r, k, λ), of the BIBD?
   (b) Show that these parameters satisfy the requirements in Theorem 8.35. ☑

The next two theorems are presented without proof. See [46] for proofs, which utilize mathematical ideas beyond the assumed background for this text.

THEOREM 8.37 ***Fisher's Inequality***

If D is a nontrivial (v, b, r, k, λ)-design, then $b \geq v$ and $r \geq k$.

THEOREM 8.38 ***Bruck–Ryser–Chowla***

If D is a symmetric balanced incomplete block design, with parameters (v, k, λ), then the following statements are true.

- If v is even, then $k - \lambda$ is a square.
- If v is odd, then the equation
$$z^2 = (k - \lambda)x^2 + (-1)^{(v-1)/2}\lambda y^2$$
has a solution in integers x, y, and z, where x, y, and z are not all zero.

The second part of the Bruck–Ryser–Chowla Theorem can be used to show that a $(43, 7, 1)$-design cannot exist.

EXAMPLE 8.23 **Fragrance Testing**

A perfume company wishes to test some new fragrances. There are six candidate aromas. They wish to test them in groups of three, with the human test subjects classifying the individual aromas in each triple of fragrances as "best", "middle", and "worst" (with no ties). They want to have every pair of fragrances matched in at least two tests. There are $\binom{6}{3} = 20$ subsets of three fragrances. These 20 subsets will match each pair of fragrances more than twice (four times to be exact). Can this be reduced without compromising the requirements?

A balanced incomplete block design is one possible mechanism to reduce the number of tests, assuming a suitable BIBD exists. What parameters are necessary? The requirements have specified that $v = 6$, since there are six fragrances to test. The number of blocks (tests) is undetermined so far. However, each test matches three fragrances, so $k = 3$. If the minimum number of pairings is used, then $\lambda = 2$. This is enough to determine b and r. From Theorem 8.35, $b \cdot 3 = 6 \cdot r$ and $r \cdot 2 = 2 \cdot 5$. Thus, $r = 5$ and $b = 10$ are required. This is a promising start; the solution values for b and r might not have been integers.

Since $10 \geq 6$ and $5 \geq 3$, Fisher's inequality holds, so it still looks promising. Since $k \neq r$, the Bruck–Ryser–Chowla theorem does not apply.

It seems likely (but not guaranteed) that a $(6, 10, 5, 3, 2)$-design exists. If one does exist, then the number of tests can be reduced from 40 to $b = 10$. This is a substantial improvement. ■

Knowing that a BIBD might exist is not the same as being able to construct one. Example 8.23 must remain incomplete for now.[23]

The next section discusses some techniques for constructing BIBDs.

8.3.1 Constructing Balanced Incomplete Block Designs

The methods for constructing BIBDs fall into two broad categories. One category consists of constructions that start with other kinds of combinatorial designs or mathematical objects and build a balanced incomplete block design.[24] The other category consists of methods to transform one BIBD into another BIBD.

New BIBDs from Old

There are a number of ways to create a new BIBD from an existing BIBD. Several will be described in this section. The first two are very general; they will work with any initial BIBD. Each of the second pair of constructions assumes that the initial BIBD is symmetric.

The first of these constructions is extremely simple: just make one or more copies of each of the blocks.

Construction 8.3 A Replicated Design:
A $(v, nb, nr, k, n\lambda)$-Design from a (v, b, r, k, λ)-Design

If D is a (v, b, r, k, λ)-design, then a $(v, nb, nr, k, n\lambda)$-design, D_n, can be created by making n copies of each block in D. Thus, the varieties of D_n are identical to the varieties in D and each block of D will appear n times in D_n.

Notice that neither v nor k will change in this construction. If a variety, u, appears in r blocks in D, then it will appear in nr blocks in D_n (all its old blocks, each repeated n times). In a similar fashion, every pair of varieties will appear in $n\lambda$ common blocks.

EXAMPLE 8.24 **A Replicated Design**

Table 8.18 shows the blocks of a (5, 10, 6, 3, 3)-design.

TABLE 8.18 The Blocks of a (5, 10, 6, 3, 3)-Design

B_1	B_2	B_3	B_4	B_5	B_6	B_7	B_8	B_9	B_{10}
1	1	1	1	1	1	2	2	2	3
2	2	2	3	3	4	3	3	4	4
3	4	5	4	5	5	4	5	5	5

A $(5, 20, 12, 3, 6)$-design can be created by taking two copies of each block (Table 8.19).

TABLE 8.19 A (5, 20, 12, 3, 6)-Design

B_1	B_2	B_3	B_4	B_5	B_6	B_7	B_8	B_9	B_{10}	B_{11}	B_{12}	B_{13}	B_{14}	B_{15}	B_{16}	B_{17}	B_{18}	B_{19}	B_{20}
1	1	1	1	1	1	2	2	2	3	1	1	1	1	1	1	2	2	2	3
2	2	2	3	3	4	3	3	4	4	2	2	2	3	3	4	3	3	4	4
3	4	5	4	5	5	4	5	5	5	3	4	5	4	5	5	4	5	5	5

■

[23] A complete solution will be given in Exercise 18 on page 472.

[24] This is similar to using a set of $n-1$ mutually orthogonal Latin squares of order n to construct a finite projective plane of order n.

The next construction is only slightly more sophisticated.

Construction 8.4 The Complement Design:
A $(v, b, b-r, v-k, b-2r+\lambda)$-Design from a (v, b, r, k, λ)-Design

If D is a (v, b, r, k, λ)-design, then a $(\overline{v}, \overline{b}, \overline{r}, \overline{k}, \overline{\lambda}) = (v, b, b-r, v-k, b-2r+\lambda)$-design, $\overline{D}$, can be created by complementing each of the blocks in D. The varieties in $\overline{D}$ are identical to those in D. Suppose the set of varieties in D is U. If B is a block in D, then the corresponding block in $\overline{D}$ is $\overline{B} = U - B$.

The numbers of varieties and blocks do not change when moving from D to $\overline{D}$, so $\overline{v} = v$ and $\overline{b} = b$. A block, B, in D contains k varieties. The corresponding block, $\overline{B}$, will contain every variety that is *not* in B. Thus $\overline{k} = v - k$. The variety, u, appears in r blocks, $B_1, B_2, \ldots, B_r$, in D but will appear in every block except $\overline{B_1}, \overline{B_2}, \ldots, \overline{B_r}$ in $\overline{D}$. Thus, $\overline{r} = b - r$. Finally, consider two varieties, u_1 and u_2. They are in λ common blocks in D. There are an additional $r - \lambda$ blocks in D that contain u_1 but not u_2, and another $r - \lambda$ different blocks in D that contain u_2 but not u_1. The varieties u_1 and u_2 will appear together in all blocks, $\overline{B} \in \overline{D}$ for which $B \in D$ does not contain either u_1 or u_2. Thus, $\overline{\lambda} = b - (\lambda + (r - \lambda) + (r - \lambda)) = b - 2r + \lambda$.

EXAMPLE 8.25 **A Complement Design**

The (5, 10, 6, 3, 3)-design in Example 8.24 has a complement that is a (5, 10, 4, 2, 1)-design. The blocks of the complement are shown in Table 8.20.

TABLE 8.20 A (5, 10, 4, 2, 1)-Design

$\overline{B_1}$	$\overline{B_2}$	$\overline{B_3}$	$\overline{B_4}$	$\overline{B_5}$	$\overline{B_6}$	$\overline{B_7}$	$\overline{B_8}$	$\overline{B_9}$	$\overline{B_{10}}$
4	3	3	2	2	2	1	1	1	1
5	5	4	5	4	3	5	4	3	2

■

The next two constructions, which transform symmetric BIBDs into new BIBDs, are more sophisticated. The following theorem guarantees a necessary condition. The proof involves more matrix algebra than is assumed for this text, so it will be omitted.

THEOREM 8.39 ***Block Intersections in Symmetric BIBDs***

If D is a symmetric (v, k, λ)-design, then every pair of blocks contains exactly λ common varieties.

Construction 8.5 Derived Designs:
A $(k, v-1, k-1, \lambda, \lambda-1)$-Design from a (v, k, λ)-Design

Let D be a symmetric (v, k, λ)-design, with blocks $\{B_0, B_1, B_2, \ldots, B_{v-1}\}$. A derived design, D', with parameters $(v', b', r', k', \lambda') = (k, v-1, k-1, \lambda, \lambda-1)$ can be constructed by selecting any block in D (assumed here to be B_0). The varieties of D' will be the set of k varieties *in* B_0. The $v-1$ blocks of D' will be formed by removing B_0 and any varieties that are *not* in B_0. That is, $B_i' = B_i \cap B_0$ for $i \neq 0$.

Since every block in D contains k varieties, there will be $v' = k$ varieties in D' (since the varieties of D' come from B_0). The blocks in D' are constructed from the $v-1$ blocks other than B_0 in D. Each variety, u, in B_0 is in $r = k$ blocks in D. That variety will still be in the corresponding blocks in D', but B_0 has no corresponding block in D'. Thus, $r' = r - 1 = k - 1$. Theorem 8.39 asserts that B_0 and B_i have exactly λ varieties in common, for $i = 1, 2, \ldots, v-1$. Thus, $k' = |B_0 \cap B_i| = \lambda$. Finally, every pair of varieties in B_0 appears in λ common blocks in D. All but one of those common appearances will be preserved in D' (the common block B_0 has no counterpart in D'). Thus, $\lambda' = \lambda - 1$ (assuming $v' \geq 2$).

EXAMPLE 8.26

A Derived Design

A symmetric (15, 7, 3)-design is listed in Table 8.21.[25]

The derived design (using the block B_0) is shown in Table 8.22. It has parameters (7, 14, 6, 3, 2).

TABLE 8.21 A Symmetric (15, 7, 3)-Design

B_0	B_1	B_2	B_3	B_4	B_5	B_6	B_7	B_8	B_9	B_{10}	B_{11}	B_{12}	B_{13}	B_{14}
0	0	0	0	0	0	0	1	1	1	1	2	2	2	2
1	1	1	3	3	5	5	3	3	4	4	3	3	4	4
2	2	2	4	4	6	6	5	6	5	6	5	6	5	6
3	7	11	7	9	7	9	7	7	8	8	8	8	7	7
4	8	12	8	10	8	10	9	10	10	9	10	9	9	10
5	9	13	11	13	13	11	11	12	11	12	12	11	12	11
6	10	14	12	14	14	12	13	14	14	13	13	14	14	13

TABLE 8.22 A (7, 14, 6, 3, 2)-Design

B'_1	B'_2	B'_3	B'_4	B'_5	B'_6	B'_7	B'_8	B'_9	B'_{10}	B'_{11}	B'_{12}	B'_{13}	B'_{14}
0	0	0	0	0	0	1	1	1	1	2	2	2	2
1	1	3	3	5	5	3	3	4	4	3	3	4	4
2	2	4	4	6	6	5	6	5	6	5	6	5	6

■

The next construction also requires a symmetric design.

Construction 8.6 Residual Designs:
A $(v-k, v-1, k, k-\lambda, \lambda)$-Design from a (v, k, λ)-Design

Let D be a symmetric (v, k, λ)-design, with blocks $\{B_0, B_1, B_2, \ldots, B_{v-1}\}$. A residual design, D^*, with parameters $(v^*, b^*, r^*, k^*, \lambda^*) = (v-k, v-1, k, k-\lambda, \lambda)$ can be constructed by selecting any block in D (assumed again to be B_0). The varieties of D^* will be the set of $v-k$ varieties that are *not* in B_0. The $v-1$ blocks of D^* will be formed by removing B_0 and any varieties that are *in* B_0: $B_i^* = B_i - B_0, i \neq 0$.

Since the k varieties in B_0 are not present in D^*, $v^* = v - k$. Also, $b^* = v - 1$ since B_0 does not transform into a block in D^*. The varieties that remain in D^* are not in B_0, so removing B_0 does not change the number of blocks they are in. Thus, $r^* = r = k$. Theorem 8.39 asserts that B_0 shares λ common varieties with B_i, for $i \neq 0$. Thus, $k^* = k - \lambda$. Finally, if u_i and u_j are in the common block, B, in D, they will still be in the common block, B^*, in D^*. Thus, $\lambda^* = \lambda$ (assuming $v^* \geq 2$).

EXAMPLE 8.27

A Residual Design

The residual design (using block B_0) for the (15, 7, 3)-design in Example 8.26 is shown in Table 8.23. It has parameters (8, 14, 7, 4, 3).

TABLE 8.23 A (8, 14, 7, 4, 3)-Design

B^*_1	B^*_2	B^*_3	B^*_4	B^*_5	B^*_6	B^*_7	B^*_8	B^*_9	B^*_{10}	B^*_{11}	B^*_{12}	B^*_{13}	B^*_{14}
7	11	7	9	7	9	7	7	8	8	8	8	7	7
8	12	8	10	8	10	9	10	10	9	10	9	9	10
9	13	11	13	13	11	11	12	11	12	12	11	12	11
10	14	12	14	14	12	13	14	14	13	13	14	14	13

■

[25]From [46, p. 128].

✔ Quick Check 8.11

The symmetric BIBD, D, in Table 8.24 has parameters (7, 4, 2).

TABLE 8.24 A Symmetric (7, 4, 2)-Design

B_0	B_1	B_2	B_3	B_4	B_5	B_6
3	2	2	1	1	1	1
4	4	3	4	3	2	2
6	5	5	5	5	6	3
7	7	6	6	7	7	4

1. Form the complement design, $\overline{D}$. What are its parameters?
2. Form the derived design, D', using block B_2. What are its parameters?
3. Form the residual design, D^*, using block B_2. What are its parameters?

✔

BIBDs from Other Mathematical Objects

The following theorem, together with Construction 8.7, provide a prime example of a construction of a BIBD from some other combinatorial design.

THEOREM 8.40 ***BIBD Iff Finite Projective Plane***

A finite projective plane of order $n \geq 2$ exists if and only if a $(n^2 + n + 1, n + 1, 1)$-design exists.

Construction 8.7 A $(n^2 + n + 1, n + 1, 1)$-Design from a Finite Projective Plane of Order n

Let $\mathcal{F}$ be a finite projective plane of order n. A symmetric balanced incomplete block design, D, can be constructed from $\mathcal{F}$ by taking the lines of $\mathcal{F}$ as the blocks of D and the points of $\mathcal{F}$ as the varieties of D. A variety will be in a block if and only if (viewed as a point in $\mathcal{F}$) it is on the line in $\mathcal{F}$ from which the block was derived. D will be a $(n^2 + n + 1, n + 1, 1)$-design.

Construction 8.8 A Finite Projective Plane of Order n from a $(n^2 + n + 1, n + 1, 1)$-Design

Let D be a symmetric balanced incomplete block design with parameters $(n^2 + n + 1, n + 1, 1)$. The lines of $\mathcal{F}$ will be the blocks of D and the points of $\mathcal{F}$ will be the varieties of D. A point will be on a line if that point (viewed as a variety) is in the block from which the line was derived.

Proof of Theorem 8.40: The proof is a verification that Constructions 8.7 and 8.8 are correct.

Construction 8.7 is valid Since $\mathcal{F}$ has n^2+n+1 lines and n^2+n+1 points, $b = v = n^2+n+1$. Since every line in $\mathcal{F}$ contains $n+1$ points and every point in $\mathcal{F}$ is on $n+1$ lines, $k = r = n + 1$. Finally, every pair of points is on exactly one common line, so $\lambda = 1$.

Construction 8.8 is valid Since D is symmetric, $b = v = n^2+n+1$ and $r = k = n+1$, so there will be $n^2 + n + 1$ points and $n^2 + n + 1$ lines. Theorem 8.26 on page 445 implies that the construction will be a finite projective plane of order n (if it is actually a finite projective plane).

Since every block contains $k = n + 1$ varieties, it is clear that every line contains $n+1$ points. Similarly, since $r = n+1$, every point will be on $n+1$ lines. Since $\lambda = 1$, every pair of points will be on exactly one common line, so FPP1 holds. Theorem 8.39 on page 465 implies that every pair of lines contains $\lambda = 1$ common point, so FPP2 holds.

It only remains to verify that FPP3 holds. Since $k = n + 1 < n^2 + n + 1 = v$, no block contains every variety. Choose any two varieties, u_1 and u_2. There is exactly one block, $B_{1\cdot 2}$, that contains both u_1 and u_2 (by Theorem 8.39). Since $k < v$, there must exist some variety, u_3, that is not in $B_{1\cdot 2}$. There are unique blocks, $B_{1\cdot 3}$ and $B_{2\cdot 3}$, such that u_1 and u_3 are both in $B_{1\cdot 3}$ and also u_2 and u_3 are both in $B_{2\cdot 3}$. The three varieties u_1, u_2, and u_3 are not all in a common block. How many distinct varieties are in the three blocks $B_{1\cdot 2}$, $B_{1\cdot 3}$, and $B_{2\cdot 3}$? There are $k = n+1$ distinct varieties in $B_{1\cdot 2}$, but only $k - 1 = n$ additional varieties in $B_{1\cdot 3}$ (since u_1 has already been counted). The block $B_{2\cdot 3}$ contributes only $k-2 = n-1$ new varieties since both u_2 and u_3 have already been counted. These three blocks therefore contain a total of $3n$ distinct varieties.[26] Since $n \geq 2$, $n^2 + n + 1 > 3n$. There must be at least one more variety, u_4, that is not in the blocks $B_{1\cdot 2}$, $B_{1\cdot 3}$, and $B_{2\cdot 3}$. Since pairs of varieties are only in one common block ($\lambda = 1$), u_4 is not in a common block with any pair among $\{u_1, u_2, u_3\}$.

Now change the viewpoint back to the projective plane interpretation. The points u_1, u_2, u_3, and u_4 are 4 points in $\mathcal{F}$ with no three among them on a common line. This means that FPP3 holds. □

✔ Quick Check 8.12

1. Create a $(7, 3, 1)$-design from the Fano plane. Sort the varieties in increasing order and list the blocks sorted by the varieties they contain.

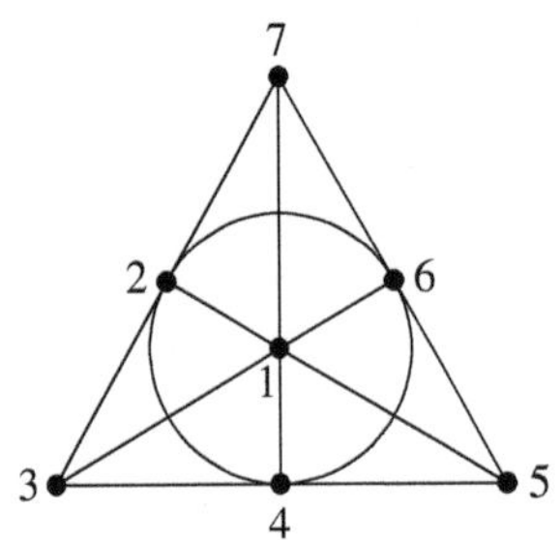

2. Show the incidence matrix for the $(7, 3, 1)$-design you just created.

The next construction uses modular arithmetic.[27]

Construction 8.9 A (13, 26, 6, 3, 1)-Design

Start with the blocks $B_1 = \begin{matrix} 1 \\ 3 \\ 9 \end{matrix}$ and $B_{14} = \begin{matrix} 2 \\ 5 \\ 6 \end{matrix}$. Create twelve additional blocks by starting with B_1 and adding 1 to the number in each of the three rows of the current block. Do the arithmetic mod 13. That is, whenever the addition produces a 13, change the value to 0 (and re-sort the block). Keep doing this until the original block appears. Discard the second copy of the original block. Repeat the process starting with B_{14}.

[26] You should convince yourself that they do not contain additional common varieties.

[27] See Definition 3.18 on page 98.

Each of the two original blocks is the source for 12 additional blocks, for a total of $b = 26$ blocks. There are 13 varieties (the numbers 0–12). Each block contains $k = 3$ varieties. You can verify that $r = 6$ and $\lambda = 1$ by examining the incidence matrix (Table 8.26). The BIBD is listed (without labels) in Table 8.25. It is split into two groupings so that the progression from the two initial blocks is apparent. The transitions from block 1 to block 2 and from block 4 to block 5 are amplified in Figure 8.11.

Figure 8.11 The transition from B_1 to B_2 and from B_4 to B_5.

TABLE 8.25 A (13, 26, 6, 3, 1)-Design

1	2	3	4	0	1	2	3	4	5	0	1	0
3	4	5	6	5	6	7	8	9	10	6	7	2
9	10	11	12	7	8	9	10	11	12	11	12	8
2	3	4	5	6	7	8	0	0	1	2	0	1
5	6	7	8	9	10	11	9	1	2	3	3	4
6	7	8	9	10	11	12	12	10	11	12	4	5

Table 8.26 shows the incidence matrix for this BIBD.

TABLE 8.26 A (13, 26, 6, 3, 1)-Design

	B_1	B_2	B_3	B_4	B_5	B_6	B_7	B_8	B_9	B_{10}	B_{11}	B_{12}	B_{13}	B_{14}	B_{15}	B_{16}	B_{17}	B_{18}	B_{19}	B_{20}	B_{21}	B_{22}	B_{23}	B_{24}	B_{25}	B_{26}
0	0	0	0	0	1	0	0	0	0	0	1	0	1	0	0	0	0	0	0	0	1	1	0	0	1	0
1	1	0	0	0	0	1	0	0	0	0	0	1	0	0	0	0	0	0	0	0	0	1	1	0	0	1
2	0	1	0	0	0	0	1	0	0	0	0	0	1	1	0	0	0	0	0	0	0	0	1	1	0	0
3	1	0	1	0	0	0	0	1	0	0	0	0	0	0	1	0	0	0	0	0	0	0	0	1	1	0
4	0	1	0	1	0	0	0	0	1	0	0	0	0	0	0	1	0	0	0	0	0	0	0	0	1	1
5	0	0	1	0	1	0	0	0	0	1	0	0	0	1	0	0	1	0	0	0	0	0	0	0	0	1
6	0	0	0	1	0	1	0	0	0	0	1	0	0	1	1	0	0	1	0	0	0	0	0	0	0	0
7	0	0	0	0	1	0	1	0	0	0	0	1	0	0	1	1	0	0	1	0	0	0	0	0	0	0
8	0	0	0	0	0	1	0	1	0	0	0	0	1	0	0	1	1	0	0	1	0	0	0	0	0	0
9	1	0	0	0	0	0	1	0	1	0	0	0	0	0	0	0	1	1	0	0	1	0	0	0	0	0
10	0	1	0	0	0	0	0	1	0	1	0	0	0	0	0	0	0	1	1	0	0	1	0	0	0	0
11	0	0	1	0	0	0	0	0	1	0	1	0	0	0	0	0	0	0	1	1	0	0	1	0	0	0
12	0	0	0	1	0	0	0	0	0	1	0	1	0	0	0	0	0	0	0	1	1	0	0	1	0	0

The next construction is similar to Construction 8.9. It will be the basis for a solution to Kirkman's schoolgirls problem.

> **Construction 8.10 A (15, 35, 7, 3, 1)-Design**
>
> Start with the blocks $B_1 = \begin{matrix} 0 \\ 1 \\ 4 \end{matrix}$ and $B_{16} = \begin{matrix} 0 \\ 2 \\ 8 \end{matrix}$ and $B_{31} = \begin{matrix} 0 \\ 5 \\ 10 \end{matrix}$. Create additional blocks by starting with B_1 and adding 1 to each row. Do the arithmetic mod 15. Keep doing this until the original block appears. Discard the second copy of the original block. Repeat the process with B_{16} and B_{31}.

The 35 blocks from Construction 8.10 are listed as columns in Table 8.27.[28]

TABLE 8.27 A (15, 35, 7, 3, 1)-Design

0	1	2	3	4	5	6	7	8	9	10	0	1	2	0
1	2	3	4	5	6	7	8	9	10	11	11	12	13	3
4	5	6	7	8	9	10	11	12	13	14	12	13	14	14

0	1	2	3	4	5	6	0	1	2	3	4	5	0	1
2	3	4	5	6	7	8	7	8	9	10	11	12	6	7
8	9	10	11	12	13	14	9	10	11	12	13	14	13	14

0	1	2	3	4
5	6	7	8	9
10	11	12	13	14

EXAMPLE 8.28

Kirkman's Schoolgirls

A schoolmistress has 15 young girls at her boarding school. Each day she lines the girls up in five rows of three girls each. She wishes to group the girls so that in the course of seven walks, each girl will have been in a row with every other girl exactly once.

This requires a (15, 35, 7, 3, 1)-design (Quick Check 8.10 on page 463). There are 80 essentially distinct BIBDs with these parameters, but only four of them are resolvable. The four resolvable BIBDs lead to seven essentially distinct solutions to the schoolgirl problem[29] [19, page 88]. Construction 8.10 produces one of the four resolvable (15, 35, 7,3, 1)-designs. The girls are numbered from 0 to 14. Table 8.28 shows these blocks (reformatted for the original problem) separated into seven disjoint groups.

TABLE 8.28 A Solution to the Original Kirkman's Schoolgirl Problem

Sunday			**Monday**			**Tuesday**			**Wednesday**		
0	1	4	0	2	8	0	3	14	0	5	10
2	9	11	1	3	9	1	6	11	1	12	13
3	10	12	4	11	13	2	7	12	2	3	6
5	7	13	5	12	14	4	5	8	4	9	14
6	8	14	6	7	10	9	10	13	7	8	11

Thursday			**Friday**			**Saturday**		
0	6	13	0	7	9	0	11	12
1	7	14	1	2	5	1	8	10
2	4	10	3	8	13	2	13	14
3	5	11	4	6	12	3	4	7
8	9	12	10	11	14	5	6	9

■

[28] The construction is from [77, p. 761].

[29] There can be more than one way to partition the BIBD's blocks into disjoint groups.

8.3.2 Exercises

The exercises marked with ★ have detailed solutions in Appendix H.

1. Let D be a (v, b, r, k, λ)-design. Prove that $r = \frac{\lambda(v-1)}{k-1}$ and $b = \frac{\lambda v(v-1)}{k(k-1)}$. This implies that it is only necessary to specify the parameters v, k, and λ. Thus, the shorthand notation "(v, k, λ)-design" need not be limited to only symmetric BIBDs.

2. Assume the inequality $b \geq v$ in Theorem 8.37 (page 463) and use Theorem 8.35 (page 462) to prove the inequality $r \geq k$.

3. Prove Lemma 8.41.

> **LEMMA 8.41**
> If D is a nontrivial (v, b, r, k, λ)-design, then $v > k$ and $r > \lambda$.

4. Can a BIBD have the following sets of parameters? If it is not possible, explain why not. Otherwise, list the parameters as "potential".
   (a) ★ $(10, 15, 6, 4, 2)$ (b) ★ $(15, 12, 5, 3, 2)$
   (c) $(8, 12, 4, 4, 1)$ (d) $(55, 99, 18, 10, 3)$

5. Can a BIBD have the following sets of parameters? If it is not possible, explain why not. Otherwise, list the parameters as "potential".
   (a) $(13, 18, 9, 3, 2)$ (b) $(27, 117, 13, 3, 1)$
   (c) $(145, 290, 20, 10, 1)$ (d) $(324, 342, 19, 18, 1)$

6. Can a BIBD have the following sets of parameters? If it is not possible, explain why not. Otherwise, list the parameters as "potential".
   (a) $(21, 70, 10, 3, 1)$ (b) $(46, 22, 5, 28, 3)$
   (c) $(9, 18, 8, 4, 3)$ (d) $(61, 305, 20, 4, 1)$

7. Can a symmetric BIBD have the following sets of parameters? If it is not possible, explain why not. Otherwise, list the parameters as "potential".
   (a) $(8, 7, 4)$ (b) ★ $(22, 7, 2)$
   (c) ★ $(15, 7, 3)$ (d) $(27, 13, 6)$

8. Can a symmetric BIBD have the following sets of parameters? If it is not possible, explain why not. Otherwise, list the parameters as "potential".
   (a) $(76, 51, 34)$ (b) $(22, 13, 4)$
   (c) $(92, 14, 2)$ (d) $(115, 19, 3)$

9. Can a symmetric BIBD have the following sets of parameters? If it is not possible, explain why not. Otherwise, list the parameters as "potential".
   (a) $(9, 8, 7)$ (b) $(188, 34, 6)$
   (c) $(23, 11, 5)$ (d) $(34, 12, 4)$

10. ★ The Bruck–Ryser–Chowla theorem can be used to show that a $(43, 7, 1)$-design cannot exist. The theorem requires a solution to an equation in x, y, and z. What is that equation for the parameters $(43, 7, 1)$? You need not attempt to show there are no integer solutions that are not all 0.

11. Prove, without using the Bruck–Ryser–Chowla theorem, that a $(43, 7, 1)$-design does not exist.

12. Construct a $(13, 4, 1)$-design.

13. Construct the incidence matrix of a $(6, 15, 10, 4, 6)$-design. You may find it helpful to think about the BIBD whose incidence matrix is shown in the following table.

	B_1	B_2	B_3	B_4	B_5	B_6	B_7	B_8
u_1	1	1	1	1	1	0	0	0
u_2	1	0	0	0	0	1	1	1
u_3	0	1	0	0	0	1	0	0
u_4	0	0	1	0	0	0	1	0
u_5	0	0	0	1	0	0	0	1
u_6	0	0	0	0	1	0	0	0

	B_9	B_{10}	B_{11}	B_{12}	B_{13}	B_{14}	B_{15}
u_1	0	0	0	0	0	0	0
u_2	1	0	0	0	0	0	0
u_3	0	1	1	1	0	0	0
u_4	0	1	0	0	1	1	0
u_5	0	0	1	0	1	0	1
u_6	1	0	0	1	0	1	1

14. A $(13, 26, 12, 6, 5)$-design.
    (a) Construct a $(13, 26, 12, 6, 5)$-design by starting with the blocks [0 1 3 6 7 11] and [0 1 2 3 7 11] and adding 1s, mod 13.
    (b) What are the parameters for the complement of the design in part (a)? List the first five blocks in the complement design.

15. The design, D, shown is a symmetric $(16, 6, 2)$-design.

B_0	B_1	B_2	B_3	B_4	B_5	B_6	B_7
a	c	d	c	a	b	b	b
b	f	f	d	e	d	c	f
c	g	i	i	h	g	h	h
d	h	k	l	i	j	k	l
e	i	m	o	m	m	n	m
f	j	n	p	o	o	o	p

B_8	B_9	B_{10}	B_{11}	B_{12}	B_{13}	B_{14}	B_{15}
b	a	a	a	a	d	c	e
e	d	c	f	b	e	e	f
g	h	g	g	i	g	j	j
i	j	l	k	j	h	k	l
k	k	m	o	n	n	m	n
l	l	n	p	p	p	p	o

   (a) ★ List the parameters for the residual design, D^*, of D, and then list the blocks ofD^*.
   (b) List the parameters for the derived design, D', of D, and then list the blocks ofD'.

16. The design, D, shown is a symmetric $(19, 9, 4)$-design.

B_0	B_1	B_2	B_3	B_4	B_5	B_6	B_7	B_8	B_9
1	2	0	0	1	2	3	4	0	1
4	5	3	1	2	3	4	5	5	6
5	6	6	4	5	6	7	8	6	7
6	7	7	7	8	9	10	11	9	10
7	8	8	8	9	10	11	12	12	13
9	10	9	9	10	11	12	13	13	14
11	12	11	10	11	12	13	14	14	15
16	17	13	12	13	14	15	16	15	16
17	18	18	14	15	16	17	18	17	18

B_{10}	B_{11}	B_{12}	B_{13}	B_{14}	B_{15}	B_{16}	B_{17}	B_{18}
0	1	0	0	0	0	1	2	0
2	3	2	1	1	1	2	3	3
7	8	4	3	2	2	3	4	4
8	9	9	5	4	3	4	5	5
11	12	10	10	6	5	6	7	6
14	15	13	11	11	7	8	9	8
15	16	16	14	12	12	13	14	10
16	17	17	17	15	13	14	15	15
17	18	18	18	18	16	17	18	16

(a) List the parameters for the residual design, D^*, of D, and then list the blocks of D^*.

(b) List the parameters for the derived design, D', of D, and then list the blocks of D'.

17. Let M be an n-by-n magic square. Create a combinatorial design by taking the individual rows, columns, and two diagonals of M as "blocks". Let the numbers 1 through n^2 be the varieties. Is this a BIBD? If so, determine the parameters. If not, explain why not.

18. Example 8.23 (on page 463) can be completed by performing the following steps.

(a) Construct a symmetric $(11, 5, 2)$-design, D, by starting with the block, B_1, whose varieties are 1, 3, 4, 5, and 9 (written as a single column) and adding 1s to the rows using mod 11 arithmetic.

(b) Form a residual design, D^*, from the design D in the previous step.

(c) Interpret the results in terms of the original example.

19. Suppose two BIBDs with common varieties and parameter sets $(v, b_1, r_1, k_1, \lambda_1)$ and $(v, b_2, r_2, k_2, \lambda_2)$ exist. What requirements are necessary for the union of their blocks to be a balanced incomplete block design? What will the parameters of the union be? (Experiment with the $(6, 10, 5, 3, 2)$-design in Exercise 18 and its complement design.)

20. Let D be a (v, b, r, k, λ)-design with incidence matrix, M. Describe the matrix MM^T.

8.4 The Knapsack Problem

The introduction to this chapter mentioned the knapsack problem as an example of a combinatorial optimization problem. Combinatorial optimization problems seek a best solution from among many contending solutions to some problem. The knapsack problem seeks to pack a knapsack with a set of items. Each item has a benefit value (or utility or cost) and a size (or weight). The goal is to obtain the maximum benefit under the constraint that the knapsack has a finite capacity. These ideas are expressed formally in the next definition.

DEFINITION 8.42 ***The Knapsack Problem***

The *knapsack problem* is concerned with a knapsack that has positive integer volume (or capacity) v. There are n distinct items that may potentially be placed into the knapsack. Item i has positive integer volume v_i and positive integer benefit b_i. In addition, there are q_i copies of item i available, where quantity q_i is a positive integer satisfying $1 \leq q_i \leq \infty$.

The integer variables $x_1, x_2, \ldots, x_n$ will determine how many copies of item i are to be placed into the knapsack. The goal is to

Maximize

$$\sum_{i=1}^{n} b_i x_i$$

Subject to the constraints

$$\sum_{i=1}^{n} v_i x_i \leq v \quad \text{and} \quad 0 \leq x_i \leq q_i$$

If $q_i = 1$ for $1 \leq i \leq n$, the problem is a *0-1 knapsack problem*. If one or more of the q_i is infinite, the problem is *unbounded*; otherwise, the problem is *bounded*.

An implicit assumption in knapsack problems is that the items are highly malleable. There will never be a conflict based on the shapes of the items. The only relevant factor is that their total volume does not exceed the capacity of the knapsack.

EXAMPLE 8.29

A 0-1 Knapsack Problem

Suppose I have a knapsack that holds 1000 cubic inches. I wish to pack several items of different sizes and different benefits. I want the greatest benefit within the constraint that the knapsack only holds 1000 cubic inches.

There are nine potential items (labeled "A," "B," "C,","D," "E," "F,","G," "H," "I"). Their volumes and benefits are shown in Table 8.29.

TABLE 8.29 Potential Items for a 1000-Cubic-Inch Knapsack

Item #	A	B	C	D	E	F	G	H	I
Benefit	13	8	11	16	4	1	1	9	10
Volume	340	210	190	450	120	20	60	120	220

I seek to maximize the total benefit:

$$\sum_{i=1}^{9} b_i x_i = 13x_1 + 8x_2 + 11x_3 + 16x_4 + 4x_5 + x_6 + x_7 + 9x_8 + 10x_9$$

subject to the constraints

$$\begin{aligned}\sum_{i=1}^{9} v_i x_i &= 340x_1 + 210x_2 + 190x_3 + 450x_4 + 120x_5 + 20x_6 \\ &\quad + 60x_7 + 120x_8 + 220x_9 \\ &\le 1000\end{aligned}$$

and

$$x_i \in \{0, 1\}, \quad \text{for } i = 1, 2, \ldots, n$$

The name "0-1 knapsack problem" arises because x_i is either 0 or 1 for all i (and so $q_i = 1$ for each i). ■

The problem in Example 8.29 seems sufficiently complex that the solution is not immediately obvious. Since there are 2^9 possible subsets of items, it would be feasible to program a computer to list all the possibilities and then identify a subset that meets the constraints and has the maximum total benefit. It is not an approach worth doing by hand. Are there any other algorithms for solving knapsack problems?

In Section 4.3, it was apparent that the obvious algorithm may not always be the most efficient. The next example shows that the obvious algorithm need not even be correct.

EXAMPLE 8.30

A Greedy Solution

Consider the 0-1 knapsack problem introduced in Example 8.29. The obvious approach is a *greedy algorithm*: Pack the item of highest utility first, then eliminate all items that will not fit. Now pack the remaining item of highest utility and eliminate any that now won't fit. Continue until no other item will fit. This approach is similar to the strategy a monkey might use to collect fruit. The monkey will most likely grab all the biggest and ripest fruit that are nearby. It will continue to take what it perceives as the most desirable pieces until it can carry no more.

A greedy algorithm may be a useful strategy for a monkey gathering fruit. However, the greedy approach fails to find an optimal (maximal) solution for this problem. To see

Item #	A	B	C
Benefit	13	8	11
Volume	340	210	190

Item #	D	E	F
Benefit	16	4	1
Volume	450	120	20

Item #	G	H	I
Benefit	1	9	10
Volume	60	120	220

this, notice that the greedy algorithm produces

Item #	Benefit	Cumulative Benefit	Cumulative Volume
D	16	16	450
A	13	29	790
C	11	40	980
F	1	**41**	1000

There is a better solution:

Item #	Benefit	Cumulative Benefit	Cumulative Volume
A	13	13	340
C	11	24	530
E	4	28	650
H	9	37	770
I	10	**47**	990

The greedy algorithm produced a series of *local optimizations* (at each step, do what seems best for the next step). The better solution was found by using an algorithm that produces a *global optimum* (make decisions that work for the total problem). ■

The greedy algorithm failed. Perhaps it was because we were focusing on the wrong property. A more sophisticated approach is to consider "benefit per unit of volume". That is, consider the ratios $\frac{b_i}{v_i}$. Perhaps if we add items with the highest benefit per unit of volume ratio first, the solution will be better.

EXAMPLE 8.31

A Sophisticated Greedy Algorithm

The items in Example 8.29 can be sorted by decreasing benefit-per-unit-of-volume order. Items can then be added to the knapsack in this order. The ratios are listed in Table 8.30.

TABLE 8.30 Ratios for a Sophisticated Greedy Algorithm

Item #	A	B	C	D	E	F	G	H	I
Benefit	13	8	11	16	4	1	1	9	10
Volume	340	210	190	450	120	20	60	120	220
B/V Ratio	.0382	.0381	.0579	.0356	.0333	.0500	.0167	.0750	.0455

Sorting by ratio produces Table 8.31.

TABLE 8.31 Items, Sorted by Benefit/Volume

B/V Ratio	.0750	.0579	.0500	.0455	.0382	.0381	.0356	.0333	.0167
Item #	H	C	F	I	A	B	D	E	G
Benefit	9	11	1	10	13	8	16	4	1
Volume	120	190	20	220	340	210	450	120	60

TABLE 8.32 "Sophisticated" Greedy Solution

Item	Benefit	Cum. Benefit	Cum. Volume
H	9	9	120
C	11	20	310
F	1	21	330
I	10	31	550
A	13	44	890
G	1	**45**	950

The "sophisticated" greedy solution produces Table 8.32. This is an improvement over the obvious greedy algorithm, but still does not produce an optimal solution. Since it does not give the maximum possible solution, it is technically an incorrect algorithm. ■

A Correct Algorithm

It is time to present an algorithm that correctly solves the knapsack problem. It will be easier to begin with a solution for completely unbounded knapsack problems ($q_i = \infty$, for $i = 1, 2, \ldots, n$). A solution to this problem will then be modified to solve bounded and 0-1 knapsack problems.

The key insight introduces recursion.[30] Suppose an optimal solution is sought for a knapsack with capacity v, with the additional requirement that at least one copy of item X must be included. Suppose that item X has volume v_x and benefit b_x. Let the optimal total benefit for a knapsack with capacity v be denoted by $B(v)$. A bit of thought will lead to the observation that the optimal solution will have total benefit $b_x + B(v - v_x)$. That is, add an item X, and then find the optimal benefit for the remaining $v - v_x$ cubic units of knapsack.

A small example will help make this concrete.

EXAMPLE 8.32

A Very Small Unbounded Knapsack Problem

Consider a knapsack with capacity $v = 13$. There are three types of items, each with an unlimited number of copies. The items and their benefits and volumes are shown in Table 8.33.

TABLE 8.33 Potential Items for a Knapsack with Capacity 13

Item	X	Y	Z
Benefit	4	3	5
Volume	6	7	8

It is easy to spot the optimal solution (two item X's, for a total benefit of 8). However, that solution will be ignored for the moment. Instead, for each of the three types of items, consider the optimal solution if at least one copy of the item must be included.

Item X The optimal benefit will be $4 + B(13 - 6) = 4 + B(7)$.

Item Y The optimal benefit will be $3 + B(13 - 7) = 3 + B(6)$.

Item Z The optimal benefit will be $5 + B(13 - 8) = 5 + B(5)$.

Consider the final case. Since no item has a volume as small as 5, $B(5)$ must be 0. Thus, 5 is the maximum benefit if at least one item Z is required.

If at least one item Y is required, it is necessary to determine the optimal benefit for a knapsack with capacity 6. Since item X is the only item that can fit into a knapsack with volume 6, $B(6) = 4$. Thus, the optimal benefit will be $3 + B(6) = 3 + 4 = 7$.

Finally, if at least one item X is required, it is necessary to determine the optimal benefit for a knapsack with capacity 7. Either item X or item Y will fit, but item X has a larger benefit. Choosing item X leads to $B(7) = 4$. Thus, the optimal benefit for a knapsack with capacity 13 for which at least one item X is needed is $4 + B(7) = 4 + 4 = 8$.

Since the knapsack will contain at least one copy of *some* item, the most beneficial solution will start with an item X. This produces a knapsack with two item X's and a total (optimal) benefit of 8. ■

The key insight [examining $b_x + B(v - v_x)$] uses recursion, but it is often easier to start with a knapsack having capacity 0 and iterate up to a knapsack with capacity v. This requires more calculations than are absolutely necessary but is simple enough that it can be accomplished with any spreadsheet program.[31]

[30] In an Operations Research course, a more efficient approach is typically introduced in a chapter on *dynamic programming*, which has no immediate connection to computer programming.

[31] That approach will not be used here because a spreadsheet solution does not adapt well to bounded and 0-1 knapsack problems.

EXAMPLE 8.33

Item	X	Y	Z
Benefit	4	3	5
Volume	6	7	8

Optimal benefit with volume w and at least one X is $b_X + B(w - v_X)$

A Tabular Implementation of the Recursion

Example 8.32 can be solved by creating a table containing all the relevant calculations. The rows of the table will be indexed by knapsack capacities (from row 0 to row v). The table will contain a column for each item type, as well as a final column for the optimal benefit for the row capacities. The first row will contain 0s. The remaining rows will be calculated by finding the optimal benefit possible by adding one more copy of an item to the current knapsack. Each item's optimum will be recorded in the proper column. The total benefit is the largest value in the row (there may be ties).

The calculations for Example 8.32 are shown in Table 8.34.

TABLE 8.34 The Calculations for Example 8.32.

w	X	Y	Z	$B(w)$	
0	0	0	0	0	
1	0	0	0	0	
2	0	0	0	0	
3	0	0	0	0	
4	0	0	0	0	
5	0	0	0	0	
6	$4 + B(0) = 4$	0	0	4	Item X is the only option here.
7	$4 + B(1) = 4$	$3 + B(0) = 3$	0	4	
8	$4 + B(2) = 4$	$3 + B(1) = 3$	$5 + B(0) = 5$	5	Item Z has the greatest benefit for this v.
9	$4 + B(3) = 4$	$3 + B(2) = 3$	$5 + B(1) = 5$	5	
10	$4 + B(4) = 4$	$3 + B(3) = 3$	$5 + B(2) = 5$	5	
11	$4 + B(5) = 4$	$3 + B(4) = 3$	$5 + B(3) = 5$	5	
12	$4 + B(6) = 8$	$3 + B(5) = 3$	$5 + B(4) = 5$	8	It is now possible to add a second item X.
13	$4 + B(7) = 8$	$3 + B(6) = 7$	$5 + B(5) = 5$	8	Two X's are better than an X and a Y.

The final row of the table indicates that the optimal benefit (for $v = 13$) is 8. To determine which items to pack, work backward. Start at the final row and find any item whose benefit matches the optimal value. Item X is the only choice for this example. Now subtract item X's volume from v and look at row 7. The optimal benefit for that row is 4, and it is achieved in the item X column. Add another item X, then subtract item X's volume from 7. The optimal benefit in row 1 is 0, so no additional items can be added. ■

✔ Quick Check 8.13

1. A knapsack can hold 10 cubic units. It is to be filled with four types of items (each with an unlimited number of copies). The item volumes and benefits are shown. Create a table like the one in Example 8.33 to determine an optimal collection of items.

Item	W	X	Y	Z
Benefit	3	4	11	1
Volume	4	3	7	2

☑

The ideas presented so far have been used to create the following algorithm for solving unbounded knapsack problems. The table will be denoted by $T_{w,i}$, where w is the current knapsack (row), and i is the current item (column). The final column, $B(w)$, will be represented as a separate list. The initial column, w, will be represented by the first subscript, rather than as an explicit column (as was done in Table 8.34).

```
 1: integer, list unboundedKnapsack(integer v, list of integers (v_1, v_2, ..., v_n),
 2:                                  list of integers (b_1, b_2, ..., b_n))
 3:    # Build the table
 4:    set T_{0,i} = 0, for i = 1, 2, ..., n
 5:    B(0) = 0
 6:    for w = 1 to v              # step through the rows
 7:       for i = 1 to n           # for each item
 8:          if v_i ≤ w            # will at least one item i fit?
 9:             T_{w,i} = b_i + B(w - v_i)
10:          else
11:             T_{w,i} = 0
12:       B(w) = max {T_{w,1}, T_{w,2}, ..., T_{w,n}}   # B(w) is the row maximum
13:    # Now step through the table (backwards) to find the set of items
14:    K = ∅
15:    w = v
16:    while B(w) > 0
17:       choose i so that T_{w,i} == B(w)
18:       K = K ∪ {i}
19:       w = w - v_i
20:    # Done
21:    return B(v), K
22: end unboundedKnapsack
```

Bounded Knapsacks and 0-1 Knapsacks

Instead of creating separate algorithms for unbounded, bounded, and 0-1 knapsack problems, an algorithm that works for all three will be developed.[32]

The feature that sets the completely unbounded knapsack problem apart is the assumption that $q_i = \infty$ for all items. Suppose that one or more of these quantities is finite. Then line 9 of **unboundedKnapsack** is inadequate. Just because the current knapsack is large enough to hold item i does not mean that there are any item i's left to recursively add to the knapsack. This can be rectified by introducing a second table to the algorithm. The new table will be denoted K, with $K_{w,i}$ recording the number of item i to optimally pack in a knapsack of capacity w.

There is still one difficulty. What value should be entered as the row w benefit for item i if there are no more copies of item i available? That is, suppose $K_{w-v_i,i} = q_i$. What should be used in place of the recursion $b_i + B(w - v_i)$? Notice that a larger capacity knapsack can always hold any set of items that will fit in a smaller knapsack. Therefore, $B(w)$ is a nondecreasing function. It may be useful to look at the knapsack with capacity $w - 1$ rather than the knapsack with capacity $w - v_i$ when no more item i are available. Thus, when $K_{w-v_i,i} = q_i$, a reasonable assignment is $T_{w,i} = B(w - 1)$ (the optimum for a slightly smaller knapsack). This makes a fairly effective heuristic, but it does not guarantee an optimal solution.

A correct algorithm can be created by using recursion (line 23 of the algorithm on the next page): set $T_{w,i}$ to b_i plus the optimal benefit for a knapsack problem with capacity $w - v_i$ and with q_i decremented by 1 (to account for the item i just used). The correct algorithm will use more memory and is potentially much slower.

[32]Recall that the 0-1 knapsack problem is a special case of the bounded knapsack problem. The 0-1 knapsack problem is often treated separately so that the special boundary conditions will lead to a more efficient algorithm. No attempt will be made here to develop a highly efficient algorithm (but see Exercise 19).

```
1: integer, list Knapsack(integer v, list of integers (v_1, v_2, ..., v_n),
2:                                   list of integers (b_1, b_2, ..., b_n),
3:                                   list of integers (q_1, q_2, ..., q_n))
4:
5: # T_{w,i} = optimal benefit if at least one i is in a knapsack of capacity w
6: # K_{w,i} = the number of item i to optimally pack in a knapsack of capacity w
7: # B(w) = the optimal benefit achievable in a knapsack with capacity w
8:
9:    # Initialize the tables
10:
11:   T_{0,i} = 0, i = 1, 2, ..., n
12:   K_{0,i} = 0, i = 1, 2, ..., n
13:   B(0) = 0
14:
15:   for w = 1 to v              # step through the rows
16:      for i = 1 to n           # for each item
17:         set R_{i,j} = 0 for j = 1,2,...,n    # clear values from any previous recursion
18:         if v_i ≤ w and q_i > 0       # the second test is for recursion
19:            if K_{w-v_i,i} < q_i      # more copies of item i available?
20:               T_{w,i} = b_i + B(w - v_i)
21:            else # add an i and use recursion to find optimum for q_i - 1
22:                 set p_k = q_k for k = 1,...,n but with p_i = q_i - 1
23:                 {rb, {R_{i,1}, R_{i,2}, ..., R_{i,n}}} =
                        Knapsack(w - v_i, (v_1, v_2, ..., v_n), (b_1, b_2, ..., b_n), (p_1, p_2, ..., p_n))
24:                 T_{w,i} = b_i + rb     # i's benefit + recursive max benefit
25:                 R_{i,i} = R_{i,i} + 1  # add back the item i removed for the recursion
26:         else
27:            T_{w,i} = 0              # no item i possible
28:
29:      # now update row w of K
30:
31:      find m such that T_{w,m} == max {T_{w,1}, T_{w,2}, ..., T_{w,n}}
32:      B(w) = T_{w,m}     # B(w) is the row maximum
33:      if v_m ≤ w and K_{w-v_m,m} < q_m # didn't run out of item m
34:         K_{w,i} = K_{w-v_m,i}, for i = 1,2,...,n
35:         K_{w,m} = K_{w,m} + 1              # added an additional item m
36:      else     # use the recursion packing
37:         K_{w,i} = R_{m,i}, for i = 1,2,...,n
38:
39:    return B(v), {K_{v,1}, K_{v,2}, ..., K_{v,n}}
40: end Knapsack
```

The availability of this new algorithm suggests a second look at Example 8.32.

EXAMPLE 8.34 **A Very Small 0-1 Knapsack Problem**

Consider a knapsack with capacity $v = 13$. There are three items that can be added to the knapsack, but only one of each is available. The items and their benefits and volumes are shown in Table 8.35.

TABLE 8.35 Potential Unique Items for a Knapsack with $v = 13$

Item	X	Y	Z
Benefit	4	3	5
Volume	6	7	8

Algorithm **Knapsack** produces Table 8.36. Pay special attention to the final two rows. These are where the restrictions $q_i = 1$ come into play.

One additional notational device is introduced. It is always the case that the first row of the K table will contain only 0s. It is helpful to overwrite these 0s with the item quantities (making it easy to see when no more copies of an item are available).[33]

TABLE 8.36 The Calculations for Example 8.34

	T				**K**		
w	X	Y	Z	$B(w)$	X	Y	Z
0	0	0	0	0	**1**	**1**	**1**
1	0	0	0	0	0	0	0
2	0	0	0	0	0	0	0
3	0	0	0	0	0	0	0
4	0	0	0	0	0	0	0
5	0	0	0	0	0	0	0
6	$4 + B(0) = 4$	0	0	4	1	0	0
7	$4 + B(1) = 4$	$3 + B(0) = 3$	0	4	1	0	0
8	$4 + B(2) = 4$	$3 + B(1) = 3$	$5 + B(0) = 5$	5	0	0	1
9	$4 + B(3) = 4$	$3 + B(2) = 3$	$5 + B(1) = 5$	5	0	0	1
10	$4 + B(4) = 4$	$3 + B(3) = 3$	$5 + B(2) = 5$	5	0	0	1
11	$4 + B(5) = 4$	$3 + B(4) = 3$	$5 + B(3) = 5$	5	0	0	1
12	$4 + B_X(6) = 4$	$3 + B(5) = 3$	$5 + B(4) = 5$	5	0	0	1
13	$4 + B_X(7) = 7$	$3 + B(6) = 7$	$5 + B(5) = 5$	7	1	1	0

The final row of the table indicates that the optimal benefit can be obtained by packing an item X and an item Y, for a total benefit of 7. The notation $B_X(6)$ and $B_X(7)$ in rows 12 and 13 serve as a visual reminder that the values come from recursions with q_x reduced by 1. This effectively removes the X columns for this example. Table 8.37 shows the details of the recursions (with rows 1–6 merged).

TABLE 8.37 The Recursions for Example 8.34

	T			**K**	
w	Y	Z	$B_X(w)$	Y	Z
0	0	0	0	**1**	**1**
1–6	0	0	0	0	0
7	$3 + B_X(0) = 3$	0	3	1	0

If the quantities were unbounded, the final three rows of the first table would be

	T				**K**		
w	X	Y	Z	$B(w)$	X	Y	Z
11	$4 + B(5) = 4$	$3 + B(4) = 3$	$5 + B(3) = 5$	5	0	0	1
12	$4 + B(6) = 8$	$3 + B(5) = 3$	$5 + B(4) = 5$	8	2	0	0
13	$4 + B(7) = 8$	$3 + B(6) = 7$	$5 + B(5) = 5$	8	2	0	0

■

[33]This expanded table (with K included) may also be used with algorithm **Knapsack** for unbounded knapsack problems. The optimal set of items to pack is readily displayed.

EXAMPLE 8.35 **More Recursions Required**

Consider a knapsack with volume 10. The benefits, volumes, and quantities of the items are listed next.

Item	X	Y	Z
Benefit	4	5	1
Volume	3	4	2
Quantity	1	1	1

The **Knapsack** algorithm produces the following table.

	T				**K**		
w	X	Y	Z	$B(w)$	X	Y	Z
0	0	0	0	0	**1**	**1**	**1**
1	0	0	0	0	0	0	0
2	0	0	$1+B(0)=1$	1	0	0	1
3	$4+B(0)=4$	0	$1+B(1)=1$	4	1	0	0
4	$4+B(1)=4$	$5+B(0)=5$	$1+B_Z(2)=1$	5	0	1	0
5	$4+B(2)=5$	$5+B(1)=5$	$1+B(3)=5$	5	1	0	1
6	$4+B_X(3)=5$	$5+B(2)=6$	$1+B(4)=6$	6	0	1	1
7	$4+B(4)=9$	$5+B(3)=9$	$1+B_Z(5)=6$	9	1	1	0
8	$4+B_X(5)=9$	$5+B_Y(4)=9$	$1+B_Z(6)=6$	9	1	1	0
9	$4+B(6)=10$	$5+B(5)=10$	$1+B(7)=10$	10	1	1	1
10	$4+B_X(7)=10$	$5+B_Y(6)=10$	$1+B(8)=10$	10	1	1	1

The optimal benefit of 10 is achieved by packing all three available items. Notice that the algorithm was forced to use recursion several times [for example $B_X(3)$, $B_Y(4)$, and $B_Z(6)$]. While completing the previous table, it was necessary to examine selected rows of the tables for B_X, B_Y, and B_Z.

Here are the required rows for B_X.

	T			**K**	
w	Y	Z	$B_X(w)$	Y	Z
0	0	0	0	**1**	**1**
1	0	0	0	0	0
2	0	$1+B_X(0)=1$	1	0	1
3	0	$1+B_X(1)=1$	1	0	1
4	$5+B_X(0)=5$	$1+B_{XZ}(2)=1$	5	1	0
5	$5+B_X(1)=5$	$1+B_{XZ}(3)=1$	5	1	0
6	$5+B_X(2)=6$	$1+B_X(4)=6$	6	1	1
7	$5+B_X(3)=6$	$1+B_X(5)=6$	6	1	1

This table requires two rows from B_{XZ} – the table for the original problem with both X and Z unavailable. Since Y has volume 4, it will not fit a knapsack with volume 2 or 3. Thus $B_{XZ}(2) = B_{XZ}(3) = 0$.

The table for B_Y is next.

w	T: X	Z	$B_Y(w)$	K: X	Z
0	0	0	0	**1**	**1**
1	0	0	0	0	0
2	0	$1 + B_Y(0) = 1$	1	0	1
3	$4 + B_Y(0) = 4$	$1 + B_Y(1) = 1$	4	1	0
4	$4 + B_Y(1) = 4$	$1 + B_{YZ}(2) = 1$	4	1	0
5	$4 + B_Y(2) = 5$	$1 + B_Y(3) = 5$	5	1	1
6	$4 + B_{YX}(3) = 5$	$1 + B_Y(4) = 5$	5	1	1

$B_{YX}(3) = 1$ because a knapsack with volume 3 and a single Z available will have benefit 1. $B_{YZ}(2) = 0$ because a knapsack with volume 2 cannot hold item X.

Finally, the table for B_Z is presented.

w	T: X	Y	$B_Z(w)$	K: X	Y
0	0	0	0	**1**	**1**
1	0	0	0	0	0
2	0	0	0	0	0
3	$4 + B_Z(0) = 4$	0	4	1	0
4	$4 + B_Z(1) = 4$	$5 + B_Z(0) = 5$	5	0	1
5	$4 + B_Z(2) = 4$	$5 + B_Z(1) = 5$	5	0	1
6	$4 + B_{ZX}(3) = 4$	$5 + B_Z(2) = 5$	5	0	1

$B_{ZX}(3) = 0$ because Y has volume 4, so it won't fit a knapsack with capacity 3. ■

✔ Quick Check 8.14

1. A knapsack can hold 10 cubic units. It is to be filled with four types of items. The item volumes, benefits, and available quantities are shown. Create tables like the ones in Example 8.34 to find an optimal collection of items.

Item	W	X	Y	Z
Benefit	3	4	5	1
Volume	4	3	7	2
Quantity	1	2	1	1

☑

An optimal solution for Example 8.29 on page 473 was found by using a computer implementation of the **Knapsack** algorithm. The solution requires 1001 rows for the tables (rows 0–1000). The discussion that follows presents a heuristic that often permits the size of the problem to be reduced before the algorithm is used.

A Heuristic Reduction Technique

The heuristic reduction technique[34] that will be developed in this section draws its inspiration from the sophisticated greedy algorithm. The essential idea is to compare the two items with the highest benefit-to-volume ratios. If the highest ranking item is sufficiently better than the next highest-ranking item, then it is possible to find an optimal solution that includes at least one copy of the highest-ranking item. If the difference is

[34] An *heuristic* technique is one that suggests a course of action that is probably correct but is not guaranteed. Often the course of action is determined after some preliminary calculations are made.

not dramatic enough, the heuristic will fail to provide a decisive answer about whether some of the highest ranking item should be packed.

The details are the subject of the following theorem.[35]

THEOREM 8.43 ***A Knapsack Reduction Heuristic***

Suppose a knapsack has volume v and there are n types of items that can be packed. Denote the item benefits by $\{b_1, b_2, \ldots, b_n\}$, the item volumes by $\{v_1, v_2, \ldots, v_n\}$, and the item quantities by $\{q_1, q_2, \ldots, q_n\}$. Assume that the items have already been ordered so that

$$\frac{b_1}{v_1} \geq \frac{b_2}{v_2} \geq \cdots \geq \frac{b_n}{v_n} \quad \text{and set} \quad \hat{q}_1 = \min\left\{q_1, \left\lfloor \frac{v}{v_1} \right\rfloor\right\}$$

If

$$b_1\hat{q}_1 \geq b_2\left(\frac{v}{v_2}\right)$$

then it is possible to find an optimal packing that includes at least one copy of item 1. If the inequality is strict or if $\frac{v}{v_2} \notin \mathbb{N}$, then every optimal packing must contain at least one item 1.

Proof: The ratio $\frac{b_i}{v_i}$ represents the benefit per unit of volume derived from packing item i. Since $\frac{b_2}{v_2} \geq \frac{b_3}{v_3} \geq \cdots \geq \frac{b_n}{v_n}$, a knapsack that does not contain any item 1s can have no greater total benefit than $b_2(\frac{v}{v_2})$. This is because item 2 offers at least as great a benefit per unit of volume as do any of the remaining items. This estimate will be an overestimate if $q_2 < \frac{v}{v_2}$ or $\left\lfloor \frac{v}{v_2} \right\rfloor < \frac{v}{v_2}$ (often $\frac{v}{v_2} \notin \mathbb{N}$); however, the estimate will be at least as large as any achievable packing using an assortment of items other than item 1.

The knapsack can hold $\hat{q}_1$ copies of item 1, for an achievable benefit of $b_1\hat{q}_1$. A packing that contains no item 1s cannot be optimal if

$$b_1\hat{q}_1 > b_2\left(\frac{v}{v_2}\right) \quad \text{or if} \quad b_1\hat{q}_1 = b_2\left(\frac{v}{v_2}\right) \text{ and } \frac{v}{v_2} \notin \mathbb{N}$$

The negation of "no item 1s" is "at least one item 1". If $b_1\hat{q}_1 = b_2\left(\frac{v}{v_2}\right)$, a packing that contains at least one item 1 will be at least as good as one that does not. (See Tables 8.38 and 8.39.) □

TABLE 8.38 Two Optimal Packings: $v = 18$.

Item	X	Y
Benefit	6	4
Volume	8	6
Quantity	2	3
B/V Ratio	.75	.67

TABLE 8.39 One Optimal Packing: $v = 18$.

Item	X	Y	Z
Benefit	6	4	1
Volume	8	6	2
Quantity	2	3	1
B/V Ratio	.75	.67	.5

Note: It may not be possible to achieve the optimal benefit by packing $\hat{q}_1$ copies of item 1. For example, consider a knapsack with volume 70 and three items with respective benefits 70, 40, and 35, respective volumes 30, 21, and 19, and $q_1 = q_2 = q_3 = 2$. Then $\hat{q}_1 = 2$ and $b_1\hat{q}_1 \geq b_2\left(\frac{v}{v_2}\right)$, but the optimal packing uses one copy of each item, for a total benefit of 145.

A few small examples will illustrate both the strengths and limitations of this theorem.

EXAMPLE 8.36 **Applying the Heuristic**

Consider a knapsack with capacity 15 and two potential item types. Table 8.40 shows the relevant information. Note that $\hat{q}_1 = \min\left\{2, \left\lfloor \frac{15}{6} \right\rfloor\right\} = 2$.

The inequality from Theorem 8.43 is $18 \cdot 2 \geq 6\left(\frac{15}{3}\right)$ or $36 \geq 30$, which is true. An optimal packing will contain at least one item X. (The optimal packing is two item X and one item Y, for a total benefit of 42.) ■

TABLE 8.40 The Heuristic at Its Best

Item	X	Y
Benefit	18	6
Volume	6	3
Quantity	2	3
B/V Ratio	3	2

[35] A simple generalization of this theorem can be found at http://www.mathcs.bethel.edu/~gossett/DiscreteMathWithProof/ .

EXAMPLE 8.37

A Second Application of the Heuristic

The knapsack in this has volume 13. The other information is shown in Table 8.41.

TABLE 8.41 The Knapsack Heuristic is Silent

Item	X	Y	Z
Benefit	12	3	5
Volume	18	7	8
Quantities	∞	∞	∞
B/V Ratio	$\frac{2}{3} \simeq .667$	$\frac{3}{7} \simeq .429$	$\frac{5}{8} = .625$

The heuristic will treat item X as the first item, followed by Z and then Y. The inequality will compare X to Z with $\hat{q}_1 = \min\left\{\infty, \left\lfloor \frac{13}{18} \right\rfloor\right\} = 0$:

$$12 \cdot 0 \not\geq 5\left(\frac{13}{8}\right)$$

$$0 \not\geq \frac{65}{8} = 8.125$$

The desired inequality fails, so the heuristic provides no recommendation about whether to pack any item X. This is correct, since item X will not fit in the knapsack. The optimal solution requires one item Z to be packed. ■

EXAMPLE 8.38

The Heuristic Is Not Perfect

Consider a knapsack with volume 9 that can contain unlimited quantities of two types of items. Table 8.42 shows the essential information.

TABLE 8.42 The Knapsack Heuristic Misses

Item	X	Y
Benefit	11	5
Volume	6	4
Quantities	∞	∞
B/V Ratio	$\frac{11}{6} \simeq 1.833$	$\frac{5}{4} = 1.25$

The inequality from Theorem 8.43 has $\hat{q}_1 = \min\left\{\infty, \left\lfloor \frac{9}{6} \right\rfloor\right\} = 1$

$$11 \cdot 1 \not\geq 5\left(\frac{9}{4}\right)$$

$$11 \not\geq 11.25$$

which is not what is needed. The heuristic has no advice to offer. However, it is easy to see that packing one item X is better than packing 2 item Y's (and it is not possible to pack one of each).

The reason that the heuristic failed to suggest packing an item X is because it is comparing packing 1 item X ($\hat{q}_1$) to 2.25 item Ys ($\left(\frac{9}{4}\right)$). However, it is not possible to pack one-quarter of an item Y. ■

The previous example might suggest that the heuristic could be improved by comparing $b_1\hat{q}_1$ to $b_2\hat{q}_2$, where $\hat{q}_2 = \min\left\{q_2, \left\lfloor \frac{v}{v_2} \right\rfloor\right\}$. However, this would invalidate the theorem, since it does not account for the possibility[36] that packing $\hat{q}_2$ item 2s and a few item 3s might be better than packing $\hat{q}_1$ item 1s. Exercise 15 on page 487 illustrates this phenomenon.

[36]This is just one among a great many alternatives that have not been accounted for by the heuristic.

✔ Quick Check 8.15

1. Consider a knapsack that holds 100 units of volume. The other details are in the adjacent table. Apply the heuristic from Theorem 8.43 as often as possible.

Item	W	X	Y	Z
Benefit	30	40	50	10
Volume	40	35	70	20
Quantity	1	3	1	1

One final example of the knapsack reduction heuristic will be considered.

EXAMPLE 8.39

Reducing a 0-1 Knapsack Problem

Consider a knapsack with a capacity of 1000 units of volume. The items can be ranked by descending benefit-to-volume ratio (Table 8.43).

TABLE 8.43 Potential Items for a 0-1 Knapsack with Capacity 1000

Item #	A	B	C	D	E
Benefit	37	11	5	4	1
Volume	520	300	240	520	160
Quantity	1	1	1	1	1
B/V Ratio	.071	.037	.021	.008	.006

Table 8.44 shows how Theorem 8.43 can be used to successively reduce the knapsack. At each stage, a smaller knapsack, having fewer potential item types, can be considered. The process will terminate as soon as the heuristic inequality becomes false.

TABLE 8.44 Applying the Knapsack Heuristic to Example 8.29

Compare Items	**Heuristic Inequality**	**Conclusion**
A and B	$37 \cdot 1 \geq 11\left(\frac{1000}{300}\right)$	pack item A
$\hat{q}_1 = \min\left\{1, \left\lfloor \frac{1000}{520} \right\rfloor\right\} = 1$	$37 \geq 36.667$	volume to fill = 480
B and C	$11 \cdot 1 \geq 5\left(\frac{480}{240}\right)$	pack item B
$\hat{q}_2 = \min\left\{1, \left\lfloor \frac{480}{300} \right\rfloor\right\} = 1$	$11 \geq 10$	volume to fill = 180
C and D	$5 \cdot 0 \not\geq 4\left(\frac{180}{520}\right)$	no recommendation
$\hat{q}_3 = \min\left\{1, \left\lfloor \frac{180}{240} \right\rfloor\right\} = 0$	$0 \not\geq 1.38$	volume to fill = 180

The sequence of reductions has lead to a knapsack that contains items A, and B, having a total benefit of 48 and a remaining volume of 180. The heuristic has been unable to decide whether to add item C.

At this point, Theorem 8.43 is no longer useful. It is now time to use the **Knapsack** algorithm. However, instead of using tables with 1001 rows and 5 columns, the algorithm will only need 181 rows and 3 columns.

We can actually avoid **Knapsack** altogether in this example by noting that neither item C nor item D will fit in the remaining space. The optimal solution is therefore to pack A, B, and E, for a total benefit of 49. ■

The heuristic reduction technique introduced in Theorem 8.43 partially answers the question "is it possible to find an optimal packing that includes some item 1s," where item 1 is the item with the largest benefit-to-volume ratio. The answer is partial because the theorem compares adding *only* item 1s against a (possibly unattainable) collection of item 2s. The heuristic is not powerful enough to consider a mixed collection that includes both item 1s and other items, nor is it powerful enough to handle every attainable collection of items 2 through n.

Exercise 18 on page 488 asks you to consider an alternative heuristic that does not use the floor function.

8.4.1 Exercises

The exercises marked with ★ have detailed solutions in Appendix H.

1. Use algorithm **unboundedKnapsack** to solve the following problems. Do not use Theorem 8.43.

(a) ★ The knapsack has volume 10.

Item	X	Y	Z
Benefit	4	5	1
Volume	3	4	2
Quantity	∞	∞	∞

(b) The knapsack has volume 10.

Item	X	Y	Z
Benefit	4	5	1
Volume	2	3	2
Quantity	∞	∞	∞

(c) The knapsack has volume 10.

Item	X	Y	Z
Benefit	4	9	2
Volume	3	6	2
Quantity	∞	∞	∞

2. Use algorithm **unboundedKnapsack** to solve the following problems. Do not use Theorem 8.43.

(a) The knapsack has volume 12.

Item	W	X	Y	Z
Benefit	8	5	7	3
Volume	6	3	5	2
Quantity	∞	∞	∞	∞

(b) The knapsack has volume 12.

Item	W	X	Y	Z
Benefit	8	5	7	3
Volume	7	4	5	3
Quantity	∞	∞	∞	∞

(c) The knapsack has volume 12.

Item	W	X	Y	Z
Benefit	7	4	6	5
Volume	7	4	6	5
Quantity	∞	∞	∞	∞

3. ★ Compare the greedy algorithm, the sophisticated greedy algorithm, and algorithm **Knapsack** for the following knapsack problem. The knapsack has capacity 7.

Item	X	Y	Z
Benefit	4	3	2
Volume	5	4	3
Quantity	1	1	1

4. Compare the greedy algorithm, the sophisticated greedy algorithm, and algorithm **Knapsack** for the following knapsack problem. The knapsack has capacity 8.

Item	X	Y	Z
Benefit	2	5	4
Volume	4	6	4
Quantity	1	1	1

5. Compare the greedy algorithm, the sophisticated greedy algorithm, and algorithm **Knapsack** for the following knapsack problem. The knapsack has capacity 12.

Item	X	Y	Z
Benefit	7	13	8
Volume	7	8	5
Quantity	1	1	1

6. Use algorithm **Knapsack** to solve the following problems. Do not use Theorem 8.43.

(a) ★ The knapsack has volume 10.

Item	X	Y	Z
Benefit	4	5	1
Volume	3	4	2
Quantity	2	2	2

(b) The knapsack has volume 10.

Item	X	Y	Z
Benefit	4	5	1
Volume	3	4	2
Quantity	1	2	1

(c) The knapsack has volume 10.

Item	X	Y	Z
Benefit	4	5	3
Volume	3	5	2
Quantity	2	2	1

7. Use algorithm **Knapsack** to solve the following problems. Do not use Theorem 8.43.

(a) The knapsack has volume 12.

Item	W	X	Y	Z
Benefit	8	5	7	3
Volume	6	3	5	2
Quantity	2	2	2	4

(b) The knapsack has volume 12.

Item	W	X	Y	Z
Benefit	8	5	7	3
Volume	6	3	5	2
Quantity	2	2	2	2

(c) The knapsack has volume 12.

Item	W	X	Y	Z
Benefit	7	4	6	4
Volume	5	2	5	3
Quantity	1	2	1	2

8. Use a computer implementation of algorithm **Knapsack** to solve the following problems. Do not use Theorem 8.43.

(a) The knapsack has volume 16.

Item	V	W	X	Y	Z
Benefit	7	6	5	7	3
Volume	4	5	6	6	3
Quantity	1	1	1	2	3

(b) The knapsack has volume 16.

Item	V	W	X	Y	Z
Benefit	3	4	6	2	2
Volume	4	3	6	2	3
Quantity	1	2	2	1	3

9. A dieter has decided that dinner must contain at most 600 calories. The food choices at a small fast-food court are limited. They include taco (180 calories), burrito (380), tostado (300), fish (170), hamburger (270), french fries (250), garden salad (100), garden salad with dressing (270), chicken nuggets (190), and hush puppy (60). The dieter has rated the choices from 1 to 10 (with 10 being best). The ratings are taco (8), burrito (10), tostado (7), fish (4), hamburger (7), french fries (6), garden salad (4), garden salad with dressing (7), chicken nuggets (6), hush puppy (2). The dieter is very hungry and has decided to ignore the health values and fat content of the various foods but still limit the total calories to at most 600. Set up a complete knapsack problem that will help the dieter find an optimal selection of foods. Optimality is determined by the ratings, not the health benefit. You need not solve the problem unless you have access to computer software that implements a knapsack algorithm.

10. ★ Suppose you are going on a shopping spree with $100 in your pocket. After arriving at the mall, you find many items that you wish to purchase. They include a watch ($30), a skirt ($20), a mug ($10), a book ($14), a swim suit ($21), a ring ($55), a pair of earrings ($12), a backpack ($28), and a box of stationary ($5). You have rated these items from 1 to 5 (with 5 being best). The ratings are watch (5), skirt (4), mug (3), book (3), swim suit (3), ring (5), earrings (2), backpack (2), and stationary (1). You will not allow yourself to use any other payment method besides cash (i.e., you will only use the money in your pocket). Set up a complete knapsack problem to help find an optimal selection of items at the mall for your shopping spree. Optimality is determined strictly by the ratings. Assume that you are willing to purchase more than one of each type of item. You need not solve the problem unless you have access to computer software that implements a knapsack algorithm.

11. Suppose that a farmer wants to maximize the profit obtained from his crops. The profits per acre for his crops have been determined and are as follows: soybeans ($100), wheat ($110), cotton ($78), hay ($69), corn ($120), barley ($71), peanuts ($88), sugar beets ($58), tomatoes ($90), and radishes ($72). The numbers of hours required to tend each acre of the specified crops are also known. They are soybeans (60), wheat (70), cotton (50), hay (45), corn (79), barley (43), peanuts (60), sugar beets (50), tomatoes (70), and radishes (66). The farmer only has 2500 hours of labor available. Set up a complete knapsack problem that will help the farmer find an optimal selection of crops to plant to maximize his profit. You need not solve the problem unless you have access to computer software that implements a knapsack algorithm.

12. As a birthday gift, you are going to make your friend a quilt. The quilt will be formed using colored squares of fabric. You cut the squares of fabric in different sizes (but all squares of the same color are the same size). The areas in square centimeters for the distinct colors are blue (144), green (121), yellow (100), purple (82), orange (144), red (81), turquoise (64), gold (48), and white (64). You have asked your friend to assign a rating to each color from 1 to 10 (with 10 being best). The results are recorded next: blue (3), green (7), yellow (8), purple (9), orange (6), red (7), turquoise (5), gold (5), and white (7). You want the quilt to be at most 1600 square centimeters in area. Set up a complete knapsack problem that will help you find an optimal selection of colored squares for the quilt. Optimality is determined by the color ratings given by your friend. You need not solve the problem unless you have access to computer software that implements a knapsack algorithm.

13. Each of the following statements is either true (always) or false (at least sometimes). Determine which option applies for each statement and provide adequate explanation for your choice.

(a) ★ Every knapsack problem has a unique optimal packing.

(b) If everything else is equal, an unbounded knapsack problem might have a higher optimal benefit than a bounded version of the same knapsack problem.

(c) The greedy algorithm will always give a less-than-optimal solution.

(d) Every unbounded knapsack problem can be solved using the bounded knapsack algorithm.

14. A young high school math teacher has just come home after an evening of parent–teacher conferences. She has 2 hours before bedtime and more items on her to-do list than can possibly fit into 2 hours. The to-do list contains the following items:

Item	Estimated Time
T: watch tape of favorite TV show (skipping over ads)	50 minutes
Q: grade a set of quizzes	20 minutes
F: phone fiancé and talk	40 minutes
L: make up lesson plans for the following week	70 minutes
C: clean the bathroom	30 minutes
E: go to the gym for a workout	70 minutes
P: phone parents and talk on the phone	20 minutes

(a) Her initial instinct is to rank the items on the to-do list on a scale of 1 to 10 and quickly solve a knapsack problem to find an optimal set of items to fill the two hours. The rankings are

T 6 Q 7 F 9 L 8 C 3 E 4 P 5

Use a software implementation of the **Knapsack** algorithm to determine the set of optimal items, the total benefit, and how much free time she will have. [*Hint*: Notice that the estimated times are in 10 minute intervals.]

(b) After looking at her list of rankings, she noticed that they are all distinct. Therefore, it is possible to list the tasks in order from 1 to 7, with 7 being the most beneficial. Her first instinct is that this should certainly provide the same solution. Her instinct is wrong. The new rankings are

T 4 Q 5 F 7 L 6 C 1 E 2 P 3

Use a software implementation of **Knapsack** to determine an optimal set of items, the total benefit, and how much free time she will have with this second set of rankings. Use the variable order T Q F L C E P.

(c) Extra credit: Explain the discrepancy between the two solutions. [*Hint*: is the solution to part (b) optimal for part (a) and vice versa?]

15. ★ Consider a knapsack with volume 9 that can contain unlimited quantities of three types of items. The following table shows the essential information:

Item	X	Y	Z
Benefit	11	5	1.1
Volume	8.5	4	1
Quantities	∞	∞	∞
B/V Ratio	$\frac{11}{8.5} \simeq 1.294$	$\frac{5}{4} = 1.25$	1.1

Show that the inequality $b_1 \left\lfloor \frac{v}{v_1} \right\rfloor \geq b_2 \left\lfloor \frac{v}{v_2} \right\rfloor$ does *not* provide a valid heuristic for deciding whether to include an item X.

16. Modify algorithm **Knapsack** so that it uses the (potentially non-optimal) heuristic described on page 477 (rather than using recursion) when no more item i are available.

17. Use Theorem 8.43 (page 482) to reduce the following knapsack problems as much as possible.

(a) The knapsack has volume 10.

Item	X	Y	Z
Benefit	4	5	1
Volume	3	4	2
Quantity	2	2	2

(b) The knapsack has volume 12.

Item	W	X	Y	Z
Benefit	8	9	7	3
Volume	6	5	5	2
Quantity	2	2	2	4

(c) The knapsack has volume 100.

Item	W	X	Y	Z
Benefit	12	4	6	8
Volume	26	15	20	35
Quantity	3	4	6	5

18. An alternative to the heuristic in Theorem 8.43 can be derived that does not use the floor function [107, page 1028]. The alternative is representative of an approach to mathematical problem solving that seeks to eliminate some messy mathematical calculations by making simplifying assumptions early in the process. What is gained in this approach is a problem that may be easier to solve. However, there is a price to pay: The solution may not be as powerful as it could be.

The alternate heuristic makes an additional simplification. The simplification starts by noticing that $\left\lfloor \frac{v}{v_1} \right\rfloor$ will truncate the fraction by some number that is less than 1 (but possibly a value very close to 1). The floor function can be eliminated by using $\frac{v}{v_1} - 1$ as an estimate for the number of copies of item 1 that can fit into the knapsack. This leads to the inequality

$$b_1 \left(\frac{v}{v_1} - 1 \right) \geq b_2 \frac{v}{v_2}$$

(a) Prove the Theorem 8.44. Note that the first inequality is strict.

THEOREM 8.44 ***An Alternative Knapsack Reduction***

Suppose a knapsack has volume v and there are n types of items that can be packed. Denote the item benefits by $\{b_1, b_2, \ldots, b_n\}$, the item volumes by $\{v_1, v_2, \ldots, v_n\}$, and the item quantities by $\{q_1, q_2, \ldots, q_n\}$. Assume that the items have already been ordered so that

$$\frac{b_1}{v_1} > \frac{b_2}{v_2} \geq \frac{b_3}{v_3} \geq \cdots \geq \frac{b_n}{v_n}$$

If

$$q_1 \geq \left(\frac{v}{v_1} - 1 \right) \quad \text{and} \quad v \geq \frac{b_1}{\frac{b_1}{v_1} - \frac{b_2}{v_2}}$$

then it is possible to find an optimal packing that includes at least one item 1.

(b) Use both heuristics to do one reduction for the following knapsacks. Then compare their suggestions with the optimal packing. Assume all problems are unbounded.

i. The knapsack has volume 10.

Item	X	Y
Benefit	50	20
Volume	5	5

ii. The knapsack has volume 10.

Item	X	Y
Benefit	40	16
Volume	10	4

iii. The knapsack has volume 10.

Item	X	Y
Benefit	16	35
Volume	4	10

iv. The knapsack has volume 10.

Item	X	Y
Benefit	30	20
Volume	6	5

v. The knapsack has volume 15.

Item	X	Y
Benefit	18	6
Volume	6	3

(c) Compare the heuristics in Theorems 8.43 and 8.44. What, if anything, has been gained by making the additional assumption? What, if anything, has been lost?

19. Algorithm **Knapsack** is not efficient. Another recursive strategy can often be more efficient. Start with an empty knapsack and keep a list of items that have not yet been added. The main loop of the algorithm looks at the remaining items one at a time. If the new item, i, fits the knapsack, recursively consider a knapsack with volume reduced by the volume of i and i removed from the set of remaining items. Add i to the optimal solution returned by the recursion and see if it is a better solution than the current best solution. Write this algorithm in pseudocode or your favorite computer language.

8.5 Error-Correcting Codes

With the advent of computer technology, information has become one of our primary commodities. This information needs to be produced, stored, transferred, and consumed. The processes of storing and transmitting this information occasionally introduce errors into the information. The mathematical field of *error-correcting codes* (often called *coding theory*) seeks to provide mechanisms to combat this problem.

Computer-compatible information is stored using a *binary* representation. This means that we use only two symbols, 0 and 1. Information is stored as *strings* of *bits*.[37]

[37] A *bit* is a **bi**nary digi**t**, that is, a 0 or a 1. Strings are defined more fully in Definition 9.4 on page 543.

For example, the letter "a" is often stored as the binary string "1100001".[38] Errors occur when a 0 gets changed to a 1, or a 1 gets changed to a 0, or bits are added or removed from the string. This can be caused by noise on a phone line, electrical interference in the atmosphere, tapes or disks getting old, electronic component failure, or other reasons.

Coding theory has combined some sophisticated theoretical mathematics with some intuitive ideas to ensure reliable transmission and storage of information. The primary intuitive idea is similar to the manner in which human languages ensure reliable information exchange: add redundancy. In English, we interpret poorly communicated words by considering the context. Teh strktr of th langage und contxtul kluus hlp us recnstrut the originla messg. In the same way, by adding cleverly chosen extra bits to a binary string before it is transmitted or stored, we can reconstruct the original form of a garbled message.

Many different techniques exist to add these redundant bits. One relatively simple technique is a 7-bit Hamming code. A preliminary definition will be helpful.

DEFINITION 8.45 ***Binary String***

A *binary string of length n* is a sequence of n symbols, where each symbol is either a "0" or a "1".

EXAMPLE 8.40 **Some Binary Strings**

The following are binary strings of length 7: 0100100 1111000 1010101 0000000. The following list contains all binary strings of length 3.

000 001 010 011 100 101 110 111 ■

PROPOSITION 8.46

There are 2^n binary strings of length n.

Proof: There are n positions in the string, each with two possible values. Since the choices are independent, the Independent Tasks Principle applies (page 227). □

8.5.1 The 7-Bit Hamming Code

A 7-bit Hamming code is a set of 16 binary strings of length 7. In each string, the first 4 bits contain an encoded message; the last 3 bits are cleverly chosen redundant bits. With this code, there are 4 bits used to carry a message, so there are 16 different messages (such as 0010, 1101, and 1001). All 16 possible messages are listed in Table 8.45. I have arbitrarily assigned meanings to each of the messages. You could assign any other meanings you wished to communicate.

TABLE 8.45 Encoding 16 Messages

0000	hello	0100	I'm hungry	1000	I'm anxious	1100	will you marry me?
0001	goodbye	0101	I'm in love	1001	yes	1101	I like snow
0010	send money	0110	I'm happy	1010	no	1110	I like rain
0011	send pizza	0111	I'm sad	1011	maybe	1111	I like sunny days

Before a message is transmitted, it will be *encoded* by adding three additional bits, called *check bits*. The encoding process starts with messages (which are 4 bits long) and adds 3 check bits (the 3 values depend on the current message), creating a 7-bit string called a *code word*. The 7-bit code word is then transmitted. The person receiving the transmission *decodes* the 7-bit string she receives, ending up with the original 4-bit

[38]Because only the symbols "0" and "1" are allowed, it is necessary to use several 0s and 1s strung together to distinguish the letter "a" from the other letters and punctuation symbols.

message (unless two or more errors have occurred). This process works because the 16 code words all differ in at least 3 places.[39] If any one of the 7 bits is changed, the received string will be closer to the original code word than to any of the other 15 code words.[40] The received message can be decoded by choosing the code word that requires the fewest changes to the received string to obtain a match.[41] The message will be the first 4 bits of that code word.

The encoding process is diagrammed as follows:

"send money" 0010 message	$\longrightarrow$ **encode**	0010 message	+	110 check bits	=	0010110 code word

The code word is then transmitted. Suppose that the first bit is accidentally changed during transmission and a different string is received. The decoding process is as follows (and is later explained in detail).

1010110 received string (not a code word)	$\longrightarrow$ **decode**	0010110 code word	$\longrightarrow$ **extract message**	"send money" 0010

If no errors occur, the decoding process will verify that the received string is actually a code word, which will then be assumed to be the code word sent.

Imagine a table that lists messages and their corresponding code words. A partial version is shown in Table 8.46. When a string is received, decoding essentially consists of finding the code word that is closest to the received string and then assuming that the corresponding message was sent.

TABLE 8.46 A Partial Listing of Messages and Corresponding Code Words

mesg	code word	mesg	code word	mesg	code word	mesg	code word
0000		0100		1000		1100	
0001		0101		1001	1001100	1101	1101001
0010	0010110	0110		1010	1010101	1110	
0011		0111		1011	1011010	1111	

As an example, suppose the message 1101 (I like snow) has been sent. The message must be changed into a code word before it is transmitted. In this case, the code word is 1101001 (the details will be presented soon). The transmitted code word will be 1101001.

Now suppose that the transmission arrives incorrectly. For example, assume that the second bit of the code word is changed from a 1 to a 0. The person receiving the transmission sees 1001001, which is not one of the 16 possible code words. (The code word for the message 1001 is 1001100.) There is only one code word of the 16 that does not differ from the received string in more than 1 bit. That code word is 1101001. The person receiving the transmission should assume that 1101001 was sent, decoding the first 4 bits as the message 1101.

This process of manually comparing received strings with a list of the possible code words is not too tedious for this code, but useful codes have many more code words (each of which is a long string of bits), so a more efficient way to encode and decode is needed. The technique presented next can be used with larger Hamming codes. In fact, a computer can be used to do the encoding and decoding.

---

[39]There are sixteen 7-bit code words, but 128 possible strings of seven 0s and 1s. The verification that the code words differ in at least three places will be presented later in this section.

[40]Closeness is measured by counting the number of places two strings differ. Strings with small total differences are considered to be close. This notion will be formalized in Definition 8.47 on page 492.

[41]For example, the received string 1010110 requires only one change to match the code word 0010110 but two changes to match the code word 1010101.

Encoding and Decoding the 7-Bit Hamming Code

TABLE 8.47 Addition mod 2

+	0	1
0	0	1
1	1	0

The encoding and decoding process can be accomplished using mod 2 arithmetic. The addition table for addition mod 2 is presented in Table 8.47.[42]

The critical observation is that $1+1=0 \pmod 2$. This can be extended to multiple additions. Thus $1+0+1+1=1 \pmod 2$. In essence, the result is 1 if there is an odd number of 1s and 0 otherwise.

Encoding: Denote the bits in the original message by x_1, x_2, x_3, and x_4. The check bits are denoted x_5, x_6, and x_7. (The subscripts indicate the position of the bit. Thus, x_3 represents the third bit.) The check bits for the code word $x_1x_2x_3x_4x_5x_6x_7$ are chosen to make the following equations true:

$$x_5 = x_2 + x_3 + x_4 \pmod 2$$
$$x_6 = x_1 + x_3 + x_4 \pmod 2$$
$$x_7 = x_1 + x_2 + x_4 \pmod 2$$

Thus the message 0010 (send money) becomes the code word 0010110, since

$$x_5 = 0+1+0 = 1 \pmod 2$$
$$x_6 = 0+1+0 = 1 \pmod 2$$
$$x_7 = 0+0+0 = 0 \pmod 2$$

Notice that the formula for x_5 adds all the message bits other than x_1, the formula for x_6 adds all the message bits except x_2, and the formula for x_7 adds all the message bits except x_3.

Decoding: To see how to decode, consider what happens to the check bits if there is an error in exactly one of the 4 original message bits. For example, if bit x_1 is changed from 0 to 1 or from 1 to 0, but x_2, x_3, and x_4 are the same, then x_5 won't change, but x_6, and x_7 will. (In the formulas for computing the check bits, the formula for x_5 does not contain x_1, but the formulas for x_6 and x_7 do contain x_1. Thus, if x_1 changes, so will x_6 and x_7.) Table 8.48 summarizes the results.

TABLE 8.48 Errors in a 7-Bit Hamming Code

an error in bit	causes changes in
x_1	x_6 and x_7
x_2	x_5 and x_7
x_3	x_5 and x_6
x_4	x_5, x_6, and x_7

When a transmitted string is received, the check bits can be recalculated. If the recalculated check bits are the same as the transmitted check bits, it is reasonable to assume the message arrived safely. If the new check bits are different, then an error has occurred. If only one check bit is different, the error occurred in sending that check bit, so the error can be ignored. If two or three check bits differ, use Table 8.42 to see which message bit was erroneously changed.

It will be helpful to introduce some additional notation. A superscript will be used to distinguish between the check bits that are received and those that are calculated from

[42] This should not be confused with the binary (or base 2) addition table, which looks like

+	0	1
0	0	1
1	1	10

where 10 is the binary representation for the number 2.

the received message bits. A superscript r will denote the received bit, and a superscript c will denote the calculated bit.

Suppose the bit string 0110110 has been received. The 4 message bits (0110) appear to be the message "I'm happy". The transmitted check bits are $x_5^r = 1$, $x_6^r = 1$, and $x_7^r = 0$. The computed check bits (using the 4 message bits 0110) are $x_5^c = 0$, $x_6^c = 1$, and $x_7^c = 1$. Thus the transmitted and computed check bits for x_5 and x_7 differ. Table 8.42 shows that x_2 is incorrect. The message actually sent was probably 0010 "send money".

Tables 8.49 and 8.50 summarize encoding and decoding for the 7-bit Hamming code. An alternative approach is presented in Problem 15 on page 501.

TABLE 8.49 Encoding a 7-Bit Hamming Code

$x_5 = x_2 + x_3 + x_4 \pmod 2$
$x_6 = x_1 + x_3 + x_4 \pmod 2$
$x_7 = x_1 + x_2 + x_4 \pmod 2$

TABLE 8.50 Decoding a 7-Bit Hamming Code

Differing Check Bits	Error in Bit
x_6 and x_7	x_1
x_5 and x_7	x_2
x_5 and x_6	x_3
x_5, x_6, and x_7	x_4
x_5	x_5
x_6	x_6
x_7	x_7

✔ Quick Check 8.16

1. Encode the following messages into code words in the 7-bit Hamming code.
   (a) 1101 (b) 0001
2. Assume that the 7-bit Hamming code is being used. Decode the following received transmissions. (Find the message that was most likely to have been sent.)
   (a) 0011001 (b) 0011011 (c) 1011001 ☑

8.5.2 A Formal Look at Coding Theory

Section 8.5.1 introduced many of the major ideas in coding theory. It is now time to formalize some of these ideas. This process will start by introducing several definitions.

DEFINITION 8.47 ***Hamming Distance***
The *Hamming distance* between two binary strings, u and v, having common length, n, is the number of positions in which u and v differ. The Hamming distance is denoted by $\text{Hd}(u, v)$.

DEFINITION 8.48 ***Hamming Weight***
The *Hamming weight* of a binary string is the number of 1s in the string. The Hamming weight of the string, u, is denoted by $\text{Hw}(u)$.

DEFINITION 8.49 ***Adding and Subtracting Binary Strings***
Let u and v be binary strings with common length, n. The *sum* and *difference* of the two strings are denoted $u + v$ and $u - v$, respectively. Both operations are defined as the bitwise (mod 2) sum. That is, the bit in position k of $u + v$ is the same as the bit in position k of $u - v$ and has the value $u_k + v_k \pmod 2$.

A simple example of the previous definition may be helpful. Let $u = 1010$ and let $v = 1100$. Then $u + v = u - v = w$, where $w_1 = 1 + 1 = 0 \pmod 2$,

$w_2 = 0 + 1 = 1 \pmod 2$, $w_3 = 1 + 0 = 1 \pmod 2$, and $w_4 = 0 + 0 = 0 \pmod 2$. Thus, $u + v = u - v = 0110$.

DEFINITION 8.50 ***Binary Error-Correcting Code***

A *binary error-correcting code* is a nonempty subset, $\mathcal{C}$, of the set of all binary strings having length, n. Let $|\mathcal{C}| = M$ and let $d = \min_{u,v \in \mathcal{C}} \text{Hd}(u, v)$. $\mathcal{C}$ is characterized by the parameters n, M, and d and is referred to as an (n, M, d) code. Parameter d is called *the minimum distance* of the code.

The binary strings that are the elements of the code are called *code words*.

An error-correcting code is *linear* if the sum (or difference) of any two code words is also a code word.

If $\mathcal{C}$ is a linear code, then the string consisting of n 0's is a member of $\mathcal{C}$. (Let $u \in \mathcal{C}$ be any code word. Then $u - u$, the string of n 0s, is also in $\mathcal{C}$.)

Note that the phrase *code word* applies to the encoded string with the extra redundancy bits. It does not refer to the embedded message.

EXAMPLE 8.41 **The 7-Bit Hamming Code Revisited**

The 7-bit Hamming code is a (7,16,3) linear code. Table 8.46 on page 490 asserts that the code contains (in part) the code words 1001100, 1010101, 0010110, and 1011010. It is easy to see that

$$\begin{aligned}
\text{Hd}(1001100, 1010101) &= 3 \\
\text{Hd}(0010110, 1001100) &= 4 \\
0010110 - 1001100 &= 1011010 \\
\text{Hw}(1011010) &= 4
\end{aligned}$$

It appears to be fairly tedious to verify that the minimum distance between any two code words is 3: There are $\binom{16}{2} = 120$ pairs of distinct code vectors, u and v, for which $\text{Hd}(u, v)$ needs to be calculated. Corollary 8.52 on page 494 provides a simpler, alternative calculation that is valid for linear codes.

To see that the 7-bit Hamming code is linear, suppose that $x = x_1x_2x_3x_4x_5x_6x_7$ and $y = y_1y_2y_2y_4y_5y_6y_7$ are code words. Then $w = x + y$ is another code word if it satisfies the three encoding equations. The value of position k of w is $w_k = x_k + y_k \pmod 2$.

Both x and y are code words. Consequently $x_5 = x_2 + x_3 + x_4 \pmod 2$ and $y_5 = y_2 + y_3 + y_4 \pmod 2$. Thus (using Proposition 3.24 on page 100)

$$\begin{aligned}
w_5 &= x_5 + y_5 \pmod 2 \\
&= ((x_2 + x_3 + x_4 \pmod 2) + (y_2 + y_3 + y_4 \pmod 2)) \pmod 2 \\
&= ((x_2 + y_2) + (x_3 + y_3) + (x_4 + y_4)) \pmod 2 \\
&= w_2 + w_3 + w_4 \pmod 2
\end{aligned}$$

Similar calculations verify that w_6 and w_7 also satisfy the encoding equations.

■

THEOREM 8.51 ***Hamming Distance and Hamming Weight***

Let u, v, and w be any binary strings having common length, n. Then

1. $\mathrm{Hd}(u, v) = \mathrm{Hw}(u - v)$
2. $\mathrm{Hd}(u, v) \leq \mathrm{Hd}(u, w) + \mathrm{Hd}(w, v)$

Proof:

1. Let u and v be any two binary strings (not necessarily distinct). The binary string $u - v$ will have a 1 in every position where u and v differ, and a 0 in every position where u and v are the same. Since $\mathrm{Hw}(u-v)$ counts the 1s, it will equal the number of positions in which u and v differ.
2. Suppose u and v differ in position k. Then either $w_k = u_k$ and $w_k \neq v_k$ or else $w_k \neq u_k$ and $w_k = v_k$. Thus, $\mathrm{Hd}(u, v)$ will have a contribution of 1 from position k, but so will exactly one of $\mathrm{Hd}(u, w)$ and $\mathrm{Hd}(w, v)$.

   On the other hand, if u and v have the same value in position k, $\mathrm{Hd}(u, v)$ will have a contribution of 0 from position k. Position k will contribute either 0 or 2 to the sum $\mathrm{Hd}(u, w) + \mathrm{Hd}(w, v)$, depending on the value of w in position k.

   In either case, position k never contributes more to $\mathrm{Hd}(u, v)$ than it does to $\mathrm{Hd}(u, w) + \mathrm{Hd}(w, v)$. Since this is true for every position, the proof is complete. □

COROLLARY 8.52 ***The Minimum Distance of a Linear Code***

The minimum distance, d, in a binary linear error-correcting code is the smallest nonzero Hamming weight among the code words.

Proof: Let u and v be distinct nonzero code words in a linear code. Then $\mathrm{Hd}(u, v) = \mathrm{Hw}(u - v)$. But $u - v$ is some other nonzero code word (since this is a linear code). Thus, finding the minimum distance is equivalent to finding the minimum weight among all nonzero code words. □

Since the 7-bit Hamming code is linear, it is easy to verify that the minimum weight of a nonzero code word is 3. (See Exercise 1 on page 500.)

The 7-bit Hamming code is not the only error-correcting code. It is time to look at an additional example.

EXAMPLE 8.42 **Repetition Codes**

There is a very simple mechanism available for adding redundancy to a message: Just repeat the message (or each bit) n times. For example, if each bit of the message is repeated three times, it is very simple to detect that a single error has occurred: One of the three received bits will differ from the other two. Decoding is also simple: Always decode to the value represented by the majority of the three bits in each bundle. Such codes are called *n-fold repetition codes*.

Suppose a 3-fold repetition code is used to encode the 16 possible 4-bit messages. If the repetition is done 1 bit at a time, then the message 1011 would be encoded as 111000111111. If 111010111111 is received, and if a single error is more likely than two or more errors, it is easy to see that bit 5 must be where the error occurred. The message must therefore be 1011. ■

The repetition code is preferable to the 7-bit Hamming code if ease of use is the main criterion for comparison. However, the 3-fold repetition code requires the transmission of 12 bits per 4-bit message, whereas the 7-bit Hamming code only requires the transmission of 7 bits.

DEFINITION 8.53 ***Error-Correcting Capability; Efficiency***

Let C be a binary error-correcting code. If it is possible to correctly decode any received string whenever t or fewer bits have been changed during transmission but not possible always to correctly decode if $t+1$ or more bits are changed, then C is said to be *t-error correcting*.

If the code words in C have length n and the messages that the code words represent have length k, then C is said to have *efficiency* $\frac{k}{n}$.

EXAMPLE 8.43 **Error-Correcting Capability and Efficiency**

Both the 7-bit Hamming code and the 3-fold repetition code in Example 8.42 are 1-error correcting. Algorithms have already been given to show that they can correct 1 error. To see that neither can correct any pattern of 2 errors, consider the message 0000. The 7-bit Hamming code encodes this as 0000000; the 3-fold repetition code encodes it as 000000000000. Now suppose the first two positions are changed during transmission.

The repetition code will be decoded under the assumption that bit 3 was changed (since the majority wins among each 3 bits). The message will therefore be incorrectly decoded as 1000.

The 7-bit Hamming code will recalculate the check bits for the received string 1100000. The calculated check bits are $x_5^c = 1$, $x_6^c = 1$, and $x_7^c = 0$. The decoding table indicates that bit 3 is in error. The incorrectly decoded message will therefore be 1110.

The efficiency of the 7-bit Hamming code is $\frac{4}{7}$, which is larger than the 3-fold repetition code's efficiency of $\frac{4}{12}$.

The 7-bit Hamming code has the highest efficiency possible among all binary error-correcting codes that transmit 4 message bits. Here is an intuitive justification for this claim. There must be some additional bits, or else there would be no redundancy and no possibility of detecting and correcting errors. Once the message has been encoded, an error could occur in any of the bits, so there must be sufficient check bits to distinguish among all the possible error positions. The 7-bit Hamming code uses 7 bits. This requires the ability to distinguish among 8 possibilities (no error, or a single error in each of the seven positions). That means at least 3 check bits are needed (since $2^3 = 8$, but $2^2 < 8$). The 7-bit Hamming code uses exactly 3 check bits, so it meets the lower bound on required extra bits. Any other code would need to distinguish between errors in the 4 message bits and one or more check bits, so more than 2 check bits would be needed. ■

The next theorem provides the key insight regarding error-correcting capability.

THEOREM 8.54 ***Error-Correcting Capability***

Let C be an (n, M, d) binary error-correcting code. If $2t+1 \le d \le 2t+2$, then C is t-error correcting.

Proof: This theorem has an implicit assumption that it is more likely to have fewer errors rather than more errors. With this assumption, the code word that is most likely to have been transmitted will always be the code word with the smallest Hamming distance to the received string (since this will be the code word for which the fewest number of bits differ from the received string). To have a unique "closest code word", it will be necessary to make sure that code words are not too close to each other.

Suppose C is a code with $d \ge 2t+1$, and that code word u was transmitted but string y was received. If at most t errors occur during transmission, then $\text{Hd}(u, y) \le t$. Now suppose that $v \ne u$ is any other code word. Then $\text{Hd}(u, v) \ge d \ge 2t+1$.

Theorem 8.51 implies $\mathrm{Hd}(u, v) \le \mathrm{Hd}(u, y) + \mathrm{Hd}(y, v)$. That is, $2t + 1 \le \mathrm{Hd}(u, v) \le \mathrm{Hd}(u, y) + \mathrm{Hd}(y, v) \le t + \mathrm{Hd}(y, v)$. Thus, $\mathrm{Hd}(y, v) \ge t + 1$. This means that u is the closest code word to y.

On the other hand, since $d \le 2t+2$ there must be some code word, w, with $2t+1 \le \mathrm{Hd}(u, w) \le 2t+2$. Suppose that $t+1$ errors occur during transmission of u, resulting in y. Then it is possible that those errors all occur in positions where u and w differ. This means that y has at least $t + 1$ bits that are identical to the corresponding bits in w. The string u has become y by modifying $t+1$ bits. Since $2t+1 \le \mathrm{Hd}(u, w) \le 2t+2$, there will be either t or $t + 1$ additional positions in which y and w differ. If $\mathrm{Hd}(y, w) = t$, then the decoding process will incorrectly assume that code word w was transmitted. On the other hand, if $\mathrm{Hd}(y, w) = t + 1$, then there is a tie for the nearest code word. Therefore, it is not possible always to decode more than t errors. □

EXAMPLE 8.44

Visualizing Theorem 8.54

Theorem 8.54 can be visualized by considering the 3-fold repetition code with 1 message bit. There are two choices for the message (0 or 1), and the message is repeated three times for transmission. The majority of the three received bits determines the decoding.

The two code words and the other 6 possible strings of length 3 can be visualized as the corners of a cube (Figure 8.12).

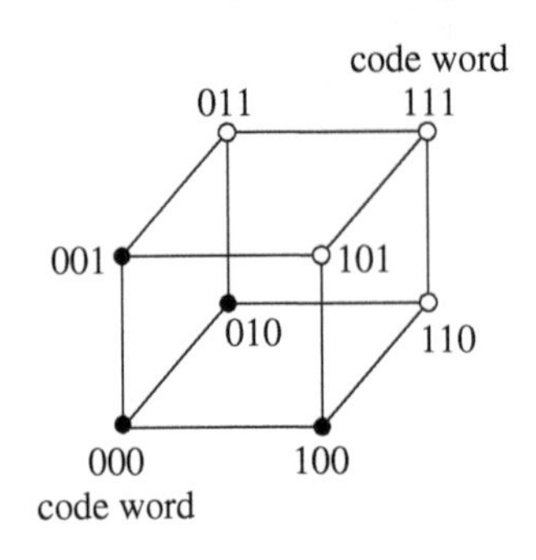

Figure 8.12. Visualizing the 3-fold repetition code.

The code word 000 and the three strings that are distance 1 from 000 are denoted by solid dots. The code word 111 and the three strings that are distance 1 from it are denoted by open circles.

Theorem 8.54 says that if it is possible to place a "sphere of radius t" around each code word ($t = 1$ in the cube diagram), and if those spheres are disjoint, then it is possible to correct t errors. In order for the spheres to be disjoint, it must be possible to change t bits of a code word and still stay inside its own sphere. One or more changes are required to move into another sphere, and then t additional changes to produce the other code word. ■

✔ Quick Check 8.17

Construct a code with 3 message bits and 3 check bits by requiring that the check bits satisfy the following equations.

$$x_4 = x_1 + x_2 \pmod 2$$
$$x_5 = x_1 + x_3 \pmod 2$$
$$x_6 = x_2 + x_3 \pmod 2$$

1. List all eight messages and the corresponding code words.
2. Determine the efficiency of this code.
3. Determine the error-correcting capability of this code. You may assume the code is linear (see Exercise 13 on page 501). ☑

The code in Quick Check 8.17 provides an additional illustration of Theorem 8.54. There are 8 code words, each having 6 bits. This means that for each code word, u, there are 6 strings with Hamming distance 1 from u. The code word u, together with the strings at distance 1 form a sphere containing 7 strings. Since there are $2^6 = 64$ distinct binary strings of length 6, the code words and their nearest neighbors do not account for all strings (there are $8 \cdot 7 = 56$ strings in the spheres). This means that there are 8 strings that are at least distance 2 from a code word. If one of those strings is received, the best we can do is declare that 2 or more errors have occurred and ask for a retransmission. (There will be several code words with Hamming distance 2 from such a string.)

8.5.3 Combinatorial Aspects of Coding Theory

The field of error-correcting codes makes extensive use of ideas found in a typical course in linear algebra. It also assumes numerous ideas from a course in algebraic structures (in particular, groups, finite fields, and polynomials over finite fields). Since these topics are not assumed for this textbook, this final section will instead look at a few combinatorial ideas that arise in the study of error-correcting codes.

Perfect Codes

Example 8.44 introduced the notion of a "sphere" around a code word. That idea is formalized in the next definition.

DEFINITION 8.55 ***The Binary Sphere Centered at a Code Word***

Let $\mathcal{C}$ be a binary error-correcting code with code words of length n, and let $\mathcal{B}_n$ be the set of all binary strings of length n. The *binary sphere of radius t centered at a code word, u*, is denoted $S_t(u)$ and is defined as $S_t(u) = \{x \in \mathcal{B}_n \mid \mathrm{Hd}(u, x) \leq t\}$.

One simple property of binary spheres of radius t is presented in the next proposition.

PROPOSITION 8.56 $|S_t(u)|$

Let $\mathcal{C}$ be a binary error-correcting code with code words of length n. Then for any code word, u,

$$|S_t(u)| = \sum_{i=0}^{t} \binom{n}{i}$$

Proof: A string whose Hamming distance from u is i must differ from u in exactly i positions. There are $\binom{n}{i}$ distinct subsets of i positions. Since a string cannot have more than one Hamming distance from u, the Mutually Exclusive Tasks Principle (page 228) completes the proof. □

Example 8.44 represents one extreme in the distribution of spheres in $\mathcal{B}_n$. In that example, every string of length 3 is in exactly one of the spheres of radius 1 centered at a code word. In contrast, the code in Quick Check 8.17 leaves 8 strings sitting outside the spheres of radius 1 centered at the code words. An attempt to include those stray strings would require the use of spheres of radius 2. This would have the unfortunate effect of placing noncode strings inside multiple spheres, negating any error-correcting capability.

The situation in Example 8.44 motivates the next definition.

DEFINITION 8.57 ***Perfect Code***

Let $\mathcal{C}$ be a binary error-correcting code whose code words have length, n. Let $\mathcal{B}_n$ be the set of all binary strings of length n and assume that $\mathcal{C}$ is a proper subset of $\mathcal{B}_n$. Then $\mathcal{C}$ is called a *perfect t-error correcting code* if, for some $t > 0$, $\mathcal{B}_n$ is a disjoint union of the spheres $S_t(u)$, for $u \in \mathcal{C}$. Such a code may also be referred to as a *perfect code*, without specifying t.

Thus, a code is perfect if every string of length n is in exactly one of the spheres of radius t about a code word.

Perfect codes are relatively rare. The next proposition introduces a necessary (but not sufficient) condition for the existence of a perfect code.

PROPOSITION 8.58 ***Binary Sphere Packing Condition***
A perfect binary t-error correcting $(n, M, 2t+1)$ code must satisfy

$$M \cdot \sum_{i=0}^{t} \binom{n}{i} = 2^n$$

Proof: Proposition 8.56 asserts that each of the M spheres contains $\sum_{i=0}^{t} \binom{n}{i}$ strings. Since the spheres are disjoint if the code is perfect, there must be $M \cdot \sum_{i=0}^{t} \binom{n}{i}$ strings. Since the spheres in a perfect code contain every string of length n, and there are 2^n distinct binary strings of length n, the equality must hold. □

The binary sphere packing condition is not a sufficient condition. For example, the parameters $(90, 2^{78}, 5)$ satisfy the theorem because $2^{78} \cdot \left(1 + 90 + \binom{90}{2}\right) = 2^{90}$, but it is known that no code with these parameters can exist [75].

During the late 1960s and early 1970s, J. H. Van Lint and other mathematicians made significant progress in determining all possible perfect codes (in a more general setting than has been presented here). In 1973, Aimo Tietäväinen successfully completed the task. The collection of perfect binary codes is presented next. (There are some additional perfect codes if strings are not restricted to just the symbols 0 and 1. One such code will be mentioned.)

Trivial Perfect Codes The are two families of trivial perfect codes.

Codes with one message
This code has parameters $(n, 1, n)$. If only one message is possible but n bits are used to transmit the message, then any error will result in an unexpected received string. The decoding process assumes that the single valid message was sent. This code can correct n errors, so the sphere about the one code word includes all of $\mathcal{B}_n$. It is not a useful code.

$(2t+1)$-fold repetition with one message bit
The code has parameters $(2t+1, 2, t)$. (Example 8.44 fits in this category.) If 1 bit is used to hold messages and the symbol in that position is repeated an odd number, $2t+1$, of times, then the code can correct any t errors using a "majority rules" decoding process. The code words are

$$\overbrace{000\cdots000}^{2t+1 \text{ times}} \quad \text{and} \quad \overbrace{111\cdots111}^{2t+1 \text{ times}}.$$

Exercise 12 on page 501 asks you to prove that this code is perfect.

The Hamming Codes (and some others)
The 7-bit Hamming code serves as a model from which to construct similar codes having parameters

$$(n, M, 2t+1) = (2^r\text{-}1, 2^{n-r}, 3) \quad \text{for } r \geq 3$$

(The 7-bit Hamming code has parameters (7, 16, 3), corresponding to $r = 3$.) Hamming codes are often referred to as 1-error correcting (n, k) codes, where $n = 2^r - 1$ and $k = n - r$. These Hamming codes are all linear. There also exist nonlinear perfect codes with the same parameters as the Hamming codes.

The 3-Error-Correcting Golay Code
This $(23, 2^{12}, 7)$ code has 4096 code words of length 23.

Quick Check 8.18

1. The 3-error-correcting Golay code satisfies the sphere-packing condition. Prove this assertion.

EXAMPLE 8.45

Nonbinary Perfect Codes

It is possible to discuss error-correcting codes that do not limit strings to just the symbols 0 and 1. In fact, there are additional perfect nonbinary trivial codes and also nonbinary Hamming codes with parameters $(n = \frac{q^r-1}{q-1}, q^{n-r}, 3)$, where $q = p^s$ for some prime number, p. If $q = 3$, the strings will contain the symbols 0, 1, and 2. The definition of Hamming distance is still valid, but the definition for Hamming weight will need a minor change. These generalized Hamming codes are all linear and perfect.

Of particular interest is the only nontrivial perfect code that is neither a Hamming code nor the previous Golay code. It is the ternary ($q = 3$) Golay code, having parameters $(11, 3^6, 5)$. Exercise 19 on page 502 investigates this code a bit more. ■

Codes and Balanced Incomplete Block Designs

An error-correcting code can be viewed as a combinatorial configuration. A balanced incomplete block design (Section 8.3) is a set of blocks that contains some (but not all) of the possible subsets of a set of objects. Similarly, a code is a subset of the set of all possible strings of length n. In both cases, there are additional requirements (for codes, the distance between code words is important).

It is often possible to use one combinatorial design to create a different kind of combinatorial design. For example, mutually orthogonal Latin squares were used to create a finite projective plane (Section 8.2.3) and a finite projective plane was used to construct a balanced incomplete block design (Section 8.3.1). This section will conclude with a description of how an error-correcting code can be constructed from a balanced incomplete block design (the construction is from [67]).

Construction 8.11 A Nonlinear $(v, b, 2k - 2)$ Error-Correcting Code from a $(v, b, r, k, 1)$-Design

Let D be a $(v, b, r, k, 1)$-design and let I_D be the v by b incidence matrix for D. The columns of I_D are binary strings having length v (the incidence matrix columns become strings by reading bits from top to bottom and writing them left to right).

Define a binary error-correcting code by taking the columns of I_D as the code words. Then each code word will have length v and there will be $M = b$ code words.

The resulting code will either have parameters $(v, b, 2k - 2)$ or $(v, b, 2k)$, with the former possibility being more likely.

The number of code words will be $b = \frac{v(v-1)}{k(k-1)}$ (Exercise 1 on page 471).

The minimum distance, d, of this code must be at least $2k - 2$. To see this, notice that every block in the BIBD contains k objects, so each code word has Hamming weight k. Also, two distinct blocks cannot contain more than one common object. This is because every pair of objects appears together in exactly 1 block ($\lambda = 1$). If two blocks each contain object i and object j, then the pair of objects would appear in more than $\lambda = 1$ common block.

Now consider two code words, u and v. It is possible that they have no common objects (considering them as blocks). In that case, they will not have any 1s in a common position, so $\text{Hd}(u, v) = 2k$ (there are k places where they differ because of the 1s in u and another k places they differ due to the 1s in v). The other possibility is that u and v have exactly one common object. In that case, they will both have a 1 in the position corresponding to that object, but will not have any other common 1s. Therefore, $\text{Hd}(u, v) = 2(k - 1)$.

EXAMPLE 8.46 **A (13, 26, 4) Code**

Construction 8.9 on page 468 presented a (13, 26, 6, 3, 1)-design. The (13, 26, 4) error-correcting code defined in Construction 8.11 contains the 26 code words listed in Table 8.51. Each code word contains 13 bits. You may verify that the Hamming distance between any two code words is either 4 or 6. Since $2t + 1 = 4$, the code is capable of correcting any $t = \left\lfloor \frac{3}{2} \right\rfloor = 1$ error.

TABLE 8.51 The 26 Code Words in a (13, 26, 4) Code

0101000001000	0010011000000
0010100000100	0001001100000
0001010000010	0000100110000
0000101000001	0000010011000
1000010100000	0000001001100
0100001010000	0000000100110
0010000101000	0000000010011
0001000010100	1000000001001
0000100001010	1100000000100
0000010000101	0110000000010
1000001000010	0011000000001
0100000100001	1001100000000
1010000010000	0100110000000

For a code of this size, it is not difficult to write a computer program that would calculate the Hamming distances between a received string and all 26 code words. The decoding would then pick the code word with the smallest Hamming distance.

A more interesting question is: Which are the message bits and which are the check bits? Suppose there are s message bits. Since there are 26 messages, s must be determined by the equation $2^s = 26$. This indicates that there should be $\log_2(26) \simeq 4.7$ message bits. For this code, perhaps it is best just to consider the entire code word as representing the message and not bother trying to identify message and check bits. Nevertheless, it is informative to estimate the efficiency of the code, which is approximately $\frac{4.7}{13} \simeq 0.36$. Contrast this with the 1-error correcting (15, 11, 3) Hamming code's efficiency of $\frac{11}{15} \simeq 0.73$. ■

8.5.4 Exercises

The exercises marked with ★ have detailed solutions in Appendix H.

1. ★ Complete the list of 16 code words for the 7-bit Hamming code. (Recall the partial table on page 490.) That is, produce a table with the 16 messages (4 bits) and their corresponding 7-bit code words.
2. Assume that the 7-bit Hamming code is being used. Decode the following received transmissions.
   (a) 1100110 (b) ★ 1110111
   (c) 1110011 (d) 1101000
3. Calculate the Hamming weight of each string in Exercise 2.
4. Assume that the 7-bit Hamming code is being used. Decode the following received transmissions.
   (a) 0101011 (b) 0100110
   (c) 1111010 (d) 1001110
5. Calculate the Hamming weight of each string in Exercise 4.
6. Calculate the Hamming distance for each pair of binary strings.
   (a) ★ 1100 and 1011
   (b) 1001110 and 1110111
   (c) 1010010001 and 1011100001
7. ★ Calculate the bitwise sums of the pairs of binary strings in Exercise 6.
8. Calculate the Hamming distance for each pair of binary strings.
   (a) 0100101 and 1100101
   (b) 0110010100 and 0010001001
   (c) 1100101001 and 0111010101
9. Calculate the bitwise sums of the binary string pairs in Exercise 8.

10. Assume that a 5-fold repetition code is being used, with repetitions done at each bit (rather than for the entire message). Decode each of the received strings.

(a) 1110000000

(b) ★ 111110000011101

(c) 110001110011001

11. Determine whether each of the following statements are true or false. Justify your answer.

(a) ★ A binary error-correcting code with parameters (n, M, d) uses n bits per code word, of which M are message bits.

(b) If the minimum distance of a code is 11, then the code can correct 6 errors.

(c) The minimum distance, d, in any binary error-correcting code is the minimum Hamming weight of a nonzero code word.

(d) There are no nontrivial perfect 2-error correcting binary codes.

(e) A binary code with parameters $(12, 32, 3)$ satisfies the binary sphere-packing condition.

12. Show that the "$(2t + 1)$-fold repetition with one message bit" code satisfies the binary sphere-packing condition. Then prove that this code is perfect. [*Hint for the first task*: Exercise 19 on page 257.]

13. Recall the error-correcting code in Quick Check 8.17 on page 496.

(a) Prove that this code is linear.

(b) Describe a decoding procedure.

(c) Determine which message was the most likely to have been intended for each received string.

i. 010001 ii. 101010

iii. 110011 iv. 111011

14. This exercise provides a concrete proof that the binary sphere-packing condition holds for the 7-bit Hamming code.

(a) ★ Create a table with 16 columns. At the top of each column, write a different code word from the 7-bit Hamming code (see Exercise 1). Under each code word, list all binary strings of length 7 that are at Hamming distance 1 from the code word at the top of the column.

(b) How many strings (including the code word) are in each column?

(c) Verify that every binary string of length 7 appears exactly once in the table.

(d) Prove that the binary sphere packing condition holds for the 7-bit Hamming code.

15. An alternative approach for encoding and decoding the 7-bit Hamming codes exists. This approach uses mod 2 matrix multiplications.

Encoding Write each 4-bit message, $x_1x_2x_3x_4$, as a column. Multiply the column times the following matrix (using mod 2 arithmetic). The resulting column will be the check bits. That is,

$$\begin{pmatrix} x_5 \\ x_6 \\ x_7 \end{pmatrix} = \begin{pmatrix} 0 & 1 & 1 & 1 \\ 1 & 0 & 1 & 1 \\ 1 & 1 & 0 & 1 \end{pmatrix} \begin{pmatrix} x_1 \\ x_2 \\ x_3 \\ x_4 \end{pmatrix}$$

Decoding Write the received string as a column and multiply it by the following matrix (using mod 2 arithmetic). If the result is a column of 0s, no error has occurred. Otherwise, the result will match the column in the matrix that corresponds to the position of the error. [For example, if the result is $\begin{pmatrix} c_1 \\ c_2 \\ c_3 \end{pmatrix} = \begin{pmatrix} 1 \\ 1 \\ 0 \end{pmatrix}$, then an error is assumed to have occurred in the third bit.] In either case, the first 4 bits in the code word are still the message bits.

$$\begin{pmatrix} c_1 \\ c_2 \\ c_3 \end{pmatrix} = \begin{pmatrix} 0 & 1 & 1 & 1 & 1 & 0 & 0 \\ 1 & 0 & 1 & 1 & 0 & 1 & 0 \\ 1 & 1 & 0 & 1 & 0 & 0 & 1 \end{pmatrix} \begin{pmatrix} x_1^r \\ x_2^r \\ x_3^r \\ x_4^r \\ x_5^r \\ x_6^r \\ x_7^r \end{pmatrix}$$

(a) Use the alternative approach to encode the following messages. (Show the corresponding code words.)

i. 1001 ii. 0111

iii. 0001 iv. 1101

(b) Use the alternative approach to decode the following received strings. (Find the most likely transmitted code word; then determine the corresponding message.)

i. 1011101 ii. 0110011

iii. 1101101 iv. 1100000

(c) Explain why this alternative approach correctly implements the 7-bit Hamming code. Don't forget to discuss the way the decoding technique determines which bit is in error.

16. Use Construction 8.11 (page 499) to create an error-correcting code from the $(5, 10, 4, 2, 1)$-design found in Example 8.25 on page 465.

(a) List the code words.

(b) What are the parameters of the code?

(c) How many errors can the code correct?

(d) Estimate the efficiency of the code.

17. Prove the following corollary to Proposition 8.58 (page 498).

COROLLARY 8.59 ***Binary Sphere Packing Bound***

A binary t-error correcting $(n, M, 2t + 1)$ code must satisfy

$$M \leq \left\lfloor \frac{2^n}{\sum_{i=0}^{t} \binom{n}{i}} \right\rfloor$$

18. Use Corollary 8.59 from Exercise 17 to determine the maximum number of code words possible in each of the following codes.

(a) ★ The code words have length $n = 14$ and the code corrects 3 errors.

(b) The code is a $(12, M, 3)$ code.

(c) The code is an $(8, M, 5)$ code.

19. The Golay ternary $(11, 3^6, 5)$ code can correct 2 errors.

(a) Extend the definitions of *Hamming distance* and *Hamming weight* to be valid for ternary codes. [*Hint*: Replace "1" with *nonzero*.]

(b) Let C be a ternary code. Define *ternary sphere centered at a code word*.

(c) Prove the following ternary version of the sphere packing condition.

> **PROPOSITION 8.60** ***Ternary Sphere Packing Condition***
>
> A perfect ternary t-error-correcting $(n, M, 2t+1)$ code with $n > t$ must satisfy
>
> $$M \cdot \sum_{i=0}^{t} 2^i \binom{n}{i} = 3^n$$

(d) Show that the ternary $(11, 3^6, 5)$ Golay code satisfies the ternary sphere packing condition.

20. A major reason for error-correcting codes is that some of the bits in a message might get changed during transmission (from a 0 to a 1 or from a 1 to a 0). Suppose the probability that a single bit is changed is p, and the probability of a change in 1 bit is independent of any changes in other bits. (This is not always a valid assumption. Some channels introduce errors in bursts.)

(a) ★

i. What is the probability (as a function of p) of exactly 1 error occurring in a transmission that is 7 bits long?

ii. What is the probability if $p = 0.1$? What is the probability if $p = 0.5$?

(b)

i. What is the probability (as a function of p) of exactly 2 errors occurring in a transmission that is 7 bits long?

ii. What is the probability if $p = 0.1$? What is the probability if $p = 0.5$?

(c)

i. What is the probability that exactly r errors occur in a transmission that is n bits long?

ii. If $n = 7$, what is the probability of no errors if $p = 0.1$ and if $p = 0.5$?

8.6 Distinct Representatives, Ramsey Numbers

This final section introduces two topics that investigate some properties of sets whose elements are other sets. The first topic, systems of distinct representatives, determines whether it is possible to choose a distinct element from each set in a collection. The second topic, Ramsey numbers, investigates the minimum size of a set necessary to guarantee the existence of a property related to some of its subsets.

8.6.1 Systems of Distinct Representatives

The notion of a *system of distinct representatives* arises when considering the *Marriage Problem.* The Marriage Problem is not the same as the Stable Marriage Problem (page 3), although there are some similarities.

The Marriage Problem

The Marriage Problem concerns a group of eligible young women and a group of unmarried young men. The two groups need not be the same size. Each young woman makes a list of acceptable mates from among the group of young men. She then checks with each man on her list to see if he is willing to marry her. She removes the name of any man on the list who is unwilling to marry her. It is assumed that any man left on the list is completely acceptable as a mate. All the lists are then handed to a neutral referee. The referee must determine whether it is possible to marry each young woman to a young man who is on her list. Bigamy, of course, is not permitted.

EXAMPLE 8.47 **Illustrating the Marriage Problem**

Suppose there are three young women: Xena, Yolanda, and Zelda. Assume also that there are four young men: Abel, Bart, Collin, and Dermot. The final lists are

Xena: Bart, Collin **Yolanda:** Collin, Dermot **Zelda:** Collin, Dermot

It is clear that Collin cannot be matched with Xena without leaving one of the other young women unattached. One possible successful matching is Xena–Bart, Yolanda–Collin, Zelda–Dermot.

Suppose that Collin wants to become a monk and therefore will never marry. The lists are now

Xena: Bart **Yolanda:** Dermot **Zelda:** Dermot

It is clearly impossible to match all three young women with acceptable mates. ■

The previous example shows that it is not always possible to find an acceptable matching for the marriage problem. The main theorem in this section will present a necessary and sufficient condition for there to be a successful matching. The theorem is stated in the more general context of sets, so the problem needs to be restated using set-theoretic terminology.

DEFINITION 8.61 ***A System of Distinct Representatives***

Let $A_1, A_2, \ldots, A_n$ be n (not necessarily distinct) subsets of a set U. A list, $r_1, r_2, \ldots, r_n$, of elements in U is called a *system of distinct representatives* for $\{A_1, A_2, \ldots, A_n\}$ if

- $r_i \in A_i$, for $i = 1, 2, \ldots, n$
- $r_i \neq r_j$, for $i \neq j$

Example 8.47 can be written in the form of this definition. The initial sets are U = {Abel, Bart, Collin, Dermot}, A_1 = {Bart, Collin}, $A_2 = A_3$ ={Collin, Dermot}. In the original version (with Collin eligible), one choice of distinct representatives is R = Bart, Collin, Dermot.

In the second version, no system of distinct representatives is possible since there is an insufficient supply of eligible males for Yolanda and Zelda. In other words, the union of the lists for the two women contains only one man.

EXAMPLE 8.48 **No System of Distinct Representatives**

Let $A_1 = \{a, b\}$, $A_2 = \{a, b\}$, $A_3 = \{c, d, e\}$ and $A_4 = \{a, b\}$. There can be no system of distinct representatives, since the set $\{r_1, r_2, r_3, r_4\} \subseteq \{a, b, c, d, e\}$ must have three distinct elements to assign as the values of r_1, r_2, and r_4, but there are only two elements, a and b, available to choose from. ■

The pattern outlined in the previous example can be generalized. Let

$$\{B_1, B_2, \ldots, B_k\} \subseteq \{A_1, A_2, \ldots, A_n\}$$

If there are fewer than k elements in $B_1 \cup B_2 \cup \cdots \cup B_k$, then it is not possible to find a system of distinct representatives for $\{B_1, B_2, \ldots, B_k\}$, and hence also not possible for $\{A_1, A_2, \ldots, A_n\}$.[43]

This pattern is central, so it is useful to give it a name.

DEFINITION 8.62 ***The Marriage Condition***

Let $A_1, A_2, \ldots, A_n$ be n (not necessarily distinct) subsets of a set U. The collection, $\{A_1, A_2, \ldots, A_n\}$, is said to satisfy the *marriage condition* if for every k with $1 \leq k \leq n$ and every choice of a size-k subcollection, $\{A_{i_1}, A_{i_2}, \ldots, A_{i_k}\}$, with $1 \leq i_1 < i_2 < \cdots < i_k \leq n$

$$|A_{i_1} \cup A_{i_2} \cup \cdots \cup A_{i_k}| \geq k$$

[43] In Example 8.48, let $B_1 = A_1$, $B_2 = A_2$, and $B_3 = A_4$.

The notation in Definition 8.62 is more complicated than the notation used in the paragraph that precedes the definition, but the idea is the same. The notation A_{i_j} is just an abstract renaming for one of the sets in $\{A_1, A_2, \ldots, A_n\}$.[44] Note that the subscripts, $i_1, i_2, \ldots, i_k$, are merely integers that form a subset of the original subscripts, $1, 2, \ldots, n$. The change in notation makes it easier to specify the requirement that each A_{i_j} represents a different A_m from the original collection. This is accomplished by requiring $1 \leq i_1 < i_2 < \cdots < i_k \leq n$.[45]

The definition says that every collection of k of the A's must contain at least k elements in their union to satisfy the marriage condition. The marriage condition covers all values of k between 1 and n: every one of the A's must have at least one element, every pair must have at least two elements in their union, and so on.

The definition of the marriage condition was motivated by the observation that a set of distinct representatives cannot be found if the marriage condition fails.

EXAMPLE 8.49 **Verifying the Marriage Condition**

Look again at Example 8.48, where $A_1 = \{a, b\}$, $A_2 = \{a, b\}$, $A_3 = \{c, d, e\}$, and $A_4 = \{a, b\}$. In order for the marriage condition to be satisfied for this example, the following must all be true.

$k = 1$ Each set must have at least one element: $|A_1| \geq 1$, $|A_2| \geq 1$, $|A_3| \geq 1$, and $|A_4| \geq 1$. All are true for this example.

$k = 2$ Each pair of sets must have at least two elements in their union: $|A_1 \cup A_2| \geq 2$, $|A_1 \cup A_3| \geq 2$, $|A_1 \cup A_4| \geq 2$, $|A_2 \cup A_3| \geq 2$, $|A_2 \cup A_4| \geq 2$, and $|A_3 \cup A_4| \geq 2$. All these conditions are also true for this example.

$k = 3$ Every collection of three of the sets must contain at least three elements in their union: $|A_1 \cup A_2 \cup A_3| \geq 3$, $|A_1 \cup A_2 \cup A_4| \not\geq 3$, $|A_1 \cup A_3 \cup A_4| \geq 3$, $|A_2 \cup A_3 \cup A_4| \geq 3$. All but one of these conditions are true. For this example, $|A_1 \cup A_2 \cup A_4| = 2$.

$k = 4$ The union of all four of the sets must contain at least four elements: $|A_1 \cup A_2 \cup A_3 \cup A_4| \geq 4$. This condition is true.

Since one of the required conditions fails, the collection $\{A_1, A_2, A_3, A_4\}$ from Example 8.48 does not satisfy the marriage condition. ■

✔ Quick Check 8.19

Determine whether the marriage condition is satisfied. If it is not satisfied, show at least one specific point of failure.

1. $A_1 = \{a, b, c\}$, $A_2 = \{a\}$, and $A_3 = \{b\}$.
2. $A_1 = \{a, b\}$, $A_2 = \{a\}$, and $A_3 = \{b\}$. ☑

The marriage condition is necessary for a system of distinct representatives to exist. The next theorem shows that it is also sufficient.

THEOREM 8.63 ***Systems of Distinct Representatives (The Marriage Theorem)***

Let $A_1, A_2, \ldots, A_n$ be n (not necessarily distinct) subsets of a set U. A system of distinct representatives exists if and only if $\{A_1, A_2, \ldots, A_n\}$ satisfies the marriage condition.

---

[44]Using this notation with Example 8.48 and with $k = 3$, there are $\binom{4}{3} = 4$ possible choices for $\{A_{i_1}, A_{i_2}, A_{i_3}\}$, having $1 \leq i_1 < i_2 < i_3 \leq 4$. They are $\{A_1, A_2, A_3\}$, $\{A_1, A_2, A_4\}$, $\{A_1, A_3, A_4\}$, and $\{A_2, A_3, A_4\}$.

[45]Try stating this using the notation $B_1, B_2, \ldots, B_k$. It is tempting to require $B_i \neq B_j$, for $i \neq j$. However, this need not be true. In Example 8.48, it is possible to have $B_1 = A_1$, and $B_2 = A_2$, but still have $B_1 = B_2$.

This theorem, sometimes called Hall's marriage theorem, was first stated and proved by Philip Hall in 1935. Many proofs exist, including some constructive proofs. The constructive proofs tend to be a bit complex but would be preferred if an efficient algorithm for producing the system of distinct representatives is desired. The proof given here uses complete induction and proof by cases [15]. The cases are illustrated by Example 8.50 and Example 8.51, which immediately follow the formal proof. You may wish to refer to them as you read the proof. The Chapter 8 section at http://www.mathcs.bethel.edu/~gossett/DiscreteMathWithProof/ has a pdf file which contains these two examples.

Proof of Theorem 8.63:

Only If The "only if" part of the theorem is logically equivalent to the following claim:

if $\{A_1, A_2, \ldots, A_n\}$ does not satisfy the marriage condition,

then a system of distinct representatives does not exist.

Thus, suppose that for some k, with $1 \leq k \leq n$, and for some set of subscripts, $i_1, i_2, \ldots, i_k$, with $1 \leq i_1 < i_2 < \cdots < i_k \leq n$, $|A_{i_1} \cup A_{i_2} \cup \cdots \cup A_{i_k}| < k$.[46]

Each of the k sets, $A_{i_1}, A_{i_2}, \ldots, A_{i_k}$, must have a different representative. However, they collectively contain fewer than k distinct elements. It is therefore impossible to find distinct representatives for these sets. Consequently, it is impossible to find distinct representatives for $\{A_1, A_2, \ldots, A_n\}$.

If This part of the theorem will be proved using complete induction on n, the number of sets. During the inductive step, it will be necessary to consider two complementary cases. The two cases are determined by a more demanding version of the marriage condition. For the purposes of this proof, it will be called **the enhanced marriage condition**:

for every k with $1 \leq k \leq n-1$ and every choice of a size-k subcollection, $\{A_{i_1}, A_{i_2}, \ldots, A_{i_k}\}$, with $1 \leq i_1 < i_2 < \cdots < i_k \leq n$

$$|A_{i_1} \cup A_{i_2} \cup \cdots \cup A_{i_k}| \geq k+1$$

The new feature is the assertion that (for $k \leq n-1$) the union of any k of the sets must contain at least $k+1$ elements. Notice that when $k = n$, the normal marriage condition will still apply.

The negation of the enhanced marriage condition is

for some m with $1 \leq m \leq n-1$ and some choice of a size-m subcollection, $\{A_{i_1}, A_{i_2}, \ldots, A_{i_m}\}$, with $1 \leq i_1 < i_2 < \cdots < i_m \leq n$

$$|A_{i_1} \cup A_{i_2} \cup \cdots \cup A_{i_m}| \leq m$$

Now that this temporary definition is in place, the proof will proceed by complete induction on n.

Base Step When $n = 1$, there is only one set, A_1. If the marriage condition is satisfied for that set, then $|A_1| \geq 1$, so the set is not empty. Choose any element in the set to be the representative.

Inductive Step The inductive hypothesis is

every collection of $p-1$ or fewer sets that satisfies the marriage condition has a set of distinct representatives.

Assume now that the collection $\{A_1, A_2, \ldots, A_p\}$ satisfies the marriage condition. The goal is to show that this collection has a set of distinct representatives. There are two complementary conditions to consider.

[46] This is the negation of the marriage condition. See Exercise 1 on page 516.

Both will use the set identity[47] $(X - Z) \cup (Y - Z) = (X \cup Y) - Z$. In particular, in the generalized form $(A_{i_1} - Z) \cup \cdots \cup (A_{i_k} - Z) = (A_{i_1} \cup \cdots \cup A_{i_k}) - Z$.

Case 1: The enhanced marriage condition is satisfied.[48]

Because the marriage condition is satisfied, A_p has at least one element. Choose any element in A_p to be its representative, r_p. Since the enhanced marriage condition is also satisfied, it should be possible to remove r_p from $A_1, A_2, \ldots, A_{p-1}$ and still have the marriage condition be satisfied for those sets. The details follow.

Let $B_i = A_i - \{r_p\}$, for $i = 1, 2, \ldots, p-1$ ($A_i \cap \{r_p\} = \emptyset$ is possible). To see that $\{B_1, B_2, \ldots, B_{p-1}\}$ satisfies the marriage condition, consider any k with $1 \le k \le p-1$ and any collection of subscripts with $1 \le i_1 < i_2 < \cdots < i_k \le p-1$. Then

$$\begin{aligned}|B_{i_1} \cup B_{i_2} \cup \cdots \cup B_{i_k}| &= |(A_{i_1} \cup A_{i_2} \cup \cdots \cup A_{i_k}) - \{r_p\}| \\ &\ge |A_{i_1} \cup A_{i_2} \cup \cdots \cup A_{i_k}| - 1 \\ &\ge (k+1) - 1 \\ &= k\end{aligned}$$

Since $\{B_1, B_2, \ldots, B_{p-1}\}$ satisfies the marriage condition, the inductive hypothesis asserts that a system, $r_1, r_2, \ldots, r_{p-1}$, of distinct representatives exists, where $r_i \in B_i$. It is also clear that $r_i \ne r_p$ for $i = 1, 2, \ldots, p-1$. Since $B_i \subseteq A_i$, it is also true that $r_i \in A_i$ for $i = 1, 2, \ldots, p-1$. Therefore, the list $r_1, r_2, \ldots, r_p$ is a system of distinct representatives for $\{A_1, A_2, \ldots, A_p\}$.

Case 2: The enhanced marriage condition is not satisfied.[49]

Since the enhanced marriage condition is not satisfied, there must be some m, with $1 \le m \le p-1$, and some collection of distinct subscripts, $\{i_1, i_2, \ldots, i_m\}$, such that $|A_{i_1} \cup A_{i_2} \cup \cdots \cup A_{i_m}| \le m$. However, the marriage condition *is* satisfied, so $|A_{i_1} \cup A_{i_2} \cup \cdots \cup A_{i_m}| \ge m$. These inequalities indicate that $|A_{i_1} \cup A_{i_2} \cup \cdots \cup A_{i_m}| = m$. To simplify the notation, rename the A's so that $|A_1 \cup A_2 \cup \cdots \cup A_m| = m$.

Since $m \le p-1$, the inductive hypothesis provides a list, $T = r_1, r_2, \ldots, r_m$, of distinct representatives for $\{A_1, A_2, \ldots, A_m\}$. Define S by $S = \{r_1, r_2, \ldots, r_m\}$. Because $|A_1 \cup A_2 \cup \cdots \cup A_m| = m$, $S = A_1 \cup A_2 \cup \cdots \cup A_m$ must be true.

In a manner similar to case 1, remove the elements of S from each set in $\{A_{m+1}, A_{m+2}, \ldots, A_p\}$. Thus, let $B_i = A_i - S$, for $i = m+1, m+2, \ldots, p$. Since $m \ge 1$, this collection of sets contains at most $p-1$ sets. If this collection satisfies the marriage condition, then the inductive hypothesis can be used a second time. A brief detour will verify that $\{B_{m+1}, B_{m+2}, \ldots, B_p\}$ does indeed satisfy the marriage condition. The set identity $(X - Z) \cup (Y - Z) = (X \cup Y) - Z$ will be used with $X = Z = S$ and $Y = (A_{i_1} \cup A_{i_2} \cup \cdots \cup A_{i_k})$. Note also that $|X - Y| = |X| - |Y|$ if $Y \subseteq X$.

Detour

Let k be any integer such that $m+1 \le k \le p$ and let $i_1, i_2, \ldots, i_k$ be indices with $m+1 \le i_1 < i_2 < \cdots < i_k \le p$. Then

$$\begin{aligned}|B_{i_1} \cup B_{i_2} \cup \cdots \cup B_{i_k}| &= |(A_{i_1} \cup A_{i_2} \cup \cdots \cup A_{i_k}) - S| \\ &= |\emptyset \cup [(A_{i_1} \cup A_{i_2} \cup \cdots \cup A_{i_k}) - S]| \\ &= |[S - S] \cup [(A_{i_1} \cup A_{i_2} \cup \cdots \cup A_{i_k}) - S]| \\ &= |[S \cup (A_{i_1} \cup A_{i_2} \cup \cdots \cup A_{i_k})] - S| \\ &= |(A_1 \cup A_2 \cup \cdots \cup A_m) \cup (A_{i_1} \cup A_{i_2} \cup \cdots \cup A_{i_k}) - S| \\ &= |A_1 \cup A_2 \cup \cdots \cup A_m \cup A_{i_1} \cup A_{i_2} \cup \cdots \cup A_{i_k}| - |S| \\ &\ge (m+k) - m \\ &= k\end{aligned}$$

---

[47] See Exercise 15(b) on page 34.

[48] See Example 8.50 on page 507 for an illustration of this case.

[49] See Example 8.51 on page 507 for an illustration of this case.

The assertion $|A_1 \cup A_2 \cup \cdots \cup A_m \cup A_{i_1} \cup A_{i_2} \cup \cdots \cup A_{i_k}| \geq m + k$ is valid since $\{A_1, A_2, \ldots, A_p\}$ satisfies the marriage condition.

End of Detour

Now, since $\{B_{m+1}, B_{m+2}, \ldots, B_p\}$ satisfies the marriage condition and the collection contains at most $p - 1$ sets, the inductive hypothesis asserts the existence of a list, $r_{m+1}, r_{m+2}, \ldots, r_p$, of distinct representatives, with $r_i \in B_i$, for $i = m + 1, m + 2, \ldots, p$. The construction of the B's ensures that $r_j \notin \{r_{m+1}, r_{m+2}, \ldots, r_p\}$ for $j = 1, 2, \ldots, m$. Since $B_i \subseteq A_i$, $r_i \in A_i$, for $i = m + 1, m + 2, \ldots, p$. Thus, $r_1, r_2, \ldots, r_m, r_{m+1}, r_{m+2}, \ldots, r_p$ is a system of distinct representatives for $\{A_1, A_2, \ldots, A_p\}$.

Conclusion The theorem is true when $n = 1$. If the theorem is true for any collection of $n = p - 1$ or fewer sets, it is also true for a collection of $n = p$ sets. The Complete Induction theorem implies that the assertion is valid for all $n \geq 1$. □

EXAMPLE 8.50

Illustrating Case 1

Let $A_1 = \{a, b\}$, $A_2 = \{b, c\}$, $A_3 = \{c, d\}$, and $A_4 = \{a, d\}$. Then $|A_i| \geq 2$ for $i = 1, 2, 3, 4$. Also, $|A_i \cup A_j| \geq 3$ whenever $1 \leq i < j \leq 4$. In addition, $|A_i \cup A_j \cup A_k| \geq 4$ for all choices of i, j, k with $1 \leq i < j < k \leq 4$. Finally, notice that $|A_1 \cup A_2 \cup A_3 \cup A_4| = 4$. Consequently, both the marriage condition and the enhanced marriage condition are satisfied.

Let $r_4 = a$. Then $B_1 = \{b\}$, $B_2 = \{b, c\}$, and $B_3 = \{c, d\}$. It is easy to verify that $\{B_1, B_2, B_3\}$ satisfies the marriage condition, so the inductive hypothesis guarantees a system of distinct representatives. In this example, there is only one such system: $r_1 = b$, $r_2 = c$, and $r_3 = d$.

The list r_1, r_2, r_3, r_4 does form a system of distinct representatives for $\{A_1, A_2, A_3, A_4\}$. ■

EXAMPLE 8.51

Illustrating Case 2

Let $A_1 = \{a, b\}$, $A_2 = \{b, c\}$, $A_3 = \{a, b\}$, and $A_4 = \{c, d, e\}$.

Since $|A_1 \cup A_3| = 2$, the enhanced marriage condition is not satisfied. However, it is easy to verify that the marriage condition *is* satisfied.

Rename the sets so that $A_1 = \{a, b\}$, $A_2 = \{a, b\}$, $A_3 = \{b, c\}$, and $A_4 = \{c, d, e\}$ (changing subscripts $i_1 = 1$ and $i_2 = 3$ to subscripts 1 and 2). The inductive hypothesis asserts the existence of a system of distinct representatives for $\{A_1, A_2\}$. One such system is $r_1 = a$ and $r_2 = b$. Notice that $m = n - 2$, so complete induction is necessary.

Removing $S = \{a, b\}$ from A_3 and A_4 results in $B_3 = \{c\}$ and $B_4 = \{c, d, e\}$. The collection $\{B_3, B_4\}$ satisfies the marriage condition and has less than 4 members, so a system of distinct representatives exists (by the inductive hypothesis). One such system is $r_3 = c$ and $r_4 = d$.

The list r_1, r_2, r_3, r_4 forms a system of distinct representatives for $\{A_1, A_2, A_3, A_4\}$. ■

You might have noticed that the inductive step in the proof of the marriage theorem provides the basis for a recursive algorithm to find a set of distinct representatives. The algorithm would start with the original collection of sets, and use one of the two cases to break the problem into one or two smaller problems (case 1 and case 2, respectively). For example, in Example 8.51, the original problem of finding distinct representatives for $\{A_1, A_2, A_3, A_4\}$ became two problems: find distinct representatives for $\{A_1, A_2\}$ and find distinct representatives for $\{A_3, A_4\}$. Each of these problems can then be solved using the case 1 recursion.

The difficulty with this algorithm is its inherent inefficiency. At each new recursion, it is necessary to decide which of the two cases applies. This requires the calculation of the number of elements in all the different unions of A's. Since there are n A's to start with, there are $2^n - 1$ such calculations for the initial step (the empty set is not considered). This means that the algorithm is at least in $\Omega(2^n)$. Except for very small problems, it is not a practical algorithm.

It should be clear from some of the examples that there may be more than one system of distinct representatives for a given collection of sets. A simple corollary to Theorem 8.63 provides a lower bound for the number of systems.

COROLLARY 8.64 ***A Lower Bound for the Number of Systems of Distinct Representatives***

Let $\{A_1, A_2, \ldots, A_n\}$ be a collection of sets that satisfies the marriage condition. Let t be the number of elements in a smallest set: $t = \min\{|A_1|, |A_2|, \ldots, |A_n|\}$. Then the number of systems of distinct representatives is at least

$$\begin{cases} t! & \text{if } t < n \\ \frac{t!}{(t-n)!} & \text{if } t \geq n \end{cases}$$

See Exercise 12 on page 517 for the proof. Note that the two cases have different values for $t!$ (depending on how they compare to n). The second case results in a larger minimum (Exercise 13 on page 517).

The following consequence of Corollary 8.64 was used in Section 8.2.1.

THEOREM 8.65 ***Lower Bound on Distinct Latin Squares***

Let $L(n)$ be the number of distinct Latin squares of order n. Then

$$L(n) \geq n! \cdot (n-1)! \cdot (n-2)! \cdots 3! \cdot 2! \cdot 1!$$

Proof: The proof constructs a Latin square a row at a time, counting the number of choices available for each new row [31].

Suppose that r of the n rows have already been specified. Each column of the partial Latin square contains some, but not all, of the numbers in $\{1, 2, \ldots, n\}$. Let A_i denote the set of numbers in $\{1, 2, \ldots, n\}$ that are *not* in the partially constructed ith column, for $i = 1, 2, \ldots, n$. Then for all i, $|A_i| = n - r$. In addition, every number in $\{1, 2, \ldots, n\}$ appears in exactly r of the partially constructed columns (since each number appears once in each of the r rows of a Latin square). Consequently, every number in $\{1, 2, \ldots, n\}$ appears in exactly $n - r$ of the A_i's.

Consider $A_{i_1} \cup A_{i_2} \cup \cdots \cup A_{i_k}$, for $1 \leq k \leq n$. If $k \leq n - r$, then any one of the A_{i_j} contains at least k elements, so $|A_{i_1} \cup A_{i_2} \cup \cdots \cup A_{i_k}| \geq k$. If $k > n - r$, then create an $n - r$ by k matrix whose pqth entry is the pth element in A_{i_q}. If there were fewer than k distinct numbers in this matrix, then some number would need to appear in more than $n - r$ of the $(n - r) \cdot k$ positions. This contradicts the already established fact that every number appears in exactly $n - r$ of the A_i's. Thus, in this case also, $|A_{i_1} \cup A_{i_2} \cup \cdots \cup A_{i_k}| \geq k$. The conclusion is that the collection $\{A_1, A_2, \ldots, A_n\}$ satisfies the marriage condition.

Since $\{A_1, A_2, \ldots, A_n\}$ satisfies the marriage condition, it has a system of distinct representatives. That system can become the next row of the Latin square.

Now count the number of possible rows at each stage of the construction. There are $n!$ possibilities for the first row. After r rows have been constructed, each of the A_i's contains $n - r$ elements, so Corollary 8.64 implies that there are at least $(n - r)!$ systems of distinct representatives. But each system of distinct representatives corresponds to a different choice for the $(r + 1)$st row.

Building row by row, there are thus at least $n! \cdot (n-1)! \cdots 2! \cdot 1!$ different Latin squares of order n. □

8.6.2 Ramsey Numbers

The following problem appeared in Exercises 5.3.3 after the pigeon-hole principle was introduced.

✔ Quick Check 8.20

1. Consider a room that contains six people. Every pair of people are either friends or are enemies. Prove that there is a subgroup of at least three of the people who are either mutual friends or are mutual enemies. That is, show that there are three people all of whom are friends or there are three people all of whom are enemies.

Hints: You may want to name them Alice, Bob, Carol, Don, Erin, and Frank. (Or you could shorten the names to A, B, C, D, E, and F.) If you are geometrically inclined, you may want to represent the guests by points. If the guests are acquainted, draw a red line between their points. If they are unacquainted, draw a blue line between their points. Now show that there exists either a red triangle or a blue triangle (or both). ☑

The puzzle with six guests has a much wider setting. Suppose I have a room full of people. I want to know if there is always a set of j mutually acquainted or a set of k mutually unacquainted people. In 1930 Frank Ramsey was able to prove that there is always a number $R(j, k)$ such that any collection of $R(j, k)$ or more people has such a set. Quick Check 8.20 was related to the *Ramsey number* $R(3, 3)$, which is 6. Thus Ramsey generalized the simple theorem about mutual-relationship sets of size three in a group of six people, into a theorem about mutual-relationship sets of any sizes j and k in some suitably larger group. Ramsey's generalizing did not stop here. He also proved such a result holds for any finite number of relationships. For example, we can expand the relationships from acquainted and unacquainted to relative, friend, and stranger. There is always a number $R(r, f, s; 2)$ such that a room with at least that many people will have either a set of r mutual relatives, or a set of f mutual friends, or a set of s mutual strangers (or a combination of such sets). The discussion that follows will develop these ideas and make one additional generalization.

The Ramsey theorems are usually stated either as statements about sets or as statements about graphs (Chapter 10). The set-theoretic versions are presented here. The discussion will frequently involve sets whose elements are other sets.

A set-theoretic example will be helpful. However, some new notation will first be introduced.

DEFINITION 8.66 $S_{\{m\}}$

If S is a set with $|S| \geq m$, then the set of all m-element subsets of S will be denoted by $S_{\{m\}}$.

The set brackets around the subscript in the notation $S_{\{m\}}$ should remind you that the m-element *subsets* of S are the focus.

EXAMPLE 8.52 $S_{\{m\}}$

Let $S = \{x, y, z\}$. Then $S_{\{1\}} = \{\{x\}, \{y\}, \{z\}\}$, $S_{\{2\}} = \{\{x, y\}, \{x, z\}, \{y, z\}\}$, and $S_{\{3\}} = \{\{x, y, z\}\}$. Notice that $S_{\{3\}}$ is the set that contains the set S.

If $|S| = n$ and $m \leq n$ then $\left|S_{\{m\}}\right| = \binom{n}{m}$ (Section 5.1.4 on page 231). ■

EXAMPLE 8.53

Six Guests

Let $S = \{a, b, c, d, e, f\}$. The collection of all two-element subsets of S is $S_{\{2\}} = \{A_1, A_2, A_3, \ldots, A_{15}\}$, where $A_1 = \{a, b\}$, $A_2 = \{a, c\}$, $A_3 = \{a, d\}$, ..., $A_{15} = \{e, f\}$. A two-element subset corresponds to a relationship (such as acquaintance) between the two elements of S.

The problem in Quick Check 8.20 can be restated as follows:

Claim:

If the two-element subsets of S are partitioned[50] into two sets, X and Y, then there is either some three-element subset, T, of S such that every 2-element subset of T is in X or there is some three-element subset, U, of S such that every two-element subset of U is in Y.

For example, let $X = \{\{a, b\}, \{a, d\}, \{a, e\}, \{a, f\}, \{b, d\}, \{b, e\}, \{b, f\}, \{d, e\}, \{d, f\}, \{e, f\}\}$ and let $Y = \{\{a, c\}, \{b, c\}, \{c, d\}, \{c, e\}, \{c, f\}\}$. Then $T = \{a, b, d\}$ has all three of its two-element subsets in X. (There are several other choices for T that also work. There are no viable choices for U with this partition.)

There are 2^{15} possible partitions of $\{A_1, A_2, A_3, \ldots, A_{15}\}$, so it is not practical to verify the claim by looking at each partition. Instead, the proof from Quick Check 8.20 can be adapted to the new setting.

Proof of the Claim:

Consider the element $a \in S$. In any partition of $S_{\{2\}}$, the collection of two-element subsets of S, a must either be in three or more of the two-element sets in X or else must be in three or more of the two-element sets in Y. (This is true because a is in five different two-element subsets of S. All five must be in $X \cup Y$.)

Suppose a is in at least three of the sets in X. With suitable renaming if necessary, it can be assumed that X contains $\{a, b\}$, $\{a, c\}$, and $\{a, d\}$. If X contains none of the sets $\{b, c\}$, $\{b, d\}$, or $\{c, d\}$, then $U = \{b, c, d\}$ satisfies the claim. Otherwise, one of the sets $\{b, c\}$, $\{b, d\}$, or $\{c, d\}$ is contained in X. Suppose $\{b, c\} \in X$ (the other possibilities are handled similarly). Then $T = \{a, b, c\}$ satisfies the claim.

On the other hand, suppose that a is in at least three of the sets in Y. Assume they are $\{a, b\}$, $\{a, c\}$, and $\{a, d\}$. If Y contains none of the sets $\{b, c\}$, $\{b, d\}$, or $\{c, d\}$, then $T = \{b, c, d\}$ satisfies the claim. Otherwise, one of the sets $\{b, c\}$, $\{b, d\}$, or $\{c, d\}$ is contained in Y. Suppose $\{b, c\} \in Y$ (the other possibilities are handled similarly). Then $U = \{a, b, c\}$ satisfies the claim. □

■

It is time for some formal definitions and theorems.

DEFINITION 8.67 ***The (j, k) Ramsey Condition***

Let S be a set with n elements. Let $j \geq 2$ and $k \geq 2$. S satisfies the (j, k) *Ramsey condition* if for every partition of $S_{\{2\}}$ into the disjoint sets, X and Y, there is either a j-element subset, T, of S such that $T_{\{2\}} \subseteq X$, or else there is a k-element subset, U, of S such that $U_{\{2\}} \subseteq Y$.

DEFINITION 8.68 $R(j, k)$

The *Ramsey number*, $R(j, k)$, is the smallest integer such that every set, S, with at least $R(j, k)$ elements satisfies the (j, k) Ramsey condition.

[50] See page 24 for the definition of *partition*.

EXAMPLE 8.54 **Determining $R(4, 2)$**

Let S be a set with n elements. Suppose the two-element subsets of S are partitioned into the sets X and Y: $S_{\{2\}} = X \cup Y$ and $X \cap Y = \emptyset$. How large must n be to guarantee that either there will be a four-element subset, T, of S such that every two-element subset of T is in X, or else there will be a two-element subset, U, of S such that every two-element subset of U will be in Y?

The key observation is that U has only one two-element subset, namely, U itself. Therefore, if Y is not empty, the required conditions can be easily reached: take U to be any one of the elements of Y.

If U cannot be found then $Y = \emptyset$. In that case, all the two-element subsets of S will be in X. To discuss four-element subsets of S requires $n \geq 4$. If $n = 4$, then $T = S$ satisfies the required conditions. If $n > 4$, let T be any four-element subset of S. All its two-element subsets will be in X, since *every* two-element subset of S is in X.

The conclusion is that the conditions are satisfied whenever S has at least four elements. Thus, $R(4, 2) = 4$. ■

Definition 8.68 defines $R(j, k)$ for all $j, k \geq 2$; However, defining the concept does not imply that such a number necessarily exists. Theorem 8.70 (soon to be presented) will provide that guarantee.

Determining the values $R(j, k)$ for all j and k is an incomplete task. In fact, very few of these numbers are known. The previous example determined that $R(4, 2) = 4$. The next example will complete the calculation of $R(3, 3)$. The subsequent proposition will determine $R(j, 2)$ and $R(2, k)$.

EXAMPLE 8.55 **$R(3, 3)$**

Example 8.53 showed that any set, S, with six elements satisfies the $(3, 3)$ Ramsey condition. The proof requires only one change if S has more than 6 elements: a will be in $n - 1$ two-element subsets of S rather than in 5. Thus, $R(3, 3) \leq 6$.

To see that $R(3, 3) > 5$, let $S = \{a, b, c, d, e\}$ and choose to partition $S_{\{2\}}$ as $X = \{\{a, b\}, \{b, c\}, \{c, d\}, \{d, e\}, \{a, e\}\}$ and $Y = \{\{a, c\}, \{a, d\}, \{b, d\}, \{b, e\}, \{c, e\}\}$. S has $\binom{5}{3} = 10$ three-element subsets. These subsets are the candidates for T and U. It is easy to verify that none of them satisfy the $(3, 3)$ Ramsey condition. (For example, $\{a, c, e\}$ has $\{a, e\} \in X$ and $\{a, c\}, \{c, e\} \in Y$.) Thus, at least one partition of $S_{\{2\}}$ fails to satisfy the $(3, 3)$ Ramsey condition when $n = 5$. The conclusion is that $R(3, 3) = 6$. ■

PROPOSITION 8.69 ***Some Elementary Properties of Ramsey Numbers***

The following properties are all valid assertions about Ramsey numbers. Let $j \geq 2$ and $k \geq 2$. Then

1. $R(j, k) = R(k, j)$
2. $R(j, 2) = j$
3. $R(2, k) = k$

Proof: The symmetry of the two parameters follows by interchanging the roles of X and Y and of T and U in the definition of the (j, k) Ramsey condition.

The proof that $R(2, k) = k$ will be left as an exercise. The proof that $R(j, 2) = j$ is similar to the proof in Example 8.54 that $R(4, 2) = 4$. The details will now be presented.

Let S be any set with $n \geq 2$ elements. Partition the two-element subsets of S into the sets X and Y. If $Y \neq \emptyset$, then any element of Y will serve as the two-element set U in the $(j, 2)$ Ramsey condition.

If $Y = \emptyset$ and $n \geq j$, then any j-element subset of S will fulfill the requirements for T in the $(j, 2)$ Ramsey condition. If $n < j$, then no such T is possible.

The conclusion is that $R(j, 2) = j$. □

THEOREM 8.70 ***R(j, k) Exists***

Let $j \geq 2$ and $k \geq 2$. Then the Ramsey number, $R(j,k)$, exists. In addition,

$$R(j,k) \leq R(j\text{-}1,k) + R(j,k\text{-}1) \quad \text{for all} \quad j,k \geq 3$$

Proof: The first assertion (existence) follows from the second, since an upper bound for $R(j,k)$ guarantees that it is a finite integer. The recursive inequality requires the values $R(2,k)$ and $R(j,2)$ as base cases. Proposition 8.69 provides those values.

The proof uses induction on j and k. The base step has already been discussed. The inductive hypothesis is as follows: For $j,k \geq 3$, $R(j\text{-}1,r)$ exists for $2 \leq r \leq k$ and $R(s,k\text{-}1)$ exists for $2 \leq s \leq j$.

Suppose now that $j \geq 3$ and $k \geq 3$ and that S is a set containing at least $R(j\text{-}1,k) + R(j,k\text{-}1)$ elements. Partition the two-element subsets into some sets X and Y. Choose an element $a \in S$.

Then a must either appear in at least $R(j\text{-}1,k)$ of the sets in X or else it must appear in at least $R(j,k\text{-}1)$ of the sets in Y.[51]

Let X' and Y' be the subsets of X and Y, respectively, formed by removing all two-element subsets of S that contain a.

Case 1: a is in at least $R(j\text{-}1,k)$ of the sets in X.

Consider the set, B, of the elements in S that appear with a in a two-element subset in X. Since there are at least $R(j\text{-}1,k)$ elements in B, the inductive hypothesis asserts the existence of either a $(j\text{-}1)$-element subset, T', of S such that all of its two-element subsets are in X' or else a k-element subset, U, of S such that all of its two-element subsets are in Y'.

In the former case, let $T = T' \cup \{a\}$. T contains j elements, and all its two-element subsets are in X. Otherwise, U already satisfies the (j,k) Ramsey condition.

Case 2: a is in at least $R(j,k\text{-}1)$ of the sets in Y.

See Exercise 18 on page 518. □

When both $j > 2$ and $k > 2$, very few values of $R(j,k)$ are known. Table 8.52 lists some of what was known at the time this text was written. Other values, such as $R(4,7)$, are not known precisely, but upper and lower bounds are known. These are indicated by a range, such as 49–61.

TABLE 8.52 Some Known Values and Bounds for $R(j,k)$

j/k	3	4	5	6	7	8	9	10
3	6	9	14	18	23	28	36	40–43
4	9	18	25	35–41	49–61	55–84	69–115	80–149
5	14	25	43–49	58–87	80–143	95–216	116–316	141–442
6	18	35–41	58–87	102–165	109–298	122–495	153–780	167–1171
7	23	49–61	80–143	109–298	205–540	216–1031	227–1713	238–2826
8	28	55–84	95–216	122-495	216–1031	282–1870	295–3583	308–6090
9	36	69–115	116–316	153–780	227–1713	295–3583	565–6625	580–12715
10	40–43	80–149	141–442	167–1171	238–2826	308–6090	580–12715	798–23854

[51]To see that this assertion is true, suppose it is false. Then a would be in at most $R(j\text{-}1,k) - 1$ sets in X and at most $R(j,k\text{-}1) - 1$ sets in Y, for a total of at most $R(j\text{-}1,k) + R(j,k\text{-}1) - 2$ sets. But a is in a two-element subset with every other element in S, so it must be in at least $R(j\text{-}1,k) + R(j,k\text{-}1) - 1$ two-element subsets. This contradiction proves the assertion.

Two generalizations will now be introduced. The first generalization considers m-element subsets of S (instead of just two-element subsets).

DEFINITION 8.71 *The $(j, k; m)$ Ramsey Condition*
Let S be a set with n elements. Let $j \geq m \geq 1$ and $k \geq m \geq 1$. S satisfies the $(j, k; m)$ *Ramsey condition* if for every partition of $S_{\{m\}}$ into the disjoint sets, X and Y, there is either a j-element subset, T, of S such that $T_{\{m\}} \subseteq X$, or else there is a k-element subset, U, of S such that $U_{\{m\}} \subseteq Y$.

Notice that $m = 1$ is permitted in this more general version.

DEFINITION 8.72 $R(j, k; m)$
The *Ramsey number*, $R(j, k; m)$, is the smallest integer such that every set, S, with at least $R(j, k; m)$ elements satisfies the $(j, k; m)$ Ramsey condition.

Notice that $R(j, k; 2) = R(j, k)$.

EXAMPLE 8.56 **$R(3, 4; 1)$**

The goal is to find a number, $R(3, 4; 1)$, such that for any set, S, with $|S| \geq R(3, 4; 1)$, whenever the one-element subsets of S are partitioned into two sets, X and Y, there will always be either a three-element subset, $T \subseteq S$, such that $T_{\{1\}} \subseteq X$, or else there will be a four-element subset, $U \subseteq S$, such that $U_{\{1\}} \subseteq Y$.

The requirement is equivalent to the claim that if the elements of S are divided into two disjoint sets, X and Y, then either $|X| \geq 3$ or $|Y| \geq 4$. If $|S| \leq 5$, it is possible to violate this condition. If $|S| \geq 6$, the pigeon-hole principle ensures that at least one of the required inequalities is true. (If $|X| \geq 3$ we are done. If $|X| < 3$, then $|Y| \geq 4$.)

Thus, $R(3, 4; 1) = 6$. ■

The previous example can be generalized.

PROPOSITION 8.73 $R(j, k; 1)$
If $j \geq 1$ and $k \geq 1$, then $R(j, k; 1) = j + k - 1$.

Proof: Let S be a set with $|S| \geq R(j, k; 1)$. The $(j, k; 1)$ Ramsey condition is equivalent to the requirement that whenever the elements of S are partitioned into the sets X and Y, either $|X| \geq j$ or $|Y| \geq k$. If $|S| \leq (j - 1) + (k - 1)$, it is possible to distribute the elements of S between X and Y in a manner that violates the Ramsey condition. If $|S| \geq (j - 1) + (k - 1) + 1$, the pigeon-hole principle[52] implies that either $|X| \geq j$ or $|Y| \geq k$. Thus, $R(j, k; 1) = j + k - 1$. □

Proposition 8.73 suggests that Ramsey numbers generalize the pigeon-hole principle.

THEOREM 8.74 ***$R(j, k; m)$ Exists***
If $j, k \geq m$ and $m \geq 2$, then $R(j, k; m)$ exists. In addition,

$$R(j, k; m) \leq R(j\text{-}1, k; m) + R(j, k\text{-}1; m) \quad \text{for all } j, k \geq 2$$

Proof: Following the proof in [72]. Since the proof uses multi-dimensional induction, it might be helpful to review section 3.5.5 starting on page 148.

[52] See Exercise 14 on page 265.

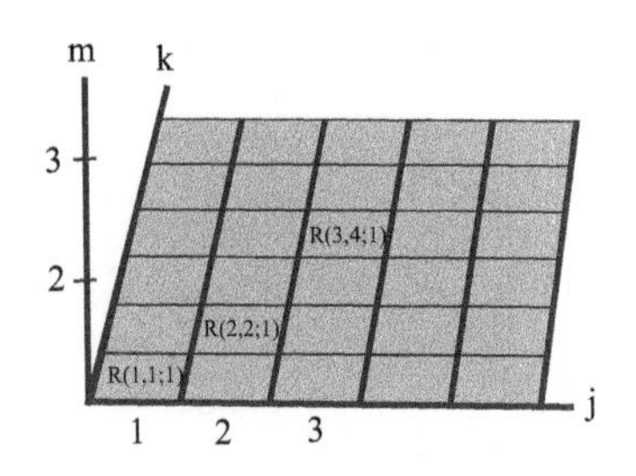

Figure 8.13. Base Case 1.

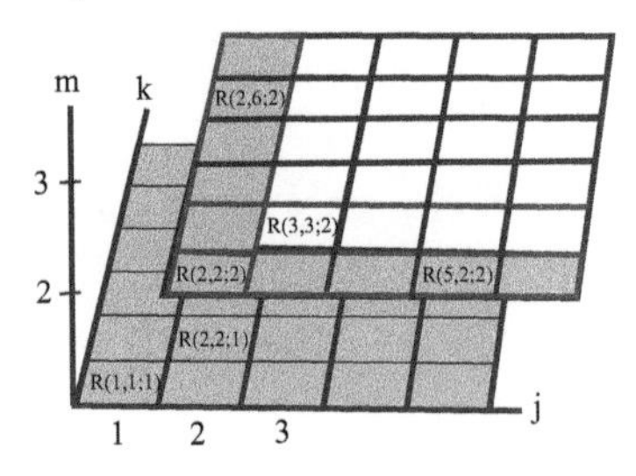

Figure 8.14. Base Case 2 for $m = 2$.

Theorem 8.74: If $j, k \geq m$ and $m \geq 2$, then $R(j, k; m)$ exists. In addition,

$$R(j, k; m) \leq R(j\text{-}1, k; m) + R(j, k\text{-}1; m) \quad \text{for all } j, k \geq 2.$$

Base Case 1 By Proposition 8.73 $R(j, k; 1) = j+k-1$, for $j, k \geq 1$. This is illustrated in Figure 8.13, where the shading indicates the known values.

Base Case 2 $R(j, m; m) = j$ and $R(m, k; m) = k$ for $j, k \geq m \geq 1$

To see that $R(j, m; m) = j$, suppose $|S| = j - 1$. Let $X = S$ and $Y = \emptyset$. Then there is no j-element subset of X nor any m-element subset of Y that satisfy the $(j, k; m)$ Ramsey condition. On the other hand, if $|S| = j$, and $S_{\{m\}} = X \cup Y$ with $X \cap Y = \emptyset$, then if $Y \neq \emptyset$ let U be any element of Y and the $(j, m; m)$ Ramsey condition is satisfied. Otherwise, every m-element subset of S is in X. Let $T = S$ and the $(j, m; m)$ Ramsey condition will be satisfied. The proof that $R(m, k; m) = k$ is similar. Figure 8.14 illustrates this for $m = 2$. $R(3, 3; 2)$ is unshaded because the base case does not ensure a finite value when $j = k = 3$. Similar bordered planes can be imagined stacking up in a receding manner for higher values of m.

Inductive Step The intuitive approach that will be taken for the inductive step is the following. We can determine that $R(3, 3; 2)$ exists because we know $R(2, 3; 2)$ and $R(3, 2; 2)$ exist and also $R(r, s; 1)$ exists for all $r, s \geq 1$. But then $R(3, 3; 2)$ can be used to show $R(3, 4; 2)$ and $R(4, 3; 2)$ exist. In this manner, successive 135° diagonals (diagonals with constant row+column sums) can be shaded until the entire level-2 plane is shaded. Now start the same process with $R(4, 4; 3)$ (using the existence of $R(3, 4; 3)$, $R(4, 3; 3)$ and $R(r, s; 2)$ for all $r, s \geq 2$) on the level-3 plane. Eventually, all planes will be shaded. The formal discussion provides the details.

Inductive Hypothesis For $j, k > m \geq 2 : R(j\text{-}1, k; m)$ and $R(j, k\text{-}1; m)$ exist. Also, for $r, s \geq m - 1 \geq 1 : R(r, s; m\text{-}1)$ exists.

The existence of $R(j, k; m)$ must now be established. By the inductive hypothesis, $r = R(j\text{-}1, k; m)$ and $s = R(j, k\text{-}1; m)$ exist so $R(r, s; m\text{-}1)$ also exists. Let $N = R(r, s; m\text{-}1) + 1$. Let $n \geq N$, and let $|S| = n$. Select an element $a \in S$, and let $P = S - \{a\}$. Partition the m-element subsets of S as $S_{\{m\}} = X \cup Y$. Now consider the $(m\text{-}1)$-element subsets of P. Create a partition of these subsets as $P_{\{m\text{-}1\}} = X' \cup Y'$ in the following manner. Let Q be an $(m\text{-}1)$-element subset of P. If $Q \cup \{a\} \in X$, then place Q in X', otherwise place Q in Y'.

By construction, there are at least $R(r, s; m\text{-}1)$ elements in P. Thus, by definition, there is either (1) a set C of r elements in P with $C_{\{m\text{-}1\}} \subseteq X'$, or (2) a set D of s elements in P with $D_{\{m\text{-}1\}} \subseteq Y'$.[53]

Case (1) We can partition $C_{\{m\}}$ as $C_{\{m\}} = \left(X \cap C_{\{m\}}\right) \cup \left(Y \cap C_{\{m\}}\right)$. Since $|C| = r = R(j\text{-}1, k; m)$, there is either:

(a) a set of j-1 elements from C whose m-element subsets are all in $X \cap C_{\{m\}}$

Then add a to the set of $j - 1$ elements, and now there is a j-element set $E \subseteq S$ with $E_{\{m\}} \subseteq X$. This is because the m-element subsets that don't include a are already in $X \cap C_{\{m\}} \subseteq X$. Also, the $(m\text{-}1)$-element subsets of C are all in X' and adding a to each of these subsets leads to an m-element subset that is in X.

or

(b) a k-element set $F \subseteq C \subset S$ whose m-element subsets are all in $Y \cap C_{\{m\}}$.

The proof of Case (1) is complete because either $E_{\{m\}} \subseteq X$ or $F_{\{m\}} \subseteq Y$.

Case (2) This is similar to the previous case and will be left as Exercise 23.

Thus $R(j, k; m)$ exists and the proof is complete. □

[53] Examples 8.58 and 8.59 on page 515 illustrate the first of these two cases.

EXAMPLE 8.57 **$R(4, 3; 3)$**

$R(4, 3; 3) = 4$. To see this, note first that if $S = \{a, b, c\}$, the three-element subsets of S can be partitioned as $X = \{\{a, b, c\}\}$ and $Y = \emptyset$. Since S has no four-element subsets, there cannot be a four-element subset $T \subseteq S$ whose three-elements subsets are all in X. On the other hand, there are no three-element subsets $U \subseteq S$ all of whose three-element subsets are in Y. Thus, $R(4, 3; 3) > 3$.

Suppose $S = \{a, b, c, d\}$. There are four three-element subsets. If any one of them is placed in Y, then that subset can be used as U. On the other hand, if every three-element subset is placed in X, then let $T = S$. In either case, the (4, 3; 3) Ramsey condition is satisfied. Since this argument works even if $|S| > 4$, it is clear that $R(4, 3; 3) \leq 4$, completing the proof. ■

EXAMPLE 8.58 **Illustrating Case (1a) in the Proof of Theorem 8.74**

Suppose the existence of $R(4, 4; 3)$ is being established in the inductive step of the proof for Theorem 8.74. The base step has established that $r = R(3, 4; 3) = 4$ and $s = R(4, 3; 3) = 4$. Table 8.52 on page 512 shows that $R(r, s; 2) = R(4, 4; 2) = 18$. So $n = N = 19$. Let $S = \{a, b, c, d, e, f, g, h, i, l, p, q, t, u, v, w, x, y, z\}$ (some letters have been skipped). $P = \{b, c, d, e, f, g, h, i, l, p, q, t, u, v, w, x, y, z\}$.

For this example, let $X = \{abc, abd, abe, bcd, acd, ace, ade\}$, where the three-element subsets have been abbreviated (for example, $bcd = \{b, c, d\}$). Let Y consist of the remaining $\binom{19}{3} - 7$ three-element subsets.

The set X' is then (using a similar abbreviation) $X' = \{bc, bd, be, cd, ce, de\}$. Notice that X' contains no two-element subset that corresponds to $bcd \in X$.

For this example, $C = \{b, c, d, e\}$ is a four-element subset of P with all its two-element subsets in X'. (The collection of two-element subsets equals X' in this example.) Thus, Case (1) applies. $C_{\{3\}} = \{bcd, bce, bde, cde\} = \{bcd\} \cup \{bce, bde, cde\}$. Note that $\{bcd\} = X \cap C_{\{3\}}$ and $\{bce, bde, cde\} = Y \cap C_{\{3\}}$.

In this example, $j - 1 = 3$. The three-element subset $\{b, c, d\}$ of C has all of its three-element subsets in $X \cap C_3$, so subcase (a) applies. Adding a creates the four-element set $E = \{a, b, c, d\}$. This set has $\binom{4}{3} = 4$ three-element subsets. The subset bcd is in $X \cap C_{\{3\}} \subset X$. Adding a to the two-element subsets of C produces $\{abc, abd, abe, acd, ace, ade\} \subset X$. (Notice again that bcd does not appear in this collection.) In addition, all the three-element subsets of E appears in one of these two subsets of X. ■

EXAMPLE 8.59 **Illustrating Case (1b) in the Proof of Theorem 8.74**

Suppose the existence of $R(4, 4; 3)$ is being established in the inductive step of the proof for Theorem 8.74. The base step has established that $r = R(3, 4; 3) = 4$ and $s = R(4, 3; 3) = 4$. It is known (Table 8.52) that $R(r, s; 2) = R(4, 4; 2) = 18$. So $n = N = 19$. Let $S = \{a, b, c, d, e, f, g, h, i, l, p, q, t, u, v, w, x, y, z\}$ (some letters have been skipped). $P = \{b, c, d, e, f, g, h, i, l, p, q, t, u, v, w, x, y, z\}$.

For this example, let $X = \{abc, abd, abe, acd, ace, ade\}$, where the three-element subsets have been abbreviated as in Example 8.58. Let Y consist of the remaining $\binom{19}{3} - 6$ three-element subsets. In this example, the set bcd has been moved from X to Y. Consequently, $X' = \{bc, bd, be, cd, ce, de\}$.

For this example, $C = \{b, c, d, e\}$ is a four-element subset of P with all its two-element subsets in X'. Thus, Case (1) applies. $C_{\{3\}} = \{bcd, bce, bde, cde\} = \emptyset \cup \{bcd, bce, bde, cde\}$. Note that $X \cap C_{\{3\}} = \emptyset$ and $Y \cap C_{\{3\}} = \{bcd, bce, bde, cde\}$.

In this example, $k = 4$. The four-element subset $\{b, c, d, e\}$ of C has all of its three-element subsets in $Y \cap C_{\{3\}}$. ■

✔ Quick Check 8.21

1. Determine the value of $R(j, 3; 3)$, where $j \geq 3$. ☑

There is one more generalization that will be presented. The proof is patterned after the proof in [72].

THEOREM 8.75 ***Ramsey's Theorem***

Let $k_1, k_2, \ldots, k_n$ and m be integers with $k_i \geq m \geq 1$ and $n \geq 2$. Then there exists a positive integer, $R(k_1, k_2, \ldots, k_n; m)$, such that if $|S| \geq R(k_1, k_2, \ldots, k_n; m)$ and $S_{\{m\}}$ is partitioned into a disjoint union of sets, $S_{\{m\}} = X_1 \cup X_2 \cup \cdots \cup X_n$, then for some i with $1 \leq i \leq n$ there is a k_i-element subset, $T \subseteq S$, such that $T_{\{m\}} \subseteq X_i$.

Even less is known about these more general Ramsey numbers. It is known that $R(2, 2, \ldots, 2; 2) = 2$ and that $R(3, 3, 3; 2) = 17$. It is also easy to prove that $R(k_1, k_2, \ldots, k_n; 1) = k_1 + k_2 + \cdots k_n - n + 1$.

Proof of Theorem 8.75: A formal proof would use induction on n. Here is an informal proof.

Exercise 22 will establish the existence of $R(k_1, k_2, \ldots, k_n; 1)$. Theorem 8.74 shows that $R(k_1, k_2; m)$ exists.

Consider the case where $n = 3$. Let $k_i \geq m \geq 1$ for $i = 1, 2, 3$. To see that $R(k_1, k_2, k_3; m)$ exists, let $v = R(k_2, k_3; m)$ and let $w = R(k_1, v; m)$. Both v and w exist by Theorem 8.74.

Let S be a set with $|S| \geq w$. Partition the m-element subsets of S as $S_{\{m\}} = X_1 \cup X_2 \cup X_3$ and set $Y = X_2 \cup X_3$. Then $S_{\{m\}} = X_1 \cup Y$, so there either exists a k_1-element subset T of S such that $T_{\{m\}} \subseteq X_1$ (and the requirements of Theorem 8.75 have been met), or else there is a v-element subset U of S such that $U_{\{m\}} \subseteq Y$.

In the second case, $|U| = v = R(k_2, k_3; m)$. Set $X'_2 = X_2 \cap U_{\{m\}}$ and $X'_3 = X_3 \cap U_{\{m\}}$. Then $U_{\{m\}} = X'_2 \cup X'_3$. Theorem 8.74 implies that there is either a k_2-element subset C of U with $C_{\{m\}} \subseteq X'_2 \subseteq X_2$ or a k_3-element subset D of U with $D_{\{m\}} \subseteq X'_3 \subseteq X_3$. In either case, the requirements of Theorem 8.75 have been met.

The case where $n = 4$ can now be proved in a similar manner by considering $v = R(k_2, k_3, k_4; m)$ and letting $w = R(k_1, v; m)$.

Continuing in this fashion will establish the theorem for every positive n. □

8.6.3 Exercises

The exercises marked with ★ have detailed solutions in Appendix H.

1. (a) ★ Rewrite the marriage condition using predicates and quantifiers. Let $M(k, i_1, i_2, \ldots, i_k)$ be the claim "$|A_{i_1} \cup A_{i_2} \cup \cdots \cup A_{i_k}| \geq k$." Let $S = \{1, 2, \ldots, n\}$ represent the set of legal subscripts for the A's.

(b) Write the negation of the expression from part (a).

(c) Write the negation of the marriage condition using a style similar to the style in Definition 8.62 on page 503.

2. Determine which of the collections of sets have a system of distinct representatives. If a system of distinct representatives does exist, show one; if no system exists, indicate clearly how the marriage condition is violated.

(a) ★ $A_1 = \{b\}, A_2 = \{b\}, A_3 = \{a, b\}$

(b) ★ $A_1 = \{a, b\}, A_2 = \{b, c\}, A_3 = \{c\}$

(c) $A_1 = \{a, b\}, A_2 = \{a, b, c\}, A_3 = \{a, b, c\}$, $A_4 = \{b, c\}$

(d) $A_1 = \{a, c\}, A_2 = \{a, e\}, A_3 = \{c, e\}, A_4 = \{b, d\}$, $A_5 = \{a, c, e\}$

3. For each collection of sets in Exercise 2, determine whether the collection satisfies the enhanced marriage condition.

4. Determine which of the collections of sets have a system of distinct representatives. If a system of distinct representatives does exist, show one; if no system exists, indicate clearly how the marriage condition is violated.

(a) $A_1 = \{x, y\}, A_2 = \{w, y\}, A_3 = \{z\}, A_4 = \{w, x\}$

(b) $A_1 = \{a, b, d\}, A_2 = \{e, f, g\}, A_3 = \{a, e, f\}$, $A_4 = \{d, g\}, A_5 = \{c, e\}$

(c) $A_1 = \{w, x, y, z\}, A_2 = \{x, y\}, A_3 = \{z\}, A_4 = \{w, x\}$, $A_5 = \{w, z\}$

(d) $A_1 = \{a, b, f\}, A_2 = \{a, c, e\}, A_3 = \{c, e\}$, $A_4 = \{b, d\}, A_5 = \{a, c, e\}, A_6 = \{c, e\}$

5. ★ For each collection of sets in Exercise 4, determine whether the collection satisfies the enhanced marriage condition.

6. A small rural high school has five extracurricular clubs. The clubs and the club members are listed in the following table.

Math	Honors	Service	Yearbook	Poetry
Don	Angus	Bart	Bart	Bart
Effie	Carla	Don	Don	Effie
	Don	Effie		

The principal wishes to form a "club council" that will contain one representative from each club. She does not want two clubs represented by the same student (otherwise the council could consist of just Bart and Don). Is this possible? If it is, show a list of five representatives; otherwise, explain why it is not possible.

7. For each collection of sets, perform the inductive step in the proof of Theorem 8.63. Indicate which case applies.

(a) ★ $A_1 = \{b, c\}$, $A_2 = \{b\}$, $A_3 = \{a, b\}$

(b) $A_1 = \{a\}$, $A_2 = \{b\}$, $A_3 = \{c\}$

(c) $A_1 = \{a, d\}$, $A_2 = \{b\}$, $A_3 = \{a, b\}$, $A_4 = \{b, c\}$

(d) $A_1 = \{b, c\}, A_2 = \{a, b\}, A_3 = \{a, c, d\}, A_4 = \{a, b, d\}$

(e) $A_1 = \{a, b, d\}$, $A_2 = \{b, e\}$, $A_3 = \{a, b\}$, $A_4 = \{b, c\}$, $A_5 = \{a, d, e\}$

8. Each of the following collections of sets has a system of distinct representatives. In each case, determine the lower bound asserted by Corollary 8.64 and then determine how many different systems of distinct representatives actually exist.

(a) ★ $A_1 = \{a, b\}$, $A_2 = \{a, b\}$

(b) ★ $A_1 = \{a, b, c\}$, $A_2 = \{b, c\}$, $A_3 = \{a, c\}$

(c) $A_1 = \{x, y, z\}$, $A_2 = \{x, y\}$

(d) $A_1 = \{a, b\}$, $A_2 = \{a, c\}$, $A_3 = \{b, c\}$

9. Each of the following collections of sets has a system of distinct representatives. In each case, determine the lower bound asserted by Corollary 8.64 and then determine how many different systems of distinct representatives actually exist.

(a) $A_1 = \{a, b, c\}$, $A_2 = \{a, b, c\}$

(b) $A_1 = \{x\}$, $A_2 = \{y\}$

(c) $A_1 = \{a, b, c, d\}$, $A_2 = \{a, b, c, d\}$

(d) $A_1 = \{a, b, c\}$, $A_2 = \{a, c, d\}$, $A_3 = \{b, c\}$, $A_4 = \{a, c\}$

10. Let n and k be positive integers. Suppose you are given $k \cdot n$ cards; each card is marked with a number from 1 to n such that each number is represented k times. You shuffle and deal k cards to each of n people. Is it possible for each person to lay down one card so that every number from 1 to n is given once?

(a) Show that for $k = 1$ or $k = 2$, the answer is yes.

(b) Is it true in general? Justify your answer.

11. Let $S = \{n \in \mathbb{Z} \mid n > 1\}$. Two integers in S are either relatively prime, or they share at least one common prime factor. How large a collection of integers in S are necessary to ensure that there exists a subset of four integers that are either mutually relatively prime, or in which every pair shares a common prime factor?

12. Prove Corollary 8.64. Relabel the A's so that $t = |A_1| \leq |A_2| \leq \cdots \leq |A_n|$. Choose any $r_1 \in A_1$ and set $B_j = A_j - \{r_1\}$, for $2 \leq j \leq n$.

(a) Show that $\{B_2, B_3, \ldots, B_n\}$ satisfies the marriage condition. [*Hint*: Suppose $|B_{i_1} \cup \cdots \cup B_{i_k}| < k$. Show that $|A_1| = |A_{i_1}| = |A_{i_2}| = \cdots = |A_{i_k}| = 1$ and derive a contradiction.]

(b) There are t choices for r_1. Since $\{B_2, B_3, \ldots, B_n\}$ satisfies the marriage condition, part (a) can be repeated. [*Hint*: Rename and reorder the B's as a new collection of A's, where $t - 1 \leq |A_1| \leq |A_2| \leq \cdots \leq |A_{n-1}|$.]

13. A note after the statement of Corollary 8.64 claimed that the second case results in a higher minimum value. Let the t values in the two cases be $t_1 < n$ and $t_2 \geq n$. Write $t_1 = n - r$ where $1 \leq r < n$ and $t_2 = n + s$ where $0 \leq s$. Prove that $t_1! < \frac{t_2!}{(t_2-n)!}$

14. Each of the following statements is either true (always) or false (at least sometimes). Determine which option applies for each statement and provide adequate explanation for your choice.

(a) ★ The sets $\{A_1, A_2, \ldots, A_n\}$ satisfy the marriage condition if for each k there exists a subcollection, $\{A_{i_1}, A_{i_2}, \ldots, A_{i_k}\} \subseteq \{A_1, A_2, \ldots, A_n\}$, such that $|A_{i_1} \cup A_{i_2} \cup \cdots \cup A_{i_k}| \geq k$.

(b) The marriage condition is a necessary and sufficient condition for a collection of sets to have a system of distinct representatives.

(c) The enhanced marriage condition is easier to satisfy than the marriage condition.

(d) $R(j, k; 1) = R(j, k)$

(e) ★ Table 8.52 indicates that $43 \leq R(5, 5) \leq 49$. This means that any set of size 49 or larger will satisfy the (5, 5) Ramsey condition, but some sets of sizes 43–48 will also satisfy the (5, 5) Ramsey condition. However, no set of size 42 or less will satisfy the (5, 5) Ramsey condition.

(f) The m in the notation $(j, k; m)$ indicates that the $(j, k; m)$ Ramsey condition is an assertion about partitions of the m-element subsets of some set, S.

15. A large union has members from 30 different industries and many companies within those industries. At the national convention, there will be hundreds of delegates from many companies and from all the associated industries. Is the following claim true or false? Give evidence for your answer.

As long as the convention contains a predetermined number of delegates, there will always be a set of 20 people such that one of the following is true:

- All 20 work at the same company.
- All 20 work at different companies in the same industry.
- All 20 work in different industries.

16. ★ The June 22, 1993 Ann Landers advice column contained a letter from a reader who had read elsewhere that

> Two professors ... finally have learned the answer to a question that has baffled scientists for 63 years. ... If you are having a party and want to invite at least four people who know each other and [the word *or* should have been used here] five who don't, how many people should you invite? The answer is 25.

The reader asked why all the time, computing resources, and money were used to answer this question instead of being used to help feed starving children around the world.

Ann Landers wrote that there must be more to this news item than just the party question. Perhaps there were more significant applications. Ann then asked any reader with more information to contact her.

If the incorrect conjunction *and* is replaced by the proper *or*, the statement proved by the mathematicians (Stanislaw Radziszowski and Brendan McKay) showed that $R(4, 5) = 25$.

Was the writer justified in criticizing these mathematicians for spending time determining a Ramsey number? Write a paragraph defending your opinion.[54]

17. Complete the proof of Proposition 8.69 (page 511).

18. Prove case 2 of Theorem 8.70 (page 512).

19. Prove the following proposition. [*Hint*: If there are no two-element subsets of S, then all of these nonexistent subsets can be considered as members of any set. Investigate the two cases: $|S| = 1$ and $|S| \geq 2$.]

> **PROPOSITION 8.76** $\boldsymbol{R(j, 1; 2)}$ ***and*** $\boldsymbol{R(1, j; 2)}$
>
> Let S be any nonempty set. If $j \geq 1$ and $k \geq 1$, then
>
> **1.** $R(j, 1; 2) = 1$
>
> **2.** $R(1, k; 2) = 1$

20. Without using Table 8.52, outline the steps in a proof that $9 \leq R(3, 4) \leq 10$.

21. Prove that $R(m, k; m) = k$ for $k \geq m \geq 1$.

22. Prove that $R(k_1, k_2, \ldots, k_n; 1) = k_1 + k_2 + \cdots k_n - n + 1$.

23. Finish Case (2) in the proof of Theorem 8.74 (page 513).

8.7 Quick Check Solutions

Quick Check 8.1

1. There are $9! = 362880$ distinct arrangements—far too many to try them all.

2. Before presenting a formal proof, the following heuristic argument shows that the proposed value is reasonable. Whatever the common sum, S, is, it should be larger than the sum of the n smallest numbers in the magic square: $S > (1+2+\cdots+n) = \frac{n(n+1)}{2}$. It should also be smaller than the sum of the n largest numbers in the magic square: $S < (n^2+(n^2-1)+\cdots+(n^2-(n-1)) = n \cdot n^2 - (0+1+\cdots(n-1)) = n^3 - \frac{(n-1)n}{2}$. It seems reasonable that S should be close to the average of these two extremes: $S \simeq \frac{1}{2} \cdot \left[\frac{n(n+1)}{2} + \left(n^3 - \frac{(n-1)n}{2}\right)\right] = \frac{n(n^2+1)}{2}$.

   Proof: Let the common sum be S. Then each row has sum S and there are n rows. Thus, adding the rows together produces a total of nS. On the other hand, adding each individual entry in the magic square will add each number in $\{1, 2, \cdots, n^2\}$ to the total: $\sum_{i=1}^{n^2} i = \frac{n^2(n^2+1)}{2}$. Equating these two totals leads to $S = \frac{n(n^2+1)}{2}$. □

3. The possible subsets are

1	5	9
1	6	8
2	4	9
2	5	8
2	6	7
3	4	8
3	5	7
4	5	6

[54] There was active discussion in the USENET newsgroups sci.math and sci.math.research about how to respond to the letter. Some mathematicians sent responses, but none were published. See the postscript at the end of [41] for one of those responses and a larger extract from the original letter to Ann Landers.

First, observe that a 3-by-3 magic square requires eight subsets of size 3 (3 rows, 3 columns, 2 diagonals). There are only eight possible subsets of size 3 that have sum 15, so all must be used in any magic square.

Next, notice that 1, 3, 7, and 9 only appear in two of the subsets. This means that they cannot be at a corner (which requires the element to be in three subsets since diagonals are important) nor in the center. Also, 2, 4, 6, and 8 each appear in three subsets, so they must each be in a corner. The number 5 is the only one that appears in four of the subsets, so it must be the center element (which is in four subsets of size 3).

4. The subset $\{1, 5, 9\}$ can appear in one of four possible ways (the second row or the second column, each in two possible orders since 5 must be the middle element). Once that subset has been placed, the subset $\{1, 6, 8\}$ has only two possible placements (perpendicular to $\{1, 5, 9\}$, with 1 at the center). There is then only one way to fill in each of the two diagonals. The remaining entries of the magic square are also then uniquely determined, so there are at most eight distinct magic squares. It is easy to see that all eight possibilities are actual magic squares.

5. The eight 3-by-3 magic squares are as follows.

```
6 7 2     8 3 4     2 7 6     4 3 8
1 5 9     1 5 9     9 5 1     9 5 1
8 3 4     6 7 2     4 3 8     2 7 6

6 1 8     8 1 6     2 9 4     4 9 2
7 5 3     3 5 7     7 5 3     3 5 7
2 9 4     4 9 2     6 1 8     8 1 6
```

Quick Check 8.2

1. Use Definition 8.1. There is only one way to partition n with only 1 summand: $n = n$.

2. Use the boundary condition $p(n, n) = 1$ to fill in the main diagonal. Then use

$$p(n, 1) = p(n\text{-}1, 0) + p(n\text{-}1, 1) = 0 + p(n\text{-}1, 1) = p(n\text{-}1, 1)$$

to fill in column 1. Now work by rows, starting with $n = 3$. The results are shown in Table 8.53.

TABLE 8.53 The Values of $p(n)$ and $p(n, k)$ for $n, k \leq 6$

	$p(n, k)$						$p(n)$
$n \backslash k$	1	2	3	4	5	6	
1	1	–	–	–	–	–	1
2	1	1	–	–	–	–	2
3	1	1	1	–	–	–	3
4	1	2	1	1	–	–	5
5	1	2	2	1	1	–	7
6	1	3	3	2	1	1	11

TABLE 8.54 The Values of $S(n, k)$ for $n, k \leq 4$

	$S(n, k)$			
$n \backslash k$	1	2	3	4
1	1	–	–	–
2	1	1	–	–
3	1	3	1	–
4	1	7	6	1

Quick Check 8.3

1. Notice that $S(n, n) = 1$ because each of the n containers must receive at least one of the n objects. This can only be accomplished by placing exactly one object in each container. This determines the main diagonal of Table 8.54. Also, $S(n, 1) = 1$ because all n objects must be placed into the single container. The remaining values can be found by listing all possible distributions. The main restrictions are that

every container must receive at least one object and containers are indistinguishable (so there is no order within distributions). The values for $S(3,2)$, $S(4,2)$, and $S(4,3)$ can be calculated by listing all distinct possibilities, as shown in the following lists.

$n \backslash k$	1	2	3	4
	$S(n,k)$			
1	1	–	–	–
2	1	1	–	–
3	1	3	1	–
4	1	7	6	1

$S(3,2)$
$\{\{o_1, o_2\}, \{o_3\}\}$
$\{\{o_1, o_3\}, \{o_2\}\}$
$\{\{o_2, o_3\}, \{o_1\}\}$

$S(4,3)$
$\{\{o_1, o_2\}, \{o_3\}, \{o_4\}\}$
$\{\{o_1, o_3\}, \{o_2\}, \{o_4\}\}$
$\{\{o_1, o_4\}, \{o_2\}, \{o_3\}\}$
$\{\{o_2, o_3\}, \{o_1\}, \{o_4\}\}$
$\{\{o_2, o_4\}, \{o_1\}, \{o_3\}\}$
$\{\{o_3, o_4\}, \{o_1\}, \{o_2\}\}$

$S(4,2)$
$\{\{o_1, o_2, o_3\}, \{o_4\}\}$
$\{\{o_1, o_2, o_4\}, \{o_3\}\}$
$\{\{o_1, o_3, o_4\}, \{o_2\}\}$
$\{\{o_2, o_3, o_4\}, \{o_1\}\}$
$\{\{o_1, o_2\}, \{o_3, o_4\}\}$
$\{\{o_1, o_3\}, \{o_2, o_4\}\}$
$\{\{o_1, o_4\}, \{o_2, o_3\}\}$

Quick Check 8.4

1. In this problem, the objects are the four CDs, which are distinguishable. The containers are the three friends, which are also distinguishable.

   (a) Since every friend will receive at least one CD, there are $3!S(4,3) = 6 \cdot 6 = 36$ ways to distribute the CDs.

   Here are some incorrect solutions:

   i. The first three CDs can be distributed in $4 \cdot 3 \cdot 2$ ways (four choices for friend A's first CD, then three choices for friend B's first CD, then two choices for friend C's first CD). There are then three choices for the friend who gets the last CD. Thus, there are $24 \cdot 3 = 72$ ways to distribute the CDs. The error is that some options are double counted. For example, I can give CD W to A, CD X to B, and CD Y to C. I can then give CD Z to A. But this same distribution could be generated by giving Z to A, X to B, and Y to C in the first round.

   ii. There are three choices for the friend who will receive two CDs, so there are three possible distributions. This is incorrect because it assumes that the CDs are indistinguishable. A correct solution could start with this observation and proceed to notice that the favored friend has $\binom{4}{2}$ ways to receive the two CDs, and then the remaining CDs can be distributed in two ways. Thus, there are $3 \cdot \binom{4}{2} \cdot 2 = 36$ ways to distribute the CDs.

   (b) This is easy: There are $3^4 = 81$ ways to distribute the CDs.

2. The objects (Sacagawea dollars) are indistinguishable, but the containers (nieces) are not. The answer depends on whether I want to risk the wrath of a niece who does not receive at least one dollar. If I choose the safe course, there will be $\binom{4-1}{3-1} = 3$ ways to distribute the coins. This makes sense: There are three ways to choose which niece will receive two coins (there is still some risk of favoritism involved).

   If I choose to ignore the possibility of hurt feelings, there will be $\binom{3+4-1}{4} = 15$ ways to distribute the coins: I can give all four to one niece (three ways), give three coins to one niece and 1 coin to another (six ways), give two coins each to two nieces (three ways), or give one niece two coins and the other nieces 1 coin each (three ways).

3. The problem statement implies that harvesting need not occur in every field on the first day. The objects (sons) are distinguishable, but the containers (fields) are indistinguishable. There are $S(4,1) + S(4,2) + S(4,3) = 1 + 7 + 6 = 14$ ways to assign fields on the first day.

Quick Check 8.5

1. There are 12 distinct Latin squares of order 3.

$$\begin{array}{ccc}1&2&3\\2&3&1\\3&1&2\end{array}\qquad\begin{array}{ccc}1&2&3\\3&1&2\\2&3&1\end{array}\qquad\begin{array}{ccc}1&3&2\\2&1&3\\3&2&1\end{array}\qquad\begin{array}{ccc}1&3&2\\3&2&1\\2&1&3\end{array}$$

$$\begin{array}{ccc}2&1&3\\1&3&2\\3&2&1\end{array}\qquad\begin{array}{ccc}2&1&3\\3&2&1\\1&3&2\end{array}\qquad\begin{array}{ccc}2&3&1\\1&2&3\\3&1&2\end{array}\qquad\begin{array}{ccc}2&3&1\\3&1&2\\1&2&3\end{array}$$

$$\begin{array}{ccc}3&1&2\\1&2&3\\2&3&1\end{array}\qquad\begin{array}{ccc}3&1&2\\2&3&1\\1&2&3\end{array}\qquad\begin{array}{ccc}3&2&1\\1&3&2\\2&1&3\end{array}\qquad\begin{array}{ccc}3&2&1\\2&1&3\\1&3&2\end{array}$$

2. There are four standardized Latin squares of order 4.

$$\begin{array}{cccc}1&2&3&4\\2&1&4&3\\3&4&1&2\\4&3&2&1\end{array}\qquad\begin{array}{cccc}1&2&3&4\\2&1&4&3\\3&4&2&1\\4&3&1&2\end{array}\qquad\begin{array}{cccc}1&2&3&4\\2&3&4&1\\3&4&1&2\\4&1&2&3\end{array}\qquad\begin{array}{cccc}1&2&3&4\\2&4&1&3\\3&1&4&2\\4&3&2&1\end{array}$$

Quick Check 8.6

Both problems are just duals of claims that have already been proved. The solutions here have interchanged the terms *point* and *line* and the terms *on* and *contains*. These new proofs are valid independent of the duality used to write them.

1. Let p be any point in the finite projective plane. Axiom FPP3 asserts the existence of four points, p_1, p_2, p_3, p_4, no three of which are on a common line. At least three of these points are distinct from p. Assume, with suitable renaming if necessary, that $p \notin \{p_1, p_2, p_3\}$. Let the common line containing both p and p_i be denoted L_i, for $i = 1, 2, 3$. If the lines L_i are all distinct, the claim is true. Otherwise, there must still be two distinct L_i (since p_1, p_2, and p_3 are on no common line). Assume that $L_1 \neq L_2$. There must be at least one other line, L'_1 containing p_1 and at least one other line, L'_2, containing p_2. They must contain a common point, p', with $p' \neq p$. Points p and p' must be on a common line, L (FPP1). If $L = L_1$, then p_1 and p' would both be on L_1 and on L'_1, contradicting FPP2. Similarly, $L \neq L_2$. Therefore, p must be on the three distinct lines L_1, L_2, and L.

2. Suppose that there are k distinct lines in $\mathcal{F}$. Since each pair of lines contains a common point, there must be $\binom{k}{2}$ such pairs. However, this collection of pairs of lines overcounts the points. In fact, each point has been counted once for each pair of lines it is common to. There are $n + 1$ lines containing each point, so each point has been counted $\binom{n+1}{2}$ times. The number of points is thus

$$\frac{\binom{k}{2}}{\binom{n+1}{2}} = \frac{k \cdot (k-1)}{(n+1) \cdot n}$$

By the duality principle, if there are k lines, then there must also be k points. Hence, $k = \frac{k \cdot (k-1)}{(n+1) \cdot n}$, and consequently, there are $k = n^2 + n + 1$ points. This proves statement 5.

Quick Check 8.7

1. **Step 1** The construction requires a set of $n - 1 = 1$ mutually orthogonal Latin squares of order 2. The one shown will work.

$$\begin{array}{cc} 1 & 2 \\ 2 & 1 \end{array}$$

The matrix A is shown next.

$$\begin{array}{cccc} 1 & 1 & 2 & 2 \\ 1 & 2 & 1 & 2 \\ 1 & 2 & 2 & 1 \end{array}$$

Step 2 Now add column labels.

1	2	3	4
1	1	2	2
1	2	1	2
1	2	2	1

Step 3 The six partial lines are shown next.

	i	
r	1	2
1	{1,2}	{3,4}
2	{1,3}	{2,4}
3	{1,4}	{2,3}

Step 4 The finite projective plane, $\mathcal{F}$, has the seven points $\{1, 2, 3, 4, \infty_1, \infty_2, \infty_3\}$ and the seven lines shown in the next table.

	i	
r	1	2
1	$\{1,2,\infty_1\}$	$\{3,4,\infty_1\}$
2	$\{1,3,\infty_2\}$	$\{2,4,\infty_2\}$
3	$\{1,4,\infty_3\}$	$\{2,3,\infty_3\}$
	$\{\infty_1, \infty_2, \infty_3\}$	

The lines and points can be shown in a diagram that should look familiar.

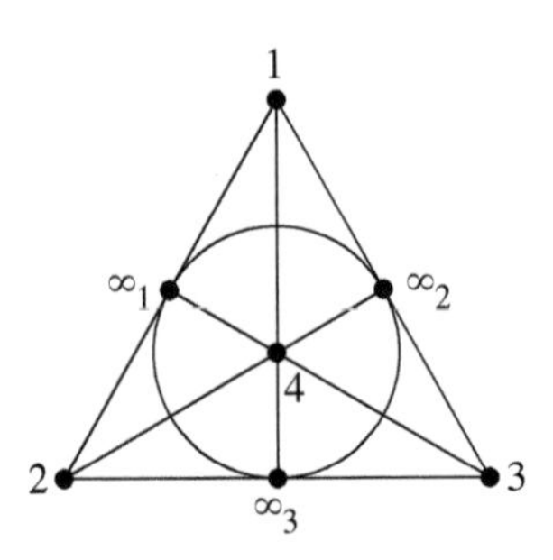

Quick Check 8.8

1. **Step 1**

5	{1, 2, 5}	{3, 4, 5}
6	{1, 3, 6}	{2, 4, 6}
7	{1, 4, 7}	{2, 3, 7}
	{5, 6, 7}	

Step 2

$T =$

1	2
{1, 2}	{3, 4}
{1, 3}	{2, 4}
{1, 4}	{2, 3}

Step 3

$A =$

1	2	3	4
1	1	2	2
1	2	1	2
1	2	2	1

Step 4

$$A' = 1 \quad 2 \quad 2 \quad 1 \qquad L^1 = \begin{matrix} 1 & 2 \\ 2 & 1 \end{matrix}$$

Quick Check 8.9

1. $v = 9, b = 12, r = 4, k = 3$, and $\lambda = 1$
2. $v = 4, b = 4, r = 3, k = 3$, and $\lambda = 2$
3. The incidence matrix is listed. Notice that every column has three 1s (since $k = 3$) and every row has three 1s (since $r = 3$). Also, every pair of rows has 1s in two common columns (since $\lambda = 2$).

	C_1	C_2	R_1	R_2
1	1	1	1	0
2	1	0	1	1
3	1	1	0	1
4	0	1	1	1

4. Denote the columns as M_1, M_2, M_3, and so on. The incidence matrix does not need to contain row and column labels. They are added here for clarity. Notice that every column has three 1s (since $k = 3$) and every row has four 1s (since $r = 4$). Also, every pair of rows has exactly one common column with a 1 (since $\lambda = 1$).

	M_1	M_2	M_3	Tu_1	Tu_2	Tu_3	Th_1	Th_2	Th_3	F_1	F_2	F_3
1	1	0	0	1	0	0	1	0	0	1	0	0
2	1	0	0	0	1	0	0	1	0	0	1	0
3	1	0	0	0	0	1	0	0	1	0	0	1
4	0	1	0	1	0	0	0	0	1	0	1	0
5	0	1	0	0	1	0	1	0	0	0	0	1
6	0	1	0	0	0	1	0	1	0	1	0	0
7	0	0	1	1	0	0	0	1	0	0	0	1
8	0	0	1	0	1	0	0	0	1	1	0	0
9	0	0	1	0	0	1	1	0	0	0	1	0

Quick Check 8.10

1. (a) The blocks are the rows of girls; the varieties are the 15 girls. There are 5 rows on each of 7 days for a total of 35 rows. Thus $v = 15$ and $b = 35$. Each row has 3 girls so $k = 3$. Each girl must walk every day, so $r = 7$. Finally, every girl is to walk in a row with every other girl once per week, so $\lambda = 1$. The BIBD needs to be a (15, 35, 7, 3, 1)-design.

   (b) $35 \cdot 3 = 15 \cdot 7$ and $7 \cdot 2 = 1 \cdot 14$

Quick Check 8.11

1. The parameters of $\overline{D}$ are (7, 3, 1).

$\overline{B_0}$	$\overline{B_1}$	$\overline{B_2}$	$\overline{B_3}$	$\overline{B_4}$	$\overline{B_5}$	$\overline{B_6}$
1	1	1	2	2	3	5
2	3	4	3	4	4	6
5	6	7	7	6	5	7

2. The parameters of D' are (4, 6, 3, 2, 1).

B'_0	B'_1	B'_3	B'_4	B'_5	B'_6
3	2	5	3	2	2
6	5	6	5	6	3

3. The parameters of D^* are (3, 6, 4, 2, 2). Notice that D^* is a twofold replication of a (3, 3, 2, 2, 1)-design.

B^*_0	B^*_1	B^*_3	B^*_4	B^*_5	B^*_6
4	4	1	1	1	1
7	7	4	7	7	4

Quick Check 8.12

1. The blocks are the columns of the following table. The varieties are the numbers 1 through 7.

B_1	B_2	B_3	B_4	B_5	B_6	B_7
1	1	1	2	2	3	5
2	3	4	3	4	4	6
5	6	7	7	6	5	7

2. The incidence matrix (with helpful labels) is unambiguous if the blocks are labeled in the sorted order used in part 1.

	B_1	B_2	B_3	B_4	B_5	B_6	B_7
1	1	1	1	0	0	0	0
2	1	0	0	1	1	0	0
3	0	1	0	1	0	1	0
4	0	0	1	0	1	1	0
5	1	0	0	0	0	1	1
6	0	1	0	0	1	0	1
7	0	0	1	1	0	0	1

Item	W	X	Y	Z
Benefit	3	4	11	1
Volume	4	3	7	2

Quick Check 8.13

1.

w	W	X	Y	Z	$B(w)$
0	0	0	0	0	0
1	0	0	0	0	0
2	0	0	0	$1+B(0)=1$	1
3	0	$4+B(0)=4$	0	$1+B(1)=1$	4
4	$3+B(0)=3$	$4+B(1)=4$	0	$1+B(2)=2$	4
5	$3+B(1)=3$	$4+B(2)=5$	0	$1+B(3)=5$	5
6	$3+B(2)=4$	$4+B(3)=8$	0	$1+B(4)=5$	8
7	$3+B(3)=7$	$4+B(4)=8$	$11+B(0)=11$	$1+B(5)=6$	11
8	$3+B(4)=7$	$4+B(5)=9$	$11+B(1)=11$	$1+B(6)=9$	11
9	$3+B(5)=8$	$4+B(6)=12$	$11+B(2)=12$	$1+B(7)=12$	12
10	$3+B(6)=11$	$4+B(7)=15$	$11+B(3)=15$	$1+B(8)=12$	15

The maximum total benefit is 15. There are two choices for the item to add in the final row: X or Y. Suppose X is chosen. Then the next item will be chosen by looking at row $10-3=7$. The only item that matches $B(7)$ is Y, so a Y is added. This leads to row $7-7=0$, so no other items are added.

If Y is chosen in row 10, the next item will be determined by row $10-7=3$. In row 3, the item that matches $B(3)$ is X, so an X is added to the knapsack. The next row to examine is row 0, so no other items are added.

Both routes lead to the same solution: one X and one Y. (Other examples may have several distinct solutions.)

Quick Check 8.14

1. The tables follow:

	T					K			
w	W	X	Y	Z	$B(w)$	W	X	Y	Z
0	0	0	0	0	0	**1**	**2**	**1**	**1**
1	0	0	0	0	0	0	0	0	0
2	0	0	0	$1+B(0)=1$	1	0	0	0	1
3	0	$4+B(0)=4$	0	$1+B(1)=1$	4	0	1	0	0
4	$3+B(0)=3$	$4+B(1)=4$	0	$1+B_Z(2)=1$	4	0	1	0	0
5	$3+B(1)=3$	$4+B(2)=5$	0	$1+B(3)=5$	5	0	1	0	1
6	$3+B(2)=4$	$4+B(3)=8$	0	$1+B(4)=5$	8	0	2	0	0
7	$3+B(3)=7$	$4+B(4)=8$	$5+B(0)=5$	$1+B_Z(5)=5$	8	0	2	0	0
8	$3+B(4)=7$	$4+B(5)=9$	$5+B(1)=5$	$1+B(6)=9$	9	0	2	0	1
9	$3+B(5)=8$	$4+B_X(6)=9$	$5+B(2)=6$	$1+B(7)=9$	9	0	2	0	1
10	$3+B(6)=11$	$4+B_X(7)=11$	$5+B(3)=9$	$1+B_Z(8)=9$	11	1	2	0	0

Item	W	X	Y	Z
Benefit	3	4	5	1
Volume	4	3	7	2
Quantity	1	2	1	1

An optimal packing is one W and two X's, for a total benefit of 11. Notice that an unbounded problem would pack three X's, for a total benefit of 12.

In row 4, $B_Z(2)$ is used in place of $B(2)$ because $K_{2,4}=1$, indicating that an item Z is already used for the optimal packing in row 2.

The recursion tables are listed next. B_{XX} means remove both copies of item X, B_{XZ} means remove one X and one Z. B_{XXZ} means remove both copies of item X and the item Z. Note that $B_{XZ}(w)=B_{ZX}(w)$.

To build the table for B_X, the only change to the knapsack problem is that there is now only one X. The capacity (10), the benefits (3, 4, 5, 1), and the volumes (4, 3, 7, 2) do not change. The highest row we need in the table for B_Z is row 8 and for B_X the highest needed is row 7.

When determining the values for K after a recursion, add one to the appropriate column of the recursion table's K row. For example, in row 10 of the table for $B(w)$, assume we decide the X column wins, the K values are determined by adding a 1 to the X entry in the K table from row 7 of the table for B_X (changing 1 1 0 0 into 1 2 0 0).

	T					**K**			
w	W	X	Y	Z	$B_X(w)$	W	X	Y	Z
0	0	0	0	0	0	**1**	**1**	**1**	**1**
1	0	0	0	0	0	0	0	0	0
2	0	0	0	$1 + B_X(0) = 1$	1	0	0	0	1
3	0	$4 + B_X(0) = 4$	0	$1 + B_X(1) = 1$	4	0	1	0	0
4	$3 + B_X(0) = 3$	$4 + B_X(1) = 4$	0	$1 + B_{XZ}(2) = 1$	4	0	1	0	0
5	$3 + B_X(1) = 3$	$4 + B_X(2) = 5$	0	$1 + B_X(3) = 5$	5	0	1	0	1
6	$3 + B_X(2) = 4$	$4 + B_{XX}(3) = 5$	0	$1 + B_X(4) = 5$	5	0	1	0	1
7	$3 + B_X(3) = 7$	$4 + B_{XX}(4) = 7$	$5 + B_X(0) = 5$	$1 + B_{XZ}(5) = 5$	7	1	1	0	0

	T				**K**		
w	W	X	Y	$B_Z(w)$	W	X	Y
0	0	0	0	0	**1**	**2**	**1**
1	0	0	0	0	0	0	0
2	0	0	0	0	0	0	0
3	0	$4 + B_Z(0) = 4$	0	4	0	1	0
4	$3 + B_Z(0) = 3$	$4 + B_Z(1) = 4$	0	4	0	1	0
5	$3 + B_Z(1) = 3$	$4 + B_Z(2) = 4$	0	4	0	1	0
6	$3 + B_Z(2) = 3$	$4 + B_Z(3) = 8$	0	8	0	2	0
7	$3 + B_Z(3) = 7$	$4 + B_Z(4) = 8$	$5 + B_Z(0) = 5$	8	0	2	0
8	$3 + B_Z(4) = 7$	$4 + B_Z(5) = 8$	$5 + B_Z(1) = 5$	8	0	2	0

	T				**K**		
w	W	X	Y	$B_{XZ}(w)$	W	X	Y
0	0	0	0	0	**1**	**1**	**1**
1	0	0	0	0	0	0	0
2	0	0	0	0	0	0	0
3	0	$4 + B_{XZ}(0) = 4$	0	4	0	1	0
4	$3 + B_{XZ}(0) = 3$	$4 + B_{XZ}(1) = 4$	0	4	0	1	0
5	$3 + B_{XZ}(1) = 3$	$4 + B_{XZ}(2) = 4$	0	4	0	1	0

w	T: W	T: Y	T: Z	$B_{XX}(w)$	K: W	K: Y	K: Z
0	0	0	0	0	**1**	**1**	**1**
1	0	0	0	0	0	0	0
2	0	0	$1 + B_{XX}(0) = 1$	1	0	0	1
3	0	0	$1 + B_{XX}(1) = 1$	1	0	0	1
4	$3 + B_{XX}(0) = 3$	0	$1 + B_{XXZ}(2) = 1$	3	1	0	0

For B_{XXZ}, rather than do the full table, we can make some simple observations. If the knapsack only has one item W and one item Y available, the knapsack must have volume at least 4 to have a nonzero benefit. If $4 \leq w \leq 6$, only a single item W will fit, for a benefit of 3. If $7 \leq w \leq 10$ the optimal choice is to add the item Y, for a benefit of 5.

Quick Check 8.15

1. The knapsack holds 100 units of volume. The benefit-to-volume ratios are shown in the next table.

Item	W	X	Y	Z
Benefit	30	40	50	10
Volume	40	35	70	20
Quantity	1	3	1	1
B/V Ratio	$\frac{30}{40} = .75$	$\frac{40}{30} \simeq 1.143$	$\frac{50}{70} \simeq .714$	$\frac{10}{20} = .5$

The heuristic will treat item X as the first item, followed by W, then Y, and finally Z. The inequality will compare X to W ($\hat{q}_2 = \min\left\{3, \left\lfloor \frac{100}{35} \right\rfloor\right\} = 2$):

$$40 \cdot 2 \geq 30\left(\frac{100}{40}\right)$$

$$80 \geq 75$$

The inequality is true, so the heuristic ensures that at least one unit of item X should be packed. The knapsack will still have 65 units of volume left to fill. A second application of the heuristic will compare X to W (notice that q_2 needs to be reduced by 1 so $\hat{q} = \min\left\{2, \left\lfloor \frac{65}{35} \right\rfloor\right\} = 1$):

$$40 \cdot 1 \not\geq 30\left(\frac{65}{40}\right) \quad \text{which reduces to } 40 \not\geq 48.75$$

The inequality is false, so the heuristic provides no additional suggestions. An optimal solution is to pack two item X's and one item Z, for a total benefit of 90. The reason that the heuristic fails to suggest packing a second item X is because Theorem 8.43 only checks to see if $\hat{q}$ item Xs is better than $\frac{v}{v_2}$ item Ws. It does not properly compensate for the fractional part of $\frac{v}{v_2}$ ($\frac{65}{40}$ in this case).

Quick Check 8.16

1. Use the three encoding equations.

   (a) The check bits are $x_5 = 1+0+1 = 0$ (mod 2), $x_6 = 1+0+1 = 0$ (mod 2), and $x_7 = 1+1+1 = 1$ (mod 2). The code word is therefore 1101001.

   (b) The check bits are $x_5 = 0+0+1 = 1$ (mod 2), $x_6 = 0+0+1 = 1$ (mod 2), and $x_7 = 0+0+1 = 1$ (mod 2). The code word is therefore 0001111.

2. Use the decoding table.

   (a) The received check bits are $x_5^r = 0$, $x_6^r = 0$, and $x_7^r = 1$. The calculated check bits are $x_5^c = 0+1+1 = 0$ (mod 2), $x_6^c = 0+1+1 = 0$ (mod 2), and $x_7^c = 0+0+1 = 1$ (mod 2). Since the two sets of check bits are the same, the transmission is assumed to be without error. The message is therefore 0011.

   (b) The received check bits are $x_5^r = 0$, $x_6^r = 1$, and $x_7^r = 1$. The calculated check bits are $x_5^c = 0+1+1 = 0$ (mod 2), $x_6^c = 0+1+1 = 0$ (mod 2), and $x_7^c = 0+0+1 = 1$ (mod 2). Since $x_6^r \neq x_6^c$, the transmission is assumed to contain an error. The decoding table indicates that bit x_6 is where the error occurred. The message bits were received unaltered, so the message is therefore 0011.

   (c) The received check bits are $x_5^r = 0$, $x_6^r = 0$, and $x_7^r = 1$. The calculated check bits are $x_5^c = 0+1+1 = 0$ (mod 2), $x_6^c = 1+1+1 = 1$ (mod 2), and $x_7^c = 1+0+1 = 0$ (mod 2). Since $x_6^r \neq x_6^c$ and $x_7^r \neq x_7^c$, the transmission is assumed to contain an error. The decoding table indicates that bit x_1 is where the error occurred. The code word that was sent is therefore assumed to have been 0011001, corresponding to the message 0011.

Quick Check 8.17

1. The eight messages and their corresponding code words are shown in the following table:

Message	Code Word	Message	Code Word
000	000000	100	100110
001	001011	101	101101
010	010101	110	110011
011	011110	111	111000

2. There are three message bits out of a total of six, so the efficiency is $\frac{3}{6} = \frac{1}{2}$.

3. Since the code is linear, the minimum distance, d, of the code is the smallest nonzero weight among the code words (Corollary 8.52). The table of code words indicates that this minimum weight is 3. Theorem 8.54 indicates that the code is 1 error-correcting ($2t + 1 = 3$, so $t = 1$).

Quick Check 8.18

1. The parameters $(23, 2^{12}, 7)$ indicate that $M = 2^{12}$, $n = 23$, and $t = 3$ errors can be corrected. Thus

$$\begin{aligned} 2^{12} \cdot \sum_{i=0}^{3} \binom{23}{i} &= 2^{12} \cdot \left(\binom{23}{0} + \binom{23}{1} + \binom{23}{2} + \binom{23}{3} \right) \\ &= 2^{12} \cdot (1 + 23 + 253 + 1771) \\ &= 2^{12} + 2^{11} = 2^{23} \end{aligned}$$

as required.

Quick Check 8.19

1. The marriage condition is satisfied.

$k = 1$ $|A_1| = 3$, $|A_2| = 1$, and $|A_3| = 1$
$k = 2$ $|A_1 \cup A_2| = 3$, $|A_1 \cup A_3| = 3$, and $|A_2 \cup A_3| = 2$
$k = 3$ $|A_1 \cup A_2 \cup A_3| = 3$

2. The marriage condition is *not* satisfied because $|A_1 \cup A_2 \cup A_3| < 3$.

$k = 1$ $|A_1| = 2$, $|A_2| = 1$, and $|A_3| = 1$
$k = 2$ $|A_1 \cup A_2| = 2$, $|A_1 \cup A_3| = 2$, and $|A_2 \cup A_3| = 2$
$k = 3$ $|A_1 \cup A_2 \cup A_3| = 2$

Quick Check 8.20

1. Single out person A. Place all A's acquaintances in a group and all those A is unacquainted with in another group. There are 5 people that are being placed in groups, so one of the groups must have at least 3 members (by the pigeon-hole principle).
   Suppose the group of acquaintances has 3 or more members. If two of them (say B and C) are also acquainted with each other, then A, B, and C form a set of three mutual acquaintances. Otherwise the 3 or more people in the group of A's acquaintances are all mutually unacquainted.
   On the other hand, suppose the group of people A is unacquainted with contains three or more people. If two of them (again named B and C) are mutually unacquainted, then A, B, and C form a set of three mutually unacquainted people. Otherwise there must be 3 people in the group of those A is unacquainted with who are all mutually acquainted.

Quick Check 8.21

1. $R(j, 3; 3) = j$.
   Suppose $|S| \leq j - 1$. The three-element subsets of S can be partitioned by placing all of them into X, with $Y = \emptyset$. Since S has no j-element subsets, there cannot be a subset $T \subseteq S$ whose three-elements subsets are all in X. On the other hand, there are no three-element subsets $U \subseteq S$ all of whose three-element subsets are in Y. Thus, $R(j, 3; 3) > j - 1$.
   Suppose $|S| = j$. If any one of the three-element subsets of S is placed in Y, then that subset can be used as U. On the other hand, if every three-element subset is placed in X, then let $T = S$. In either case, the $(j, 3; 3)$ Ramsey condition is satisfied. Since this argument works even if $|S| > j$, $R(j, 3; 3) \leq j$.
   Combining the inequalities shows that $R(j, 3; 3) = j$.

8.8 CHAPTER REVIEW

8.8.1 Summary

This chapter provides an introduction to combinatorics by exploring several representative topics. These topics can be broadly organized by using three categories that encompass much of combinatorics: existence, enumeration, and optimization. Many of the topics in this chapter have subtopics in more than one of these broad categories.

The chapter starts with some enumeration topics (Section 8.1). More specifically, it starts with partitions of an integer and occupancy problems. In the course of the discussion, the sets of Stirling numbers of the first and second kinds are introduced. Occupancy problems can be subdivided into eight categories. The categories add new kinds of counting problems to the categories of permutations and combinations (with and without repetition) encountered in Chapter 5.

Section 8.2 discusses two apparantly unrelated topics: Latin squares and finite projective planes. Existence is important in both topics. In fact, the two topics are related because constructions exist that create sets of Latin squares from finite projective planes and that create finite projective planes from an appropriate collection of mutually orthogonal Latin squares. Enumeration also shows up in this section (estimating the number of mutually orthogonal Latin squares of order n). You may find some of the material in this section to be challenging.

Existence is also a key concern in Section 8.3. Balanced incomplete block designs are examples of combinatorial designs (as are Latin squares and finite projective planes). One of the early theorems in the section provides another opportunity to use a combinatorial proof. The basic definitions and some standard properties of BIBDs are presented, as well as some applications in the design of experiments. The rest of the section presents a number of constructions. This section illustrates two distinct aspects of existence. Some of the theorems can be used to show BIBDs with certain parameters cannot exist. There are also constructions that actually create BIBDs with other sets of parameters.

Section 8.4 describes a topic from the combinatorial optimization category. Knapsack problems have optimal solutions, but exhaustively checking all possible arrangements is not feasible in most cases. Rather than providing an analytic determination of an optimal packing, an algorithm is presented that produces an optimal solution.[55]

Section 8.5 is about error-correcting codes. Much of the work in error-correcting codes has been accomplished by using ideas found in courses about linear algebra and algebraic structures. The material presented here is more combinatorial in nature. The basic definitions and some introduction to the notion of perfect codes are perhaps the most significant portions of the section.

The final section in the chapter (Section 8.6) contains material that is about sets of sets. The section contains two topics. The first topic, systems of distinct representatives, determines whether it is possible to choose a distinct element from each set in a collection. The second topic, Ramsey numbers, investigates the minimum size of a set necessary to guarantee the existence of a property related to some of its subsets. This section contains some fairly advanced material.

This chapter contains some fairly abstract, theoretical material. There is also some fairly concrete material, either in the form of identifying the proper formula or in the form of constructions, often small enough to list in a table or draw in a diagram. You will need to spend extra time reviewing the more abstract material. Gaining an intuitive feel for the definitions and theorems will help you master the material. You may find it very helpful to form a study group to discuss this chapter. You will benefit from the opportunity to express ideas verbally to others and will also benefit from their insights.

[55] An example of analytically determining an optimal value is the familiar max-min problems in calculus. A function is found that describes the problem, and then the first derivative is used to find the critical points. There are several other derivative-based tests that can be used to determine which of the critical points might provide an optimal solution to the original problem.

8.8.2 Notation

Notation	Page	Brief Description
$p(n)$	419	the number of partitions of n
$p(n, k)$	419	the number of partitions of n that contain exactly k summands
O C D I $\emptyset$ $\neg\emptyset$	423	in occupancy problems: O—object, C—container, D—distinguishable, I—indistinguishable, $\emptyset$—containers may be empty, $\neg\emptyset$—containers may not be empty
$S(n, k)$	424	Stirling number of the second kind
$s(n, k)$	432	Stirling number of the first kind
$(x)_n$	430	the falling factorial
L^j	438	one of a set of mutually orthogonal Latin squares
p_j	442	a point in a finite projective plane
$L_j = \{p_1, p_2, \ldots, p_k\}$	442	a line in a finite projective plane
$m(n)$	439	the maximum number of mutually orthogonal Latin squares of order n
$\mathcal{F}$	444	a finite projective plane
∞_i	448	a point at infinity in a finite projective plane
(v, b, r, k, λ)	461	the parameters of a balanced incomplete block design
(v, k, λ)	462	the parameters of a symmetric balanced incomplete block design
$\overline{D}$	465	the complement design of the BIBD, D
D'	465	the derived design of the BIBD, D
D^*	466	the residual design of the BIBD, D

Notation	Page	Brief Description
v, v_i, b_i, q_i	472	parameters for a knapsack problem
$B(v)$	475	optimal total benefit for a knapsack with capacity, v
$\mathrm{Hd}(u, v)$	492	the Hamming distance between binary strings u and v
$\mathrm{Hw}(u)$	492	the Hamming weight of binary string, u
(n, M, d)	493	the parameters of a binary error-correcting code
B_n	497	the set of all binary strings of length n
$S_t(u)$	497	the binary sphere of radius t about u
$S_{\{m\}}$	509	the collection of m-element subsets of S
$R(j, k)$	510	a two-way relationship Ramsey number
$R(j, k; m)$	513	a Ramsey number
$R(k_1, k_2, \ldots, k_n; m)$	516	a generalized Ramsey number

8.8.3 The Fano Plane

The smallest finite projective plane is known as the Fano plane. It contains seven points and seven lines. Every line contains three points. The points are a, b, c, d, e, f, and g. The lines are listed at the top of the figure. Line L_4 is conventionally depicted as a circle for aesthetic reasons.

FPP1 Any two distinct points are on one and only one common line.

FPP2 Any two distinct lines contain one and only one common point.

FPP3 There exist four distinct points, no three of which are on a common line.

FPP3′ There exist four distinct lines, no three of which contain a common point.

$L_1 = \{a, b, d\}$ $\quad L_2 = \{a, c, e\}$ $\quad L_3 = \{b, c, f\}$ $\quad L_4 = \{d, e, f\}$
$L_5 = \{b, e, g\}$ $\quad L_6 = \{c, d, g\}$ $\quad L_7 = \{a, f, g\}$

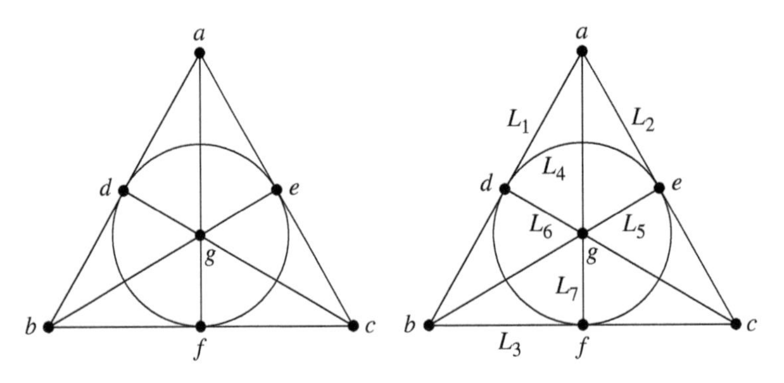

8.8.4 Occupancy Problems

TABLE 8.3 The Number of Ways to Place *n* Objects into *k* Containers

Objects		Containers: Distinguishable	Containers: Indistinguishable
Distinguishable	∅:	k^n	$\sum_{i=1}^{k} S(n, i)$
Distinguishable	¬∅:	$k!S(n, k)$	$S(n, k)$
Indistinguishable	∅:	$\binom{k+n-1}{n}$	$\sum_{i=1}^{k} p(n, i)$
Indistinguishable	¬∅:	$\binom{n-1}{k-1}$	$p(n, k)$

∅: containers may be empty
¬∅: containers must contain at least one object

8.8.5 Additional Review Material

Go to http://www.mathcs.bethel.edu/~gossett/DiscreteMathWithProof/review.xhtml for additional review material, including a sample chapter exam, with solutions.

CHAPTER

9

Formal Models in Computer Science

I don't want to play anymore. – Helen (an artificial intelligence)

Richard Powers — Galatea 2.2

This chapter presents several formal models that capture some important aspects of computing. The chapter begins with a mathematical model for information. Sections 9.2 and 9.6 present models of increasingly more powerful machines that capture the nature of a computer. Section 9.3 introduces the notion of a formal language, which is a topic that arises when compilers[1] are examined at a nontrivial level. Section 9.4 presents the very useful notion of a regular expression. Section 9.5 shows that finite automata (Section 9.2), regular grammars (Section 9.3), and regular expressions (Section 9.4) are distinct models that have identical expressive power.

9.1 Information

In 1948, Claude E. Shannon provided a mathematical description of the engineering aspects of communication. In that year, the *Bell System Technical Journal* published Shannon's paper "A Mathematical Theory of Communication" [81]. The paper provided a clear answer to the question, How much information can a communications channel carry? The answer requires a mathematical definition of *information*, a description of channel capacity, a discussion of channel noise, and an introduction to the notions of encoding and decoding messages.

The primary goal in this section is the presentation of Shannon's definition of *information*. The discussions of channel capacity, noise, and the encoding/decoding process involve background ideas that are not assumed for readers of this text.

In the short introduction, Shannon set the agenda for the paper with the following paragraph.

> The fundamental problem of communication is that of reproducing at one point either exactly or approximately a message selected at another point. Frequently the messages have *meaning*; that is they refer to or are correlated according to some system with certain physical or conceptual entities. These semantic aspects of communication are irrelevant to the engineering problem. The significant aspect is that the actual message is one *selected from a set* of possible messages. The system must be designed to operate for each possible selection, not just the one which will actually be chosen since this is unknown at the time of design.

[1] A *compiler* is a computer program that translates human-readable computer code into machine-executable binary code.

The key points for our purposes are as follows:

- The claim that meaning is irrelevant in this context.
- The notion that actual messages have been chosen from a set of potential messages.

The irrelevance of meaning in this context arises from the need to describe mathematically the process of transmitting a message from one point to another. The semantic content of the message is not unimportant; it is just very difficult to describe mathematically. Fortunately, the questions related to how much information can be sent across a channel can be answered without needing to worry about the meaning of the information. All that is required is that the original message arrives at the destination in an uncorrupted form.

The second key point will be discussed in Section 9.1.2.

9.1.1 A General Model of Communication

Shannon's introduction contains a broadly applicable schematic model of a communication system. This model provides the appropriate context for his definition of information.

The model is outlined in Figure 9.1 and then the various components of the model are described.

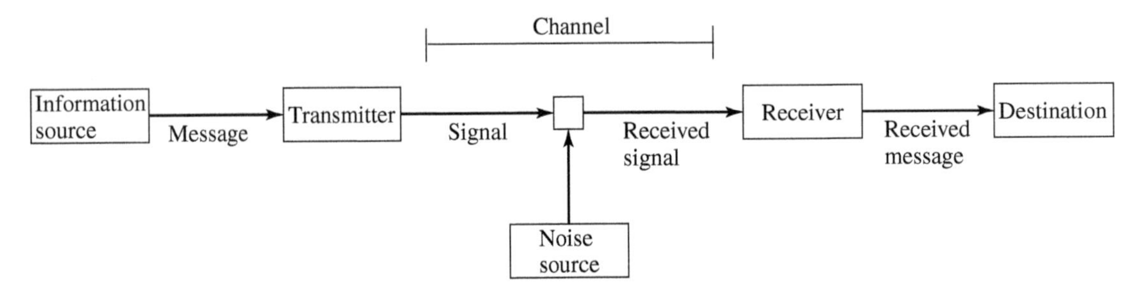

Figure 9.1 Shannon's model of a communication system.

Information source The information source may be a person who is speaking to another person, or a computer sending instructions to another computer over an ethernet cable. The information source selects the message that is to be communicated from among all possible messages that might be sent. The model assumes that there are only a finite (but possibly very large) number of possible messages.

Destination The destination is the other end of the communication system. It could be the person in a conversation who is currently listening, or it could be a computer receiving a file sent by another computer.

Message The message implicitly carries the desired information. It is one among a set of possible messages.

Transmitter The message often needs to be *encoded* to be successfully sent over the channel. For example, a person may think of a message to send to another person. The message is changed into sound waves as the person speaks. The larynx and mouth are thus parts of the transmitter. If an ATM machine is sending a message (perhaps a request for cash) to a central bank computer, the transmitter may need to transform the request into digital signals or light waves that are carried over a phone line or optical cable. The transmitter performs the encoding and sends the result over the channel.

Channel The channel is the medium used to carry the signal. It might be an optical cable, the air, or a piece of paper.

Signal The signal is an encoded output of the transmitter.

Noise source The channel is not always able to deliver the transmitted signal to the receiver in unmodified form because the channel is usually subjected to noise. If the channel is the air in a crowded restaurant where two people are conversing, the noise of other patrons talking, of dishes being loaded onto a cart, of cell phones ringing, and so on, may make the conversation difficult. If the channel is a telephone line near a lightning storm, there may be bursts of static on the line.

Received signal The received signal is the original signal with the changes caused by the channel noise.

Receiver The receiver takes the received signal and decodes it into a message that is compatible with the destination. For example, it might be some kind of analog-to-digital converter that changes a light wave from an optical cable back into a stream of bits for a computer. It might be the human ear changing a sound wave into a form that the brain can comprehend.

Received message The received message is the output of the receiver. The goal is for it to correspond properly to the original message.

The model assumes that the message's meaning is of significance to the information source and the destination, but meaning is not an explicit part of the model.

If the message and the signal both consist of a sequence of discrete symbols, the system is said to be a *discrete communication system*. If the message and signal are both represented by continuous functions, the system is said to be a *continuous communication system*. Otherwise, the system is said to be a *mixed communication system*.

A computer sending a stream of bits over a pulse-modulated wire to another computer would be an example of a discrete system. Two people conversing by means of sound waves would be an example of a mixed system (the words in the message are discrete, but the signal is continuous).

9.1.2 A Mathematical Definition of Information

There are several intuitive ideas that motivate the mathematical definition of information. Several of these insights will be discussed before attempting a complete definition.

Initial Insights

Insight 1: The length of a message is not necessarily related to the amount of information conveyed by the message.

If your classmate tells you "all the dining centers on campus are open for lunch today", you have not (under normal circumstances) learned anything new or unexpected. On the other hand, if the classmate (who you know only on a casual basis) says to you "I love you", you are likely to consider this valuable new information. The second message is shorter, but carries (under normal circumstances) more information.

The next insight is closely related to the first.

Insight 2: Information is related to the "surprise factor" in the message.

Suppose you meet a friend in the hallway between classes and have the following conversation:

You: "Good morning. How are you?"
Friend: "Fine. How are you?"

This conversation carries very little information. You could have shortened it to one word per person and achieved the same effect.

You: "Hi."
Friend: "Hi."

On the other hand, consider the following conversation:

> You: "Good morning. How are you?"
> Friend: "I am not doing well. I found out yesterday that I have leukemia."

The unexpected announcement that your apparently healthy friend actually has a serious disease carries much more information than does the routine response "Fine. How are you?"

The notion of surprise will be revisited after some other ideas are introduced. The next collection of insights leads toward a quantification of information. To that end, it is useful to consider the communication system in probabilistic terms.[2] The finite collection of possible messages can be thought of as a sample space. The message that is actually transmitted is the outcome of the random experiment.[3]

Insight 3: If the sample space has only one outcome, no information is gained by receiving that message. An information value of 0 seems appropriate.

Insight 4: A sample space consisting of two equally likely pieces of news is an obvious choice for a standard nontrivial random experiment. It is reasonable to assign this situation an information value of 1.

Insight 5: A message containing two pieces of news should have the same information content as two messages that each contain only one piece of news.

This is analogous to the effective equivalence of the two experiments:

- Flip two coins.
- Flip one coin twice.

This insight can be extended:

Insight 5′: Let $\{S_1, S_2, \ldots, S_k\}$ be identical, finite sample spaces of messages, each to be delivered on k distinct occasions (so there will be new information, even if a received message repeats an earlier received message). The random experiment "select one message from S_i, $i = 1, 2, \ldots, k$" should provide the same amount of information as the experiment "select k messages, with repetition,[4] from S_1".

If the amount of information in selecting a message from S_1 is $I(S_1)$, then the composite experiments in insight 5′ should have information value $k \cdot I(S_1)$.

Insight 6: Suppose S_1 is a sample space having n_1 equally likely messages and S_2 is a sample space with n_2 equally likely messages. Assume also that $S_1 \cap S_2 = \emptyset$. Form the sample space $T = S_1 \times S_2 = \{(a, b) \mid a \in S_1, b \in S_2\}$ and assign equal probabilities to each of the $n_1 n_2$ outcomes in T. Then the random experiments "select a message from S_1 and another from S_2" and "select an ordered pair of messages from T" should have the same information value.

The information values for the experiments in insight 6 should be related as $I(T) = I(S_1) + I(S_2)$.

It is time to make some preliminary suggestions for a function that will map sample spaces of messages to information values. Table 9.1 indicates that there is one obvious choice that meets the intuitive requirements.

---

[2]See Section 6.1.1 for the required definitions.

[3]These conditions implicitly indicate that a discrete communication system is under discussion.

[4]See Section 5.1 for a brief discussion of repetition.

TABLE 9.1 A Function that Matches the Intuitive Insights

Properties of $\log_2(n)$	Intuitive Insights About $I(S)$ (equally likely messages)
$\log_2(1) = 0$	insight 3: if $\|S\| = 1$ then $I(S) = 0$
$\log_2(2) = 1$	insight 4: if $\|S\| = 2$ then $I(S) = 1$
$\log_2(x^k) = k \log_2(x)$	insight 5': if S^k represents k copies of S, then $I(S^k) = k \cdot I(S)$
$\log_2(xy) = \log_2(x) + \log_2(y)$	insight 6: if $T = S_1 \times S_2$ then $I(T) = I(S_1) + I(S_2)$

Table 9.1 should lead to the following tentative definition:

Tentative Definition of Information

If S is a sample space consisting of n equally likely messages, then the information value of selecting/receiving a message from S is $I(S) = \log_2(n)$.

Validating the Tentative Definition

The tentative definition for $I(S)$ made an intuitive leap from claims of the form $I(T) = I(S_1) + I(S_2)$ to claims of the form $I(S_1) = \log_2(n_1)$, $I(S_2) = \log_2(n_2)$, and $I(T) = \log_2(n_1 n_2)$. That is, the sample spaces S_1, S_2, and T have been replaced by logarithmic equations involving the sizes of the sample spaces. That intuitive leap will now be justified.

Suppose then that S has n equally likely messages. A message from S has been transmitted, but you do not yet know which message. How can you find out? One option is to play a version of "20 questions". That is, ask a series of yes–no questions until you narrow the options to a single message.

EXAMPLE 9.1 **$|S| = 2$**

If the sample space contains messages that reveal the results of a coin flip, then it will only take a single question to determine the message. For example, "was the coin flip a head?" If the answer is yes, you know the heads message was sent. Otherwise, you know the tails message was sent.

We would assign $I(S) = 1$ question in this example. ■

EXAMPLE 9.2 **$|S| = 4$**

If the sample space contains messages that reveal the suit of a card chosen from a standard deck, then it will take two questions to determine the message. For example, "Was the card red?" If the answer is yes, you could ask "Was it a heart?" Otherwise, you could ask "was it a spade?" In any case, you can determine the actual message sent with two questions.

Notice that you might ask "Was it a heart?" first and actually get the answer "Yes". However, you are not guaranteed to get yes for an answer. You might actually need three questions with this approach.

The first approach suggests assigning $I(S) = 2$ questions for this example. ■

✔ Quick Check 9.1

1. Construct a minimal sequence of questions for the purpose of determining which of the equally likely messages was sent.
   (a) The sample space consists of the eight compass directions $\{N, S, E, W, NW, SW, NE, SE\}$.
   (b) The sample space consists of the numbers 1–6. The experiment is to roll a die. ☑

The "rolling a die" example raises a question about the tentative definition. The quick check solution indicated that three questions were needed. How does this relate to the tentative definition? The definition asserts $I(\text{roll a die}) = \log_2(6) \simeq 2.585$. It appears that the number of questions approach is related to the logarithmic approach by the equation

$$\text{number of questions} = \lceil \log_2(n) \rceil$$

when the sample space has n equally likely messages.[5] A few other experiments should convince you that this equation is valid.

There is one other profitable perspective from which to view the tentative definition. Recall that

$$\log_2(n) = x \quad \text{if and only if} \quad 2^x = n$$

There is a simple connection between $\log_2(n)$ and the number of bits needed to represent the nonnegative integer n as a binary (base 2) number. Let $\log_2(n) = x$. If x is an integer, then $x+1$ bits are needed to represent n (a bit position for each of $2^0, 2^1, \ldots, 2^x$). If x is not an integer, then $\lceil x \rceil$ bits are needed to represent n as a binary number. This follows since $2^0 + 2^1 + \cdots + 2^{\lceil x \rceil - 1} = 2^{\lceil x \rceil} - 1 \geq n$ (see Exercise 16 on page 542). Thus, the binary representation of n requires only powers of 2 up through at most $2^{\lceil x \rceil - 1}$.

The amount of information in selecting a message from a sample space of n equally likely messages is closely related to the number of bits needed to represent the number n. The previous discussion implies that only $k = \lceil \log_2(n) \rceil$ bits are needed to represent the n numbers $\{0, 1, 2, \ldots, n-2, n-1\}$.

The General Case: Nonequally Likely Messages

At this point you should be convinced that the tentative definition on page 537 seems to properly capture (except for the ceiling function modification) the amount of information in an experiment with n equally likely messages. How can this be extended to cover sample spaces where the messages are not equally likely? The following example will indicate the proper direction.

EXAMPLE 9.3 **Victim at the Board**

Suppose a math class has n_1 female students and n_2 male students. There are $n = n_1 + n_2$ students in the class.

The instructor wants to randomly choose a student to go to the board and write the solution to a homework problem. The instructor wishes to be fair, so all the names have been placed on slips of paper and placed in a hat. One name is randomly chosen. Since the names are equally likely to be chosen, when the lucky student's name is read, the amount of information conveyed is $\log_2(n)$.

The instructor could choose instead to pick the student by using a two-step process: first choose whether the lucky student will be female or male, and then choose from among all students of the chosen gender. Let the amount of information in knowing the gender be denoted by I. (Clearly there is value in knowing the gender; if you are female and the instructor chooses "male" in step 1, you know you will not be at the board. However, knowing the gender is not the complete information.) Once the gender is known, the information from step 2 is needed. The information in choosing a female name, once it has been decided to choose a female, is $\log_2(n_1)$. Similarly, the information in a male name in step 2 is $\log_2(n_2)$.

How should the information from the two steps be combined? Notice that the probability of choosing a female is $\frac{n_1}{n_1+n_2} = \frac{n_1}{n}$. Similarly, the probability of choosing a male is $\frac{n_2}{n}$. If the female and male information values are apportioned by probability, then the information for the combined steps should be $\frac{n_1}{n}(I + \log_2(n_1)) + \frac{n_2}{n}(I + \log_2(n_2))$.

[5]Recall that $\lceil x \rceil$ is the smallest integer that is not smaller than x. See page 99.

That is, $\frac{100n_1}{n}$ percent of the time, a female will be chosen, with information value $I + \log_2(n_1)$. The rest of the time, a male will be chosen, with information value $I + \log_2(n_2)$.

The two distinct processes for choosing the student should have the same information value.[6] Thus

$$\log_2(n) = \frac{n_1}{n}(I + \log_2(n_1)) + \frac{n_2}{n}(I + \log_2(n_2)) = I + \frac{n_1}{n}\log_2(n_1) + \frac{n_2}{n}\log_2(n_2)$$

This indicates that

$$I = \log_2(n) - \frac{n_1}{n}\log_2(n_1) - \frac{n_2}{n}\log_2(n_2)$$ ■

The interesting aspect of the previous example is the determination of the information value, I, in deciding which of two unequally likely genders was chosen. There is a probability of $p_1 = \frac{n_1}{n}$ that a female is chosen and a probability of $p_2 = \frac{n_2}{n}$ that a male is chosen. It is not too difficult to express I in terms of the probabilities p_1 and p_2.

$$\begin{aligned}
I &= \log_2(n) - \frac{n_1}{n}\log_2(n_1) - \frac{n_2}{n}\log_2(n_2) \\
&= \frac{n_1 + n_2}{n}\log_2(n) - \frac{n_1}{n}\log_2(n_1) - \frac{n_2}{n}\log_2(n_2) \\
&= \frac{n_1}{n}(\log_2(n) - \log_2(n_1)) + \frac{n_2}{n}(\log_2(n) - \log_2(n_2)) \\
&= \frac{n_1}{n}\log_2\left(\frac{n}{n_1}\right) + \frac{n_2}{n}\log_2\left(\frac{n}{n_2}\right) \\
&= \frac{n_1}{n}\log_2\left(\left(\frac{n_1}{n}\right)^{-1}\right) + \frac{n_2}{n}\log_2\left(\left(\frac{n_2}{n}\right)^{-1}\right) \\
&= -\frac{n_1}{n}\log_2\left(\frac{n_1}{n}\right) - \frac{n_2}{n}\log_2\left(\frac{n_2}{n}\right) \\
&= -p_1\log_2(p_1) - p_2\log_2(p_2)
\end{aligned}$$

✔ Quick Check 9.2

1. Calculate the information, I, in knowing the gender in Example 9.3 if
   (a) $n_1 = 10$ and $n_2 = 20$
   (b) $n_1 = 10$ and $n_2 = 1$
2. Show that the expression $I = -p_1\log_2(p_1) - p_2\log_2(p_2)$ reduces to $\log_2(2)$ (the old tentative definition) when the genders are equally likely. ☑

There are two significant observations that Quick Check 9.2 should lead to. First, the information, I, is positive in each example examined so far. This is because the probabilities, p_i, are positive numbers less than 1. Thus, their logarithms are negative, cancelling the existing minus signs in $I = -p_1\log_2(p_1) - p_2\log_2(p_2)$.

The second observation is that the case where the probabilities are equally likely has the highest information. Before discussing why this is true in general, consider the following intuitive justification. If one of the genders is much less likely to be chosen, then the other gender will usually be picked. Thus, most of the time, there is little surprise value to the experiment. That means that, on average, the amount of surprise (information) is lower than in a case where there is more uncertainty.

[6] Let M represent the event "Mary was chosen" and let F represent the event "*female* was chosen at step 1." In process 1, $P(M) = \frac{1}{n_1+n_2}$. In process 2, $P(M) = P(M|F) \cdot P(F) = \frac{1}{n_1} \cdot \frac{n_1}{n_1+n_2} = \frac{1}{n_1+n_2}$ (see Section 6.2). Since the two probabilities are equal, it is reasonable to assume both processes have the same information value.

A proof that I reaches its maximum when the two probabilities are equal uses calculus to find the optimal value of p_1. Note that $p_2 = 1 - p_1$, so we can rewrite I as

$$I = -p_1 \log_2(p_1) - (1 - p_1)\log_2(1 - p_1)$$

Thus the derivative, I', can be found as a function of p_1:

$$I'(p_1) = \log_2(1 - p_1) - \log_2(p_1)$$

Consequently, $I'(p_1) = 0$ when $\log_2(1 - p_1) = \log_2(p_1)$. Thus, $2^{\log_2(1-p_1)} = 2^{\log_2(p_1)}$, that is, $1 - p_1 = p_1$. The only critical point is at $p_1 = \frac{1}{2}$. Since $I''(\frac{1}{2}) < 0$, the second derivative shows there is a maximum at $p_1 = p_2 = \frac{1}{2}$.

Figures 9.2 shows how the value of I varies as p_1 varies from 0 to 1.

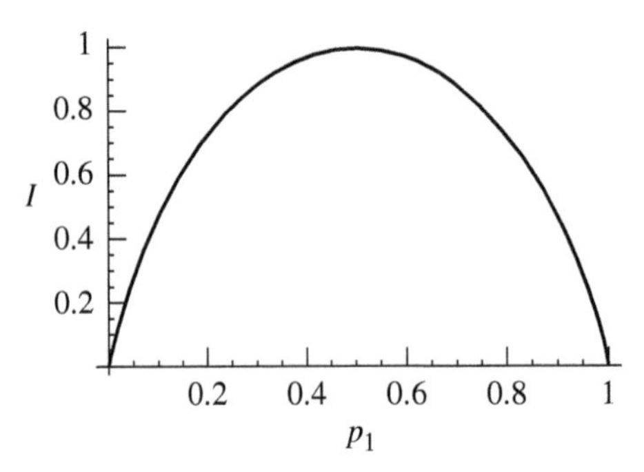

Figure 9.2 How I varies as a function of p_1.

The process that was used to justify the definition $I = -p_1 \log_2(p_1) - p_2 \log_2(p_2)$ can be easily extended into the final definition of information.

DEFINITION 9.1 ***Information***

Let $S = \{m_1, m_2, \ldots, m_n\}$ be a finite collection of messages, having respective nonzero probabilities $\{p_1, p_2, \ldots, p_n\}$. Then the average *information* in a randomly selected message from S is

$$I(S) = -\sum_{k=1}^{n} p_k \log_2(p_k)$$

The definition should remind you of the definition of expected value (page 304). That similarity provides additional insight into why this definition is the proper measure for the average information for a message in S. An easy homework exercise shows that this definition reduces to the tentative definition of *information* (page 537) if the respective probabilities are equal.

THEOREM 9.2 ***Maximizing Information***

Let $S = \{m_1, m_2, \ldots, m_n\}$ be a finite collection of messages, having respective probabilities $\{p_1, p_2, \ldots, p_n\}$. Then $I(S)$ is maximized when $p_1 = p_2 = \cdots = p_n = \frac{1}{n}$.

The proof uses ideas from multivariable calculus, so it will not be presented here.

9.1.3 A Summary of Other Ideas in Shannon's Paper

Shannon's definition of information has already been discussed. Some of the other highlights of the paper are very briefly presented here.[7]

[7] Warren Weaver has provided a more complete, fairly nontechnical summary of Shannon's paper [81].

Channel capacity is described in terms of the amount of information the channel can carry per unit time (rather than the number of symbols it can carry). It does not matter whether the channel is carrying successive symbols from a single message that are being transmitted very rapidly, or signals from multiple messages (as a typical multiplexed phone line does). The channel capacity is a measure of either the maximum number of bits per second that can be physically transmitted over the channel without error or the maximum data rate for a specified error rate.

One of the key notions in Shannon's paper is that of *encoding*. It is possible to encode the message (at the transmitter) in a manner that utilizes the channel very efficiently. Assume, for the moment, that there is no noise in the channel. Suppose the channel has a capacity of C bits per second, and the information source transmits the message at I bits per symbol. A major theorem Shannon proved states that it is theoretically possible to find an encoding that enables the information in a discrete communication system to be transmitted across the channel at an average rate that is as close to $\frac{C}{I}$ symbols per second as we desire. The theorem also states that it is impossible to transmit at a rate faster than $\frac{C}{I}$ symbols per second, no matter how clever the encoding.[8]

If the channel contains noise sources, the situation becomes more complex. If the channel capacity, C, exceeds the rate, I, that the information source is producing the message, then it is always possible to encode and transmit the message in such a manner that the non-zero probability of having errors in the received message is as small as desired. However, if $I > C$, then no matter how cleverly we encode the message, there will always be uncertainty that the received message matches the original message. This uncertainty has a lower bound related to $I - C$.

Devising practical encoding/decoding algorithms to be used when transmitting over a noisy channel has developed into a separate branch of mathematics and engineering called *coding theory* (see Section 8.5). Some very sophisticated techniques use ideas typically found in courses taught at the junior-senior level to mathematics majors.

How does the situation change when a continuous channel is involved? It turns out that the analysis is not too different. This is because it is practically possible to convert a continuous system into a discrete system that is effectively as capable. The reason for this is that people have limited capacities of perception. We are unable to distinguish sound waves that are beyond certain frequencies. We can only "see" light between certain frequencies (and cannot distinguish frequencies that are too close together). It is therefore possible to *sample* a continuous function and use the finite (discrete) collection of samples to represent the function. As long as the sampling occurs sufficiently often per second, the samples constitute a faithful representation of the original function. (Sampling of this kind is used in audio CD and video DVD technology.)

9.1.4 Exercises

The exercises marked with ★ have detailed solutions in Appendix H.

1. Consider the communication system formed by an instructor giving a lecture and a student listening to the lecture. Describe each component in Shannon's model of a communication system for this setting.

2. Insight 1 asserts that the length of a message is not necessarily an indication of the amount of information the message carries. Give an example of a long message with little information and also an example of a short message with a large amount of information.

3. ★ Insight 2 is about surprise. Discuss how this relates to the final, probabilistic definition of information. Incorporate Theorem 9.2 into your discussion.

4. How many bits are needed to uniquely encode a collection of messages if there are n messages, where

   (a) ★ $n = 14$ (b) $n = 512$ (c) $n = 1000$

5. Design a minimal sequence of questions to determine whether a student is a freshman, sophomore, junior, senior, or graduate student.

[8]There is a price for being able to transmit at near the optimum rate: More nearly optimum transmission rates are achieved by using more complex (and hence time-consuming) encoding techniques.

6. During the balcony scene in *Romeo and Juliet*, Juliet utters the famous phrase, "O Romeo, Romeo! wherefore art thou Romeo?", lamenting the fact that Romeo has the wrong name (he is a Montague and should therefore be her enemy since she is a Capulet).

   Suppose Juliet has sent her nurse to deliver a message to Romeo. The nurse encounters a group of young men in the town square. The young men decide to make it difficult for her to determine which of them is Romeo, so they agree to only answer yes-no questions. If there are eleven young men in the group (including Romeo), what is the minimum number of questions the nurse can ask and be guaranteed to determine which young man is Romeo?

7. Find $I''(x)$ if $I(x) = -x\log_2(x) - (1-x)\log_2(1-x)$. Then calculate $I''(\frac{1}{2})$.

8. ★ Show that

$$\lim_{x\to 0^+}\left[-x\log_2(x) - (1-x)\log_2(1-x)\right] = 0$$

   [*Hint*: A standard application of L'Hôpital's rule with indeterminate forms shows that $\lim_{x\to 0^+} x\ln(x) = 0$.]

9. Suppose that the probability of conceiving a boy is 0.512 and the probability of conceiving a girl is 0.488. If a nurse is using an ultrasound to determine an unborn baby's gender, what is the (average) information associated with the gender announcement?

10. ★ Recall the candy jar in Example 6.13 (page 279). It contains four kinds of candy, (L, C, B, W), having respective probabilities $\frac{3}{16}$, $\frac{1}{4}$, $\frac{1}{4}$, and $\frac{5}{16}$. What is the average information associated with randomly choosing a piece of candy from the jar?

11. According to Henry Wadsworth Longfellow's famous poem *The Midnight Ride of Paul Revere*, a signal would be sent to warn the Americans about how the British soldiers would attack. If the British would attack by land, a single lantern would be placed in the bell tower of North Church. If the British launched their attack by sea (across Boston Bay), two lanterns would be placed in the tower (hence the well-known line: "one if by land and two if by sea").
    (a) What is the average information if the two options were equally likely (and the event had been repeated randomly many times)?
    (b) What is the average information if a sea attack was twice as likely as a land attack?
    (c) Suppose that there was a third option: no attack. If the probabilities were equally likely for the three options, what is the average information?
    (d) If a sea attack had probability 0.3, a land attack had probability 0.2, and no attack had probability 0.5, what is the average information in observing the bell tower of North Church? (Again, assume that the historical event had been repeated with random results many times.)

12. The definition of (average) *information* was compared to the definition of *expected value*. In the information context, what corresponds to the random variable, X? What corresponds to the value, x, of the random variable? What corresponds to $P(x)$?

13. Prove that Definition 9.1 reduces to the tentative definition of information (page 537) if the probabilities, p_k, are equal.

14. Each of the following statements is either true (always) or false (at least sometimes). Determine which option applies for each statement and provide adequate explanation for your choice.
    (a) The expression for average information, $I(S)$, in Definition 9.1 is always a value which is less than zero.
    (b) If the message set, S, has only one message, than the average information of a randomly selected message is 1.
    (c) In Shannon's model of a communication system, it is possible for the message and the received message to differ.
    (d) ★ Shannon primarily based his definition of information on the "surprise factor" in the meaning (semantics) of a message.

15. Design message spaces and message probabilities for which the average information is
    (a) 2 (b) 3 (c) 4 (d) n, where $n \in \mathbb{N}$

16. Let n be a natural number and let $\log_2(n) = x$.
    (a) If $x \notin \mathbb{N}$, prove that

$$\sum_{i=0}^{\lceil x\rceil - 1} 2^i \geq n$$

    (b) If $x \in \mathbb{N}$, prove that

$$\sum_{i=0}^{\lceil x\rceil - 1} 2^i = n - 1$$

9.2 Finite-State Machines

Section 9.1 presented a model of information. This section presents two models of very simple computers. Both models are examples of *finite-state machines*.

At the heart of the notion of a finite-state machine is the concept of a *state*. Consider a system of some kind (perhaps a DVD player or a cafeteria line) that has a number of distinct conditions that it can be in. For example, a DVD player can be off, paused, playing, fast-forwarding, and so on. A student in a cafeteria line may be at the salad bar, at the main dish station, getting a drink, and so on. The distinguishable conditions are called the *states* of the system. We will be concerned with systems that have a finite number of associated states.

EXAMPLE 9.4

A Simple DVD Player

Consider a very simple DVD player. It has a power button, a DVD tray that slides in and out to load or eject a DVD, a play button, a stop button, and a pause button. A natural collection of states (with symbolic descriptors) for this machine is

Off off with no DVD in the tray

OffDVD off with a DVD in the tray

On on with no DVD in the tray

Stop stopped with a DVD in the tray

Play playing

Pause paused ■

Finite-state machines are formal models of machines or systems with a finite number of states. There is more than one variety of finite-state machine. We will start with one of the simplest kinds: a finite-state machine that models the changes of state within a system that occur until the machine achieves one of a collection of desired states.

9.2.1 Finite Automata

DEFINITION 9.3 ***Finite Automaton***

A *finite automaton*, A, is a model that consists of

- A finite set of states, $\mathcal{S}$.
- A set of input values, Σ.
- A transition function, $t(s, i)$, that maps state–input pairs to states.
- A special state called the *start state* (generically named s_0).
- A subset, $\mathcal{F} \subseteq \mathcal{S}$, of *final (or accepting) states.*

This can be represented as $A = (\mathcal{S}, \Sigma, t, s_0, \mathcal{F})$.

The plural of *finite automaton* is *finite automata.*

The transition function specifies the actions of the system. Suppose the system is in state s_1 and the system receives input i. If $t(s_1, i) = s_2$, then the system will change to state s_2.

We are usually interested in the state changes that occur as a result of a sequence of input symbols. If the input symbols are all single characters, the input sequence is often called an *input string*.

DEFINITION 9.4 ***String***

A finite sequence of characters that are concatenated[a] together is called a *string.* A finite sequence of 1s and 0s that are concatenated together is often called a *binary string.*

[a]The word *concatenate* means to "connect or link in a series or chain". The words *dog* and *house* can be concatenated to form the word *doghouse*.

When an input string causes the finite automaton to land in a final state, the string is said to be *recognized* (or *accepted*).

There are two common mechanisms used for describing a finite automaton. The first uses a table called a *state table*; the second uses a diagram called a *state diagram.* The table is easier to present in a typeset document but the diagram is easier for most people to understand.

EXAMPLE 9.5

A Finite Automaton for a DVD Player

Consider the DVD player introduced in Example 9.4. We already know the states: {Off, OffDVD, On, Stop, Play, Pause}. The inputs are all related to button pushes:

PowerB power button has been pushed

PlayB play button has been pushed

StopB stop button has been pushed

PauseB pause button has been pushed

EjectB eject button has been pushed

For this example, I will assume that the start state is Off and $\mathcal{F}$ = {Off, OffDVD}. It remains to specify the transition function. This will be done in the next two examples using the two standard mechanisms. ■

To complete the definition of the model of a simple DVD player, we need to specify for each state–input pair what the next state will be. There are 6 states and 5 inputs, so there are 30 ordered pairs that need to be specified. In many cases, there will be no transition to a new state (for example, pushing Play when the state is already Play). However, we need to specify the new state for each ordered pair.

EXAMPLE 9.6

DVD Player via State Table

Table 9.2 defines the function $t(s, i)$ for the simple DVD player. Notice the assumptions built into the function. In particular, the assumption that after pushing the eject button while in either state On or state Off, someone will always change whether the tray is empty or not (if the player is currently stopped – with a DVD in the tray – and the eject button is pushed, the tray will extend, the DVD will be removed and the tray will retract, leaving the player in state On with an empty tray.). Also, many buttons assume a particular state before they will activate (real DVD players are more flexible).

TABLE 9.2 The Transition Function for a DVD Player

	Input				
State	**PowerB**	**PlayB**	**StopB**	**PauseB**	**EjectB**
Off	On	Off	Off	Off	Off
OffDVD	Stop	OffDVD	OffDVD	OffDVD	OffDVD
On	Off	On	On	On	Stop
Stop	OffDVD	Play	Stop	Stop	On
Play	Play	Play	Stop	Pause	Play
Pause	Pause	Play	Stop	Pause	Pause

Read the table as $t(\text{row}, \text{column}) = \text{intersection}$. Thus, $t(\text{Play}, \text{PauseB}) = \text{Pause}$. That is, if the state is currently Play and the pause button is pushed, the state will become Pause. ■

EXAMPLE 9.7

DVD Player via State Diagram

To display the transition function on a diagram, several conventions need to be established.

- States are shown as circles.
- A final state is a denoted by a pair of concentric circles.
- The transition $t(s_1, i) = s_2$ is represented by an arrow that starts at s_1 and ends (has the arrow head at) s_2. The arrow bears the label i.
- The start state is shown by having an unlabeled arrow ending at that state (and with no state as its beginning).

The simple DVD player can be represented using the state diagram in Figure 9.3.

Figure 9.3 A state diagram for a DVD player.

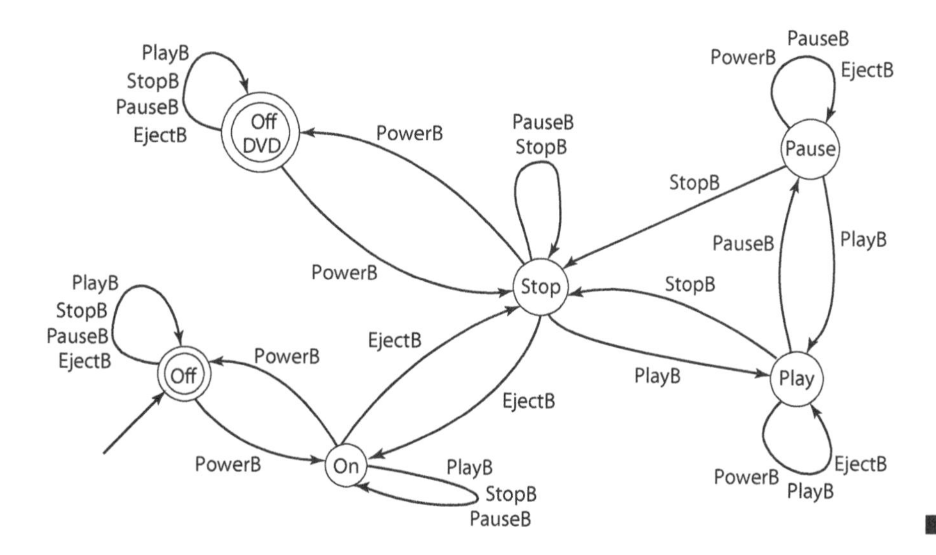

■

Many authors follow a convention of leaving out any transition that results in a loop back to the same state. This tends to keep the diagram less cluttered. However, by making sure that every state has a transition for every input, you are less likely to neglect an important transition. For that reason, I would strongly recommend that you follow the examples in this text and show every transition.[9]

✔ Quick Check 9.3

1. Create a finite automaton that accepts strings of 0s and 1s as input and recognizes (accepts) any string with an odd number of 1s. Clearly label the states and transitions. Indicate the starting state and any final states.
   (a) Use a state table to represent the transition function.
   (b) Use a state diagram to represent the transition function. ☑

EXAMPLE 9.8

Two 0s

Suppose we want a finite automaton that accepts strings of 0s and 1s as input and recognizes any input string that contains at least two adjacent 0s.

A simple automaton that performs this task will consist of three states—Zero, One, and Two—indicating, respectively, that the most recent substring has contained 0, 1, or 2 zeros in a row. The state Two also indicates that at least two adjacent 0s have been encountered somewhere in the input string.

The start state is Zero and the only final state is Two (Figure 9.4).

Figure 9.4 Two 0s state diagram.

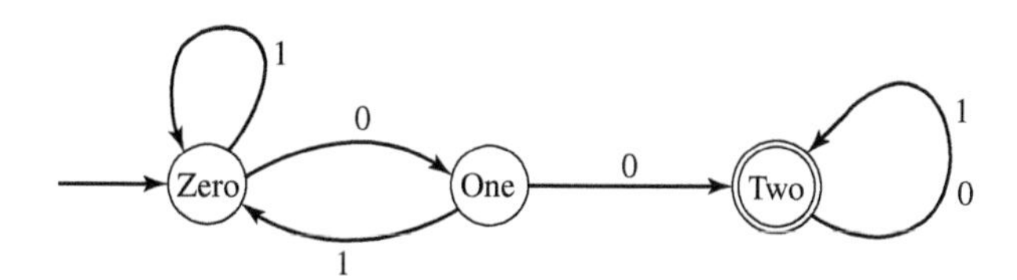

Notice the "black hole" nature of state Two. Once the finite automaton transitions to this state it never leaves. This is a mechanism used to "remember" that two adjacent zeros have been encountered somewhere in the input string. ■

[9] You should ask your instructor what his/her preference is before you start any homework in this section.

It is often helpful to simulate finite-state machines. One of the available simulations is Matt Chapman's Java application.[10] With such simulations, you may define the automaton and then watch the state transitions as a sequence of input values are processed. His diagrams merge pairs of arrows between two states into a single two-headed arrow.[11] The input label for a transition is written next to the arrowhead (closer to the new state). It is not possible to label states using this simulation. Figure 9.5 shows a simulation for Example 9.8 after it has processed the string 1101001.

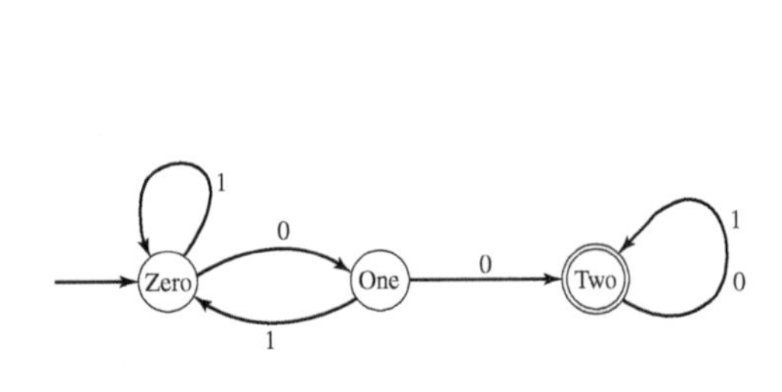

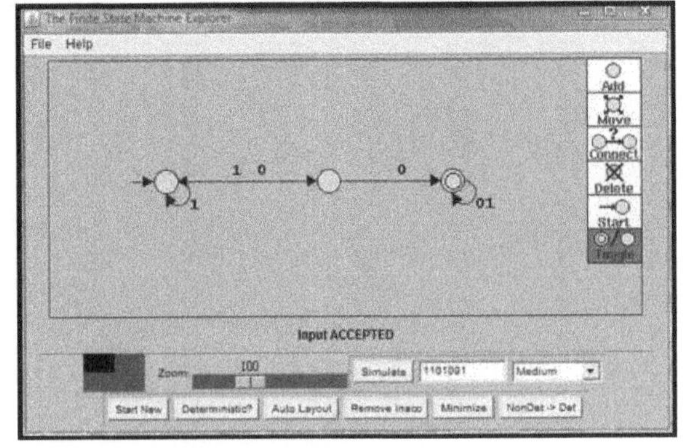

Figure 9.5 The finite-state machine explorer.

EXAMPLE 9.9

The Wizard of Oz

In the 1939 version of the movie *The Wizard of Oz*, Glinda the Good Witch tells young Dorothy to "click your heels together three times and say, 'there's no place like home'." Dorothy will then find herself magically transported back to her home in Kansas.

Suppose for this example that the directions mean that Dorothy should first click her heels three times and then say "there's no place like home" once. Any deviation from this formula will leave Dorothy forever in Oz. That will include making additional heel clicks or repeating the magic phrase once she is back in Kansas. Figure 9.6 represents this using a finite-state machine. Notice that even though Kansas is a final state,[12] it is

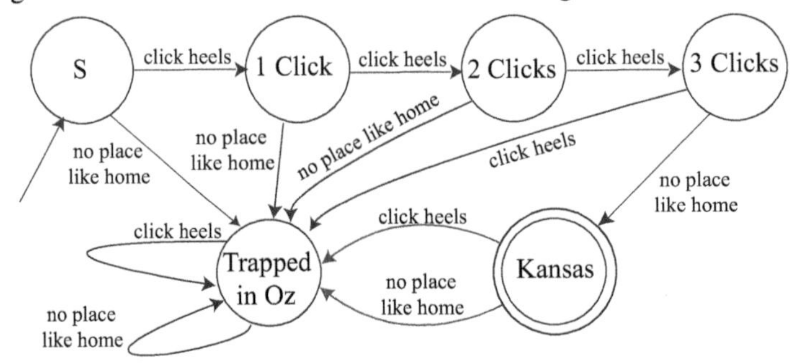

Figure 9.6 "Click your heels together three times and say, 'There's no place like home.' "

still possible to transition out of that state. It is *final* in the sense that Dorothy "wins" if she remains in that state. ■

Note: If from some state s_k an input i is impossible, then no transition corresponding to that input should be added to the transition table or state diagram. Your diagrams *should* contain transitions for *all* possible inputs at each state.

---

[10] Available at http://www.mathcs.bethel.edu/~gossett/DiscreteMathWithProof/ or at http://www.belgarath.org/java/fsme.html as an applet.

[11] This is not ideal. In particular, it makes it harder to determine which input symbols go with which direction. However, for a student project, this simulation is impressive.

[12] In the finite-state machine sense.

9.2.2 Finite-State Machines with Output

A second type of finite-state machine is one for which each transition has an associated output. The output might provide feedback about the current state of the machine (Example 9.11) or it may serve as a stream of information that the machine was created to produce (Example 9.12). Notice that in this model there are no final states.

DEFINITION 9.5 ***Finite-State Machine with Output***

A *finite-state machine with output*, M, is a model that consists of

- A finite set of states, $\mathcal{S}$.
- A finite set of input values, Σ.
- A transition function, $t(s, i)$, that maps state–input pairs to states.
- A finite set of output values, Γ.
- An output function, $g(s, i)$, that maps state–input pairs to output values.
- A special state called the *start state* (generically denoted s_0).

This is represented as $M = (\mathcal{S}, \Sigma, t, \Gamma, g, s_0)$.

There are still two mechanisms commonly used to represent the associated functions. In the tabular presentation, additional columns are added to indicate the output values. In a state diagram, the transition labels are represented as ordered pairs, (i, j), containing the input and the output. If more than one input generates the same output, the label will be in the form $(i_1\, i_2\, i_3, j)$. Notice that there is only one comma in the label (it is an ordered pair). If several inputs result in distinct outputs but cause a transition to the same state, the arrow will need several labels (for example, the arrow from state "25" to state "0" in Figure 9.8 on page 549).

EXAMPLE 9.10 **Red Pill or Blue Pill?**

In the movie *The Matrix*, the hero, Neo, is told to choose one of two pills. The red pill will lead him into the unsettling true world. The blue pill will lead him back to a comfortable illusory world. We can model Neo's decision.

There are three states: Ready to Decide, Real World, and Illusion. The inputs are the two choices: red pill and blue pill. The outputs are Learn the Truth, Return to Ignorance, Remain the Same. I have made the assumption that neither pill has any effect after the initial decision. Table 9.3 and Figure 9.7 show the transition and output functions.

TABLE 9.3 Neo's State Table

	Transition Function		**Output Function**	
	Input		**Input**	
State	Red Pill	Blue Pill	Red Pill	Blue Pill
Ready to Decide	Real World	Illusion	Learn the Truth	Return to Ignorance
Real World	Real World	Real World	Remain the Same	Remain the Same
Illusion	Illusion	Illusion	Remain the Same	Remain the Same

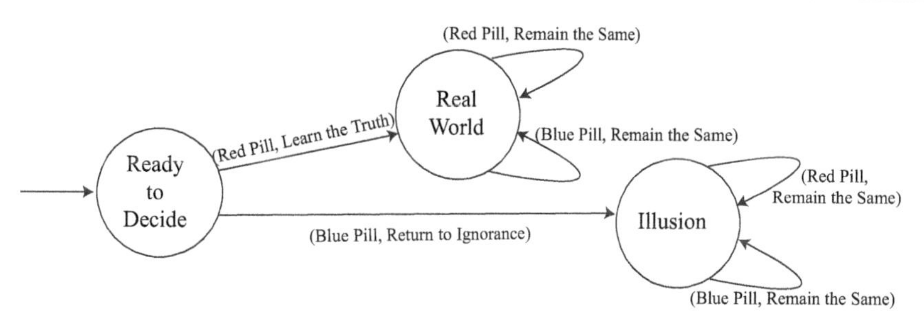

Figure 9.7 State diagram for Neo's Decision.

EXAMPLE 9.11

A Simple Vending Machine

Suppose we want to model a simple vending machine that dispenses pencils and erasers. Pencils cost \$0.25 and erasers cost \$0.15. The vending machine contains a small message screen that shows how much money has been deposited so far. The vending machine only accepts nickels, dimes and quarters. It does not give change, nor does it keep track of extra money. If you deposit \$0.30 and request a pencil, you will not have the extra \$0.05 returned, nor will it count as part of the next purchase.

The vending machine has two buttons to select the item to buy, but they only work after sufficient funds to make the purchase have been deposited.

One choice for the set of states is to provide one state for each possible sum of money deposited (stopping at \$0.25). The start state is 0. The input set can be represented as $\{0, 5, 10, 25, \text{P}, \text{E}\}$, where P represents pushing the pencil button and E represents pushing the eraser button. The output set is defined as $\{0, 5, 10, 15, 20, 25, \text{DP}, \text{DE}\}$, where DP (DE) indicates a pencil (eraser) has been delivered.

The transition and output functions are represented in Table 9.4.

TABLE 9.4 The State Table for a Vending Machine

	Transition Function						Output Function					
	Input						Input					
State	0	5	10	25	P	E	0	5	10	25	P	E
0	0	5	10	25	0	0	0	5	10	25	0	0
5	5	10	15	25	5	5	5	10	15	25	5	5
10	10	15	20	25	10	10	10	15	20	25	10	10
15	15	20	25	25	15	0	15	20	25	25	15	DE
20	20	25	25	25	20	0	20	25	25	25	20	DE
25	25	25	25	25	0	0	25	25	25	25	DP	DE

The output function looks almost the same as the transition function in this example. They differ only when there is enough money accumulated to purchase an item. For example, if the machine is in state 15 and the E button is pushed, the transition function shows that the next state is 0, whereas the output function emits DE, indicating that an eraser was delivered.

The model can also be represented with a state diagram (Figure 9.8).

The next example has a more interesting output function.

EXAMPLE 9.12

Delaying Input

Suppose I want to construct a model for a circuit that reads a binary string. It reads one bit every microsecond. The circuit outputs the same input string, but each bit is delayed by 2 microseconds.

Assume that the input string contains at least 2 bits. The finite-state machine will require three transitional states that allow for the initial period where there is no output. It then needs to keep track of what bits have arrived over the previous two time periods. This requires four more states.

The start state, denoted $s_\emptyset$, will be a transitional state indicating that no input has yet been received. The other transitional states, denoted s_0 and s_1, indicate that the first bit was a 0 or 1, respectively. The remaining states will be denoted s_{00}, s_{01}, s_{10}, and s_{11}, where s_{ij} indicates that the previous two characters were i and j, in that order.

The transition and output functions in tabular form are shown in Table 9.5. The symbol "—" indicates that there is no output.

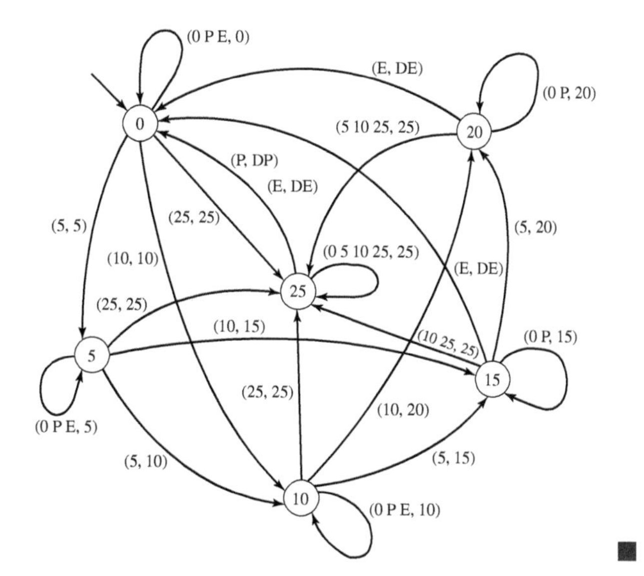

Figure 9.8 State diagram for a vending machine.

■

TABLE 9.5 The State Table for a Delay Circuit

	Transition		Output	
	Input		Input	
State	0	1	0	1
$s_\emptyset$	s_0	s_1	—	—
s_0	s_{00}	s_{01}	—	—
s_1	s_{10}	s_{11}	—	—
s_{00}	s_{00}	s_{01}	0	0
s_{01}	s_{10}	s_{11}	0	0
s_{10}	s_{00}	s_{01}	1	1
s_{11}	s_{10}	s_{11}	1	1

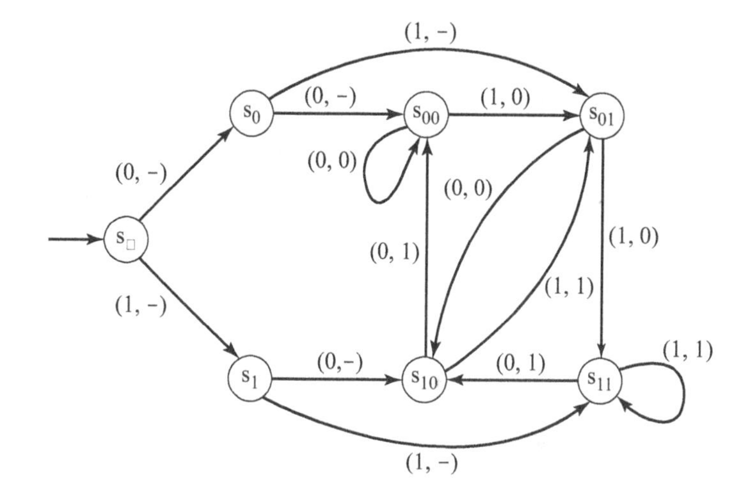

Figure 9.9 A state diagram for a delay circuit.

The state diagram in Figure 9.9 is an alternative representation of this finite-state machine.

The state diagram makes it easy to trace the state changes and output for an input string, such as 110100. You should trace this string yourself. If you do not generate the output — — 1101 and end at state s_{00}, you probably need to reread this example.

Notice that the model does not flush the final two input bits. This deficiency needs to be corrected. ■

✔ Quick Check 9.4

1. Correct the oversight in Example 9.12. To flush the final characters, you will need to add an additional character to the input set, namely —, which can be appended to the end of any input string. You may assume that the input symbol — will never occur except at the very end of an input string, and that every input string terminates with — — (two —s). You may add additional states. ☑

One more example should be sufficient to demonstrate finite-state machines with output.

EXAMPLE 9.13

A Binary Adder

It is possible to design a simple finite-state machine for adding two positive binary integers. Suppose the two numbers are $x_n x_{n-1} \cdots x_2 x_1 x_0$ and $y_n y_{n-1} \cdots y_2 y_1 y_0$, where $x_i, y_i \in \{0, 1\}$ for $i \in \{0, 1, 2, \ldots, n\}$. If one number has fewer bits than the other, prepend some zeros so that they are the same length. (For example, the numbers might be 1101 and 0011.)

The machine will read pairs of corresponding bits from the two numbers, reading right-to-left. Output will also be produced in right-to-left order. If $x_i = 0$ and $y_i = 1$, the input symbol will be "01." The set of input symbols is $I = \{00, 01, 10, 11\}$.

You are aware that the addition of two bits might result in a carry. How can this be represented in a finite-state machine? The simplest mechanism is to use the states. We could create two states, c_0 and c_1, representing a carry of 0 or 1 on the previous addition. The bit that would normally be written under the column as the result of the addition for that column will be the output symbol. Thus, $\Gamma = \{0, 1\}$.

Since the addition starts with no previous carry, the start state will be c_0.

Notice that when the current state is c_1 (previous addition required a carry), an input of 01 will cause another carry: $0 + 1 +$ previous-carry $= 0 + 1 + 1 = 10_2$, where 10_2 is binary for "2". The output will be 0, and the next state will again be c_1 since we just carried again. On the other hand, if the current state is c_0 and the input is 01, no carry will be generated: $0 + 1 +$ previous-carry $= 0 + 1 + 0 = 1_2$. The output will be 1 and the next state will be c_0.

Before reading the state table, try to determine what the output and next state will be if the current state is c_1 and the input is 11. The state table and state diagrams are shown in Table 9.6 and Figure 9.10.

TABLE 9.6 The State Table for a Binary Adder

	Transition				Output			
	Input				Input			
State	00	01	10	11	00	01	10	11
c_0	c_0	c_0	c_0	c_1	0	1	1	0
c_1	c_0	c_1	c_1	c_1	1	0	0	1

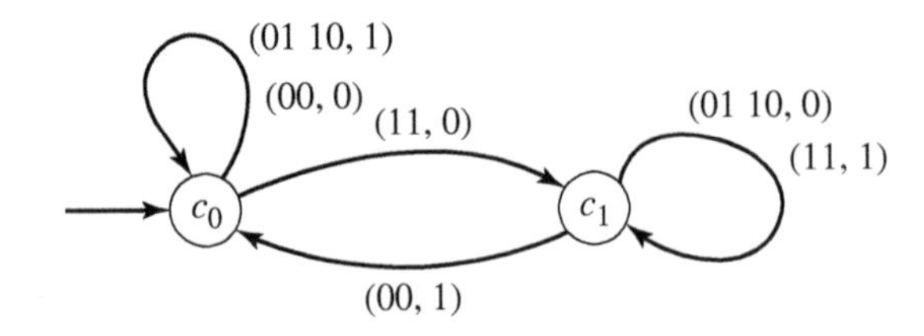

Figure 9.10 The state diagram for a binary adder.

It is informative to trace through a simple example. Suppose we want to add 1101 and 1011. To get the machine to output the final carry bit, it will be necessary to change the numbers to 01101 and 01011.

The input pairs will be 11 01 10 11 00. Table 9.7 shows the machine's actions.

TABLE 9.7 Adding 1101 and 1011

Current State	Input	Next State	Output
c_0	11	c_1	0
c_1	01	c_1	0
c_1	10	c_1	0
c_1	11	c_1	1
c_1	00	c_0	1

The sum $1101_2 + 1011_2 = 11000_2$. Converting to decimal gives $13 + 11 = 24$, which agrees with the binary result. ■

9.2.3 Exercises

The exercises marked with ★ have detailed solutions in Appendix H.

1. List the sequence of states assumed by the automaton in Example 9.8 on page 545 when it is given each input string below. Also, indicate whether the string is recognized.

(a) ★ 01101001 (b) 1101110
(c) 00001111 (d) 10110100

2. Create a state diagram for the finite automaton whose state table is shown. S_0 is the start state; S_3 is the only final state.

State	Input			
	a	b	c	d
S_0	S_1	S_1	S_2	S_0
S_1	S_2	S_0	S_1	S_1
S_2	S_1	S_0	S_3	S_3
S_3	S_3	S_3	S_1	S_2

3. Create a state table for the finite automaton whose state diagram is shown.

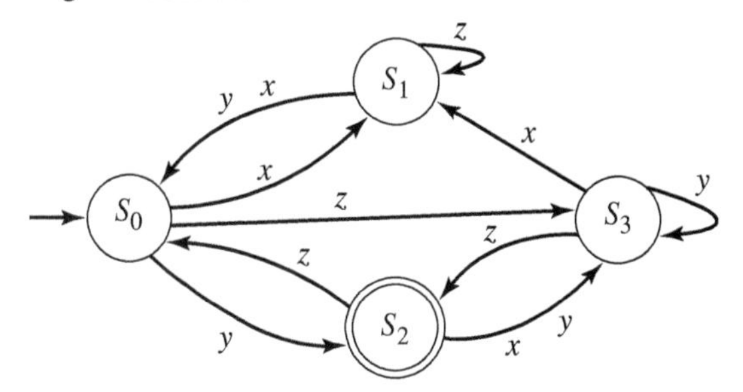

4. Describe the set of strings that the following finite automaton recognizes.

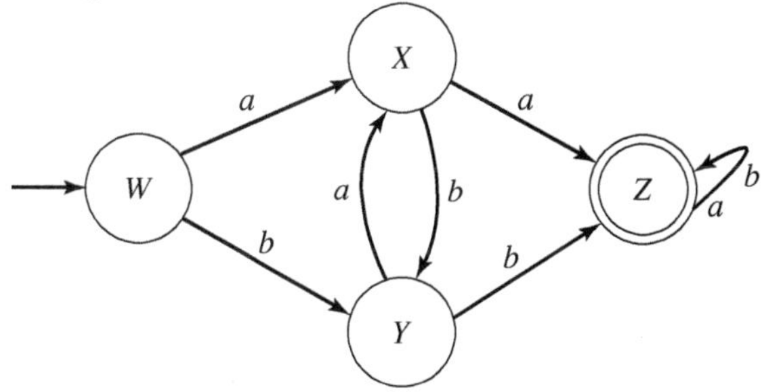

5. Create finite automata with the following characteristics. Clearly label the states and transitions. Indicate the starting state and any final states. In each case, indicate the transition function using both a state table and a state diagram. Each automaton will accept input strings of 0s and 1s.

(a) Create a finite automaton that recognizes any string with an even number of 1s.

(b) Create a finite automaton that recognizes any input string that contains two 0s as the final characters.

(c) ★ Create a finite automaton that recognizes any input string that contains at least one 1 and at least one 0.

(d) Create a finite automaton that recognizes any input string that contains no 1s.

(e) Create a finite automaton that recognizes any input string that contains either all 1s or all 0s (but must contain at least one bit).

6. Create finite automata with the following characteristics. Clearly label the states and transitions. Indicate the starting state and any final states. Use a state diagram to represent the transition function. Each automaton will accept input strings containing the letters a, b, and c.

(a) Create a finite automaton that recognizes any string that contains the substring abc.

(b) Create a finite automaton that recognizes any string that *does not* contain the substring abc.

(c) Create a finite automaton that recognizes any string that contains at least one a and at least one b, but no cs.

(d) Create a finite automaton that recognizes any string that contains at least one each of two distinct letters, but does not contain the third letter. (A string of all as is not recognized. The string $aaabaaa$ is recognized.)

7. The campus "Young Moguls" club has three requirements for membership:

- You must be a full-time student.
- You must have earned at least $100,000 through your own investments.
- You must be nominated by a current member.

The full-time student requirement must be verified first. The other two requirements can occur in either order. Consider the input symbols F, H, and N to represent, respectively, verified full-time student status, verified earnings of at least $100,000, and a nomination from a current member in good standing.

Create a finite automaton to model the membership acceptance process. Clearly describe and label the model. For this problem, you may part with the usual convention of including *all* inputs for each state. You only need to include inputs that result in a state change.

8. An arcade game consists of three raised cylinders, labeled, A, B, and C, respectively. The object of the game is to push down the cylinders in the proper sequence. A cylinder that is pushed down out of sequence will stay down, but the other two cylinders will pop up. When a cylinder is pushed down in its proper position in the sequence, all previous cylinders in the sequence will also stay down. The proper sequence is BCA. Design a finite automaton that models this arcade game. [*Hint*: Use the states to represent which cylinders are down. There is only one final state.]

9. The game of Monopoly® is a celebration of unrestrained capitalism, but it does allow for a player to "go to jail". There are four ways to get out of jail when it is your turn: (1) hand in a "get out of jail free" card, (2) pay a $50 fine, (3) roll doubles with a pair of dice – any other roll means you remain in jail, or (4) after remaining in jail for three turns you must pay the $50 fine and must then leave on your next turn. Suppose you are in jail. Create a finite automaton that models getting out of jail. The final state should be named *Free*.

10. ★ A somewhat devious contest that can be presented to your friends is the following. The contestant needs to choose (in order) two numbers from the set $\{1, 2, 3, 4\}$. The contestant is (truthfully) told that 25% of the possible pairs are winning pairs. The winning pairs happen to be $(1, 1)$, $(2, 2)$, $(3, 3)$, and $(4, 4)$, so less than 25% of the contestants are likely to win (because people tend to avoid double numbers unless explicitly told that double numbers are legal).

    Design a finite automaton that models this contest. How many winning states are needed?

11. Design a finite automaton that accepts binary strings of length five as input. Let the number of 1s in the string be n and the number of 0s in the string be m. The finite automaton should accept the binary strings for which $|n - m| = 1$.

12. Show the output string and state changes if the following pairs of numbers are used as the numbers to add using the finite-state machine in Example 9.13 on page 550.
    (a) ★ 01101 and 00011
    (b) 11011 and 01010
    (c) 00111 and 00001
    (d) 11111 and 00001

13. Create a state diagram for the finite-state machine whose state table is shown below. S_0 is the start state.

	Transition Function			Output Function		
	Input			Input		
State	a	b	c	a	b	c
S_0	S_0	S_1	S_2	x	x	y
S_1	S_2	S_3	S_0	y	z	x
S_2	S_3	S_4	S_4	x	x	z
S_3	S_2	S_4	S_2	y	z	x
S_4	S_1	S_2	S_3	z	z	z

14. Create a state table for the finite-state machine whose state diagram is shown.

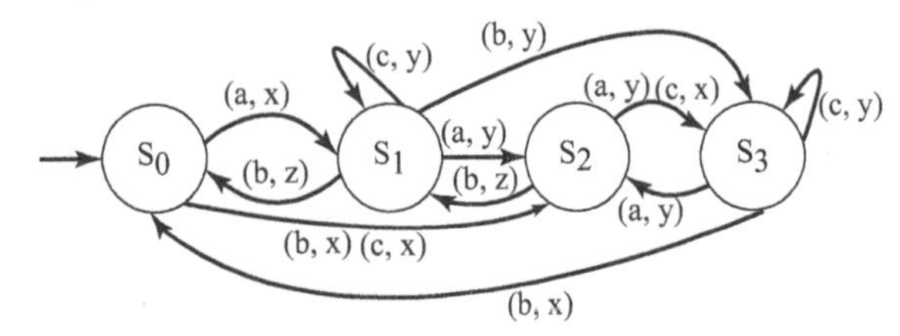

15. Describe the set of possible output strings for the following finite-state machine with output.

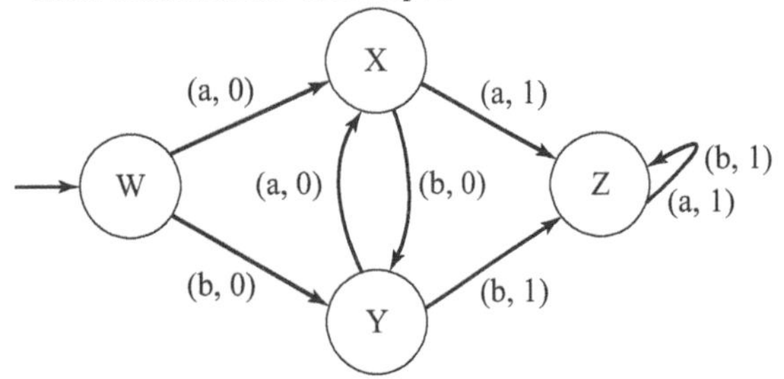

16. The finite-state machines in this problem all accept strings of 0s and 1s and generate output strings of 0s and 1s. Clearly label the states and transitions. Indicate which state is the start state. Produce both a state table and a state diagram.
    (a) Create a finite-state machine that complements its input string; 0s are turned into 1s and 1s are turned into 0s. (Thus, 001 as input will result in 110 as output.)
    (b) Create a finite-state machine that delays its input by one time unit. Use a minimum number of states.
    (c) Create a finite-state machine that copies bits in odd positions of the input string into the corresponding position of the output string. It should complement bits in even positions of the input string. Thus, 1011001 should produce the output 1110011.
    (d) Create a finite-state machine that delays the input by one time unit. The output will be a 0 if the two most recent input bits are different, and a 1 if the two most recent input bits are the same. (The input string 0110001 will result in the output −010110.) It will not need a − to flush the last bit because its output is comparing the two most recent bits. After the final bit is received, the output has already compared it to its predecessor.

17. The finite-state machines in this problem all accept strings whose characters are in $\{a, b, c\}$ and produce output strings of Ts and Fs. Clearly label the states and transitions. Indicate which state is the start state. You should produce a state diagram, not a state table.
    (a) Create a finite-state machine that outputs a T for every b in the input string, and an F for any other character.
    (b) ★ Create a finite-state machine that outputs Ts once the character pair ab (must be adjacent) is encountered. Before then, it should output Fs.
    (c) Create a finite-state machine that outputs a T if the two most recent input characters were ab (in that order). It should output an F in all other cases.
    (d) Create a finite-state machine that outputs a T if the most recent pair of input characters are identical. It should output an F in all other cases.

18. In the game of Monopoly®, players roll a pair of dice when it is their turn to play. The player moves a marker forward on the playing board by the sum of the numbers showing on the dice, and then completes his or her turn. However, if the roll of the dice results in "doubles" (the same number on each die), the player gets to roll the dice again (effectively taking another turn without waiting for the other players to have a turn). The same rules apply to the second roll of the dice. It is possible to get a third successive turn by rolling two doubles in a row. However, if on the third turn doubles result again, the player must "go to jail", which ends the turn—a bad outcome early in the game but a good outcome late in the game.

    Design a finite-state machine with output that models a single player's turn at Monopoly.

    Let Σ = {doubles, not doubles} and Γ = {turn done, roll again, go to jail}. You do not need to have transitions from any state that signals the end of the player's turn.

19. ★ The *twos complement* of an n-bit binary number is another n-bit binary number that acts like the negation of the binary number. One way to form the twos complement is to change all 0s to 1s and simultaneously change all 1s to 0s. Then add 1 (using binary addition with carries as necessary), keeping only the least significant n bits. For example, to find the 6-bit twos complement of the 6-bit binary number 000000, convert it to 111111 and then add 1, producing 1000000. Now discard the left-most bit, resulting in 000000 (as should be expected for the negation of zero).

An alternative algorithm for calculating the twos complement is to read the number from right to left. As long as the bits are 0s, output a 0. Output a 1 for the first 1 that is read. After that, output $(1 - b)$ for each bit, b, that is read. For example, the twos complement of the 6-bit binary number 100100 is 011100. Note that $100100+011100 = 000000$ as long as the result is truncated to 6 bits.

Create a finite-state machine with output that takes a binary integer as input (in right-to-left order) and outputs the twos complement (also in right-to-left order). Specify the formal model and create a state diagram.

20. Two middle school students have devised a clever code for passing notes during class. The code ensures the privacy of their communications if the teacher intercepts a note. The code works by using every third letter in the message to carry the real content. The other symbols (including spaces and punctuation) are to be ignored. The true message starts with the first letter in the note. For example, the note might read

Ice milk in kopeks alms at tache.

The secret message is "I like math".

(a) Design a finite-state machine with output that decodes these secret messages. Use the Greek letter λ to represent an empty output.

(b) Suppose the code is changed so that all letters up to and including the first lowercase "e" are ignored. The next letter, and every third letter thereafter, will constitute the secret message. Modify the finite-state machine in part (a) to decode these new messages.

21. A nonnegative integer, written in binary form, can be divided by 2 simply by deleting the bit in the 0s place and shifting all other bits one place to the right. Note that integer division truncates the fractional part (so dividing the integer, n, by 2 results in $\lfloor \frac{n}{2} \rfloor$). Design a finite-state machine with output which outputs $\lfloor \frac{n}{2} \rfloor$ when given the nonnegative binary integer, n, as input.

9.3 Formal Languages

The motivation for much of this section is the need to understand and formalize the translation (compilation) of human-readable computer programs into machine-executable form. The material will be presented in a more general context because some of the ideas have been borrowed from the research of linguists.

The initial concepts are derived from some of the features of languages such as English. The language can be expressed using words. The words, in turn, are composed by concatenating symbols from an alphabet of symbols or characters.[13]

DEFINITION 9.6 ***Symbols; Alphabet; Word; Null String***

An *alphabet* is a finite, nonempty set, Σ, of elements (called *symbols*). In this context, any finite string of symbols from Σ will be called a *word*. The *null string* (also called the *empty string*) will be denoted by λ. It is the zero-length string that contains no characters.

Some authors use *vocabulary* in place of *alphabet*. They may also use *sentence* instead of *word*. This alternative terminology will not be used here, but you are likely to encounter it elsewhere.

You should think of the null string as an actual entity; λ is a valid word (a string of 0 symbols from Σ). It is not the same as having no word at all.[14]

It is easy to concatenate symbols from the alphabet together into arbitrary words. It is more useful to consider some subset of all the possible words. For example, the English language does not include the word *abbaswxzwp* but does include the other space-separated strings in this sentence.

[13]Concatenating the symbols π, μ, and ϵ results in $\pi\mu\epsilon$.

[14]If you are familiar with the mechanism used by UNIX systems to store strings in memory, this will be easy to understand. UNIX systems always use a special bit pattern consisting of all 0s to mark the end of a string. That bit pattern is called the "null character". The null character uses the same size memory location as every other character, so the empty string takes one memory location to store. On the other hand, a nonexistent string uses no memory at all.

DEFINITION 9.7 *Language*

The set of all finite-length words using symbols from Σ will be denoted Σ^*. A *language over* Σ is a subset of Σ^*.

Notice that *any* subset of Σ^* is a language. We will need additional mechanisms to sort through all the possible languages and focus on the useful ones.

You should note well that Σ is a finite set and also observe that all elements in Σ^* are finite strings of symbols. However, Σ^* is an infinite set.

EXAMPLE 9.14 **A Simple Language**

Let $\Sigma = \{a, b, c\}$. Then $\Sigma^* = \{\lambda, a, b, c, aa, ab, ac, ba, bb, bc, \ldots\}$. One interesting language over this Σ is the subset of words that start with an a, end with a c, and have zero or more bs in the middle: $L = \{ac, abc, abbc, abbbc, \ldots\}$. ■

How do we specify a language? We need some mechanism that simultaneously excludes the unwanted words and describes the words that are in the language. In the previous example, I used a simple description in English. Other, more precise mechanisms are available.

The mechanisms that will be presented in this chapter are quite diverse. One mechanism will involve the notion of a *grammar* and will use rules called productions to specify the words we want in the language. Another option will use finite-state machines. Finally, a recursive technique will be introduced for languages that have a sufficiently nice structure.

As a preview, consider the finite automaton in the next example.

EXAMPLE 9.15 **Recognizing Words Using a Finite Automaton**

The finite automaton shown in Figure 9.11 will take input strings consisting of as, bs, and cs. It will recognize any string that is in the language, L, defined in Example 9.14. The state P indicates that the input string is at least partially compatible with the language. The final state recognizes the string, the state *No* indicates that the string should be rejected as a member of L.

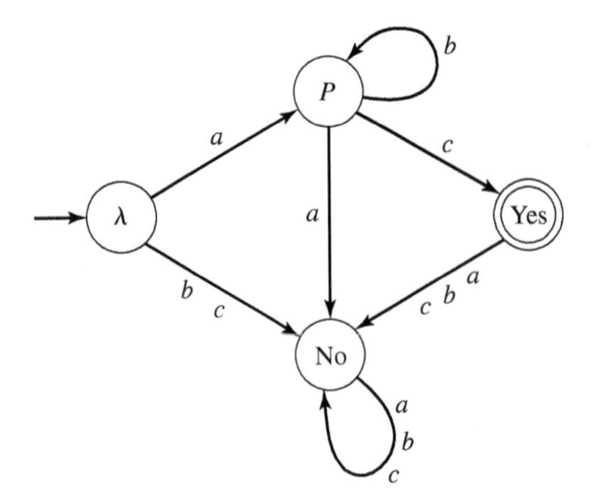

Figure 9.11 A finite automaton to recognize strings in L.

■

9.3.1 Regular Grammars

A grammar is one mechanism for specifying a language. A grammar consists of some symbols and some rules for transforming symbols into other symbols, eventually arriving at a word in a language. The nature of the transformation rules will determine the nature of the resulting language. In this section, a class of languages called *regular languages* will be discussed. More complex classes of languages will be discussed in Section 9.6.1.

DEFINITION 9.8 ***Regular Grammar***

A *regular grammar*, $\mathcal{G} = (\Sigma, \Delta, S, \Pi)$, consists of

- An alphabet, Σ, also called the set of *terminal symbols*.
- A set, Δ, of *nonterminal symbols*.
- A nonterminal symbol, S, called the *start symbol*.
- A set, Π, of replacement rules called *productions*.

The productions are all in the form

$$N \to \nu$$

where N is a nonterminal and $\nu \in (\Sigma \cup \Delta)^*$. The string ν satisfies the following conditions:

- It must contain at most one nonterminal symbol.
- If ν contains a nonterminal, it must be the rightmost symbol.
- If the only terminal symbol is λ, then there can be no nonterminal symbol.
- The string ν must contain at least one terminal symbol.

Notice some conventions in notation: nonterminals are represented by uppercase letters, whereas terminal symbols are represented by lowercase letters. Any string that is a mixture of terminal and nonterminal symbols will be represented by a lowercase Greek letter.

It is possible to drop the third assumption and allow productions of the form $N_1 \to N_2$ for nonterminals N_1 and N_2. In a course with only one chapter (rather than a full semester) devoted to this topic, it is better to add this restriction and avoid a few complications that arise if we allow such productions.

EXAMPLE 9.16

A Grammar for Example 9.14

Let $\Sigma = \{a, b, c\}$, $\Delta = \{S, B\}$, and let S be the start symbol. The productions are $\Pi = \{S \to aB, B \to bB, B \to c\}$.

The productions are used by performing a sequence of substitutions. For example, the word *abbc* can be obtained by the following substitutions:

- S is replaced by aB (using the production $S \to aB$).
- aB is replaced by abB (using the production $B \to bB$ to replace the B).
- abB is replaced by $abbB$ (again using the production $B \to bB$ to replace the B).
- $abbB$ is replaced by $abbc$ (using the production $B \to c$).

This is often abbreviated as

$$S \Rightarrow aB \Rightarrow abB \Rightarrow abbB \Rightarrow abbc$$

The word *ac* can be obtained by

$$S \Rightarrow aB \Rightarrow ac$$

■

DEFINITION 9.9 *Derivation*

The process of transforming the start symbol into a word over Σ is called a *derivation*. More generally, if v_1 and v_2 are strings in $\Sigma \cup \Delta$ such that there is a production in Π that enables v_1 to be replaced by v_2, we say that v_2 is *directly derivable* from v_1 and denote the fact by $v_1 \Rightarrow v_2$. If v_n can be obtained from v_1 by a sequence of substitutions using productions in Π, we say that v_n is *derivable* from v_1 and denote this as $v_1 \stackrel{*}{\Rightarrow} v_n$.

DEFINITION 9.10 *Language Generated by a Grammar*

Let $\mathcal{G} = (\Sigma, \Delta, S, \Pi)$ be a grammar. The *language generated by* $\mathcal{G}$ is the set of all words over Σ that can be derived from S. This set is denoted $L(\mathcal{G})$. Thus,

$$L(\mathcal{G}) = \{w \in \Sigma^* \mid S \stackrel{*}{\Rightarrow} w\}$$

If the grammar, $\mathcal{G}$, is a regular grammar, then $L(\mathcal{G})$ is called a *regular language*.

EXAMPLE 9.17 **A Simple Language**

The language generated by the grammar in Example 9.16 is precisely the language described verbally in Example 9.14. ■

EXAMPLE 9.18

Let $\Sigma = \{a, b, \lambda\}$ and $\Delta = \{S, X\}$. Let Π contain the productions

- $\mathrm{S} \rightarrow \lambda$
- $\mathrm{S} \rightarrow abX$
- $\mathrm{X} \rightarrow abX$
- $\mathrm{X} \rightarrow \lambda$

Let $\mathcal{G} = (\Sigma, \Delta, S, \Pi)$. Then

$$L(\mathcal{G}) = \{\lambda, ab, abab, ababab, \ldots\}$$ ■

The description of the language in the previous example can be simplified with the introduction of some useful notational conventions. The concept of concatenation is prominent. If $x =$"*abc*" and $y =$"*defg*", then the concatenation of x and y is the string "*abcdefg*".

Notation

Let x and y be strings. Then

- xy is the string formed by concatenating the two strings.
- $x\lambda = \lambda x = x$.
- $x^0 = \lambda$.
- $x^n = xx^{n-1}$ for $n \geq 1$.
- x^* represents zero or more copies of x. That is, $x^* \in \{x^n \mid n \geq 0\}$.
- x^+ represents one or more copies of x. That is, $x^+ \in \{x^n \mid n \geq 1\}$.

With this notation, the language in the previous example can be described as

$$L(\mathcal{G}) = \{(ab)^n \mid n \geq 0\}$$

String concatenation is associative: If x, y, and z are strings, then $(xy)z = x(yz)$. One additional set of notation will be useful.

Notation

Let A and B be sets of symbols or sets of strings. Then

- $AB = \{ab \mid a \in A \text{ and } b \in B\}$.
- $A^0 = \{\lambda\}$.
- $A^n = AA^{n-1}$ for $n \geq 1$.
- The *Kleene closure* of A is $A^* = \{\lambda\} \cup A \cup A^2 \cup A^3 \cup \cdots = \cup_{i=0}^{\infty} A^i$.
- $A^+ = A \cup A^2 \cup A^3 \cup \cdots = \cup_{i=1}^{\infty} A^i$.

EXAMPLE 9.19

Notation Practice

Let $x =$ "ab" and $y =$ "c". Let $A = \{r, s, t\}$ and $B = \{u, v\}$. (*Note*: I have used the quotation marks to emphasize that x and y are strings. In this context, the quotation marks are often omitted.)

- $x^3 =$ "$ababab$"
- $x^2y =$ "$ababc$"
- $(x^2y)^3 =$ "$ababcababcababc$"
- $x^* \in \{\lambda,$ "ab", "$abab$", "$ababab$", $\ldots\}$
- $x^* \in \{\lambda\} \cup x^+$
- $AB = \{ru, rv, su, sv, tu, tv\}$
- $B^3 = \{uuu, uuv, uvu, uvv, vuu, vuv, vvu, vvv\}$ ■

✔ Quick Check 9.5

Let $x = ab$ and let $y = cde$. Let $A = \{r, s\}$ and $B = \{t, u\}$.

1. Is "$abba$" in $\{x^n \mid n \geq 0\}$?
2. What is xy^2?
3. Are x^0 and A^0 the same?
4. What is AB?
5. How many elements are in AB^2A? ☑

The following examples show how grammars can be constructed to produce some well-defined languages.

EXAMPLE 9.20

$\{a^i b^j c^k \mid i, j, k \geq 0\}$

How do we construct a grammar for which $L(\mathcal{G}) = \{a^i b^j c^k \mid i, j, k \geq 0\}$?

The set consists of zero or more as followed by zero or more bs followed by zero or more cs. We need productions that will let us generate a bunch of as then move on to generate a bunch of bs and finally generate some cs, then quit. The null string also needs to be derivable. The grammar can be formally specified as $\Sigma = \{\lambda, a, b, c\}$, $\Delta = \{S, A, B, C\}$, and Π contains

$S \to A$	move to the a section
$A \to aA$	generate another a
$A \to B$	move to the b section
$B \to bB$	generate another b
$B \to C$	move to the c section
$C \to cC$	generate another c
$C \to \lambda$	quit

The string $aaccc$ can be derived by

$$S \Rightarrow A \Rightarrow aA \Rightarrow aaA \Rightarrow aaB \Rightarrow aaC \Rightarrow aacC \Rightarrow aaccC \Rightarrow aacccC \Rightarrow aaccc$$

Everything seems fine. However, the grammar produced is *not* a regular grammar since it contains the productions $S \to A$, $A \to B$, and $B \to C$.

Π : first attempt.

$S \to A$

$A \to aA$

$A \to B$

$B \to bB$

$B \to C$

$C \to cC$

$C \to \lambda$

Here is an alternative set of productions that do conform to the rules of regular grammars.

$S \to aA$	move to the *a* section
$S \to bB$	move directly to the *b* section
$S \to cC$	move directly to the *c* section
$S \to \lambda$	generate the null string and quit
$A \to aA$	generate another *a*
$A \to bB$	move to the *b* section
$A \to cC$	move to the *c* section (skipping the *b* section)
$A \to \lambda$	quit
$B \to bB$	generate another *b*
$B \to cC$	move to the *c* section
$B \to \lambda$	quit
$C \to cC$	generate another *c*
$C \to \lambda$	quit

■

The listing of the productions in the previous example is a bit long (and this is just a simple example). A shorthand notation is available to simplify the specification of productions. The shorthand notation uses the "pipe" symbol, |, to represent the word *or*. The productions

- $A \to aA$
- $A \to bB$

can both be written on one line as:

$$A \to aA \mid bB$$

The grammar in Example 9.20 can be written succinctly as $\Sigma = \{\lambda, a, b, c\}$, $\Delta = \{S, A, B, C\}$, with start symbol S, and Π contains

- $S \to aA \mid bB \mid cC \mid \lambda$
- $A \to aA \mid bB \mid cC \mid \lambda$
- $B \to bB \mid cC \mid \lambda$
- $C \to cC \mid \lambda$

✔ Quick Check 9.6

1. Produce a regular grammar that generates the language
$$\{a^i b^j c^k \mid i, k \geq 0, j \geq 1\}$$
Use at most four nonterminals. Write productions using the shorthand notation with |.
2. Derive the string *bbc* using the grammar from 1.

EXAMPLE 9.21

No Double Letters

Let $T = \{a, b\}$. How would we specify a grammar that generates the language of all non-null strings in T^* that contain no double letters? (The string *ababbab* would be rejected since it contains *bb*.)

On first thought, it might be possible to make every production always contain either an *ab* or a *ba*, depending on what the previous letter was (perhaps represented by the nonterminal we are replacing). However, this will not allow acceptable strings such as *a* or *aba* to be generated.

What is needed are productions that produce a terminal symbol, followed by a nonterminal that "remembers" which terminal it came with. Then we ensure that the nonterminal only leads to the other character. Formally, $\mathcal{G} = (\Sigma, \Delta, S, \Pi)$, where $\Sigma = \{a, b, \lambda\}$, $\Delta = \{S, A, B\}$ and Π contains the six productions

$S \to aA \mid bB$	can start with either letter
$A \to bB \mid \lambda$	switch letters or quit
$B \to aA \mid \lambda$	switch letters or quit

The string aba would be derived as

$$S \Rightarrow aA \Rightarrow abB \Rightarrow abaA \Rightarrow aba$$

■

Recall that the definition of a regular grammar specified that a production can have at most one nonterminal on the right-hand side and any such nonterminal must be the rightmost symbol. This restriction makes regular grammars easy to work with and causes regular languages to have a very simple structure. However, regular grammars are not always as expressively powerful as we need.

EXAMPLE 9.22 **A Language that Is Not Regular**

Let $\Sigma = \{a, b\}$. The language (subset of Σ^*) specified by $L = \{a^n b^n \mid n \geq 0\}$ cannot be generated by a regular grammar (and so is not a regular language).

The problem centers on the need to remember the number of as that were in the first part of the string while producing the final half of the string. You may have observed that having the only nonterminal in a production's right-hand side as the final symbol causes derivations to always expand from left to right. That means that all the as will have been generated (and "forgotten") before any of the bs can be produced. The productions will have no way to remember how many as were generated (or even that multiple as *were* produced).

A more formal justification for the claim that this language is not regular uses a result called the *pumping lemma*. Details can be found in [59]. ■

9.3.2 Exercises

The exercises marked with ★ have detailed solutions in Appendix H.

1. Which of the following sets of productions can be from a regular grammar?
   (a) $\{S \to aX \mid \lambda,\ X \to \lambda\}$
   (b) ★ $\{S \to aX \mid bY,\ X \to bY,\ Y \to aX \mid Y \mid b\}$
   (c) $\{S \to aX \mid bY,\ X \to bYa \mid b,\ Y \to bX \mid a\}$
   (d) $\{S \to aX \mid YZ,\ X \to aZ,\ Y \to b,\ Z \to aX \mid a\}$
2. Prove that string concatenation is associative.
3. ★ Show that $\{uvw\} \cup \{vw\} = \{u, \lambda\}\{vw\}$.
4. Prove that if A and B are sets of strings, then
   $$(AB) \cup B = (A \cup \{\lambda\})B$$
5. Let $\Sigma = \{0, 1\}$, $\Delta = \{S, X, Y\}$, and
   $$\Pi = \{S \to 01X \mid 1Y,\ X \to 0X \mid 1,\ Y \to 00Y \mid 1\}$$
   and let S be the start symbol. Which of the following strings are derivable from S? If the string *can* be derived, show a derivation; otherwise give some reason why it cannot.
   (a) 011 (b) 101 (c) ★ 100001
   (d) 010001 (e) 0101
6. Let $\Sigma = \{a, b, c\}$, $\Delta = \{S, W, X, Y\}$, and
   $$\Pi = \{S \to bW \mid cX \mid aY,\ W \to bW \mid aY,\ X \to aX \mid b,\ Y \to aX \mid cY \mid c\}$$
   and let S be the start symbol. Which of the following strings are derivable from S? If the string *can* be derived, show a derivation; otherwise give some reason why it cannot.
   (a) $abba$ (b) $aaab$ (c) $bbacab$
   (d) $bbaacb$ (e) $caabb$
7. Describe the language generated by the grammar $\mathcal{G} = (\Sigma, \Delta, S, \Pi)$, where $\Sigma = \{a, b, c\}$, $\Delta = \{S, X, Y\}$ and Π is
   (a) ★ $\{S \to aX \mid bY,\ Y \to bY \mid b,\ X \to aX \mid c\}$
   (b) $\{S \to aX \mid bY,\ Y \to bY \mid c,\ X \to aY \mid a\}$
   (c) $\{S \to aX,\ X \to aX \mid bY,\ Y \to cY \mid b\}$
8. Describe the language generated by the grammar $\mathcal{G} = (\Sigma, \Delta, S, \Pi)$, where $\Sigma = \{0, 1, \lambda\}$, $\Delta = \{S, X, Y, Z\}$ and Π is
   (a) $\{S \to 0X \mid 0Y \mid 1Z,\ X \to 1Z,\ Y \to 0Y \mid 0,\ Z \to 1Z \mid 1\}$
   (b) $\{S \to 0X \mid 1Y,\ X \to 1Y \mid 1Z,\ Y \to 0X \mid 0Z,\ Z \to 0\}$
   (c) $\{S \to 0X \mid 1Y,\ X \to 1Y \mid 1Z,\ Y \to 0X \mid 0Z,\ Z \to 0 \mid \lambda\}$

9. Let $\Sigma = \{a, b\}$. Create a regular grammar that generates each of the following languages.

(a) ★ $L = \{w \in \Sigma^* \mid w \neq \lambda$ and w does not contain $bb\}$

(b) $L = \{a^n baa \mid n \geq 0\}$

(c) $L = \{a^k bba^m \mid k \geq 1$ and $m \geq 0\}$

10. Let $\Sigma = \{0, 1, \lambda\}$. Create a regular grammar that generates each of the following languages.

(a) $L = \{(01)^n \mid n \geq 1\}$ (b) $L = \{(01)^n \mid n \geq 0\}$

(c) $L = \{001^n 00 \mid n \geq 2\}$

11. Each of the following statements is either true (always) or false (at least sometimes). Determine which option applies for each statement and provide adequate explanation for your choice.

(a) If Σ is a finite set of symbols, then Σ^* is also a finite set.

(b) A production in a regular grammar is not permitted to contain a terminal symbol on its left-hand side.

(c) ★ Let G be a regular grammar. Then $\lambda \in L(\mathcal{G})$.

(d) Let x and y be symbols. Then $(xy)^+$ represents one or more copies of the string "xy," concatenated together.

12. Let Σ be a set of symbols.

(a) Prove that $(\Sigma^*)^* = \Sigma^*$.

(b) Prove that $(\Sigma^+)^+ = \Sigma^+$.

13. ★ Let Σ be a set of symbols. Prove that $(\Sigma^+)^* = (\Sigma^*)^*$.

14. Is $(\Sigma^*)^+ = (\Sigma^+)^*$ always true? If it is always true, provide a proof; otherwise, provide a counterexample.

15. Find a grammar whose language has more words of length three than it has words of length 4.

16. Find sets A and B such that $(A \cup B)^* \neq (A^* \cup B^*)$.

17. Let $A = \{x^n y^m \mid n \geq 1, m \geq 0\}$. Use the $^+$ and * operators to express A more succinctly.

18. Let A and B be any two (not necessarily disjoint) sets of finite strings over a common set, Σ, of symbols. (That is, A and B are two languages over Σ). Prove that $|AB| \leq |A| \cdot |B|$.

19. Let A be a nonempty set such that $A^2 = A$.

(a) Prove that $A^+ = A$.

(b) Prove that $\lambda \in A$.
[*Hint* [44]: Consider the cases $|A| = 1$ and $|A| > 1$. For the second case, think about a non-null string in A with minimal length.]

(c) Prove that $A^* = A$.

20. Prove that it is impossible to have a language, $L \subseteq \Sigma^*$, where L contains at least two elements that are not λ, and with the additional property:

> Let $x, y \in L$, with $x \neq \lambda$ and $y \neq \lambda$. Then the concatenation xy is in L if and only if $x \neq y$.

(This problem is from [18].)

9.4 Regular Expressions

One of the most immediately useful topics in this chapter is the notion of regular expressions. Many computer programs and operating systems support the use of regular expressions to specify a pattern to find in a file. For example, if your word processor has opened a large document and you wish to find the phrase "separate the novices from the experienced" but cannot remember whether the first word is spelled "separate" or "seperate" and also cannot recall whether "novice" or "experienced" comes first, a regular expression would allow you to search and find the phrase in any of the possible combinations. One regular expression that would achieve this goal (and will be explained shortly) is "sep[ae]rate.+novice".

9.4.1 Introduction to Regular Expressions

DEFINITION 9.11 ***An Informal Definition of Regular Expressions***
Let Σ be an alphabet. A *regular expression* over Σ is a mechanism for building or recognizing or matching a subset of Σ^*. The subset is called the *regular set* generated by the regular expression. A regular expression serves as an abstract pattern that specifies which strings in Σ^* belong to the corresponding regular set.

EXAMPLE 9.23 **A Very Simple Regular Expression**

Let Σ be the ASCII character set. The regular expression `mom` matches the strings

Hi mom!

The moment I saw you, I fell in love.

mom

but not the strings

I love math!

Mom said I could have it!

Plant the mums over there.

since "mom" occurs as a substring of each string in the first group but not in any string in the second group. The strings in the first group are all part of the regular set specified by the regular expression "mom". ■

If directly specifying a substring were all there was to regular expressions, they would not be very useful or important. Fortunately, there are more powerful mechanisms for specifying patterns.

Many UNIX programs use some form of regular expressions. A simple application would be to locate all lines in a document that contain the letter q followed by any letter *except* u. A regular expression that would identify all such lines is

```
q[^u]
```

(an explanation of the role played by `[` and `]` and `^` will be given shortly).

We need to use several special characters (called *metacharacters*) to help specify the more interesting regular expressions. These characters will be presented next.

Details

Unfortunately, there is no standard set of rules for building regular expressions. In this section, a close approximation to a minimal set of generic rules (at least for UNIX systems) will be presented.

The simplest rule can be approximated by stating:

Most characters match themselves.

In the first example, the characters in the string "mom" matched themselves in the first group of strings. There are a few characters (the metacharacters) that do not match themselves. Instead, they serve other roles.

Metacharacters

The characters on the next line have special meanings when they appear in a regular expression:

```
$   ^   [   ]   |   (   )   \   .   "   *   ?   +
```

Their meanings are described in the following discussion. A summary description can be found on page 598.

- The `$` character matches the end of a line. For example, the regular expression

  ```
  mom$
  ```

  will only match strings for which "mom" occurs as the last characters on the line (followed by the newline character).
- The `^` character matches the beginning of a line. (There is another usage of the `^` metacharacter if it is used with a `[]` pair. This will be described in the next bullet.) The regular expression

  ```
  ^mom$
  ```

  will only match strings for which "mom" occurs as the only characters on the line.
- The `[` character initiates a `[]` pair. A regular expression consisting of a pair `[]` with a set of characters inside matches any one character from the set. A regular expression consisting of `[^]` with a set of characters after the `^` matches any one character that *doesn't* occur inside the `[]` pair.

EXAMPLE 9.24 **Using [] Pairs**

The regular expression

```
[02468]
```

matches any string with at least one even digit.

The regular expression

```
[^0123456789]
```

matches any string with at least one character that is not a digit.

The regular expression

```
q[^u]
```

will match any string that contains a q followed by any character other than u.

The regular expression

```
[0-9]
```

matches any string with at least one digit. The "–" character acts as a range operator.

If the "–" character is one of the characters you want as a choice in the [] pair, it needs to be the first character so that it is not interpreted as a range operator. The regular expression `[-+*/]` matches any one of the four arithmetic operators. A common method for specifying a single alphabetic character is `[a-zA-Z]`. ■

- The metacharacter | indicates an alternative. A regular expression containing a | matches any string that contains either the left or the right alternative. For example,

  ```
  a|b
  ```

  matches any string that contains either an a or a b. The regular expression

  ```
  a|b|c|d
  ```

  matches any string that contains one of the first four lowercase letters.
- The `()` metacharacters are used to group characters into subpatterns of the regular expression. For example, the regular expression

  ```
  s(a|e|i)t
  ```

  matches any string that contains *sat*, *set*, or *sit* as a substring, but does not match *seat* or *seit*.
- The \ metacharacter is used to convert a metacharacter back into a normal symbol (so it can match itself). For example, the regular expression

  ```
  \$5
  ```

  will match any string containing the monetary unit `$5`.
  A DOS path name can be specified using

  ```
  C:\\Classes\\CWC
  ```

  which matches any string containing "`C:\Classes\CWC`" as a substring.

- The . metacharacter matches any single character except a newline.
- The " metacharacter is used to surround a regular expression. Everything between a pair of " characters is treated as the regular expression. The main reason for this is for use with software that accepts regular expressions as command-line arguments. The " characters keep the operating system from trying to evaluate the regular expression as a part of a command.
  Notice that you need a straight quote character, ", on each end. Do not use the matching quotes, " and ".

The role played by the remaining metacharacters will be defined in the next subsection.

Before attempting to create regular expressions, it is helpful to locate computer software that you can use to check your own answers. If you have access to a UNIX operating system, the standard utilities *grep* and *egrep* can be used.[15]

An easier approach is to use the simple utility at

http://www.mathcs.bethel.edu/~gossett/cgi-bin/test-regex.pl

to test your regular expressions.

To use the utility to validate a regular expression for finding a line with at least one even digit, use the following regular expression:

```
[02468]
```

Then type the target string to see if it matches the regular expression. The string "`I am 3 years old.`" does not match, but "`My lunch cost $8.46 today.`" does match. The screen grabs in Figures 9.12 and 9.13 show the testing for the second string. Notice that only the first possible match is echoed.

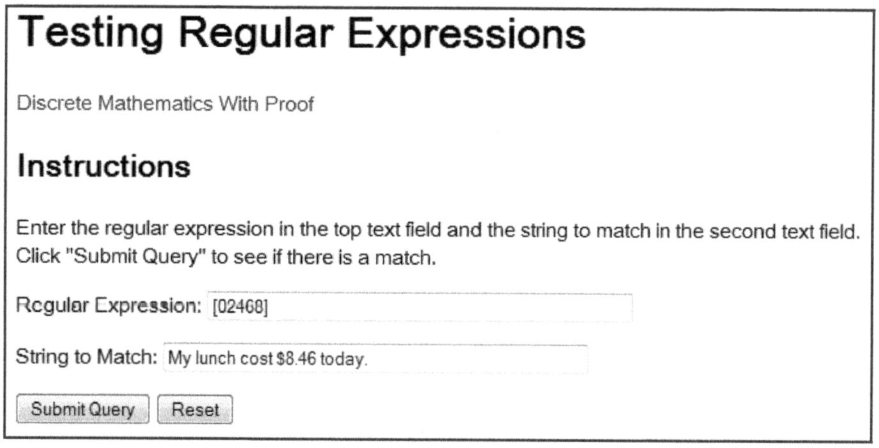

Testing Regular Expressions

Discrete Mathematics With Proof

Instructions

Enter the regular expression in the top text field and the string to match in the second text field. Click "Submit Query" to see if there is a match.

Regular Expression: [02468]

String to Match: My lunch cost $8.46 today.

Submit Query Reset

Figure 9.12 Testing a regular expression.

Results

Regular Expression	[02468]
The String	My lunch cost $8.46 today.
Match?	8

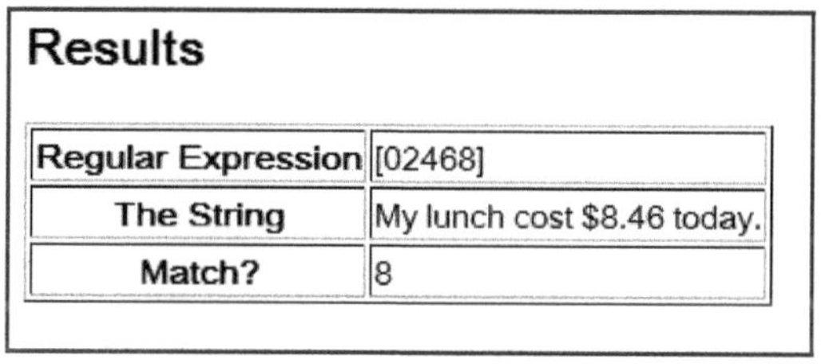

Figure 9.13 The result from testing a regular expression.

[15]For details on use, see the man pages: man grep. You may find it helpful to use the metacharacters ^ and $ at the beginning and end of your pattern.

Quick Check 9.7

1. Write a regular expression that matches the words *ear* and *eat*.
2. Write a regular expression that matches any word with an *i* that is *not* followed by an *e* or a *z*.

Repetition Specification

The real power of regular expressions comes from the ability to use concatenation, alternation, and Kleene closure (page 557). Kleene closure provides the ability to repeat a subpattern 0 or more times. As a convenience, most regular expression specifications add the ability to specify 0 or 1 occurrences and 1 or more occurrences.

- The `*` metacharacter matches zero or more copies of the immediately preceding character or subpattern.

EXAMPLE 9.25

Using *

The regular expression

```
(a|b)*cd*
```

will match any string that contains any finite (possibly empty) sequence of *a*s and *b*s, followed by exactly one *c*, followed by zero or more *d*s.

The following strings are in the regular set generated by "`(a|b)*cd*`".

c
aabac
xyzbbacdddnm
bbbbcd
abdcd (Think carefully about this one.)

The string *abdd* will not be matched. ■

- The `?` metacharacter matches either zero or one copy of the immediately preceding character or subpattern. The regular expression

```
wan?d
```

will match strings containing either *wad* or *wand*, but not the string *wannd*.

- The `+` metacharacter matches one or more copies of the immediately preceding character or subpattern. Thus, the regular expression

```
a.+z
```

will match any string containing an *a* and a *z*, in sequence, with at least one other character between them. The strings *a*8*z* and *a*   *z* would fit this criterion.

Quick Check 9.8

1. Write a regular expression that matches any positive integer.
2. Write a regular expression that matches any integer.

Precedence

As usual, the major regular expression operators have an agreed-on precedence. Subexpressions that are inside parentheses are evaluated first. Kleene closure, X^*, has the next highest priority, followed by concatenation, XY, and then alternative, $X \mid Y$.

EXAMPLE 9.26

Precedence

The regular expression $ab^*|c$ matches substrings that either begin with a single a, followed by zero or more bs or else contain exactly one c. That is, it generates the regular set $\{ab^n \mid n \geq 0\} \cup \{c\}$.

The regular expression $a(b^*|c)$ matches all substrings that begin with a single a, followed by either zero or more bs or else followed by one c. The associated regular set is thus $\{ab^n \mid n \geq 0\} \cup \{ac\}$.

The regular expression $(ab)^*|c$ matches all substrings that begin with any collection of zero or more abs or else exactly one c. The associated regular set is thus $\{(ab)^n \mid n \geq 0\} \cup \{c\}$. ■

The Formal Definition

DEFINITION 9.12 ***A Formal Definition of Regular Expressions***

Let Σ be an alphabet. A *regular expression* over Σ is defined recursively by the following:

- The empty set, $\emptyset$, is a regular expression.
- The null string, λ, is a regular expression.
- The symbol, a, is a regular expression for every symbol $a \in \Sigma$.
- If the symbols or strings, A and B, are regular expressions, then their concatenation, AB, is also a regular expression.
- If the symbols or strings, A and B, are regular expressions, then $A|B$ is also a regular expression.
- If the symbol or string, A, is a regular expression, then A^* is also a regular expression.

The metacharacters introduced in the previous subsections provide convenient mechanisms for implementing this definition. For example, the regular expression

```
^$
```

implements the regular expression λ. The `[]` pair is an alternative (more compact) mechanism for implementing | alternation when the subpatterns in the alternative are single symbols of Σ. The `?` and `+` metacharacters could be eliminated by using `*` and | in combination, but it is convenient to keep them.

EXAMPLE 9.27

Novices and Experts

Recall the example used to introduce the topic of regular expressions:

> Your word processor has opened a large document and you wish to find the phrase "separate the novices from the experienced" but cannot remember whether the first word is spelled "separate" or "seperate" and also cannot recall whether "novice" or "experienced" comes first.

A regular expression that would achieve this goal is "`sep[ae]rate.+novice`". This will match the initial letters "sep", then match either an a or an e (depending on which actually appears in the document). It will then match an arbitrary nonempty string followed by the letters *novice*. The arbitrary string will be either "the" or "the experienced from the" (again depending on the order in which the words *novice* and *experienced* actually appear in the document).

Of course, the regular expression, "`sep[ae]rate.+novice`", will also match strings such as "separate the young novices before they start a fight". However, the context assumes that any other such match is unlikely, so it is not worth the effort to create a more precise regular expression (but see Exercise 6 on page 568 for an improved version). ■

9.4.2 Perl Extensions

The programming language Perl has a number of useful extensions to the regular expression constructors presented so far. A few of these extensions are presented in Table 9.8. Perl uses regular expressions within a very powerful pattern matching operator that allows you to specify whether the match should be case sensitive or not and also allows the use of variables within the pattern. See a Perl text for the complete details.

TABLE 9.8 Perl Extensions for Regular Expressions

Symbol	Represents
`\n`	newline
`\r`	carriage return
`\t`	tab
`\f`	formfeed
`\d`	a digit, same as [0-9]
`\D`	a nondigit
`\w`	a (single) word character (alphanumeric), same as `[_0-9a-zA-Z]`
`\W`	a nonword character
`\b`	a word boundary (between `\w` and `\W` in some order)
`\B`	not a word boundary
`\s`	a whitespace character, same as `[ \t\n\r\f]`
`\S`	a nonwhitespace character

The following examples illustrate more complex uses of regular expressions (with Perl extensions).

EXAMPLE 9.28

Finding an HTML Tag

Hyper-Text Markup Language (HTML) is used to define the logical structure of a Web page. One of the features that is often overlooked is the tag that defines a title to display on the top border of the Web browser, in a list of bookmarks, or in the list some search engines return.

The HTML tag that defines this title looks like

```
<title>Your Title Here</title>
```

The word *title* in the pairs of brackets could be in upper case in the old HTML standard. The regular expression listed next can be used to scan an HTML file to make sure it has a valid *title* tag. A regular expression that will locate a valid HTML title is:

```
(<title>|<TITLE>)(.*)(</title>|</TITLE>)
```

Figure 9.14 shows the result from searching a substring that contains a match (for the lowercase option).

Regular Expression	(<title>|<TITLE>)(.*)(</title>|<\TITLE>)
The String	<html><head><title>My Web Page</title></head>
Match?	<title>My Web Page</title>

Figure 9.14 Searching for an HTML title tag.

■

EXAMPLE 9.29 **E-mail Addresses**

The following Perl regular expression will allow a file to be scanned for an email address in the form `username@bethel.edu`.[16] It also allows additional characters between `@` and `bethel.edu`; for example, `username@homer.acs.bethel.edu`. Period, hyphen, and underscore characters are allowed in the username.[17]

```
((\w|\.|-|_)+)[@](\S*)bethel[.]edu
```

The regular expression requires one or more characters from the alphanumeric set (with periods, hyphens, and underscores also allowed). Then it must match an `@` character, followed by 0 or more nonwhitespace characters, followed by `bethel.edu`. Notice that the period needs to either be enclosed in brackets, `[.]`, or else backslash-escaped, `\.`, so that it is not interpreted as the match-almost-everything metacharacter. ■

The final example uses a rather complicated regular expression.

EXAMPLE 9.30 **Class Lists**

At one time, it was possible for an instructor at my university to receive an e-mail from the Administrative Computing Center that contained a list of all students enrolled in a class. However, the list contained lots of information (student name, student ID number, post office box number, class rank, major code, grading type, number of credits, student status, academic advisor). A typical list might look like the following fragment:

```
1 Anderson, Erik Vincent        12345   181 JR   COMP   LT   3.0   CR TurnquistB
2 Berget, Neil Jonathan         67890   375 SO   COMS   LT   3.0   CR Gossett,EJ
3 Crownhart, Brian Scott        24680   392 SO   COMP   LT   3.0   CR Gossett,EJ
```

Suppose I want to extract the name and PO number but ignore the rest of the information. I want a list that looks like

```
1 Anderson, Erik Vincent       181   |
2 Berget, Neil Jonathan        375   |
3 Crownhart, Brian Scott       392   |
```

that can be used for recording homework scores by hand.

The critical portion of a Perl script that accomplishes this task is as follows. The only part you need to understand is the regular expression that appears inside `m/ /` inside the `if` line.[18]

```
if (m/(\s*\d+\s\w+,\s\w+\s\D*)(\d+)(\s)(\s*\d+)/) {
  print "$1$4   |\n";
}
```

The regular expression looks for one or more leading digits (perhaps preceded by some whitespace), then a comma separated name: `(\s*\d+\s\w+,\s\w+\s\D*)`. It calls that part of the match $1. Then it tries to also match another digit (the student ID, which will be discarded): `(\d+)`, and then some more whitespace: `(\s)`, followed by the PO number (named $4): `(\s*\d+)` . If a successful match is made, the name (with leading number) and the PO are printed (with a trailing "|" to make things look nice). ■

---

[16]See the Chapter 9 section of http://www.mathcs.bethel.edu/~gossett/DiscreteMathWithProof/ for a fully general email-matching regular expression.

[17]To place the string `username@homer.acs.bethel.edu` directly into a Perl script, it would be necessary to backslash escape the @ symbol: \@.

[18]The m// characters tell Perl to match the regular expression inside the / pair. The variables $1, $2, and so on, match the subregular expressions inside the parentheses.

9.4.3 Exercises

The exercises marked with ★ have detailed solutions in Appendix H.

1. Show that the + operator (metacharacter) is not necessary. That is, show that the regular expression `x+` can be replaced by an equivalent regular expression that only uses concatenation and the | and * operators.
2. Write a regular expression that matches telephone numbers in the form `###-###-####`.
3. ★ Write a regular expression that matches words that start and end in vowels (the standard English vowels aeiou). Assume that only lowercase letters are being used. Assume that the word must have at least two letters.
4. Write a regular expression that matches any line of characters that *does not* contain any periods, commas, colons, or semicolons.
5. Create regular expressions that match single words with the following characteristics:
   (a) Containing at least one `q`.
   (b) ★ Containing a double lowercase vowel. A double lowercase vowel would be the same vowel twice, as in *book*.
   (c) Containing at least two double lowercase vowels. They can be the same double lowercase vowel, as in `beekeeper`.
6. Write a regular expression that will match only the eight phrases: "*X* the *Y* from the *Z*", where *X* is either "separate" or "seperate", and *Y* and *Z* are both either "novices" or "experienced", with $Y \neq Z$.
7. Write a regular expression that matches any non-null binary string.
8. Write a regular expression that matches a valid Pascal identifier. A Pascal identifier consists of a letter, followed by 0 or more letters and/or digits.
9. Write a regular expression that matches a C identifier. A C identifier consists of a letter or underscore, followed by 0 or more letters, digits, and/or underscores.
10. A hexadecimal constant is usually written as a 0, followed by either an *x* or an *X* and then a mixture of digits and letters from the set $\{a, b, c, d, e, f, A, B, C, D, E, F\}$ (for example, `0X3B6`). Write a regular expression that matches any hexadecimal constant.
11. Write a regular expression that matches the string "`\begin{definition}`".
12. Write a regular expression that matches a complete question. The sentence should start with a capital letter and end with a question mark. Assume that the only permissible non-alphabetic characters are commas and spaces, but two commas may not be adjacent. Allow capital letters in the middle.
13. Write a regular expression that matches a formal name. The name may have an optional title or honorific (such as Dr., Doctor, Miss, Mr.), a first name, an optional middle name or initial (initials will always have a period), a last name, followed by an optional period-free designation (such as Junior or IV). The designation should be preceded by a comma. All parts of the name should begin with uppercase letters. You may assume that all words, except a middle initial, contain at least two letters.
14. ★ Write a regular expression that matches a credit card number that contains an expiration date. The number may either be in the form "dddd dddd dddd dddd mm/yy" or the form "dddd-dddd-dddd-dddd mm/yy" or the form "dddddddddddddddd mm/yy," where "d," "m," and "y" represent digits.
15. Write a regular expression that matches a letter grade. Assume that letter grades must be in the following set: {A, A−, B+, B, B−, C+, C, C−, D+, D, F, I, W}. You may not use the regular expression "`(A|A-|B+|B|B-|C+|C|C-|D+|D|F|I|W)`", even if you fix the error.
16. Write a regular expression that matches any single lowercase word. For this problem, a word must contain one or more consonants and at least one vowel. (Assume that *a*, *e*, *i*, *o*, and *u* are the only vowels, so the word *why* will not be matched.)
17. Write a regular expression that matches a single line of text (one delimited by a beginning and end of line, with arbitrary strings in the middle).
18. ★ Is it possible to write a regular expression that matches any line of text that does not contain the string "Percival"? Explain your answer (possibly by producing such a regular expression).
19. Write a regular expression that matches any complete English word that contains at least two double letters (such as "bookkeeper"). Assume all letters are lowercase.
20. Write a regular expression that matches any complete English word that contains at least two adjacent vowels (such as "eager"). Assume all letters are lowercase.
21. Write a regular expression that matches a real number in standard decimal notation (with optional decimal point). A leading minus sign is optional; leading plus signs need not be matched. Leading digits before the decimal point are optional, as are trailing digits after the decimal point. Multiple leading 0s are acceptable for this problem.
22. An infix expression is written in the form *exp op exp*, where *exp* is any expression and *op* is a binary operator. Assume for now that expressions are either integers or one-letter variables. Assume that operators are one of the four standard arithmetic operators: $\{+, -, *, /\}$. Write a regular expression that matches infix expressions with these restrictions.
23. Extend the previous exercise to allow expressions of the form (*exp*) *op* (*exp*). The parentheses are optional, but the expression inside can now consist of either a real number or a one-letter variable. [*Hint*: First write a regular expression that matches just (*exp*) (parentheses optional) and uses integers or one-letter variables for the expression. Next, add real numbers. Finally, reintroduce the operators and two expressions. *Note*: The final expression will be long.]

9.5 The Three Faces of Regular

The main result of this section is that the input strings recognized by finite automata, the languages generated by regular grammars, and the sets generated by regular expressions are all the same.

A bit of review may be helpful. Suppose that we have an alphabet Σ. The symbols in Σ might be used as the input symbols for a finite automaton.[19] In that case, we are interested in which strings in Σ^* lead to a final state. On the other hand, if we are given a grammar, $\mathcal{G} = (\Sigma, \Delta, S, \Pi)$, with Σ as its set of terminal symbols, we are interested in which words in Σ^* belong to $L(\mathcal{G})$.[20] Finally, if we construct a regular expression over Σ, we are interested in the regular set (a subset of Σ^*) generated by the regular expression.[21]

We will show that for corresponding choices of the finite automaton, regular grammar, and regular expression, the subset of Σ^* will be the same in each case. This means that the three mechanisms are equivalent in expressive power; any subset of Σ^* that can be specified using one of the mechanisms can be expressed using the other two as well.

Figure 9.15 shows the relationships that will be proved.

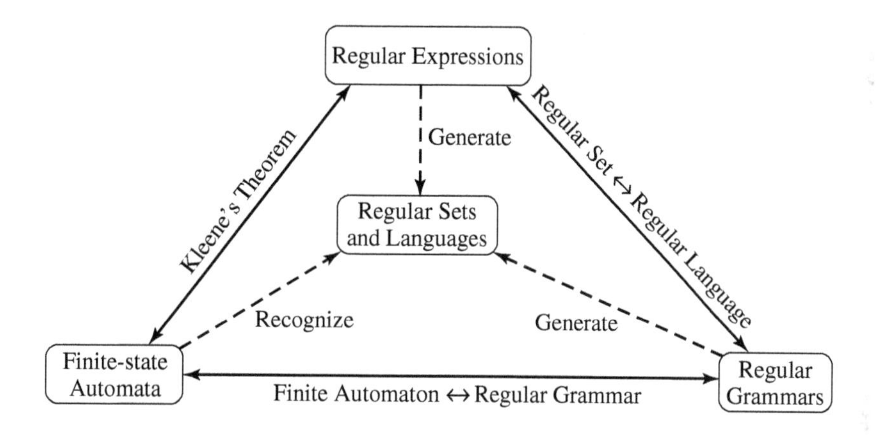

Figure 9.15 The three faces of *regular*.

The proofs of two of the theorems in the diagram are greatly simplified by introducing nondeterministic finite automata.

Definition 9.3
Finite Automaton

A *finite automaton*, A, is a model that consists of

- A finite set of states, $\mathcal{S}$.
- A set of input values, Σ.
- A transition function, $t(s, i)$, that maps state–input pairs to states.
- A special state called the *start state* (generically named s_0).
- A subset, $\mathcal{F} \subseteq \mathcal{S}$, of *final (or accepting) states*.

This can be represented as $A = (\mathcal{S}, \Sigma, t, s_0, \mathcal{F})$.

DEFINITION 9.13 ***Nondeterministic Finite Automaton***

A *nondeterministic finite automaton* is a finite-state machine that relaxes three requirements in the definition of a finite automaton.

- It is permissible to move between states without any input symbol to trigger the transition. Such transitions are called *λ-transitions* and denoted by labeling the transition with the null string.
- States do not need transitions associated with every input symbol.
- States may have more than one transition associated with the same input symbol.

The adjective *deterministic* is often used to emphasize that a finite automaton is *not* nondeterministic.

---

[19]Definition 9.3 on page 543.

[20]Definitions 9.7 and 9.10 on pages 554 and 556.

[21]Definition 9.12 on page 565.

EXAMPLE 9.31

A Nondeterministic Finite Automaton

Figure 9.16 illustrates the previous definition. Notice that

- There is a λ-transition from the start state to state x.
- There are many omitted transitions. For instance, there is no b transition from state x and there are no transitions at all from state z.
- There are two distinct a transitions leaving state x.

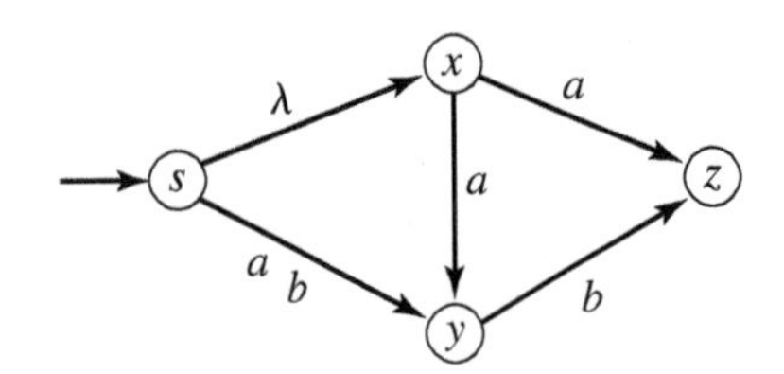

Figure 9.16 A nondeterministic finite automaton.

■

There seems to be significant inherent ambiguity associated with nondeterministic finite automata (hence the adjective *nondeterministic*). However, the next theorem indicates that the ambiguity can be removed without changing the functionality of the automaton.

THEOREM 9.14

Any nondeterministic finite automaton can be transformed into a deterministic finite automaton that recognizes the same set of strings.

A constructive proof can be found in either [49] or [59]. The key idea is to have the states in the new (deterministic) finite automaton represent subsets of states in the original (nondeterministic) finite automaton.

It is now time to start proving that regular grammars, regular expressions, and finite automata are three equally expressive mechanisms for specifying sets of strings (languages).

THEOREM 9.15 ***Finite Automaton → Regular Grammar***

Let Σ be the set of input symbols for a finite automaton $A = (\mathcal{S}, \Sigma, t, s_0, \mathcal{F})$. Let R be the subset of Σ^* that is recognized by A. Then there is a regular grammar $\mathcal{G} = (\Sigma, \Delta, S, \Pi)$ such that $L(\mathcal{G}) = R$.

The theorem will be illustrated with an example that will motivate the formal proof.

EXAMPLE 9.32

Converting Example 9.8 to a Grammar

Example 9.8 constructed a finite automaton that recognizes any binary string that contains two adjacent 0s. The automaton's state diagram is repeated in Figure 9.17.

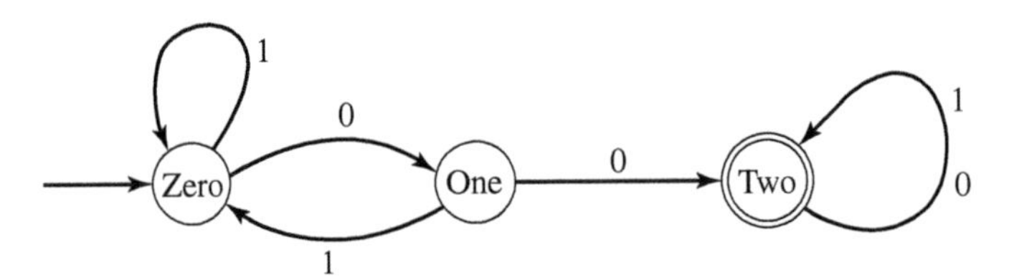

Figure 9.17 Two 0s state diagram.

The automaton can be formally specified by the following:[22]

[22]Look at Definition 9.3 on page 543.

- $\mathcal{S} = \{Zero, One, Two\}$ is the set of states.
- $\Sigma = \{0, 1\}$ is the set of input values.
- The transition function t is described by the state diagram.
- The start state is *Zero*.
- $\mathcal{F} = \{Two\}$ is the set of final states.

We need to convert this to a grammar, $\mathcal{G}$, that has a corresponding nature. The grammar can be specified as follows:[23]

- $\Sigma = \Sigma$ (the set of terminal symbols in $\mathcal{G}$ is the set of input values for A).
- $\Delta = \{Z, E, T\}$ (one nonterminal in $\mathcal{G}$ for each state in A—notice that E has replaced the notationally less suitable O).
- The start symbol is Z.
- Π is constructed from the transition function t. Transitions to nonfinal states will generate one production; transitions to final states will generate two productions.

Transition	First Production	Second Production
$t(Zero, 0) = One$	$Z \to 0E$	
$t(Zero, 1) = Zero$	$Z \to 1Z$	
$t(One, 0) = Two$	$E \to 0T$	$E \to 0$
$t(One, 1) = Zero$	$E \to 1Z$	
$t(Two, 0) = Two$	$T \to 0T$	$T \to 0$
$t(Two, 1) = Two$	$T \to 1T$	$T \to 1$

The pattern used is $t(s_k, i) = s_m$ becomes $S_k \to iS_m$. If s_m is a final state, also add $S_k \to i$.

Does $\mathcal{G}$ generate the strings that A recognizes? A few samples will raise confidence that it does. A will recognize 0101001. This string can be derived as

$$Z \Rightarrow 0E \Rightarrow 01Z \Rightarrow 010E \Rightarrow 0101Z \Rightarrow 01010E \Rightarrow 010100T \Rightarrow 0101001$$

A does not recognize the string 11. The only production that can begin at the start symbol, Z and produce 1s is $Z \to 1Z$. There is no way to eliminate the nonterminal after generating just two 1s, so $\mathcal{G}$ cannot generate the string 11. ■

The formal proof follows the construction hinted at in the previous example.

Proof of Theorem 9.15: We need to construct a grammar with the required properties.

We can begin by using the set, Σ, of input values for A as the set, Σ, of terminal symbols for the grammar (but this set may eventually need to be expanded to include λ). The set, Δ, of nonterminal symbols can be created from the set, $\mathcal{S}$, of states in A. For each state, $s_i \in \mathcal{S}$, create a nonterminal symbol, $S_i \in \Delta$. The nonterminal, S_0, that is created to correspond to the start state, s_0, will be the start symbol for the grammar.

Finally, we need to create a set of productions from which we can derive any string that A can recognize. There will be either one or two productions created for each transition specified by t.

Suppose that $t(s_k, i) = s_m$. Then the production $S_k \to iS_m$ will be added to Π. If $s_m \in \mathcal{F}$, then the production $S_k \to i$ will also be added to Π. If the start state is a final state, we also add $S_0 \to \lambda$ to Π and add λ to Σ if it is not already there.

Can every string recognized by A be derived from the start symbol in $\mathcal{G}$?

[23] See Definition 9.8 on page 555.

Let $i_1 i_2 i_3 \cdots i_j$ be an input string that is recognized by A, corresponding to the state changes $r_1 r_2 r_3 \cdots r_j$, where r_k is one of the states in $\mathcal{S}$, for each $k \in \{1, 2, \ldots, j\}$. Thus, the input symbol i_1 moves A from state s_0 to r_1. Then the input symbol i_2 moves A from r_1 to r_2. This continues until the symbol i_j moves A from r_{j-1} to $r_j \in \mathcal{F}$.

We can derive the string $i_1 i_2 i_3 \cdots i_j$ using the grammar. Begin with the start symbol, S_0, and apply the production $S_0 \rightarrow i_1 R_1$ (where R_1 is really one of the states, $S_k \in \Delta$, previously defined). Then replace R_1 using the production $R_1 \rightarrow i_2 R_2$. Such a production must exist because the transition $t(r_1, i_2) = r_2$ has been used by A. Continue in this fashion until the string $i_1 i_2 i_3 \cdots i_{j-1} R_{j-1}$ has been generated. The final substitution uses the production $R_{j-1} \rightarrow i_j$, which is in Π since r_j is a final state. In summary, we can derive $i_1 i_2 i_3 \cdots i_j$ as

$$S_0 \Rightarrow i_1 R_1 \Rightarrow i_1 i_2 R_2 \Rightarrow i_1 i_2 i_3 R_3 \Rightarrow \cdots \Rightarrow i_1 i_2 i_3 \cdots i_{j-1} R_{j-1} \Rightarrow i_1 i_2 i_3 \cdots i_j$$

Thus, any string recognized by A can be generated by $\mathcal{G}$ (so $R \subseteq L(\mathcal{G})$) . Will $\mathcal{G}$ generate any strings that A will not recognize?

Notice that each production of $\mathcal{G}$ follows a transition that exists in A. If a production permits the replacement of some nonterminal S_k by $i S_m$, then the input symbol i will move A from state s_k to state s_m. The only productions that enable the derivation to end are those that correspond to a final state in A. So a replacement of the form $S_k \rightarrow i$ corresponds to a transition from s_k to some final state. Therefore, $\mathcal{G}$ will never generate a string that A does not recognize [so $L(\mathcal{G}) \subseteq R$].

Since $R \subseteq L(\mathcal{G})$ and $L(\mathcal{G}) \subseteq R$, it must be the case that $L(\mathcal{G}) = R$. □

THEOREM 9.16 ***Regular Grammar → Finite Automaton***

Let $L(\mathcal{G})$ be the language generated by a regular grammar, $\mathcal{G} = (\Sigma, \Delta, S, \Pi)$. Then there exists a finite automaton, $A = (\mathcal{S}, \Sigma_A, t, s_0, \mathcal{F})$, which recognizes $L(\mathcal{G})$.

The following example demonstrates the major ideas in the formal proof.

EXAMPLE 9.33

Grammar to Finite Automaton

Let $\Sigma = \{a, b, \lambda\}$ and $\Delta = \{S, X\}$. Let $\Pi = \{S \rightarrow \lambda \mid abX,\ X \rightarrow bX \mid a\}$. Let $\mathcal{G} = (\Sigma, \Delta, S, \Pi)$. Then

$$L(\mathcal{G}) = \{\lambda, aba, abba, abbba, \ldots\}$$

The finite automaton, A, will be defined to have input values $\Sigma_A = \Sigma - \{\lambda\} = \{a, b\}$. The set, $\mathcal{S}$, of states will be built incrementally. The start symbol, S, for $\mathcal{G}$ will correspond to the start state, s_0, of A. Every other nonterminal symbol in Δ will correspond to a lowercase version of itself as a state in $\mathcal{S}$. The initial version of $\mathcal{S}$ for this example is therefore $\mathcal{S} = \{s_0, x\}$.

Because there is a production, $X \rightarrow a$, that has no nonterminals on the right-hand side (and also does not have λ as the right-hand side), we will add a new state f to $\mathcal{S}$. So we now have $\mathcal{S} = \{s_0, x, f\}$.

At this point it is possible to identify the final states. Any nonterminal symbol that has a production of the form $Y \rightarrow \lambda$ will be a final state. If the state f has been added, it will also be a final state. For this example, $\mathcal{F} = \{s_0, f\}$.

It remains to specify the transition function, t. In the process of specifying t it will be necessary to add additional states. The transition function will be specified by converting productions to transitions.

The production $S \rightarrow \lambda$ has already been accounted for by making s_0 a final state. The production $X \rightarrow a$ will become the transition $t(x, a) = f$. The production $X \rightarrow bX$ will become the transition $t(x, b) = x$.

The remaining production, $S \rightarrow abX$, will take more effort. The problem is the presence of multiple terminal symbols on the right-hand side. A finite automaton must accept single input values and specify a transition for each value. This can be accommodated by introducing a new intermediate state, s_1, that fills the gap between the input symbols a and b in the right-hand side, abX. Thus, $\mathcal{S} = \{s_0, x, f, s_1\}$. We now specify a transition on either side of s_1: $t(s_0, a) = s_1$ and $t(s_1, b) = x$.

We are almost done. It might be helpful to examine the partial finite automaton constructed so far (Figure 9.18).

Figure 9.18 Grammar to partial finite automaton.

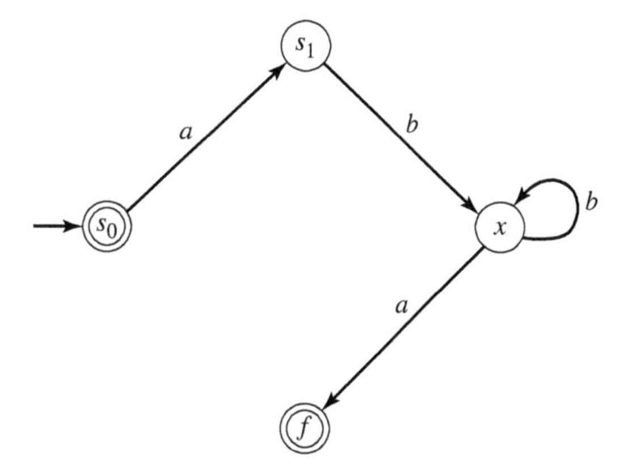

How should the transition function treat the missing input values at each state? Any other input will lead to a string that the grammar cannot generate. Thus, a state bh (which acts like a black hole) will be added. All other input symbols will drive A to bh and the string will *not* be recognized (Figure 9.19).

Figure 9.19 Grammar to finite automaton.

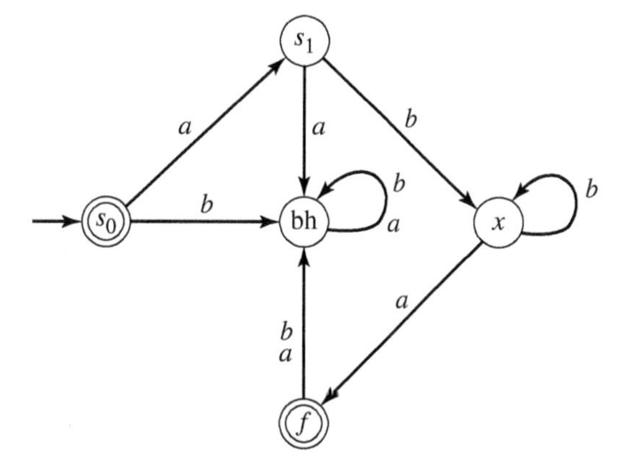

If you try a few strings, both in and out of $L(\mathcal{G})$, you should become convinced that this finite automaton recognizes exactly the set of strings in $L(\mathcal{G})$. ■

The previous example is a bit misleading. It has been carefully constructed to avoid a pair of messy possibilities (λ-transitions and multiple transitions with the same input symbol). In particular, a regular grammar may contain pairs of productions $X \rightarrow aY$ and $X \rightarrow aZ$. The previous example would indicate that the partial finite automaton in Figure 9.20 would be constructed. This leads to an indeterminate automaton.

Figure 9.20 Multiple transitions with the same input.

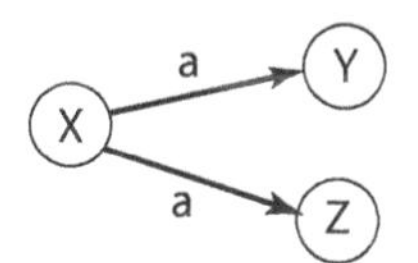

The next example introduces some of the messiness that Example 9.33 suppressed.

EXAMPLE 9.34 A Grammar That Adds Nondeterminacy

Let $\Sigma = \{a, b\}$ and $\Delta = \{S, X, Y\}$. Let $\Pi = \{S \to a, S \to aX, X \to bY, Y \to a, Y \to bY\}$. Then $L(\mathcal{G}) = \{a\} \cup \{ab^n a \mid n \geq 1\}$.

The process outlined in the previous example produces (Figure 9.21) a nondeterministic finite automaton (the start state has two a transitions).

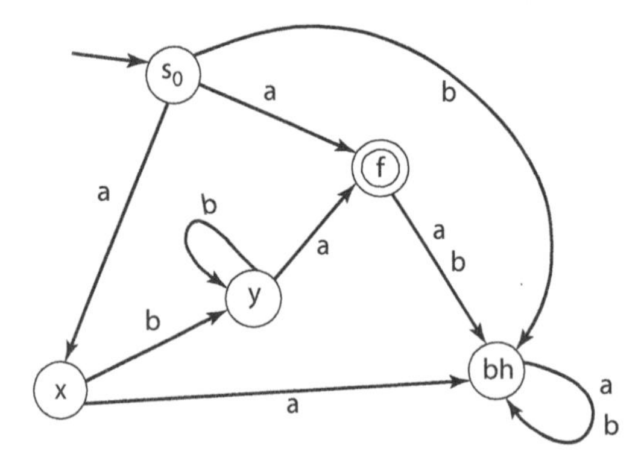

Figure 9.21 Grammar to a nondeterministic finite automaton.

It is easy (for this example) to eliminate the nondeterminacy by adding a new state (Figure 9.22) and then adding λ to Σ. Even if there were more than two transitions from a state, a sequence of λ-transitions to new states would remove the nondeterminacy. For example, $W \to aX|aY|aZ$ would produce Figure 9.23.

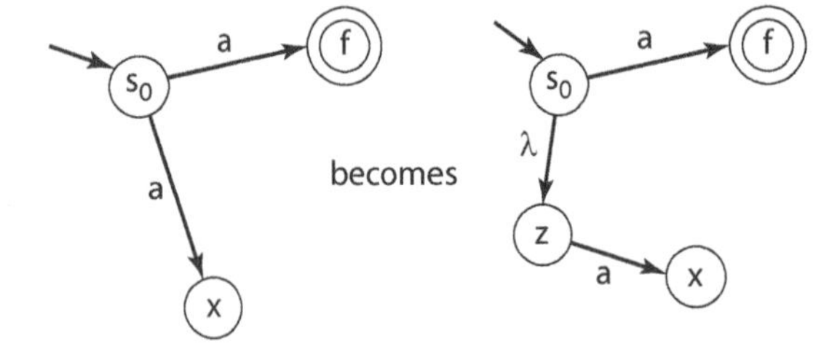

Figure 9.22 Adding a new state with a λ-transition.

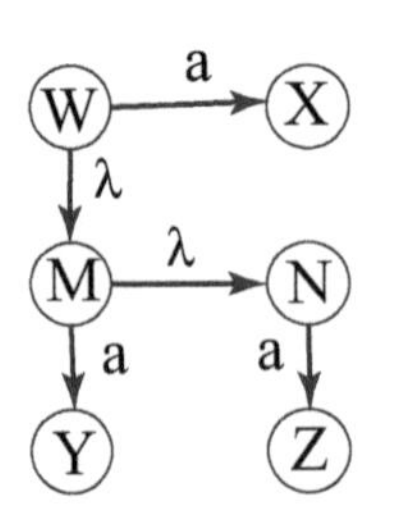

Figure 9.23. Three a transitions.

However, if there are multiple λ-transitions from some state, the previous trick will fail. An additional step (not shown here) would be required to convert this to a deterministic finite automaton. The full algorithm from Theorem 9.14 would be needed. ■

The formal proof follows the pattern of the previous two examples.

Proof of Theorem 9.16: Suppose we are given the grammar, $\mathcal{G} = (\Sigma, \Delta, S, \Pi)$. We need to construct a deterministic finite automaton, $A = (\mathcal{S}, \Sigma_A, t, s_0, \mathcal{F})$, which recognizes $L(\mathcal{G})$. The process begins by constructing a nondeterministic finite automaton, N, which recognizes $L(\mathcal{G})$.

The nondeterministic finite automaton N will have input values $\Sigma_N = \Sigma - \{\lambda\}$. Note that $\Sigma_N = \Sigma$ if $\lambda \notin \Sigma$. The set, $\mathcal{S}$, of states will be built incrementally. The start symbol, S, for $\mathcal{G}$ will correspond to the start state, s_0, of N. Every nonterminal symbol in Δ will correspond to a lowercase version of itself as a state in $\mathcal{S}$.

If there is a production whose right-hand side contains neither λ nor a nonterminal, we will add a new state f to $\mathcal{S}$.

At this point it is possible to identify the final states. Any nonterminal symbol that has a production of the form $Y \to \lambda$ will cause state y to be a final state. If the state f has been added, it will also be a final state.

It remains to specify the transition function, t. In the process of specifying t it may be necessary to add additional states. The transition function will be specified by converting productions to transitions.

Any production in the form $Y \to \lambda$ has already been accounted for by making y a final state. Any production $Y \to i$, where $i \neq \lambda$ is a single terminal symbol, will become the transition $t(y, i) = f$.

Productions of the form $X \to iY$ will become the transition $t(x, i) = y$.

The remaining productions are all in the form, $X \to i_1 i_2 \cdots i_k Y$, with $k \geq 2$. For each pair of adjacent terminal symbols, $i_j i_{j+1}$, in the production, introduce a new state, s_j (thus adding states $s_1, s_2, \ldots, s_{k-1}$). We now specify the transitions $t(x, i_1) = s_1, t(s_{k-1}, i_k) = y$, and $\{t(s_{j-1}, i_j) = s_j \mid j = 2, \ldots, k-1\}$.

At this point, N has been fully specified. The fairly routine verification that the set of strings recognized by N is exactly $L(\mathcal{G})$ will be left to the reader.

Using the constructive algorithm from the proof of Theorem 9.14, the nondeterministic finite automaton, N, can be converted to a deterministic finite automaton, A, that recognizes the language generated by $\mathcal{G}$, completing the proof. □

The finite automaton initially specified by the previous constructive proof is usually non-deterministic. (1) If productions of the form $X \to \lambda Y$ were allowed, then the finite automaton will contain λ-transitions. (2) There will usually be many transitions that the productions do not specify. (The formal proof has omitted the creation of the "black hole" state presented in Example 9.33.) (3) Finally, if there are productions of the form $X \to aY$ and $X \to aZ$, then there will be two a transitions leaving state x.

✔ Quick Check 9.9

1. (a) Convert the finite automaton defined by the following state diagram into an equivalent regular grammar.
   (b) Determine $L(\mathcal{G})$.

2. Create a finite automaton that recognizes $L(\mathcal{G})$ [and only $L(\mathcal{G})$], where the regular grammar $\mathcal{G}$ is defined by $\Sigma = \{a, b\}$, the start symbol is S, $\Delta = \{S\}$, $\Pi = \{S \to aS \mid baa\}$.

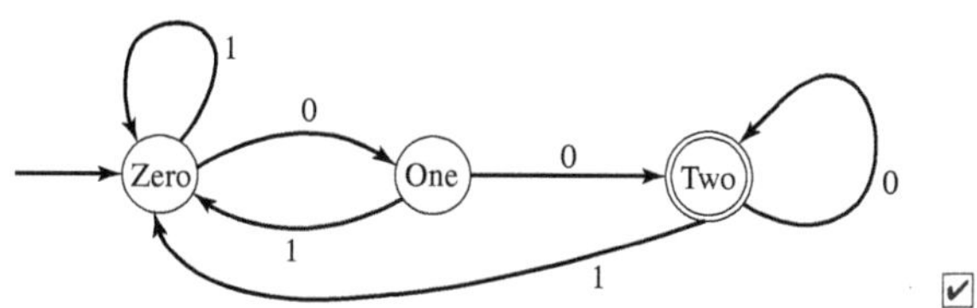

Recall that "regular set" is the name for the collection of strings generated by a regular expression. The following theorem shows that regular expressions and finite automata are equally expressive; any set of strings generated by a regular expression can be recognized by a finite automaton (without recognizing any additional strings), and vice versa.

THEOREM 9.17 ***Kleene's Theorem***

A set is regular if and only if it is recognized by a finite automaton.

Proof: The theorem will be proved by appealing to a pair of constructive proofs, each presented as a lemma.

The first lemma will show that any regular set can be recognized by a finite automaton. This will be accomplished by starting with the regular expression that generates the regular set and then using the recursive definition (page 565) of *regular expression* to *construct recursively* a nondeterministic finite automaton that recognizes exactly those strings in the regular set. The nondeterministic finite automaton could then be converted to a deterministic finite automaton by using Theorem 9.14.

The second lemma will show that it is possible to construct a regular expression that matches exactly those strings recognized by a finite automaton. □

Definition 9.12
Regular Expressions
Let Σ be an alphabet. A *regular expression* over Σ is defined recursively by:
- The empty set, $\emptyset$, is a regular expression.
- The null string, λ, is a regular expression.
- The symbol, a, is a regular expression for every symbol $a \in \Sigma$.
- If the symbols or strings, A and B, are regular expressions, then their concatenation, AB, is also a regular expression.
- If the symbols or strings, A and B, are regular expressions, then $A|B$ is also a regular expression.
- If the symbol or string, A, is a regular expression, then A^* is also a regular expression.

Before presenting the lemmas that prove Kleene's theorem, there is an easy corollary to Kleene's theorem that fills in the final equivalence on the diagram from page 569.

COROLLARY 9.18 ***Regular Set If and Only If Regular Language***
A subset $R \subseteq \Sigma^*$ is a regular set if and only if it is a regular language.

Proof: Suppose R is a regular set. Then Kleene's theorem implies that there is a finite automaton, A, that recognizes R. Theorem 9.15 asserts that there is a regular grammar that generates R. Since R can be generated by a regular grammar, R is a regular language (Definition 9.10).

If $R = L(\mathcal{G})$ for some regular grammar, $\mathcal{G}$, then Theorem 9.16 implies that there is a finite automaton that recognizes R. Kleene's theorem asserts that R is a regular set. □

9.5.1 Optional: Completing the Proof of Kleene's Theorem

LEMMA 9.19 ***Converting Regular Expressions to Nondeterministic Finite Automata***
Given any regular expression, R, it is possible to find a nondeterministic finite automaton that recognizes all strings, and only those strings, in the regular set generated by R.

Proof: The main idea is to go back to the recursive definition of regular expressions (repeated in the left margin) and build a finite automaton for each piece of the definition. If the finite automaton and the regular expression generate/recognize the same set of strings, and if the construction shows how to hook the pieces together, then the lemma will be proved. To that end, each part of Definition 9.12 will be discussed in order.

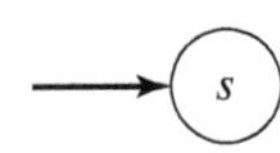

Figure 9.24. A finite automaton for $\emptyset$.

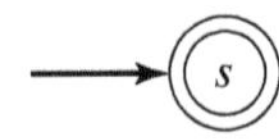

Figure 9.25. A finite automaton for λ.

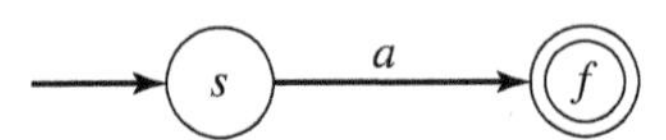

Figure 9.26. A finite automaton for symbol a.

$\emptyset$ The regular expression $\emptyset$ does not match any strings. Its regular set is the empty set. A nondeterministic finite automaton with a single, nonfinal state recognizes the empty set (Figure 9.24).

λ The regular expression λ matches the null string. A nondeterministic finite automaton with a single, final state recognizes only the null string (Figure 9.25).

a Every symbol, $a \in \Sigma$, is a regular expression that matches only itself. A nondeterministic finite automaton with a nonfinal start state and one final state will match that one string (Figure 9.26).

The remaining three parts of the definition require recursion. To that end, it is necessary to introduce some diagrammatic notation. Suppose that X is a regular expression for which a corresponding nondeterministic finite automaton, A_X, exists. Figure 9.27 will represent A_X. There may be no final states, exactly one final state, or several final states in A_X. The generic diagram shows two. The diagram does not show the intermediate transitions except to indicate that they exist. The dotted inner circle on the start state is to indicate that the start state may or may not be a final state. The subscripts indicate that this nondeterministic finite automaton corresponds to the regular expression X.

Figure 9.27 A_X.

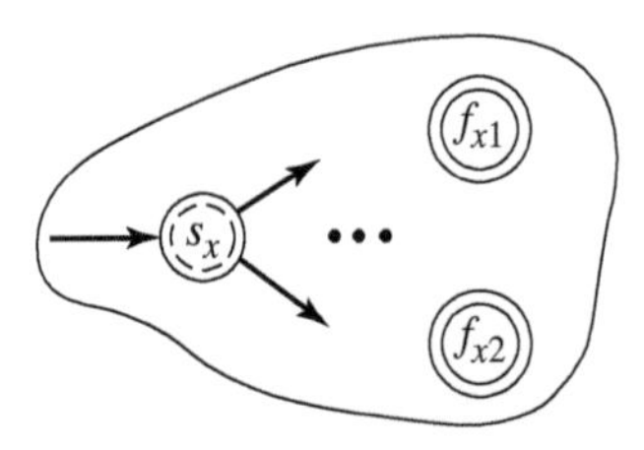

XY Suppose X and Y are regular expressions having corresponding nondeterministic finite automata A_X and A_Y. The concatenation, XY, is also a regular expression. A nondeterministic finite automaton that recognizes exactly those strings in the regular set generated by XY can be built from A_X and A_Y.

The following steps will create the new nondeterministic finite automaton. First, create a λ-transition from each final state in A_X to the start state of A_Y. Second, transform all final states in A_X into nonfinal states. The λ-transitions are needed so that once the string X is recognized, the new machine continues on to look for the string Y. The removal of final states from the A_X machine is needed so that the new machine does *not* recognize strings that match X but do not have a Y appended (Figure 9.28).

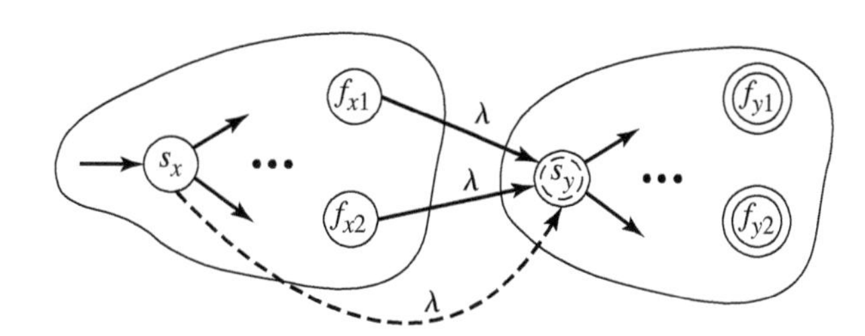

Figure 9.28 A nondeterministic finite automaton for XY.

X | Y Suppose X and Y are regular expressions having corresponding nondeterministic finite automata A_X and A_Y. The alternative, $X \mid Y$, is also a regular expression. A nondeterministic finite automaton that recognizes exactly those strings in the regular set generated by $X \mid Y$ can be built from A_X and A_Y.

First, a new start state, $s_{x|y}$, will be created. Then a λ-transition will be created from $s_{x|y}$ to the start states in A_X and A_Y. The λ-transitions allow the machine to either move to a submachine that recognizes X, or to a submachine that recognizes Y (Figure 9.29).

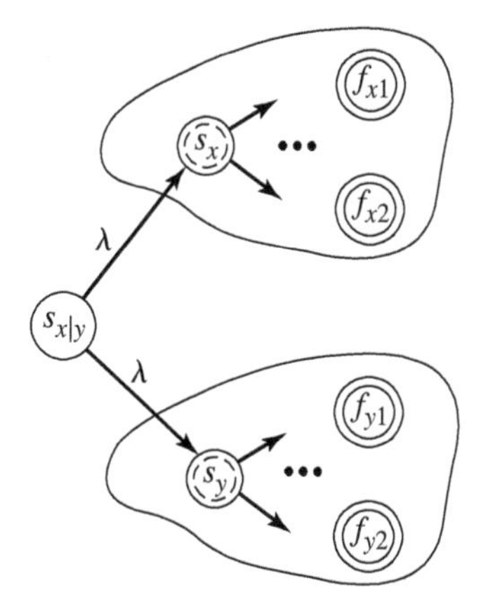

Figure 9.29. A nondeterministic finite automaton for $X \mid Y$.

X* Suppose X is a regular expression having corresponding nondeterministic finite automata A_X. The Kleene closure, X^*, is also a regular expression. A nondeterministic finite automaton that recognizes exactly those strings in the regular set generated by X^* can be built from A_X.

First, make the start state of A_X a final state (so that λ can be recognized). Then make a λ-transition from every final state in A_X to the start state. The λ-transitions enable the machine to look for an additional copy of X, once a valid copy is recognized in the input string (Figure 9.30).

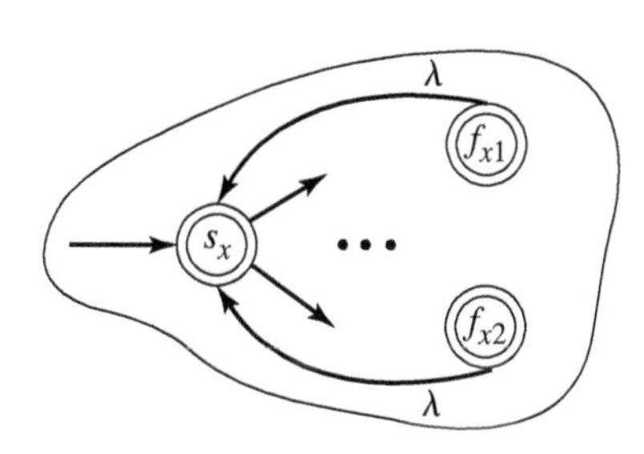

Figure 9.30 A nondeterministic finite automaton for X^*.

Because it is possible to provide a recursive construction that corresponds exactly to the recursive definition of *regular expression*, the construction completes the proof of the lemma. □

Before proceeding to the second lemma, it will be helpful to look at an example that illustrates the algorithm inherent to the constructive proof of Lemma 9.19.

EXAMPLE 9.35

Transforming a Regular Expression into a Nondeterministic Finite Automaton

The construction used to prove Lemma 9.19 can be used to convert the regular expression $ab^*a \mid c^*$ into a nondeterministic finite automaton that recognizes the regular set, $\{ab^n a \mid n \geq 0\} \cup \{c^m \mid m \geq 0\}$, generated by the regular expression.

The process begins by creating finite automata for the three regular expressions a, b, and c (Figure 9.31).

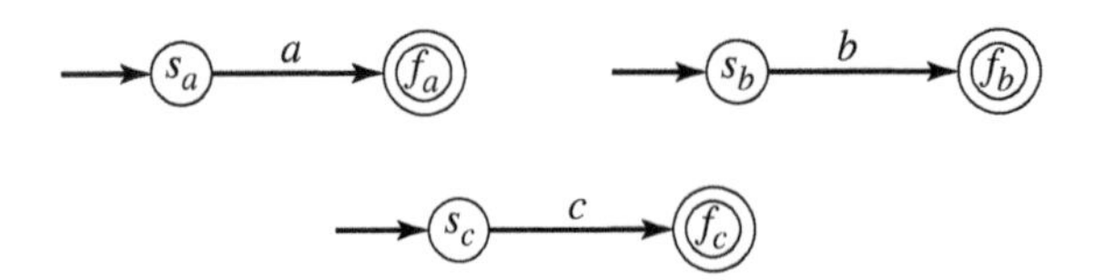

Figure 9.31 Finite automata for a, b, and c.

Next, the nondeterministic finite automata corresponding to b^* and c^* can be formed from the finite automata for b and c using the rule for Kleene closure (Figure 9.32).

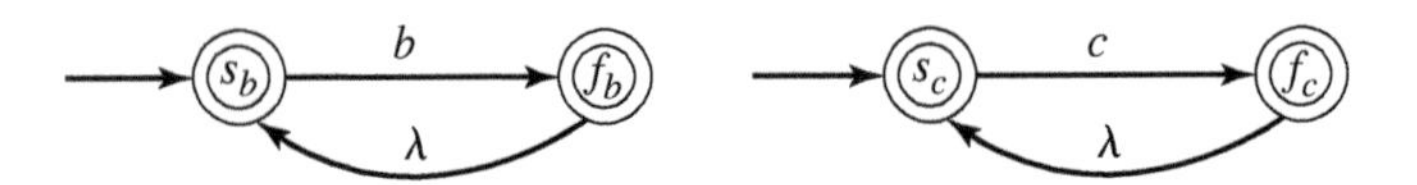

Figure 9.32 Nondeterministic finite automata for b^* and c^*.

The concatenation rule can be used to construct a non-deterministic finite automaton for the regular expression ab^* and then for ab^*a (Figure 9.33).

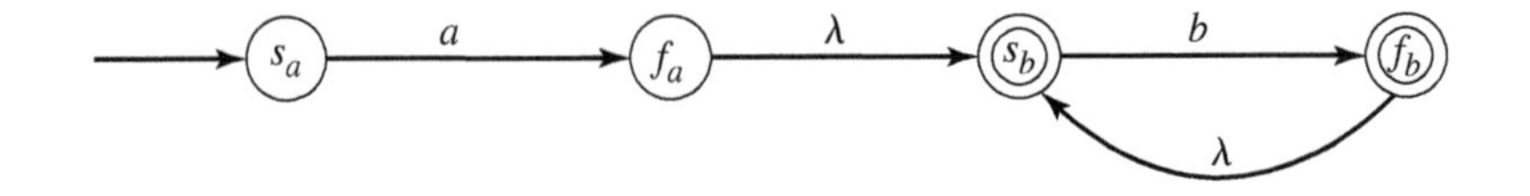

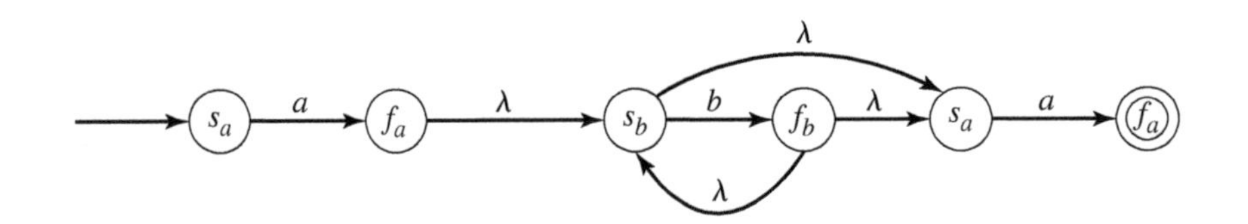

Figure 9.33 Nondeterministic finite automata for ab^* and ab^*a.

Finally, the rule for alternative can be used to complete the construction (Figure 9.34).

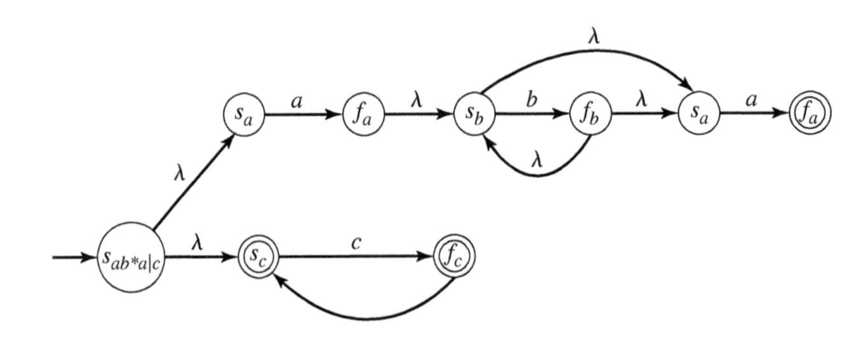

Figure 9.34 A nondeterministic finite automata that recognizes $\{ab^n a \mid n \geq 0\} \cup \{c^m \mid m \geq 0\}$.

■

> **LEMMA 9.20** ***Converting Nondeterministic Finite Automata to Regular Expressions***
>
> Given any nondeterministic finite automaton, A, it is possible to find a regular expression that generates all strings, and only those strings, recognized by A.

Proof Highlights: The translation from nondeterministic finite automaton to regular expression is fairly straightforward. The key observations are captured in Figures 9.35 – 9.37.

If the nondeterministic finite automaton contains subsets A_X, A_Y, and so on. that can be translated into regular expressions X, Y, and so on, then a configuration like Figure 9.35 can be converted to the equivalent regular expression $aX \mid bY \mid cZ$.

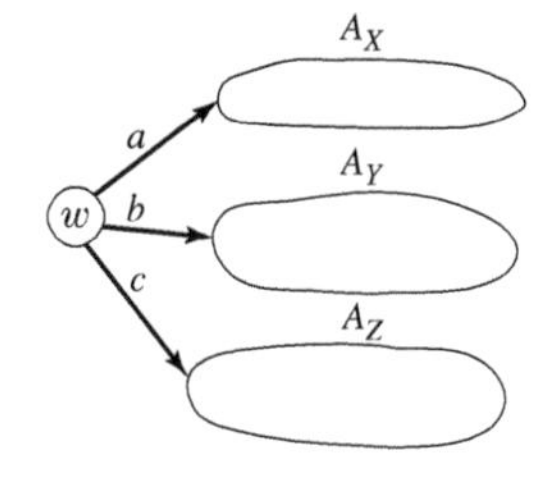

Figure 9.35 Parallel subdiagrams.

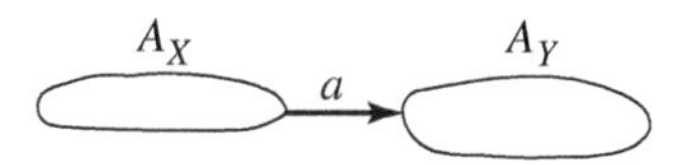

Figure 9.36. Sequential subdiagrams.

If the nondeterministic finite automaton contains subsets, A_X and A_Y, that can be translated into regular expressions, X and Y, then a configuration like Figure 9.36 can be converted to the equivalent regular expression XaY.

Finally, if the nondeterministic finite automaton contains a subset, A_X, that corresponds to the regular expression, X, and if a well-defined loop transition is added to A_X (as in Figure 9.37), then the new configuration corresponds to the regular expression $(Xa)^*$. Note that $a = \lambda$ is permitted. □

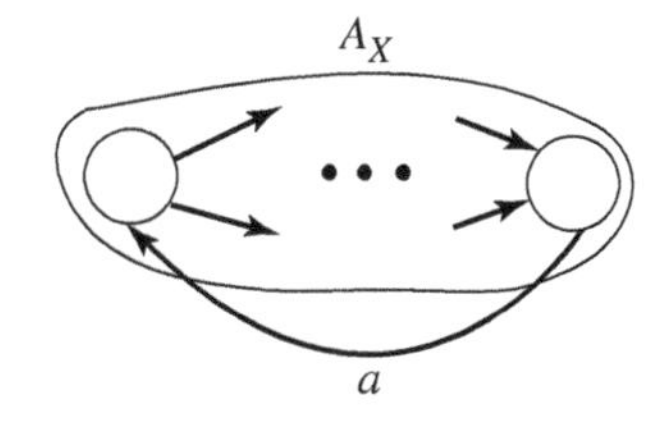

Figure 9.37. Loops.

Rather than formalizing these insights into a proof, an algorithm that uses them will be presented. The algorithm follows that found in [49, pp. 592, 593].

The main task is to reduce the finite automaton to a finite-state machine that has two states and one transition. The transition will be labeled with the regular expression that corresponds to the original finite automaton.

There are four preliminary steps to the algorithm:

- Create a new start state, s_0, and connect it to the old start state via a λ-transition.
- Create a new final state, f_0, and connect all the old final states to it via λ-transitions.
- Eliminate any nonfinal state that has no transitions that move from the state to some different state. (That is, remove nonfinal black hole states. A string that moves to such a state will never be recognized.)
- Eliminate multiple transitions. For each pair of states, x and y, that have more than one transition from x to y, replace those transitions by a single transition whose label is the regular expression formed by using the alternative operator, |, with the labels on the former transitions.

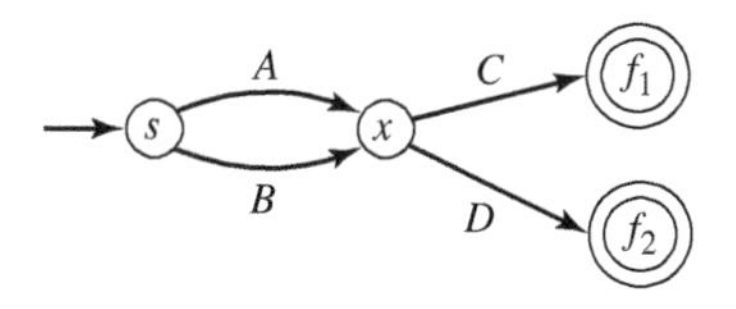

Figure 9.38. The initial finite automaton.

If the finite automaton looks like the state diagram in Figure 9.38, then the preliminary steps would produce the finite-state machine in Figure 9.39.

Figure 9.39 Finite-state machine after preliminary steps.

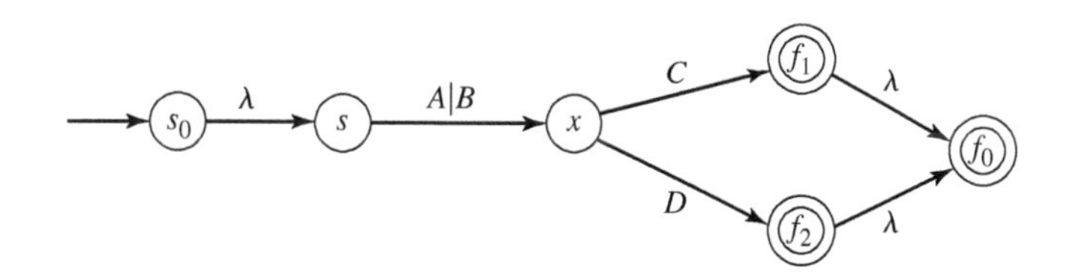

The rest of the algorithm is an iterative process that reduces the number of states (and ultimately, the number of transitions). More precisely, while there are still more than two states,

- Eliminate some state, $y \notin \{s_0, f_0\}$.

The algorithm will be complete once the state-elimination procedure is described. Figure 9.40 shows a state, y, and some of the possible transitions connected to it.

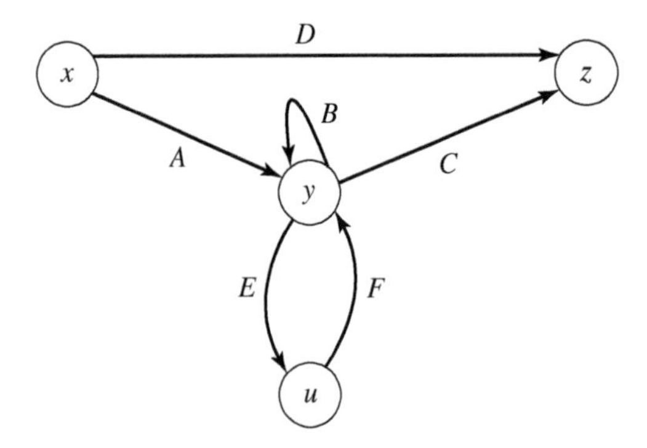

Figure 9.40 State y is ready to eliminate.

Note that it is possible to move from state x to state z by either a string that is matched by the regular expression D, or else by a string that is matched by the regular expression A, followed by zero or more strings that are matched by B, followed by a string that is matched by C. Thus, if y is eliminated, a new transition from x to z with label $D \mid AB^*C$ will need to replace the transitions from x to z, and from x to y to z. Similarly, the state u will need a loop with label FB^*E.

More formally, let $< x, z >$ represent the transition from state x to state z. Let $Lo(x, z)$ denote the current (old) label on that transition and $Ln(x, z)$ denote the new label, after eliminating some state which is different from x and z. Define $Lo(x, z) = \emptyset$ if there is no transition $< x, z >$.

To eliminate a state, y, consider all pairs of transitions $< x, y >$ and $< y, z >$ such that $x \neq y$ and $z \neq y$. (Note that $x = z$ is permissible.) For each such pair, create a new label for the transition $< x, z >$. The new label is the regular expression, $Ln(x, z)$, where $Ln(x, z) = Lo(x, z)|Lo(x, y)Lo(y, y)^*Lo(y, z)$.

After applying this process, Figure 9.40 would become the finite-state machine shown in Figure 9.41.

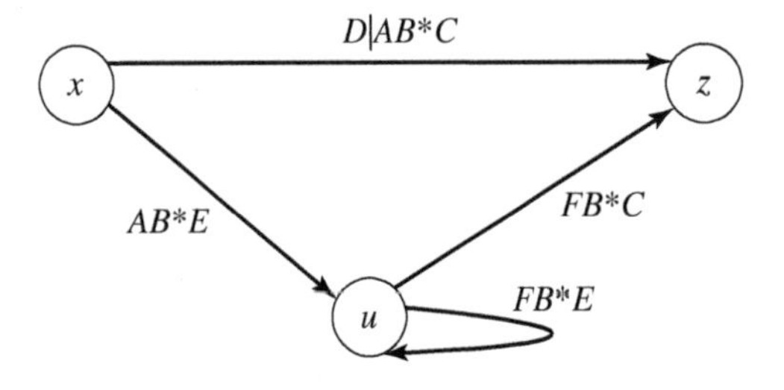

Figure 9.41 State y has been eliminated.

EXAMPLE 9.36

Transforming a Finite Automaton into a Regular Expression

Suppose the finite automaton from Example 9.15 on page 554 needs to be converted to a regular expression. The state diagram is repeated in Figure 9.42.

Note that there are really multiple transitions between several of the states (the diagram uses some shorthand notation). For example, there are really three transitions between states Yes and No.

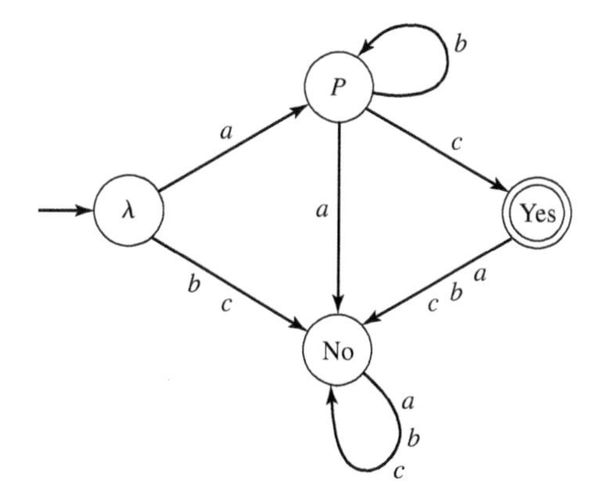

Figure 9.42 A finite automaton to recognize strings in $L = \{ab^*c\}$.

Three of the four preliminary steps produce the state diagram in Figure 9.43.

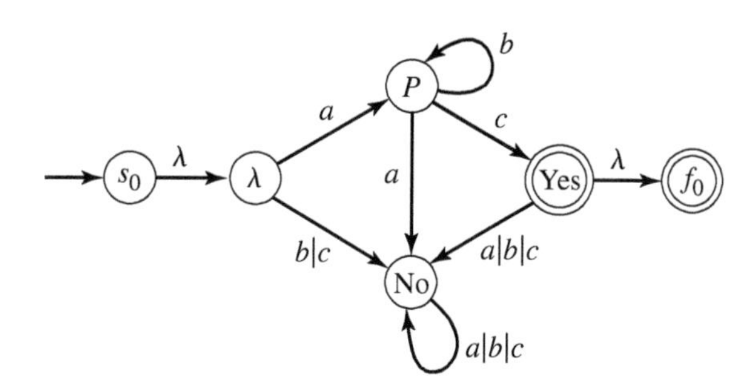

Figure 9.43 State diagram after preliminary steps.

The iteration can start with any of the original states. It is easiest to start with state No because there are no pairs of transitions $< x, \text{No} >$ and $< \text{No}, z >$ with x and z not the same as No. The relabeling rule for any existing transition becomes $Ln(x, z) = (Lo(x, z) \mid \emptyset) = Lo(x, z)$. This effectively eliminates the nonfinal black hole state, which could have been done during the preliminary phase (Figure 9.44).

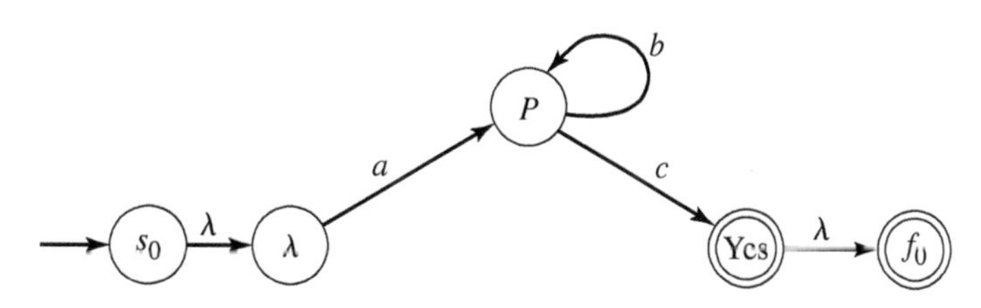

Figure 9.44 Eliminate state No.

The states λ and Yes can be eliminated next, producing the finite-state machine in Figure 9.45. Note that the regular expressions $\emptyset \mid \lambda\emptyset^*a$ and a generate the same regular set.[24] The simpler expression has been used as a label in each case.

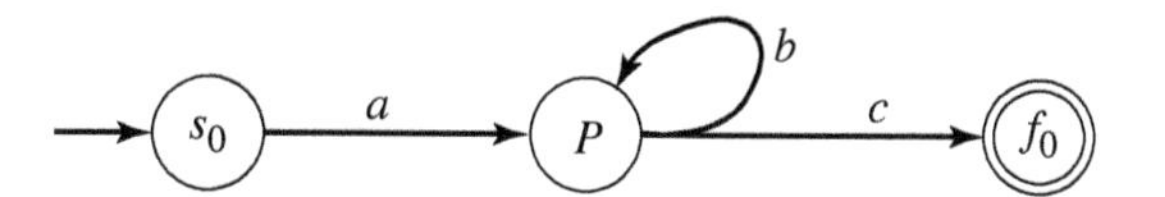

Figure 9.45 Eliminate states λ and Yes.

The final iteration removes the state p (Figure 9.46).

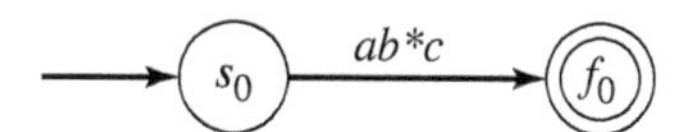

Figure 9.46 Eliminate state p.

A regular expression that generates the set of strings recognized by the original finite automaton is ab^*c, which is consistent with the original example on page 554. ■

[24]The regular expression $\emptyset^*$ generates the regular set $\{\lambda\}$.

✔ Quick Check 9.10

1. Use the pseudocode notation from Chapter 4 to express the algorithm for converting from a finite automaton to a regular expression. You may use the notation already introduced ($<x, y>$, $Lo(x, y)$, and $Ln(x, y)$). ☑

9.5.2 Exercises

The exercises marked with ★ have detailed solutions in Appendix H.

1. Use the construction in Theorem 9.15 (page 570) to convert the following finite automata to regular grammars.
   (a) ★ The finite automaton from Quick Check 9.3 on page 545.

   (b) Let A be defined by $\mathcal{S} = \{s_0, s_1, s_2\}$, $\Sigma = \{0, 1\}$, the start state is s_0, $\mathcal{F} = \{s_0, s_2\}$, and t is defined by the following state table.

State	Input 0	1
s_0	s_1	s_2
s_1	s_1	s_2
s_2	s_1	s_2

   (c) The following finite automaton has the set of input symbols $\Sigma = \{a, b\}$ and recognizes any string with two adjacent as or two adjacent bs. The start state is $s_\emptyset$. $\mathcal{S} = \{s_\emptyset, A, B, DBL\}$ and $\mathcal{F} = \{DBL\}$. The transition function is implicitly defined by the state diagram.

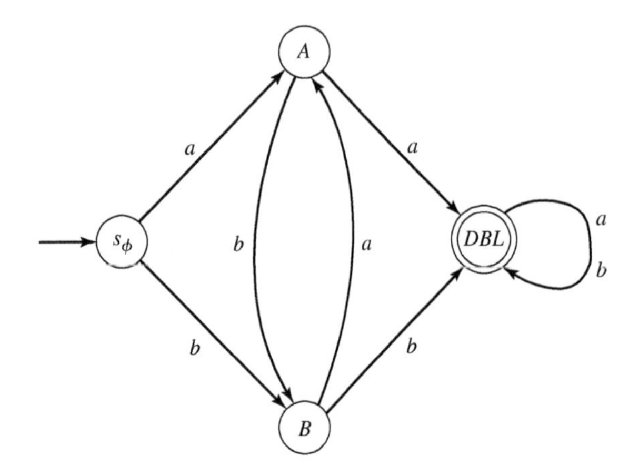

   (d) The following automaton has the set of input symbols $\Sigma = \{a, b, c\}$ and recognizes any nonempty string with no cs. The start state is $s_\emptyset$. $\mathcal{S} = \{s_\emptyset, A, B, C\}$ and $\mathcal{F} = \{A, B\}$. The transition function is implicitly defined by the state diagram.

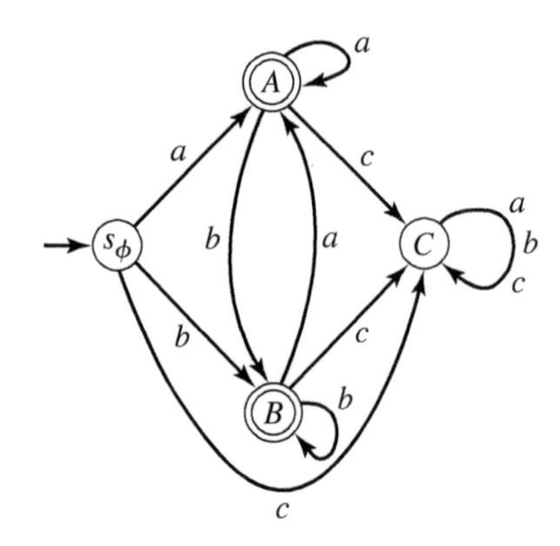

2. The following finite automaton helps you pick petals off of daisies. It has the set of input symbols $\Sigma = \{l, n\}$, where l = "loves me" and n = "loves me not". The start state is P ("pick" the poor helpless daisy from the ground). The states are $\mathcal{S} = \{P, L, N\}$. The final state is L. The transition function is just what you expect. It is represented by the following table.

State	Input l	n
P	L	N
L	L	N
N	L	N

   (a) The finite automaton just described creates a social problem. It enables (in the psychological sense) someone to cheat because it accepts strings of the form $lllll$. In proper "loves me, loves me not" form, the ls and ns should alternate and the string should start with an l. Design a finite automaton that accepts any nonempty string of ls and ns in proper "loves me, loves me not" form and rejects all other strings. Clearly define Σ, $\mathcal{S}$, $\mathcal{F}$, and t. Clearly label the start state.

   (b) Modify the finite automaton in part (a) so that it only accepts properly formed strings that end with l.

   (c) Create a grammar that recognizes all strings recognized by the finite automaton in part (b).

3. Use the construction from Theorem 9.16 (page 572) to convert the following regular grammars to finite automata.
   (a) ★ The grammar in Example 9.16 on page 555. Use uppercase letters for states in this exercise.
   (b) The grammar in Example 9.21 on page 558. Use uppercase letters for states in this exercise.
   (c) The grammar in Example 9.20 (summarized on page 558). Use uppercase letters for states in this exercise. Also, explain why the original, nonregular grammar would be harder to convert to a finite automaton.
   (d) The grammar in Problem 1 of Quick Check 9.6 on page 558. Use uppercase letters for states in this exercise.
   (e) The grammar in Example 9.18 on page 556. Use lowercase letters for states in this exercise.
4. The recursive definition of *regular expression* does not discuss the "+" operator. Suppose X is a regular expression having corresponding nondeterministic finite automaton, A_X. Design a nondeterministic finite automaton that recognizes X^+.
5. Use the construction in Lemma 9.19 (page 576) to convert the following regular expressions to nondeterministic finite automata.
   (a) ab^*c (b) ★ $(ab \mid cd)e$
   (c) $((a \mid b)(c \mid d))*$
   (d) $(a + c) \mid b$ (This part requires the result from Exercise 4.)
6. Use the construction in Lemma 9.20 (page 579) to convert each of the following finite automata to an equivalent regular expression.
   (a) The finite automaton from Example 9.8 on page 545
   (b) ★ The finite automaton from part (b) of Exercise 1 of this set of exercises
   (c) The finite automaton from part (c) of Exercise 1 of this set of exercises
   (d) The finite automaton from part (d) of Exercise 1 of this set of exercises
7. Each of the following statements is either true (always) or false (at least sometimes). Determine which option applies for each statement and provide adequate explanation for your choice.
   (a) Suppose, in the nondeterministic finite automaton in Example 9.31 on page 570, a loop is added at state x having transition label, a. Then the new state diagram still represents a nondeterministic finite automaton.
   (b) Any finite automaton will recognize the language generated by a regular grammar, G.
   (c) Suppose that a set of strings is recognized by a finite automaton. Then the set must be regular.
   (d) ★ A nondeterministic finite automaton differs from a finite automaton in that there will always be at least one state that does not have transitions associated with every input symbol.
8. Each of the following statements is either true (always) or false (at least sometimes). Determine which option applies for each statement and provide adequate explanation for your choice.
   (a) In the conversion process from a finite automaton to a regular grammar, it is always valid to let the set of terminal symbols for the grammar be the set of input values for the finite automaton.
   (b) An important idea in this section is that the input strings recognized by finite automata, the languages generated by regular grammars, and the sets generated by regular expressions exhibit strong connections.
   (c) A finite automaton is not nondeterministic, and thus it cannot recognize the same set of strings as a nondeterministic finite automaton.
   (d) In the conversion process from a regular grammar to a finite automaton, it is always valid to let the set of input values for the finite automaton be the alphabet of the regular grammar.

The following definition and theorems are motivated by the idea that a language is a subset of Σ^*.

DEFINITION 9.21 *The Complement of a Language*

Let L be a language over Σ. The *complement* of L is denoted by $\overline{L}$ and consists of all strings in Σ^* that are not in L.

9. Prove Theorem 9.22.

THEOREM 9.22 *The Complement of a Regular Language*

If L is a regular language, then $\overline{L}$ is also a regular language.

[*Hint*: Use Definition 9.21 and Theorems 9.15 and 9.16. Think about final and nonfinal states.]

10. Prove Theorem 9.23.

THEOREM 9.23 *The Intersection of Regular Languages*

If L_1 and L_2 are both regular languages over the same alphabet, Σ, then $L_1 \cap L_2$ is also a regular language.

[*Hint*: Use De Morgan's laws, Theorem 9.22, and most of the other theorems in this section.]

9.6 A Glimpse at More Advanced Topics

As is true with many chapters in this textbook, this chapter could easily be extended into a full-semester course. In particular, regular grammars and finite automata are not the only useful grammars or models of computation. This section gives a brief glimpse at some other important ideas that extend what has already been presented.

9.6.1 Context-Free Languages and Grammars

Recall Example 9.22, where it was mentioned that the set $\{a^n b^n \mid n \geq 0\}$ cannot be specified by a regular grammar. That does *not* mean we are unable to find a grammar that can specify this language. This section specifies a hierarchy that provides a collection of increasingly more expressive grammars. The price for greater expressive power is a grammar that is harder to use. For example, a regular grammar can be modeled by a finite automaton. The more expressive grammars require more complex computational models.

The hierarchy of grammars is called the Chomsky hierarchy. It was first introduced in 1959 by the linguist Noam Chomsky. The grammars are all in the form $\mathcal{G} = (\Sigma, \Delta, S, \Pi)$. The only point of difference is the nature of productions in Π. In all cases, productions look like $\alpha \rightarrow \beta$, where α and β are strings in $(\Sigma \cup \Delta)^*$. Table 9.9 shows how the productions are defined for each grammar in the Chomsky hierarchy.[25]

TABLE 9.9 The Chomsky Hierarchy of Grammars

Type	Grammar	Productions $\alpha \rightarrow \beta$	Recognized by
0	Phrase structured (unrestricted)	$\alpha \in (\Sigma \cup \Delta)^+$, $\beta \in (\Sigma \cup \Delta)^*$ (*Note*: $\alpha \neq \lambda$, but $\beta = \lambda$ is ok)	Turing machine
1	Context sensitive	$\alpha_1 N \alpha_2 \rightarrow \alpha_1 \gamma \alpha_2$ for $\alpha_1, \alpha_2, \gamma \in (\Sigma \cup \Delta)^*$, $\gamma \neq \lambda$ and $N \in \Delta$	Turing machine
2	Context free	$N \rightarrow \beta$ where $N \in \Delta$, $\beta \in (\Sigma \cup \Delta)^*$	Pushdown automaton
3	Regular	$N_1 \rightarrow \beta_1 N_2$ or $N \rightarrow \beta$ where $N_1, N_2, N \in \Delta$ and $\beta, \beta_1 \in \Sigma^*$, $\beta_1 \neq \lambda$	Finite automaton

A context-sensitive grammar cannot generate the null string. With a suitable adjustment, the grammars form a hierarchy:

$$\begin{aligned}\{\text{regular grammars}\} - \lambda &\subseteq \{\text{context-free grammars}\} - \lambda \\ &\subseteq \{\text{context-sensitive grammars}\} \\ &\subseteq \{\text{phrase-structured grammars}\}\end{aligned}$$

Pushdown automata and Turing machines will be discussed in the next section.

The productions $\alpha_1 N \alpha_2 \rightarrow \alpha_1 \gamma \alpha_2$ in a context-sensitive grammar allow the non-terminal N on the left to be replaced by the non-null string γ whenever N appears in the context of α_1, α_2. Notice that the context, $\alpha_1 \cdots \alpha_2$, is preserved.

[25]Computer scientists are also interested in other grammars. I have expressed some of the grammars described here by using normal forms. That is, the original definitions are not exactly what is shown in the table, but any grammar following the more general rules can be transformed into one that conforms to the normal form.

EXAMPLE 9.37

A Phrase Structured Language

Define the grammar $\mathcal{G}$ as $\Sigma = \{a, b, c\}$, $\Delta = \{S, A, B\}$, with S as the start symbol. The productions in Π are listed next. The left-hand sides (for example Ab are the N

$$\begin{aligned} S &\to abc \mid aAbc \\ Ab &\to bA \\ Ac &\to Bbcc \\ bB &\to Bb \\ aB &\to aa \mid aaA \end{aligned}$$

This grammar generates the language L = $\{a^n b^n c^n \mid n \geq 1\}$.

Notice that the nonterminal A can only be replaced if it is in the context of having either a b or a c as the symbol on its right. This language can also be generated by a context sensitive grammar (but requiring more productions). ■

Context-free languages have some of the nice properties of regular grammars (a single nonterminal on the left side of every production) but are more expressive. The next example shows a grammar for the nonregular language $\{a^n b^n \mid n \geq 0\}$.

EXAMPLE 9.38

A Context-Free Language

Let $\mathcal{G}$ be defined by having $\Sigma = \{a, b, \lambda\}$, $\Delta = \{S\}$, with S as the start symbol, and $\Pi = \{S \to \lambda \mid aSb\}$.

With two simple productions, this grammar produces a language that is not possible using a regular grammar. The cost is the need for something more complex than a finite automaton if we want a machine to recognize strings in this language. The more complex machine is called a *pushdown automaton* and combines a finite automaton with a stack. More on this in the next section. ■

✔ Quick Check 9.11

1. Define a context-free grammar that specifies the language $L_1 = \{a^n b^{2n} \mid n \geq 0\}$.
2. Define a context-free grammar that specifies the language $L_2 = \{a^n b c^n \mid n \geq 1\}$. Use $\Sigma = \{a, b, c, \lambda\}$. ☑

9.6.2 Turing Machines

The goal of this section is to provide informal descriptions of pushdown automata and Turing machines and then to comment briefly on their significance.

Pushdown Automata

The previous section alluded to pushdown automata as computational models that can recognize context-free languages. A pushdown automaton is a finite automaton with some limited-access memory. That is, in addition to the states and input symbols found in a finite automaton, there is also a place to store symbols for later use. This limited-access memory is called a stack.

A *stack* consists of a (potentially very large) collection of storage bins and two operations that move items in and out of bins. The operations are called *push* and *pop*. Push and pop enforce a "last-in, first-out" discipline on accessing the memory bins. Push places a new item (input symbol, string, etc.) on the top of the stack. Pop removes the item at the top of the stack. No other access to the stack's bins is permitted.

A stack can be informally visualized as a pile of clean dinner plates in a spring-loaded serving cylinder. Only the top plate is available for the next customer to grab. As the pile of plates becomes almost empty, the serving staff can add new plates to the top (one at a time in this visualization).

Transitions in a pushdown automaton are determined by looking at the current state, the current input symbol, and the item on the top of the stack. As a transition is performed, it is permissible to pop the top item off the stack, push a new item onto the stack, or leave the stack alone.

EXAMPLE 9.39

Recognizing $\{a^n b^n \mid n \geq 0\}$

Here is an outline of how a pushdown automaton can be used to recognize the nonregular language $\{a^n b^n | \geq 0\}$.

The start state, s_0, is a final state and the stack is initially empty. If the first input symbol is an a, push some special marker symbol, ϵ, onto an empty stack and move to nonfinal state s_1. Then, as long as input symbol a is encountered, remain in state s_1 and push a second marker, δ, onto the stack.

As soon as the first b is encountered in the input, and as long as the top of the stack is a δ, move to nonfinal state s_2 and pop the top item off the stack. If a b is encountered and the top of the stack is an ϵ, move to final state s_3.

Any other combination of state, input, and top of stack will lead to a black hole nonfinal state s_4.

This pushdown automaton will terminate in a final state for exactly those strings in the form, $a^n b^n$, for some $n \geq 0$. ■

The major result about pushdown automata is summarized in the next theorem. Pushdown automata are also important tools when thinking about the design of compilers[26] for computer programs. In particular, pushdown automata are appropriate models for the process of parsing the source code of the program that needs compiling.

THEOREM 9.24 ***Context-Free Grammars and Pushdown Automata***

Any language generated by a context-free grammar can be recognized by a pushdown automaton. Any language recognized by a pushdown automaton can be generated by a context-free grammar.

Context-free grammars and pushdown automata are therefore equally expressive mechanisms.

Turing Machines

Pushdown automata are more expressive than finite automata. However, there are easily defined languages that cannot be recognized by them. A simple example is the language $\{a^n b^n c^n \mid n \geq 0\}$. A more powerful class of finite-state machines, called Turing machines, can recognize this language, and in fact, any language generated by a phrase-structured grammar.

Turing machines were first formally described in a paper by Alan Turing that was published in 1936—a decade before the first electronic computer. There are many computationally equivalent descriptions of Turing machines available today. The one presented here is similar to the original presentation.

A Turing machine consists of a *control unit* and an infinite *data tape*. The data tape is divided into *cells* that are arranged in a line that extends forever in both directions. Each cell may either be blank (contain the symbol λ) or contain one symbol from some finite alphabet.[27] The control unit has the ability to read the contents of a cell, write a new symbol in a cell, and move the Turing machine one cell to the left or right.

[26]A *compiler* is a program that translates human-readable computer *source code* into a form that the computer can run. *Parsing* is the process of breaking the source code into strings that are meaningful components of the grammar for the programming language.

[27]The data tape is similar in spirit to the stack used by a pushdown automaton, but it is a more powerful mechanism for external data storage.

The control unit also contains a finite set of *states* together with a set of *instructions*. The instructions are 5-tuples that specify the current state, the symbol in the current cell, and the response the control unit should exhibit. The response consists of three parts: which symbol to write in the current cell; whether to move left, right, or remain over the same cell; and, finally, which state should become the current state. The 5-tuples can be written in the form

<current state, symbol in current cell, symbol to write in current cell, direction of motion on data tape, next state>

or, more succinctly,

<current state, read, write, move, next state>

Adopting the shorthand notation L = move left, R = move right, S = stay at the same cell, a typical instruction might be

$$< x, a, b, L, y >$$

where x and y are states, and a and b are symbols in the input/output alphabet.

A few other conventions are needed before a simple example is possible.

- The tape will always start with only a finite number of nonblank cells (and the nonblank cells will be contiguous).
- The control unit always starts at the leftmost nonblank cell.
- There is a single start state, s_0, and a single final state, f.
- If the control unit is in a state, x, and the current cell contains a symbol, a, such that there is no instruction that begins $< x, a, \ldots >$, the machine will halt in state x.
- If the machine halts in state f, the Turing machine is said to recognize the string that was initially written on the data tape.
- If the Turing machine halts in any state other than f or if the machine runs forever without halting, the initial string on the data tape is not recognized.

Notice that Turing machines do not use black hole states to reject strings; rejection is accomplished by failing to specify transitions and actions for dead-end state–input pairs.

EXAMPLE 9.40

A Turing Machine That Recognizes $\{a^n b^n c^n \mid n \geq 0\}$

A Turing machine that recognizes exactly the strings $W = \{a^n b^n c^n \mid n \geq 0\}$ will need to have an input/output alphabet that contains the symbols $\{a, b, c, \lambda\}$. We need to use the data tape to "count" as and then bs and finally cs. The tape must initially contain a string for which the Turing machine must determine membership in W.

Label the start state s_0 and the final state f. One easy-to-determine instruction is needed to recognize the empty string (where $n = 0$). That instruction is $< s_0, \lambda, \lambda, S, f >$. It really doesn't matter whether another blank is written to the cell and no motion to another cell is made. I have chosen those two components of the 5-tuple so as to minimize activity.

How can the data tape and states be used to ensure that any symbols come in runs having the same number of elements? That is not apparent yet. However, it seems reasonable to start consuming as, moving to the right after each new a. If a c is encountered (where another a or else the first b is expected), there should be no instruction; the machine should halt in some state other than f. Similarly, during a run of bs, encountering a blank cell or an a when more bs or the first c is expected should also cause a halt in a nonfinal state.

It is clear that we can't use the states to count the number of as because there could potentially be a billion as. Perhaps the control unit should move to state s_1 as soon as an a is read and stay in that state until a b is read.

The key insight is that the machine needs some additional output symbols. Perhaps it can use A, B, and C to indicate that the cell previously contained the lower case version of the symbol, and also that the cell has already been processed. With these symbols, the machine can use a series of left-to-right scans that replace a single a-b-c non-adjacent sequence by A-B-C, effectively counting one of each input symbol. Then the machine can return to the left edge of the input string and look for another matched set a-b-c. Keep this up until a return sweep recognizes that all cells have been processed or until an unexpected input symbol is found.

With a bit of thought, the following instructions can be developed:

$< s_0, \lambda, \lambda, S, f >$	Recognize the empty string
$< s_0, a, A, R, s_1 >$	The first a on this sweep
$< s_1, a, a, R, s_1 >$	Skip over other as on this sweep
$< s_1, b, B, R, s_2 >$	The first b on this sweep
$< s_2, b, b, R, s_2 >$	Skip over other bs this sweep
$< s_2, c, C, L, s_3 >$	The first c this sweep; start moving left
$< s_3, b, b, L, s_3 >$	Ignore any bs when moving left
$< s_3, B, B, L, s_3 >$	Ignore any Bs when moving left
$< s_3, a, a, L, s_3 >$	Ignore any as when moving left
$< s_3, A, A, R, ???>$	Past active region; start moving right

After using these instructions once, in sequence, the machine is ready to start sweeping right again. It seems inefficient to add additional states (as might be tempting for the incomplete final instruction). The only difference from the initial sweep right is that now there are As, Bs, and Cs on the tape. Additional instructions are needed to skip over them (in the proper order, of course).

How does the machine know it is done? It is done when a right sweep encounters only B and C (but no lowercase b or c). If the right sweep starts in state s_0 and never leaves that state unless an a is encountered, then the instruction $< s_0, \lambda, \lambda, S, f >$ will properly halt the machine at the first empty cell to the right of the string, after first verifying that all original symbols match in number and are in the proper order.

Here is the full list of instructions.

$< s_0, \lambda, \lambda, S, f >$	Recognize a valid string
$< s_0, a, A, R, s_1 >$	The first a on this sweep
$< s_0, B, B, R, s_0 >$	Should be the final sweep
$< s_0, C, C, R, s_0 >$	Should be the final sweep
$< s_1, a, a, R, s_1 >$	Skip over other as on this sweep
$< s_1, B, B, R, s_1 >$	Skip over any Bs on this sweep
$< s_1, b, B, R, s_2 >$	The first b on this sweep
$< s_2, b, b, R, s_2 >$	Skip over other bs this sweep
$< s_2, C, C, R, s_2 >$	Skip over any Cs on this sweep
$< s_2, c, C, L, s_3 >$	The first c this sweep; start moving left
$< s_3, C, C, L, s_3 >$	Ignore any Cs when moving left
$< s_3, b, b, L, s_3 >$	Ignore any bs when moving left
$< s_3, B, B, L, s_3 >$	Ignore any Bs when moving left
$< s_3, a, a, L, s_3 >$	Ignore any as when moving left
$< s_3, A, A, R, s_0 >$	Past active region; start moving right

■

✔ Quick Check 9.12

1. For each input string (shown on the data tape), determine whether the Turing machine in Example 9.40 halts. If it does halt,
   - Indicate the state at which the Turing machine halts.
   - Show the data tape after halting.
   - Put parentheses around the input symbol that caused the machine to halt.
   - Indicate whether or not the string is recognized.

   (a) $\cdots\lambda aabbcc\lambda\cdots$

   (b) $\cdots\lambda aabcc\lambda\cdots$

   (c) $\cdots\lambda ab\lambda\cdots$

   (d) $\cdots\lambda abbc\lambda\cdots$ ☑

The Church–Turing Thesis

Turing machines are important because they are as expressive as any phrase-structured grammar. They are important for an additional reason. A famous assertion, called the Church–Turing thesis,[28] claims that a Turing machine is as expressive as any possible model of computation. The assertion is called a thesis (rather than a theorem) because it compares the precise, formal notion of a Turing machine to the imprecise, intuitive notion of computability.

> **Church–Turing Thesis**
>
> A Turing machine is capable of performing any computable algorithm.

This thesis asserts that a Turing machine is capable of performing any computation that a massively parallel supercomputer can do. Of course, the supercomputer may be capable of completing some tasks in a few seconds that a Turing machine would need billions of years to complete.

Turing Machines and Neural Nets

There is an additional significant issue related to Turing machines that should be mentioned.

A Turing machine is a model for machine computation. In the early 1940s, McCulloch and Pitts formulated a model of the human brain called a *neural net*. The basic component in their model is a *neuron*. Neurons have *excitatory inputs* and *inhibitory inputs*. If the excitatory signal minus the inhibitory signal is greater than some threshold value, the neuron fires, sending a signal to one of another neuron's inputs. Threshold values can vary from neuron to neuron. Neurons are connected to other neurons in nets. There are also inputs to the net from outside the net and outputs that leave the net. It is possible to hook neurons together to form simple logic circuits.

Pitts showed that a neural net can express the same computations as a Turing machine.

If Pitts's observation is coupled with the Church–Turing thesis, you arrive at the conclusion that there exist expressively equivalent *models* for computers and for human brains. It is tempting to carry this one step further and assert that there is no essential difference between humans and intelligent machines. Many people have made just this claim.

In fact, they go even further. Recall Shannon's definition of *information*. That definition focused on pattern and probability and excluded any notion of *meaning*. If information is just pattern and if humans and intelligent machines are essentially the same, it is not hard to generate a description (or metamodel) that describes both humans and machines: *information processor*. The information may be stored in different ways, and the manipulations may be done using different mechanisms, but both are capable of performing the same transformations on some collection of data.

[28]Named for Alonzo Church and Alan Turing.

This equating of human and machine is not without its critics. One simple observation is that the equation is extremely reductionistic. Many details of what humans and machines really are have been omitted to produce the simple models. Those omitted details may be quite significant in defining the very real differences. A second line of criticism asserts that humans are more intimately connected to their bodies than the simple models admit. These critics would assert that human intelligence cannot be defined (or developed) apart from the human body.

This is not the place to carry this discussion any further. It has been mentioned to indicate how seemingly academic models of computation have entered the discussion of what it means to be human or intelligent. These models are no longer just theoretical curiosities for professors to play with; they have become a frequently-mentioned artifact in our attempt to understand ourselves.

9.6.3 Exercises

The exercises marked with ★ have detailed solutions in Appendix H.

1. ★ Design a context-free grammar that specifies the language $L = \{a^n bc^n d \mid n \geq 1\}$. Use $\Sigma = \{a, b, c, d\}$ and $\Delta = \{S, X\}$ (you really do need at least two nonterminals).
2. Design a context-free language that produces all binary strings that have two zeros at both the beginning and end of the string. (The string 00 of length 2 qualifies.)
3. Define a grammar (of any type) that specifies the language $L = \{a^n b^m c^n \mid n, m \geq 0\}$.
4. A context-free grammar is defined by $\Sigma = \{a, b\}$, $\Delta = \{S, X, Y\}$, and $\Pi = \{S \to aXb, X \to bYa \mid ba, Y \to aXb \mid ab\}$. The start symbol is S.
   (a) Describe the language generated by this grammar.
   (b) Is there a regular grammar that generates the same language? If there is, describe one; otherwise, show why there is none.
   (c) Speculate about the answer to the following assertion: "A nonregular grammar can generate a regular language".
5. For each set of productions, determine the most restrictive type of grammar that the productions can be part of. (A higher grammar type in the set {0, 1, 2, 3} indicates a more restrictive grammar.) Assume that uppercase letters represent nonterminal symbols and lowercase letters represent terminal symbols.
   (a) ★ $\{abN \to abc, M \to cN \mid a\}$
   (b) $\{N \to \lambda \mid aN, M \to cMa \mid aN\}$
   (c) $\{aNb \to aMc \mid acc \mid M, M \to cN \mid \lambda\}$
   (d) $\{N \to aaNM \mid bb, M \to aM\}$
6. For each set of productions, determine the most restrictive type of grammar that the productions can be part of. (A higher grammar type in the set {0, 1, 2, 3} indicates a more restrictive grammar.) Assume that uppercase letters represent nonterminal symbols and lowercase letters represent terminal symbols.
   (a) ★ $\{aaNa \to bN \mid cNc \mid c, \lambda \to bM \mid N\}$
   (b) $\{N \to bNb \mid cMc, dMe \to dbNbbe \mid dace\}$
   (c) $\{N \to aN \mid \lambda, M \to aabM \mid bcN \mid a\}$
   (d) $\{ccN \to aNMa \mid bM, M \to \mid Ma \mid bc\}$
7. Design a grammar, of any type, that generates all strings of *a*s and *b*s that contain an equal number of each letter. Thus, the grammar should generate *abaababbba* and λ, but not *aba*. Show the derivation for *abaababbba* using your grammar.
8. Design a regular language that generates all strings of *a*s and *b*s that contain exactly one *a*. Use $\Sigma = \{a, b\}$ (so λ should not be used).
9. Design a context-free grammar that generates the language $\{a^n b^{n+1} \mid n \geq 0\}$.
10. Describe a pushdown automaton that recognizes all the strings in the language $\{a^n b^{n+1} \mid n \geq 0\}$.
11. Each of the following statements is either true (always) or false (at least sometimes). Determine which option applies for each statement and provide adequate explanation for your choice.
    (a) ★ Any language recognized by a Turing machine can be recognized by a pushdown automaton.
    (b) A stack has two operations that move items in and out of bins. *Push* and *pop* have the jobs of removing the item at the top of the stack and placing a new item on the top of the stack, respectively.
    (c) Given any computational model, a Turing machine will be at least as expressive.
    (d) Turing machines do not use black hole states to reject strings. Instead, Turing machines simply do not specify transitions and actions for dead-end state–input pairs.
12. Each of the following four statements is either true (always) or false (at least sometimes). Determine which option applies for each statement and provide adequate explanation for your choice.
    (a) A language can be generated by a context-free grammar if and only if it can be recognized by a pushdown automaton.
    (b) It is a universal requirement for the grammars specified in this text that at least one nonterminal appears on the left side of a production.
    (c) It is permissible for a cell in the data tape for a Turing machine to be blank.
    (d) Context-free grammars cannot generate the null string.

13. Describe a pushdown automaton that recognizes all the strings in the language, $\{a^n b^m \mid n \geq m \geq 0\}$.

14. For each input string (shown on the data tape), determine whether the Turing machine in Example 9.40 halts. If it does halt,

- Indicate the state at which the Turing machine halts.
- Show the data tape after halting.
- Put parentheses around the input symbol that caused the machine to halt.
- Indicate whether or not the string is recognized.

(a) ★ $\cdots \lambda aaabbcc\lambda \cdots$

(b) $\cdots \lambda aabbcc\lambda \cdots$

(c) $\cdots \lambda abbbccc\lambda \cdots$

(d) $\cdots \lambda aaabbbcccc\lambda \cdots$

15. Design a Turing machine that recognizes the language $\{0^n 1 0^n \mid n \geq 1\}$.

9.7 Quick Check Solutions

Quick Check 9.1

1. Alternative sequences of questions are valid. The *number* of questions should not vary.

(a) Three questions suffice. One possible sequence might be

- Does the compass reading contain two letters?—Suppose the answer is yes.
- Does the compass reading contain an N?—Suppose the answer is no.
- Does the compass reading contain an E?—If the answer is yes, the message was *SE*, otherwise the message was SW.

Similar sequences of questions will always determine the message as long as the number of options is cut in half each time.

This experiment should be assigned information value "three questions".

(b) Three questions are guaranteed to work. Sometimes two will suffice. One set of questions would be

- Is the value less than 4? Suppose it is.
- Is the value an even number? If the answer is yes, then a 2 was rolled. Suppose the answer is no (an odd was rolled).
- Is the number a 1? The answer will determine whether the message is informing us that a 1 or a 3 was rolled.

Since sometimes three questions are needed, the information value should be "three questions."

Quick Check 9.2

1. The change of base formula (page 187) is needed to calculate base 2 logarithms.

(a) $I = -\frac{10}{30}\log_2\left(\frac{10}{30}\right) - \frac{20}{30}\log_2\left(\frac{20}{30}\right) \simeq -\left(\frac{1}{3}\right)(-1.585) - \left(\frac{2}{3}\right)(-0.585) = 0.918$

(b) $I = -\frac{10}{11}\log_2\left(\frac{10}{11}\right) - \frac{1}{11}\log_2\left(\frac{1}{11}\right) \simeq -\left(\frac{10}{11}\right)(-.1375) - \left(\frac{1}{11}\right)(-3.459)$
$= 0.439$

2. In this case, $p_1 = p_2 = \frac{1}{2}$. Thus,

$$I = -\frac{1}{2}\log_2(2^{-1}) - \frac{1}{2}\log_2(2^{-1}) = \frac{1}{2}\log_2(2) + \frac{1}{2}\log_2(2) = 1.$$

Quick Check 9.3

1. This can be accomplished very efficiently using only two states: Odd (an odd number of 1s so far), and Even (an even number of ones so far). The start state is Even, the final state is Odd. The inputs are {0, 1}.

   (a) The state table is

	Input	
State	**0**	**1**
Even	Even	Odd
Odd	Odd	Even

   (b) The state diagram is

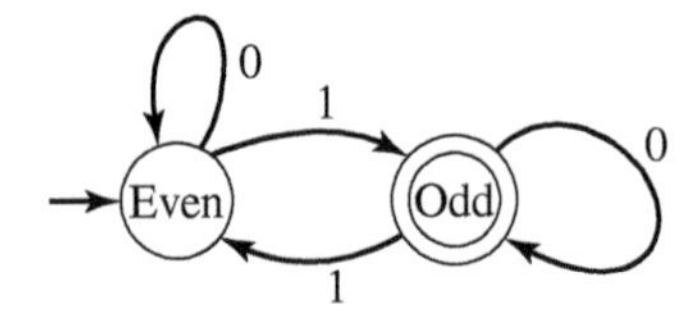

Quick Check 9.4

1. The modified finite-state machine will require three additional transitional states. These states keep track of the bits that are being flushed at the end of the input string. I will call these new states f_0 (still have one 0 to flush), f_1 (a final 1 to flush), and $f_\emptyset$ (all done).

   To be really complete, I should add one more state (an "error state") that would indicate an error in the input string (a "—" that is in the wrong place). The error state would be a black hole state. I have made the state $f_\emptyset$ serve in this additional role.

   The transition and output functions in tabular form are shown next. (A state diagram follows.) The symbol "—" now has dual roles (input and output symbol).

	Transition			Output		
	Input			Input		
State	**0**	**1**	**—**	**0**	**1**	**—**
$s_\emptyset$	s_0	s_1	$f_\emptyset$	—	—	—
s_0	s_{00}	s_{01}	$f_\emptyset$	—	—	—
s_1	s_{10}	s_{11}	$f_\emptyset$	—	—	—
s_{00}	s_{00}	s_{01}	f_0	0	0	0
s_{01}	s_{10}	s_{11}	f_1	0	0	0
s_{10}	s_{00}	s_{01}	f_0	1	1	1
s_{11}	s_{10}	s_{11}	f_1	1	1	1
f_0	$f_\emptyset$	$f_\emptyset$	$f_\emptyset$	0	0	0
f_1	$f_\emptyset$	$f_\emptyset$	$f_\emptyset$	1	1	1
$f_\emptyset$	$f_\emptyset$	$f_\emptyset$	$f_\emptyset$	—	—	—

   A state diagram can also be used to represent this finite-state machine.

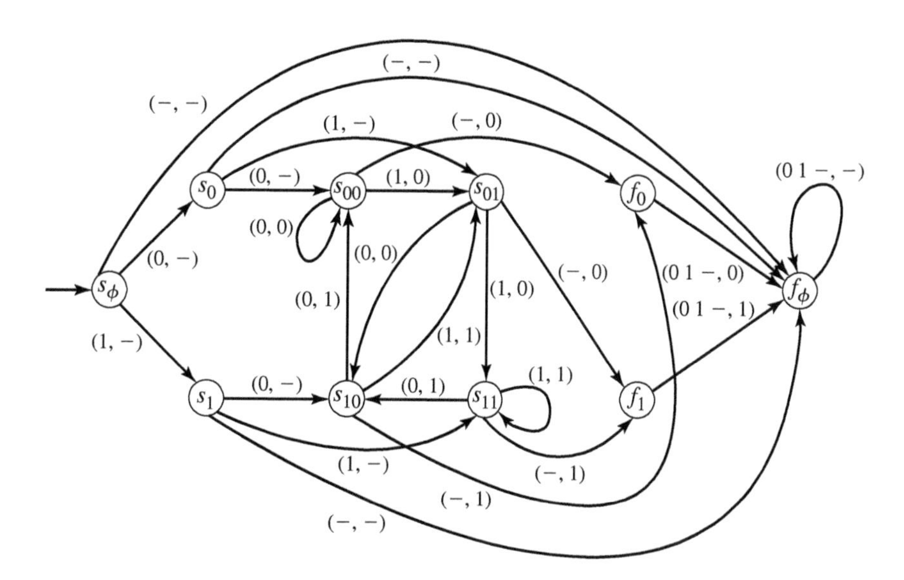

Quick Check 9.5

1. No. The concatenations preserve the order of the characters in the strings. The string *abab is* in $\{x^n \mid n \geq 0\}$, but *abba* is not.
2. *abcdecde*
3. No, $x^0 = \lambda$, the null *string*; $A^0 = \{\lambda\}$, the *set* containing the null string as its only element.
4. $AB = \{rt, ru, st, su\}$
5. 16. Each successive concatenation pairs every element of the left-hand set with every element of the right-hand set. AB will have four elements. Concatenating with another B will cause the set to double (since $|B| = 2$). Finally, concatenating with another A will double the set again.

Quick Check 9.6

1. The major change from Example 9.20 is that the transition to the b section of the string must guarantee that at least one b is generated. The productions $S \to cC$, $S \to \lambda$, $A \to cC$, and $A \to \lambda$ must be dropped.

$$\mathcal{G} = (\Sigma, \Delta, S, \Pi), \text{ where}$$
$$\Sigma = \{\lambda, a, b, c\}, \quad \Delta = \{S, A, B, C\}, \quad \text{and } \Pi \text{ contains}$$

- $S \to aA \mid bB$
- $A \to aA \mid bB$
- $B \to bB \mid cC \mid \lambda$
- $C \to cC \mid \lambda$

2. $S \Rightarrow bB \Rightarrow bbB \Rightarrow bbcC \Rightarrow bbc$

Quick Check 9.7

1. Two simple solutions are possible:
   - `ea(r|t)`
   - `ea[rt]`

   Notice that `ear|t` is *not* a valid solution (try it).
2. `i[^ez]`

Quick Check 9.8

1. `[0-9]+` is incorrect (it allows 0). The solution is `[1-9][0-9]*` (which excludes leading 0s).

2. We need to allow for a leading − sign. `(-)?[0-9]+` will almost work (but allows −0 and 00). A better solution is `^(0|-?[1-9][0-9]*)$` (either the number is 0, or else start with an optional minus sign followed by a nonzero digit, followed by zero or more digits).

Quick Check 9.9

1. (a) Use the construction from the proof of Theorem 9.15.

   Let $\Sigma = \{0, 1\}$ and $\Delta = \{Z, E, T\}$ (where state One has become nonterminal E). The start symbol will be Z.

   The productions are $\Pi = \{Z \rightarrow 1Z,\ Z \rightarrow 0E,\ E \rightarrow 0T,\ E \rightarrow 0,\ E \rightarrow 1Z,\ T \rightarrow 1Z, T \rightarrow 0T,\ T \rightarrow 0\}$.

   (b) $L(\mathcal{G})$is the set of all bit strings that end in two zeros.

2. Use the construction from the proof of Theorem 9.16.

   The initial part of the construction leads to $\Sigma = \{a, b\}$, the start state is s_0, and since there is a production $s \rightarrow baa$ that contains neither a nonterminal nor λ on the right-hand side, $\mathcal{S}$ initially contains s_0 and f. We also know that $\mathcal{F} = \{f\}$.

   From the production $S \rightarrow aS$ we have the transition $t(s_0, a) = s_0$. The production $S \rightarrow baa$ will introduce two more states, which can be named s_1 and s_2 for this example. The transitions $t(s_0, b) = s_1,\ t(s_1, a) = s_2,$ and $t(s_2, a) = f$.

   Finally, adding the state bh and the transitions $t(s_1, b) = bh$, $t(s_2, b) = bh$, $t(f, a) = bh$, $t(f, b) = bh$, $t(bh, a) = bh$, and $t(bh, b) = bh$ completes the finite automaton.

   A is therefore defined by $\mathcal{S} = \{s_0, f, s_1, s_2, bh\}$, with start state s_0. The input values are $\Sigma = \{a, b\}$ and the final states are $\mathcal{F} = \{f\}$. The transition function is summarized in the following state diagram.

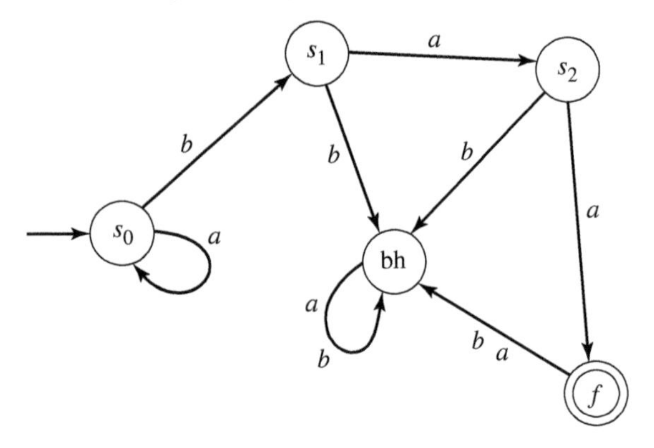

Quick Check 9.10

1. This is quite straightforward.

```
1. Create a new start state, s0, and connect it to the old
   start state via a λ-transition.
2. Create a new final state, f0, and connect all the old
   final states to it via λ-transitions.
3. Eliminate any nonfinal state that has no transitions that
   move from the state to some different state (that is,
   remove all nonfinal black holes).
4. For each pair of states, x and y, that have more than one
   transition from x to y, replace those transitions by a
   single transition whose label is the regular expression
   formed by using the alternative operator, |, with the
   labels on the former transitions.
5. While there are still more than two states
     A. Choose some state, y ∉ {s0, f0}, to eliminate.
     B. For each pair of transitions, <x,y> and <y,z>,
        with x≠y, z≠y:
          α. Save Lo(x,z) and eliminate the transition
                         <x,z> if it exists.
          β. Create a new transition, <x,z>, with
                   Ln(x,z) = Lo(x,z)|Lo(x,y)Lo(y,y)*Lo(y,z).
```

Quick Check 9.11

1. Let $\mathcal{G}$ be defined by having $\Sigma = \{a, b, \lambda\}$, $\Delta = \{S\}$, with S as the start symbol, and $\Pi = \{S \to \lambda \mid aSbb\}$.

2. One solution is $\Sigma = \{a, b, c, \lambda\}$, $\Delta = \{S\}$, with S as the start symbol, and $\Pi = \{S \to aSc \mid b\}$.

Quick Check 9.12

1. The input symbol that caused the halt is in parentheses.

(a) Halts in state f (recognized). Tape $= \cdots \lambda AABBCC(\lambda) \cdots$

(b) Halts in state s_1 (not recognized). Tape $= \cdots \lambda AABC(c)\lambda \cdots$

(c) Halts in state s_2 (not recognized). Tape $= \cdots \lambda AB(\lambda) \cdots$

(d) Halts in state s_0 (not recognized). Tape $= \cdots \lambda AB(b)C\lambda \cdots$

9.8 CHAPTER REVIEW

9.8.1 Summary

This chapter presents some mathematical models that seek to describe several aspects of computer science. The chapter starts with Shannon's models of information and of a communication system. The model of a communication system is simple, but it is very useful. The development of error-correcting codes (Section 8.5) is a response to the presence of noise in the communication channel. Shannon's model of information is not as simple, but it is based on some intuitive ideas. It is worth remembering that Shannon's definition of information does not consider "meaning." It is an attempt to form an abstraction that concentrates on the engineering aspects of reliably transmitting data. Notions such as the "average information" in a randomly selected message help to discuss ideas such as "how much information per second can be transmitted over this channel?".

The remainder of the chapter presents several models that are abstractions created to help theorists understand computers. Finite-state machines (Section 9.2) are capable of representing the notion that at any given instant, a computer has a definite state (the contents of the memory, registers, program counter, etc.). In addition, state changes (and perhaps output) can be triggered by new input, but the action will also depend on the current state.

Section 9.3 is motivated by the effort to understand compilers. Human languages are too ambiguous and complex to make the process of translating a human-oriented program into machine-oriented object code that can run on a central processing unit. Formal languages remove some of the complexity and ambiguity of human languages (as well as some of the expressiveness). What is gained is conformity to a grammar (similar to a set of axioms). This regularity and concrete set of transformation rules (productions) enables the creation of compilers (programs whose task is to translate other programs into machine language).

One very simple formal language that is used extensively is the language of regular expressions (Section 9.4). Regular expressions enable people to do very sophisticated pattern matching for text processing. The patterns can be very general (such as matching everything that looks like a phone number, or everything that looks like a zip code, or an e-mail address).

Section 9.5 explores the connections between finite automata (Section 9.2), regular grammars (Section 9.3), and regular expressions (Section 9.4). The main theorem in that section demonstrates that they are distinct models that have identical expressive power.

The material in Section 9.6 extends the simple finite-state machine models into a series of increasingly more powerful models (and associated formal languages). This section is intended as a brief preview of ideas that you may encounter in future courses. It does not contain enough details to serve as a solid introduction.

Except for (perhaps) the model of information and the proof of Kleene's theorem, the material in this chapter is mostly of a concrete nature. There are many algorithmic manipulations and there are fewer definitions and theorems than some other chapters contain. There is, however, a large amount of new notation.

This chapter contains some very useful models, especially for students who plan to do more advanced work in computer science. Instructors in future courses expect students to be comfortable with these models and capable of using them effectively. As you review, don't concentrate on the theory alone; spend some time reviewing how to use the models to solve practical problems.

9.8.2 Notation

Notation	Page	Brief Description
$I(S)$	537	the information value (or average information) of selecting/receiving a message from sample space S
$A = (\mathcal{S}, \Sigma, t, s_0, \mathcal{F})$	543	a finite automaton
$\mathcal{S}$	543	the set of states for a finite-state machine
Σ	543	the set of input values for a finite-state machine (alternative usage listed below)
$t(s, i)$	543	a transition function that maps state–input pairs to states
s_0	543	the start state in a finite-state machine
$\mathcal{F}$	543	the set of final (or accepting) states in a finite automaton
$M = (\mathcal{S}, \Sigma, t, \Gamma, g, s_0)$	547	a finite-state machine with output
Γ	547	the set of output values for a finite-state machine with output
$g(s, i)$	547	an output function that maps state–input pairs to output values
Σ	553	an alphabet in a formal language (alternative usage listed above)
λ	553	the null string
Σ^*	554	the set of all finite-length words over Σ
$\mathcal{G} = (\Sigma, \Delta, S, \Pi)$	555	a regular grammar
Δ	555	the set of nonterminal symbols in a regular grammar
S	555	the start symbol in a regular grammar
Π	555	the set of productions in a regular grammar
$\nu_1 \Rightarrow \nu_2$	556	ν_2 is directly derivable from ν_1
$\nu_1 \overset{*}{\Rightarrow} \nu_n$	556	ν_n is derivable from ν_1
$L(\mathcal{G})$	556	the language generated by grammar, $\mathcal{G}$
$\overline{L}$	583	the complement of the language, L

String Concatenation

(See page 556.) Let x and y be strings. Then

- xy is the string formed by concatenating the two strings.
- $x\lambda = \lambda x = x$
- $x^0 = \lambda$
- $x^n = xx^{n-1}$ for $n \geq 1$
- x^* represents zero or more copies of x. That is, $x^* \in \{x^n \mid n \geq 0\}$.
- x^+ represents one or more copies of x. That is, $x^+ \in \{x^n \mid n \geq 1\}$.

Kleene Operators

(See page 557.) Let A and B be sets of symbols or sets of strings. Then

- $AB = \{ab \mid a \in A \text{ and } b \in B\}$
- $A^0 = \{\lambda\}$
- $A^n = AA^{n-1}$ for $n \geq 1$
- The *Kleene closure* of A is $A^* = \{\lambda\} \cup A \cup A^2 \cup A^3 \cup \cdots = \cup_{i=0}^{\infty} A^i$.
- $A^+ = A \cup A^2 \cup A^3 \cup \cdots = \cup_{i=1}^{\infty} A^i$.

Regular Expression Metacharacters

(See page 561.)

- The $ character matches the end of a line.
- The ^ character matches the beginning of a line.
- The [character initiates a [] pair. A regular expression consisting of a pair [] with a set of characters inside matches any one character from the set. A regular expression consisting of [^] with a set of characters after the ^ matches any one character that *doesn't* occur inside the [] pair. The "—" character acts as a range operator. If the "—" character is one of the characters you want as a choice in the [] pair, it needs to be the first character so that it is not interpreted as a range operator.
- The metacharacter | indicates an alternative. A regular expression containing a | matches any string that contains either the left or the right alternative.
- The () metacharacters are used to group characters into subpatterns of the regular expression.
- The \ metacharacter is used to convert a metacharacter back into a normal symbol (so it can match itself).
- The . metacharacter matches any single character except a newline.
- The " metacharacter is used to surround a regular expression. Everything between a pair of " characters is treated as the regular expression. The main reason for this is for use with software that accepts regular expressions as command-line arguments. The " characters keep the operating system from trying to evaluate the regular expression as a part of a command.
- The * metacharacter matches zero or more copies of the immediately preceding character or subpattern.
- The ? metacharacter matches either zero or one copy of the immediately preceding character or subpattern.
- The + metacharacter matches one or more copies of the immediately preceding character or subpattern.

9.8.3 Additional Review Material

Go to http://www.mathcs.bethel.edu/~gossett/DiscreteMathWithProof/review.xhtml for additional review material, including a sample chapter exam, with solutions.

CHAPTER 10

Graphs

The origins of graph theory are humble, even frivolous.
N. Biggs, E. K. Lloyd, and R. J. Wilson — Graph Theory: 1736 – 1936

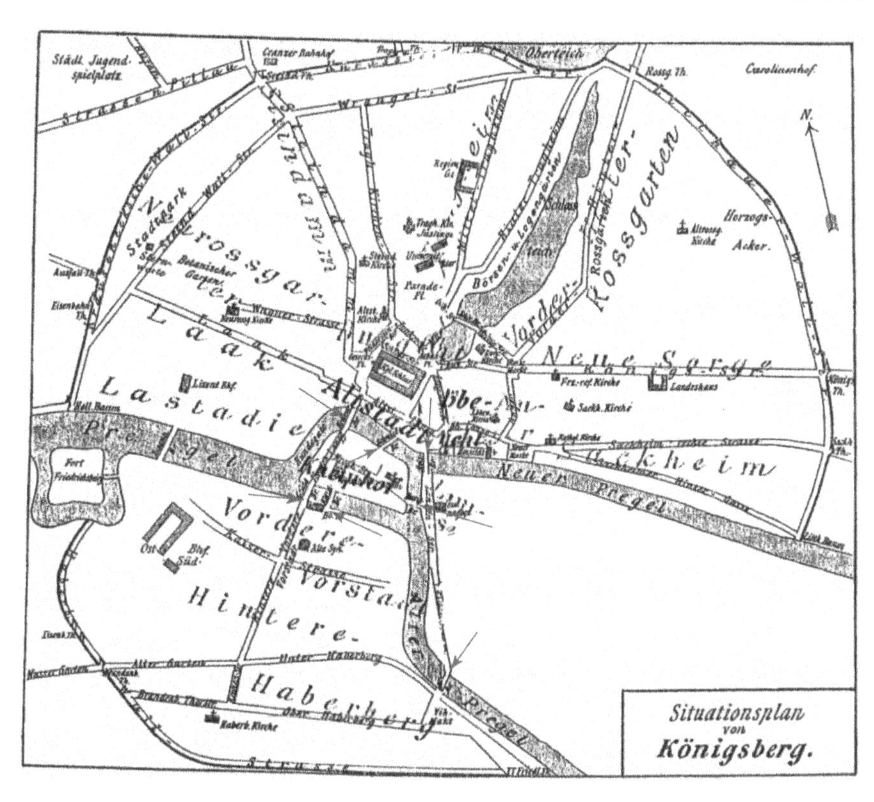

Figure 10.1 The town of Königsberg.

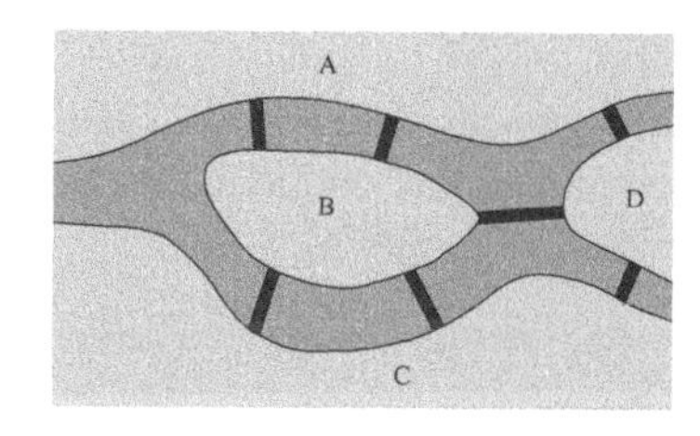

Figure 10.2. The bridges of Königsberg.

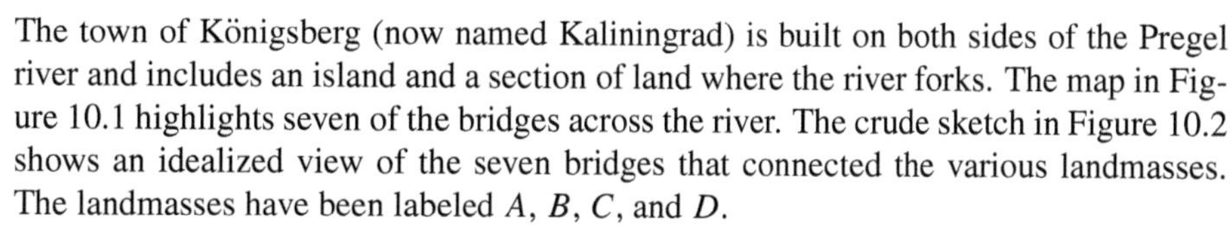

The town of Königsberg (now named Kaliningrad) is built on both sides of the Pregel river and includes an island and a section of land where the river forks. The map in Figure 10.1 highlights seven of the bridges across the river. The crude sketch in Figure 10.2 shows an idealized view of the seven bridges that connected the various landmasses. The landmasses have been labeled A, B, C, and D.

A popular puzzle in the town was to produce a tour of the town that crossed each of the bridges exactly once. A first-rate solution would design a tour that began and ended at the same place. Many people tried (and failed) to find a solution. In 1736 Leonhard Euler provided a solution to this problem, and also to every other bridge-crossing problem you might invent. His solution has been credited as the origin of *graph theory*.

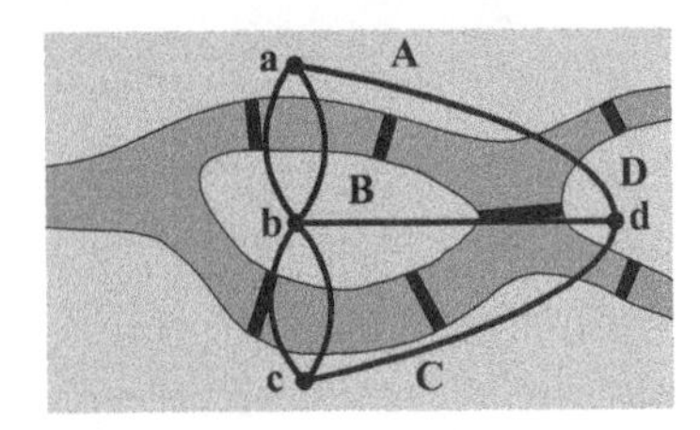

Figure 10.3. The Königsberg abstraction.

Euler cleverly abstracted the really important parts of the problem and removed the nonessential features. To this end, he shrank the landmasses to points, totally removed the river, and kept the bridges as lines and arches joining the points representing the landmasses.

The diagrams in Figures 10.3 and 10.4 show the abstract representation.

A tour is now represented by starting at a dot and crossing arches until every arch has been crossed. Euler's solution will be presented after some terminology and necessary background concepts have been established.

10.1 Terminology

It is unfortunate that mathematicians use the name *graph* for two quite distinct categories of visual representations. You are most familiar with the kind of graph that represents a function. For example, the cosine function can be represented as shown in Figure 10.5.

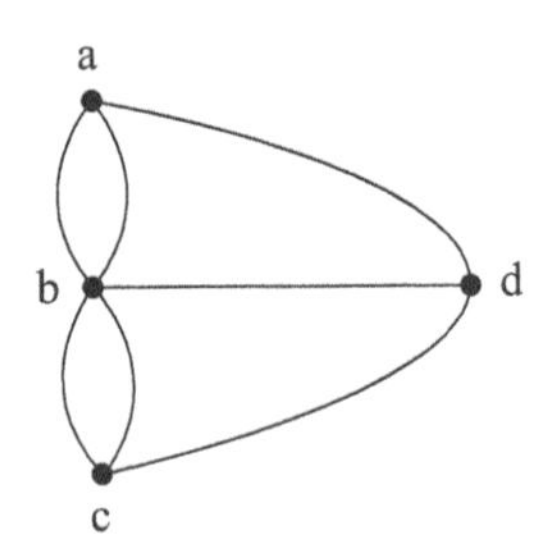

Figure 10.4. The Königsberg graph.

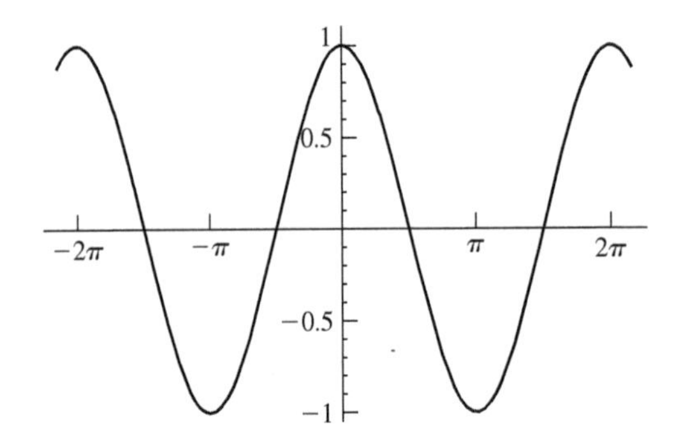

Figure 10.5 The cosine function.

The other category of graph is the topic of this chapter. Intuitively, a graph consists of a collection of dots (called *vertices*) and a collection of lines or arches (called *edges*). Each edge connects two vertices. The Königsberg graph (Figure 10.4) is an example.

A formal approach is more in the style of axiomatic mathematics. In this approach, vertices are taken as undefined terms and the relationship between vertices and edges is specified. As you read the definitions that follow, it does not hurt to keep a mental model of dots and lines lurking in the background.

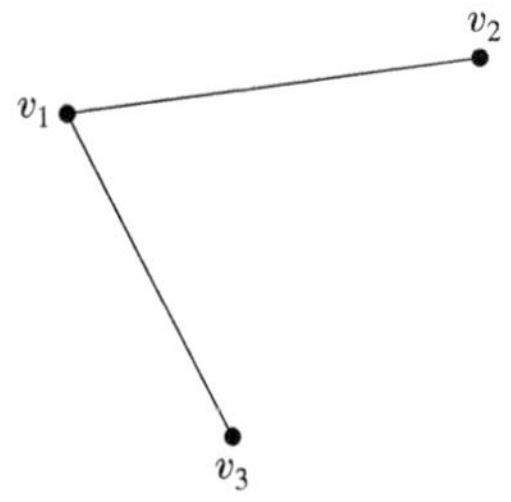

DEFINITION 10.1 ***Simple Graph—Preliminary Definition***
A *simple graph*, G, consists of a nonempty, finite set of *vertices*, V, and a set of *edges*, E, where the elements of E are unordered pairs of distinct elements from V. Thus $G = (V, E)$.

The singular form of *vertices* is *vertex*.[1] The terms *vertex* and *edge* have been borrowed from plane geometry (recall the manner in which we speak about triangles). Intuitively, the unordered pairs in E specify the two vertices that are joined by the edge. The requirement that V be a finite set would be removed from the definition in a more advanced course on graph theory, and some graph theorists would allow $V = \emptyset$.

Figure 10.6. Informal view of G.

EXAMPLE 10.1 **A Small Graph**

Define the graph $G = (V, E)$ by setting $V = \{v_1, v_2, v_3\}$ and $E = \{v_1v_2, v_1v_3\}$. Notice that we could specify the first edge in E as either v_1v_2 or as v_2v_1 since the edges are unordered. The less formal representation of G is shown in Figure 10.6. ■

[1] The terms *vertice* and *vertexes* are both incorrect.

There are three important generalizations of the definition. These generalizations permit loops,[2] multiple edges between vertices, and directed edges. The description of these extensions will require some additional thought about notation. For example, suppose we want to define a loop as an edge that joins a single vertex back to itself. The notation $\{v, v\}$ is unacceptable because set notation does not allow duplicates. The notation (v, v) has traditionally denoted an *ordered* pair, so that notation is also out. The notation used for simple graphs works here: a loop at v can be written vv.

Multiple edges between the same vertices introduce another problem. If E is a set of pairs of vertices, there would be no proper method for listing the duplicate pairs. That is, vw cannot be listed more than once in $E = \{vw, \dots\}$ using proper set notation. What is often done is to define E as a set of edges, $\{e_1, e_2, \dots, e_k\}$, and use a function, ϕ, from E to unordered pairs of elements in V. It is then possible to have $\phi(e_1) = vw$ and $\phi(e_2) = vw$.

Directed graphs will require ϕ to map to *ordered* pairs of vertices: $\phi(e_1) = (v, w)$. This extension will be discussed in Section 10.6.

Because some of these extensions require the introduction of ϕ, it is more consistent to revise the definition of a simple graph.

DEFINITION 10.2 ***Simple Graph***

A *simple graph*, $G = (V, E, \phi)$, consists of a nonempty, finite set of *vertices*, V, a finite set of *edges*, E, and a one-to-one incidence function, ϕ, that maps edges to unordered pairs of distinct vertices.

Recall that a one-to-one function is one that does not map two distinct elements of the domain to the same element of the range. The definition therefore eliminates multiple edges joining the same two vertices. The requirement that ϕ maps edges to pairs of *distinct* vertices eliminates loops.

EXAMPLE 10.2 **A Small Graph Revised**

The graph in Example 10.1 can be rewritten as $G = (V, E, \phi)$, where $V = \{v_1, v_2, v_3\}$, $E = \{e_1, e_2\}$ and $\phi(e_1) = v_1v_2$, $\phi(e_2) = v_1v_3$.

The visual representation has not changed (Figure 10.7). ■

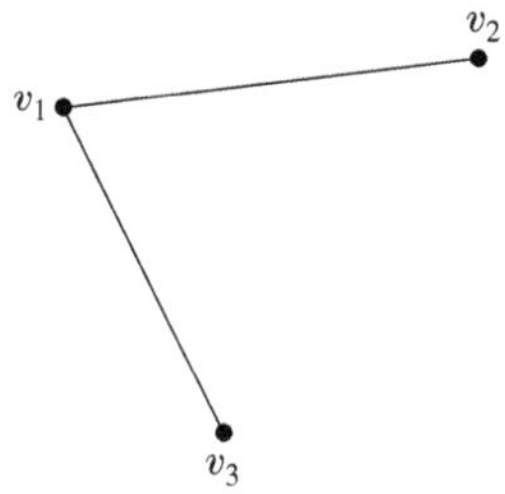

v_1 v_2 v_3

Figure 10.7. Visually the same as Figure10.6.

DEFINITION 10.3 ***Graph***

A *graph*, $G = (V, E, \phi)$, consists of a nonempty, finite set of *vertices*, V, a finite set of *edges*, E, and an incidence function, ϕ, that maps edges to unordered pairs of vertices. If the two vertices in the unordered pair are the same vertex, the edge is called a *loop*.

Two changes have been made to Definition 10.2: The phrases *one-to-one* and *distinct* have been dropped from the definition of ϕ. These omissions allow for multiple edges between vertices and loops, respectively.

If the context permits, we may sometimes be lazy and write edges directly as pairs of vertices instead of formally referring to the function ϕ. Thus, we might write $e_1 = v_1v_2$ rather than the more awkward statement "e_1 is an edge with $\phi(e_1) = v_1v_2$".

If we want to emphasize that loops and/or multiple edges between vertices are allowed (but not required), the graph may be called a *multigraph*.

The letter n will often be used to denote the number of vertices in a graph G: $n = |V|$. Similarly, the Greek letter *epsilon* will be often be used to denote the number of edges in the graph: $\epsilon = |E|$.

[2] A loop in a diagram would be an edge with both ends connected to the same vertex.

If G is a multigraph, it is easy to form a related simple graph, called the *underlying simple graph of G*. The simple graph is formed by deleting all loops and then deleting all but one edge from each set of edges that map to the same unordered pair of vertices. Notice that there is some ambiguity in this definition since it does not precisely specify which edges from each multiple set should be eliminated. However, suppose S_1 and S_2 are two different simple graphs resulting from different choices of edges to eliminate. Then, in all important respects, S_1 and S_2 will have identical properties.[3]

DEFINITION 10.4 ***Adjacent; Incident***

Two vertices, v and w, are said to be *adjacent* if there is an edge $e \in E$ for which $\phi(e) = vw$. In this case, the vertices v and w are said to be *incident* with e (and vice versa).

DEFINITION 10.5 ***Degree of a Vertex; Regular***

The *degree* of the vertex $v \in V$, denoted $\deg(v)$, is the number of edges that are incident with v. Loops are counted twice (once for each end of the loop).

If every vertex in G has the same degree, the graph is said to be *regular*.

EXAMPLE 10.3 **Degree and Adjacency**

Consider the graph in Figure 10.8 (in visual form).

The degrees and adjacencies are summarized in Table 10.1.

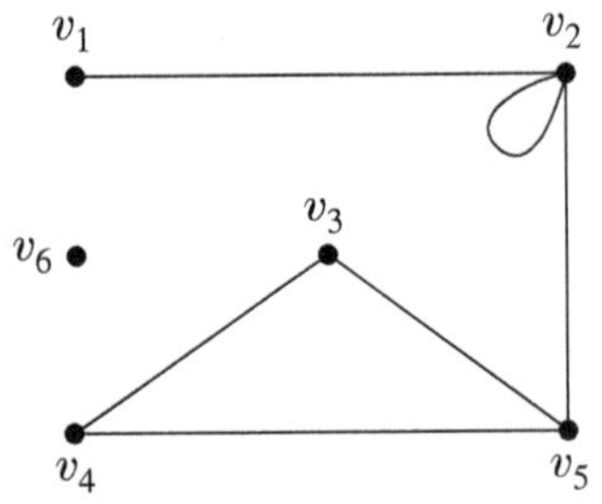

Figure 10.8. Degrees and adjacency.

TABLE 10.1 Degrees and Adjacency

Vertex	Degree	Adjacent Vertices
v_1	1	v_2
v_2	4	v_1 v_2 v_5
v_3	2	v_4 v_5
v_4	2	v_3 v_5
v_5	3	v_2 v_3 v_4
v_6	0	

■

With these few definitions it is possible to state and prove a simple, but useful theorem. Before you read the proof, you should verify the truth of the theorem for the previous example and for the Königsberg graph (Figure 10.9).

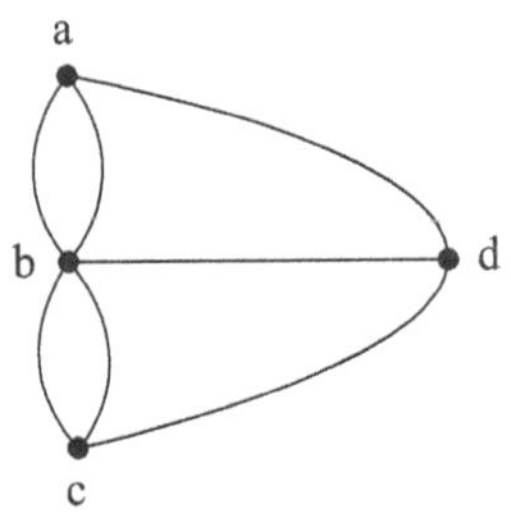

Figure 10.9. The Königsberg graph.

THEOREM 10.6 ***The Handshake Theorem***

Let $G = (V, E, \phi)$ be a graph with $\epsilon = |E|$ edges. Then

$$\sum_{v \in V} \deg(v) = 2\epsilon$$

Proof: For any vertex v, $\deg(v)$ is the number of edges incident with v. Every edge is incident with two vertices, so every edge is counted twice in the sum $\sum_{v \in V} \deg(v)$. □

[3] This statement could be made more precise once *graph isomorphism* is introduced in Definition 10.36 on page 629.

✔ Quick Check 10.1

1. List the degree for each vertex in the following graph.

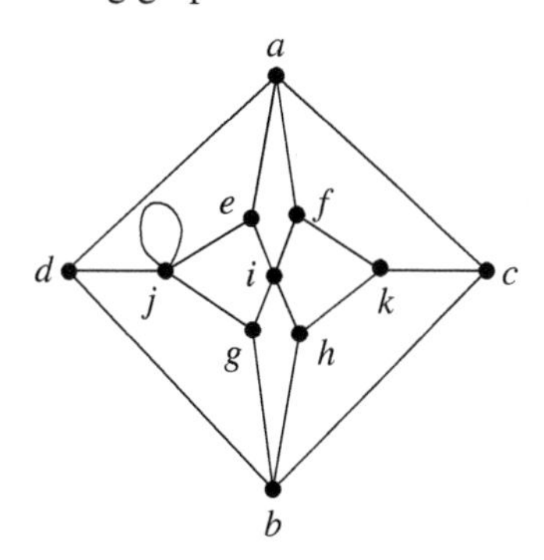

2. The handshake theorem has not been arbitrarily named. Suppose you enter a room containing 12 people. After asking each person to report the number of handshakes he or she participated in (before you entered), you learn two facts:
   - The sum of the reported numbers is 36.
   - Every person participated in the same number, h, of handshakes.

   What is h? ☑

10.1.1 New Graphs from Old

There are mechanisms for constructing a new graph from an old one. The most prominent such mechanisms are defined next.

DEFINITION 10.7 ***Subgraph***

Let $G = (V, E, \phi)$ be a graph. If $V' \subseteq V$ and $E' \subseteq E$ is such that no edge in E' is incident to a vertex that is not in V', we say the $G' = (V', E', \phi)$ is a *subgraph* of G.

If $V' \subseteq V$, the subgraph formed by taking all vertices in V', together with all edges that join two vertices in V', is called the *subgraph induced by* V'.

Informally, we construct a subgraph by removing zero or more edges. We may also remove some vertices and all edges incident with the removed vertices.

The prime notation has no special meaning. It is just a convenient way to denote a second graph.

EXAMPLE 10.4 **A Subgraph**

By removing the vertex j and the edges ac, bc, and kc from the graph in Figure 10.10 we obtain the subgraph shown in Figure 10.11.

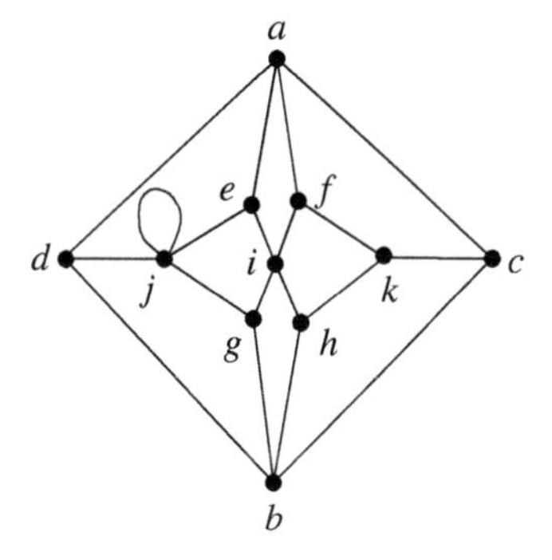

Figure 10.10 Original graph.

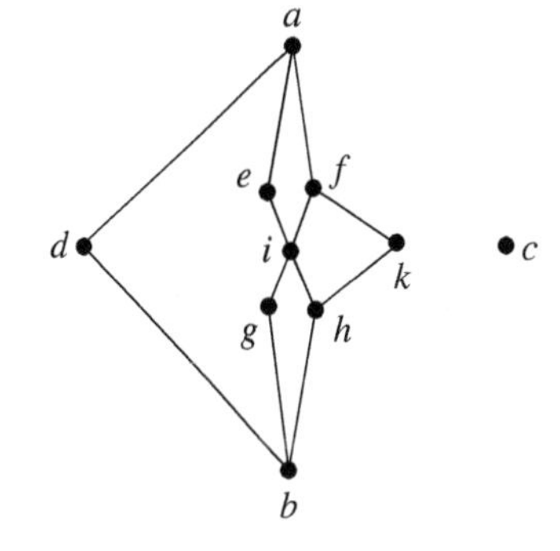

Figure 10.11 A subgraph.

■

DEFINITION 10.8 ***Complement of a Simple Graph***

Let $G = (V, E, \phi)$ be a simple graph. The *complement of* G is the simple graph $\overline{G} = (V, \overline{E}, \overline{\phi})$ where $\overline{E}$ contains an edge $\overline{e}$ with $\overline{\phi}(\overline{e}) = vw$ if and only if E does *not* contain an edge e with $\phi(e) = vw$.

The complement $\overline{G}$ has an edge joining the members of every pair of vertices that are not joined in G, and vice versa.

EXAMPLE 10.5

A Complement

Figure 10.12 shows a simple graph and its complement.

In $\overline{G}$, the intersection of edges v_1v_5 and v_2v_4 is *not* a vertex. That crossing is merely an artifact of the manner in which the graph was represented visually.

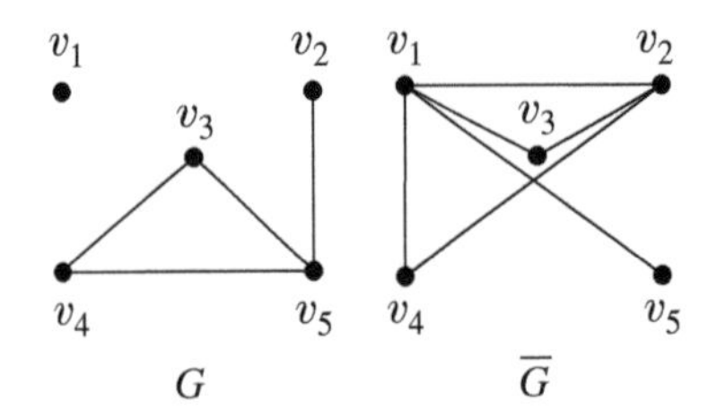

Figure 10.12 A graph and its complement.

■

DEFINITION 10.9 ***The Line Graph***

Let $G = (V, E, \phi)$ be a graph with no loops. The *line graph*, G_L is the graph $G_L = (E, F, \zeta)$ whose set of vertices is the set E of edges of G. Two vertices in E are adjacent in G_L if and only if they are both incident (as edges in G) with a common vertex in V. There will be one edge in F joining e_i and e_j for each shared vertex in V.

EXAMPLE 10.6

A Line Graph

The Königsberg graph can be shown with the edges labeled (Figure 10.13).

The line graph is shown in Figure 10.14.

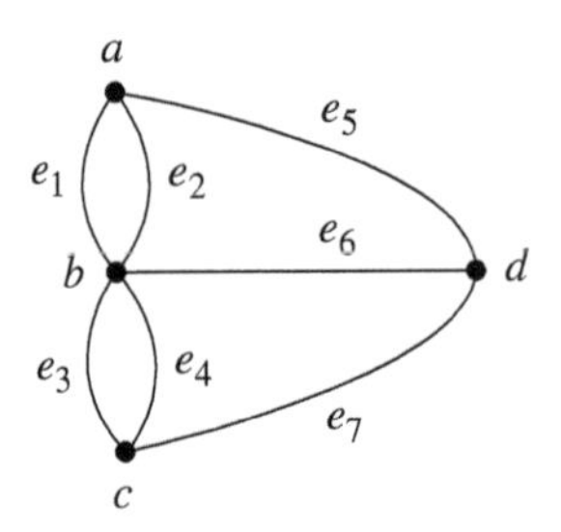

Figure 10.13 The Königsberg graph with labeled edges.

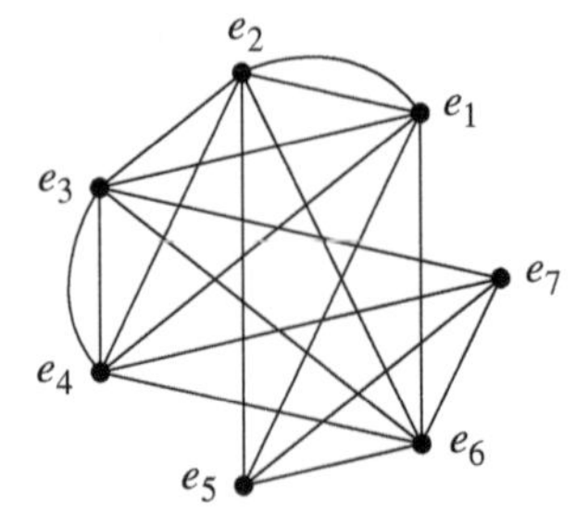

Figure 10.14 The line graph of the Königsberg graph.

■

One other technique for creating a new graph from an existing graph will be presented later (page 643).

10.1.2 Special Graph Families

There are a number of families of graphs that occur often enough to be given names. Several will be presented next.

The Family of Complete Graphs

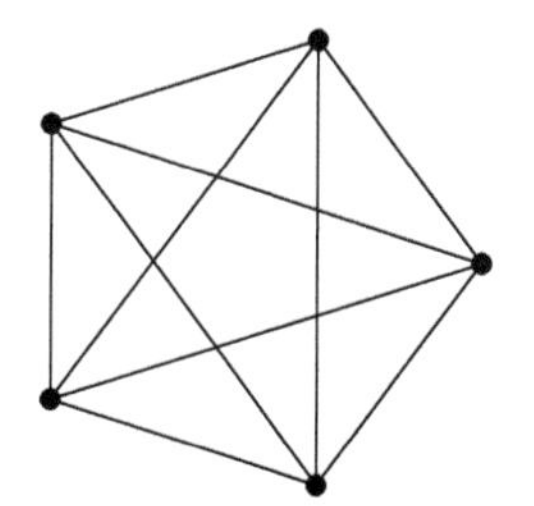

Figure 10.15. K_5.

The *complete graph on n vertices* is the simple graph for which there is a common edge incident with both members in every pair of distinct vertices. The complete graph on n vertices is denoted K_n. The graph K_5 is shown in Figure 10.15.

The Family of Complete Bipartite Graphs

DEFINITION 10.10 ***Bipartite Graph***
A graph $G = (V, E, \phi)$ is *bipartite* if $E \neq \emptyset$ and if V can be partitioned into two disjoint subsets V_α and V_β such that for every edge $e \in E$, $\phi(e) = v_1v_2$, where $v_1 \in V_\alpha$ and $v_2 \in V_\beta$.

A graph is bipartite if there are no edges that join vertices in V_α to other vertices in V_α and also no edges that join vertices in V_β to other vertices in V_β, but there is at least one edge. If G is a bipartite graph, then we can write G as $G = (V_\alpha \cup V_\beta, E, \phi)$, where $V_\alpha \cap V_\beta = \emptyset$ and all edges in E connect a vertex in V_α to a vertex in V_β. Notice that the definition essentially says "*if* there is an edge, then its two incident vertices must not be from the same subset in the partition". It does *not* say that a randomly chosen vertex from V_α must be on a common edge with every (or any) vertex in V_β.

EXAMPLE 10.7 **A Bipartite Graph**

The graph in Figure 10.16 is bipartite because the vertices can be partitioned as $V_\alpha = \{a, b, c\}$, and $V_\beta = \{d, e\}$. ■

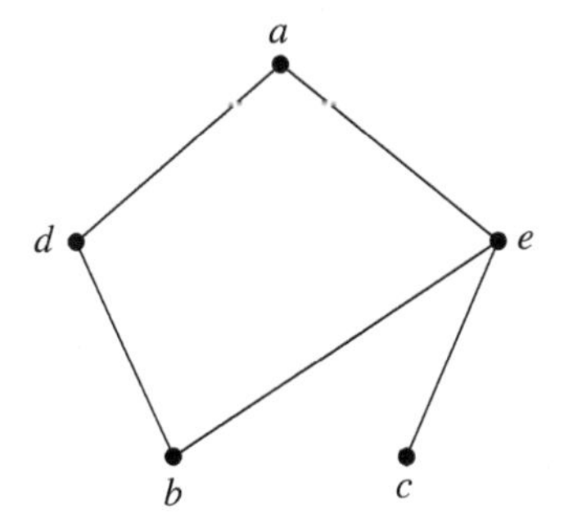

Figure 10.16. A bipartite graph.

The *complete bipartite graph*, $K_{n,m}$, is a bipartite simple graph $G = (V_\alpha \cup V_\beta, E, \phi)$ with $n + m$ vertices, where $|V_\alpha| = n$, $|V_\beta| = m$, and every possible edge joining a vertex in V_α to a vertex in V_β is in E, but E contains no other edges.

The complete bipartite graph $K_{3,4}$ is shown in Figure 10.17.

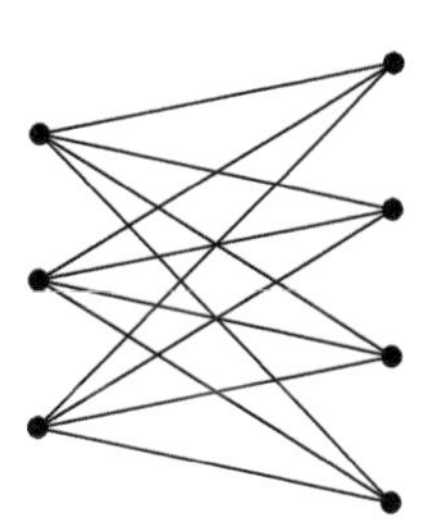

Figure 10.17 $K_{3,4}$.

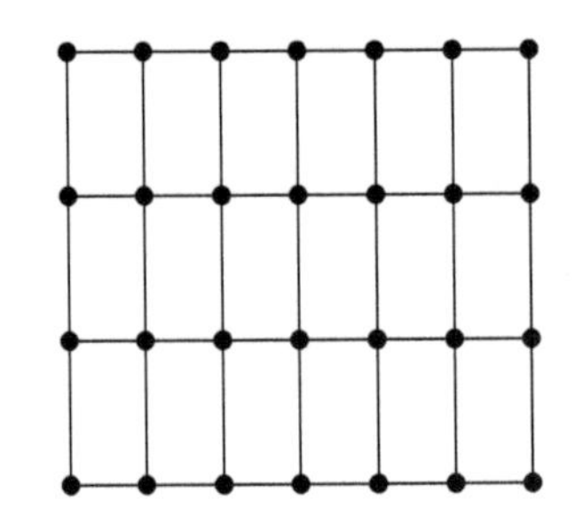

Figure 10.18. $G_{4,7}$.

Grids, Cycles, Wheels, and Stars

The *grid graph*, $G_{n,m}$, is a simple graph with vertices arranged in an n by m grid. Edges join vertices that are vertically or horizontally adjacent. Figure 10.18 represents $G_{4,7}$.

The *cycle graph on* $n \geq 3$ *vertices*, C_n, is a simple graph with all vertices having degree 2. The graph may be represented in a circular arrangement, as shown in Figure 10.19 for C_{30}.

The *wheel graph on* $n \geq 4$ *vertices*, W_n, is a simple graph with all but one vertex having degree 3. The remaining vertex has degree $n-1$ (so it is adjacent to all other vertices). The graph may be represented as a wheel with spokes, as shown in Figure 10.20 for W_{16}.

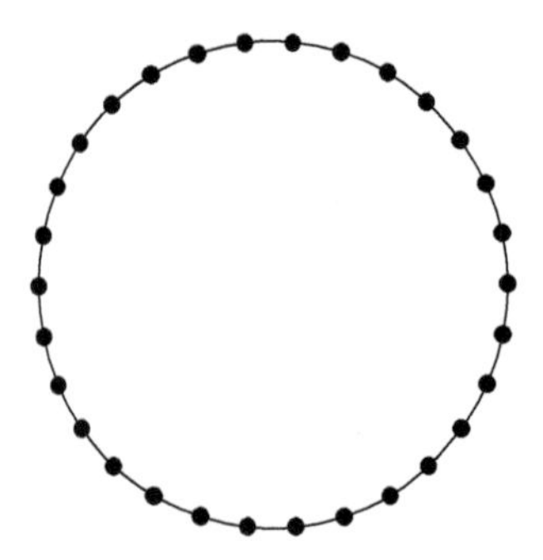

Figure 10.19 C_{30}.

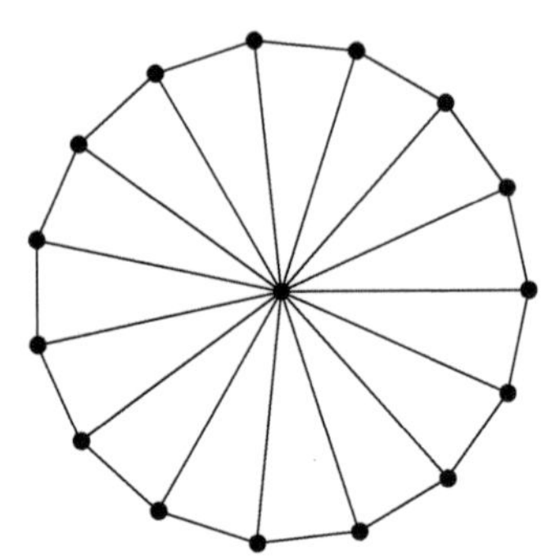

Figure 10.20 W_{16}.

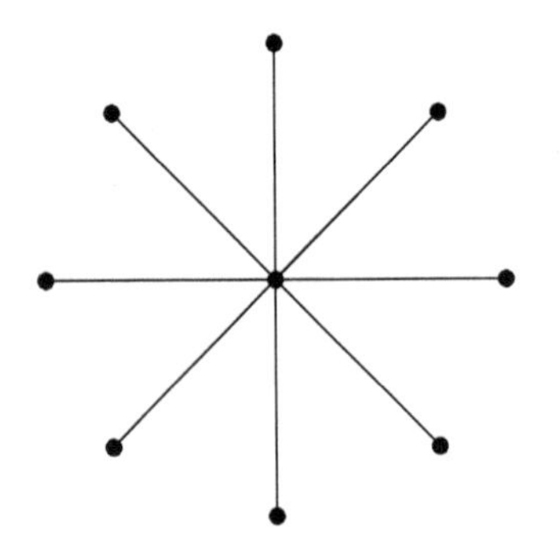

Figure 10.21. S_9.

The *star graph on* $n \geq 1$ *vertices*, S_n, is a simple graph with all but one vertex having degree 1. The remaining vertex has degree $n-1$. The graph may be represented as a central vertex that is adjacent to $n-1$ orbiting vertices, as shown in Figure 10.21 for S_9.

The Family of Hypercubes

The *hypercube on n vertices*, H_n, represents an n dimensional cube. One way to think of a hypercube is to label the vertices using binary strings with length n. Two vertices are adjacent if their labels differ in exactly one position. For example, in H_3, the vertices labeled 101 and 111 would be adjacent but 101 and 110 would not.

Figure 10.22 represents H_3. Figure 10.23 represents H_4.

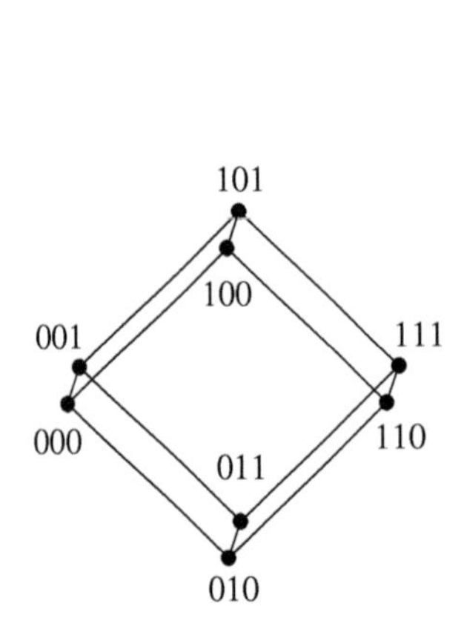

Figure 10.22 H_3.

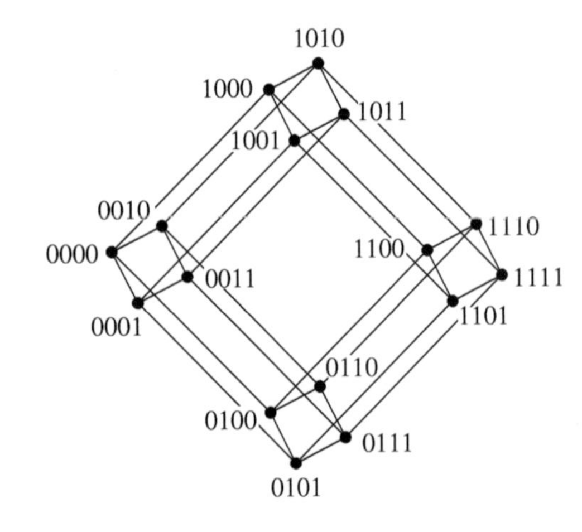

Figure 10.23 H_4.

Quick Check 10.2

1. List two important subgraphs of the wheel graph W_n.
2. Draw the graph $\overline{G_{3,2}}$.

Software Tools for Manipulating Graphs

It is very useful to have access to a software tool that enables you to easily create and manipulate graphs. Simple drawing programs are of some value, but they require you to place each vertex, edge, and label as separate items. They also do not allow you to perform operations like forming the complement of a graph or calculating the degrees of vertices.

Mathematica contains a collection[4] of functions that can manipulate and perform operations on graphs. The collection is named *Combinatorica* and was first developed by Steven Skiena [84]. Its major deficiency is the lack of ability to manipulate the visual representation. In particular, it would be useful to click and drag a vertex and have all adjacent edges adjust themselves to follow.

There have been several projects started to create software with this click-and-drag capability. One such project is Sandbox. It provides both an applet and an application user interface.[5] The application version can import and export Combinatorica graph files.

The screen shot in Figure 10.24 shows the application in action.

The screen shot in Figure 10.25 shows the graph in Figure 10.24 after vertex a has been dragged to the right and vertex b is part way through being repositioned.

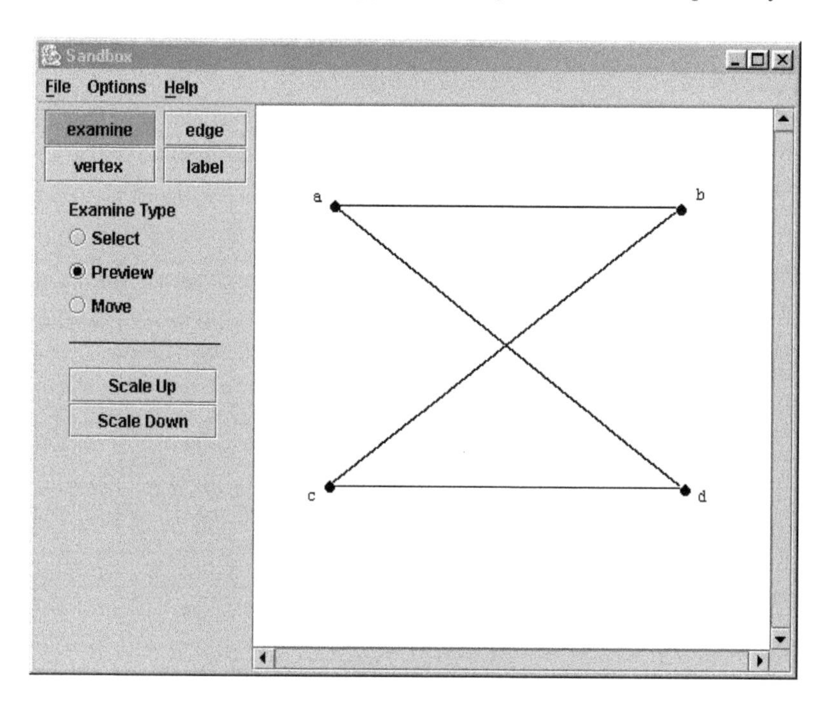

Figure 10.24 Sandbox.

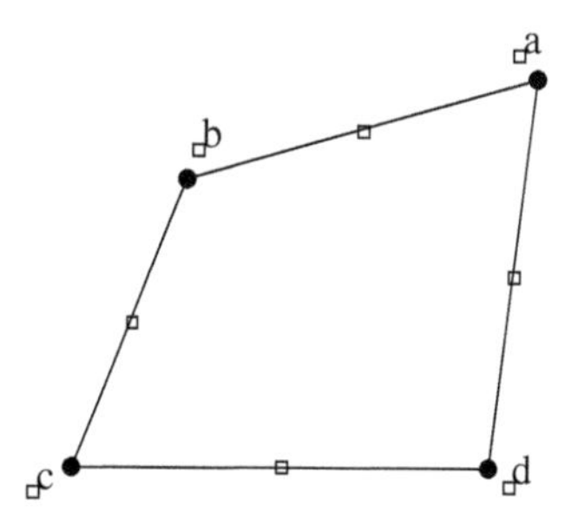

Figure 10.25 The graph in Figure 10.24 after dragging.

[4]The collection is called a *package* in *Mathematica* terminology.

[5]The software can be run in a Web browser or downloaded for free. Both options are available at http://www.mathcs.bethel.edu/~gossett/DiscreteMathWithProof/ .

10.1.3 Exercises

The exercises marked with ★ have detailed solutions in Appendix H.

1. Produce the underlying simple graph for the following graph.

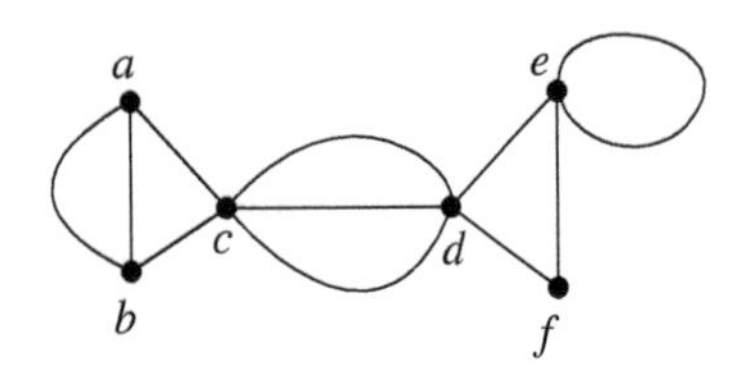

2. For which values of n and m are the following graphs bipartite? (No formal proofs are required for this problem.)
   (a) ★ $G_{n,m}$ (b) K_n (c) S_n (d) C_n (e) W_n

3. Draw each of the following complement graphs.
   (a) $\overline{W_5}$ (b) $\overline{S_5}$ (c) $\overline{G_{2,2}}$

4. Draw the complement graphs for each of the following three graphs.
   (a) The graph in Figure 10.16 on page 605.
   (b) ★

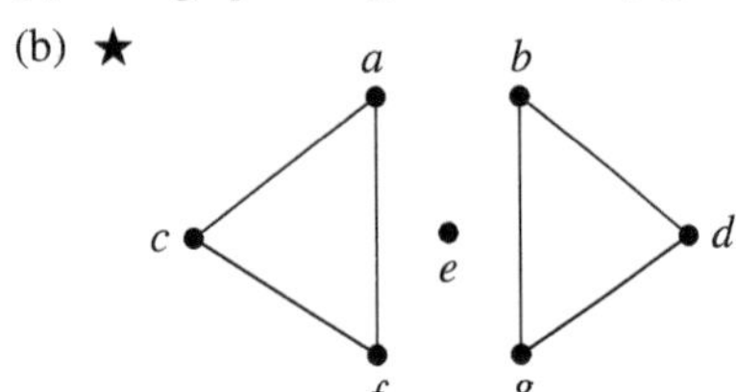

   (c)

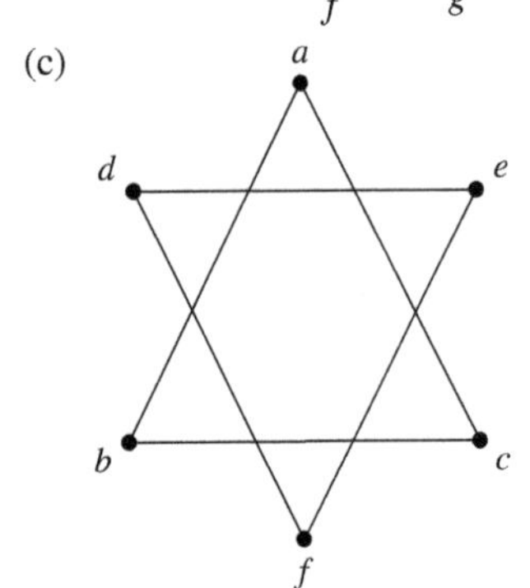

5. For which values of n and m are the following graphs regular?
   (a) K_n (b) ★ S_n (c) $G_{n,m}$

6. Draw the line graph for each of the following graphs:
   (a) K_3 (b) ★ C_5 (c) K_4
   (d) W_4 (e) S_5 (f) $G_{2,3}$

7. ★ How many edges does K_n have? Prove your result.

8. How many edges does $K_{n,m}$ have? Prove your result.

9. Prove that no graph can have an odd number of vertices with odd degrees. [*Hint*: Look at Exercise 10 on page 162.]

10. Suppose G is a regular graph in which every vertex has odd degree, k. Prove that the number of edges in G is a multiple of k.

11. In a class with nine students, each student sends a valentine to three other students. Is it possible for every student to only receive valentines from the three classmates he or she sent cards to? (From [100].)

12. Show that if G is a simple graph with n vertices and ϵ edges, then $\epsilon \le \binom{n}{2}$.

13. Is it possible for a graph to have half its vertices having degree a and half its vertices having degree $a + 1$? If it is possible, provide a necessary condition and then give an example. Otherwise, provide a proof that no such graph exists.

14. For which values of n can there be a multigraph with $|V| = n$ and having a vertex with each degree in $\{1, 2, \ldots, n\}$? Prove your claim. Draw visual representations for the two smallest values of n for which this is possible.

15. ★ Let G be a simple graph with n vertices and ϵ edges. How many edges are in $\overline{G}$?

16. Let $G = (V, E, \phi)$ be a graph with n vertices and ϵ edges. Show that

$$\min_{w \in V} \deg(w) \le \frac{2\epsilon}{n} \le \max_{w \in V} \deg(w)$$

17. If G has 10 edges and $\overline{G}$ has 5 edges, how many vertices does G have?

18. If G has j edges and $\overline{G}$ has k edges, where $2(j + k) = m(m + 1)$ for some positive integer, m, how many vertices does G have? Prove your claim.

19. Each of the following statements is either true (always) or false (at least sometimes). Determine which option applies for each statement and provide adequate explanation for your choice.
   (a) If there is an edge joining two vertices in a graph, we say that these two vertices are incident.
   (b) ★ It is permissible to call some graph a *multigraph* to emphasize that loops and/or multiple edges between vertices are not permitted.
   (c) If there is no edge joining two of the vertices in a simple graph, G, then there must be an edge joining these two vertices in $\overline{G}$.
   (d) The underlying simple graph of a multigraph can be formed easily by deleting all loops and then deleting all edges from each set of edges that map to the same unordered pair of vertices.
   (e) The complement, $\overline{K_{m,n}}$, of $K_{m,n}$ is also a bipartite graph.

20. Let G be a simple graph with at least two vertices. Prove that there must exist two vertices in G that have the same degree.

21. (a) Prove that the cycle graph, C_n, is bipartite if and only if n is even.
   (b) Prove that if a graph, G, is bipartite, then it does not contain a cycle having an odd number of edges.

22. Draw a visual representation of a regular simple graph $G = (V, E, \phi)$ where $|V| = 10$ and $\deg(v) = 3$ for all $v \in V$. [*Hints*:
(1) How many edges will the graph need?
(2) You can get 10 vertices by starting with two copies of C_5.]

> **DEFINITION 10.11** ***Clique; Independent Set***
> A *clique* in a graph, $G = (V, E, \phi)$, is a subset, C, of V such that every pair of vertices in C is joined by an edge in E.
> An *independent set* in a graph, $G = (V, E, \phi)$, is a subset, I, of V such that no pair of vertices in I is joined by an edge in E.

23. Let G be a simple graph with n vertices and suppose that G contains an independent set of size k (Definition 10.11).
    (a) What is an upper bound on the number of edges in G?
    (b) Create a simple graph with $n = 5$ vertices and an independent set of size $k = 3$ that attains the upper bound.

24. Prove that a maximal clique in a graph, G, is a maximal independent set in $\overline{G}$.

25. ★ Definition 10.11 has nice visual interpretations. If C is a clique in a graph, G, then a diagram of the subgraph of G induced by C looks like a diagram of K_n, for some $n \geq 1$. An independent set induces a subgraph that contains no edges.
For each of the graphs in Exercise 4, find a maximal clique and a maximal independent set.

26. Prove Theorem 10.12 and Corollary 10.13. [*Hints*: For the theorem, use Exercise 21 and Proposition 11.41 on page 725. For the corollary, see Exercise 24.]

> **THEOREM 10.12** ***Bipartite Graphs and Independent Sets***
> A graph, G, is bipartite if and only if every subgraph, H, of G contains an independent set consisting of at least half the vertices in H.

> **COROLLARY 10.13** ***Bipartite Graphs and Cliques***
> Let G be a graph. Then $\overline{G}$ is bipartite if and only if every subgraph, H, of G contains a clique consisting of at least half the vertices in H.

10.2 Connectivity and Adjacency

Paths and connectivity are generalized forms of adjacency. These topics are the substance of this section.

10.2.1 Connectivity

The goal in this section is to define and then explore what it means for a graph to be connected.

You may already have an intuitive idea of what it means for a graph to be "connected". One way to test your intuitive notion is to make up several graphs that you think should be connected and several that you think should not be called connected. Check your definition on each example. Your goal is to put your definition through a torture test to see if it is really pure gold or still contains impurities. Try to be devious; make your definition fail if possible. Then create a better definition and start over.

✔ Quick Check 10.3

1. Carry out the process just outlined. ☑

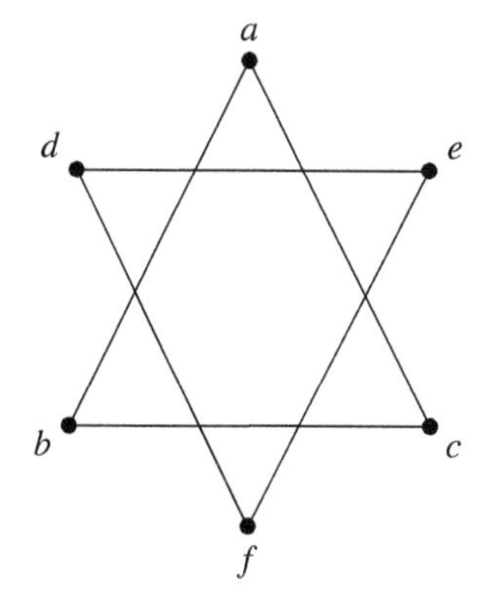

Figure 10.26. Is this connected?

Here is a final test for your definition. Should the graph in Figure 10.26 be considered to be connected?

Notice that the vertices $V_\alpha = \{a, b, c\}$ and the vertices $V_\beta = \{d, e, f\}$ are not connected by any edges. A definition that uses terms such as "without lifting your pencil from the paper" fails on two counts. First, it would imply that the previous graph is connected. Second, a definition in terms of the visual representation is inadequate.

It seems necessary that "getting from one vertex to any other" be part of the final definition. This concept needs to be expressed in terms of the formal definition $G = (V, E, \phi)$. To that end, some new auxiliary definitions are in order. These definitions are presented as a basic concept (a walk), together with some additional refinements (can edges and vertices be repeated?, does the walk start and end at the same place?).

DEFINITION 10.14 ***Walk; Trail; Path; Closed Walk; Circuit; Cycle***

Let $G = (V, E, \phi)$ be a graph.

The Basic Concept

A *walk* of length k is a nonempty sequence of alternating vertices and edges, $v_0 e_1 v_1 e_2 v_2 \cdots e_k v_k$, such that $\phi(e_i) = v_{i-1}v_i$, for $i = 1, 2, \ldots, k$. The vertices, v_0 and v_k, are called the *endpoints* of the walk.

Excluding Repeated Edges and Vertices

- A *trail* is a walk with no repeated edges.
- A *path* is a walk with no repeated vertices.

Starting and Ending at the Same Vertex

- A walk is *closed* if it has length of at least one and its endpoints are the same vertex.
- A *circuit* is a closed trail.
- A *cycle* is a circuit in which the endpoints are the only repeated vertices.

Alternative Notation for Walks in Simple Graphs

If G is a simple graph, or a simple graph with loops, then the edge joining two vertices is uniquely determined. Consequently, the walk $v_0 e_1 v_1 e_2 v_2 \cdots e_k v_k$ can be denoted unambiguously by $v_0 v_1 v_2 \cdots v_k$.

You may have wondered why there is no special name for a closed path, especially since there are two special names for different kinds of closed trails. A moment's reflection provides the answer. To be a closed walk, the initial vertex must be repeated. The presence of a repeated vertex means that the walk cannot be a path.

EXAMPLE 10.8 **Walks and Circuits**

The graph in Figure 10.27 contains many walks. Since it is a simple graph with loops, it is not necessary to include edges when listing a walk. Several walks of various kinds are listed next. To keep the graph from becoming cluttered, the edges have not been labeled. The edges will be designated by an e with subscripts that indicate the incident vertices. For example, the edge with endpoints p and r will be denoted by e_{pr}.

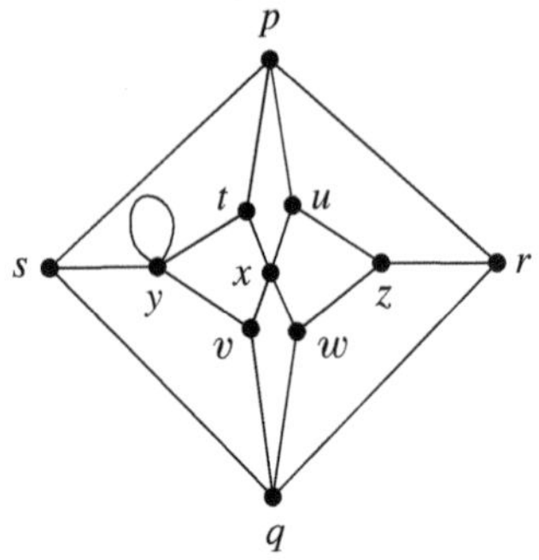

p t u s r x y z v w q

Figure 10.27. Walks and circuits.

- The walk $pe_{pr}re_{rz}ze_{zw}w$ (or $przw$) is both a trail and a path.
- The loop $ye_{yy}y$ is a circuit of length one.
- The walk $syytx$ is not a path (the vertex y is repeated), but it is a trail.
- The walk $ptxupsq$ is a trail, is not a path, and is not a circuit (even though it contains a subwalk that *is* a circuit).
- The walk s has length zero.
- The walk $prqsptxup$ is a circuit.
- The walk $tpuxt$ is a cycle. ■

EXAMPLE 10.9 **Sorting Out Notation**

The fact that the definition of a walk contains a list of subscripted v's, in numeric order, might be confusing when applied to a graph in which the vertices have been labeled using subscripted v's. Figure 10.28 shows such a graph. The definition should be interpreted as showing a general pattern. The subscripts there help to show the ordering of the elements in the walk and also help to determine the length of the walk. However, they are not to be taken too literally.

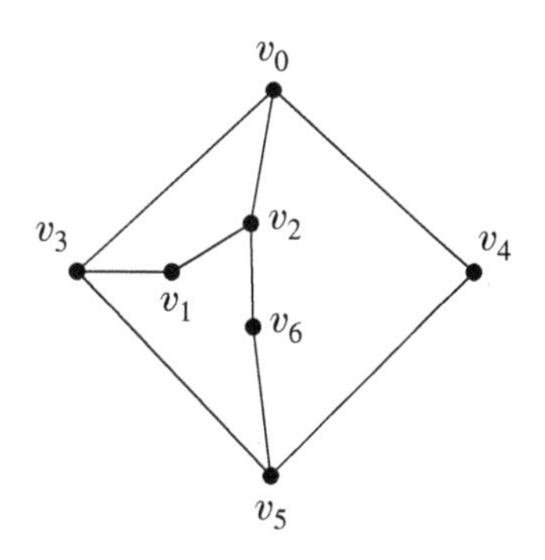

Figure 10.28. Vertices labeled as subscripted v's.

In Figure 10.28, the walk $v_4v_5v_3v_1v_2$ has length 4. The walk $v_4v_5v_3v_1v_2v_0v_4$ is a cycle with length 6. The walk $v_3v_1v_3$ is a closed walk with length 2. ■

It is finally time to define *connected.*

DEFINITION 10.15 *Connected*

A graph $G = (V, E, \phi)$ is *connected* if $|V| = 1$ or if there is a walk in G between every pair of distinct vertices.

The set V of vertices in G can always be partitioned into a disjoint union $V = V_1 \cup V_2 \cup \cdots V_k$ such that $v, w \in V_j$ for a common j if and only if there is a walk in V from v to w. The subgraphs induced by the vertex subsets V_j are the *components* (or *connected components*) of G. A connected graph has one component.

EXAMPLE 10.10

Components

This graph has three components.

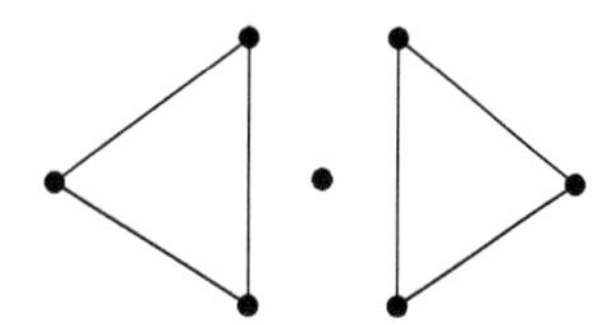

■

Consider the three graphs in Figure 10.29.

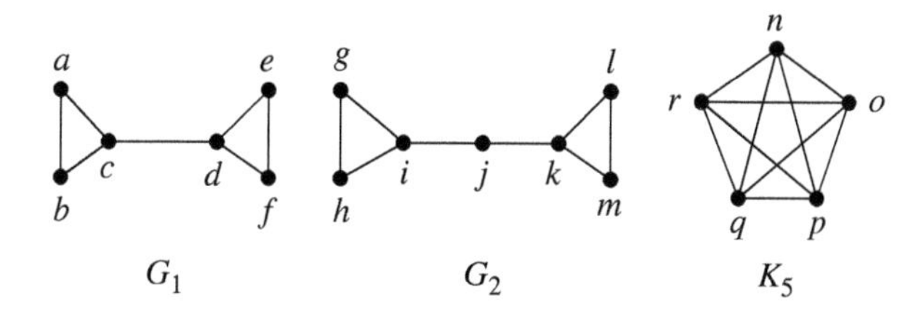

Figure 10.29 Levels of connectedness.

The rightmost graph (K_5) seems in some sense more connected than the other two graphs. This intuitive notion can be made concrete in several ways. First, notice that graphs G_1 and G_2 can be disconnected by removing a single edge. Graph G_2 can be disconnected by removing the vertex j (and the edges ij and jk). However, K_5 can have any vertex and its incident edges removed and still stay connected. In fact, to disconnect K_5, at least four edges must be removed.

DEFINITION 10.16 *Cut Vertex; Cut Edge*

Let v be a vertex in a component, G_1, of a graph G. The vertex v is called a *cut vertex* if removing v and all edges incident with v causes G_1 to become disconnected.

Let e be an edge in a component, G_1, of a graph G. The edge e is called a *cut edge* if removing e causes G_1 to become disconnected.

EXAMPLE 10.11

Cut Vertices and Cut Edges

The graph G_1 in Figure 10.29 has cut edge cd and cut vertices c and d. The graph G_2 has cut edges ij and jk and cut vertices i, j, and k. The graph K_5 has no cut edges and no cut vertices. ■

A second way to make the notion of "more strongly connected" concrete is to consider the minimum number of vertices that need to be removed before a connected graph becomes disconnected.

DEFINITION 10.17 ***Connectivity***
A graph $G = (V, E, \phi)$ is said to have *connectivity* k if k is the size of the smallest subset of vertices whose removal from V either causes the graph to become disconnected or reduces V to a single vertex. G is *k-connected* if the connectivity of G is at least k.

Notice that a connected graph with more than one vertex is at least 1-connected. Disconnected graphs and graphs with a single vertex are 0-connected.

EXAMPLE 10.12 **Vertex Connectivity**

The graphs G_1 and G_2 in Figure 10.29 are 1-connected. The graph K_5 is 4-connected. ■

The amount of connectedness in a graph can also be measured in terms of the edges.

DEFINITION 10.18 ***Edge Connectivity***
A connected graph $G = (V, E, \phi)$ is said to have *edge connectivity* k if k is the size of the smallest subset of edges whose removal from E causes the graph to become disconnected. G is *k-edge-connected* if the connectivity of G is at least k.

If G is not connected, it has edge connectivity 0.

EXAMPLE 10.13 **Edge Connectivity**

The graph G_1 in Figure 10.30 is 1-edge-connected. The graph G_3 is 2-edge-connected. The graph K_5 is 4-edge-connected.

Figure 10.30 Edge connectivity.

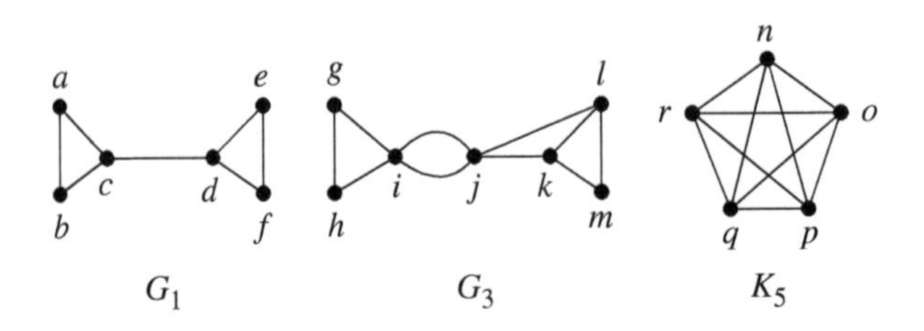

■

✔ Quick Check 10.4

1. Determine the vertex and edge connectivity of the following graphs.

(a)

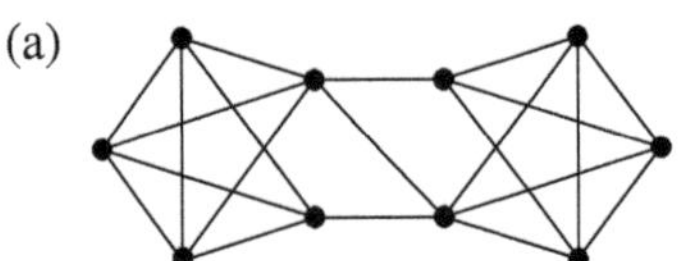

(b)

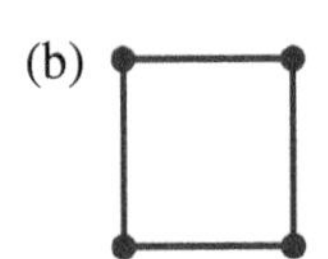

(c)

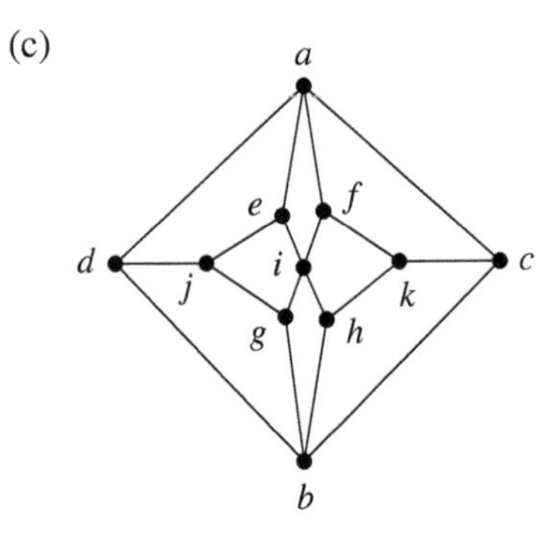

☑

Connectivity will not be developed much further in this text. I will conclude the discussion with one theorem that relates connectivity and edge connectivity. The proof can be found in [12].

THEOREM 10.19 ***Connectivity; Edge Connectivity; Minimal Degree***

Let G be a graph with minimal vertex degree $\delta = \min_{w \in V} \deg(w)$. If G has connectivity k and edge connectivity m, then

$$k \le m \le \delta$$

The graph in Problem 1(a) of Quick Check 10.4 shows that we can have $k < m < \delta$. The graph $G_{2,2}$ in Problem 1(b) of Quick Check 10.4 shows that $k = m = \delta$ is also possible.

10.2.2 The Adjacency Matrix

It is frequently fruitful to apply ideas from one branch of mathematics to the study of a different branch. Your study of calculus relied heavily on the use of analytic geometry. Analytic geometry introduced coordinates to geometry, allowing the algebraic descriptions of functions you are familiar with.

Graph theory similarly benefits from the use of ideas from linear algebra. Most of these ideas require you to have a good background in linear algebra.[6] Since this text does not assume the reader has taken a linear algebra class yet, most of these tools are beyond the scope of this course. However, some tools used to represent graphs algebraically are appropriate here.

This section assumes that the student is familiar with matrix multiplication. Appendix E provides a quick introduction for students who are not familiar with this simple concept.

DEFINITION 10.20 ***Adjacency Matrix of a Simple Graph***

Let $G = (V, E, \phi)$ be a simple graph with $V = \{v_1, v_2, \ldots, v_n\}$. The *adjacency matrix*, $A = (a_{ij})$, is the n by n matrix whose rows and columns are indexed by the elements of V (in the same order). The element a_{ij} is defined by

$$a_{ij} = \begin{cases} 1 & \text{if } v_i \text{ and } v_j \text{ are adjacent} \\ 0 & \text{otherwise} \end{cases}$$

EXAMPLE 10.14 **An Adjacency Matrix for a Simple Graph**

The graph in Figure 10.31 has the adjacency matrix in Table 10.2.

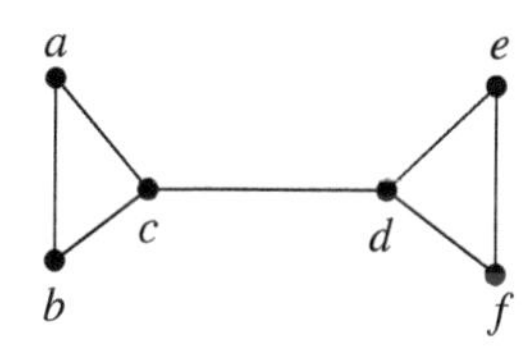

Figure 10.31 A simple graph.

TABLE 10.2 The Adjacency Matrix for the Graph in Figure 10.31

	a	b	c	d	e	f
a	0	1	1	0	0	0
b	1	0	1	0	0	0
c	1	1	0	1	0	0
d	0	0	1	0	1	1
e	0	0	0	1	0	1
f	0	0	0	1	1	0

■

[6]Concepts, such as eigenvalues and eigenvectors, play a key role in algebraic graph theory.

Definition 10.20 can be extended to include multigraphs.

DEFINITION 10.21 ***Adjacency Matrix***

Let $G = (V, E, \phi)$ be a graph with $V = \{v_1, v_2, \ldots, v_n\}$. The *adjacency matrix*, $A = (a_{ij})$, is the n by n matrix whose rows and columns are indexed by the elements of V (in the same order). The element a_{ij} is defined by

$$a_{ij} = \begin{cases} k & \text{if } v_i \text{ and } v_j \text{ are joined by } k \text{ common edges} \\ 0 & \text{otherwise} \end{cases}$$

A loop counts as a single edge.

EXAMPLE 10.15 **An Adjacency Matrix for a Multigraph**

The graph in Figure 10.32 has the adjacency matrix in Table 10.3.

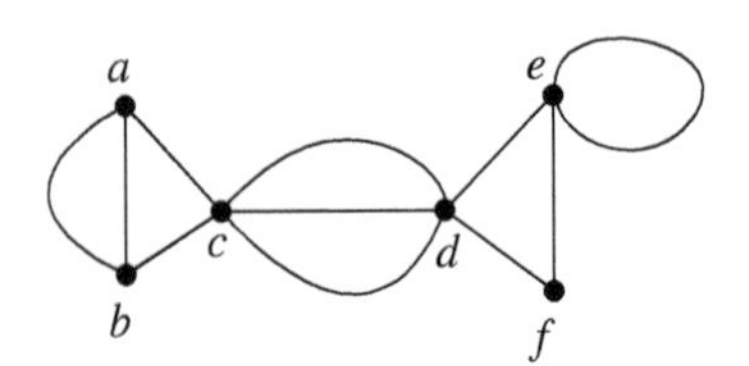

Figure 10.32 A multigraph.

TABLE 10.3 The adjacency matrix for the graph in Figure 10.32

	a	b	c	d	e	f
a	0	2	1	0	0	0
b	2	0	1	0	0	0
c	1	1	0	3	0	0
d	0	0	3	0	1	1
e	0	0	0	1	1	1
f	0	0	0	1	1	0

The sum along any row or column is the number of incident edges (loops count as 1). ■

The final theorem in this section is a very nice application of the adjacency matrix.

THEOREM 10.22 ***Number of Walks in a Connected Graph***

Let $G = (V, E, \phi)$ be a connected graph with adjacency matrix A. Let $V = \{v_1, v_2, \ldots, v_n\}$ and let A be indexed using the natural ordering of these vertices. Then the ijth entry of A^k, for $k \geq 0$, is the number of distinct walks of length k joining v_i and v_j.

Proof: Since the theorem has a claim about A^k for $k \geq 0$, it is reasonable to use a proof by mathematical induction. Thus, assume that G is as described in the statement of the theorem.

Base Step

By definition, the matrix $A^0 = I$, the identity matrix. Every entry is zero except the entries a_{ii} along the main diagonal. Each of these entries is 1. The interpretation is that every vertex has 1 walk of length 0 connecting it to itself, and no walks of length zero to any other vertex.

This is sufficient for the base step. However, notice that $A^1 = A$ does correctly list the number of walks of length 1 from vertex v_i to vertex v_j (because of how a_{ij} was defined in Definition 10.21). With this understanding, the inductive step can use mathematical induction rather than complete induction.

Inductive Step

Assume that the entries of A^k correctly count the number of walks of length k between the vertices of G.

We need to examine the ijth entry of A^{k+1}. Some new notation is needed first. Temporarily denote the ijth entry of A^k as b_{ij} and the ijth entry of A^{k+1} as c_{ij}. Since

$A^{k+1} = AA^k$, using the definition of matrix multiplication (see Appendix E), we can write c_{ij} as

$$c_{ij} = a_{i1}b_{1j} + a_{i2}b_{2j} + \cdots + a_{in}b_{nj}$$

Now think about how a walk from v_i to v_j must progress. It must start at v_i and move to an adjacent vertex, say v_m. There are a_{im} edges along which that may occur. Once the adjacent vertex v_m has been reached, the walk can proceed to v_j in any of b_{mj} ways (since that is how many walks of length k exist between v_m and v_j). Since these choices are independent,[7] the number of walks from v_i to v_j that first pass through v_m is $a_{im}b_{mj}$.

Finally, since moving from v_i to v_m excludes the possibility of moving from v_i to v_s (with $s \neq m$) as the second vertex in the walk, the choices for second vertex are mutually exclusive.[8] Thus it is necessary to add the numbers of ways to reach v_j by passing through each of the possible second vertices in the walk.

This is exactly the form of the entry c_{ij} in A^{k+1}. Thus, A^{k+1} properly counts the number of walks of length $k+1$ between any two vertices in G. □

COROLLARY 10.23 ***Test for Connectedness***

Let the graph G have the n by n adjacency matrix A. Then G is connected if and only if $\sum_{k=0}^{n-1} A^k$ has only nonzero entries.

Proof: If $\sum_{i=0}^{n-1} A^i$ has a zero in the ij^{th} entry, then there is no walk between v_i and v_j that has length less than or equal to $n-1$. By the contrapositive of Exercise 13 on page 617, there is no walk between v_i and v_j, so G is disconnected.

If $\sum_{i=0}^{n-1} A^i$ has a nonzero value in the ij^{th} position, then for at least one value of i, there is a walk of length i between v_i and v_j. If this is true for all choices of i and j, G must be connected. □

10.2.3 Exercises

The exercises marked with ★ have detailed solutions in Appendix H.

1. Prove that if two distinct vertices in a graph are joined by a walk, then they are also joined by a path. (Bullet 4 in Example 10.8 is a walk from p to q having a repeated vertex.)
2. Prove: If G is a graph, then either G or $\overline{G}$ (or both) is connected. [*Hint*: Suppose G is not connected. Let $v, w \in V$ and consider two cases: v and w in the same component of G and v and w in different components of G.]
3. For each walk listed, determine which of the following are true: It is a trail, it is a path, it is a closed walk, it is a circuit, it is a cycle.

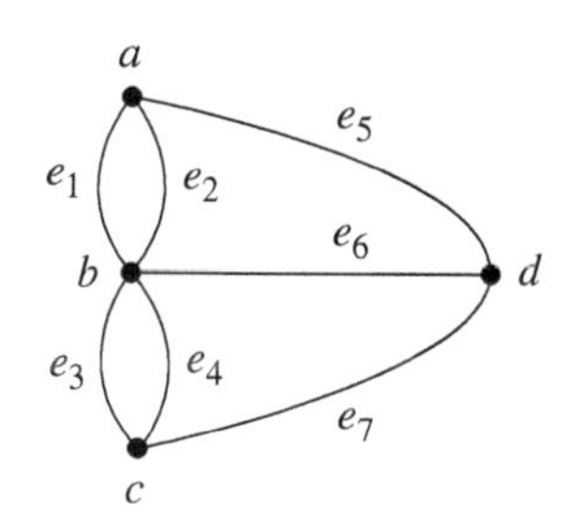

(a) ★ $be_3ce_7de_5a$ (b) ★ $de_6be_4ce_7d$
(c) $ae_1be_2ae_5de_6be_2a$ (d) $ce_7de_5ae_1be_3ce_7d$

4. For each walk listed, determine which of the following are true: It is a trail, it is a path, it is a closed walk, it is a circuit, it is a cycle.

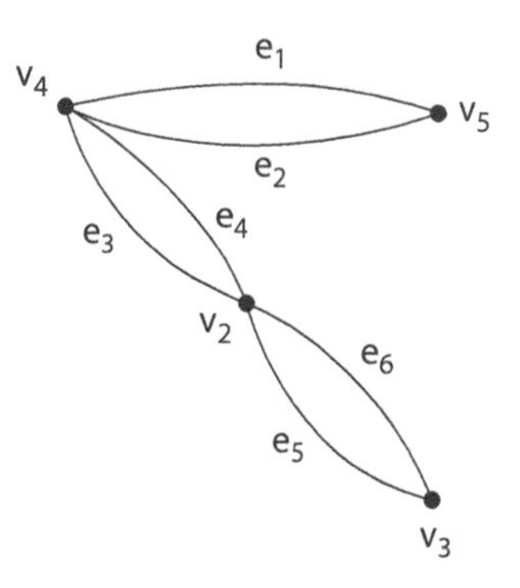

(a) $v_4e_1v_5e_2v_4$ (b) $v_5e_1v_4e_3v_2e_6v_3$
(c) $v_3e_5v_2e_4v_4e_3v_2$ (d) $v_2e_3v_4e_2v_5e_1v_4e_4v_2$
(e) $v_3e_5v_2e_4v_4e_3v_2e_5v_3$ (f) $v_4e_3v_2e_6v_3$

[7] See the Independent Tasks Principle on page 227.
[8] See the Mutually Exclusive Tasks Principle on page 228.

5. What is the connectivity and edge connectivity for the following graphs?

(a) ★

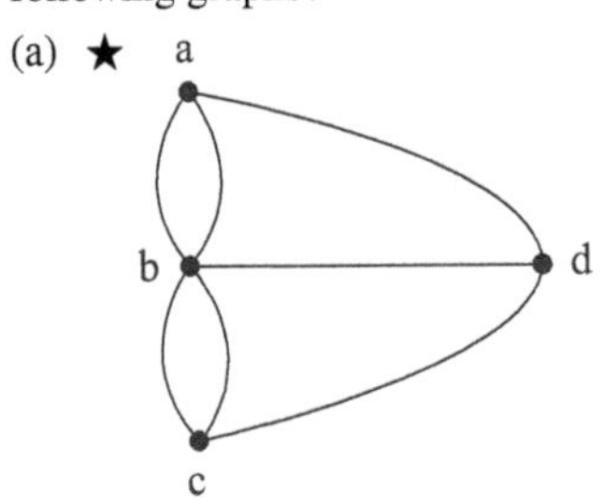

(b)

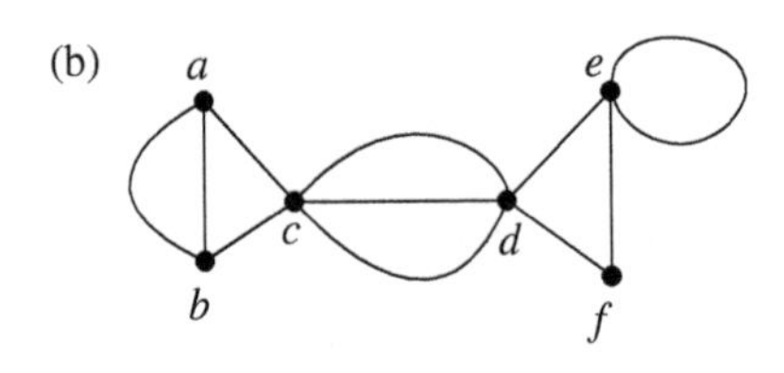

(c)

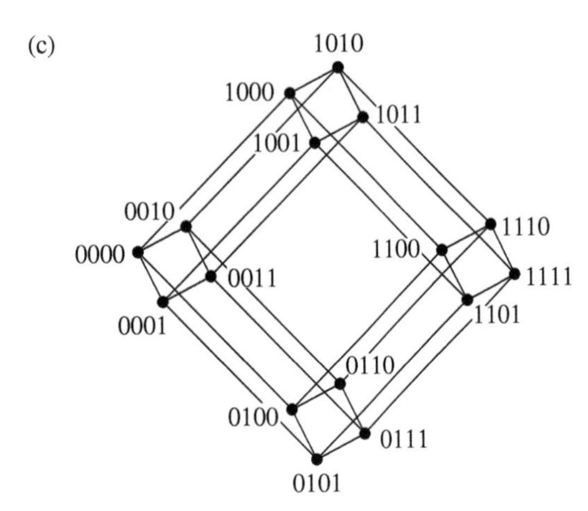

6. What is the connectivity and edge connectivity for the following graphs?

(a)

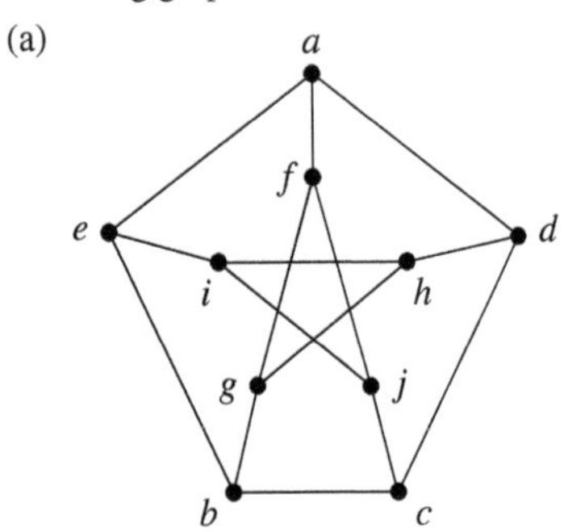

(b)

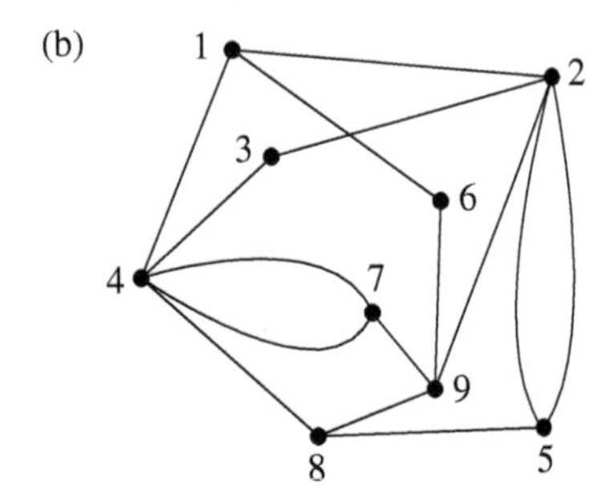

(c)

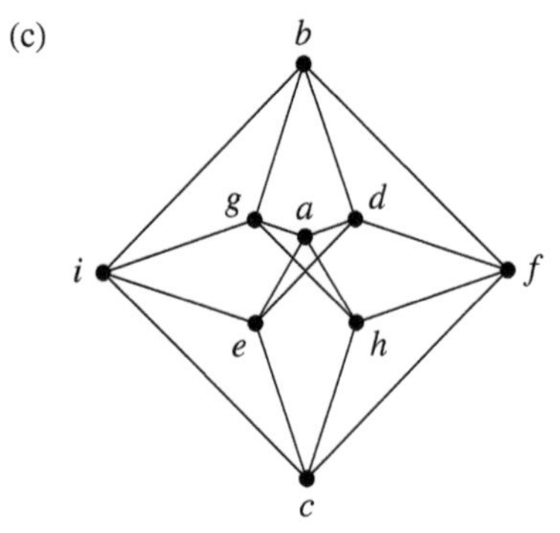

7. Produce the adjacency matrix for the following graphs. Use the natural ordering for the vertices.

(a) ★

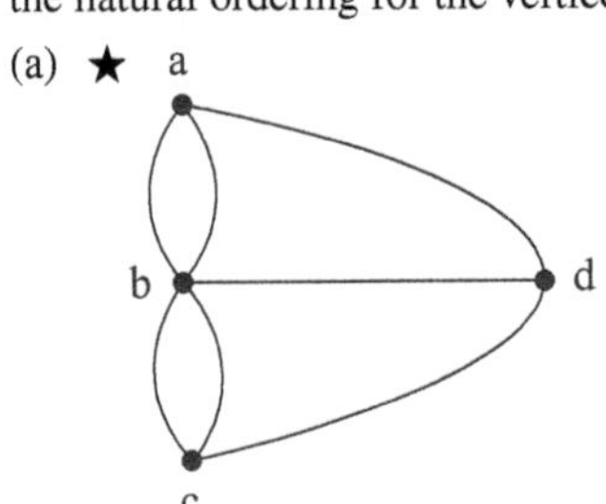

(b)

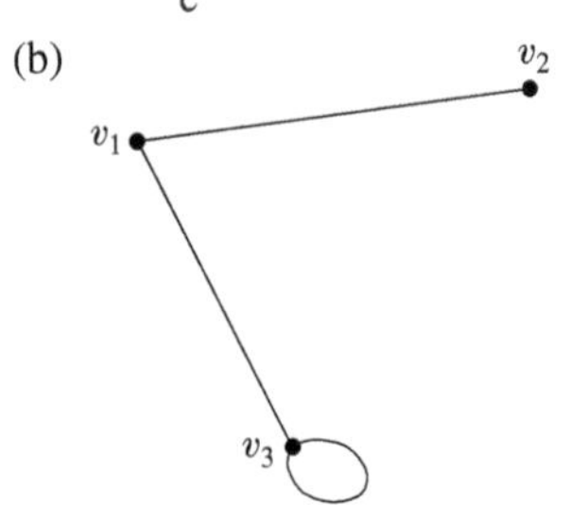

(c)

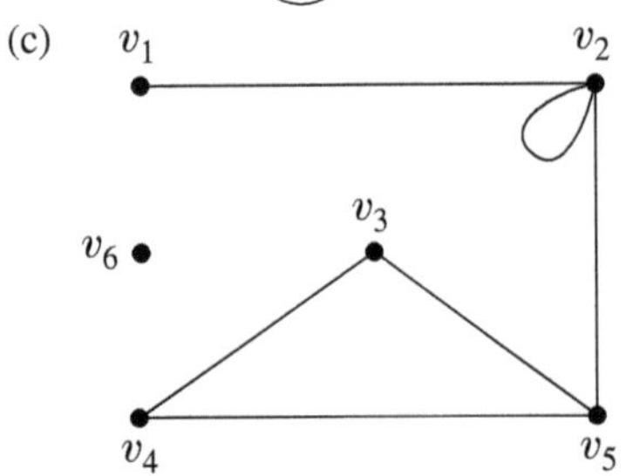

(d)

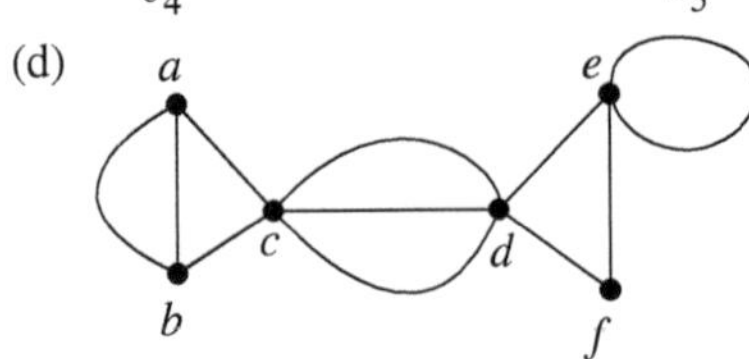

(e)

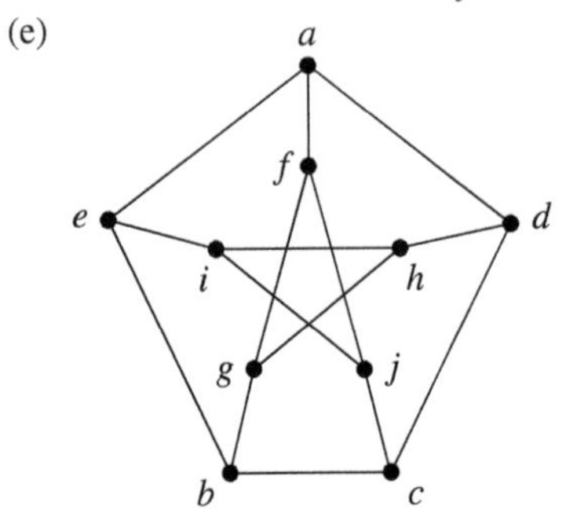

8. Let A be the adjacency matrix for the following graph.

(a) Calculate A and A^2.

(b) Let $k = 2$, $i = 1$, and $j = 6$. Trace through the inductive step of Theorem 10.22. That is, verify that the claims about c_{16} are valid for this example.

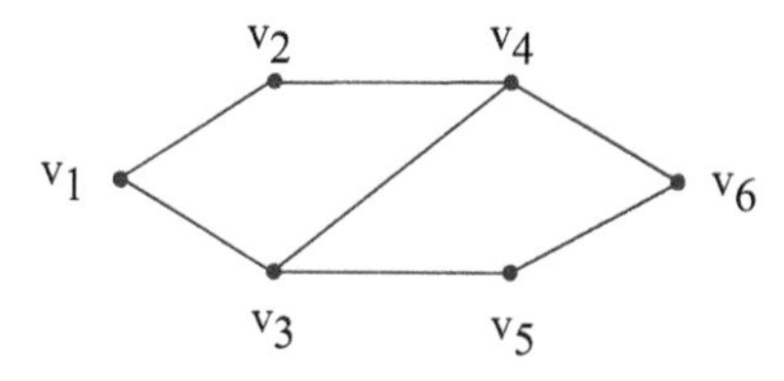

9. ★ Let A be the adjacency matrix for Exercise 7(a).

(a) Calculate A^3.

(b) How many walks of length 3 are there between vertex a and vertex d? List them (with alternating vertices and edges).

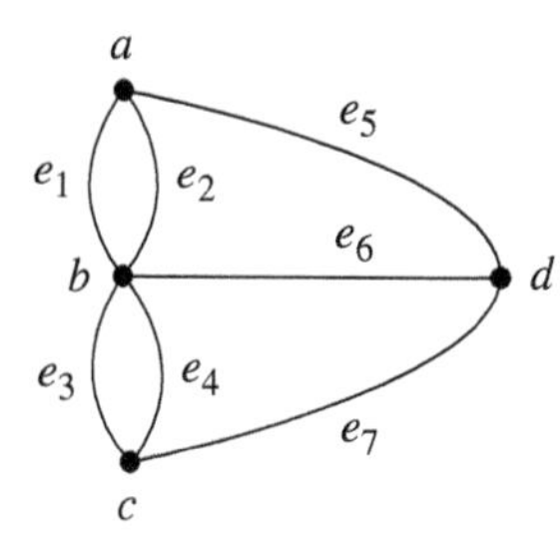

10. Let A be the adjacency matrix for Exercise 7(b).

(a) Calculate A^3.

(b) How many walks of length 3 are there between vertex v_3 and vertex v_3? List them.

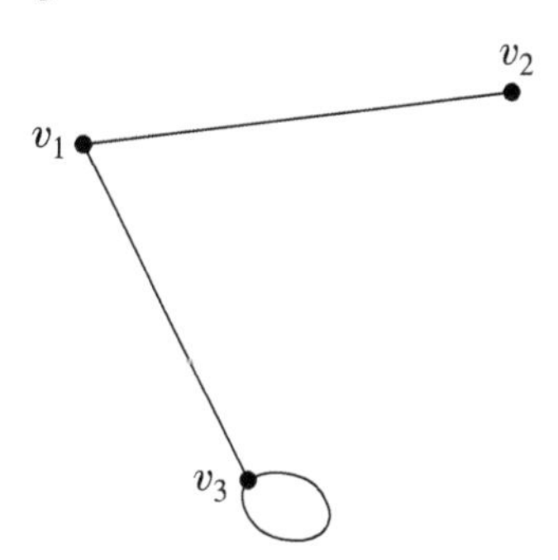

11. Let A be the adjacency matrix for Exercise 7(c).

(a) Calculate A^4.

(b) How many walks of length 4 are there between vertex v_1 and vertex v_3? You should use a calculator or computer software that can multiply matrices. List the walks.

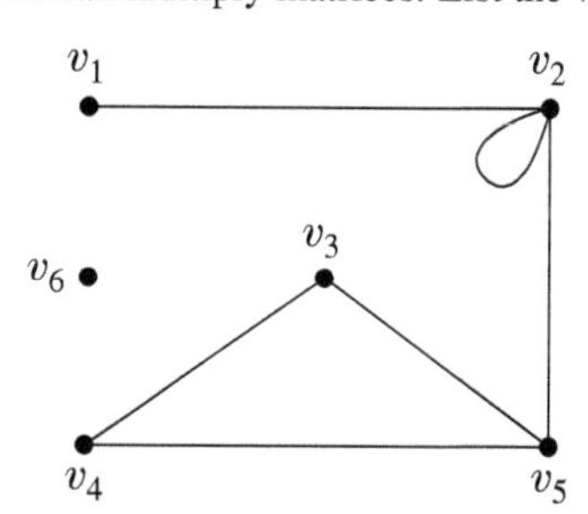

12. Let A be the adjacency matrix for Exercise 7(d).

(a) Calculate A^4.

(b) How many walks of length 4 are there between vertex a and vertex e? You should use a calculator or computer software that can multiply matrices. List the walks as a sequence of alternating vertices and edges.

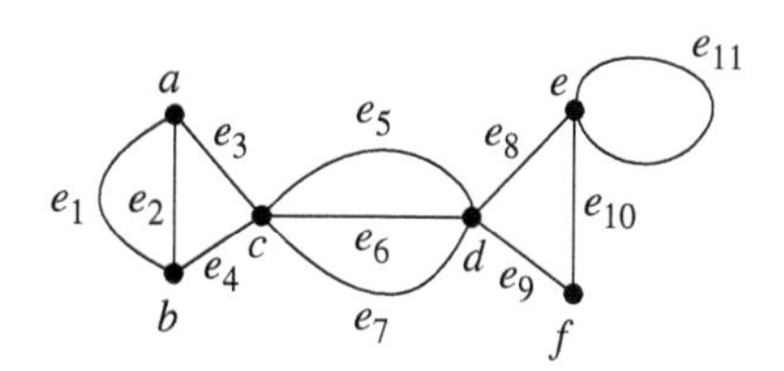

13. Let $G = (V, E, \phi)$ be a graph with $|V| = n$. Let $a, b \in V$ be vertices that are in the same component of G. Prove there is a walk of length at most $n - 1$ that joins a and b.

14. Use Corollary 10.23 on page 615 to show that the graph in Exercise 7(b) is connected.

15. Each of the following statements is either true (always) or false (at least sometimes). Determine which option applies for each statement and provide adequate explanation for your choice.

(a) ★ Every trail in a graph is also a path.

(b) Walks can exist in a graph with a single vertex, but neither paths nor trails can exist.

(c) The adjacency matrix of a graph may include exactly one nonzero entry.

(d) A circuit can have repeated vertices.

(e) A loop at vertex v_i will add 2 to the iith entry of the adjacency matrix.

16. Let v be a vertex that is part of a cycle in a graph, G. Can v be a cut vertex? If it can, provide an example; if not, provide a proof. (The cycle need not contain every vertex in G.)

17. Let G be a simple graph with cut vertex, v. Prove that $\overline{G} - \{v\}$ is connected.

18. Let G be a connected graph. Let $p_1 = v_1v_2\cdots v_n$ and $p_2 = w_1w_2\cdots w_n$ be two paths in G of maximal length. Prove that $v_i = w_j$ for some $i, j \in \{1, 2, \ldots, n\}$.

19. Prove Proposition 10.24. [*Hint*: Prove the contrapositive.]

> **PROPOSITION 10.24** ***Path Implies Trail***
> Every path in a graph is also a trail.

20. Let G be a simple graph which contains a circuit, C.

(a) Prove that C contains a cycle.

(b) Prove that if C is not a cycle, then C must contain at least two cycles.

21. Let G be a simple disconnected graph with n vertices having a maximal number of edges.

(a) Prove that G has exactly two components and each component is a complete graph.

(b) Prove that G has $\frac{(n-1)(n-2)}{2}$ edges. [*Hint*: Calculus will be handy. Maximums do not always occur at a critical point.]

10.3 Euler and Hamilton

10.3.1 Euler Circuits and Euler Trails

It is time to finally present Euler's solution to the Bridges of Königsberg problem. Euler used a strategy mentioned in Section 3.6.2: Prove a more general theorem.

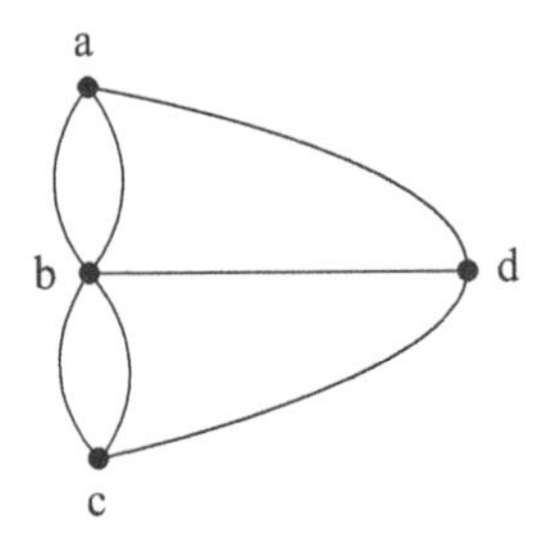

Figure 10.33. The Königsberg graph.

> **DEFINITION 10.25** *Euler Trail; Euler Circuit*
> Let $G = (V, E, \phi)$ be a graph without loops. An *Euler trail in G* is a walk that contains every edge in E exactly once. An *Euler circuit in G* is a circuit in which every edge appears exactly one time.

The Königsberg bridge puzzle seeks an Euler trail or an Euler circuit in the Königsberg graph (the multigraph in Figure 10.33).

Euler's key insight was the following:

> Any vertex that is not the beginning or the end of the Euler trail must have an even number of incident edges if such a trail exists.

This is because vertices in the interior portion of the trail are visited on an in-out basis. One edge is followed into the vertex and a different edge is immediately followed out. Thus, the edges are consumed in pairs. If there is an odd number of edges, the final edge will be traveled to arrive at the vertex, but there will be no new edge available to leave the vertex.

On the other hand, if all but two vertices have an even number of incident edges, it may be possible to start at a vertex with odd degree, pass multiple times through the vertices with even degrees, and finally arrive at the other vertex with odd degree.

It should now be clear that if the graph has one vertex with odd degree or else three or more vertices with odd degree, there can be no Euler trail. So the Königsberg graph has no Euler trail (and the original puzzle is impossible). Even more can be said.

> **THEOREM 10.26** *Euler Circuits*
> Let G be a connected multigraph. Then G has an Euler circuit if and only if there are no vertices in G with odd degree.

Proof: The need for the assumption that G is connected should be obvious. Most of the work needed to prove the "only if" part has already been done.

Only If (first proved by Leonhard Euler in 1736)
Suppose G has a vertex v with odd degree. If v is the initial vertex in the circuit, one edge will be crossed as the first leg of the circuit is traveled. There may be other visits to v with one new edge being crossed on the way in and one new edge being crossed on the way out. Eventually, the final edge is crossed. That final edge will always be crossed while traveling away from v. There will be no edge left to make the final trip back to v to complete the circuit. Hence, no circuit exists.

If v is not the initial vertex in the circuit, edges will be used in pairs until the final edge. That edge will be crossed and then there will be no edge left to leave v. However, v is not the initial vertex so it is not the end of the circuit. Again, no circuit can exist.

If (stated by Euler, first printed proof by Carl Hierholzer in 1873)
A constructive proof will be given.[9] Pick any vertex, v_1, as the initial vertex. Start traveling along edges until arriving at a vertex that has no new edges to travel out over. That vertex must be v_1 because the trail would otherwise have stopped at a vertex having

[9]Example 10.16 on page 619 illustrates this process.

an odd number of edges eliminated (so the vertex would have at least one more edge to travel out on). If every edge has been crossed, the construction is complete.

Otherwise there must be some other vertex, v_2, in the circuit[10] that still has incident edges that have not been crossed. In fact, v_2 must have an even number of unused incident edges. The current circuit looks like $v_1 \cdots v_2 \cdots v_1$. Since it is a circuit, we can rotate it so that v_2 is the initial vertex. The v_1's in the previous listing will merge and v_2 will now be the initial and final symbols in the listing of the circuit: $v_2 \cdots v_1 \cdots v_2$. The final v_2 in this list has more edges, so we can extend the circuit. Because every vertex still has an even number of unused edges, this extension must terminate at v_2. The extension has used additional edges. If all edges have been used, we are done.

Otherwise there is a vertex, v_3, in the circuit which still has unused incident edges. Make v_3 the new initial vertex and extend the circuit again. Every time this process is repeated, more edges are consumed. Since the graph has only a finite number of edges, the process must eventually use them all. At each step, a circuit exists. The final circuit uses all edges and so it is an Euler circuit. □

EXAMPLE 10.16

Constructing an Euler Circuit

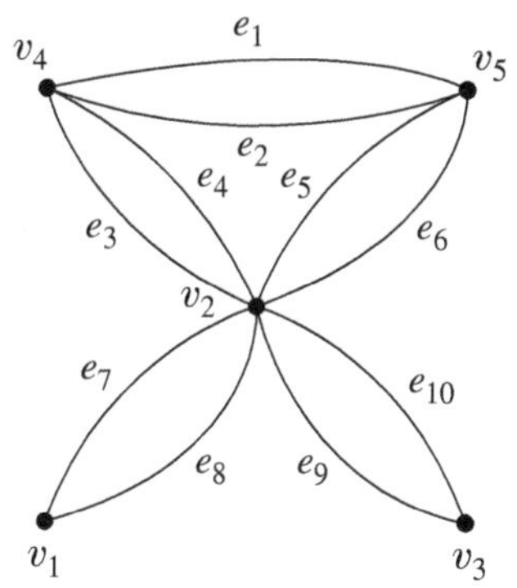

The graph shown in Figure 10.34 has no vertices with odd degree. Therefore, it must have an Euler circuit. The construction from Theorem 10.26 will be used to find one.

Suppose we start at v_1 and form the circuit $v_1e_7v_2e_6v_5e_5v_2e_8v_1$. Figure 10.35 shows the remaining edges and vertices.

Figure 10.34. This must have an Euler circuit.

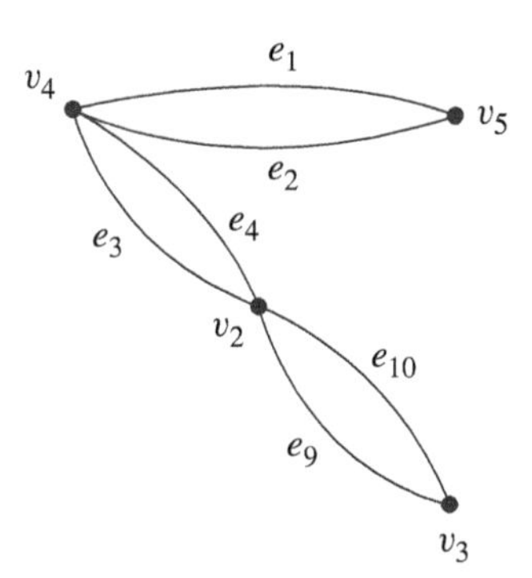

Figure 10.35 The edges and vertices that are not in $v_1e_7v_2e_6v_5e_5v_2e_8v_1$.

Either v_2 or v_5 could serve as the new initial vertex of the circuit $v_1e_7v_2e_6v_5e_5v_2e_8v_1$. Suppose we start it at v_2: $v_2e_6v_5e_5v_2e_8v_1e_7v_2$. The new extension might be $v_2e_3v_4e_4v_2e_{10}v_3e_9v_2$, making the larger circuit $v_2e_6v_5e_5v_2e_8v_1e_7v_2e_3v_4e_4v_2e_{10}v_3e_9v_2$.

The edges and vertices that are left are shown in Figure 10.36.

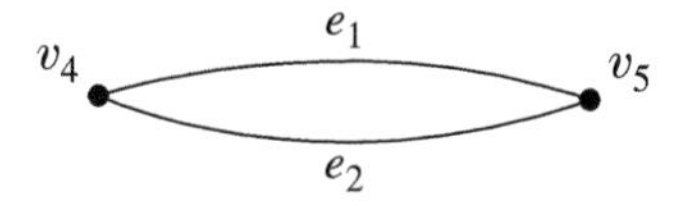

Figure 10.36 The edges and vertices that are not in $v_2e_6v_5e_5v_2e_8v_1e_7v_2e_3v_4e_4v_2e_{10}v_3e_9v_2$.

The only remaining vertex that is connected to the current circuit is v_4, so we rotate the circuit and write it as $v_4e_4v_2e_{10}v_3e_9v_2e_6v_5e_5v_2e_8v_1e_7v_2e_3v_4$. The circuit can now be extended by adding $v_4e_1v_5e_2v_4$. This uses all the edges and the construction is complete. The Euler circuit is $v_4e_4v_2e_{10}v_3e_9v_2e_6v_5e_5v_2e_8v_1e_7v_2e_3v_4e_1v_5e_2v_4$. ■

[10] If the only vertex with remaining edges in not in the circuit, the graph would not be connected.

THEOREM 10.27 ***Euler Trails***

Let G be a connected multigraph. Then G has an Euler trail (but not an Euler circuit) if and only if there are exactly two vertices with odd degree.

Proof:

Only If

If there are no vertices with odd degree, then an Euler circuit exists. If only one vertex has odd degree,[11] it must either be the initial or the final vertex in the trail, but it can't be both, so no Euler trail can exist. Finally, if three or more vertices have odd degree, each needs to be either the initial or the final vertex, which is impossible.

If

Let the two vertices with odd degree be v_1 and v_2. Temporarily add an edge, e, joining v_1 and v_2 (there may already be other edges joining them). Every vertex in the new graph now has even degree, so Theorem 10.26 guarantees the existence of an Euler circuit. We can list the circuit in the form $v_1 \cdots v_2 e v_1$. Now remove the temporary edge, e and the final v_1. The walk $v_1 \cdots v_2$ is an Euler trail. □

✔ Quick Check 10.5

1. Find an Euler circuit if possible; otherwise find an Euler trail. If neither is possible, explain why not.

   (a) The Herschel graph.

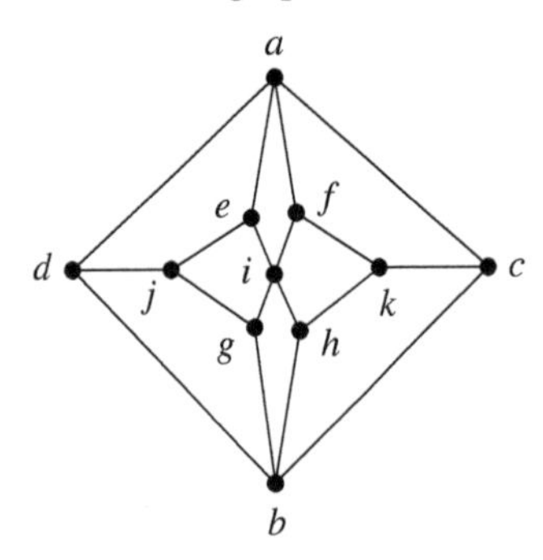

   (b)

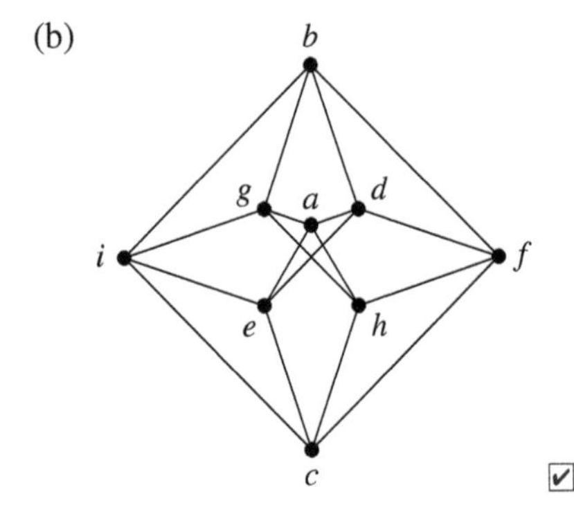

☑

10.3.2 Hamilton Cycles and Hamilton Paths

After Euler's elegant solution to the problem of crossing every edge exactly once, it seems natural that people would investigate walks and circuits that visit each *vertex* exactly once. That question was raised by the mathematician William Rowan Hamilton in 1856.

DEFINITION 10.28 ***Hamilton Path; Hamilton Cycle***

Let $G = (V, E, \phi)$ be a multigraph. A *Hamilton path in* G is a walk that contains every vertex in V exactly once. A *Hamilton cycle in* G is a cycle that contains every vertex.

Hamilton sent a letter to a friend in which he outlined a game played on a model of the dodecahedron. Figure 10.37 shows a dodecahedron next to a dodecahedron with one of its surface pentagons removed, and the dodecahedron stretched about the hole and squashed onto a plane.

[11] Actually, the handshake theorem implies that no graph can have an odd number of vertices with odd degree.

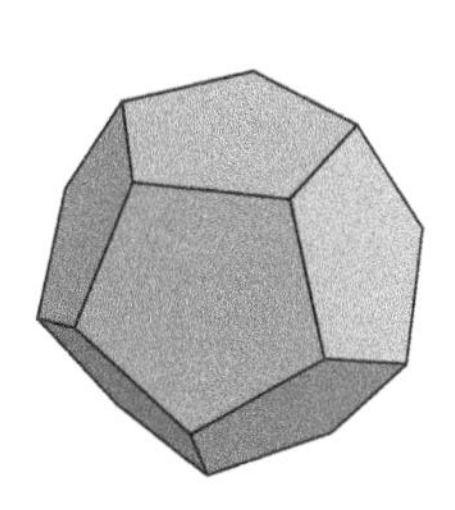
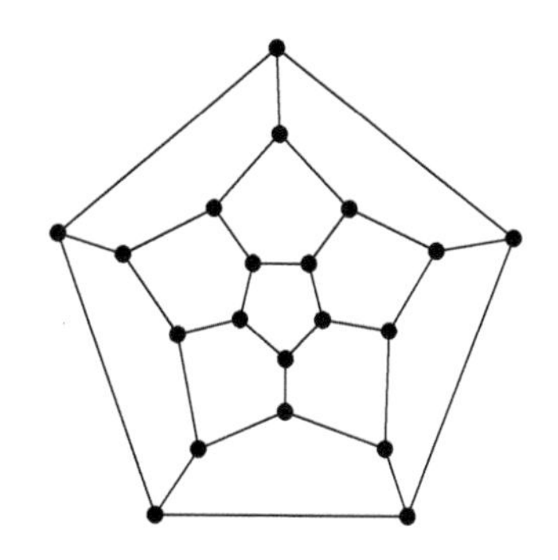

Figure 10.37 The dodecahedron and its two-dimensional embedding.

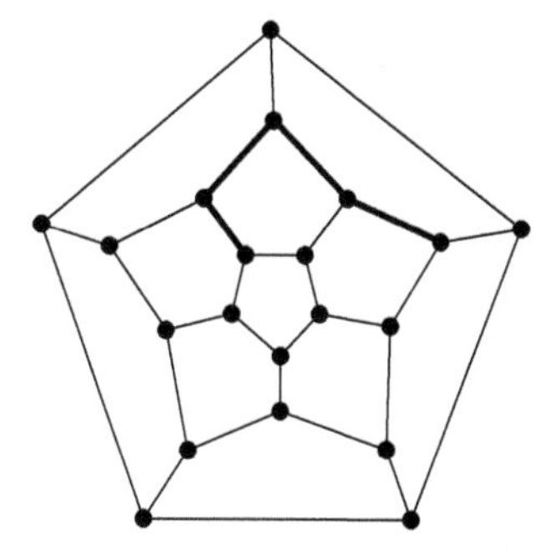

Figure 10.38. The first five vertices have been chosen.

The puzzle was a challenge: The first person would use a string to connect five successively adjacent vertices. The second person needed to extend the path to a Hamilton path on the dodecahedron.

Figure 10.38 depicts the puzzle after the first person has chosen the five vertices.

Hamilton named his puzzle the "Icosian game" and sold the idea to a dealer in games and puzzles. An instruction leaflet (written mainly by Hamilton) containing several games was included. The game was marketed in 1859 but was not a commercial success.[12]

✔ Quick Check 10.6

1. Extend the path in the previous graph into a Hamilton cycle. ☑

Unfortunately, there is no known, all-inclusive, "if and only if" theorems that characterize Hamilton cycles and paths.[13] This does not mean that there are no meaningful theorems about Hamilton cycles. A few results will be presented here. Some omitted theorems require concepts that are more appropriately introduced in a full-semester course in graph theory.

Since vertices are visited only once in a Hamilton path, removing multiple edges between the same pair of vertices will not change whether or not a graph has a Hamilton path.

PROPOSITION 10.29 ***Hamilton Paths in Multigraphs***

Let G be a multigraph and let G' be its underlying simple graph. Then G contains a Hamilton path if and only if G' contains one.

PROPOSITION 10.30 ***Hamilton Cycles in Complete Graphs***

A complete graph with at least three vertices has a Hamilton cycle.

Proof: Start at any vertex. There will be an edge to every other vertex. Pick one. There will be an edge leading from this second vertex to any of the remaining vertices. We can continue finding available edges until the last vertex has been reached. The edge connecting the initial and final vertices has not been used yet. Crossing that edge completes the cycle. □

[12] See [10] for more details and a look at the instructions.

[13] Many people would disagree with the word *unfortunately*. Since no grand theorem exists, there is lots of opportunity for mathematicians to find and prove new theorems about Hamilton cycles. This is considered to be a good thing.

THEOREM 10.31 ***A Sufficient Condition for a Hamilton Cycle***

Let G be a simple graph with $n \geq 3$ vertices. Let $\delta = \min_{v \in V} \deg(v)$. If $\delta \geq \frac{n}{2}$, then G has a Hamilton cycle.

EXAMPLE 10.17 **Illustrating Theorem 10.31**

The graph in Figure 10.39 has $n = 6$ and $\delta = 3 \geq \frac{6}{2}$. It must therefore have a Hamilton cycle. It is easy to find one.

Notice that a graph, such as the one in Figure 10.40, *does not* satisfy the hypotheses of Theorem 10.31, yet it certainly contains a Hamilton cycle. The theorem is not an "if and only if" assertion, so this does not contradict the theorem.

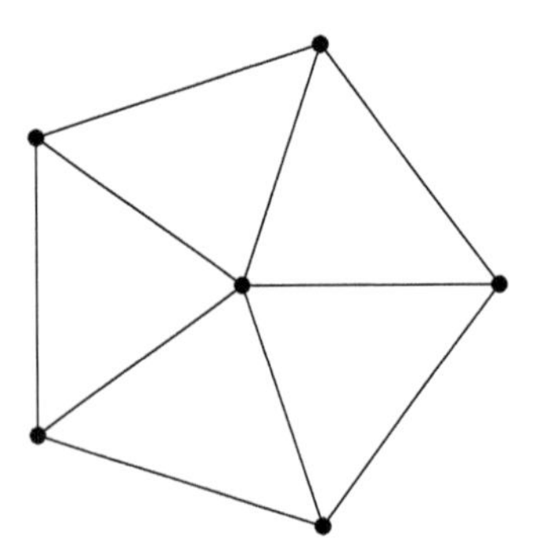

Figure 10.39 This has a Hamilton cycle.

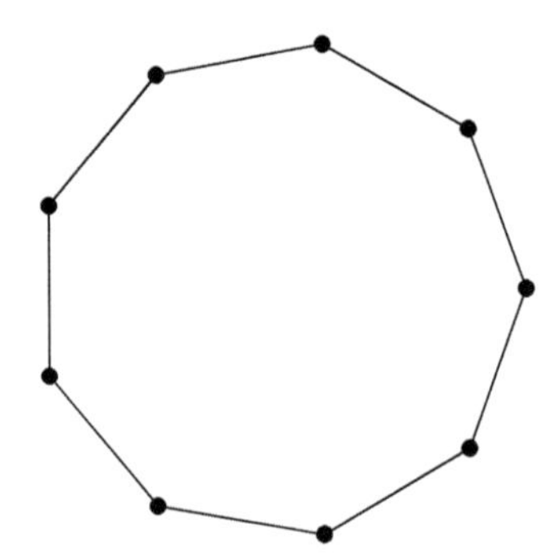

Figure 10.40 This has a Hamilton cycle even though $\delta < \frac{n}{2}$.

■

Proof of Theorem 10.31: The intuitive motivation for a theorem of this type is the following observation. When the minimum degree is large enough in relation to the number of vertices, there should be enough edges available to travel from any vertex to a sufficient number of other vertices to enable a path to keep progressing.

The existence of a Hamilton cycle implicitly assumes that G is connected. The theorem is still true when G is not connected. To see this, suppose that G is not connected, but the hypotheses of the theorem are true for G. Since G is disconnected, there are vertices, v_1 and v_2, that are in different components of G. Let v_1 be in the component, C_1, and let v_2 be in the component, C_2. Let C_1 contain n_1 vertices and C_2 contain n_2 vertices. Since $\deg(v_i) \geq \frac{n}{2} \geq \frac{3}{2} > 1$ for $i = 1, 2$, $n_1 > 1$ and $n_2 > 1$. Thus, each component contains at least two vertices. In fact, C_i contains v_i and at least $\deg(v_i)$ other vertices. Consequently, $n_i \geq 1 + \deg(v_i) \geq 1 + \frac{n}{2}$. This implies that

$$n \geq n_1 + n_2 \geq \left(1 + \frac{n}{2}\right) + \left(1 + \frac{n}{2}\right) = 2 + n > n$$

a clear contradiction. Therefore, if the hypotheses of the theorem are valid for G, then G must be connected.

Assume for the rest of the proof that G is connected. A proof by contradiction will be given. Thus, assume that the theorem is false for some graphs that satisfy the hypotheses. Suppose the theorem fails for at least one graph having $n \geq 3$ vertices. There may be more than one graph with n vertices and $\delta \geq \frac{n}{2}$ for which the theorem fails. Among this group, choose one with a maximal number of edges.[14]

We are therefore assuming that we have a simple, connected graph, $G = (V, E, \phi)$, having n vertices and minimum degree $\delta \geq \frac{n}{2}$ that does not contain a Hamilton cycle.

[14]The word *maximal* is used to indicate that the largest number of edges may occur on more than one graph. It won't matter which one of these we choose.

We are also assuming that any simple, connected graph with n vertices and more edges than G *does* contain a Hamilton cycle.

Proposition 10.30 indicates G is not a complete graph, so there must be at least two vertices that are not joined by an edge in E. By suitably renaming vertices if necessary, we may assume that the first vertex, w_1, and the last vertex, w_n, are both in V, but the edge w_1w_n is not in E.[15] Let $E' = E \cup \{w_1w_n\}$ and $G' = (V, E', \phi)$. G' is the graph G with the edge w_1w_n added to G.

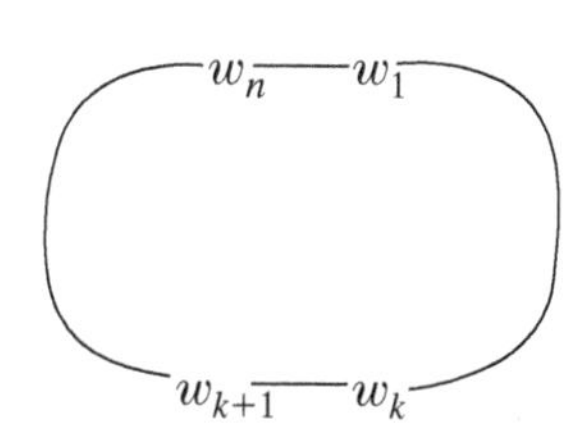

Figure 10.41. The Hamilton cycle in G'.

By the maximality of G (with respect to the number of edges), G' must contain a Hamilton cycle. That cycle must contain the edge w_1w_n (or else the cycle would be entirely in G). We may write this cycle as $w_1w_2 \cdots w_{n-1}w_nw_1$ (perhaps after renaming the vertices in the middle of the cycle). Figure 10.41 shows the Hamilton cycle in G' in outline form.

To reach the desired contradiction, we will show that $\deg(w_1) + \deg(w_n) < n$. If that is true, $2\,\delta = \delta + \delta \leq \deg(w_1) + \deg(w_n) < n$, and so $\delta < \frac{n}{2}$, contradicting the original hypothesis.

Some background work is needed to show $\deg(w_1) + \deg(w_n) < n$. To that end, define two subsets, B and F, of V.

$$B = \{w_i \in V \mid w_iw_n \in E\} \qquad F = \{w_i \in V \mid w_1w_{i+1} \in E\}$$

B is the set of vertices that are adjacent to w_n. F is the set of vertices in the Hamilton cycle whose next neighbor in the cycle is adjacent to w_1. Since every vertex is in the cycle, the number of elements in F is the number of vertices adjacent (in G) to w_1.

Notice that $w_n \notin B$ since a simple graph has no loops. Also, $w_n \notin F$ because there is no loop at w_1. This means that

$$|B \cup F| < |V| = n$$

Suppose that $B \bigcap F \neq \emptyset$. Then there is a vertex, w_k such that $w_k \in B$ and $w_k \in F$. Since $w_k \in B$, the edge w_kw_n is in E. Since $w_k \in F$, the edge $w_{k+1}w_1$ is in E. That means a new Hamilton cycle can be created: insert the edges w_kw_n and $w_{k+1}w_1$ and remove the edges w_1w_n and w_kw_{k+1}. Figure 10.42 shows this.

Figure 10.42 $B \cap F \neq \emptyset$ results in a contradiction.

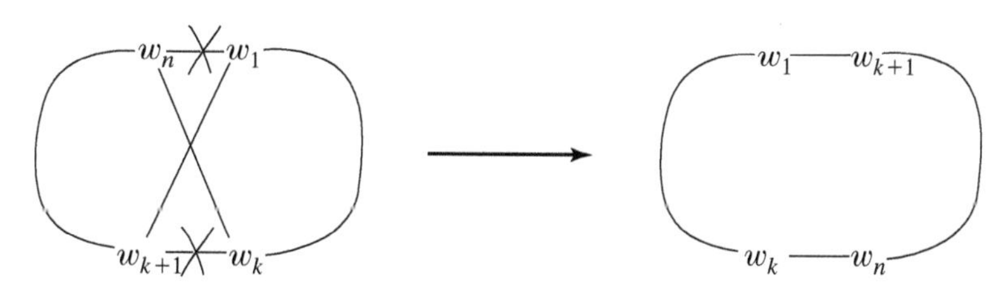

Because the new cycle no longer contains the edge w_1w_n, the cycle exists entirely in G. But that is impossible since G has no Hamilton cycle. Therefore, $B \bigcap F = \emptyset$. Consequently,

$$|B \cap F| = 0$$

Finally, since $|B \cup F| < n$ and $|B \cap F| = 0$,

$$2\delta \leq \deg(w_1) + \deg(w_n) = |F| + |B| = |B \cup F| + |B \cap F| < n$$

contradicting[16] the original hypothesis $\delta \geq \frac{n}{2}$.

The assumption that there exists a simple, connected graph G with $n \geq 3$ vertices and $\delta \geq \frac{n}{2}$ that does not have a Hamilton cycle must be an incorrect assumption. The theorem must be true. □

[15] Notice the informal notation. It is more precise to say that there is not an $e_j \in E$ with $\phi(e_j) = w_1w_n$.

[16] Simple inclusion–exclusion has been used: $|X \cup Y| = |X| + |Y| - |X \cap Y|$.

10.3.3 Exercises

The exercises marked with ★ have detailed solutions in Appendix H.

1. By adding at most four edges, modify the graph below so that it will contain an Euler circuit. The new graph should remain a simple graph.

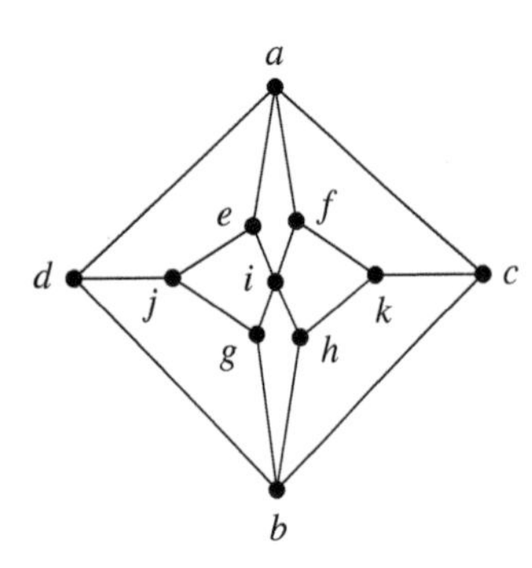

2. Use the construction from Theorem 10.26 to produce an Euler circuit for the following graph.

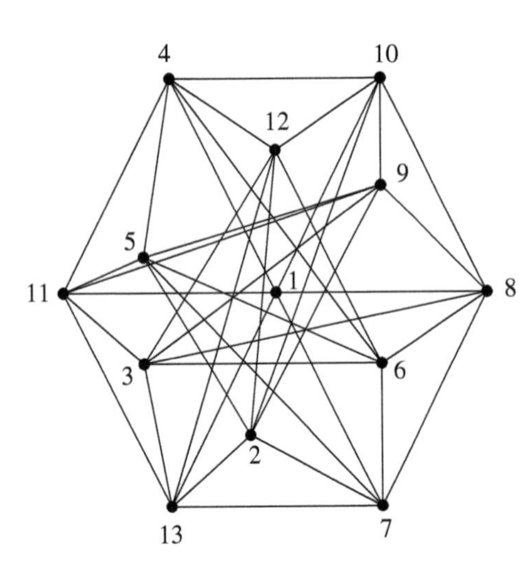

3. ★ Proposition 10.30 on page 621 states that a complete graph on $n \geq 3$ vertices has a Hamilton cycle. Why is this proposition false for $n = 2$?

4. Let G be a simple graph with at least two vertices. Suppose G contains a vertex w with $\deg(w) < 2$. Prove that G cannot contain a Hamilton cycle.

5. Find a Hamilton cycle in the icosahedron graph.

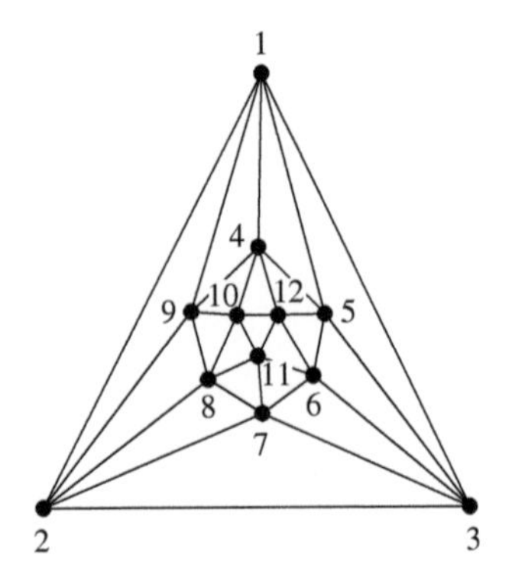

6. ★ Consider H_3.

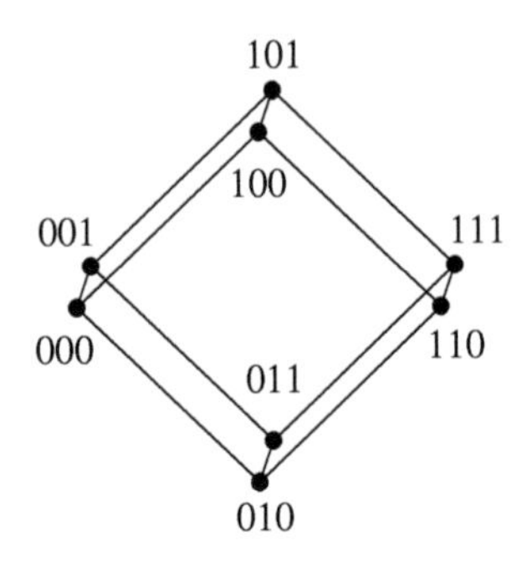

   (a) If an Euler circuit exists, describe one; otherwise, explain why there is no Euler circuit.

   (b) If a Hamilton cycle exists, describe one; otherwise, find a Hamilton path.

7. Consider the Petersen graph.

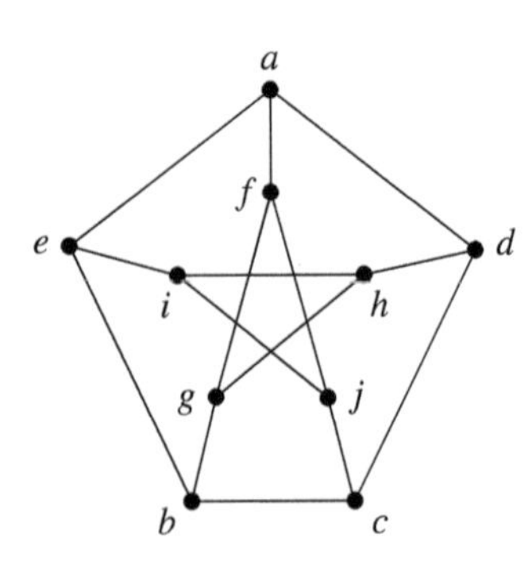

   (a) If an Euler circuit exists, describe one; otherwise, explain why there is no Euler circuit.

   (b) If a Hamilton cycle exists, describe one; otherwise, find a Hamilton path.

8. Consider the Herschel graph.

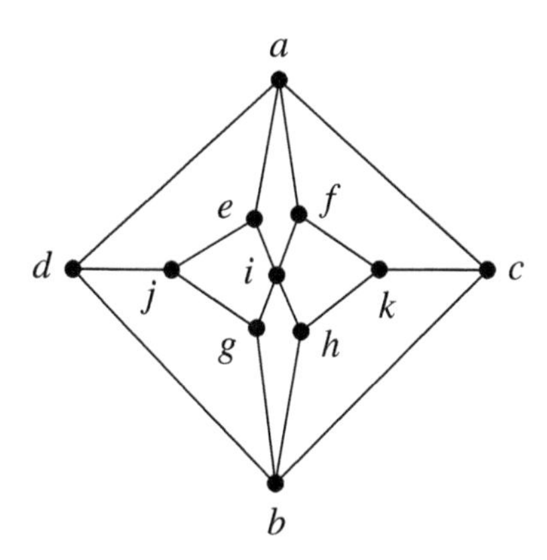

(a) Prove that a Hamilton cycle does not exist in this graph. [*Hint*: (from [44]) Consider the set of vertices $\{a, b, i, j, k\}$. A Hamilton cycle will contain exactly two of the edges incident with each of these vertices. The other edges cannot be in the Hamilton cycle. Now consider the numbers of edges and vertices in a Hamilton cycle.]

(b) If a Hamilton path exists, describe one; otherwise, prove that a Hamilton path does not exist.

9. Prove Theorem 10.32. [*Hint*: Look at Exercise 8. You may also need to look at Definition 10.11 on page 609.]

THEOREM 10.32 ***Hamilton Cycles and Independent Sets***

Let I be an independent set in a graph, G. Let G have ϵ edges and n vertices. Then G does not contain a Hamilton cycle if

$$\epsilon - \sum_{v \in I} \deg(v) + 2|I| < n$$

10. ★ Prove that no Hamilton path (and hence no Hamilton cycle) can exist in the following graph:

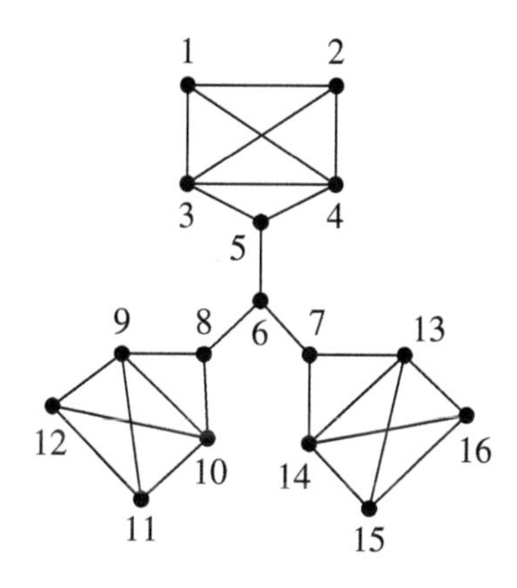

11. Make a photocopy of the Tutte graph[17] (Figure 10.43). There are no Hamilton cycles in this graph. Find a Hamilton path.

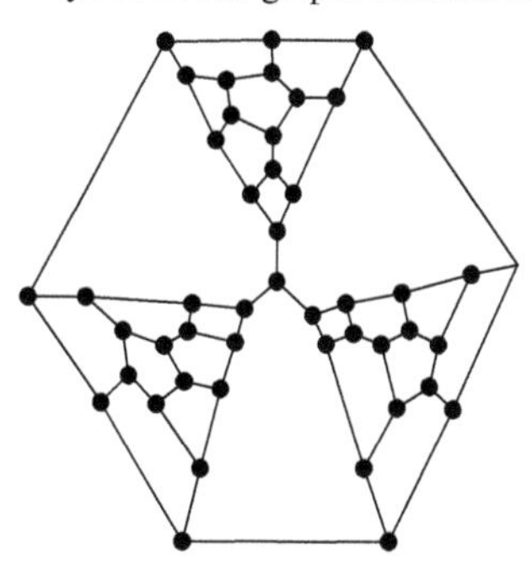

Figure 10.43 The Tutte graph.

12. Each of the following statements is either true (always) or false (at least sometimes). Determine which option applies for each statement and provide adequate explanation for your choice.

(a) Any connected multigraph with exactly two vertices of odd degree will not have an Euler circuit.

(b) Suppose that the minimal degree of the vertices in a connected simple graph with 10 vertices is 2. Then this graph cannot contain a Hamilton cycle.

(c) A multigraph that contains a Hamilton cycle also contains a Hamilton path.

(d) ★ A graph, G, has an Euler circuit if and only if there are no vertices with odd degree in G.

13. (a) For which values of m and n does $K_{m,n}$ have an Euler circuit? Provide adequate justification for your answer.

(b) For which values of m and n does $K_{m,n}$ have an Euler trail but not an Euler circuit? Provide adequate justification for your answer.

14. Is it possible to trace the following representation of an open envelope without lifting your pencil from the paper and without tracing a line more than once? Is this familiar puzzle related to Euler circuits or to Hamilton cycles?

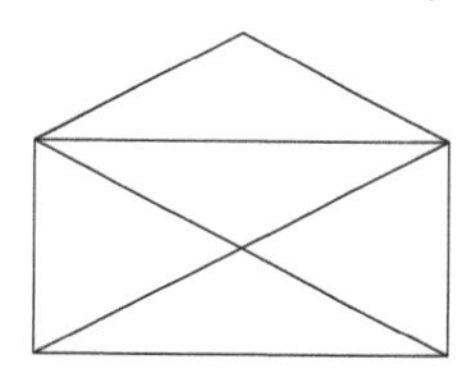

15. Let G be a connected bipartite graph, with vertices partitioned by the vertex sets, V_α and V_β.

(a) Prove that G cannot have a Hamilton cycle if $|V_\alpha| \neq |V_\beta|$.

(b) Prove that if G has a Hamilton path, then $|V_\alpha| = |V_\beta| \pm 1$.

(c) Give an example of a connected bipartite simple graph with $|V_\alpha| = |V_\beta|$ but in which there is no Hamilton cycle.

[17] William Tutte's last name is pronounced the same as that of the abbreviated version of the ancient Egyptian, King Tut. A pdf file with a larger copy of the graph can be found at http://www.mathcs.bethel.edu/~gossett/DiscreteMathWithProof/ (or turn to page 672).

16. Suppose that in a group of $2n$ people, each person has at least n friends, where $n > 1$. The host wishes to seat everyone in a circle such that everyone in the circle is seated next to at least one friend. Is this possible? If it is, provide a proof; if it is not possible, provide a counterexample.

17. ★ Can a simple graph with an Euler circuit have an even number of vertices and an odd number of edges? If it is possible, provide an example; otherwise, provide a proof that it is not possible.

18. Suppose the graph, G, has an Euler circuit. If e_1 and e_2 are two edges in G that share a common endpoint, v, is there an Euler circuit that contains e_1 and e_2 as consecutive edges? (The desired Euler circuit will look like either $\cdots e_1 v e_2 \cdots$ or $\cdots e_2 v e_1 \cdots$.) If it is possible, provide a proof; otherwise, provide a counterexample.

19. Prove Theorem 10.33 (from [74]). [*Hints*: Prove the contrapositive. Start adding edges (but not vertices) to G until a graph, H, is reached such that no Hamilton cycle exists in H, but adding any new edge to H will produce a graph that contains a Hamilton cycle.]

THEOREM 10.33 ***Another Sufficient Condition for a Hamilton Cycle***

Let G be a graph with $n \geq 3$ vertices such that for all nonadjacent vertices, u and v, with $u \neq v$,

$$\deg(u) + \deg(v) \geq n$$

Then G contains a Hamilton cycle.

20. Prove that Theorem 10.33 is not a necessary condition.

10.4 Representation and Isomorphism

There are many ways to represent a graph. The different choices have been developed to provide natural and convenient representations for differing purposes. For example, the "dot and line" diagrams that have already been used extensively provide a very intuitive visual representation of a graph. The adjacency matrix introduced in Definition 10.21 on page 614 is an excellent representation for calculating the number of walks between two vertices (Theorem 10.22 on page 614).

Several other common representations will be presented next. Then the question "when are two graphs really the same?" will be addressed.

10.4.1 Representation

A graph is formally defined as an ordered triple (V, E, ϕ), where V is a set of vertices (an undefined term), E is a set of edges (defined only by their incidence relations with vertices), and the function ϕ that maps edges to unordered pairs of vertices.

A representation of a graph is a mechanism to make this definition more concrete. The dot and line diagrams are visually appealing but can lead to visual dis-information if we are not careful.[18]

Visual representations are often referred to as *embeddings* since they embed the graph in the plane. To store such an embedding on a computer, we need to specify the graph and the locations of the vertices. The *Mathematica* package Combinatorica (prior to Mathematica version 4.2) represented graphs in the form `Graph[A,coords]`, where A is an adjacency matrix for the graph and `coords` is a list of ordered xy-pairs specifying the xy-coordinates of the vertices.[19]

EXAMPLE 10.18 **An Embedding**

The graph embedding shown in Figure 10.44 was represented by Combinatorica as

```
Graph[{{0, 1, 1, 0, 0, 0}, {1, 0, 1, 0, 0, 0},
{1, 1, 0, 1, 0, 0}, {0, 0, 1, 0, 1, 1}
{0, 0, 0, 1, 0, 1}, {0, 0, 0, 1, 1, 0}},
{{-5, 4}, {-5, -4}, {-2, 0}, {4, 0}, {7, 4}, {7, -4}}]
```
■

Figure 10.44. An embedding in the plane.

[18] If the vertices are placed in unfortunate places, some edges may be hidden. It is also possible to have apparent edges that do not really exist. See the solution to Quick Check 10.2 (page 665) for an example.

[19] The current version essentially replaces the adjacency matrix by a list of all the edges (as vertex pairs).

Adjacency matrices are one of the most useful nonvisual representations. A few other common representations will now be presented.

DEFINITION 10.34 ***Incidence Matrix***

Let $G = (V, E, \phi)$ be a graph with no loops. Let $V = \{v_1, v_2, \ldots, v_n\}$ and $E = \{e_1, \ldots, e_\epsilon\}$. The *incidence matrix* $B = (b_{ij})$, is the n by ϵ matrix defined by

$$b_{ij} = \begin{cases} 1 & \text{if } v_i \text{ is incident with } e_j \\ 0 & \text{otherwise} \end{cases}$$

The rows of the incidence matrix are indexed by the vertices and the columns of the incidence matrix are indexed by the edges.

EXAMPLE 10.19

An Incidence Matrix

The Königsberg graph with labeled edges (Figure 10.45) has the incidence matrix shown in Table 10.4.

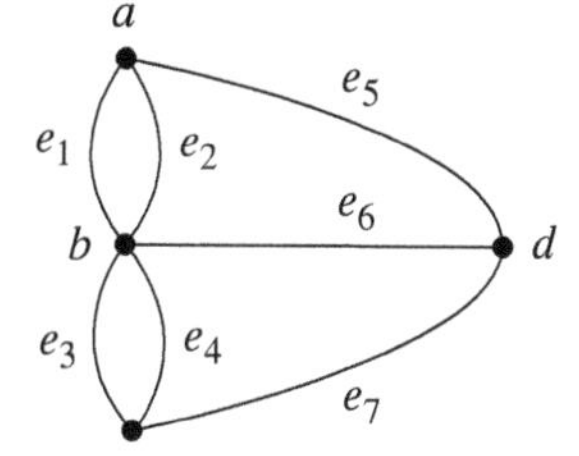

Figure 10.45. The Königsberg graph.

TABLE 10.4 The Incidence Matrix for the Graph in Figure 10.45

	e_1	e_2	e_3	e_4	e_5	e_6	e_7
a	1	1	0	0	1	0	0
b	1	1	1	1	0	1	0
c	0	0	1	1	0	0	1
d	0	0	0	0	1	1	1

Notice that every column sum should be 2 and row i's sum should be $\deg(v_i)$. ■

One application for which the incidence matrix is a very useful representation is the following. The superscript, T, denotes the transpose of the matrix B. The matrix I_ϵ is the ϵ by ϵ identity matrix. Matrix transpose and identity matrices are described in Appendix E.

THEOREM 10.35 ***The Line Graph***

Let G be a graph with no loops having n by ϵ incidence matrix B. Then the adjacency matrix, A_L, of the line graph, G_L, is

$$A_L = B^T B - 2I_\epsilon$$

Proof: Define $\delta_{rs} = \begin{cases} 1 & \text{if } r = s \\ 0 & \text{otherwise} \end{cases}$

Let $A_L = (a'_{ij})$ be defined by $A_L = B^T B - 2I_\epsilon$. Then, since the entry in row i and column k of B^T is b_{ki},

$$a'_{ij} = \sum_{k=1}^{n} b_{ki} b_{kj} - 2\delta_{ij}$$

The summation counts (in the original graph) the number of vertices v_k that edges e_i and e_j are both incident with (including multiple incidences). The term $2\delta_{ij}$ removes the artificial common incidences of an edge (in the original graph) with its two endpoints.

Since adjacency in the line graph is defined in terms of common incidences in the original graph, the expression properly calculates the entry in the ith row and jth column of the adjacency matrix A_L. □

EXAMPLE 10.20

Producing a Line Graph

The incidence matrix (repeated in the margin) for the Königsberg graph (shown in Figure 10.46 with its line graph) was found in Example 10.19.

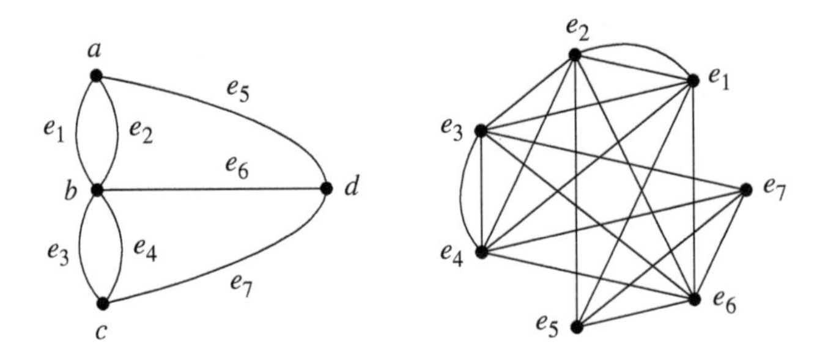

Figure 10.46 The Königsberg graph and its line graph.

	e_1	e_2	e_3	e_4	e_5	e_6	e_7
a	1	1	0	0	1	0	0
b	1	1	1	1	0	1	0
c	0	0	1	1	0	0	1
d	0	0	0	0	1	1	1

Theorem 10.35 implies that the line graph of the Königsberg graph has adjacency matrix $A_L = B^T B - 2I_\epsilon$:

$$\begin{pmatrix} 1 & 1 & 0 & 0 \\ 1 & 1 & 0 & 0 \\ 0 & 1 & 1 & 0 \\ 0 & 1 & 1 & 0 \\ 1 & 0 & 0 & 1 \\ 0 & 1 & 0 & 1 \\ 0 & 0 & 1 & 1 \end{pmatrix} \begin{pmatrix} 1 & 1 & 0 & 0 & 1 & 0 & 0 \\ 1 & 1 & 1 & 1 & 0 & 1 & 0 \\ 0 & 0 & 1 & 1 & 0 & 0 & 1 \\ 0 & 0 & 0 & 0 & 1 & 1 & 1 \end{pmatrix} - \begin{pmatrix} 2 & 0 & 0 & 0 & 0 & 0 & 0 \\ 0 & 2 & 0 & 0 & 0 & 0 & 0 \\ 0 & 0 & 2 & 0 & 0 & 0 & 0 \\ 0 & 0 & 0 & 2 & 0 & 0 & 0 \\ 0 & 0 & 0 & 0 & 2 & 0 & 0 \\ 0 & 0 & 0 & 0 & 0 & 2 & 0 \\ 0 & 0 & 0 & 0 & 0 & 0 & 2 \end{pmatrix}$$

$$= \begin{pmatrix} 0 & 2 & 1 & 1 & 1 & 1 & 0 \\ 2 & 0 & 1 & 1 & 1 & 1 & 0 \\ 1 & 1 & 0 & 2 & 0 & 1 & 1 \\ 1 & 1 & 2 & 0 & 0 & 1 & 1 \\ 1 & 1 & 0 & 0 & 0 & 1 & 1 \\ 1 & 1 & 1 & 1 & 1 & 0 & 1 \\ 0 & 0 & 1 & 1 & 1 & 1 & 0 \end{pmatrix}$$

This adjacency matrix is in agreement with the line graph (Figure 10.46) produced in Example 10.6 on page 604. ■

✔ Quick Check 10.7

1. What is the incidence matrix of the graph to the right?

2. Produce the adjacency matrix for the line graph of the graph in Question 1. Then, draw the line graph.

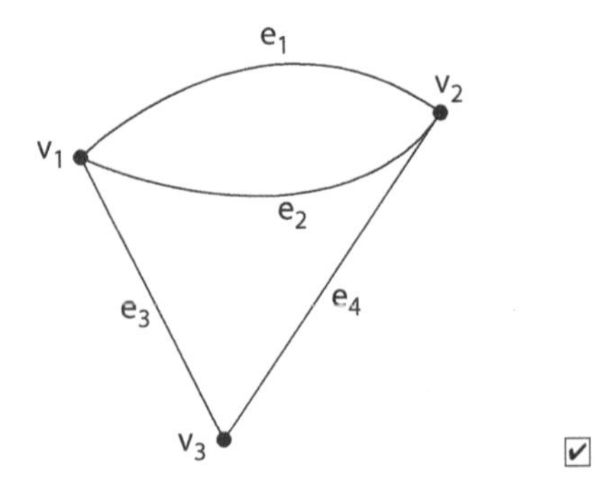

☑

Some other graph representations are as follows:

Edge lists An edge list is a table with rows indexed by the vertices. Each row contains a list of all edges incident with the row's indexing vertex.

Adjacency lists An adjacency list is a table with rows indexed by the vertices. Each row contains a list of all vertices adjacent with the row's indexing vertex.

10.4.2 Isomorphism

Since graphs can have multiple representations (including multiple embeddings), how can we tell if two representations are describing the same graph? A related question is, How can we tell if two graphs (different sets of vertices and edges, different incidence functions) are representing graphs that should really be considered the same? Before the questions can be answered, the notion "the same" needs to be refined.

> **DEFINITION 10.36** ***Isomorphic***
> The graphs $G = (V, E, \phi)$ and $G' = (V', E', \phi')$ are *isomorphic* if there exists a one-to-one and onto function, ζ, from V to V' such that v_i and v_j are adjacent in G if and only if $\zeta(v_i)$ and $\zeta(v_j)$ are adjacent in G'.
> In a multigraph, the adjacency condition needs to be enhanced by also requiring a one-to-one and onto function, ξ, between the edge sets such that $\phi(e) = v_i v_j$ if and only if $\phi'(\xi(e)) = \zeta(v_i)\zeta(v_j)$.

Notice that for G and G' to be isomorphic we require

- $|V| = |V'|$ (the one-to-one and onto requirement).
- $|E| = |E'|$ (the adjacency if and only if requirement).

The function ζ is really a renaming function. The graphs are isomorphic if we can rename the vertices of G so that they correspond to the vertices of G' and by so doing, the adjacencies are the same.

EXAMPLE 10.21 **A Simple Isomorphism**

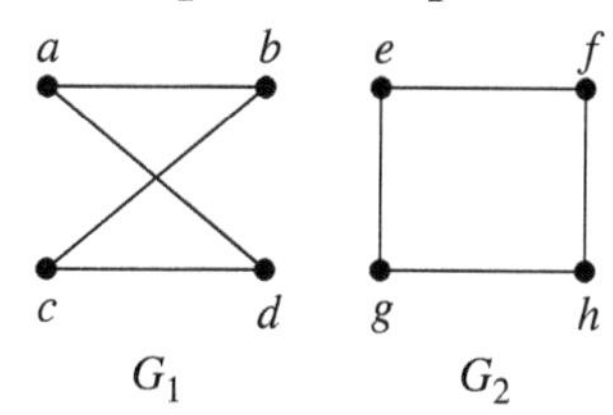

Figure 10.47 Are these the same graph?

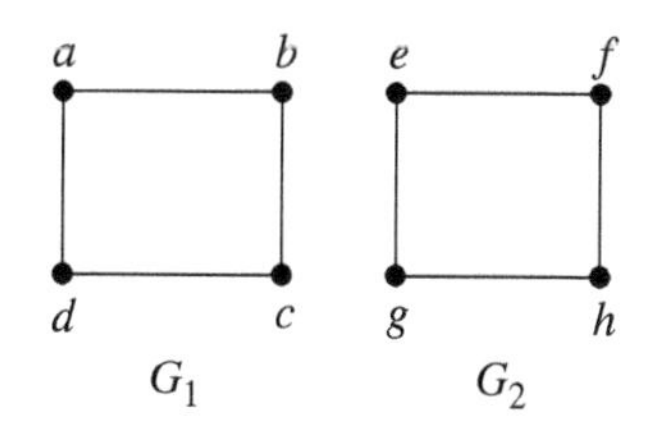

Figure 10.48. Vertices c and d have been embedded in new positions.

Using a graph visualization tool such as Sandbox (page 607) it is possible to do some click-and-drag manipulations to establish what ζ should be. For example, consider the graphs in Figure 10.47. Swapping the positions of vertices c and d produces the embedding in Figure 10.48.

The function ζ can be defined as follows:

V	$V' = \zeta(V)$
a	e
b	f
c	h
d	g

■

It should be clear from Definition 10.36 that if G and G' are isomorphic graphs and if $v_1 \in V$ has degree k, then $\zeta(v_1)$ must also have degree k. A formal way to state this is as follows:

Vertex degree is isomorphism invariant.

This observation leads naturally to the next definition.

> **DEFINITION 10.37** ***Degree Sequence***
> Let $G = (V, E, \phi)$ be a graph with $V = \{v_1, v_2, \ldots, v_n\}$. The *degree sequence* of G is the set $\{\deg(v_1), \deg(v_2), \ldots, \deg(v_n)\}$ rearranged as a sequence in decreasing order. A sequence of decreasing numbers $i_1, i_2, \ldots, i_n$ is said to be *graphic* if there is a simple graph whose degree sequence is $i_1, i_2, \ldots, i_n$.

EXAMPLE 10.22 **Degree Sequences**

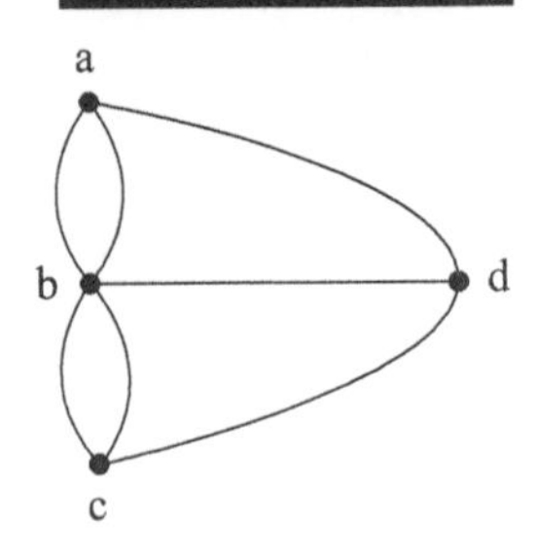

Figure 10.49. The Königsberg graph.

The Königsberg graph (Figure 10.49) has degree sequence 5, 3, 3, 3.

The sequence 5, 3, 2, 2, 1 is not graphic since the sum of the degrees is odd, contradicting the handshake theorem. ■

Because the degree of a vertex is isomorphism invariant, if two graphs are isomorphic then their degree sequences must be the same. The contrapositive of this statement is quite useful.

> **PROPOSITION 10.38** ***Degree Sequences***
> If G and G' are two graphs with distinct degree sequences, then G and G' are *not* isomorphic.

The converse of the statement "if two graphs are isomorphic then their degree sequences must be the same" is not true in general. That is, two graphs can have identical degree sequences but still not be isomorphic.

EXAMPLE 10.23 **Identical Degree Sequences Does Not Imply Isomorphic**

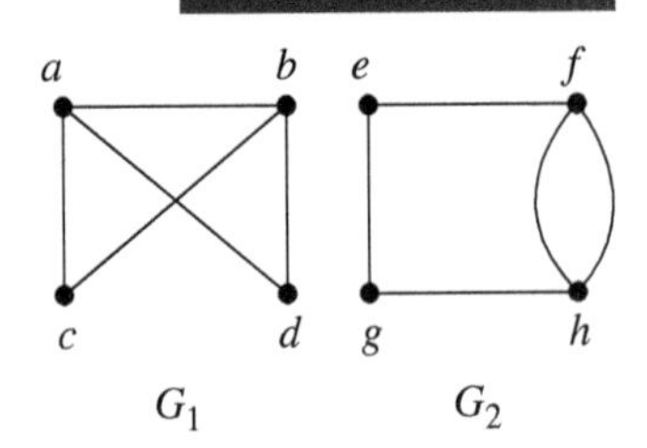

Figure 10.50. Same degree sequence but not isomorphic.

The graphs G_1 and G_2 in Figure 10.50 both have degree sequence 3, 3, 2, 2. However, they are not isomorphic.

Notice that the two vertices with degree 2 are adjacent in G_2 but not in G_1. An isomorphism would need to map c and d to nonadjacent vertices in G_2. However, it would also need to map both of them to vertices of degree 2. Since it is impossible to meet both these requirements, the two graphs cannot be isomorphic. ■

✔ Quick Check 10.8

1. List the degree sequence of the following graph.

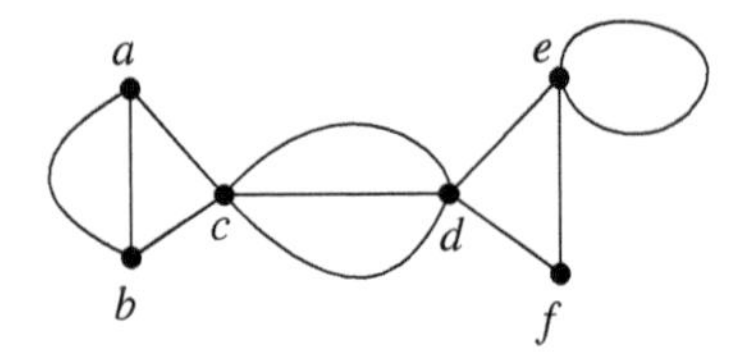

2. Which of the following graphs are isomorphic? Show the function ζ if they are isomorphic, or explain why they are not isomorphic.

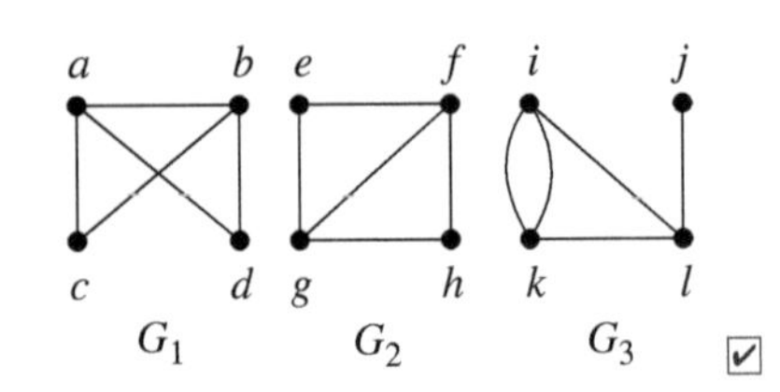

☑

Comparing the numbers of vertices and edges and comparing the degree sequences are useful tools if we are trying to show that two graphs are not isomorphic. If any one of these tests fails, the graphs cannot be isomorphic. However, we have seen that even if these all match, the graphs may still not be isomorphic. Another useful test is to examine walks and circuits.

Walks and circuits are also isomorphism invariant. That is, if G and G' are isomorphic graphs and $v_0 e_1 v_2 e_2 \cdots e_n v_n$ is a walk (circuit) in G, then $\zeta(v_1) f_1 \zeta(v_2) \cdots f_n \zeta(v_n)$

must be a walk (circuit) in G', for some edges, $f_1, \ldots, f_n$. Graphs G_2 and G_3 in Quick Check 10.8 cannot be isomorphic because G_3 contains a circuit of length 2, but G_2 does not. (An isomorphism from G_3 to G_2 would require vertices i and k to map to a pair of degree-3 vertices in G_2 that are part of a length 2 circuit.)

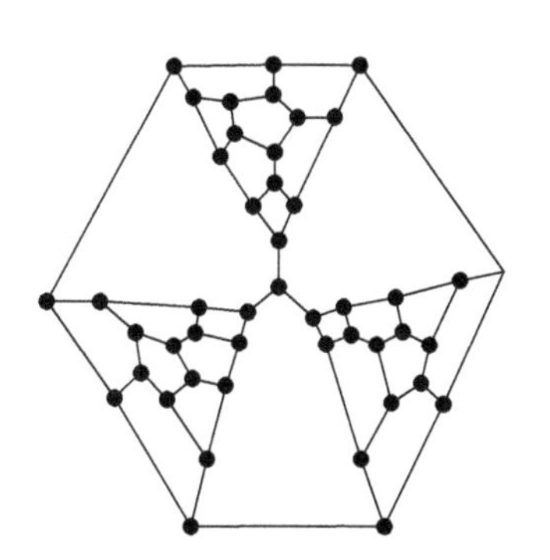

Figure 10.51. The Tutte graph.

Showing that two graphs *are* isomorphic is even more difficult in general. Consider trying to find the one-to-one function ζ if one of the graphs is the Tutte graph (Figure 10.51).

Since the set of vertices in a graph is a finite set, it is clearly possible to create an algorithm that would manipulate computer representations of the two graphs and determine whether they were isomorphic or not. At the very worst, every possible mapping could be tried and then corresponding adjacencies could be tested. As soon as an adjacency test failed, the next possible mapping ζ could be tried.

How many possible one-to-one functions ζ are there if the vertex sets each have n elements? The sad answer is that there are $n!$ distinct one-to-one functions. Even ignoring the adjacency testing, such an algorithm would have a complexity of at least $\Omega(n!)$. This is a useless algorithm for even moderately sized graphs.

Are there any efficient approaches? The answer in general is no. There is no known algorithm that has a polynomial complexity measure that will test any two graphs for isomorphism. Most researchers believe a polynomial time algorithm does not exist.[20] There are a few algorithms that are fairly efficient in practice, but will slow down dramatically on some graphs. Combinatorica provides the function `IsomorphicQ`$[G_1, G_2]$ that implements one such algorithm [84].

10.4.3 Exercises

The exercises marked with ★ have detailed solutions in Appendix H.

1. For each graph, produce
   - The adjacency matrix.
   - The incidence matrix.

(a)

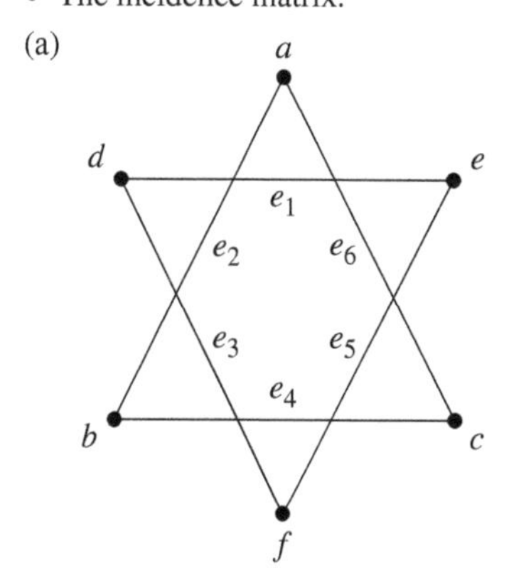

(b) ★

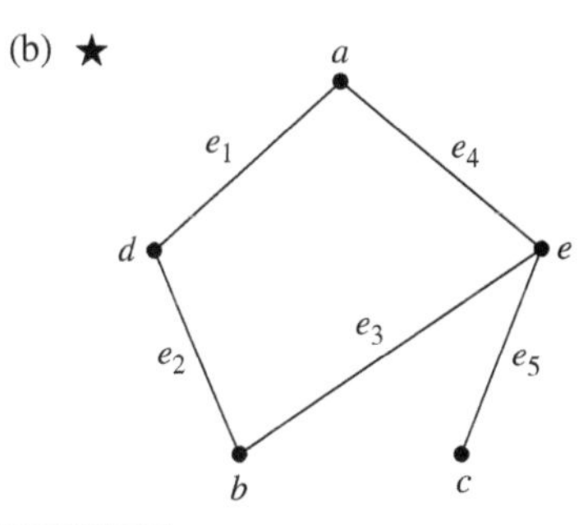

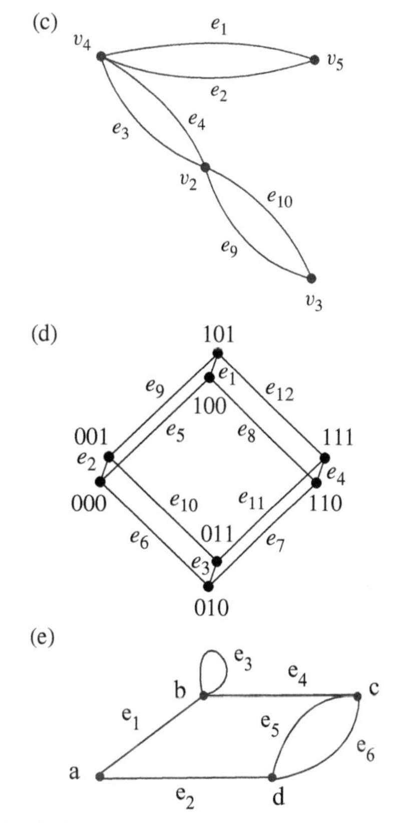

2. ★ For each graph in Exercise 1 (parts a–d), use Theorem 10.35 to produce the adjacency matrix for its line graph.

[20] The notions of P, NP, and NP-complete algorithms are needed to make this more precise. These notions are presented in more advanced courses.

3. Draw the line graph for each graph in Exercise 1 (parts a–d). For part (d), you may want use a pipe cleaner model, or at least switch from the embedding

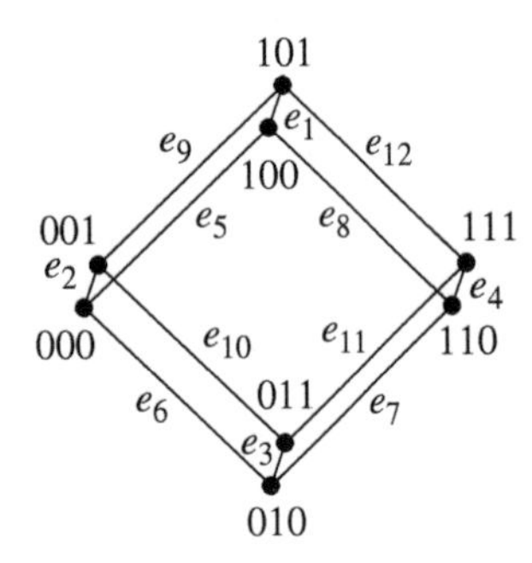

to the embedding

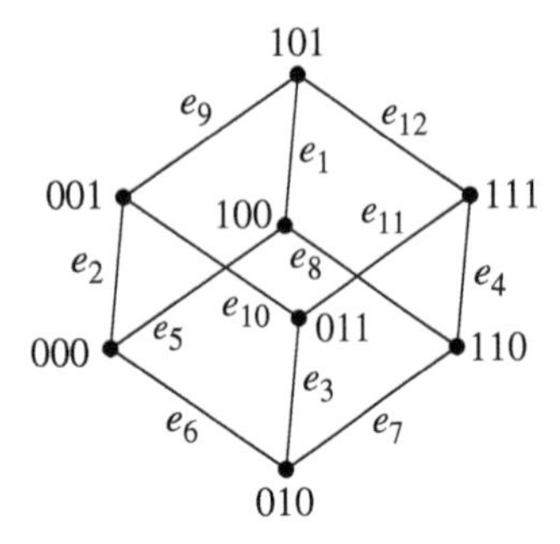

4. For each matrix, A, determine whether it can be the adjacency matrix for a simple graph. If it cannot, explain why not; otherwise, draw a diagram for a graph whose adjacency matrix (under the natural sorting of vertex labels) is A.

(a) ★

$$\begin{array}{c} \\ a \\ b \\ c \\ d \end{array} \begin{array}{c} \begin{array}{cccc} a & b & c & d \end{array} \\ \begin{pmatrix} 0 & 1 & 1 & 0 \\ 1 & 0 & 0 & 1 \\ 1 & 1 & 0 & 0 \\ 0 & 1 & 0 & 0 \end{pmatrix} \end{array}$$

(b)

$$\begin{array}{c} \\ a \\ b \\ c \\ d \\ e \\ f \end{array} \begin{array}{c} \begin{array}{cccccc} a & b & c & d & e & f \end{array} \\ \begin{pmatrix} 0 & 1 & 1 & 0 & 0 & 1 \\ 1 & 0 & 1 & 0 & 0 & 0 \\ 1 & 1 & 0 & 1 & 0 & 0 \\ 0 & 0 & 1 & 0 & 1 & 1 \\ 0 & 0 & 0 & 1 & 0 & 1 \\ 1 & 0 & 0 & 1 & 1 & 0 \end{pmatrix} \end{array}$$

(c)

$$\begin{array}{c} \\ a \\ b \\ c \\ d \\ e \\ f \end{array} \begin{array}{c} \begin{array}{cccccc} a & b & c & d & e & f \end{array} \\ \begin{pmatrix} 1 & 1 & 1 & 0 & 0 & 0 \\ 1 & 0 & 1 & 0 & 0 & 0 \\ 1 & 1 & 0 & 1 & 0 & 0 \\ 0 & 0 & 1 & 0 & 1 & 1 \\ 0 & 0 & 0 & 1 & 0 & 1 \\ 0 & 0 & 0 & 1 & 1 & 1 \end{pmatrix} \end{array}$$

5. For each matrix, B, determine whether it can be the incidence matrix for a simple graph. If it cannot, explain why not; otherwise, draw a diagram for a graph whose incidence matrix (under the natural sorting of vertex and edge labels) is B.

(a)

$$\begin{array}{c} \\ a \\ b \\ c \\ d \end{array} \begin{array}{c} \begin{array}{cccccc} e_1 & e_2 & e_3 & e_4 & e_5 & e_6 \end{array} \\ \begin{pmatrix} 1 & 1 & 0 & 0 & 1 & 0 \\ 1 & 1 & 1 & 1 & 0 & 1 \\ 0 & 0 & 1 & 0 & 0 & 0 \\ 0 & 0 & 0 & 0 & 1 & 1 \end{pmatrix} \end{array}$$

(b)

$$\begin{array}{c} \\ a \\ b \\ c \\ d \end{array} \begin{array}{c} \begin{array}{ccccccc} e_1 & e_2 & e_3 & e_4 & e_5 & e_6 & e_7 \end{array} \\ \begin{pmatrix} 1 & 1 & 0 & 0 & 1 & 0 & 0 \\ 1 & 1 & 1 & 1 & 0 & 1 & 0 \\ 0 & 0 & 1 & 1 & 0 & 0 & 1 \\ 0 & 0 & 0 & 0 & 1 & 1 & 1 \end{pmatrix} \end{array}$$

(c)

$$\begin{array}{c} \\ a \\ b \\ c \\ d \end{array} \begin{array}{c} \begin{array}{ccccc} e_1 & e_2 & e_3 & e_4 & e_5 \end{array} \\ \begin{pmatrix} 1 & 1 & 0 & 0 & 0 \\ 1 & 0 & 1 & 1 & 0 \\ 0 & 0 & 1 & 0 & 1 \\ 0 & 1 & 0 & 1 & 1 \end{pmatrix} \end{array}$$

6. Prove that K_3 and $K_{1,3}$ have isomorphic line graphs but are not isomorphic to each other.

7. ★ Produce the degree sequence for each graph in Exercise 1.

8. Create a graph with degree sequence 4, 3, 2, 1.

9. Produce two graphs with degree sequence 4, 3, 3, 3, 3 that are not isomorphic.

10. Produce an isomorphism for the pair of graphs, or explain why the two graphs are not isomorphic. A click-and-drag tool, such as Sandbox, will greatly simplify the task.

(a) ★

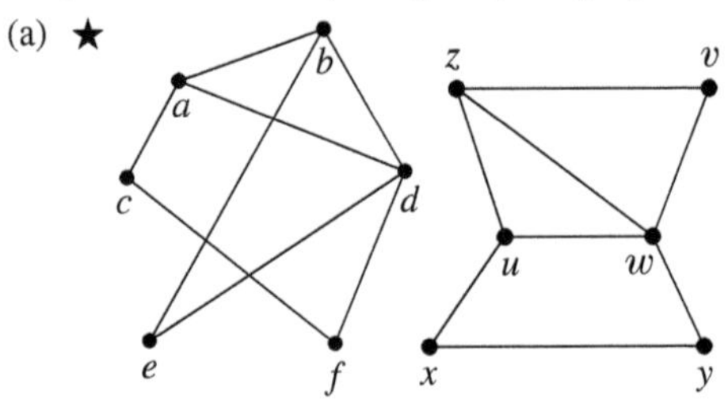

(b)

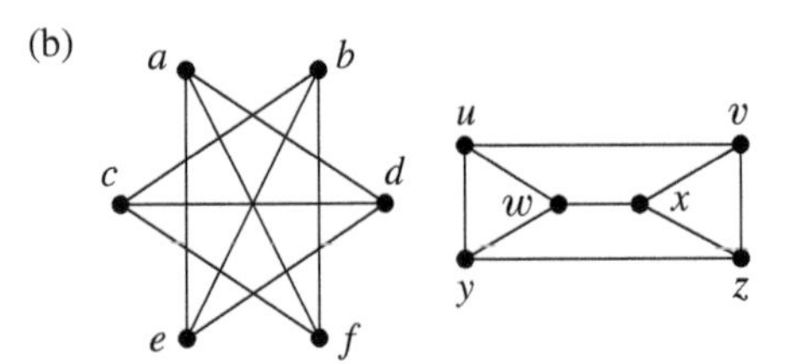

(c)

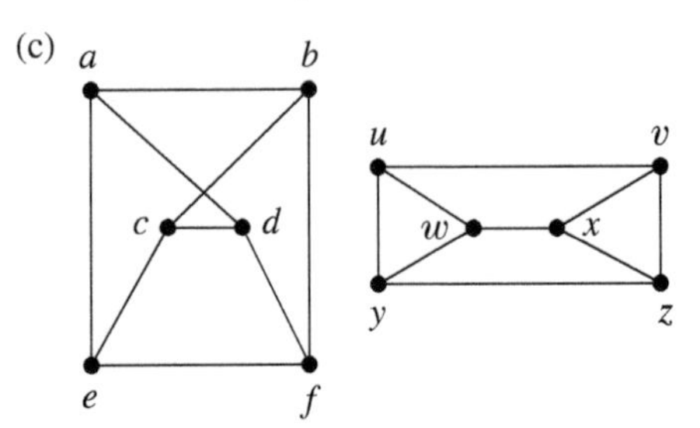

11. Produce an isomorphism for the pair of graphs, or explain why the two graphs are not isomorphic. A click-and-drag tool such as Sandbox will greatly simplify the task.

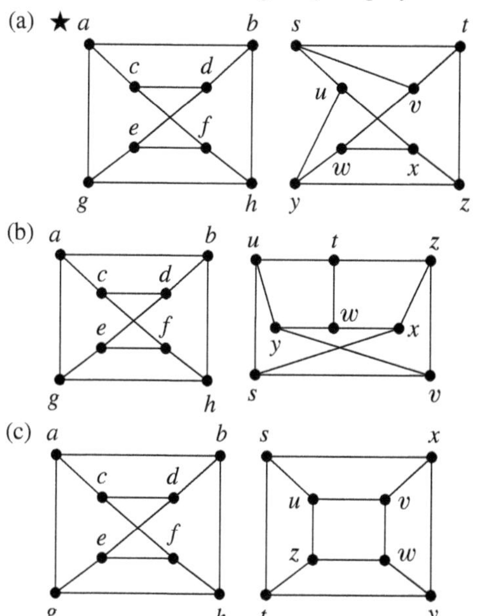

12. Each of the following statements is either true (always) or false (at least sometimes). Determine which option applies for each statement and provide adequate explanation for your choice.

   (a) The incidence matrix of a simple graph, G, with n vertices, $\{v_1, v_2, \cdots, v_n\}$, and ϵ edges, $\{e_1, e_2, \cdots, e_\epsilon\}$, is the n by ϵ matrix, $B = (b_{ij})$, defined by

$$b_{ij} = \begin{cases} 1 & \text{if } v_i \text{ is incident with } e_j \\ 0 & \text{otherwise} \end{cases}$$

   (b) ★ If ζ is an isomorphism from a simple graph, G, to a multigraph, G', which is not a simple graph, then the degree sequences of G and G' are identical.

   (c) Given any sequence of natural numbers arranged in decreasing order, there may not be a *simple graph* with this degree sequence. However, it will still be possible to construct a *graph* with this degree sequence.

   (d) If two graphs have identical degree sequences, then they are isomorphic.

   (e) Suppose B is the n by ϵ incidence matrix for a graph, G. Then the adjacency matrix, A, of the graph G is $A = B^T B - 2I_\epsilon$.

   (f) ★ If an isomorphism exists between two graphs without loops, then vertex degree will be isomorphism invariant.

13. (a) For which values of k can there be a graph with degree sequence $2k-1, 2k-3, \ldots, 5, 3, 1$? Prove your claim. [*Hint*: question 2 in Quick Check 3.10 on page 135.]

   (b) For the two smallest values of k for which a graph exists, create a visual representation of a graph with the required degree sequence.

14. Are K_6 and $K_{3,3}$ isomorphic? Give sufficient evidence for your answer.

15. Can two graphs have identical incidence matrices and not be isomorphic? Explain your answer.

16. Suppose G and H are both regular simple graphs with $n > 2$ vertices, where every vertex in each graph has degree 2.

   (a) Are G and H isomorphic? If they are, provide a proof; if they need not be isomorphic, provide a counterexample.

   (b) Suppose G and H are both connected graphs. Are G and H isomorphic? If they are, provide a proof; if they need not be isomorphic, provide a counterexample.

17. Use adjacency matrices to show that the graphs, G_1 and G_2, from Quick Check 10.8 are isomorphic.

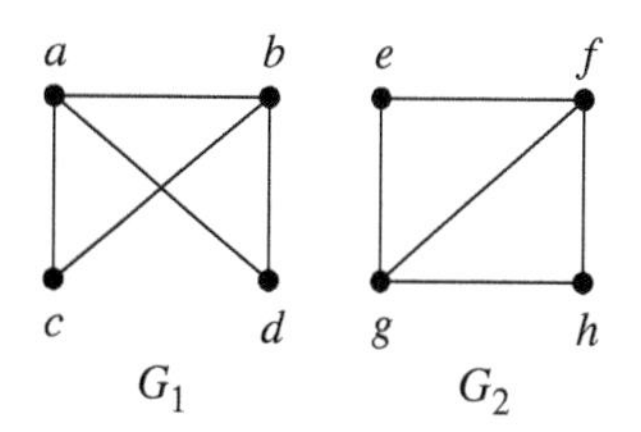

   (a) Write a vertex-labeled adjacency matrix for each graph.

   (b) Now transform the adjacency matrix for G_2 into the same matrix (ignoring the row and column labels) as the adjacency matrix for G_1. Do this by choosing two rows and swapping them, then swapping the corresponding columns. Then swap other pairs of rows (and corresponding columns) until the adjacency matrices are identical.

   (c) The function, ζ, can be defined by matching row labels for the two adjacency matrices. What is ζ?

   (d) Prove that each swap (rows and corresponding columns) in part (b) will produce an adjacency matrix for a graph that is isomorphic to the graph corresponding to the pre-swap adjacency matrix.

18. Can two graphs be isomorphic if one is bipartite and the other is not? Support your answer.

19. Without actually constructing the complement graph, show the following graph and its complement are not isomorphic.

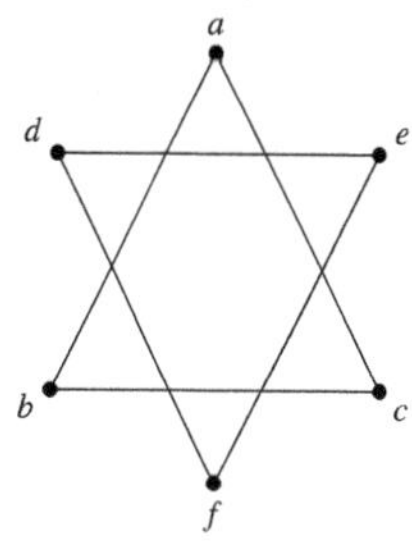

20. Let G and H be two simple graphs. Prove that G and H are isomorphic if and only of $\overline{G}$ and $\overline{H}$ are isomorphic.

21. Let G be a simple graph with n vertices. If G has degree sequence, $d_1, d_2, \ldots, d_n$, what is the degree sequence of $\overline{G}$?

22. Find an example of a graph, G, that is isomorphic to its complement and that has

   (a) ★ Four vertices. (b) Five vertices.

   (c) Five vertices and is not isomorphic to the part (b) example.

   [*Hint*: Use Exercise 21.]

10.5 The Big Theorems: Planarity, Euler, Polyhedra, Chromatic Number

There are a few theorems in graph theory that stand apart from the rest; they are part of the lore that every educated person should know. The goal of this section is to present four significant theorems. The first theorem answers the question, When can a graph be embedded in the plane without any of the edges crossing another edge? The second theorem is the famous Euler formula, which will then be used to prove the third theorem, an important result from antiquity: There are only five regular solids. Finally, a simple problem about coloring maps of countries, posed by a high school student in the mid-1800s, led to the difficult-to-prove four-color theorem.

10.5.1 Planarity

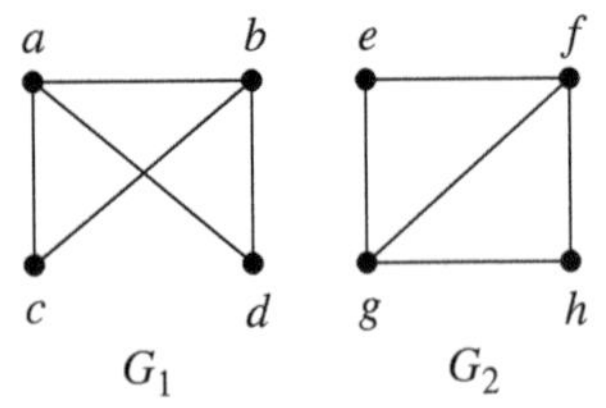

Quick Check 10.8 showed that graphs G_1 and G_2 in Figure 10.52 are isomorphic. This can be interpreted in another manner. The graph G_1 can be represented using the alternate embedding shown in Figure 10.53.

Figure 10.52. G_1 and G_2 are isomorphic.

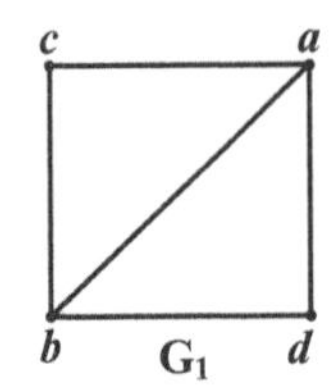

Figure 10.53 An alternate embedding for G_1.

This alternate embedding is preferred because it avoids the unnecessary crossing of the edges *ad* and *bc* in the previous embedding. This section examines the question, When can a graph be embedded in the plane without any edge crossings?

DEFINITION 10.39 *Planar Graph*

A graph, G, is *planar* if G has an embedding in the plane that has no edge crossings.

EXAMPLE 10.24 **The Utility Graph**

Imagine a small neighborhood with three houses. The city wants to connect the three houses to three utilities: electricity, gas, and telephone. Figure 10.54 shows that this can be modeled using the bipartite complete graph $K_{3,3}$. In this context, $K_{3,3}$ is often called the *utility graph*.

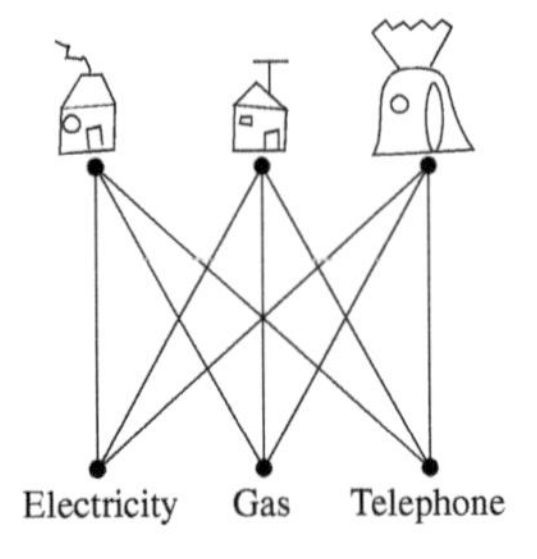

Figure 10.54 The utility graph.

■

✔ Quick Check 10.9

1. Is $K_{3,3}$ a planar graph? That is, is it possible to redraw the connections from utilities to houses so that there are no edge crossings? ☑

There are two graphs that play an essential role in the quest to determine which graphs are planar. These two nonplanar graphs are K_5 and $K_{3,3}$ (Figure 10.55). The reason they are so important will be apparent when Kuratowski's theorem is presented later in this section.

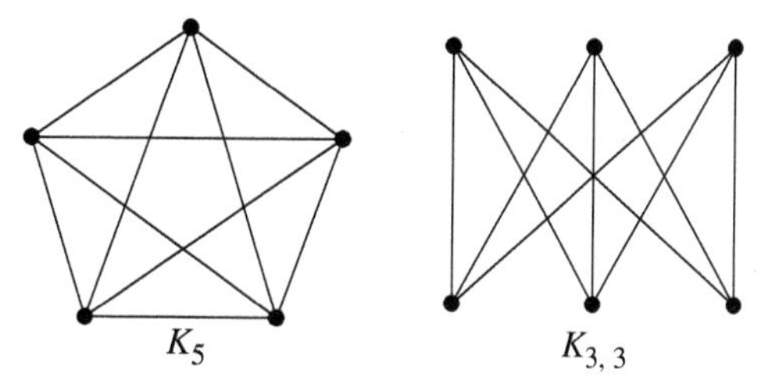

Figure 10.55 K_5 and $K_{3,3}$ are nonplanar.

THEOREM 10.40 ***K_5 is Nonplanar***

The complete graph, K_5, is nonplanar.

Proof: See Exercises 1 and 2 on page 648. □

THEOREM 10.41 ***$K_{3,3}$ is Nonplanar***

The complete bipartite graph, $K_{3,3}$, is nonplanar.

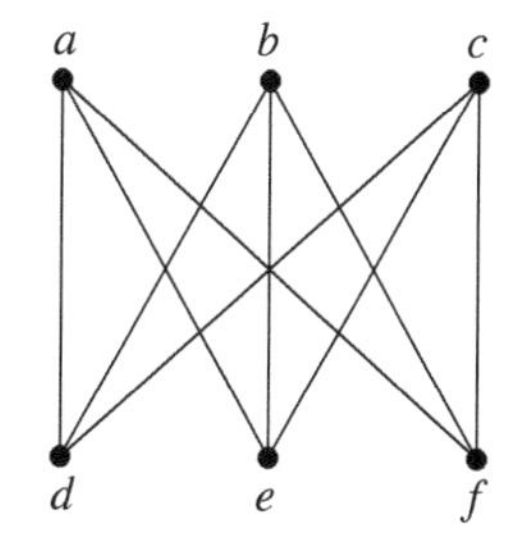

Figure 10.56. $K_{3,3}$ with vertex labels.

Proof: Label the vertices of $K_{3,3}$ as a, b, c, d, e, and f (Figure 10.56).

Now attempt to produce a planar embedding. We can start by just adding the vertices a, b, d, and e. Since a is connected to both d and e and b is also connected to these vertices, we must have an embedding that is essentially that of Figure 10.57 if we insist on a planar embedding. The key feature is that a planar representation will divide the plane into two regions, r_1 and r_2.

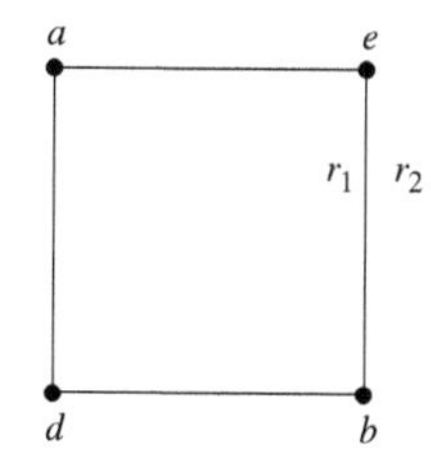

Figure 10.57 Attempting a planar embedding of $K_{3,3}$.

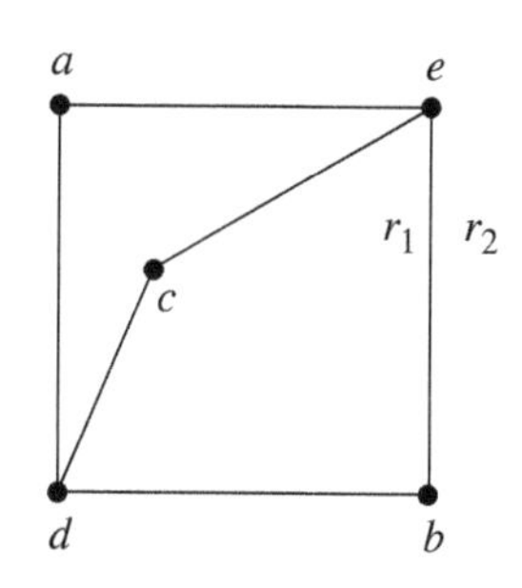

Figure 10.58. Case 1: c is in r_1.

We must now add vertex c. There are two possibilities. Either c must be placed in region r_1 or it must be placed in region r_2.

Case 1: c is in r_1

Since c must be connected to both d and e, a planar embedding must look similar to Figure 10.58.

There is a difficulty. We need to add the vertex f in such a way that it is adjacent to a, b, and c, but without any edge crossings. This is impossible. (The vertex f must be placed in one of the three regions; in each case, one of a, b, or c cannot be reached without crossing an edge.)

Case 2: c is in r_2

Since c must be connected to both d and e, a planar embedding must look similar to Figure 10.59.

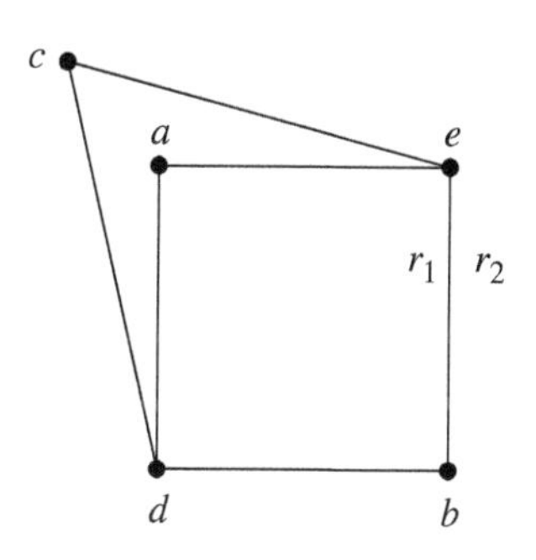

Figure 10.59. Case 2: c is in r_2.

It should be clear that vertex f cannot be added without creating edge crossings. □

The graphs K_5 and $K_{3,3}$ are fairly small and uncomplicated graphs. It would appear that there will be many nonplanar graphs. In such a myriad of nonplanar graphs, it seems like wishful thinking to expect that there is any pattern or order to what makes one graph planar and another nonplanar. Contrary to expectation, the main theorem of this section (Kuratowski's theorem) provides a surprisingly simple property that all nonplanar graphs share. Before the main theorem can be presented, a few new ideas need to be presented.

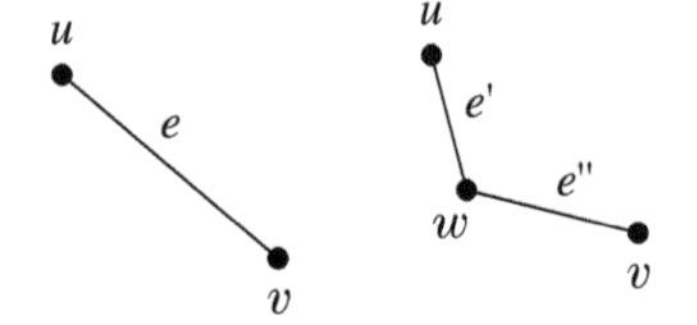

Figure 10.60. An elementary subdivision of the edge, uv.

DEFINITION 10.42 ***Elementary Subdivision of an Edge***
Let e be an edge in a graph that is incident with vertices u and v. An *elementary subdivision* of e is a path of length two from u to v that replaces e.

As Figure 10.60 indicates, an elementary subdivision of an edge appears visually as if a new vertex was placed in the middle of an edge.

EXAMPLE 10.25 **Elementary Subdivisions**

The graph at the top of Figure 10.61 is the Petersen graph.

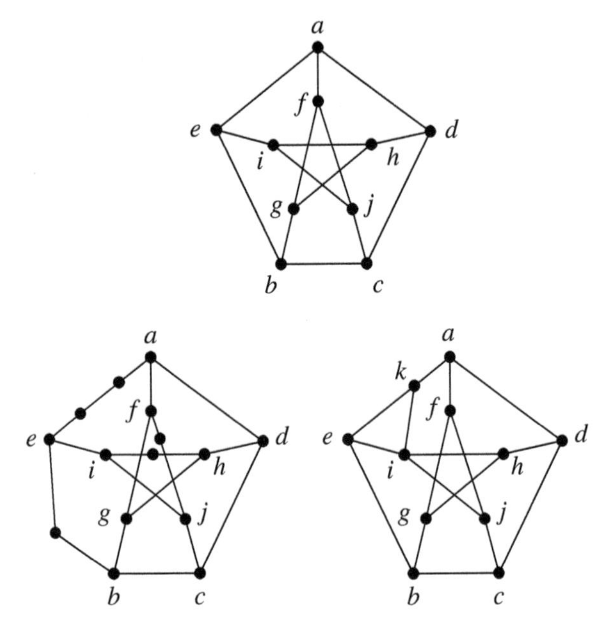

Figure 10.61 The Petersen graph and two derived graphs.

On the bottom left is a graph that has been derived from the Petersen graph by a sequence of five elementary subdivisions of edges. The graph on the bottom right has also been derived from the Petersen graph, but it has *not* been produced using only elementary subdivisions. The edge ik has been added after an elementary subdivision of the edge ae. ■

DEFINITION 10.43 ***Homeomorphic***
Two graphs are said to be *homeomorphic* if both can be derived from a common ancestor using only a sequence of elementary subdivisions.

Notice that homeomorphism is a property of the graphs and not just of the visual representation.

EXAMPLE 10.26 **Homeomorphic Graphs**

The two graphs at the bottom of Figure 10.62 are homeomorphic because both have been produced by sequences of elementary subdivisions of the graph at the top.

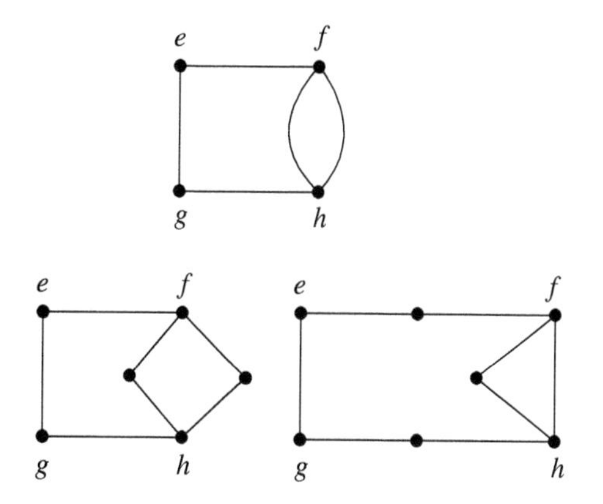

Figure 10.62 Homeomorphic graphs.

Determining whether two graphs are homeomorphic is generally much harder if the common ancestor is not present. ■

✔ Quick Check 10.10

1. The two graphs on the right are homeomorphic. Find the common ancestor.

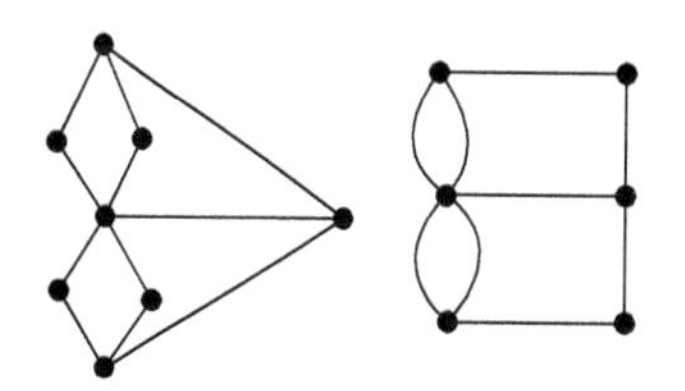

☑

It is finally time to state the main result on planarity, first proved by the Polish mathematician Kazimierz Kuratowski in 1930. The proof is too involved for a one-chapter survey of graph theory. However, it is not too difficult to present to undergraduates in a semester course in graph theory.

THEOREM 10.44 ***Kuratowski's Theorem***

A graph, G, is nonplanar if and only if it contains a subgraph that is homeomorphic to either K_5 or $K_{3,3}$.

Pay close attention to the statement of the theorem. It does not say that G must contain K_5 or $K_{3,3}$ as a subgraph. Nor does it say that G itself must be homeomorphic to one of K_5 or $K_{3,3}$.

Graph G is nonplanar if we can find a subgraph of G (i.e., delete some edges and vertices) that is homeomorphic to K_5 or to $K_{3,3}$.

This theorem is remarkable in its conceptual simplicity.[21] It says that at least one of the graphs K_5 or $K_{3,3}$ is hiding (perhaps as a homeomorphic descendant) in every nonplanar graph. In addition, if neither K_5 nor $K_{3,3}$ is hidden in G, then G is planar.

[21] It is not so simple technically, since the notions of elementary subdivisions and homeomorphism are needed.

EXAMPLE 10.27

Applying Kuratowski's Theorem

The graph at the top of Figure 10.63 is the line graph of the Königsberg graph. The subgraph on the bottom left is K_5 (and so is trivially homeomorphic to K_5). It was obtained by deleting vertices e_5 and e_7, as well as deleting the two curved edges e_1e_2 and e_3e_4. The subgraph on the right is $K_{3,3}$. It was obtained by deleting vertex e_2 and all remaining edges e_3e_4, e_1e_6, and e_6e_7.

The line graph of the Königsberg graph is clearly nonplanar. ■

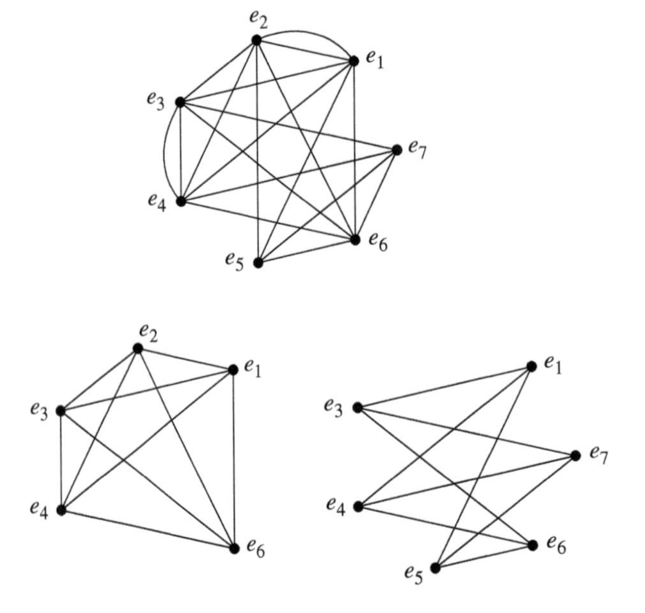

Figure 10.63 The line graph of the Königsberg graph contains subgraphs homeomorphic to K_5 and to $K_{3,3}$.

EXAMPLE 10.28

A More Complex Application of Kuratowski's Theorem

The graph on the left of Figure 10.64 appears to be nonplanar. To be certain, it is helpful to use Kuratowski's theorem. The graph at the right is a minor re-embedding of a subgraph. It is clearly homeomorphic to K_5 (with K_5 itself as the common ancestor). The graph on the left is therefore nonplanar.

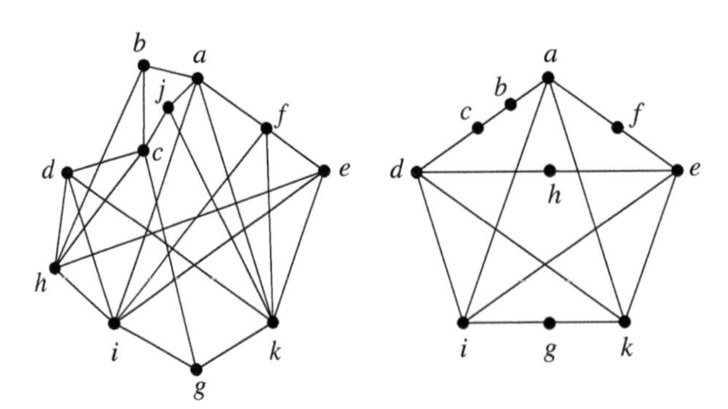

Figure 10.64 Verifying nonplanarity.

Notice that many edges and one vertex have been deleted. It is not possible to remove any more vertices. For example, in the original graph, vertices d and e are not adjacent. The path of length two passing through vertex h is needed to "reproduce" the hidden copy of K_5.

To find a subgraph that is homeomorphic to K_5 or $K_{3,3}$, it is very helpful to have a program that allows click-and-drag manipulations of graphs. ■

10.5.2 The Regular Polyhedra

DEFINITION 10.45 ***Polyhedron***

A *polyhedron* is a simple closed surface in space whose outer boundary is composed of polygons. The shape is determined by the vertices, edges, and *faces* and by the incidence relationships among them. Edges and vertices are related exactly as they are in graphs. (In fact, a polyhedron *is* a graph if we ignore the faces.) Faces are bounded by edges.

The plural form of *polyhedron* is *polyhedra.*

EXAMPLE 10.29 **A Polyhedron**

Figure 10.65 shows a three-dimensional rendering of a polyhedron. It has 14 faces, 24 vertices, and 36 edges. ■

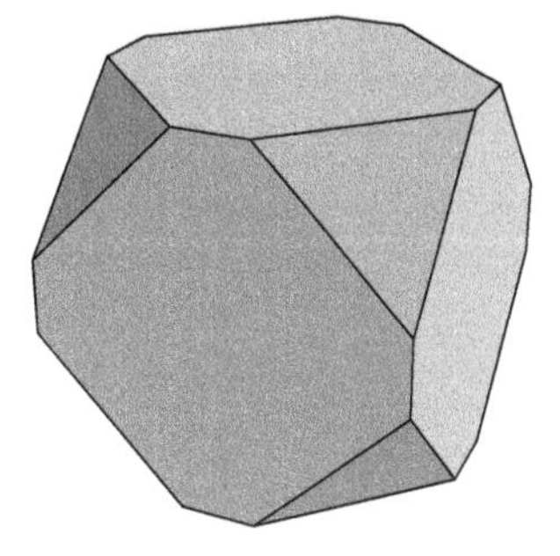

Figure 10.65. A truncated cube.

Recall that a graph is called *regular* if every vertex has the same degree.

DEFINITION 10.46 ***Regular Polyhedron***

A polyhedron is *regular* if

- Every vertex has the same degree.
- The faces are all congruent.

Since the faces are all congruent, every face will have the same number of bounding edges. A cube is an example of a regular polyhedron: Every vertex has degree 3 and every face is a square (with four bounding edges).

Notice that a regular polyhedron can be embedded in the plane by removing one face and stretching. The missing face can be thought of as the area outside the rest of the graph that is formed. When a polyhedron is embedded in the plane, the faces are usually called *regions*. The regions will not be congruent polygons.

EXAMPLE 10.30 **A Planar Embedding of the Cube**

The cube and a planar embedding are shown in Figure 10.66. Imagine removing the top face of the cube, then stretching the sides outward and down.

Figure 10.66 The cube and a planar embedding.

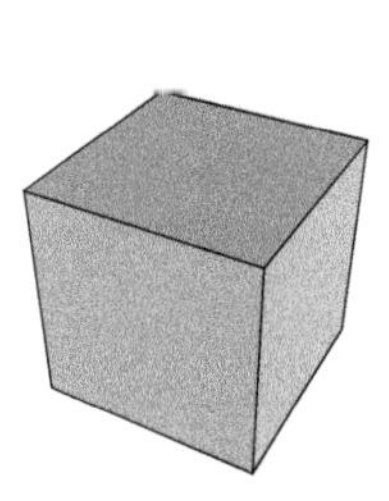

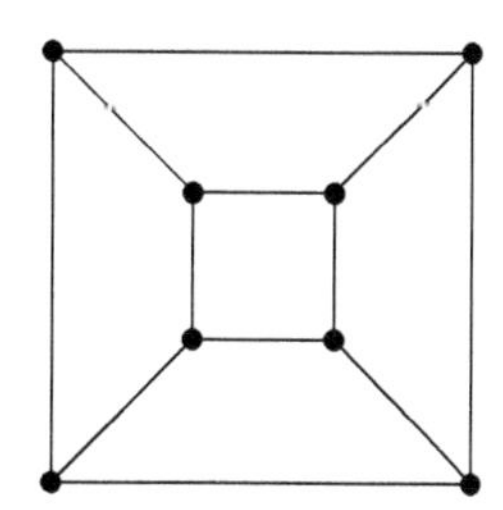

The polyhedron on the left has six faces. The graph on the right has six regions. The sixth region is the area outside the larger square in the graph. ■

THEOREM 10.47 ***Euler's Formula***

Let G be a connected, planar graph with ϵ edges and n vertices. Let ρ be the number of regions in a planar embedding of G (including the region on the outside). Then

$$\rho = \epsilon - n + 2$$

Before reading the proof, you should verify that the formula is correct for the cube. Also, notice the adjectives in the hypothesis. Both are essential.

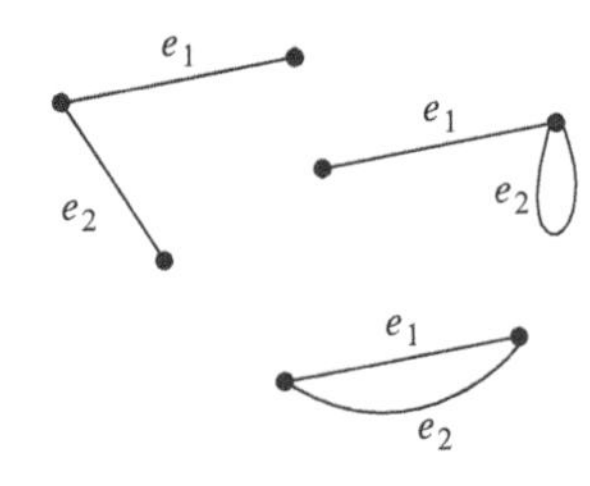

Figure 10.67. G_2

Proof of Theorem 10.47: $\rho = \epsilon - n + 2$

Label the edges of G as $e_1, e_2, \ldots, e_\epsilon$. We can always assign labels so that the subgraph, G_k, containing only edges $e_1, e_2, \ldots, e_k$, is a connected graph, for $1 \le k \le \epsilon$. Let G_k have ϵ_k edges, n_k vertices, and ρ_k regions. Notice that $G = G_\epsilon$.

Now consider building G one edge at a time, using the order just established. The first stage, G_1, will have one edge, e_1, two vertices and one region. Thus $\rho_1 = 1, \epsilon_1 = 1$, and $n_1 = 2$. Clearly, $\rho_1 = \epsilon_1 - n_1 + 2$.

After adding e_2 to G_1, the subgraph G_2 will look like one of the graphs in Figure 10.67. The relationship $\rho_2 = \epsilon_2 - n_2 + 2$ is true.

Any additional edges that are added will fall into one of two cases. Suppose that the most recent edge added is e_k.

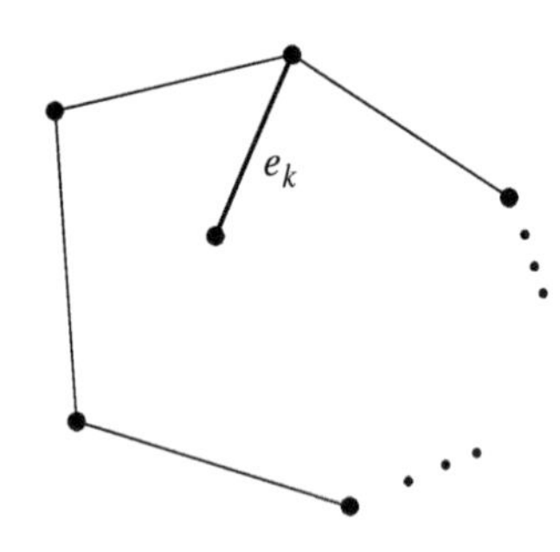

Figure 10.68. A new vertex is added.

A new vertex is introduced.
The next edge might add one new vertex. (It cannot add two vertices because each subgraph G_k is connected.) In this case, adding e_k introduces one new edge and one new vertex (Figure 10.68). However, it does not change the number of regions. The new edge and the new vertex will cancel in the formula $\rho_k = \epsilon_k - n_k + 2$.

No new vertices are added. If the new edge joins vertices that are already in G_{k-1}, the edge must divide a current region into two regions (even if the region is the outside region). This is true because G_{k-1} is connected and G is planar. The connectedness of G_{k-1} ensures that the new edge completes a circuit. The planarity of G means that adding e_k will not create an edge crossing, which would introduce multiple new regions. Thus, one new edge and one new region will be added, but no new vertices will be added (Figure 10.69). Thus, $\rho_k = \epsilon_k - n_k + 2$.

In either case, if $\rho_{k-1} = \epsilon_{k-1} - n_{k-1} + 2$, then $\rho_k = \epsilon_k - n_k + 2$. Since this is true for all k $(1 \le k \le \epsilon)$, the graph G itself must satisfy $\rho = \epsilon - n + 2$. □

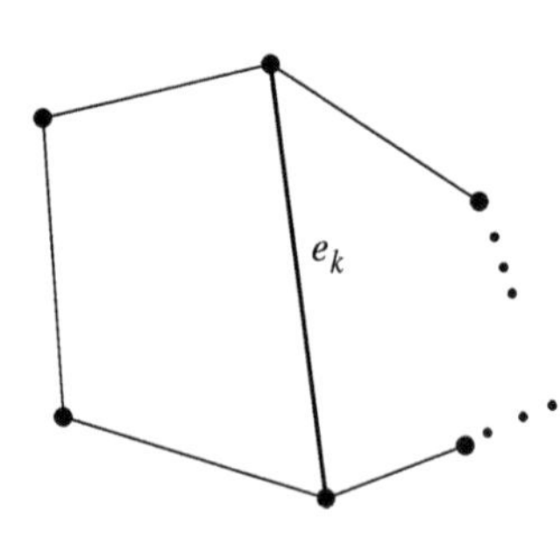

Figure 10.69. No new vertex is added.

Since all polyhedra can be represented as planar graphs, every polyhedron must satisfy Euler's formula (with faces considered as the regions).

The Five Regular Polyhedra

We would like to find all possible regular polyhedra. There are five values that characterize such geometric figures:

- The number of edges: ϵ.
- The number of vertices: n.
- The number of regions (faces): ρ.
- The common degree of the vertices: d.
- The number of edges bounding each face: b.

THEOREM 10.48 ***There are exactly five regular polyhedra***

The only regular polyhedra are listed in the next table.

Polyhedron	Face Shape	Faces	Vertices	Edges
Tetrahedron	Equilateral triangles	4	4	6
Cube	Squares	6	8	12
Octahedron	Equilateral triangles	8	6	12
Dodecahedron	Equilateral pentagons	12	20	30
Icosahedron	Equilateral triangles	20	12	30

The five regular polyhedra are also called the *five Platonic solids*. They are shown in a three-dimensional rendering and in a planar embedding in Figures 10.70 – 10.74.

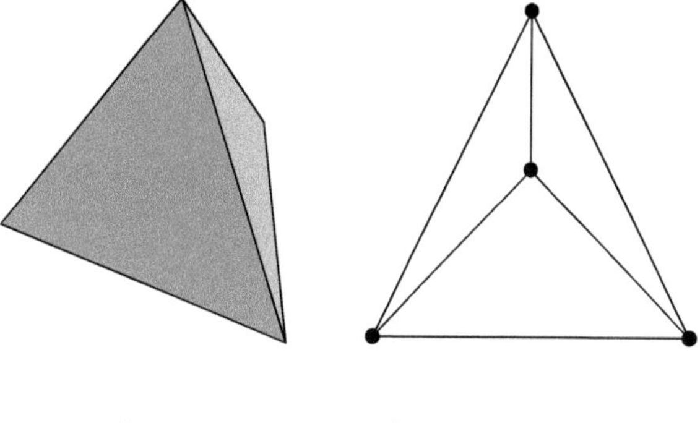

Figure 10.70 The tetrahedron: 4 faces, 4 vertices, 6 edges.

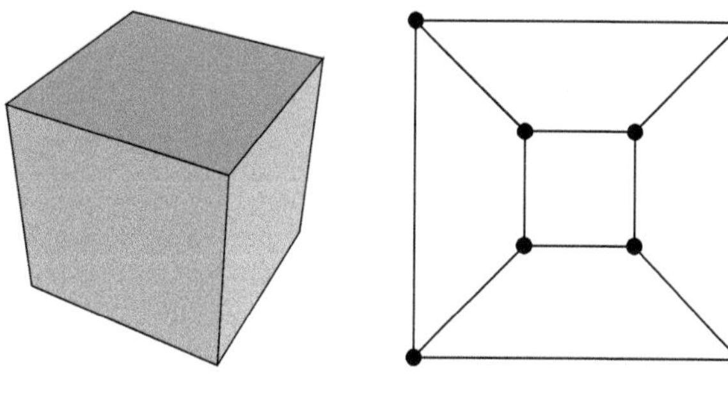

Figure 10.71 The cube: 6 faces, 8 vertices, 12 edges.

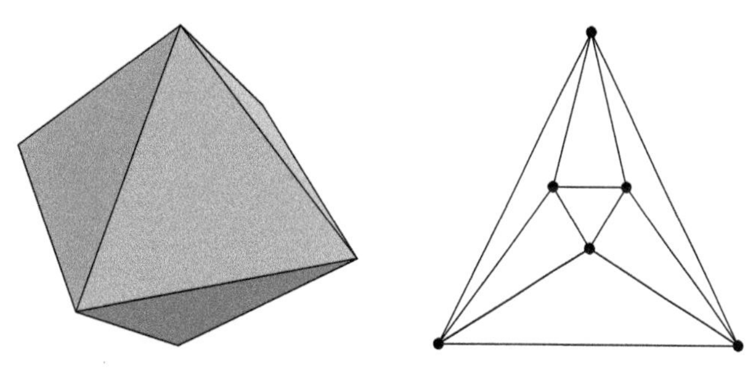

Figure 10.72 The Octahedron: 8 faces, 6 vertices, 12 edges.

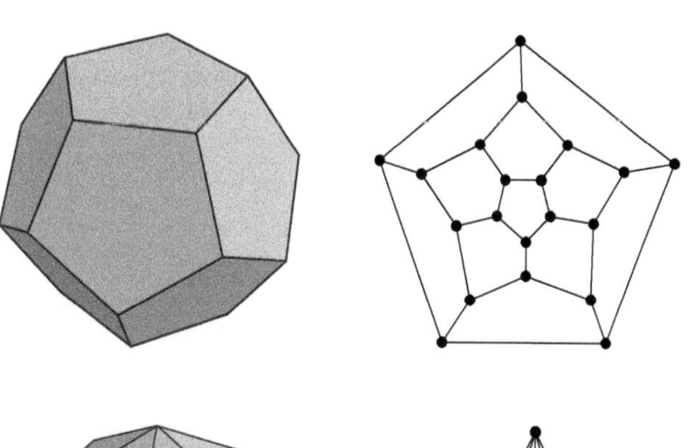

Figure 10.73 The Dodecahedron: 12 faces, 20 vertices, 30 edges.

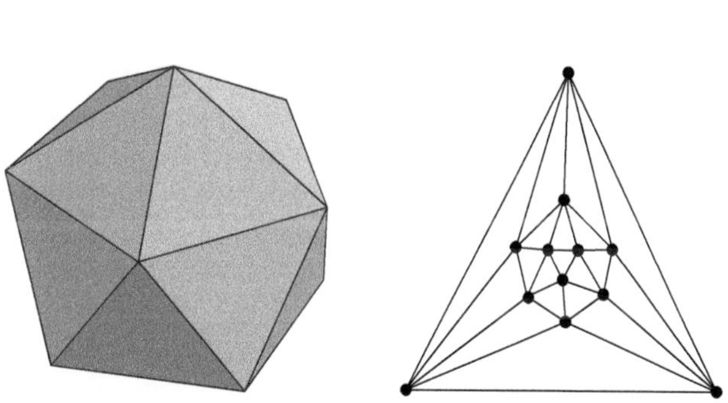

Figure 10.74 The Icosahedron: 20 faces, 12 vertices, 30 edges.

- Number of edges: ϵ.
- Number of vertices: n.
- Number of faces: ρ (faces = regions).
- Vertex degree: d.
- Number of edges bounding each face: b.

$\rho = \epsilon - n + 2$

Proof of Theorem 10.48: We can write n and ρ in terms of ϵ, d, and b.

Using the handshake theorem,

$$n = \frac{2\epsilon}{d}.$$

Also, since each face is bounded by b edges but each edge is then counted twice,

$$\epsilon = \frac{\rho \cdot b}{2} \quad \text{thus} \quad \rho = \frac{2\epsilon}{b}$$

Substituting into Euler's formula,

$$\frac{2\epsilon}{b} = \epsilon - \frac{2\epsilon}{d} + 2$$

dividing by 2ϵ, and rearranging the terms, we have

$$\frac{1}{b} + \frac{1}{d} = \frac{1}{2} + \frac{1}{\epsilon}$$

Notice that if b and d get too large, the left-hand side will be less than or equal to $\frac{1}{2}$, making the equality impossible. In fact, with $b = d = 4$, the equality fails.

It is not possible to form a polyhedron unless $b \geq 3$ and $d \geq 3$. Since they cannot both be 4 or greater, at least one of b and d must be 3. Since $\frac{1}{3} + \frac{1}{6} = \frac{1}{2}$, there are only six viable cases:

d	$b = 3$
3	Tetrahedron
4	Octahedron
5	Icosahedron

b	$d = 3$
3	Tetrahedron
4	Cube
5	Dodecahedron

The cases are easily checked, producing the table in the theorem statement. □

10.5.3 Chromatic Number

One of the most interesting problems in graph theory arose from a student's question. Around 1850, Francis Guthrie was coloring a map of the counties of England. He of course wanted to use different colors for adjacent counties. He began to wonder how many colors would be needed. It is not hard to create a map that requires four colors, but he could not design one that needed more than four colors. Francis asked his brother Frederick for help. Frederick was at that time a student of the famous mathematician Augustus De Morgan. In 1852 De Morgan sent a letter to his friend and fellow mathematician William Rowan Hamilton. Hamilton apparently was not interested. Finally, in 1878, Arthur Cayley announced the problem to the London Mathematical Society, indicating that he had failed to prove that only four colors were enough to color any map.

The next year, a lawyer and amateur mathematician named Alfred Kempe[22] published what he claimed was a proof that four colors sufficed. He was soon elected as a Fellow of the Royal Society. He had about a decade to revel in his fame. In 1890, Percy Heawood found an error in Kempe's "proof".

Heawood was able to modify Kempe's ideas to complete a valid proof that five colors were sufficient to color any map. However, no one was able to create a map that required more than four colors. The four-color conjecture became one of the most famous unsolved problems in mathematics.

Finally, in 1976, Kenneth Appel and Wolfgang Haken used over 1000 hours on an IBM 360 computer to investigate over 2000 configurations for a graph that would require five colors. These configurations represented the critical configuration for every

[22]The pronunciation of "Kempe" rhymes with "hemp".

potential case where five colors might be needed. The search determined that in each case, four colors were all that was needed. The four-color theorem was finally proved—at least in the minds of some mathematicians.

Appel and Haken's proof created some controversy within the mathematics community. Was a "proof" that relied on a computer program and that could not be carried out by hand by a human really a valid proof? Lots of discussion ensued.

What progress has been made? In 1996, Robertson, Sanders, Seymour, and Thomas reduced the number of cases that must be investigated to 633. However, their proof still requires a computer to complete the investigation. The majority of mathematicians believe the proof is valid. However, anyone who produces a proof that does not require a computer will become famous.

Modern discussions of the four-color theorem are usually presented as a problem about coloring the vertices of a graph. To see that this is really not a different problem, it is necessary to investigate the notion of a dual graph.

DEFINITION 10.49 ***The Dual Graph***

Let $G = (V, E, \phi)$ be a planar graph. Choose a particular planar embedding of G. The *dual graph*, G^* is the planar graph $G^* = (V^*, E^*, \phi^*)$ formed by creating a vertex in V^* for each region in the embedding of G (including the exterior region). Two vertices in V^* are adjacent if and only if they represent regions in the embedding of G that share a common bounding edge.

Notice that the dual graph is relative to the particular embedding chosen for G. Notice also that the dual graph itself may have multiple planar embeddings.

EXAMPLE 10.31

Some Dual Graphs

Figure 10.75 shows two different planar embeddings of the same graph.

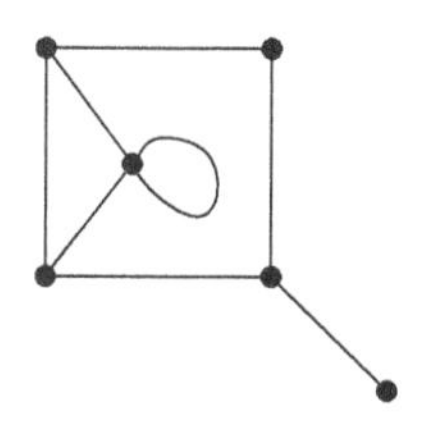

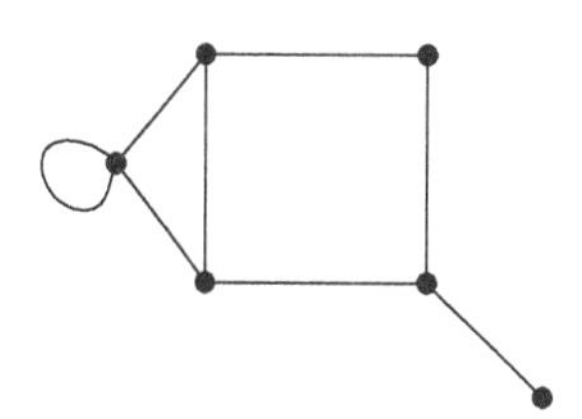

Figure 10.75 Two planar embeddings of the same graph.

Figure 10.76 shows the respective dual graphs (each using one of the many possible planar embeddings). Notice that the degree of a vertex in the dual is equal to the number of bounding edges for the corresponding region in the original embedding. The respective degree sequences are: 6, 6, 3, 1 and 8, 4, 3, 1. The dual graphs are *not* isomorphic.

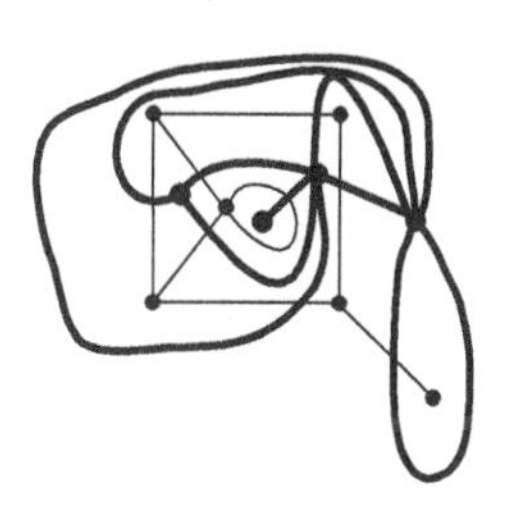

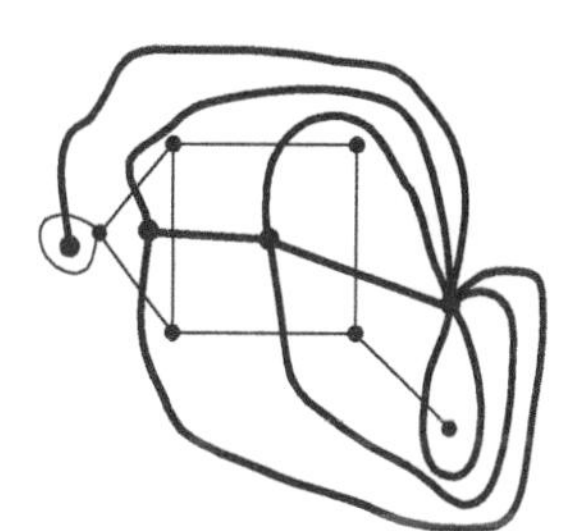

Figure 10.76 The dual graphs for Figure 10.75.

■

An interesting relationship exists between the five regular solids and their dual graphs. It can be shown (quite easily with pipe cleaner models) that the tetrahedron is its own dual, the cube and octahedron are duals of each other, and the dodecahedron and icosahedron are duals of each other.

✔ Quick Check 10.11

1. Consider the following map, M (with vertices emphasized).

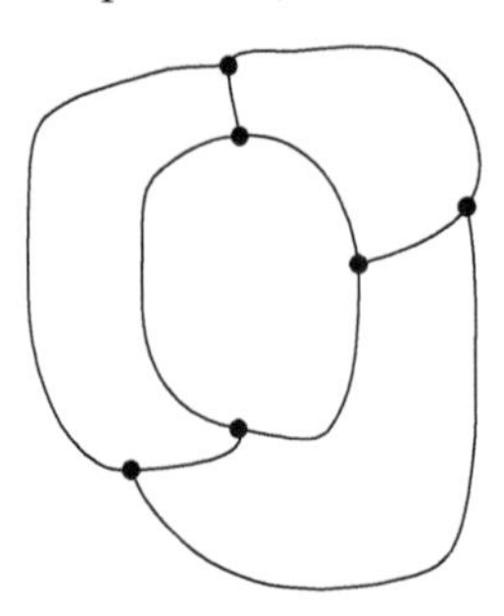

   (a) Draw a planar embedding of the dual graph of M.

   (b) Draw the map, M, using polygonal regions (no curved edges).

2. Show that the following graph is isomorphic to its dual graph:

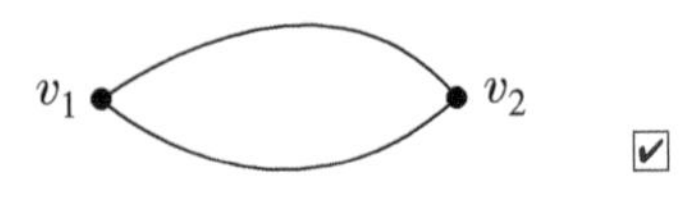

☑

A bit of reflection on the graphs in the previous example should indicate that if it is possible to color the regions in the original graphs with at most k colors so that no bordering regions have the same color, then those same colors can be used to color the vertices in the dual graphs so that no adjacent vertices have the same color (and vice versa). This observation provides the transition from the original problem about coloring regions of maps to the modern problem of coloring vertices.

DEFINITION 10.50 ***Chromatic Number***

A *coloring* of a graph $G = (V, E, \phi)$ is an assignment of a color to each vertex in V. The coloring is a *proper coloring* if the two vertices in every pair of distinct adjacent vertices have different colors.

The *chromatic number* of G is the minimum number of colors in any proper coloring. The chromatic number of G is denoted by $\chi(G)$, or just by χ if the choice of G is clear from the context.

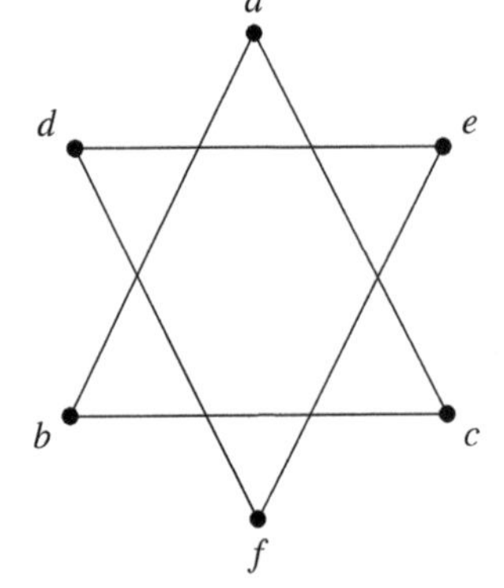

Figure 10.77. A graph with $\chi = 3$.

THEOREM 10.51 ***Chromatic Number of a Bipartite Graph***

The chromatic number of any connected bipartite graph with at least two vertices is 2.

Proof: Let $G = (V, E, \phi)$ be a bipartite graph with at least two vertices. Then we can choose subsets V_α and V_β of V with $V = V_\alpha \cup V_\beta$ and $V_\alpha \cap V_\beta = \emptyset$. In addition, every edge connects one vertex from V_α to one vertex from V_β.

We can clearly color every vertex in V_α with a single color and every vertex in V_β with a second color. That means that $\chi(G) \leq 2$.

However, E is not empty. That means that V_α and V_β must be colored with different colors. Thus, $\chi(G) = 2$. □

EXAMPLE 10.32 **Chromatic Number by Experiment**

The graph in Figure 10.77 has $\chi = 3$.

This is true because once a is assigned a color, a new color must be used for b. Then c must be assigned a color that is different from those assigned to both a and b. Thus $\chi \geq 3$. However, the same three colors can be used for d, e, and f. Thus, $\chi \leq 3$. Combining these two inequalities leads to the conclusion that $\chi = 3$. ■

Note: An actual assignment of colors to a graph will provide an upper bound for χ. If your assignment uses k colors, then you know that $\chi \leq k$. It may be possible, however, for someone else to find an assignment that uses even fewer colors. Consequently, an assignment of colors using k colors does not constitute a proof that $\chi = k$.

✔ Quick Check 10.12

1. What is the chromatic number of the following graph? Give evidence for your answer.

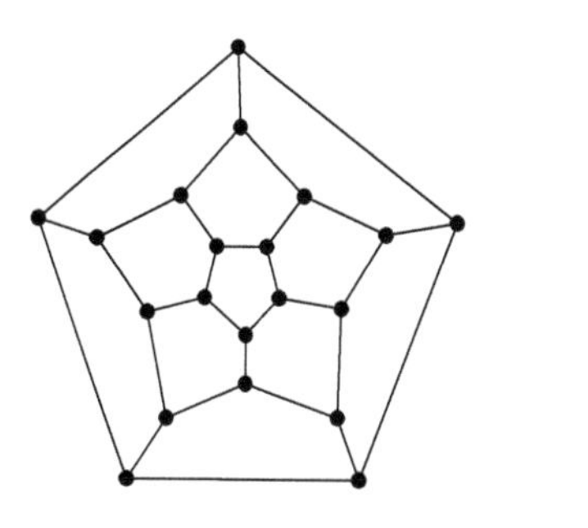

The big theorem is the four-color theorem. The proof has already been discussed.

THEOREM 10.52 ***The Four-Color Theorem***

The chromatic number of a planar graph is at most 4.

Warning! The graph K_5 shows that graphs can have a chromatic number greater than 4, as long as we do not insist on planarity.

The Five-Color Theorem

Although the proof of the four-color theorem is quite complex, a proof of the five-color theorem can be presented in just a few pages. The proof follows that found in [94] or [48] or [24].

THEOREM 10.53 ***The Five-Color Theorem***

The chromatic number of a planar graph is at most 5.

Before launching into the proof, it will be helpful to simplify the problem with some preliminary observations. Notice first that if a planar graph is disconnected, then each component can be colored independently, so it is only necessary to prove the theorem for connected graphs. Also, a graph can be properly colored with n colors if and only if its underlying simple graph can be colored with n colors. The proof of the five-color theorem can therefore be restricted to a proof about connected, planar, simple graphs.

The proof depends heavily on the following consequence of Euler's formula.

THEOREM 10.54 ***Connected, Planar, Simple Graphs Contain a Vertex with Small Degree***

Let $G = (V, E, \phi)$ be a connected, planar, simple graph. Then G has at least one vertex, w, with $\deg(w) \leq 5$.

Proof: Let G have ϵ edges and n vertices. If $n \leq 6$, there is nothing to prove because every vertex in the simple graph, G, has degree less than 6.

Assume that $n > 6$ and suppose, by way of contradiction, that $\deg(w) \geq 6$ for every vertex, w, in V. Then the handshake theorem implies that

$$2\epsilon = \sum_{w \in V} \deg(w) \geq 6n$$

Proposition 10.55 on page 648 implies that $\epsilon \leq 3n - 6$. Combining these results leads to

$$3n \leq \epsilon \leq 3n - 6$$

an obvious contradiction. The contradiction arose from assuming that $\deg(w) \geq 6$ for every vertex in V. Consequently, there is at least one vertex with $\deg(w) \leq 5$. □

Proof of Theorem 10.53: The proof will use complete induction.

Base Steps
If the graph has $n \le 5$ vertices, then $\chi \le 5$, since there cannot be more colors than vertices.

Inductive Step
Assume that every connected, planar, simple graph with n or fewer vertices has $\chi \le 5$. Let G be a connected, planar, simple graph with $n + 1$ vertices.

Theorem 10.54 (page 645) asserts the existence of a vertex, v, in G with $\deg(v) \le 5$. The graph, $G' = G - \{v\}$ (formed by removing v and its incident edges from G), has n vertices. The inductive hypothesis implies that every component of G can be properly colored with at most five colors. It will now be shown that this five-coloring of G' can always be transformed into a five-coloring of G. There are several cases to examine.

$\deg(v) < 5$

In the coloring of G', at most four colors have been used to color the vertices that are adjacent to v in G. There is thus at least one free color to use for v. Therefore, G can be five-colored.

$\deg(v) = 5$ and $\chi(G') < 5$

Since at most four colors have been used to color G', v can be colored with the unused fifth color, producing a five-coloring for G.

$\deg(v) = 5$ and $\chi(G') = 5$

Let $E_v = \{v_1, v_2, v_3, v_4, v_5\}$ be the set of five vertices in G that are adjacent to v. If they have been colored (in G') with fewer than five colors, then there is a color available for v and thus G can be five-colored. Otherwise, assume that v_i has the color, c_i, for $i = 1, 2, 3, 4, 5$ where $c_i \ne c_j$ if $i \ne j$. Assume also that the planar embedding has the vertices in E_v oriented about v in increasing subscript order.

Now consider v_1. Start a subgraph of G' that includes v_1 and keep adding vertices to the subgraph by adhering to the following rules:

- The new vertex must be adjacent to some vertex that is already in the subgraph (but need not be incident with the most recently added edge).
- The vertex must have either color c_1 or color c_3.

Quit when no additional vertices can be added to the subgraph. The subgraph that results will be called the c_1–c_3 chain anchored at v_1. It is possible for the chain to also end with v_1, producing a subgraph with only one vertex.

The c_1–c_3 chain is a maximal connected subgraph of G' that contains v_1 and only vertices with colors c_1 and c_3. Since G' has been five-colored, the vertices in any walk in the chain must alternate between the two colors.

There are two subcases to consider.

- **The c_1–c_3 chain anchored at v_1 does not include v_3.**

  In this case, the c_1–c_3 chain anchored at v_1 must end before it reaches v_3 (the left-hand side of Figure 10.78 shows such a chain for a specific graph). If the vertices in the c_1–c_3 chain starting at v_1 all swap colors, the subgraph will still be a c_1–c_3 chain (the right-hand side of Figure 10.78). However, v_1 will now be colored using c_3, the same color as v_3. (Note that the color swap has not caused any adjacent vertices in G' to have the same color. This is because every vertex not in the chain but adjacent to some vertex in the chain must have a color other than either c_1 or c_3—otherwise that vertex would have been a member of the chain.)

  Since no vertex in E_v is now colored with c_1, that color can be used for v, and G can be five-colored.

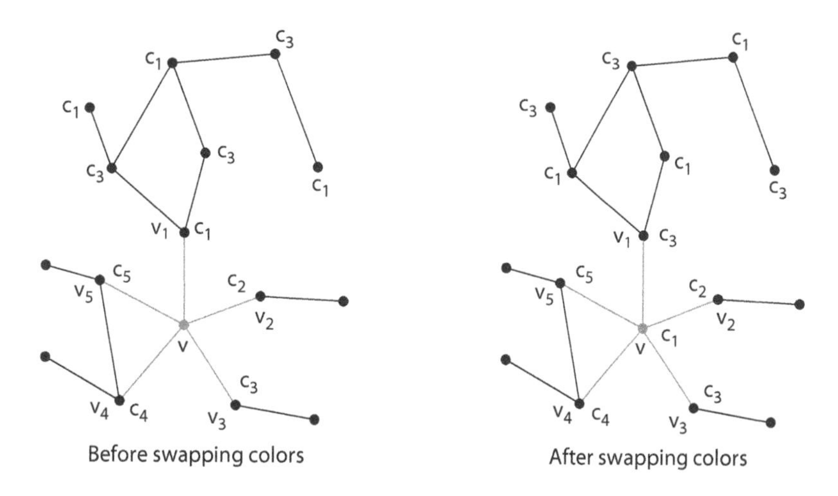

Figure 10.78 The c_1–c_3 chain anchored at v_1 before and after the color swap.

- **The c_1–c_3 chain anchored at v_1 includes v_3.**
  The c_1–c_3 chain anchored at v_1 must[23] contain a path, p, which includes v_1 and v_3 as it endpoints (Figure 10.77 shows such a path for a specific graph).

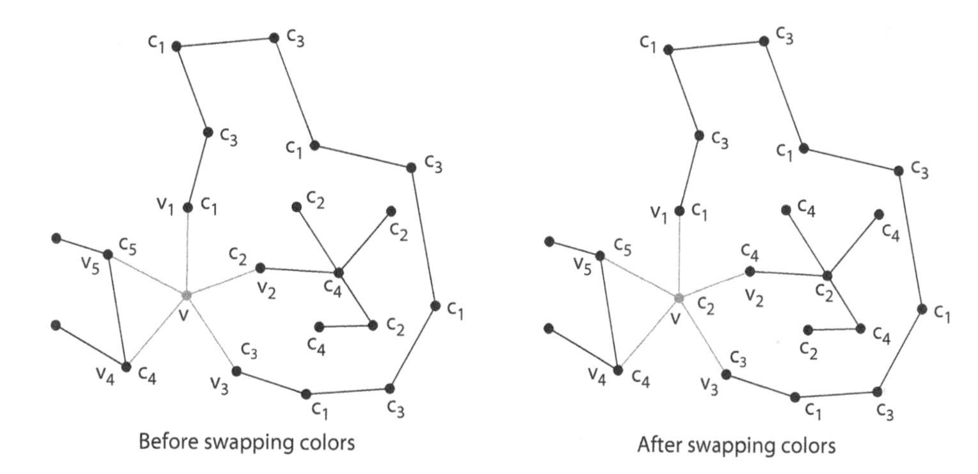

Figure 10.79 The c_2–c_4 chain anchored at v_2 before and after the color swap.

Now consider the c_2–c_4 chain anchored at v_2 in G' (defined in a manner similar to the earlier construction of the c_1–c_3 chain). That chain cannot include v_4 because it would need to include one of the vertices in the c_1–c_3 path, p, but each of those vertices have been colored with either c_1 or c_3. That is, the c_2–c_4 chain would need to cross[24] p if it were to include v_4 (since v is not part of G'). But crossing p requires the walk to include a vertex in p. That vertex would then need to have two distinct colors. Thus, the c_2–c_4 chain anchored at v_2 in G' does not include v_4.

It is therefore possible to swap the colors of all vertices in the c_2–c_4 chain anchored at v_2 (Figure 10.79). This will result in v_2 and v_4 both having color c_4. The color, c_2, can now be used to color v, producing a five-coloring of G.

Conclusion Every connected, planar, simple graph with fewer than six vertices can be five-colored. If every connected, planar, simple graph with n or fewer vertices can be five-colored, then so can every connected, planar, simple graph with $n+1$ vertices. The theorem of complete induction implies that every connected, planar, simple graph can be five-colored. This is the same as the assertion that $\chi(G) \leq 5$ for every connected, planar, simple graph. □

[23]Exercise 1 in Exercises 10.2.3 on page 615.

[24]This intuitively appealing assertion is actually not easy to prove. It is a consequence of the Jordan curve theorem. A proof of a restricted version of that theorem can be found in [100].

10.5.4 Exercises

The exercises marked with ★ have detailed solutions in Appendix H.

1. This exercise will walk you through a proof of the following proposition.

> **PROPOSITION 10.55** ***Edges in a Connected, Planar, Simple Graph***
>
> Let G be a connected, planar, simple graph with ϵ edges and $n \geq 3$ vertices. Then $\epsilon \leq 3n - 6$.

Define the degree of a region as follows:

> **DEFINITION 10.56** ***Degree of a Region***
>
> The *degree of a region*, r, in a planar graph is denoted $\deg(r)$ and is equal to the number of bounding edges. An edge terminating in a degree 1 vertex will be counted twice (once per side).

(a) What is the degree of the region inside a loop? What is the degree of a region between two edges that have the same pair of vertices at their ends (as in the following graph)?

(b) What is a lower bound on the degree of the regions in a connected, planar, simple graph with $n \geq 3$? (Be sure to include the outer region.)

(c) Prove that the sum of the degrees of the regions is twice the number of edges. (Recall the handshake theorem.)

(d) Combine parts (b) and (c) to show that if there are ρ regions, then $\rho \leq \frac{2\epsilon}{3}$.

(e) Use Euler's formula to complete the proof of the proposition.

2. Use Proposition 10.55 (from Exercise 1) to prove that K_5 is not a planar graph (thus proving Theorem 10.40).

3. Show that the Petersen graph (Example 10.25 on page 636) contains a subgraph that is homeomorphic to $K_{3,3}$ (and so is nonplanar). Click-and-drag graph software is recommended.

4. Does the left-hand graph in Example 10.28 on page 638 contain a subgraph that is homeomorphic to $K_{3,3}$? If so, show one. If not, give some justification for why it doesn't.

5. Prove:

(a) ★ Every proper subgraph of K_5 is planar.

(b) Every proper subgraph of $K_{3,3}$ is planar.

6. Is the following graph planar? Support your answer.

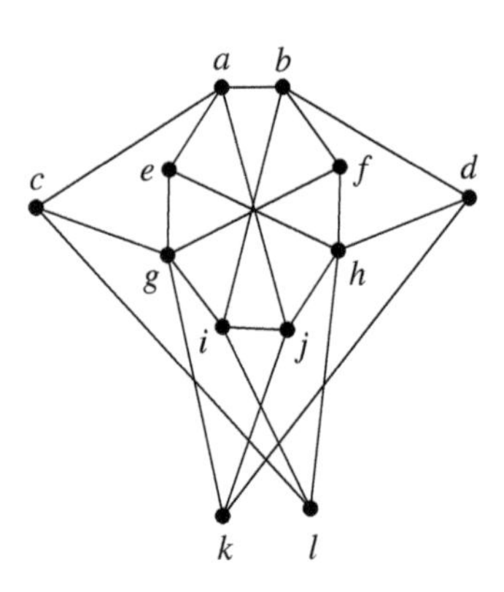

7. Is the following graph planar? Support your answer.

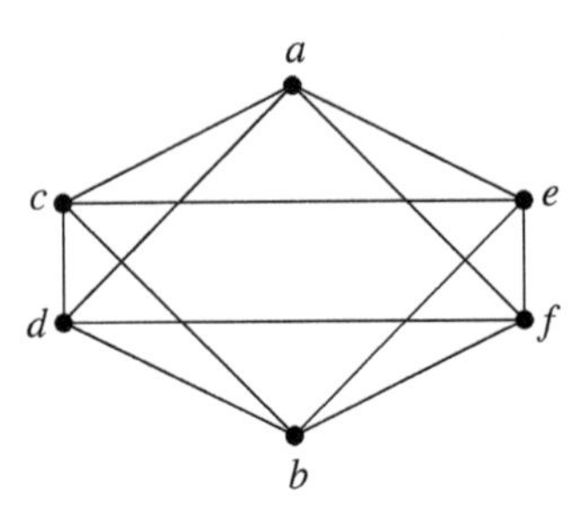

8. Is the following graph planar? Support your answer.

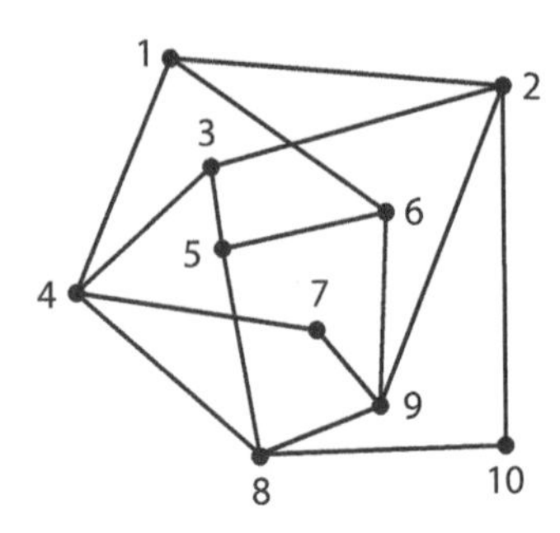

9. ★ The Euler formula does not work for this graph. Why not?

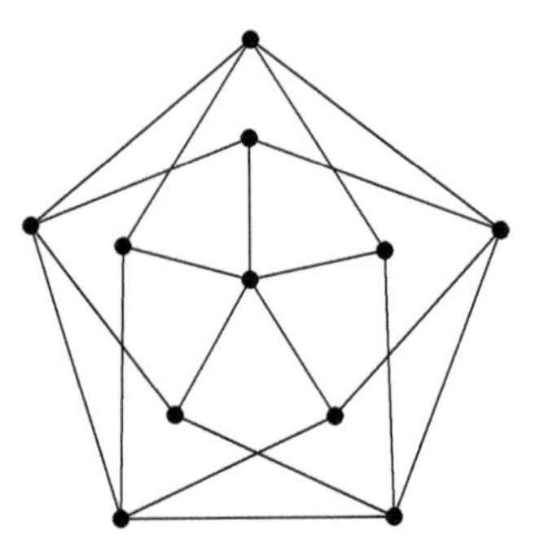

10. What is the chromatic number of each of the following graphs? Give sufficient supporting evidence for your answer.

(a) ★ The icosahedron graph.

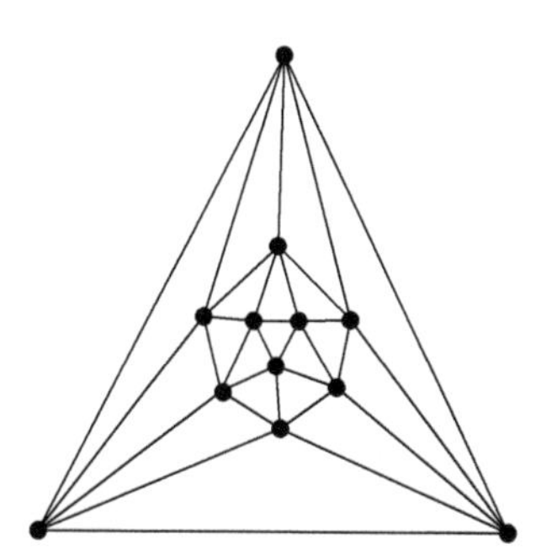

(b) The Heawood graph.

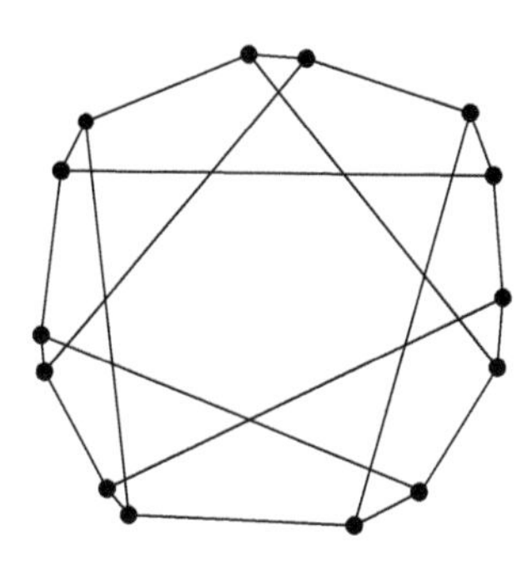

(c) The Walther graph.

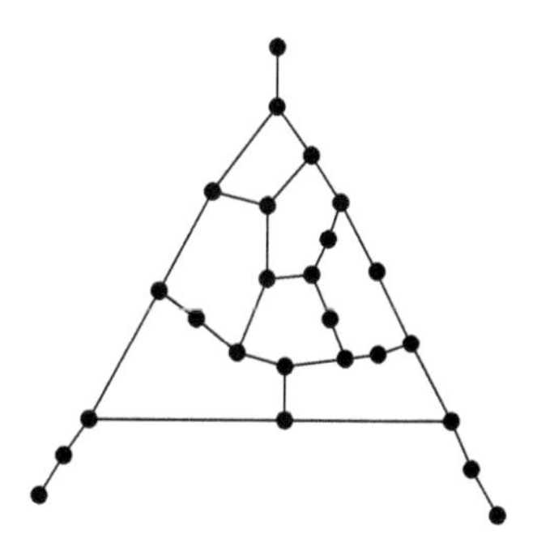

(d) The "alien spaceship" graph.

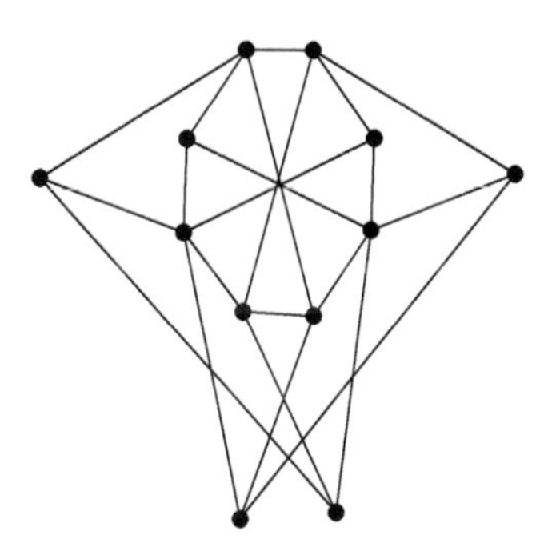

(e) The Grotztsch graph.

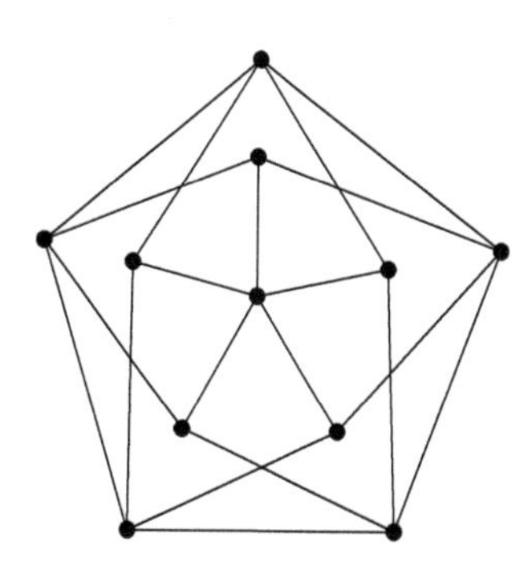

11. What is the chromatic number of the following graphs? Support your answer.

(a) The cycle graph, C_n.

(b) The wheel graph, W_n.

(c) The star graph, S_n.

12. The dean of a small college wants to increase enrollment in summer school. She has decided to offer classes during three time periods: morning, afternoon, and evening. Students may enroll in one or two courses. She has found 10 professors who are willing to offer summer courses. The dean has asked students to indicate the courses they are interested in so that she can decide which courses should be scheduled in the three time slots to maximize the number of students who will be able to enroll in two courses (should they desire to do so).

The dean used her survey to produce the following graph. Each vertex represents one of the 10 courses. There is an edge between courses if at least one student wants to enroll in both courses.

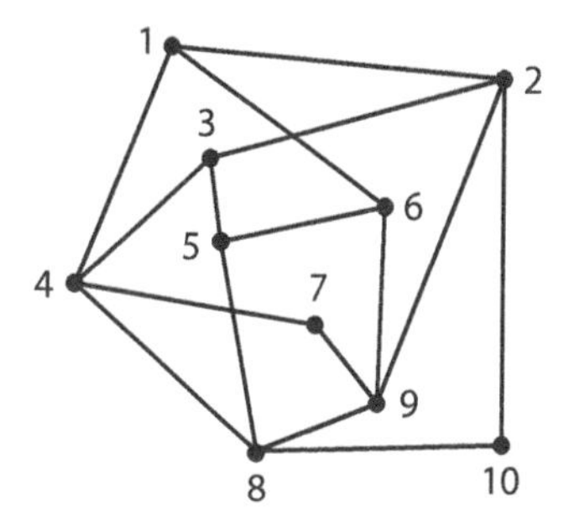

(a) Is it possible to schedule the courses so that there are no student time conflicts? If it is possible, show a set of time assignments; if it is not possible, explain why not.

(b) Is it possible to eliminate the morning session without causing any time conflicts? If it is possible, show a set of time assignments; if it is not possible, explain why not.

13. Find the dual of the following graphs. Show the superimposed dual, and then show a simplified version of the dual by itself.

(a) ★

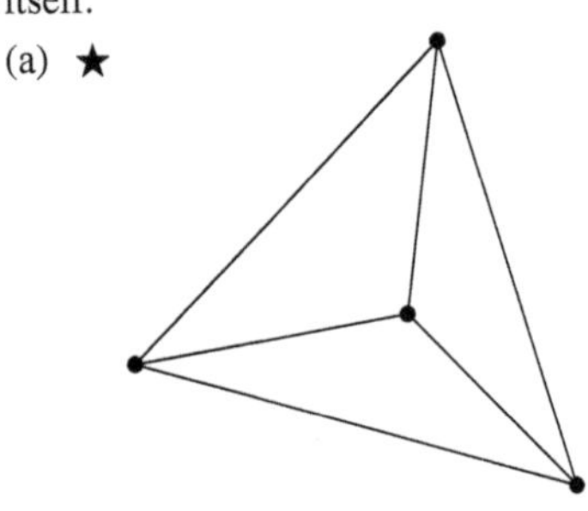

(b)

(c)

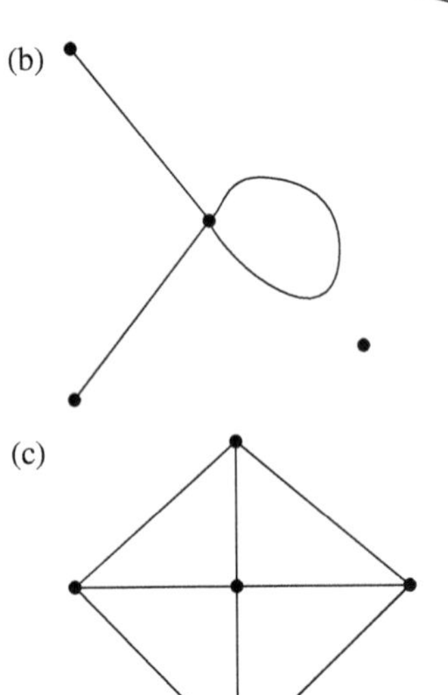

14. Prove that the dual of a planar graph is planar. [*Hint*: Suppose the graph has regions, r_1 and r_2, with common edge, e. The dual will have vertices, x_1 and x_2, corresponding to the regions. Instead of considering the edge in the dual that joins x_1 and x_2 through e, consider the "half-edges" that start at x_i and stop at a common point on e.]

15. (a) Is the dual of a connected planar graph connected?
(b) Is the dual of a disconnected planar graph disconnected?

16. Show that the graphs, G and H, in the following diagram are isomorphic but their dual graphs are not isomorphic. (This exercise is from [12].)

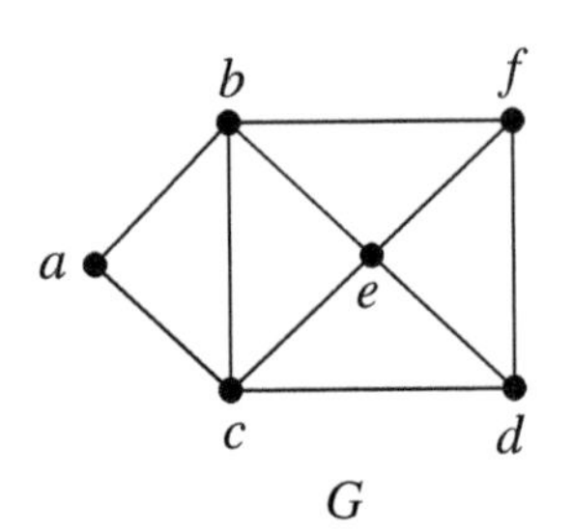

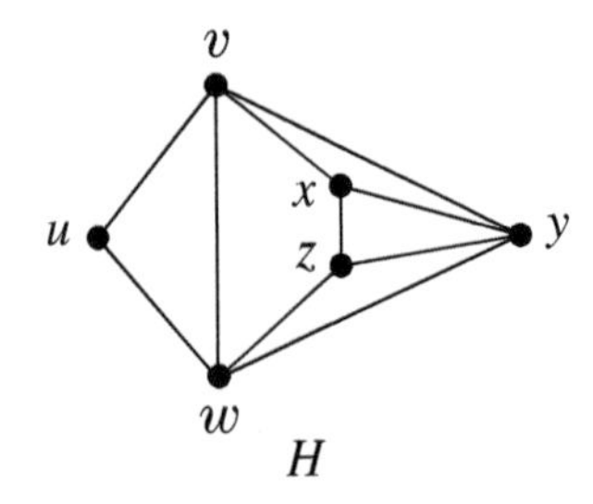

17. Find the dual of the graphs, G, depicted. Then find the dual of the dual. After completing both parts, make an observation about the differences that arise.

(a)

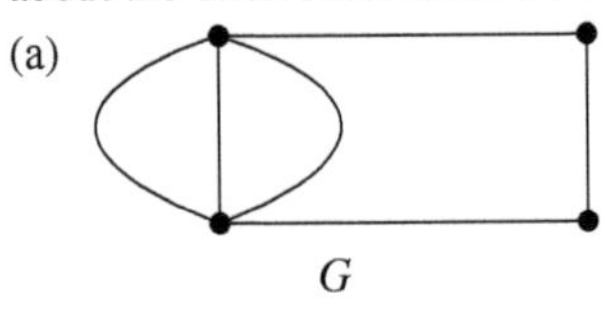

(b)

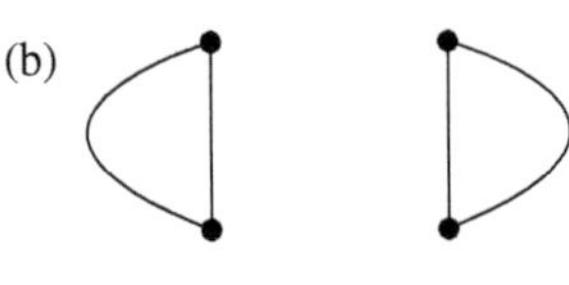

18. Each of the following statements is either true (always) or false (at least sometimes). Determine which option applies for each statement and provide adequate explanation for your choice.

(a) Just as was defined for graphs, a polyhedron is regular if every vertex has the same degree.
(b) A main result on planarity is that a graph is nonplanar if it does not contain a subgraph that is homeomorphic to K_5.
(c) ★ The chromatic number of a graph, G, is the number of colors in any proper coloring of G.
(d) If two graphs are homeomorphic, then they are isomorphic.

19. Let G be a connected, planar, simple graph with n vertices and ϵ edges. Prove: if every region has at least four bounding edges, then

$$\epsilon \leq 2n - 4$$

20. Let G be a connected, planar, simple graph with n vertices and ϵ edges. Prove: if every region is isomorphic to C_k, then

$$\epsilon = \frac{k(n-2)}{k-2}$$

21. Find the chromatic number of the following graph. Provide adequate justification for your answer. A pdf file with a larger version of this graph can be found at www.mathcs.bethel.edu/~gossett/DiscreteMathWithProof/ (or turn to page 672).

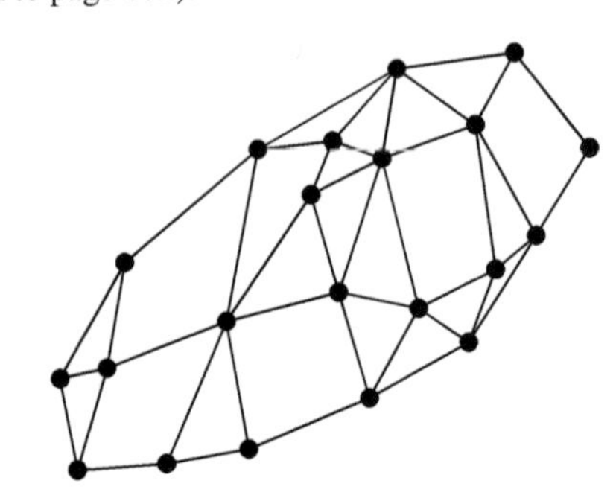

10.6 Directed Graphs and Weighted Graphs

Simple graphs were defined in the first section of this chapter. The more general notion of a multigraph allows loops and multiple edges. Two additional generalizations of "simple graph" are frequently encountered. The first variation permits edges to have a direction; the second permits edges to have numeric labels called weights.

When edges have a direction, the graph is called a *directed graph*; when the edges have weights, the graph is a *weighted graph*.

10.6.1 Directed Graphs

The primary definitions will follow the previous definitions fairly closely. The changes will capture the notion of a directed edge. In addition, it is common to call these directed edges arcs. You should compare the details of Definition 10.2 on page 601 to the details in the next definition.

DEFINITION 10.57 ***Simple Directed Graph***

A *simple directed graph* (also called a *simple digraph*), $D = (V, E, \phi)$, consists of a nonempty, finite set of *vertices*, V; a finite set of *arcs*, E; and a one-to-one incidence function, ϕ, that maps arcs to ordered pairs of distinct vertices.

If $\phi(e) = (v_i, v_f)$, the vertex v_i is called the *initial vertex* of arc e, and the vertex v_f is called the *terminal vertex* of e. The arc e may be informally referred to as $v_i v_f$.

EXAMPLE 10.33 **A Simple Digraph**

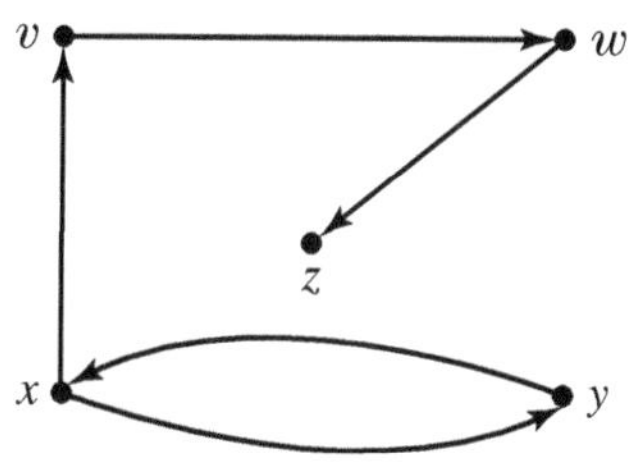

Figure 10.80 contains an example of a simple directed graph. Notice that the arcs are represented by arrows. The arrowhead is at the terminal vertex. The arcs xy and yx are distinct because the ordered pairs (x, y) and (y, x) are distinct. If this were an undirected graph, the (undirected, no arrowhead) edges xy and yx would be multiple edges and the graph would not be simple.

The arcs are like one-way streets. It is possible to traverse from vertex x to vertex z, but not possible to traverse from z to x in this example. ■

Figure 10.80. A simple digraph.

As in Definition 10.3 (page 601), directed graphs can have loops and multiple edges.

DEFINITION 10.58 ***Directed Graph***

A *directed graph*, $D = (V, E, \phi)$, consists of a nonempty, finite set of *vertices*, V; a finite set of *arcs*, E; and an incidence function, ϕ, that maps arcs to ordered pairs of vertices. If the two vertices in the ordered pair are the same vertex, the arc is called a *loop*.

Directed graphs are often called *digraphs*. A directed graph may be called a *directed multigraph* to emphasize the presence of loops and/or multiple arcs.

EXAMPLE 10.34 **A Digraph**

Figure 10.81 is an example of a directed graph. Notice the multiple arcs between vertices x and v and the loop at vertex z. ■

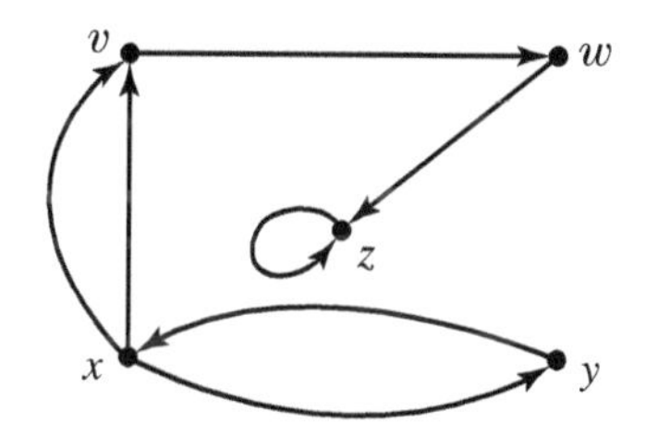

Figure 10.81. A digraph.

Suppose D is a directed graph. Form a graph, G, by replacing every arc in D by an undirected edge. The graph G is called the *underlying graph* for D. The underlying graph is what you would obtain by erasing all the arrowheads in a diagram of the directed graph.

The first theorem that was presented in this chapter was the handshake theorem (page 602). The theorem related the sum of the degrees of the vertices of a graph to the number of edges. The notion "degree of a vertex" needs to be modified somewhat in the context of digraphs.

DEFINITION 10.59 ***Indegree; Outdegree***

Let v be a vertex in a directed graph, D. The *indegree* of v, denoted $\deg^-(v)$, is the number of arcs whose terminal vertex is v. The *outdegree* of v, denoted $\deg^+(v)$, is the number of arcs whose initial vertex is v. The *degree* of v is $\deg^-(v) + \deg^+(v)$.

Notice that the definition of degree is identical to the degree of v in the underlying graph. Notice also that loops contribute to both the indegree and the outdegree of the same vertex.

The handshake theorem for directed graphs is quite straightforward and so will be left as an exercise. The next theorem is also quite simple.

THEOREM 10.60 ***Sums of Indegrees and Outdegrees***

The sum of the indegrees of all vertices in a directed graph is equal to the sum of the outdegrees of all vertices.

Proof: The sum in each case is the total number of arcs in the digraph. □

The Adjacency Matrix

The adjacency matrix of a graph (page 614) is always a symmetric matrix. That is, $a_{ij} = a_{ji}$, where a_{ij} is the number in row i and column j of the adjacency matrix, A. The adjacency matrix of a digraph need not be symmetric.

DEFINITION 10.61 ***Adjacency Matrix of a Directed Graph***

Let $D = (V, E, \phi)$ be a digraph with $V = \{v_1, v_2, \ldots, v_n\}$. The *adjacency matrix*, $A = (a_{ij})$, is an n-by-n matrix whose rows and columns are indexed by the elements of V. The matrix entry a_{ij} is the number of arcs in E with initial vertex v_i and terminal vertex v_j.

EXAMPLE 10.35

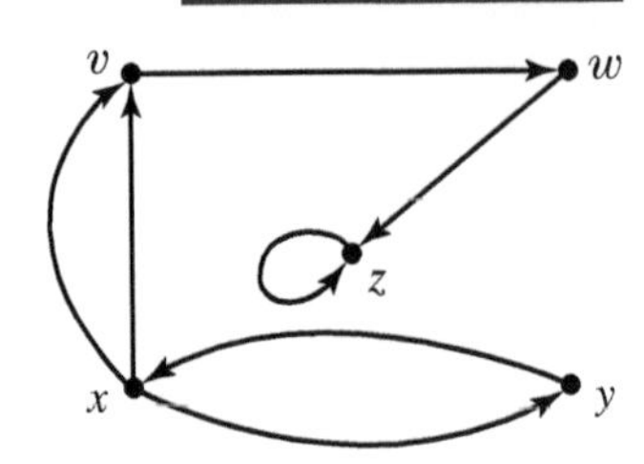

Figure 10.81. A digraph.

An Adjacency Matrix for a Digraph

Using the natural ordering for the vertices in Example 10.34 (Figure 10.81), the adjacency matrix is shown in Table 10.5.

TABLE 10.5 The Adjacency Matrix for the Digraph in Example 10.34

	v	w	x	y	z
v	0	1	0	0	0
w	0	0	0	0	1
x	2	0	0	1	0
y	0	0	1	0	0
z	0	0	0	0	1

Notice that each row sum is the outdegree for the corresponding vertex, and each column sum is the indegree.

The adjacency matrix in this example is not symmetric. One example of nonsymmetry is found by comparing the number in row 1 column 2 with the number in row 2 column 1. ■

Theorem 10.22 used the adjacency matrix of a graph to count the number of walks of length k joining pairs of vertices. A similar theorem is available for digraphs. However, the notion of connectivity needs to be refined a bit first.

Directed Walks and Connectivity

The notion of a walk in a directed graph should respect the directedness of the arcs. This requires modified definitions of *walk* and *circuit*.

DEFINITION 10.62 ***Directed Walk; Directed Circuit***

Let $D = (V, E, \phi)$ be a directed graph.

The Basic Concept

A *directed walk* of length k is a sequence of alternating vertices and arcs, $v_0e_1v_1e_2v_2 \cdots e_kv_k$, such that $\phi(e_i) = (v_{i-1}, v_i)$, for $i = 1, 2, \ldots, k$. The vertices, v_0 and v_k, are called the *endpoints* of the walk.

Excluding Repeated Arcs and Vertices

- A *directed trail* is a directed walk with no repeated arcs.
- A *directed path* is a directed walk with no repeated vertices.

Starting and Ending at the Same Vertex

- A directed walk is *closed* if it has length at least one and its endpoints are the same vertex.
- A *directed circuit* is a closed directed trail.
- A *directed cycle* is a directed circuit in which the endpoints are the only repeated vertices.

Alternative Notation for Walks in Simple Digraphs

If G is a simple digraph, or a simple digraph with loops, then the arc joining two vertices is uniquely determined. Consequently, the directed walk $v_0e_1v_1e_2v_2 \cdots e_kv_k$ can be unambiguously denoted by $v_0v_1v_2 \cdots v_k$.

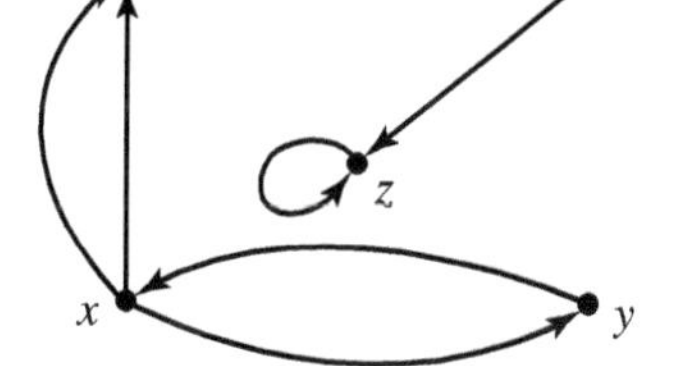

Figure 10.81. A digraph.

The digraph in Example 10.34 on page 651 (repeated in the margin) appears (visually) to be connected. However, Definition 10.15 (page 611) requires a walk from any vertex to any other vertex in order for the graph to be connected. That is not true for this digraph (when *walk* is replaced by *directed walk*). This suggests that there are two levels of "connectivity" in digraphs for which the underlying undirected graph is connected.

DEFINITION 10.63 ***Weakly Connected***

A digraph is *weakly connected* if the underlying graph is connected.

DEFINITION 10.64 ***Strongly Connected***

A digraph, $D = (V, E, \phi)$, is *strongly connected*if there is a directed walk from every vertex $w \in V$ to every vertex $z \in (V - \{w\})$.

The graph in Example 10.34 is weakly connected but not strongly connected.

✔ Quick Check 10.13

1. What is the minimum number of additional arcs needed to make the digraph in Figure 10.81 strongly connected?
2. Let G be a digraph that is not just "a single vertex with no arcs". Devise a very simple test that uses the adjacency matrix of G to determine that G has an isolated vertex (and hence is not even weakly connected). ☑

It is now possible to state and prove the digraph version of Theorem 10.22.

THEOREM 10.65 ***Number of Directed Walks in a Weakly Connected Digraph***

Let $D = (V, E, \phi)$ be a digraph with adjacency matrix A. Let $V = \{v_1, v_2, \ldots, v_n\}$ and let A be indexed using the natural ordering of these vertices. Then the ijth entry of A^k, for $k \geq 0$, is the number of distinct directed walks of length k from v_i to v_j.

Before reading this proof, you should look at the proof of the undirected version (page 614). Try to write the new proof yourself; then compare your proof to this proof.

Proof:

The proof will be done using mathematical induction.

Base Step

The matrix $A^0 = I$ is the identity matrix. Every entry is zero except the entries a_{ii} along the main diagonal. Each of these entries is 1. The interpretation is that every vertex has 1 directed walk of length 0 connecting it to itself, and no directed walks of length zero to any other vertex.

Notice also that $A^1 = A$ does correctly list the number of directed walks of length 1 from vertex v_i to vertex v_j.

Inductive Step

Assume that the entries of A^k correctly count the number of directed walks of length k between the vertices of D. That is, the ijth entry correctly counts the number of directed walks from v_i to v_j.[25]

We need to examine the ijth entry of A^{k+1}. Some new notation is needed first. Temporarily denote the ijth entry of A^k as b_{ij} and the ijth entry of A^{k+1} as c_{ij}. Since $A^{k+1} = AA^k$, using the definition of matrix multiplication (see Appendix E), we can write[26] c_{ij} as

$$c_{ij} = a_{i1}b_{1j} + a_{i2}b_{2j} + \cdots + a_{in}b_{nj}$$

Now think about how a directed walk from v_i to v_j must progress. It must start at v_i and move to an adjacent vertex, v_m, that is the terminal vertex of a common arc. There are a_{im} arcs along which that may occur. Once the adjacent vertex v_m has been reached, the walk can proceed to v_j in any of b_{mj} ways (since the inductive hypothesis asserts that there are that many directed walks of length k between v_m and v_j). Since these choices are independent, the number of directed walks from v_i to v_j that first pass through v_m is $a_{im}b_{mj}$.

Finally, since moving from v_i to v_m excludes the possibility of moving from v_i to v_s as the second vertex in the directed walk, the choices for second vertex are mutually exclusive. Thus it is proper to add the number of ways to reach v_j by passing through each of the vertices as second in the directed walk.

This is exactly the form of the entry c_{ij} in A^{k+1}. Thus, A^{k+1} properly counts the number of directed walks of length $k + 1$ between any two vertices in D. □

COROLLARY 10.66 ***Test for Strong Connectedness***

Let the digraph D have the n by n adjacency matrix A. Then D is strongly connected if and only if $\sum_{i=0}^{n-1} A^i$ has only nonzero entries.

---

[25] Notice that the ji^{th} entry does *not*, in general, count the number of directed walks from v_i to v_j.

[26] It is conceptually important to write the product as AA^k and not as A^kA, since neither A nor A^k is symmetric and matrix multiplication is not commutative. However, the associativity of matrix multiplication guarantees that the two products will be identical in this setting.

✔ Quick Check 10.14

1. The following graph is strongly connected. Calculate the number of directed walks of length 3 between vertices z and y. List all of them.

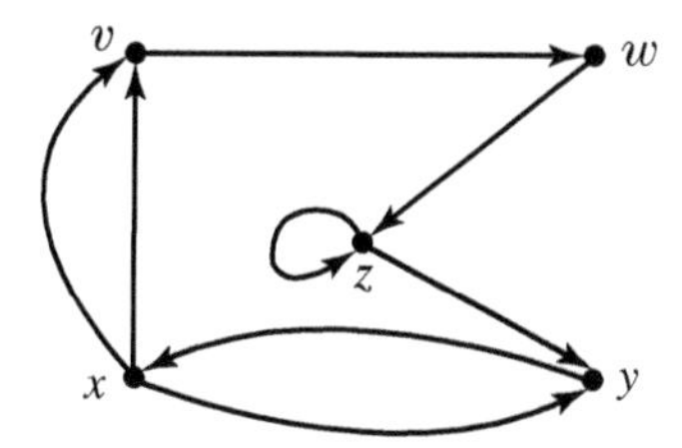

2. Use Corollary 10.66 to show that the following graph (from Example 10.34) is not strongly connected.

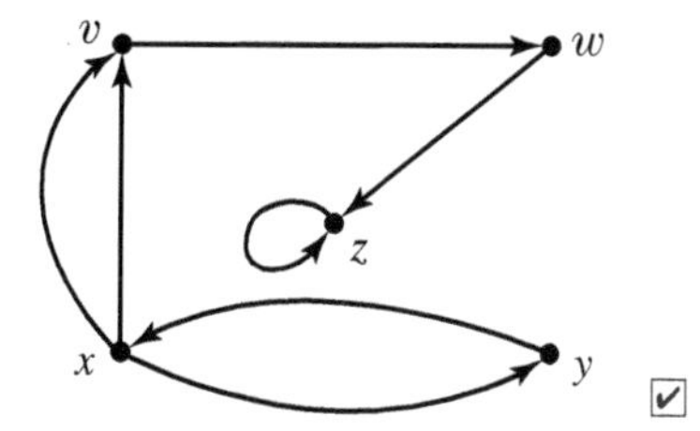

☑

10.6.2 Weighted Graphs and Shortest Paths

A *weighted graph* is a graph whose edges contain numeric labels. The numeric labels typically represent costs, benefits, capacities, distances, and so on. For example, the graph may represent a network of highways and the edge labels may represent distances between cities (represented by vertices).

Models of this nature are often used to represent the key data in an optimization problem. Thus, in a complex network of highways, we might be interested in finding the shortest route between two of the cities. If the weighted graph represents capacities of a natural gas pipeline, we might be interested in determining the combined pathways that will allow the maximum volume of gas to be moved per minute.

One of these problems will be addressed next.

Dijkstra's Shortest Path Algorithm

Suppose that $G = (V, E, \phi)$ is a weighted simple graph containing vertices a and z.[27] If the edge weights in G represent distances, what is the shortest path from a to z?

The following algorithm is a repackaging of an algorithm developed by Edsger Dijkstra and is known as *Dijkstra's shortest path algorithm.*

Some notation is needed. If x and y are vertices in G, define the *edge weight function*, w, by

$$w(x, y) = \begin{cases} \text{the nonnegative weight on edge } xy & \text{if edge } xy \text{ exists} \\ \infty & \text{otherwise} \end{cases}$$

The algorithm creates a subset, B, of vertices that starts with only vertex a and adds additional vertices in order of their distance from a. A vertex, v, is not added to B until the shortest distance from a to v is known. Eventually, either B will include z, and the distance from a to z will have been successfully and efficiently calculated, or else a point will be reached where no additional vertices can be added to B (indicating that z is in a different component from a).

The version of the algorithm presented next will provide a constructive mechanism to produce the shortest path, assuming that a and z are in the same component of G.

The key idea in the algorithm is to find the vertex that has the shortest distance to a among all vertices that are adjacent to B (but not in B).[28] This is accomplished by assigning every vertex a tentative distance from a, and then updating these tentative

---

[27] The discussion that follows is also valid for directed graphs, if *edge* is replaced by *arc* and *path* is replaced by *directed path*.

[28] Think of B as "the blob" (the main "character" in a 1950s science fiction movie). The blob is a large, mobile pile of jello that wishes to subsume everything around it. The blob constantly searches its perimeter looking for the tastiest new item to eat. Being lazy, it only subsumes items that are already next to it. In this way, it may expand in seemingly random directions.

distances each time a new vertex is added to B. The tentative distance from a to v at *iteration* n will be denoted $d_n(v)$. The actual distance from a to v is denoted by $d(a, v)$.

The algorithm is shown in Table 10.6. As is usual, the line numbers are not part of the algorithm but are provided to make discussion easier.

TABLE 10.6 Djikstra's Shortest Path Algorithm for Connected Graphs

```
 1: {d(a,z), shortest path} shortestPath(weighted connected simple graph G, vertex a, vertex z)

 2:    # Initialization

 3:    B = {a}
 4:    n = 0     # initial iteration
 5:    r = a     # the most recent vertex added to B
 6:    d(a,a) = 0   # the distance from a to a is known
 7:    for each vertex v in G - {a}
 8:      d0(v) = ∞

 9:    # Start the main loop

10:    while z ∉ B
11:       n = n + 1
12:       A becomes the set of vertices in V - B which are adjacent to r
13:       for each vertex, u in A              # a shorter estimate may be possible
14:          dn(u) = min{dn-1(u), d(a,r) + w(r,u)}
15:          if dn(u) ≠ dn-1(u)
16:             p(u) = r            # u is currently best reached by passing through r
17:       for each vertex, v ∈ (V - A)     # no change in the estimate
18:          dn(v) = dn-1(v)
19:       x = a vertex in V-B with minimum value for dn(u) among vertices u ∈ V-B
20:       d(a,x) = dn(x)     # the true distance from a to x is now known
21:       add x to B
22:       r = x                          # x becomes the most recently added vertex

23:    # z has been reached, now construct the path

24:    P = an ordered list with z as its only element   # start building the path
25:    r = z                              # the most recently added vertex
26:    while r ≠ a
27:       x = p(r)              # r can be reached by passing through x
28:       prepend x to P        # add next vertex to the front of P
29:       r = x

30:    return {d(a,z), P}
31: end shortestPath
```

At line 14 of the algorithm, the vertex most recently added to B is r. It may now be possible to re-evaluate the tentative distances from a to all vertices, u, that are not in B but are adjacent to r. We need only compare the current estimate, $d_{n-1}(u)$, to the newly possible path created by starting with the known shortest path from a to r and then following an edge from r to u (Figure 10.82).

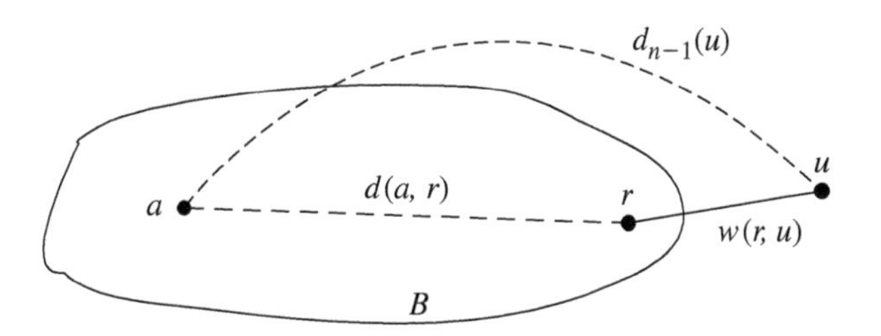

Figure 10.82 Re-evaluating the tentative distances.

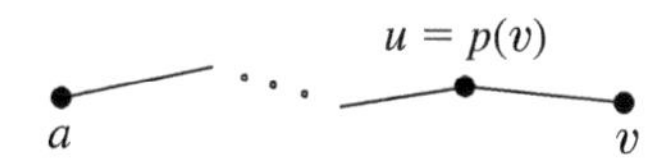

Figure 10.83. $p(v)$.

The function $p(v)$ will record, for each vertex, v with $v \neq a$, the vertex, u, which is adjacent to v and is in a shortest path from a to v, as depicted in Figure 10.83.

The value of $p(v)$ will remain undefined until vertex v becomes adjacent to some vertex in B.

Ties at line 19 can be broken in an arbitrary manner.

It will be useful to introduce a tabular format that will aid in using the algorithm by hand. The table indicates, for each iteration, the values of B, r, A, d_n, and p. It will typically look like Table 10.7. The vertices need not be in any particular order. A simple example will provide an overview before a more detailed example is given.

EXAMPLE 10.36 A Small Example of Dijkstra's Algorithm

Find the shortest path from vertex a to vertex e in the weighted graph in Figure 10.84. Trace through the algorithm while referring to Table 10.7.

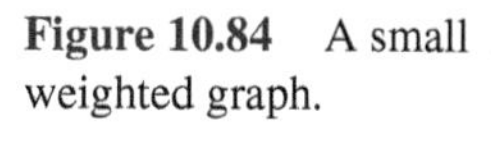

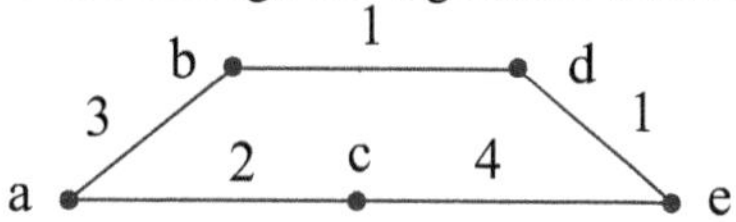

Figure 10.84 A small weighted graph.

TABLE 10.7 Shortest Path in a Small Weighted Graph

				d_n					p				
n	B	r	A	a	b	c	d	e	a	b	c	d	e
0	$\emptyset$			[0]	∞	∞	∞	∞					
1	$\{a\}$	a	$\{b, c\}$		3	[2]	∞	∞		a	a		
2	$\{a, c\}$	c	$\{e\}$		[3]		∞	6					c
3	$\{a, b, c\}$	b	$\{d\}$				[4]	6				b	
4	$\{a, b, c, d\}$	d	$\{e\}$					[5]					d
5	$\{a, b, c, d, e\}$												

Once a vertex is added to B, the distance is known, so that column in the table can be frozen (indicated by a box around the true distance). Entries in a column need only be updated when the value is recalculated or when the vertex in that column is chosen for the next r.

In the first iteration, adding a to B allows b and c to be reached, so their tentative distances are updated. Since c has the smaller tentative distance to a, it is added to B and the tentative distance becomes the actual distance. Also, a is recorded as the current best vertex from which to move to b or c.

In the second iteration, the tentative distance to b is unchanged, but e is now reachable. However, the tentative distance from a to b is shorter, so b is added to B. Notice that c is tentatively recorded as the best vertex from which to approach e, but is not recorded for b since the tentative distance to b did not change.

Once e has been included into B, the shortest path can be constructed. Start in the e column of p and look for the lowest entry. That entry, d, is prepended onto e, producing de. Now look at the lowest entry in the d column. Prepending gives bde. From the b column we find the final entry in the shortest path. That path is $abde$. ■

EXAMPLE 10.37 **Using Dijkstra's Algorithm**

Find the shortest path from vertex a to vertex z in the weighted graph displayed in Figure 10.85.

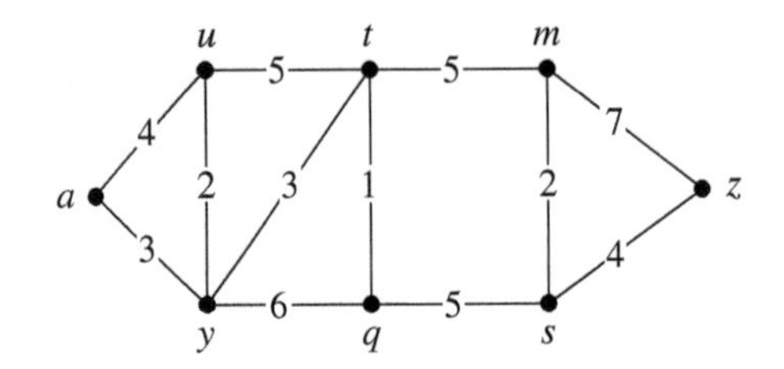

Figure 10.85 Using Dijkstra's algorithm.

TABLE 10.8 Dijkstra's Algorithm for Example 10.37

				d								p							
n	B	r	A	a	q	y	s	t	m	u	z	a	q	y	s	t	m	u	z
0				$\boxed{0}$	∞	∞	∞	∞	∞	∞	∞								
1	$\{a\}$	a	$\{u, y\}$			$\boxed{3}$				4				a				a	
2	$\{a, y\}$	y	$\{q, u, t\}$		9			6		$\boxed{4}$			y			y			
3	$\{a, u, y\}$	u	$\{t\}$					$\boxed{6}$											
4	$\{a, t, u, y\}$	t	$\{q, m\}$		$\boxed{7}$				11				t				t		
5	$\{a, q, t, u, y\}$	q	$\{s\}$				12		$\boxed{11}$						q				
6	$\{a, m, q, t, u, y\}$	m	$\{s, z\}$				$\boxed{12}$				18								m
7	$\{a, m, q, s, t, u, y\}$	s	$\{z\}$								$\boxed{16}$								s

Table 10.8 organizes the algorithm steps.[29] Once a vertex is added to B, the distance is known, so that column in the table can be frozen (indicated by a box around the true distance). Entries in a column need only be updated when the value is recalculated or when the vertex in that column is chosen for the next r. An example of updating when the value has not been recalculated would be step 5 in Table 10.8. Vertex m has been chosen as the next r but has not been updated. Nevertheless, its value (11) has been copied again to highlight its choice (with a box).

At the start of the first iteration ($n = 1$), $B = \{a\}$ and $r = a$. Thus, $A = \{u, y\}$. At line 14 of Dijkstra's algorithm (Table 10.6), the old distance estimates for u and y (∞ in both cases) are replaced by the newer estimates (4 and 3, respectively). This is because

$$\min\{d_0(u), d(a, a) + w(a, u)\} = \min\{\infty, 0 + 4\} = 4$$

and

$$\min\{d_0(y), d(a, a) + w(a, y)\} = \min\{\infty, 0 + 3\} = 3$$

These weights have been updated, and the vertex in A with the smaller distance (y) has been chosen as the new vertex to add to B (temporarily denoted x at line 19 of the algorithm). The values for $p(y)$ and $p(u)$ can then be updated (line 16). This is reflected in the y and u columns in the p region of the table. The vertex y is then promoted to most-recently-added status and listed in the r column for the next iteration ($n = 2$).

[29]Go to the Chapter 10 section of http:// www.mathcs.bethel.edu/ gossett/DiscreteMathWithProof/ to find a pdf file with Tables 10.6 (Dijkstra's Algorithm) and 10.8 on a single page.

When $n = 2$, the set $A = \{q, u, t\}$. Notice that it is now possible to move from a to u by more than one path. The distance for the new path, ayu, will be compared to the previous distance estimate for the shortest path from a to u. That comparison is in line 14: $\min\{d_1(u), d(a, y) + w(y, u)\} = \min\{4, 3 + 2\} = 4$. The old estimate is still smaller, so it is kept. Similar calculations are completed for q and t:

$$d_2(q) = \min\{d_1(q), d(a, y) + w(y, q)\} = \min\{\infty, 3 + 6\} = 9$$

and

$$d_2(t) = \min\{d_1(t), d(a, y) + w(y, t)\} = \min\{\infty, 3 + 3\} = 6$$

These values are reflected in the $n = 2$ row of the d region in the table. The shortest distance among all vertices in $V - B$ is 4, associated with the vertex u, so u is added to B and becomes the new r. The values for $p(q)$ and $p(t)$ can then be updated (line 16), but since $d_2(u) = d_1(u)$, the value of $p(u)$ does not change.

When $n = 3$, the set of finished vertices is $B = \{a, u, y\}$ and $r = u$ is the most recently added vertex. The set of vertices whose estimated distance might potentially change due to adding $r = u$ to B is $A = \{t\}$. The updated distance estimate for t is $d_3(t) = \min\{d_2(t), d(a, u) + w(u, t)\} = \min\{6, 4 + 5\} = 6$. The estimate has not changed, so $p(t)$ also does not change. The candidates for the new vertex to add to B are q and t, so t is chosen (line 19). Although the distance estimate for q has not been updated, since line 18 updates this value it is still permissible to copy that number into the $n = 3$ row of the table, as was done in Example 10.36.

During the $n = 4$ iteration, the estimated distances for q and m change, and q is added to B. When $n = 5$, the estimated distance for s changes and m is added to B (the estimated distance for m did not change, but it is convenient to copy that value to the $n = 5$ row so that the box can be added).

During the $n = 6$ iteration, the vertices that need updates are s and z. The calculations are $d_6(s) = \min\{d_5(s), d(a, m) + w(m, s)\} = \{12, 11 + 2\} = 12$ and $d_6(z) = \min\{d_5(z), d(a, m) + w(m, z)\} = \{\infty, 11 + 7\} = 18$. The vertex s is chosen to add to B and $p(z)$ is updated.

The final iteration ($n = 7$) has $r = s$ and $A = \{z\}$. The estimated distance to z is updated: $d_7(z) = \min\{d_6(z), d(a, s) + w(s, z\} = \min\{18, 12 + 4\} = 16$. Since $d_7(z) \neq d_6(z)$, the value of $p(z)$ is updated, and z is chosen (as the vertex with smallest estimated distance) to add to B.

The length of the shortest path is now known: $d(a, z) = 16$. It only remains to construct the path. Line 24 initializes $P = z$. It then prepends $p(z) = s$ (the final value in the z column) to P, producing $P = sz$. Next, $p(s) = q$ and then $p(q) = t$ are prepended, so that $P = tqsz$. Then $p(t) = y$ and $p(y) = a$ are prepended and the shortest path is complete: $P = aytqsz$. ■

✔ Quick Check 10.15

1. Use Dijkstra's algorithm to find a shortest path between vertices a and f for the following graph. List both the shortest distance and the shortest path.

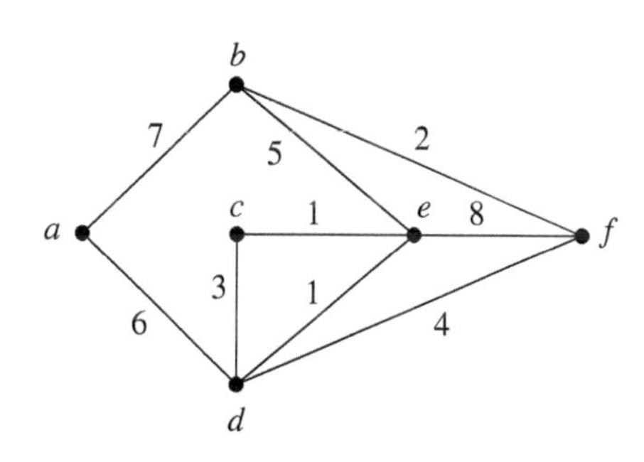

☑

THEOREM 10.67 ***Dijkstra's Algorithm is Correct***

If G is a weighted connected simple graph with nonnegative weights, then Dijsktra's algorithm correctly calculates $d(a, z)$ for any pair of vertices, a and z, in G.

Proof: The proof is by mathematical induction. The inductive hypothesis is fairly complex. The reason for the use of $B - \{r\}$ in the first inductive assertion will become clear during the base and inductive steps.

Inductive hypothesis At the end of iteration n,

- $\forall u \in (V - B)$, $d_n(u)$ is the shortest distance from a to u using paths of the form: $av_2v_3v_4 \cdots v_k u$ for which $v_i \in (B - \{r\})$ for $i = 1, 2, \ldots, k$.
- If $v \in B$ and $u \notin B$, then $d(a, v) \leq d(a, u)$.
- $\forall v \in B, d(a, v) = d_n(v)$.

The first assertion essentially claims that among all vertices, u, that are immediate neighbors of B, the estimate $d_n(u)$ represents the length of a shortest path from a to u that stays inside the previous B until just before reaching u. The second assertion states that B grows by adding vertices in the order of their distance from a. Vertices that are closer to a are added earlier. The final assertion claims that once a vertex is added to B, the distance of the shortest path has been calculated correctly.

The reason for such an elaborate inductive hypothesis is to make the inductive step easier. The third inductive assertion is the one that leads to the desired conclusion. The other two are necessary facts that lead to that conclusion. Rather than attempting some convoluted proof of the third assertion without the aid of these extra hypotheses, it is simpler to prove them in the early part of the inductive step, and then use them to prove the third assertion.

All three inductive assertions need to be verified in both the base step and the inductive step.

Base Step Lines 3–8 complete the iteration with $n = 0$.

- Since $B - \{r\} = B - \{a\} = \emptyset$, the first assertion in the inductive hypothesis is valid because there are no paths with the given requirements. Thus the estimated distance $d_0(u) = \infty$ is valid for all $u \neq a$.
- Since $d(a, a) = 0$ and all weights are nonnegative, it is clear that $d(a, a) \leq d(a, u)$ for any other vertex, u.
- After this iteration, $B = \{a\}$ and so $d(a, a) = 0$ has been calculated correctly for the only element in B.

Inductive Step Assume that after iteration $n - 1$ the inductive hypothesis holds.

Iteration n is completed by performing lines 11–22. The inductive hypothesis asserts that prior to performing those lines of the algorithm, $d(a, v)$ is known for every vertex, v, in B. It also asserts that $d_{n-1}(u)$ is the length of the shortest path from a to $u \in (V - B)$ among all paths having nonfinal vertices in B and with r excluded from the path. In addition, all vertices, v, in B have paths from a to v with lengths that are not longer than paths from a to vertices outside B.

- The addition of vertex r to B during iteration $n - 1$ permits additional paths from a to u that pass through r. The set A in line 12 contains the vertices that are adjacent to r but are not in B. The new path will go from a to r and then use an edge from r to u. The length of the new path will be $d(a, r) + w(r, u)$. Line 14 assigns to $d_n(u)$ the smaller of $d_{n-1}(u)$ and $d(a, r) + w(r, u)$. At the end of iteration n, a new value for r will exist and the current r will be permitted in candidate paths, so this newly calculated value for $d_n(u)$ will represent the shortest distance from a to u among all paths of the form $av_2v_3 \cdots v_k u$ where each v_i is in $B - \{\text{new } r\}$.

- The estimated lengths, $d_n(u)$, are constructed by successively adding actual edge weights (line 14), so $d_n(u)$ represents the distance for an actual path. Thus, $d_n(u)$ is an upper bound for $d(a, u)$. The choice of x in line 19 ensures that the vertices in B will continue to have smaller distances from a than do vertices outside of B, even after x is added. This is true since $d(a, x) \leq d_n(x) \leq d_n(u)$ for all $u \in (V - B)$.
- The choice of x in line 19 will result in a new vertex being added to B and a new choice for r. The first inductive assertion (already verified) indicates that $d_n(x)$ is the shortest distance from a to x using a path of the form $av_2 \cdots v_k x$ and $v_i \in (B - \{x\})$ for $i > 1$.

  It is not possible for a path that includes a vertex v_j with $v_j \notin (B - \{x\})$ to be shorter. This is because once x has been added to B, the second inductive assertion (already verified) indicates that $d(a, x) \leq d(a, v_j)$ if $v_j \notin (B - \{x\})$. Thus, a path from a to x that includes v_j cannot be a shorter path.[30]

  Because a shortest path from a to x will contain only vertices in B and since $d_n(x)$ is the length of the shortest such path, $d(a, x) = d_n(x)$.

Conclusion Each iteration preserves the inductive hypothesis. Eventually, vertex z will be added to B (since there are only a finite number of vertices and a new vertex is added to B during each iteration). Suppose z is added to B during iteration n. The inductive hypothesis asserts (among other things) that $d(a, z) = d_n(z)$. The algorithm has therefore correctly calculated $d(a, z)$. □

Representing Weighted Graphs

Which of the previously encountered representations for a graph would be easiest to extend to become a suitable representation for weighted graphs? Certainly a visual embedding is suitable; several such diagrams have already been used. What about some of the nonvisual representations?

The adjacency matrix seems like a possibility: Just place the edge weight, $w(v_i, v_j)$, in the ijth entry if there is an edge connecting v_i and v_j. There are two problems with this approach. The first relates to multiple edges. The adjacency matrix has already been extended to have entries other than 0 or 1 to count the number of distinct edges connecting two vertices. The proposed extension for weighted graphs would not work for weighted multigraphs. The second problem arises from the nature of the edge weights. Since edge weights are only restricted to be nonnegative, an entry of 0 in the matrix would be ambiguous. Does the 0 represent an edge with weight 0 or the absence of an edge?

The incidence matrix (page 627) is another possibility. This representation has a separate column for each edge, so edge multiplicities are not an issue. However, the ambiguity from edges with weight 0 still remains.

These difficulties indicate that it is not always possible to redefine one of the previous representations. The alternative is to use a separate listing of the edge weights along with an adjacency or incidence matrix.[31] However, if only simple graphs are of interest and if all edge weights are positive, then either a modified adjacency matrix or a modified incidence matrix might be acceptable.

[30] It is possible that $d(a, v_j) = d(a, x)$ and the other connecting edges all have weight 0.

[31] An array with as many entries as there are edges would be easy to use with graphs that are manipulated from within a computer program.

10.6.3 Exercises

The exercises marked with ★ have detailed solutions in Appendix H.

1. Which of the following directed graphs are strongly connected? Which are weakly connected?

   (a) ★

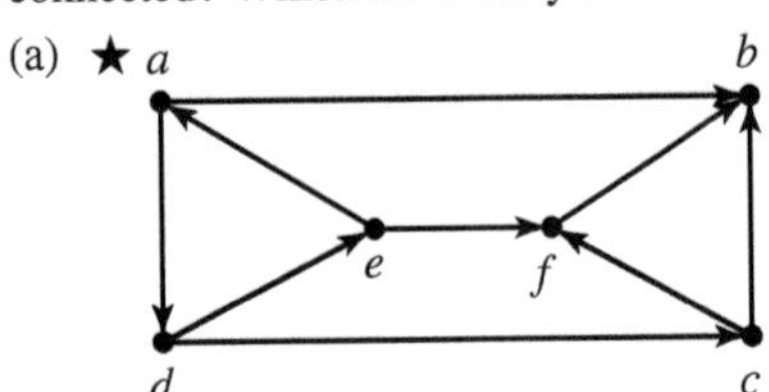

   (b)

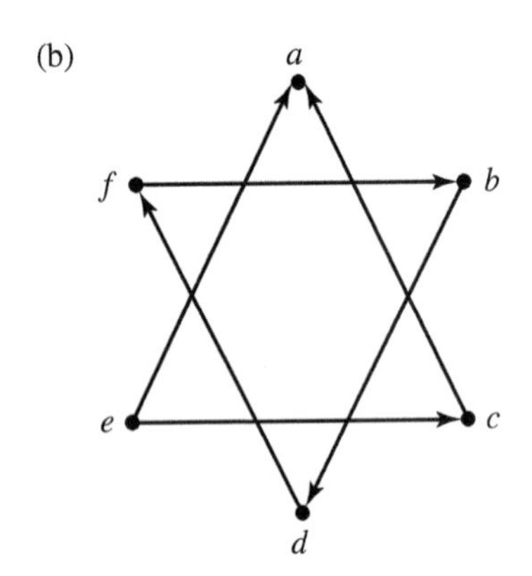

   (c)

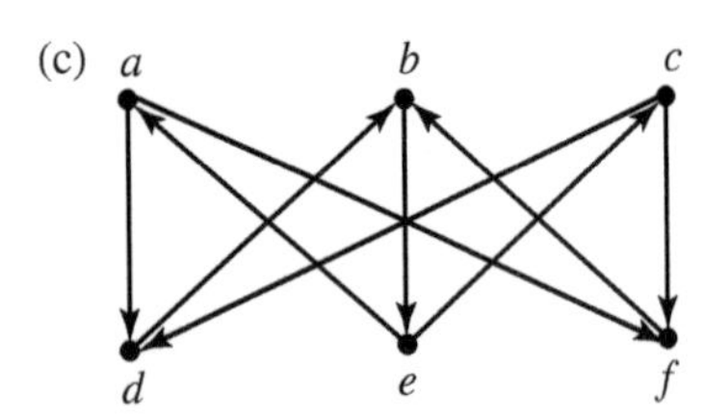

   (d)

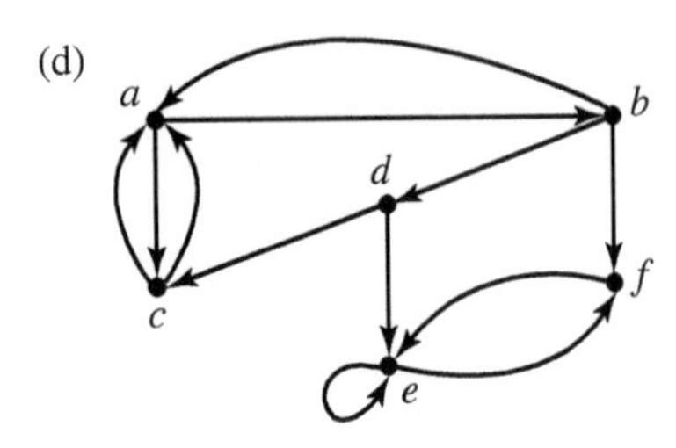

2. ★ Produce the adjacency matrix for each of the directed graphs in Exercise 1.

3. Which of the four following directed graphs are strongly connected? Which are weakly connected?

   (a)

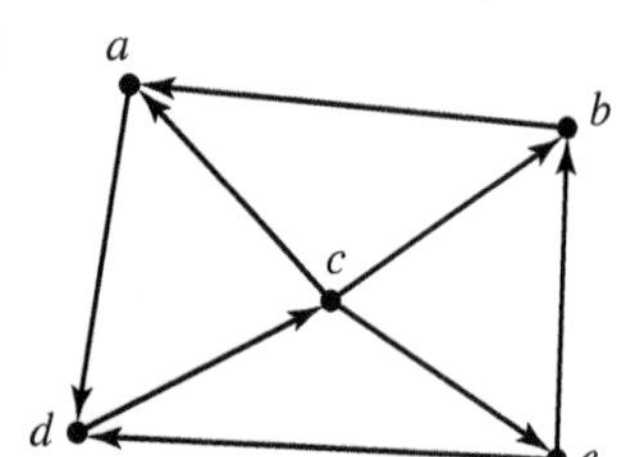

   (b)

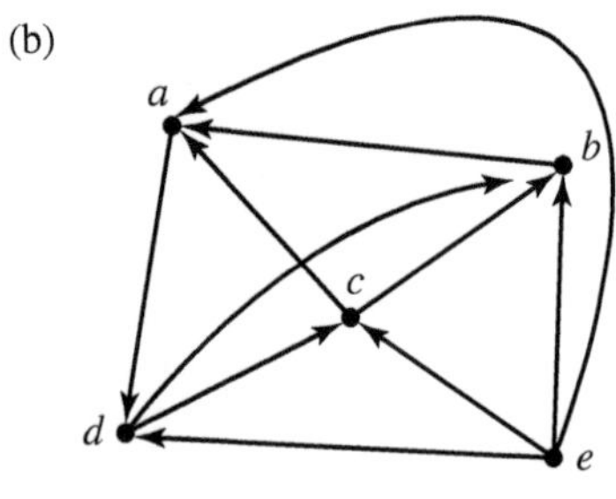

   (c)

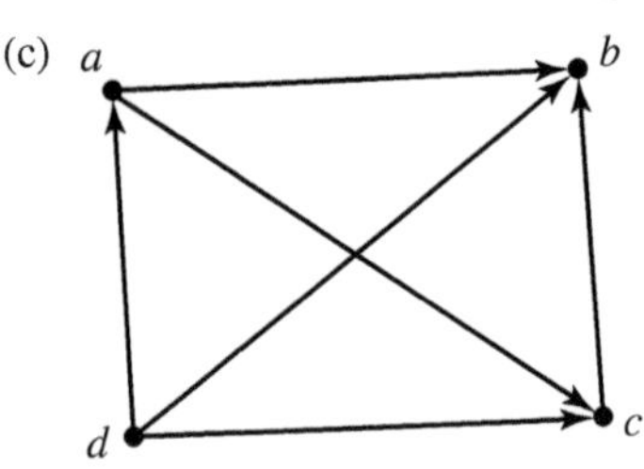

   (d)

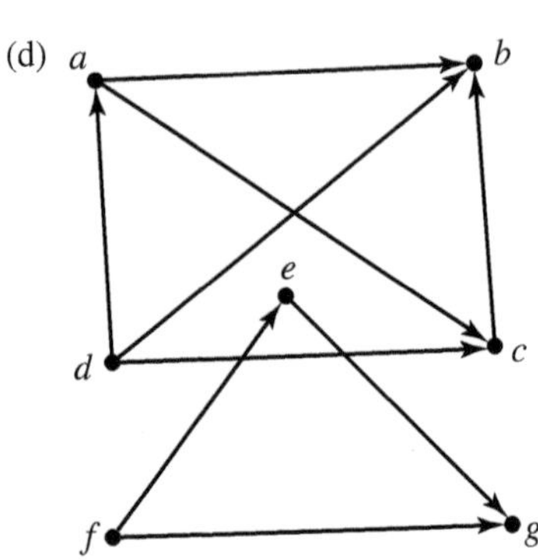

4. Produce the adjacency matrix for each of the directed graphs in Exercise 3.

5. Consider the directed "utility" graph in Exercise 1(c). Use the adjacency matrix to answer the following questions.
   (a) ★ Determine the number of directed walks of length 3 from vertex b to vertex d.
   (b) Determine the number of directed walks of length 3 that start from vertex b. Explain how you arrived at your answer.
   (c) Determine the number of directed walks of length 3 that end at vertex c. Explain how you arrived at your answer.

6. Consider the directed multigraph in Exercise 1(d). Use the adjacency matrix to answer the following questions:
   (a) Determine the number of directed walks of length 4 from vertex a to vertex d.
   (b) Determine the number of directed walks of length 4 that start from vertex c. Explain how you arrived at your answer.
   (c) Determine the number of directed walks of length 4 that end at vertex d. Explain how you arrived at your answer.

7. State and prove a "directed graph" version of the handshake theorem on page 602.

8. Define *directed Euler circuit*, and then prove Theorem 10.68.

> **THEOREM 10.68 Directed Euler Circuits in Digraphs**
>
> Let D be a weakly connected digraph without loops. Then the following assertions are equivalent.
>
> 1. D contains a directed Euler circuit.
> 2. D is strongly connected and if v is a vertex in D, then $\deg^-(v) = \deg^+(v)$.

9. Let $D = (V, E, \phi)$ be a directed graph with $|V| = n$. Let $a, b \in V$ be vertices such that there is a directed walk from a to b in D and $a \neq b$. Prove that there is a directed path of length at most $n - 1$ from a to b.

10. ★ Use Exercise 9 to prove Corollary 10.66 on page 654.

11. Each of the following assertions is either always true or is false in some or all cases. Determine which case applies; then provide some justification for your answer.
    (a) Every strongly connected digraph is a weakly connected digraph.
    (b) ★ If e_1 and e_2 are distinct arcs in a digraph, $D = (V, E, \phi)$, such that $\phi(e_1) = (v_i, v_f)$ and $\phi(e_2) = (v_f, v_i)$, then e_1 and e_2 are multiple arcs.
    (c) If v is a vertex in a digraph, then $\deg^-(v) = \deg^+(v)$.
    (d) If v is a vertex in a digraph, then $\deg^-(v) + \deg^+(v)$ is the degree of v as a vertex in the underlying graph. That is, $\deg(v) = \deg^-(v) + \deg^+(v)$.
    (e) If A is the adjacency matrix of a digraph, D, then a_{km} is the number of distinct arcs with terminal vertex v_k and initial vertex v_m.

12. Use Dijkstra's algorithm to find a shortest path for each weighted graph. Also, find the length of the shortest path.
    (a) ★ From a to f

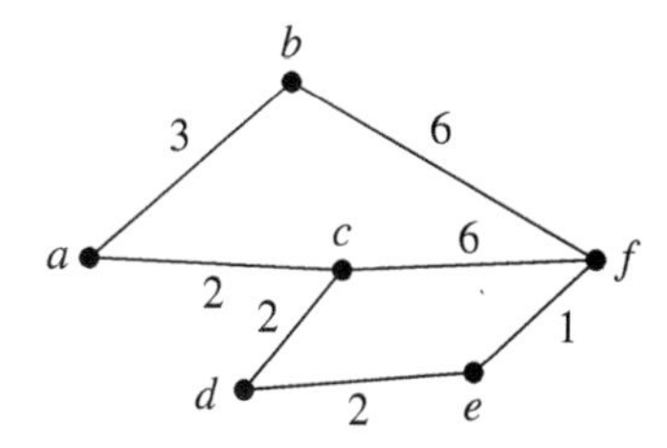

    (b) From a to f

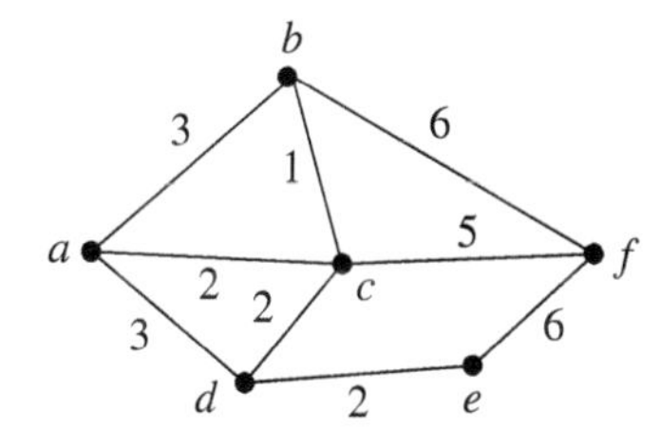

    (c) From a to e

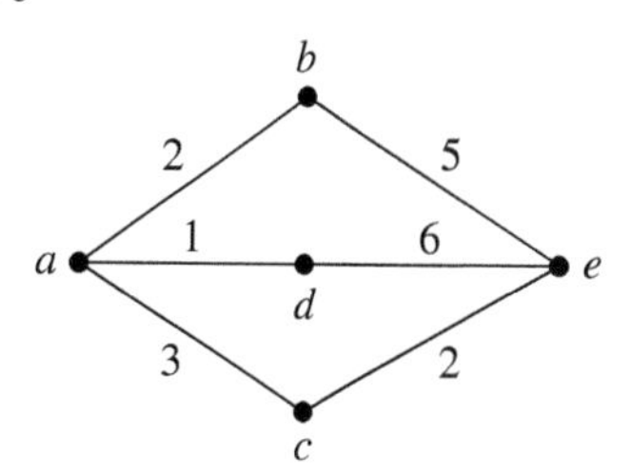

    (d) From a to z

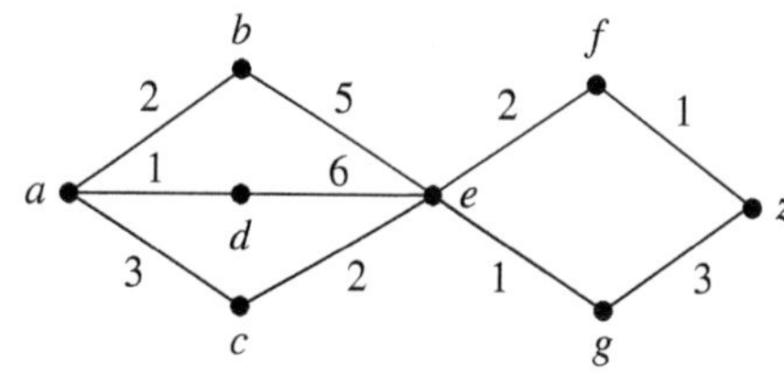

13. Use Dijkstra's algorithm to find a shortest path for each weighted graph. Also, find the length of the shortest path.
    (a) From a to f

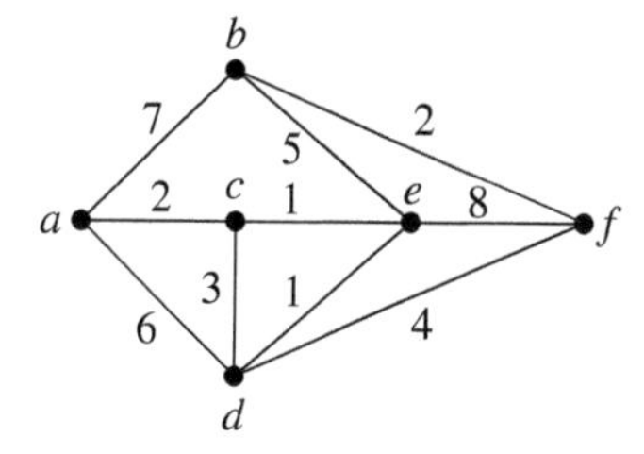

    (b) From a to f

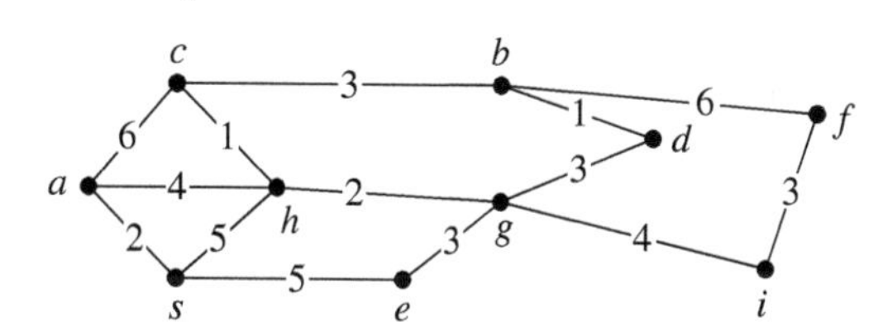

    (c) From a to z

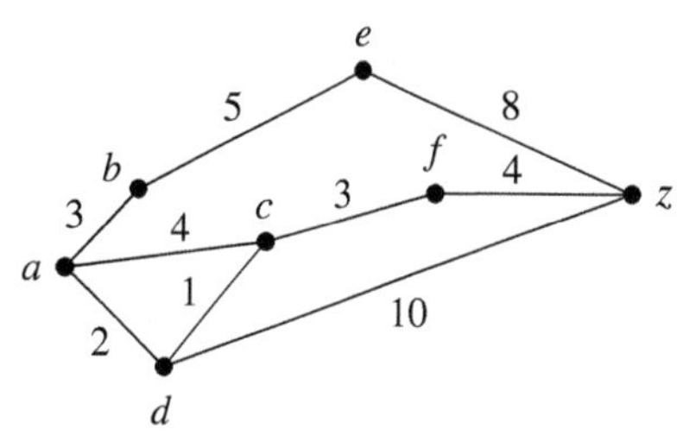

    (d) From a to z

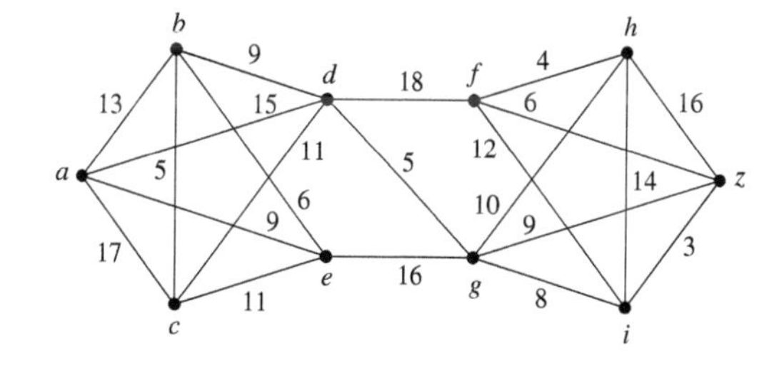

14. Design a definition for the adjacency matrix of a weighted simple graph with positive weights. Then use an adjacency matrix to represent each weighted graph in Exercise 13.

15. Design a definition for the incidence matrix of a weighted simple graph with positive weights. Then use an incidence matrix to represent each weighted graph in Exercise 13.

16. Prove the following theorem.

THEOREM 10.69 ***Dijkstra's Algorithm is in $\mathcal{O}(n^2)$***

If the weighted, connected, simple graph, G, has n vertices, then Dijkstra's algorithm for finding a shortest path is in $\mathcal{O}(n^2)$ in the worst case.

17. Modify Dijkstra's shortest path algorithm so that it will work with weighted simple graphs that may not be connected.

18. Define the *indegree sequence* to be the sequence of the indegrees of the vertices in a directed graph, written in decreasing order. Similarly, define the *outdegree sequence* to be the sequence of the outdegrees of the vertices in a directed graph, written in decreasing order. Theorem 10.60 asserts that the sum of the indegrees must equal the sum of the outdegrees. Must the indegree sequence always equal the outdegree sequence? If the answer is yes, provide a proof; otherwise, produce a counterexample.

DEFINITION 10.70 ***Tournament Graph; Transitive Tournament Graph***

A simple directed graph, $D = (V, E, \phi)$, is called a *tournament graph* if for all vertices, u and v, in V with $u \neq v$ there is either an arc, e_{uv}, such that $\phi(e_{uv}) = (u, v)$, or there is an arc, e_{vu}, such that $\phi(e_{vu}) = (v, u)$, but not both.

A tournament graph is a *transitive tournament graph* if whenever there is an arc from u to v and an arc from v to w, then there is also an arc from u to w.

19. ★ According to Definition 10.70, which of the digraphs in Exercise 3 are tournament graphs? Which are transitive tournament graphs?

20. Prove Theorem 10.71 (related to Definition 10.70). [*Hint*: Use mathematical induction on the number of vertices in the directed graph. Don't forget to use the defining characteristic of a tournament graph somewhere in the proof.]

THEOREM 10.71 ***Tournament Graphs***

Every tournament graph contains a Hamilton path.

21. Prove Theorem 10.72.

THEOREM 10.72 ***Transitive Tournament Graphs***

A tournament graph contains a unique Hamilton path if and only if it is a transitive tournament graph.

[*Hints* (from [74]):

if Show that if a tournament graph is transitive and there exist two distinct Hamilton paths, then there must be two vertices, u and v, that appear in opposite relative order in the two paths. Show that this leads to a contradiction.

only if Assume that a tournament graph has a unique Hamilton path, $v_1 v_2 \cdots v_n$. Use complete induction to prove the following claim, and then use the claim to prove transitivity.

- **Claim:** If $j > i$, then the arc joining v_i and v_j must have v_i as its initial vertex.

]

22. ★ One consequence of Theorem 10.71 is that every tournament graph provides a linear ranking of the vertices. Thus, tournament graphs can be used to rank the participants in a contest where every pair of participants are matched and a winner is recorded (and no ties are allowed). The participants can be represented by vertices in a directed graph. The winner in each pair will be the initial vertex of a directed arc and the loser will be the terminal vertex. A Hamilton path provides a linear ranking of the participants.

The following table indicates the results in a ping pong tournament. The table lists the winner in each pair. Use a tournament graph to answer the following questions.

(a) List at least two Hamilton paths in the tournament graph.

(b) Count the number of games each player won. Is the list of numbers of wins consistent with the ranking of the people using the Hamilton paths?

	Assumpta	**Bob**	**Carmen**	**Donald**	**Eurydice**	**Faramir**
Assumpta		Assumpta	Assumpta	Donald	Eurydice	Assumpta
Bob	Assumpta		Bob	Bob	Eurydice	Faramir
Carmen	Assumpta	Bob		Carmen	Carmen	Faramir
Donald	Donald	Bob	Carmen		Donald	Faramir
Eurydice	Eurydice	Eurydice	Carmen	Donald		Faramir
Faramir	Assumpta	Faramir	Faramir	Faramir	Faramir	

10.7 Quick Check Solutions

Quick Check 10.1

1.

vertex	a	b	c	d	e	f	g	h	i	j	k
degree	4	4	3	3	3	3	3	3	4	5	3

2. Solve the equation $\sum_{v \in V} h = 36$ for h. The common number of handshakes is 3, since there are 12 vertices in V.

Quick Check 10.2

1. The star graph S_n and the cycle graph C_{n-1} (notice the change in subscript).
2. The graphs $G_{3,2}$ and $\overline{G_{3,2}}$ are

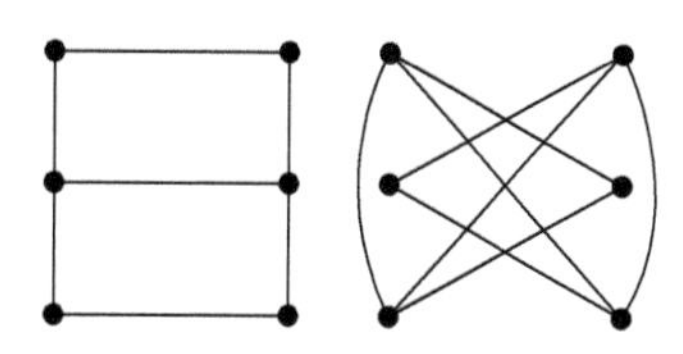

The following is an unacceptable representation of $\overline{G_{3,2}}$. (It contains apparent incidences that do not actually exist in the graph.)

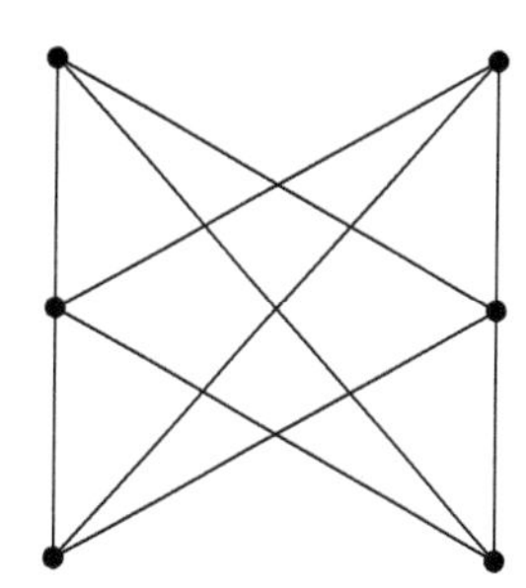

Quick Check 10.3

1. There are many initial definitions, many sets of examples, and many ways to refine the initial definition. You will benefit from doing the work; don't look here for an easy answer.

Quick Check 10.4

1. (a) This graph has connectivity 2 (remove the two left-most vertices from among the four central vertices) and edge connectivity 3 (remove the three central edges).

   (b) This has connectivity 2 (remove opposite vertices) and edge connectivity 2 (remove edges that are incident with a common vertex).

   (c) This graph has connectivity 3 (remove a, b, and j) and edge connectivity 3 (remove the edges ad, dj, and bd).

Quick Check 10.5

1. (a) There is neither an Euler circuit nor an Euler trail because there are more than two vertices with odd degree (c, d, e, f, g, h, j, and k).

   (b) One Euler circuit is $b\ i\ c\ f\ b\ g\ i\ e\ c\ h\ f\ d\ e\ a\ h\ g\ a\ d\ b$.

Quick Check 10.6

1. One solution is shown.

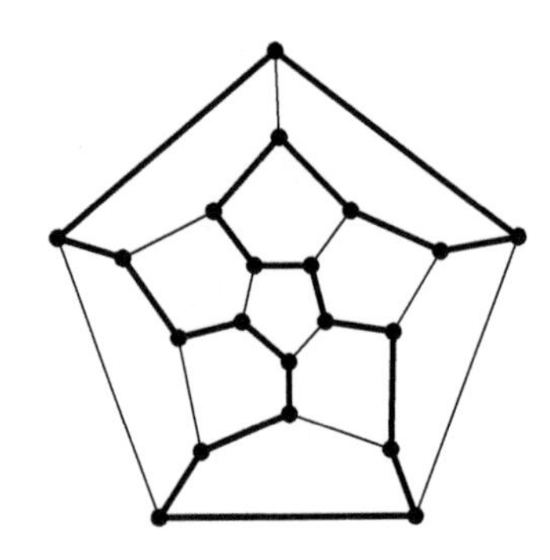

Quick Check 10.7

1. The incidence matrix is shown.

$$\begin{array}{c} \\ v_1 \\ v_2 \\ v_3 \end{array} \begin{array}{c} \begin{array}{cccc} e_1 & e_2 & e_3 & e_4 \end{array} \\ \begin{pmatrix} 1 & 1 & 1 & 0 \\ 1 & 1 & 0 & 1 \\ 0 & 0 & 1 & 1 \end{pmatrix} \end{array}$$

2. Using Theorem 10.35,

$$A_L = \begin{pmatrix} 1 & 1 & 0 \\ 1 & 1 & 0 \\ 1 & 0 & 1 \\ 0 & 1 & 1 \end{pmatrix} \begin{pmatrix} 1 & 1 & 1 & 0 \\ 1 & 1 & 0 & 1 \\ 0 & 0 & 1 & 1 \end{pmatrix} - \begin{pmatrix} 2 & 0 & 0 & 0 \\ 0 & 2 & 0 & 0 \\ 0 & 0 & 2 & 0 \\ 0 & 0 & 0 & 2 \end{pmatrix} = \begin{pmatrix} 0 & 2 & 1 & 1 \\ 2 & 0 & 1 & 1 \\ 1 & 1 & 0 & 1 \\ 1 & 1 & 1 & 0 \end{pmatrix}.$$

The graph is shown in the next diagram.

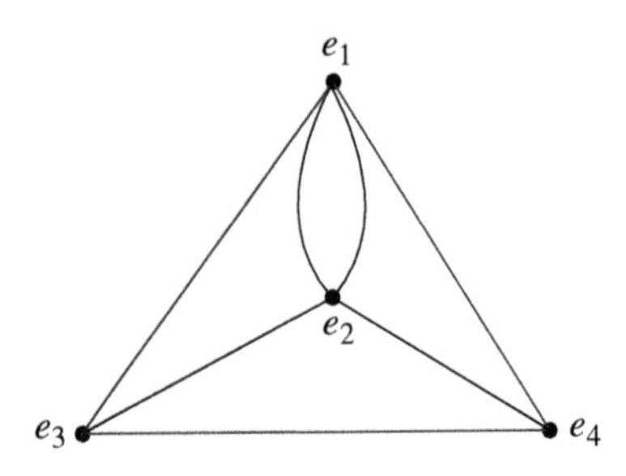

Quick Check 10.8

1. The degree sequence is 5, 5, 4, 3, 3, 2.
2. G_1 and G_2 are isomorphic. Any of the mappings shown in the following table will work for ζ.

V	ζ_1	ζ_2	ζ_3	ζ_4
a	f	f	g	g
b	g	g	f	f
c	e	h	e	h
d	h	e	h	e

G_3 is not isomorphic to either G_1 or G_2 since the degree sequence for G_3 is 3, 3, 3, 1, but the other two graphs have degree sequence 3, 3, 2, 2.

Quick Check 10.9

1. $K_{3,3}$ is *not* a planar graph. Theorem 10.41 on page 635 provides one proof of this important fact.

Quick Check 10.10

1. The common ancestor is the Königsberg graph.

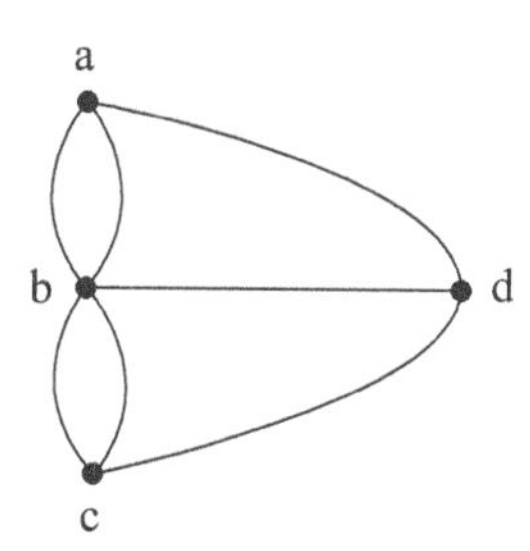

Quick Check 10.11

1. (a) The dual graph has the following planar embedding.

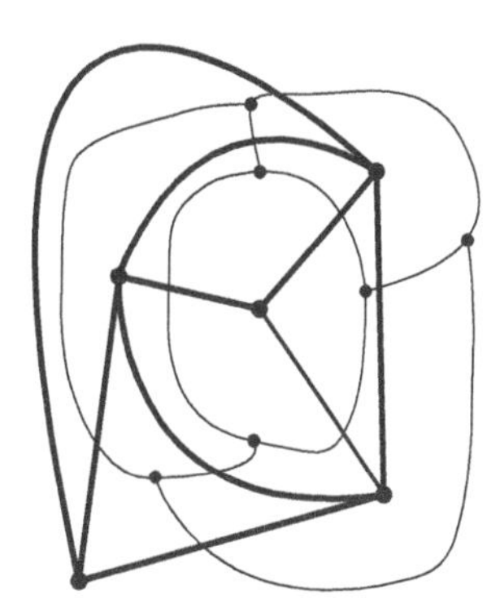

(b) Any polygonal embedding looks something like the following:

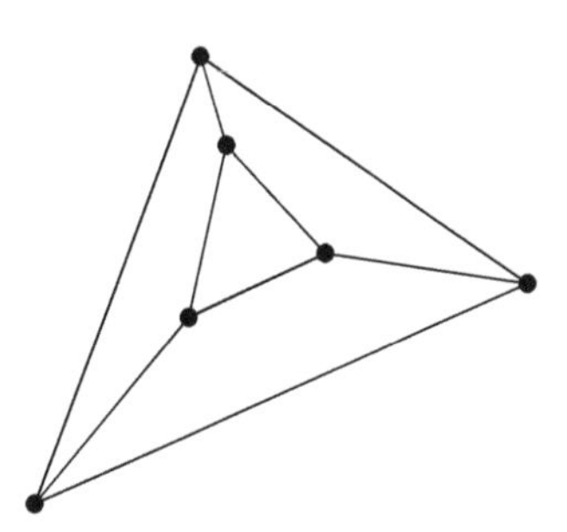

2. The mapping $\zeta(v_1) = r_1$ and $\zeta(v_2) = r_2$ establishes the isomorphism.

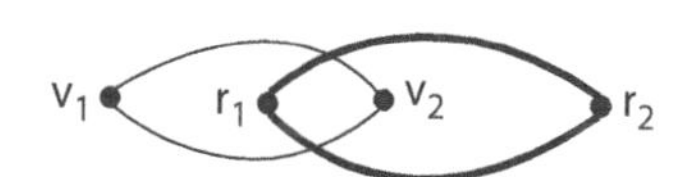

Quick Check 10.12

1. $\chi = 3$

   To see that $\chi \geq 3$, color the top vertex red. The two adjacent vertices on the outside need to be a different color, so make them green (trying to get by with only 2 colors). The final two outer vertices cannot be green and they cannot both be red. Thus at least three colors are needed.

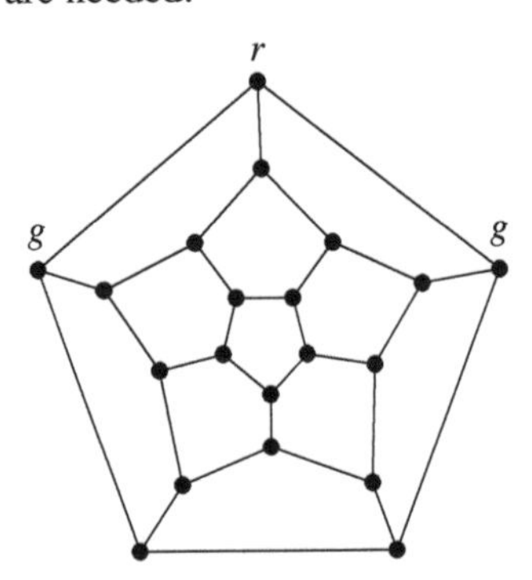

   To see that only three colors are needed, it is sufficient to assign colors to the vertices so that only three colors are used and the coloring is proper. I have used red, green, and blue.

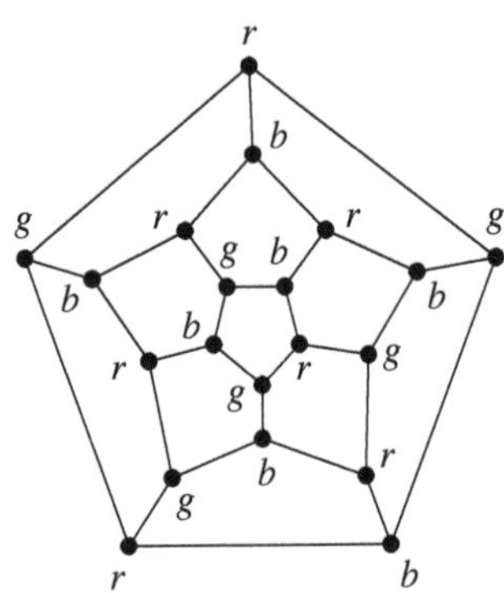

Quick Check 10.13

1. Only one additional arc is needed: either from z to x or from z to y.

2. If G consists of a single vertex with no arcs, then it is weakly connected. It is also a graph with an isolated vertex. The assumption that G does not consist of only a single vertex with no arcs eliminates this special case.

   If the sum of row i and the sum of column i are both 0, the digraph cannot be weakly connected. This is because if a row sum is 0, then starting at the vertex corresponding to the row index, it is impossible to reach any other vertex. If a column sum is 0, then it is not possible to start at any vertex other than the one corresponding to the column index and reach that vertex. If both are true the digraph has an isolated vertex and hence cannot be weakly connected.

Quick Check 10.14

1. The adjacency matrix is

$$A = \begin{array}{c} \\ v \\ w \\ x \\ y \\ z \end{array} \begin{array}{c} \begin{array}{ccccc} v & w & x & y & z \end{array} \\ \begin{pmatrix} 0 & 1 & 0 & 0 & 0 \\ 0 & 0 & 0 & 0 & 1 \\ 2 & 0 & 0 & 1 & 0 \\ 0 & 0 & 1 & 0 & 0 \\ 0 & 0 & 0 & 1 & 1 \end{pmatrix} \end{array}$$

Thus,

$$A^3 = \begin{array}{c} \\ v \\ w \\ x \\ y \\ z \end{array} \begin{array}{c} \begin{array}{ccccc} v & w & x & y & z \end{array} \\ \begin{pmatrix} 0 & 0 & 0 & 1 & 1 \\ 0 & 0 & 1 & 1 & 1 \\ 2 & 0 & 0 & 1 & 2 \\ 0 & 2 & 1 & 0 & 0 \\ 2 & 0 & 1 & 2 & 1 \end{pmatrix} \end{array}$$

Since $a_{54} = 2$, there are two directed walks of length 3 from z to y. They are $zyxy$ and $zzzy$.

2. The adjacency matrix is

$$A = \begin{array}{c} \\ v \\ w \\ x \\ y \\ z \end{array} \begin{array}{c} \begin{array}{ccccc} v & w & x & y & z \end{array} \\ \begin{pmatrix} 0 & 1 & 0 & 0 & 0 \\ 0 & 0 & 0 & 0 & 1 \\ 2 & 0 & 0 & 1 & 0 \\ 0 & 0 & 1 & 0 & 0 \\ 0 & 0 & 0 & 0 & 1 \end{pmatrix} \end{array}$$

Thus, $\sum_{i=0}^{n-1} A^i$ is

$$\begin{pmatrix} 1 & 0 & 0 & 0 & 0 \\ 0 & 1 & 0 & 0 & 0 \\ 0 & 0 & 1 & 0 & 0 \\ 0 & 0 & 0 & 1 & 0 \\ 0 & 0 & 0 & 0 & 1 \end{pmatrix} + \begin{pmatrix} 0 & 1 & 0 & 0 & 0 \\ 0 & 0 & 0 & 0 & 1 \\ 2 & 0 & 0 & 1 & 0 \\ 0 & 0 & 1 & 0 & 0 \\ 0 & 0 & 0 & 0 & 1 \end{pmatrix} + \begin{pmatrix} 0 & 0 & 0 & 0 & 1 \\ 0 & 0 & 0 & 0 & 1 \\ 0 & 2 & 1 & 0 & 0 \\ 2 & 0 & 0 & 1 & 0 \\ 0 & 0 & 0 & 0 & 1 \end{pmatrix}$$

$$+ \begin{pmatrix} 0 & 0 & 0 & 0 & 1 \\ 0 & 0 & 0 & 0 & 1 \\ 2 & 0 & 0 & 1 & 2 \\ 0 & 2 & 1 & 0 & 0 \\ 0 & 0 & 0 & 0 & 1 \end{pmatrix} + \begin{pmatrix} 0 & 0 & 0 & 0 & 1 \\ 0 & 0 & 0 & 0 & 1 \\ 0 & 2 & 1 & 0 & 2 \\ 2 & 0 & 0 & 1 & 2 \\ 0 & 0 & 0 & 0 & 1 \end{pmatrix}$$

which simplifies to

$$\begin{array}{c} \\ v \\ w \\ x \\ y \\ z \end{array} \begin{array}{c} \begin{array}{ccccc} v & w & x & y & z \end{array} \\ \begin{pmatrix} 1 & 1 & 0 & 0 & 3 \\ 0 & 1 & 0 & 0 & 4 \\ 4 & 4 & 3 & 2 & 4 \\ 4 & 2 & 2 & 3 & 2 \\ 0 & 0 & 0 & 0 & 5 \end{pmatrix} \end{array}$$

This matrix contains 0s, so the original digraph is not strongly connected.

Quick Check 10.15

1. The shortest path has length 9 and is *abf*. The following table shows the details of Dijkstra's algorithm. Note that when $n = 3$, the estimated distance for vertex b is an update candidate, but the value does not change since the old estimate is smaller than the length of the new path through vertex e. Thus, $p(b)$ does not change.

				d						***p***					
n	B	r	A	a	b	c	d	e	f	a	b	c	d	e	f
0				$\boxed{0}$	∞	∞	∞	∞	∞						
1	$\{a\}$	a	$\{b, d\}$		7		$\boxed{6}$				a		a		
2	$\{a, d\}$	d	$\{c, e, f\}$			9		$\boxed{7}$	10			d		d	d
3	$\{a, d, e\}$	e	$\{b, f\}$		$\boxed{7}$										
4	$\{a, b, d, e\}$	b	$\{f\}$						$\boxed{9}$						b

10.8 CHAPTER REVIEW

10.8.1 Summary

This chapter provides a survey of some foundational material in graph theory. Graphs provide a rich assortment of theoretically interesting ideas and theorems as well as some very effective tools for practical applications. Part of the attractiveness of graph theory is the intuitively appealing visualization of graphs as diagrams composed of dots and lines. However, you should not lose sight of the fact that the diagram is merely a visualization; the graph itself is an abstract mathematical entity.

Section 10.1 introduces the fundamental definitions. It is here that the definition of a graph is stated. That definition presents graphs as sets of vertices and edges, together with an incidence function. That section also contains the elegant handshake theorem.

Section 10.2 introduces the notion of connectivity, defined in terms of walks in the graph. Extra effort is required to keep the definitions of *walk*, *trail*, *path*, *circuit*, and *cycle* clear in your thinking. This section also introduces the adjacency matrix of a graph. The adjacency matrix enables important algebraic tools from linear algebra to be applied to the study of graphs. The primary application in this book is the determination of the number of walks of a predetermined length that exist between two vertices. There are other, more advanced, applications, but they assume a deeper knowledge of linear algebra than this book requires.

Section 10.3 discusses the Königsberg bridge problem, which is credited as the problem that inspired Euler to begin the field of graph theory. The existence of Euler circuits and Hamilton cycles appear to be closely related topics. However, there is a simple theorem that completely characterizes which graphs contain Euler circuits, but no such theorem for characterizing which graphs contain Hamilton cycles.

Section 10.4 starts with descriptions of several alternative ways to represent graphs. Many of them (such as the incidence matrix) are tailored toward algorithms that can be programmed for a computer. One of the reasons that there are multiple representations is because some problems are easier to solve using one representation but are more difficult to solve with another.[32] The section continues by introducing a more advanced notion: graph isomorphism. Determining whether two graphs are isomorphic is, in general, a difficult problem. This section presented one way to determine that two graphs are *not* isomorphic: Observe that they have different degree sequences. However, graphs can have identical degree sequences and still not be isomorphic.

Section 10.5 presents several famous theorems from graph theory. The first is Kuratowski's theorem, which characterizes all graphs that have planar embeddings. The key determination hinges on the graphs K_5 and $K_{3,3}$. The next major result is Euler's formula, which relates the numbers of vertices, edges, and regions in a planar graph. Euler's formula is then used to determine that there are exactly five regular polyhedra. The section concludes with a discussion of the four-color theorem. That discussion introduces the chromatic number of a graph and also contains a proof of the five-color theorem.

Section 10.6 introduces graphs with directed edges and graphs with weighted edges. Most of the previous discussion survives the change with only minor revisions. Once those revisions have been presented, the section discusses an important application of weighted graphs: determining a shortest path. In particular, Dijkstra's shortest path algorithm is presented.

Graph theory is a rich field of mathematical exploration. It contains ideas that have immediate intuitive appeal and are comprehensible to mathematically unsophisticated

[32]Compare this to the ease of adding complex numbers that are written in Cartesian form and the ease of finding powers of complex numbers that are written in trigonometric form. The opposite representations for the same tasks are not as simple to perform.

people. It also contains ideas and theorems that challenge even the greatest of mathematicians. There is abundant material that lies between these extremes. Graph theory also has numerous practical applications. For example, the connections in a computer network can be viewed using a graph.

As you review the material in this chapter, you should again seek a balance between reinforcing your understanding of definitions and theorems and your proficiency with the algorithmic parts of the chapter.

10.8.2 Notation

Notation	Page	Brief Description
$G = (V, E, \phi)$	601	a graph, with vertex set, V, edge set, E, and incidence function, ϕ
$\deg(v)$	602	the degree of vertex v
$\overline{G}$	604	the complement graph of the simple graph, G
G_L	604	the line graph of the graph, G
K_n	605	the complete graph on n vertices
$K_{n,m}$	605	the complete bipartite graph on n and m vertices
$G_{n,m}$	606	a grid graph
C_n	606	a cycle graph with n vertices
W_n	606	a wheel graph with n vertices
S_n	606	a star graph with n vertices
H_n	606	the hypercube on n vertices
$v_0 e_1 v_1 e_2 v_2 \cdots e_k v_k$	610	a walk in a graph
$v_0 v_1 v_2 \cdots v_k$	610	a walk in a simple graph
$A = (a_{ij})$	613	the adjacency matrix of a graph
$B = (b_{ij})$	627	the incidence matrix of a graph
G^*	643	the dual graph of the graph, G
$\chi(G)$	644	the chromatic number of the graph, G
$\deg^-(v)$	652	the indegree of the vertex, v, in a directed graph
$\deg^+(v)$	652	the outdegree of the vertex, v, in a directed graph
$w(x, y)$	655	the edge weight on the edge joining vertices, x and y, in a weighted graph
$d(a, v)$	656	the distance between vertices, a and v, in a weighted graph
$d_n(v)$	656	the tentative distance from a to v at *iteration* n of Dijkstra's algorithm
$p(v)$	657	the vertex, u, which is adjacent to v and is in a shortest path from a to v

10.8.3 Additional Review Material

Go to http://www.mathcs.bethel.edu/~gossett/DiscreteMathWithProof/review.xhtml for additional review material, including summaries of notation, definitions, and theorems from this chapter. There is also a sample chapter exam, with solutions.

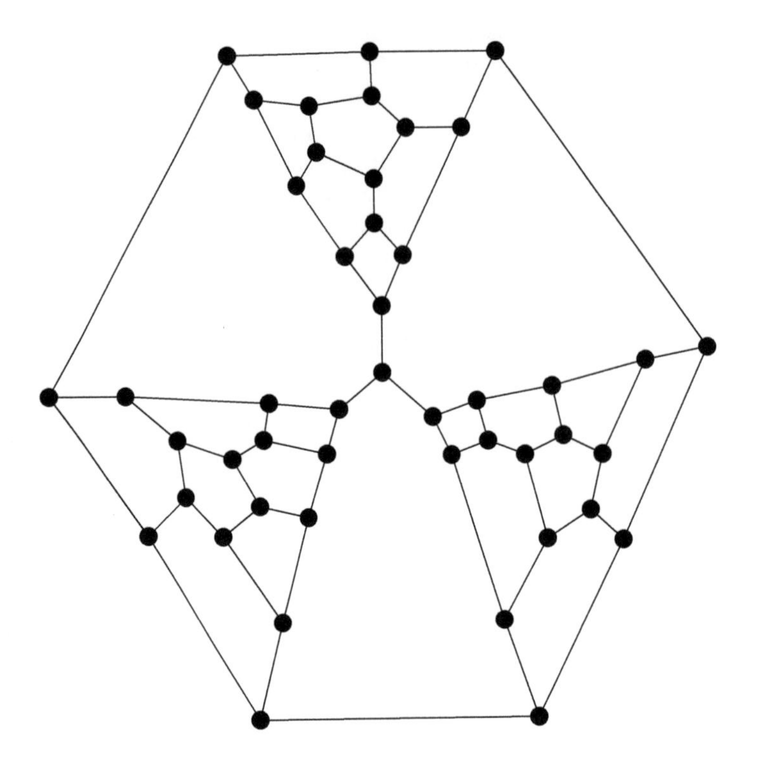

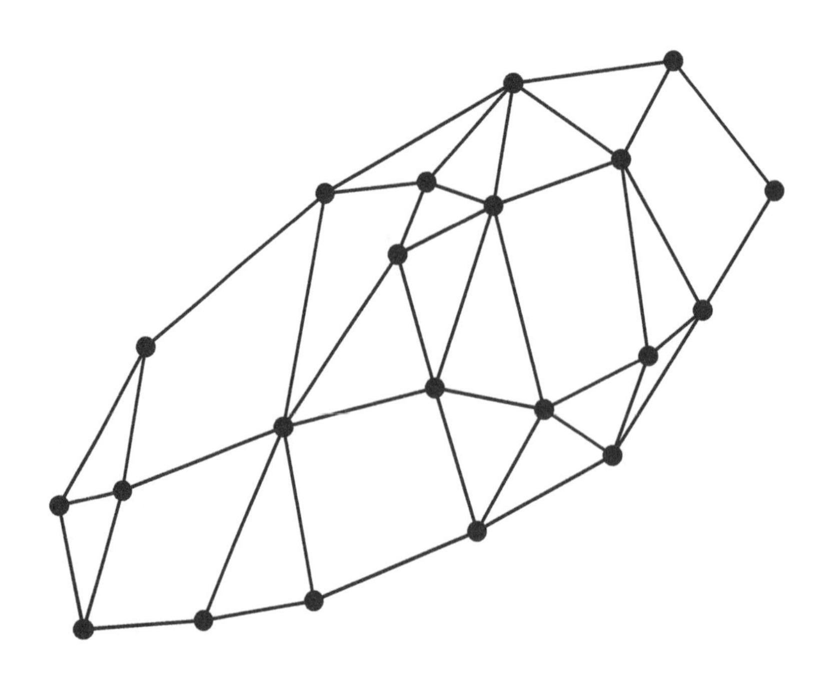

CHAPTER

11

Trees

For in the true nature of things, if we rightly consider, every green tree is far more glorious than if it were made of gold and silver.

Martin Luther (1483 – 1546)

One of the most useful structures for storing information on a computer is called a tree. The mathematical model for such a storage structure is also called a tree. The mathematical tree is just a special kind of graph. However, it is such a useful subclass of graph that trees merit a chapter of their own.

11.1 Terminology, Counting

DEFINITION 11.1 ***Tree***

A connected graph with no cycles is called a *tree*.

Notice that the exclusion of cycles implies that a tree cannot have multiple edges or loops.

EXAMPLE 11.1 **A Few Trees**

The graph in Figure 11.1 is a tree. It is shown in two different embeddings.

The right-hand embedding is the more traditional form. Notice the top-to-bottom hierarchical nature of the right-hand embedding. It is deliberate. ■

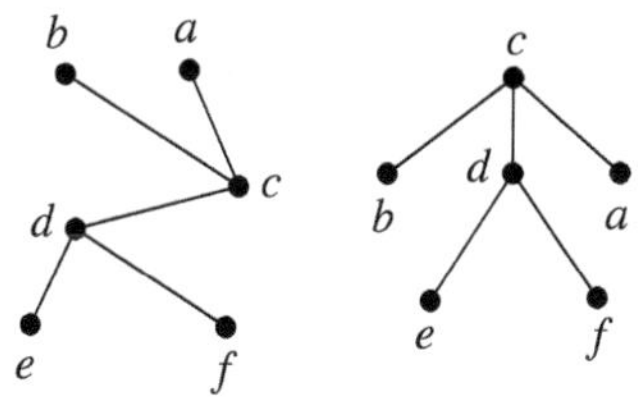

Figure 11.1. Two embeddings for a small tree.

DEFINITION 11.2 ***Node***

The vertices of a tree (or a graph) are often called *nodes* by computer scientists.

The exclusion of cycles imposes some fairly strong structure on trees, as the next theorem shows.

THEOREM 11.3 ***A Characterization of Trees***

A connected simple graph is a tree if and only if it contains a unique path between any two vertices.

Proof: Trees are always connected simple graphs, so the hypotheses "connected" and "simple" are understood to be in effect during both halves of the proof.

If: Suppose that the connected simple graph G contains a unique path between any two of its vertices. Then G cannot contain any cycles, since a cycle would imply two paths between any two of the vertices in the cycle. G is therefore a connected, simple graph with no cycles. Thus, G is a tree.

Only If: Suppose that the connected simple graph T is a tree. Then T is connected so there is a walk between any two vertices (nodes). Exercise 1 on page 615 implies that we may assume there is a path between any pair of vertices. Finally, if there were a second path between two vertices, the two paths could be joined into a cycle (Exercise 1 on page 680). Since trees do not contain cycles, this is impossible. Hence, the path is unique. □

DEFINITION 11.4 ***Root; Rooted Tree***

It is possible to designate one of the vertices as the *root* (often called the *root node*). A tree with a root is called a *rooted tree*.

The previous theorem leads naturally to the next definition, which is a property of individual nodes rather than of the entire tree.

DEFINITION 11.5 ***Level***

The *level* of a node in a rooted tree is the length of the unique path from the root to that node. An alternate name for this number is the *depth* of the node. The root node is at level 0.

Choosing a root node therefore imposes a hierarchical structure on a tree. It is traditional to use the language of family relationships to discuss nodes with this imposed hierarchy.

DEFINITION 11.6 ***Parent; Child; Ancestor; Descendant; Sibling***

Node p in a rooted tree is said to be the *parent* of node c if p and c are adjacent and the path length from p to the root is 1 less than the path length from c to the root. The node, c, is called a *child* of p.

Every node on the path from c to the root (excluding c but including the root) is an *ancestor* of c. The node, c, is said to be a *descendant* of every one of its ancestors. Nodes c_1 and c_2 are *siblings* if they have a common parent.

Note that a node can have many children but at most one parent (so the analogy with human family trees breaks down at this point).

DEFINITION 11.7 ***Subtree***

Any node, x, in a rooted tree can be chosen as the root of a *subtree* consisting of x and all its descendants.

✔ Quick Check 11.1

1. Let r be the root of the following tree (shown in an embedding that displays the imposed hierarchy in a natural manner). Identify the parent of node d. Identify the children of node b. What is the level of each node? Which nodes are in the subtree rooted at c?

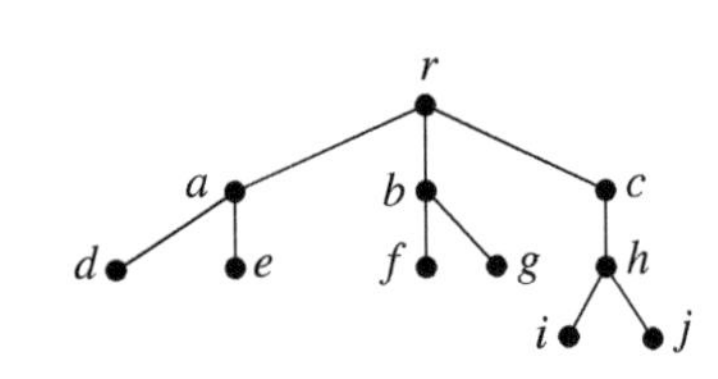

2. Prove that siblings are always at the same level in a tree. ☑

There are several other definitions that you need to become familiar with. Some enable us to talk about special regions of a tree. Others aid in discussing the general shape of the tree.

DEFINITION 11.8 ***Leaf Node; Interior Node***

A node with no children is called a *leaf* (or a *leaf node*). A node that has one or more children is called an *interior node.*

DEFINITION 11.9 ***Height; Balanced Trees***

The *height* of a rooted tree is the length of the longest path from the root to a leaf node. A tree of height h is *balanced* if all leaves appear at levels $h - 1$ and h.

Notes:

1. A tree with exactly one node (the root) has height 0.
2. *Height* is a feature of a tree (or a subtree). *Level* is a feature of an individual node.
3. If the tree has height h and node a is at level k and is on a maximal length path in the tree, then node a is the root of a subtree of height $h - k$.

The hierarchy imposed on a tree by choosing a root can be extended by imposing an ordering on the nodes.

DEFINITION 11.10 ***Ordered Rooted Tree***

A rooted tree is *ordered* if all groups of sibling nodes are assigned a relative order. For binary trees, the ordering is specified by declaring a child node as either the *left child* or as the *right child*.

EXAMPLE 11.2

Ordered Trees

The embedding of the tree in Figure 11.2 implicitly defines an ordering (due to the left-to-right alphabetical placement of the nodes). Thus, the node a precedes its siblings b and c.

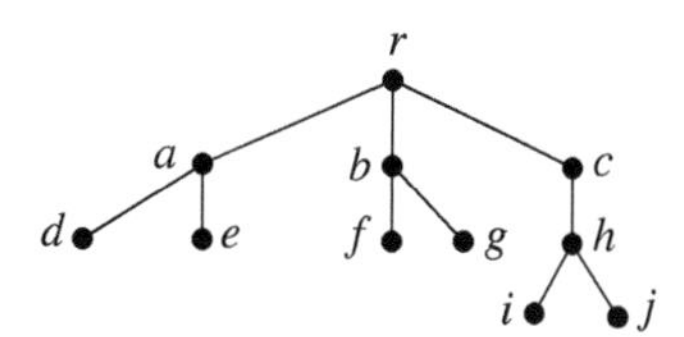

Figure 11.2 An implicit ordering.

Even though there is no alphabetical ordering to the nodes in Figure 11.3, this binary tree still has a clear ordering due to its embedding. Node g is the left sibling of node e. Node c is the right sibling of a (and the right child of d). The embedding also implies that node f is the right child of node c.

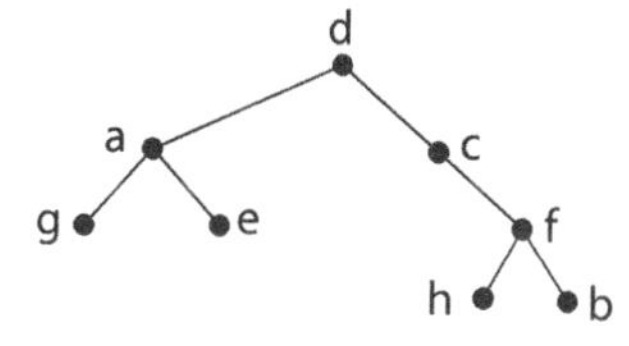

Figure 11.3 An embedding-imposed ordering.

■

It is useful to classify trees according to the maximum number of children each node may have.

DEFINITION 11.11 ***m-ary Tree; Binary Tree; Ternary Tree; Full Tree; Complete Tree; Maximal Complete Tree***

A rooted tree in which every node has at most m children is called an *m-ary* tree ($m \geq 2$). If $m=2$, we call the tree a *binary tree.* If $m=3$, it is called a *ternary tree.*

An m-ary tree in which every interior node has exactly m children is called a *full m-ary tree.*

An m-ary tree of height h is called *complete* if it is a balanced tree having all levels filled except level h. If the tree is an ordered tree, all nodes in level h are in the left-most positions. A complete and full tree for which leaves appear only in the final level is called a *maximal complete tree.*

EXAMPLE 11.3 **Some Trees**

The tree from Quick Check 11.1 (Figure 11.4) is a ternary tree since node r has three children, but no node has more than three children. The leaves are nodes d, e, f, g, i, j. The tree has height 3 and it is a balanced tree because all leaves appear at levels 2 and 3.

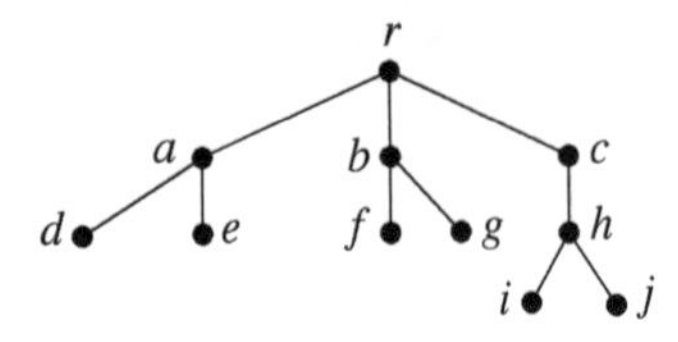

Figure 11.4. A balanced ternary tree.

The tree in Figure 11.5 is a height 3 binary tree that is balanced, but neither full nor complete.

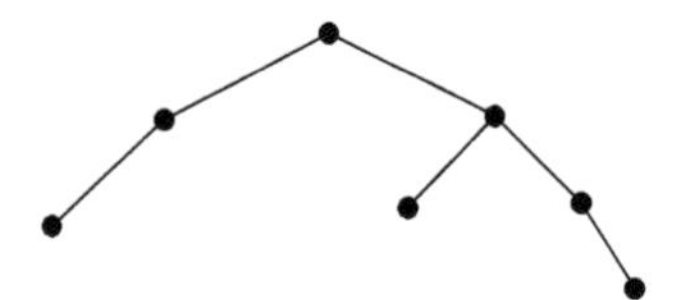

Figure 11.5 A balanced height 3 binary tree.

Figure 11.6 shows a full ternary tree with 13 nodes. It is not balanced, nor is it complete.

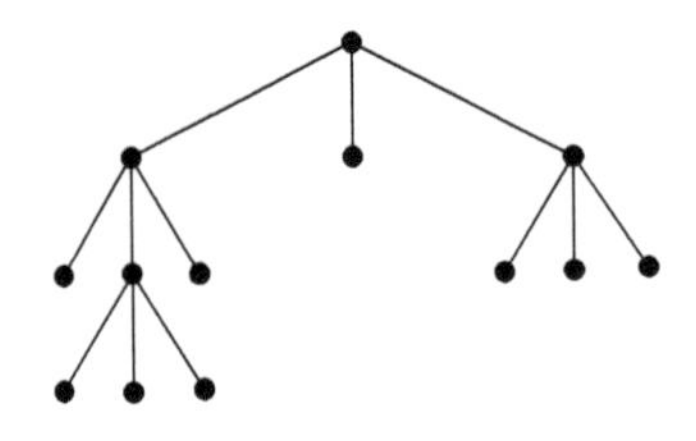

Figure 11.6 A full ternary tree.

The tree in Figure 11.7 is a balanced, ternary tree.

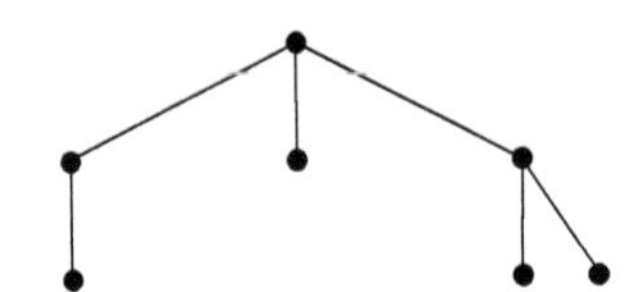

Figure 11.7 A balanced, ternary tree.

The tree in Figure 11.8 is a complete binary tree with 12 nodes.

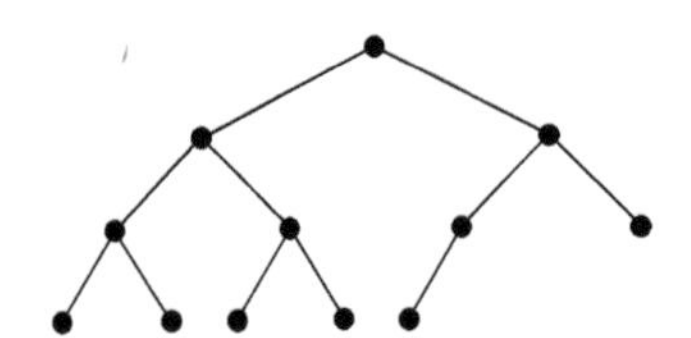

Figure 11.8 A complete binary tree.

The final tree (Figure 11.9) is a maximal complete binary tree with 15 nodes. It is full and balanced and has the maximum number of nodes for a binary tree with height 3.

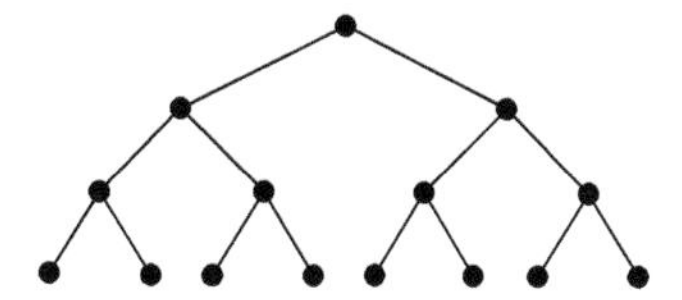

Figure 11.9 A maximal complete binary tree.

Notice that complete ordered trees are balanced and are either full or fail to be full at just one interior node. ■

✔ Quick Check 11.2

1. Draw a complete ordered binary tree with eight nodes.
2. Draw a maximal complete ternary tree of height 2.
3. How many nodes are in
   (a) a maximal complete binary tree of height h?
   (b) a maximal complete ternary tree of height h?
4. How many leaf nodes are there in
   (a) a maximal complete binary tree of height h?
   (b) a maximal complete ternary tree of height h? ☑

This section will conclude with some counting theorems. These are often of interest in practice, since the usefulness and efficiency of trees as computer storage models is directly related to the number and placement of nodes in a tree.

THEOREM 11.12 ***The Number of Edges***

A tree with n nodes has $n - 1$ edges.

Proof: The proof is quite simple and elegant. Observe that every node except the root has exactly one adjacent edge on the unique path joining the node to the root. Every edge has two incident nodes. Partner each edge with its incident node that is farther from the root (determined by the respective levels). There is then a one-to-one pairing between edges in the tree and nonroot nodes. Since there are $n - 1$ nodes that are not the root, there must be $n - 1$ edges. □

THEOREM 11.13 ***Full m-ary Trees***

Let T be a full m-ary tree with n nodes, l leaves, and i interior nodes. Then

- $n = l + i$
- $n = mi + 1$

Proof: All nodes are either interior nodes or leaves (but not both). Thus $n = l + i$.

The second relationship is valid since all nodes are either the root node or else have a parent. The root node is the only node that is not a child. Every interior node is a parent of m children, so there must be mi children. Adding the root node leads to $n = mi + 1$. □

The next theorem expresses the number of nodes, interior nodes, and leaves in a maximal complete m-ary tree as functions of m and the height h of the tree.

THEOREM 11.14 ***Maximal Complete m-ary Trees***

Let T be a maximal complete m-ary tree of height h. Then T has

- $n = \frac{m^{h+1}-1}{m-1}$ nodes
- $l = m^h$ leaves
- $i = \frac{m^h-1}{m-1}$ interior nodes

Proof: At level 0 there is $m^0 = 1$ node. At the next level there will be $m^1 = m$ nodes, since every interior node in a full tree has m children. At level 2 there will be m^2 nodes since there are m nodes at level 1, each having m children. In general, there will be m^j nodes at level j. In particular, there will be m^h leaves, since all leaves in a maximal complete m-ary tree of height h are at level h.

The total number of nodes will be $\sum_{i=0}^{h} m^i = \frac{m^{h+1}-1}{m-1}$. Since $n = i + l$, the number of interior nodes is

$$i = n - l = \frac{m^{h+1}-1}{m-1} - m^h = \frac{m^h-1}{m-1}$$

□

COROLLARY 11.15 ***Balanced m-ary Trees***

Let T be a complete m-ary tree of height h having n nodes. Then

$$\frac{m^h-1}{m-1} < n \leq \frac{m^{h+1}-1}{m-1}$$

Proof: The nodes in levels 0 through h-1 form a maximal complete m-ary tree of height h-1. Theorem 11.14 indicates that there are $\frac{m^h-1}{m-1}$ nodes in levels 0 through $h-1$. There must be at least 1 and at most m^h more nodes on the final level. □

Binary trees are often used to store information in a computer. The height of the tree determines how quickly information can be retrieved.

THEOREM 11.16 ***The Height of a Binary Tree***

Suppose T is a binary tree with n nodes and height h. Then

$$\lceil \log_2(n+1) - 1 \rceil \leq h \leq n - 1$$

Proof: The proof, which uses Corollary 11.15, is left as an exercise. □

One rather interesting idea related to m-ary trees is to consider the nodes that are missing. These nodes are formally defined in the next definition.

DEFINITION 11.17 ***External Nodes***

Suppose a node in an m-ary tree has k children, with $0 \leq k \leq m$. Then the node is said to have $m - k$ *external nodes*. The external nodes represent the node's missing children. External nodes are often depicted as squares, connected by dashed edges.

EXAMPLE 11.4 **Some External Nodes**

The trees in Figure 11.10 are a binary and a ternary tree (respectively). They each have nine external nodes.

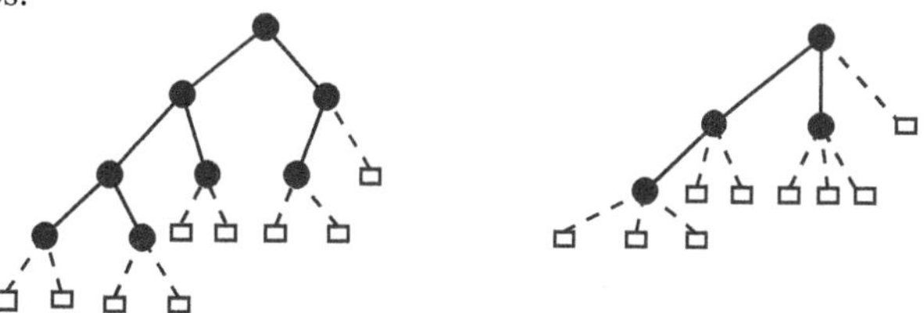

Figure 11.10 A binary tree and a ternary tree.

■

✔ Quick Check 11.3

1. There are five distinct ordered binary trees with three nodes. In each case, count the number of external nodes.
2. Draw four or five distinct binary trees having four nodes. Determine the number of external nodes in each case.
3. Make a conjecture about the number of external nodes there are for a binary tree with n nodes.

THEOREM 11.18 ***Maximal Complete Binary Trees***

The sum of the heights of the subtrees in a maximal complete binary tree with n nodes and height h is $n - h - 1$.

Proof: The proof uses a clever counting argument (from [99]). The next few sentences present the main idea. For each subtree of height k in the tree, color k of the edges. Do this in a manner that never colors an edge twice. After all the coloring has been completed, count the number of colored edges.

The choice of edges to color is easy: Start with the edge leading to the left child. After that, always take the right branch, continuing until the bottom level is reached. (It might be helpful to read Example 11.5 on page 680 before continuing with this proof.) The number of edges on a path from any node to the bottom level will always correspond to that subtree's height because all paths to the bottom have the same length in a maximal complete binary tree.

To see that the colored edges will not overlap, suppose node x is an ancestor of node y. It may happen that node y appears on the path that is being colored from x down toward the bottom level. The path "owned" by y will branch to the left, but the path "owned" by x will branch to the right. The two paths will therefore be disjoint.

Which edges will be colored? Let x be any node. If x has a left child, the edge joining x and its left child will always be colored. Now suppose the node, x, has a parent, y. Consider the edge leading to x's right child (if there is one). That edge will be colored as part of the path associated with y if x is a left child of y. If x is a right child of y, the edge toward x's right child will be colored as part of the path associated with x's grandparent (assuming there *is* a grandparent).

After the coloring is complete, the only edges that are not colored are those along the rightmost path from the root to the bottom level. There are exactly h such edges. Theorem 11.12 asserts that there are $n - 1$ edges in all, so the number of colored edges is $(n-1) - h = n - 1 - h$. The colored edges are in one-to-one correspondence with the heights of the subtrees, so the sum of the heights is the number of colored edges. □

EXAMPLE 11.5 **Coloring Edges**

Consider the maximal complete graph with 15 nodes (Figure 11.11).

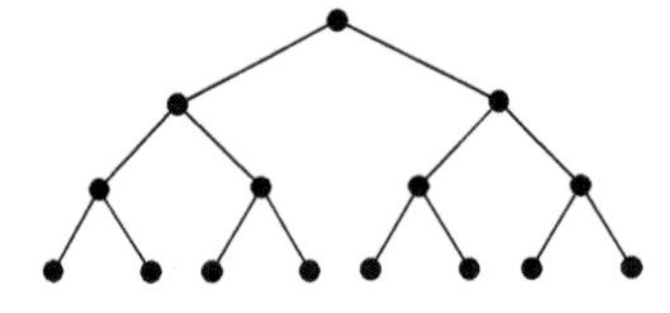

Figure 11.11 A maximal complete graph with 15 nodes.

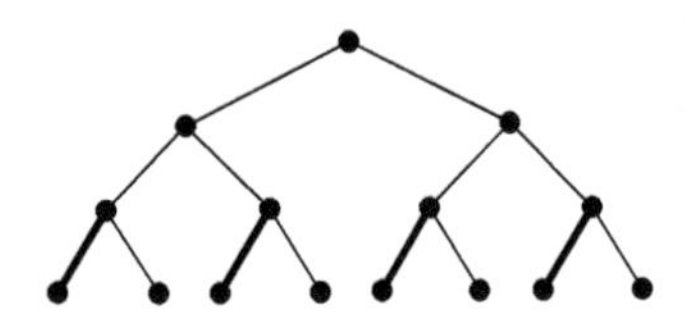

Figure 11.12. Coloring height 1.

Every node on the bottom level (the leaves) is the root of a subtree with height 0. No edges need to be colored to count the sum of these heights. The next level from the bottom consists of nodes that are the roots of subtrees with height 1. One edge should be colored for each of these nodes. Figure 11.12 colors the edge incident with the left child of each of these nodes.

Moving up one level, each subtree needs two edges colored. The newly colored edges are shown in Figure 11.13.

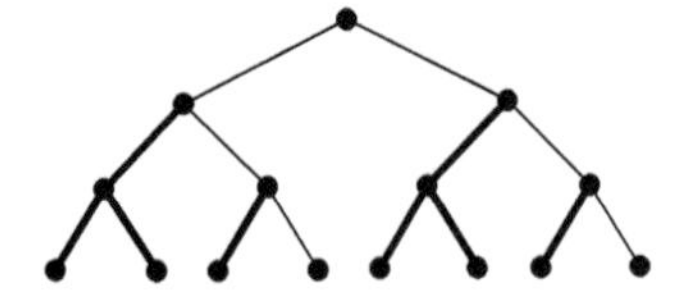

Figure 11.13. Coloring height 2.

Finally, the root node needs three more edges colored. Figure 11.14 shows the final, fully colored tree.

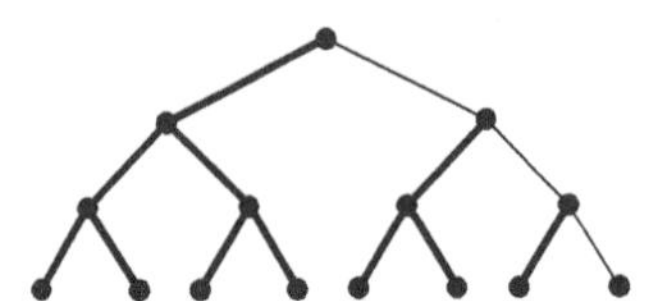

Figure 11.14 The fully colored tree.

The only edges that have not been colored are the three edges along the rightmost path in the tree. ■

COROLLARY 11.19 ***Average Height in Maximal Complete Binary Trees***

The average height of subtrees in a maximal complete binary tree with n nodes is

$$1 - \frac{\log_2(n+1)}{n}$$

Proof: The previous theorem implies that the average height is $\frac{n-h-1}{n} = 1 - \frac{h+1}{n}$. Since the tree is a maximal complete binary tree, $n = 2^{h+1} - 1$ (Theorem 11.14), or $h = \log_2(n+1) - 1$. Thus, the average height of the subtrees is $1 - \frac{\log_2(n+1)}{n}$. □

The corollary may seem unintuitive at first because it asserts that the average height is less than 1. However, there are $2^{h+1} - 1$ nodes, and 2^h of them (just over half), are roots of subtrees with height 0. The pyramid structure of the tree ensures that very few subtrees exist with much larger heights.

11.1.1 Exercises

The exercises marked with ★ have detailed solutions in Appendix H.

1. Let v_a and v_b be distinct vertices in a simple graph. If $v_a v_1 v_2 \cdots v_k v_b$ and $v_a w_1 w_2 \cdots w_n v_b$ are distinct paths from v_a to v_b, prove that the walk $v_a v_1 v_2 \cdots v_k v_b w_n \cdots w_2 w_1 v_a$ contains a cycle.

2. Prove Proposition 11.20.

PROPOSITION 11.20

Every tree with two or more nodes has at least two nodes of degree 1. Every tree with more than two nodes has at least one node with degree greater than 1.

3. Let T be a full m-ary tree. Prove the following:
   (a) ★ If T has n nodes, then there are $i = \frac{n-1}{m}$ interior nodes and $l = \frac{n(m-1)+1}{m}$ leaves.
   (b) If T has i interior vertices, then there are $n = mi + 1$ nodes and $l = i(m-1) + 1$ leaves.
   (c) If T has l leaves, then there are $n = \frac{ml-1}{m-1}$ nodes and $i = \frac{l-1}{m-1}$ interior nodes.
4. Suppose a con artist starts a chain letter asking people to send one dollar to the name at the top of the list, cross that name off and add their own name at the bottom, and then mail the letter to four others. The con artist places five bogus names on the list, all with PO boxes that she is renting. She mails the letter to four victims. Suppose everybody that receives the letter slavishly follows the instructions. How much money will the con artist receive?
5. Suppose you want to organize a single elimination tournament. If there are 500 contestants, what is the minimum number of rounds that will be needed?
6. ★ Suppose 10,250 items need to be stored in the nodes of an m-ary tree of height five (one item per node). What is the smallest value of m for which this is possible if
   (a) the items can be stored in all nodes?
   (b) the items can only be stored in leaf nodes?
7. Use a tree to represent the table of contents for this chapter.
8. The following graph is not balanced. Choose a new root which will produce a balanced embedding. Show the embedding.

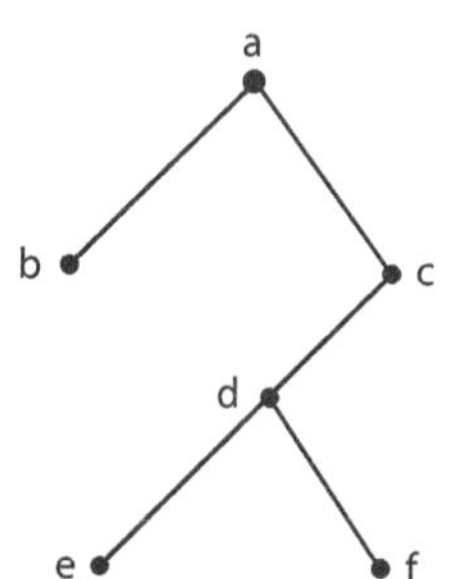

9. It is possible to create ordered, rooted trees that have a Fibonacci structure. The Fibonacci trees are defined recursively. Let F_0 and F_1 be trees with a single node. Define F_n as the tree in which the root's left subtree is F_{n-1} and the root's right subtree is F_{n-2}.
   (a) Draw F_2, F_3, F_4, and F_5.
   (b) Write a recurrence relation for the number of nodes in F_n. You need not solve the recurrence relation.
   (c) (Difficult: Requires Section 7.4) Solve the recurrence relation.
10. ★ Write a recursive algorithm that determines the height of a tree. The algorithm should begin

```
integer height(x)
```

where x is a node. The algorithm should give the correct answer if it is applied to the root node.

11. ★ Suppose T is a binary tree with n nodes and height h. Prove that $\lceil \log_2(n+1) - 1 \rceil \le h \le n - 1$.
12. Prove that adding a new edge to a tree without adding any new nodes will create a graph that is not a tree.
13. Prove that every binary tree with $n \ge 1$ nodes has $n + 1$ external nodes.
14. Find a function of n and m for the number of external nodes in every m-ary tree with $n \ge 1$ nodes. Then prove the formula is correct. [*Hint*: Let the formula for an m-ary tree with n nodes be a_n. Write an expression for a_n in terms of a_{n-1} and m.]
15. Prove that a simple, connected graph with $n > 1$ vertices and $n - 1$ edges must contain a degree 1 vertex.
16. ★ Use Exercise 15 to prove that a connected, simple graph with n vertices and $n - 1$ edges is a tree.
17. Use Exercise 23 on page 152 to provide an alternative proof of Theorem 11.18 (page 679).
18. Each of the following statements is either true (always) or false (at least sometimes). Determine which option applies for each statement and provide adequate explanation for your choice.
   (a) ★ A node's level is equal to the height of the subtree for which that node is the root.
   (b) A tree can contain a cycle.
   (c) The leaves in a balanced tree appear only at the two levels that are farthest from the root.
   (d) A tree with one node has height 1.
19. Each of the following statements is either true (always) or false (at least sometimes). Determine which option applies for each statement and provide adequate explanation for your choice.
   (a) A complete tree is a balanced tree of height, h, having all its leaves at level h.
   (b) In a full m-ary tree, every node is either a leaf or it has exactly m children.
   (c) A complete tree is always a full tree.
   (d) A full tree is always a complete tree.
20. Let a_n represent the number of distinct ordered rooted binary trees with n nodes.
   (a) Write a recurrence relation for a_n.
   (b) Fill in the final two entries in the following table.

n	a_n
0	1
1	1
2	2
3	5
4	
5	

> **DEFINITION 11.21** ***Eccentricity; Center***
> The *eccentricity* of a vertex, v, in a tree is the maximum length of a path starting at v. A vertex, v, in a tree is a *center* if its eccentricity is less than or equal to the eccentricity of all other vertices in the tree.

21. For each tree, use Definition 11.21 to calculate the eccentricities of each vertex and then identify any centers.

(a) ★

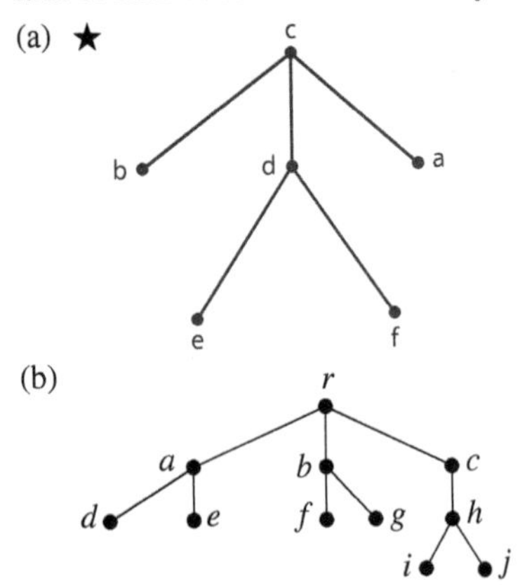

(b)

(c)

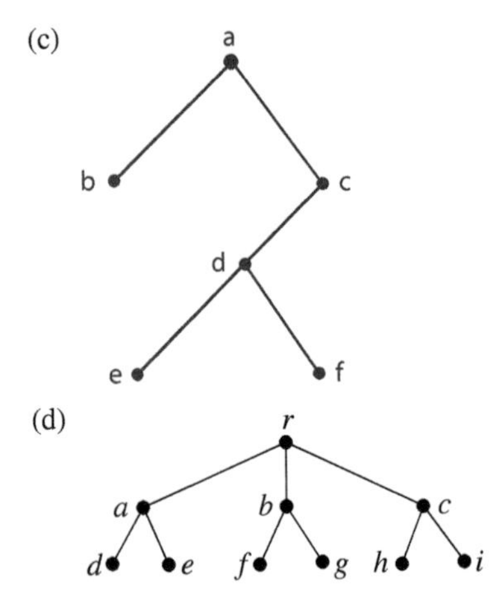

(d)

22. Repeat Exercise 21, but first remove all the nodes of degree 1 from each tree. [So, for example, the tree in part (a) will have only nodes c and d left.]

23. Prove that every nonempty tree has either one center or else two adjacent centers. [*Hint*: Use complete induction, after first looking at Exercises 21 and 22.]

11.2 Traversal, Searching, and Sorting

Since one of the major applications of trees is to provide a repository for storing information for later retrieval and processing, it is important to devise appropriate mechanisms for visiting all the nodes in a tree. In this context, it is assumed that any tree is an ordered rooted tree (Definition 11.10 on page 675). With this assumption, it is possible to impose a precedence order upon the nodes. This ordering can then be used to determine the sequence in which the nodes will be visited.

> **DEFINITION 11.22** ***Traversal***
> The process of visiting every node of a tree in a systematic manner is called a *traversal* of the tree.

For the remainder of this section, the discussion will be limited to binary trees.

11.2.1 Traversing Binary Trees

In an ordered binary tree, there is a natural (embedding dependent) ordering of a node's children. The children are denoted as the left child and the right child. It is traditional to regard the left child as the first child and the right child as the second child.

Trees are inherently recursive (since every node can be considered the root of its subtree). The simplest mechanism for systematically visiting every node is also recursive. Once the commitment has been made to use a recursive traversal and once the left child has been given precedence over the right child, there are only three kinds of traversal of interest.

The three kinds of traversal are *preorder*, *inorder*, and *postorder*. The traversals are completely determined by the relative order in which a parent node and its two children are visited.[1]

[1] If one or more children are missing, the visit can still be understood to have occurred, but no processing will occur at that missing node.

DEFINITION 11.23 ***Preorder; Inorder; Postorder***

In a *preorder traversal*, the parent node is processed before the children. The nodes are visited in the order parent, left child, right child.

An *inorder traversal* processes the left child first. The nodes are visited in the order left child, parent, right child.

In a *postorder traversal*, the parent node is processed after the children. The nodes are visited in the order left child, right child, parent.

Notice that the names of the traversals focus on the position that the parent node takes, relative to the left and right children. The left child always precedes the right child.

EXAMPLE 11.6 **Traversals**

The tree in Figure 11.15 is an ordered rooted binary tree.

A preorder traversal will visit the nodes in the order

$$g \quad h \quad c \quad b \quad e \quad a \quad d \quad f$$

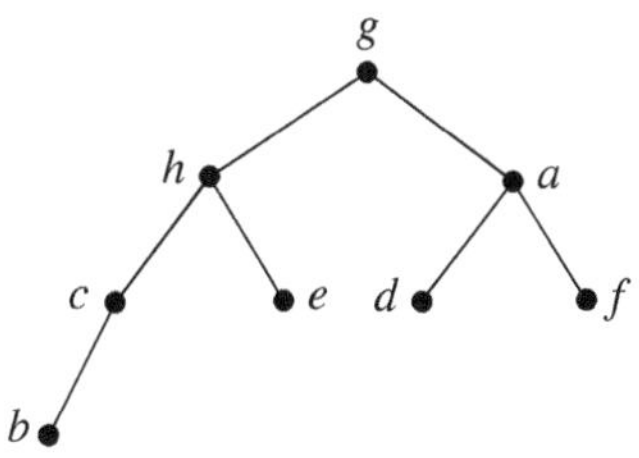

The traversal starts at the root node and processes that node. It then looks at the left child, h, and processes h before looking at the children of h. The left child of h is c, so c (as the root/parent of its own subtree) is processed next. The left child, b, of c is next. It has no children, so the traversal goes back to the right child of c ($c \rightarrow$ left child of $c \rightarrow$ right child of c). There is no right child of c, so the traversal looks for the right child of h ($h \rightarrow$ left child of $h \rightarrow$ right child of h). The node e is processed. It has no children. Since both children of h have now been processed, the traversal looks for the right child of g. The node a is processed, then its left child, d, then its right child, f, completing the traversal.

Figure 11.15. An ordered rooted binary tree.

An inorder traversal will visit the nodes in the order

$$b \quad c \quad h \quad e \quad g \quad d \quad a \quad f$$

In this case, the traversal again begins at the root, but before processing g it must process h, the left child of g. Before processing h it must process h's left child, c. Before processing c, it must process c's left child, b. Since b has no left child, the node b will be the first node actually processed. The right child of b should be next, but there is no right child. The node c can now be processed. If c had a right child, it would be processed next. Since there is no right child of c, the next node will be h. The right child, e, of h comes next. It has no children so the root node, g, is finally processed (because the entire left subtree has been completed). The node a is next but cannot be processed until its left child, d, is visited. Next, a is processed, followed by its right child, f. The traversal is complete.

A postorder traversal will visit the nodes in the order

$$b \quad c \quad e \quad h \quad d \quad f \quad a \quad g$$

You should work carefully through the postorder traversal to verify the result. ■

✔ Quick Check 11.4

1. List the nodes in the order they will be visited in a
   (a) Postorder traversal.
   (b) Preorder traversal.
   (c) Inorder traversal.

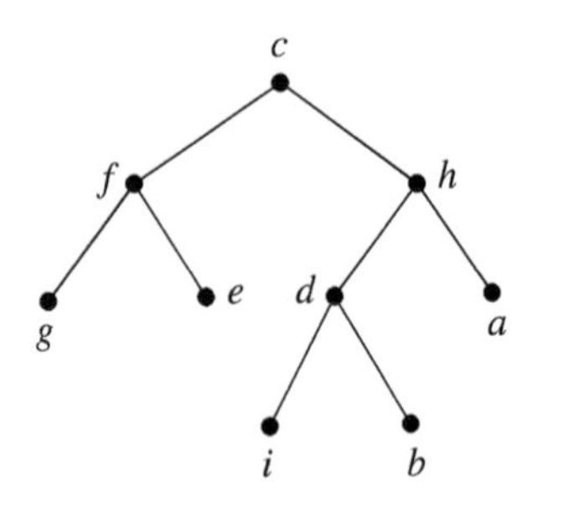

The following example is a simple application of tree traversals. It provides a mechanism for translating from *infix* notation (normal algebraic notation) to *postfix* notation (also called *reverse Polish* notation). With infix notation, the arithmetic operator is written between the two operands. For example, to add the numbers 3 and 4, we write $3+4$ using infix notation. Postfix notation writes the two operands followed by the operator: $3\,4\,+$.

The advantage of postfix notation is that there is never any need to use parentheses. With infix notation, ambiguities can arise unless we establish precedence conventions. To write an expression that changes the convention, we need to use parentheses. For example, the expression $3 + 4 * 5$ should evaluate to 23 using the normal rule that multiplication is done before addition.[2] We would need to write $(3+4)*5$ if we wanted 35 as the result. Using postfix notation, the expression $3\ 4\ 5\ *\ +$ will evaluate to 23 and the expression $3\ 4\ +\ 5\ *$ will evaluate to 35. Parentheses are unnecessary in either case.

EXAMPLE 11.7

Converting Infix to Postfix

An infix expression can be converted to an equivalent postfix expression by creating a binary tree for which each operator is the parent of a subtree with its left subtree containing the (possibly) parenthesized expression that appeared to the left of the operator, and the right subtree containing the (possibly) parenthesized expression that appeared to the right of the operator in the original infix expression.

For example, the expression $((4 + 5) \cdot 3)^2$ could be written as ((4 + 5) * 3)^2 and converted to the tree in Figure 11.16 (using the standard operator precedence rules).

Figure 11.16. The tree for $((4+5)*3)^2$.

Notice that the interior nodes are all operators and the leaf nodes are all operands. This will always be true.

The postfix expression can be obtained by performing a postorder traversal of the tree. The processing consists of writing down the label of the node.

For the preceding tree, the postorder traversal produces 4 5 + 3 * 2 ^. The expression (in either format) simplifies to 729. ■

The previous example did not show how to treat the convention for working with operators having the same precedence. For example, the expression $a * b * c + d$ is commonly interpreted to mean $((a * b) * c) + d$. The addition is done last because multiplication has higher precedence than addition. The two multiplications have the same precedence, so they are evaluated in a left-to-right manner. In general, any sequence of additions or subtractions (possibly mixed) are to be evaluated in left-to-right order. Similarly, any sequence of multiplications or divisions (possibly mixed) are to be evaluated in left-to-right order.

The exception is a sequence of exponentiations. They should be evaluated in right-to-left order. Exercise 10 on page 691 extends this discussion to *unary minus*.

[2] The symbol * is a commonly used keyboard symbol that denotes multiplication.

EXAMPLE 11.8

A More Complex Infix to Postfix Conversion

The infix expression $abc + d^{e^f}$ should be interpreted as $((ab)c) + d^{(e^f)}$. The associated binary tree is displayed in Figure 11.17.

The postfix expression is $a\ b * c * d\ e\ f$ ^ ^ +. ■

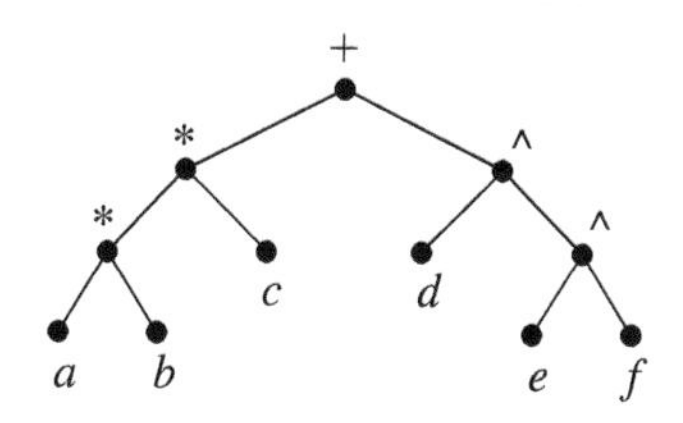

Figure 11.17. The tree for $abc + d^{e^f}$.

11.2.2 Binary Search Trees

One very clever way to exploit the structure in an ordered rooted binary tree is to create what is known as a *binary search tree.*

> **DEFINITION 11.24** ***Binary Search Tree***
> A *binary search tree* is an ordered rooted binary tree used to organize and store data that can be lexicographically ordered. It is defined by the property that the data in a node's left child must precede the node's data in lexicographical order and the data in the right child must lexicographically follow the node's data.

EXAMPLE 11.9

A Binary Search Tree

The tree in Figure 11.18 is a small binary search tree. The data are "stored" as the labels on the nodes. In this example, the data are a collection of alphabet letters. The lexicographical order is the standard ordering of the letters in the alphabet. ■

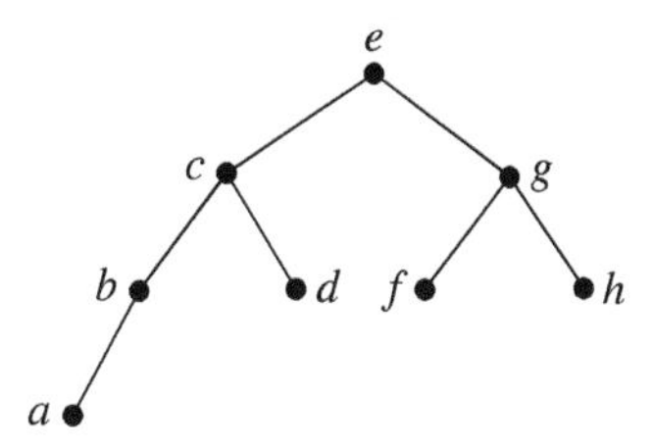

Figure 11.18. A binary search tree.

How are binary search trees formed? There is a very simple process. Line the data up in any order. The first data item will become the root node. After that, each new data item is compared to the root's data. If the new item comes before the root's data lexicographically, then move to the left. If the new item comes after the root's data, then move to the right. At each new child node, make the same decision: Move left if the new item is smaller, move right if the new item is larger. When there is no edge to follow, move in the proper direction by creating a new node. Make the new node's data equal to the new item. Each new item starts at the root and moves down the tree in the manner just described until an empty spot on the tree is found.

EXAMPLE 11.10

Constructing a Binary Search Tree

The previous example was constructed from the data being added in the order

$$e\ g\ c\ f\ h\ b\ d\ a$$

Figure 11.19 shows the intermediate steps.

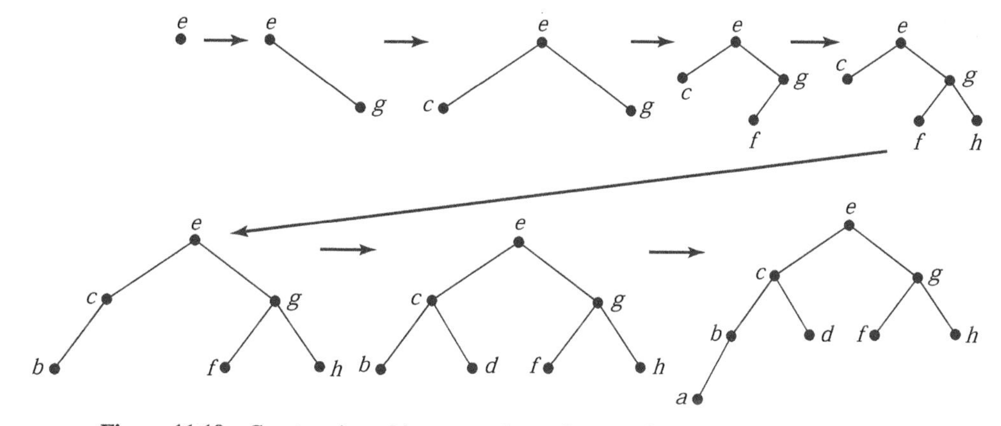

Figure 11.19 Constructing a binary search tree for *e g c f h b d a*. ■

A binary search tree can be used to access the data once the tree has been built. The process that was used to build the tree can be used essentially unchanged to look for an item of information that may already be in the tree. Just follow the path until the item is found or until the path ends without finding the item. If the path ends without finding the item, we conclude that the item of information is not present.

Quick Check 11.5

1. Construct a binary search tree using the data: dog cat monkey aardvark kangaroo.
2. List the path followed to find the data item "lemming" in the binary search tree in question 1.
3. List the path followed to find the data item d in the binary search tree

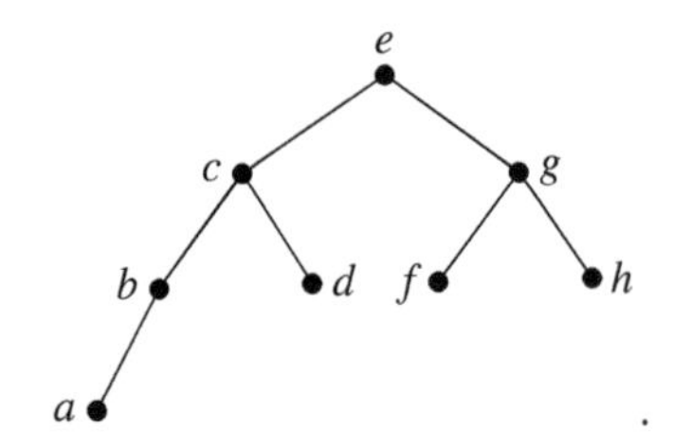

If the data are already in sorted order, the binary search tree will be a chain because each new insertion will be a right child. In this worst-case situation, constructing the tree requires a comparison of the kth data item to the $k - 1$ previously constructed nodes. The construction thus requires $0 + 1 + \cdots + (n-1) = \frac{(n-1)n}{2}$ comparisons.

The following theorem examines the effort needed to construct a binary search tree in a best case situation.

THEOREM 11.25 ***Binary Search Trees***

Looking for an item in a complete binary search tree with n nodes is in $\Theta(\log_2 n)$.

Constructing a binary search tree with n nodes requires at least $\Theta(n \log_2 n)$ comparisons. This lower bound is achieved if the tree is complete at each stage of the construction.

Proof:

Searching for an Item

Consider a complete binary tree. Corollary 11.15 on page 678 indicates that the height, h, of the tree satisfies $2^h - 1 < n \leq 2^{h+1} - 1$. Adding 1 to all three parts and taking base 2 logarithms leads to $h < \log_2(n+1) \leq h + 1$. This can be rewritten as

$$\log_2(n+1) - 1 \leq h < \log_2(n+1)$$

Notice that $\log_2(n) < \log_2(n+1) \leq \log_2(n+n) = \log_2(2n) = 1 + \log_2(n)$ and so

$$\log_2(n) - 1 < \log_2(n+1) - 1 \quad \text{and} \quad \log_2(n+1) \leq \log_2(n) + 1$$

Combining this with the inequalities for h leads to

$$\log_2(n) - 1 < h < \log_2(n) + 1$$

Therefore, $h \in \Theta(\log_2 n)$.

Finally, in a search tree, the search path is bounded by the height of the tree, so the search is in $\Theta(\log_2 n)$.

Constructing the Tree

Assume the data are arranged so that new nodes are added one level at a time. Each new level is filled before the next level begins. This is clearly a best case situation since there is no other way to keep the number of comparisons as small at each stage.

Placing the first data item in the root node does not require any comparisons. The two items in level 1 will require one comparison apiece. The $4 = 2^2$ items in level 2 will each require two comparisons. The $8 = 2^3$ items in level 3 each require three comparisons. In general, the 2^k items that will become the data for level k each require k comparisons to determine their proper place in the evolving tree.

The process will end when the nth node has been placed. That node will appear at level h, where n is bounded by the expressions below (read from the inside out).

$$2^h - 1 = \frac{2^h - 1}{2 - 1} = \sum_{k=0}^{h-1} 2^k < n \leq \sum_{k=0}^{h} 2^k = \frac{2^{h+1} - 1}{2 - 1} = 2^{h+1} - 1$$

(The left-hand sum, $\sum_{k=0}^{h-1} 2^k$, counts the number of nodes in the subtree that excludes the final level. The right-hand sum, $\sum_{k=0}^{h} 2^k$, counts the number of nodes if the final level is totally filled.)

Adding 1 to all parts of $2^h - 1 < n \leq 2^{h+1} - 1$ and taking base 2 logarithms leads to

$$h < \log_2(n+1) \leq h + 1$$

Thus, $h = \lceil \log_2(n+1) \rceil - 1$.

The total number of comparisons, T, has a contribution of $k \cdot 2^k$ from level k for $k < h$. Level h may not be full. Thus

$$\sum_{k=0}^{h-1} k \cdot 2^k < T \leq \sum_{k=0}^{h} k \cdot 2^k$$

Exercise 24 on page 152 implies T is between the lower and upper bounds T_l and T_u:

$$T_l = 2 + (h-2)2^h < T \leq 2 + (h-1)2^{h+1} = T_u$$

One extreme case will occur when the final level is totally filled ($n = 2^{h+1} - 1$). In this case, $h = \log_2(n+1) - 1$.

So when the final level is filled,

$$T_u = 2 + (\log_2(n+1) - 2)2^{\log_2(n+1)} = (n+1)\log_2(n+1) - 2n$$

The other extreme is when the final level has only one node. We can make this even more extreme by removing that item. In that case, we would have a tree with one less level. Substituting $h - 1$ for h in the previous calculation,

$$T_l = 2 + (\log_2(n+1) - 3)2^{\log_2(n+1)-1} = 2 + \frac{(n+1)}{2}(\log_2(n+1) - 3)$$

The relationship $\log_2(n+1) \in \Theta(\log_2 n)$ has already been established. The expressions for T_l and T_u imply that $T \in \Theta(n \log_2(n))$. □

A major topic in a data structures course is how to build binary search trees that are as short as possible (hence the desire for balanced search trees). That topic will not be explored in this textbook.

The Average Case for a Binary Search

In Section 4.2.4, it was shown that algorithm **binarySearch** is in $\Theta(1)$ in the best case and in $\Theta(\log_2(n))$ in the worst case. It was also suggested that the average case is likely to be in $\Theta(\log_2(n))$. That assertion can now be proved.

A binary search uses a sorted list and keeps halving the list until the target item is found or the next sublist is empty. By properly placing the list items in a binary search tree, the steps of any particular binary search with the list can be exactly duplicated as a search in the binary search tree. In particular, since the binary search always cuts the list in half until the final step, the corresponding binary search tree must be balanced and all interior nodes at levels 0 through $h - 2$ will have two children.

A recursive algorithm that builds the desired binary search tree is not hard to produce. It mimics that halving done in the **binarySearch** algorithm to find the center element. The expression $\left\lfloor \frac{\text{low} + \text{high}}{2} \right\rfloor$ becomes the equivalent $\left\lfloor \frac{\text{len} - 1}{2} \right\rfloor$, where "len" is the number of elements in the list. The value at that center location in the list becomes the value of a node. The node's children are then recursively created to correspond to the center elements in the sublists to the left and right of the central element. The notation assumes that the sorted list, L, contains elements denoted as L_i, where i starts at 0. The algorithm assumes that sublist positions are always renumbered, starting at 0.

```
1: tree node center(sorted list L)
2:     create a new tree node, temporarily named: tnode
3:     len = number of elments in L
4:     cen = ⌊(len - 1)/2⌋
5:     set tnode's value to L_cen
6:     set tnode's left child = center(sublist of L to the left of L_cen)
7:     set tnode's right child = center(sublist of L to the right of L_cen)
8:     return tnode
9: end center
```

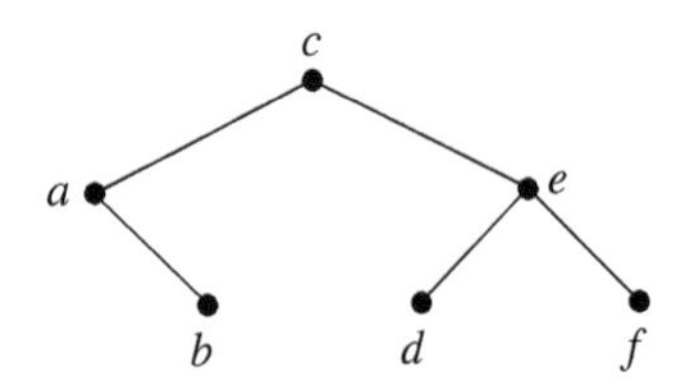

Figure 11.20. A binary search tree for $\{a, b, c, d, e, f\}$.

You can verify that the list $\{a, b, c, d, e, f\}$ corresponds to the binary search tree in Figure 11.20. If you perform a binary search for each item in the list and examine the set of intermediate comparisons, you will see that they are identical to the path used to search for the items in the tree.

The correspondence between binary searches in a sorted list and searches in a carefully constructed binary search tree means that the average behavior of the binary search algorithm will be the same as the average behavior of the search in the binary search tree. This second average is not hard to calculate. The next theorem is the key.

THEOREM 11.26 ***Average Level of a Maximal Complete Binary Tree***

The average level for nodes in a maximal complete binary tree having height h is

$$\frac{(h-1)2^{h+1}+2}{2^{h+1}-1} = (h-1) + \frac{h+1}{2^{h+1}-1}$$

Proof: At level i, there are 2^i nodes. The sum of the levels is thus $\sum_{i=0}^{h} i2^i = (h-1)2^{h+1} + 2$ using the result of Exercise 24 on page 152. A maximal complete binary tree has $n = 2^{h+1} - 1$ nodes. The average level is thus

$$\frac{(h-1)2^{h+1}+2}{2^{h+1}-1} = \frac{(h-1)\left(2^{h+1}-1\right)+(h-1)+2}{2^{h+1}-1} = (h-1) + \frac{h+1}{2^{h+1}-1} \qquad \square$$

COROLLARY 11.27 ***Average Case Binary Search***

A successful binary search is in $\Theta(\log_2(n))$ on average.

Proof: Given a sorted list with n items, algorithm **center** will create a balanced binary search tree with all interior nodes at levels 0 through $h-2$ having two children. Other than having all nodes at level h left justified, the tree is essentially a complete tree. Thus, Corollary 11.15 asserts that $2^h - 1 < n \le 2^{h+1} - 1$.

The average level will therefore be between the maximal complete trees at these extremes. That is,

$$(h-2) + \frac{h}{2^h - 1} < \text{average level} \le (h-1) + \frac{h+1}{2^{h+1}-1}$$

At these extremes, $h = \log_2(n+1) - 1$. So

$$\begin{aligned} \log_2(n+1) - 3 + \frac{2\log_2(n+1) - 2}{n-1} &< \text{average level} \\ &\le \log_2(n+1) - 2 + \frac{\log_2(n+1)}{n} \end{aligned}$$

The expressions on both sides of the inequality are in $\Theta(\log_2(n))$, completing the proof. □

11.2.3 Sorting

Binary search trees can be used to sort a collection of data. The procedure is simple. Build a search tree from the data, then use an inorder traversal to list the data. Building the tree is in $\Theta(n \log_2 n)$ on average,[3] whereas traversing the tree is in $\Theta(n)$, so the average case combined procedure is in $\Theta(n \log_2 n)$.[4]

The inorder traversal will list the items in lexicographical order since the left child always precedes the parent node, and the right node always follows the parent node.

EXAMPLE 11.11 **Sorting**

The tree in Figure 11.21 is a binary search tree.

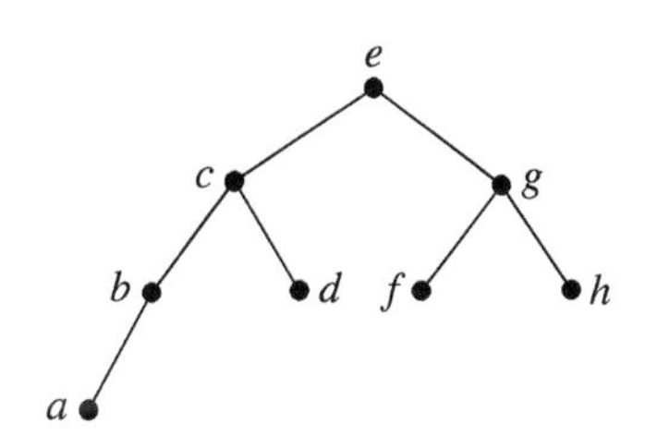

Figure 11.21 Sorting via a binary search tree.

An inorder traversal produces $a\ b\ c\ d\ e\ f\ g\ h$. ■

[3] See [99], page 506 for a proof.

[4] Theorem 4.14 on page 192.

11.2.4 Exercises

The exercises marked with ★ have detailed solutions in Appendix H.

1. For each tree, list the order that the nodes are processed using a preorder, inorder, and postorder traversal. Clearly label each traversal.

   (a) ★

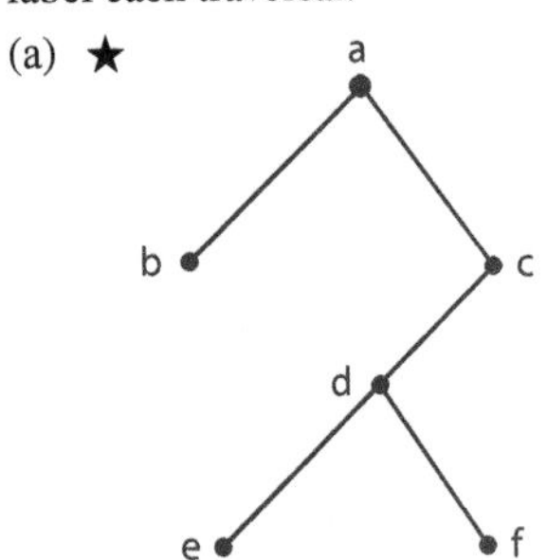

   (b)

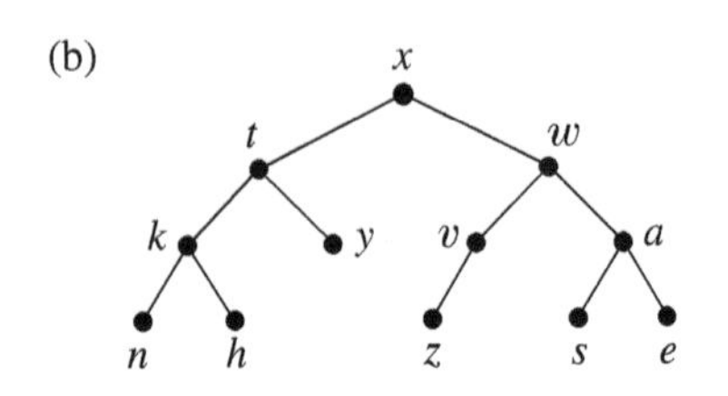

   (c)

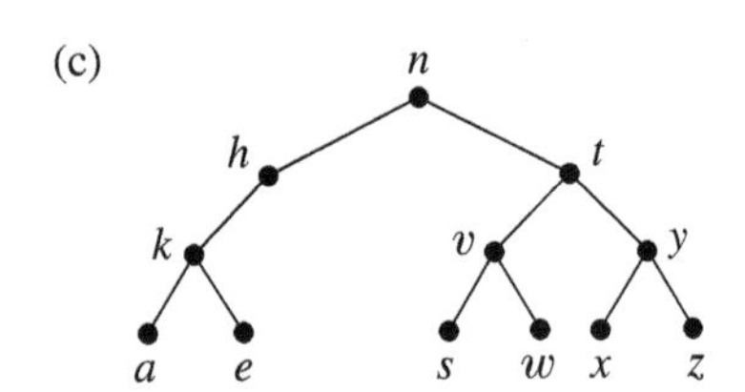

2. For each tree, list the order that the nodes are processed using a preorder, inorder, and postorder traversal. Clearly label each traversal.

   (a)

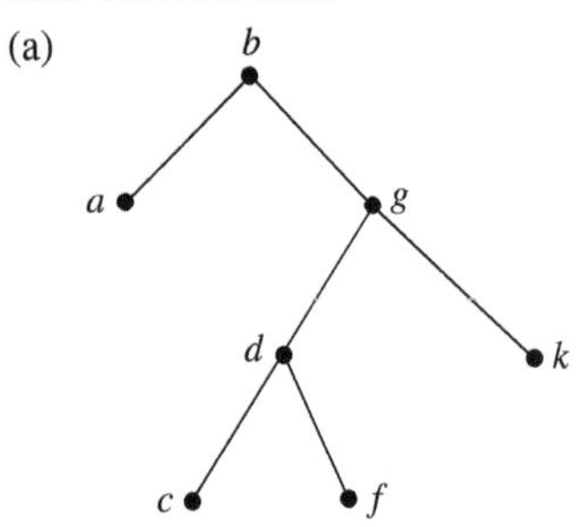

   (b)

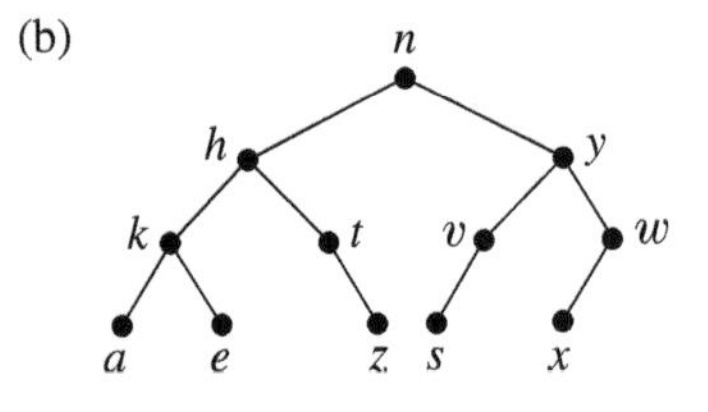

   (c)

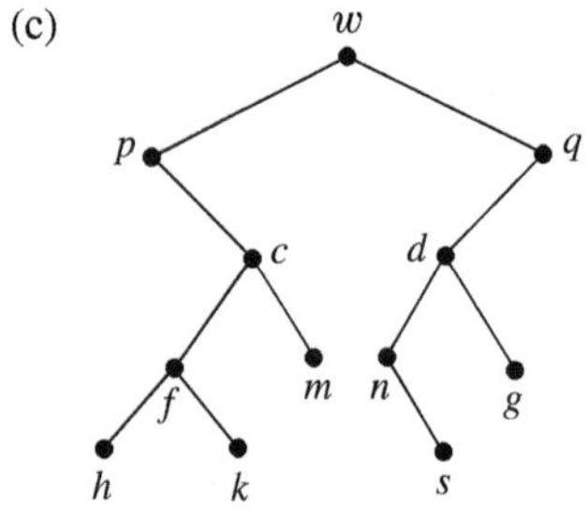

3. Create binary search trees for each data sequence. (Add the nodes in the order specified.)
   (a) ★ grape apple peach kiwi orange grapefruit tangerine starfruit mango cherry
   (b) *x r e f m s t d a b*
   (c) red blue green yellow orange violet pink fuchsia
4. Create binary search trees for each data sequence. (Add the nodes in the order specified.)
   (a) *d a h v r f g u m e c b*
   (b) oak redwood palm elm pine maple apple dogwood birch aspen cottonwood eucalyptus
   (c) Minneapolis Tucson San Diego Saint Paul San Francisco Albuquerque Trenton Detroit Fargo Bismark Saint Louis
5. List the order in which the nodes will be traversed in the following binary search tree when searching for the presence of *e*, which is not in the tree.

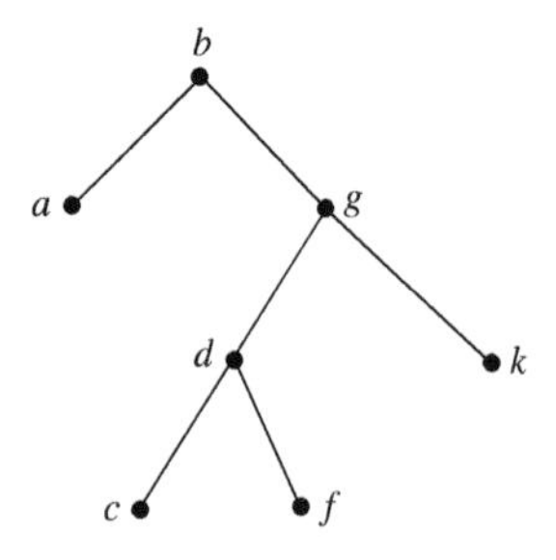

6. Is this a binary search tree? Fully explain your answer.

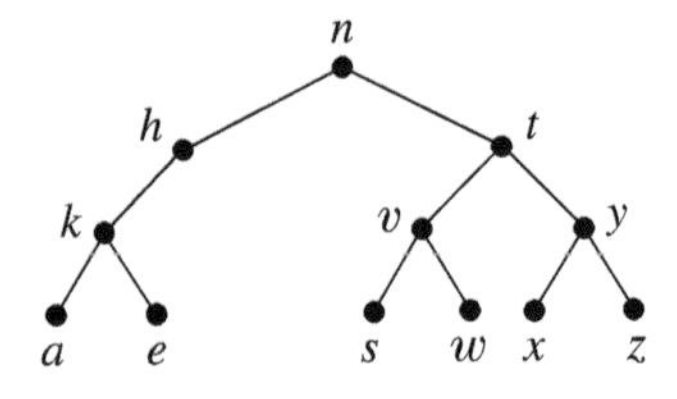

7. Some data sets contain duplicate items. It is possible to modify the definition of a binary search tree so that the left child must precede or be equal to the node. It is also possible to make the nodes in a binary search tree store data and also a count of the number of matching data items. An alternative to using a binary search tree to store such data is to use a ternary search tree. Create a definition for *ternary search tree*. Then build a ternary search tree for each list.
   (a) *c b d a c b a a d* (b) *d c e c a b e f f d*

8. Use binary trees to convert the following infix expressions into algebraically equivalent postfix expressions. Show the binary tree.

(a) $(a + b) + c \cdot d$

(b) $z \cdot (x \cdot y + w) - v$

(c) $a^x \cdot b^y + (c + d)^z$

(d) $(a \cdot b + c)^d + e$

9. Use binary trees to convert the following infix expressions into algebraically equivalent postfix expressions. Show the binary tree.

(a) ★ $x^{(y^z)} + (x^y)^z$

(b) $(a + b) \cdot (c + d)^n$

(c) $\left(a + (b + c)^d\right) + e \cdot f$

(d) $x^n + y \cdot z^m + w$

10. The infix-to-prefix algorithm can be extended to include unary minus as follows. First, create a new unary minus operator for the postfix expression. We will use $\sim$ for this symbol. The infix expression $-x$ would then be written in postfix as $x \sim$. Second, write any unary minus as a node (labeled by $\sim$) with no right child. The left child will be the tree for the expression that is being negated. For example, $-(x + y)$ becomes the following tree.

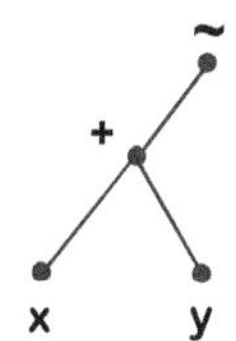

Use binary trees to convert the following infix expressions into algebraically equivalent postfix expressions. Show the binary tree.

(a) $-(-x + y)$

(b) $(-x)^c \cdot (-y)$

(c) $y^{(-x)^z}$

(d) $-x^n + y \cdot z^m + w$

11. Use the associative property of the real numbers to *completely* parenthesize the following expression with every possible ordering of operations (there are five of them). Then use trees to translate each of them into postfix.

$$w \cdot x \cdot y \cdot z$$

12. Use trees to sort the following lists of letters. Provide adequate explanation for your work.

(a) ★ *b r t w q f c s e*

(b) *p f g k n e w x a*

(c) *z x w q m k f a*

13. Fill in the missing details in the proof of Corollary 11.27.

(a) Show that $\log_2(n + 1) \in \Theta(\log_2(n))$.

(b) Show that $1 + \frac{1}{n} \in \Theta(1)$.

(c) Show that $\log_2(n + 1) - 2 + \frac{\log_2(n+1)}{n} \in \Theta(\log_2(n))$.

DEFINITION 11.28 ***Binary Heap***
An ordered rooted binary tree in which every parent is lexicographically less than or equal to both its children is called a *binary heap*.

14. Which of the following trees are heaps (using the standard alphabetical ordering)? Explain your answers.

(a) ★

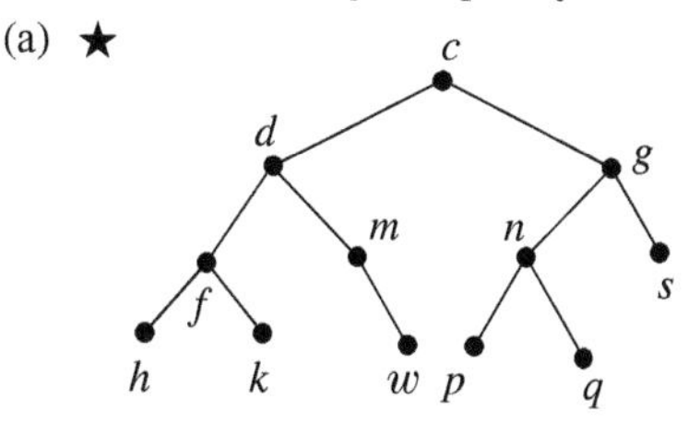

(b)

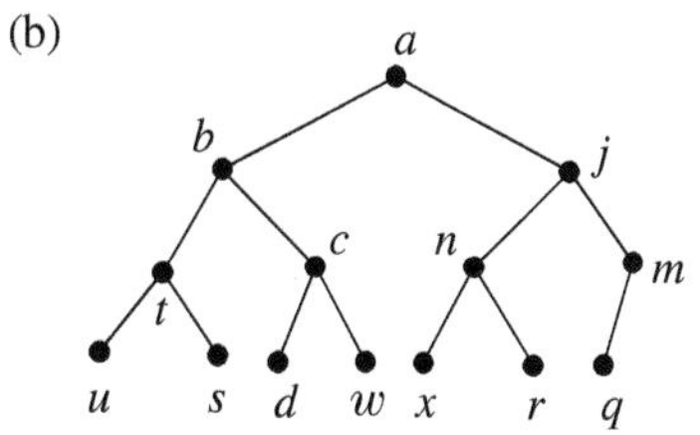

(c)

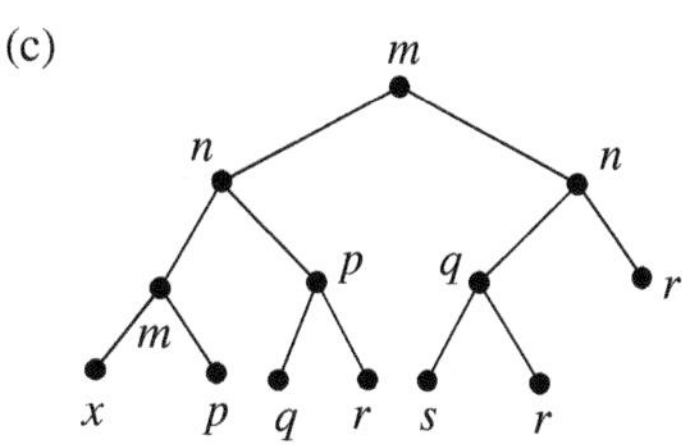

15. According to Definition 11.28, which binary search trees are also binary heaps?

16. ★ A binary heap can always be converted to a binary heap which is also a complete binary tree. (An efficient algorithm exists to accomplish this. See [99] for the details.) Assume that H is a complete binary heap with n nodes.

(a) What important property does the root node have (in relation to the other nodes in the heap)?

(b) Suppose the root node is removed and replaced by the node in the rightmost position of the final level (and that position eliminated from the tree). The new tree will probably not be a heap. Devise a $\Theta(\log_2(n))$ algorithm to restore the heap property.

(c) The algorithm that builds a complete binary heap is in $\mathcal{O}(n)$ (it is also in $\Theta(n)$). Devise a heap-based $\mathcal{O}(n \log_2(n))$ algorithm for sorting a list containing n items. (This algorithm is called *heap sort.*)

17. Use algorithm **center** on page 688 to create binary search trees that correspond to the following sorted lists.

(a) $a\ b\ c\ d\ e\ f$

(b) $a\ b\ c\ d\ e\ f\ g\ h$

(c) $a\ b\ c\ d\ e\ f\ g\ h\ i$

18. An alternative strategy for traversing binary trees is a level-order traversal. In a level-order traversal, nodes are processed in a left-to-right order, level by level, starting at the root and ordering the levels in increasing order. This is quite easy to accomplish by hand but requires more effort if done on a computer. The following pseudocode algorithm is an incorrect attempt to write a recursive computer algorithm for level order. Where is the error?

The algorithm can be used by invoking **failed-level-order** with the root node as the initial input value.

```
 1: void failed-level-order(node current)
 2:     # the root node needs special
          handling
 3:     if current does not have a parent
 4:        process(current)
 5:     # process everything at the
          descendant level
 6:     if current has a left child
 7:        process(leftChild)
 8:     if current has a right child
 9:        process(rightChild)
10:     # process everything at the
          next (grandchild) level
11:     if current has a left child
12:        failed-level-order(leftChild)
13:     if current has a right child
14:        failed-level-order(rightChild)
15:     return
16: end failed-level-order
```

19. Exercise 18 contains a failed attempt to create an algorithm for a level-order traversal. Such an algorithm is possible with the aid of a simple data structure called a *queue*. A queue is a data list in which new data are added to the back of the list and data are removed from the front of the list (so it is a first-in, first-out structure). A queue has two operations of interest here: *enqueue*, which adds a data item to the end of the queue, and *dequeue*, which removes the item at the front of the list.

Use a queue to write a nonrecursive pseudocode algorithm for a level-order traversal. Use the notation `enqueue(x)` to add node `x` to the back of the queue and `y = dequeue()` to remove the item at the front of the list and store it in `y`. The first line of the algorithm should look like the following sample.

```
void level-order(node rootNode)
```

[*Hint*: Since sibling nodes are not directly connected in a tree, it is best to collect the nodes at a level by using the links from their parent nodes.]

20. Use pseudocode to write recursive algorithms for preorder, inorder, and postorder traversals of a binary tree. Use the notation "leftChild" and "rightChild" to refer to the current node's left and right child, respectively. Use the notation "process(x)" to represent a command to process the information at node x.

21. Use pseudocode to write a recursive algorithm to search an unordered binary tree for the presence of a node whose value equals the input parameter. The algorithm should return "true" if the desired node is present and return "false" otherwise. The algorithm's first line should look like the following line:

```
boolean searchTree(node n, value x)
```

11.3 More Applications of Trees

This section contains three additional applications of trees. They may be read in any order (or one or more may be skipped).

11.3.1 Parse Trees

It is possible to capture the details of a derivation (Section 9.3.1) in a tree. Recall that derivations consist of a sequence of substitutions. A nonterminal symbol is replaced at each step using some production. This can be illustrated by Example 9.21 on page 558, which specified the grammar $\mathcal{G} = \{\Sigma, \Delta, S, \Pi\}$, where $\Sigma = \{a, b\}$, $\Delta = \{S, A, B\}$, and $\Pi = \{S \to aA \mid bB,\ A \to bB \mid \lambda,\ B \to aA \mid \lambda\}$.

The string *aba* was derived as

$$S \Rightarrow aA \Rightarrow abB \Rightarrow abaA \Rightarrow aba$$

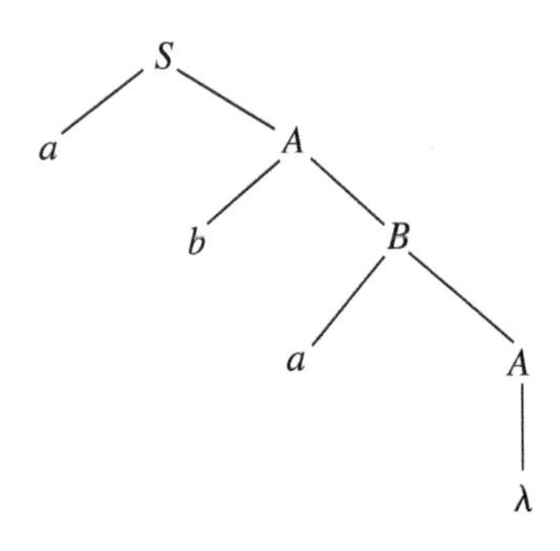

Figure 11.22. A parse tree for deriving *aba*.

The tree in Figure 11.22 captures this derivation. The tree is constructed in the following manner. The start symbol will be the root. When a production is used, the terminal symbol that is replaced will become a parent node. Its children are the symbols on the right-hand side of the production that was used, placed in left-to-right order. (This implies that a context-free grammar is being used.) The derived string is read from the leaves (all of which will be terminal symbols).

More formally,

> **DEFINITION 11.29** ***Parse Tree***
>
> A *parse tree* is a representation of a derivation in a context-free grammar. The representation is a tree whose root is the start symbol of the grammar. Nodes represent symbols in $\Sigma \cup \Delta$. All leaves are symbols in Σ, and symbols in Σ appear only as leaf nodes. All interior nodes are symbols in Δ. The children of an interior node, N, are the symbols (in left-to-right order) from the right-hand side of a production having N as its left-hand side.

EXAMPLE 11.12

A Parse Tree

Let $\mathcal{G} = \{\Sigma, \Delta, S, \Pi\}$, where $\Sigma = \{a, b, \lambda\}$, $\Delta = \{S, B\}$, and $\Pi = \{S \to aSa \mid B \mid \lambda,\ B \to bB \mid b\}$.

The language generated is $\{a^n b^k a^n \mid n \geq 0, k \geq 0\}$. The string *aabbbaa* can be derived as

$$S \Rightarrow aSa \Rightarrow aaSaa \Rightarrow aaBaa \Rightarrow aabBaa \Rightarrow aabbBaa \Rightarrow aabbbaa.$$

The parse tree is displayed in Figure 11.23.

Figure 11.23 A parse tree for *aabbbaa*.

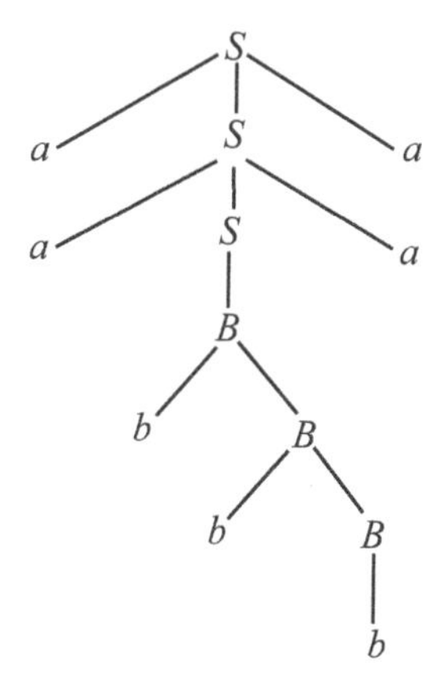

Notice that the leaves are read in left-to-right order, traveling counterclockwise around the outside of the tree. ■

Although moving from a derivation to a parse tree will always produce the same tree, if you try to convert from a parse tree to a derivation, it is possible that there will be many derivations for the string produced by the tree.

EXAMPLE 11.13

Ambiguity

Let $\mathcal{G} = \{\Sigma, \Delta, S, \Pi\}$, where $\Sigma = \{a, b\}$, $\Delta = \{S, A, B\}$, and $\Pi = \{S \to AB,\ A \to aA \mid a,\ B \to bB \mid b\}$.

This grammar generates the language $\{a^k b^m \mid k, m \geq 1\}$.

Consider the parse tree in Figure 11.24.

There are several derivations of *aabbb* that are consistent with the parse tree. Here are three.

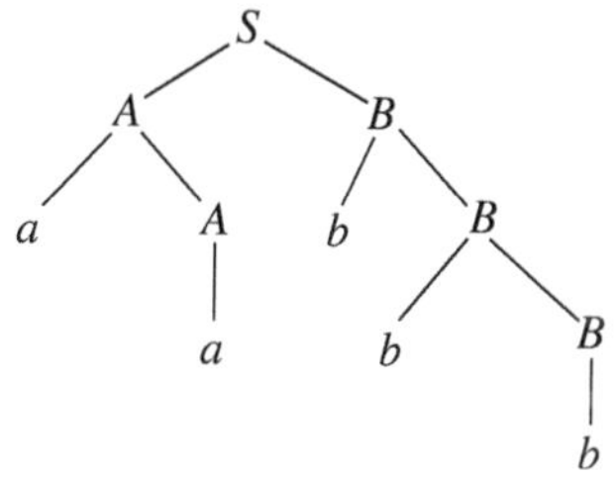

$$S \Rightarrow AB \Rightarrow aAB \Rightarrow aaB \Rightarrow aabB \Rightarrow aabbB \Rightarrow aabbb$$
$$S \Rightarrow AB \Rightarrow AbB \Rightarrow AbbB \Rightarrow Abbb \Rightarrow aAbbb \Rightarrow aabbb$$
$$S \Rightarrow AB \Rightarrow aAB \Rightarrow aAbB \Rightarrow aAbbB \Rightarrow aabbB \Rightarrow aabbb$$

■

Figure 11.24. An ambiguous parse tree.

✔ Quick Check 11.6

1. Let $\Sigma = \{0, 1\}$, $\Delta = \{S, X, Y\}$, and $\Pi = \{S \rightarrow 01X \mid 1Y, X \rightarrow 0X \mid 1, Y \rightarrow 00Y \mid 1\}$, and let S be the start symbol. Produce parse trees for the following strings:
   (a) 100001
   (b) 01001 ☑

11.3.2 Huffman Compression

The American National Standards Institute (ANSI) character set uses an eight bit encoding for each of the 256 characters in the set. Thus, a document having 1000 characters requires 1000 bytes of storage.[5] The key insight that motivates Huffman compression is that these characters do not appear in real documents with equal frequencies. It should therefore be possible to use a nonuniform encoding to reduce the number of bits needed to store a long message composed from these nonuniformly encoded characters. In particular, if characters that appear frequently were encoded using only 2 or 3 or 4 bits, we would save more than half a byte per occurrence.

EXAMPLE 11.14 **Using Variable-Length Encoding**

We might decide to let $e = 11$, $a = 10$, $s = 110$, $i = 01$, and $t = 111$, with other bit patterns for the other characters. We would then encode the word *sea* as 1101110. This requires 7 bits instead of the normal 24 bits using the ANSI encoding. ■

There are two issues that need to be addressed before this scheme will work.

- With variable-length encoding, how can we reconstruct the characters from the stream of bits?
- Can we still use eight or fewer bits per character with a variable-length encoding?

The answer to the second question will depend directly on the practical solution to the first question.

Prefix Codes

If you think carefully about Example 11.14, you may notice a problem. The next example illustrates the issue.

EXAMPLE 11.15 **Ambiguity**

Suppose we start with the encoded bit string 1101110 and seek to transform it back into characters.

We have several choices:

110	11	10
s	*e*	*a*
11	01	110
e	*i*	*s*
110	111	0
s	*t*	??

The problem with this encoding is that there is ambiguity in the decoding process. ■

What is needed is an encoding that has the prefix property.

[5] A *byte* is 8 bits.

DEFINITION 11.30 *The Prefix Property*
An encoding has the *prefix property* if the code for any character never occurs as the initial bits in the encoding for another character.

EXAMPLE 11.16 **The Prefix Property Fails**

In Example 11.14, the encoding for *e* (11) appears as the initial bits in the encodings for *s* (110) and *t* (111). The prefix property does not hold for this example. ■

By insisting on an encoding that has the prefix property, we must allow some characters to require more than 8 bits so that it is possible to find suitable representations for all 256 ANSI characters. We can arrange to have the infrequently occurring characters be the ones with more bits. We hope that the savings in bits from short characters will outweigh the loss in bits for long characters.

The Fundamental Algorithm

The essential idea of the Huffman compression algorithm is to build a tree having the various characters as leaves, but placing the frequently used characters as close to the root as possible and the infrequently used ones as far away as necessary. A corresponding character (requiring 8 bits in ASCII or ANSI)[6] would then be encoded as a 0/1-sequence by noting whether we move to a left child (0) or the right child (1) as we traverse the tree toward the character. If the tree looks like the one in Figure 11.25, the encoding for the letter *e* would be 00. The full collection of encodings is listed in Table 11.1.

TABLE 11.1 The Encodings for Figure 11.25

a	e	i	m	q	z
10	00	11	010	0111	0110

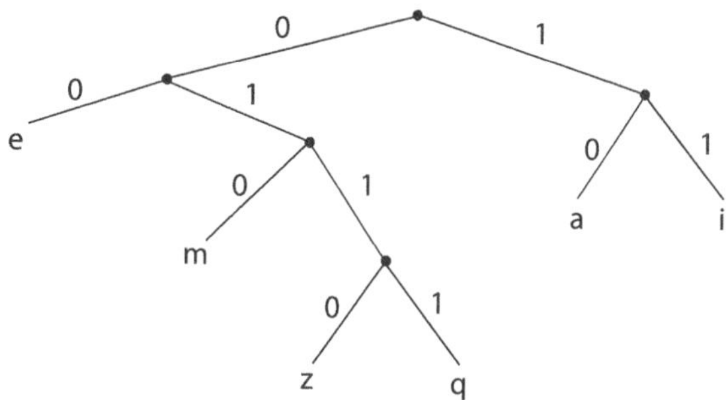

Figure 11.25 A Huffman encoding tree.

Notice that the character encodings have the prefix property. This is guaranteed by the placement of characters as leaves.

If the frequently used characters (in an encoding that covers the full ANSI set of characters) are close to the top, then we should not need anywhere near 8 levels in the tree to locate them. The infrequently used characters, though probably requiring more than 8 bits, are used so rarely that the overall effect is to reduce the total number of bits needed for the original character sequence, sometimes by as much as 40% in real applications. As an extreme example, a message that contains 5 as, 4 es, 3 ms, and 1 q would require $13 \cdot 8 = 104$ bits using the standard ANSI encoding. With the encoding in Table 11.1 it would require $5 \cdot 2 + 4 \cdot 2 + 3 \cdot 3 + 1 \cdot 4 = 31$ bits, for a 70% reduction in size.

[6]The American Standard Code for Information Interchange (ASCII) character encoding scheme is a collection of 128 characters, originally encoded using 7 bits per character. An 8-bit version (using a parity check bit) was later introduced. The ANSI encoding uses 8 bits to encode a collection of 256 characters. The 8-bit ASCII set is identical to the first 128 characters in the ANSI set.

Details The tree is built using the following procedure:

1. Scan the document, recording the frequency of occurrence for each character.
2. Place the entries from the frequency table into a priority queue[7] with the lowest frequencies ready to exit first. (So the queue is sorted in increasing order.)
3. At this step, the tree will be built from the bottom up. The leaves correspond to the characters in the ANSI set. Each leaf node will have a unique character associated with it (perhaps by writing the character inside the leaf node). Each node will also have a label (a frequency if it is a leaf node, a sum of frequencies in its subtree if it is an interior node). To build a tree, start removing objects from the queue. Begin by creating a parent node for the first two characters in the queue. Make the first character the left child and the second character the right child. Label the parent node with the sum of the labels of the two children. Now insert the new three-node tree into the queue at the place where its root label is in proper order. In case other items in the queue have the same label, place the new subtree closer to the back of the queue. (The two characters that were removed to create this subtree are not directly placed back in the queue.)
   Now repeat the following until there is only a single tree left.
   a. Remove the first two elements from the queue.
   b. Create a new parent node with the two chosen objects as its children. Let the first object be the left child.
   c. Label the new parent node with the sum of the labels on the children.
   d. Place the new tree into the queue at the position where its root node's label will be in proper order (as close to the back of the queue as possible without violating the label ordering).
4. Relabel the tree. (This step produces the encoding.)
   a. Go down the tree, placing a zero on left branches and a one on right branches.
   b. Now traverse the tree to each leaf, collecting the branch labels into the encoding for the character attached to the leaf node. For example, a character for which we traverse (from the root) right, right, left would receive the encoding 110.

EXAMPLE 11.17

Compressing a Brief Message

The string "mammals amaze mama" contains 18 characters and requires 144 bits using the ANSI encoding. The Huffman algorithm first builds the table of character frequencies (Table 11.2).[8]

TABLE 11.2 Character Frequencies for Example 11.17

␣	*a*	*e*	*l*	*m*	*s*	*z*
2	6	1	1	6	1	1

The characters then are placed into a priority queue. For this textbook, the agreement will be that the front of the queue is on the left, and the back of the queue is on the right. The initial queue for this example is shown in Figure 11.26. Note that there is some ambiguity when ties occur in the frequency values. The order in which ties are placed in the queue will effect the ultimate character encodings. For consistency, place ties in alphabetical order.

---

[7] A *queue* is a data collection for which new items are added to the back and old items are removed from the front, much like the line at a bank window. A *priority queue* is a queue in which items are sorted by some priority value. When a new item is added, it is added to the back and then moved forward until it either bumps into an item with the same priority or reaches the front of the queue.

[8] The character "␣" represents a space, which comes before the alphabet in the ASCII and ANSI encodings.

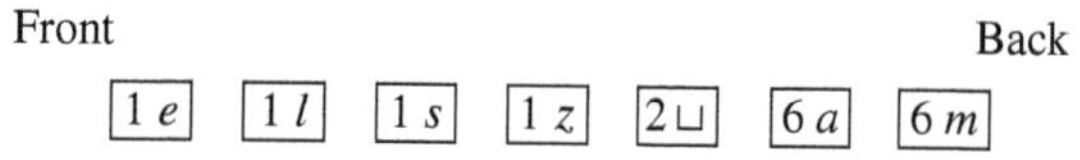

Figure 11.26 The initial queue.

Front Back

1 *e* 1 *l* 1 *s* 1 *z* 2 ⊔ 6 *a* 6 *m*

The next step is to start pulling pairs of subtrees from the front of the queue, joining them into a single tree, and then reinserting the new tree into the queue. The first pair will be the leaf nodes containing *e* and *l*. The joined tree will have a frequency value of 2 (so will be inserted after the node with the space character, as in Figure 11.27).

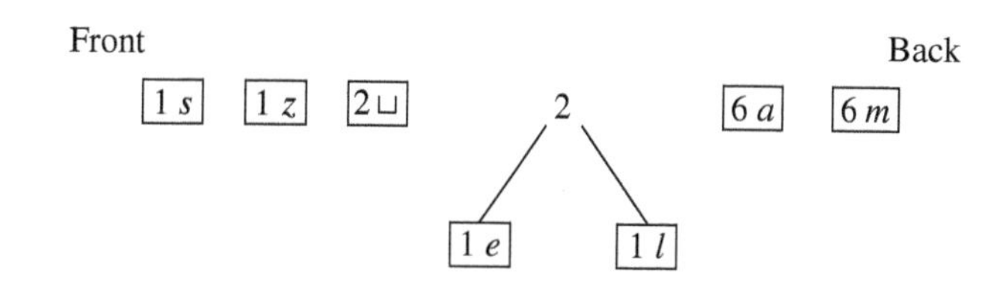

Figure 11.27 The first iteration.

The *s* and *z* nodes come off next. The new queue is shown in Figure 11.28.

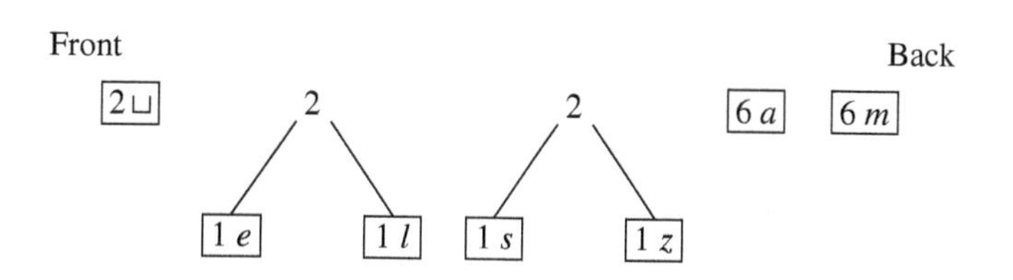

Figure 11.28 The second iteration.

The space character and the *e*-*l* tree come off next. The newly joined tree has frequency value 4 and will be inserted in front of the *a* node (Figure 11.29).

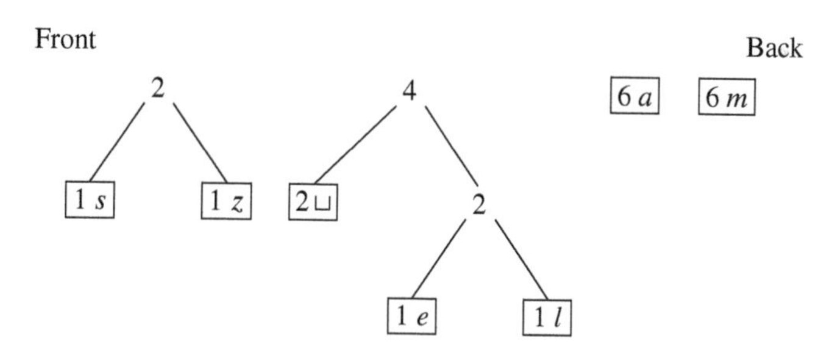

Figure 11.29 The third iteration.

Joining the two trees at the front of the queue produces the three-element queue shown in Figure 11.30.

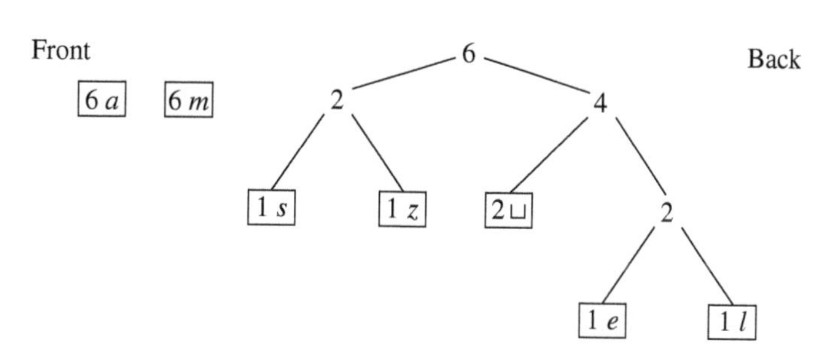

Figure 11.30 The fourth iteration.

The *a* and *m* nodes are then joined to become part of a frequency 12 tree (Figure 11.31 on page 698).

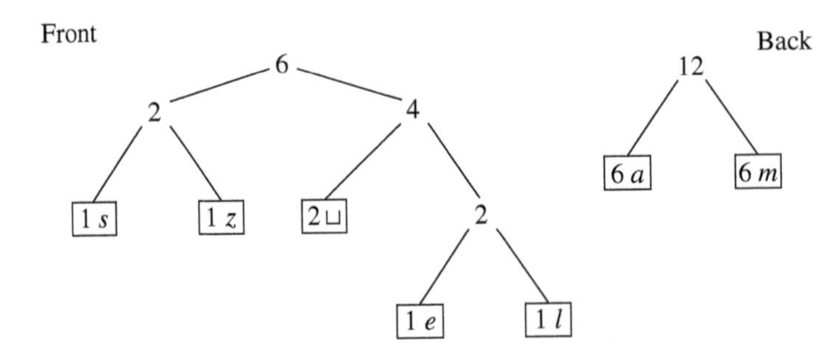

Figure 11.31 The fifth iteration.

This phase of the algorithm is then completed by joining the two remaining trees in the queue. The result, after removing it from the queue, is the tree in Figure 11.32.

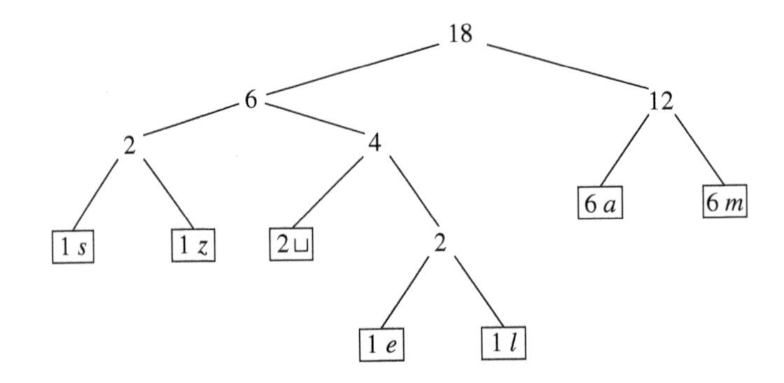

Figure 11.32 The last iteration.

The final phase of the algorithm is where the tree's edges are given 0/1 labels, and then the leaves are assigned their encodings. Figure 11.33 shows the edge labeling.

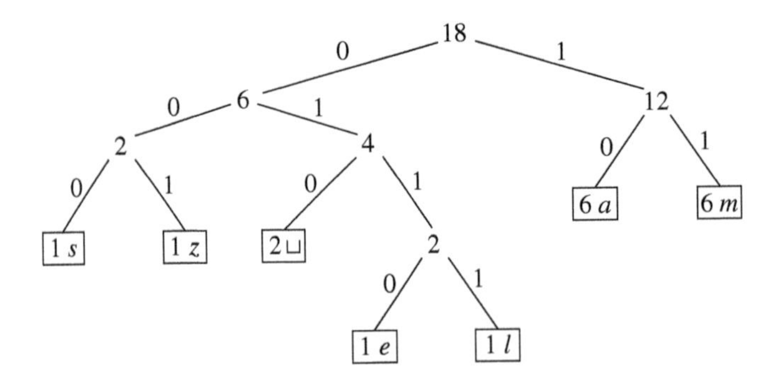

Figure 11.33 The edge labels.

Now the leaves can be labeled (Figure 11.34). For each leaf, follow the path from the root to the leaf, building the encoding from the edge labels (in the order they are encountered).

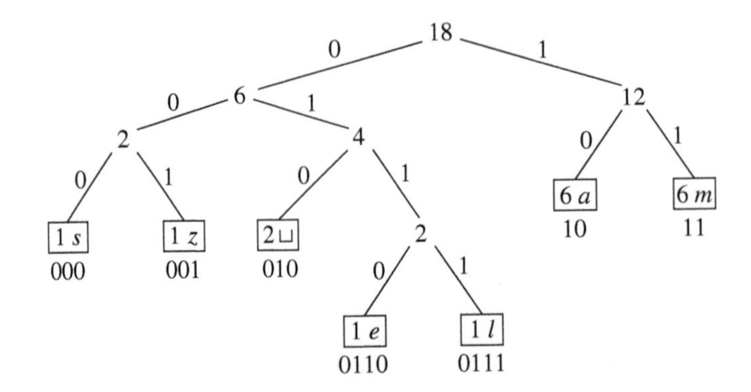

Figure 11.34 The leaf labels.

The encoding is summarized in Table 11.3.

TABLE 11.3 The Final Encoding

␣	a	e	l	m	s	z
010	10	0110	0111	11	000	001

The encoded version of the string, "mammals amaze mama", requires $6 \cdot 2 = 12$ bits for the m's, 12 bits for the a's, $2 \cdot 3 = 6$ bits for the spaces, 4 bits each for the e and l, and 3 bits each for the s and z. The total for the encoded string is 44 bits—a 100-bit savings over the original ANSI encoding.

The comparison with the number of bits required by the ANSI encoding is not really fair. With only seven characters, it would be possible to use a 3-bit-per-character encoding. The constant-sized encoding would then need $18 \cdot 3 = 54$ bits. The savings is therefore only 10 bits over what a constant-width encoding could attain. ■

Practicalities Once the tree has been formed, the original file needs to be read character by character and a new file with the bit patterns (from the tree) written out. This will probably require some memory buffering of the bits since the operating system typically needs bits written in 8-bit packets. Any left-over spaces in the last byte to be written will need to somehow be filled in a way that the decoding algorithm can ignore.

A second practical issue is to provide a mechanism for the decoding algorithm to know what the encoding tree was (since this depends on the original document). One solution is to store the tree at the beginning of the compressed file. The decoding algorithm would start by reading the tree and then the compressed document. This means the algorithm could actually produce a larger file if the original file contains only a short message. For instance, the 100-bit savings in Example 11.17 would be almost entirely used up storing the encoding table.

✔ Quick Check 11.7

1. Consider the DNA string
   GGGACTGAAAAC
   (a) How many bits are needed to encode this string using a constant width encoding with 2 bits per character?
   (b) Produce the final labeled tree and the encoding table for a Huffman encoding of this string.
   (c) How many bits are needed for the Huffman encoding if the encoding table is not included in the file? ☑

11.3.3 XML

You are almost certainly aware of HyperText Markup Language (HTML), the World Wide Web markup language. It provides a collection of *tags* to wrap around text to instruct a Web browser about how to format a Web document for display.

HTML has some critical limitations that prevent it from being of use for any purpose other than Web pages. First, although it was designed to present the *logical* structure of a Web document, its main purpose is to suggest to a Web browser the manner in which the document is to be displayed. In fact, HTML contains a number of tags that are purely for formatting purposes (such as font style and size).

A second problem that limits the use of HTML is that the only tags that can be used are those that have been defined as part of the HTML standard.

A newer markup language, called XML (eXtensible Markup Language), removes these limitations. XML provides a mechanism for the document's creator to define the tags that provide the logical structure of the document. There is no need for any formatting information to be provided in the document. XML provides other mechanisms that allow very precise control over how the document will be displayed. In fact, those mechanisms allow for different formatting to be applied based on the display medium. One version could be used for Web pages and another for a laser printer. XML is structured in a manner that allows computer programs to manipulate, modify, or process the document. In addition, it does not contain any proprietary document markup (as is used with word processors like Microsoft Word, or with spreadsheet programs). XML documents are just text documents that can be read and edited with any text editor.

The connection between XML and trees is simple: An XML document can be directly translated into a tree that accurately represents the document. The discussion that follows will present an overview of XML. Only the most basic parts will be discussed. The topic is much richer than this brief introduction can indicate.

XML Building Blocks

An XML document is created using a small number of building blocks. The most important are presented here. They are tags, elements, attributes, strings (called "character data" in most discussions about XML), and processing instructions.

A short example will provide a context in which to introduce these components.

EXAMPLE 11.18 **A Very Simple XML Document**

The document shown is a complete XML document that contains examples of all the building blocks mentioned previously. Line numbers have been added to aid the discussion, but the actual document does not contain them.

```
 1: <?xml version="1.0"?>
 2: <letter>
 3:     <greeting>My dearest darling:</greeting>
 4:     <body>
 5:         <paragraph>
 6:             How I have missed you! The knowledge that I will see
 7:             you again is the only reason ...
 8:         </paragraph>
 9:         <paragraph>
10:             The good news is that my Discrete Mathematics class is
11:             really interesting. It is a lot of work, but I am doing well ...
12:         </paragraph>
13:     </body>
14:     <closing>Your faithful admirer,</closing>
15:     <signature>Name suppressed to protect the privacy of the author</signature>
16: </letter>
```

■

An XML document contains the following kinds of components:[9]

Processing instructions Line 1 of Example 11.18 contains a processing instruction. Its purpose is to inform any program that is reading the document which version of XML is being used. Processing instructions are not part of the tree that represents the document. You can copy this line exactly as the first line of any XML documents you create as homework.

Attributes The only attribute in this example is in line 1. It is

```
version="1.0"
```

and is part of the processing instruction. The string "version" is the attribute *name*, and the quoted string "1.0" is the attribute *value*. Attributes will be discussed in more detail on page 704.

[9]Other components exist but are not necessary for this overview.

Tags Tags are the mechanism used by XML (and HTML) to create the logical structure of the document. Tags come in pairs. The opening tag consists of a pair of angle brackets, <>, enclosing a string that identifies the *element* the tag is creating, and possibly some optional attributes. The closing tag[10] consists of an opening angle bracket and a slash, </, followed by the same element string and a closing angle bracket, >. There are no attributes in closing tags. Lines 2 and 16 of Example 11.18 contain a pair of tags for the "letter" element. Line 3 contains a pair of tags for the "greeting" element. The enclosed string, "My dearest darling:", is not part of the tags.

Elements Elements are what give an XML document its logical structure. They are one of the two building blocks that make up the tree nodes in the tree representation. In Example 11.18, the elements are "letter", "body", "greeting", "paragraph", "closing", "signature". Element *names* are case sensitive. Thus, "letter" and "Letter" are considered to be distinct elements.

The element "letter" at line 2 is called the *root element*. It will be the root of the tree representation.

Everything between the element's opening and closing tags is considered part of the element. An element can be composed of any mixture of other elements and character data. It can also be an *empty element* and have nothing between the two tags. In Example 11.18, the "letter" element is composed of the "greeting", "body", "closing", and "signature" elements. The "greeting" element is composed of character data.

Character data The strings that appear between matching tags are called character data. They are the second of the building blocks that become nodes in the associated tree. The string "My dearest darling:" on line 3 is an example, as are the longer strings between the "paragraph" elements.

There are a few symbols that cannot be used as part of character data. They are the two angle brackets, < and >, the ampersand, &, and single and double quotes, ' and ". They cannot be used because they have special meanings in XML. For example, using an angle bracket, <, inside character data will probably cause a program that is reading the document to mistakenly start processing a closing tag. These illegal symbols *can* be represented by using the special *entities* listed in Table 11.4. The following XML fragment demonstrates the use of these entities:

TABLE 11.4 Some XML Entities

Symbol	XML Entity
<	<
>	>
&	&
'	'
"	"

```
<theorem>If x &lt; 4, then x squared &lt; 16.</theorem>
```

On a Web browser that can display XML, this would appear as either

```
<theorem>If x < 4, then x squared < 16.</theorem>
```

or as

```
If x < 4, then x squared < 16.
```

Well-Formed XML Documents

There are some requirements about the manner in which these building blocks may be composed. One requirement is that element and attribute names are case sensitive. Another significant fact is that multiple adjacent whitespace characters[11] are usually treated as a single space. The other requirements are summarized in the next definition.

[10]Opening and closing tags are also called *start* and *end* tags.

[11]See Section 9.4.2 on page 566 for a brief description of whitespace characters.

DEFINITION 11.31 ***Well-formed***

An XML document is *well-formed* if the following are all true:

- There is a single root element.
- Each element has a matching opening and closing tag.
- All attribute values are in single or double quotes.
- Elements are properly nested.

The final requirement in the previous definition needs some additional explanation. An element can be nested inside another element. For example, in Example 11.18, there are two "paragraph" elements nested inside the "body" element. The indentation used in that example helps to highlight the nesting.

Two elements are properly nested if both the opening and closing tags of the second element are entirely contained inside the two tags of the first element.

```
<first>
   <second>
   </second>
</first>
```

Improper nesting occurs when the opening and closing tags are not both inside the other element's tags.

```
<first>
   <second>
</first>
   </second>          INCORRECT!
```

EXAMPLE 11.19

Nesting

The following fragment of an XML document shows proper nesting with parallel element and character data.

```
<paragraph>
   One of my favorite books is <book-title>Lord of
   the Rings</book-title> by
   <author>J. R. R. Tolkien</author>.
</paragraph>
```

This could also be formatted in the following manner (because extra whitespace is not significant). The period must be adjacent to `</author>` or a space would precede it.

```
<paragraph>
   One of my favorite books is
   <book-title>Lord of the Rings</book-title>
   by
   <author>J.R.R. Tolkien</author>.
</paragraph>
```

This clearly shows that this "paragraph" is composed of a sequence of five items (three character data sections, including the period, and two elements). The "book-title" and "author" elements consist of character data. ■

There are a number of software programs that can determine if an XML document is well formed. Perhaps the most accessible is to use a current version of a web browser. Saving Example 11.18 in a file (perhaps named "letter.xml") and opening the file with a web browser will produce a display similar to Figure 11.35.[12] If the XML document had not been well formed, an error message would have been displayed instead.

[12] Firefox omits the line containing `<?xml version="1.0"?>`.

```
<?xml version="1.0" ?>
- <letter>
    <greeting>My dearest darling:</greeting>
  - <body>
      <paragraph>How I have missed you! The knowledge that I will see you again is the only reason ...</paragraph>
      <paragraph>The good news is that my Discrete Mathematics class is really interesting. It is a lot of work, but I
        am doing well ...</paragraph>
    </body>
    <closing>Your faithful admirer,</closing>
    <signature>Name suppressed to protect the privacy of the author</signature>
  </letter>
```

Figure 11.35 Opening letter.xml in Internet Explorer.

The Tree Representation of an XML Document

The nesting of elements leads directly to a tree representation for an XML document. The root element becomes the root of the tree. Because every element is composed of a sequence of other elements and character data, it is easy to determine the children of each element. Just take each item in the sequence and create a new child node.

EXAMPLE 11.20

A Subtree

Example 11.19 contained a properly nested XML fragment. A subtree that represents that fragment is shown in Figure 11.36. The element nodes and character data nodes have been visually distinguished, but that is not necessary.

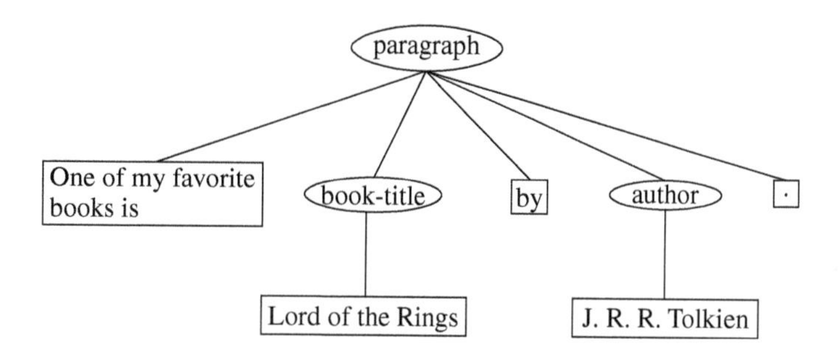

Figure 11.36 A tree for Example 11.19.

■

EXAMPLE 11.21

The Tree Representation for Example 11.18

A tree for the XML document in Example 11.18 (page 700) is shown in Figure 11.37.

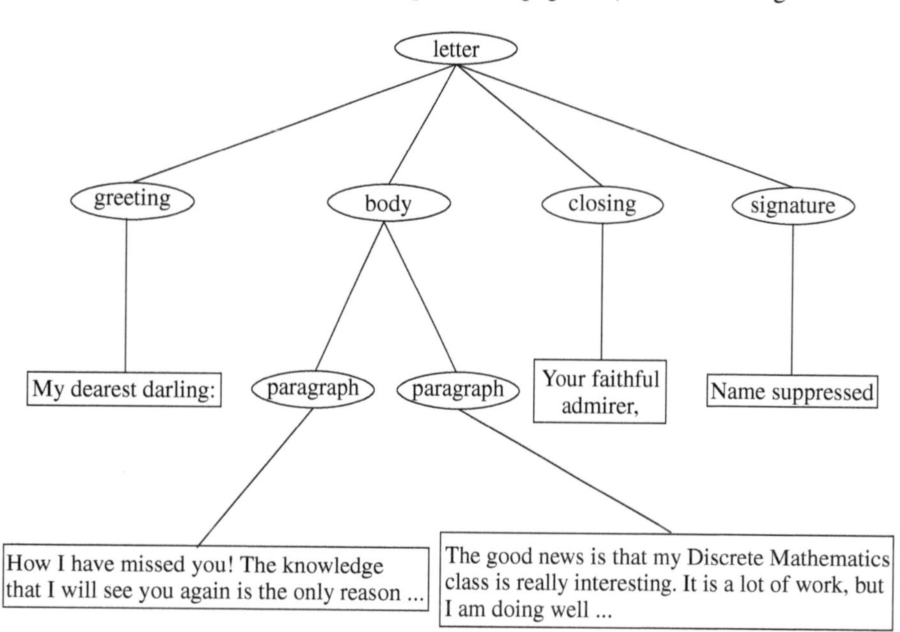

Figure 11.37 A tree for Example 11.18.

■

✔ Quick Check 11.8

1. Produce the tree representation for this well-formed XML document.

```
<?xml version="1.0" ?>
<classlist>
   <course>Discrete Mathematics</course>
   <enrolled>2</enrolled>
   <student>
      <name>Jennifer Adams</name>
      <cohort>Sophomore</cohort>
      <major>Mathematics</major>
   </student>
   <student>
      <name>William Zembrot</name>
      <cohort>Junior</cohort>
      <major>Computer Science</major>
   </student>
</classlist>
```

✔

Attributes

The "classlist" element in Quick Check 11.8 contains two elements that appear once each ("course" and "enrolled"). The other elements are all of the same type ("student"). These elements differ in other ways as well. The "course" and "enrolled" elements contain information *about* the "classlist" whereas the "student" elements are what the "classlist" is composed from.

The XML standard provides an alternate mechanism for recording information about an element: attributes. Attributes are listed inside the opening element tag. They appear in the following format:

```
AttributeName = "Attribute Value"
```

The attribute name must not contain any whitespace, but the attribute value may. The attribute value must be enclosed by either single or double quotes. The whitespace around the equal sign is optional.

EXAMPLE 11.22

A Revised Class List

The classlist document in Quick Check 11.8 can be revised using attributes.

```
<?xml version="1.0" ?>
<classlist course = "Discrete Mathematics" enrolled = "2">
   <student>
      <name>Jennifer Adams</name>
      <cohort>Sophomore</cohort>
      <major>Mathematics</major>
   </student>
   <student>
      <name>William Zembrot</name>
      <cohort>Junior</cohort>
      <major>Computer Science</major>
   </student>
</classlist>
```

The attributes can be considered to be an integral part of the element. The tree representation (Figure 11.38) can show this by making the attributes part of the element's node.

Figure 11.38 A tree with attributes.

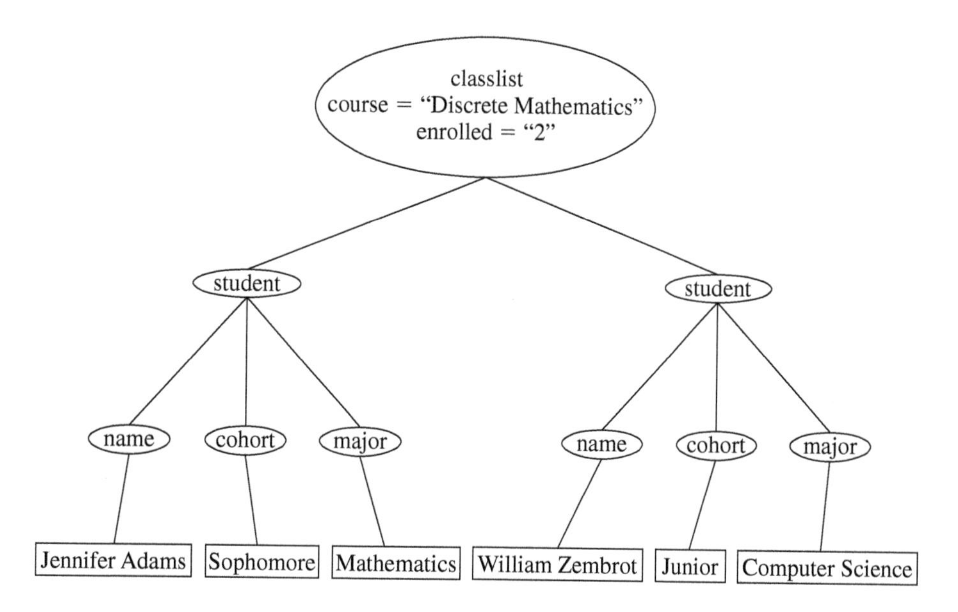

■

The decision about whether some information related to an element should be represented as an attribute or as a nested element or as nested character data is complex. Part of the decision relates to how the document will be manipulated by the software it is intended to be used by. There are no rules that always apply. This introduction to XML is much too brief to worry about those issues.

Validation

So far, XML appears to be a complicated way to create a document. What are the benefits? There are several.

1. The first benefit is that XML can be used instead of the less flexible HTML standard for creating Web pages.
2. A second advantage is that XML documents are ideal for sharing electronically, and more important, are designed so that software can automatically perform complex manipulations. For example, a program could print a report about the class list in Example 11.22. The report might sort students by cohort and then by major. The report could also very easily count the number of students in each major or cohort.
3. One of the other significant advantages is the ability to define in advance what a particular class of documents should look like. Documents that do not conform to the user-defined standard can be rejected.

There are several mechanisms available to create a user-defined standard for a class of documents. The two most common are Document Type Definitions (DTDs) and Schemas. A Schema is written as a well-formed XML document, so it can also be represented as a tree. DTDs use a format that does not translate into a tree. DTDs will be briefly presented here because they are a bit easier to describe and because they provide another application of simple regular expression notation.[13]

The next example will provide a context to introduce some of the major kinds of DTD rules.

[13] See Section 9.4.

EXAMPLE 11.23 **A Classlist DTD**

The following DTD defines the structure for "class list" XML documents. The line numbers are not part of the DTD, but are included to aid the discussion.

```
1: <!ELEMENT classlist (student)+>
2:    <!ATTLIST classlist course CDATA #REQUIRED>
3:    <!ATTLIST classlist enrolled CDATA #REQUIRED>
4: <!ELEMENT student (name, cohort, major)>
5: <!ELEMENT name (#PCDATA)>
6: <!ELEMENT cohort (#PCDATA)>
7: <!ELEMENT major (#PCDATA)>
```

■

DTDs use a fairly rich set of rules. For the purposes of this introduction, only two kinds are discussed (and neither in its fullest form).

1. Element declarations An element's structure is defined by a declaration of the following form.

```
<!ELEMENT ElementName ConsistsOfRule>
```

The word "ELEMENT" that appears after the exclamation point *must* be all uppercase. The "ElementName" will be the name of the type of element being specified. The "ConsistsOfRule" portion uses regular expression notation to indicate exactly what other elements and character data are permitted to be nested inside the element.

For instance, in line 1 of Example 11.23, the "classlist" element is composed of one or more "student" elements. Line 4 specifies that a "student" element consists of a "name" followed by a "cohort" followed by a "major". Those three elements must appear in the order listed. Line 5 specifies that a "name" consists of character data.[14]

The specification of what the element consists of will always be inside parentheses. It will typically use combinations of the following regular-expression-inspired constructors.

Repeat factors The regular expression symbols "+" and "*" and "?" retain their meanings here (one or more, zero or more, zero or one, respectively).

Notice the convention that if there is a single repeated element, the repeat factor is placed outside the parentheses, as was done in line 1 of Example 11.23.

Sequence Element names separated by commas must appear in the order specified. For example,

```
<!ELEMENT workday (meetings+, lunch, email?, meetings*)>
```

indicates that a "workday" consists of one or more meetings, followed by a single lunch, possibly followed by an e-mail session, and then perhaps some more meetings.

Alternative The character "|" is used to indicate that only one of a collection of elements will appear. For example,

```
<!ELEMENT wallcovering (paint | wallpaper | paneling)>
```

indicates that "wallcovering" will contain exactly one nested element. That element will be either a "paint" element or a "wallpaper" element or a "paneling" element.

---

[14]The DTD specification requires you to use the declaration "#PCDATA" to specify character data nested inside an element. It stands for "parsed character data." The distinction between this and the "CDATA" declaration in line 2 is best left for a more complete look at DTDs. You will have no trouble as long as you just mimic their use here. Note that both "#PCDATA" and "CDATA" must be all uppercase.

It is also possible to use "#PCDATA" in place of an element name. The one special requirement for "#PCDATA" is that it must be the first option in an alternative rule that ends with an asterisk, and no other rules may be nested inside. For example, there may be some predefined elements representing the language(s) used to write a document. Perhaps only the most common languages are listed with special elements. Less commonly used languages will just be written as character data.

```
<!ELEMENT WrittenIn (#PCDATA | English | French |
Spanish | Russian | Chinese | Japanese)*>
```

These constructors can be freely mixed (with the help of parentheses), as shown in the next rule.

```
<!ELEMENT segment (SegmentHeader?, (paragraph+ |
(poem, interpretation)+))>
```

The rule indicates that a segment has an optional segment header and then either one or more paragraphs, or else one or more poem-with-interpretation pairs.

2. Attribute declarations Lines 2 and 3 of Example 11.23 show how attributes can be specified. The general form (for the purposes of this overview) looks like the following pattern.

```
<!ATTLIST OwningElementName AttributeName CDATA #REQUIRED>
```

The string "OwningElementName" is the name of the element that the attribute is attached to (e.g, "classlist"). The string "AttributeName" is the name of the attribute (e.g., "enrolled"). The remaining two items are two common responses for much more general categories of information. The "CDATA" string indicates that the value of the attribute will be character data. The string "#REQUIRED" indicates that this attribute is not optional. If the string "#IMPLIED" were used instead, the attribute could be left out of any particular tag of type "OwningElementName."

✔ Quick Check 11.9

1. Create a DTD for letters, using the XML document in Example 11.18 as a concrete example of what a letter should look like. ☑

Any program that reads an XML document for the purpose of manipulation or processing must use a component called a *parser*. A parser makes sense of the structure inherent in the document. If the document is not well formed, the parser will issue an error message, and the controlling program will be unable to complete its task. However, a well-formed document may still cause the task to fail. For example, if the program's task is to count the number of mathematics majors in a class list, it will have problems if some of the "student" elements don't contain a "major" element. The DTD in Example 11.23 contains all the information necessary to enable the parser to issue an error message if the XML document doesn't conform to the requirements in the DTD. A parser with this ability is called a *validating parser*.[15]

One additional processing instruction is needed in the XML document to inform the parser how to find the associated DTD. The document in Example 11.22 (page 704) can be modified to include this new kind of processing instruction.

[15]Look at http://www.mathcs.bethel.edu/~gossett/DiscreteMathWithProof/ for a link to *rxp*, a free validating parser. It is run from a command line: rxp -V classlist.xml where "classlist.xml" may be replaced by some other XML document name. The DTD file will need to be in the same directory as the XML file.

EXAMPLE 11.24 **A Class List with DTD**

The class list XML document in Example 11.22 can be associated with the DTD in Example 11.23 by adding the processing instruction shown below at line 2. The strings "DOCTYPE" and "SYSTEM" must be all uppercase. Between them the name of the root element will be listed. At the end, in quotes, will be the name of the file that contains the DTD ("class.dtd" in this case).

```
 1: <?xml version="1.0" ?>
 2: <!DOCTYPE classlist SYSTEM "class.dtd">
 3: <classlist course = "Discrete Mathematics" enrolled = "2">
 4:    <student>
 5:       <name>Jennifer Adams</name>
 6:       <cohort>Sophomore</cohort>
 7:       <major>Mathematics</major>
 8:    </student>
 9:    <student>
10:       <name>William Zembrot</name>
11:       <cohort>Junior</cohort>
12:       <major>Computer Science</major>
13:    </student>
14: </classlist>
```

■

This completes the brief introduction to XML. There is much more to learn if you wish to become proficient. For example, one of the more advanced XML technologies, called DOM, is strongly tied to this chapter. *DOM*, which stands for the document object model, provides a collection of software components that support XML document processing. DOM components are used to parse an XML document and create a tree that represents the document. The DOM components provide mechanisms to traverse the tree and either gather information or modify the tree. You can even specify which kind of tree traversal you want to use.

11.3.4 Exercises

The exercises marked with ★ have detailed solutions in Appendix H.

1. Let $\mathcal{G} = \{\Sigma, \Delta, S, \Pi\}$, where

$$\Sigma = \{a, b\}$$
$$\Delta = \{S, A, B, C\}$$

and

$$\Pi = \{S \to aABb \mid C,\ A \to a,\ B \to b,\ C \to AabB\}$$

(a) ★ Produce two distinct parse trees for the string *aabb* in $L(\mathcal{G})$.

(b) Produce all distinct parse trees for this grammar.

2. Consider the grammar in Example 11.13 on page 693.

(a) Create a parse tree for the string *aaab*.

(b) Produce three different derivations that are consistent with the following parse tree.

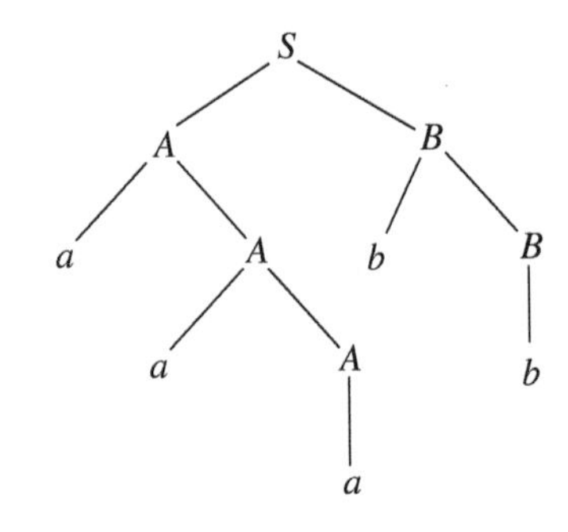

3. ★ A language is generated by the grammar $\mathcal{G} = \{\Sigma, \Delta, S, \Pi\}$, where

$$\Sigma = \{a, b, c\}$$
$$\Delta = \{S, X, Y\}$$

and

$$\Pi = \{S \to aX \mid bY,\ Y \to bY \mid b,\ X \to aX \mid c\}$$

Is the following parse tree consistent with this grammar? Give reasons for your answer.

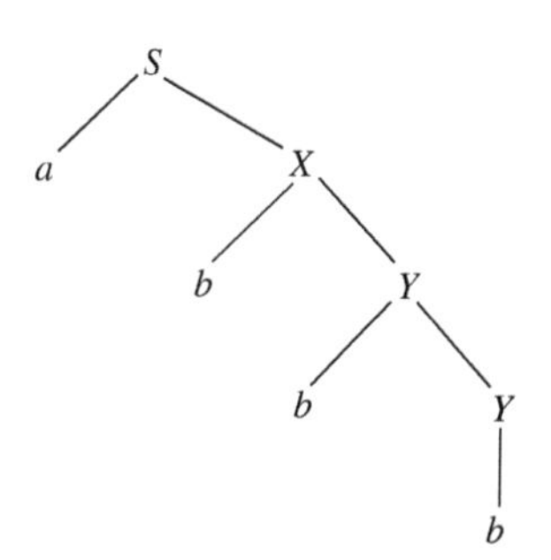

4. A language is generated by the grammar $\mathcal{G} = \{\Sigma, \Delta, S, \Pi\}$, where $\Sigma = \{a, b, c\}$, $\Delta = \{S, X, Y\}$, and $\Pi = \{S \to aXc \mid bY, Y \to bYX \mid c, X \to XaY \mid a\}$. Create a parse tree for the string *aaabcac*.
5. A language is generated by the grammar $\mathcal{G} = \{\Sigma, \Delta, S, \Pi\}$, where $\Sigma = \{a, b, c\}$, $\Delta = \{S, X, Y, Z\}$, and $\Pi = \{S \to Xb \mid YZ, Y \to bZ \mid b, X \to aZ \mid a, Z \to c\}$. Draw all parse trees for this grammar.
6. Let $\mathcal{G} = \{\Sigma, \Delta, S, \Pi\}$, where $\Sigma = \{a, b, c, (,), +, *\}$, $\Delta = \{S, E, V\}$, and Π contains the following productions.

$$S \to E$$
$$E \to (E) \mid E + E \mid E * E \mid V$$
$$V \to a \mid b \mid c$$

$\mathcal{G}$ is a simple grammar for algebraic expressions in the three variables a, b, and c and the operators $+$ and $*$, with properly paired parentheses permitted.

(a) Create the parse tree corresponding to the following derivations. Note that "(" and ")" and "+" and "*" are each separate terminal symbols and will consequently appear as separate nodes in the parse tree.

i. $S \Rightarrow E \Rightarrow (E) \Rightarrow (E*E) \Rightarrow (V*E) \Rightarrow (a*E) \Rightarrow (a*V) \Rightarrow (a*c)$

ii. $S \Rightarrow E \Rightarrow E+E \Rightarrow V+E \Rightarrow a+E$
$\Rightarrow a+(E) \Rightarrow a+(E+E) \Rightarrow a+(E+V)$
$\Rightarrow a+(E+c) \Rightarrow a+(V+c) \Rightarrow a+(b+c)$

iii. $S \Rightarrow E \Rightarrow E*E \Rightarrow (E)*E \Rightarrow (E)*(E)$
$\Rightarrow (E+E)*(E) \Rightarrow (E+E)*(E+E)$
$\Rightarrow (V+E)*(E+E) \Rightarrow (a+E)*(E+E)$
$\Rightarrow (a+V)*(E+E) \Rightarrow (a+b)*(E+E)$
$\Rightarrow (a+b)*(V+E) \Rightarrow (a+b)*(a+E)$
$\Rightarrow (a+b)*(a+V) \Rightarrow (a+b)*(a+c)$

(b) Show that the unparenthesized expression $a+b*c$ has two distinct parse trees using this grammar.

(c) Do the expressions $(a+b)*c$ and $a+(b*c)$ have unique parse trees? Support your answer.

7. ★ Produce a Huffman encoding for the phrase "pied piper." Assume that the encoding alphabet consists of the letters d, e, i, p, r, and ␣.
8. Produce a Huffman encoding for the phrase "eager beaver". Assume that the encoding alphabet consists of the letters a, b, e, g, r, v, and ␣.
9. Assume that the encoding alphabet consists of the letters b, i, o, t, y, and ␣. Produce a Huffman encoding for the phrase "bibbity bobbity boo".
10. For each DNA string, complete the following tasks:
    - Produce the final labeled tree and the encoding table for a Huffman encoding of this string.
    - Determine how many bits are needed to encode this string using a constant width encoding with 2 bits per character.
    - Determine how many bits are needed for the Huffman encoding if the encoding table is not included in the file.

    (a) ★ ACGTACGACA

    (b) ACTGGTACCCAGTTAACCCG

    (c) TATAACACATATTGTTGACTTACTTTATAACACAT-ATTGTTGACTTACTT
11. I am currently using the following constant-width encoding to store certain characters as binary strings on a computer.

␣	f	i	l	m	n	o	s
101	110	001	100	011	000	010	111

I wish to store the string "millions of minions".

(a) Produce the final labeled Huffman tree and the table of Huffman encodings.

(b) How many bits will the constant-width encoding require to store the string?

(c) How many bits will the Huffman encoding require to store the string?

12. I am currently using the following constant-width encoding to store certain characters as binary strings on a computer.

␣	a	d	e	i	m	n	o
111	000	001	010	011	100	101	110

I wish to store the string "mamie minded momma".[16]

(a) Produce the final labeled Huffman tree and the table of Huffman encodings.

(b) How many bits will the constant-width encoding require to store the string?

(c) How many bits will the Huffman encoding require to store the string?

---

[16] The original is from a bit of verse by Walt Kelly from the *Pogo* comic strip [60, p. 90]. The context is

O, Mamie minded momma
'Til one day in Singapore
A sailorman from Turkestan
Came knocking at the Door.

13. Produce the tree representation for the following well-formed XML document.

```
<?xml version = "1.0" ?>
<VegetableGarden year = "2001">
   <PlantingDate>May 28</PlantingDate>
   <planted>
      <corn num = "20">
        Yellow Majesty
      </corn>
      <tomato num = "2">
        Big Boy
      </tomato>
      <tomato num = "3">
        Cherry Delight
      </tomato>
   </planted>
</VegetableGarden>
```

14. ★ The "letter" DTD in Quick Check 11.9 (page 707) could be improved by adding an element that indicates that the enclosed character data requires emphasis. This element might be used, for example, by a program that prints the document on a laser printer.

The following fragment is from an XML document that conforms to the new "emphasis" element.

```
<?xml version = "1.0" ?>
<!DOCTYPE letter SYSTEM "letter2.dtd">
<letter>
  <greeting>Dear mom and dad,</greeting>
  <body>
     <paragraph>
       I <emphasis>really</emphasis>
       need money! Please send
       some <emphasis>soon!</emphasis>
       Otherwise ...
     </paragraph>
```

Modify the letter DTD on page 728 by including this new element.

15. Consider the following DTD, which describes various modes of transportation. As usual, the line numbers are not part of the DTD.

```
 1: <!ELEMENT transportation (land,
    water)>
 2:   <!ATTLIST transportation
       updated CDATA #REQUIRED>
 3: <!ELEMENT land (machine-powered+,
      human-powered+)>
 4: <!ELEMENT water (machine-powered+,
    human-powered+, air-powered+)>
 5: <!ELEMENT machine-powered
      (description, power-source)>
 6:   <!ATTLIST machine-powered
        capacity CDATA #IMPLIED>
 7: <!ELEMENT human-powered
      (description)>
 8: <!ELEMENT air-powered (description)>
 9: <!ELEMENT power-source (#PCDATA)>
10: <!ELEMENT description (#PCDATA)>
```

The XML document in Figure 11.39 fails to conform to the DTD. Determine all the ways in which it fails to follow the DTD, and then modify the document so that it conforms to the DTD while still keeping all the intended information. The line numbers are not part of the XML document.

```
 1: <?xml version="1.0"?>
 2: <!DOCTYPE transportation SYSTEM
    "transportation.dtd">
 3: <transportation updated="6-08-2009">
 4:   <land>
 5:     <machine-powered capacity="5">
 6:       <description>automobile</description>
 7:       <power-source>gasoline
           engine</power-source>
 8:       <description>semi tractor
          </description>
 9:       <power-source>diesel
          engine</power-source>
10:     </machine-powered>
11:     <human-powered>
12:       <description>bicycle</description>
13:     </human-powered>
14:     <human-powered>
15:       <description>feet</description>
16:     </human-powered>
17:     <water>
18:       <machine-powered>
19:         <description>submarine<description>
20:         <power-source>nuclear
            reactor</power-source>
21:       </machine-powered>
22:       <air-powered>
23:         <description>sail boat</description>
24:       </air-powered>
25:       <human-powered>
26:         <description>canoe</description>
27:         <description>rowboat</description>
28:       </human-powered>
29:     </water>
30:   </land>
31: </transportation>
```

Figure 11.39 An invalid XML document.

16. A computer program that allows the display and manipulation of graphs also has the ability to read and write the essential information as a disk file. The disk file uses XML to represent the graph.

The displayed graph consists of a nonempty collection of vertices, edges, and labels (in any order). The element names are "vertex", "edge", and "label", respectively.

Vertices have three attributes: a diameter (in points), an x-coordinate, and a y-coordinate. Edges have four attributes:

the x- and y-coordinates of the two vertices at the endpoints of the edge.[17] Labels have four attributes: a font, a (font) size (in points), and the x- and y-coordinates of the top left corner of the label. A label also must enclose some character data.

(a) Construct the tree representation for the following well-formed XML document.

```
<?xml version = "1.0" ?>
<!DOCTYPE graph SYSTEM "graph.dtd">
<graph>
   <label font="Times"
     size = "12" x = "20" y = "25">
   Two vertices, one edge.
   </label>
   <vertex diameter = "4" x = "40"
        y = "40"></vertex>
   <edge x1 = "40" y1 = "40"
         x2 = "80"
         y2 = "50"></edge>
   <vertex diameter = "4" x = "80"
         y = "50"></vertex>
</graph>
```

(b) Create a DTD for graph documents. Elements with no enclosed items can be specified by replacing the parenthesized "ConsistsOfRule" by the string "EMPTY" (all uppercase). The vertex rule will be written in the following manner.

```
<!ELEMENT vertex EMPTY>
```

17. ★ Suppose an e-mail letter has one or more "to" elements, zero or more "cc" elements, one "replyto" element, one "subject", and a "body" with zero or more paragraphs.

(a) Create a DTD for an e-mail letter.

(b) Create a nontrivial conforming XML document.

18. A "CD-library" will contain one or more "CD" elements. A "CD" element has two attributes and four sub-elements. The attributes are "TotalTime" and "IDnum". A "CD" will contain a "CD-name" followed by a "CD-label" (the recording company), followed by an "artist-composer" element followed by a "contents" element. An "artist-composer" element contains a mixture of one or more "artist" and "composer" elements. The "contents" element contains one or more "track" elements. Each track contains a "TrackTime" attribute. The elements "artist", "composer", and "track" contain character data.

(a) Create a DTD for a "CD-library" XML document.

(b) Create a nontrivial conforming XML document.

11.4 Spanning Trees

A typical connected graph will have more edges than are necessary to ensure connectivity. It is usually possible to remove one or more edges and still maintain a connected graph with the same set of vertices. In fact, it is always possible to find a subgraph that contains all the vertices and is a connected subgraph with no cycles. Such a subgraph must be a tree.

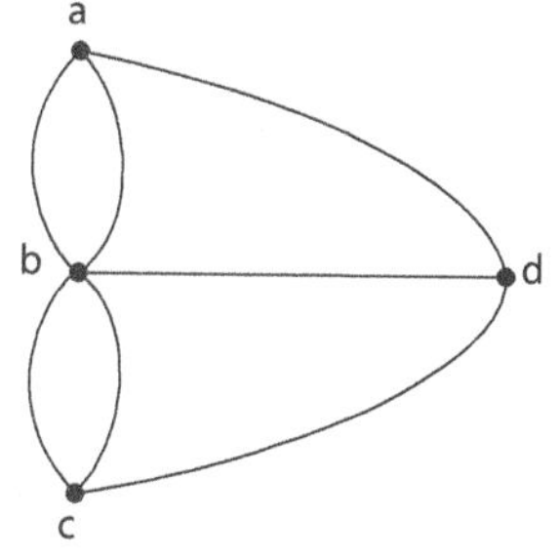

Figure 11.40. The Königsberg graph.

11.4.1 Spanning Trees in Unweighted Graphs

> **DEFINITION 11.32** ***Spanning Tree***
> A subgraph, $T = \{V_T, E_T, \phi\}$, of a connected graph, $G = \{V, E, \phi\}$, is called a *spanning tree* if T is a tree and $V_T = V$.

A spanning tree is therefore a tree that contains every vertex in G. Since T is a tree, it is possible (Theorem 11.3 on page 673) to get from any vertex in G to any other vertex by crossing edges in T. In that sense, T "spans" G.

EXAMPLE 11.25 **Some Spanning Trees**

Recall the Königsberg graph (Figure 11.40). This graph has numerous spanning trees. Two of them are shown in Figure 11.41. ■

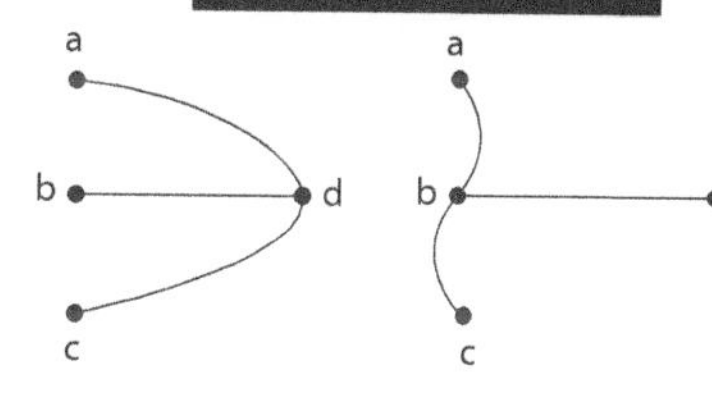

Figure 11.41. Two spanning trees for the Königsberg graph.

The next theorem asserts that spanning trees are the largest possible trees in a graph in the sense that adding *any* other edge will create a cycle, causing the new subgraph to cease being a tree. (So there are no edges that are isolated from a spanning tree.)

[17] There is a better way (using the ID and IDREF attributes) to do this in XML.

THEOREM 11.33 ***Spanning Trees Are Maximal Trees***

Let G be a connected graph with spanning tree T. Let T' be the graph formed by adding any other edge in G to T. The graph T' must contain a cycle.

Proof: The proof is left as an exercise (exercise 4 on page 723). □

The introduction to this section claimed that every connected graph has a spanning tree. The next theorem provides the formal details.

THEOREM 11.34 ***Connected If and Only If Spanning Tree***

A graph is connected if and only if it contains a spanning tree.

Proof: Suppose that a graph, G, has a spanning tree, T. Then every vertex in G is also in T. Theorem 11.3 on page 673 ensures that there is a path in T (and hence in G) between any two distinct vertices. Thus, G is connected.

Now suppose that G is a connected graph. A spanning tree can be constructed as follows. First, remove any loops. If there are multiple edges connecting two vertices, remove all but one of them. Denote the resulting connected subgraph as G'. Notice that every vertex in G is still in G'. If G' is a tree, then the proof is complete. Otherwise, there must be a cycle in G'. Remove any edge in the cycle. The new subgraph is still connected and contains every vertex in G. If it is a tree, the proof is complete. Otherwise, keep removing edges from cycles until there are no cycles left. (This process will terminate because the graph has only a finite number of edges.)

Each time an edge is removed from a cycle, its two incident vertices will not be removed because they are each still connected to at least one other edge (or else they would not be in a cycle). Thus, the new subgraph will still contain every vertex of G. Also, removing an edge from a cycle will not disconnect the graph (edges in cycles are never cut-edges).

The process of removing edges terminates when there are no longer any cycles. The resulting subgraph still contains all the vertices of G and is still connected. It is therefore the desired spanning tree. □

EXAMPLE 11.26 **Producing a Spanning Tree**

The constructive algorithm in the proof of Theorem 11.34 for producing a spanning tree can be used to find a spanning tree for the connected multigraph in Figure 11.42.

The first step is to remove the loop and one of the multiple edges (Figure 11.43).

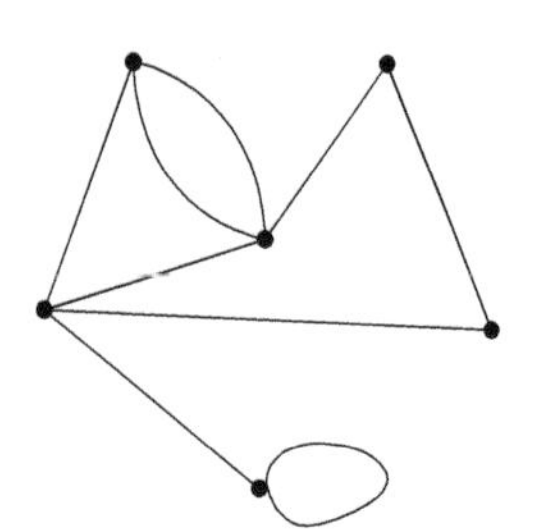

Figure 11.42 A connected multigraph.

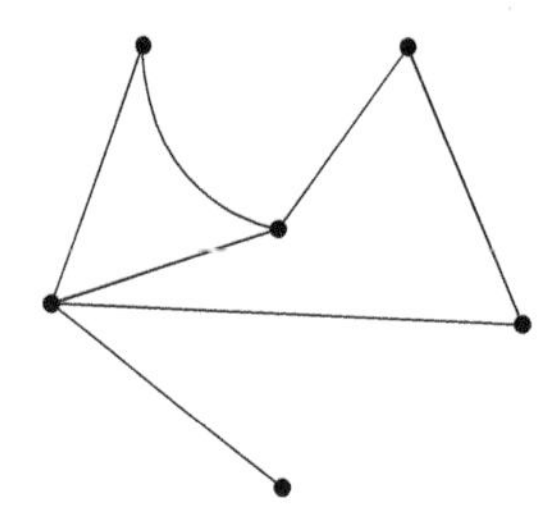

Figure 11.43 Drop the loop and a multiple edge.

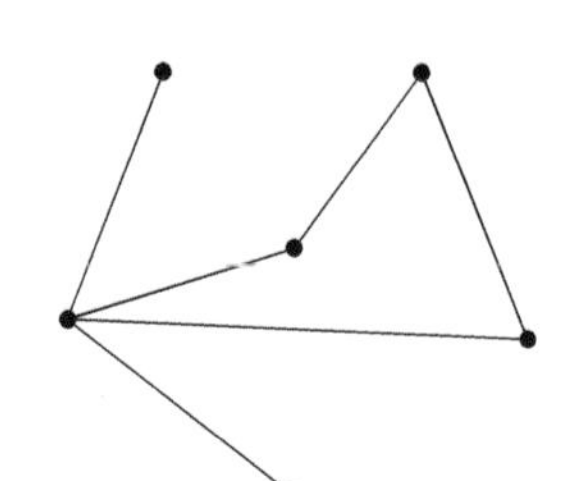

Figure 11.44 Remove an edge in a cycle.

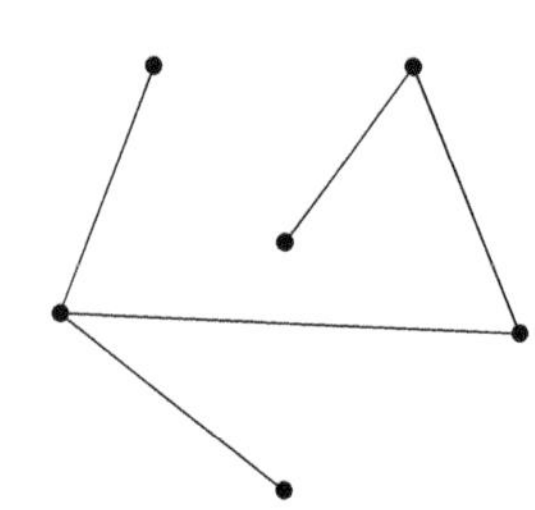

Figure 11.45 Remove another edge in a cycle.

Since the resulting subgraph is not a tree, any edge from one of the cycles is chosen to be removed (Figure 11.44). The resulting subgraph is still not a tree so an edge from the remaining cycle is chosen to be removed (Figure 11.45), producing a spanning tree. ■

The second part of the previous proof is constructive but it is not a particularly efficient construction. This is because it is not always easy to identify a cycle in a very large graph. Two alternate algorithms will be briefly outlined. Both avoid the need to detect existing cycles and both use a very simple rule to avoid creating cycles.

Depth First

Suppose that G is a connected simple graph. One approach for finding a spanning tree is to use a *depth-first* algorithm. The algorithm arbitrarily chooses a vertex to be the root node of the tree. It then starts adding new edges that satisfy the following criteria:

- Each new edge must be incident with the last vertex added to the tree.
- Each new edge must be incident with a vertex that is not currently in the tree.

The first criterion ensures that the tree stays connected. The second criterion ensures that no cycles are added.

Once it is not possible to add another edge, the algorithm backs up to the second most recently added vertex and tries to extend a new edge that meets the two criteria. Each time it is impossible to add a new edge, the algorithm backs up to a previous vertex. This continues until the root node is reached (while backing up) and there are no new edges possible. The resulting graph is connected and does not contain any cycles, so it is a tree. In addition, it must contain every vertex of G or else at some point it would have been possible to add another edge.

EXAMPLE 11.27 **Depth First**

The depth-first algorithm can be used to construct a spanning tree for the Walther graph (Figure 11.46). The edge labels indicate the order in which edges are added to the tree. The edges that are not part of the spanning tree have been left in with a lighter coloring.

Notice that new edges are added up through edge 8 with no difficulty. The algorithm will then back up to the vertex shared by edges 7 and 8, but there are no new edges to add, so it will back up to the vertex shared by edges 6 and 7. Note also that after adding edge 22, the algorithm must back up to the vertex between edges 19 and 20 before it can find a new edge that meets the two criteria.

The final spanning tree is shown in Figure 11.47, isolated from the rest of the graph.

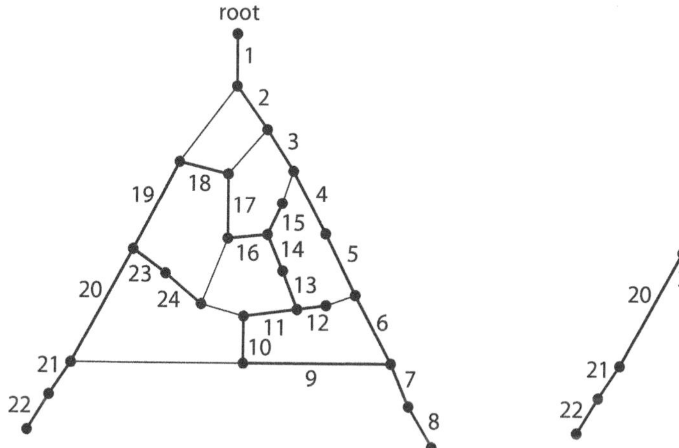

Figure 11.46 The Walther graph.

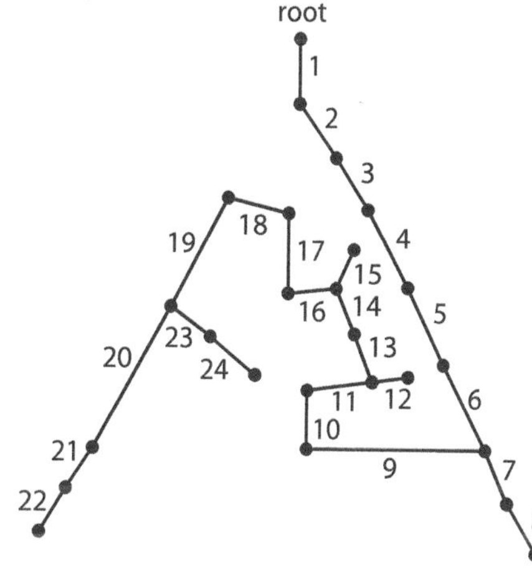

Figure 11.47 A depth-first spanning tree for the Walther graph.

The resulting tree is far from unique; many other trees could have been constructed using this algorithm. ■

Breadth First

Suppose that G is a connected simple graph. Another approach for finding a spanning tree is to use a *breadth-first* algorithm. The algorithm arbitrarily chooses a vertex to be the root node of the tree. That vertex also becomes the current vertex. It then adds all edges that satisfy the following criteria:

- Every new edge must be incident with the current vertex.
- Every new edge must be incident with a vertex that is not currently in the tree.

The first criterion ensures that the tree stays connected. The second criterion ensures that no cycles are added.

As an edge is added to the tree, a reference to it is also placed at the back of a list of pending items.[18] Once all the edges that satisfy the criteria have been added, the item at the front of the list is removed. The current vertex becomes the vertex (incident with the newly removed edge) that is farthest from the root. The process then repeats: Add all new edges that satisfy the two criteria, placing references at the back of the list as edges are added to the tree.

The process continues until the list of pending items is empty.

EXAMPLE 11.28 **Breadth First**

The breadth-first algorithm can also be used to construct a spanning tree for the Walther graph (repeated as Figure 11.48). The edge labels indicate the order in which edges are added to the tree. The edges that are not part of the spanning tree have been left in with a lighter coloring.

The final spanning tree is shown in Figure 11.49, isolated from the rest of the graph.

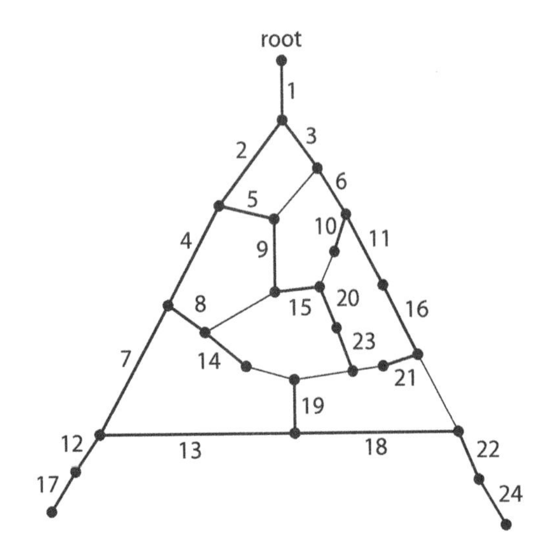

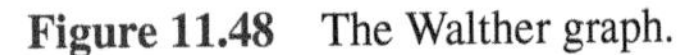

Figure 11.48 The Walther graph.

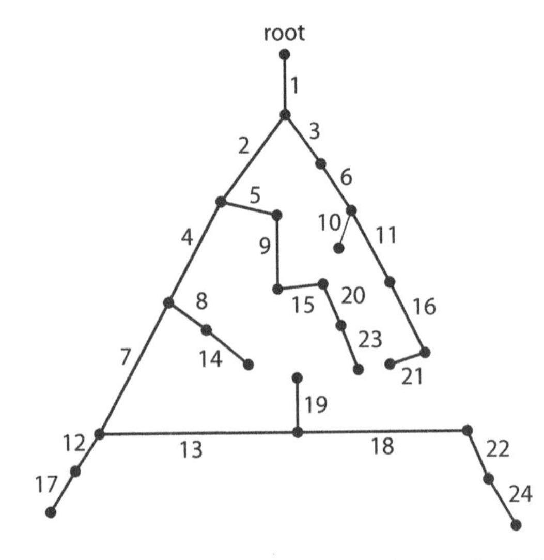

Figure 11.49 A breadth-first spanning tree for the Walther graph.

The resulting tree does have some sense of uniqueness once the root node is chosen. To ensure uniqueness, some agreement would need to be made about the order in which to add the edges incident to a vertex (perhaps counterclockwise, as was done in this example). ■

---

[18]Computer science majors will recognize this list as a queue.

Counting Spanning Trees (Optional: requires knowledge of determinants and cofactors)

It would appear to be quite difficult to determine the number of distinct spanning trees contained in an arbitrary connected graph. In fact, it *is* quite difficult to determine the number of *non-isomorphic* spanning trees. However, there is a straightforward procedure for counting the number of distinct trees in a connected loopless graph. The procedure is presented in the next theorem.

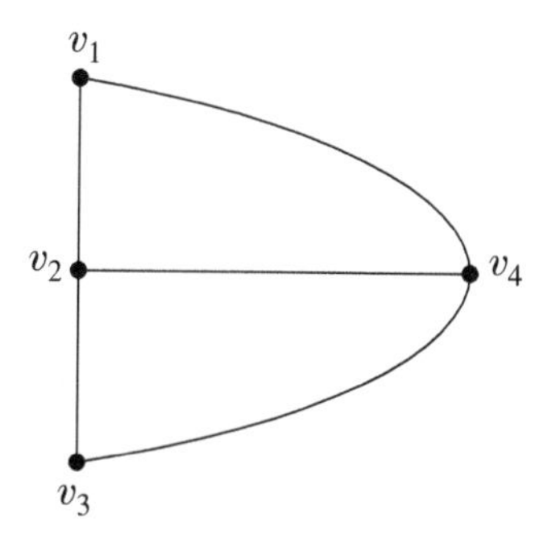

Figure 11.50. Underlying Königsberg simple graph.

THEOREM 11.35 ***Kirchoff's Matrix-Tree Theorem***

Let $G = (V, E, \phi)$ be a connected, loopless graph with $V = \{v_1, v_2, \ldots, v_n\}$. Let A be the adjacency matrix (ordered the same as V). Finally, let K be the matrix whose entries are defined by

$$k_{ij} = \begin{cases} -a_{ij} & i \neq j \\ \deg(v_i) - a_{ii} & i = j \end{cases}$$

Then the number of distinct spanning trees in G is the value of any cofactor of K.

A proof may be found in [100, p. 67].

EXAMPLE 11.29 **Counting Spanning Trees**

The goal of this example is to count the number of distinct spanning trees in the Königsberg graph. The verification will be simpler by first counting the number of spanning trees in the underlying simple graph (Figure 11.50).

The adjacency matrix for this graph is (using the natural vertex ordering)

$$A = \begin{pmatrix} 0 & 1 & 0 & 1 \\ 1 & 0 & 1 & 1 \\ 0 & 1 & 0 & 1 \\ 1 & 1 & 1 & 0 \end{pmatrix}$$

The matrix K from Theorem 11.35 is

$$K = \begin{pmatrix} 2 & -1 & 0 & -1 \\ -1 & 3 & -1 & -1 \\ 0 & -1 & 2 & -1 \\ -1 & -1 & -1 & 3 \end{pmatrix}$$

Choosing a convenient cofactor,

$$C_{13} = (-1)^{1+3} \det A_{13} = \begin{vmatrix} -1 & 3 & -1 \\ 0 & -1 & -1 \\ -1 & -1 & 3 \end{vmatrix} = 8$$

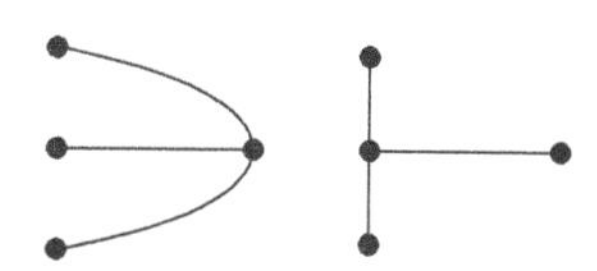
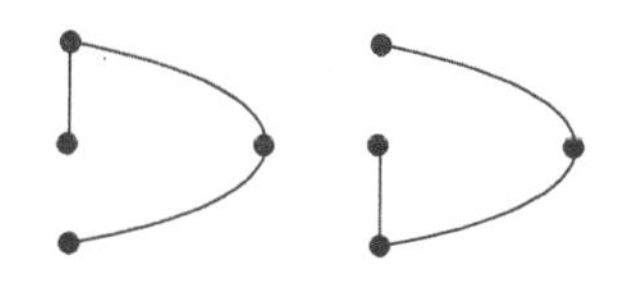
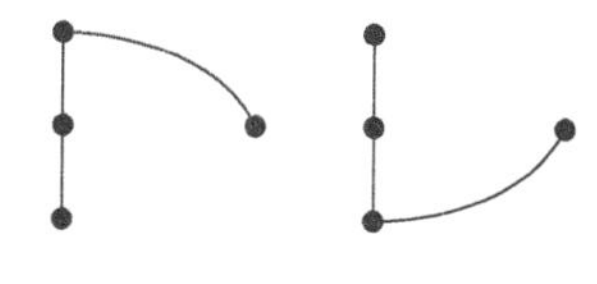
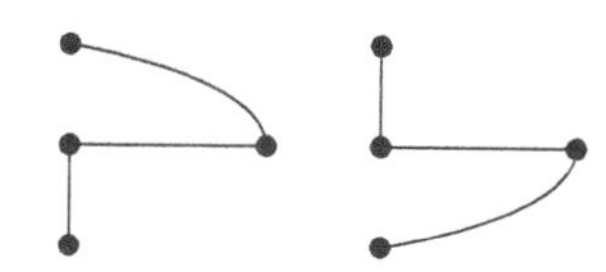

Figure 11.51. Distinct spanning trees for Figure 11.50.

To see that there are indeed eight distinct spanning trees, the collection of trees in Figure 11.51 is an exhaustive listing of all of them.

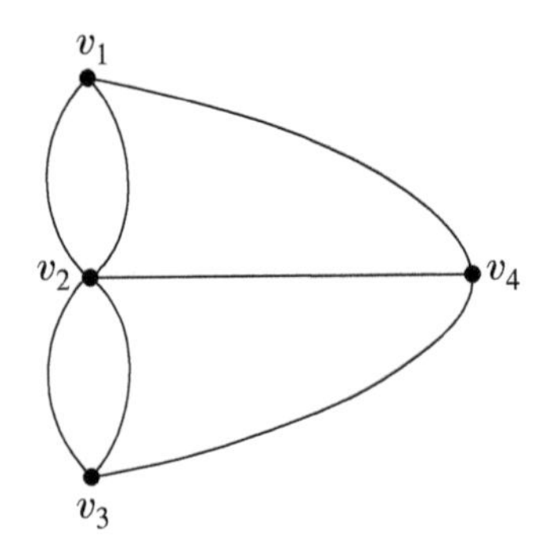

Figure 11.52. The Königsberg graph.

It is now a simple matter to examine the complete Königsberg graph (repeated in Figure 11.52).

The adjacency matrix for this graph is (again using the natural vertex ordering)

$$A = \begin{pmatrix} 0 & 2 & 0 & 1 \\ 2 & 0 & 2 & 1 \\ 0 & 2 & 0 & 1 \\ 1 & 1 & 1 & 0 \end{pmatrix}$$

The matrix K from Theorem 11.35 is

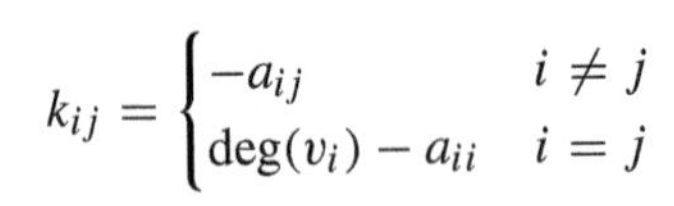

$$k_{ij} = \begin{cases} -a_{ij} & i \neq j \\ \deg(v_i) - a_{ii} & i = j \end{cases}$$

$$K = \begin{pmatrix} 3 & -2 & 0 & -1 \\ -2 & 5 & -2 & -1 \\ 0 & -2 & 3 & -1 \\ -1 & -1 & -1 & 3 \end{pmatrix}$$

Choosing a convenient cofactor,

$$C_{13} = (-1)^{1+3}\det A_{13} = \begin{vmatrix} -2 & 5 & -1 \\ 0 & -2 & -1 \\ -1 & -1 & 3 \end{vmatrix} = 21$$

Rather than exhaustively listing all 21 of the distinct spanning trees in the Königsberg graph, notice that the previous listing contains all the essential information. There are now two possibilities for each of the vertical edges in the previous listing. Thus, every tree in the second and fourth rows will become two trees in the list for the complete Königsberg graph. Also, all but the first tree in the first and third rows will have four variations. The total number of trees will therefore be 21, as predicted. ■

THEOREM 11.36 ***Cayley's Formula***

There are n^{n-2} distinct trees on the set $\{v_1, v_2, \ldots, v_n\}$ of n nodes.

The proof (exercise 6 on page 723) counts the number of distinct spanning trees in the complete graph, K_n.

EXAMPLE 11.30 **Counting Trees**

Cayley's formula predicts that there are $3^{3-2} = 3$ distinct trees on three nodes. (Note well: These need not be non-isomorphic.) It is quite easy to list them all (Figure 11.53). Notice that they are all isomorphic to each other.

Figure 11.53 All distinct trees with 3 nodes.

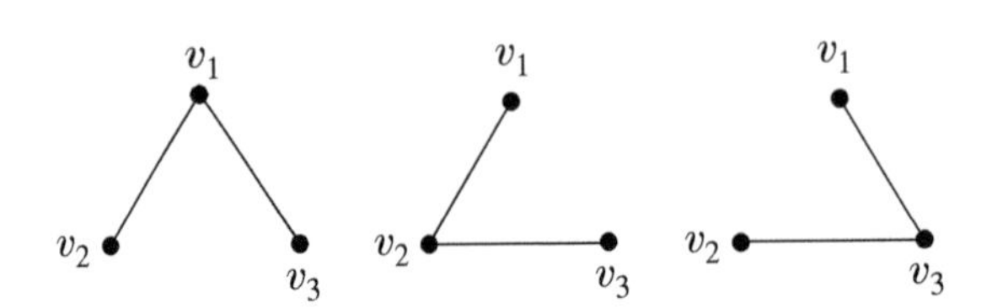

■

11.4.2 Minimal Spanning Trees in Weighted Graphs

If G is a weighted graph, the spanning trees of interest change in nature. The weighted graph is often a model of some problem, and the weights will often be costs associated with traversing edges. The spanning trees that are of interest will be ones that keep the cost as low as possible. This leads naturally to the next definition.

DEFINITION 11.37 ***Minimal Spanning Tree***

Let G be a connected, weighted, simple graph. A *minimal spanning tree* for G is a spanning tree, T, for which the sum of the weights on the edges of T is less than or equal to the sums of the weights on the edges of any other spanning tree for G.

The definition of *minimal spanning tree* is associated with simple graphs. This is not a significant restriction because any spanning tree in a multigraph will exclude all loops. In addition, any minimal spanning tree will always choose an edge of minimal weight from among any set of multiple edges. Therefore, to find a minimal spanning tree for a multigraph, first remove all loops and then, from each set of multiple edges, remove all but an edge with lowest weight. Any minimal spanning tree for the resulting simple graph will also be a minimal spanning tree for the original multigraph.

For example, the simple graph in Figure 11.55 is the proper choice for the underlying weighted simple graph of the weighted multigraph in Figure 11.54.

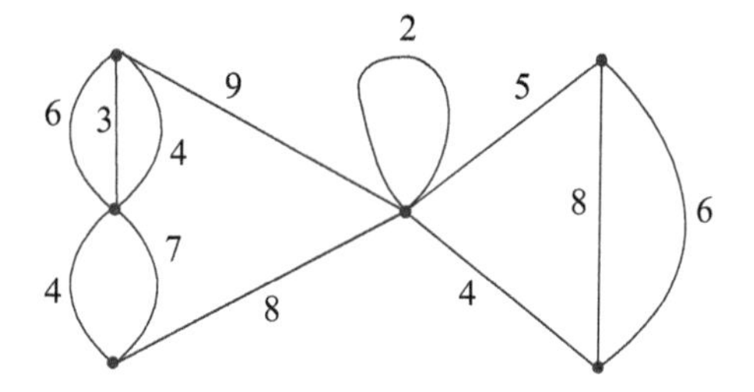

Figure 11.54 A multigraph.

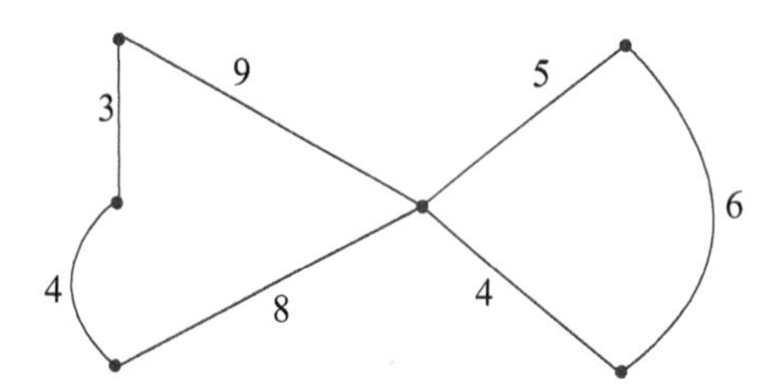

Figure 11.55 The correct underlying simple graph.

It is not hard to verify (by exhaustively looking at all possible spanning trees) that the spanning tree in Figure 11.56 is a minimal spanning tree for the weighted graphs in Figures 11.54 and 11.55. The total weight of the spanning tree is 24.

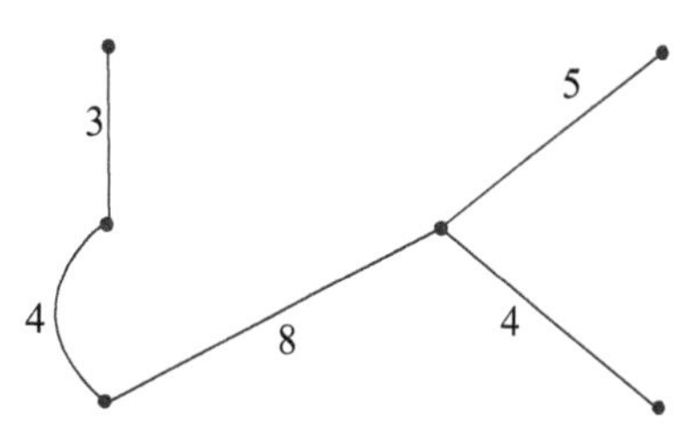

Figure 11.56 A minimal spanning tree.

It turns out to be quite easy to construct minimal spanning trees. Two closely related algorithms will be presented here. Both algorithms will be described and examples of their use will be presented before proving that the algorithms always work.

Both algorithms are examples of *greedy algorithms*.[19] They seek to achieve a global optimum by making locally optimal choices at each stage of the algorithm. This strategy falls short of the intended goal for many tasks, but does work in this setting.

[19] See Example 8.30 on page 473 for more on greedy algorithms.

Prim's Algorithm for Constructing Minimal Spanning Trees

Prim's algorithm for producing a minimal spanning tree, T, for a graph, G, starts by choosing any edge with smallest weight. That edge becomes the initial edge in T. At each successive step, a new edge is chosen that satisfies the following criteria:

- The new edge must share a vertex with some edge that is already in T.
- The new edge does not create a cycle in T.
- The new edge has the smallest possible weight without violating one of the other criteria.

The algorithm stops when the number of edges in T is one less than the number of vertices in G.

If at some step there are multiple viable edges with the same weight, the algorithm may not produce a unique minimal spanning tree. Different trees might emerge depending on which of the viable options are chosen at various steps.

It is easier to discuss the algorithm if the following notation is adopted: Define T_k to be the subgraph of T that exists after step k of the algorithm. Thus, the initial step produces T_1, having one edge. The final step produces $T_{n-1} = T$ (assuming G has n vertices).

EXAMPLE 11.31 **Prim's Algorithm**

Prim's algorithm has been used to construct a minimal spanning tree for the weighted graph in Figure 11.57. The edges have been labeled in the order they were added to the spanning tree. The first edge, e_1, was arbitrarily chosen from among the two edges of weight 2.

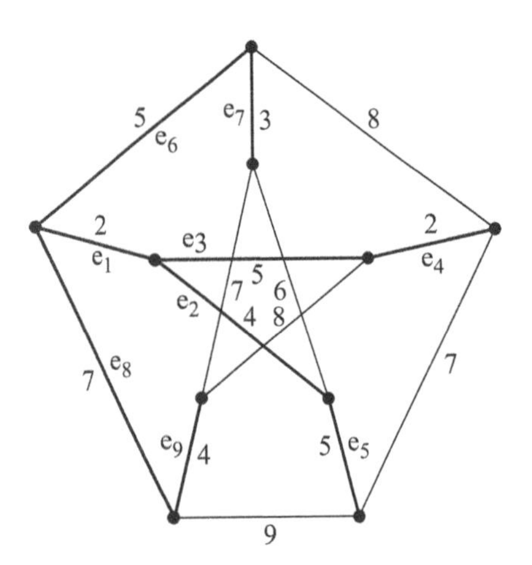

Figure 11.57 The Petersen graph with edge weights.

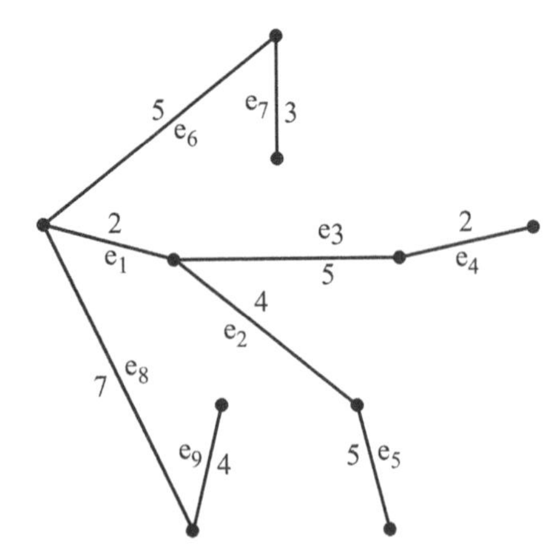

Figure 11.58 A spanning tree via Prim's algorithm.

The edges that share a vertex with e_1 have weights 4, 5, and 7. The weight 4 edge was added next. There are three weight 5 edges that could be used for e_3. The one chosen allowed the other weight 2 edge to be added as e_4. You should convince yourself that adding one of the other weight 5 edges at this stage would not have resulted in a spanning tree with a smaller total weight.

The final spanning tree is shown in Figure 11.58. ■

Kruskal's Algorithm for Constructing Minimal Spanning Trees

Kruskal's algorithm for producing a minimal spanning tree, T, for a graph, G, differs in only one detail from Prim's algorithm. That detail will be noted after the algorithm is presented.

Kruskal's algorithm starts by choosing any edge with smallest weight (as does Prim's algorithm). That edge becomes the initial edge in T. At each successive step, a new edge is chosen that satisfies the following criteria:

- The new edge does not create a cycle in T.
- The new edge has the smallest possible weight without violating the other criterion.

The algorithm stops when the number of edges in T is one less than the number of vertices in G.

The one detail that is different is that Kruskal's algorithm does not require T_k to be connected for $1 < k < n - 1$.

EXAMPLE 11.32 **Kruskal's Algorithm**

Kruskal's algorithm has been used to construct a minimal spanning tree for the weighted graph in the Prim example (repeated as Figure 11.59). The edges have again been labeled in the order they were added to the spanning tree.

There were two choices for the initial edge. In this example, they were added as e_1 and e_2 in the order shown. The weight 3 edge was added next. All weight 4 and 5 edges were also added without any complications. The final edge could not be the weight 6 edge since adding it would create a cycle. Two of the three weight 7 edges were viable candidates. One of them was chosen arbitrarily.

The final spanning tree is shown in Figure 11.60.

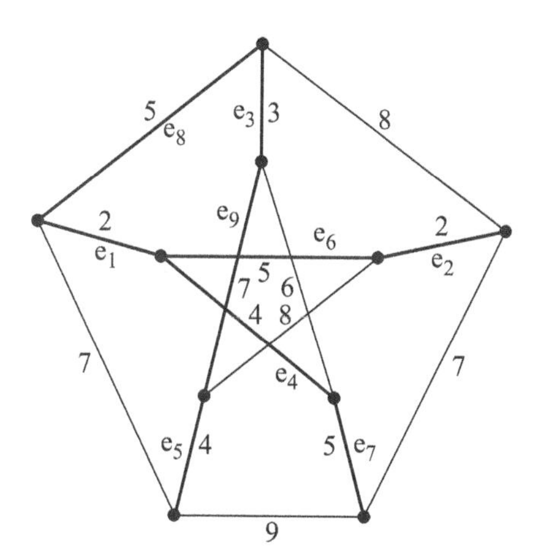

Figure 11.59 The Petersen graph with edge weights.

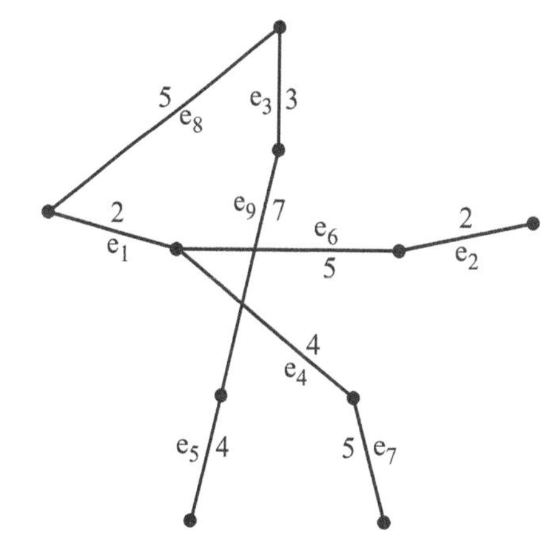

Figure 11.60 A spanning tree via Kruskal's algorithm. ■

Proving that the Prim and Kruskal Algorithms are Correct

The proofs that the Prim and Kruskal algorithms always produce minimal spanning trees are quite interesting. The strategy is to pretend to be temporarily omniscient and compare the tree, T, produced by the algorithm to all possible minimal spanning trees for the underlying graph. A proof by contradiction will show that T must be a member of the set of all minimal spanning trees.

The proof of Theorem 11.38 is a bit abstract, so an example would be helpful. However, the proof is at its core a proof by contradiction. Therefore, an example of the difficult part of the proof can never really exist. However, it is possible to create an example that does illustrate some aspects of the difficult section. Example 11.33 attempts that feat. You may wish to read it in parallel with the proof of the theorem.

THEOREM 11.38 ***Prim's Algorithm***

Let G be a connected, weighted, simple graph. Then Prim's algorithm will always produce a minimal spanning tree for G.

Proof: Suppose G has n vertices. Prim's algorithm adds edges in some order $e_1, e_2, \ldots, e_{n-1}$, forming a tree, T.

Consider the finite set of all minimal spanning trees for G. Choose any one of them, M, which contains the longest sequence $e_1, e_2, \ldots, e_k$ of the initial edges in T.[20]

If $T = M$ ($k = n - 1$), we then know that T is a minimal spanning tree for G and the proof is complete.

Otherwise, let T_k be the (nonspanning) tree formed by the edges $e_1, e_2, \ldots, e_k$ with $k < n - 1$. Since M is a spanning tree, adding another edge to M will produce a cycle (Theorem 11.33 on page 712). In particular, adding e_{k+1} to M will produce a cycle in M. Since e_{k+1} is in the cycle, and since it shares a vertex with an edge in T_k, at least one of the vertices in T_k is part of the cycle. Since T is a tree it cannot contain a cycle, so there must be some edge, $\hat{e}$, in the cycle that is part of M, is not in T, but does have a vertex connected to T_k.

$M \cup \{e_{k+1}\}$ has one too many edges to be a tree. Remove $\hat{e}$ to form M'. (Notice that both vertices of $\hat{e}$ are still "spanned.") M' is a spanning tree. It also has no larger weight than M, since by Prim's construction the weight on e_{k+1} is less than or equal to the weight on $\hat{e}$.[21] But M' contains a longer sequence of edges than does M. This contradicts the maximality of M.

What assumption created the contradiction? It was the assumption that $k < n - 1$. Thus, it must always be true that $k = n - 1$ and T is one of the minimal spanning trees for G. □

EXAMPLE 11.33 **Illustrating the Proof of Theorem 11.38**

Let the graph, G, be the graph in Example 11.31 on page 718 and let T be the minimal spanning tree produced in that example. Both the graph (Figure 11.61) and the spanning tree (T in Figure 11.62) are repeated for easy reference.

Even though there will never be a minimal spanning tree, M, which contains a maximal initial sequence, $e_1, e_2, \ldots, e_k$, of edges in $\{e_1, e_2, \ldots, e_9\}$ for which $k < 9$, pretend for the moment that there is one. For concreteness, let M be the minimal spanning tree produced by Kruskal's algorithm in Example 11.32 on page 719. However, forget about the fact that that tree was produced by an algorithm. Figure 11.62 shows the Prim tree on the left (T in the proof of Theorem 11.38) and the hypothetical M on the right.

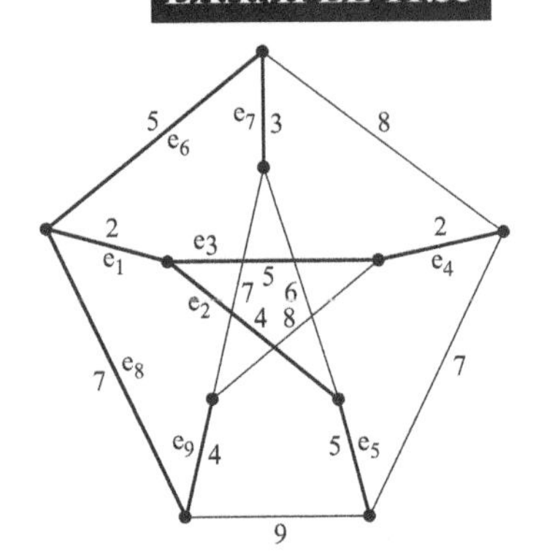

Figure 11.61. The weighted Petersen graph.

[20] These edges are the first k edges that Prim's algorithm chooses. How they happened to be in M is not important, nor is the order in which they were added to M a relevant concept. The key idea is that no other minimal spanning tree contains a longer sequence of initial edges from T than does M.

[21] The edge $\hat{e}$ cannot have a smaller weight than e_{k+1} because Prim's algorithm always picks an edge with smallest weight among the remaining edges that don't form a cycle. Starting at T_k, the algorithm chose e_{k+1} instead of $\hat{e}$. This means that either $\hat{e}$ has a weight that is equal or larger than that of e_{k+1} or else adding $\hat{e}$ to T_k would form a cycle. This second option is not true, because every edge in T_k and the edge $\hat{e}$ are all in the tree M.

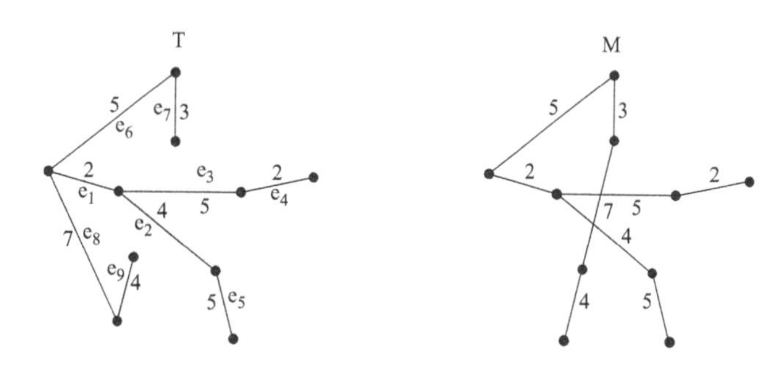

Figure 11.62 The Prim spanning tree (T) and a hypothetical spanning tree (M).

This M contains the initial sequence, $e_1, e_2, \ldots, e_7$, of edges in common with T. Adding e_8 to M will create a graph, $M \cup \{e_8\}$, with a cycle, as shown in Figure 11.63.

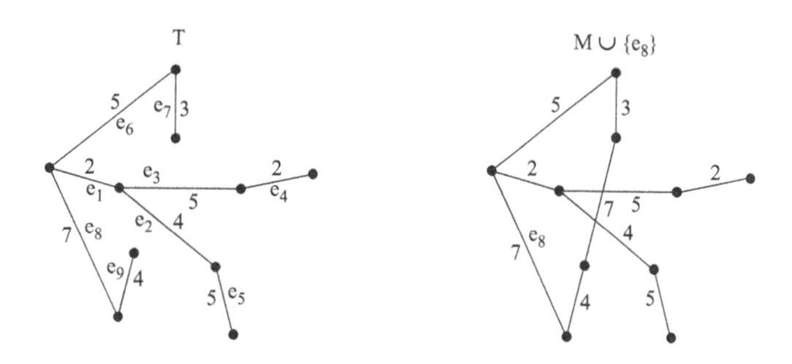

Figure 11.63 T and $M \cup \{e_8\}$.

Figure 11.64 shows T_7 on the left. The edge, $\hat{e}$, in $M \cup \{e_8\}$ must satisfy the following conditions: It is in the four-sided cycle, it is not in T, but it is incident with a vertex in T_7. Such an edge exists. There is only one choice for the vertex that $\hat{e}$ shares with T_7. It has been labeled "v".

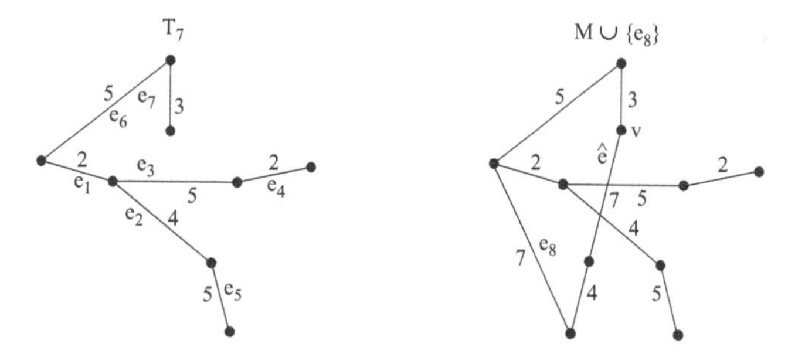

Figure 11.64 T_7 and $M \cup \{e_8\}$.

Removing $\hat{e}$ from $M \cup \{e_8\}$ and leaving e_8 in does not change the total weight. The new graphs, M' and T_8, are shown in Figure 11.65. M' is a minimal spanning tree for G that contains a larger initial sequence of edges from T, completing the contradiction.

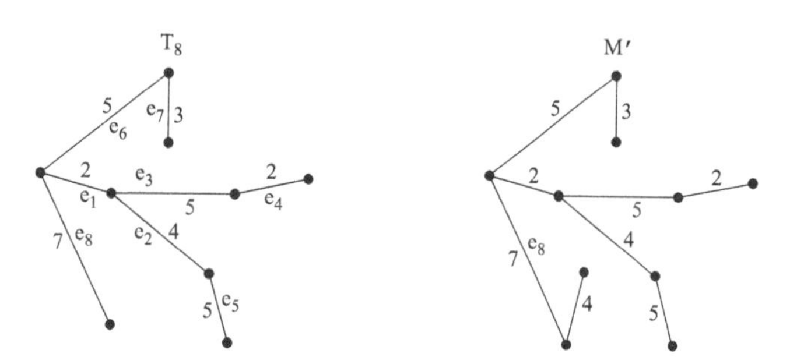

Figure 11.65 T_8 and M'.

■

THEOREM 11.39 ***Kruskal's Algorithm***

Let G be a connected, weighted, simple graph. Then Kruskal's algorithm will always produce a minimal spanning tree for G.

Proof: Kruskal's algorithm drops the requirement that T_k be connected for $1 < k < n - 1$. This makes the algorithm slightly easier to use. However, you might expect the proof to become more difficult since there is less structure in T_k to exploit.

The details of the proof are left as Exercise 9 on page 723. □

✔ Quick Check 11.10

1. How many distinct spanning trees exist for the underlying unweighted graph?
2. Use Prim's algorithm to find a minimal spanning tree. Label the edges in the order they are added.
3. Use Kruskal's algorithm to find a minimal spanning tree. Label the edges in the order they are added.

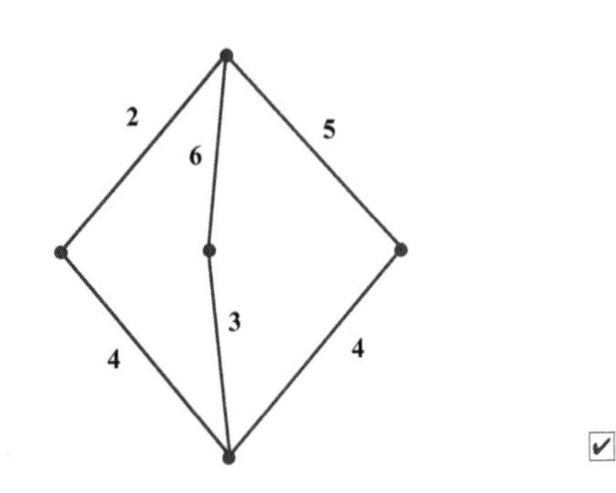

☑

Maximal Spanning Trees

Suppose that G is a weighted graph. It is certainly possible to develop modified Prim and Kruskal algorithms that will produce a maximal spanning tree. Here is a clever alternative.

Assume that the edges of G are labeled with the nonnegative numbers $\{n_1, n_2, \ldots, n_k\}$ and that the largest number in the set of labels is M.[22] Form a graph, G', that is identical to G except that all the edge weights will be replaced. In particular, the weight n_i will be replaced by $M - n_i$. The weights on the edges of G' will still be nonnegative and will still maintain the relative differences in size. That is, if $|n_i - n_j| = d$, then $|(M - n_i) - (M - n_j)| = d$ is still true. However, the relative magnitudes have reversed. Thus, if $n_i \geq n_j$, then $(M - n_i) \leq (M - n_j)$.

Use either the Prim or the Kruskal algorithm to find a minimal spanning tree, T', for G'. Now create a new tree, T, that is identical to T', except that the edge weights will be replaced by subtracting each weight from M. Since $M - (M - n_i) = n_i$, the weights on edges in T will be the same as the weights on the corresponding edges in G. Thus, T will be a maximal spanning tree for G.

11.4.3 Exercises

The exercises marked with ★ have detailed solutions in Appendix H.

1. ★ Find spanning trees for each of the following graphs. Use three different algorithms:
   - The algorithm from the proof of Theorem 11.34, page 712.
   - The depth-first algorithm (page 713).
   - The breadth-first algorithm (page 714).

   For depth first and breadth first, label the root node and number the edges in the order that they were added to the spanning tree.

(a)

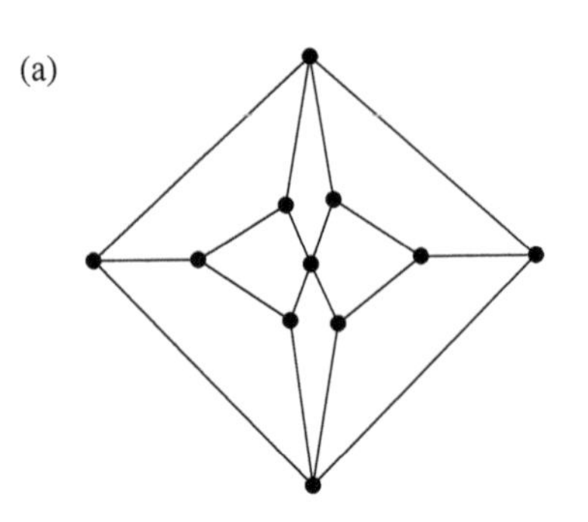

[22]Note that M may appear more than once in the list of weights.

(b)

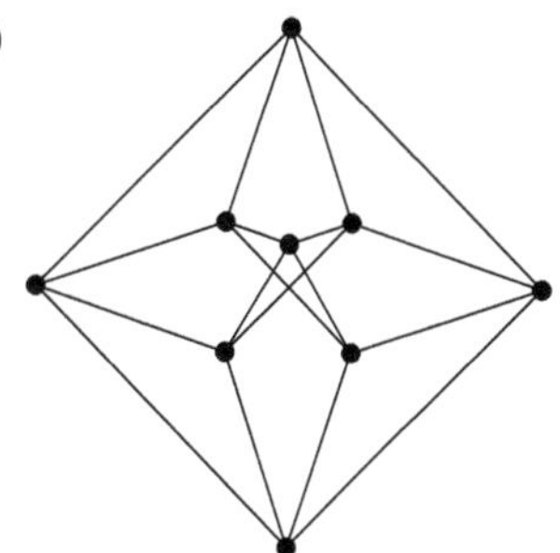

(c)

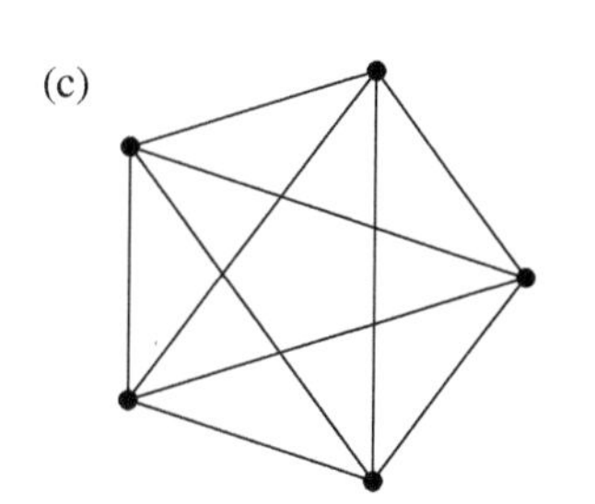

(d)

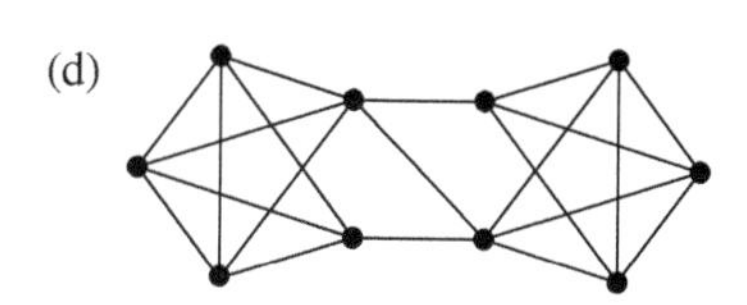

2. Find spanning trees for each of the following graphs. Use three different algorithms:
   - The algorithm from the proof of Theorem 11.34, page 712.
   - The depth-first algorithm (page 713).
   - The breadth-first algorithm (page 714).

   For depth first and breadth first, label the root node and number the edges in the order that they were added to the spanning tree.

   (a)

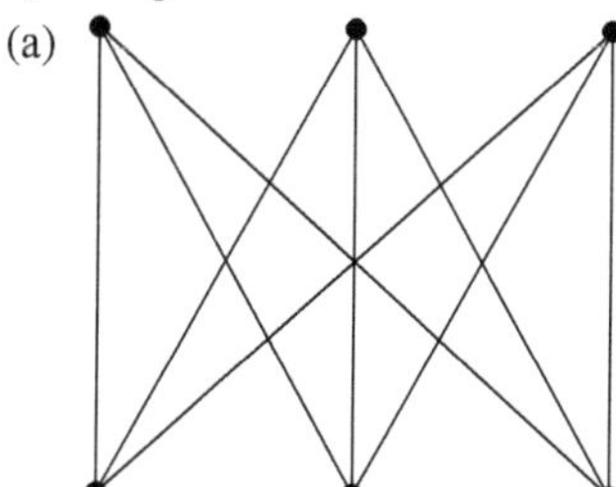

(b)

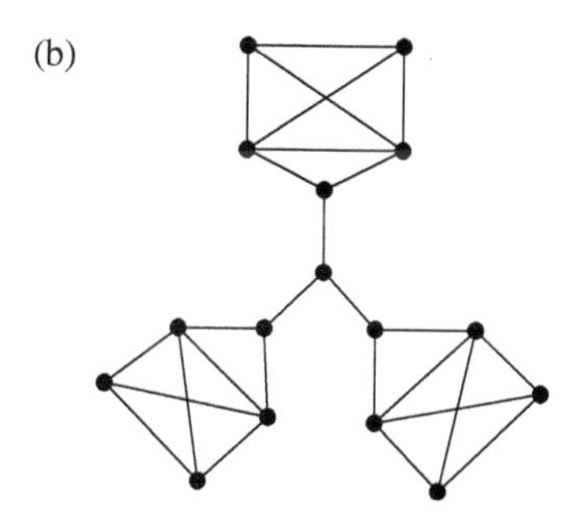

(c)

(d)

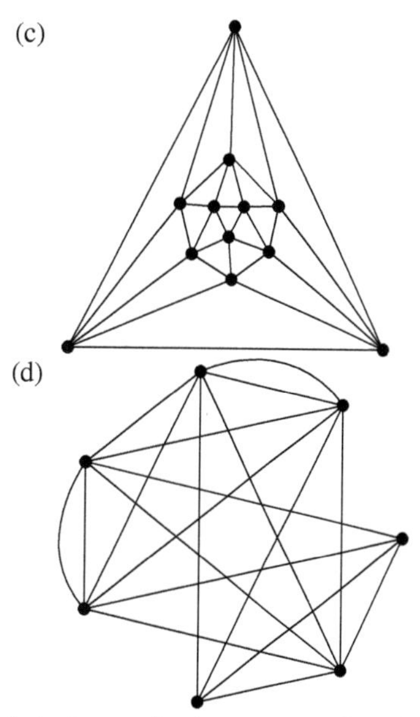

3. Let e be an edge in a connected graph, G, such that e is not in any cycle in G. Prove that e must be an edge in any spanning tree of G.
4. Prove Theorem 11.33 on page 712.
5. Use the pseudocode notation from Section 4.1 to express the Prim and Kruskal algorithms.
6. Prove Theorem 11.36 on page 716.
7. Prove Proposition 11.40.

> **PROPOSITION 11.40** ***The Number of Edges in a Spanning Tree***
>
> Let G be a connected graph with n vertices. Then every spanning tree for G has $n - 1$ edges.

8. Suppose a weighted graph does not contain any duplicate weights. Can there be more than one minimal spanning tree? If the answer is yes, provide an example; if the answer is no, provide a proof.
9. Prove that Kruskal's algorithm produces a minimal spanning tree by completing the following steps.
   (a) Identify where the proof of Prim's algorithm would fail as a proof for Kruskal's algorithm.
   (b) What changes need to be made to correct the failure?
   (c) Now write a complete proof that Kruskal's algorithm produces a minimal spanning tree.
10. Consider trees with the four nodes $\{v_1, v_2, v_3, v_4\}$.
    (a) How many distinct trees are there with four nodes?
    (b) Draw the distinct trees.
    (c) Draw a complete set of non-isomorphic trees with four nodes.
    (d) How many non-isomorphic trees with four nodes are there?
11. Although there are $5^{5-2} = 125$ distinct trees on five vertices, there are only three non-isomorphic trees on five vertices. Draw three non-isomorphic trees on five vertices. Explain why they are non-isomorphic.

12. ★ Produce minimal spanning trees for the following weighted graphs:

- Use both Prim's and Kruskal's algorithms.
- Label the edges by order of inclusion into the spanning tree.
- In each case, clearly record the sum of the weights on the edges of the minimal spanning tree.

(a)

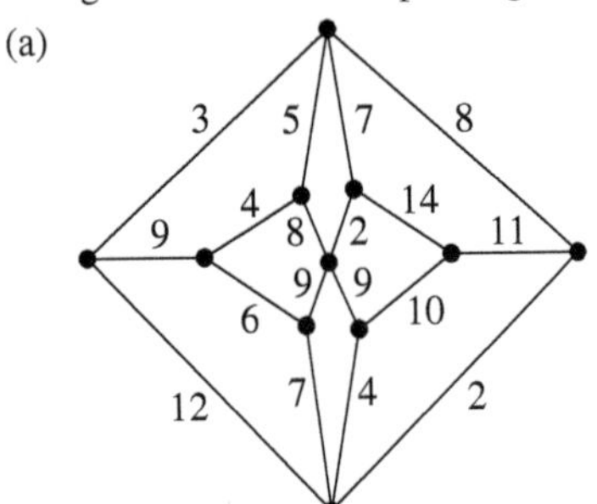

(b)

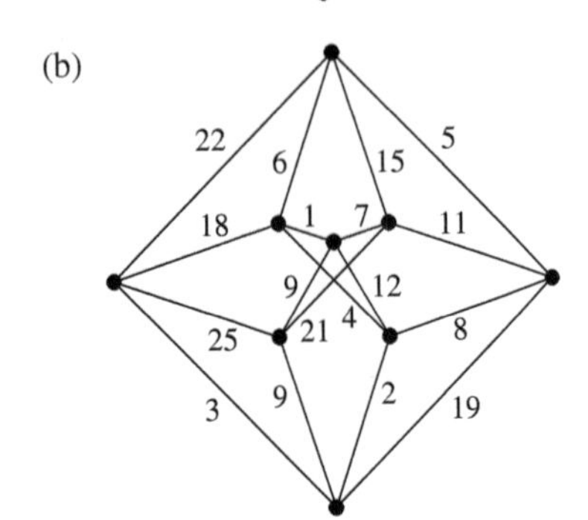

(c)

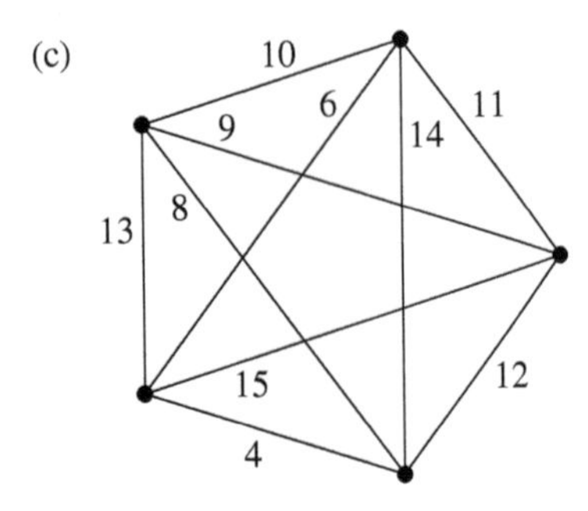

(d)

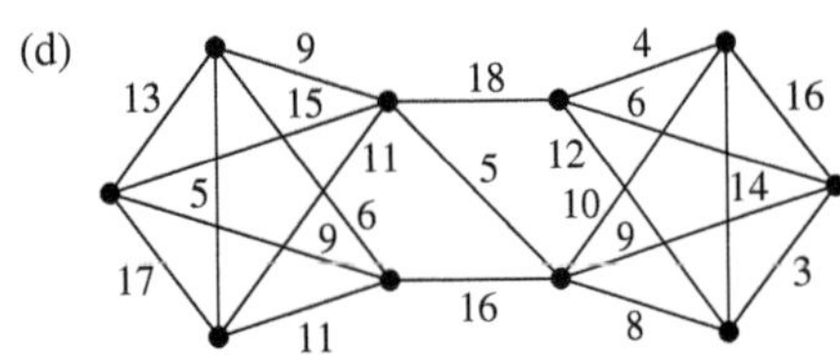

13. Produce minimal spanning trees for the following four weighted graphs:

- Use both Prim's and Kruskal's algorithms.
- Label the edges by order of inclusion into the spanning tree.
- In each case, clearly record the sum of the weights on the edges of the minimal spanning tree.

(a)

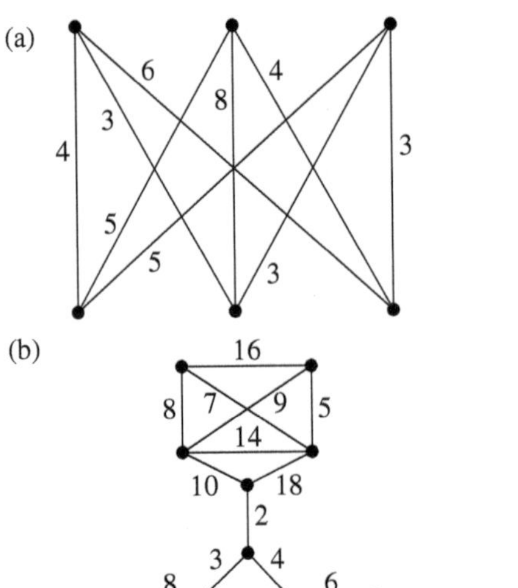

(b)

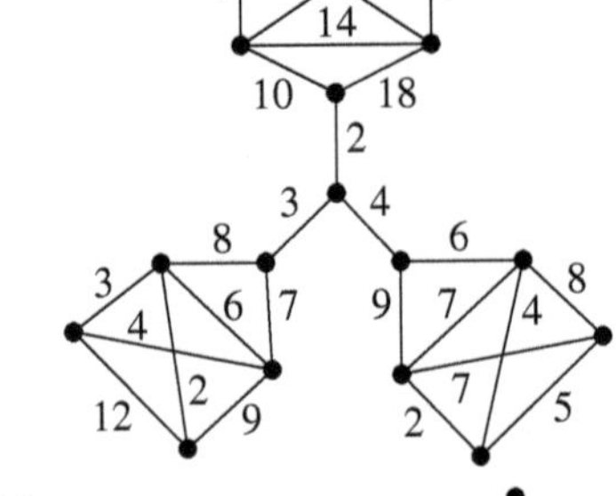

(c)

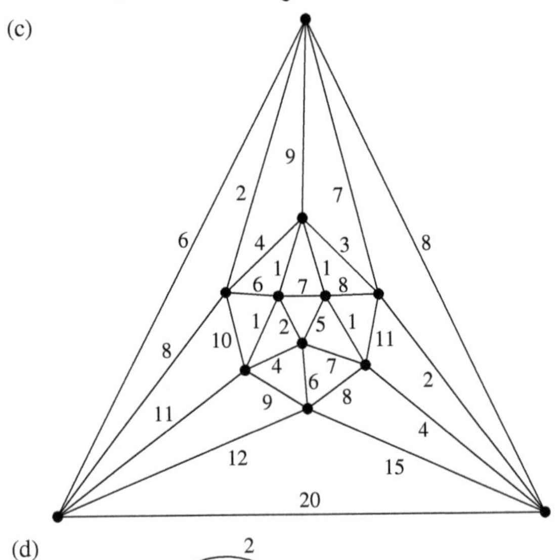

(d)

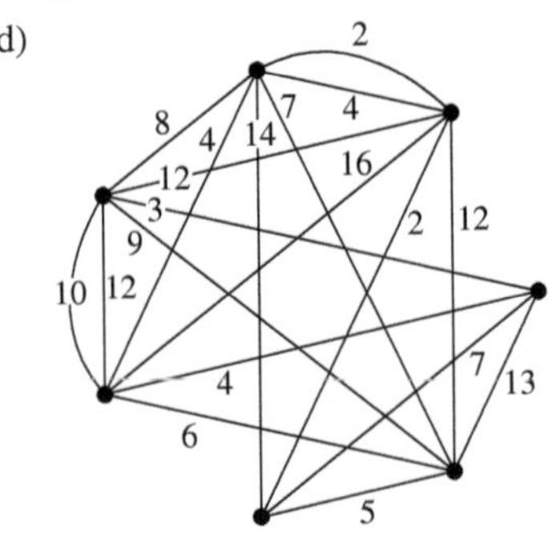

14. ★ Use the algorithm on page 722 to produce a maximal spanning tree for each graph in Exercise 12. Indicate the sum of the weights on the edges of the maximal spanning tree.

15. Use the algorithm on page 722 to produce a maximal spanning tree for each graph in Exercise 13. Indicate the sum of the weights on the edges of the maximal spanning tree.

16. Use Theorem 11.35 on page 715 to count the number of distinct spanning trees for each of the following graphs. Also, list all the distinct spanning trees. (This exercise requires knowledge of linear algebra or access to software or to a graphing calculator that can calculate determinants.)

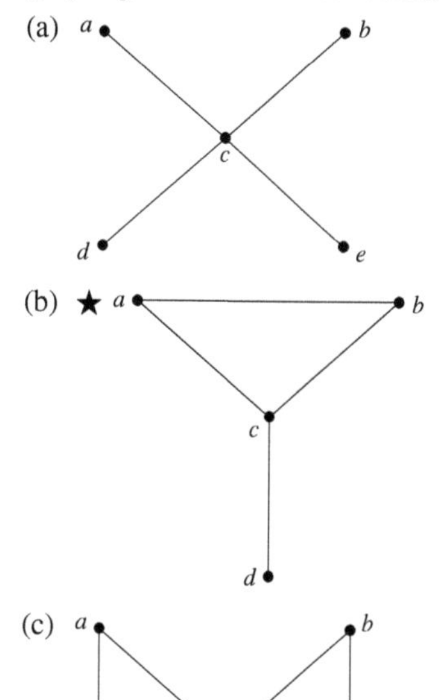

(c)

17. Use Theorem 11.35 on page 715 to count the number of distinct spanning trees for each of the following graphs. (This exercise requires knowledge of linear algebra or access to software or a graphing calculator that can calculate determinants.)

(a) ★ C_5 (b) K_5 (c) $K_{3,3}$

18. Prove Proposition 11.41. [*Hints*: "only if" is Exercise 21(b) on page 608. Use a spanning tree for "if".]

> **PROPOSITION 11.41** ***Bipartite Graphs and Cycles***
>
> Let G be a graph. G is bipartite if and only if it does not contain a cycle with an odd number of edges.

19. A small company produces video games. The company has been unable to find a suitable building to hold all the employees, so it has rented space in five different places within the same city. Since the video game industry is so competitive, the CEO does not want to risk using email or telephones to discuss projects. Instead, he has contracted with a small courier service to make daily runs from headquarters to each of the four satellite offices. The courier service charges solely by the mile; the amount of material delivered does not matter. The courier service works on a "pony express" model. They send a courier from one office location to another, at which point the materials to distribute are passed to one or more other couriers to deliver.

Thus, the CEO might be able to send the packets for offices A, B, and C in a bundle to office B, at which point the packets for offices A and C will be handed to new couriers and the packet for B delivered to the waiting employees. An alternative option is to have three couriers start at the main office and each head for a different satellite office.

Assume that the main office is M, and the four satellite offices are A, B, C, and D. What is an optimal delivery route if the CEO wants to send identical packets to each of the satellite offices, given the distances in the following table? Describe the route and also explain how you arrived at this route. If the courier service charges \$10 per mile, what is the cost for this delivery?

	A	**B**	**C**	**D**	**M**
A		2.5			5
B	2.5		1		4
C		1		3	4.5
D			3		6
M	5	4	4.5	6	

20. A small town in northern Minnesota gets large amounts of snow in the winter. The town plows the streets, but during and right after a heavy snow, it is only possible to keep some major streets open. The town has decided that during a snowstorm, access to the church, the post office, the grocery store, the fire department, and the school must all be available. There are multiple routes between several of these locations. However, some locations do not have direct routes joining them. For example, the post office is directly between the school and the grocery store. Therefore, if the routes between the school and the post office and between the post office and the grocery store are kept open, it will also be possible to travel between the school and the grocery store.

The following table indicates the distance, in blocks, between different locations. Alternate routes are separated by semicolons. All routes are nonintersecting and also nonoverlapping.

Assuming that snow-plowing costs are directly proportional to the number of blocks plowed, determine a least expensive collection of routes to keep access available between these locations.

	Church	**Post office**	**Grocery store**	**Fire department**	**School**
Church			25	3	3; 5
Post office			3	2	8; 14
Grocery store	25	3		4	
Fire department	3	2	4		4
School	3; 5	8; 14		4	

11.5 Quick Check Solutions

Quick Check 11.1

1. The parent of d is a. The children of b are f and g. Level 0: r Level 1: a, b, c Level 2: d, e, f, g, h Level 3: i, j. The subtree rooted at c contains the nodes c, h, i, j.
2. Let the sibling nodes be c_1 and c_2, with common parent p. Suppose p is at level k. Since c_1 is a child of p, it must be at level $k+1$. This is also true for c_2. Thus, both c_1 and c_2 are at level $k+1$.

Quick Check 11.2

1.

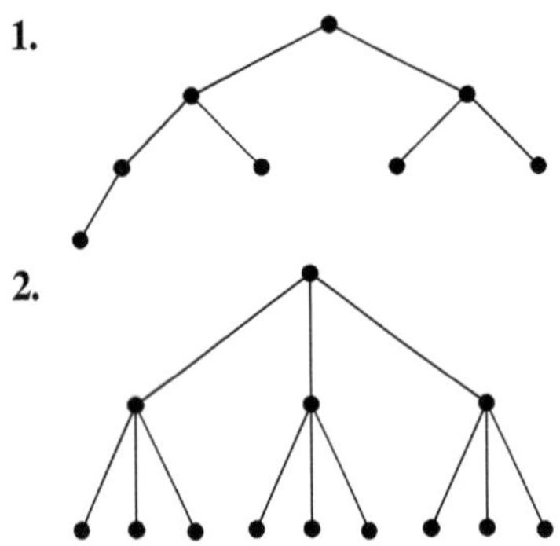

2.

3. (a) Count by levels: $2^0 + 2^1 + 2^2 + \cdots + 2^h = \frac{2^{h+1}-1}{2-1} = 2^{h+1} - 1$.

   (b) Count by levels again: $3^0 + 3^1 + 3^2 + \cdots + 3^h = \frac{3^{h+1}-1}{3-1} = \frac{3^{h+1}-1}{2}$.
4. (a) All leaves in a maximal complete tree are on the final level. There are 2^h nodes in the final level.

   (b) There are 3^h leaves.

Quick Check 11.3

1. There are four external nodes in each case.
2. You should find five external nodes in every binary tree with four nodes.
3. Conjecture: A binary tree with n nodes has $n+1$ external nodes. The proof of the conjecture is a homework problem.

Quick Check 11.4

1. (a) Postorder: $g\,e\,f\,i\,b\,d\,a\,h\,c$

   (b) Preorder: $c\,f\,g\,e\,h\,d\,i\,b\,a$

   (c) Inorder: $g\,f\,e\,c\,i\,d\,b\,h\,a$

Quick Check 11.5

1. The final tree is

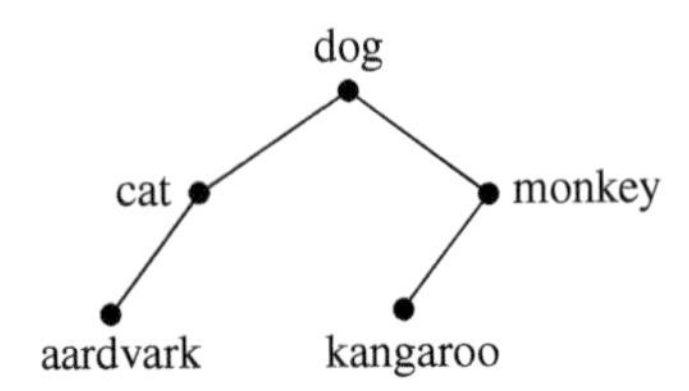

2. dog monkey kangaroo. Attempt to go right; conclude that lemming is not present.
3. $e\,c\,d$

Quick Check 11.6

1.

S
1 Y
0 0 Y
0 0 Y
1

(a)

S
0 1 X
0 X
0 X
1

(b)

Quick Check 11.7

1. (a) The 2-bit-per-character encoded string uses $12 \cdot 2 = 24$ bits. One encoding that uses 2 bits per character is shown in the next table.

A	C	G	T
00	01	10	11

(b) The frequency table is

A	C	G	T
5	2	4	1

so the initial queue looks like the following diagram.

Front			Back
1 T	2 C	4 G	5 A

The final labeled tree is shown.

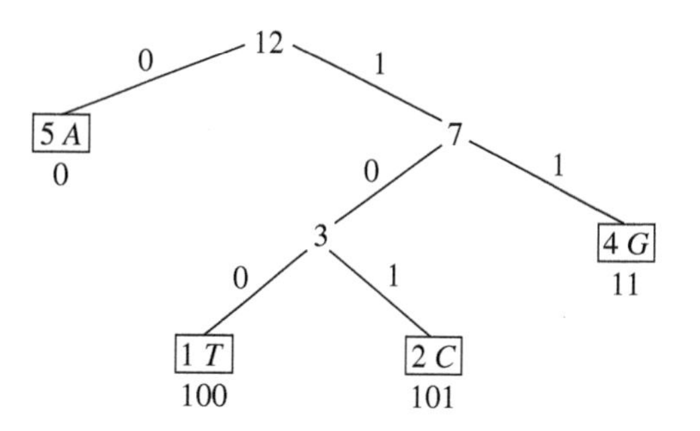

The encoding table is

A	C	G	T
0	101	11	100

(c) The Huffman encoded string requires $5 \cdot 1 + 2 \cdot 3 + 4 \cdot 2 + 1 \cdot 3 = 22$ bits.

Quick Check 11.8

1.

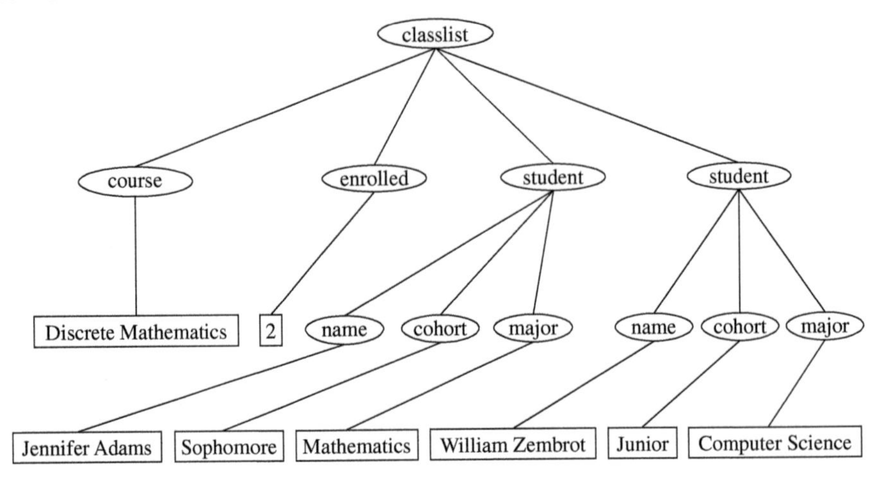

Quick Check 11.9

1. This is a fairly simple DTD.

```
<!ELEMENT letter (greeting, body, closing, signature)>
<!ELEMENT greeting (#PCDATA)>
<!ELEMENT body (paragraph)+>
<!ELEMENT paragraph (#PCDATA)>
<!ELEMENT closing (#PCDATA)>
<!ELEMENT signature (#PCDATA)>
```

Note that the order in which the elements are listed is not important. The following is also a valid DTD for the letter example.

```
<!ELEMENT letter (greeting, body, closing, signature)>
<!ELEMENT signature (#PCDATA)>
<!ELEMENT greeting (#PCDATA)>
<!ELEMENT closing (#PCDATA)>
<!ELEMENT paragraph (#PCDATA)>
<!ELEMENT body (paragraph)+>
```

Quick Check 11.10

1. Since there are 5 vertices, Theorem 11.36 on page 716 indicates that there will be $5^3 = 125$ distinct spanning trees.

2. and 3.

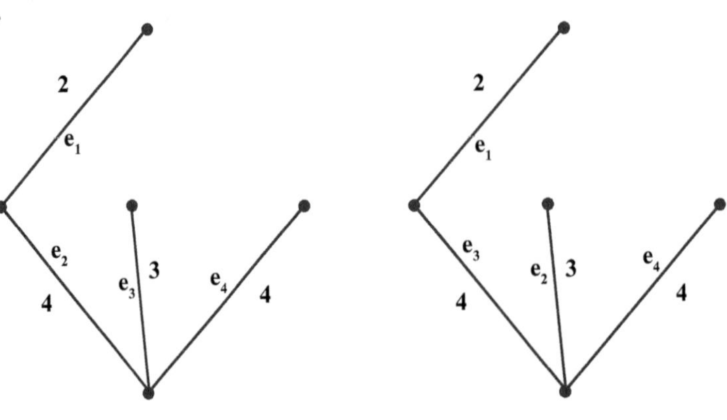

Prim

Kruskal
(e_3 and e_4 can appear in either order)

11.6 CHAPTER REVIEW

11.6.1 Summary

A tree is a connected graph that contains no cycles. These two restrictions (connected, no cycles) create a sufficiently rich structure to support an extensive body of research. Trees also play a prominent role in computer science as structures for organizing data. The presentation in this chapter is strongly influenced by various applications of trees in computer science.

Section 11.1 presents the basic terminology and also presents several theorems that relate to the numbers of vertices (nodes) in trees having some additional properties.

Section 11.2 discusses some recursive algorithms for traversing a tree. Several applications of trees are also presented. One application is a tree-based algorithm for converting infix expressions to postfix form. The important notion of a search tree is introduced in this section. These trees can be used for searching and sorting with collections of data. Search trees also provide a setting to answer a question which was only partially resolved in Chapter 4: What is the average case behavior of a binary search?

Section 11.3 presents several more-detailed applications of trees: parse trees, Huffman compression, and XML. Parse trees are often used as a data structure by compilers. Huffman compression is a simple form of data compression. There are much more effective compression algorithms available, but they are also more complex. Huffman compression is a very nice introduction to compression algorithms. The third application in this section is a relatively new application of trees, but its present and future prominence is assured. XML documents are organized as trees that reflect the logical structure of the document.

Section 11.4 shifts back to a more mathematics-oriented topic: spanning trees. A spanning tree is a subgraph of another graph that is typically not itself a tree. The section presents several algorithms for finding spanning trees. If the original graph is a weighted graph, then the goal is usually to find a spanning tree which has either minimal or maximal weight among all spanning trees.

Much of this chapter is algorithmic in nature. That material is best learned through practice. The theorems are best learned while gaining an understanding of the ideas in the proofs.

11.6.2 Notation

Notation	Page	Brief Description
m	677	the maximum number of children a node may have (in counting theorems)
n	677	the number of nodes (in counting theorems)
l	677	the number of leaves (in counting theorems)
i	677	the number of interior nodes (in counting theorems)
h	677	the height of a tree (in counting theorems)
<a string>	701	an opening XML tag
</a string>	701	a closing XML tag
<!ELEMENT ··· >	706	an element declaration in a DTD
<!ATTLIST ··· >	707	an attribute declaration in a DTD

11.6.3 Additional Review Material

Go to http://www.mathcs.bethel.edu/~gossett/DiscreteMathWithProof/review.xhtml for additional review material, including a sample chapter exam, with solutions.

CHAPTER 12

Functions, Relations, Databases, and Circuits

A function is a well-behaved relation.
Elizabeth Stapel – purplemath.com

This chapter begins with the familiar, but fundamental, concept of a *function* and develops that concept in two directions.

One direction is the introduction of a generalization of a *function*, called a *relation*. Relations with some additional properties are examined in the section on *equivalence relations*. An important application of relations is briefly introduced in the section that discusses *relational databases*.

The other direction that is developed is the notion of *Boolean functions*. This notion is then applied to the construction of logic circuits in digital electronics.

12.1 Functions and Relations

This section will begin with a definition of *function* that can be naturally extended to the more general notion of *relation*. Basic properties and associated definitions will also be presented in each case.

12.1.1 Functions

The process of stating, and then refining, the definition of a *function* can be traced back[1] to Leonard Euler [27]. Euler's mathematics textbook *Introductio in analysin infinitorum*, published in 1748, shaped what we would now call the "precalculus" curriculum. Euler defined a function in this (no longer used) manner:

> A *function* of a variable quantity is an analytic expression composed in any way whatsoever of the variable quantity and numbers or constant quantities [32].

This obsolete definition emphasizes the algebraic expression rather than the ideas of mapping or association. However, it implicitly contains the notion of an algebraic expression that can transform some variable into an associated value.

Euler's textbook introduced the standard collection of functions that you have studied in previous courses: polynomials, trigonometric functions, exponential functions, and logarithms.[2]

The modern definition of a function did not finally appear until it was formulated by Peter Dirichlet in 1837: "y is a function of x when to each value of x in a given interval there corresponds a unique value of y" [62, p. 950].

---

[1] For a brief but more extensive view of this development, see "Where Do Functions Come From?" by Leigh Atkinson [5].

[2] Euler provided the transition from log *tables*, which are not used much at present, to log *functions*, which are essential in much of modern mathematics. Euler was also the originator of the natural log function, $\ln(x)$, and its associated exponential function, e^x.

Two essentially equivalent definitions are most often used at present. Both are restatements of Dirichlet's definition. The first defines a function to be a mapping from a domain to a range for which no element in the domain is mapped to more than one element of the range. The second definition, presented next, deemphasizes the mapping idea and highlights the association between elements in two sets. It might be helpful to review the definition of Cartesian product on page 25.

DEFINITION 12.1 ***Function; Domain; Range***

A *function* from the nonempty set $\mathcal{D}$ into the nonempty set $\mathcal{R}$ is a subset, $\mathcal{F}$, of the Cartesian product $\mathcal{D} \times \mathcal{R}$ such that every element of $\mathcal{D}$ appears as the first coordinate in one and only one ordered pair in $\mathcal{F}$.

The set, $\mathcal{D}$, is called the *domain* and the set, $\mathcal{R}$, is called the *range*. The *image* of the function is the subset of $\mathcal{R}$ consisting of elements that actually appear in the right-hand side of at least one ordered pair in $\mathcal{F}$.

The assertion "every element of $\mathcal{D}$ appears in one and only one ordered pair in $\mathcal{F}$" means (by the definition of the Cartesian product $\mathcal{D} \times \mathcal{R}$) that no two ordered pairs in $\mathcal{F}$ have the same first element. It also implies that every element of $\mathcal{D}$ appears in some ordered pair in $\mathcal{F}$.[3]

This may seem to be a completely different idea from the mapping version of a function. However, the similarities in the definitions are intuitively compelling. For example, to move from the mapping version to the set version, think about building a (possibly infinite) table similar to those used to graph functions by hand.

x	$f(x)$
x_1	$f(x_1)$
x_2	$f(x_2)$
$\vdots$	$\vdots$

It is easy to see the corresponding set of ordered pairs: $\{(x_1, f(x_1)), (x_2, f(x_2)), \ldots\}$.

Definition 12.1 does not require the specification of an algebraic expression or an algorithm for producing the association of elements in $\mathcal{D}$ and $\mathcal{R}$. If such an algebraic expression, f, is known, it is easy to see that the subset, $\mathcal{F}$, of ordered pairs can be viewed as the result of mapping elements of $\mathcal{D}$ to elements of $\mathcal{R}$ using the expression f. Even if such an expression is not known, there is an implicit "mapping" available: Associate each element of $\mathcal{D}$ with the element of $\mathcal{R}$ with which it appears in an ordered pair of $\mathcal{F}$.

There are some additional definitions that will allow more precise discussions about functions.

DEFINITION 12.2 ***Onto and One-to-One Functions***

Let $\mathcal{F}$ be a function from $\mathcal{D}$ into $\mathcal{R}$. Then $\mathcal{F}$ is called *onto* if every element of $\mathcal{R}$ appears as a second coordinate in at least one ordered pair in $\mathcal{F}$. If no element of $\mathcal{R}$ appears as a second coordinate in more than one ordered pair in $\mathcal{F}$, then $\mathcal{F}$ is called *one-to-one* (also abbreviated as 1-1).[a]

[a]Some more advanced texts refer to onto functions as *surjective* functions and one-to-one functions as *injective* functions. Functions that are both one-to-one and onto are referred to as *bijective* functions. This terminology is awkward and will not be used here.

[3]Note that changing the domain also changes the function. Thus, using the mapping view of functions for the moment, $f(x) = x$ with domain $\mathbb{Z}$ is not the same function as $f(x) = x$ with domain $\mathbb{R}$. The second function is defined at π, whereas the first is not.

Note that it is possible for a function to be one-to-one without being onto, for example $\{(x, e^x)\} \subseteq \mathbb{R} \times \mathbb{R}$. It is also possible to be onto without being one-to-one, for example $\{(x, x(x-2)(x+2))\} \subseteq \mathbb{R} \times \mathbb{R}$. It is also possible to have both properties, $\{(x, x)\} \subseteq \mathbb{R} \times \mathbb{R}$, or neither property, $\{(x, x^2)\} \subseteq \mathbb{R} \times \mathbb{R}$.

EXAMPLE 12.1

The Sine Function

Most of the functions you have encountered prior to this course were functions with an infinite domain and an infinite range. For instance, the familiar sine function, $\sin(x)$, has the entire set of real numbers as its domain and as its range, but the image is the closed interval $[-1, 1]$.

The function is neither onto nor one-to-one. It is not onto because the image is a proper subset of the range. It is not one-to-one because the set of ordered pairs $\{(n\pi, 0) \mid n \in \mathbb{Z}\}$ is a subset of the sine function.[4]

For the purposes of this textbook, the alternate, mapping-oriented terminology is acceptable as an informal mode for describing functions. Thus, the sine function could be informally described as a mapping that "sends" radian angles to real numbers in $[-1, 1]$. For a particular radian measure, x, we would denote the value it is mapped to as $\sin(x)$.

More formally, the sine function can be described by defining $\sin(x)$ to be the y-coordinate of the point, $(x, \sin(x))$, determined by the standard wrapping function from trigonometry.[5] The sine function is then defined by

$$\sin = \{(x, \sin(x)) \mid x \in \mathbb{R}\}$$ ■

Many functions of interest in a discrete mathematics course have a finite or countably infinite domain and a finite or countably infinite range. The next example describes some functions with a finite domain and range.

EXAMPLE 12.2

Marriage Assignments

The six possible marriage assignments in Example 1.1 on page 4 all represent one-to-one, onto functions with domain $\mathcal{D} = \{A, B, C\}$ and range $\mathcal{R} = \{X, Y, Z\}$.

The assignment labeled "female 1st choice" is associated with the function $\mathcal{F}_1 = \{(A, Y), (B, Z), (C, X)\}$. ■

✔ Quick Check 12.1

1. Theorem 3.54 on page 134 implicitly defines a function, $\mathcal{S}$, with a countably infinite domain and range. Use the formal definition of *function* to define explicitly the function, its domain, and its range. Also, indicate whether the function is one-to-one and if it is onto.

2. Formally describe the parity function, $\mathcal{P}$, corresponding to the informal notion that an integer is either even or odd. Describe the function and its domain and range, and then determine whether the function is one-to-one and if it is onto. ☑

A Word about Notation Up to this point, the exposition about functions has mostly used an uppercase, script font for the name of a function. (The sine function is too steeped in tradition to use anything other than "sin" as its name.) This is the same as the font used to represent the domain and range. The font choice emphasizes their commonality as sets.

---

[4] It may seem strange to say that a set of ordered pairs is a subset of the sine function, but this is the proper terminology when using Definition 12.1.

[5] That is, start at the point $(1, 0)$ and move counterclockwise around the unit circle until a distance of x radians has been traveled. The y-coordinate of the final point is then labeled as $\sin(x)$.

Once the nature of functions as "sets of ordered pairs" is understood, it is more in keeping with standard notational practice to use lowercase letters (such as f) to represent functions. It is then easy to move from the more formal notation to the standard notation. For example, if $f = \{(x, x^2)\} \subseteq \mathbb{R} \times \mathbb{R}$, then we can write $f(2) = 4$, or more generally, $f(x) = x^2$.

Functions that are both one-to-one and onto have a naturally associated partner function, called the inverse function. If the function is informally denoted as f, the inverse function is often denoted as f^{-1}. Familiar examples are sin (with a suitably restricted domain, such as $[-\frac{\pi}{2}, \frac{\pi}{2}]$) and its inverse, $\sin^{-1}$ (also denoted by "arcsin"), and the pair of inverse functions whose second coordinates are denoted by e^x and $\ln(x)$.

Informally, we say that functions f and g are inverses if the following conditions are met:

- The domain, $\mathcal{D}_f$, of f is the same as the range, $\mathcal{R}_g$, of g.
- The domain, $\mathcal{D}_g$, of g is the same as the range, $\mathcal{R}_f$, of f.
- $\forall x \in \mathcal{D}_f, g(f(x)) = x$.
- $\forall y \in \mathcal{D}_g, f(g(y)) = y$.

This can be expressed formally by the following definition:

DEFINITION 12.3 ***Inverse Function***

Let $\mathcal{F}$ be a one-to-one and onto function with domain $\mathcal{D}_\mathcal{F}$ and range $\mathcal{R}_\mathcal{F}$. A function, $\mathcal{G}$, whose domain is $\mathcal{R}_\mathcal{F}$ and whose range is $\mathcal{D}_\mathcal{F}$ is called the *inverse of* $\mathcal{F}$ if the following conditions hold:

- If $(x, y) \in \mathcal{F}$, then $(y, x) \in \mathcal{G}$.
- If $(y, x) \in \mathcal{G}$, then $(x, y) \in \mathcal{F}$.

It is easy to show that $\mathcal{G}$ exists and is unique as long as $\mathcal{F}$ is one-to-one and onto. $\mathcal{G}$ is also one-to-one and onto. (See Exercise 16 on page 738.)

The function $\mathcal{F}_1 = \{(A, Y), (B, Z), (C, X)\}$, from Example 12.2, has inverse function $\mathcal{G}_1 = \{(Y, A), (Z, B), (X, C)\}$. The ordered pairs in $\mathcal{F}_1$ indicate the spouse for each female, whereas the ordered pairs in $\mathcal{G}_1$ indicate the spouse for each male.

The familiar notion of composition of functions can also be described using the set-oriented definition of *function*. Recall that if f is informally defined as a function that maps elements, x, from X to Y, and if g is a function that maps elements, y, from Y to Z, then $g \circ f$ is the function that maps elements from X to Z using the rule $(g \circ f)(x) = g(f(x))$.

DEFINITION 12.4 ***Composition of Functions***

Let $\mathcal{F}$ be a function whose domain is $\mathcal{X}$ and whose range is $\mathcal{Y}$. Let $\mathcal{G}$ be a function whose domain is $\mathcal{Y}$ and whose range is $\mathcal{Z}$. The composition of $\mathcal{G}$ and $\mathcal{F}$ is denoted by $\mathcal{G} \circ \mathcal{F}$ and is defined by

$$\mathcal{G} \circ \mathcal{F} = \{(x, z) \mid \exists y \in \mathcal{Y} \text{ with } (x, y) \in \mathcal{F} \text{ and } (y, z) \in \mathcal{G}\}$$

EXAMPLE 12.3

A Simple Composition

Let $\mathcal{F} = \{(1, 2), (2, 4), (3, 6), (4, 8), (5, 4)\}$ and $\mathcal{G} = \{(2, -6), (4, -12), (6, -24), (8, -24), (10, -100)\}$. Then

$$\mathcal{G} \circ \mathcal{F} = \{(1, -6), (2, -12), (3, -24), (4, -24), (5, -12)\}$$

■

12.1.2 Relations

Functions have been defined as subsets of a Cartesian product for which there is no duplication in the first elements of the ordered pairs. A relation is also a subset of a Cartesian product. However, there are no restrictions.

DEFINITION 12.5 ***Relation***

A *relation* between the set $\mathcal{A}$ and the set $\mathcal{B}$ is a subset, $\mathcal{R}$, of the Cartesian product $\mathcal{A} \times \mathcal{B}$.

If $(a, b) \in \mathcal{R}$, it is common to write $a\mathcal{R}b$ and to say that a is related to b. If $\mathcal{A} = \mathcal{B}$, the relation is said to be a *relation on* $\mathcal{A}$.

The set $\mathcal{A}$ is called the *domain* of the relation and the set $\mathcal{B}$ is called the *range*.

The definitions of *one-to-one* and *onto* do not change.

DEFINITION 12.6 ***Onto and One-to-One Relations***

Let $\mathcal{R}$ be a relation between the sets $\mathcal{A}$ and $\mathcal{B}$. Then $\mathcal{R}$ is called *onto* if every element of $\mathcal{B}$ appears as a second coordinate in at least one ordered pair in $\mathcal{R}$. If no element of $\mathcal{B}$ appears as second coordinate in more than one ordered pair in $\mathcal{R}$, then $\mathcal{R}$ is called *one-to-one*.

Why would an unrestricted subset of a Cartesian product be of interest, and how is it related to the notion of a function? Perhaps a simple example will begin to answer these questions.

EXAMPLE 12.4

A Relation between Large Discrete Sets

Let $\mathcal{A}$ be the set of all people who were alive at some point between 1000 A.D. and 2000 A.D., and let $\mathcal{B}$ represent the set of all roles that might characterize a person at some time in life during that millenium. It is easy to imagine a relation, $\mathcal{R}$, between $\mathcal{A}$ and $\mathcal{B}$ that correctly captures all the roles each person assumed. Some entries might be as follows:

- (Shirley Temple Black, daughter), (Shirley Temple Black, child actor), (Shirley Temple Black, wife), (Shirley Temple Black, mother), (Shirley Temple Black, U.S. Ambassador), ...
- (Hayao Miyazaki, son), (Hayao Miyazaki, animator), (Hayao Miyazaki, husband), (Hayao Miyazaki, father), ...
- (Leonhard Euler, son), (Leonhard Euler, mathematician), (Leonhard Euler, teacher), (Leonhard Euler, husband), (Leonhard Euler, father), ...
- and so on

The relation $\mathcal{R}$ is clearly not a function, since Shirley Temple Black appears as first coordinate in multiple ordered pairs. It is also not one-to-one, since Leonhard Euler and Hayao Miyazaki are both associated with the role of father. The relation is onto if, and only if, every role in $\mathcal{B}$ has at least one person associated with it.

Although the uniqueness of second coordinates has been lost, this relation clearly retains some of the flavor of a function. The ordered pairs serve to associate the first coordinates with their (multiple) second coordinates. ■

Many functions do not have an inverse; every relation does.

DEFINITION 12.7 ***Inverse Relation***
Let $\mathcal{R}$ be a relation between the sets $\mathcal{A}$ and $\mathcal{B}$. The *inverse relation* of $\mathcal{R}$ is denoted $\mathcal{R}^{-1}$ and is a subset of the Cartesian product $\mathcal{B} \times \mathcal{A}$. More precisely,

$$\mathcal{R}^{-1} = \{(b, a) \in \mathcal{B} \times \mathcal{A} \mid (a, b) \in \mathcal{R}\}$$

If $\mathcal{R}$ is the relation defined in Example 12.4, then $\mathcal{R}^{-1}$ contains (among many other ordered pairs):

(daughter, Shirley Temple Black), (son, Hayao Miyazaki), (son, Leonhard Euler), (U.S. Ambassador, Shirley Temple Black), (animator, Hayao Miyazaki), and (mathematician, Leonhard Euler).

Composition of relations will be used extensively in Section 12.2.

DEFINITION 12.8 ***Composition of Relations***
Let $\mathcal{R}$ be a relation whose domain is $\mathcal{A}$ and whose image is $\mathcal{B}$. Let $\mathcal{S}$ be a relation whose domain contains $\mathcal{B}$ and whose range is $\mathcal{C}$. The *composition* of $\mathcal{S}$ and $\mathcal{R}$ is a subset of $\mathcal{A} \times \mathcal{C}$. It is denoted by $\mathcal{S} \circ \mathcal{R}$ and is defined as

$$\mathcal{S} \circ \mathcal{R} = \{(a, c) \mid \exists b \in \mathcal{B} \text{ with } (a, b) \in \mathcal{R} \text{ and } (b, c) \in \mathcal{S}\}$$

The range of $\mathcal{R}$ need not be contained in the domain of $\mathcal{S}$. All that is required is Image($\mathcal{R}$) $\subseteq$ Domain($\mathcal{S}$).

The definition can be intuitively visualized by placing ordered pairs next to each other, with elements of $\mathcal{R}$ on the left and elements of $\mathcal{S}$ on the right:

$$(a, b)(b, c) \to (a, c)$$

However, the notation, $\mathcal{S} \circ \mathcal{R}$, reverses this ordering.[6]

EXAMPLE 12.5

A Small Relation

Let $\mathcal{A} = \{1, 2, 3, 4\}$. Define a relation, $\mathcal{R}$, on $\mathcal{A}$ by[7]

$$x\mathcal{R}y \text{ if and only if } x \mid y$$

Define a second relation, $\mathcal{S}$, on $\mathcal{A}$ by

$$x\mathcal{S}y \text{ if and only if } x \text{ and } y \text{ have the same parity and } x \neq y$$

Then

- $\mathcal{R} = \{(1, 1), (1, 2), (1, 3), (1, 4), (2, 2), (2, 4), (3, 3), (4, 4)\}$
- $\mathcal{S} = \{(1, 3), (2, 4), (3, 1), (4, 2)\}$
- $\mathcal{S} \circ \mathcal{R} = \{(1, 3), (1, 4), (1, 1), (1, 2), (2, 4), (2, 2), (3, 1), (4, 2)\}$
- $\mathcal{R} \circ \mathcal{S} = \{(1, 3), (2, 4), (3, 1), (3, 2), (3, 3), (3, 4), (4, 2), (4, 4)\}$

The composition $\mathcal{S} \circ \mathcal{R}$ can be determined systematically by starting with the ordered pairs in $\mathcal{R}$ and finding all ordered pairs in $\mathcal{S}$ whose first coordinate matches the second coordinate of the ordered pair in $\mathcal{R}$: $(a, \mathbf{b})(\mathbf{b}, c) \to (a, c)$. For example, the ordered pair $(1, 2)$ in $\mathcal{R}$ can be matched with the ordered pair $(2, 4)$ in $\mathcal{S}$. The ordered pair $(1, 4)$ is thus a member of $\mathcal{S} \circ \mathcal{R}$ (it appears as the second element of $\mathcal{S} \circ \mathcal{R}$ in the preceding list). ■

[6]One way to keep this straight is to read "$\mathcal{S} \circ \mathcal{R}$" as "$\mathcal{S}$ follows $\mathcal{R}$." That is, first apply the association in $\mathcal{R}$ $(a \to b)$, and then apply the association in $\mathcal{S}$ $(b \to c)$.

[7]See Definition 3.8 on page 93 for a refresher on "x divides y".

✔ Quick Check 12.2

Let $\mathcal{A} = \{1, 2, 3\}$, $\mathcal{B} = \{a, b, c, d\}$, and $\mathcal{C} = \{x, y, z\}$.

1. Let $\mathcal{R} = \{(1, a), (2, b), (2, c), (3, c)\}$ be a relation in $\mathcal{A} \times \mathcal{B}$. What is the inverse relation, $\mathcal{R}^{-1}$?
2. If $\mathcal{S} = \{(a, x), (a, z), (b, z), (c, y), (d, y)\}$ is a relation in $\mathcal{B} \times \mathcal{C}$, what is $\mathcal{S} \circ \mathcal{R}$? ☑

The next theorem indicates a simple algebraic property of composition.

THEOREM 12.9 ***Composition of Relations is Associative***

Let $\mathcal{R}$ be a relation whose domain is $\mathcal{A}$ and whose image is $\mathcal{B}$. Let $\mathcal{S}$ be a relation whose domain contains $\mathcal{B}$ and whose image is $\mathcal{C}$. Finally, let $\mathcal{T}$ be a relation whose domain contains $\mathcal{C}$ and whose range is $\mathcal{D}$. Then

$$(\mathcal{T} \circ \mathcal{S}) \circ \mathcal{R} = \mathcal{T} \circ (\mathcal{S} \circ \mathcal{R})$$

A simple consequence of this theorem is that the notation $\mathcal{T} \circ \mathcal{S} \circ \mathcal{R}$ is unambiguous.

Proof: This theorem is claiming the equality of two sets. Before choosing a proof strategy, it might be helpful to see if the two sets are subsets of the same Cartesian product. If they are not, then the claim in the theorem must be false.

The set $\mathcal{T} \circ \mathcal{S}$ is a subset of Domain($\mathcal{S}$) $\times \mathcal{D}$, so $(\mathcal{T} \circ \mathcal{S}) \circ \mathcal{R}$ is a subset of $\mathcal{A} \times \mathcal{D}$. In addition, the set $\mathcal{S} \circ \mathcal{R}$ is a subset of $\mathcal{A} \times \mathcal{C}$, so $\mathcal{T} \circ (\mathcal{S} \circ \mathcal{R})$ is a subset of $\mathcal{A} \times \mathcal{D}$. Consequently, the claim is viable.

Section 2.1.3 provides some proof strategy options. A natural strategy for this theorem is to show that each set is a subset of the other.

***Part 1:* $(\mathcal{T} \circ \mathcal{S}) \circ \mathcal{R} \subseteq \mathcal{T} \circ (\mathcal{S} \circ \mathcal{R})$**

It is useful to recall the previously determined Cartesian products:

$$\mathcal{T} \circ \mathcal{S} \subseteq \text{Domain}(\mathcal{S}) \times \mathcal{D} \qquad \mathcal{S} \circ \mathcal{R} \subseteq \mathcal{A} \times \mathcal{C}$$
$$(\mathcal{T} \circ \mathcal{S}) \circ \mathcal{R} \subseteq \mathcal{A} \times \mathcal{D} \qquad \mathcal{T} \circ (\mathcal{S} \circ \mathcal{R}) \subseteq \mathcal{A} \times \mathcal{D}$$

Let $(a, d) \in (\mathcal{T} \circ \mathcal{S}) \circ \mathcal{R}$. If (a, d) is also an element in $\mathcal{T} \circ (\mathcal{S} \circ \mathcal{R})$, then the claim will be verified since (a, d) represents any element in the left-hand set.[8]

Since $(a, d) \in (\mathcal{T} \circ \mathcal{S}) \circ \mathcal{R}$, there must be[9] an element $b \in \mathcal{B}$ such that $(a, b) \in \mathcal{R}$ and $(b, d) \in \mathcal{T} \circ \mathcal{S}$. But if $(b, d) \in \mathcal{T} \circ \mathcal{S}$, then there exists an element $c \in \mathcal{C}$ such that $(b, c) \in \mathcal{S}$ and $(c, d) \in \mathcal{T}$.

Consequently, since $(a, b) \in \mathcal{R}$ and $(b, c) \in \mathcal{S}$, it is clear that[10] $(a, c) \in \mathcal{S} \circ \mathcal{R}$. Since $(a, c) \in \mathcal{S} \circ \mathcal{R}$ and $(c, d) \in \mathcal{T}$, it follows that $(a, d) \in \mathcal{T} \circ (\mathcal{S} \circ \mathcal{R})$ is true.

This completes the verification that $(\mathcal{T} \circ \mathcal{S}) \circ \mathcal{R} \subseteq \mathcal{T} \circ (\mathcal{S} \circ \mathcal{R})$.

***Part 2:* $\mathcal{T} \circ (\mathcal{S} \circ \mathcal{R}) \subseteq (\mathcal{T} \circ \mathcal{S}) \circ \mathcal{R}$**

Exercise 18 on page 738. □

12.1.3 Exercises

The exercises marked with ★ have detailed solutions in Appendix H.

1. Which of the following sets of ordered pairs represent functions? (Recall that $\mathbb{Z}^+$ is the set of positive integers.)
   (a) ★ $\{(1, 1), (2, 4), (3, 3), (3, 4), (4, 5), (5, 5)\} \subseteq \{1, 2, 3, 4, 5\} \times \{1, 2, 3, 4, 5\}$
   (b) $\{(1, 4), (2, 4), (3, 4), (4, 4), (5, 4)\} \subseteq \{1, 2, 3, 4, 5\} \times \{1, 2, 3, 4, 5\}$
   (c) $\{(x, y) \in \mathbb{Z}^+ \times \mathbb{Z}^+ \mid \gcd(x, y) = 2\}$
   (d) $\{(x, y) \in \mathbb{N} \times \{-1, 0, 1\} \mid y = \sin(x\frac{\pi}{2})\}$
2. ★ Which of the relations in Exercise 1 are onto?
3. ★ Which of the relations in Exercise 1 are one-to-one?

---

[8] See Section 2.1.3 if you need a quick review on this proof strategy.
[9] See Definition 12.8 (concept → properties) with $\mathcal{T} \circ \mathcal{S}$ in the role of the definition's $\mathcal{S}$.
[10] Definition 12.8 (properties → concept).

4. ★ Describe the inverse relation for each relation in Exercise 1.

5. Which of the following sets of ordered pairs represent functions? (Recall that $\mathbb{Z}^+$ is the set of positive integers.)
   (a) $\{(a, a), (b, c), (c, x), (d, y), (x, z), (y, d), (d, b)\} \subseteq \{a, b, c, d, x, y, z\} \times \{a, b, c, d, x, y, z\}$
   (b) $\{(x, y) \in \{1, 2, 3, 4, 5, 6, 7, 8\} \times \{1, 2, 3, 4, 5, 6, 7, 8\} \mid \text{lcm}(x, y) > 8\}$
   (c) $\{(x, y) \in \mathbb{Z}^+ \times \mathbb{N} \mid y = \lceil \log_2(x) \rceil\}$
   (d) Let $\mathcal{P}$ be the set of all people and let $\mathcal{W}$ be the set of all human women. The set of ordered pairs is $\{(p, w) \in \mathcal{P} \times \mathcal{W} \mid w \text{ is the biological mother of } p\}$.

6. Which of the relations in Exercise 5 are onto?

7. Which of the relations in Exercise 5 are one-to-one?

8. Describe the inverse relation for each relation in Exercise 5.

9. Determine the ordered pairs in $\mathcal{S} \circ \mathcal{R}$ (if they are defined), where $\mathcal{R}, \mathcal{S} \subseteq \mathbb{Z}^+ \times \mathbb{Z}^+$.
   (a) $\mathcal{R} = \{(1, 5), (1, 6), (2, 6), (2, 8)\}$ and $\mathcal{S} = \{(1, 2), (5, 4), (6, 5), (7, 8), (8, 9), (8, 10)\}$
   (b) $\mathcal{R} = \{(1, 1), (2, 1), (3, 2), (4, 1)\}$ and $\mathcal{S} = \{(1, 1), (1, 2), (2, 3), (2, 4)\}$
   (c) ★ $\mathcal{R} = \{(2a, 2a) \mid a \in \mathbb{Z}^+\}$ and $\mathcal{S} = \{(3b, 3b + 1) \mid b \in \mathbb{Z}^+\}$
   (d) $\mathcal{R} = \{(2a, 6a) \mid a \in \mathbb{Z}^+\}$ and $\mathcal{S} = \{(3b, 3b + 1) \mid b \in \mathbb{Z}^+\}$
   (e) $\mathcal{R} = \{(n, 2n + 1) \mid n \in \mathbb{Z}^+\}$ and $\mathcal{S} = \{(2k - 1, k^2) \mid k \in \mathbb{Z}^+\}$

10. Determine which (if any) of the compositions, $\mathcal{S} \circ \mathcal{R}$ and $\mathcal{R} \circ \mathcal{S}$, are defined. If the composition is defined, list the resulting ordered pairs. In each case, $\mathcal{R}, \mathcal{S} \subseteq \mathbb{N} \times \mathbb{N}$.
   (a) $\mathcal{R} = \{(0, 5), (1, 8), (9, 6), (9, 7)\}$ and $\mathcal{S} = \{(0, 0), (0, 9), (8, 1), (8, 9)\}$
   (b) $\mathcal{R} = \{(2, 5), (5, 3), (6, 6), (6, 8)\}$ and $\mathcal{S} = \{(1, 2), (5, 4), (6, 5), (7, 8)\}$
   (c) $\mathcal{R} = \{(1, 1), (2, 4), (3, 9), (4, 16)\}$ and $\mathcal{S} = \{(1, 1), (4, 2), (9, 3), (16, 4)\}$
   (d) $\mathcal{R} = \{(1, 2), (1, 3), (1, 4), (1, 5)\}$ and $\mathcal{S} = \{(1, 2), (2, 4), (3, 5), (4, 6), (5, 7)\}$

11. Let $\mathcal{R} = \{(1, 5), (1, 6), (2, 6), (2, 8), (3, 1), (4, 7)\}$ and $\mathcal{S} = \{(1, 2), (5, 4), (6, 5), (7, 8), (8, 9), (8, 10)\}$ (both are subsets of $\mathbb{N} \times \mathbb{N}$).
   (a) Determine the ordered pairs in $\mathcal{R}^{-1}$.
   (b) Determine the ordered pairs in $\mathcal{S}^{-1}$.
   (c) Determine the ordered pairs in $(\mathcal{S} \circ \mathcal{R})^{-1}$.
   (d) Is there any connection between parts (a), (b), and (c)?

12. Let $\mathcal{R} = \{(0, 2), (0, 5), (0, 9), (1, 9), (1, 12), (1, 15), (2, 2)\}$ and $\mathcal{S} = \{(2, 0), (2, 6), (5, 6), (9, 8), (12, 1), (12, 7), (15, 4)\}$ (both are subsets of $\mathbb{N} \times \mathbb{N}$).
   (a) Determine the ordered pairs in $\mathcal{R}^{-1}$.
   (b) Determine the ordered pairs in $\mathcal{S}^{-1}$.
   (c) Determine the ordered pairs in $(\mathcal{S} \circ \mathcal{R})^{-1}$.
   (d) Is there any connection between parts (a), (b), and (c)?

13. Determine the ordered pairs in $\mathcal{T} \circ \mathcal{S} \circ \mathcal{R}$. Assume that $\mathcal{R} \subseteq \{GA, MI, PA, SD\} \times \{CA, ND, WI\}$, $\mathcal{S} \subseteq \{CA, ND, WI\} \times \{NC, SC, WA\}$, and $\mathcal{T} \subseteq \{NC, SC, WA\} \times \{IL, KA, MA, MN, RI\}$.
   (a) ★ $\mathcal{R} = \{(GA, CA), (MI, CA)\}$, $\mathcal{S} = \{(CA, NC), (ND, WA)\}$, $\mathcal{T} = \{(NC, IL), (NC, KA)\}$
   (b) $\mathcal{R} = \{(MI, ND), (PA, ND), (PA, WI), (SD, CA)\}$, $\mathcal{S} = \{(CA, NC), (ND, NC), (ND, SC), (ND, WA), (WI, WA)\}$, $\mathcal{T} = \{(NC, MN), (SC, MA), (SC, RI), (WA, IL)\}$
   (c) $\mathcal{R} = \{(GA, ND), (MI, WI), (PA, CA), (SD, CA)\}$, $\mathcal{S} = \{(CA, SC), (ND, NC), (ND, SC), (WI, WA)\}$, $\mathcal{T} = \{(NC, KA), (SC, MA), (SC, RI), (WA, IL)\}$

14. Determine the ordered pairs in $\mathcal{T} \circ \mathcal{S} \circ \mathcal{R}$. Assume that $\mathcal{R} \subseteq \{1, 2, 3\} \times \{a, b, c\}$, $\mathcal{S} \subseteq \{a, b, c\} \times \{\alpha, \beta, \gamma\}$, and $\mathcal{T} \subseteq \{\alpha, \beta, \gamma\} \times \{\heartsuit, \diamondsuit, \clubsuit, \spadesuit\}$.
   (a) $\mathcal{R} = \{(1, a), (1, b)\}$, $\mathcal{S} = \{(a, \alpha), (b, \beta)\}$, $\mathcal{T} = \{(\alpha, \diamondsuit), (\beta, \diamondsuit)\}$
   (b) $\mathcal{R} = \{(1, b), (1, c), (2, a), (3, a)\}$, $\mathcal{S} = \{(a, \alpha), (a, \gamma), (b, \alpha), (c, \beta), (c, \gamma)\}$, $\mathcal{T} = \{(\alpha, \heartsuit), (\beta, \spadesuit), (\beta, \diamondsuit), (\gamma, \heartsuit)\}$
   (c) $\mathcal{R} = \{(1, a), (1, b), (2, a), (3, b)\}$, $\mathcal{S} = \{(a, \alpha), (b, \beta), (c, \beta)\}$, $\mathcal{T} = \{(\alpha, \heartsuit), (\alpha, \clubsuit), (\beta, \spadesuit), (\gamma, \spadesuit)\}$

15. Each of the following statements is either true (always) or false (at least sometimes). Determine which option applies for each statement and provide adequate explanation for your choice.
   (a) If a relation is one-to-one and onto, then it is a function.
   (b) Composition of relations is associative for all relations. That is, $(\mathcal{T} \circ \mathcal{S}) \circ \mathcal{R} = \mathcal{T} \circ (\mathcal{S} \circ \mathcal{R})$, where $\mathcal{R}$, $\mathcal{S}$, and $\mathcal{T}$ are any relations.
   (c) A relation that is not one-to-one can still have an inverse.
   (d) ★ If the function $\mathcal{F}$ is defined by $\mathcal{F} = \{(a, 1), (b, 2), (c, 3), (d, 1), (e, 2)\}$, then the range of $\mathcal{F}$ is $\{1, 2, 3\}$.
   (e) Let $\mathcal{A}$ and $\mathcal{B}$ be sets. Then it is possible for a subset of the Cartesian product $\mathcal{A} \times \mathcal{B}$ to be both a function and a relation.

16. Let $\mathcal{F}$ be a one-to-one and onto function with domain $\mathcal{D}_{\mathcal{F}}$ and range $\mathcal{R}_{\mathcal{F}}$. Prove the following.
   (a) An inverse function, $\mathcal{G}$, must exist.
   (b) The inverse function is unique.
   (c) The inverse function is also one-to-one and onto.

17. The informal definition of the *identity function*, $\mathcal{I}$, requires $\mathcal{I}(x) = x$ to be true for all x in the domain (so the range must contain the domain). Produce a formal definition for the identity function on $\mathcal{D} \times \mathcal{D}$.

18. Complete the proof of Theorem 12.9 on page 737.

19. Let $\mathcal{R}$ and $\mathcal{S}$ be two relations for which $\mathcal{R} \circ \mathcal{S}$ and $\mathcal{S} \circ \mathcal{R}$ are both defined. Is it always true that $\mathcal{R} \circ \mathcal{S} = \mathcal{S} \circ \mathcal{R}$? If it is true, provide a proof; otherwise find a counterexample.

20. Let $\mathcal{R}$ be a relation between $\mathcal{A}$ and $\mathcal{B}$, and let $\mathcal{S}$ be a relation between $\mathcal{B}$ and $\mathcal{C}$ with Image($\mathcal{R}$) = Domain($\mathcal{S}$). Prove that

$$(\mathcal{S} \circ \mathcal{R})^{-1} = \mathcal{R}^{-1} \circ \mathcal{S}^{-1}$$

12.2 Equivalence Relations, Partially Ordered Sets

There are a number of special properties that can be used to describe relations. These properties will be defined in the next section. A special bundle of properties, called an *equivalence relation*, will be explored in more detail in Section 12.2.2.

12.2.1 Properties that Characterize Relations

Before the properties are formally defined, a few examples will be given.

EXAMPLE 12.6

Some Properties that Describe Relations

Let $\mathcal{A} = \mathbb{Z}^+$. Define the following relations on $\mathcal{A}$.

- $\mathcal{R}_1 = \{(x, y) \in \mathbb{Z}^+ \times \mathbb{Z}^+ \mid (x - y) \geq 0\}$
- $\mathcal{R}_2 = \{(x, y) \in \mathbb{Z}^+ \times \mathbb{Z}^+ \mid \gcd(x, y) = 1\}$
- $\mathcal{R}_3 = \{(x, y) \in \mathbb{Z}^+ \times \mathbb{Z}^+ \mid x < y\}$

The relation, $\mathcal{R}_1$, contains (among many other ordered pairs) all the ordered pairs (x, x) for $x \in \mathbb{Z}^+$. Any relation such that $a\mathcal{R}a$ for *every* $a \in \mathcal{A}$ is called *reflexive*.

If the ordered pair (x, y) is in $\mathcal{R}_2$, it is also true that $(y, x) \in \mathcal{R}_2$. Relations with this feature are called *symmetric*.

Relation $\mathcal{R}_3$ is *transitive*. That is, whenever $(x, y) \in \mathcal{R}_3$ and $(y, z) \in \mathcal{R}_3$, then $(x, z) \in \mathcal{R}_3$. Using the traditional notation, this can be expressed as: if $x < y$ and $y < z$, then $x < z$.

Observe that $\mathcal{R}_1$ is not symmetric [since $(4, 2) \in \mathcal{R}_1$ but $(2, 4) \notin \mathcal{R}_1$]. However, it *is* transitive.

Neither $\mathcal{R}_2$ nor $\mathcal{R}_3$ is reflexive.[11] Finally, $\mathcal{R}_3$ is not symmetric and $\mathcal{R}_2$ is not transitive (since $\gcd(4, 5) = 1$ and $\gcd(5, 8) = 1$, but $\gcd(4, 8) \neq 1$). ■

DEFINITION 12.10 ***Reflexive; Symmetric; Transitive***

Let $\mathcal{R}$ be a relation on a set, $\mathcal{A}$.

- $\mathcal{R}$ is *reflexive* if and only if $(x, x) \in \mathcal{R}$ for *all* $x \in \mathcal{A}$.
- $\mathcal{R}$ is *symmetric* if and only if for all $x, y \in \mathcal{A}$, $(x, y) \in \mathcal{R}$ implies $(y, x) \in \mathcal{R}$.
- $\mathcal{R}$ is *transitive* if and only if for all $x, y, z \in \mathcal{A}$, $((x, y) \in \mathcal{R}) \wedge ((y, z) \in \mathcal{R})$ implies $(x, z) \in \mathcal{R}$.

Relations that are simultaneously reflexive, symmetric, and transitive are of special interest. It is useful to attach a name to relations with this bundle of properties.

DEFINITION 12.11 ***Equivalence Relation***

Let $\mathcal{R}$ be a relation on a set, $\mathcal{A}$. If $\mathcal{R}$ is reflexive, symmetric, and transitive, then it is called an *equivalence relation*.

✔ Quick Check 12.3

For each relation listed, determine whether it is reflexive, symmetric, or transitive. Also, determine which relations are equivalence relations.

1. $\mathcal{R}_1 = \{(x, y) \in \mathbb{Z}^+ \times \mathbb{Z}^+ \mid x \leq y\}$
2. $\mathcal{R}_2 = \{(x, y) \in \mathbb{Z}^+ \times \mathbb{Z}^+ \mid x \mid y\}$
3. $\mathcal{R}_3 = \{(x, y) \in \mathbb{Z}^+ \times \mathbb{Z}^+ \mid \gcd(x, y) = 2\}$
4. $\mathcal{R}_4 = \{(x, y) \in \mathbb{N} \times \mathbb{N} \mid (x \bmod 5) = (y \bmod 5)\}$
5. $\mathcal{R}_5 = \{(x, y) \in \{-1, 1\} \times \{-1, 1\} \mid y = x^2\}$ ☑

[11] Even though $(1, 1) \in \mathcal{R}_2$, $\mathcal{R}_2$ is not reflexive because $(x, x) \in \mathcal{R}_2$ does not hold for *all* $x \in \mathbb{Z}^+$.

The next definition introduces some additional properties that relations can exhibit.

DEFINITION 12.12 ***Antireflexive; Antisymmetric; Asymmetric***

Let $\mathcal{R}$ be a relation on a set, $\mathcal{A}$.

- $\mathcal{R}$ is *antireflexive* if and only if for all $x, y \in \mathcal{A}$, $(x, y) \in \mathcal{R}$ implies $x \neq y$.
- $\mathcal{R}$ is *antisymmetric* if and only if for all $x, y \in \mathcal{A}$ with $x \neq y$, $(x, y) \in \mathcal{R}$ implies $(y, x) \notin \mathcal{R}$.
- $\mathcal{R}$ is *asymmetric* if and only if for all $x, y \in \mathcal{A}$, $(x, y) \in \mathcal{R}$ implies $(y, x) \notin \mathcal{R}$.

The strict inequality relation, $\{(x, y) \in \mathbb{N} \times \mathbb{N} \mid x < y\}$, is the prototypical antireflexive relation.

Several additional comments are in order. First, *antireflexive* does not mean "not reflexive" and *antisymmetric* does not mean "not symmetric".[12] Second, the definition of *antisymmetric* could have been stated as "a relation, $\mathcal{R}$, is antisymmetric if $(x, y) \in \mathcal{R}$ and $(y, x) \in \mathcal{R}$ implies $x = y$."[13]

✔ Quick Check 12.4

For each relation listed below, determine whether it is antireflexive, antisymmetric, or asymmetric.

1. $\mathcal{R}_1 = \{(x, y) \in \mathbb{Z}^+ \times \mathbb{Z}^+ \mid x \leq y\}$
2. $\mathcal{R}_2 = \{(x, y) \in \mathbb{Z}^+ \times \mathbb{Z}^+ \mid x \mid y\}$
3. $\mathcal{R}_3 = \{(x, y) \in \mathbb{Z}^+ \times \mathbb{Z}^+ \mid \gcd(x, y) = 2\}$
4. $\mathcal{R}_4 = \{(x, y) \in \mathbb{N} \times \mathbb{N} \mid (x \bmod 5) = (y \bmod 5)\}$
5. $\mathcal{R}_5 = \{(x, y) \in \{-1, 1\} \times \{-1, 1\} \mid y = x^2\}$ ☑

For the material in this book, the antisymmetric property is more important than the antireflexive and asymmetric properties. The next definition incorporates that property.

DEFINITION 12.13 ***Partial Ordering; Poset***

A relation, $\mathcal{R}$, on a set, $\mathcal{A}$, is a *partial ordering* if it is reflexive, antisymmetric, and transitive. The pair, $(\mathcal{A}, \mathcal{R})$, is called a *partially ordered set*, or *poset*.

EXAMPLE 12.7 **Two Posets**

The relation, $\mathcal{R}_1 = \{(x, y) \in \mathbb{N} \times \mathbb{N} \mid x \leq y\}$, is a partial ordering on $\mathbb{N}$. Thus, the pair, $(\mathbb{N}, \mathcal{R}_1)$, is a poset. It is also common to use the notation "$(\mathbb{N}, \leq)$ is a poset".

The name *partially ordered set* indicates that the elements of the set can be ordered but that it may not be possible to specify an ordering for every pair of elements. The relation just presented actually creates a "complete ordering" on $\mathbb{N}$, since every pair of natural numbers satisfies either $x \leq y$ or $y \leq x$ (or both inequalities).[14]

The next relation is a partial ordering that is *not* a complete ordering.

Let $\mathcal{R}_2 = \{(x, y) \in \mathbb{Z}^+ \times \mathbb{Z}^+ \mid x \mid y\}$. Quick Checks 12.3 and 12.4 have shown that $\mathcal{R}_2$ is a partial ordering. Since neither $4 \mid 7$ nor $7 \mid 4$, it is not a complete ordering. The pair, $P_2 = (\mathbb{Z}^+, \mathcal{R}_2)$, *is* a poset. ■

[12] Exercise 4 on page 747 asks you to find examples to validate these claims.

[13] Exercise 5 on page 747 asks you to verify this claim.

[14] Note that every complete ordering is also a partial ordering.

Posets are often visually represented using an oriented graph, called a *Hasse diagram*, that is reminiscent of the wooden lattices used in formal gardens for trellises. Suppose the poset is $P = (\mathcal{A}, \mathcal{R})$. The Hasse diagram satisfies three properties.

- There is a vertex for every element of $\mathcal{A}$.
- Element y appears higher in the graph than element x if $x\mathcal{R}y$ and $x \neq y$.
- If $x\mathcal{R}y$ and there is no $z \in \mathcal{A}$ with both $x\mathcal{R}z$ and $z\mathcal{R}y$, then there is an edge joining the vertices representing x and y.

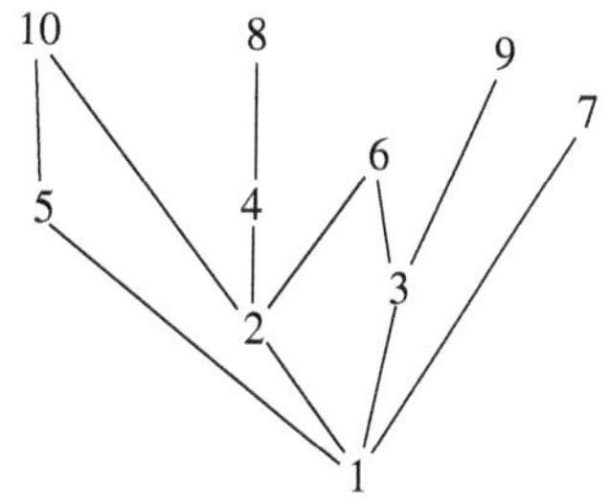

Reflexivity is assumed, so it is not explicitly shown on the diagram.

The lower portion of a Hasse diagram for poset P_2 in Example 12.7 is shown in Figure 12.1.

10 8 9 7 6 5 4 3 2 1

Figure 12.1. A Hasse diagram for poset P_2.

A common notion in higher mathematics is that of the *closure of a set*. The key idea is that the original set, $\mathcal{S}$, does not necessarily satisfy some property. The deficiency might be corrected by adding additional elements to $\mathcal{S}$. The goal is to find the smallest set that contains $\mathcal{S}$ and also satisfies the desired property. The following definitions illustrate this idea (with a relation in the role of the set $\mathcal{S}$).

DEFINITION 12.14 ***Closures of Relations***

Let $\mathcal{R}$ be a relation on a set, $\mathcal{A}$.

- The *reflexive closure* of $\mathcal{R}$ is the smallest relation, $\overline{\mathcal{R}}_r$, such that $\mathcal{R} \subseteq \overline{\mathcal{R}}_r$ and $\overline{\mathcal{R}}_r$ is reflexive on $\mathcal{A}$.
- The *symmetric closure* of $\mathcal{R}$ is the smallest relation, $\overline{\mathcal{R}}_s$, such that $\mathcal{R} \subseteq \overline{\mathcal{R}}_s$ and $\overline{\mathcal{R}}_s$ is symmetric on $\mathcal{A}$.
- The *transitive closure* of $\mathcal{R}$ is the smallest relation, $\overline{\mathcal{R}}_t$, such that $\mathcal{R} \subseteq \overline{\mathcal{R}}_t$ and $\overline{\mathcal{R}}_t$ is transitive on $\mathcal{A}$.

All three of these closures exist, since they will never be larger than $\mathcal{A} \times \mathcal{A}$. You should convince yourself that these closures are also unique. For example, there cannot be two distinct reflexive closures of some relation.

EXAMPLE 12.8

Some Closures

Let $\mathcal{A} = \{1, 2, 3, 4\}$ and let $\mathcal{R} = \{(1, 1), (1, 2), (1, 3), (2, 2), (2, 3), (3, 3), (3, 4)\}$. It is easy to verify that

- $\overline{\mathcal{R}}_r = \{(1, 1), (1, 2), (1, 3), (2, 2), (2, 3), (3, 3), (3, 4), \mathbf{(4, 4)}\}$
- $\overline{\mathcal{R}}_s = \{(1, 1), (1, 2), (1, 3), \mathbf{(2, 1)}, (2, 2), (2, 3), \mathbf{(3, 1)}, \mathbf{(3, 2)}, (3, 3), (3, 4), \mathbf{(4, 3)}\}$

where the boldface ordered pairs are the pairs added for the closure. Determining the transitive closure usually takes more effort. In this case,

$$\overline{\mathcal{R}}_t = \{(1, 1), (1, 2), (1, 3), \mathbf{(1, 4)}, (2, 2), (2, 3), \mathbf{(2, 4)}, (3, 3), (3, 4)\}$$

One systematic way to arrive at this is to look at every possible matched pair of the form $(x, z) \leftrightarrow (z, y)$ and then make sure that the ordered pair (x, y) is either in the relation or is added to the relation. Once that is done, it is sometimes necessary to make a second pass (or a third, or a fourth, or . . .).

The first pass in this example requires the addition of $(1, 4)$ and $(2, 4)$. The second pass yields no new ordered pairs, so we can quit. ■

The heuristic approach for finding the transitive closure that was used in Example 12.8 can be justified by the next theorem. A new definition is needed to state the theorem concisely. The definition provides a standard notation for a sequence of compositions involving the same relation.

DEFINITION 12.15 $\mathcal{R}^n$

Let $\mathcal{R}$ be a relation on a set, $\mathcal{A}$. Then

- $\mathcal{R}^0 = \mathcal{I} = \{(x, x) \in \mathcal{A} \times \mathcal{A} \mid x \in \mathcal{A}\}$.
- $\mathcal{R}^1 = \mathcal{R}$.
- $\mathcal{R}^n = \mathcal{R} \circ \mathcal{R}^{n-1}$ for all $n \geq 1$.

THEOREM 12.16 ***Transitive Relations***

Let $\mathcal{R}$ be a relation on a set, $\mathcal{A}$. Then $\mathcal{R}$ is transitive if and only if

$$\mathcal{R}^n \subseteq \mathcal{R}, \quad \forall n \in \mathbb{Z}^+$$

Proof:

Part 1: Suppose $\mathcal{R}^n \subseteq \mathcal{R}$, $\forall n \in \mathbb{Z}^+$.

Let (x, z) and (z, y) be two ordered pairs in $\mathcal{R}$. Then $(x, y) \in \mathcal{R} \circ \mathcal{R} = \mathcal{R}^2 \subseteq \mathcal{R}$. Thus, $(x, y) \in \mathcal{R}$ and so $\mathcal{R}$ is transitive.

Part 2: Suppose $\mathcal{R}$ is transitive.

If $\mathcal{R}$ is transitive, then a simple induction will show that $\mathcal{R}^n \subseteq \mathcal{R}$, $\forall n \in \mathbb{Z}^+$.

The base case is straightforward: $\mathcal{R}^1 = \mathcal{R} \subseteq \mathcal{R}$. For the inductive step, suppose that $\mathcal{R}^k \subseteq \mathcal{R}$ for some $k \in \mathbb{Z}^+$. Let (x, y) be any element of $\mathcal{R}^{k+1} = \mathcal{R} \circ \mathcal{R}^k$. Then, using the definition of composition of relations, there must be an element, $z \in \mathcal{A}$, such that $(x, z) \in \mathcal{R}^k$ and $(z, y) \in \mathcal{R}$. By the inductive hypothesis, $(x, z) \in \mathcal{R}$. Since $\mathcal{R}$ is transitive and since both $(x, z) \in \mathcal{R}$ and $(z, y) \in \mathcal{R}$, it follows that $(x, y) \in \mathcal{R}$. Thus, every element of $\mathcal{R}^{k+1}$ is also in $\mathcal{R}$, so $\mathcal{R}^{k+1} \subseteq \mathcal{R}$, completing the induction. □

12.2.2 Equivalence Relations and Partitions

Equivalence relations merit additional exposition. One notable application of equivalence relations occurs when the chain of familiar number systems, $\mathbb{N} \subseteq \mathbb{Z} \subseteq \mathbb{Q} \subseteq \mathbb{R} \subseteq \mathbb{C}$, is carefully constructed. In particular, when moving from the integers, $\mathbb{Z}$, to the set of rational numbers, $\mathbb{Q}$, it is common to define the elements of $\mathbb{Q}$ by starting with ordered pairs of integers such that the second coordinate is not 0. The first coordinate represents the numerator and the second coordinate represents the denominator. One complication is that the ordered pairs $(1, 2)$ and $(2, 4)$ really represent the same rational number (the fraction $\frac{1}{2}$). Equivalence relations provide a mechanism for combining the infinitely many representations for the same number into a single element of the new set, $\mathbb{Q}$. Additional details will be presented later in this section.

A previous example from a Quick Check will motivate the major new idea.

EXAMPLE 12.9 **Congruence Classes mod 5**

Quick Check 12.3 on page 739 introduced the equivalence relation

$$\mathcal{R}_4 = \{(x, y) \in \mathbb{N} \times \mathbb{N} \mid (x \bmod 5) = (y \bmod 5)\}$$

It is easy to see that if $(x, y) \in \mathcal{R}_4$, then $x - y = 5n$, for some integer, n. Definition 3.19 on page 98 provides an alternate description: $(x \bmod 5) = (y \bmod 5)$ if and only if $x \equiv y \pmod 5$.

There is a natural partitioning[15] of $\mathbb{N}$ into the *congruence classes* mod 5:

Remainder 0 $\{0, 5, 10, 15, 20, 25, \ldots\}$

Remainder 1 $\{1, 6, 11, 16, 21, 26, \ldots\}$

Remainder 2 $\{2, 7, 12, 17, 22, 27, \ldots\}$

Remainder 3 $\{3, 8, 13, 18, 23, 28, \ldots\}$

Remainder 4 $\{4, 9, 14, 19, 24, 29, \ldots\}$

Every pair of natural numbers from the same congruence class is in the relation, $\mathcal{R}_4$. Any two numbers from different congruence classes are not in the relation. ■

The phenomenon of being able to partition the elements of a set into natural subsets defined by an equivalence relation is not unique to congruences in $\mathbb{N}$. It happens with every equivalence relation.

DEFINITION 12.17 ***Equivalence Class***

Let $\mathcal{R}$ be an equivalence relation on a set, $\mathcal{A}$, and let $x \in \mathcal{A}$. The *equivalence class* of x is denoted by $[x]$ and is defined as

$$[x] = \{a \in \mathcal{A} \mid (x, a) \in \mathcal{R}\}$$

The set $\{[x] \mid x \in \mathcal{A}\}$ is referred to as the set of equivalence classes *induced by* R on A. The element, x, that appears in the notation "$[x]$" is called a *class representative*.

✔ Quick Check 12.5

Let $\mathcal{A}$ be the set of all students who reside on campus at a particular college or university. Let $\mathcal{R} \subseteq \mathcal{A} \times \mathcal{A}$ be the relation defined by $(x, y) \in \mathcal{R}$ if and only if x and y live in the same residence hall (dorm).

1. Show that $\mathcal{R}$ is an equivalence relation.
2. Determine the equivalence classes of $\mathcal{R}$. ☑

PROPOSITION 12.18 ***Equivalence Classes Are Disjoint***

Let $\mathcal{R}$ be an equivalence relation on a set, $\mathcal{A}$. If x and y are two elements in $\mathcal{A}$, then either $[x] = [y]$ or else $[x] \cap [y] = \emptyset$.

Proof:

***Case 1:* $(x, y) \in \mathcal{R}$**

If $(x, y) \in \mathcal{R}$, then $y \in [x]$ (by Definition 12.17). Since $\mathcal{R}$ is symmetric, it is also true that $(y, x) \in \mathcal{R}$, so $x \in [y]$. $\mathcal{R}$ is reflexive, so $x \in [x]$. Thus, $[x] \cap [y] \neq \emptyset$.

Suppose that $a \in [x]$. Then $(x, a) \in \mathcal{R}$. Since $\mathcal{R}$ is symmetric, $(a, x) \in \mathcal{R}$ is also true. But then the transitivity of $\mathcal{R}$, combined with $(a, x) \in \mathcal{R}$ and $(x, y) \in \mathcal{R}$ implies that $(a, y) \in \mathcal{R}$ and (by symmetry) $(y, a) \in \mathcal{R}$. Consequently, $a \in [y]$ and $[x] \subseteq [y]$. A similar argument shows that $[y] \subseteq [x]$. The conclusion is that $[x] = [y]$ whenever $x\mathcal{R}y$.

***Case 2:* $(x, y) \notin \mathcal{R}$**

Suppose that $(x, y) \notin \mathcal{R}$ but that $a \in [x]$ and $a \in [y]$. Then $(x, a) \in \mathcal{R}$ and $(y, a) \in \mathcal{R}$. Using the symmetry and transitivity of $\mathcal{R}$, it is then true that $(x, y) \in \mathcal{R}$, a contradiction. The contradiction arose by assuming that $[x]$ and $[y]$ had a common element. The conclusion is that $[x] \cap [y] = \emptyset$. □

[15]See Defintion 2.13 on page 24.

Since an equivalence relation is reflexive, $(x, x) \in \mathcal{R}$ for every $x \in \mathcal{A}$. Consequently, $x \in [x]$. The partitioning phenomenon that was observed in Example 12.9 is fully developed in the next theorem.

One consequence of Proposition 12.18 is that any one of the elements in an equivalence class may be unambiguously used as the class representative.

THEOREM 12.19 ***Equivalence Relations and Partitions***

Let $\mathcal{A}$ be a set.

- If $\mathcal{R}$ is an equivalence relation on $\mathcal{A}$, then the equivalence classes of $\mathcal{R}$ form a partition of $\mathcal{A}$.
- Every partition of $\mathcal{A}$ determines an equivalence relation on $\mathcal{A}$.

Proof:

Equivalence Relation Implies Partition

Let $\mathcal{R}$ be an equivalence relation on $\mathcal{A}$. Proposition 12.18 implies that the equivalence classes induced by $\mathcal{R}$ are disjoint. Since every element, x, in $\mathcal{A}$ is in an equivalence class, $[x]$, every element of $\mathcal{A}$ is in some equivalence class. Therefore, the equivalence classes form a partition of $\mathcal{A}$; they are disjoint and their union is all of $\mathcal{A}$.

Partition Implies Equivalence Relation

Let $P_{\mathcal{A}} = \{\mathcal{A}_i \subseteq \mathcal{A} \mid i \in \Upsilon\}$ for some set of indices, Υ.[16] Assume also that $\mathcal{A}_i \cap \mathcal{A}_j = \emptyset$ if $i \neq j$ and $\mathcal{A} = \cup_{i \in \Upsilon} \mathcal{A}_i$. $P_{\mathcal{A}}$ is a partition of $\mathcal{A}$.

Define a relation, $\mathcal{R}$, on $\mathcal{A}$ by

$$(x, y) \in \mathcal{R} \text{ if and only if } x \in \mathcal{A}_i \text{ and } y \in \mathcal{A}_i, \text{ for a common } i \in \Upsilon$$

It is an easy exercise to show that $\mathcal{R}$ is an equivalence relation whose equivalence classes are the members of $P_{\mathcal{A}}$. □

It is now time to provide the missing details for deriving the rational numbers from the integers.

EXAMPLE 12.10 **Extending the Integers to the Rational Numbers—Part 1**

Assume that the integers, $\mathbb{Z}$, have been rigorously defined, along with the operations of addition and multiplication in $\mathbb{Z}$.

The intuitive model for the rational numbers is the set of all fractions: $\left\{\frac{p}{q} \mid p, q \in \mathbb{Z}, q \neq 0\right\}$. The previously mentioned problem is that $\frac{1}{2}$ and $\frac{2}{4}$ are different fractions, but really represent the same number.

The formal approach starts with the set of ordered pairs whose second coordinate is not 0 (merely a change in notation from the set of fractions). Let $\mathcal{F} = \{(p, q) \in \mathbb{Z} \times \mathbb{Z} \mid q \neq 0\}$. Define a relation, $\mathcal{R}$, on $\mathcal{F}$ by

$$((p_1, q_1), (p_2, q_2)) \in \mathcal{R} \text{ if and only if } p_1 q_2 = p_2 q_1 \tag{12.1}$$

A routine proof shows that $\mathcal{R}$ is an equivalence relation.[17]

The set of rationals can now be defined as the set of equivalence classes induced by $\mathcal{R}$ on $\mathcal{F}$. More formally,

$$\mathbb{Q} = \{[(p, q)] \mid (p, q) \in \mathcal{F}\}$$

This definition ensures that different representations for the same number resolve to the same element of $\mathbb{Q}$. For example, $(1, 3)\mathcal{R}(4, 12)$ since $1 \cdot 12 = 4 \cdot 3$, so $[(1, 3)] = [(4, 12)]$.

[16] The set Υ will be finite or countably infinite for most examples in this text. The proof does not make this assumption. See page 23 for a bit more about index sets.

[17] Even though the proof is routine, it is a healthy activity for you, the student, to engage in. See Exercise 14 on page 747.

Since we want $\mathbb{Z} \subseteq \mathbb{Q}$, we can identify the integer, n, with the equivalence class $[(n, 1)]$.[18] This effectively embeds the set of integers inside the newly defined set of rational numbers.

The process can be completed by defining the operations of addition and multiplication for the new set of numbers. This will be done by using the already defined operations on the integers. Example 12.12 will provide the details. ■

Before completing the derivation of the rational numbers, an additional important concept needs to be introduced. This is necessitated by the following complication. The elements of $\mathbb{Q}$ have been defined as sets (equivalence classes). How should the operations of addition and multiplication be defined when the operations are applied to sets? The operations will be defined using equivalence class representatives. But this leads to the very real possibility that a different choice of class representative might lead to a different result.

More specifically, suppose that $\mathcal{R}$ is an equivalence relation on a set, $\mathcal{A}$. Suppose that there is a binary operation, $\odot$, defined on $\mathcal{A}$. Thus, for all $x, y \in \mathcal{A}$, $x \odot y \in \mathcal{A}$, but $x \mathcal{R} y$ may or may not be true. Let $\oslash$ be a binary operation that is to be defined for the set, $P_{\mathcal{A}}$, of equivalence classes that $\mathcal{R}$ induces on $\mathcal{A}$. We require $[x] \oslash [y] \in P_{\mathcal{A}}$ for all $[x], [y] \in P_{\mathcal{A}}$.

Suppose that the new operation is defined as $[x] \oslash [y] = [x \odot y]$. This seems reasonable: $\oslash$ takes two equivalence classes and produces another equivalence class. There is a major land mine lurking just below the surface. Suppose for the moment that $[x] = [x']$ for $x \neq x'$ and $[y] = [y']$ for $y \neq y'$. Since $[x] = [x']$ and $[y] = [y']$, we clearly want $([x] \oslash [y]) = ([x'] \oslash [y'])$. But there is no guarantee that $[x \odot y] = [x' \odot y']$.

EXAMPLE 12.11

A Failed Class Operation

Consider the relation $\mathcal{R}_4$ from Example 12.9 on page 742. Suppose we define $x \odot y = \gcd(x, y) \bmod 5$. If $\oslash$ is defined as $[x] \oslash [y] = [x \odot y]$, the binary operation $\oslash$ will not be unambiguously defined. For example, $[2] = [12]$ and $[3] = [8]$. However, $[2] \oslash [3] = [2 \odot 3] = 1$, but $[12] \oslash [8] = [12 \odot 8] = 4$. ■

The problem originated because the new operation, $\oslash$, was defined using class representatives. What is needed is a definition for the operation that is independent of which class representative is chosen. If that independence holds, the operation is said to be *well-defined*.

DEFINITION 12.20 ***Well-Defined***

Let $\mathcal{R}$ be an equivalence relation on a set, $\mathcal{A}$, and let $\odot$ be a binary operation on $\mathcal{A}$. Let $\mathcal{E}$ be the set of equivalence classes that $\mathcal{R}$ induces on $\mathcal{A}$. Finally, let $\oslash$ be a binary operation on $\mathcal{E}$ that is defined by $[x] \oslash [y] = [x \odot y]$, for $x, y \in \mathcal{A}$.

If $([x] \oslash [y]) = ([x'] \oslash [y'])$ whenever $[x] = [x']$ and $[y] = [y']$, then $\oslash$ is said to be *well-defined*.

Note: Notice that $\oslash$ takes two equivalence classes and returns a third equivalence class. The definition ensures that $\oslash$ is independent of which class representatives are used to define the equivalence classes.

[18] That is, think of the fraction, $\frac{n}{1}$, and the integer, n, as being the same number.

EXAMPLE 12.12 **Extending the Integers to the Rational Numbers—Part 2**

The derivation of $\mathbb{Q}$ will be complete if well-defined versions of addition and multiplication can be produced. Since we want these operations to correspond to the usual operations on fractions, it is suggestive to use the standard addition and multiplication of fractions as the model for the new operations. Thus, we should make the following definitions.[19]

$$[(p_1, q_1)] \cdot [(p_2, q_2)] = [(p_1 \cdot p_2, q_1 \cdot q_2)]$$
$$[(p_1, q_1)] + [(p_2, q_2)] = [(p_1 \cdot q_2 + p_2 \cdot q_1, q_1 \cdot q_2)]$$

It remains to show that these operations are well-defined. To that end, assume that $[(p_1, q_1)] = [(p_1', q_1')]$ and $[(p_2, q_2)] = [(p_2', q_2')]$.

Notice that $[(p_1, q_1)] = [(p_1', q_1')]$ implies $p_1 \cdot q_1' = p_1' \cdot q_1$ and also the equality $[(p_2, q_2)] = [(p_2', q_2')]$ implies $p_2 \cdot q_2' = p_2' \cdot q_2$.[20]

Multiplication

Since $p_1 \cdot q_1' = p_1' \cdot q_1$ and $p_2 \cdot q_2' = p_2' \cdot q_2$, multiplying the integers on the left-hand and right-hand sides of the equations (respectively) leads to $p_1 \cdot q_1' \cdot p_2 \cdot q_2' = p_1' \cdot q_1 \cdot p_2' \cdot q_2$. This can be rearranged as $(p_1 \cdot p_2) \cdot (q_1' \cdot q_2') = (p_1' \cdot p_2') \cdot (q_1 \cdot q_2)$.

This implies that $((p_1 \cdot p_2, q_1 \cdot q_2), (p_1' \cdot p_2', q_1' \cdot q_2')) \in \mathcal{R}$ and thus $[(p_1 \cdot p_2, q_1 \cdot q_2)] = [(p_1' \cdot p_2', q_1' \cdot q_2')]$. Consequently, multiplication on $\mathbb{Q}$ is well-defined.

Addition

Multiplying (as integers) both sides of $p_1 \cdot q_1' = p_1' \cdot q_1$ by $q_2 \cdot q_2'$ leads to

$$(p_1 \cdot q_2) \cdot (q_1' \cdot q_2') = (p_1' \cdot q_2') \cdot (q_1 \cdot q_2)$$

Similarly, multiplying both sides of $p_2 \cdot q_2' = p_2' \cdot q_2$ by $q_1 \cdot q_1'$ leads to

$$(p_2 \cdot q_1) \cdot (q_1' \cdot q_2') = (p_2' \cdot q_1') \cdot (q_1 \cdot q_2)$$

Adding these equations produces

$$(p_1 \cdot q_2 + p_2 \cdot q_1) \cdot (q_1' \cdot q_2') = (p_1' \cdot q_2' + p_2' \cdot q_1') \cdot (q_1 \cdot q_2)$$

Equation 12.1 on page 744 then implies that

$$[(p_1 \cdot q_2 + p_2 \cdot q_1, q_1 \cdot q_2)] = [(p_1' \cdot q_2' + p_2' \cdot q_1', q_1' \cdot q_2')]$$

showing that addition is also well-defined. ■

12.2.3 Exercises

The exercises marked with ★ have detailed solutions in Appendix H.

1. For each relation in the following set
   (a) Determine which of the following properties apply: reflexive, symmetric, transitive, antireflexive, antisymmetric, asymmetric.
   (b) Decide whether the relation is an equivalence relation.
   (c) Decide whether the relation is a partial order.

   Organize your work in a table with four columns with one row for each relation.

Relation	(a) Properties	(b) Equivalence relation?	(c) Partial order?

   Let $\mathcal{H}$ be the set of all living humans.

   R₁: ★ $\{(x, y) \in \mathcal{H} \times \mathcal{H} \mid x$ and y are siblings with the same biological mother and father$\}$

   R₂: $\{(x, y) \in \mathcal{H} \times \mathcal{H} \mid y$ is a biological parent of $x\}$

   R₃: $\{(x, y) \in \mathcal{H} \times \mathcal{H} \mid x$ and y are cousins$\}$

   R₄: $\{(x, y) \in \mathbb{N} \times \mathbb{N} \mid x$ and y have the same parity (both even or both odd)$\}$

   R₅: $\{(x, y) \in \mathbb{Z}^+ \times \mathbb{Z}^+ \mid \text{lcm}(x, y) = 10\}$

2. Let $\mathcal{R}_2 = \{(x, y) \in \mathbb{Z}^+ \times \mathbb{Z}^+ \mid x \mid y\}$. Prove (using Definition 3.8 on page 93) that $\mathcal{R}_2$ is transitive.

[19] Notice that when + and · appear inside []'s, they refer to integer operations. When they appear outside []'s, they refer to the newly defined operations on elements of $\mathbb{Q}$.

[20] See the definition of $\mathcal{R}$ (Equation 12.1) in Example 12.10 on page 744.

3. For each relation in the following set
   (a) Determine which of the following properties apply: reflexive, symmetric, transitive, antireflexive, antisymmetric, asymmetric.
   (b) Decide whether the relation is an equivalence relation.
   (c) Decide whether the relation is a partial order.

   Organize your work in a table with four columns with one row for each relation.

	(a)	(b)	(c)
Relation	Properties	Equivalence relation?	Partial order?

R$_6$: $\{(x, y) \in \mathbb{Z} \times \mathbb{Z} \mid |x| = |y|\}$

R$_7$: $\{(x, y) \in \mathbb{N} \times \mathbb{N} \mid x$ has the same number of prime factors as $y\}$

R$_8$: $\{(x, y) \in \mathbb{N} \times \mathbb{N} \mid x$ does not have more prime factors than $y\}$

R$_9$: $\{(x, y) \in \mathbb{R} \times \mathbb{R} \mid y = x^2\}$

R$_{10}$: $\{(x, y) \in \mathbb{R} \times \mathbb{R} \mid y = x^2 + 1\}$

4. Verify that the following pairs of properties are independent. [*Hint*: In each part, make up examples of relations that have, respectively, only one of the two properties, then both properties, and then neither property. Explain the conclusion.]
   (a) ★ Antireflexive and not reflexive.
   (b) Antisymmetric and not symmetric.
5. Show that the alternative definition of *antisymmetric* (below Definition 12.12 on page 740) is equivalent to the formal definition.
6. Investigate the definitions of *antisymmetric* and *asymmetric*.
   (a) Can you find a relation that is antisymmetric but not asymmetric?
   (b) Can you find a relation that is asymmetric but not antisymmetric?
   (c) Formulate (and prove) a theorem that properly relates the two properties.
7. For each relation on $\mathcal{A} = \{a, b, c, d\}$ listed, find the reflexive, symmetric, and transitive closures.
   (a) ★ $\mathcal{S}_1 = \{(a, b), (a, c), (b, c), (c, c)\}$
   (b) $\mathcal{S}_2 = \{(a, b), (a, d), (b, d), (c, c), (d, a), (d, c)\}$
   (c) $\mathcal{S}_3 = \{(a, a), (b, a), (b, c), (c, b), (c, d), (d, b), (d, d)\}$
8. For each relation on $\mathcal{H}$, the set of all living humans, listed, find the reflexive, symmetric, and transitive closures.
   (a) $\mathcal{R}_1 = \{(x, y) \in \mathcal{H} \times \mathcal{H} \mid x$ and y are siblings with the same biological mother and father$\}$
   (b) $\mathcal{R}_2 = \{(x, y) \in \mathcal{H} \times \mathcal{H} \mid y$ is a biological parent of $x\}$
   (c) $\mathcal{R}_3 = \{(x, y) \in \mathcal{H} \times \mathcal{H} \mid x$ and y are cousins$\}$
9. For each relation listed, find the reflexive, symmetric, and transitive closures.
   (a) $\mathcal{R}_4 = \{(x, y) \in \mathbb{N} \times \mathbb{N} \mid x$ and y have the same parity (both even or both odd)$\}$
   (b) $\mathcal{R}_5 = \{(x, y) \in \mathbb{Z}^+ \times \mathbb{Z}^+ \mid \text{lcm}(x, y) = 10\}$
   (c) $\mathcal{R}_9 = \{(x, y) \in \mathbb{Z} \times \mathbb{Z} \mid y = x^2\}$
   (*Extra credit*: Draw graphs of the closures.)
10. Show that each relation listed is a partial order. Then draw a Hasse diagram for the poset.
    (a) ★ Let $\mathcal{F}$ be the set of all integers, x, such that $\{-10 \le x \le 10\}$. Let $\mathcal{R}_1$ be defined by $(x, y) \in \mathcal{R}_1$ if and only if $x = y$.
    (b) Let $\mathcal{A}$ be the power set of $\{1, 2, 3\}$ (Definition 2.15 on page 25). Let $\mathcal{R}_2$ be defined by $(\mathcal{S}_1, \mathcal{S}_2) \in \mathcal{R}_2$ if and only if $\mathcal{S}_1 \subseteq \mathcal{S}_2$.
    (c) Let $\mathcal{S}$ be the set of strings {"reflexive", "transitive", "antisymmetric", "A relation is a partial ordering if it is reflexive, antisymmetric, and transitive", "A relation", "if it is reflexive, antisymmetric, and transitive", "reflexive, antisymmetric"}. Let an ordered pair of strings, (s_1, s_2), be in the relation, $\mathcal{R}_3$, if and only if s_1 is a substring of s_2.
    (d) Let $\mathcal{B}$ be the set of all bit strings of length 4. Let $\mathcal{R}_4$ be defined by $(s_1, s_2) \in \mathcal{R}_4$ if and only if $s_1 = s_2$ or the number of 1s in s_1 is less than the number of 1s in s_2.
11. Let $G = \{V, E, \phi\}$ be a graph (Definition 10.3). Define two symmetric relations on V as follows:

    Adjacency $\mathcal{A} = \{(v, w) \in V \times V \mid \phi(v, w) \in E\}$

    Connectedness $\mathcal{C} = \{(v, w) \in V \times V \mid v$ and w are connected by a walk in $G\}$.

    (a) What property must hold in G in order for $\mathcal{A}$ to be a reflexive relation?
    (b) Is $\mathcal{A}$ transitive? Explain your answer.
    (c) Is $\mathcal{C}$ transitive? Explain your answer.
    (d) What is the transitive closure of A?
    (e) What is the transitive closure of C?
12. ★ Prove the claim, made in Example 12.9, that $(x \bmod 5) = (y \bmod 5)$ if and only if $x - y = 5n$ for some integer, n. [*Hint*: Use the Quotient–Remainder theorem.]
13. Complete the proof of Theorem 12.19 (page 744): Show that $\mathcal{R}$ is an equivalence relation whose equivalence classes are the members of $P_{\mathcal{A}}$.
14. Show that the relation, $\mathcal{R}$, in Example 12.10 on page 744 is an equivalence relation. Do not use any algebraic manipulations involving fractions.
15. ★ Define a relation, $\mathcal{S}$, on $\mathbb{Z} \times \mathbb{Z}$ by

    $$(a, b) \in \mathcal{S} \text{ if and only if } a + b \text{ is even}$$

    (a) Show that $\mathcal{S}$ is an equivalence relation.
    (b) Describe the equivalence classes.
16. Define a relation, $\mathcal{S}$, on $\mathbb{Z} \times \mathbb{Z}$ by

    $$(a, b) \in \mathcal{S} \text{ if and only if } a = b \text{ or } a = -b$$

    (a) Show that $\mathcal{S}$ is an equivalence relation.
    (b) Describe the equivalence classes.
17. Define a relation, $\mathcal{D}$, on $\mathbb{R} \times \mathbb{R}$ by

    $$(a, b) \in \mathcal{D} \text{ if and only if } a - b \text{ is an integer}$$

    (a) Show that $\mathcal{D}$ is an equivalence relation.
    (b) Describe the equivalence classes.

18. Let $\mathbb{R}^2 = \mathbb{R} \times \mathbb{R}$ be the familiar Cartesian plane. Define a relation, $\mathcal{D}$, on $\mathbb{R}^2 \times \mathbb{R}^2$ by

$$((x_1, y_1), (x_2, y_2)) \in \mathcal{D}$$
$$\text{if and only if } (x_1^2 + y_1^2) = (x_2^2 + y_2^2)$$

(a) Show that $\mathcal{D}$ is an equivalence relation.

(b) Describe the equivalence classes.

19. The empty set is a relation, because it is a subset of every Cartesian product. Consider the six properties: reflexive, symmetric, transitive, antireflexive, antisymmetric, asymmetric. Which properties characterize the empty set as a relation on $\mathbb{N}$? Which properties characterize the empty set as a relation on $\emptyset$?

20. ★ Let $\mathcal{A} = \{(x, y) \in \mathbb{R} \times \mathbb{R} \mid x - \lfloor x \rfloor = y - \lfloor y \rfloor\}$. That is, x and y are in $\mathcal{A}$ if and only if they have the same expansion to the right of the decimal. Define an operation, $\oslash$, on the equivalence classes $\mathcal{A}$ induces on $\mathbb{R}$ by

$$[x] \oslash [y] = [x + y]$$

Show that $\oslash$ is well-defined.

21. Let $\mathcal{A} = \{(x, y) \in \mathbb{Z} \times \mathbb{Z} \,\big|\, |x| = |y|\}$. Define an operation, $\odot$, on the equivalence classes $\mathcal{A}$ induces on $\mathbb{Z}$ by

$$[x] \odot [y] = [\,|\max(x, y)|\,]$$

(Take the absolute value of the maximum of x and y. The result will be the equivalence class representative.) Show that $\odot$ is not well-defined.

22. Let $\mathcal{C}_5 = \{(x, y) \in \mathbb{N} \times \mathbb{N} \mid x \equiv y \pmod 5\}$. Define an operation, $\oslash$, on the equivalence classes $\mathcal{C}_5$ induces on $\mathbb{N}$ by

$$[x] \oslash [y] = [xy]$$

(Multiply x and y, and then look at the remainder, mod 5. The equivalence class of that remainder will be the desired element.) Show that $\oslash$ is well-defined.

23. Let $\mathcal{C}_5 = \{(x, y) \in \mathbb{N} \times \mathbb{N} \mid x \equiv y \pmod 5\}$. Define an operation, $\odot$, on the equivalence classes $\mathcal{C}_5$ induces on $\mathbb{N}$ by

$$[x] \odot [y] = \left[\left\lceil \frac{x}{y+1} \right\rceil\right]$$

(Divide x by $y + 1$, round up to the nearest integer, and then look at the remainder, mod 5. The equivalence class of that remainder will be the desired element.) Show that $\odot$ is not well-defined.

12.3 *n*-ary Relations and Relational Databases

The relations that were examined in the previous sections were relations in $\mathcal{A} \times \mathcal{B}$, for some pair of sets, $\mathcal{A}$ and $\mathcal{B}$. These relations can be referred to as *binary relations*, since they involve two sets.

An obvious generalization is that of n-ary relations, which are relations in a Cartesian product involving n sets. This section formally introduces n-ary relations and also introduces relational databases, which are an important application of n-ary relations. The section ends with a brief look at *normal forms*, which aid in avoiding some undesirable anomalies in databases.

12.3.1 *n*-ary Relations

The definition of an n-ary relation is a simple extension of the definition of a (binary) relation.

DEFINITION 12.21 *n-ary Relation*
An *n-ary relation* in (or on) the sets $\mathcal{A}_1, \mathcal{A}_2, \ldots, \mathcal{A}_n$ is a subset, $\mathcal{R}$, of the Cartesian product $\mathcal{A}_1 \times \mathcal{A}_2 \times \cdots \times \mathcal{A}_n$.

If the context makes the value of n clear, it is convenient to refer to $\mathcal{R}$ simply as a "relation", rather than use the more formal phrase "n-ary relation".

When $n = 2$, it is common to refer to the relation as a *binary relation*. A 3-ary relation (sometimes called a *ternary relation*) may be viewed as a generalization of a function of two variables. If the relation is a subset of $\mathcal{X} \times \mathcal{Y} \times \mathcal{Z}$, then $\mathcal{X} \times \mathcal{Y}$ would be the domain and $\mathcal{Z}$ would be the range of the relation. This interpretation can be emphasized by viewing the relation as a subset of $(\mathcal{X} \times \mathcal{Y}) \times \mathcal{Z}$, having elements in the form $((x, y), z)$.

EXAMPLE 12.13 **Family Relations**

Let $\mathcal{H}$ be the set of all humans. Define $\mathcal{B} \in \mathcal{H} \times \mathcal{H} \times \mathcal{H}$ to be the set of all ordered triples (f, m, c), where f is the biological father of c and m is the biological mother of c, and f and m are currently married.[21] Since it is possible (in fact, common) for parents to have more than one child, this relation cannot be interpreted as a function of two variables.

Let $\mathcal{A} \in \mathcal{H} \times \mathcal{H} \times \mathcal{H}$ be the set of all ordered triples (f, m, c), where f is the adoptive father of c and m is the adoptive mother of c, and f and m are currently married.[22]

The relations $\mathcal{B}$ and $\mathcal{A}$ are disjoint.[23] ■

12.3.2 Relational Databases

One of the most significant applications of n-ary relations is to provide the theoretical framework for relational databases.[24] A database is a (typically large) collection of related information, generally managed through the use of computers. A relational database consists of a collection of n-ary relations (with multiple concurrent values for n). The relational database model was originally proposed in 1970 by E. F. Codd. The theoretical model is clearly seen in the name *relational* database. Nevertheless, the terminology used in discussions of relational databases includes words that are more influenced by the need to use computers to implement the database model and less influenced by the mathematical underpinnings.

Databases are used to store and manipulate information about some real-world entity (such as a university or a business). In order to accomplish this, it is necessary to create a model for the information that needs to be stored, including constraints on values and on how various pieces of information are permitted to interact. This model is called a *schema*. When a database is created and actual information is entered, the result is called an *instance*. A schema is a description that will usually have a very large number of possible instances that conform to the model. At any given time, there will be only one instance that is stored in the database. So, it is helpful to keep in mind the distinction between the real-world entity, the model of the entity (a schema), and an actual collection of stored information (an instance).

The next few definitions introduce the terminology needed here.

DEFINITION 12.22 ***Schema; Instance***

A *database schema* is a formal description of the information that will be contained in a database. The schema defines how the information is grouped and also contains constraints on acceptable values and on how different pieces of information can be associated.

A *database instance* is an actual collection of information that conforms to a predefined schema.

[21] The marriage requirement helps simplify subsequent use of this example.

[22] Notice that this definition excludes situations where a single parent adopts a child, and also excludes cases where one parent is a biological parent and the other is an adoptive parent. These omissions serve to keep the example simple.

[23] You may consider the assertion that they are disjoint to be an explicit assumption. Bizarre twists in human relations might make an amusing Gilbert and Sullivan operetta, but will only detract from an example that seeks to introduce concepts about n-ary relations.

[24] There are two older frameworks for databases: the *hierarchical* model and the *network* model. There are also more recent frameworks, such as *object* databases. See [61] or [73] for details.

DEFINITION 12.23 *Relational Database; Attributes; Tuples*

A *relational database* is defined by a schema, which includes a collection, $\{\mathcal{T}_1, \mathcal{T}_2, \ldots, \mathcal{T}_k\}$, of relations, where $\mathcal{T}_j$ is an n_j-ary relation, for $j = 1, 2, \ldots, k$. The schema (and hence the relational database) also includes constraints on the data in the relations and on how relations interact. A relational database is made manifest by creating instances for its collection of relations.

The coordinate positions in the n_j-tuples of $\mathcal{T}_j$ are called *attributes* and must be single-valued. Each attribute has an *attribute name*. The set, $\mathcal{A}$, of all attribute names will be called the *attribute set*.

The individual ordered n_j-tuples in the relation, $\mathcal{T}_j$, are simply called *tuples* when the value of n_j is understood.

Note: The tuples in $\mathcal{T}_j$ are ordered tuples. Consequently, $\mathcal{T}_j$ imposes an ordering on the subset of $\mathcal{A}$ that corresponds to attributes in $\mathcal{T}_j$. Because, for $j \neq k$, $\mathcal{T}_j$ and $\mathcal{T}_k$ can have common attributes, this imposed ordering is *not* an ordering on $\mathcal{A}$ itself.

The requirement that attributes be single valued means that an attribute cannot be a set. This keeps the database model simple and eliminates some potential anomalies and inconsistencies in the database.

The previous definitions emphasize the mathematical origins. There are two other systems of designations that are commonly used. One emphasizes the matrix-like structure that can be easily imposed on the relation: a two-dimensional *table* with the ordered n-tuples as *rows* and with the ordered coordinate positions aligned as *columns*. The second alternative emphasizes the need to store the database on a computer. In this view, each relation is stored as a separate *file*.[25] The tuples are considered to be *records* and the attributes are called *fields*. To add to the confusion, it is not uncommon to mix the three systems of terminology. For example, you may encounter a reference to a table along with its constituent records and attributes.

The correspondences among the three systems of terminology are shown in Table 12.1.

TABLE 12.1 Three Systems of Terminology for Relational Databases

Mathematical View	Two-Dimensional View	Computer Storage View
relation	table	file
tuple	row	record
attribute	column	field

EXAMPLE 12.14

More Family Relations

Recall Example 12.13. A relational database model for that example would consist of two ternary relations. We could call the first relation "Biological" and the second relation "Adoptive". The attributes of the Biological relation can be named "Father", "Mother", and "Biological Child", respectively. The attributes of the "Adoptive" relation can be named "Father", "Mother", and "Adoptive Child", respectively. A tuple in the "Adoptive" relation might then be (John Petersen, Mary Petersen, Chae-Ok Petersen).

Astute readers may have noticed a potential problem with this relational database: It is likely that there will be multiple identical tuples in the relation. Mathematically, this is not possible, since relations are themselves sets. However, this is a practical likelihood. For example, there may be a tuple (John Smith, Kathy Smith, Tom Smith)

[25] This is not necessarily true, but combining distinct relations within a single file is an implementation detail that is not important when considering the theoretical model.

that represents a family in America and another (identical) tuple that represents a family in England. Additional terminology is needed to resolve this problem. ■

12.3.3 Functional Dependence; Models and Instances

DEFINITION 12.24 ***Functional Dependence (Informal)***
Let $\mathcal{T}$ be a relation and A be the attribute set of $\mathcal{T}$. Let $B \in \mathsf{A}$ and suppose that $\{A_1, A_2, \ldots, A_j\} \subseteq \mathsf{A}$ with $j \geq 1$. The attribute B is *functionally dependent* on $\{A_1, A_2, \ldots, A_j\}$ if every distinct choice of values for the attributes $A_1, A_2, \ldots, A_j$ uniquely determines the value of B.

A more formal definition of functional dependence will be given on page 756.

In Example 12.14, the values of "Father" and "Mother" in the relation "Biological" do *not* uniquely determine the value of "Biological Child" (since parents can have several biological children). Therefore, "Biological Child" is *not* functionally dependent on {Father, Mother}. On the other hand, both "Father" and "Mother" are functionally dependent on "Biological Child".

EXAMPLE 12.15 **Functional Dependence**

Let $\mathcal{T}$ be a relation with attribute set {Student, GPA, Honors}, where "Honors" has values in the set {none, cum laude, magna cum laude, summa cum laude}. Since honors rankings are determined by the value of a student's GPA, the attribute "Honors" is functionally dependent on the attribute "GPA". If a GPA of 3.6 is in the cum laude range, then *every* student with a GPA of 3.6 will have cum laude as the value of the third attribute.

Note that "GPA" is not functionally dependent on "Honors", since knowing that a student will graduate cum laude does not determine whether the GPA is 3.6 or 3.69 or some other value in the range determined by the college for cum laude. ■

The final sentence prior to Example 12.15 makes the assertion that both "Father" and "Mother" are functionally dependent on "Biological Child". The basis for that assertion is our understanding of human biology: each human has a unique biological father and mother. Selecting a specific human makes it possible (in theory) to determine the actual biological father and mother.

However, the tuple (John Smith, Kathy Smith, Tom Smith) in Example 12.14 indicates that additional thought may be needed to ensure that actual instances of relations conform to our schema. In this case, the name, Tom Smith, does not sufficiently distinguish between two distinct biological entities. The solution to this particular problem will be introduced in Section 12.3.4.

The important idea at present is the distinction between a theoretical model (the schema) for how attributes in a relation interact and an actual table (an instance) containing tuples for that relation.

In Example 12.15, the functional dependence of "Honors" on "GPA" is established by our knowledge of the inner workings of academia. It is not because in a particular table containing student names, GPA values, and honors designations it is always possible to associate a particular honors designation to a fixed range of GPA values.

Functional Dependence is a Property of a Schema

The notion of functional dependence is determined by the schema, not by an instance of a relation.

Consider Example 12.15 again. Even though "Honors" is functionally dependent on "GPA", the exact nature of that dependence will vary from year to year and from school to school. That is, the GPA range for "cum laude" may be 3.60 – 3.74 this year and 3.58 – 3.76 next year. This information must be imposed on the schema to reflect constraints determined by some real-world entity (the Registrar's Office in this case).

In the same way, corporation A may decide that "salary" is functionally dependent upon "job title", "years of service", "highest academic degree", whereas corporation B may determine that salary is functionally dependent upon "job title", "highest academic degree", and "performance review".

Sometimes functional dependencies in a schema are easy to determine from the policies of the institution that is implementing the database (such as the GPA and salary examples just discussed). In other cases, it is helpful to build some experimental tables to build insight into the kinds of tuples that might actually be encountered. From these experiments, some additional functional dependencies might be inferred. You might also rule out some functional dependencies that you think ought to exist.

For instance, it might at first seem reasonable that a book's title and author(s) should be functionally dependent on the book's ISBN number. However, this presents a problem if you are to start populating a relation with this information: many books were published before ISBN numbers were invented. Those books indicate that the simple schema must be revised.

12.3.4 Keys; Operations on Relations

Recall the tuple (John Smith, Kathy Smith, Tom Smith) in Example 12.14. Two distinct families both generate that tuple, but a relation can only contain one copy of a tuple. One way to correct this problem is to add one or more additional attributes. Those attributes will cause the two tuples to be distinct. We could perhaps add a "location" attribute. However, there is a better approach. Add some attributes which will guarantee that every distinct biological parent–child instance will generate a distinct tuple. The additional attributes are called a *key*.

DEFINITION 12.25 ***Key; Primary Key; Alternate Key; Nonkey Attribute***
Let $\mathcal{T}$ be a relation and A be the attribute set of $\mathcal{T}$. A nonempty subset, $\mathsf{K} = \{A_1, A_2, \ldots, A_j\}$, of A is called a *key* for $\mathcal{T}$ if

1. All attributes in the set difference $\mathsf{A} - \mathsf{K}$ are functionally dependent on K.
2. No nonempty proper subset of K has property 1.

If a key consists of only one attribute, B, it is customary to speak of the key B instead of the key $\{B\}$. If there is more than one key, one of them is chosen to be the *primary key*, and the other choices are demoted to the status of *alternate keys*. An attribute that is not part of the primary key is called a *nonkey attribute*.

Note: A key is one of the constraints specified by the schema. It must apply to every possible instance of the relation.

In Example 12.14, Biological Child should be a primary key for the Biological relation. In Example 12.15, the only potential primary key would be the Student attribute. If this attribute takes on unique values (such as a student ID) rather than values that might appear in other tuples (such as a name), then it will be a primary key; otherwise, there is no primary key for the relation. The next example hints at the kinds of problems that the absence of a primary key creates.

EXAMPLE 12.16 **Driver's Licenses**

Let $\mathcal{T}$ be a relation that has attribute set {Last, First, Middle, Birthdate, Driver's License Number, Street Address, City, State, Zip}. It is tempting to choose Driver's License Number as the primary key. This would be a mistake in many states. The algorithms used to determine driver's license numbers do not always assign a unique number to each person.[26] Imagine the trouble you might encounter if you share a driver's license number with someone who has a tendency to drive recklessly while intoxicated. If your name appears earlier in the state's database, the tickets might be sent to you instead of the real offender.

In states that do not have unique driver's license numbers, at least one other attribute must be added to form the primary key. The Birthdate attribute might be a good choice. It is not mathematically certain that {Driver's License Number, Birthdate} is truly a primary key, but the probability of failure is extremely low.

A better solution would be to use an algorithm that *does* generate unique driver's license numbers. ■

It is now possible to resolve the potential practical problem encountered in Example 12.14. This practical problem was caused by real-life databases possibly having identical tuples that represent distinct entities in the database. The solution is simple. When a relational database schema is chosen, it must be designed so that there is a primary key. If no such key arises naturally, some kind of unique ID must be assigned to each entity represented in the relation.

In Example 12.14, a unique ID number could be assigned to each person. Then the names could be replaced by those ID numbers. A second relation with attributes {ID, LastName, FirstName} could be added so that reports could be generated that are easier for humans to read. If computer storage space is not an issue, the ID number could be a numeric representation of each person's DNA sequence. Uniqueness would be guaranteed (but at a high cost - both in storage and in the cost of obtaining DNA sequences and in ensuring that the information was entered correctly into the database).

In practice, unique ID numbers are often generated by first predicting how many digits will allow a sufficient number of IDs for the future needs of the database. Then the database software will use the next available number in a sequence that starts at the lowest allowed number. So, for example, if a small business determines that at most 10,000 customers will ever be encountered, they could use ID numbers in either the range 0–9,999 or the range 10,000–19,999.

The next list summarizes the features of relational databases that have been presented so far. The two-dimensional terminology is used as the primary vocabulary. This choice may help reinforce the mathematical terminology given in Definition 12.23.

Features of a Relational Database

- A relational database is a collection of relations, each of which can be visualized as a two-dimensional table.
- The columns in a table represent attributes.
- The rows in a table are ordered tuples of attributes that represent a unique entity.
- There are no duplicate rows.
- The entry in each row-column intersection is single valued.
- The order of the rows is unimportant.
- Each table has a primary key, consisting of one or more attributes.

[26]See [40] for more complete details. Some related information can be found at the "Driver's License" link at http://www.mathcs.bethel.edu/~gossett/DiscreteMathWithProof/ .

✔ Quick Check 12.6

Assume that the following table represents a relation with an unknown schema.

Variety	Vegetable	Germination	Harvest
Bush Champion	Cucumber	7–14 days	55 days
Cool Kitty	Cucumber	8–12 days	60 days
Nantes Half Long	Carrot	7–21 days	70 days
What's Up Doc?	Carrot	7–14 days	75 days
Cherry Belle	Radish	7–14 days	22 days
Cherry Bomb	Radish	7–14 days	20 days

1. Determine the attribute set.
2. Determine all functional dependencies consistent with the information in the table.
3. What is a good choice for the primary key (using the available information)? ✔

There are several very useful operations that may be applied to relations in a relational database. The operations that are most relevant for this discussion are introduced next.

DEFINITION 12.26 ***Projection; Join***

Let $\mathcal{T}_1$ be an n_1-ary relation with attribute set A_1 and $\mathcal{T}_2$ be an n_2-ary relation with attribute set A_2. If $\{B_1, B_2, \ldots, B_j\} \subseteq \mathsf{A}_1$, then the *projection* of $\mathcal{T}_1$ onto $\{B_1, B_2, \ldots, B_j\}$ is the relation obtained by

1. removing from each tuple in $\mathcal{T}_1$ the components that do not correspond to an attribute in $\{B_1, B_2, \ldots, B_j\}$,
2. and then removing any duplicate tuples (keeping one copy).

The projection of a relation, $\mathcal{T}$, onto attributes $\mathsf{B} = \{B_1, B_2, \ldots, B_j\}$ is denoted by $\mathcal{T}[B_1, B_2, \ldots, B_j]$ or $\mathcal{T}[\mathsf{B}]$. Similar notation is used to denote the projection of a single tuple in $\mathcal{T}$ onto the attribute set $\{B_1, B_2, \ldots, B_j\}$.

The *join*, $\mathcal{T}_1 * \mathcal{T}_2$, of $\mathcal{T}_1$ and $\mathcal{T}_2$ is a relation having attribute set $\mathsf{B} = \mathsf{A}_1 \cup \mathsf{A}_2$, with an ordering imposed on B. Assume that the ordered attributes in the union, B, are $B_1, B_2, \ldots, B_n$. Then

$$\mathcal{T}_1 * \mathcal{T}_2 = \{r \in B_1 \times B_2 \times \cdots \times B_n \mid \exists r_1 \in \mathcal{T}_1 \text{ and } \exists r_2 \in \mathcal{T}_2 \text{ with } r[\mathsf{A}_1] = r_1 \text{ and } r[\mathsf{A}_2] = r_2\}$$

Joins and projections are of practical benefit when applied to relation instances. When applied to relation instances, the resulting relations will be new instances. When applied to instances, the Cartesian product, $B_1 \times B_2 \times \cdots \times B_n$, in the definition of *join* is constructed using the values of each attribute that actually appear in the relation instances, $\mathcal{T}_1$ and $\mathcal{T}_2$, and not the set of all potential values for the attributes. The result will a finite table (but often a very large table).

The join defined in Definition 12.26 is also called the *natural join*. A simple algorithmic procedure can create a join:

- Form a Cartesian product from $\mathcal{T}_1$ and $\mathcal{T}_2$. That is, make a table with one row for each distinct pair of rows in $\mathcal{T}_1$ and $\mathcal{T}_2$. Each new row will consist of a tuple from $\mathcal{T}_1$ concatenated with a tuple from $\mathcal{T}_2$. If $\mathcal{T}_1$ has n rows and $\mathcal{T}_2$ has m rows, then the new table will have nm rows. There may be columns with duplicate attribute names.

- Remove all rows in the new table for which the duplicate columns do not have identical values. (This step corresponds to the requirement $\exists r_1 \in \mathcal{T}_1$ and $\exists r_2 \in \mathcal{T}_2$ with $r[\mathsf{A}_1] = r_1$ and $r[\mathsf{A}_2] = r_2$ in Definition 12.26. That definition does not start with duplicate columns.)
- Form the attribute set (as a true set, with no duplicates). Project onto the attribute set (thus eliminating one copy of each duplicate column).

A short example should help clarify the definitions of projection and join.

EXAMPLE 12.17 **Projection and Join with Family Relations**

The relations in Example 12.13 on page 749 contain a very large number of tuples. To make this example simpler, suppose that the two relations, "Biological" and "Adoptive", contain only the tuples shown in Tables 12.2 and 12.3.

TABLE 12.2 Biological

Father	Biological Mother	Biological Child
John Smith	Jane Smith	William Smith
John Smith	Jane Smith	Susan Smith
Esteban Rodriguez	Anita Rodriguez	Pablo Rodriguez
Walter Leblanc	Miranda Leblanc	Wanda Leblanc
Robert Westlund	Virginia Westlund	Derwin Westlund
Robert Westlund	Virginia Westlund	Darwin Westlund

TABLE 12.3 Adoptive

Father	Adoptive Mother	Adoptive Child
John Smith	Jane Smith	Carmen Smith
John Smith	Jane Smith	Polly Smith
Esteban Rodriguez	Anita Rodriguez	Tran-minh Rodriguez
Isaac Levitz	Helen Levitz	Aaron Levitz
Isaac Levitz	Helen Levitz	Hanna Levitz
Bob Jones	Betty Jones	Samantha Jones

The projection of Biological onto {Father, Mother} is the relation containing all couples who are (jointly) biological parents (Table 12.4).

TABLE 12.4 Biological[Father, Mother]

Biological[Father,Mother]	
Father	**Mother**
John Smith	Jane Smith
Esteban Rodriguez	Anita Rodriguez
Walter Leblanc	Miranda Leblanc
Robert Westlund	Virginia Westlund

The join Biological * Adoptive is essentially the relation containing all pairs of biological and adoptive children having the same parents (Table 12.5).

TABLE 12.5 Biological * Adoptive

Biological * Adoptive			
Father	**Mother**	**Biological Child**	**Adoptive Child**
John Smith	Jane Smith	William Smith	Carmen Smith
John Smith	Jane Smith	William Smith	Polly Smith
John Smith	Jane Smith	Susan Smith	Carmen Smith
John Smith	Jane Smith	Susan Smith	Polly Smith
Esteban Rodriguez	Anita Rodriguez	Pablo Rodriguez	Tran-minh Rodriguez

More precisely, the join is the set of all tuples in Father × Mother × Biological Child × Adoptive Child for which the projection onto {Father, Mother, Biological Child} is a tuple in Biological, and the projection onto {Father, Mother, Adoptive Child} is a tuple in Adoptive. This rules out tuples such as (Isaac Levitz, Helen Levitz, Wanda Leblanc, Aaron Levitz) because the projection, (Isaac Levitz, Helen Levitz, Wanda Leblanc), onto {Father, Mother, Biological Child} is not in Biological. ■

✔ Quick Check 12.7

Use the relation instances defined in Example 12.17 to complete the following tasks.

1. Form the projection Adoptive[Mother, Adoptive Child].
2. Use the alternative procedure (outlined immediately after Definition 12.26) to produce the join, Biological * Adoptive.
3. The join in part (2) of this Quick Check is not terribly useful. However, it is an intermediate step for producing a more useful relation. Form (Biological * Adoptive)[Father, Mother]; that is, first form the join, and then project onto {Father, Mother}. Also, describe what this new relation represents. ☑

Projections can be used to provide a more precise definition of functional dependence.

DEFINITION 12.27 *Functional Dependence (Formal)*
Let $\mathcal{T}$ be a relation and A be the attribute set of $\mathcal{T}$. Let $B \in \mathsf{A}$ and $\{A_1, A_2, \ldots, A_j\} \subseteq \mathsf{A}$ with $j \geq 1$. The attribute, B, is *functionally dependent* on $\{A_1, A_2, \ldots, A_j\}$ if for every pair of tuples, r_1 and r_2 in $\mathcal{T}$,

$$r_1[A_1, A_2, \ldots, A_j] = r_2[A_1, A_2, \ldots, A_j] \quad \text{implies} \quad r_1[B] = r_2[B].$$

The notation

$$\{A_1, A_2, \ldots, A_j\} \to B$$

indicates that B is functionally dependent on $\{A_1, A_2, \ldots, A_j\}$.

If $\{A_1, A_2, \ldots, A_j\} \to B_i$ for $i = 1, 2, \ldots, n$ is a collection of functional dependencies on a common set of attributes, the shorthand notation

$$\{A_1, A_2, \ldots, A_j\} \to \{B_1, B_2, \ldots, B_n\}$$

may be used.

12.3.5 Normal Forms

If care is not taken in the formation of a relational database schema, some serious deficiencies can be introduced. In particular, if too much redundancy is designed into the relations, there can be problems if a piece of information needs to be modified or deleted. The following example illustrates such a problem.

EXAMPLE 12.18 **Anomalies in a Relational Database Schema**

The relation in Table 12.6 might arise if the registrar's office is attempting to keep track of course assignments together with useful information to help someone contact the instructor.

TABLE 12.6 Schedule

Schedule					
Course	**Section**	**Semester**	**Instructor**	**Office**	**Phone**
MAT241	1	Fall	Gossett	CC 224	x6131
MAT241	2	Fall	Gossett	CC 224	x6131
MAT124M	1	Fall	Pederson	CC 225	x6348
MAT124M	2	Fall	Kinney	CC 229	x6532
MAT124M	3	Fall	Pederson	CC 225	x6348

Suppose Professor Pederson decides to change her last name to Conrath. It is possible that the person who modifies the database might be in a hurry and forget to change every occurrence of the name. This will result in inconsistencies in the database.

Note also that some other information is repeated multiple times (for example, the fact that Professor Gossett is in CC 224 and has phone extension 6131). This redundancy requires the database to take more space than is necessary.

There is another potential problem with this design. Suppose section 2 of MAT124M does not have a sufficient number of students, so the registrar's office decides to cancel section 2. There is some possibility (depending on the other relations in the database) that deleting this tuple may result in the loss of the information placing Professor Kinney in CC 229 with phone extension 6532. ■

The solution to the kinds of problems introduced in Example 12.18 is to cleverly decompose the database into a collection of relations that exhibit some desirable properties. These properties are codified in a sequence of *normal forms*. Before introducing some normal forms, the decomposition process needs to be discussed.

DEFINITION 12.28 ***Decomposition; Lossless Decomposition***

Let $\mathcal{T}$ be a relation with attribute set $\mathsf{A} = \mathsf{D}_1 \cup \mathsf{D}_2 \cup \ldots \cup \mathsf{D}_n$, where the subsets, D_i, of attributes are not necessarily disjoint. The set of relations $\{\mathcal{T}[\mathsf{D}_1], \mathcal{T}[\mathsf{D}_2], \ldots, \mathcal{T}[\mathsf{D}_n]\}$ is a *decomposition* of $\mathcal{T}$. If, in addition, $\mathcal{T} = \mathcal{T}[\mathsf{D}_1] * \mathcal{T}[\mathsf{D}_2] * \cdots * \mathcal{T}[\mathsf{D}_n]$, the decomposition is called a *lossless decomposition*.

EXAMPLE 12.19

A Lossless Decomposition

Suppose the relation, Schedule, defined in Table 12.6 from Example 12.18 is decomposed as

{Schedule[Course, Section, Semester, Instructor],
Schedule[Instructor, Office, Phone]}

Tables 12.7 and 12.8 show these projections.

TABLE 12.7 Schedule[Course, Section, Semester, Instructor]

Schedule[Course, Section, Semester, Instructor]			
Course	**Section**	**Semester**	**Instructor**
MAT241	1	Fall	Gossett
MAT241	2	Fall	Gossett
MAT124M	1	Fall	Pederson
MAT124M	2	Fall	Kinney
MAT124M	3	Fall	Pederson

TABLE 12.8 Schedule[Instructor, Office, Phone]

Schedule[Instructor, Office, Phone]		
Instructor	**Office**	**Phone**
Gossett	CC 224	x6131
Pederson	CC 225	x6348
Kinney	CC 229	x6532

This decomposition is lossless because

Schedule = Schedule[Course, Section, Semester, Instructor] *
Schedule[Instructor, Office, Phone] ■

EXAMPLE 12.20

A Lossy Decomposition

Suppose the relation, Schedule, in Table 12.6 is decomposed as
{Schedule[Course, Section, Semester], Schedule[Course, Instructor, Office, Phone]}.

The two projections are shown in Tables 12.9 and 12.10.

TABLE 12.9 Schedule[Course, Section, Semester]

Schedule[Course, Section, Semester]		
Course	**Section**	**Semester**
MAT241	1	Fall
MAT241	2	Fall
MAT124M	1	Fall
MAT124M	2	Fall
MAT124M	3	Fall

TABLE 12.10 Schedule[Course, Instructor, Office, Phone]

Schedule[Course, Instructor, Office, Phone]			
Course	**Instructor**	**Office**	**Phone**
MAT241	Gossett	CC 224	x6131
MAT241	Gossett	CC 224	x6131
MAT124M	Pederson	CC 225	x6348
MAT124M	Kinney	CC 229	x6532
MAT124M	Pederson	CC 225	x6348

This decomposition is lossy[27] because

$$\begin{aligned}\text{Schedule} \neq\ & \text{Schedule[Course, Section, Semester]} * \\ & \text{Schedule[Course, Instructor, Office, Phone]}\end{aligned}$$

In fact, the join (Table 12.11) contains tuples that should be excluded. For instance, Professor Kinney is *not* scheduled to teach section 1 of MAT124M.

TABLE 12.11 Schedule[Course, Section, Semester] * Schedule[Course, Instructor, Office, Phone]

Schedule[Course, Section, Semester] * Schedule[Course, Instructor, Office, Phone]					
Course	**Section**	**Semester**	**Instructor**	**Office**	**Phone**
MAT241	1	Fall	Gossett	CC 224	x6131
MAT241	2	Fall	Gossett	CC 224	x6131
MAT124M	1	Fall	Pederson	CC 225	x6348
MAT124M	2	Fall	Pederson	CC 225	x6348
MAT124M	3	Fall	Pederson	CC 225	x6348
MAT124M	1	Fall	Kinney	CC 229	x6532
MAT124M	2	Fall	Kinney	CC 229	x6532
MAT124M	3	Fall	Kinney	CC 229	x6532

■

The primary goal for the rest of this section is to introduce first, second, and third normal forms. Brief mention will be made of Boyce Codd normal form. The next definition starts the process. It uses notation introduced in Definition 12.27 on page 756.

DEFINITION 12.29 ***First, Second, and Third Normal Forms***
Let $\mathcal{T}$ be a relation with attribute set A.

First Normal Form $\mathcal{T}$ is in *first normal form* if every attribute in $\mathcal{T}$ is single valued.

Second Normal Form $\mathcal{T}$ is in *second normal form* if it is in first normal form and if for all subsets of attributes $\mathsf{D} \subseteq \mathsf{A}$ and for all attributes $B \notin \mathsf{D}$ with B not in any key:

$\mathsf{D} \to B$ implies that D is not properly contained in any key of $\mathcal{T}$.

Third Normal Form $\mathcal{T}$ is in *third normal form* if it is in first normal form and if for all subsets of attributes $\mathsf{D} \subseteq \mathsf{A}$ and for all attributes $B \notin \mathsf{D}$ with B not in any key:

$$\mathsf{D} \to B \quad \text{implies that } \mathsf{D} \text{ contains some key of } \mathcal{T}.$$

Notice that Definition 12.23 requires relations in a relational database to be in first normal form. Exercise 10 on page 771 asserts that any relation that is in third normal form is also in second normal form.

The intuitive motivation for second normal form is the observation that having a functional dependency on a proper subset of a key is undesirable. The requirement for third normal form just strengthens this insight by eliminating dependencies on attributes that are not keys or supersets of keys.

[27] The term *lossy* does not mean that some tuples in the original relation may be lost. It means that some of the *information* in the relation is lost. In this case, information about who is teaching various sections has been lost.

EXAMPLE 12.21 **First Normal Form**

Table 12.12 represents a relation instance from a database design that is not in first normal form.

TABLE 12.12 ScheduleA

ScheduleA					
Course	**Section**	**Semester**	**Instructor**	**Office**	**Teaching Assistant**
MAT241	1	Fall	Gossett & Turnquist	CC 224 & CC 226	Nielsen
MAT241	2	Fall	Gossett	CC 224	Nielsen
MAT124M	1	Fall	Conrath	CC 225	Dowdey
MAT124M	2	Fall	Kinney	CC 229	Ness
MAT124M	1	Spring	Conrath	CC 225	Dowdey
MAT422	2	Fall	Kinney	CC 229	Ness

The problem, of course, is that section 1 of MAT241 is team taught, resulting in attributes that are not single valued.

The solution is simple: Break tuples with multivalued attributes into a set of tuples. For this example, one additional tuple suffices (Table 12.13).

TABLE 12.13 ScheduleB

ScheduleB					
Course	**Section**	**Semester**	**Instructor**	**Office**	**Teaching Assistant**
MAT241	1	Fall	Gossett	CC 224	Nielsen
MAT241	1	Fall	Turnquist	CC 226	Nielsen
MAT241	2	Fall	Gossett	CC 224	Nielsen
MAT124M	1	Fall	Conrath	CC 225	Dowdey
MAT124M	2	Fall	Kinney	CC 229	Ness
MAT422	2	Fall	Kinney	CC 229	Ness
MAT124M	1	Spring	Conrath	CC 225	Dowdey

■

The discussion in the remainder of this section will benefit from an informal definition. Let $\mathcal{T}$ be a relation. If $\{\mathsf{A}_1, \mathsf{A}_2, \mathsf{A}_3\} \to \mathsf{A}_4$ is a functional dependence for $\mathcal{T}$, then clearly a functional dependence of the form $\{\mathsf{A}_1, \mathsf{A}_2, \mathsf{A}_3, \mathsf{A}_4\} \to \mathsf{B}$ can be reduced to $\{\mathsf{A}_1, \mathsf{A}_2, \mathsf{A}_3\} \to \mathsf{B}$. This is because $\{\mathsf{A}_1, \mathsf{A}_2, \mathsf{A}_3\} \to \mathsf{A}_4$ ensures that if $r_1, r_2 \in \mathcal{T}$, then $r_1[A_1, A_2, A_3] = r_2[A_1, A_2, A_3]$ implies $r_1[A_4] = r_2[A_4]$. Thus $r_1[A_1, A_2, A_3, A_4] = r_2[A_1, A_2, A_3, A_4]$ if and only if $r_1[A_1, A_2, A_3] = r_2[A_1, A_2, A_3]$.

DEFINITION 12.30 ***Essential Dependencies***

Let $\mathcal{M}$ be the schema for a relation and let $\mathcal{F}$ be the set of all functional dependencies determined by $\mathcal{M}$. Let $\mathcal{E} \subseteq \mathcal{F}$ be the subset of functional dependencies of the form $\{A_1, A_2, \ldots, A_j\} \to B$ where for $i = 1, 2, \ldots, j$, A_i is *not* functionally dependent on $\{A_1, A_2, \ldots, A_j\} - \{A_i\}$. The set $\mathcal{E}$ is called the set of *essential dependencies*.

EXAMPLE 12.22

Second Normal Form

The relation in Example 12.18 on page 757 was shown to result in update and deletion anomalies. One indicator that such problems will occur for a relation is that the relation is not in second normal form.

Consider the relation, ScheduleB, from Example 12.21 (Table 12.13). That relation is in first normal form, but not in second normal form. To see this, consider the functional dependencies that can be inferred from the table. The essential dependencies are shown in Table 12.14.[28]

TABLE 12.14 Essential Dependencies Inferred from ScheduleB

{Course, Section, Semester, Instructor} → Office

{Course, Section, Semester, Instructor} → Teaching Assistant

{Course, Section, Semester, Office} → Instructor

{Course, Section, Semester, Office} → Teaching Assistant

Instructor → Office

Instructor → Teaching Assistant

Office → Teaching Assistant

Office → Instructor

To verify that {Course, Section, Semester, Instructor} → Office is an essential dependency, try removing any one of the attributes on the left and show that it is *not* functionally dependent on the rest.

The dependencies indicate that {Course, Section, Semester, Instructor} can be chosen as the primary key. There is one alternate key: {Course, Section, Semester, Office}.

Notice that D = {Instructor} is properly contained in a key, but D → Teaching Assistant is true. Thus, this relation is *not* in second normal form.

You may have questioned the dependencies Instructor → Teaching Assistant and Office → Teaching Assistant. There are two possibilities: the table reflects a deliberate policy that each professor is assigned only one teaching assistant, or else the table is too small to demonstrate that a professor may indeed have multiple teaching assistants. ■

If a relation is not in second normal form (but is in first normal form), there are algorithms that will transform it into an equivalent relation that *is* in second normal form. See [70] or most other database textbooks for details. The algorithms typically create a lossless decomposition to achieve their goal. The approach in this book will be to introduce an algorithm that will generate a decomposition whose relations are in third normal form.

EXAMPLE 12.23

Third Normal Form

Suppose the registrar's office has decreed that team-taught sections will no longer be permitted. In addition, an instructor is allowed to have different teaching assistants for different sections or semesters of a course. However, teaching assistants may only work

[28]There are other dependencies, such as {Instructor, Office} → Teaching Assistant, which are not listed. They have been omitted because they are not essential dependencies. For instance, Instructor → Office, so {Instructor, Office} → Teaching Assistant should be omitted. The inclusion of both {Course, Section, Semester, Instructor} → Office and Instructor → Office is because neither has any internal dependencies on the left-hand side.

for one instructor. The following table conforms to these policies.

TABLE 12.15 ScheduleC

ScheduleC

Course	Section	Semester	Instructor	Office	Teaching Assistant
MAT241	1	Fall	Gossett	CC 224	Nielsen
MAT241	2	Fall	Gossett	CC 224	Nielsen
MAT124M	1	Fall	Conrath	CC 225	Dowdey
MAT124M	2	Fall	Kinney	CC 229	Ness
MAT422	2	Fall	Kinney	CC 229	Ness
MAT124M	1	Spring	Conrath	CC 225	Berg

The essential dependencies in Table 12.16 conform to this relation instance.

TABLE 12.16 Essential Dependencies for ScheduleC

{Course, Section, Semester} → Instructor
{Course, Section, Semester} → Office
{Course, Section, Semester} → Teaching Assistant
Instructor → Office
Teaching Assistant → Instructor
Teaching Assistant → Office
Office → Instructor

The primary (and only) key is {Course, Section, Semester}. This relation is in second normal form. This follows since the left-hand side of every dependency is either the entire primary key (and hence not a proper subset of the primary key), or is a nonkey attribute.

This relation is not, however, in third normal form. To see this, set D = {Teaching Assistant}. Notice that D → Instructor is true, but D does not contain any key in the relation. ■

✔ Quick Check 12.8

For each relation listed, determine whether it is in first, second, or third normal form (focus on the relation instance, not the unspecified schema). Assume that there are never players with the same name in the same position on the same team and that job grades have predetermined salaries.

1.

League Rosters

Team	Player	Position	Captain
Mud Hens	Casey	first base	yes
Mud Hens	O'Reilly	pitcher	no
Mud Hens	Issacson	pitcher	no
Mud Hens	Johnson	catcher	no
Mud Hens	Johnson	right field	no
⋮	⋮	⋮	⋮
Prairie Chickens	Svenson	pitcher	no
Prairie Chickens	Johnson	shortstop	yes
Prairie Chickens	Johnson	catcher	no
Prairie Chickens	Hidalgo	left field	no
⋮	⋮	⋮	⋮

2.

Salaries

EmployeeID	Name	Job Grade	Salary
1214	John Chen	GS-11	\$50,000
1225	Mary Thompson	GS-10	\$45,000
1309	Sue Witkowski	GS-11	\$50,000
1356	Ahmed Mosse	GS-12	\$55,000
1443	John Chen	GS-9	\$40,000
1455	Yolanda Roberts	GS-9.5	\$40,000
⋮	⋮	⋮	⋮

✔

An algorithm is needed that will convert a relation (in first or second normal form) into a lossless decomposition of relations that are each in third normal form. The corollary to the next theorem (from [38]) will be useful.

THEOREM 12.31 ***A Lossless Decomposition***

Let $\mathcal{T}$ be a relation with attribute set A, and let $\mathsf{X} \subseteq \mathsf{A}$ and $\mathsf{Y} \subseteq \mathsf{A}$ with $\mathsf{X} \cap \mathsf{Y} = \emptyset$. Set $\mathsf{Z} = \mathsf{A} - (\mathsf{X} \cup \mathsf{Y})$. If $\mathsf{X} \to \mathsf{Y}$, then the two projections, $\mathcal{T}[\mathsf{X} \cup \mathsf{Y}]$ and $\mathcal{T}[\mathsf{X} \cup \mathsf{Z}]$, form a lossless decomposition of $\mathcal{T}$.

Proof: In the expressions that follow, x, y, z are tuples of values for the sets of attributes, $\mathsf{X}, \mathsf{Y}, \mathsf{Z}$, respectively. Assume that the attributes in A are ordered with all attributes in X appearing first, all attributes in Y appearing in the middle, and all attributes in Z appearing at the end. Then

$$\mathcal{T}[\mathsf{X} \cup \mathsf{Y}] * \mathcal{T}[\mathsf{X} \cup \mathsf{Z}] = \{(x, y, z) \mid x \in \mathsf{X}, y \in \mathsf{Y}, z \in \mathsf{Z} \text{ and } (x, y) \in \mathcal{T}[\mathsf{X} \cup \mathsf{Y}] \text{ and } (x, z) \in \mathcal{T}[\mathsf{X} \cup \mathsf{Z}]\}$$

Since x, y, and z are chosen independently, it is possible that $\mathcal{T}[\mathsf{X} \cup \mathsf{Y}] * \mathcal{T}[\mathsf{X} \cup \mathsf{Z}]$ may contain tuples that are not in $\mathcal{T}$. It is certainly true that $\mathcal{T} \subseteq \mathcal{T}[\mathsf{X} \cup \mathsf{Y}] * \mathcal{T}[\mathsf{X} \cup \mathsf{Z}]$.

Suppose that $(x, y, z) \in \mathcal{T}[\mathsf{X} \cup \mathsf{Y}] * \mathcal{T}[\mathsf{X} \cup \mathsf{Z}]$. It is not certain that $(x, y, z) \in \mathcal{T}$. However, it is true that $(x, z) \in \mathcal{T}[\mathsf{X} \cup \mathsf{Z}]$. But (x, z) is a projection of some tuple in $\mathcal{T}$, so there is some $y' \in \mathsf{Y}$ such that $(x, y', z) \in \mathcal{T}$. But then $(x, y') \in \mathcal{T}[\mathsf{X} \cup \mathsf{Y}]$.

Thus, $(x, y) \in \mathcal{T}[\mathsf{X} \cup \mathsf{Y}]$ and $(x, y') \in \mathcal{T}[\mathsf{X} \cup \mathsf{Y}]$. In addition, $\mathsf{X} \to \mathsf{Y}$. But then Definition 12.27 implies that $y = (x, y)[\mathsf{Y}] = (x, y')[\mathsf{Y}] = y'$ (because $(x, y)[\mathsf{X}] = (x, y')[\mathsf{X}]$). Consequently, $(x, y, z) = (x, y', z) \in \mathcal{T}$. Thus, $\mathcal{T}[\mathsf{X} \cup \mathsf{Y}] * \mathcal{T}[\mathsf{X} \cup \mathsf{Z}] \subseteq \mathcal{T}$.

The two subset inclusions establish the validity of the theorem. □

COROLLARY 12.32

Let $\mathcal{T}$ be a relation with attribute set A, and let $\mathsf{X} \subseteq \mathsf{A}$ and $\mathsf{Y} \subseteq \mathsf{A}$. Set $\mathsf{Z} = \mathsf{A} - (\mathsf{X} \cup \mathsf{Y})$. If $\mathsf{X} \to \mathsf{Y}$, then the two projections, $\mathcal{T}[\mathsf{X} \cup \mathsf{Y}]$ and $\mathcal{T}[\mathsf{X} \cup \mathsf{Z}]$, form a lossless decomposition of $\mathcal{T}$.

Proof: The corollary does not assume $\mathsf{X} \cap \mathsf{Y} = \emptyset$. Thus, suppose $\mathsf{X} \cap \mathsf{Y} \neq \emptyset$. Set $\mathsf{Y}' = \mathsf{Y} - \mathsf{X}$. Then $\mathsf{X} \cap \mathsf{Y}' = \emptyset$, $\mathsf{X} \cup \mathsf{Y}' = \mathsf{X} \cup \mathsf{Y}$, and $\mathsf{Z} = \mathsf{A} - (\mathsf{X} \cup \mathsf{Y}) = \mathsf{A} - (\mathsf{X} \cup \mathsf{Y}')$.

The theorem asserts that $\mathcal{T}[\mathsf{X} \cup \mathsf{Y}]$ $(= \mathcal{T}[\mathsf{X} \cup \mathsf{Y}'])$ and $\mathcal{T}[\mathsf{X} \cup \mathsf{Z}]$ form a lossless decomposition of $\mathcal{T}$. □

Converting Relations to Collections in Third Normal Form

Suppose $\mathcal{T}$ is a relation in a relational database that is not in third normal form. Since it is in a relational database, it is already in first normal form.

Let A be the attribute set of $\mathcal{T}$. Since $\mathcal{T}$ is not in third normal form, there must be a subset, $\mathsf{D} \subseteq \mathsf{A}$, and an attribute, $B \notin \mathsf{D}$, which is not part of any key, such that $\mathsf{D} \to B$, but D does not contain any key.

Let $\mathsf{X} = D$ and $\mathsf{Y} = \{B\}$ and $\mathsf{Z} = \mathsf{A} - (\mathsf{D} \cup \{B\})$. With these identifications, $\mathsf{D} \cup \mathsf{Z} = \mathsf{D} \cup (\mathsf{A} - (\mathsf{D} \cup \{B\})) = \mathsf{A} - \{B\}$, so Corollary 12.32 implies that the relations $\mathcal{T}[\mathsf{D} \cup \{B\}]$ and $\mathcal{T}[\mathsf{A} - \{B\}]$ form a lossless decomposition of $\mathcal{T}$.

Notice that $\mathsf{D} \cup \{B\} \neq \mathsf{A}$ because otherwise D would be a key for $\mathcal{T}$. Thus, both $\mathcal{T}[\mathsf{D} \cup \{B\}]$ and $\mathcal{T}[\mathsf{A} - \{B\}]$ are relations with fewer attributes than $\mathcal{T}$ has. If both these new relations are in third normal form, the goal has been achieved. Otherwise, the process can be repeated. Since the number of attributes in each new partition is strictly fewer than the number of attributes in the parent relation, the process must terminate after a finite number of steps.

The process is summarized in the following algorithm.

Convert to Third Normal Form

```
1. convert tables into first normal form (as necessary) to form a relational database
2. while there are relations in the database that are not in third normal form
   a. choose a relation, T, with attribute set A, which is not in third normal form
   b. find a subset, D, of attributes in T and an attribute,
        B ∉ D, where B is not in any key, but D → B
   c. replace T with T[D ∪ {B}] and T[A − {B}]
```

The following theorem formally summarizes the main consequence of the process just outlined:

THEOREM 12.33 ***Any Relational Database can be Converted to Third Normal Form***

Let $\{\mathcal{T}_1, \mathcal{T}_2, \ldots, \mathcal{T}_n\}$ be a relational database. Then the relations, $\mathcal{T}_i$, can be losslessly decomposed into a collection of relations that are each in third normal form.

Note: The decomposition algorithm can (and should) be used before any tuples reside in tables. It is a process that should be applied to the schema.

However, there is pedagogical value in using the algorithm with relation instances. That approach will be used in the next few examples.

EXAMPLE 12.24 **Converting First Normal Form to Third Normal Form**

The relation, ScheduleB, is in first normal form. It is repeated in Table 12.17. The essential dependencies were found in Example 12.22 and are repeated in Table 12.18.

TABLE 12.17 ScheduleB

ScheduleB					
Course	**Section**	**Semester**	**Instructor**	**Office**	**Teaching Assistant**
MAT241	1	Fall	Gossett	CC 224	Nielsen
MAT241	1	Fall	Turnquist	CC 226	Nielsen
MAT241	2	Fall	Gossett	CC 224	Nielsen
MAT124M	1	Fall	Conrath	CC 225	Dowdey
MAT124M	2	Fall	Kinney	CC 229	Ness
MAT422	2	Fall	Kinney	CC 229	Ness
MAT124M	1	Spring	Conrath	CC 225	Dowdey

TABLE 12.18 Essential Dependencies for ScheduleB

{Course, Section, Semester, Instructor} → Office

{Course, Section, Semester, Instructor} → Teaching Assistant

{Course, Section, Semester, Office} → Instructor

{Course, Section, Semester, Office} → Teaching Assistant

Instructor → Office

Instructor → Teaching Assistant

Office → Teaching Assistant

Office → Instructor

The algorithm for converting to third normal form starts by finding a subset, D, of attributes that does not contain a key and another attribute, $B \notin \mathsf{D}$, which is not in any key and which is functionally dependent on D. One such choice is D = {Instructor} and B = Teaching Assistant.

The next step is to create the lossless decomposition {ScheduleB[Instructor, Teaching Assistant], ScheduleB[Course, Section, Semester, Instructor, Office]} (Tables 12.19 and 12.20).

TABLE 12.19 ScheduleB[Instructor, Teaching Assistant]

ScheduleB[Instructor, Teaching Assistant]	
Instructor	**Teaching Assistant**
Gossett	Nielsen
Turnquist	Nielsen
Conrath	Dowdey
Kinney	Ness

TABLE 12.20 ScheduleB[Course, Section, Semester, Instructor, Office]

ScheduleB[Course, Section, Semester, Instructor, Office]				
Course	**Section**	**Semester**	**Instructor**	**Office**
MAT241	1	Fall	Gossett	CC 224
MAT241	1	Fall	Turnquist	CC 226
MAT241	2	Fall	Gossett	CC 224
MAT124M	1	Fall	Conrath	CC 225
MAT124M	2	Fall	Kinney	CC 229
MAT422	2	Fall	Kinney	CC 229
MAT124M	1	Spring	Conrath	CC 225

Both of these new relations are in third normal form. The essential dependencies in ScheduleB[Course, Section, Semester, Instructor, Office] are as follows:

{Course, Section, Semester, Instructor} → Office

{Course, Section, Semester, Office} → Instructor

Instructor → Office

Office → Instructor.

The keys are {Course, Section, Semester, Instructor} and {Course, Section, Semester, Office}. There is no choice for B that is not in some key.

Notice that Instructor → Office still exists in this table. That can be eliminated by moving to Boyce–Codd normal form (Definition 12.34 on page 768 and Exercise 17 on page 772). ■

✔ **Quick Check 12.9**

1. Show that the relation ScheduleB[Instructor, Office, Teaching Assistant] is in third normal form. (Showing another way to losslessly decompose ScheduleB into two third normal form relations.) ☑

In the final decomposition in Example 12.24, Instructor is the primary key for the relation ScheduleB[Instructor, Teaching Assistant]. It also appears in the relation ScheduleB[Course, Section, Semester, Instructor, Office] as a *foreign key*.[29] Modifying or deleting an instructor's name will require changes to both tables. However, since Instructor always appears as either a primary key or as a foreign key, software implementations of relational databases can typically update all instances automatically.

EXAMPLE 12.25

Converting Second Normal Form to Third Normal Form

The relation, ScheduleC, (reproduced as Table 12.21) is already in second normal form. In Example 12.23 it was demonstrated that $\mathsf{D} = \{\text{Teaching Assistant}\}$ and $B = \text{Instructor}$ show that ScheduleC is not in third normal form. The conversion algorithm suggests that ScheduleC can be decomposed into ScheduleC$[\mathsf{D} \cup \{B\}]$ and ScheduleC$[\mathsf{A} - \{B\}]$.

TABLE 12.21 ScheduleC

ScheduleC					
Course	**Section**	**Semester**	**Instructor**	**Office**	**Teaching Assistant**
MAT241	1	Fall	Gossett	CC 224	Nielsen
MAT241	2	Fall	Gossett	CC 224	Nielsen
MAT124M	1	Fall	Conrath	CC 225	Dowdey
MAT124M	2	Fall	Kinney	CC 229	Ness
MAT422	2	Fall	Kinney	CC 229	Ness
MAT124M	1	Spring	Conrath	CC 225	Berg

Thus, the algorithm suggests the decomposition into ScheduleC[Teaching Assistant, Instructor] (Table 12.25 on page 767) and ScheduleC[Course, Section, Semester, Office, Teaching Assistant] (Table 12.22).

TABLE 12.22 ScheduleC[Course, Section, Semester, Office, Teaching Assistant]

ScheduleC[Course, Section, Semester, Office, Teaching Assistant]				
Course	**Section**	**Semester**	**Office**	**Teaching Assistant**
MAT241	1	Fall	CC 224	Nielsen
MAT241	2	Fall	CC 224	Nielsen
MAT124M	1	Fall	CC 225	Dowdey
MAT124M	2	Fall	CC 229	Ness
MAT422	2	Fall	CC 229	Ness
MAT124M	1	Spring	CC 225	Berg

ScheduleC[Course, Section, Semester, Office, Teaching Assistant] is not in third normal form. To see this, list the inferred essential dependencies, which would not be present if we worked with a schema instead of a relation instance.

$$\{\text{Course, Section, Semester}\} \rightarrow \text{Office}$$
$$\{\text{Course, Section, Semester}\} \rightarrow \text{Teaching Assistant}$$
$$\text{Teaching Assistant} \rightarrow \text{Office}$$

[29] A *foreign key* is an attribute (or set of attributes) that is a primary key in some other relation in the relational database.

The functional dependency Teaching Assistant → Office shows that this relation is not in third normal form.

Create two new relations, ScheduleC[Course, Section, Semester, Teaching Assistant] (Table 12.23) and ScheduleC[Teaching Assistant, Office] (Table 12.24) .

TABLE 12.23 ScheduleC[Course, Section, Semester, Teaching Assistant]

ScheduleC[Course, Section, Semester, Teaching Assistant]			
Course	**Section**	**Semester**	**Teaching Assistant**
MAT241	1	Fall	Nielsen
MAT241	2	Fall	Nielsen
MAT124M	1	Fall	Dowdey
MAT124M	2	Fall	Ness
MAT422	2	Fall	Ness
MAT124M	1	Spring	Berg

Both of these new relations are in third normal form. The primary key for the first is {Course, Section, Semester}. There are no other choices for primary key and also no functional dependencies with D not the primary key. The second new relation has only one functional dependency (Teaching Assistant → Office) whose left-hand side is its primary key. They must therefore be in third normal form.

TABLE 12.24 ScheduleC[Teaching Assistant, Office]

ScheduleC[Teaching Assistant, Office]	
Teaching Assistant	**Office**
Nielsen	CC 224
Dowdey	CC 225
Ness	CC 229
Berg	CC 225

The relation, ScheduleC[Teaching Assistant, Instructor] in Table 12.25 has only one functional dependency: Teaching Assistant → Instructor. The primary key is therefore Teaching Assistant and this relation is in third normal.

TABLE 12.25 ScheduleC[Teaching Assistant, Instructor]

ScheduleC[Teaching Assistant, Instructor]	
Teaching Assistant	**Instructor**
Nielsen	Gossett
Dowdey	Conrath
Ness	Kinney
Berg	Conrath

The three relations,

ScheduleC[Course, Section, Semester, Teaching Assistant] (Table 12.23)
ScheduleC[Teaching Assistant, Office] (Table 12.24) and
ScheduleC[Teaching Assistant, Instructor] (Table 12.25)

form a third normal form decomposition for ScheduleC.

If we had focused on a schema (instead of an instance), the decomposition would more likely use Instructor rather than Teaching Assistant as the foreign key. ■

Other Normal Forms

Although third normal form prevents many potential anomalies from entering a relational database, it does not prevent all such problems. One undesirable property is having an attribute be functionally dependent on something less than a key. Table 12.20 on page 765 presents an example of this.

EXAMPLE 12.26

Beyond Third Normal Form

Quick Check 12.8 introduced a relation (Table 12.26) with team rosters. Suppose that instead of listing whether a player is the captain, it lists the player's Contract ID. League Contract ID's are tied to both team and status, where status is either rookie or veteran.

TABLE 12.26 League rosters

	League Rosters		
Team	**Player**	**Position**	**Contract ID**
Mud Hens	Casey	first base	mhr
Mud Hens	O'Reilly	pitcher	mhv
Mud Hens	Issacson	pitcher	mhv
Mud Hens	Johnson	catcher	mhv
Mud Hens	Johnson	right field	mhr
⋮	⋮	⋮	⋮
Prairie Chickens	Svenson	pitcher	pcr
Prairie Chickens	Johnson	shortstop	pcv
Prairie Chickens	Johnson	catcher	pcr
Prairie Chickens	Hidalgo	left field	pcv
⋮	⋮	⋮	⋮

The essential functional dependencies are as follows.

$$\{\text{Team, Player, Position}\} \to \text{Contract ID}$$
$$\text{Contract ID} \to \text{Team}$$

This relation is in third normal form because Team is not a nonkey attribute. However the functional dependency Contract ID → Team may be considered undesirable since Contract ID is a nonkey attribute. ■

The undesirable feature in this example can be removed by decomposing the relation into {(League Rosters)[Contract ID, Team], (League Rosters)[Player, Position, Contract ID]}. The primary keys are Contract ID and {Player, Position, Contract ID}, respectively. The relations in this decomposition are in what is called Boyce–Codd normal form.

DEFINITION 12.34 ***Boyce–Codd Normal Form***

Let $\mathcal{T}$ be a relation in a relational database with attribute set A. Let $\mathsf{D} \subseteq \mathsf{A}$ and $B \notin \mathsf{D}$. Then $\mathcal{T}$ is in *Boyce–Codd normal form* if $\mathsf{D} \to B$ implies that D contains some key of $\mathcal{T}$.

The change from third normal form is the elimination of the requirement that B not be an attribute in any key.

For more information on Boyce–Codd normal form, see [73] or [38].

It is tempting to assume that the sequence of normal forms will ultimately reach a form that guarantees that all undesirable features have been eliminated from the database schema. Unfortunately, this is not the case. The higher normal forms remove some additional problems but eventually lead to problems with preserving functional dependencies. There are also trade-offs that must be considered. For example, some database designers might choose to give up third normal form for some relations in the database to reduce the number of relations. For instance, it is not uncommon to leave both Zip Code and City as attributes in a larger relation, even though City is functionally dependent on Zip Code, and Zip Code may not be a key in the relation.

Both theoretical and practical research continues in this area. See [38] for a mathematically oriented overview.

12.3.6 Exercises

The exercises marked with ★ have detailed solutions in Appendix H.

1. Which of the following tables could be instances of relations in a relational database? If the table could represent a relation in a relational database, list the attribute set.

(a) ★ **Cast List**

Character	Actor	Understudy
Hamlet	Helmut Weiner	John Garner, Sam Ranier
Claudius	Richard Gunther	Bob Searle
Ophelia	Suzanne Bonner	Sally Richards
Gertrude	Virginia Smith	no understudy

(b) **Teaching Assignments**

Course	Section	Semester	Instructor
MAT241	1	Fall	Gossett
MAT241	2	Fall	Gossett
MAT222	1	Spring	Kinney
MAT124M	1	Fall	Conrath
MAT124M	2	Fall	Kinney
MAT124M	3	Fall	Conrath

(c) **My Movie Log**

Movie	Where Watched
Ponette	Video at home
Fellowship of the Ring	Ritz Theater
Secret of Roan Inish	Theatre Leo
Fellowship of the Ring	Ritz Theater
Princess Mononoke	DVD at home

2. Prove Proposition 12.35.

PROPOSITION 12.35
Let $\mathcal{T}$ be a relation with attribute set A. If $\{A_1, A_2, \ldots, A_j\} \subseteq \mathsf{A}$ and $B \in \{A_1, A_2, \ldots, A_j\}$, then B is functionally dependent on $\{A_1, A_2, \ldots, A_j\}$.

3.
- Determine the functional dependencies in the following relations. You will need to infer the schema. Ignore trivial dependencies, such as Proposition 12.35. Do not include dependencies with more attributes than necessary. For example, if B is functionally dependent on $\{A_1, A_2\}$, then it is also functionally dependent on $\{A_1, A_2, C\}$, for any attribute C. List only the smaller set of attributes.
- Then list a good choice for the primary key.

Only the attribute sets are provided in some cases.

(a) {Course, Section, Semester, Year, Instructor} (Assume that the database only stores information for current or previous time periods and that there is only one instructor per section.)

(b) {Book Title, Author, Publisher, Edition} (You may assume there is only a single author in each tuple.)

(c) {Composition, Composer, Original Instrumentation, Composition Date, Original Performance Date} (Assume that there will never be two composers with the same name and composition.)

(d) Partial information from *The Lady Bird Johnson Wildflower Center Master Plant List* has been used for this table (current Web address available at www.mathcs.bethel.edu/~gossett/DiscreteMathWithProof/).

The Lady Bird Johnson Wildflower Center Master Plant List

Genus/ Species	Common Name	Family
Abutilon fruticosum	Indian Mallow	Malvaceae
Abutilon incanum	Indian Mallow	Malvaceae
Abutilon incanum	Pelotazo	Malvaceae
⋮	⋮	⋮
Aesculus pavia	Red Buckeye	Hippocastanaceae
Aesculus pavia var. pavia	Red Buckeye	Hippocastanaceae
⋮	⋮	⋮

4. ★ Form the projection (Teaching Assignments)[Course, Instructor].

Teaching Assignments

Course	Section	Semester	Instructor
MAT241	1	Fall	Gossett
MAT241	2	Fall	Gossett
MAT222	1	Spring	Kinney
MAT124M	1	Fall	Conrath
MAT124M	2	Fall	Kinney
MAT124M	3	Fall	Conrath

5. Form the following projections for the truncated version of Chores.
   (a) Chores[Day, Person]
   (b) Chores[Task, Person]—Sort the result by Task, and within identical tasks sort by Person.

Chores

Day	Task	Person
Sunday	cook	Sue
Sunday	dishes	Franka
Monday	vacuum	Beth
Monday	cook	Franka
Monday	dishes	Beth
Monday	shop	Sue
Tuesday	dust	Beth
Tuesday	cook	Sue
Tuesday	dishes	Sue

6. Form the join for the following pairs of relations.
   (a) ★

Yearbook

Student	Task	Homeroom
Joe	Advertisements	H4
Martha	Activities	G3
Kim	Teacher Photos	H4
Rosa	Student Photos	D5

Newspaper

Student	Feature	Cohort
Amelia	News	Junior
Wesley	City Page	Sophomore
Kim	Photos	Senior
Rosa	Editorials	Junior
Bob	Sports	Sophomore

   (b)

Employees

Employee ID	Name	Salary Level
21457	Said Sachdev	S3
21490	Millie Volk	S1
21688	Dee Delacroix	S2
22000	June Hebert	S4

Salaries

Salary Level	Yearly Pay
S1	$20,000
S2	$25,000
S3	$30,000
S4	$90,000

7. Form the join for the following pairs of relations.
   (a)

Teaching Assignments

Course	Section	Instructor
MAT241	1	Gossett
MAT241	2	Gossett
MAT223	1	Wetzell
MAT124M	1	Conrath
MAT124M	2	Kinney
MAT124M	3	Conrath

Room Assignments

Course	Section	Room
MAT241	1	CC325
MAT241	2	CC325
MAT223	1	CC431
MAT124M	1	CLC109
MAT124M	2	RC424
MAT124M	3	AC203

   (b)

Creation

Composer	Composition	Date
John Newton	Amazing Grace	c. 1770
Ludwig van Beethoven	Moonlight Sonata	1801
Johann Sebastian Bach	Brandenburg Concertos	1721

Performance

Artist/ Orchestra	Composer	Composition
Charlotte Church	John Newton	Amazing Grace
Judy Collins	John Newton	Amazing Grace
Alan Schiller	Ludwig van Beethoven	Für Elise
Amsterdam Baroque	Johann Sebastion Bach	Brandenburg Concertos

8. Prove the following proposition about the composition of projections.

> **PROPOSITION 12.36**
> Let $\mathcal{T}$ be a relation with attribute set $\{A_1, \ldots, A_j, B_1, \ldots, B_k, C_1, \ldots, C_n\}$, $j, k, n \geq 1$.
> Set $\mathcal{T}' = \mathcal{T}[A_1, \ldots, A_j, B_1, \ldots, B_k]$.
> Then $\mathcal{T}'[A_1, \ldots, A_j] = \mathcal{T}[A_1, \ldots, A_j]$.

9. Prove the following proposition about the natural join. [*Hint*: use Proposition 12.36 for associativity.]

> **PROPOSITION 12.37**
> Let $\mathcal{T}_1$, $\mathcal{T}_2$, and $\mathcal{T}_3$ be relations with attribute sets A_1, A_2, A_3, respectively. Impose orderings on $\mathsf{A}_1 \cup \mathsf{A}_2$, $\mathsf{A}_2 \cup \mathsf{A}_3$, and $\mathsf{A}_1 \cup \mathsf{A}_2 \cup \mathsf{A}_3$. Then
> - The join operator is commutative: $\mathcal{T}_1 * \mathcal{T}_2 = \mathcal{T}_2 * \mathcal{T}_1$.
> - The join operator is associative: $(\mathcal{T}_1 * \mathcal{T}_2) * \mathcal{T}_3 = \mathcal{T}_1 * (\mathcal{T}_2 * \mathcal{T}_3)$.
> - If $\mathsf{A}_1 \cap \mathsf{A}_2 = \emptyset$, then $\mathcal{T}_1 * \mathcal{T}_2 = \mathcal{T}_1 \times \mathcal{T}_2$.

10. Prove that any relation that is in third normal form is also in second normal form.

11. For each relation,

- List the essential dependencies. (You will need to infer the schema.)
- Determine whether it is in first, second, or third normal form. Your answer should reflect the highest form that applies. Give adequate reasons to justify your answer.

(a) Assume that "Katia and Marielle Labèque" is a single-valued entry for this relational database.

CD-1

UPC	Title	Artist
743215911227	The Ultimate Recorder Collection	Michala Petri
093624742623	A Day Without Rain	Enya
075992677424	Watermark	Enya
038146202125	Kalevala: Dream of the Salmon Maiden	Ruth MacKenzie
074645256825	Encore!	Midori
074644838121	Encore!	Katia and Marielle Labèque
⋮	⋮	⋮

(b)

CD-2

Title	Artists
The Daemon Lover	Custer LaRue, The Baltimore Consort
The Mad Buckgoat	The Baltimore Consort
Trio	Dolly Parton, Linda Ronstadt, Emmylou Harris
Crouching Tiger Hidden Dragon	Tan Dun, Yo-Yo Ma
⋮	⋮

(c) ★ Assume that all widgets are stored in Warehouse W and all flanges are stored in Warehouse F.

Widgets and Flanges

Part ID	Part Name	Part Location
W1256	Widget (metric)	Warehouse W
W1257	Widget (metric)	Warehouse W
W2276	Widget (English)	Warehouse W
F4	Flange (4 inch)	Warehouse F
F6	Flange (6 inch)	Warehouse F
⋮	⋮	⋮

(d) Assume that there will never be two composers with the same name.

Compositions

Title	Composer	Composer's Birthdate
Scheherazade	Nikolai Rimsky-Korsakov	1844
Scheherazade	Maurice Ravel	1875
Concierto de Aranjuez	Joaquín Rodrigo	1902
Fantasia para un gentilhombre	Joaquín Rodrigo	1902
Concerto for four harpsichords	Johann Sebastian Bach	1685
Messiah	George Frederic Handel	1685
⋮	⋮	⋮

12. Using the relations in Exercise 11 as the base relations, which of the following decompositions are lossless? Justify your answers.

(a) {CD-1[UPC, Title], CD-1[Title, Artist]}
(b) {CD-1[UPC, Title], CD-1[UPC, Artist]}
(c) ★ {(Widgets and Flanges)[Part ID], (Widgets and Flanges)[Part Name, Part Location]}
(d) {Compositions[Title, Composer], Compositions[Composer, Composer's Birthdate}

13. Convert the relation, Salaries, from Quick Check 12.8 (page 762) into third normal form. Use the algorithm on page 764.

14. ★ Use the algorithm on page 764 to convert each relation in Exercise 11 into a set of relations in third normal form.

15. Suppose your mathematics department wishes to create a relational database that lists students who are willing to tutor other students. The database will contain the following information: the tutor's first and last names, the tutor's student ID number, the courses the tutor is qualified to tutor, and the tutor's hourly fee. A tutor may be qualified to tutor multiple courses and may charge different fees for different courses.
   (a) Design a good schema for this relational database. All relations should be in third normal form.
   (b) List the essential dependencies and keys.
   (c) Prove that each relation is in third normal form.

16. Consider the following relation.

Members

ID	Name	Initials
44	Joe Smith	JS
51	Carl Carlson	CC
52	Betty Boop	BB
64	Carl Carlson	CC
75	Bob Burquist	BB

   (a) List the inferred essential dependencies.
   (b) What are the possible primary keys?
   (c) Which normal form is this table in? (List the highest form.)
   (d) If the relation is *not* in third normal form, convert it to a collection of tables in third normal form whose join is the original table. If it *is* in third normal form, attempt to convert it into a collection of tables in Boyce–Codd normal form whose join is the original table.

17. Example 12.24 on page 764 introduced the projection ScheduleB[Course, Section, Semester, Instructor, Office]. This relation is in third normal form.
   (a) Show that it is not in Boyce–Codd normal form.
   (b) Show that the projections ScheduleB[Course, Section, Semester, Instructor] and ScheduleB[Instructor, Office] *are* in Boyce–Codd normal form.

18. A more realistic course registration database would need to allow multiple professors to have the same name. Suppose that the registrar's office has chosen to use Name and Office to uniquely identify professors (rather than the more sensible decision to assign unique ID numbers). Assume also that offices and phone numbers can be shared, and that more than one phone can be assigned to an office, but that professors with the same name are never assigned to the same office.

Instructor Info

Name	Office	Phone	Teaching Assistant
Gossett	CC 224	x6131	Nielsen
Conrath	CC 224	x6131	Ness
Kinney	CC 224	x6532	Nygren
Brown	CC 224	x6532	Ness
Jones	HC 414K	x6335	Anderson
Jones	AC 123	x6312	Nelson

   (a) Show that Instructor Info is in third normal form.
   (b) Choose {Name, Office} as primary key. Show that there is an attribute in the primary key that is functionally dependent on a nonkey attribute.
   (c) Convert this relation to a pair of relations in Boyce–Codd normal form.

19. Each of the following statements is either true (always) or false (at least sometimes). Determine which option applies for each statement and provide adequate explanation for your choice.
   (a) It is always better to use a higher normal form than to use a lower normal form.
   (b) ★ A relation, $\mathcal{T}$, can always be recovered from a lossless decomposition (of $\mathcal{T}$).
   (c) Redundancy in a relational database is usually a desirable feature.
   (d) ★ If nonempty relations, $\mathcal{R}$ and $\mathcal{T}$, have no common attributes, then the join, $\mathcal{R} * \mathcal{T}$, is empty.

20. Prove that functional dependence is transitive on attributes. That is, show that $X \to Y$ and $Y \to Z$ implies that $X \to Z$ for attributes X, Y, Z.

21. Let $\mathcal{T}$ be a relation that is in third normal form. If the primary key is the only key in the relation, prove that $\mathcal{T}$ is also in Boyce–Codd normal form.

12.4 Boolean Functions and Boolean Expressions

The main goal of this section is to develop some of the primary theory relating to functions whose domain and range is a Boolean algebra. This development will culminate in a canonical form, called *disjunctive normal form*, for Boolean functions. A secondary goal is to develop an algorithm for converting a Boolean expression into disjunctive normal form.

One application of Boolean functions is minimization of logic circuits. This will be introduced in Section 12.5.

It is easy to create functions whose domain is a Boolean algebra. Just start with some variables whose values are elements of the Boolean algebra whose associated set is $\{0, 1\}$ and form a Boolean expression. For example, the Boolean expression

$(x + y) \cdot (\overline{x} + y)$ becomes a Boolean function of the two variables, x and y, if the values 0 and 1 are substituted for x and y. If this function is denoted as $f(x, y)$, then $f(0, 1) = (0 + 1) \cdot (\overline{0} + 1) = 1$.

12.4.1 Boolean Functions

DEFINITION 12.38 ***Single-Variable Boolean Function***
Let $\mathbb{B}$ be a Boolean algebra with associated set, B. A *single-variable Boolean function* is a function whose domain is B and whose range is $\{0, 1\}$.

EXAMPLE 12.27 **A Single-Variable Boolean Function**

Let $\mathbb{B}$ be the Boolean algebra defined in Example 2.30 on page 61. Recall that the associated set is $B = \{\emptyset, \{a\}, \{b\}, \{a, b\}\}$. A single-variable Boolean function on $\mathbb{B}$ can be specified by showing its action for each element in B.

x	$f(x)$
$\emptyset$	0
$\{a\}$	0
$\{b\}$	1
$\{a, b\}$	0

■

One easy-to-answer question is, How many single-variable Boolean functions are there with domain B? The answer depends on the size of the associated set.

THEOREM 12.39 ***The Number of Single-Variable Boolean Functions on*** $\mathbb{B}$
Let $\mathbb{B}$ be a Boolean algebra with associated set B. If $|B| = m$, then there are 2^m distinct single-variable Boolean functions on $\mathbb{B}$.

Proof: Two functions are distinct if they disagree on at least one element in B. A function can independently assign either 0 or 1 to each element in B. Since there are m elements, each with two possible values, there are 2^m distinct functions. □

EXAMPLE 12.28 **All Single-Variable Boolean Functions**

Table 12.27 shows all 16 possible Boolean functions for the Boolean algebra in Example 2.30.

TABLE 12.27 All 16 Boolean Functions for the Boolean Algebra with Associated Set $B = \{\emptyset, \{a\}, \{b\}, \{a, b\}\}$

x	f_1	f_2	f_3	f_4	f_5	f_6	f_7	f_8	f_9	f_{10}	f_{11}	f_{12}	f_{13}	f_{14}	f_{15}	f_{16}
$\emptyset$	0	0	0	0	0	0	0	0	1	1	1	1	1	1	1	1
$\{a\}$	0	0	0	0	1	1	1	1	0	0	0	0	1	1	1	1
$\{b\}$	0	0	1	1	0	0	1	1	0	0	1	1	0	0	1	1
$\{a, b\}$	0	1	0	1	0	1	0	1	0	1	0	1	0	1	0	1

■

Single-variable Boolean functions are not as interesting as multivariable Boolean functions, which are defined next.

DEFINITION 12.40 ***Multivariable Boolean Function***

Let $\mathbb{B}$ be a Boolean algebra with associated set, B. An *n-variable Boolean function* is a function whose domain is $\overbrace{B \times B \times \cdots \times B}^{n \text{ times}}$ and whose range is $\{0, 1\}$.

EXAMPLE 12.29

A Multivariable Boolean Function

Let $\mathbb{B}$ be the Boolean algebra defined in Example 2.30. A two-variable Boolean function on $\mathbb{B}$ can be specified by showing its action for each pair of elements in B. Table 12.28 shows one such function. ■

TABLE 12.28 A Two-Variable Boolean Function on $B = \{\emptyset, \{a\}, \{b\}, \{a, b\}\}$

x	y	$f(x, y)$
$\emptyset$	$\emptyset$	0
$\{a\}$	$\emptyset$	0
$\{b\}$	$\emptyset$	1
$\{a, b\}$	$\emptyset$	0
$\emptyset$	$\{a\}$	0
$\{a\}$	$\{a\}$	1
$\{b\}$	$\{a\}$	1
$\{a, b\}$	$\{a\}$	0
$\emptyset$	$\{b\}$	1
$\{a\}$	$\{b\}$	0
$\{b\}$	$\{b\}$	1
$\{a, b\}$	$\{b\}$	1
$\emptyset$	$\{a, b\}$	0
$\{a\}$	$\{a, b\}$	1
$\{b\}$	$\{a, b\}$	1
$\{a, b\}$	$\{a, b\}$	0

The proof of the following theorem is left as an exercise.

THEOREM 12.41 ***The Number of Multivariable Boolean Functions on $\mathbb{B}$***

Let $\mathbb{B}$ be a Boolean algebra with associated set, B. If $|B| = m$, then there are $2^{m^n} = 2^{(m^n)}$ distinct n-variable Boolean functions on $\mathbb{B}$.

The most important case is when the Boolean algebra is the one defined in Example 2.29 on page 60. For that example, $B = \{0, 1\}$. Since the remainder of this section will mainly be interested in this case, another definition will be helpful.

DEFINITION 12.42 ***Binary Function***

A *binary function of order n* is an n-variable Boolean function on the Boolean algebra whose associated set is $B = \{0, 1\}$.

The counting theorem for this definition is a corollary of Theorem 12.41.

COROLLARY 12.43 ***The Number of Binary Functions***

There are $2^{2^n} = 2^{(2^n)}$ distinct binary functions of order n.

The collection of all possible binary functions of order 2 is shown in the next example.

EXAMPLE 12.30

All Binary Functions of Order 2

Table 12.29 shows all 16 possible binary functions of order 2. Compare this table with Table 12.27 on page 773.

TABLE 12.29 All 16 Binary Functions of Order 2

x	y	f_1	f_2	f_3	f_4	f_5	f_6	f_7	f_8	f_9	f_{10}	f_{11}	f_{12}	f_{13}	f_{14}	f_{15}	f_{16}
0	0	0	0	0	0	0	0	0	0	1	1	1	1	1	1	1	1
0	1	0	0	0	0	1	1	1	1	0	0	0	0	1	1	1	1
1	0	0	0	1	1	0	0	1	1	0	0	1	1	0	0	1	1
1	1	0	1	0	1	0	1	0	1	0	1	0	1	0	1	0	1

■

12.4.2 Binary Functions and Disjunctive Normal Form

Up to this point, there has been no attempt to produce algebraic expressions that define a Boolean function. It is time to consider this for binary functions. Recall Definition 2.27 on page 59. This definition can be restricted to the case of current interest.

DEFINITION 12.44 ***Binary Variable; Binary Expression***
A *binary variable* is one whose possible values are either 0 or 1. A *binary expression* is an algebraic expression that is composed using the symbols 0, 1, +, ·, $\overline{}$, and binary variables.

EXAMPLE 12.31

A Binary Expression

The expression $0 + 1 \cdot x \cdot \overline{y}$ is a binary expression in two variables. It is instructive to evaluate this expression for each possible value of x and y.

x	y	$\mathbf{0+1\cdot x\cdot \overline{y}}$
0	0	0
0	1	0
1	0	1
1	1	0

Note that this binary expression corresponds to the function, f_3, in Example 12.30. ■

When you studied functions over the real numbers, you may have encountered functions that have no corresponding algebraic expression. For example, a function that is important in numerical analysis and probability theory is the *error function*, commonly denoted by erf(x), and defined by

$$\operatorname{erf}(x) = \frac{2}{\sqrt{\pi}} \int_0^x e^{-\frac{1}{2}t^2}\, dt$$

Even if the most common trigonometric or transcendental functions [such as e^x and $\ln(x)$] are used, there is still no simple expression that defines this function; it has no easy-to-express integration-free antiderivative.[30]

The situation for binary functions is much simpler. It will soon be shown that *every* binary function (no exceptions) can be expressed as a binary expression. Thus, the correspondence between f_3 and the binary expression in Example 12.31 was not an accident. The formal proof will depend upon the notion of a *minterm*.

DEFINITION 12.45 ***Minterm***
Let $x_1, x_2, \ldots, x_n$ be n binary variables. A *minterm* is a binary expression in the form

$$\hat{x}_1 \cdot \hat{x}_2 \cdots \hat{x}_n$$

where $\hat{x}_i$ is either x_i or $\overline{x_i}$, for $i = 1, 2, \ldots, n$. There is an assumed ordering of the variables.

In other words, a minterm is a product, in order, of the n binary variables, each appearing in either complemented or uncomplemented form. There are no summations in the expression.

[30]For more information, see the Chapter 12 section at http://www.mathcs.bethel.edu/~gossett/DiscreteMathWithProof/ .

EXAMPLE 12.32 **Some Minterms**

The following are all minterms in the 4 binary variables, x_1, x_2, x_3, x_4.

$$x_1 \cdot \overline{x_2} \cdot x_3 \cdot x_4$$
$$x_1 \cdot x_2 \cdot x_3 \cdot x_4$$
$$\overline{x_1} \cdot \overline{x_2} \cdot x_3 \cdot \overline{x_4}$$

The binary expressions in the next list are *not* minterms in x_1, x_2, x_3, x_4.

$x_1 \cdot x_3 \cdot x_4$	Missing x_2
$x_1 \cdot x_2 + x_3 \cdot x_4$	Cannot include summations
$x_2 \cdot \overline{x_1} \cdot x_3 \cdot x_4$	Variables are out of order
$\overline{x_1 \cdot x_2} \cdot x_3 \cdot \overline{x_4}$	Complement should only involve single variables ■

DEFINITION 12.46 ***Disjunctive Normal Form***

A Boolean expression is in *disjunctive normal form* if it is either a sum of distinct minterms or it is the expression, 0.

✔ Quick Check 12.10

Assume that all expressions in this Quick Check may be constructed using the binary variables, x_1, x_2 and x_3.

1. Which of the following binary expressions are minterms in x_1, x_2, x_3? If an expression is not a minterm, explain why it is not.
   (a) $1 \cdot x_1 \cdot \overline{x_2} \cdot \overline{x_3}$
   (b) 0
   (c) $\overline{x_1} \cdot \overline{x_2} \cdot \overline{x_3}$
2. Which of the following binary expressions in x_1, x_2, x_3 are in disjunctive normal form? If an expression is not in disjunctive normal form, explain why it is not.
   (a) $\overline{x_1} \cdot \overline{x_2} \cdot \overline{x_3} + x_1 \cdot \overline{x_2} \cdot \overline{x_3}$
   (b) $0 + x_1 \cdot \overline{x_2} \cdot x_3$
   (c) $\overline{x_1} \cdot \overline{x_2} \cdot \overline{x_3}$ ☑

It is now time to start proving the main results in this section. This will be done by proving a sequence of intermediate propositions. The following familiar ideas will be used.

- Two binary expressions are *equivalent* if one can be transformed into the other using the axioms and fundamental properties of Boolean algebras.
- Two binary functions are *equal* if they have identical values at each element of their common domain.

The major results that will be proved are summarized as follows:

1. Every binary function can be represented as a binary expression in disjunctive normal form.
2. Every binary expression is equivalent to a binary expression in disjunctive normal form.

These results indicate that disjunctive normal form is a *canonical form* for binary functions and binary expressions. In mathematics, a canonical form is a standard form into which all members of some class of mathematical objects can be placed. Classes of mathematical objects that have a canonical form demonstrate an underlying regularity and order that is considered desirable.

PROPOSITION 12.47 ***Evaluating Minterms***

Let $\hat{x}_1 \cdot \hat{x}_2 \cdots \hat{x}_n$ be a minterm in the binary variables, $x_1, x_2, \ldots, x_n$. Define an n-variable binary function, f, as

$$f(x_1, x_2, \ldots, x_n) = \hat{x}_1 \cdot \hat{x}_2 \cdots \hat{x}_n$$

Then f has the value 1 at only one element in its domain; it has the value 0 at all other elements of its domain. The n-tuple at which f has the value 1 is determined by setting

$$x_i = \begin{cases} 1 & \text{if } \hat{x}_i = x_i \\ 0 & \text{if } \hat{x}_i = \overline{x_i} \end{cases}$$

for $i = 1, 2, \ldots, n$.

An example will demonstrate how simple the proof will be.

EXAMPLE 12.33

A Minterm as a Function

Let $f(x_1, x_2, x_3) = x_1 \cdot \overline{x_2} \cdot x_3$. It is not hard to list the value of f at each of the 8 ordered triples in its domain. However, a simple observation will achieve the same goal. Notice that f must evaluate to 0 if any of the three factors, x_1, $\overline{x_2}$, x_3, is 0. The only way to make each of these factors evaluate to 1 is to set $x_1 = 1$, $x_2 = 0$, and $x_3 = 1$. ■

Proof of Proposition 12.47: The function, f, is defined in the n factors, $\hat{x}_1, \hat{x}_2, \ldots, \hat{x}_n$. If any one of those factors evaluates to 0, then f will also evaluate to 0. The only way to make each of the factors evaluate to 1 is to make the assignments specified in the statement of the proposition. □

The critical observation in the proof of the first main result essentially reverses Proposition 12.47. If the ordered n-tuples at which a Boolean function, f, equals 1 are known, then a set of minterms that evaluate to 1 at those n-tuples can be constructed. If these minterms are properly combined, a Boolean expression that represents f will result.[31]

THEOREM 12.48 ***Every Binary Function can be Represented in Disjunctive Normal Form***

Let f be a binary function in the binary variables, $x_1, x_2, \ldots, x_n$. Then there is a binary expression in disjunctive normal form that is equal to f when viewed as a binary function.

The proof will be easier to follow if an example is available.

EXAMPLE 12.34

From Binary Function to Binary Expression

TABLE 12.30 A Binary Function of x_1, x_2, and x_3

x_1	x_2	x_3	$f(x_1, x_2, x_3)$
0	0	0	1
0	0	1	0
0	1	0	0
0	1	1	1
1	0	0	0
1	0	1	0
1	1	0	1
1	1	1	0

Let f be a binary function in the binary variables, x_1, x_2, x_3, with values defined by Table 12.30.

There are three ordered triples at which f has the value 1. Using the main idea of Proposition 12.47, it should be clear that the three minterms, $\overline{x_1} \cdot \overline{x_2} \cdot \overline{x_3}$, $\overline{x_1} \cdot x_2 \cdot x_3$, and $x_1 \cdot x_2 \cdot \overline{x_3}$, each have the value 1 at an ordered triple where f has the value 1. Adding these minterms will produce a Boolean expression that evaluates to the same value as f at every triple in the domain. This occurs because the ordered triples, (0, 0, 0), (0, 1, 1), and (1, 1, 0), will each cause one of the terms to evaluate to 1 and the other two terms to evaluate to 0. Any other ordered triple will cause each of the minterms to evaluate to 0.

[31] The constructive proof that follows is very similar in spirit to the standard construction of a Lagrange interpolating polynomial. Consult a numerical analysis text for details.

We can thus write (informally blurring the distinction between f and the sum of minterms)

$$f(x_1, x_2, x_3) = \overline{x_1} \cdot \overline{x_2} \cdot \overline{x_3} + \overline{x_1} \cdot x_2 \cdot x_3 + x_1 \cdot x_2 \cdot \overline{x_3}$$

This expression is in disjunctive normal form. ■

Proof of Theorem 12.48: If f evaluates to 0 at every ordered n-tuple in its domain, then $f(x_1, x_2, \ldots, x_n) = 0$ is a representation of f as a Boolean expression in disjunctive normal form.

Otherwise, for each ordered n-tuple at which f evaluates to 1, define a corresponding minterm, $\hat{x}_i \cdots \hat{x}_n$, by

$$\hat{x}_i = \begin{cases} x_i & \text{if the } i\text{th coordinate of the } n\text{-tuple is 1} \\ \overline{x_i} & \text{if the } i\text{th coordinate of the } n\text{-tuple is 0} \end{cases}$$

for $i = 1, 2, \ldots, n$.

The sum of these minterms is a Boolean expression in disjunctive normal form. It has been constructed so that exactly one of the minterms in the sum will evaluate to 1 at each ordered n-tuple where f evaluates to 1. At any other ordered n-tuple, each minterm will evaluate to 0, so the sum will also evaluate to 0. □

✔ Quick Check 12.11

1. Let f be the Boolean function (with an 8-element domain) defined by the following table. Construct a Boolean expression in disjunctive normal form that represents a Boolean function that is equal to f.

x_1	x_2	x_3	$f(x_1, x_2, x_3)$
0	0	0	0
0	0	1	0
0	1	0	1
0	1	1	1

x_1	x_2	x_3	$f(x_1, x_2, x_3)$
1	0	0	0
1	0	1	0
1	1	0	1
1	1	1	1

☑

12.4.3 Binary Expressions and Disjunctive Normal Form

The sequence of propositions leading up to the second major result is patterned after the sequence that is used in [66]. It may be helpful to review the axioms and fundamental properties for Boolean algebras (see Section 2.5).

PROPOSITION 12.49 ***Moving Complements onto Single Variables***
Every binary expression is equivalent to a binary expression in which the only occurrences of the complement operator, $\overline{}$, are to complement single variables.

Proof: Every binary expression is equivalent to itself, so if every complement operator in the original expression already involves only a single variable (or if no complement operators appear), there is nothing to prove.

If either $\overline{0}$ or $\overline{1}$ appear in the expression, an equivalent expression can be obtained by using the substitutions $\overline{0} \to 1$ and $\overline{1} \to 0$.

The involution property can be used to eliminate multiple complements of the same subexpression.

Any other complements must involve subexpressions containing one (or both) of the operators, $+$ and $\cdot$. Associativity and De Morgan's laws can be used to reduce these to subexpressions in which the complements involve strictly smaller subexpressions.

These transformations can be applied recursively, generating a sequence of equivalent binary expressions. Eventually, an expression in the form asserted by the proposition must be reached because binary expressions are finite, and each transformation makes progress toward the stated goal. □

EXAMPLE 12.35

Moving Complements onto Single Variables

Consider the binary expression $\overline{1 \cdot x_1 \cdot x_2} + \overline{\overline{x_2 \cdot (x_3 + \overline{x_4})}}$. One possible sequence of transformations is shown below.

$$
\begin{aligned}
\overline{1 \cdot x_1 \cdot x_2} + \overline{\overline{x_2 \cdot (x_3 + \overline{x_4})}} &= \overline{1 \cdot x_1 \cdot x_2} + x_2 \cdot (x_3 + \overline{x_4}) && \text{involution} \\
&= \overline{1 \cdot (x_1 \cdot x_2)} + x_2 \cdot (x_3 + \overline{x_4}) && \text{associativity} \\
&= \left(\overline{1} + \overline{x_1 \cdot x_2}\right) + x_2 \cdot (x_3 + \overline{x_4}) && \text{De Morgan} \\
&= (0 + \overline{x_1 \cdot x_2}) + x_2 \cdot (x_3 + \overline{x_4}) && \text{complement of 1} \\
&= (0 + \overline{x_1} + \overline{x_2}) + x_2 \cdot (x_3 + \overline{x_4}) && \text{De Morgan}
\end{aligned}
$$

■

PROPOSITION 12.50 ***Transforming to a Sum of Products***

Let E be a binary expression in which all occurrences of the complement operator involve only single variables. Then E is equivalent to a binary expression that is a sum of products, where each product contains factors that are either one of the constants, 0 or 1, or are complemented or uncomplemented single variables. A product can consist of a single such factor.

Proof: The associativity and distributivity properties are all that is needed. Suppose there is a factor that is not one of the four acceptable entities.

If the factor is a product of acceptable entities, then the associativity property allows the factor to be ungrouped into a product of factors that are acceptable.

If the factor is the only factor in the term, and the addition operator appears one or more times, the associativity property can be used to convert the term into a sum of strictly simpler terms.

Otherwise, the factor must contain at least one addition operator and there must be at least one other factor. The distributivity property can be used to transform the term into two strictly simpler terms.

These transformations can be applied recursively, generating a sequence of equivalent binary expressions. Eventually, an expression in the form asserted by the proposition must be reached because binary expressions are finite, and each transformation makes progress toward the stated goal. □

EXAMPLE 12.36

Transforming to a Sum of Products

The final binary expression in Example 12.35 can be transformed into an equivalent binary expression that is a sum of products.

$$
\begin{aligned}
(0 + \overline{x_1} + \overline{x_2}) + x_2 \cdot (x_3 + \overline{x_4}) &= 0 + \overline{x_1} + \overline{x_2} + x_2 \cdot (x_3 + \overline{x_4}) && \text{associativity} \\
&= 0 + \overline{x_1} + \overline{x_2} + (x_2 \cdot x_3) + (x_2 \cdot \overline{x_4}) && \text{distributivity} \\
&= 0 + \overline{x_1} + \overline{x_2} + x_2 \cdot x_3 + x_2 \cdot \overline{x_4} && \text{associativity}
\end{aligned}
$$

■

PROPOSITION 12.51 ***Transforming Sums of Products to Disjunctive Normal Form***

Let E be a binary expression that is a sum of products in which each factor is either 0, 1, or a complemented or uncomplemented single variable. Then E is equivalent to a binary expression in disjunctive normal form.

Note that disjunctive normal form implies a preestablished ordering among the variables so that it is possible to discuss minterms.

Proof: If a term includes 0 as a factor, then the term can be replaced by the term containing only 0. Then all terms consisting of only the factor, 0, may be removed (using the identity axiom). The exception will be an expression that only contains a sum of terms that are each 0. In that case, all but one 0 may be removed.

If a term includes 1 as a factor and also contains other factors, the factor, 1, can be removed (using the identity axiom). If a term consists only of the constant, 1, and there are other terms, then all terms except the term consisting of only 1 may be removed (using the domination axiom).

At this point, either the entire expression is the constant, 0, and the expression is already in disjunctive normal form, or the entire expression is the constant, 1, or else the constants 0 and 1 do not appear in the expression.

Let the set of variables be $\{x_1, x_2, \ldots, x_n\}$. For the remainder of this proof, a term will be said to contain x_i if either x_i or $\overline{x_i}$ is a factor of the term. Assume also that 0 does not appear in the expression.

The expression will be transformed into disjunctive normal form using the following replacement algorithm.

```
for i = 1, 2, ..., n
  while E contains a term, t, which does not contain x_i
    replace t with t · x_i + t · x̄_i
```

The validity of the replacement step is proved below.

$$\begin{aligned} t &= t \cdot 1 && \text{identity axiom} \\ &= t \cdot (x_i + \overline{x_i}) && \text{complement axiom} \\ &= (t \cdot x_i) + (t \cdot \overline{x_i}) && \text{distributivity axiom} \\ &= t \cdot x_i + t \cdot \overline{x_i} && \text{associativity} \end{aligned}$$

At each step of the algorithm, there is a smaller total number of missing variables. Since there is a finite number of terms in the original expression and a finite number of missing variables, the process must eventually terminate with an expression in disjunctive normal form. Note that the expression, 1, is missing all n variables, but the algorithm still applies (starting with the replacement $1 = x_1 + \overline{x_1}$).

When the algorithm terminates, every term will contain all n variables. The commutativity axiom can then be used to sort the factors in each term, using the preestablished ordering of the variables. If a term contains multiple copies of x_1 (or of $\overline{x_i}$), the idempotence property can be used to remove all but one copy. If a term contains both x_i and $\overline{x_i}$, the complement axiom implies that the term is equivalent to the term, 0. If there are multiple 0 terms, all but one can be eliminated using the identity axiom.

Finally, if 0 is the only term, the expression is in disjunctive normal form. Otherwise, the resulting expression is in disjunctive normal form. □

EXAMPLE 12.37 **Transforming Sums of Products to Disjunctive Normal Form**

The final expression in Example 12.36 can be transformed into an equivalent binary expression in disjunctive normal form using the following steps.

$$0 + \overline{x_1} + \overline{x_2} + x_2 \cdot x_3 + x_2 \cdot \overline{x_4}$$

$$= \overline{x_1} + \overline{x_2} + x_2 \cdot x_3 + x_2 \cdot \overline{x_4} \qquad \text{identity}$$

$$= \overline{x_1} + \overline{x_2} \cdot x_1 + \overline{x_2} \cdot \overline{x_1} + x_2 \cdot x_3 + x_2 \cdot \overline{x_4} \qquad \text{replacement algorithm } (i = 1)$$

$$= \overline{x_1} + \overline{x_2} \cdot x_1 + \overline{x_2} \cdot \overline{x_1} + x_2 \cdot x_3 \cdot x_1 + x_2 \cdot x_3 \cdot \overline{x_1} + x_2 \cdot \overline{x_4} \qquad \text{replacement algorithm } (i = 1)$$

$$= \overline{x_1} + \overline{x_2} \cdot x_1 + \overline{x_2} \cdot \overline{x_1} + x_2 \cdot x_3 \cdot x_1 + x_2 \cdot x_3 \cdot \overline{x_1} + x_2 \cdot \overline{x_4} \cdot x_1 + x_2 \cdot \overline{x_4} \cdot \overline{x_1} \qquad \text{replacement algorithm } (i = 1)$$

$$= \overline{x_1} \cdot x_2 + \overline{x_1} \cdot \overline{x_2} + \overline{x_2} \cdot x_1 + \overline{x_2} \cdot \overline{x_1} + x_2 \cdot x_3 \cdot x_1 + x_2 \cdot x_3 \cdot \overline{x_1} + x_2 \cdot \overline{x_4} \cdot x_1 + x_2 \cdot \overline{x_4} \cdot \overline{x_1} \qquad \text{replacement algorithm } (i = 2)$$

$$= \overline{x_1} \cdot x_2 \cdot x_3 + \overline{x_1} \cdot x_2 \cdot \overline{x_3} + \overline{x_1} \cdot \overline{x_2} + \overline{x_2} \cdot x_1 + \overline{x_2} \cdot \overline{x_1} + x_2 \cdot x_3 \cdot x_1 + x_2 \cdot x_3 \cdot \overline{x_1} + x_2 \cdot \overline{x_4} \cdot x_1 + x_2 \cdot \overline{x_4} \cdot \overline{x_1} \qquad \text{replacement algorithm } (i = 3)$$

$$= \overline{x_1} \cdot x_2 \cdot x_3 + \overline{x_1} \cdot x_2 \cdot \overline{x_3} + \overline{x_1} \cdot \overline{x_2} \cdot x_3 + \overline{x_1} \cdot \overline{x_2} \cdot \overline{x_3} + \overline{x_2} \cdot x_1 + \overline{x_2} \cdot \overline{x_1} + x_2 \cdot x_3 \cdot x_1 + x_2 \cdot x_3 \cdot \overline{x_1} + x_2 \cdot \overline{x_4} \cdot x_1 + x_2 \cdot \overline{x_4} \cdot \overline{x_1} \qquad \text{replacement algorithm } (i = 3)$$

$$= \overline{x_1} \cdot x_2 \cdot x_3 + \overline{x_1} \cdot x_2 \cdot \overline{x_3} + \overline{x_1} \cdot \overline{x_2} \cdot x_3 + \overline{x_1} \cdot \overline{x_2} \cdot \overline{x_3} + \overline{x_2} \cdot x_1 \cdot x_3 + \overline{x_2} \cdot x_1 \cdot \overline{x_3} + \overline{x_2} \cdot \overline{x_1} + x_2 \cdot x_3 \cdot x_1 + x_2 \cdot x_3 \cdot \overline{x_1} + x_2 \cdot \overline{x_4} \cdot x_1 + x_2 \cdot \overline{x_4} \cdot \overline{x_1} \qquad \text{replacement algorithm } (i = 3)$$

$$= \overline{x_1} \cdot x_2 \cdot x_3 + \overline{x_1} \cdot x_2 \cdot \overline{x_3} + \overline{x_1} \cdot \overline{x_2} \cdot x_3 + \overline{x_1} \cdot \overline{x_2} \cdot \overline{x_3} + \overline{x_2} \cdot x_1 \cdot x_3 + \overline{x_2} \cdot x_1 \cdot \overline{x_3} + \overline{x_2} \cdot \overline{x_1} \cdot x_3 + \overline{x_2} \cdot \overline{x_1} \cdot \overline{x_3} + x_2 \cdot x_3 \cdot x_1 + x_2 \cdot x_3 \cdot \overline{x_1} + x_2 \cdot \overline{x_4} \cdot x_1 + x_2 \cdot \overline{x_4} \cdot \overline{x_1} \qquad \text{replacement algorithm } (i = 3)$$

$$
\begin{aligned}
&= \overline{x_1} \cdot x_2 \cdot x_3 + \overline{x_1} \cdot x_2 \cdot \overline{x_3} \\
&\quad + \overline{x_1} \cdot \overline{x_2} \cdot x_3 + \overline{x_1} \cdot \overline{x_2} \cdot \overline{x_3} \\
&\quad + \overline{x_2} \cdot x_1 \cdot x_3 + \overline{x_2} \cdot x_1 \cdot \overline{x_3} \\
&\quad + \overline{x_2} \cdot \overline{x_1} \cdot x_3 + \overline{x_2} \cdot \overline{x_1} \cdot \overline{x_3} \\
&\quad + x_2 \cdot x_3 \cdot x_1 + x_2 \cdot x_3 \cdot \overline{x_1} \\
&\quad + x_2 \cdot \overline{x_4} \cdot x_1 \cdot x_3 + x_2 \cdot \overline{x_4} \cdot x_1 \cdot \overline{x_3} \\
&\quad + x_2 \cdot \overline{x_4} \cdot \overline{x_1} \qquad \text{replacement algorithm } (i = 3)
\end{aligned}
$$

$$
\begin{aligned}
&= \overline{x_1} \cdot x_2 \cdot x_3 + \overline{x_1} \cdot x_2 \cdot \overline{x_3} \\
&\quad + \overline{x_1} \cdot \overline{x_2} \cdot x_3 + \overline{x_1} \cdot \overline{x_2} \cdot \overline{x_3} \\
&\quad + \overline{x_2} \cdot x_1 \cdot x_3 + \overline{x_2} \cdot x_1 \cdot \overline{x_3} \\
&\quad + \overline{x_2} \cdot \overline{x_1} \cdot x_3 + \overline{x_2} \cdot \overline{x_1} \cdot \overline{x_3} \\
&\quad + x_2 \cdot x_3 \cdot x_1 + x_2 \cdot x_3 \cdot \overline{x_1} \\
&\quad + x_2 \cdot \overline{x_4} \cdot x_1 \cdot x_3 + x_2 \cdot \overline{x_4} \cdot x_1 \cdot \overline{x_3} \\
&\quad + x_2 \cdot \overline{x_4} \cdot \overline{x_1} \cdot x_3 + x_2 \cdot \overline{x_4} \cdot \overline{x_1} \cdot \overline{x_3} \qquad \text{replacement algorithm } (i = 3)
\end{aligned}
$$

Another 10 steps (with $i = 4$) of the replacement algorithm yields the final expression in disjunctive normal form.

$$
\begin{aligned}
&\overline{x_1} \cdot x_2 \cdot x_3 \cdot x_4 + \overline{x_1} \cdot x_2 \cdot x_3 \cdot \overline{x_4} + \overline{x_1} \cdot x_2 \cdot \overline{x_3} \cdot x_4 + \overline{x_1} \cdot x_2 \cdot \overline{x_3} \cdot \overline{x_4} \\
&+ \overline{x_1} \cdot \overline{x_2} \cdot x_3 \cdot x_4 + \overline{x_1} \cdot \overline{x_2} \cdot x_3 \cdot \overline{x_4} + \overline{x_1} \cdot \overline{x_2} \cdot \overline{x_3} \cdot x_4 + \overline{x_1} \cdot \overline{x_2} \cdot \overline{x_3} \cdot \overline{x_4} \\
&+ \overline{x_2} \cdot x_1 \cdot x_3 \cdot x_4 + \overline{x_2} \cdot x_1 \cdot x_3 \cdot \overline{x_4} + \overline{x_2} \cdot x_1 \cdot \overline{x_3} \cdot x_4 + \overline{x_2} \cdot x_1 \cdot \overline{x_3} \cdot \overline{x_4} \\
&+ \overline{x_2} \cdot \overline{x_1} \cdot x_3 \cdot x_4 + \overline{x_2} \cdot \overline{x_1} \cdot x_3 \cdot x_4 + \overline{x_2} \cdot \overline{x_1} \cdot \overline{x_3} \cdot x_4 + \overline{x_2} \cdot \overline{x_1} \cdot \overline{x_3} \cdot \overline{x_4} \\
&+ x_2 \cdot x_3 \cdot x_1 \cdot x_4 + x_2 \cdot x_3 \cdot x_1 \cdot \overline{x_4} + x_2 \cdot x_3 \cdot \overline{x_1} \cdot x_4 + x_2 \cdot x_3 \cdot \overline{x_1} \cdot \overline{x_4} \\
&+ x_2 \cdot \overline{x_4} \cdot x_1 \cdot x_3 + x_2 \cdot \overline{x_4} \cdot x_1 \cdot \overline{x_3} + x_2 \cdot \overline{x_4} \cdot \overline{x_1} \cdot x_3 + x_2 \cdot \overline{x_4} \cdot \overline{x_1} \cdot \overline{x_3}
\end{aligned}
$$

The individual terms can now be sorted.

$$
\begin{aligned}
&\overline{x_1} \cdot x_2 \cdot x_3 \cdot x_4 + \overline{x_1} \cdot x_2 \cdot x_3 \cdot \overline{x_4} + \overline{x_1} \cdot x_2 \cdot \overline{x_3} \cdot x_4 + \overline{x_1} \cdot x_2 \cdot \overline{x_3} \cdot \overline{x_4} \\
&+ \overline{x_1} \cdot \overline{x_2} \cdot x_3 \cdot x_4 + \overline{x_1} \cdot \overline{x_2} \cdot x_3 \cdot \overline{x_4} + \overline{x_1} \cdot \overline{x_2} \cdot \overline{x_3} \cdot x_4 + \overline{x_1} \cdot \overline{x_2} \cdot \overline{x_3} \cdot \overline{x_4} \\
&+ x_1 \cdot \overline{x_2} \cdot x_3 \cdot x_4 + x_1 \cdot \overline{x_2} \cdot x_3 \cdot \overline{x_4} + x_1 \cdot \overline{x_2} \cdot \overline{x_3} \cdot x_4 + x_1 \cdot \overline{x_2} \cdot \overline{x_3} \cdot \overline{x_4} \\
&+ \overline{x_1} \cdot \overline{x_2} \cdot x_3 \cdot x_4 + \overline{x_1} \cdot \overline{x_2} \cdot x_3 \cdot x_4 + \overline{x_1} \cdot \overline{x_2} \cdot \overline{x_3} \cdot x_4 + \overline{x_1} \cdot \overline{x_2} \cdot \overline{x_3} \cdot \overline{x_4} \\
&+ x_1 \cdot x_2 \cdot x_3 \cdot x_4 + x_1 \cdot x_2 \cdot x_3 \cdot \overline{x_4} + \overline{x_1} \cdot x_2 \cdot x_3 \cdot x_4 + \overline{x_1} \cdot x_2 \cdot x_3 \cdot \overline{x_4} \\
&+ x_1 \cdot x_2 \cdot x_3 \cdot \overline{x_4} + x_1 \cdot x_2 \cdot \overline{x_3} \cdot \overline{x_4} + \overline{x_1} \cdot x_2 \cdot x_3 \cdot \overline{x_4} + \overline{x_1} \cdot x_2 \cdot \overline{x_3} \cdot \overline{x_4}
\end{aligned}
$$

The remaining steps in the proof are not needed for this example. ■

The previous example indicates that Proposition 12.51 does not accomplish everything we would like. In particular, the replacement algorithm adds unnecessary redundancy to the expression that is in disjunctive normal form. For instance, the final expression begins with the term $\overline{x_1} \cdot x_2 \cdot x_3 \cdot x_4$ but later contains another copy of that term. By using the commutativity axiom and the idempotence property, the multiple copies of $\overline{x_1} \cdot x_2 \cdot x_3 \cdot x_4$ can be combined into a single term, $\overline{x_1} \cdot x_2 \cdot x_3 \cdot x_4$. This is true in general.

THEOREM 12.52 ***Every Binary Expression is Equivalent to a Unique Expression in Disjunctive Normal Form***

Every binary expression is equivalent to a unique expression in disjunctive normal form. The uniqueness requires a preestablished lexicographical ordering of the variables.

Proof: The previous three propositions establish the assertion that a binary expression, E, is equivalent to a binary expression, D, which is in disjunctive normal form.

Proposition 12.51 assumes that some lexicographical ordering of variables exists (if the variables are subscripted, the natural subscript order can be used). The (possibly complemented) variables in each term of D are in lexicographical order. These terms are all minterms. Make the additional assumption that the uncomplemented variable, x_i, comes before its own complement, $\overline{x_i}$. With this additional ordering, the collection of minterms can be sorted.

Now use the idempotence property to remove all but one copy of any identical terms. Call the final expression C. Since there are no duplicate terms in C, each minterm has an imposed internal sorting, and there is a well-defined sort order among the minterms, C is unique. □

EXAMPLE 12.38

A Unique Expression

The overly long expression at the end of Example 12.37 can be placed into the unique form specified by the previous theorem using the natural ordering imposed by the subscripts. That expression is repeated here.

$$\begin{aligned}
&\overline{x_1}\cdot x_2\cdot x_3\cdot x_4+\overline{x_1}\cdot x_2\cdot x_3\cdot \overline{x_4}+\overline{x_1}\cdot x_2\cdot \overline{x_3}\cdot x_4+\overline{x_1}\cdot x_2\cdot \overline{x_3}\cdot \overline{x_4}\\
&+\overline{x_1}\cdot \overline{x_2}\cdot x_3\cdot x_4+\overline{x_1}\cdot \overline{x_2}\cdot x_3\cdot \overline{x_4}+\overline{x_1}\cdot \overline{x_2}\cdot \overline{x_3}\cdot x_4+\overline{x_1}\cdot \overline{x_2}\cdot \overline{x_3}\cdot \overline{x_4}\\
&+x_1\cdot \overline{x_2}\cdot x_3\cdot x_4+x_1\cdot \overline{x_2}\cdot x_3\cdot \overline{x_4}+x_1\cdot \overline{x_2}\cdot \overline{x_3}\cdot x_4+x_1\cdot \overline{x_2}\cdot \overline{x_3}\cdot \overline{x_4}\\
&+\overline{x_1}\cdot \overline{x_2}\cdot x_3\cdot x_4+\overline{x_1}\cdot \overline{x_2}\cdot x_3\cdot x_4+\overline{x_1}\cdot \overline{x_2}\cdot \overline{x_3}\cdot x_4+\overline{x_1}\cdot \overline{x_2}\cdot \overline{x_3}\cdot \overline{x_4}\\
&+x_1\cdot x_2\cdot x_3\cdot x_4+x_1\cdot x_2\cdot x_3\cdot \overline{x_4}+\overline{x_1}\cdot x_2\cdot x_3\cdot x_4+\overline{x_1}\cdot x_2\cdot x_3\cdot \overline{x_4}\\
&+x_1\cdot x_2\cdot x_3\cdot \overline{x_4}+x_1\cdot x_2\cdot \overline{x_3}\cdot \overline{x_4}+\overline{x_1}\cdot x_2\cdot x_3\cdot \overline{x_4}+\overline{x_1}\cdot x_2\cdot \overline{x_3}\cdot \overline{x_4}
\end{aligned}$$

The minterms can be sorted, using the convention that x_i comes before $\overline{x_i}$.

$$\begin{aligned}
&x_1\cdot x_2\cdot x_3\cdot x_4+x_1\cdot x_2\cdot x_3\cdot \overline{x_4}+x_1\cdot x_2\cdot x_3\cdot \overline{x_4}+x_1\cdot x_2\cdot \overline{x_3}\cdot \overline{x_4}\\
&+x_1\cdot \overline{x_2}\cdot x_3\cdot x_4+x_1\cdot \overline{x_2}\cdot x_3\cdot \overline{x_4}+x_1\cdot \overline{x_2}\cdot \overline{x_3}\cdot x_4+x_1\cdot \overline{x_2}\cdot \overline{x_3}\cdot \overline{x_4}\\
&+\overline{x_1}\cdot x_2\cdot x_3\cdot x_4+\overline{x_1}\cdot x_2\cdot x_3\cdot x_4+\overline{x_1}\cdot x_2\cdot x_3\cdot \overline{x_4}+\overline{x_1}\cdot x_2\cdot x_3\cdot \overline{x_4}\\
&+\overline{x_1}\cdot x_2\cdot x_3\cdot \overline{x_4}+\overline{x_1}\cdot x_2\cdot \overline{x_3}\cdot x_4+\overline{x_1}\cdot x_2\cdot \overline{x_3}\cdot \overline{x_4}+\overline{x_1}\cdot x_2\cdot \overline{x_3}\cdot \overline{x_4}\\
&+\overline{x_1}\cdot \overline{x_2}\cdot x_3\cdot x_4+\overline{x_1}\cdot \overline{x_2}\cdot x_3\cdot x_4+\overline{x_1}\cdot \overline{x_2}\cdot x_3\cdot x_4+\overline{x_1}\cdot \overline{x_2}\cdot x_3\cdot \overline{x_4}\\
&+\overline{x_1}\cdot \overline{x_2}\cdot \overline{x_3}\cdot x_4+\overline{x_1}\cdot \overline{x_2}\cdot \overline{x_3}\cdot x_4+\overline{x_1}\cdot \overline{x_2}\cdot \overline{x_3}\cdot \overline{x_4}+\overline{x_1}\cdot \overline{x_2}\cdot \overline{x_3}\cdot \overline{x_4}
\end{aligned}$$

Finally, remove duplicate minterms.

$$\begin{aligned}
&x_1\cdot x_2\cdot x_3\cdot x_4+x_1\cdot x_2\cdot x_3\cdot \overline{x_4}+x_1\cdot x_2\cdot \overline{x_3}\cdot \overline{x_4}+x_1\cdot \overline{x_2}\cdot x_3\cdot x_4\\
&+x_1\cdot \overline{x_2}\cdot x_3\cdot \overline{x_4}+x_1\cdot \overline{x_2}\cdot \overline{x_3}\cdot x_4+x_1\cdot \overline{x_2}\cdot \overline{x_3}\cdot \overline{x_4}+\overline{x_1}\cdot x_2\cdot x_3\cdot x_4\\
&+\overline{x_1}\cdot x_2\cdot x_3\cdot \overline{x_4}+\overline{x_1}\cdot x_2\cdot \overline{x_3}\cdot x_4+\overline{x_1}\cdot x_2\cdot \overline{x_3}\cdot \overline{x_4}+\overline{x_1}\cdot \overline{x_2}\cdot x_3\cdot x_4\\
&+\overline{x_1}\cdot \overline{x_2}\cdot x_3\cdot \overline{x_4}+\overline{x_1}\cdot \overline{x_2}\cdot \overline{x_3}\cdot x_4+\overline{x_1}\cdot \overline{x_2}\cdot \overline{x_3}\cdot \overline{x_4}
\end{aligned}$$

A good way to check to see if any mistakes were made in this long process is to compare this result with the original expression, thought of as a Boolean function. The minterms in the final expression indicate that the final expression will evaluate to 1 at the 15 ordered 4-tuples: $\{(1,1,1,1),\ (1,1,1,0),\ (1,1,0,0),\ (1,0,1,1),\ (1,0,1,0),\ (1,0,0,1),\ (1,0,0,0),\ (0,1,1,1),\ (0,1,1,0),\ (0,1,0,1),\ (0,1,0,0),\ (0,0,1,1),\ (0,0,1,0),\ (0,0,0,1),\ (0,0,0,0)\}$.

Table 12.31 shows values for the function defined by the original expression. Note that whenever either x_1 or x_2 is 0, the term $\overline{1\cdot x_1\cdot x_2}$ will evaluate to 1, but then the domination property indicates that the function must evaluate to 1. This leaves only four additional ordered 4-tuples to examine. It turns out that the only way the expression can have the value 0 is to have $x_1 = x_2 = 1$ and $x_3 = 0$ and $x_4 = 1$. The ordered 4-tuple $(1,1,0,1)$ is the only possible binary 4-tuple missing from the previous list. Thus, the final disjunctive normal form expression was calculated correctly. ■

TABLE 12.31 The Function Defined by $\overline{1\cdot x_1\cdot x_2}+\overline{\overline{x_2\cdot(x_3+\overline{x_4})}}$

x_1	x_2	x_3	x_4	$\overline{1\cdot x_1\cdot x_2}+\overline{\overline{x_2\cdot(x_3+\overline{x_4})}}$
0	0	0	0	1
0	0	0	1	1
0	0	1	0	1
0	0	1	1	1
0	1	0	0	1
0	1	0	1	1
0	1	1	0	1
0	1	1	1	1
1	0	0	0	1
1	0	0	1	1
1	0	1	0	1
1	0	1	1	1
1	1	0	0	1
1	1	0	1	0
1	1	1	0	1
1	1	1	1	1

✔ Quick Check 12.12

1. Use the process developed in Propositions 12.49, 12.50, 12.51, and Theorem 12.52 to transform the following Boolean expression in x_1, x_2 and x_3 into disjunctive normal form. Use the natural lexicographical order imposed by the subscripts.

$$\overline{(0+x_1)\cdot\overline{x_2}+x_1\cdot(x_2+\overline{x_3})}$$ ☑

12.4.4 Exercises

The exercises marked with ★ have detailed solutions in Appendix H.

1. Count the number of Boolean functions with domain equal to the Boolean algebra defined by setting $B = \mathcal{P}(S)$, where $S = \{a, b, c\}$ and $\mathbb{B}$ is defined using the standard rules presented on page 60 for creating a Boolean algebra from a set.

2. Prove Theorem 12.41 on page 774.

3. ★ How many distinct binary functions of order 3 are there?

4. How many distinct functions of order n are there with domain $\{0, 1, 2\}$ and range $\{0, 1, 2\}$?

5. Assume that all expressions in this problem may be constructed using the binary variables, w, x, y, z. Which of the following binary expressions are minterms in w, x, y, z? If an expression is not a minterm, explain why it is not.
   (a) $0\cdot w\cdot x\cdot y\cdot\overline{z}$ (b) ★ 1 (c) $\overline{w}\cdot\overline{x}\cdot\overline{y}$
   (d) $w\cdot\overline{x}\cdot y\cdot z+\overline{w}\cdot x\cdot\overline{y}\cdot z$ (e) $\overline{w}\cdot x\cdot\overline{y}\cdot\overline{z}$

6. A binary function evaluates to 1 at the following ordered 5-tuples: $(0, 0, 1, 1, 1)$, $(1, 0, 1, 0, 1)$, $(1, 1, 0, 0, 1)$. It evaluates to 0 at all other 5-tuples. Write a binary expression in disjunctive normal form that represents this function.

7. Assume that all expressions in this problem may be constructed using the binary variables, w, x, y, z. Which of the following binary expressions in w, x, y, z are in disjunctive normal form? If an expression is not in disjunctive normal form, explain why it is not.
   (a) $\overline{w}\cdot\overline{x}\cdot\overline{y}\cdot z+x\cdot\overline{y}\cdot\overline{z}$ (b) ★ 1
   (c) $\overline{w}\cdot\overline{y}\cdot\overline{x}\cdot z+w\cdot x\cdot y\cdot\overline{z}$
   (d) $w\cdot x\cdot\overline{y}\cdot z+w\cdot x\cdot\overline{y}\cdot z$
   (e) $0+0$ (f) $w\cdot w\cdot x\cdot y\cdot z$

8. Write a binary expression in disjunctive normal form for each binary function of order 2 in Example 12.30 on page 774.

9. ★ Create a binary expression that represents a function of two binary variables that evaluates to 1 when exactly one of the variables is 1 and evaluates to 0 otherwise. (XOR)

10. Create a binary expression that represents a function of two binary variables that evaluates to 1 when the variables have the same value and evaluates to 0 otherwise. (The biconditional function.)

11. Use the process developed in Proposition 12.49 to transform the following binary expressions in x_1, x_2, and x_3 into an expression for which any complement operator is acting on only a single variable.
    (a) ★ $\overline{x_1+\overline{x_2}\cdot(x_1+\overline{x_3})}$ (b) $\overline{x_2\cdot\overline{x_1\cdot x_3}}$
    (c) $\overline{1\cdot x_3}+\overline{0+x_1}$

12. Use the process developed in Proposition 12.50 to transform the following binary expressions in x_1, x_2, and x_3 into a sum of products where every factor is either 0, 1, or a complemented or uncomplemented single variable.
    (a) ★ $(\overline{x_1}\cdot 1+x_2\cdot x_3)\cdot x_1$
    (b) $(x_1+x_2\cdot x_3)\cdot(x_2+\overline{x_3})$
    (c) $(1+x_1)\cdot(x_2\cdot\overline{x_3}+0)$

13. Use the process developed in Proposition 12.51 to transform the following binary expressions in x_1, x_2, and x_3 into an expression in disjunctive normal form.
    (a) ★ $x_1\cdot x_3\cdot\overline{x_1}$ (b) $x_1\cdot\overline{x_2}$ (c) $0+x_1\cdot x_2\cdot x_1$

14. Use the process developed in Propositions 12.49, 12.50, 12.51, and Theorem 12.52 to transform the following binary expressions in x_1, x_2, and x_3 into disjunctive normal form. Use the natural lexicographical order imposed by the subscripts.
    (a) $x_1\cdot x_2+x_2\cdot\overline{x_3}$ (b) ★ 1
    (c) $x_2\cdot\overline{x_3}+\overline{x_1}\cdot x_2\cdot x_1$
    (d) $\overline{1+x_1}+x_2\cdot\overline{x_3\cdot(x_1+\overline{x_2})}$

15. Use the process developed in Propositions 12.49, 12.50, 12.51, and Theorem 12.52 to transform the following binary expressions in x_1, x_2, and x_3 into disjunctive normal form. Use the natural lexicographical order imposed by the subscripts.
    (a) $x_3+x_1\cdot\overline{\overline{x_1}+x_2+\overline{x_3}}\cdot 1$ (b) $\overline{x_2}\cdot x_3+\overline{x_1\cdot x_3}$
    (c) $(x_1+x_2)\cdot(x_2+\overline{x_3})$ (d) $\overline{\overline{x_1}\cdot\overline{x_2}\cdot\overline{x_3}}$

16. Use the process developed in Propositions 12.49, 12.50, 12.51, and Theorem 12.52 to transform the following binary expressions in x_1, x_2, x_3, and x_4 into disjunctive normal form. Use the natural lexicographical order imposed by the subscripts.
    (a) $\overline{x_1+x_2+x_3+x_4}$
    (b) $(x_1\cdot x_2+x_3\cdot\overline{x_4})\cdot\overline{x_2+x_3+x_4}$
    (c) $1\cdot x_2\cdot\overline{x_3}+x_1\cdot x_2\cdot x_1\cdot x_3$
    (d) $x_1\cdot x_2\cdot x_3+x_1\cdot x_2\cdot x_4+x_1\cdot x_3\cdot x_4+x_2\cdot x_3\cdot x_4$

17. Use the process developed in Propositions 12.49, 12.50, 12.51, and Theorem 12.52 to transform the following binary expressions in x_1, x_2, x_3, and x_4 into disjunctive normal form. Use the natural lexicographical order imposed by the subscripts.
    (a) $x_2\cdot x_3\cdot x_4+\overline{1+x_3}+\overline{x_1+x_2}$
    (b) ★ $\overline{x_1\cdot x_4+x_1\cdot\overline{x_2}}\cdot x_2$
    (c) $x_3\cdot(\overline{x_1}+\overline{x_4}\cdot x_3)+\overline{x_2}\cdot(\overline{x_3}\cdot x_4+x_1)$
    (d) $\overline{x_2\cdot(x_3+x_4)}+\overline{x_2+x_3+\overline{x_4}}$

18. ★ An alternative canonical form.

(a) Use the definition of a minterm as a model from which to create a definition for a *maxterm*.

(b) Define *conjunctive normal form*. Recall that the logic operator $\vee$ is also called the disjunction operator and that $\wedge$ is also called the conjunction operator. Also recall that Boolean algebras that are created from collections of propositions make the associations: $\vee$ with $+$ and $\wedge$ with $\cdot$.

(c) State and prove a maxterm-based proposition patterned after Proposition 12.47. Then,

(d) State (but do not prove) a maxterm-based theorem patterned after Theorem 12.48

(e) State (but do not prove) a maxterm-based theorem patterned after Theorem 12.52.

19. Use part (c) of Exercise 18 to write a binary expression in conjunctive normal form for each binary function of order 2 in Example 12.30 on page 774.

20. Use any method you wish to convert the binary expressions in Exercise 15 into conjunctive normal form. (See Exercise 18 for the definition of conjunctive normal form.)

21. Each pair of binary expressions contains an expression in disjunctive normal form and one in conjunctive normal form. In each case, decide whether the two binary expressions are equivalent.

(a) $x_1 \cdot \overline{x_2}$; and $(x_1 + x_2) \cdot (x_1 + \overline{x_2}) \cdot (\overline{x_1} + \overline{x_2})$

(b) $x_1 \cdot \overline{x_2} + \overline{x_1} \cdot x_2$; and $(x_1 + \overline{x_2}) \cdot (\overline{x_1} + x_2)$

(c) $x_1 \cdot x_2 + \overline{x_1} \cdot \overline{x_2}$; and $(\overline{x_1} + \overline{x_2}) \cdot (x_1 + x_2)$

(d) $x_1 \cdot x_2 + \overline{x_1} \cdot x_2 + x_1 \cdot \overline{x_2}$; and $(x_1 + x_2)$

(e) $x_1 \cdot x_2 + \overline{x_1} \cdot x_2 + x_1 \cdot \overline{x_2}$; and $(x_1 + x_2) \cdot (\overline{x_1} + x_2) \cdot (x_1 + \overline{x_2})$

12.5 Combinatorial Circuits

Section 12.4 introduced a technique for converting binary functions into binary expressions in disjunctive normal form. Disjunctive normal form is very nice as a canonical representation, but it is often not ideal if the goal is to have a simple binary expression with as few products and sums as possible. This section will present a technique for simplifying binary expressions. This will be beneficial because Section 12.5.2 will establish a connection between binary expressions and the design of combinatorial circuits. Since circuits are built from real components, it is desirable to minimize how many components are needed to accomplish the circuit's task.

12.5.1 Minimizing Binary Expressions

The process of simplifying a binary expression in disjunctive normal form relies heavily on the four axioms for a Boolean algebra. It also uses the absorbtion properties.

The major simplification rule can be justified by the following set of Boolean equivalences.

$$e \cdot f + \overline{e} \cdot f = (e + \overline{e}) \cdot f = 1 \cdot f = f \cdot 1 = f$$

Both e and f can be binary expressions, instead of merely single binary variables. The most common usage will have e as a single binary variable and f as a product of binary variables. A typical simplification might be

$$x_1 \cdot x_2 \cdot \overline{x_3} \cdot x_4 + x_1 \cdot \overline{x_2} \cdot \overline{x_3} \cdot x_4 = x_1 \cdot \overline{x_3} \cdot x_4$$

The expanded version of this simplification is shown next.

$$\begin{aligned}
& x_1 \cdot x_2 \cdot \overline{x_3} \cdot x_4 + x_1 \cdot \overline{x_2} \cdot \overline{x_3} \cdot x_4 \\
&\quad = x_2 \cdot x_1 \cdot \overline{x_3} \cdot x_4 + \overline{x_2} \cdot x_1 \cdot \overline{x_3} \cdot x_4 && \text{commutativity (twice)} \\
&\quad = x_2 \cdot (x_1 \cdot \overline{x_3} \cdot x_4) + \overline{x_2} \cdot (x_1 \cdot \overline{x_3} \cdot x_4) && \text{associativity (twice)} \\
&\quad = (x_2 + \overline{x_2}) \cdot (x_1 \cdot \overline{x_3} \cdot x_4) && \text{distributivity} \\
&\quad = 1 \cdot (x_1 \cdot \overline{x_3} \cdot x_4) && \text{complement} \\
&\quad = (x_1 \cdot \overline{x_3} \cdot x_4) \cdot 1 && \text{commutativity} \\
&\quad = (x_1 \cdot \overline{x_3} \cdot x_4) && \text{identity} \\
&\quad = x_1 \cdot \overline{x_3} \cdot x_4 && \text{associativity}
\end{aligned}$$

Binary Expression Simplification Rule

Let e and f be binary expressions. Then

$$e \cdot f + \overline{e} \cdot f = f$$

The binary expression simplification rule (with perhaps some commutativity) can be used iteratively to reduce an expression in disjunctive normal form into a much simpler (but equivalent) expression.

EXAMPLE 12.39 **Simplifying the Binary Expression in Quick Check 12.11**

The binary function defined in Quick Check 12.11 on page 778 was found to be equivalent to the binary expression $\overline{x_1} \cdot x_2 \cdot \overline{x_3} + \overline{x_1} \cdot x_2 \cdot x_3 + x_1 \cdot x_2 \cdot \overline{x_3} + x_1 \cdot x_2 \cdot x_3$. The binary expression simplification rule can be used several times.

$$\begin{aligned}
&\overline{x_1} \cdot x_2 \cdot \overline{x_3} + \overline{x_1} \cdot x_2 \cdot x_3 + x_1 \cdot x_2 \cdot \overline{x_3} + x_1 \cdot x_2 \cdot x_3 \\
&\quad = (\overline{x_1} \cdot x_2 \cdot \overline{x_3} + \overline{x_1} \cdot x_2 \cdot x_3) + x_1 \cdot x_2 \cdot \overline{x_3} + x_1 \cdot x_2 \cdot x_3 \\
&\quad = (\overline{x_1} \cdot x_2) + x_1 \cdot x_2 \cdot \overline{x_3} + x_1 \cdot x_2 \cdot x_3 \\
&\quad = \overline{x_1} \cdot x_2 + (x_1 \cdot x_2 \cdot \overline{x_3} + x_1 \cdot x_2 \cdot x_3) \\
&\quad = \overline{x_1} \cdot x_2 + (x_1 \cdot x_2) \\
&\quad = (\overline{x_1} \cdot x_2 + x_1 \cdot x_2) \\
&\quad = x_2
\end{aligned}$$

Consequently, the function can be expressed as either $f(x_1, x_2, x_3) = \overline{x_1} \cdot x_2 \cdot \overline{x_3} + \overline{x_1} \cdot x_2 \cdot x_3 + x_1 \cdot x_2 \cdot \overline{x_3} + x_1 \cdot x_2 \cdot x_3$, or as $f(x_1, x_2, x_3) = x_2$. A quick look at the table (on page 778) that defines f should convince you that $f(x_1, x_2, x_3) = x_2$ is a valid representation. For most purposes, this shorter expression is the preferred representation. ■

Doing hand simplifications (as in the previous example) works well for small problems. It becomes less suitable when the number of variables increases. There are two fairly elementary methods that can turn the process into an algorithm. The first method is an algorithm that uses a matrix, called a Karnaugh map, to represent the minterms. It proceeds by circling pairs of minterms that are candidates for simplification. This method becomes difficult to use when the number of variables gets larger than five or six. The most common alternative is the Quine–McCluskey algorithm. This algorithm uses tables to organize the work. It is also easier to translate into a computer algorithm.

The Quine–McCluskey algorithm will be introduced by an example before a formal description is given.

EXAMPLE 12.40 **Illustrating Quine–McCluskey**

Consider the binary function $f(x, y, z) = x \cdot y \cdot z + x \cdot y \cdot \overline{z} + x \cdot \overline{y} \cdot \overline{z} + \overline{x} \cdot y \cdot z + \overline{x} \cdot y \cdot \overline{z} + \overline{x} \cdot \overline{y} \cdot \overline{z}$, which is in disjunctive normal form.

TABLE 12.32

1	111
2	110
3	011
4	100
5	010
6	000

The Quine–McCluskey algorithm associates a bit string with each minterm. If a variable appears in uncomplemented form, the bit string contains a 1 in the corresponding position; otherwise, the bit string contains a 0 in the corresponding position. It will be helpful to sort the minterms according to the number of 1s in their associated bit strings. The minterms can also be numbered for future reference (Table 12.32).

An application of the binary expression simplification rule to two binary expressions corresponds to a merging of the two associated bit strings. For instance, the simplification, $x \cdot y \cdot z + x \cdot y \cdot \overline{z} = x \cdot y$, corresponds to merging 111 and 110 into the expression 11–, where the hyphen represents the elimination of z. Notice that bit strings cannot be merged unless they differ by exactly one in the number of 1s they contain. However, this condition doesn't guarantee a merge is possible: For example, 1100 and 0001 cannot be merged.

The algorithm extends the table by making a new column for all the possible merged bit-hyphen expressions. In addition, a column is added to place a check mark next to each minterm that became part of a merged expression, and another column to keep track of which minterms were used to create each merged expression. The check mark indicates that a minterm has been "covered" by a merged expression. For instance, the minterms $x \cdot y \cdot z$ and $x \cdot y \cdot \overline{z}$ are covered by $11-$.

1	111	✓	1, 2	1 1 −
2	110	✓	1, 3	− 1 1
3	011	✓	2, 4	1 − 0
4	100	✓	2, 5	− 1 0
5	010	✓	3, 5	0 1 −
6	000	✓	4, 6	− 0 0
			5, 6	0 − 0

The newly merged expressions can now be merged with each other. The only new requirement is that hyphens must match another hyphen in the same position. New columns are introduced to keep track of which bit-hyphen expressions have been covered and also which of the original minterms have been covered by the new bit-hyphen expression.

1	111	✓	1, 2	1 1 −	✓	1, 2, 3, 5	− 1 −
2	110	✓	1, 3	− 1 1	✓	2, 4, 5, 6	− − 0
3	011	✓	2, 4	1 − 0	✓		
4	100	✓	2, 5	− 1 0	✓		
5	010	✓	3, 5	0 1 −	✓		
6	000	✓	4, 6	− 0 0	✓		
			5, 6	0 − 0	✓		

Notice that the expression $-1-$ can be obtained by merging $11-$ and $01-$ (leading to a covering of 1, 2, 3, 5) and also by merging -11 and -10 (leading to the same covering). The new expression is listed only once.

No additional merging is possible, so the algorithm is ready to start phase 2. A matrix is created by listing the original minterms as column labels and listing all unchecked bit-hyphen expressions as row labels. (In this example, all unchecked bit-hyphen expressions are in the final section of the table, but this need not be true in general.) An $\times$ is placed in row i, column j if the bit-hyphen expression in row i covers the minterm in column j.

	1 $x \cdot y \cdot z$	**2** $x \cdot y \cdot \overline{z}$	**3** $\overline{x} \cdot y \cdot z$	**4** $x \cdot \overline{y} \cdot \overline{z}$	**5** $\overline{x} \cdot y \cdot \overline{z}$	**6** $\overline{x} \cdot \overline{y} \cdot \overline{z}$
$-1-$	×	×	×		×	
$--0$		×		×	×	×

Every minterm in the original binary expression must be covered by the reduced expression. The final goal is to pick a smallest set of rows so that every column has at least one X among the chosen rows. For this example, both rows are needed. The final reduced binary expression is therefore $y + \overline{z}$. You can evaluate both representations at each of the eight ordered triples in the domain to verify that

$$f(x, y, z) = x \cdot y \cdot z + x \cdot y \cdot \overline{z} + x \cdot \overline{y} \cdot \overline{z} + \overline{x} \cdot y \cdot z + \overline{x} \cdot y \cdot \overline{z} + \overline{x} \cdot \overline{y} \cdot \overline{z} = y + \overline{z}$$

■

The Quine–McCluskey Algorithm

Phase 1:

```
Sort the minterms according to the number of 1's in their
  associated bit-hyphen strings.
Number the sorted bit-hyphen strings.
while the current section contains bit-hyphen strings that
  can be merged
    List all new merged bit-hyphen strings in a new column.
    Make a check next to each bit-hyphen string in the
       current section that was used in a merge.
    Create a new column that lists the original minterms
       covered by the newest bit-hyphen strings.
```

Phase 2:

```
Create a matrix that has the original minterms as column
   labels and all unchecked bit-hyphen expressions
   as row labels.
Place an X in row i, column j if the bit-hyphen expression
   in row i covers the minterm in column j.
Pick a smallest set, S, of rows so that every column has at
   least one X among the chosen rows.
Convert the bit-hyphen expressions in S into their
   corresponding binary expressions and add them.
```

EXAMPLE 12.41

A Four-Variable Quine–McCluskey

The Quine–McCluskey algorithm can be used to minimize the binary expression

$$w \cdot \overline{x} \cdot y \cdot z + w \cdot \overline{x} \cdot \overline{y} \cdot \overline{z} + \overline{w} \cdot x \cdot y \cdot \overline{z} + \overline{w} \cdot x \cdot \overline{y} \cdot z$$
$$+ \overline{w} \cdot \overline{x} \cdot y \cdot z + \overline{w} \cdot \overline{x} \cdot y \cdot \overline{z} + \overline{w} \cdot \overline{x} \cdot \overline{y} \cdot z + \overline{w} \cdot \overline{x} \cdot \overline{y} \cdot \overline{z}$$

Phase 1:

1	1011	√	1, 4	−011		4, 6, 7, 8	00−−
2	0110	√	2, 6	0−10			
3	0101	√	3, 7	0−01			
4	0011	√	4, 6	001−	√		
5	1000	√	4, 7	00−1	√		
6	0010	√	5, 8	−000			
7	0001	√	6, 8	00−0	√		
8	0000	√	7, 8	000−	√		

Phase 2:

	1 1011	2 0110	3 0101	4 0011	5 1000	6 0010	7 0001	8 0000
−011	×			×				
0−10		×				×		
0−01			×				×	
−000					×			×
00−−				×		×	×	×

Notice that if the term $\overline{w} \cdot \overline{x}$ (00−−) is chosen, then each of the other four terms must also be chosen. However, the other four terms are sufficient to cover all columns. The equivalent minimized binary expression is thus[32]

$$\overline{x} \cdot y \cdot z + \overline{w} \cdot y \cdot \overline{z} + \overline{w} \cdot \overline{y} \cdot z + \overline{x} \cdot \overline{y} \cdot \overline{z}$$

No proof of the correctness of the Quine–McCluskey algorithm will be given in this textbook. However, a bit of justification seems appropriate for phase 2 of this example.

How can the exclusion of $\overline{w} \cdot \overline{x}$ be justified? One approach is to notice that

$$\begin{aligned}
\overline{w} \cdot \overline{x} &= \overline{w} \cdot \overline{x} \cdot y + \overline{w} \cdot \overline{x} \cdot \overline{y} \\
&= \overline{w} \cdot \overline{x} \cdot y \cdot z + \overline{w} \cdot \overline{x} \cdot y \cdot \overline{z} + \overline{w} \cdot \overline{x} \cdot \overline{y} \\
&= \overline{w} \cdot \overline{x} \cdot y \cdot z + \overline{w} \cdot \overline{x} \cdot y \cdot \overline{z} + \overline{w} \cdot \overline{x} \cdot \overline{y} \cdot z + \overline{w} \cdot \overline{x} \cdot \overline{y} \cdot \overline{z}
\end{aligned}$$

Thus,

$$\begin{aligned}
\overline{x} \cdot y \cdot z &+ \overline{w} \cdot y \cdot \overline{z} + \overline{w} \cdot \overline{y} \cdot z + \overline{x} \cdot \overline{y} \cdot \overline{z} + \overline{w} \cdot \overline{x} \\
&= \overline{x} \cdot y \cdot z + \overline{w} \cdot y \cdot \overline{z} + \overline{w} \cdot \overline{y} \cdot z + \overline{x} \cdot \overline{y} \cdot \overline{z} \\
&\quad + (\overline{w} \cdot \overline{x} \cdot y \cdot z + \overline{w} \cdot \overline{x} \cdot y \cdot \overline{z} + \overline{w} \cdot \overline{x} \cdot \overline{y} \cdot z + \overline{w} \cdot \overline{x} \cdot \overline{y} \cdot \overline{z}) \\
&= (\overline{x} \cdot y \cdot z + \overline{w} \cdot \overline{x} \cdot y \cdot z) + (\overline{w} \cdot y \cdot \overline{z} + \overline{w} \cdot \overline{x} \cdot y \cdot \overline{z}) \\
&\quad + (\overline{w} \cdot \overline{y} \cdot z + \overline{w} \cdot \overline{x} \cdot \overline{y} \cdot z) + (\overline{x} \cdot \overline{y} \cdot \overline{z} + \overline{w} \cdot \overline{x} \cdot \overline{y} \cdot \overline{z}) \\
&= \overline{x} \cdot y \cdot z + \overline{w} \cdot y \cdot \overline{z} + \overline{w} \cdot \overline{y} \cdot z + \overline{x} \cdot \overline{y} \cdot \overline{z}
\end{aligned}$$

The final equivalence follows from four applications of the absorption property. For example,

$$\begin{aligned}
\overline{x} \cdot y \cdot z + \overline{w} \cdot \overline{x} \cdot y \cdot z &= (\overline{x} \cdot y \cdot z) + \overline{w} \cdot (\overline{x} \cdot y \cdot z) \\
&= (\overline{x} \cdot y \cdot z) + (\overline{x} \cdot y \cdot z) \cdot \overline{w} \\
&= (\overline{x} \cdot y \cdot z)
\end{aligned}$$

This confirms the conclusion of the algorithm: $\overline{w} \cdot \overline{x}$ is not needed. ■

✔ Quick Check 12.13

1. Use the Quine–McCluskey algorithm to minimize the binary expression $x \cdot y \cdot \overline{z} + x \cdot \overline{y} \cdot z + \overline{x} \cdot y \cdot z + \overline{x} \cdot y \cdot \overline{z}$. ☑

12.5.2 Combinatorial Circuits and Binary Expressions

During the late 1930s, Claude Shannon noticed a useful connection between electronic circuits and binary expressions. That connection is the focus of this section.

The two circuit diagrams in Figures 12.2 and 12.3 will help motivate Shannon's observations. The first diagram (Figure 12.2) shows part of an electrical circuit. In order for current to flow from the left end to the right end, both of the connections, x and y, need to be closed. It is common to call the connections *gates* or *logic gates* (perhaps since they look like an open gate in a fence). The gates are in a *serial* arrangement.

Figure 12.2. Two serial gates.

If either gate is open, no current will flow. The behavior is the same as the familiar logic operator, AND. That is, current will flow if and only if $x \wedge y = \mathbf{T}$. This can also be expressed using binary expression notation: Current will flow if and only if $x \cdot y = 1$.

The second diagram (Figure 12.3) shows two gates in a *parallel* arrangement. In this diagram, current can flow if either (or both) of the gates is closed. Thus, current will flow if and only if $x \vee y = \mathbf{T}$. Using binary expression notation: Current will flow if and only if $x + y = 1$.

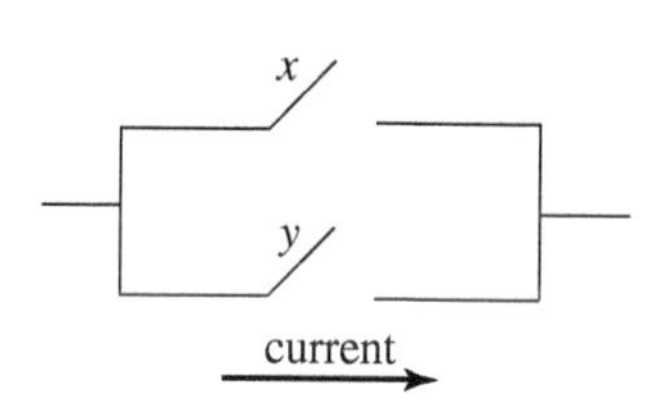

Figure 12.3. Two parallel gates.

[32]The final expression can be verified using the QuineMcCluskey application/applet available at http://www.mathcs.bethel.edu/~gossett/DiscreteMathWithProof/.

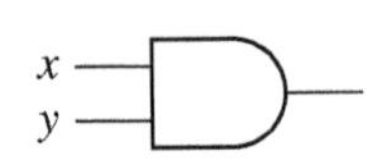

Figure 12.4. The standard symbol for an AND gate.

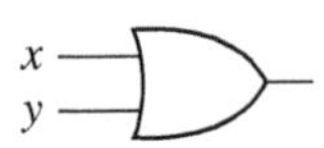

Figure 12.5. The standard symbol for an OR gate.

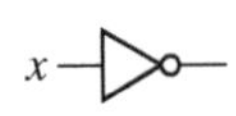

Figure 12.6. The standard symbol for a NOT gate.

It is now customary to abstract the logical structure of circuits and use special notation for the gates. The standard symbol for an AND gate is shown in Figure 12.4. If both inputs, x, y, are 1, then the output will be 1. Otherwise, the output will be 0.

The standard symbol for an OR gate is shown in Figure 12.5. If either input, x, y, is 1, then the output will be 1. Otherwise, the output will be 0.

The final symbol (Figure 12.6) represents a NOT gate (or an *inverter*). This gate changes an input value of 1 to a 0 and a 0 to a 1.

These three kinds of logic gates can be used to design many useful circuits. The circuits of interest here are called combinatorial circuits. A *combinatorial circuit* is a circuit in which there are no delay elements. This is in contrast to *sequential circuits*, in which delay elements exist. (One example would be a finite-state automaton, with its built-in memory of the current state. Finite-state automata can accept a stream of input symbols; combinatorial circuits do not.)

A general process can be used to create combinatorial circuits.

1. Decide how many inputs the circuit will need.
2. Create a binary function that represents the desired output for each combination of input values.
3. Create a binary expression, in disjunctive normal form, which represents the function.
4. Use the Quine–McCluskey algorithm to create a simpler (but equivalent) binary expression.
5. Use logic gates to build a diagram that matches the binary expression.

EXAMPLE 12.42

A Very Simple Combinatorial Circuit

Suppose a combinatorial circuit is needed that accepts three inputs, x, y, z, and generates a 1 if $x = 1$ and $y = z$, but generates a 0 in all other cases. The function is specified by Table 12.33.

TABLE 12.33 $f(x, y, z)$

x	y	z	$f(x, y, z)$
0	0	0	0
0	0	1	0
0	1	0	0
0	1	1	0
1	0	0	1
1	0	1	0
1	1	0	0
1	1	1	1

This can be represented by the binary expression $x \cdot \overline{y} \cdot \overline{z} + x \cdot y \cdot z$. Since the numbers of 1s in the two minterms differ by more than one, the Quine–McCluskey algorithm will not result in any simplification. Figure 12.7 on page 791 shows how the binary expression can be translated into a circuit diagram. Notice the use of the associative property; the binary expression that has been diagramed is actually $(x \cdot \overline{y}) \cdot \overline{z} + (x \cdot y) \cdot z$

There are two alternative depictions for this diagram. In the first alternative (Figure 12.8), the three source inputs are shown as two distinct clusters on the left. This is merely a convenience to help keep the diagram simple.

The second alternative (Figure 12.9) uses multiple-input AND and OR gates. These serve to simplify the diagram. They also reflect a valid option. Many commercial AND and OR gates are capable of multiple inputs. ■

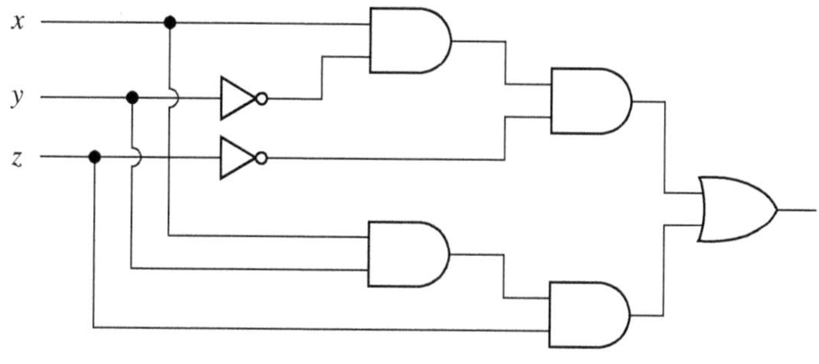

Figure 12.7 A combinatorial circuit for $(x \cdot \overline{y}) \cdot \overline{z} + (x \cdot y) \cdot z$.

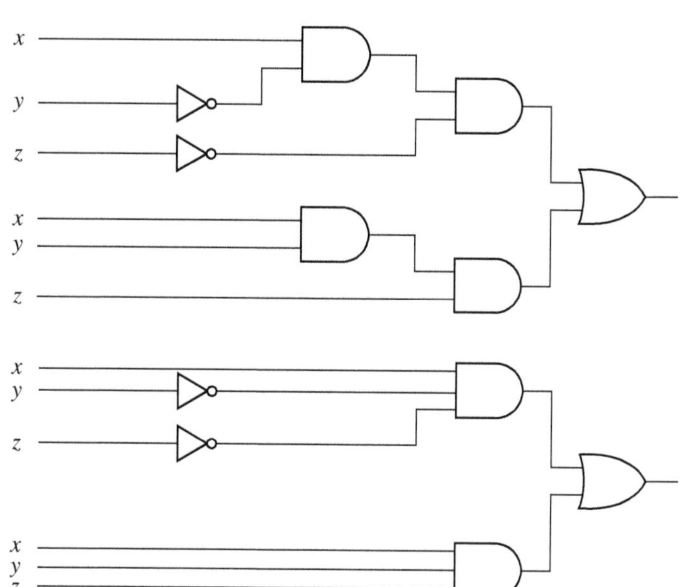

Figure 12.8 A combinatorial circuit for $(x \cdot \overline{y}) \cdot \overline{z} + (x \cdot y) \cdot z$ that duplicates the inputs.

Figure 12.9 A combinatorial circuit for $x \cdot \overline{y} \cdot \overline{z} + x \cdot y \cdot z$ that uses multiple-input gates.

✔ Quick Check 12.14

1. Design a combinatorial circuit that has three inputs. Its output should be the same value as the majority of its inputs. ☑

The basic strategy presented for designing combinatorial circuits can be extended to create more versatile circuits. The next example introduces one possible extension: a circuit with multiple outputs.

EXAMPLE 12.43

Paper, Rock, Scissors

The rules for the game paper, rock, scissors were explained in Quick Check 6.1 on page 275. It is fairly easy to design a combinatorial circuit that will determine the winner (if any) of a game of paper, rock, scissors.

There will be six inputs: p_1, r_1, s_1, and p_2, r_2, s_2, where the subscript indicates the player and the letter indicates which object the player chose. If player 1 chooses scissors, then $s_1 = 1$ and $p_1 = r_1 = 0$. (Illegal input will be considered soon.) Similarly, if player 2 chooses rock, then $r_2 = 1$ and $p_2 = s_2 = 0$ should be the input values.

It is tempting to design the circuit so that a single output variable is assigned the value 1 if player 1 wins and the value 0 if player 2 wins. The defect with this strategy is that many paper, rock, scissors games end in a tie. A binary variable cannot represent a three-valued outcome.

One solution is to design a circuit with three output variables: p, r, s. At most one of these variables will be assigned the value 1 (representing a win by the player who chose that object). A tie will result in all three output variables being assigned the value 0.

If player 1 chooses scissors and player 2 chooses rock, then player 2 will win. Consequently, $r = 1$ and $p = s = 0$ are the appropriate output values.

With six inputs, there are $2^6 = 64$ values in the domain of the function. However, some of these values represent illegal inputs for this game. For example, $p_1 = r_1 = s_1 = 1$ would mean that player 1 tried to show all three objects simultaneously. The circuit can prevent illegal input from giving player 1 an unfair advantage by making the outcome a tie. Since illegal input and legitimate ties all result in an output of three 0s, there are not that many input sets that will result in a non-tie output. Only those values are shown in Table 12.34.

TABLE 12.34 The Input 6-Tuples that Result in an Output Value Being Set to 1

p_1	r_1	s_1	p_2	r_2	s_2	p	r	s
0	0	1	0	1	0	0	1	0
0	0	1	1	0	0	0	0	1
0	1	0	0	0	1	0	1	0
0	1	0	1	0	0	1	0	0
1	0	0	0	0	1	0	0	1
1	0	0	0	1	0	1	0	0

There will be three expressions to simplify:

$$p(p_1, r_1, s_1, p_2, r_2, s_2) = \overline{p_1} \cdot r_1 \cdot \overline{s_1} \cdot p_2 \cdot \overline{r_2} \cdot \overline{s_2} + p_1 \cdot \overline{r_1} \cdot \overline{s_1} \cdot \overline{p_2} \cdot r_2 \cdot \overline{s_2}$$
$$r(p_1, r_1, s_1, p_2, r_2, s_2) = \overline{p_1} \cdot \overline{r_1} \cdot s_1 \cdot \overline{p_2} \cdot r_2 \cdot \overline{s_2} + \overline{p_1} \cdot r_1 \cdot \overline{s_1} \cdot \overline{p_2} \cdot \overline{r_2} \cdot s_2$$
$$s(p_1, r_1, s_1, p_2, r_2, s_2) = \overline{p_1} \cdot \overline{r_1} \cdot s_1 \cdot p_2 \cdot \overline{r_2} \cdot \overline{s_2} + p_1 \cdot \overline{r_1} \cdot \overline{s_1} \cdot \overline{p_2} \cdot \overline{r_2} \cdot s_2.$$

Since both minterms in each expression have the same number of 1s, the Quine–McCluskey algorithm will not result in simpler expressions. Figure 12.10 shows the desired combinatorial circuit.

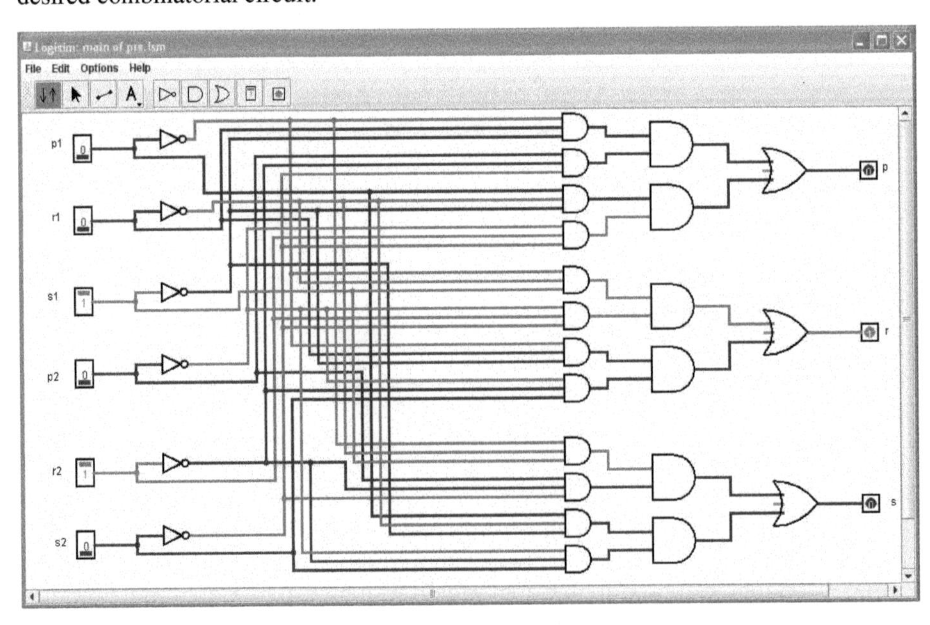

Figure 12.10 A combinatorial circuit for Paper, Rock, Scissors.

An alternative approach to a paper, rock, scissors circuit is presented in Exercise 12 on page 796. ■

The diagram in Figure 12.10 was created with a program called Logisim, which has been designed to simulate combinatorial circuits.[33] The program can be downloaded

[33] Combinatorial circuits are also called *logic circuits*; hence the name Logisim (logic circuit simulator).

free.[34] Logisim provides drag-and-drop templates for the logic gates and the connections. It also lets you simulate the circuit. It provides "input switches" (the boxes on the left-hand side of Figure 12.10) and "output LEDs" (the boxes on the right-hand side of Figure 12.10). The input switches can be clicked to toggle between 0 and 1. The output LEDs show the resulting output. The circuit in Figure 12.10 shows player 1 choosing scissors and player 2 choosing rock. The output, r, is 1, signifying that rock (player 2) is the winner. The program uses green lines when the connecting line carries a 1 and red lines when the value is 0.

Figure 12.10 is a bit messy in the middle, but it is possible to follow the lines properly because junctions (where a line splits) are indicated by a dot. All other crossings are artificial and assumed to be insulated so that no accidental bit changes arise. The sets of smaller AND gates were used because the large AND gates in Logisim are limited to 5 inputs.

The general process for constructing combinatorial circuits with multiple output functions is summarized next.

Creating Combinatorial Circuits

Step 1 Decide how many inputs and how many outputs the circuit will need. Create a binary variable for each input and each output.

Step 2 For each output, create a binary function that represents the desired output for each combination of input values.

Step 3 Create binary expressions, in disjunctive normal form, which represent each of the binary functions.

Step 4 Use the Quine–McCluskey algorithm to create simpler (but equivalent) binary expressions.

Step 5 Use logic gates to build a diagram that matches the set of binary expressions.

12.5.3 Functional Completeness

Propositional logic was introduced in Section 2.3. Five primary logic operators were used to construct propositions (statements). These operators were NOT, AND, OR, implication, and biconditional.

The implication and biconditional logical equivalences on page 53 can be used to show that implication and biconditional are not really necessary. In particular,

$$(P \rightarrow Q) \Leftrightarrow [\neg(P \wedge (\neg Q))] \Leftrightarrow [(\neg P) \vee Q]$$

asserts that implication can be replaced with either NOT and AND or with NOT and OR. Similarly,

$$(P \leftrightarrow Q) \Leftrightarrow [(P \rightarrow Q) \wedge (Q \rightarrow P)]$$

asserts that the biconditional can be replaced by AND and implication, which can then be replaced by AND, NOT, and OR or by AND and NOT. Thus, any statement in propositional logic can be expressed using only NOT, AND, and OR.

DEFINITION 12.53 ***Functionally Complete***
A collection of logic operators in propositional logic is *functionally complete* if every compound statement is logically equivalent to some statement that contains only operators from the collection.

[34] Go to http://www.mathcs.bethel.edu/~gossett/DiscreteMathWithProof/ and look for the Logisim link. The data file, prs.lsm, for the paper, rock, scissors circuit can also be found there.

The discussion just prior to the definition indicates that $\{\neg, \wedge, \vee\}$ is a functionally complete set of logic operators. Since De Morgan's laws and the law of double negation imply that

$$(a \vee b) \Leftrightarrow \neg((\neg a) \wedge (\neg b))$$

every statement that contains OR can be replaced with a logically equivalent statement that only contains NOT and AND. Hence, $\{\neg, \wedge\}$ is also a functionally complete set of logic operators.

The notion of functional completeness is relevant to designing combinatorial circuits. Each gate in a circuit represents a kind of component that needs to be manufactured. If the number of distinct components can be reduced, it may be possible to reduce the cost of the circuit.

One simple mechanism for reducing the number of components arises from introducing two new kinds of gates (rather than eliminating kinds of gates). These gates are NAND gates and NOR gates.

A NAND gate replaces an AND gate followed by a NOT gate. The new symbol is shown in Figure 12.11. Similarly, a NOR gate replaces an OR gate followed by a NOT gate (Figure 12.11).[35]

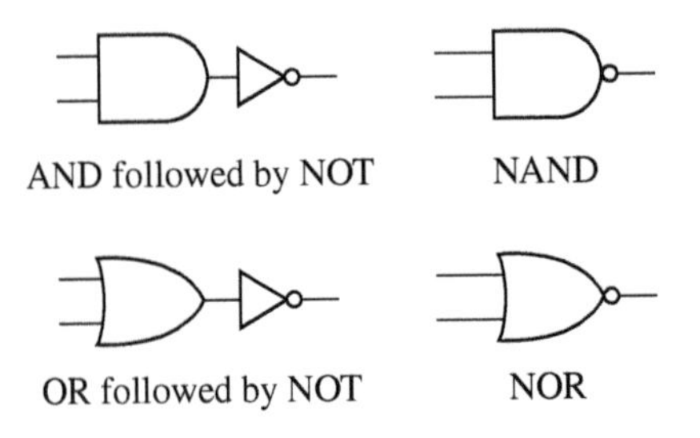

Figure 12.11 The symbols for NAND and NOR gates.

From a mathematical viewpoint, NAND and NOR should be defined by truth tables. These truth tables are presented in Tables 12.35 and 12.36. New notation has also been introduced: NAND is denoted by $\uparrow$ and NOR is denoted by $\downarrow$.

TABLE 12.35 The Truth Table for NAND

P	Q	$P \uparrow Q$
T	T	F
T	F	T
F	T	T
F	F	T

TABLE 12.36 The Truth Table for NOR

P	Q	$P \downarrow Q$
T	T	F
T	F	F
F	T	F
F	F	T

THEOREM 12.54 ***NAND Is Functionally Complete***

The set $\{\uparrow\}$ is a functionally complete collection of logic operators.

Proof: The proof strategy is to show that each of the operators, $\neg, \wedge, \vee$, can be expressed using only $\uparrow$. Then any statement that uses these three operators can be replaced by a logically equivalent statement that uses only $\uparrow$. Since $\{\neg, \wedge, \vee\}$ is a functionally complete set of operators, $\{\uparrow\}$ must also be functionally complete.

The details of the strategy are straightforward. Three easy exercises with truth

[35] Think of NAND as "NOT AND" and NOR as "NOT OR".

tables show that

$$\begin{aligned} \neg P &\leftrightarrow (P \uparrow P) \\ (P \wedge Q) &\leftrightarrow [(P \uparrow Q) \uparrow (P \uparrow Q)] \\ (P \vee Q) &\leftrightarrow [(P \uparrow P) \uparrow (Q \uparrow Q)] \end{aligned}$$

□

There is a price for reducing the number of kinds of logic operators to allow. The next example illustrates this.

EXAMPLE 12.44 **Converting NOT, AND, OR to NAND**

The statement $(x \wedge (\neg y)) \vee z$ contains the three logic operators, $\neg$, $\wedge$, $\vee$. It is logically equivalent to one that contains only the operator, $\uparrow$. The logical equivalences introduced in the proof of Theorem 12.54 can be used to produce the new statement.

$$\begin{aligned} (x \wedge (\neg y)) \vee z &\Leftrightarrow ((x \uparrow (\neg y)) \uparrow (x \uparrow (\neg y))) \vee z \\ &\Leftrightarrow (((x \uparrow (\neg y)) \uparrow (x \uparrow (\neg y))) \uparrow \\ &\qquad ((x \uparrow (\neg y)) \uparrow (x \uparrow (\neg y)))) \uparrow (z \uparrow z) \\ &\Leftrightarrow (((x \uparrow (y \uparrow y)) \uparrow (x \uparrow (y \uparrow y))) \uparrow \\ &\qquad ((x \uparrow (y \uparrow y)) \uparrow (x \uparrow (y \uparrow y)))) \uparrow (z \uparrow z) \end{aligned}$$

The final statement contains only one kind of logic operator. However, the original statement contained only three operators in total; the new statement contains thirteen. ■

✔ **Quick Check 12.15**

For each statement, find a logically equivalent statement that contains only the NAND operator. Provide adequate justification for your answer.

1. $\neg(x \wedge y)$ **2.** $a \rightarrow b$ 

12.5.4 Exercises

The exercises marked with ★ have detailed solutions in Appendix H.

1. Use the Quine–McCluskey algorithm to simplify the following binary expressions. Show all the details.

(a) $w \cdot x \cdot \overline{y} \cdot z + w \cdot x \cdot \overline{y} \cdot \overline{z}$

(b) ★ $x \cdot y \cdot z + x \cdot y \cdot \overline{z} + x \cdot \overline{y} \cdot z + x \cdot \overline{y} \cdot \overline{z} + \overline{x} \cdot y \cdot z$

(c) $x \cdot y \cdot \overline{z} + x \cdot \overline{y} \cdot z + \overline{x} \cdot y \cdot z + \overline{x} \cdot y \cdot \overline{z} + \overline{x} \cdot \overline{y} \cdot \overline{z}$

(d) $x_1 \cdot x_2 \cdot x_3 \cdot x_4 + x_1 \cdot x_2 \cdot x_3 \cdot \overline{x_4} + x_1 \cdot x_2 \cdot \overline{x_3} \cdot x_4 + x_1 \cdot \overline{x_2} \cdot \overline{x_3} \cdot x_4$

2. Use the Quine–McCluskey algorithm to simplify the following binary expressions. Show all the details.

(a) $x_1 \cdot \overline{x_2} \cdot x_3 \cdot \overline{x_4} + \overline{x_1} \cdot x_2 \cdot x_3 \cdot x_4 + \overline{x_1} \cdot x_2 \cdot \overline{x_3} \cdot \overline{x_4} + \overline{x_1} \cdot \overline{x_2} \cdot \overline{x_3} \cdot \overline{x_4}$

(b) $x \cdot y \cdot z + x \cdot y \cdot \overline{z} + x \cdot \overline{y} \cdot z + x \cdot \overline{y} \cdot \overline{z} + \overline{x} \cdot y \cdot \overline{z} + \overline{x} \cdot \overline{y} \cdot z$

(c) $x \cdot y \cdot z + x \cdot \overline{y} \cdot z + x \cdot \overline{y} \cdot \overline{z} + \overline{x} \cdot y \cdot z + \overline{x} \cdot \overline{y} \cdot \overline{z}$

(d) $a \cdot b \cdot c \cdot \overline{d} + a \cdot \overline{b} \cdot c \cdot d + \overline{a} \cdot b \cdot c \cdot \overline{d} + \overline{a} \cdot \overline{b} \cdot c \cdot \overline{d}$

3. There are sixteen distinct binary functions of order 2. (See Example 12.30 on page 774 and Exercise 8 on page 784.) Use the binary expression simplification rule (and perhaps other Boolean axioms and properties) to simplify the associated binary expressions.

4. Simplify the following binary expressions. Show the details.

(a) ★ $w \cdot \left(\overline{x \cdot y + x}\right)$

(b) $(x + y) \cdot (x + \overline{y}) \cdot (y + z)$

(c) $(x + y) \cdot (x \cdot z) + (x + y)$

5. Simplify the following binary expressions. Show the details.

(a) $w \cdot \overline{(\overline{w} \cdot y)} \cdot \overline{(x + z)}$

(b) $\overline{x \cdot w \cdot \overline{z}} \cdot (\overline{x} + z) + w$

(c) $x \cdot \overline{y} + (y \cdot z + z) \cdot (\overline{z} + y)$

6. Create a combinatorial circuit diagram that represents each of the binary functions. Do not simplify the binary expressions.

(a) $f(x, y) = \overline{x + x \cdot \overline{y}}$

(b) ★ $f(x, y) = (\overline{x} + \overline{y}) \cdot \left(\overline{x + y}\right)$

(c) $f(x, y, z) = (x + y) \cdot (\overline{y} + z) + x \cdot y \cdot \overline{z}$

(d) $f(x, y, z) = (x \cdot y + y \cdot \overline{z}) + \overline{x \cdot \overline{y}}$

7. For each binary expression in Exercise 6,

i. Convert the binary expression to an equivalent expression in disjunctive normal form.

ii. Use the Quine–McCluskey algorithm to minimize the expression in part (i).

iii. Create a combinatorial circuit diagram that represents the binary function defined by the expression in part (ii).

8. ★ Design a combinatorial circuit for the simplified version of each expression in Exercise 1.

9. Design a combinatorial circuit for the simplified version of the expression in Exercise 2, part(d). Use only two-input AND and OR gates.

10. Design a combinatorial circuit that simulates a pair of wall switches that control a single light. The light will be on if both switches are in the "up" position or if both switches are in the "down" position. If one switch is "up" and the other switch is "down", the light will be off.

11. I once bought a computer version of a popular TV quiz show. The players were each assigned a key on the keyboard to press if they knew the answer to the current question. The first contestant to press a key was supposed to be chosen to answer the question (and gain points if a correct answer was given). Unfortunately, the program was poorly written. It seems that the program would show the question, then start a timer. It apparently gathered all key presses during a time window lasting a few seconds. It would then look through the list of keys pressed and choose the first pressed key. The list was always scanned in the same order, so player 1 had an unfair advantage.

Suppose, for example, that players 1, 2, 3, and 4 used keys a, b, c, and d, respectively. If keys a, c, and d were pressed at about the same time, the program would always choose player 1. If keys b and d were pressed at nearly the same time, the program always choose player 2. Thus, player 1 was favored over 2, 3, and 4, player 2 was favored over 3 and 4, and player 3 was favored over player 4.

Design a combinatorial circuit to simulate this defective program.

12. Recall the strategy for designing a combinatorial circuit to simulate a game of Paper, Rock, Scissors with output variables p, r, s, each designating a potential winning object. An alternative strategy is to use output variables, w_1, w_2, t, which represent player 1 wins, player 2 wins, a tie, respectively. All three output variables will be set to 0 for illegal input values.

(a) Create binary expressions to represent the binary functions which specify the appropriate behavior of the output variables, w_1, w_2, t.

(b) Design a combinatorial circuit to simulate this new strategy.

13. ★ A single mom has made an agreement with her kids, Annabelle and Boris, about eating out on Saturdays. If Mom wants to eat at home, they will not eat out. However, if Mom wants to eat out, at least one of the kids must also want to eat out or else they will eat at home. Design a combinatorial circuit that will correctly decide whether they eat out on Saturday. Show details for the entire process.

14. Mom spends a lot of time chauffeuring her three children. The children all like mom, so they fight over who gets to sit in the front of the car with her. Mom has devised a fair way to determine which kid gets the front seat. Each child will flip a coin. If all three are heads or all three are tails, they must all flip again. Otherwise, the child with the odd face gets the front seat. (No, not the kid who makes the funniest face!) Thus, if Sharayah's coin shows a head, Adelyn's coin shows a tail, and Landon's coin shows a head, Adelyn gets the front seat.

Follow the process outlined in this section to devise a combinatorial circuit that simulates mom's seating process.

15. Design a combinatorial circuit that will add two 2-bit binary numbers, perhaps resulting in a 3-bit result. Show details for the entire process.

16. Design a combinatorial circuit that will multiply two 2-bit binary numbers, resulting in a 4-bit result. Show details for the entire process.

17. Three friends meet each week for dinner. One friend drives everyone to the restaurant, on a rotating basis. They decide (by secret ballot) on the night of the dinner which of four favorite restaurants to visit. If at least two friends agree on a restaurant, then that is where they will go. If they all have different choices, the driver's choice is where they will eat. Designate the friends as x, y, and z, with x being the driver.

Design a combinatorial circuit that will correctly specify the restaurant. Show the function definition table(s) and the simplified functions. Building the circuit from the simplified functions is optional.

[*Hint*: Four restaurants can be specified with 2 bits, so use variables x_1, x_2, y_1, y_2, and z_1, z_2. You will need two output functions, f_1 and f_2. You may use the Quine–McCluskey applet at http://www.mathcs-bethel.edu/~gossett/DiscreteMathWithProof/ to create the minimized binary expressions. You can even rename the column headers as x_1, x_2, y_1, y_2, z_1, z_2 and just click to specify the function.]

18. Show that $\{\neg, \vee\}$ is a functionally complete set of logic operators.

19. Prove that $\{\downarrow\}$ is a functionally complete set of logic operators.

20. Use truth tables to prove that the following are tautologies.

(a) $\neg P \leftrightarrow (P \uparrow P)$

(b) ★ $(P \wedge Q) \leftrightarrow [(P \uparrow Q) \uparrow (P \uparrow Q)]$

(c) $(P \vee Q) \leftrightarrow [(P \uparrow P) \uparrow (Q \uparrow Q)]$

21. Find a logically equivalent statement that uses only the NAND operator for each statement. Provide adequate justification that your answer is logically equivalent to the original statement.

(a) $\neg(P \vee Q)$

(b) ★ $P \vee Q \vee R$

(c) $(\neg P) \wedge (\neg Q)$

(d) $(P \wedge Q) \vee R$

12.6 Quick Check Solutions

Quick Check 12.1

1. The domain and range are most naturally defined to be the set, $\mathbb{Z}^+$, of positive integers. The function can then be represented as

$$\mathcal{S} = \{(n, \sum_{j=1}^{n} j) \mid n \in \mathbb{Z}^+\}$$

Notice that the second coordinates in $\mathcal{S}$ are all distinct, since each one is larger than the previous second coordinate. The function is therefore one-to-one. Observe that the first few ordered pairs are $\{(1, 1), (2, 3), (3, 6), \ldots\}$. It is clear that many natural numbers (for example, 2, 4, and 5) do not appear as second coordinates. The function is therefore not onto.

2. This function has a countably infinite domain (the integers) and a finite range (the set of words {even, odd}). The function is not one-to-one because it contains distinct ordered pairs, (2, even) and (4, even), which have the same second coordinate. It is clearly onto since the ordered pairs (2, even) and (3, odd) are in the function. The function might be described as

$$\{\ldots, (-3, \text{odd}), (-2, \text{even}), (-1, \text{odd}), (0, \text{even}), (1, \text{odd}), (2, \text{even}), (3, \text{odd}), \ldots\}$$

Quick Check 12.2

1. $\mathcal{R}^{-1} = \{(a, 1), (b, 2), (c, 2), (c, 3)\}$
2. $\mathcal{S} \circ \mathcal{R} = \{(1, x), (1, z), (2, z), (2, y), (3, y)\}$

Quick Check 12.3

1. $\mathcal{R}_1$ is reflexive, is not symmetric, is transitive, and is not an equivalence relation.
2. $\mathcal{R}_2$ is reflexive, is not symmetric, is transitive (see Exercise 2 on page 746) and is not an equivalence relation.
3. $\mathcal{R}_3$ is not reflexive, is symmetric, is not transitive [since $\gcd(4, 6) = 2$ and $\gcd(6, 20) = 2$, but $\gcd(4, 20) \neq 2$], and is not an equivalence relation.
4. $\mathcal{R}_4$ is reflexive, symmetric, and transitive. It is therefore an equivalence relation.
5. $\mathcal{R}_5 = \{(-1, 1), (1, 1)\}$. It is not reflexive (since $(-1, -1) \notin \mathcal{R}_5$). It is not symmetric (since $(1, -1) \notin \mathcal{R}_5$). It *is* transitive, since $x = -1, y = 1, z = 1$ and $x = 1, y = 1, z = 1$ are the only choices for x, y, and z in the definition of *transitivity*.

Quick Check 12.4

1. $\mathcal{R}_1$ is not antireflexive ($x \le y$ does not imply $x \ne y$), it *is* antisymmetric (if $x \le y$ and $x \ne y$, then in fact, $x < y$, so $y \not\le x$), but it is not asymmetric (since $4 \le 4$).
2. $\mathcal{R}_2$ is not antireflexive ($4|4$), it *is* antisymmetric (if $x \mid y$ and $x \ne y$, then in fact, x is a proper factor of y, so y cannot be a divisor of x), and it is not asymmetric ($4 \mid 4$ again).
3. $\mathcal{R}_3$ is not antireflexive ($\gcd(2, 2) = 2$) and is neither antisymmetric nor asymmetric [both fail because $\gcd(2, 6) = \gcd(6, 2) = 2$ but $2 \ne 6$].
4. $\mathcal{R}_4$ is neither antireflexive, nor antisymmetric, nor asymmetric.
5. $\mathcal{R}_5 = \{(-1, 1), (1, 1)\}$. This relation is not antireflexive because $(1, 1)$ is in $\mathcal{R}_5$. It *is* antisymmetric, but is *not* asymmetric [since $(1, 1)$ is in $\mathcal{R}_5$].

Quick Check 12.5

1. This is quite easy to see intuitively. However, it is helpful to attempt a carefully worded proof.

   Reflexivity holds because every student lives in the same dorm as his or her own self. (This may seem a bit artificial, but it is valid.) If student x lives in the same dorm as student y, then changing the order of their names does not cause them to live elsewhere; y and x live in the same dorm. Thus, symmetry holds.

   Suppose that students x and y live in the same dorm. For the moment, assume the dorm is named "Marshall Hall". Suppose also that students y and z live in the same dorm. We already know that y lives in Marshall Hall, so z also lives in Marshall Hall. But then x and z both live in Marshall Hall, so they live in the same dorm. This establishes the transitivity property.

   Since the relation is reflexive, symmetric, and transitive, it is an equivalence relation.

2. There is one equivalence class per dorm. The equivalence classes consist of the sets of students living in a common dorm.

Quick Check 12.6

1. The attribute set is {Variety, Vegetable, Germination, Harvest}.

2. It is currently consistent with the table that every other attribute is functionally dependent on "Variety" (and also on "Harvest"). However, if the table is extended, it is quite possible to have a variety name repeated. For example, Bush Champion is a likely name for a variety of green bean (in addition to the current cucumber variety). If such a variety is added to the table, then knowing the variety name does not uniquely determine the rest of the tuple. It is almost certain to have repeated Harvest values, so the other attributes will not be functionally dependent on Harvest.

   A reasonable model would require the attributes "Germination" and "Harvest" to be functionally dependent on {Variety, Vegetable}.

3. The discussion in part (2) shows that the primary key should be {Variety, Vegetable}.

Quick Check 12.7

1. **Adoptive[Mother, Adoptive Child]**

Mother	**Adoptive Child**
Jane Smith	Carmen Smith
Jane Smith	Polly Smith
Anita Rodriguez	Tran-minh Rodriguez
Helen Levitz	Aaron Levitz
Helen Levitz	Hanna Levitz
Betty Jones	Samantha Jones

2. The three steps are listed below.

   - Form the Cartesian product (Table 12.37 on page 799). (This would normally all be done by computer.) Some of the first names have been abbreviated to fit the tuples on the page.

TABLE 12.37 The Cartesian Product Biological * Adoptive

Father	Mother	Biological Child	Father	Mother	Adoptive Child
John Smith	Jane Smith	William Smith	John Smith	Jane Smith	Carmen Smith
John Smith	Jane Smith	William Smith	John Smith	Jane Smith	Polly Smith
John Smith	Jane Smith	William Smith	E. Rodriguez	A. Rodriguez	T. Rodriguez
John Smith	Jane Smith	William Smith	Isaac Levitz	Helen Levitz	Aaron Levitz
John Smith	Jane Smith	William Smith	Isaac Levitz	Helen Levitz	Hanna Levitz
John Smith	Jane Smith	William Smith	Bob Jones	Betty Jones	Samantha Jones
John Smith	Jane Smith	Susan Smith	John Smith	Jane Smith	Carmen Smith
John Smith	Jane Smith	Susan Smith	John Smith	Jane Smith	Polly Smith
John Smith	Jane Smith	Susan Smith	E. Rodriguez	A. Rodriguez	T. Rodriguez
John Smith	Jane Smith	Susan Smith	Isaac Levitz	Helen Levitz	Aaron Levitz
John Smith	Jane Smith	Susan Smith	Isaac Levitz	Helen Levitz	Hanna Levitz
John Smith	Jane Smith	Susan Smith	Bob Jones	Betty Jones	Samantha Jones
E. Rodriguez	A. Rodriguez	Pablo Rodriguez	John Smith	Jane Smith	Carmen Smith
E. Rodriguez	A. Rodriguez	Pablo Rodriguez	John Smith	Jane Smith	Polly Smith
E. Rodriguez	A. Rodriguez	Pablo Rodriguez	E. Rodriguez	A. Rodriguez	T. Rodriguez
E. Rodriguez	A. Rodriguez	Pablo Rodriguez	Isaac Levitz	Helen Levitz	Aaron Levitz
E. Rodriguez	A. Rodriguez	Pablo Rodriguez	Isaac Levitz	Helen Levitz	Hanna Levitz
E. Rodriguez	A. Rodriguez	Pablo Rodriguez	Bob Jones	Betty Jones	Samantha Jones
W. Leblanc	M. Leblanc	Wanda Leblanc	John Smith	Jane Smith	Carmen Smith
W. Leblanc	M. Leblanc	Wanda Leblanc	John Smith	Jane Smith	Polly Smith
W. Leblanc	M. Leblanc	Wanda Leblanc	E. Rodriguez	A. Rodriguez	T. Rodriguez
W. Leblanc	M. Leblanc	Wanda Leblanc	Isaac Levitz	Helen Levitz	Aaron Levitz
W. Leblanc	M. Leblanc	Wanda Leblanc	Isaac Levitz	Helen Levitz	Hanna Levitz
W. Leblanc	M. Leblanc	Wanda Leblanc	Bob Jones	Betty Jones	Samantha Jones
R. Westlund	V. Westlund	Derwin Westlund	John Smith	Jane Smith	Carmen Smith
R. Westlund	V. Westlund	Derwin Westlund	John Smith	Jane Smith	Polly Smith
R. Westlund	V. Westlund	Derwin Westlund	E. Rodriguez	A. Rodriguez	T. Rodriguez
R. Westlund	V. Westlund	Derwin Westlund	Isaac Levitz	Helen Levitz	Aaron Levitz
R. Westlund	V. Westlund	Derwin Westlund	Isaac Levitz	Helen Levitz	Hanna Levitz
R. Westlund	V. Westlund	Derwin Westlund	Bob Jones	Betty Jones	Samantha Jones
R. Westlund	V. Westlund	Darwin Westlund	John Smith	Jane Smith	Carmen Smith
R. Westlund	V. Westlund	Darwin Westlund	John Smith	Jane Smith	Polly Smith
R. Westlund	V. Westlund	Darwin Westlund	E. Rodriguez	A. Rodriguez	T. Rodriguez
R. Westlund	V. Westlund	Darwin Westlund	Isaac Levitz	Helen Levitz	Aaron Levitz
R. Westlund	V. Westlund	Darwin Westlund	Isaac Levitz	Helen Levitz	Hanna Levitz
R. Westlund	V. Westlund	Darwin Westlund	Bob Jones	Betty Jones	Samantha Jones

- Remove any tuples for which the duplicate attributes {Father, Mother} do not have identical values (Table 12.38).

TABLE 12.38 The Tuples with Identical Values for Duplicate Attributes

Father	Mother	Biological Child	Father	Mother	Adoptive Child
John Smith	Jane Smith	William Smith	John Smith	Jane Smith	Carmen Smith
John Smith	Jane Smith	William Smith	John Smith	Jane Smith	Polly Smith
John Smith	Jane Smith	Susan Smith	John Smith	Jane Smith	Carmen Smith
John Smith	Jane Smith	Susan Smith	John Smith	Jane Smith	Polly Smith
E. Rodriguez	A. Rodriguez	Pablo Rodriguez	E. Rodriguez	A. Rodriguez	T. Rodriguez

- Project onto {Father, Mother, Biological Child, Adoptive Child} (Table 12.39).

TABLE 12.39 The Projection onto {Father, Mother, Biological Child, Adoptive Child}

Father	Mother	Biological Child	Adoptive Child
John Smith	Jane Smith	William Smith	Carmen Smith
John Smith	Jane Smith	William Smith	Polly Smith
John Smith	Jane Smith	Susan Smith	Carmen Smith
John Smith	Jane Smith	Susan Smith	Polly Smith
E. Rodriguez	A. Rodriguez	Pablo Rodriguez	T. Rodriguez

3. The projection (Biological * Adoptive)[Father, Mother] contains all couples who have both biological and adoptive children.

Father	Mother
John Smith	Jane Smith
E. Rodriguez	A. Rodriguez

Quick Check 12.8

1. This relation is in first normal form because there are no attributes with sets as values. The inferred essential dependencies are listed next.

$$\{\text{Team, Player, Position}\} \rightarrow \text{Captain}$$

The only key is {Team, Player, Position}, so it is the primary key. This relation is in third normal form because the only functional dependency has a key for the left-hand side.

2. This relation is in first normal form because there are no attributes with sets as values. The inferred essential dependencies are listed next.

$$\text{EmployeeID} \rightarrow \text{Name}$$
$$\text{EmployeeID} \rightarrow \text{Job Grade}$$
$$\text{EmployeeID} \rightarrow \text{Salary}$$
$$\text{Job Grade} \rightarrow \text{Salary}$$

Thus, the primary (and only) key is EmployeeID. Since there are no proper subsets of the primary key, this relation is in second normal form. However, the functional dependency Job Grade $\rightarrow$ Salary prevents this from being in third normal form.

Quick Check 12.9

1. The relation ScheduleB[Instructor, Office, Teaching Assistant] is in third normal form because either Instructor or Office could be chosen as primary key. (The most likely choice for primary key is Instructor, with Office as alternate key.) Teaching Assistant is the only choice for B in Definition 12.29. However, there is no choice for D that does not contain a key.

   The decomposition {ScheduleB[Course, Section, Semester, Instructor}, ScheduleB[Instructor, Office, Teaching Assistant]} is a viable alternative decomposition to the decomposition to {ScheduleB[Instructor, Teaching Assistant}, ScheduleB[Course, Section, Semester, Instructor, Office]}.

Quick Check 12.10

1. (a) This is not a minterm because it contains the constant, 1.
   (b) This is not a minterm, but it *is* in disjunctive normal form. A minterm must contain each of the candidate binary variables in either complemented or uncomplemented form.
   (c) This is a minterm in x_1, x_2, x_3.
2. (a) This binary expression is in disjunctive normal form.
   (b) This is not in disjunctive normal form because it is not a sum of minterms (0 is not a minterm) and it is not just 0.
   (c) This binary expression is in disjunctive normal form. (The summation may contain just one term.)

Quick Check 12.11

1. There are four ordered triples at which f evaluates to 1. The corresponding minterms are listed.

triple	minterm
(0,1,0)	$\overline{x_1} \cdot x_2 \cdot \overline{x_3}$
(0,1,1)	$\overline{x_1} \cdot x_2 \cdot x_3$
(1,1,0)	$x_1 \cdot x_2 \cdot \overline{x_3}$
(1,1,1)	$x_1 \cdot x_2 \cdot x_3$

   Thus, it is possible to write

$$f(x_1, x_2, x_3) = \overline{x_1} \cdot x_2 \cdot \overline{x_3} + \overline{x_1} \cdot x_2 \cdot x_3 + x_1 \cdot x_2 \cdot \overline{x_3} + x_1 \cdot x_2 \cdot x_3$$

Quick Check 12.12

1. **Move all complements onto single variables**

$$\overline{(0+x_1)}\cdot\overline{x_2}+x_1\cdot(x_2+\overline{x_3}) = \left(\overline{0}\cdot\overline{x_1}\right)\cdot\overline{x_2}+x_1\cdot(x_2+\overline{x_3}) \text{ De Morgan}$$
$$= (1\cdot\overline{x_1})\cdot\overline{x_2}+x_1\cdot(x_2+\overline{x_3}) \text{ complement of 0}$$

Transform to a sum of products

$$(1\cdot\overline{x_1})\cdot\overline{x_2}+x_1\cdot(x_2+\overline{x_3}) = (1\cdot\overline{x_1})\cdot\overline{x_2}+x_1\cdot x_2+x_1\cdot\overline{x_3} \text{ distributivity}$$
$$= 1\cdot\overline{x_1}\cdot\overline{x_2}+x_1\cdot x_2+x_1\cdot\overline{x_3} \quad \text{associativity}$$

Transform into disjunctive normal form

$$\begin{aligned}
&1\cdot\overline{x_1}\cdot\overline{x_2}+x_1\cdot x_2+x_1\cdot\overline{x_3} \\
&\quad = 1\cdot(\overline{x_1}\cdot\overline{x_2})+x_1\cdot x_2+x_1\cdot\overline{x_3} && \text{associativity} \\
&\quad = (\overline{x_1}\cdot\overline{x_2})\cdot 1+x_1\cdot x_2+x_1\cdot\overline{x_3} && \text{commutativity} \\
&\quad = \overline{x_1}\cdot\overline{x_2}+x_1\cdot x_2+x_1\cdot\overline{x_3} && \text{identity} \\
&\quad = \overline{x_1}\cdot\overline{x_2}+x_1\cdot x_2 \\
&\qquad +x_1\cdot\overline{x_3}\cdot x_2+x_1\cdot\overline{x_3}\cdot\overline{x_2} && \text{replacement algorithm } (i=2) \\
&\quad = \overline{x_1}\cdot\overline{x_2}\cdot x_3+\overline{x_1}\cdot\overline{x_2}\cdot\overline{x_3}+x_1\cdot x_2 \\
&\qquad +x_1\cdot\overline{x_3}\cdot x_2+x_1\cdot\overline{x_3}\cdot\overline{x_2} && \text{replacement algorithm } (i=3) \\
&\quad = \overline{x_1}\cdot\overline{x_2}\cdot x_3+\overline{x_1}\cdot\overline{x_2}\cdot\overline{x_3}+x_1\cdot x_2\cdot x_3 \\
&\qquad +x_1\cdot x_2\cdot\overline{x_3}+x_1\cdot\overline{x_3}\cdot x_2 \\
&\qquad +x_1\cdot\overline{x_3}\cdot\overline{x_2} && \text{replacement algorithm } (i=3)
\end{aligned}$$

Now sort within terms to produce minterms.

$$\overline{x_1}\cdot\overline{x_2}\cdot x_3+\overline{x_1}\cdot\overline{x_2}\cdot\overline{x_3}+x_1\cdot x_2\cdot x_3+x_1\cdot x_2\cdot\overline{x_3}+x_1\cdot x_2\cdot\overline{x_3}+x_1\cdot\overline{x_2}\cdot\overline{x_3}$$

Sort and reduce into a unique disjunctive normal form

Sort the minterms.

$$x_1\cdot x_2\cdot x_3+x_1\cdot x_2\cdot\overline{x_3}+x_1\cdot x_2\cdot\overline{x_3}+x_1\cdot\overline{x_2}\cdot\overline{x_3}+\overline{x_1}\cdot\overline{x_2}\cdot x_3+\overline{x_1}\cdot\overline{x_2}\cdot\overline{x_3}$$

Remove duplicate minterms.

$$x_1\cdot x_2\cdot x_3+x_1\cdot x_2\cdot\overline{x_3}+x_1\cdot\overline{x_2}\cdot\overline{x_3}+\overline{x_1}\cdot\overline{x_2}\cdot x_3+\overline{x_1}\cdot\overline{x_2}\cdot\overline{x_3}$$

Check the answer.

x_1	x_2	x_3	$\overline{(0+x_1)}\cdot\overline{x_2}+x_1\cdot(x_2+\overline{x_3})$
0	0	0	1
0	0	1	1
0	1	0	0
0	1	1	0
1	0	0	1
1	0	1	0
1	1	0	1
1	1	1	1

The minterms in the expression $x_1\cdot x_2\cdot x_3+x_1\cdot x_2\cdot\overline{x_3}+x_1\cdot\overline{x_2}\cdot\overline{x_3}+\overline{x_1}\cdot\overline{x_2}\cdot x_3+\overline{x_1}\cdot\overline{x_2}\cdot\overline{x_3}$ correspond to the ordered triples where the original expression evaluates to 1.

Quick Check 12.13

1. The two phases are shown.

Phase 1:

1	110	✓	1, 4 −10
2	101		3, 4 01−
3	011	✓	
4	010	✓	

Phase 2:

	1 **110**	**2** **101**	**3** **011**	**4** **010**
101		×		
−10	×			×
01−			×	×

To cover every column, all three of the rows are required. The minimized binary expression is thus $x \cdot \overline{y} \cdot z + y \cdot \overline{z} + \overline{x} \cdot y$.

Quick Check 12.14

1. The first step is to create a binary function that matches the desired circuit.

x	y	z	$f(x, y, z)$
0	0	0	0
0	0	1	0
0	1	0	0
0	1	1	1
1	0	0	0
1	0	1	1
1	1	0	1
1	1	1	1

A binary expression in disjunctive normal form that represents this function is

$$\overline{x} \cdot y \cdot z + x \cdot \overline{y} \cdot z + x \cdot y \cdot \overline{z} + x \cdot y \cdot z.$$

Now use Quine–McCluskey to simplify the expression.

Phase 1:

1	011	✓	1, 4 −11
2	101	✓	2, 4 1−1
3	110	✓	3, 4 11−
4	111	✓	

Phase 2:

	1 $\overline{x} \cdot y \cdot z$	**2** $x \cdot \overline{y} \cdot z$	**3** $x \cdot y \cdot \overline{z}$	**4** $x \cdot y \cdot z$
−11	×			×
1−1		×		×
11−			×	×

All three of the rows are needed. The simplified function is

$$f(x, y, z) = y \cdot z + x \cdot z + x \cdot y$$

A circuit that implements this function is shown in Figure 12.12.

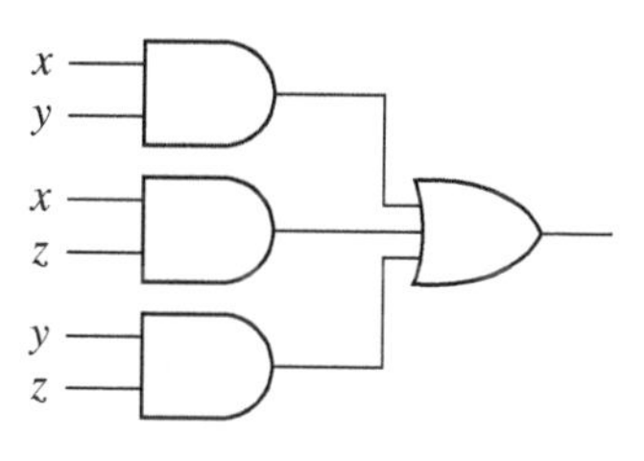

Figure 12.12. A combinatorial circuit for a three-way majority.

Quick Check 12.15

1. $\neg(x \wedge y) \Leftrightarrow \neg((x \uparrow y) \uparrow (x \uparrow y))$

$\Leftrightarrow ((x \uparrow y) \uparrow (x \uparrow y)) \uparrow ((x \uparrow y) \uparrow (x \uparrow y))$

2. $a \to b \Leftrightarrow (\neg a) \vee b$

$\Leftrightarrow (a \uparrow a) \vee b$

$\Leftrightarrow ((a \uparrow a) \uparrow (a \uparrow a)) \uparrow (b \uparrow b)$

12.7 CHAPTER REVIEW

12.7.1 Summary

This chapter is about functions and relations, together with some important applications. You have been working with functions for many years. Relations, which are a generalization of functions, may be less familiar. The material in this chapter demonstrates that this is a profitable generalization.

The chapter begins with a review of the notion of a function (Section 12.1). The review presents functions as special subsets of a Cartesian product. This definition is equivalent to the more familiar notion of a mapping, but it provides a simpler transition to the notion of a relation. The section includes discussions of the important notions of one-to-one, onto, composition, and inverse.

Section 12.2 discusses several properties that may apply to a particular relation. Three of these properties are especially noteworthy: reflexive, symmetric, and transitive. A relation for which all three of these properties apply is called an equivalence relation. Equivalence relations play an important role in algebraic structures courses. Examples 12.10 and 12.12 give a glimpse of that role.

Section 12.3 generalizes the notion of a relation to that of an n-ary relation. Perhaps the most visible application of n-ary relations is their use as the foundation for relational databases. Some of the power and flexibility of relational databases derives from the firm theoretical foundation that is available from the mathematical theory of relations. This section does not attempt to provide a well-rounded introduction to relational databases. Instead, it focuses on the concept of normal forms. By converting the relations in a relational database to an appropriate normal form, some potential inconsistencies in the database can be avoided.

Section 12.4 switches back from relations to functions. The section starts with a brief look at functions whose domain is a Boolean algebra but concentrates on functions whose domain is the set $\{0, 1\}$. These binary functions are often expressed by using binary expressions. The study of binary expressions is aided by introducing minterms and disjunctive normal form. Binary functions have much more inherent structure than do functions whose domain is the real numbers. The major result of this section is a proof that every binary function can be expressed in disjunctive normal form.

Section 12.5 builds on the foundation laid in Section 12.4. It begins with the Quine–McCluskey algorithm for minimizing a binary expression which is in disjunctive normal form. A direct application is in the design of combinatorial circuits. A binary function can be constructed that specifies the input–output behavior of the desired circuit. That function can be expressed in disjunctive normal form, which can then be minimized using the Quine–McCluskey algorithm. The minimized expression can then be converted into a circuit design with fewer components than would have been the case for the original version of the function.

This chapter effectively illustrates that there is great depth to even apparently simple mathematical ideas. The notion of a function is a familiar part of the modern mathematical landscape. However, even that familiar idea can be extended in interesting and useful ways.

Most of the material in this chapter is very straightforward and can be mastered with a reasonable investment of time and effort. The material on normal forms in relational databases is perhaps the hardest part of the chapter. You might want to spend extra time reviewing that section.

12.7.2 Notation

Notation	Page	Brief Description
$\mathcal{G} \circ \mathcal{F}$	734	the composition of the functions (or relations), $\mathcal{F}$ and $\mathcal{G}$
$\mathcal{R}^{-1}$	736	the inverse relation of $\mathcal{R}$
$a\mathcal{R}b$	735	an alternative way to indicate that $(a, b) \in \mathcal{R}$
$\overline{\mathcal{R}}_r$	741	the reflexive closure of the relation, $\mathcal{R}$
$\overline{\mathcal{R}}_s$	741	the symmetric closure of the relation, $\mathcal{R}$
$\overline{\mathcal{R}}_t$	741	the transitive closure of the relation, $\mathcal{R}$
$\mathcal{R}^n$	742	The composition of the relation, $\mathcal{R}$, with itself n times
$[x]$	743	the equivalence class of element, x
$\mathcal{T}[B_1, B_2, \ldots, B_j]$	754	the projection of a relation, $\mathcal{T}$, onto attributes $\{B_1, B_2, \ldots, B_j\}$
$\mathcal{T}_1 * \mathcal{T}_2$	754	the join of the relations, $\mathcal{T}_1$ and $\mathcal{T}_2$
$\{A_1, A_2, \ldots, A_j\} \rightarrow B$	756	B is functionally dependent on $\{A_1, A_2, \ldots, A_j\}$
$\overline{x}$	775	the complement of the binary variable, x (interchanges 0 and 1)
$\hat{x}$	775	generic notation for either x or $\overline{x}$
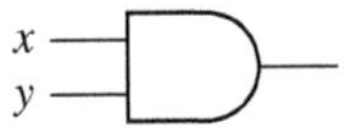 (AND gate symbol with inputs x, y)	790	the standard symbol for an AND gate
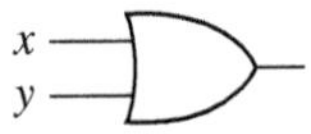 (OR gate symbol with inputs x, y)	790	the standard symbol for an OR gate
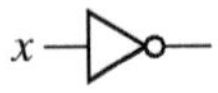 (NOT gate symbol with input x)	790	the standard symbol for a NOT gate
$\uparrow$	794	the NAND logic operator
$\downarrow$	794	the NOR logic operator

12.7.3 Additional Review Material

Go to http://www.mathcs.bethel.edu/~gossett/DiscreteMathWithProof/review.xhtml for additional review material, including a sample chapter exam, with solutions.

APPENDIX A

Number Systems

This appendix contains brief descriptions of and interesting facts about the commonly used number systems. It is not intended to be a formal or complete introduction.

Informally, a *number system* consists of a set of mathematical abstractions called *numbers*, together with the operations *addition* and *multiplication*.

A number system therefore consists of more than just the numbers. The operations (which take pairs of numbers and produce another number) are an essential ingredient. *Subtraction* and *division* can be defined by using addition and multiplication, together with the notion of an *inverse* (defined later).

A.1 The Natural Numbers

The simplest number system is the set of *natural numbers*:

$$\mathbb{N} = \{0, 1, 2, 3, 4, \dots\}$$

These numbers arise naturally in the context of counting. Very young children can easily grasp the concept of attaching a number (or number name) to a quantity, irrespective of what kind of objects are being considered (4 balls, 4 dolls, 4 houses, etc.).

This system of numbers has a limited mathematical structure. There is a notion of addition but only a limited notion of subtraction ($4 - 6$ is no longer a natural number). Multiplication of natural numbers is defined, but division generally does not work ($7 \div 3$ is not a natural number).

Every natural number has a uniquely defined *successor* (5 is the successor of 4). Every natural number except 0 also has a uniquely defined *predecessor* (8 is the predecessor of 9). This is not the case with the rational, real, and complex numbers. The notions of successor and predecessor can be used to define the terms *greater than* and *less than* for elements of $\mathbb{N}$. This leads to the seemingly trivial but actually significant well-ordering principle:

AXIOM 3.1 ***The Well-Ordering Principle***
Every nonempty set of natural numbers has a smallest element.

The inclusion of the number 0 permits discussion of the absence of some object. It also is a significant part of a place value representation of numbers. For example, the number 203 means "two one hundreds, no tens, and three units". Common use of a special symbol (such as "0") to represent the notion of "none" was a fairly late occurrence in the history of mathematics.

One simple property[1] of the natural numbers is that whenever a product of two such numbers is zero, at least one of those numbers must also be zero:

$$ab = 0 \quad \text{implies that} \quad a = 0 \text{ or } b = 0 \text{ (or both are zero)}$$

This property is sometimes called *the zero product principle.*

You should note that some authors define the natural numbers to be the set $\{1, 2, 3, \dots\}$. That is, they exclude zero.

A.2 The Integers

The natural numbers do not allow sufficient flexibility for the kinds of mathematical concepts and calculations we commonly need. In particular, subtraction is not always meaningful. For example, there is no natural number to represent $4 - 17$. The *integers* correct this deficiency by introducing some negative numbers:

$$\mathbb{Z} = \{\dots, -4, -3, -2, -1, 0, 1, 2, 3, 4, \dots\}$$

The letter $\mathbb{Z}$ is traditionally used to denote the integers.[2] Two important subsets of $\mathbb{Z}$ are the set of positive integers, $\mathbb{Z}^+ = \{1, 2, 3, \dots\}$, and the set of nonnegative integers (another name for the set of natural numbers).

The introduction of negative numbers allows convenient mathematical manipulation of notions such as "a deficit of \$1000" or "5 degrees below 0". The Chinese were comfortably using negative numbers in financial calculations somewhere between 500 B.C. and 250 A.D., probably closer to the earlier date [93].

The integers have a well-defined addition, subtraction, and multiplication. There is still no fully developed division, however. A significant feature of the integers is the notion of prime numbers and factorization. Recall that a prime number is a positive integer $n > 1$ that is not evenly divisible by any other positive integers except itself and 1.

THEOREM 3.15 ***The Fundamental Theorem of Arithmetic***

Every integer n, with $n \geq 2$, can be written uniquely as a product of primes in ascending order.

Factorization provides a mechanism to partially overcome the lack of a full division. You are familiar with "short division" from elementary school. The key ideas are listed in the next theorem.

THEOREM 3.9 ***The Quotient–Remainder Theorem***

Let a and b be integers with $b \neq 0$. Then there exist unique integers q and r such that $a = bq + r$ and $0 \leq r < |b|$.

The integer q is called the *quotient* and r is called the *remainder.*

A.3 The Rational Numbers

The set of *rational numbers*, denoted $\mathbb{Q}$, extends the set of integers in a manner that provides a fully developed notion of division. That is, if p and q are two rational numbers, then $p \div q$ with $q \neq 0$ is completely determined and is itself a rational number.

$$\mathbb{Q} = \left\{ \frac{p}{q} \mid p \in \mathbb{Z}, q \in \mathbb{Z}, \text{ and } q \neq 0 \right\}$$

[1] A property shared by the other number systems presented here, but not by all algebraic systems.

[2] The German word for "numbers" is *zahlen.*

Thus, $\mathbb{Q}$ is the familiar set of all fractions.[3]

The set of rationals is the first of the number systems listed here that satisfies all the *field axioms*:

> A *field* is a set **F** of elements, together with two operations on those elements. The operations will be symbolically denoted as "+" and "·". Each operation takes two elements (or two copies of one element) and produces another (not necessarily distinct) element. There are at least two elements, denoted 0 and 1, with $0 \neq 1$. The elements and operations obey the following axioms.
>
> **Closure** Given any two elements x and y in **F**, both $x + y$ and $x \cdot y$ are also in **F**.
>
> **Associativity** For any x, y, and z in **F**,
>
> $$(x + y) + z = x + (y + z)$$
>
> and
>
> $$(x \cdot y) \cdot z = x \cdot (y \cdot z)$$
>
> **Commutativity** For any x and y in **F**,
>
> $$x + y = y + x$$
>
> and
>
> $$x \cdot y = y \cdot x$$
>
> **Identities** There is an element of **F**, denoted 0, such that for any element x in **F**,
>
> $$x + 0 = 0 + x = x$$
>
> There is an element of **F**, denoted 1, such that for any element x in **F**,
>
> $$x \cdot 1 = 1 \cdot x = x$$
>
> **Inverses** For any element x in **F** there is an element, denoted $-x$, such that
>
> $$x + (-x) = (-x) + x = 0$$
>
> For any element y in **F**, except 0, there is an element, denoted y^{-1}, such that
>
> $$y \cdot y^{-1} = y^{-1} \cdot y = 1$$
>
> **Distributivity** For any elements x, y, and z in **F**,
>
> $$x \cdot (y + z) = (x \cdot y) + (x \cdot z)$$

Subtraction and absolute value can be used to define the distance between two numbers: If a and b are any two rational numbers, define $|a - b|$ as the distance between them.

The set of rational numbers has the property of being *dense*. That is, given any two distinct rational numbers, it is possible to find another rational number that is between the original two.[4] One simple consequence of this is that given any rational number a, and a tiny distance d, it is possible to find another rational number b with $|a - b| < d$.

It would be tempting to think that rational numbers are all the numbers that we need. However, many of the numbers we need are not included. For example, if the Pythagorean theorem is applied to a triangle whose perpendicular sides both have length

[3] It might make more sense to use the letter $\mathbb{R}$ to denote the *rationals*, but that letter is reserved for the *real* numbers. The letter $\mathbb{Q}$ should remind you of the word *quotient*.

[4] A simple example would be the average of the two rational numbers.

1, the hypotenuse must have length $\sqrt{2}$. It is fairly easy[5] to show that $\sqrt{2}$ cannot be expressed in the form $\frac{a}{b}$, where a and b are integers. Thus $\sqrt{2}$ is not a rational number.

The ratio of the circumference of a circle to its diameter is commonly denoted as π. In 1767, Johann Heinrich Lambert showed that π is not a rational number.

Numbers such as $\sqrt{2}$ and π are called *irrational numbers*. There are many more irrational numbers than there are rational numbers (see [26] for more detail).

A.4 The Real Numbers

The set of *real numbers*, $\mathbb{R}$, consists of all the rational and all the irrational numbers. There are two visualizations that are helpful.

We can consider the set of real numbers to be all the possible numeric decimal expansions, where the digits to the right of the decimal point are allowed to go on forever. The set of rational numbers can be thought of as all the expansions for which the right-hand side of the decimal either terminates (ends in all 0s) or eventually repeats. Thus $\frac{1}{8} = 0.125$, and $\frac{811}{700} = 1.158\overline{571428}$ are rational.[6] All other decimal expansions are irrational. Hence, $\pi = 3.1415926535897932\ldots$ never terminates or repeats.

The other common visualization is to identify the set of real numbers with the set of points on a line (Figure A.1). This is the familiar number line that composes the x-axis of many of the function graphs in this text.

−4 −2 0 2 4

Figure A.1. The real number line.

The set of real numbers forms a field but also has two additional kinds of axioms: order axioms and a completeness axiom. The first collection of axioms specifies an *order* to the real numbers. One of these order axioms requires that for any two real numbers x and y, either $x > y$, or $y > x$, or $x = y$. The final axiom specifies that if a sequence of real numbers (such as the sequence $\{.1, .11, .111, .1111, \ldots\}$) gets closer and closer to some value ($.11111\ldots$ gets close to $\frac{1}{9}$ in this example), then the value they approach is also a real number. In more formal language, the set of real numbers contains all *limit points* of sequences of real numbers. This notion is called *completeness*. The set of real numbers is thus a *complete ordered field*.

The order axioms can be applied to the integers and rational numbers. However, there are sequences of rational numbers whose limit point is not a rational number. For example, the sequence of rational numbers

$$\{3, 3.1, 3.14, 3.141, 3.1415, 3.14159, \ldots\}$$

approaches the value π, which is not rational. Thus, the set of rational numbers is not complete.

The set of real numbers is also dense: Between any two distinct real numbers, there is always another (distinct) real number.

A.5 The Complex Numbers

The set of real numbers might seem to be a large enough set of numbers to answer all our mathematical questions adequately. However, there are some natural mathematical questions that have no solution if answers are restricted to be real numbers.

In particular, many simple equations have no solution in the real numbers. For example,

$$x^2 + 1 = 0$$

A solution would require a number whose square is -1. For many centuries, mathematicians were content with the answer, "there is no such number". Eventually, it became acceptable to allow the existence of a number, denoted i, such that $i^2 = -1$. Once the

[5] See Proposition 3.35 on page 104.

[6] The line over the digits 571428 indicates that the group repeats forever.

proper definitions of addition, subtraction, multiplication, and division were found, a new field, called the complex numbers, was available.[7]

$$\mathbb{C} = \{a + bi \mid a, b \in \mathbb{R}\}$$

That is, the set of complex numbers contains all expressions of the form $a + bi$, where a and b are real numbers, and $i = \sqrt{-1}$. The set of real numbers is a subset of $\mathbb{C}$, since $a + 0i = a$ is a real number for every a.

Addition and multiplication of complex numbers are defined by

$$\begin{aligned}(a + bi) + (c + di) &= (a + c) + (b + d)i \\ (a + bi) \cdot (c + di) &= ac + adi + bci + bdi^2 \\ &= ac + adi + bci - bd \\ &= (ac - bd) + (ad + bc)i\end{aligned}$$

The set of complex numbers is often visualized as the set of all points in the plane (Figure A.2). The complex number $a + bi$ is identified with the point (a, b) in the plane. The old x-axis is renamed the "real axis" and the old y-axis is renamed the "imaginary axis".[8]

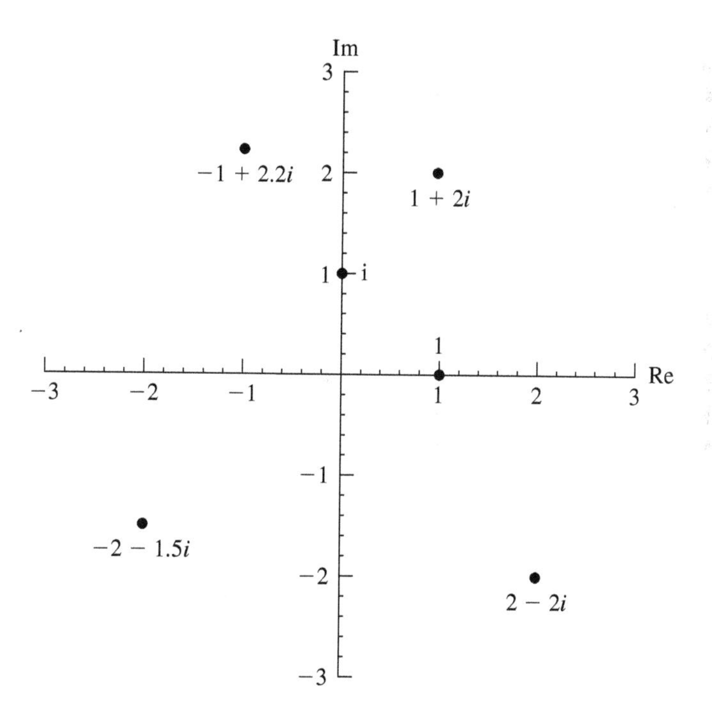

Figure A.2 The complex plane.

The set of complex numbers forms the final number system in the chain of number systems presented. The system of complex numbers is a complete field. However, the notion of order needs to be relaxed. For example, in Figure A.2, neither $i > 1$, nor $1 > i$, nor $i = 1$. The distance (in the complex plane) from a complex number to the origin can be used to obtain a weaker form of order. This distance is denoted by the familiar absolute value symbol. Thus $|i| = |1|$ and $|1 + 2i| > |i|$, as can be seen in Figure A.2.

One of the nice algebraic properties of complex numbers is the following theorem about polynomials with complex coefficients.

---

[7] Although square roots of negative numbers appeared in computations at least as early as 50 A.D., general acceptance of complex numbers as a meaningful and valid concept only occurred during the last half of the nineteenth century [96].

[8] The unfortunate term *imaginary* is a historical relic we are stuck with.

THEOREM A.1 ***The Fundamental Theorem of Algebra***

The polynomial

$$a_n x^n + a_{n-1} x^{n-1} + \cdots + a_2 x^2 + a_1 x + a_0$$

with complex coefficients $a_n, a_{n-1}, \ldots, a_1, a_0$, factors into a product of n linear factors with complex coefficients $a_n, c_n, \ldots, c_1$:

$$a_n(x + c_n)(x + c_{n-1}) \cdots (x + c_2)(x + c_1)$$

If the coefficients a_k are real numbers, there may be no real solutions to the equation

$$a_n x^n + a_{n-1} x^{n-1} + \cdots + a_2 x^2 + a_1 x + a_0 = 0$$

However, the fundamental theorem of algebra ensures that there will always be n complex solutions. Recall that the set of real numbers is a subset of the set of complex numbers, so some or all of the solutions may be real numbers. Also, some solutions may be counted more than once. Thus

$$x^2 + 4x + 4 = (x + 2)(x + 2) = 0$$

has the two solutions $x = -2$ and $x = -2$. That is, -2 is a multiple solution. The equation

$$x^2 + 1 = (x - i)(x + i) = 0$$

has as solutions $x = i$ and $x = -i$.

A.6 Other Number Systems

Mathematicians are aware of other "number" systems. The simplest alternative number system is called *arithmetic mod 2*. This system of numbers is defined by

$$\mathbb{Z}_2 = \{0, 1\}$$

with the following addition and multiplication tables:

+	0	1
0	0	1
1	1	0

·	0	1
0	0	0
1	0	1

This number system can be shown to be a field. It is used in error-correcting codes, generally with polynomials whose coefficients are in $\mathbb{Z}_2$.

Similar fields can be defined for any number p that is a prime:

$$\mathbb{Z}_p = \{0, 1, 2, 3, \ldots, p-1\}$$

with addition and multiplication defined *mod p*. That is, first do the addition or multiplication in the integers. Call the answer a. The mod p answer is the remainder upon dividing a by p. For example, in $\mathbb{Z}_5$

$$3 + 4 = 7 \pmod 5 = 2$$

and

$$2 \cdot 4 = 8 \pmod 5 = 3$$

A more formal name for these algebraic systems is *the integers mod p*. Arithmetic in the number systems $\mathbb{Z}_p$ is often called *clock arithmetic* due to the way numbers seem to wrap around (Figure A.3).

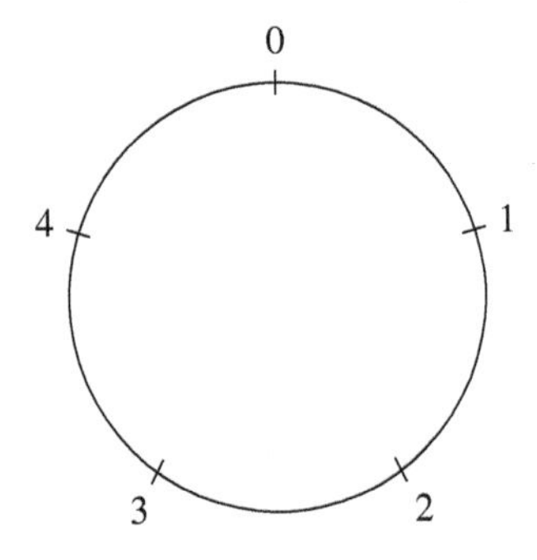

Figure A.3. Clock arithmetic.

For example, the sum 3 + 4 in $\mathbb{Z}_5$ can be envisioned by starting at 3 and moving 4 ticks clockwise, ending at 2, which agrees with the preceding example.

A.7 Representation of Numbers

You are familiar with the decimal representation of numbers. The key features of this system are that it uses base 10 and is a place value system. Thus, the digits in the number 123.45 represent

- 1 100 (10^2)
- 2 10s (10^1)
- 3 1s (10^0)
- 4 .1s (10^{-1})
- 5 .01s (10^{-2})

Ten is not the only base that has been used to represent numbers. The ancient Mayans used a base 20 representation, and the ancient Babylonians used a base 60 system. Our conventions that there are 60 minutes in an hour and 360 degrees in a circle are derived from the Babylonian representation. Perhaps one reason that 60 was chosen as a base is that 60 has a large number of prime factors:

$$60 = 2^2 \cdot 3 \cdot 5$$

The net effect is that numbers can be easily divided into halves, thirds, fourths, fifths, sixths, tenths, twelfths, fifteenths, twentieths, thirtieths, and sixtieths. Thus, fractional arithmetic and mercantile operations were simplified.

The most important modern alternatives to base 10 (decimal) are base 2 (binary), base 8 (octal), and base 16 (hexadecimal). These alternatives are important because computers use a binary representation for storing and manipulating numbers.[9] Octal and hexadecimal are used as convenient shorthand notations for binary numbers because binary numbers are difficult for humans to keep in short-term memory.

Binary numbers use the symbols {0, 1} and a place value notation. The symbols are called *bits*, an acronym for *bi*nary digi*ts*. Since the base is 2, each place represents a power of 2 (as opposed to a power of 10 in decimal representation). Thus the binary number 101.11 represents

- 1 4 (2^2)
- 0 2s (2^1)
- 1 1 (2^0)
- 1 $\frac{1}{2}$ (2^{-1})
- 1 $\frac{1}{4}$ (2^{-2})

If the subscripts $_2$ and $_{10}$ are understood to indicate the representation's base, then the number five and three-fourths can be expressed as 5.75_{10} and as 101.11_2. Both are valid representations of the same number. The binary representation is easier to store in a computer because we only need some medium that can exist in two states (such as magnetized-demagnetized, on–off).[10] The decimal form requires fewer digits to represent a number ($65,535_{10} = 1111111111111111_2$) and is generally easier for humans to use.

[9]So much arithmetic is currently done on computers or calculators that binary representations may be more important than the more familiar decimal representation.

[10]The manner in which computers actually store the binary number five and three-fourths is slightly more complex than "101.11." It is more like a binary version of scientific notation.

Table A.1 shows the binary, octal, and hexadecimal representations of the natural numbers from 0–31.

TABLE A.1 The Numbers 0 to 31 in Decimal, Binary, Octal, and Hexadecimal

Decimal	Binary	Octal	Hexadecimal	Decimal	Binary	Octal	Hexadecimal
0	0	0	0	16	10000	20	10
1	1	1	1	17	10001	21	11
2	10	2	2	18	10010	22	12
3	11	3	3	19	10011	23	13
4	100	4	4	20	10100	24	14
5	101	5	5	21	10101	25	15
6	110	6	6	22	10110	26	16
7	111	7	7	23	10111	27	17
8	1000	10	8	24	11000	30	18
9	1001	11	9	25	11001	31	19
10	1010	12	A	26	11010	32	1A
11	1011	13	B	27	11011	33	1B
12	1100	14	C	28	11100	34	1C
13	1101	15	D	29	11101	35	1D
14	1110	16	E	30	11110	36	1E
15	1111	17	F	31	11111	37	1F

You might observe common patterns in the two binary columns. One reason for the similarity is that 16_{10} is represented by 10000_2 in binary notation. Since, for example, $23_{10} = 16_{10} + 7_{10}$, the representation in binary is $10000_2 + 111_2 = 10111_2$. What should the binary, octal, and hexadecimal representations be for the number 32_{10}?

Octal representation uses the symbols {0, 1, 2, 3, 4, 5, 6, 7} and powers of 8. Thus the number five and three-fourths is 5.6_8; that is, one five and six eighths.

Hexadecimal representation uses the symbols

$$\{0, 1, 2, 3, 4, 5, 6, 7, 8, 9, \text{ A, B, C, D, E, F}\}$$

The symbol A is the hexadecimal symbol for the number 10. The symbol F represents the number 15. Hexadecimal representation uses powers of 16 for the place values. Five and three-fourths is $5.\text{C}_{16}$; that is, five and twelve sixteenths. Notice that $65{,}535_{10} = 177777_8 = \text{FFFF}_{16}$.

Because both 8 and 16 are powers of 2, it is easy to convert numbers between binary representation and octal or hexadecimal representation.

To convert from binary to octal, group the bits into groups of three (moving out from the binary point). Replace each group by its octal value:

$$101110.010_2$$
$$101\ \ 110.\ \ 010_2$$
$$5\ \ \ 6\ .\ \ 2_8$$
$$56.2_8$$

To convert from binary to hexadecimal, group the bits into groups of four (again, moving out from the binary point) and replace each group by its hexadecimal value:

$$101110.010_2$$
$$0010\ \ 1110.\ \ 0100_2$$
$$2\ \ \ \text{E}\ .\ \ 4_{16}$$
$$2\text{E}.4_{16}$$

To convert from octal or hexadecimal to binary, just reverse the process:

$$47.15_8$$
$$4 \quad 7\,. \quad 1 \quad 5_8$$
$$100 \quad 111. \quad 001 \quad 101_2$$
$$100111.001101_2$$

$$1C.B3_{16}$$
$$1 \quad C\,. \quad B \quad 3_{16}$$
$$0001 \quad 1100. \quad 1011 \quad 0011_2$$
$$11100.10110011_2$$

To convert the binary number 101110.010_2 to its decimal equivalent, just expand the place value representation using base 10 arithmetic.

$$2^5 + 2^3 + 2^2 + 2^1 + 2^{-2} = 32 + 8 + 4 + 2 + \frac{1}{4} = 46.25_{10}$$

Converting from decimal to binary is the most tedious of the common conversions. Start subtracting powers of 2 (in descending order) until the remainder is zero. It is helpful to memorize a small set of powers of 2 (Table A.2).

TABLE A.2 Some Useful Powers of 2

n	-3	-2	-1	0	1	2	3	4	5	6	7	8	9	10
2^n	$\frac{1}{8}$	$\frac{1}{4}$	$\frac{1}{2}$	1	2	4	8	16	32	64	128	256	512	1024

For example, the powers of 2 in 117.25 can be determined:

Current Remainder	**2^n**	**New Remainder**
117.25	64	53.25
53.25	32	21.25
21.25	16	5.25
5.25	4	1.25
1.25	1	0.25
0.25	$\frac{1}{4}$	0

The binary representation is built using the powers of 2 that were used (with 0s for the powers that were skipped). Thus, $117.25_{10} = 1110101.01_2$. That is,

$$117.25 = 1 \cdot 2^6 + 1 \cdot 2^5 + 1 \cdot 2^4 + 0 \cdot 2^3 + 1 \cdot 2^2 + 0 \cdot 2^1 + 1 \cdot 2^0 + 0 \cdot 2^{-1} + 1 \cdot 2^{-2}$$

APPENDIX

B Summation Notation

Suppose there are n numbers $a_1, a_2, \ldots, a_n$ that are to be added together. Writing the summation using the $\cdots$ notation is awkward:

$$a_1 + a_2 + \cdots + a_{n-1} + a_n$$

If additional calculations need to be done, the notation becomes unwieldy. Mathematicians have developed a shorthand notation for sums that has a compact form. This notation is called *summation notation*. They have also established some simple rules for manipulating expressions in summation notation.

The basic notation involves the Greek uppercase symbol sigma: $\sum$. There are four other significant components. The first is the set of numbers to be added. In the preceding example, these would be the numbers $a_1, a_2, \ldots, a_n$. The second component is an *index* variable (typically one of the letters i, j, k, l, m, or n). This variable allows us to refer generically to one of the numbers to be added. Thus, we might write a_i to refer to the ith number. The final two components are the starting and ending values for the index. The pieces fit together as follows:

$$\sum_{i=k}^{n} a_i$$

which is read as "the sum of the numbers a_i, for i between k and n, inclusive". Thus, the notational shorthand is defined as follows:

DEFINITION B.1 ***Summation Notation***

$$\sum_{i=k}^{n} a_i = a_k + a_{k+1} + a_{k+2} + \cdots + a_{n-1} + a_n$$

Notice that the choice of index variable does not change the sum:

$$\sum_{i=k}^{n} a_i = \sum_{j=k}^{n} a_j$$

Changing the starting or ending value for the index *does* change the sum.

The properties listed below are direct consequences of this definition. Notice that the symbol c represents a number that does not change with the index. You will improve your understanding if you make up simple examples to test the properties. For example,

$$\sum_{i=1}^{4} (c \cdot a_i) = c \cdot \sum_{i=1}^{4} a_i$$

is merely an extended version of the distributive property[1] of the real numbers:

$$ca_1 + ca_2 + ca_3 + ca_4 = c(a_1 + a_2 + a_3 + a_4)$$

There are additional properties that are not listed here. These additional properties are useful with manipulations that are more complex than those needed for this text.

1. $\sum_{i=1}^{n} c = nc$

2. $\sum_{i=k}^{n}(a_i + b_i) = \sum_{i=k}^{n} a_i + \sum_{i=k}^{n} b_i$

3. $\sum_{i=k}^{n}(a_i - b_i) = \sum_{i=k}^{n} a_i - \sum_{i=k}^{n} b_i$

4. $\sum_{i=k}^{n} (c \cdot a_i) = c \cdot \sum_{i=k}^{n} a_i$

Property 4 has already been illustrated. It might be helpful to describe property 1 in a sentence. Property 2 is also easy to illustrate (as is property 3).

$$\begin{aligned}
\sum_{i=3}^{6}(a_i + b_i) &= (a_3 + b_3) + (a_4 + b_4) + (a_5 + b_5) + (a_6 + b_6) \\
&= (a_3 + a_4 + a_5 + a_6) + (b_3 + b_4 + b_5 + b_6) \\
&= \sum_{i=3}^{6} a_i + \sum_{i=3}^{6} b_i
\end{aligned}$$

These properties can be combined. For example,

$$\sum_{i=1}^{n}(a_i + c) = \sum_{i=1}^{n} a_i + nc$$

combines properties 1 and 2 (set each number $b_i = c$).

Property 1 can be used twice to establish a more general form of that property:

$$\sum_{i=m}^{n} c = \sum_{i=1}^{n} c - \sum_{i=1}^{m-1} c = nc - (m-1)c = (n - m + 1)c$$

For example,

$$\sum_{i=4}^{8} 7 = (8 - 4 + 1)7 = 35 = 7_{i=4} + 7_{i=5} + 7_{i=6} + 7_{i=7} + 7_{i=8}$$

An important notational convention:

$$\sum_{i=1}^{n} a_i + b_i = \left(\sum_{i=1}^{n} a_i\right) + b_i \quad \textbf{not} \quad \sum_{i=1}^{n} (a_i + b_i) \quad \text{(often the erroneous intent)}$$

---

[1] See page A3.

APPENDIX

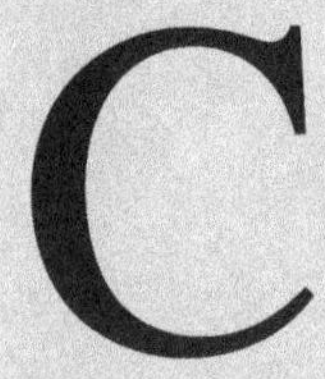

Logic Puzzles and Analyzing Claims

C.1 Logic Puzzles

C.1.1 Logic Puzzles about AND, OR, and NOT

The most enjoyable way that I know to become comfortable with the logic operators is to solve logic puzzles. Raymond Smullyan is a master at creating interesting logic puzzles [87, 88]. Consider the following examples.[1]

EXAMPLE C.1

The Island of Knights and Knaves

On a certain island, every inhabitant is either a knight, who *always* tells the truth, or a knave, who *always* lies. Suppose you meet two such inhabitants, A and B. A makes the following statement: "At least one of us is a knave." What are A and B? (It helps to know that the opposite of *at least one* is *none*.)

We can use a variation on truth tables to help solve this puzzle. If we let T represent knight, and F represent knave, we see that there are four possible answers to this puzzle: both knights, both knaves, A a knight and B a knave, or A a knave and B a knight. These possibilities are listed in the following table.

A	B
T	T
T	F
F	T
F	F

We need to check each of these possibilities for consistency with the statement by A. The results are presented in the following table.

A	B	A: "At least one of us is a knave"
T	T	*Inconsistent*
T	F	*Consistent*
F	T	*Inconsistent*
F	F	*Inconsistent*

[1] All puzzle statements are reprinted from *What is the name of this book?*, c 1978 by R. M. Smullyan, by permission of Collier Associates, P.O. Box 20149, West Palm Beach, FL 33416 USA.

Both being knights is inconsistent because A should be telling the truth, but neither is a knave. The second row (A a knight and B a knave) is consistent; A should be telling the truth, and one of them, B, is a knave. The third row is inconsistent since A should be lying, but one of them, A, is a knave. The final row is similar to the third row. A should be lying, but at least one (actually both) is a knave. The only consistent entry is when A is a knight and B is a knave.

Lurking behind the previous table is a collection of compound statements. For example, the first row can be expanded to

$$(\text{A is a knight}) \wedge (\text{B is a knight}) \wedge (\text{At least one of A, B is a knave})$$

This simplifies to $T \wedge T \wedge F = F$, which I have listed as inconsistent.

Similarly, the third row can be expanded to

$$(\text{A is a knave}) \wedge (\text{B is a knight}) \wedge (\neg(\text{At least one of A, B is a knave}))$$[2]

This simplifies to $T \wedge T \wedge (\neg T) = F$, which I have listed as inconsistent.[3] ■

EXAMPLE C.2

How to Choose a Bride

Suppose you are a visitor to the island of knights and knaves. Every female there is either a knight or a knave. You fall in love with one of the females there—a girl named Elizabeth—and are thinking of marrying her. However, you want to know just what you are getting into; you do not wish to marry a knave. If you were allowed to question her, there would be no problem, but an ancient taboo of the island forbids a man to hold speech with any female unless he is already married to her. However, Elizabeth has a brother Arthur who is also a knight or a knave (but not necessarily the same as his sister). You are allowed to ask the brother just one question, but the question must be answerable by "yes" or "no".

The problem is for you to design a question such that upon hearing the answer, you will know for sure whether Elizabeth is a knight or a knave. What question would you ask?

I shall again let T represent knight and F represent knave. There are four combinations possible for Arthur and Elizabeth. To avoid overlooking possibilities, I will again use a modified truth table. What would happen if I ask the obvious question, *Is your sister a knight?* The next table shows Arthur's answer in each possible case.

A	E	Is your sister a knight?
T	T	Yes
T	F	No
F	T	No
F	F	Yes

Clearly, the obvious question will not work. I would ask the question and receive an answer. The answer would narrow the possibilities from 4 to 2, but not in a useful way. For example, if Arthur answers "yes", Arthur and Elizabeth are represented by either the first or the fourth row. In the first row, Elizabeth is a knight, but in the fourth row, Elizabeth is a knave. I have no way of knowing which of the two rows is correct. If Arthur answers "No", I face the same obstacle. I need a different question. Convince yourself that *Is your sister a knave?* will not work either.

[2] In this case, since A is a knave we should negate his statement.

[3] The statement "At least one of A, B is a knave" is true in this case. Its negation is thus false, causing the compound statement to be false.

Perhaps I can try the question *Is either of you a knight?* The results are summarized as follows:

A	E	**Is either of you a knight?**
T	T	Yes
T	F	Yes
F	T	No
F	F	Yes

This appears to be better. If Arthur answers "no", I know that Elizabeth is a knight. But what if he answers "yes"? I still would not know what type Elizabeth is.

Rather than randomly choosing another question, it may be profitable to analyze the preceding tables. Perhaps we can gain some insight into what we want the question to produce. The problem with my first question was that the common answers did not have a common knight/knave value for Elizabeth. The problem with the last question was that it only worked sometimes. What I really need is a question for which the first and third rows have one answer and the second and fourth rows have the opposite answer.

My goal is a table that looks like this:

A	E	???
T	T	Yes
T	F	No
F	T	Yes
F	F	No

Perhaps I can use one of the truth tables from Section 2.3.2. Remember that if Arthur is a knave, he will negate the value in a truth table. The truth tables for AND and OR, with Arthur's filtered view, are given as follows:

A	E	$A \wedge E$	**Are you both knights?**
T	T	T	Yes
T	F	F	No
F	T	F	Yes
F	F	F	Yes

A	E	$A \vee E$	**Is at least one of you a knight?**
T	T	T	Yes
T	F	T	Yes
F	T	T	No
F	F	F	Yes

Neither does what I want.[4] However, the $(A \wedge E)$ table comes close. The fourth row is the only entry that is not what I want. Perhaps a similar question would work. In fact, the question I want is *Are you both the same type?*

A	E	**Are you both the same type?**
T	T	Yes
T	F	No
F	T	Yes
F	F	No

[4]The second is really the same question as *Is either of you a knight?*

Unfortunately, the single question will not help determine what kind of brother-in-law you might gain.[5] ■

The final puzzle by Smullyan that I wish to present in this section involves a character from one of Shakespeare's more controversial[6] plays. One of the concepts hiding below the surface is that you can't believe everything you read. A statement is not true merely because it claims to be true.

EXAMPLE C.3

Portia's Caskets

Portia is a wealthy young woman whose father has specified in his will that any suitor must pass a test. The suitor must agree, before taking the test, that if he fails he will remain a bachelor for the rest of his life. Here is Smullyan's puzzle, based on Act II, Scene VII of *The Merchant of Venice*.

In Shakespeare's *The Merchant of Venice* Portia had three caskets[7]—gold, silver, and lead—inside one of which was Portia's portrait. The suitor was to choose one of the caskets, and if he was lucky enough (or wise enough) to choose the one with the portrait, then he could claim Portia as his bride.

On the lid of each casket was an inscription to help the suitor choose wisely:

This first of gold, who this inscription bears:
"Who chooseth me shall gain what many men desire."
The second silver, which this promise carries:
"Who chooseth me shall get as much as he deserves."
This third dull lead, with warning all as blunt:
"Who chooseth me must give and hazard all he hath."

Suppose instead that Portia wished to choose her husband not on the basis of virtue, but simply on the basis of intelligence. She had new inscriptions put on the caskets (Figure C.1).

Figure C.1 The inscriptions on Portia's caskets.

GOLD	SILVER	LEAD
The portrait is in this casket	The portrait is not in this casket	The portrait is not in the gold casket

TABLE C.1 Possible Truth Values for the Casket Inscriptions

Gold	Silver	Lead
T	T	T
T	T	F
T	F	T
T	F	F
F	T	T
F	T	F
F	F	T
F	F	F

Portia explained to the suitor that of the three statements, at most one was true. Which casket should the suitor choose?

In this problem there are more than two people or things to combine. We can consider the possible combinations of the truthfulness of the messages on the three caskets. Table C.1 lists the possible combinations.

To solve the problem, we need to find a row that is consistent with the inscriptions. Portia has told us that we may eliminate rows 1, 2, 3, and 5. The consistency checks are presented in Table C.2.

[5] That would require a single question with four possible answers, or else two questions.
[6] Controversial in our era.
[7] A casket (in this context) is a small chest or box, not a coffin.

TABLE C.2 Checking the Consistency of the Inscriptions

Gold	Silver	Lead
The portrait is in this casket	The portrait is not in this casket	The portrait is not in the gold casket.

Gold	Silver	Lead	Consistency
T	T	T	—
T	T	F	—
T	F	T	—
T	F	F	*Inconsistent*
F	T	T	—
F	T	F	*Inconsistent*
F	F	T	*Consistent*
F	F	F	*Inconsistent*

In the fourth row, the gold inscription is true, so the portrait is in the gold casket. However, the silver casket should have a false inscription. But the inscription on the silver casket correctly states that the portrait is not in the silver casket. This is an inconsistency. You should take the time to explain the remaining entries in the table.

According to the table, the seventh row is the proper row. The gold inscription is false, so the portrait is not in the gold casket. The lead inscription is true but provides no new information since we already know the portrait is not in the gold casket. The silver inscription is false, so the portrait must be in the silver casket. The suitor should choose the silver casket. ■

✔ Quick Check C.1

1. Suppose in Example C.2 you were allowed to ask Elizabeth one yes/no question (rather than asking Arthur). What question could you ask her to determine whether she is a knight or a knave? ☑

C.1.2 Logic Puzzles about Implication, Biconditional, and Equivalence

Here are a few more of Raymond Smullyan's delightful logic puzzles. The ones presented next help explore implication, the biconditional, and equivalence.

EXAMPLE C.4

More Knights and Knaves

We have two people, A, B, each of whom is either a knight or a knave. Suppose A makes the following statement: "If I am a knight, then so is B." Can it be determined what A and B are?

Once again, I will let T represent knight and F represent knave. Table C.3 records the consistency of the statement by A.

TABLE C.3 Consistency Check for Example C.4

A	B	If I am a knight, then so is B
T	T	*Consistent*
T	F	*Inconsistent*
F	T	*Inconsistent*
F	F	*Inconsistent*

In the first row, since A is a knight, she speaks truthfully. B is also a knight, so the statement is consistent with the types of A and B. In the second row, B is a knave, contradicting the truthfulness of A's statement. The final two rows are inconsistent since A, being a knave, can never speak truthfully. But the implication "If I am a knight, then so is B" *is true* (A is a knave, so the "hypothesis" is false, making the implication true). We conclude that they are both knights. ■

EXAMPLE C.5

Romance among the Knights and Knaves

This problem, though simple, is a bit surprising.

Suppose it is given that I am either a knight or a knave. I make the following two statements:

1. I love Linda.
2. If I love Linda, then I love Kathy.

Am I a knight or a knave?

We need to decide if I can consistently make the two statements. Suppose I am a knave; every statement I make is false. Then (1) is actually false; so I do not love Linda. The hypothesis of the implication (2) is therefore false, so the implication is true. But then I could never say the true implication (2). The assumption that I am a knave makes the two statements inconsistent. No logical inconsistency occurs if we assume I am a knight. Therefore, I must be a knight.

Notice that since (1) is true, the hypothesis to (2) is also true. But the entire implication (2) is also true. We are led to conclude[8] that I must also love Kathy. Although the logic of the puzzle ensures that I am a knight, some might be tempted to label me a two-timing knave! ■

EXAMPLE C.6

Is There Gold on this Island?

On a certain island of knights and knaves, it is rumored that there is gold buried on the island. You arrive on the island and ask one of the natives, A, whether there is gold on this island. He makes the following response: "There is gold on this island if and only if I am a knight."

Our problem has two parts:

a. Can it be determined whether A is a knight or a knave?
b. Can it be determined whether there is gold on the island?

I will let T and F serve multiple duty for this example. The symbol T will represent A being a knight, gold being on the island, and the biconditional (A's response) being true. The symbol F will represent A being a knave, there being no gold on the island, and the biconditional being false. For each combination, we will check for consistency (Table C.4).

TABLE C.4 Consistency Check for Example C.6

A	Gold	Gold $\leftrightarrow$ Knight	Consistency
T	T	T	*Consistent*
T	F	F	*Inconsistent*
F	T	F	*Consistent*
F	F	T	*Inconsistent*

In the first row, A tells the truth, so the claim that his knightly nature is identical with the presence of gold is consistent. In the second row, the biconditional is false. A could never make the claim that the biconditional is true. In the third row, the biconditional is false, but A is a knave and would tell us that it was true. In the final row, the knave A would never tell us that the biconditional was true, which it is in that row.

We cannot answer (a), but we do know that the island contains gold. ■

[8] Modus ponens on page 54.

✔ Quick Check C.2

1. Suppose you are on the island of knights and knaves. Inhabitant A makes the following two statements:
   (a) If I am a knave, then B is a knight.
   (b) I am a knave if and only if B is a knave.

   Do you know what A is? Do you know what B is? ☑

C.1.3 Exercises

The exercises marked with ★ have detailed solutions in Appendix H. The first five problems concern two identical twins, Ebenezum and Jedediah. Ebenezum always tells the truth. Jedediah tells the truth on some days and lies on others. Since the twins are identical, I cannot tell which brother is speaking merely by looking or listening. I must consider the content of their statements.

1. ★ One day the twins approached me and made the following statements:
   - I always tell the truth.
   - My brother is lying.

   Is Jedediah telling the truth today, or is this one of his lying days? (Remember, the twins are identical, so I cannot tell which is which by looking or listening.)
2. I once asked them what their mother's name was. They replied,
   - Mother's name is either Ann or Betty.
   - Mother's name is Ann.

   Can I determine their mother's name?
3. The twins have a sister whose name is either Maud or Gerty. I asked them the following question: *Does Ebenezum ever lie or is your sister's name Maud?* They both gave the same yes/no answer. What did I learn?
4. Suppose I ask Ebenezum and Jedediah *Are you lying today?* What possible pairs of answers can I get?
5. If I ask the question *Is your brother lying today?* can I determine if Jedediah is telling the truth or lying?
6. King Nebuchadnezzar employed three personal physicians. The physicians tended to be jealous of one another. Since he was a despot, Nebuchadnezzar only allowed high officials or personal servants one big mistake or two small mistakes. Each physician had already made one small mistake, so they were all eagerly looking for opportunities to cause their rivals to stumble.

   One morning, Nebuchadnezzar (hereafter called "the king") woke up feeling ill. He called in physician A, who prescribed a diet of chicken soup and black bread for three days, promising that this would cure him.

   Physician B then came rushing in and exclaimed, *Don't listen to A, he and C are both liars.* The king then called in C to get his opinion. C was so frightened that all he could say was, *One of the others is lying.*

   The king thought for a minute, then had one of the physicians executed immediately. Which doctor was executed, and why? What assumptions did you need to make to solve this problem? (You may not assume that the king is stupid; nor may you assume that he acted arbitrarily. Within his worldview, he made a rational decision based on the facts and so should you.)
7. ★ Suppose you arrive on the island of knights and knaves. You want to visit the capital city. As you travel, you come to a fork in the road. There is a local inhabitant standing at the fork. She informs you that she will answer only one question. You do not know whether she is a knight or a knave. Can you think of a single question to ask whose answer is guaranteed to provide the proper route to the capital city? (You need to be a bit devious.)

C.2 Analyzing Claims

C.2.1 Analyzing Claims that Contain Implications

Many arguments in daily life explicitly or implicitly contain implications. By extending the analysis to the forms of the implications, it is possible to improve upon the guidelines for informal logic that were presented in Section 2.2.

Recall the brief discussion of *syllogistic logic* that was presented in the introduction to Chapter 2. Webster's dictionary defines syllogism as

> a deductive scheme of a formal argument consisting of a major and a minor premise and a conclusion (as in "every virtue is laudable; kindness is a virtue; therefore kindness is laudable").

The example from the dictionary can be modified[9] and rewritten as

major premise:	If kindness is a virtue, then kindness is laudable.
minor premise:	Kindness is a virtue.
	Kindness is laudable. VALID

In essence, we state an implication (the major premise) with the implicit assumption that it is a true implication, and we also affirm the truth of the hypothesis (via the minor premise). Then we assert the truth of the conclusion. The astute student should notice that this is merely a cosmetic repackaging of *modus ponens* (page 54).

The term *valid* is used to describe the *form* of reasoning used. The previous example is valid since it is based on *modus ponens*. Our study of derived implications (Section 2.3.8) shows that an argument form based on the contrapositive of a true implication also leads to a valid argument. Consider the implication "If selfishness is a virtue, then selfishness is laudable." The contrapositive can be expressed as

If selfishness is not laudable, then selfishness is not a virtue.
Selfishness is not laudable.
Selfishness is not a virtue. VALID

This is really the same form of reasoning. We state an implication (which incidentally happens to be the contrapositive of another implication); then we affirm its hypothesis.

The reasoning based on the contrapositive can be presented in a visually different manner:

If selfishness is a virtue, then selfishness is laudable.
Selfishness is not laudable.
Selfishness is not a virtue. VALID

In this form, we state the original implication and affirm the negation of the conclusion, thus affirming the negation of the hypothesis. Study this table to see that it contains essentially the same information as the previous table. Notice that the minor premise in this example is related to the negation of the conclusion of the implication.

Our study of derived implications should warn us that the following are not valid forms of reason:

If academic excellence is a virtue, then academic excellence is laudable.
Academic excellence is laudable.
Academic excellence is a virtue. INVALID

If athletic prowess is a virtue, then athletic prowess is laudable.
Athletic prowess is not a virtue.
Athletic prowess is not laudable. INVALID

The first of these invalid forms is based on the converse of the original implication. It is often called *the fallacy of affirming the consequent*. The second invalid form is based on the inverse of the original implication. It has been called *the fallacy of denying the antecedent*.

[9]I have simplified the more general "every virtue" to the more specific "kindness" and also expressed the phrase "every virtue is laudable" as an implication. Syllogistic reasoning does not always need to have an implication for the major premise.

We thus have two valid and two invalid forms of reason based on an implication. They are summarized next:

If A, then B.
A

B VALID—direct syllogistic reason

If A, then B.
¬B

¬A VALID—syllogistic reason using contrapositive

If A, then B.
B

A INVALID—fallacy of affirming the consequent

If A, then B.
¬A

¬B INVALID—fallacy of denying the antecedent

To emphasize an important idea introduced at the beginning of this section, I will introduce some additional definitions. We will label a valid argument as *sound* if both the major and minor premise are true. A valid argument form will be labeled *unsound* if either premise is not true. A valid argument form will be called *incomplete* if either of its premises has not yet been shown to be true or untrue. Notice that invalid argument forms are neither sound nor unsound nor incomplete. All the previous examples of valid argument forms have been sound. Table C.5 summarizes these definitions.

TABLE C.5 Classifying Argument Forms

		Premises		
		true	unknown	false
Form	valid	sound	incomplete	unsound
	invalid	—	—	—

If an invalid form is used, the argument provides no evidence for the truth (or lack of truth) of the conclusion. If a valid form is used and the premises are true, then the conclusion is also true.

The next set of examples all contain errors in reasoning.

EXAMPLE C.7

Ending Drug Abuse

An argument some people are making as a solution to the problem of drug abuse is *If all the drug dealers were given mandatory life sentences, there would be no more drug abuse.* We can translate this into the following argument form:

If all the drug dealers were given mandatory life sentences,
then there would be no more drug abuse.
We should give life sentences to convicted drug dealers.

The drug abuse problem will be solved. INCOMPLETE

No one has established the implication that mandatory life sentences will stop drug abuse. One possible reason that the implication may be false is that new dealers may replace the old dealers as fast as the old ones are locked up. The proposed solution also ignores the influence of the drug consumer.

It *may* be possible to prove that harsh sentences prevent some people from becoming drug dealers. The matter needs more study. The politician should make a more tentative statement: *Giving convicted drug dealers long sentences may help slow the abuse of drugs.* This is a more logically palatable argument but is poor political rhetoric. ■

EXAMPLE C.8

Everybody Is Doing It

Mom, can I put peanut butter up my nose? Everybody is doing it. This argument is a common one. Ignoring the standard response of *If everybody jumped off a cliff, would you?* consider the implied argument form:

If everybody else is doing A, then I should do A.
Everybody else is doing A.

I should do A. Usually UNSOUND

It is not true that *everybody* else is putting peanut butter up their nose. Even if we modify this to "everybody that I think is important" or "Susie Jones and Tom Smith", the major premise is still of dubious general value. If the claim were "everybody is breathing", the argument might be considered sound; but stuffing peanut butter up your nose needs additional justification. ■

EXAMPLE C.9

Pets and Longevity

There is some statistical evidence that people who have a pet live longer after the loss of a spouse than those who do not have pets. Consider the claim *If you have a pet, then you will probably live longer. My friend died soon after his wife died. He must not have had a pet.* We can diagram the argument as follows:

If you have a pet after your spouse dies,
  then you are more likely to live longer.
You do not live long after your spouse dies.

You did not have a pet INVALID

The error here is a bit subtle. The argument form appears to be the valid form based on the contrapositive. The error is that the minor premise is *not* the negation of the conclusion. The implication (based on statistical data) only claims you are more *likely* to live longer. It does not claim a guaranteed longer life. The argument just presented really has the following form:

If A, then B.
$\neg$C

$\neg$A INVALID—using unrelated information ■

EXAMPLE C.10

Conditional Love

Have you ever done poorly in a class? Perhaps you thought to yourself, *I flunked this class. My parents are going to kill me!* Your conversation to yourself might be toned down and rephrased as

If I get good grades, then my parents will love me.
I did not get good grades.

My parents do not love me. INVALID—fallacy of denying the antecedent

Almost all parents are less conditional in their love than this invalid syllogism implies. Your parents may be disappointed, but will not cease to love you because of poor grades. Perhaps the tendency to commit the preceding logical fallacy is encouraged by the phrasing of the major premise. Love can hardly be genuine if it is so narrowly conditional. We should be very cautious about oversimplifying complex realities. ■

EXAMPLE C.11

Music Lovers

Suppose your friend says to you *I know you love music, so you will love this new CD.* Your friend's statement is based on the following argument form.

If you love music, then you will love this CD.
You love music.

You will love this CD. UNSOUND

This is a valid *form* of reason. However, even granting the assumption that you love music, the major premise is not a universally true statement. There is no compelling reason to believe that a love of music will entail a love of all genres of music. Even if your tastes are very eclectic and you *do* like the musical selections on the CD, you still may be disappointed in the quality of the performance. ■

EXAMPLE C.12

Logic and Problem Solving

Consider the statements *If I study logic I will be able to solve problems. I can solve problems. Therefore I have studied logic.* The argument form is

If I study logic, then I will be able to solve problems.
I can solve problems.

I have studied logic. INVALID—fallacy of affirming the consequent ■

✔ Quick Check C.3

1. Analyze the following arguments.
   (a) If I practice piano for six hours a day for at least one year, then I will become famous. I have practiced six hours a day for the past two years. Someday I will become famous.
   (b) Productive employees at this company get raises. Ichabod received a raise. Therefore, Ichabod is a productive employee. (For this problem, assume that the major premise *is* true.)
   (c) Using a seat belt has been shown to decrease the chance of injury in an accident. I wear a seat belt, so I am less likely to be injured in an accident. ✔

C.2.2 Analyzing Claims that Contain Quantifiers

Many syllogisms contain major or minor premises that include quantifiers. For example, the major premise of this syllogism contains a universal quantifier:

All people are mortal.
I am a person.

I am mortal. VALID

The next syllogism contains an existential quantifier in the major premise:

Some people can ride a bicycle.
George can ride a bicycle.

George is a person. INVALID

There is a simple technique that is sometimes useful when analyzing syllogisms that contain quantifiers. It involves the use of the familiar Venn diagram.[10]

[10]This application of Venn diagrams goes back at least as far as the mid-1700s, when Leonard Euler used them for this purpose [33].

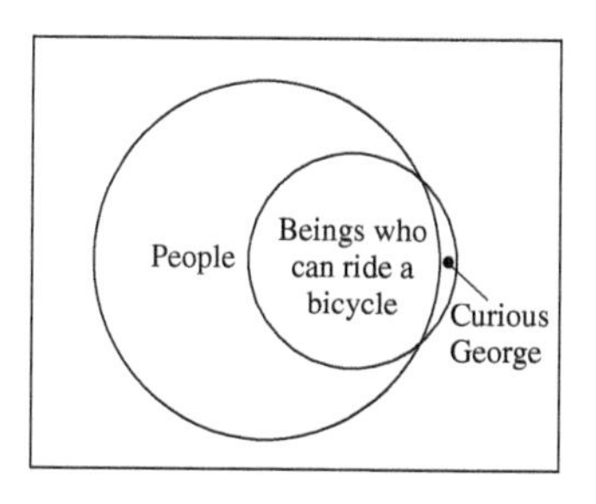

Figure C.2. Bicycle riders.

To see that the previous syllogism is invalid, consider the Venn diagram in Figure C.2. Since an existential quantifier was used, we may not assert that *all* people can ride bicycles. We also do not know whether all bicycle riders are people. George's full name may be Curious George. A chimpanzee may know how to ride a bicycle, but a chimpanzee is not a person.

The discussion on page 26 in Section 2.1.1 should serve as a warning that this intuitive approach using Venn diagrams is risky.

✔ Quick Check C.4

1. Use a collection of Venn diagrams to analyze informally the possibilities for the following syllogism, then comment on the validity of the syllogism.

All A are B.
Some B are C, but not all B are C.
Some C are not B.

Some A are C.

☑

C.2.3 Exercises

The exercises marked with ★ have detailed solutions in Appendix H.

1. What must we do to show that an implication is true?
2. What must we do to show that a biconditional is true?
3. The two previous questions make the assumption that an implication or a biconditional might be false. Give a reason for this assumption.
4. Write the original implication, the contrapositive, the converse, and the inverse for each of the following statements. Indicate which of the four implications in each set are true.
   (a) ★ If I am pregnant, then I am a female.
   (b) If I am a Republican, then I am not a Democrat.
   (c) If I am a student, then I live in a dorm and I go to class.
   (d) If I have an X chromosome, then I am male.
5. Analyze the following implication-based claims.
   (a) ★ If Socrates was a man, then he was mortal. Socrates was mortal. Therefore, Socrates was a man.
   (b) If Euclid proved the theorem, then Euclid was a great mathematician. Euclid proved the theorem. Therefore, Euclid was a great mathematician.
   (c) If Hippocrates was an ancient Greek scholar, then he was very wise. Hippocrates was an ancient Greek. Therefore, he was very wise.
   (d) If I do not get enough sleep, then I am crabby. I am not crabby. Therefore, I got enough sleep. (Assume that I am telling the truth in my major and minor premises.)
   (e) When I am sad, I like to take a walk in the forest. I am walking in the forest right now. Therefore, I am sad. (Assume that I am telling the truth in my major and minor premises.)
   (f) If I am angry, then I shout. I am not angry. Therefore, I am not shouting. (Assume that I am telling the truth in my major and minor premises.)
   (g) • Major premise: Sixty men can do a piece of work sixty times as quickly as one man.
   • Minor premise: One man can dig a posthole in sixty seconds.
   • Conclusion: Sixty men can dig a posthole in one second.[11]
6. Analyze the following claims. Draw Venn diagrams where appropriate in each case (even if it seems trivial to solve without a diagram). If multiple diagrams might apply, show them.
   (a) All teachers are boring. Dr. Beeblebrox is a teacher. Therefore, Dr. Beeblebrox is boring.
   (b) ★ Some students live at home. Zelda is a student. Therefore, Zelda lives at home.
   (c) Some students are athletes. Some athletes are students. Every person is either a student or an athlete.
   (d) Some students are math majors. Every math major is a student. Every math major does homework.[12] Therefore, some students do homework.
   (e) Some students are musicians. Every musician practices. Therefore, some musician is not a student.
   (f) Every rose is a flower. Some flowers are perennials. Therefore, every rose is a perennial.
7. Use the concepts and terminology of derived implications to explain your answers to the following questions.
   (a) Assume that the implication

   If it is raining, then I use an umbrella.

   is true. Suppose I am not using an umbrella. Can I conclude that it is not raining?
   (b) Assume that the implication

   If my car is out of gas, then it will not start.

   is also true. My car started. Can I conclude that I am not out of gas?
   (c) Suppose in part (b) that my car won't start. Can I conclude that it is out of gas?

[11] Ambrose Bierce, *The Devil's Dictionary*.
[12] Assume true for this problem. Math majors who do not do homework usually switch majors eventually.

8. (a) Write a true implication whose converse is false. Then write a true implication whose converse is also true. You may not use an example found in this text.

(b) Rewrite the second pair as an equivalence.

The next set of questions relates to a small college in Frostbite Falls, Minnesota. There are only two kinds of students on the campus, scholars and dunces. Scholars never make a mistake. Every statement they make is true. Dunces never get anything completely correct. In any conversation, some of what they say may be true, but no matter how hard they try, they always say at least one thing that is false.

9. ★ One of the students made the statement *If I am a dunce, then I am a scholar.* Can we determine which kind of student he is?

10. Another student said *If I am a scholar, then I am not a dunce.* What is she?

11. Her roommate stated *If I am a dunce, then I am not a scholar.* What is she?

12. One day when I was very thirsty, a dunce that I knew handed me a cup of juice. He told me two things.

(a) I am a scholar.

(b) I have put poison in the juice.

Even though he was incorrect in his first statement, I declined the juice. Why?

13. Two very popular math majors, A and B, were roommates. One day they made the following statements:

A: If I am a dunce, then B is a dunce.

B: I am a dunce if and only if A is a dunce.

What kind of student is A? [*Hint*: Check the consistency of both students' statements.]

14. I asked a student what kind he was. He replied, *If I am wrong, then I am a dunce.* Is he a scholar or a dunce?

15. One student had an oral exam. She made the following three statements:

- All implications are either true or false.
- Every statement is either an implication or a biconditional.
- The statement $P \leftrightarrow Q$ is equivalent to the compound statement $(P \to Q) \wedge (Q \to P)$.

Is she a scholar or a dunce?

C.3 Quick Check Solutions

Quick Check C.1

1. Suppose you ask her the same question that would work with Arthur. Her response would be as follows:

A	E	Are you both the same type?
T	T	Yes
T	F	Yes
F	T	No
F	F	No

This tells us what kind of resident Arthur is but nothing about Elizabeth.

The goal is to find a question with a response:

A	E	???
T	T	Yes
T	F	No
F	T	Yes
F	F	No

One simple question is, *Do you have a brother?* The key is to find a question that is always true.

Quick Check C.2

1. You need to keep the truth tables for implication and equivalence in mind for this puzzle. As usual, let T represent knight and F represent knave.

Consider the first statement. If both are knights, then the hypothesis is false and the conclusion is true, so the implication is true. Since A tells the truth, the statement is one that could be made by A. The second row follows by an almost identical argument. In the third row, both the hypothesis and conclusion are true, so the implication is true. But A is a knave and could never make a true statement. Thus, row 3 is inconsistent. In the fourth row, the hypothesis is true and the conclusion is false, so the implication is false. A could make this false statement.

A	B	If I am a knave, then B is a knight.
T	T	*Consistent*
T	F	*Consistent*
F	T	*Inconsistent*
F	F	*Consistent*

Now consider the second statement. If both are knights, the statement (an equivalence) is true. Since A is a knight, A could make the statement. If both are knaves, the statement is also true. But A, being a knave, could not make a true statement. Thus, row 4 is inconsistent. If A is a knight but B is a knave, then the statement is false. A (a knight) would not make such a statement. Thus, row 2 is inconsistent. If A is a knave and B is a knight, the statement is false. The knave A *could* utter such a statement. Row 3 is consistent.

A	B	I am a knave if and only if B is a knave.
T	T	*Consistent*
T	F	*Inconsistent*
F	T	*Consistent*
F	F	*Inconsistent*

Row 1 is the only row in which both statements are consistent. Both A and B are knights.

Quick Check C.3

1. (a) The argument can be diagrammed:

If I practice piano for 6 hours a day for at least 1 year,
then I will become famous.
I have practiced 6 hours a day for the past 2 years.
———————————————————
Someday I will become famous. UNSOUND

The major premise is not true. I may have no talent (my 6 hours a day may be pure noise). Even if I have talent, I may never give a public performance, so no one else will ever know how good I am. There are other reasons why I may never be "discovered".

(b) The diagram is

If you are a productive employee, then you will receive a raise.
You receive a raise.
———————————————————
You are a productive employee. INVALID—fallacy of affirming the consequent

It may be true that Ichabod is a productive employee and deserves a raise. However, just knowing that Ichabod received a raise does not mean that Ichabod is productive. It may be that Ichabod is the boss's nephew. Alternately, Ichabod may be blackmailing the boss. Perhaps a clerical error was made.

An invalid argument form means that the claim has not been proved to be either true or false. Ichabod's productivity needs to be established by some means other that the attempted logic used here.

(c) This is a sound argument. The major premise *has* been shown to be true (notice the somewhat weasely way it was stated). The problem implies that the minor premise is also true.

If I wear a seat belt, then I will be less likely to be injured.
I wear a seat belt.
____
I will be less likely to be injured. SOUND

Quick Check C.4

1. The first premise (all A are B) means that A and B are related as in this diagram:

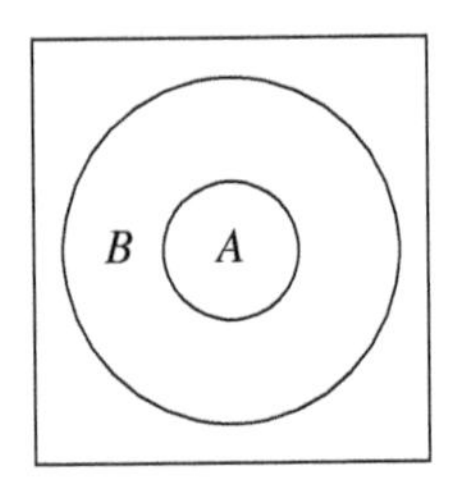

The next two premises require B and C to be related as

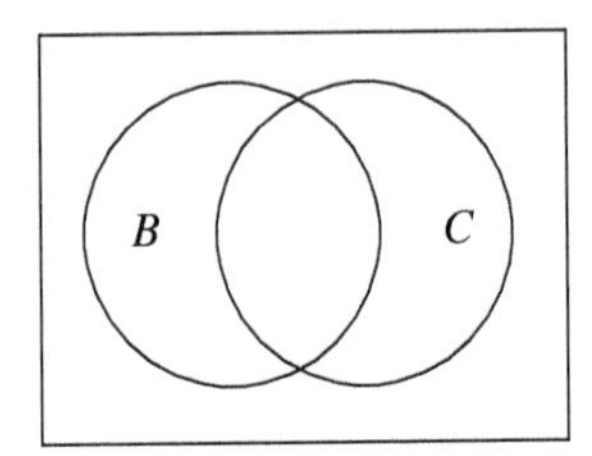

Thus the diagram must be either

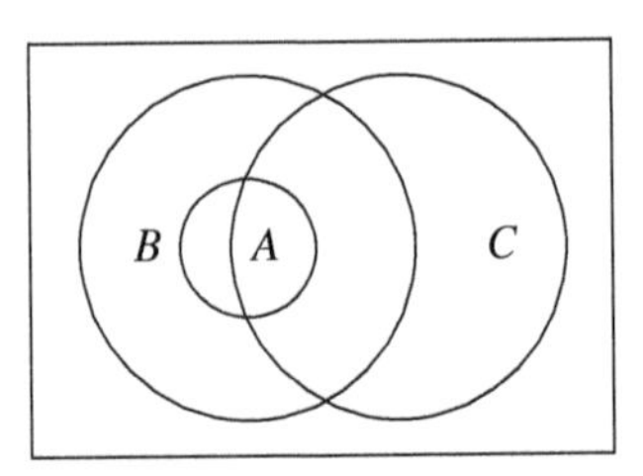

or

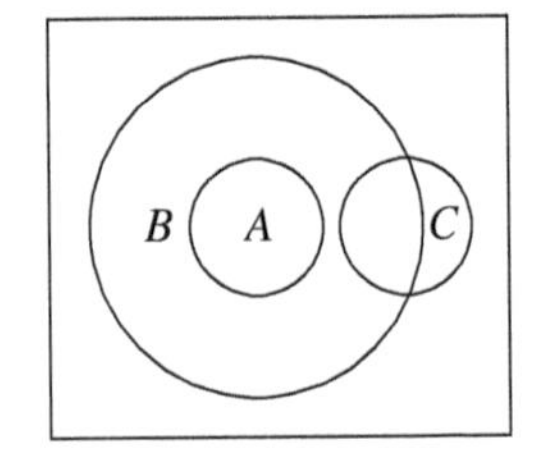

Since we can't tell which of these possibilities might occur, the syllogism is invalid (no A is C in the final diagram).

APPENDIX

D The Golden Ratio

Is it possible to quantify an aesthetic notion? Some people have believed it is possible. A number, named the *golden ratio* (also called the *golden section* or the *divine proportion*), is the result of one such attempt. This number derives from a geometric ratio that has its roots in antiquity. Euclid called the process of creating the ratio a "division in extreme and mean ratio". The notion reappeared in Fra Luca Pacioli's book, *De Divina Proportione* (written in 1509 and illustrated by Leonardo DaVinci). The adjective *golden* seems to have first appeared in the 1800s in Germany. The name *golden section* first appeared in English in the 1875 *Encyclopedia Britannica.*

Some (but not all) mathematicians use the Greek letter Φ to represent this number (in honor of the ancient Greek sculptor Phidias). The number (Euclid's "division in extreme and mean ratio") can be derived by considering the problem:

> Given a line segment, divide it into two pieces, S and L, having respective lengths, s and l, so that the number of copies of S that will fit in L is the same as the number of copies of L that will fit inside the original line segment.

Another way to state the problem is to require *the ratio of the longer segment's length to the shorter segment's length* to be the same as *the ratio of the entire line segment's length to the longer segment's length*:

$$\frac{l}{s} = \frac{l+s}{l}$$

Making the substitution $\Phi = \frac{l}{s}$, we seek a solution to the equation

$$\Phi = 1 + \frac{1}{\Phi}$$

or

$$\Phi^2 - \Phi - 1 = 0$$

The possible solutions are $\frac{1+\sqrt{5}}{2}$ and $\frac{1-\sqrt{5}}{2}$. Since $\frac{1-\sqrt{5}}{2}$ is negative, the only possibility is

$$\Phi = \frac{1+\sqrt{5}}{2}$$

In terms of the original question, the line should be divided so that $l = s\Phi$ (Figure D.1).

Figure D.1. Dividing a line by the golden ratio.

It has been asserted numerous times during the previous century that the golden ratio can also be used to determine the length and width of an aesthetically pleasing rectangle, called the *golden rectangle* (Figure D.2). The ratio of the width to the length of the sides is the golden ratio.

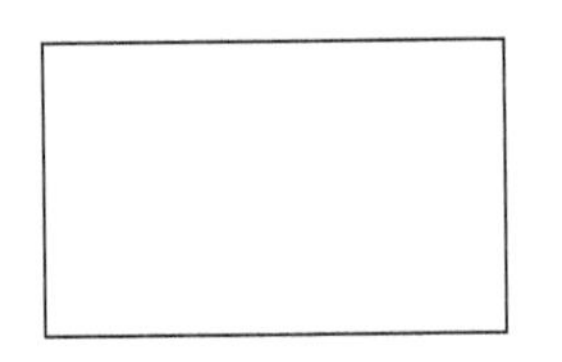

Figure D.2. A golden rectangle.

This assertion has been extended to suggest that golden rectangles can be found in ancient architecture and in many famous paintings and sculptures.[1] Perhaps the most noteworthy claim of this sort involves the front face of the Parthenon (Figure D.3), a temple in Athens dedicated to Athena. It was completed around 447 B.C. The width and height (including the peak) are said to form a golden rectangle.

Figure D.3 The Parthenon in Nashville, Tennessee, USA is a full-scale replica of the original Greek Parthenon [58]. Photo by Ryan Kaldari.

Many of these claims seem to be valid as rough approximations. However, serious doubt has been expressed about (a) the mathematical accuracy of such claims and (b) the validity of claims that the architects, painters, and sculptures actually intended to use the golden ratio as a basis for their work. For more information about these objections (and some other dubious claims regarding the golden ratio), see *Misconceptions about the Golden Ratio*, by George Markowsky [69].

[1] For example, the face in the *Mona Lisa*.

APPENDIX

E Matrices

Matrices appear in many contexts in mathematics. They are most fully explored in linear algebra courses. They appear in some of the chapters of this book, so a very brief introduction will be helpful for students who have not previously studied them in a linear algebra course or an upper-level high school course.

DEFINITION E.1 ***Matrix; Square Matrix; Main Diagonal***
An m-by-n *matrix* is a numeric table containing m rows and n columns. One common notational device is to abbreviate the matrix A by writing $A = (a_{ij})$. This is interpreted as "A is the matrix whose entry in row i and column j is the number a_{ij}".

If $n = m$, we say the matrix is a *square matrix*. The *main diagonal* is the set of entries $\{a_{11}, a_{22}, a_{33}, \ldots, a_{nn}\}$.

For example, a 3-by-4 matrix will have the form

$$\begin{pmatrix} a_{11} & a_{12} & a_{13} & a_{14} \\ a_{21} & a_{22} & a_{23} & a_{24} \\ a_{31} & a_{32} & a_{33} & a_{34} \end{pmatrix}$$

DEFINITION E.2 ***Transpose of a Matrix***
Every m-by-n matrix, A, has an associated n-by-m matrix called its *transpose* that is denoted A^T. The entry in the ith row and jth column of A^T is the entry in the jth row and ith column of A. That is, if $A = (a_{ij})$ and $A^T = (b_{ij})$, then $b_{ij} = a_{ji}$. This is often abbreviated as $A^T = (a_{ji})$.

The transpose effectively interchanges rows and columns of A.

The transpose of the 3-by-4 matrix shown previously is

$$\begin{pmatrix} a_{11} & a_{21} & a_{31} \\ a_{12} & a_{22} & a_{32} \\ a_{13} & a_{23} & a_{33} \\ a_{14} & a_{24} & a_{34} \end{pmatrix}$$

One common use for matrices is to compactly organize the important information in a system of linear equations. One matrix that arises in that context is called a coefficient matrix.

DEFINITION E.3 ***Coefficient Matrix***

Let

$$\begin{aligned} a_{11}x_1 + a_{12}x_2 + \cdots + a_{1n}x_n &= b_1 \\ a_{21}x_1 + a_{22}x_2 + \cdots + a_{2n}x_n &= b_2 \\ \vdots \quad & \quad \vdots \\ a_{m1}x_1 + a_{m2}x_2 + \cdots + a_{mn}x_n &= b_m \end{aligned}$$

be a system of linear equations. Then the matrix $A = (a_{ij})$ is called the *coefficient matrix* for the system of equations.

EXAMPLE E.1

A Coefficient Matrix

The system of linear equations

$$\begin{aligned} -5w + 2x + 3y + 4z &= 7 \\ 4w - 6x + 9y \quad &= 8 \\ w + 3x + 4y - z &= -6 \end{aligned}$$

has coefficient matrix

$$\begin{pmatrix} -5 & 2 & 3 & 4 \\ 4 & -6 & 9 & 0 \\ 1 & 3 & 4 & -1 \end{pmatrix}$$

■

Matrices are frequently used in applied settings. They also provide interesting examples of algebraic systems with properties that differ from some of the standard field properties that characterize the rational numbers. Only some simple algebraic manipulations of matrices will be presented here.

Just as we can add, subtract and multiply numbers, it is also possible to define analogous operations on matrices.

There are two special families of matrices that correspond to the numbers 0 and 1.

DEFINITION E.4 ***Zero Matrix***

A *zero matrix* is a matrix for which every entry is the number 0.

The 3-by-4 zero matrix is as follows:

$$\begin{pmatrix} 0 & 0 & 0 & 0 \\ 0 & 0 & 0 & 0 \\ 0 & 0 & 0 & 0 \end{pmatrix}$$

Once matrix addition is defined, it is easy to show that adding any m-by-n matrix, A, to an m-by-n zero matrix will result in A. Compare this to $a + 0 = a$ for any real number, a.

DEFINITION E.5 ***Identity Matrix***

An *identity matrix* is an n-by-n (square) matrix with

$$a_{ij} = \begin{cases} 0 & \text{if } i \neq j \\ 1 & \text{if } i = j \end{cases}$$

The n-by-n identity matrix is denoted I_n.

The 3-by-3 identity matrix is shown below.

$$I_3 = \begin{pmatrix} 1 & 0 & 0 \\ 0 & 1 & 0 \\ 0 & 0 & 1 \end{pmatrix}$$

Once matrix multiplication is defined, it will be easy to show that multiplying any m-by-n matrix, A, by an appropriate identity matrix will result in A. In fact, $AI_n = A$ and $I_m A = A$ will both be true. Compare this to $a \cdot 1 = a$ and $1 \cdot a = a$ for any real number, a.

Two m-by-n matrices can be added (and subtracted) by adding (subtracting) the corresponding entries.

DEFINITION E.6 ***Matrix Addition***

Let $A = (a_{ij})$ be an m-by-n matrix and let $B = (b_{ij})$ be another m-by-n matrix. Then the sum, $C = A + B$, is also an m-by-n matrix. Moreover, $C = (c_{ij})$, where

$$c_{ij} = a_{ij} + b_{ij}$$

EXAMPLE E.2 **Matrix Addition**

$$\begin{pmatrix} 3 & 0 & 11 & -4 \\ 12 & 5 & 1 & -2 \\ 3 & -5 & 8 & 2 \end{pmatrix} + \begin{pmatrix} 2 & 3 & 0 & 4 \\ -9 & 4 & 7 & -2 \\ 1 & 10 & -6 & 7 \end{pmatrix} = \begin{pmatrix} 5 & 3 & 11 & 0 \\ 3 & 9 & 8 & -4 \\ 4 & 5 & 2 & 9 \end{pmatrix}$$

■

At first it may seem reasonable to define an entry-by-entry multiplication (similar to what was done for addition). However, most applications of matrices favor a more complex definition. Under this definition, matrix multiplication is not commutative (in general). The product AB may be defined, but BA may not exist. Even if AB and BA both exist, it is usually the case that $AB \neq BA$.

DEFINITION E.7 ***Matrix Multiplication***

Let $A = (a_{ij})$ be an m-by-n matrix and let $B = (b_{ij})$ be an n-by-p matrix. Then the product, $C = AB$, is an m-by-p matrix. Moreover, $C = (c_{ij})$, where

$$c_{ij} = \sum_{k=1}^{n} a_{ik} b_{kj}$$

Notice that the number of columns of A must be the same as the number of rows in B. Observe also that the product matrix has the same number of rows as A and the same number of columns as B.

One visualization of this process is as follows. The entry in the ith row and jth column of the product is the result of grabbing the jth column of B, rotating 90° counterclockwise, and dropping it on the ith row of A. We then multiply numbers that occupy the same position and add the results:

$$c_{ij} = a_{i1}b_{1j} + a_{i2}b_{2j} + \cdots + a_{ik}b_{kj} + \cdots + a_{im}b_{mj}.$$

$$\begin{pmatrix} \vdots & \vdots & \vdots & \vdots & \vdots \\ a_{i1} & a_{i2} & \cdots & a_{i(m-1)} & a_{im} \\ \vdots & \vdots & \vdots & \vdots & \vdots \end{pmatrix} \begin{pmatrix} \cdots & b_{1j} & \cdots \\ \cdots & b_{2j} & \cdots \\ \vdots & \vdots & \vdots \\ \cdots & b_{(m-1)j} & \cdots \\ \cdots & b_{mj} & \cdots \end{pmatrix} = \begin{pmatrix} \vdots & \vdots & \vdots \\ \cdots & c_{ij} & \cdots \\ \vdots & \vdots & \vdots \end{pmatrix}$$

EXAMPLE E.3

Matrix Multiplication

Make sure you can reproduce this calculation.

$$\begin{pmatrix} 3 & 0 & 1 & -4 \\ 2 & 5 & 1 & -2 \\ 3 & -5 & 8 & 2 \end{pmatrix} \begin{pmatrix} 2 & 3 \\ -1 & 4 \\ 1 & 7 \\ 0 & 5 \end{pmatrix} = \begin{pmatrix} 7 & -4 \\ 0 & 23 \\ 19 & 55 \end{pmatrix}$$

■

Note that matrix division is *not* defined. For square matrices, it is common to define the notion of an inverse matrix and gain some of the benefits of division.

DEFINITION E.8 ***Matrix Inverse***

If A is an n-by-n matrix, then A has an *inverse* if there is a matrix, A^{-1}, such that $AA^{-1} = A^{-1}A = I_n$.

Many square matrices do not have an inverse.

The definition of matrix multiplication can be extended to cover nonnegative integer exponents for square matrices.

DEFINITION E.9 A^k

Let A be an n-by-n matrix. The nonnegative powers of A are

$$A^0 = I_n$$
$$A^1 = A$$
$$A^k = AA^{k-1}$$

EXAMPLE E.4

Matrix Powers

Let $A = \begin{pmatrix} 1 & 2 & 1 \\ 2 & 1 & 2 \\ 1 & 0 & 1 \end{pmatrix}$. Then

$$A^2 = AA = \begin{pmatrix} 6 & 4 & 6 \\ 6 & 5 & 6 \\ 2 & 2 & 2 \end{pmatrix} \quad \text{and} \quad A^3 = AA^2 = \begin{pmatrix} 20 & 16 & 20 \\ 22 & 17 & 22 \\ 8 & 6 & 8 \end{pmatrix}$$

■

APPENDIX

F The Greek Alphabet

The following Greek letters are commonly used in mathematics.

Name	Lowercase	Uppercase	Pronunciation
alpha	α	A	
beta	β	B	
gamma	γ	Γ	
delta	δ	Δ	
epsilon	ϵ	E	
zeta	ζ	Z	"e" like the "a" in "ate"
eta	η	H	"e" like the "a" in "ate"
theta	θ	Θ	"e" like the "a" in "ate"
iota	ι	I	"i" like the "e" in "he"
kappa	κ	K	
lambda	λ	Λ	
mu	μ	M	like "moo"
nu	ν	N	like "new"
xi	ξ	Ξ	"x" like in "vex," "i" like the "e" in "he"
omicron	o	O	
pi	π	Π	
rho	ρ	P	like "row"
sigma	σ	Σ	
tau	τ	T	
upsilon	υ	Υ	"ups" like "oops"
phi	ϕ	Φ	with a long "i"
chi	χ	X	with a hard "ch" (like a "k")
psi	ψ	Ψ	
omega	ω	Ω	

The Hebrew letter aleph, $\aleph$, is also used in discussions about the cardinality of infinite sets.

APPENDIX

G Writing Mathematics

The following suggestions will help you to submit properly written homework solutions. The suggestions are appropriate for both computational exercises and for proofs.

The goal of any writing is to clearly communicate ideas to another person. (That other person may even be your future self.) When you write for another person, you will need to include ideas that may be in your mind but omitted when you are writing a rough draft on scratch paper. If you keep your intended audience in mind, you will produce higher quality work. For a course in mathematics, the intended audience is usually your instructor or student grader. This implies that your task is to show that you thoroughly understand your solution. Consequently, you should routinely include more of the details.

1. **Use Sentences** The feature that best distinguishes between a properly written mathematical exposition and a piece of scratch paper is the use (or lack) of sentences. Properly written mathematics can be read in the same manner as properly written sentences in any other discipline. Sentences force a linear presentation of ideas. They provide the connections between the various mathematical expressions you use. This linearity will also keep you from handing in a page with randomly scattered computations with no connections. The sentences may contain both words and mathematical expressions. The following extract from a Quick Check solution illustrates these ideas.

   Let n be odd. Then Definition 3.10 indicates that there does not exist an integer, k, such that $n = 2k$. That is, n is not divisible by 2. The Quotient–Remainder theorem asserts that n can be uniquely expressed in the form $n = 2q + r$, where r is an integer with $0 \leq r < 2$. Thus, $r \in \{0, 1\}$. Since n is not divisible by 2, the only admissible choice is $r = 1$. Thus, $n = 2q + 1$, with q an integer.

2. **Read out loud** The sentences you write should read well out loud. This will help you to avoid some common mistakes. Avoid sentences like:

   Suppose the graph has n number of vertices.
   The piggy bank contains n amount of coins.

   If you substitute an actual number for n (such as 4 or 6) and read these out loud they will sound wrong (because they *are* wrong). The variable n is already a numeric variable so it should be read just like an actual number. The correct versions are:

   Suppose the graph has n vertices.
   (Read this as: "Suppose the graph has en vertices".)
   The piggy bank contains n coins.

3. **= is not a conjunction** The mathematical symbol $=$ is an assertion that the expression on its left and the expression on its right are equal. Do not use it as a connection between steps in a series of calculations. Use words for this purpose.

Here is an example that misuses the $=$ symbol ($6 = \frac{3x}{3}$ is false):

Incorrect! $$3x = 6 = \frac{3x}{3} = \frac{6}{3} = x = 2$$

One proper way to write this is

$3x = 6$. Dividing both sides by 3 leads to $\frac{3x}{3} = \frac{6}{3}$, which simplifies to $x = 2$.

4. **Do not Merge Steps** Suppose you need to calculate the final price for a \$20 item with 7% sales tax. One strategy is to first calculate the tax, then add the \$20. Here is an incorrect way to *write* this.

Incorrect! $$20 \cdot 0.07 = 1.4 + 20 = \$21.4.$$

The main problem (besides the magically-appearing dollar sign at the end) is that $20 \cdot 0.07 \neq 1.4 + 20$. The writer has taken the result of the multiplication (1.4) and merged directly into the addition step, creating a lie (since $1.4 \neq 21.4$). The calculations could be written as:

$$\$20 \cdot 0.07 = \$1.40 \quad \text{so the total price is} \quad \$1.40 + \$20 = \$21.40$$

5. **Use a vertical format for long calculations** If you have a long sequence of calculations, consider writing it vertically instead of horizontally. Line things up on some connecting symbol such as $=$ or $\leq$. Whenever possible, make the vertical format read in a linear fashion. The first line will have expressions on both sides of the connecting symbol, but subsequent lines will only have expressions on the right. For example, the sequence of simplifications

$$2 \cdot \left(n \cdot \frac{n}{2} + \frac{n}{2}\right) = 2 \cdot \left(\frac{n^2}{2} + \frac{n}{2}\right) = 2 \cdot \left(\frac{n^2 + n}{2}\right) = n^2 + n$$

can be written (notice the empty left-hand sides in rows 2 and 3):

$$\begin{aligned} 2 \cdot \left(n \cdot \frac{n}{2} + \frac{n}{2}\right) &= 2 \cdot \left(\frac{n^2}{2} + \frac{n}{2}\right) \\ &= 2 \cdot \left(\frac{n^2 + n}{2}\right) \\ &= n^2 + n \end{aligned}$$

6. **Avoid ambiguity** When in doubt, repeat a noun rather using unspecific words like "it" or "the". For example, in the sentences

Let G be a simple graph with $n \geq 2$ vertices that is not complete and let $\overline{G}$ be its complement. Then it must contain at least one edge.

there is some ambiguity about whether "it" refers to G or to $\overline{G}$. The second sentence is better written as "Then $\overline{G}$ must contain at least one edge".

An expanded version of this appendix can be found at
http://www.mathcs.bethel.edu/DiscreteMathWithProof/ .

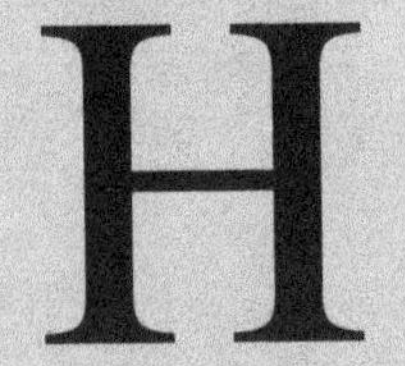

Solutions to Selected Exercises

H.1 Introduction

Exercises 1.4

4. The women must both rate the one man as their top choice, so it is not possible for such an assignment to be stable. The only way for an assignment to be stable in this context is for the man to be paired with his first choice. For if he is paired with anyone at a lower preference rating, he will elope with his first choice.

9. **(d)** True. See the definition of viable (Definition 1.5 on page 5).

10. **(c)** True. See the "female 1st choice" diagram in Example 1.1 on page 4.

11. **(a)** discrete
    (b) continuous

13.

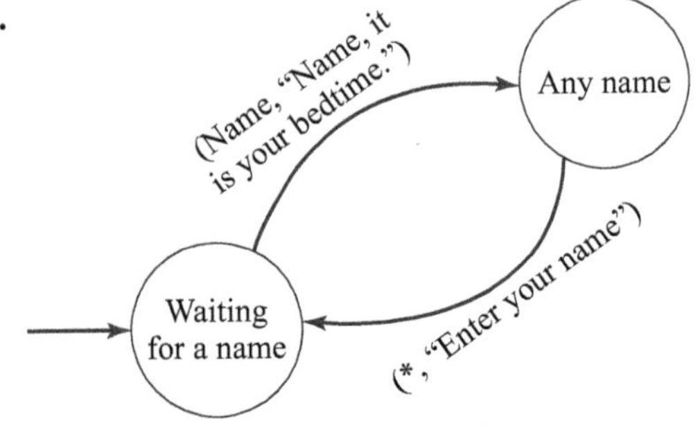

H.2 Sets, Logic, and Boolean Algebras

Exercises 2.1.2

2. **(a)**

$d \notin A$	$A \not\subseteq B$	$B \not\subseteq A$	$A \neq B$
no, $A \cap B \neq \emptyset$	$\|A\| = 6$	$\|B\| = 4$	$\overline{A} = \{d, e, 4, 5\}$
$A \cup B = \{a, b, c, d, 1, 2, 3\}$	$A \cap B = \{c, 1, 3\}$	$A - B = \{a, b, 2\}$	$B - A = \{d\}$

(b)

$d \notin A$	$A \subseteq B$	$B \not\subseteq A$	$A \neq B$
no, $A \cap B \neq \emptyset$	$\|A\| = 5$	$\|B\| = 7$	$\overline{A} = \{b, d, e, 4, 5\}$
$A \cup B = \{a, c, d, 1, 2, 3, 4\} = B$	$A \cap B = \{a, c, 1, 2, 3\} = A$	$A - B = \emptyset$	$B - A = \{d, 4\}$

6. **(b)**

$CT \notin X$	$X \subseteq Y$	$Y \not\subseteq X$	$X \neq Y$
yes, $X \cap Y = \emptyset$	$\|X\| = 0$	$\|Y\| = 4$	$\overline{X} = U$
$X \cup Y = \{NC, NH, NJ, NY\} = Y$	$X \cap Y = \emptyset = X$	$X - Y = \emptyset$	$Y - X = \{NC, NH, NJ, NY\} = Y$

7. **(c)** $\{\ldots, -12, -9, -6, -3, 0, 3, 6, 9, 12, \ldots\}$

(d) These are the ordered pairs of integers on or inside a circle centered at the origin with radius 2.

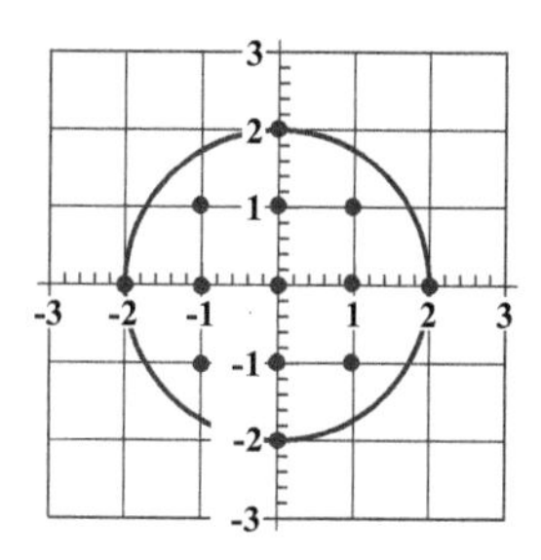

$\{(0, 2), (-1, 1), (0, 1), (1, 1), (-2, 0), (-1, 0), (0, 0), (1, 0), (2, 0), (-1, -1), (0, -1), (1, -1), (0, -2)\}$

8. **(b)** $\{n \in \mathbb{Z} \mid n = 2k \text{ for some integer } k\}$

9. **(d)** $\{n \in \mathbb{Z} \mid n = 6k \text{ for some integer } k \leq 3\}$

10. **(a)** The set of odd integer multiples of 3.

11. **(a)** $(A \cap C) - B = \emptyset$ and $A \cup (B \cap C) = \{a, b, c, f\}$

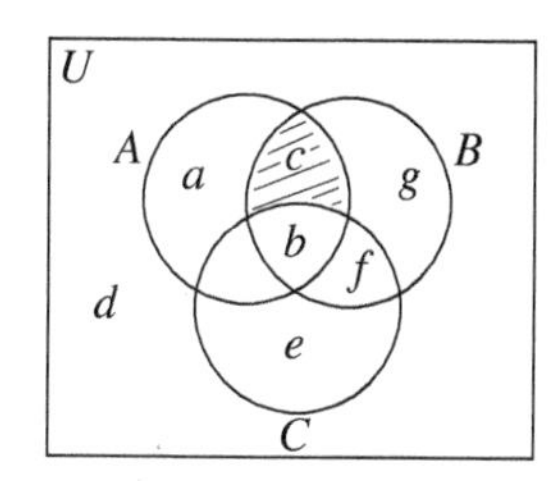

12. **(a)** i. There is no "proper value" because set difference is not associative. The expression $B - C - A$ is not well-formed. Notice that $(B - C) - A = \{2, 3, 4, z\} - A = \{2, 3, 4\}$, whereas $B - (C - A) = B - \{1\} = \{2, 3, 4, z\}$. These sets are not equal.

ii. $\overline{A \cap B} = \{w, x, y, 1, 2, 3, 4\}$ and $(A \cap C) \cup B = \{w, x, z, 1, 2, 3, 4\}$

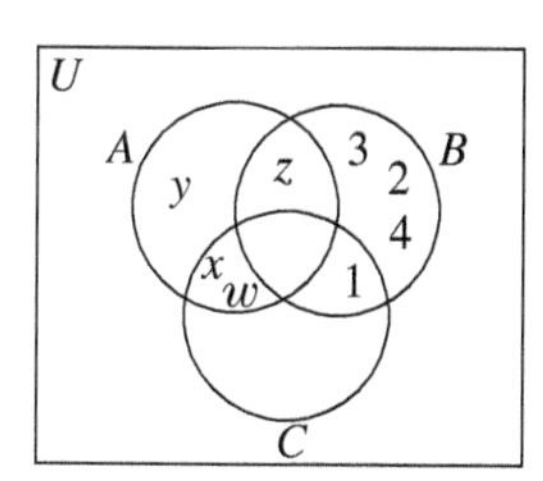

14.

problem	$A - B$	$B - A$	$A \triangle B$
(a)	$\{2, 4\}$	$\{3\}$	$\{2, 3, 4\}$
(d)	$\{2, 4\}$	$\emptyset$	$\{2, 4\}$

16. **(a)** $\{(a, 1), (a, 2), (a, 3), (a, 4), (b, 1), (b, 2), (b, 3), (b, 4)\}$

(b) $\{(a, 1, \alpha), (a, 1, \beta), (a, 1, \gamma), (a, 2, \alpha), (a, 2, \beta), (a, 2, \gamma), (b, 1, \alpha), (b, 1, \beta), (b, 1, \gamma), (b, 2, \alpha), (b, 2, \beta), (b, 2, \gamma)\}$

18. **(a)** $\{\emptyset\}$ Note well: $\emptyset \neq \{\emptyset\}$

(c) $\{\emptyset, \{\alpha\}, \{\beta\}, \{\alpha, \beta\}\}$

20. **(c)** On the left, $A - B$ is denoted by positive slopes, $A - C$ by negative slopes. The union is the region with either shading. On the right, $B \cap C$ is represented by negative sloped lines. The difference $A - (B \cap C)$ is represented by the lines with positive slope.

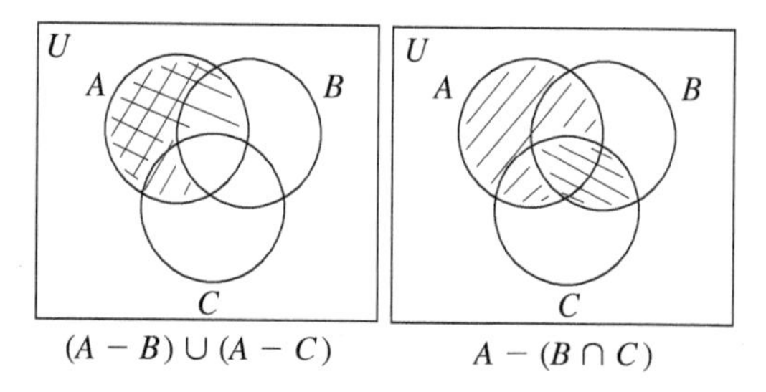

$(A - B) \cup (A - C)$ $\qquad$ $A - (B \cap C)$

22. **(d)** True. The first set contains exactly one element (the empty set). The empty set is defined as the set with no elements.

(e) True. $\overline{A} = U - A$, the set of all elements not in A. The set of elements that are not in $\overline{A}$ are those elements that are in $U - \overline{A} = U - (U - A)$. To obtain $U - A$, we remove, from U, all elements of A (but no others). What is in U that is not in $U - A$? Precisely the elements of A.

23. **(d)** True. Since $A \subseteq B$, every element in A is also in B. This fact and the assumptions that $|A| = |B|$ and A and B are finite sets, imply that B cannot contain any elements that are not in A. Thus, $B \subseteq A$ and so $A = B$. This claim is false if A and B are infinite (for example, A the set of integers and B the set of rationals).

(e) True. By the definition of *Cartesian product*, $\emptyset \times A = \{(a_1, a_2) \mid a_1 \in \emptyset \text{ and } a_2 \in A\} = \emptyset$. Similarly, $A \times \emptyset = \{(a_1, a_2) \mid a_1 \in A \text{ and } a_2 \in \emptyset\} = \emptyset$. Thus, $\emptyset \times A = A \times \emptyset = \emptyset$.

24. **(a)** $A = B$

26. **(a)** {{Fall, Spring, Summer, Winter}},
{{Fall}, {Spring, Summer, Winter}},
{{Spring}, {Fall, Summer, Winter}},
{{Summer}, {Fall, Spring, Winter}},
{{Winter}, {Fall, Spring, Summer}},
{{Fall, Spring}, {Summer, Winter}},
{{Fall, Summer}, {Spring, Winter}},
{{Fall, Winter}, {Spring, Summer}},
{{Fall}, {Spring}, {Summer, Winter}},
{{Fall}, {Summer}, {Spring, Winter}},
{{Fall}, {Winter}, {Spring, Summer}},
{{Spring}, {Summer}, {Fall, Winter}},
{{Spring}, {Winter}, {Fall, Summer}},
{{Summer}, {Winter}, {Fall, Spring}},
{{Fall}, {Spring}, {Summer}, {Winter}}

27. **(a)** This collection of subsets does not partition the integers. The union of these sets is not equal to the set of integers. In particular, the integer zero is not in this union because it is neither positive nor negative.

Exercises 2.1.4

2. Let $x \in C$. Since $C \subseteq B$, $x \in B$. But then $x \notin (B - C)$. On the other hand, $x \in C$ and $C \subseteq A$ (Quick Check 2.3 Problem 1) indicate that $x \in A$. Since $x \in A$ and $x \notin (B - C)$, the definition of *set difference* leads to the conclusion that $x \in A - (B - C)$. Since x was a generic element of C, the proof is complete.

6. (c) For this counterexample, let $A = \{4\}$, $B = \{3, 4, 5, 6, 7\}$, and $C = \{1, 3, 5, 9\}$. Then $A \subseteq B$, $A \cap C = \emptyset$, but $B \cap C = \{3, 5\} \neq \emptyset$.

7. (a) For this counterexample, let $A = \{4\}$, $B = \{w, x, 1, 2\}$, and $C = \{1, 2, y, z\}$. $A - C = \{4\} = A - B$ but $C \neq B$.

10. Idempotence

$x \in (A \cup A)$ iff $(x \in A)$ or $(x \in A)$	definition of *union*
iff $x \in A$	identical requirements
$x \in (A \cap A)$ iff $(x \in A)$ and $(x \in A)$	definition of intersection
iff $x \in A$	identical requirements

11. (a)

$(A \cap B) - C = (A \cap B) \cap \overline{C}$	Proposition 2.20
$= A \cap (B \cap \overline{C})$	associativity
$= A \cap (B - C)$	Proposition 2.20

(b)

$\overline{A} \cap (A \cup B) = (\overline{A} \cap A) \cup (\overline{A} \cap B)$	distributivity
$= (A \cap \overline{A}) \cup (\overline{A} \cap B)$	commutativity
$= \emptyset \cup (\overline{A} \cap B)$	complement
$= (\overline{A} \cap B) \cup \emptyset$	commutativity
$= \overline{A} \cap B$	identity
$= B \cap \overline{A}$	commutativity
$= B - A$	Proposition 2.20

14. (a)

$A - \emptyset = A \cap \overline{\emptyset}$	Proposition 2.20
$= A \cap U$	complement
$= A$	identity

$\emptyset - A = \emptyset \cap \overline{A}$	Proposition 2.20
$= \overline{A} \cap \emptyset$	commutativity
$= \emptyset$	domination

(b) Clearly, if $\mathcal{P}(A) = \mathcal{P}(B)$, then $|\mathcal{P}(A)| = |\mathcal{P}(B)|$. Thus, $|\mathcal{P}(A)| = |\mathcal{P}(B)| = 2^n$ for some n and therefore $|A| = |B| = n$. Suppose $x \in A$. Then x is in at least one subset of A. Since $\mathcal{P}(A) = \mathcal{P}(B)$, x is in at least one subset of B. Thus, $x \in B$ and $A \subseteq B$. Similarly, it can be shown that for any $x \in B$, it is also true that $x \in A$ and so $B \subseteq A$. Thus, $A = B$.

15. (a)

$\overline{A} \cup B = \overline{\overline{\overline{A}} \cap \overline{B}}$	De Morgan in reverse
$= \overline{A \cap \overline{B}}$	involution
$= \overline{A - B}$	Proposition 2.20

17. (e)

$A \triangle \overline{A} = (A - \overline{A}) \cup (\overline{A} - A)$	definition of *symmetric difference*
$= (A \cap \overline{\overline{A}}) \cup (\overline{A} \cap \overline{A})$	Proposition 2.20 (twice)
$= (A \cap A) \cup (\overline{A} \cap \overline{A})$	involution
$= A \cup \overline{A}$	idempotence (twice)
$= U$	complement

Exercises 2.2.4

1. (b) Appeal to authority.

(c) This is unclear without additional information. If John Rockjaw has demonstrated the ability to solve such problems in some other city, there may be no problem. However, if there is no evidence that John Rockjaw can solve these problems, the statement is a bit of a nonsequitur. If crime, drugs, and rebellious teens are not as rampant as claimed, the statement is guilty of using manipulative emotionalism.

(d) There is no fallacy or deception here. The advertisement is communicating facts about the seating arrangements available. Not all claims are deceptive or fallacious.

Exercises 2.3.4

1. The following systematic list demonstrates that there are $2^4 = 16$ combinations.

A	*B*	*C*	*D*	*A*	*B*	*C*	*D*
T	T	T	T	F	T	T	T
T	T	T	F	F	T	T	F
T	T	F	T	F	T	F	T
T	T	F	F	F	T	F	F
T	F	T	T	F	F	T	T
T	F	T	F	F	F	T	F
T	F	F	T	F	F	F	T
T	F	F	F	F	F	F	F

4. In common usage, they might. However, using the definitions presented in the text, they do not. In the first case, I might want bread without butter (or butter without bread). In the second case, I do not want either.

6. (a) Inclusive OR: Not enough tickets being sold or one of the singers getting sick, or both, will cause the concert to be canceled.
Exclusive OR: Not enough tickets being sold or one of the singers getting sick, but not both, will cause the concert to be cancelled.
An "inclusive OR" is most likely intended.

9. (a) A ∧ B ∧ C ∧ (D ∨ E) ∧ F ∧ G

(b) No, your GPA in the major is too low.

13. (c) False. They differ when P and Q have different truth values.

(d) True: build a truth table (see page A40).

Exercises 2.3.9

1. The final columns are the same.

P	Q	$P \to Q$
T	T	T
T	F	F
F	T	T
F	F	T

P	Q	$\neg Q$	$\neg P$	$(\neg Q) \to (\neg P)$
T	T	F	F	T
T	F	T	F	F
F	T	F	T	T
F	F	T	T	T

4. (a) Look carefully at the inequalities.

Original	If $n > 1$, then $\sqrt{n} < n$.
Contrapositive	If $\sqrt{n} \geq n$, then $n \leq 1$
Converse	If $\sqrt{n} < n$, then $n > 1$.
Inverse	If $n \leq 1$, then $\sqrt{n} \geq n$.

5. (a) "On every day that is sunny, we go to the beach" can be written as "If it is a sunny day, then we go to the beach".

Original	If it is a sunny day, then we go to the beach.
Contrapositive	If we do not go to the beach, then it is not a sunny day.
Converse	If we go to the beach, then it is a sunny day.
Inverse	If it is not a sunny day, then we do not go to the beach.

7. (a)

$P \to [(\neg P) \to Q]$ *is* a tautology.

P	Q	$\neg P$	$(\neg P) \to Q$	$P \to [(\neg P) \to Q]$
T	T	F	T	T
T	F	F	T	T
F	T	T	T	T
F	F	T	F	T

7. (b)

$(P \wedge Q) \vee Q$ is *not* a tautology.

P	Q	$P \wedge Q$	$(P \wedge Q) \vee Q$
T	T	T	T
T	F	F	F
F	T	F	T
F	F	F	F

9. (a) These are *not* equivalent.

P	Q	$P \wedge Q$	$\neg P \vee \neg Q$
T	T	T	F
T	F	F	T
F	T	F	T
F	F	F	T

13. (a)

P	Q	$P \wedge \neg Q$	$P \vee Q$	$(P \wedge \neg Q) \to (P \vee Q)$
T	T	F	T	T
T	F	T	T	T
F	T	F	T	T
F	F	F	F	T

13. (b)

This is a tautology because the final column is all **T**'s. That is, for all possible values of P and Q, the statement is true.

15. (a)

original	$(\neg P) \to (\neg Q)$	If a man hasn't discovered something he will die for, then he isn't fit to live.
contrapositive	$Q \to P$	If a man is fit to live, then has discovered something he will die for.
converse	$(\neg Q) \to (\neg P)$	If a man isn't fit to live, then he hasn't discovered something he will die for.
inverse	$P \to Q$	If a man has discovered something he will die for, then he is fit to live.

(b) If the original is true, we know that the contrapositive is also true. We also know that the converse and inverse have the same truth value.
On less firm ground, it is in general not the case that the converse and inverse are true statements. (Someone might choose something very evil that he is willing to die for.)

21. (a) False. To be a contradiction, *every* entry in the final column would need to be **F**.

Exercises 2.4.3

2. Original $(P \wedge Q) \rightarrow (R \vee S)$
Contrapositive $(\neg R \wedge \neg S) \rightarrow (\neg P \vee \neg Q)$
Converse $(R \vee S) \rightarrow (P \wedge Q)$
Inverse $(\neg P \vee \neg Q) \rightarrow (\neg R \wedge \neg S)$

4. (c) Associativity The two sides of the biconditional have the same truth table values.

P	Q	R	$P \vee Q$	$Q \vee R$	$[(P \vee Q) \vee R]$	$[P \vee (Q \vee R)]$
T	T	T	T	T	T	T
T	T	F	T	T	T	T
T	F	T	T	T	T	T
T	F	F	T	F	T	T
F	T	T	T	T	T	T
F	T	F	T	T	T	T
F	F	T	F	T	T	T
F	F	F	F	F	F	F

P	Q	R	$P \wedge Q$	$Q \wedge R$	$[(P \wedge Q) \wedge R]$	$[P \wedge (Q \wedge R)]$
T	T	T	T	T	T	T
T	T	F	T	F	F	F
T	F	T	F	F	F	F
T	F	F	F	F	F	F
F	T	T	F	T	F	F
F	T	F	F	F	F	F
F	F	T	F	F	F	F
F	F	F	F	F	F	F

(g) Distributivity ($\wedge$ over $\vee$) The two sides of the biconditional have the same truth table values.

P	Q	R	$P \wedge Q$	$P \wedge R$	$Q \vee R$	$P \wedge (Q \vee R)$	$(P \wedge Q) \vee (P \wedge R)$
T	T	T	T	T	T	T	T
T	T	F	T	F	T	T	T
T	F	T	F	T	T	T	T
T	F	F	F	F	F	F	F
F	T	T	F	F	T	F	F
F	T	F	F	F	T	F	F
F	F	T	F	F	T	F	F
F	F	F	F	F	F	F	F

P	Q	R	$P \vee Q$	$P \wedge R$	$Q \wedge R$	$(P \vee Q) \wedge R$	$(P \wedge R) \vee (Q \wedge R)$
T	T	T	T	T	T	T	T
T	T	F	T	F	F	F	F
T	F	T	T	T	F	T	T
T	F	F	T	F	F	F	F
F	T	T	T	F	T	T	T
F	T	F	T	F	F	F	F
F	F	T	F	F	F	F	F
F	F	F	F	F	F	F	F

5. **(a) Implication** All sides of the biconditional have the same truth table.

P	Q	$\neg P$	$\neg Q$	$P \wedge (\neg Q)$	$P \to Q$	$\neg(P \wedge (\neg Q))$	$(\neg P) \vee Q$
T	T	F	F	F	T	T	T
T	F	F	T	T	F	F	F
F	T	T	F	F	T	T	T
F	F	T	T	F	T	T	T

9. **(a)**

P	Q	$(P \to Q)$	$[(P \to Q) \wedge Q]$	$[(P \to Q) \wedge Q] \to P$
T	T	T	T	T
T	F	F	F	T
F	T	T	T	F
F	F	T	F	T

10. **(b)** $(P \to [(\neg P) \to Q])$

$\Leftrightarrow P \to (\neg(\neg P) \vee Q)$ implication

$\Leftrightarrow P \to (P \vee Q)$ double negation

$\Leftrightarrow T$ law of addition

(c) $[P \to (Q \wedge (\neg Q))] \to (\neg P)$

$\Leftrightarrow [P \to F] \to (\neg P)$ law of contradiction

$\Leftrightarrow ((\neg P) \vee F) \to (\neg P)$ implication

$\Leftrightarrow \neg P \to \neg P$ identity

$\Leftrightarrow \neg(\neg P) \vee (\neg P)$ implication

$\Leftrightarrow P \vee \neg P$ double negation

$\Leftrightarrow T$ law of the excluded middle

12. **(f)** $\neg(P \to Q) \to P$

$\Leftrightarrow \neg(\neg(P \to Q)) \vee P$ implication

$\Leftrightarrow (P \to Q) \vee P$ double negation

$\Leftrightarrow ((\neg P) \vee Q) \vee P$ implication

$\Leftrightarrow P \vee ((\neg P) \vee Q)$ commutativity

$\Leftrightarrow (P \vee (\neg P)) \vee Q$ associativity

$\Leftrightarrow T \vee Q$ law of the excluded middle

$\Leftrightarrow Q \vee T$ commutativity

$\Leftrightarrow T$ domination

Exercises 2.5.3

2. Prove one of these; then use the duality principle for the other.

$(a + (\overline{a} \cdot (\overline{b} + b))) \cdot b$

$= (a + (\overline{a} \cdot (b + \overline{b}))) \cdot b$ commutativity

$= (a + (\overline{a} \cdot 1)) \cdot b$ complement

$= (a + \overline{a}) \cdot b$ identity

$= 1 \cdot b$ complement

$= b \cdot 1$ commutativity

$= b$ identity

$(a \cdot (\overline{a} + (\overline{b} \cdot b))) + b$

$= (a \cdot (\overline{a} + (b \cdot \overline{b}))) + b$ commutativity

$= (a \cdot (\overline{a} + 0)) + b$ complement

$= (a \cdot \overline{a}) + b$ identity

$= 0 + b$ complement

$= b + 0$ commutativity

$= b$ identity

5. Prove one of these; then use the duality principle.

$a + \overline{a} \cdot b = (a + \overline{a}) \cdot (a + b)$ distributivity

$= 1 \cdot (a + b)$ complement

$= (a + b) \cdot 1$ commutativity

$= a + b$ identity

$b + a \cdot \overline{b} = (b + a) \cdot (b + \overline{b})$ distributivity

$= (b + a) \cdot 1$ complement

$= (a + b) \cdot 1$ commutativity

$= a + b$ identity

9. This is not a Boolean algebra. The obvious assignments for 0 and 1 are $0 = 0$ and $1 = n$. The identity property holds, but the complement axiom fails. For example, if $n = 11$, then $\overline{3} = 8$. However, $3 + \overline{3} = \max\{3, 8\} = 8 \neq 11$ and $3 \cdot \overline{3} = \min\{3, 8\} = 3 \neq 0$.

12. This follows from two applications of the absorption lemma.

(a) $x \cdot (x + y) = x = x + x \cdot y$ absorbtion (twice)

(b) The dual is the same equation (with the sides swapped).

14. **(d)** This is *not* true by the duality principle. The reason is that $x \cdot y = 0$ is not a theorem. It is true sometimes but false other times.

Here is a counterexample. Let $S = \{a, b, c\}$ and let $B = \mathcal{P}(S)$ with the usual associations. Then $x = \{a\}$ and $y = \{b\}$ have $x \cdot y = 0$ but $x + y \neq 1$.

15. **(b)** True. This follows because the third element, x, in the Boolean algebra ($x \neq 0$ and $x \neq 1$) would have no complement. It was already proven that $\overline{0} = 1$ and $\overline{1} = 0$ (see Quick Check 2.10). Thus, it would have to be the case that $x = \overline{x}$ due to the uniqueness of complements in a Boolean algebra. Then, $x = x + x = x + \overline{x} = 1$, but it was assumed in the beginning that $x \neq 1$. Thus, there must be at least one other element in the Boolean algebra.

17. **(a)** Let $x = \{a\}$, $y = \{b\}$. Then $x + y = \{a, b\}$, but $x \cdot y = \emptyset$.

18. **(a)** Let $x = \{a, b\}$, $y = \{c\}$, $z = \{a, c\}$. Then $x \cdot (y + z) = \{a\}$ but $x + y \cdot z = \{a, b, c\}$.

22. Let $S = \{\alpha, \beta, \gamma\}$ and $B = \mathcal{P}(S)$. Now let $a = \{\alpha, \beta\}$ and $b = \{\alpha, \beta, \gamma\}$. The set x could be any one of the sets $\{\gamma\}$, $\{\alpha, \gamma\}$, $\{\beta, \gamma\}$, $\{\alpha, \beta, \gamma\}$. This means that x cannot be uniquely determined from the equation $x + a = b$, so the equation cannot be solved uniquely for x.

Exercises 2.6.2

1. **(a)** The universe of discourse is the set of all dogs. Let $H(d)$ mean that dog d goes to heaven. $\forall d, H(d)$

(b) The universe of discourse is $\mathbb{Z}$. $\forall x \in \mathbb{Z}, \forall y \in \mathbb{Z}, (x + y \in \mathbb{Z})$

2. **(b)** The universe of discourse is the set of people in the small group of friends. Let $T(p)$ mean that person p has talked out of turn during a conversation. Let $B(p)$ mean that person p has broken a promise. $\exists p, [T(p) \vee B(p)]$

3. **(b)** For all real numbers, x and y, the equation $x \cdot y = 4x$ is true. False (let $x = 1$, $y = 9$).

(c) For all real numbers, x, there is another real number, y, such that $x \cdot y = 4x$ is true. True (set $y = 4$).

5. **(a)** There is an integer, x, and an integer, y which is greater than 3, such that the equation $x^2 + y^2 = 5$ has a solution. False ($y^2 > 9$).

8. **(a)** $\forall x \in U, [(\neg Q(x)) \wedge P(x)]$ because

$$\neg[\exists x \in U, [Q(x) \vee (\neg P(x))]]$$
$$\Leftrightarrow \forall x \in U, \neg[Q(x) \vee (\neg P(x))]$$
$$\Leftrightarrow \forall x \in U, [(\neg Q(x)) \wedge \neg(\neg P(x))]$$
$$\Leftrightarrow \forall x \in U, [(\neg Q(x)) \wedge P(x)]$$

9. **(b)** $\forall x, \exists y, \exists z, [(\neg F(x, y) \vee (\neg G(x, z)) \vee H(y, z)]$

$$\neg[\exists x, \forall y, \forall z, [(F(x, y) \wedge G(x, z)) \rightarrow H(y, z)]]$$
$$\Leftrightarrow \forall x, \neg[\forall y, \forall z, [(F(x, y) \wedge G(x, z)) \rightarrow H(y, z)]]$$
$$\Leftrightarrow \forall x, \exists y, \neg[\forall z, [(F(x, y) \wedge G(x, z)) \rightarrow H(y, z)]]$$
$$\Leftrightarrow \forall x, \exists y, \exists z, \neg[(F(x, y) \wedge G(x, z)) \rightarrow H(y, z)]$$
$$\Leftrightarrow \forall x, \exists y, \exists z, [(F(x, y) \wedge G(x, z)) \wedge \neg H(y, z)]$$
$$\Leftrightarrow \forall x, \exists y, \exists z, [F(x, y) \wedge G(x, z) \wedge (\neg H(y, z))]$$

13. The statements are first translated into (hopefully) obvious symbolic statements. The universes of discourse are implied.

Original	Symbolic	Symbolic Neg.	Negation
(a) All Cretans are liars.	$\forall c, L(c)$	$\exists c, \neg L(c)$	There is at least one Cretan who is not a liar.
(b) There are no good men.	$\forall m, \neg G(m)$	$\exists m, G(m)$	There is at least one good man.

17. **(a)**

$$\neg[\forall s, [(\exists d, [P(s, d)]) \vee Q(s)]]$$
$$\Leftrightarrow \exists s, \neg[(\exists d, [P(s, d)]) \vee Q(s)]$$
$$\Leftrightarrow \exists s, [(\forall d, [\neg P(s, d)]) \wedge (\neg Q(s))]$$

(b) **Original:** Every Bethel University student either lives in a dorm or rents a book locker (or both).
Negation: There is a Bethel University student who neither lives in a dorm nor rents a book locker.

18. **(b)** False. The negation is "some good boy does not do fine (well)".

(c) True. It is true if x and y may be complex numbers, but false if they are restricted to the real numbers.

20. Let the universe of discourse be $\mathbb{R}$ and let $P(x, y)$ mean $x^2 = y$. Then by choosing $y_0 = x^2$, the proposition $\forall x, \exists y, (x^2 = y)$ is always true. However, setting $y = -1$ shows that $\forall y, \exists x, (x^2 = y)$ is not always true. Since the hypothesis is always true but the conclusion is often false, the implication is also often false.

H.3 Proof

Exercises 3.1.3

3. Theorem W becomes unnecessary. Nevertheless, Theorem Y is still valid. However, Theorem Y contradicts the new Axiom 4. The axioms are therefore inconsistent.

4. **(a)** True. See the discussion just before the section entitled "Propositions, Theorems, Lemmas, and Corollaries" on page 89.

(e) False. The intended audience needs to verify that you, the student, truly understand the proof. That requires complete details in most cases.

5. It is not possible to define every term. Each definition needs to use already defined terms, which in turn need other previously defined terms. The options are to either eventually circle back to using terms that depend on what is currently being defined (as is done in dictionaries) or else stop the process and declare some terms as undefined.

This is not a "second best" approach because well-chosen axioms provide all the information needed to manipulate the undefined terms.

6. Suppose i is another identity element. Since e is an identity, $a = a \diamond e$ for all $a \in G$. This must therefore hold for $a = i$. But i is also an identity element, so $i \diamond b = b$ for all $b \in G$. In particular, this is true when $b = e$. Combining these identities we have $i = i \diamond e = e$. This means there is really only one identity element in G.

Exercises 3.2.6

1. (e) True. Definition 3.19.
2. (d) False. Let $m = 3$ and $n = 5$. Then both m and n are odd integers, but $m = 3 = 2 \cdot 1 + 1$ and $n = 5 = 2 \cdot 2 + 1$. Since $1 \neq 2$, it is incorrect to say that both m and n are equal to $2k + 1$ for some integer, k. It is valid, however, to say that $m = 2j + 1$ and $n = 2k + 1$ for some integers, j and k.
3.

	x	y	q	r
(a)	2961	987	$q = 3$	$r = 0$

4.

	x	x (as a product of primes)
(a)	2548	$x = 2^2 \cdot 7^2 \cdot 13$

5. Factoring x and y helps to determine $\gcd(x, y)$.

	x	y	$\gcd(x, y)$
(a)	$688 = 2^4 \cdot 43$	$108 = 2^2 \cdot 3^3$	$2^2 = 4$

6. Factoring x and y helps to determine $\operatorname{lcm}(x, y)$.

	x	y	$\operatorname{lcm}(x, y)$
(a)	$999 = 3^3 \cdot 37$	$93 = 3 \cdot 31$	$3^3 \cdot 31 \cdot 37 = 30969$

7.

	x	y	x mod y
(a)	57	701	57

Exercises 3.3.10

1. (a) Notice that $n^3 - n = n(n^2 - 1) = n(n - 1)(n + 1)$. Either n or $n - 1$ is even, so $n^3 - n$ has a factor of 2 and is thus even.
3. There are no even primes, so the hypothesis is false, making the implication true.
9.

$$\begin{aligned}
&(a+b)^2 \\
&= (a+b)(a+b) && \text{definition of } \textit{squared} \\
&= (a+b)a + (a+b)b && \text{distributivity} \\
&= a(a+b) + b(a+b) && \text{commutativity (twice)} \\
&= [aa+ab] + [ba+bb] && \text{distributivity (twice)} \\
&= [aa+ab] + [ab+bb] && \text{commutativity} \\
&= ([aa+ab]+ab) + bb && \text{associativity} \\
&= (aa+[ab+ab]) + bb && \text{associativity} \\
&= (aa+2ab) + bb && \text{properties of addition and multiplication} \\
&= (a^2+2ab) + b^2 && \text{definition of } \textit{squared} \text{ (twice)} \\
&= a^2+2ab+b^2 && \text{associativity}
\end{aligned}$$

17. An indirect proof can be used. If c is even, then there is an integer, k, such that $c = 2k$. Thus, $c^5 + 7 = (2k)^5 + 7 = 32k^5 + 7 = 32k^5 + 6 + 1 = 2(16k^5 + 3) + 1$ and so $c^5 + 7$ is odd, which is the negation of the hypothesis.
22. Suppose that $0 \leq \frac{a+b}{2} < \sqrt{ab}$. Then $\frac{a+b}{2} \cdot \frac{a+b}{2} < \sqrt{ab} \cdot \sqrt{ab}$, so $(a+b)^2 < 4ab$. Thus, $a^2 + 2ab + b^2 < 4ab$, so $a^2 - 2ab + b^2 < 0$. But then $(a-b)^2 < 0$, a contradiction, since no square is negative.
24. The cases are: n even and n odd.
    Even: If n is even, then $n = 2k$ for some $k \in \mathbb{Z}$. Thus $n^2 + n = 4k^2 + 2k = 2(2k^2 + k)$ is even.
    Odd: If n is odd, then $n = 2k + 1$ for some $k \in \mathbb{Z}$. Thus $n^2 + n = (4k^2 + 4k + 1) + (2k + 1) = 2(2k^2 + 3k + 1)$ is even.
30. Let A be the set of negative integers and let B be the set of nonnegative integers. (Evens and odds is another easy solution.)
35. Suppose no such prime exists. Then every prime is in $\{2, 3, 4, 5, \ldots, n\}$, contradicting Theorem 3.38.
40. $4 = (1+1)^2 \neq 1^2 + 1^2 = 2$
46. This can be proved as two implications.
    If a is composite, then the sum of the positive divisors of a is greater than $a + 1$.
    Since a is composite, a is not prime. Thus, there exists a positive integer, q, that is a positive divisor of a, but is not equal to 1 and is not equal to a. Additionally, 1 and a are positive divisors of any positive integer, a. Therefore, the sum of the positive divisors of a is minimally $a + 1 + q > a + 1$ since $q > 1$.
    If the sum of the positive divisors of a is greater than $a + 1$, then a is composite.
    Let S be the set of the positive divisors of a. Clearly 1 and a are included in S since these are positive divisors of any positive integer, a. It is also obvious that any element, x, in S satisfies $1 \leq x \leq a$. Since the sum of all the elements in S is greater than $a + 1$, there must exist a positive integer, r, in S that is not equal to 1 and is not equal to a. This implies that $1 < r < a$. Thus, the set of positive integer divisors of a includes more that just 1 and a, and so a is composite.

Exercises 3.4.7

1. (b) 346 and 1056

$$\begin{aligned}
1056 &= 346 \cdot (3) + 18 \\
346 &= 18 \cdot (19) + 4 \\
18 &= 4 \cdot (4) + 2 \\
4 &= 2 \cdot (2) + 0
\end{aligned}$$

So $\gcd(346, 1056) = 2$.

$$\begin{aligned}
2 &= 18 + 4 \cdot (-4) \\
&= 18 + (346 + 18 \cdot (-19)) \cdot (-4) && \text{substitute} \\
&= 346 \cdot (-4) + 18 \cdot (77) && \text{simplify} \\
&= 346 \cdot (-4) + (1056 + 346 \cdot (-3)) \cdot (77) && \text{substitute} \\
&= 1056 \cdot (77) + 346 \cdot (-235)
\end{aligned}$$

10. Use the Chinese Remainder Theorem.
    $x = (4 \cdot 385 \cdot 1 + 21 \cdot 66 \cdot 26 + 6 \cdot 210 \cdot 1) \pmod{2310} = 38836 \pmod{2310} = 1876$.
    The following calculations were used to find modular inverses.
    $6 \mid (385 - 1)$ so $385 \cdot 1 \equiv 1 \pmod{6}$
    $11 \mid (210 - 1)$ so $210 \cdot 1 \equiv 1 \pmod{11}$
    The final inverse (for 66, mod 35) takes a bit more

work (using the Euclidean Algorithm).

$$\begin{aligned} 66 &= 35 \cdot 1 + 31 \\ 35 &= 31 \cdot 1 + 4 \\ 31 &= 4 \cdot 7 + 3 \\ 4 &= 3 \cdot 1 + 1 \\ 3 &= 1 \cdot 3 + 0 \end{aligned} \qquad \begin{aligned} 1 &= 4 - 3(1) \\ &= 4 - (31 - 4(7))(1) \\ &= 4(8) - 31 \\ &= (35 - 31)(8) - 31 \\ &= 35(8) + 31(-9) \\ &= 35(8) + (66 - 35(1))(-9) \\ &= 66(-9) + 35(17) \end{aligned}$$

One inverse is -9, but that is congruent to $35 - 9 = 26$, mod 35.

16. **(a)** Notice that $48 = 2^4 \cdot 3$ so $5x \equiv 24 \pmod{48}$ has the same solution set as

$$5x \equiv 24 \pmod{16}$$
$$5x \equiv 24 \pmod{3}$$

The next step requires an inverse for 5, mod 16. Instead of trial and error, a single step of the Euclidean Algorithm enables us to express $1 = \gcd(5, 16)$ as a linear combination: $16 = 5 \cdot (3) + 1$ so $16 \cdot (1) + 5 \cdot (-3) = 1$ and $5 \cdot (-3) - 1 = 16 \cdot (-1)$. This shows that -3 is an inverse for 5, mod 16. We can also notice that $-3 + 16 = 13$, so $-3 \equiv 13 \pmod{16}$. This indicates that 13 is another inverse for 5, mod 16. A similar calculation shows that $3 \cdot 11 \equiv 1 \pmod{16}$. An inverse for 5, mod 3 is easier: $5 \cdot 2 \equiv 1 \pmod{3}$. The two inverses for 5 can be used to convert the system of linear congruences to

$$x \equiv 312 \equiv 8 \pmod{16}$$
$$x \equiv 48 \equiv 0 \pmod{3}$$

Thus, $x = (8 \cdot 3 \cdot 11 + 0 \cdot 16 \cdot 1) \bmod 48 = 264 \bmod 48 = 24$.

23. **(a)** work converts to the number sequence 23 15 18 11.

$$23^5 \bmod 91 = 6436343 \bmod 91 = 4$$
$$15^5 \bmod 91 = 759375 \bmod 91 = 71$$
$$18^5 \bmod 91 = 1889568 \bmod 91 = 44$$
$$11^5 \bmod 91 = 161051 \bmod 91 = 72$$

(c)

$$31^5 \bmod 91 = 28629151 \bmod 91 = 5$$
$$33^5 \bmod 91 = 39135393 \bmod 91 = 24$$
$$1^5 \bmod 91 = 1 \bmod 91 = 1$$
$$13^5 \bmod 91 = 371293 \bmod 91 = 13$$

Converting back to letters yields exam.

Exercises 3.5.6

1. **(b)** $2^m > m$ for $m \geq 1$

Set $P(m) : 2^m > m$.

Base Step

$P(1)$ is true because $2^1 = 2 > 1$.

Inductive Step

Assume that $P(m)$ is true. Then

$$\begin{aligned} 2^{m+1} &= 2 \cdot 2^m \\ &> 2 \cdot m && \text{by the inductive hypothesis} \\ &= m + m && \text{the hardest step to see} \\ &\geq m + 1 && \text{since } m \geq 1 \end{aligned}$$

Thus $2^{m+1} > (m+1)$ and $P(m+1)$ is also true.

Conclusion

The Mathematical Induction theorem implies that $P(m)$ is true for all $m \geq 1$.

5. To guess the formula, we need to experiment a bit.

n	$\sum_{k=1}^{n} \frac{1}{2^k}$
1	$\frac{1}{2}$
2	$\frac{3}{4}$
3	$\frac{7}{8}$
4	$\frac{15}{16}$

It seems very likely that $\sum_{k=1}^{n} \frac{1}{2^k} = \frac{2^n - 1}{2^n}$ for $n \geq 1$.

Proof:

Let $P(n)$ be the equation $\sum_{k=1}^{n} \frac{1}{2^k} = \frac{2^n - 1}{2^n}$.

Base Step

$P(1)$ is true because $\sum_{k=1}^{1} \frac{1}{2^k} = \frac{1}{2} = \frac{2^1 - 1}{2^1}$.

Inductive Step

Assume that $P(n)$ is true. Then

$$\begin{aligned} \sum_{k=1}^{n+1} \frac{1}{2^k} &= \left(\sum_{k=1}^{n} \frac{1}{2^k} \right) + \frac{1}{2^{n+1}} \\ &= \frac{2^n - 1}{2^n} + \frac{1}{2^{n+1}} && \text{by the inductive hypothesis} \\ &= \frac{(2^{n+1} - 2) + 1}{2^{n+1}} \\ &= \frac{2^{n+1} - 1}{2^{n+1}} \end{aligned}$$

so $P(n+1)$ is also true.

Conclusion

The theorem of mathematical induction implies that $P(n)$ is true for all $n \geq 1$.

10. The exercise is much easier if we use a previous theorem and write it as:

$$\sum_{k=1}^{n} \left(\sum_{i=1}^{k} i \right) = \sum_{k=1}^{n} \frac{k(k+1)}{2} = \frac{n(n+1)(n+2)}{6} \quad \text{for } n \geq 1$$

Let $P(n)$ be given by

$$P(n) : \sum_{k=1}^{n} \frac{k(k+1)}{2} = \frac{n(n+1)(n+2)}{6}$$

Base Step
$P(1)$ is true because
$\sum_{k=1}^{1} \frac{k(k+1)}{2} = \frac{2}{2} = 1 = \frac{1(1+1)(1+2)}{6}$.
Inductive Step Assume $P(n)$ is true for some $n \geq 1$. Then $P(n+1)$ is also true because

$$\begin{aligned}\sum_{k=1}^{n+1} \frac{k(k+1)}{2} &= \left(\sum_{k=1}^{n} \frac{k(k+1)}{2}\right) + \frac{(n+1)(n+2)}{2} \\ &= \frac{n(n+1)(n+2)}{6} + \frac{(n+1)(n+2)}{2} \\ &\quad \text{by the inductive hypothesis} \\ &= \frac{(n(n+1)(n+2)) + 3\,((n+1)(n+2))}{6} \\ &= \frac{(n+1)(n+2)(n+3)}{6}\end{aligned}$$

Conclusion The theorem of mathematical induction implies that $P(n)$ is true for all $n \geq 1$.

13. Let $P(n)$ be the claim that $(2n+1)^2 - 1$ is divisible by 8 ($n \geq 0$). Note that by verifying $P(n)$ is true for all $n \geq 0$, this will prove that $x^2 - 1$ is divisible by 8 when x is any positive odd integer.
Base Step
$P(0)$ is true because $(2 \cdot 0 + 1)^2 - 1 = 1 - 1 = 0$, and 0 is divisible by 8 ($0 = 8 \cdot 0$).
Inductive Step
Assume $P(n)$ is true. We need to show that $P(n+1)$ is true under this assumption.

$$P(n+1) : (2(n+1)+1)^2 - 1 \text{ is divisible by } 8$$

$$\begin{aligned}(2(n+1)+1)^2 - 1 &= (2n+3)^2 - 1 \\ &= (4n^2 + 12n + 9) - 1 \\ &= 4n^2 + 12n + 8 \\ &= 4n^2 + 4n + 8n + 8 \\ &= 4n^2 + 4n + 8(n+1) \\ &= 4n^2 + 4n + 1 - 1 + 8(n+1) \\ &= ((2n+1)^2 - 1) + 8(n+1)\end{aligned}$$

Clearly $8(n+1)$ is divisible by 8, and $((2n+1)^2 - 1)$ is divisible by 8 by the inductive hypothesis. Thus, $P(n+1)$ is true when $P(n)$ is true.
Conclusion
The Mathematical Induction theorem implies that $P(n)$ is true for all $n \geq 0$. Thus, $x^2 - 1$ is divisible by 8 for any positive odd integer, x.

20. Since all the expressions in the assertion are positive, the assertion is equivalent to the claim

$$\frac{4n^2 + 4n + 1}{4n^2 + 8n + 4} \leq \frac{n+1}{n+2}$$

which is in turn equivalent to the assertion

$$(n+1)(4n^2 + 8n + 4) - (n+2)(4n^2 + 4n + 1) \geq 0$$

This reduces to the equivalent assertion

$$3n + 2 \geq 0$$

Since $n \geq 0$, the final assertion is clearly true.

23. **Base Step** $h = 0$
$\sum_{i=0}^{0} (h-i)2^i = (0-0)2^0 = 0 = 2^{0+1} - 0 - 2$
Inductive Step
Assume that for some $h \geq 0$,
$\sum_{i=0}^{h} (h-i)2^i = 2^{h+1} - h - 2$. Consider
$\sum_{i=0}^{h+1} ((h+1) - i)2^i$.

$$\begin{aligned}\sum_{i=0}^{h+1} (h+1-i)2^i &= \sum_{i=0}^{h} (h+1-i)2^i \\ &\quad + ((h+1) - (h+1))2^{h+1} \\ &= \sum_{i=0}^{h} [(h-i)2^i + 2^i] \\ &= \sum_{i=0}^{h} (h-i)2^i + \sum_{i=0}^{h} 2^i \\ &= [2^{h+1} - h - 2] + \sum_{i=0}^{h} 2^i \\ &\quad \text{by the inductive hypothesis} \\ &= [2^{h+1} - h - 2] + 2^{h+1} - 1 \\ &= 2 \cdot 2^{h+1} - h - 3 \\ &= 2^{h+2} - (h+1) - 2\end{aligned}$$

Conclusion
The assertion is true for $h = 0$ and whenever it is true for h, it is also true for $h+1$. The Mathematical Induction theorem indicates that the assertion is true for all nonnegative integers, h.

24. **Base Step** $h = 0$
$\sum_{i=0}^{0} i2^i = 02^0 = 0 = (0-1)2^{0+1} + 2$
Inductive Step
Assume that for some $h \geq 0$, $\sum_{i=0}^{h} i2^i = (h-1)2^{h+1} + 2$.
Consider $\sum_{i=0}^{h+1} i2^i$.

$$\begin{aligned}\sum_{i=0}^{h+1} i2^i &= \sum_{i=0}^{h} i2^i + (h+1)2^{h+1} \\ &= [(h-1)2^{h+1} + 2] + (h+1)2^{h+1} \\ &\quad \text{by the inductive hypothesis} \\ &= [(h-1) + (h+1)]2^{h+1} + 2 \\ &= (2h) \cdot 2^{h+1} + 2 \\ &= h2^{h+2} + 2\end{aligned}$$

Conclusion The assertion is true for $h = 0$ and whenever it is true for h, it is also true for $h+1$. The Mathematical Induction theorem indicates that the assertion is true for all nonnegative integers, h.

27. (b) $f_0 + f_2 + \cdots + f_{2n} = f_{2n+1}$ for $n \geq 0$

Set $P(n): \ f_0 + f_2 + \cdots + f_{2n} = f_{2n+1}$

Base Step $P(0)$ is true because $f_0 = 1 = f_1 = f_{2\cdot 0+1}$.

Inductive Step Assume that $P(n)$ is true. Then

$$\begin{aligned} & f_0 + f_2 + \cdots + f_{2n} + f_{2n+2} \\ &= (f_0 + f_2 + \cdots + f_{2n}) + f_{2n+2} \\ &= f_{2n+1} + f_{2n+2} \quad \text{by the inductive hypothesis} \\ &= f_{2n+3} \quad \text{using the definition of the Fibonacci sequence} \end{aligned}$$

so $P(n+1)$ is also true.

Conclusion The Mathematical Induction theorem implies that $P(n)$ is true for all $n \geq 0$.

Exercises 3.6.3

2. (a) $\sum_{k=1}^{2m+1} k = \left(\sum_{k=1}^{2m} k\right) + (2m+1) = m(2m+1) + (2m+1) = (m+1)(2m+1)$

(b) If n is even, then there is an integer m such that $n = 2m$. Thus, using Exercise 1, $\sum_{k=1}^{n} k = \sum_{k=1}^{2m} k = m(2m+1) = \left(\frac{n}{2}\right) \cdot (n+1) = \frac{n(n+1)}{2}$.
If n is odd, then there is an integer m such that $n = 2m + 1$. Thus, using part (a), $\sum_{k=1}^{n} k = \sum_{k=1}^{2m+1} k = (m+1)(2m+1) = \left(\frac{n-1}{2} + 1\right) \cdot n = \frac{n(n+1)}{2}$.

5. Some useful examples: $2^4 - 1 = 15 = 3 \cdot 5$, $2^6 - 1 = 63 = 7 \cdot 9$, $2^9 - 1 = 511 = 7 \cdot 73$, $2^{10} - 1 = 1023 = 3 \cdot 11 \cdot 31 = 31 \cdot 33$, $2^{15} - 1 = 32767 = 7 \cdot 31 \cdot 151 = 7 \cdot 4681$.

Observation 1: If $n = 2m$, then $2^{2m} - 1 = (2^m - 1)(2^m + 1)$

Observation 2: If $n = ab$, then $2^{ab} - 1 = (2^a - 1)\left(\sum_{k=1}^{b} 2^{a(b-k)}\right)$

Thus, if $n = ab$ is composite, then $2^n - 1 = 2^{ab} - 1 = (2^a - 1)\left(\sum_{k=1}^{b} 2^{a(b-k)}\right)$ is also composite.

11. Let the rational numbers be $\frac{a}{b}$ and $\frac{c}{d}$, with $b \neq 0$ and $d \neq 0$. Their sum is $\frac{a}{b} + \frac{c}{d} = \frac{ad+bc}{bd}$. Since a, b, c, and d are integers, so are $ad + bc$ and bd. Also, $b \neq 0$ and $d \neq 0$, imply that $bd \neq 0$. Thus $\frac{ad+bc}{bd}$ meets the requirements of the definition of a rational number.

Note: The previous three sentences that exhibit the "properties → concept" part of the definition are important. The proof is not complete without this verification.

21. A proof using four cases will work. The key idea is the definition of absolute value:

$$|a| = \begin{cases} a & a \geq 0 \\ -a & a < 0 \end{cases}$$

Note that $|0| = 0 = -0$.

Case 1: $x \geq 0$, $y \geq 0$

$xy \geq 0$ so $|xy| = xy = |x| \cdot |y|$

Case 2: $x \geq 0$, $y < 0$

$xy \leq 0$ so $|xy| = -(xy) = x(-y) = |x||y|$

Case 3: $x < 0$, $y \geq 0$

$xy \leq 0$ so $|xy| = -(xy) = (-x)y = |x||y|$

Case 4: $x < 0$, $y < 0$

$xy > 0$ so

$|xy| = xy = (-1)(-1)(xy) = (-x)(-y) = |x||y|$

26. Since $0 < a \leq b$, the Quotient–Remainder theorem implies the existence of unique integers, q and r, such that $b = aq + r$. In addition, $b \bmod a = r$.

Let $d = \gcd(a, b)$. Then there exist integers, x and y such that $b = dx$ and $a = dy$.

Clearly, d is a common divisor of both a and r since $r = b - aq = d(x - yq)$. Suppose that c is any other common divisor of r and a. Notice that c is also a divisor of b $(b = aq + r)$. Thus, c is a common divisor of a and b. This means that $c|d$, since d is the greatest common divisor of a and b. Since any common divisor of r and a divides d and d is a common divisor of r and a, d must be the greatest common divisor of r and a.

Here is an incorrect argument for the final part of the proof.

Suppose that there exists some common divisor, c, of r and a for which $c \nmid d$. Since $a = dy$ and $c|a$, it is tempting to conclude that $c|y$. Then, since $c|r$,

$b - aq = b - dczq = cw$. Thus, $b = c(w + dzq)$ and so $c|b$. Thus, since $c \nmid d$, it must be that $c|x$. But then $y = cs$ and $x = ct$ for some integers, s and t. Then dc is a divisor of both a and b that is larger than d, a contradiction. Thus, $c|d$ must be true.

Exercise 42 on page 112 should make us cautious about the general validity of this approach. (There can be no such c, so it is not possible to make up a counterexample to illustrate the error.)

33. This works well by induction.

Base Step

Let S be a set with 0 elements. Then $S = \emptyset$ is the only subset. So $|\mathcal{P}(S)| = 2^0 = 1$.

As a bit of paranoia, we could check the case $n = 1$: let $S = \{a\}$. The subsets are $\emptyset$ and $\{a\}$. Thus $|\mathcal{P}(S)| = 2^1 = 2$.

Inductive Step

Let $n \geq 0$ be some integer. Assume that every set with n elements has exactly 2^n subsets. Let $|S| = n + 1$. Single out any element and denote it x. Then $S - \{x\}$ has n elements, so by the inductive hypothesis, it has 2^n subsets (none of which contain x).

The subsets of S can be partitioned into two disjoint groups: all the subsets that contain x and all the subsets that do not contain x. Every subset T that contains x can be uniquely paired with a subset $T - \{x\}$ from the subsets of $S - \{x\}$. The two groups thus have the same number of elements. Since the two groups are disjoint and their union equals the set of all subsets of S, there must be $2^n + 2^n = 2^{n+1}$ subsets of S.

Conclusion

The theorem of mathematical induction implies that a set with n elements has 2^n subsets, for $n \geq 0$.

37. Try to find a counterexample to the claim. Then we need both a and b to have remainder 1 or 2 when divided by 3. By Exercise 36, a must be odd and b must be even (or vice versa). Also, c must be odd.

The easiest way to coordinate remainder mod 3 and even and odd is to classify numbers by their remainder mod 6. A bit of thought and experimentation shows that we must have $a \in \{6k+1, 6k+5\}$, $b \in \{6m+2, 6m+4\}$, and $c = 2n+1$ if we want to make sure that 3 divides neither a nor b.

There are four cases to investigate to find a counterexample to the claim.

$a = 6k + 1$ and $b = 6m + 2$ $a^2 + b^2 = c^2$ becomes
$36k^2 + 12k + 1 + 36m^2 + 24m + 4 = 4n^2 + 4n + 1$
which simplifies to
$3(3k^2 + k + 3m^2 + 2m + 1) + 1 = n(n+1)$

$a = 6k + 1$ and $b = 6m + 4$ $a^2 + b^2 = c^2$ becomes
$36k^2 + 12k + 1 + 36m^2 + 48m + 16 = 4n^2 + 4n + 1$
which simplifies to
$3(3k^2 + k + 3m^2 + 4m + 5) + 1 = n(n+1)$

$a = 6k + 5$ and $b = 6m + 2$ $a^2 + b^2 = c^2$ becomes
$36k^2 + 60k + 25 + 36m^2 + 24m + 4 = 4n^2 + 4n + 1$
which simplifies to
$3(3k^2 + k + 3m^2 + 2m + 2) + 1 = n(n+1)$

$a = 6k + 5$ and $b = 6m + 4$ $a^2 + b^2 = c^2$ becomes
$36k^2 + 60k + 25 + 36m^2 + 48m + 16 = 4n^2 + 4n + 1$
which simplifies to
$3(3k^2 + k + 3m^2 + 4m + 13) + 1 = n(n+1)$

Exercise 35 shows that each of the final equations is impossible.

H.4 Algorithms

Exercises 4.1.3

1. Other algorithms are possible.

```
{integer, integer, integer, integer}
change(real price)
   amount = 1.0 − price
   q = 0
   while amount ≥ .25
      q = q + 1
      amount = amount − .25
   d = 0
   while amount ≥ .10
      d = d + 1
      amount = amount − .10
   n = 0
   while amount ≥ .05
      n = n + 1
      amount = amount − .05
   p = amount * 100
   return {q, d, n, p}
end change
```

6. There are several possible algorithms for this exercise. A simple one is shown here. Exercise 7 on page 350 provides another alternative.

```
void greatestCommonDivisor(natural
number a, natural number b)
   if a == 0 and b == 0
      display "invalid input" message
  else
      if (a < b and a ≠ 0) or (b == 0)
         d = a
      else
         d = b

      # Cycle through possible gcds.
      # d will eventually become 1 if no
      # other common divisor can be found.

      while d > 0
         if a is divisible by d and
            b is divisible by d
             display d
             d = 0
         else
             d = d − 1
end greatestCommonDivisor
```

9.

```
real absoluteValueOfAverage(integer n,
                 real {x1, x2, ..., xn})
   sum = 0
   average = 0
   for i = 1 to n
       sum = sum + xi
   average = sum ÷ n
   if average < 0
       average = −average
   return average
end absoluteValueOfAverage
```

15. There are more efficient ways to do this. This algorithm is simple and straightforward. The goal at this point is familiarity with the algorithm notation, not efficiency.

```
boolean relativelyPrime(integer a,
integer b)
  if (a > 1) and (a == b)
    return false
  else
    # do a brute-force check for
    # common factors

    c = min(a, b)
    for i = 2 to c
      if (i | a) and (i | b)
         return false

    # no common factors

    return true
end relativelyPrime
```

Exercises 4.2.3

1. **(c)** $f_3(n) = 3n\log_2(n)$ $c = 1$ and $n_0 = 10$ work ($n_0 > 9.93954$).

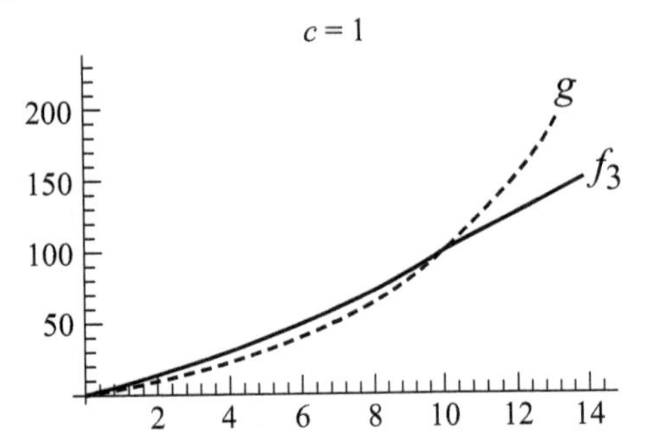

(d) $f_4(n) = \frac{n^3}{8}$ This function is *not* in $\mathcal{O}(n^2)$. Be careful: it may seem from a poorly chosen graph that it is (see the first graph). However, if n gets large enough (second graph), $\frac{n^3}{8}$ will always get larger (no matter what you choose for c).

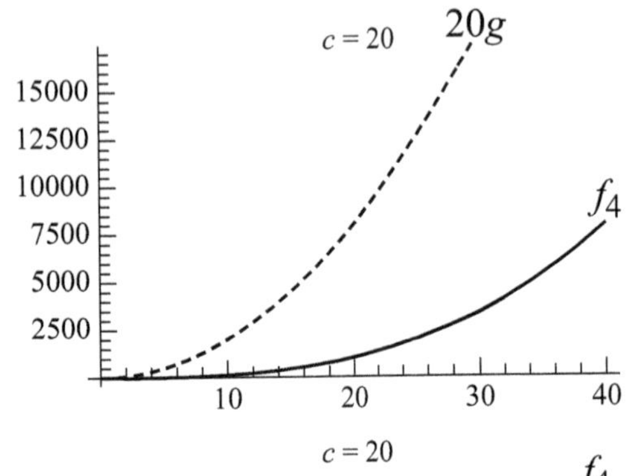

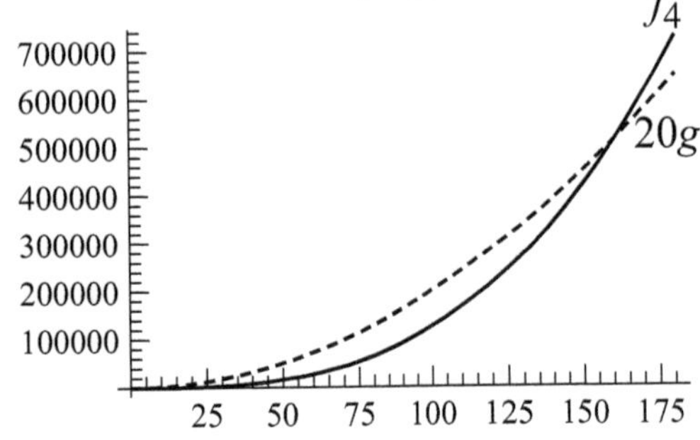

2. **(d)** $f_4(n) = \frac{n^3}{8}$ $c = 2$ and $n_0 = 16$ work

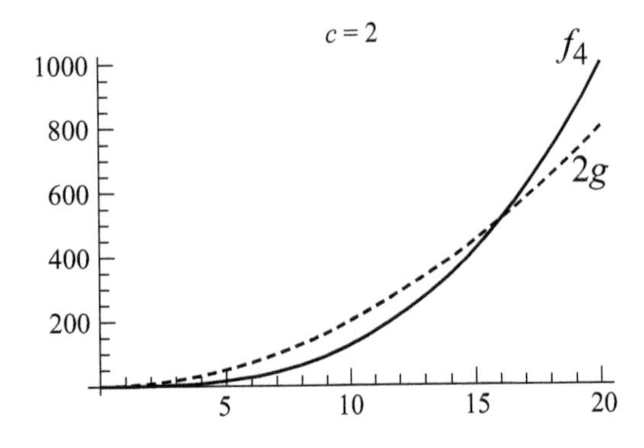

4. **(c)** True. Since $n > 0$ for $\log(n)$ to be defined, we can drop the absolute value signs.

$$3n\log_2(n) \le 3n \cdot n = 3n^2 \quad \text{for } n > 0$$

since $\log_2(n) < n$ for $n > 0$. Thus $|3n\log_2(n)| \le 3|n^2|$ for $n \ge 1$.

(d) False. Assume $n > 0$ and drop the absolute value signs. Suppose that there do exist constants c and n_0 such that $\frac{n^3}{8} \le cn^2$ for all $n > n_0$. Then $n^2(\frac{n}{8} - c) \le 0$ for all $n > n_0$. This is clearly a contradiction, since for all $n > 8c$ both factors are positive. Consequently, there cannot exist a c and n_0 such that $|\frac{n^3}{8}| \le c|n^2|$ for all $n > n_0$.

5. **(d)** True. Assume $n > 0$ and drop the absolute value signs.

$$\frac{n^3}{8} \ge \frac{n^2}{8} \quad \text{for } n \ge 1$$

Let $c = \frac{1}{8}$ and $n_0 = 1$.

8. **(b)** $f(n) = \frac{n^2+4}{n+4}$; $g(n) = n$

Note that $f(n) = \frac{n^2+4}{n+4} = n - 4 + \frac{20}{n+4}$.

Is $f \in \mathcal{O}(g)$?

Assume $n > 0$ and drop the absolute value signs. Then

$$\begin{aligned}\frac{n^2+4}{n+4} &= n - 4 + \frac{20}{n+4}\\ &\le n + \frac{20}{n+4}\\ &\le n + 4 \quad \text{for } n \ge 1\\ &\le n + 4n \quad \text{for } n \ge 1\\ &= 5n\end{aligned}$$

Let $n_0 = 1$ and $c = 5$. Then $|\frac{n^2+4}{n+4}| \le 5|n|$ for $n \ge 1$ and so $f \in \mathcal{O}(g)$.

Is $f \in \Omega(g)$?

Assume $n > 0$ and drop the absolute value signs. Then

$$\begin{aligned}\frac{n^2+4}{n+4} &\ge \frac{n^2+4}{n+n} \quad \text{for } n \ge 4\\ &= \frac{n^2+4}{2n}\\ &= \frac{1}{2}n + \frac{2}{n}\\ &\ge \frac{1}{2}n \quad \text{for } n > 0\end{aligned}$$

Let $n_0 = 4$ and $c = \frac{1}{2}$. Then $|\frac{n^2+4}{n+4}| \ge \frac{1}{2}|n|$ for $n \ge 4$ and so $f \in \Omega(g)$.

Is $f \in \Theta(g)$?

Since it was shown that $f \in \mathcal{O}(g)$ and $f \in \Omega(g)$, it is valid to conclude that $f \in \Theta(g)$.

(c) $f(n) = n!$; $g(n) = n^n$
Is $f \in \mathcal{O}(g)$?
Assume $n > 0$ and drop the absolute value signs. Then

$$\begin{aligned} n! &= n \cdot (n-1) \cdot (n-2) \cdots 3 \cdot 2 \cdot 1 \\ &\le n \cdot n \cdot n \cdots n \cdot n \cdot n \quad \text{if } n \ge 1 \\ &= 1 \cdot n^n \end{aligned}$$

Let $n_0 = 1$ and $c = 1$. Then, $|n!| \le 1|n^n|$ for $n \ge 1$ and so $f \in \mathcal{O}(g)$.
Is $f \in \Omega(g)$?
Assume $n > 0$ and drop the absolute value signs. Suppose that there do exist positive constants c and n_0 such that $n! \ge cn^n$ for all $n > n_0$. Then $\frac{n!}{n^n} \ge c$ for all $n > n_0$. Since $\lim_{n\to\infty} \frac{n!}{n^n} = 0$, there must be a positive integer, n_1, which can be assumed to be greater than n_0, such that $\frac{n!}{n^n} < c$ for all $n > n_0$. This contradicts the previous assertion that $\frac{n!}{n^n} \ge c$ for all $n > n_0$. Therefore, there do *not* exist constants c and n_0 such that $|n!| \ge c|n^n|$ for all $n > n_0$ and so $f \notin \Omega(g)$.
Is $f \in \Theta(g)$?
Although $f \in \mathcal{O}(g)$, $f \notin \Omega(g)$ and so $f \notin \Theta(g)$.

9. If $x \ge 0$, we can drop the absolute value signs. Since $\lfloor x \rfloor \le x$, it is clear that $\lfloor x \rfloor \in \mathcal{O}(x)$ ($c = 1$ and $n_0 = 0$).
Since $\lfloor x \rfloor \ge x - 1$ it is reasonable to expect $\lfloor x \rfloor \in \Omega(x)$. More formally, assume $x \ge 0$ so that we can drop the absolute value signs. Then $\lfloor x \rfloor \ge x - 1 \ge \frac{1}{2}x$ for $x \ge 2$. Let $c = \frac{1}{2}$ and $n_0 = 2$.
Since $\lfloor x \rfloor \in \mathcal{O}(x)$ and $\lfloor x \rfloor \in \Omega(x)$, we conclude that $\lfloor x \rfloor \in \Theta(x)$.

19. **(c)** $121(\log_2(n) + n)(n + 3n\log_2(n)) + 6n^2$
The factor $\log_2(n) + n \in \Theta(n)$ by Theorem 4.14 and the factor $n + 3n\log_2(n) \in \Theta(n\log_2(n))$ by the same theorem. The initial term $(121(\log_2(n) + n)(n + 3n\log_2(n)))$ is in $\Theta(n^2\log_2(n))$ by Theorem 4.12. A final application of Theorem 4.14 leads to
$121(\log_2(n)+n)(n+3n\log_2(n))+6n^2 \in \Theta(n^2\log_2(n))$.

20. **(a)** $(2n^2 + 7)\log_2(n) + 2^n(n^3 + 4)$
The factor $2n^2 + 7 \in \Theta(n^2)$ by Theorem 4.14 and so $(2n^2 + 7)\log_2(n) \in \Theta(n^2\log_2(n))$ by Theorem 4.12. The factor $n^3 + 4 \in \Theta(n^3)$ by Theorem 4.14 and so the term $2^n(n^3 + 4) \in \Theta(2^n \cdot n^3)$ by Theorem 4.12. A final application of Theorem 4.14 and use of the standard reference functions leads to the conclusion that $(2n^2 + 7)\log_2(n) + 2^n(n^3 + 4) \in \Theta(2^n \cdot n^3)$.

22. $n!$ grows faster. A graph clearly shows this. The algebraic justification is

$$2^n = \overbrace{2 \cdot 2 \cdots 2}^{n \text{ times}} \le 2 \cdot 2 \cdot 3 \cdots n = 2n!$$

This is not quite what is needed. A small adjustment will work. Notice that $2 \cdot 2 \cdot 2 \cdot 2 < 1 \cdot 2 \cdot 3 \cdot 4$. Thus, it is in fact the case (for $n \ge 4$)

$$2^n = \overbrace{2 \cdot 2 \cdots 2}^{n \text{ times}} \le 1 \cdot 2 \cdot 3 \cdots n = n!$$

25. **(a)** Let $f(n) = 3n$ and $g(n) = n$. Clearly, $f \in \mathcal{O}(g)$, but $\lim_{n\to\infty} \left|\frac{f(n)}{g(n)}\right| = 3$.

27. **(c)** Let $f(n) = 121(\log_2(n) + n)(n + 3n\log_2(n)) + 6n^2$ and $g(n) = n^2\log_2(n)$. Then clearly $g(n) \ne 0$ for $n \ge 2$. It is evident that

$$\begin{aligned} &\lim_{n\to\infty} \left|\frac{f(n)}{g(n)}\right| \\ &= \lim_{n\to\infty} \left|\frac{121(\log_2(n) + n)(n + 3n\log_2(n)) + 6n^2}{n^2\log_2(n)}\right| \\ &= \lim_{n\to\infty} \left|\frac{127n^2 + 363n(\log_2(n))^2 + 363n^2\log_2(n) + 121n\log_2(n)}{n^2\log_2(n)}\right| \\ &= \lim_{n\to\infty} \left|\frac{127}{\log_2(n)} + \frac{363\log_2(n)}{n} + 363 + \frac{121}{n}\right| = 363 \end{aligned}$$

Note the assumption made that the term $\frac{363\log_2(n)}{n}$ approaches 0 as $n \to \infty$. This can be verified by one application of L'Hospital's Rule. (Recall that $\frac{d}{dn}\log_2(n) = \frac{1}{n\ln(2)}$.) Since 363 is a real number with $363 \ne 0$, Theorem 4.15 implies that $f \in \Theta(g)$.

Exercises 4.2.5

1. **(c)**

$$\sum_{i=k}^{m} i = \sum_{i=1}^{m} i - \sum_{i=1}^{k-1} i = \frac{m(m+1)}{2} - \frac{(k-1)k}{2}$$

3. We can assume that the list has n items and successfully search for each item, followed by n searches for items that are not present. The total number of comparisons can then be divided by $2n$, the total number of searches.

$$\begin{aligned} &\frac{(1 + 2 + 3 + \cdots + n) + (n \cdot n)}{2n} \\ &= \frac{1 + 2 + 3 + \cdots + n}{2n} + \frac{n \cdot n}{2n} \\ &= \frac{n+1}{4} + \frac{n}{2} = \frac{3n}{4} + \frac{1}{4} \in \Theta(n) \end{aligned}$$

5. The binary search algorithm will still work correctly. The use of the floor function in the original is for the purpose of rounding down the quotient to an integer. The ceiling function can serve the same "rounding to an integer" purpose, even though a different integer may result. In other words, the main goal in line 5 is to assign `mid` with an integer value, regardless of whether it is a rounded up or rounded down version of the average of `low` and `high`.

The efficiency of the binary search algorithm with the new line 5 may depend on various factors, such as whether or not the target character is in the list, the position of the target character if it is in the list, and the size of the list. As one example, consider the list {a, b, c, d}, and let x = c. Then it is obviously more efficient to employ the ceiling function. However, it would be better to use the floor function when x = b.

7. (a) The exact count c is represented by

$$c = \sum_{i=1}^{20}\left(\sum_{j=1}^{i} 1\right) = \sum_{i=1}^{20} i = \frac{20(21)}{2} = 210$$

This is *not* in $\Theta(1)$! Replacing 20 with n leads to the proper estimate. The count would be $c = \frac{n(n+1)}{2}$ which we know is in $\Theta(n^2)$.

8. (a) The exact count c is represented by

$$c = \sum_{i=0}^{n-1}\left(\sum_{j=1}^{30} 2\right) = \sum_{i=0}^{n-1} 60 = 60n$$

Warning: This is *not* in $\Theta(n)$. Replacing 30 with n leads to the proper estimate. The count would be $c = 2n^2$, which we know is in $\Theta(n^2)$.

Exercises 4.3.4

1. The only match occurs at character 146.

Algorithm	Exam/Char	Chars
Obvious	1.025	4244
KMP	1.018	4234
BM with shift	1.024	4261
BM with last	0.104	448
Full BM	0.104	469

4. (b)

m	i	m	m	m	i
1	1	3	2	3	5

6. (b)

i	m
6	5

9. (b)

m	i	m	i	m	i
2	2	4	4	6	1

12. (a) obvious

Text and Pattern										Hits	Misses
f	e	m	f	u	f	o	f	u	m		
f	**u**	m								1	1
	f	u	m							0	1
		f	u	m						0	1
			f	u	**m**					2	1
				f	u	m				0	1
					f	**u**	m			1	1
						f	u	m		0	1
							f	u	m	3	0
										—	—
									Total	7	7

(b) KMP
The shift table is

f	u	m
1	1	2

Text and Pattern										Hits	Misses
f	e	m	f	u	f	o	f	u	m		
f	**u**	m								1	1
	f	u	m							0	1
		f	u	m						0	1
			f	u	**m**					2	1
					f	**u**	m			1	1
						f	u	m		0	1
							f	u	m	3	0
										—	—
									Total	7	6

(c) BM with last table
The last table is

e	f	m	o	u
0	1	3	0	2

Text and Pattern										Hits	Misses	$j - L[T_{i+j-1}]$
f	e	m	f	u	f	o	f	u	m			
f	**u**	m								1	1	$2-0$
		f	u	**m**						0	1	$3-2$
			f	u	**m**					0	1	$3-1$
					f	u	**m**			0	1	$3-1$
							f	u	m	3	0	
										–	–	
									Total	4	4	

(d) BM with shift table
The shift table is

f	u	m
3	3	1

Text and Pattern										Hits	Misses
f	e	m	f	u	f	o	f	u	m		
f	**u**	m								1	1
			f	u	**m**					0	1
				f	u	**m**				0	1
					f	u	**m**			0	1
						f	u	**m**		0	1
							f	u	m	3	0
										–	–
									Total	4	5

(e) full BM
The tables are

f	u	m
3	3	1

e	f	m	o	u
0	1	3	0	2

Text and Pattern										Hits	Misses	Win	$j - L[T_{i+j-1}]$
f	e	m	f	u	f	o	f	u	m				
f	**u**	m								1	1	S	$2-0$
			f	u	**m**					0	1	L	$3-1$
					f	u	**m**			0	1	L	$3-1$
							f	u	m	3	0		
										–	–		
									Total	4	3		

H.5 Counting

Exercises 5.1.6

2. These are mutually exclusive tasks, so the Mutually Exclusive Tasks Principle implies I can spend the evening in one of $4 + 3 + 3 = 10$ ways.

3. These are independent tasks, so the Independent Tasks Principle applies. I can while away the evening in any one of $12 \cdot 4 \cdot 3 = 144$ ways.

6. Choosing a doctor and choosing a dentist are mutually exclusive tasks, so the Mutually Exclusive Tasks Principle implies that there are $70 + 23 = 93$ choices for the speaker.

11. The assignment of the features to the badges involves independent choices, so the Independent Tasks Principle applies. Noting that there are 26 possible uppercase letters and 10 single digits, there are $7 \cdot 8 \cdot 26 \cdot 4 \cdot 10 = 58240$ different ways to assign features to a badge. Thus, the largest number of distinct badges is 58240.

15. The problem statement is ambiguous as to whether order is important. There is no repetition. I can visit in $C(5,3) = \frac{5!}{3! \cdot 2!} = 10$ ways if order is not important. Otherwise, I can visit in $P(5,3) = 60$ ways.

18. Using permutations with repetition, there are $2^7 = 128$ possible characters.

19. I will first assume that a phone number can begin with any digit (including 0). I would then need the smallest exponent n such that $10^n \geq 9$ million. This is first true for $n = 7$. So 7 digits are needed. If instead, we wish to reserve the digits 0 and 1 for long distance and operator access, then we need the smallest k for which $8 \cdot 10^k \geq 9$ million. This first occurs for $k = 7$. Thus, 8 digits are needed with this assumption.

22. This is a combination with repetition having $n = 2$ and $r = 11$. There are $C(2+11-1, 11) = C(12, 11) = 12$ ways to send the money. An alternate approach is to decide how many dollar bills to send. I have choices $0, 1, 2, \ldots, 11$, for a total of 12 options. Once I decide how many dollar bills to send, the number of coins is already determined.

30. The claims are very basic.

(a) Each pair of workers needs to communicate. With n workers, there are $C(n,2) = \frac{n!}{2!(n-2)!} = \frac{n(n-1)}{2} = \frac{n^2-n}{2}$ distinct pairs of workers.

(b) There are 2^n subsets of a set of size n. The empty set is one of them, but is not a viable team. There are thus $2^n - 1 \simeq 2^n$ possible teams.

34. (a) Note that the answer is *not* $P(4,4) = 4!$ because this counts many of the same arrangements twice. For instance, since the chairs are identical, there is not a new permutation if every person shifts over one chair, two chairs, or three chairs in a counterclockwise direction. However, there would be four permutations included in the count of $P(4,4) = 4!$ for the one distinct arrangement. Instead, it is necessary to consider one person fixed. The other three people can be arranged in $P(3,3) = 3! = 6$ ways. Thus, there are six ways to situate 4 people in identical chairs at a circular table.

(b) There are $(n-1)!$ ways to situate n people in identical chairs at a circular table. Proof: Fix one of the people. The number of ways to arrange the other $n-1$ people is $P(n-1, n-1) = (n-1)!$.

36. First choose a face value, then choose four cards of this face value out of the four possible cards of this kind in the deck, and finally choose the remaining three cards, keeping in mind that no 5s can be selected. Thus, using the Independent Tasks Principle, the number of 7-card crazy eights hands that contain no 5s but have four cards of the same kind is

$$C(12,1) \cdot C(4,4) \cdot C(44,3) = \frac{12!}{1! \cdot 11!} \cdot \frac{4!}{4! \cdot 0!} \cdot \frac{44!}{3! \cdot 41!} = 158{,}928$$

42. The question is asking, In how many ways can I pick an empty set? Most people feel that $C(n,0)$ should be either 0 or 1. Mathematicians assign the value 1, which agrees with the combinations formula if we define $0! = 1$.

The expression $C(n,n)$ denotes the number of ways to choose all n of the objects in a set of size n. The intuitive answer is 1, which agrees with the combinations counting formula.

Exercises 5.1.8

1. None of the counting principles or formulas applies directly. It might be helpful to think of breakfast and lunch as markers about which the other three activities need to be placed (keeping breakfast before lunch and placing at least one activity between the markers). Order is important and there is no repetition. A semi-exhaustive listing (as in the following table) is the easiest way to finish the problem. Each row shows the number of activities that occur in each of the three available positions (relative to breakfast and lunch). The final column totals the number of distinct arrangements for each pattern (row).

In the first row, there are three ways to pick the before-breakfast activity, with two choices remaining for the between activity. The after-lunch activity is then determined. There are therefore $3 \cdot 2 = 6$ ways this pattern can be achieved. In fact, if you notice that the order for the three activities is important in every row, you will see that each pattern must have $3! = 6$ distinct ways to occur. The total number of distinct arrangements is thus 36.

	B		L		#
1	B	1	L	1	6
2	B	1	L		6
	B	1	L	2	6
1	B	2	L		6
	B	2	L	1	6
	B	3	L		6
		Total =			36

2. (b) There is some ambiguity in the English description. I will give both reasonable answers. (Recall that part of the problem solving process is to define the problem.)

i. (Either six letters or two digits) followed by four uppercase letters: $(52^6 + 10^2) \cdot 26^4 = 9{,}034{,}694{,}167{,}513{,}664$

ii. Either six letters or (two digits followed by four uppercase letters): $52^6 + (10^2 \cdot 26^4) = 19{,}816{,}307{,}264$

4. multinomial$(5,2,7) = 72{,}072$

9. $C(7,3) \cdot C(5,2) = \frac{7!}{3!4!} \frac{5!}{2!3!} = 350$ (this is not a multinomial problem)

11. multinomial$(3,2,2,1) = 1{,}680$

17. There are six mutually exclusive ways to distribute the two empty seats. The following table lists the number of ways to arrange the students in each case. Assume Van 1 has 2 seats, Van 2 has 3 seats, and Van 3 has 4 seats.

Number of Nonempty Seats			Ways to Arrange Students
Van 1	Van 2	Van 3	
0	3	4	multinomial(3, 4) = 35
2	1	4	multinomial(2, 1, 4) = 105
2	3	2	multinomial(2, 3, 2) = 210
1	2	4	multinomial(1, 2, 4) = 105
1	3	3	multinomial(1, 3, 3) = 140
2	2	3	multinomial(2, 2, 3) = 210
	Total		805

19. Don't neglect the three workers who get to stand around and watch the others work.
multinomial$(5, 3, 4, 3) = 12,612,600$

31. **(a)** The number of ways to assign the seven ticket booth work sessions with Laura not included is $C(14+7-1, 7)$. The number of ways to assign the seven ticket booth work sessions with Laura included is $C(10+0-1, 0) + C(10+1-1, 1)+$ $C(10+2-1, 2) + C(10+3-1, 3)+$ $C(10+4-1, 4) + C(10+5-1, 5)+$ $C(10+6-1, 6)$ (using the Mutually Exclusive Tasks Principle). In this last expression, a term such as $C(10+5-1, 5)$ means that Laura is working two ticket booth sessions and so five others are needed for the remaining ticket booth sessions. The "10" comes from the fact that there are 10 committee members who are not Laura, John, Sue, Richard, or Emily. Since choosing to include Laura and choosing not to include Laura as a worker in the ticket booth are mutually exclusive tasks, the Mutually Exclusive Tasks Principle implies that there are $C(14+7-1, 7) + C(10+0-1, 0)+$ $C(10+1-1, 1) + C(10+2-1, 2)+$ $C(10+3-1, 3) + C(10+4-1, 4)+$ $C(10+5-1, 5) + C(10+6-1, 6)$ $= 77,520 + 1 + 10 + 55 + 220 + 715 + 2,002+$ $5,005 = 85,528$ ways you can assign the seven ticket booth work sessions.

Exercises 5.2.3

2. They are the same. (Expand and add in reverse order.)

$$\begin{aligned}
&\sum_{j=n+1}^{2n+1} \binom{2n+1}{j} \\
&= \binom{2n+1}{n+1} + \binom{2n+1}{n+2} + \cdots + \binom{2n+1}{2n} + \binom{2n+1}{2n+1} \\
&= \binom{2n+1}{2n+1} + \binom{2n+1}{2n} + \cdots + \binom{2n+1}{n+2} + \binom{2n+1}{n+1} \\
&= \sum_{i=0}^{n} \binom{2n+1}{(2n+1)-i} \\
&= \sum_{i=0}^{n} \binom{2n+1}{i} \text{ by Proposition 5.9}
\end{aligned}$$

4. $\binom{6}{2} = 15$

9. **(a)** If $n = 0$, then $n^2 = 0 = 2 \cdot 0 + 0 = 2 \cdot \binom{n}{2} + \binom{n}{1}$. If $n = 1$, then $n^2 = 1 = 2 \cdot 0 + 1 = 2 \cdot \binom{n}{2} + \binom{n}{1}$.
For $n \geq 2$,

$$\begin{aligned}
2\binom{n}{2} + \binom{n}{1} &= 2 \cdot \frac{n!}{(n-2)! \cdot 2!} + \frac{n!}{(n-1)! \cdot 1!} \\
&= n(n-1) + n = n^2
\end{aligned}$$

15. There will be a committee consisting of from 1 to n people. Different sized committees are mutually exclusive, so one counting strategy is to count the number of ways to have a committee of size k, for each k in $\{1, 2, \ldots, n\}$ and then add. There are $\binom{n}{k}$ ways to pick the k committee members and there are then k choices for the driver. There are consequently $k\binom{n}{k}$ ways to pick a committee (with driver) of size k. The Mutually Exclusive Tasks Principle asserts that there are $\sum_{k=1}^{n} k\binom{n}{k}$ ways to choose a shopping committee.

An alternative strategy is to choose a driver first. This can be done in n ways. There are $n-1$ people left. From 0 to $n-1$ of them can be chosen to add to the committee. That is, any subset of the $n-1$ remaining people can be added to the shopping committee. Corollary 5.12 implies that there are 2^{n-1} subsets, so there are $n2^{n-1}$ ways to choose a shopping committee with driver.

Equating these two counts establishes the desired identity.

Exercises 5.3.3

2. **(b)** 37 people (use the pigeon-hole principle: Find the smallest n such that $\lceil \frac{n}{12} \rceil = 4$.)

5. 16 names (Formally: Make 15 boxes—one per marriage. Now try to fill them without having 2 entries in a box.)

11. The set $\{1, 2, 3, \ldots, 2n\}$ contains n even numbers and n odd numbers. We can get the maximum number of elements that differ by more than 1 by choosing all the even numbers (odd would also work) . This collection still leaves out one integer. The final value must therefore be odd and hence will differ by 1 from at least one neighbor.

17. Let R be the set of students whose resumés meet the company's standards. Let D be the set of students whose clothing was acceptable and let S be the set of students who are willing to accept the salary range offered.

(a) Using the inclusion–exclusion theorem

$$\begin{aligned}
65 &= (5 + 30 + 44 + 28) \\
&\quad - (18 + 15 + 16) + |R \cap D \cap S|
\end{aligned}$$

so $|R \cap D \cap S| = 7$.

(b) A Venn diagram that uses the information found in part (a) makes this easy to answer. There are four students with acceptable resumés but unacceptable clothing and who were unwilling to work at the salary offered.

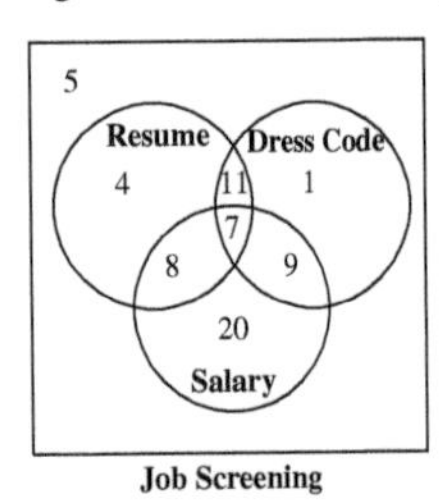

22. Let A_1 be the set of all bit strings of length 7 with 1s in the first 5 positions. Let A_2 be the set of bit strings of length 7 with 1s in positions 2–6, and let A_3 be the set of bit strings of length 7 with 1s in positions 3–7. We want to count the number, x, of bit strings of length 7 that are not in $A_1 \cup A_2 \cup A_3$.

There are $2^7 = 128$ bit strings of length 7. $|A_1| = |A_2| = |A_3| = 2^2 = 4$ since there are only two unspecified bits in each case. $|A_1 \cap A_2 \cap A_3| = 1$ since there is only one element in that set: 1111111. Finally, $|A_1 \cap A_2| = |A_2 \cap A_3| = 2$ (for example, $A_1 \cap A_2 = \{1111110, 1111111\}$), but $|A_1 \cap A_3| = 1$.

$$\begin{aligned} x &= 128 - |A_1 \cup A_2 \cup A_3| \\ &= 128 - \big[(|A_1| + |A_2| + |A_3|) - (|A_1 \cap A_2| \\ &\quad + |A_1 \cap A_3| + |A_2 \cap A_3|) + |A_1 \cap A_2 \cap A_3|\big] \\ &= 128 - [(4+4+4) - (2+2+1) + 1] \\ &= 120 \end{aligned}$$

The bit strings with five consecutive 1s are:

1111100, 1111101, 1111110, 1111111,
0111110, 01111111, 0011111, 1011111

H.6 Finite Probability Theory

Exercises 6.1.3

1. **(b)** Let the sample space be denoted by R for this problem. $R = \{L_1, L_2, S\}$. This sample space uses equally likely outcomes. It is also possible to use the sample space $R = \{L, S\}$ with probabilities $\{\frac{2}{3}, \frac{1}{3}\}$, but this sample space is often harder to use properly.

(c) Since there are three distinct physical coins, it is best to use ordered triples for the model:

$$S = \{(T,T,T), (T,T,H), (T,H,T), (T,H,H), (H,T,T), (H,T,H), (H,H,T), (H,H,H)\}$$

2. **(b)** i. With equally likely outcomes, each has theoretical probability $\frac{1}{3}$.

ii. $\mathbf{P}(L) = \frac{2}{3}$, $\mathbf{P}(S) = \frac{1}{3}$. (These are theoretical with unequally likely outcomes.)

(c) Again, using equally likely outcomes, the theoretical probability is $\frac{1}{8}$ for each outcome.

6. **(c)** A number card or an ace that is, in either case, a diamond or a spade or a club. (This is a long description, but is easier than listing all 30 cards.)

7. **(b)** The set of irrational integers is empty. The complement is the set of all integers.

8. **Picking a card**

(a) (a ten, a diamond) **NOT mutually exclusive**

(b) (a red card, a black card) **mutually exclusive**

10. **Picking a card**

(b) **P**(the sum is 8) = $\frac{5}{36}$, since the dice can show (2, 6), (6, 2), (3, 5), (5, 3), or (4, 4).

11. A suitable sample space for this random experiment is $\{1, 2, 3, 4, 5, 6\}$. This is a sample space with unequally likely outcomes. The numbers in the set $\{1, 4\}$ fit the criteria for x in the problem statement. The numbers in this set are twice as likely to occur than any of the numbers $\{2, 3, 5, 6\}$. Thus, assign a probability of $\frac{2}{8} = \frac{1}{4}$ to each of the outcomes 1 and 4, and assign a probability of $\frac{1}{8}$ to the remaining outcomes.

(a) **P**(subtracting 5 from the number produces a negative number) = $\frac{3}{4}$, since four numbers fit this criteria, two of which have been assigned a probability of $\frac{1}{4}$ and two of which have been assigned probability of $\frac{1}{8}$.

12. There are two possible outcomes for the flip of the first coin, six possible outcomes for the roll of the die, and then two possible outcomes for the second flip of a coin. The Independent Tasks Principle from Chapter 5 implies that there should be $2 \cdot 6 \cdot 2 = 24$ outcomes in the sample space for this random experiment. A typical outcome may look like $\{H, 3, T\}$, indicating that the first coin flip is a head, the roll of the dice gives a 3, and the second coin flip gives a tail.

(b) **P**(the number is 2 and at least one head) = $\frac{3}{24} = \frac{1}{8}$, since the experiment can yield $\{H, 2, H\}$, $\{H, 2, T\}$, or $\{T, 2, H\}$.

Exercises 6.2.3

2. This can be calculated directly from the reduced sample space. An alternative is to use probability principle 4 in the form

$$\mathbf{P}(B \mid A) = \frac{\mathbf{P}(A \cap B)}{\mathbf{P}(A)}$$

I have given both solutions.

(b) $\mathbf{P}(4 \mid \text{face card}) = \frac{0}{12} = 0$, or $\frac{(\frac{0}{52})}{(\frac{12}{52})} = 0$.

(c) $\mathbf{P}(4 \mid \text{a red card}) = \frac{2}{26} = \frac{1}{13}$, or $\frac{(\frac{2}{52})}{(\frac{26}{52})} = \frac{1}{13}$.

3. This can be calculated directly from the reduced sample space. An alternative is to use probability principle 4 in the form

$$\mathbf{P}(B \mid A) = \frac{\mathbf{P}(A \cap B)}{\mathbf{P}(A)}$$

I have given both solutions.

(a) $\mathbf{P}(\text{the third is a tail} \mid \text{the second is a tail}) = \frac{2}{4} = \frac{1}{2}$, or $\frac{(\frac{2}{8})}{(\frac{4}{8})} = \frac{1}{2}$.

5. Using principle 1, **P**(at least one vowel is chosen) = **P**(first letter is a vowel) + **P**(second letter is a vowel) - **P**(both vowels) = $\frac{5}{26} + \frac{5}{26} - \frac{25}{676} = \frac{235}{676}$.

6. Using principles 1 and 6, **P**(no trip in August) = $1 -$ **P**(trip in August) $= 1 -$ (**P**(trip in August this year) + **P**(trip in August next year) − **P**(trip in August both years)) = $1 - (\frac{1}{12} + \frac{1}{12} - \frac{1}{144}) = 1 - (\frac{23}{144}) = \frac{121}{144}$.

8. (a) **P**(an even sum | one die is a 3) = $\frac{5}{11}$
In the 6-by-6 table of possible rolls, the column and row indexed by 3 constitute the restricted sample space. There are 11 outcomes in the restricted sample space, since (3, 3) is only counted once. Only five of these outcomes have an even sum.

9. Even numbers are twice as likely as odd numbers. If **P**(even) $= x$ and **P**(odd) $= y$, then $x = 2y$ and $3y + 3x = 1$. Thus, $y = \frac{1}{9}$ and $x = \frac{2}{9}$.
(a) **P**(1,3) $= \frac{1}{9} \cdot \frac{1}{9} = \frac{1}{81}$ by principle 5

11. **Picking a card**
(a) (a ten, a diamond) **independent**
P(a ten) $= \frac{4}{52} = \frac{1}{13}$
P(a ten | a diamond) $= \frac{1}{13}$
(b) (a red card, a black card) **not independent**
P(a red card) $= \frac{26}{52} = \frac{1}{2}$
P(a red card | a black card) $= \frac{0}{26} = 0$

13. (a) Two events that are mutually exclusive but not independent:
Picking a card, with events "a red card" and "a black card".

16. Use probability principle 1.

$$\begin{aligned}\mathbf{P}(\text{spade} \cup \{3\}) &= \mathbf{P}(\text{spade}) + \mathbf{P}(3) - \mathbf{P}(3 \text{ of spades}) \\ &= \frac{13}{52} + \frac{4}{52} - \frac{1}{52} = \frac{4}{13}\end{aligned}$$

19. (a) Using the following table, there are two ways, (4, 6) and (6, 4), to get a sum of 10 with even numbers. The probability is $\frac{2}{36} = \frac{1}{18}$.

	1	**2**	**3**	**4**	**5**	**6**
1	2	3	4	5	6	7
2	3	4	5	6	7	8
3	4	5	6	7	8	9
4	5	6	7	8	9	10
5	6	7	8	9	10	11
6	7	8	9	10	11	12

(b) i. **P**(sum is 10 ∩ both even) = **P**(sum is 10) · **P**(both even | sum is 10) $= \frac{3}{36} \cdot \frac{2}{3} = \frac{1}{18}$
ii. **P**(sum is 10 ∩ both even) = **P**(both even) · **P**(sum is 10 | both even) $= \frac{9}{36} \cdot \frac{2}{9} = \frac{1}{18}$

24. (a) {HHH, HHT, HTH, HTT, THH, THT, TTH, TTT}
(b) The flips are determined by the weighting of the coin, not the order of the flip. Successive flips are independent.
(c) I used probability principle 5 and theoretical probabilities with unequally likely outcomes. For example, **P**(HTH) $= 0.8 \cdot 0.2 \cdot 0.8 = 0.128$.

Outcome	**Probability**
HHH	0.512
HHT	0.128
HTH	0.128
HTT	0.032
THH	0.128
THT	0.032
TTH	0.032
TTT	0.008

Exercises 6.3.1

2. (a) The different flips of the coin and the roll of the die are all independent. Thus, the Independent Tasks Principle implies that there are $2 \cdot 2 \cdot 2 \cdot 6 = 48$ distinct outcomes in the sample space.
(b) In order for the product of the number of heads and the digit on the die to be greater than 10, the digit on the die must be at least 4. Consider the three mutually exclusive cases: The digit on the die is 4, the digit on the die is 5, the digit on the die is 6. If the digit on the die is 4, then the number of heads must be 3. Since there is only one way to have three heads, there is one possible coin/die combination in this case (HHH4). Similarly, if the digit is a 5, there is only one possible coin/die combination (HHH5). If the digit on the die is 6, then the number of heads must be at least 2. Thus, there are 3 + 1 = 4 possible coin/die combinations in this case (HHT6, HTH6, THH6, HHH6). Therefore, the probability that the number of heads multiplied by the digit on the die is greater than 10 is $\frac{1+1+4}{48} = \frac{1}{8}$. To answer the question in the problem statement, the probability that Jake's parents get to choose the vacation spot is $1 - \frac{1}{8} = \frac{7}{8}$.

4. (c) The possible combinations for the exact numbers of heads and tails such that they differ by at most two is shown in the table below.

Number of Heads	**Number of Tails**
4	4
3	5
5	3

The three cases in the table are mutually exclusive. The number of ways to have exactly four heads is $C(8, 4)$ (choose four of the eight coins which will be heads). Similarly, the number of ways to have exactly three and exactly five heads are $C(8, 3)$ and $C(8, 5)$, respectively. The probability that the number of heads and the number of tails differ by at most two is $\frac{C(8,4)+C(8,3)+C(8,5)}{256} = \frac{70+56+56}{256} = \frac{91}{128} \simeq 0.711$.

6. **(a)** There are $C(90, 7) = 7{,}471{,}375{,}560$ ways that you can select seven distinct integers.

(b) There are $C(7, 6)$ ways that you can select six of the seven winning integers. There are then $C(83, 1)$ ways that you can select the remaining incorrect integer. Since these choices are independent, the probability that you will win a prize at the fair is

$$\frac{C(7,6)\cdot C(83,1)}{7,471,375,560} = \frac{581}{7,471,375,560} \simeq 0.0000000778$$

9. **(a)** There are six different spins and the repetition of colors spun is permitted. Consequently, the number of spin combinations possible is $4^6 = 4096$.

(b) The probability that you will win two stuffed animals is 1 - **P**(less than 2 stuffed animals). How many ways can you win less than two stuffed animals? Well, the number of ways to win no stuffed animals is 3^6. The number of ways to win exactly one stuffed animal is $C(6, 1) \cdot 3^5$ because the winning spin number must be chosen and then the outcomes of the other five spins must be independently determined. Thus, the number of ways to win less than two stuffed animals is $3^6 + C(6, 1) \cdot 3^5$. Consequently, the probability that you will win two or more stuffed animals (i.e., the wheel pointer lands in the orange section at least twice) is

$$1 - \frac{3^6 + C(6,1)\cdot 3^5}{4096} = 1 - \frac{729 + 1458}{4096}$$
$$= 1 - \frac{2,187}{4096} = \frac{1909}{4096}$$

(c) The number of ways to spin the wheel such that you will win exactly one cookie and exactly one book is $P(6, 2) \cdot 2^4$, since the particular spins that you win one cookie and one book need to be chosen (with order) among the distinct spins and the outcomes of the remaining four spins need to be independently determined. Therefore, the probability that you will win exactly one cookie and exactly one book (i.e., the wheel pointer lands in the red section exactly once and in the yellow section exactly once) is

$$\frac{P(6,2)\cdot 2^4}{4096} = \frac{30 \cdot 16}{4096} = \frac{15}{128} \simeq 0.117$$

11. **(a)** We could add the probabilities of exactly 3 eights and exactly 4 eights. We could also subtract the probability of 2 or fewer eights from 1. Since we already know the the number of hands containing at most one eight (Example 5.31 on page 241), the second alternative is simpler. The number of crazy eights hands with exactly two eights is

$$C(4, 2) \cdot C(48, 5) = 10{,}273{,}824$$

and so

$$\mathbf{P}(\text{exactly two eights}) = \frac{10{,}273{,}824}{133{,}784{,}560}$$

The probability of at least three eights is therefore

$$1 - \frac{73{,}629{,}072 + 49{,}086{,}048 + 10{,}273{,}824}{133{,}784{,}560} \simeq 0.006$$

The alternative approach is to add the probabilities of exactly 3 eights and exactly 4 eights:

$$\frac{C(4,3)\cdot C(48,4)}{133{,}784{,}560} + \frac{C(4,4)\cdot C(48,3)}{133{,}784{,}560}$$
$$= \frac{778{,}320 + 17{,}296}{133{,}784{,}560} \simeq 0.006$$

15. **(b)** All 5 cards are the same suit: pick a suit, then the 5 cards

$$\binom{4}{1}\binom{13}{5} = 5,148$$

The probability is $\frac{5,148}{2,598,960} \simeq 0.002$

19. **(a)** First count the number of 8-card hands containing exactly two 2s, exactly three 3s, and exactly three cards with a third common face value. There are four choices from which to select the two cards with a face value of 2. Similarly, there are four choices from which to pick the three cards with a face value of 3. There are then $13 - 2 = 11$ face values left over, one of which we may assign to the remaining three cards. Finally, once this last face value is chosen, there are four choices from which we may select the remaining three cards for the hand. There are consequently (using the Independent Tasks Principle)

$$C(4,2)\cdot C(4,3)\cdot C(11,1)\cdot C(4,3)$$
$$= \frac{4!}{2!\cdot 2!}\cdot\frac{4!}{3!\cdot 1!}\cdot\frac{11!}{1!\cdot 10!}\cdot\frac{4!}{3!\cdot 1!} = 1,056$$

8-card hands containing exactly two 2s, exactly three 3s, and exactly three cards with a third common face value. This implies that the probability that exactly two 2s, exactly three 3s, and exactly three cards with a third common face value are chosen is

$$\frac{1056}{C(52,8)} = \frac{1056}{752,538,150} \simeq 0.0000014$$

20. **(a)** For the difference between the digit on a die and the integer 3 to be positive, a 4, 5, or 6 must show up on at least one of the dice. There are $3^2 = 9$ ways for both digits on the dice to be among 4, 5, and 6. There are $2 \cdot C(3, 1) \cdot C(3, 1) = 18$ ways for there to be exactly one of 4, 5, and 6 on a die, since exactly one of these three numbers can be on either of the two dice, and there are three choices for the last die. Thus, the Mutually Exclusive Tasks Principle implies that there are $9 + 18 = 27$ ways for the difference between the digit on a die and the integer 3 to be positive. This implies that the probability that this event will occur is $\frac{27}{36} = \frac{3}{4}$. Consequently, player A more likely to win. An alternative approach is to let X be the event: "$x > 3$" and Y be the event "$y > 3$". Then

$$\mathbf{P}(X \cup Y) = \mathbf{P}(X) + \mathbf{P}(Y) - \mathbf{P}(X \cap Y)$$
$$= \frac{1}{2} + \frac{1}{2} - \frac{1}{2}\cdot\frac{1}{2} = \frac{3}{4}$$

25. **(a)** Taking into account that you must buy a pair of jeans if you buy any T-shirt(s), consider the four mutually exclusive cases: buy exactly one item, buy exactly two items, buy exactly three items, buy exactly four items. Within each of these last three cases, there are two subcases: Buy a T-shirt, do not buy a T-shirt. The number of ways to buy exactly one item is $C(23, 1)$ (a T-shirt may not be purchased alone). The number of ways to buy exactly two items when one is a T-shirt is 1 (a T-shirt and jeans). The number of ways to buy exactly two items when neither is a T-shirt is $C(23+2-1, 2) = C(24, 2)$. The number of ways to buy exactly three items when at least 1 is a T-shirt is 24 (one T-shirt and one pair of jeans leaves 24 choices for the third item). The number of ways to buy exactly three items when none are T-shirts is $C(23+3-1, 3) = C(25, 3)$. The number of ways to buy exactly four items when at least one is a T-shirt is $C(24+2-1, 2)$ (one T-shirt and one pair of jeans, then any two items), while the number of ways to buy exactly four items when none are T-shirts is $C(23+4-1, 4) = C(26, 4)$. Consequently, the Mutually Exclusive Tasks Principle implies that the number of acceptable distinct purchases is $C(23, 1)+(1+C(24, 2))+(24+C(25, 3))+(C(25, 2)+C(26, 4)) = 23 + 277 + 2,324 + 15,250 = 17,874$.

Exercises 6.4.1

1. The ratios express the approximate odds of winning (including the possibility of a free ticket) and winning cash (no free tickets included). These approximate probabilities are, respectively, $\frac{1}{4.15} \simeq 0.241$ and $\frac{1}{8.26} \simeq 0.121$. These can also be computed by adding all but the first (respectively, first two) rows in the probability column of the table computed in the text. Thus the probability of winning something is approximately $1 - 0.7589 = 0.2411$ and the probability of winning a cash prize is approximately $0.2411 - 0.12 = 0.1211$.

5. **(a)**

X	$\mathbf{P}(X)$	$X \cdot \mathbf{P}(X)$
2	0.35	0.70
5	0.40	2.00
8	0.25	2.00
$\sum x \cdot \mathbf{P}(x) \simeq$		4.70

7. There are two prizes: a \$500 car and a hearty wish for better luck next time. The probability of winning is $\frac{1}{2,000} = 0.0005$. The probability of losing is $\frac{1,999}{2,000} = 0.9995$. The expected value is thus

$$500 \cdot \frac{1}{2000} + 0 \cdot \frac{1999}{2000} = \$0.25.$$

The average loss is \$2.00 − \$0.25 = \$1.75 per ticket.

15. The expected profit per shirt is

$$\begin{aligned}\mathbf{E}(X) &= -2 \cdot 0.30 + 0 \cdot 0.23 + 1 \cdot 0.19 \\ &\quad + 2 \cdot 0.10 + 5 \cdot 0.11 + 7 \cdot 0.07 \\ &= \$0.83\end{aligned}$$

Thus, the estimated profit on the 5000 shirts is $5000 \cdot \$0.83 = \$4,150$.

17. **(a)** $\mathbf{E}(D) = 28 \cdot \frac{1}{12} + 30 \cdot \frac{4}{12} + 31 \cdot \frac{7}{12} \simeq 30.417$

(b) There are 3 months that start with a "J": January, June, and July. There are three months that count double and 9 months that count single. A denominator of 15 is appropriate. January and July have 31 days; June has 30. The expression below counts the "J" months first.

$$\begin{aligned}\mathbf{E}(D) &= 2\left(31 \cdot \frac{2}{15}\right) + 1\left(30 \cdot \frac{2}{15}\right) + 1\left(28 \cdot \frac{1}{15}\right) \\ &\quad + 3\left(30 \cdot \frac{1}{15}\right) + 5\left(31 \cdot \frac{1}{15}\right) \\ &\simeq 30.467\end{aligned}$$

22. I will assume that exactly 1 million people participate, so that every possible number is chosen (there are 10^6 numbers using permutations with repetition). I will assume that every number is equally likely to be chosen. The probability of losing is

$$\mathbf{P}(\text{Losing}) = \frac{999,999}{1,000,000} = .999999$$

The expected value is

$$\mathbf{E}(X) = 5,000,000 \cdot \frac{1}{1,000,000} + 0 \cdot \frac{999,999}{1,000,000} = \$5$$

This is a fair game.

Exercises 6.5.1

1. **(a)**

0	$\binom{6}{0}0.2^0 0.8^6$	$= \frac{4096}{15625}$	$= 0.262144$
1	$\binom{6}{1}0.2^1 0.8^5$	$= \frac{6144}{15625}$	$= 0.393216$
2	$\binom{6}{2}0.2^2 0.8^4$	$= \frac{768}{3125}$	$= 0.24576$
3	$\binom{6}{3}0.2^3 0.8^3$	$= \frac{256}{3125}$	$= 0.08192$
4	$\binom{6}{4}0.2^4 0.8^2$	$= \frac{48}{3125}$	$= 0.01536$
5	$\binom{6}{5}0.2^5 0.8^1$	$= \frac{24}{15625}$	$= 0.001536$
6	$\binom{6}{6}0.2^6 0.8^0$	$= \frac{1}{15625}$	$= 0.000064$

6. **(a)**

$$\begin{aligned}\mathbf{P}(X \le 2) &= \mathbf{P}(X=0) + \mathbf{P}(X=1) + \mathbf{P}(X=2) \\ &= \frac{1}{32} + \frac{5}{32} + \frac{5}{16} = \frac{1}{2} \\ &= 0.03125 + 0.15625 + 0.3125 = 0.5\end{aligned}$$

17. Convergence can be shown by using the ratio test since $0 < p < 1$ and each term $k(1-p)^k$ is positive.

$$\lim_{k\to\infty} \frac{(k+1)(1-p)^{k+1}}{k(1-p)^k} = \lim_{k\to\infty}\left(1 + \frac{1}{k}\right)(1-p) = 1-p < 1$$

To calculate the sum, notice that if we set $z = 1 - p$ then $0 < z < 1$ so Corollary 3.60 implies that $\sum_{k=0}^{\infty} z^k = \frac{1}{1-z}$. Thus

$$\left(\frac{1}{1-z}\right)^2 = \left(\sum_{k=0}^{\infty} z^k\right)\cdot\left(\sum_{k=0}^{\infty} z^k\right)$$
$$= \left(1+z+z^2+\cdots\right)\cdot\left(1+z+z^2+\cdots\right)$$
$$= \sum_{k=0}^{\infty}\left(\sum_{j=0}^{k} 1\cdot 1\right) z^k$$
$$= \sum_{k=0}^{\infty}(k+1)z^k$$
$$= \sum_{k=0}^{\infty} kz^k + \sum_{k=0}^{\infty} z^k$$
$$= \sum_{k=0}^{\infty} kz^k + \frac{1}{1-z}$$

Solving for $\sum_{k=0}^{\infty} kz^k$ leads to

$$\sum_{k=0}^{\infty} kz^k = \frac{1}{(1-z)^2} - \frac{1}{1-z} = \frac{z}{(1-z)^2}$$

Now substitute $1-p$ for z.

Exercises 6.6.1

1. **(a)** $\mathbf{P}(\overline{T} \mid W)$

$$= \frac{\mathbf{P}(\overline{T})\cdot\mathbf{P}(W \mid \overline{T})}{\mathbf{P}(\overline{T})\cdot\mathbf{P}(W \mid \overline{T}) + \mathbf{P}(T)\cdot\mathbf{P}(W \mid T)}$$
$$= \frac{(0.998)(0.15)}{(0.998)(0.15) + (0.002)(0.775)}$$
$$\simeq 0.9898$$

or

$$\mathbf{P}(\overline{T} \mid W) = 1 - \mathbf{P}(T \mid W) \simeq 1 - 0.0102 = 0.9898$$

(b) $\mathbf{P}(\overline{T} \mid \overline{W})$

$$= \frac{\mathbf{P}(\overline{T})\cdot\mathbf{P}(\overline{W} \mid \overline{T})}{\mathbf{P}(\overline{T})\cdot\mathbf{P}(\overline{W} \mid \overline{T}) + \mathbf{P}(T)\cdot\mathbf{P}(\overline{W} \mid T)}$$
$$= \frac{(0.998)(0.85)}{(0.998)(0.85) + (0.002)(0.225)} \simeq 0.9995$$

or

$$\mathbf{P}(\overline{T} \mid \overline{\mathbf{W}}) = 1 - \mathbf{P}(T \mid \overline{W}) \simeq 1 - 0.0005 = 0.9995$$

3. $\mathbf{P}(\mathrm{M}) = \frac{1{,}709{,}919}{2{,}167{,}071} \simeq 0.7890$, $\quad \mathbf{P}(\mathrm{F}) = \frac{457{,}152}{2{,}167{,}071} \simeq 0.2110$

$\mathbf{P}(U18 \mid M) = \frac{516{,}494}{1{,}709{,}919} \simeq 0.3021$,

$\mathbf{P}(U18 \mid F) = \frac{124{,}911}{457{,}152} \simeq 0.2732$

(a) $\mathbf{P}(M \mid U18)$

$$= \frac{\mathbf{P}(M)\cdot\mathbf{P}(U18 \mid M)}{\mathbf{P}(M)\cdot\mathbf{P}(U18 \mid M) + \mathbf{P}(F)\cdot\mathbf{P}(U18 \mid F)}$$
$$= \frac{(0.7890)(0.3021)}{(0.7890)(0.3021) + (0.2110)(0.2732)}$$
$$\simeq 0.8053$$

6. $\mathbf{P}(G \mid B) = 0.4$, $\quad \mathbf{P}(G \mid C) = 0.25$, $\quad \mathbf{P}(G \mid R) = 0.2$,
$\mathbf{P}(G \mid M) = 0.1$, $\quad \mathbf{P}(B) = 0.3$, $\quad \mathbf{P}(C) = 0.4$,
$\mathbf{P}(R) = 0.2$, $\quad \mathbf{P}(M) = 0.1$

The generalized Bayes's formula should be used.

(b) $$\mathbf{P}(C|G) = \frac{\mathbf{P}(C)\cdot\mathbf{P}(G \mid C)}{\mathbf{P}(B)\cdot\mathbf{P}(G \mid B) + \mathbf{P}(C)\cdot\mathbf{P}(G \mid C) + \mathbf{P}(R)\cdot\mathbf{P}(G \mid R) + \mathbf{P}(M)\cdot\mathbf{P}(G \mid M)}$$
$$= \frac{(.4)(.25)}{(.3)(.4) + (.4)(.25) + (.2)(.2) + (.1)(.1)}$$
$$\simeq 0.3704$$

The next part uses the values $\mathbf{P}(D \mid X) = 1 - \mathbf{P}(G \mid X)$.

(c) $$\mathbf{P}(R \mid D) = \frac{\mathbf{P}(R)\cdot\mathbf{P}(D \mid R)}{\mathbf{P}(B)\cdot\mathbf{P}(D \mid B) + \mathbf{P}(C)\cdot\mathbf{P}(D \mid C) + \mathbf{P}(R)\cdot\mathbf{P}(D \mid R) + \mathbf{P}(M)\cdot\mathbf{P}(D \mid M)}$$
$$= \frac{(.2)(.8)}{(.3)(.6) + (.4)(.75) + (.2)(.8) + (.1)(.9)}$$
$$\simeq 0.2192$$

11. Note that S will represent T-shirt; T will represent ticket.
$\mathbf{P}(C) = 0.33$, $\quad$ so $\mathbf{P}(A) = 0.67$
$\mathbf{P}(P \mid A) = 0.04$, $\quad \mathbf{P}(S \mid A) = 0.21$, $\quad \mathbf{P}(T \mid A) = 0.65$,
$\mathbf{P}(V \mid A) = 0.10$, $\quad \mathbf{P}(P \mid C) = 0.09$, $\quad \mathbf{P}(S \mid C) = 0.28$,
$\mathbf{P}(T \mid C) = 0.32$, $\quad \mathbf{P}(V \mid C) = 0.31$

(a) $$\mathbf{P}(A \mid V) = \frac{\mathbf{P}(A)\cdot\mathbf{P}(V \mid A)}{\mathbf{P}(A)\cdot\mathbf{P}(V \mid A) + \mathbf{P}(C)\cdot\mathbf{P}(V \mid C)}$$
$$= \frac{(0.67)(0.10)}{(0.67)(0.10) + (0.33)(0.31)} \simeq 0.3957$$

(b) $$\mathbf{P}(C \mid T) = \frac{\mathbf{P}(C)\cdot\mathbf{P}(T \mid C)}{\mathbf{P}(A)\cdot\mathbf{P}(T \mid A) + \mathbf{P}(C)\cdot\mathbf{P}(T \mid C)}$$
$$= \frac{(0.33)(0.32)}{(0.67)(0.65) + (0.33)(0.32)} \simeq 0.1952$$

(c) $$\mathbf{P}(A \mid P) = \frac{\mathbf{P}(A)\cdot\mathbf{P}(P \mid A)}{\mathbf{P}(A)\cdot\mathbf{P}(P \mid A) + \mathbf{P}(C)\cdot\mathbf{P}(P \mid C)}$$
$$= \frac{(0.67)(0.04)}{(0.67)(0.04) + (0.33)(0.09)} \simeq 0.4743$$

(d) $$\mathbf{P}(C \mid S) = \frac{\mathbf{P}(C)\cdot\mathbf{P}(S \mid C)}{\mathbf{P}(A)\cdot\mathbf{P}(S \mid A) + \mathbf{P}(C)\cdot\mathbf{P}(S \mid C)}$$
$$= \frac{(0.33)(0.28)}{(0.67)(0.21) + (0.33)(0.28)} \simeq 0.3964$$

19. $\mathbf{P}(\text{White}) = 0.781 \quad \mathbf{P}(\text{Low income} \mid \text{White}) = 0.078$
$\mathbf{P}(\text{Black}) = 0.103 \quad \mathbf{P}(\text{Low income} \mid \text{Black}) = 0.278$
$\mathbf{P}(\text{Other}) = 0.116 \quad \mathbf{P}(\text{Low income} \mid \text{Other}) = 0.192$

(a) $$\mathbf{P}(W \mid L) = \frac{\mathbf{P}(W)\cdot\mathbf{P}(L \mid W)}{\mathbf{P}(W)\cdot\mathbf{P}(L \mid W) + \mathbf{P}(B)\cdot\mathbf{P}(L \mid B) + \mathbf{P}(O)\cdot\mathbf{P}(L \mid O)}$$
$$= \frac{(.781)(.078)}{(0.781)(0.078) + (0.103)(0.278) + (0.116)(0.192)}$$
$$\simeq 0.5448$$

(b) $$\mathbf{P}(B \mid L) = \frac{\mathbf{P}(B)\cdot\mathbf{P}(L \mid B)}{\mathbf{P}(W)\cdot\mathbf{P}(L \mid W) + \mathbf{P}(B)\cdot\mathbf{P}(L \mid B) + \mathbf{P}(O)\cdot\mathbf{P}(L \mid O)}$$
$$= \frac{(.103)(.278)}{(0.781)(0.078) + (0.103)(0.278) + (0.116)(0.192)}$$
$$\simeq 0.2561$$

(c) $\mathbf{P}(O \mid L) = \dfrac{\mathbf{P}(O)\cdot\mathbf{P}(L \mid W)}{\mathbf{P}(W)\cdot\mathbf{P}(L \mid W)+\mathbf{P}(B)\cdot\mathbf{P}(L \mid B)+\mathbf{P}(O)\cdot\mathbf{P}(L \mid O)}$

$= \dfrac{(.116)(.192)}{(0.781)(0.078)+(0.103)(0.278)+(0.116)(0.192)}$

$\simeq 0.1992$

(d) Many people do not expect over half the poor to be white. The high probability of being white offsets the low probability of being poor, given white.

H.7 Recursion

Exercises 7.1.7

1. (a)
```
integer recursiveSum(integer n)
   if n ≤ 1
      return 1
   else
      return n + recursiveSum(n-1)
end recursiveSum
```
(b) It does use tail-end recursion.

(c)
```
integer loopSum(integer n)
   sum = 0
   i = 1
   while i ≤ n
      sum = sum + i
      i = i + 1
   return sum
end loopSum
```

5. (a) The complete diagram still has two invocations that lead to a sum of 4. However, there are more invocations than were done in Quick Check 7.1.

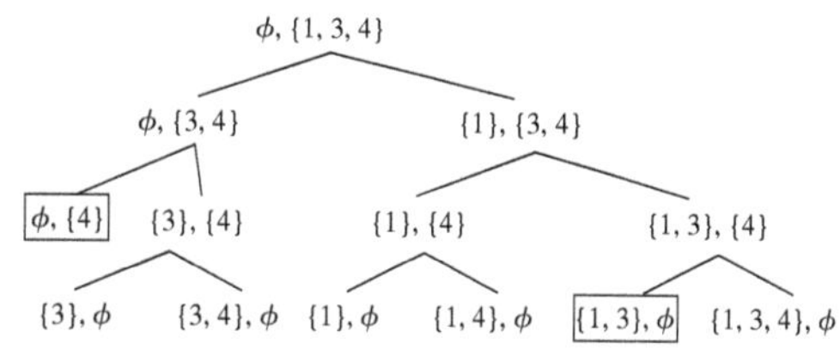

(b) Choosing the larger element is more efficient. There are more opportunities to eliminate some recursions at lines 5 or 7, since the recursive invocations at line 15 are producing sets with small sums faster than if small elements were eliminated.

7. (a) The following base conditions are sufficient:
- $\gcd(0, n) = n$
- $\gcd(1, n) = 1$, where $n \geq 1$

(b)
```
integer gcd(integer a, integer b)
   if (a > b)
      swap a and b
   if a == 0
      return b
   if a == 1
      return 1
   return gcd(b mod a, a)
```

15.
```
real aToAPowerOf2(real a, natural
number n)
   if n == 0
      return a
   else
      return aToAPowerOf2(a, n − 1)·
             aToAPowerOf2(a, n − 1)
end aToAPowerOf2
```

21. **Page 1: [1,10]** $\tau = .01$, $S(f, 1, 10) \simeq 2.74091$, $S(f, 1, 5.5) \simeq 1.80944$, and $S(f, 5.5, 10) \simeq 0.59846$, so $|\text{whole} - \text{left} - \text{right}| = 0.333008 > 10 \cdot 0.01$.

Page 2: [1,5.5] Page 3: [5.5,10] is pending. $\tau = 0.005$, $S(f, 1, 5.5) \simeq 1.80944$, $S(f, 1, 3.25) \simeq 1.19627$, and $S(f, 3.25, 5.5) \simeq 0.526424$, so $|\text{whole} - \text{left} - \text{right}| = 0.08675 > 10 \cdot 0.005$.

Page 4: [1,3.25] $\tau = 0.0025$, $S(f, 1, 3.25) \simeq 1.19627$, $S(f, 1, 2.125) \simeq 0.755735$, and $S(f, 2.125, 3.25) \simeq 0.424997$, so $|\text{whole} - \text{left} - \text{right}| = 0.0155343 < 10 \cdot 0.0025$. The algorithm returns 0.755735 + 0.424997 = 1.18073 as the value of $\int_1^{3.25} \frac{1}{x}\,dx$.

Page 5: [3.25,5.5] $\tau = 0.0025$, $S(f, 3.25, 5.5) \simeq 0.526424$, $S(f, 3.5, 4.375) \simeq 0.297271$, and $S(f, 4.375, 5.5) \simeq 0.228847$, so $|\text{whole} - \text{left} - \text{right}| = 0.000306028 < 10 \cdot 0.0025$. The algorithm returns 0.297271 + 0.228847 = 0.526118 as the value of $\int_{3.25}^{5.5} \frac{1}{x}\,dx$.

Back to Page 2 Page 2 returns $\int_1^{3.25} \frac{1}{x}dx + \int_{3.25}^{5.5} \frac{1}{x}dx = 1.18073 + 0.526118 = 1.706848$.

Page 3: [5.5,10] $\tau = 0.005$, $S(f, 5.5, 10) \simeq 0.59846$, $S(f, 5.5, 7.75) \simeq 0.342984$, and $S(f, 7.75, 10) \simeq 0.254901$, so $|\text{whole} - \text{left} - \text{right}| = 0.00057522 < 10 \cdot 0.005$. The algorithm returns 0.342984 + 0.254901 = 0.597885 as the value of $\int_{5.5}^{10} \frac{1}{x}\,dx$.

Back to Page 1 The final result is $1.706848 + 0.597885 = 2.30473$, using the intervals [1,3.25], [3.25,5.5], and [5.5,10].

The actual value is 2.30259, with an error in the adaptive quadrature estimate of approximately −0.00214.

Exercises 7.2.5

2.

```
integer Fibonacci(integer n)
   if (n == 0) or (n == 1)
      return 1
   else
      return Fibonacci(n – 1) +
             Fibonacci(n – 2)
 end Fibonacci
```

This is not an efficient algorithm; it is an example of redundant recursion.

3. (b)
$$\begin{aligned} a_n &= n \cdot a_{n-1} \\ &= n \cdot ((n-1) \cdot a_{n-2}) \quad \text{substitute} \\ &= (n(n-1)) \cdot a_{n-2} \quad \text{simplify} \\ &= (n(n-1)) \cdot ((n-2)a_{n-3}) \quad \text{substitute} \\ &= (n(n-1)(n-2)) \cdot a_{n-3} \quad \text{simplify} \\ &\ \ \vdots \qquad\qquad \vdots \\ &= (n(n-1)(n-2)\cdots(n-k+1)) \cdot a_{n-k} \\ &\ \ \vdots \qquad\qquad \vdots \\ &= (n(n-1)(n-2)\cdots 2 \cdot 1) \cdot a_0 \\ &= n! \quad \text{for } n \geq 0 \end{aligned}$$

(c)
$$\begin{aligned} a_n &= 2a_{n-2} \\ &= 2(2a_{n-4}) \quad \text{substitute} \\ &= 2^2 a_{n-4} \quad \text{simplify} \\ &= 2^2(2a_{n-6}) \quad \text{substitute} \\ &= 2^3 a_{n-6} \quad \text{simplify} \\ &\ \ \vdots \qquad\qquad \vdots \\ &= 2^k a_{n-2k} \end{aligned}$$

If n is even, this terminates when $k = \frac{n}{2}$. If n is odd, this terminates when $k = \frac{n-1}{2}$.

$$= \begin{cases} 2^{\frac{n}{2}} a_0 & \text{if } n \text{ is even} \\ 2^{\frac{n-1}{2}} a_1 & \text{if } n \text{ is odd} \end{cases}$$

$$= \begin{cases} -(2^{\frac{n}{2}}) & \text{if } n \text{ is even, } n \geq 0 \\ 3 \cdot 2^{\frac{n-1}{2}} & \text{if } n \text{ is odd, } n \geq 1 \end{cases}$$

The first few values of $\{a_n\}$ are $\{-1, 3, -2, 6, -4, 12, -8, 24, \ldots\}$. The recurrence relation essentially interleaves two distinct sequences.

4. (c)
$$\begin{aligned} a_n &= 5a_{n-1} + n \\ &= 5(5a_{n-2} + (n-1)) + n \quad \text{substitute} \\ &= 5^2 a_{n-2} + (5(n-1) + n) \quad \text{simplify} \\ &= 5^2(5a_{n-3} + (n-2)) \\ &\quad +(n-2)) + (5(n-1) + n) \quad \text{substitute} \\ &= 5^3 a_{n-3} + (5^2(n-2) + 5(n-1) + n) \quad \text{simplify} \\ &\ \ \vdots \qquad\qquad \vdots \\ &= 5^k a_{n-k} + (5^{k-1}(n-k+1) + 5^{k-2}(n-k+2) \\ &\quad + \cdots + 5(n-1) + n) \\ &\ \ \vdots \qquad\qquad \vdots \\ &= 5^n a_0 + (5^{n-1} \cdot 1 + 5^{n-2} \cdot 2 + \cdots + 5 \cdot (n\text{-}1) + n) \\ &= \textstyle\sum_{k=1}^{n} k \cdot 5^{n-k} \quad \text{the next step is not trivial} \\ &= \frac{5^{n+1} - 4n - 5}{16} \quad \text{for } n \geq 0 \text{ (more details follow)} \end{aligned}$$

Some hints for the final simplification:

$$\sum_{k=1}^{n} k \cdot 5^{n-k} = 5^n \sum_{k=1}^{n} \frac{k}{5^k}$$

and

$$\frac{k}{5^k} = \underbrace{\frac{1}{5^k} + \cdots + \frac{1}{5^k}}_{k \text{ times}}$$

Now rearrange terms to get

$$5^n \left(\sum_{k=1}^{n} \frac{1}{5^k} + \sum_{k=2}^{n} \frac{1}{5^k} + \cdots + \sum_{k=n}^{n} \frac{1}{5^k} \right)$$

This can be expressed in terms of a collection of geometric series:

$$5^n \left(\sum_{k=1}^{n} \frac{1}{5^k} + \frac{1}{5} \sum_{k=1}^{n-1} \frac{1}{5^k} + \frac{1}{5^2} \sum_{k=1}^{n-2} \frac{1}{5^k} + \cdots + \frac{1}{5^n} \sum_{k=1}^{1} \frac{1}{5^k} \right)$$

Using the formula for the sum of a geometric series, together with some algebraic simplification leads to the final result.

11. (a) The characteristic equation is $x^2 + 2x - 15 = 0$, having roots $r_1 = 3$ and $r_2 = -5$ This has a general solution in the form $a_n = \theta_1 3^n + \theta_2(-5)^n$. The system of linear equations is

$$\begin{aligned} \theta_1 + \theta_2 &= 2 \\ 3\theta_1 - 5\theta_2 &= -2 \end{aligned}$$

Using the substitution $\theta_2 = 2 - \theta_1$, it is easy to find $\theta_1 = 1$, and hence $\theta_2 = 1$. The solution is therefore

$$a_n = 3^n + (-5)^n \quad \text{for } n \geq 0$$

12. (c) The characteristic equation is $x^3 - 6x^2 + 11x - 6 = 0$. Possible rational roots are $\pm 1, \pm 2, \pm 3, \pm 6$. It is easy to verify that 1 is a root, so we can factor this as $(x-1)(x^2 - 5x + 6) = 0$ [using a simple polynomial division $(x-1)\overline{)x^3 - 6x^2 + 11x - 6}$]. An easy factorization now produces $(x-1)(x-2)(x-3) = 0$. The roots are therefore $r_1 = 1, r_2 = 2$, and $r_3 = 3$. The general solution is $a_n = \theta_1 1^n + \theta_2 2^n + \theta_3 3^n = \theta_1 + \theta_2 2^n + \theta_3 3^n$. The system of linear equations is

$$\begin{aligned} \theta_1 + \theta_2 + \theta_3 &= 3 \\ \theta_1 + 2\theta_2 + 3\theta_3 &= 4 \\ \theta_1 + 4\theta_2 + 9\theta_3 &= 6 \end{aligned}$$

Subtract the first equation from the second and subtract the first equation from the third equation. The resulting reduced system is

$$\theta_2 + 2\theta_3 = 1$$
$$3\theta_2 + 8\theta_3 = 3$$

Subtracting 3 times the first equation (of the reduced system) from the second allows us to determine that $\theta_3 = 0$. Substituting this result into the second equation (of the reduced system) allows us to determine that $\theta_2 = 1$. Substituting these results into any of the original equations results in $\theta_1 = 2$. The solution is

$$a_n = 2 + 2^n \quad \text{for } n \geq 0$$

17. (a) Let B_n stand for the number of distinct bit strings of length n that do not contain three consecutive 1s.
$B_0 = 1$ since there is only one way to have a bit string with no bits. Additionally, $B_1 = 2$, since both the bit strings 0 and 1 do not have three consecutive 1s.
$B_2 = 4$, since the bit strings 00, 01, 10, and 11 do not have three consecutive 1s.
If $n \geq 3$, the number of distinct bit strings of length n that do not contain three consecutive 1s is equivalent to the sum of the number of bit strings of length n that do not contain three consecutive 1s that end with a 0 and the number of bit strings of length n that do not contain three consecutive 1s that end with a 1.

Bit strings of length n that do not contain three consecutive 1s that end with a 0 are just the bit strings of length $n - 1$ with no three consecutive 1s with a 0 added at the end. There are B_{n-1} such bit strings.
The number of bit strings of length n that do not contain three consecutive 1s and end in 1 is equivalent to the sum of the number of bit strings of length n that do not contain three consecutive 1s that end with a 01 and the number of bit strings of length n that do not contain three consecutive 1s that end with a 11. Bit strings of length n that do not contain three consecutive 1s that end with a 01 are just the bit strings of length $n - 2$ with no three consecutive 1s with a 01 added at the end. There are B_{n-2} such bit strings. Bit strings of length n that do not contain three consecutive 1s that end with a 11 must have 0 as their $(n - 2)$nd bit. Thus, these are just the bit strings of length $n - 3$ with no three consecutive 1s with a 011 added at the end. There are B_{n-3} such bit strings.
It is now possible to create the recurrence relation.

- $B_0 = 1,\ B_1 = 2,\ B_2 = 4$
- $B_n = B_{n-1} + B_{n-2} + B_{n-3} \quad \text{for } n \geq 3$

(b) The characteristic equation is

$$x^3 - x^2 - x - 1 = 0$$

The rational roots theorem indicates that the possible rational roots are ± 1. It is easy to verify that neither is a root. In more advanced courses you might learn how to determine that this polynomial has one real and two complex roots. It is still possible to apply the linear homogeneous solution process, but the algebraic details are pretty nasty.

If you try back substitution, you might find it difficult to see any patterns emerge.

Techniques to solve this recurrence relation are presented in Section 7.4.

18. (b) The characteristic equation is $x^3 - 8x^2 + 21x - 18 = 0$. The rational roots theorem indicates that the possible rational roots are $\pm 1, \pm 2, \pm 3, \pm 6, \pm 9, \pm 18$. It is easy to verify that ± 1 are not roots. However, $r_1 = 2$ *is* a root. So

$$x^3 - 8x^2 + 21x - 18 = (x - 2)(x^2 - 6x + 9)$$

which easily factors completely as $(x - 2)(x - 3)^2$. Thus, $r_2 = 3$ has multiplicity $\nu_2 = 2$ and $\nu_1 = 1$. The general solution has the form

$$a_n = \alpha_0 2^n + (\beta_0 + \beta_1 n)\, 3^n$$

The linear system is

$$\alpha_0 + \beta_0 = 1$$
$$2\alpha_0 + 3\beta_0 + 3\beta_1 = 2$$
$$4\alpha_0 + 9\beta_0 + 18\beta_1 = 1$$

The solution is $\alpha_0 = -2,\ \beta_0 = 3,\ \beta_1 = -1$, so the explicit formula for the sequence is

$$a_n = -2 \cdot 2^n + (3 - n)\, 3^n = -2^{n+1} + (3 - n)\, 3^n,\ n \geq 0$$

Exercises 7.3.1

3. Since $h \in \mathcal{O}(g)$, there are constants, c and n_1, such that, for all $n > n_1$, $|h(n)| \leq c|g(n)|$. Thus, for all $n > \max(n_0, n_1)$, $|f(n)| \leq |h(n)| \leq c|g(n)|$. Consequently, $f \in \mathcal{O}(g)$.

5. Since $\log_y(z) - 1 \leq \left\lfloor \log_y(z) \right\rfloor \leq \log_y(z)$ for $z > 0,\ y > 1$,

$$x^{\log_y(z)-1} \leq x^{\lfloor \log_y(z) \rfloor} \leq x^{\log_y(z)}$$

Let $f(z) = x^{\lfloor \log_y(z) \rfloor}$. Then Definition 4.3 or Exercise 3 implies that $f \in \mathcal{O}\left(x^{\log_y(z)}\right)$. Definition 4.4 or Exercise 4 implies that $f \in \Omega\left(x^{\log_y(z)-1}\right)$.

8. Use Theorem 7.17.

Part	a	b	c	d	v	a vs b^v	Θ
(a)	3	2	4	5	2	$3 < 2^2$	$\Theta(n^2)$

9. Use Theorem 7.17.

Part	a	b	c	d	v	a vs b^v	Θ
(b)	3	2	$\frac{4}{3}$	4	1	$3 > 2^1$	$\Theta(n^{\log_2(3)})$

13. (a) $a = 1, b = c = 2$, and $d = 1$. Theorem 7.19 implies that $f \in \Theta\left(\left[\log_2(n)\right]^2\right)$.

16. `shuffle`

(a) There are no data copies in the base case, so $f(1) = 0$ ($d = 0$). There are three recursions, each processing one third of the original list, so $a = b = 3$. There are n data copies after the recursions, so $c = v = 1$. Thus,

- $f(n) = 3f\left(\frac{n}{3}\right) + n$
- $f(1) = 0$

(b) Theorem 7.17 (with $a = 3 = 3^1 = b^v$) implies that $f \in \Theta(n \log_3(n))$.

Exercises 7.4.1

2. (a) The product is

$$\sum_{k=0}^{\infty}\left(\sum_{j=0}^{k} 1 \cdot (k-j)\right) z^k = \sum_{k=0}^{\infty}\left(\sum_{j=0}^{k} j\right) z^k = \sum_{k=0}^{\infty} \frac{k(k+1)}{2} z^k$$

4. Assume $n \geq 1$.

$$\begin{aligned}
a_n &= (n-1)a_{n-1} + 1 \\
&= (n-1)[(n-2)a_{n-2} + 1] + 1 \quad \text{substitute} \\
&= (n-1)(n-2)a_{n-2} + (n-1) + 1 \quad \text{simplify} \\
&= (n-1)(n-2)[(n-3)a_{n-3} + 1] \\
&\quad + (n-1) + 1 \quad \text{substitute} \\
&= (n-1)(n-2)(n-3)a_{n-3} + (n-1)(n-2) \\
&\quad + (n-1) + 1 \quad \text{simplify} \\
&\vdots \qquad \vdots \\
&= (n-1)(n-2)\cdots(n-(n-1))a_1 + (n-1)(n-2) \\
&\quad \cdots(n-(n-2)) + \cdots + (n-1)(n-2) \\
&\quad + (n-1) + 1 \\
&= (n-1)(n-2)\cdots(n-(n-1))a_1 + 1 + (n-1) \\
&\quad + (n-1)(n-2) + \cdots \\
&\quad + (n-1)(n-2)\cdots(n-(n-2)) \\
&= P(n-1, n-1) + \sum_{r=0}^{n-2} P(n-1, r) \\
&= \sum_{r=0}^{n-1} P(n-1, r) \quad \text{for } n \geq 1
\end{aligned}$$

5. (b)

$$\begin{aligned}
(1+3z)^{-2} &= \sum_{k=0}^{\infty}\binom{-2}{k} 3^k z^k \\
&= \sum_{k=0}^{\infty}(-1)^k\binom{k+1}{k} 3^k z^k \\
&= \sum_{k=0}^{\infty}(-1)^k (k+1) 3^k z^k
\end{aligned}$$

The series starts as

$$1 - 6z + 27z^2 - 108z^3 + 405z^4 - 1458z^5 + \cdots$$

9. (b)

$$\begin{aligned}
\sum_{n=1}^{\infty} a_n z^n &= \sum_{n=1}^{\infty} 3a_{n-1} z^n + \sum_{n=1}^{\infty} 7z^n \quad \text{so} \\
&= 3z \sum_{n=1}^{\infty} a_{n-1} z^{n-1} + 7 \sum_{n=1}^{\infty} z^n \\
&= 3z \sum_{n=0}^{\infty} a_n z^n + 7\left(\frac{1}{1-z} - 1\right)
\end{aligned}$$

$$A(z) - a_0 = 3zA(z) + \frac{7}{1-z} - 7$$

Solving for $A(z)$:

$$A(z) = -\frac{6}{1-3z} + \frac{7}{(1-z)(1-3z)}$$

The first term expands to $-6\sum_{k=0}^{\infty} 3^k z^k$ (using the table of useful generating functions). The second term requires a use of Theorem 7.22.

$$\begin{aligned}
7\frac{1}{(1-z)(1-3z)} &= 7\left(\sum_{k=0}^{\infty} z^k\right)\left(\sum_{k=0}^{\infty} 3^k z^k\right) \\
&= 7\sum_{k=0}^{\infty}\left(\sum_{j=0}^{k} 3^j\right) z^k \\
&= \sum_{k=0}^{\infty} \frac{7}{2}\left(3^{k+1} - 1\right) z^k
\end{aligned}$$

Combining

$$\begin{aligned}
A(z) &= \left(\sum_{k=0}^{\infty} \frac{7}{2}\left(3^{k+1} - 1\right) z^k\right) - \left(\sum_{k=0}^{\infty} 6 \cdot 3^k z^k\right) \\
&= \sum_{k=0}^{\infty}\left(\frac{9}{2} \cdot 3^k - \frac{7}{2}\right) z^k
\end{aligned}$$

Thus

$$a_n = \frac{9}{2} \cdot 3^n - \frac{7}{2} \quad \text{for } n \geq 0$$

10. (d) Multiply by z^n and sum from $n = 2$:

$$\begin{aligned}
\sum_{n=2}^{\infty} a_n z^n &= -5\sum_{n=2}^{\infty} a_{n-1} z^n + 36\sum_{n=2}^{\infty} a_{n-2} z^n \\
&= -5z\sum_{j=1}^{\infty} a_j z^j + 36z^2 \sum_{j=0}^{\infty} a_j z^j
\end{aligned}$$

This leads to

$$A(z) - a_0 - a_1 z = -5z(A(z) - a_0) + 36z^2 A(z)$$

so

$$A(z)(1 + 5z - 36z^2) = a_0 + (5a_0 + a_1)z = 3 + 3z$$

Thus

$$\begin{aligned}
&A(z) \\
&= (3+3z)\left(\frac{1}{(1+9z)(1-4z)}\right) \\
&= (3+3z)\left(\sum_{k=0}^{\infty}\left(\sum_{j=0}^{k}(-9)^j 4^{k-j}\right) z^k\right)
\end{aligned}$$

$$= (3+3z)\left(\sum_{k=0}^{\infty} 4^k \left(\sum_{j=0}^{k}\left(-\frac{9}{4}\right)^j\right) z^k\right)$$

$$= (3+3z)\left(\sum_{k=0}^{\infty} 4^k \left(\frac{\left(-\frac{9}{4}\right)^{k+1}-1}{\left(-\frac{9}{4}\right)-1}\right) z^k\right)$$

$$= (3+3z)\left(\sum_{k=0}^{\infty} \frac{1}{13}\left(-(-9)^{k+1}+4^{k+1}\right) z^k\right)$$

$$= \frac{3}{13}(1+z)\left(\sum_{k=0}^{\infty} \left(-(-9)^{k+1}+4^{k+1}\right) z^k\right)$$

$$= \frac{3}{13}\left(\sum_{k=0}^{\infty} \left(-(-9)^{k+1}-(-9)^k+4^{k+1}+4^k\right) z^k\right)$$

$$= \frac{3}{13}\left(\sum_{k=0}^{\infty} \left(8\cdot(-9)^k+5\cdot 4^k\right) z^k\right)$$

$$= \left(\sum_{k=0}^{\infty} \left(\frac{24}{13}(-9)^k+\frac{15}{13}\cdot 4^k\right) z^k\right)$$

Consequently,

$$a_n = \left(\frac{24}{13}(-9)^n+\frac{15}{13}\cdot 4^n\right) \quad \text{for } n \ge 0$$

12. The solution is the coefficient of z^{38} in the expansion of

$$\left(\sum_{i=0}^{38} z^i\right)\left(\sum_{j=0}^{7} z^{5j}\right)\left(\sum_{k=0}^{3} z^{10k}\right)\left(\sum_{m=0}^{1} z^{25m}\right)$$

Mathematica was used to determine that the desired coefficient is 24. The valid part of the expansion is

$$\begin{aligned}
&1+z+z^2+z^3+z^4+2z^5+2z^6+2z^7\\
&+2z^8+2z^9+4z^{10}+4z^{11}+4z^{12}+4z^{13}\\
&+4z^{14}+6z^{15}+6z^{16}+6z^{17}+6z^{18}\\
&+6z^{19}+9z^{20}+9z^{21}+9z^{22}+9z^{23}\\
&+9z^{24}+13z^{25}+13z^{26}+13z^{27}+13z^{28}\\
&+13z^{29}+18z^{30}+18z^{31}+18z^{32}+18z^{33}\\
&+18z^{34}+24z^{35}+24z^{36}+24z^{37}+24z^{38}
\end{aligned}$$

18. **(a)** Since $(1-z-6z^2) = (1-3z)(1+2z)$,

$$\frac{3z}{1-z-6z^2} = \frac{A}{1-3z}+\frac{B}{1+2z}$$

$$= \frac{\frac{3}{5}}{1-3z}-\frac{\frac{3}{5}}{1+2z}$$

$$= \left(\frac{3}{5}\right)\frac{1}{1-3z}-\left(\frac{3}{5}\right)\frac{1}{1+2z}$$

By using Table 7.10, this can be written as

$$\left(\frac{3}{5}\right)\left(\sum_{n=0}^{\infty} 3^n z^n\right)-\left(\frac{3}{5}\right)\left(\sum_{n=0}^{\infty}(-1)^n 2^n z^n\right)$$

The summations can be combined, leading to the generating function

$$\frac{3}{5}\sum_{n=0}^{\infty}\left[3^n+(-1)^n 2^n\right]z^n$$

$$= \sum_{n=0}^{\infty}\left[\frac{3}{5}\cdot 3^n+(-1)^n\cdot\frac{3}{5}\cdot 2^n\right]z^n$$

Exercises 7.5.1

4. **(a)** Look at the diagonal elements in the table:

n	**Elimination Sequence**	j_n
1	—	1
2	2	1
3	2 1	3
4	2 4 3	1
5	2 4 1 5	3
6	2 4 6 3 1	5
7	2 4 6 1 5 3	7
8	2 4 6 8 3 7 5	1
9	2 4 6 8 1 5 9 7	3

They represent the position of the second-to-last person. It appears that a similar pattern is developing. Notice that $n = 2$ is a special case (do not eliminate anyone), so $p_2 = 2$ is not part of the general pattern. The base value is $p_3 = 1$. Extend the table:

n	3	4	5	6	7	8	9	10	11	12
p_n	1	3	5	1	3	5	7	9	11	1

The divisions in the pattern appear to be where $n = 3\cdot 2^m$. In analogy with the solution to the modified Josephus problem, we might conjecture that if $n > 2$ and $n = 3\cdot 2^m + i$, where $0 \le i < 3\cdot 2^m$, then $p_n = 2i+1$. This fits the table developed so far.

H.8 Combinatorics

Exercises 8.1.4

2. **(b)** $p(8,5) = p(7,4) + p(3,5) = p(6,3) + p(3,4) + 0 = 3 + 0 + 0 = 3$

3. **(a)** $p(5) = 7$

$k = 1$	$k = 2$	$k = 3$	$k = 4$	$k = 5$
5=5	5=4+1	5=3+1+1	5=2+1+1+1	5=1+1+1+1+1
	5=3+2	5=2+2+1		

4. **(a)** $p(n, k)$ satisfies the recurrence relation

$$p(n, k) = \sum_{i=0}^{k} p(n-k, i) \quad 0 < k \leq n$$

because there is a one-to-one correspondence between partitions of n into exactly k summands and partitions of $n-k$ into k or fewer summands. To see this, note that any partition of n into exactly k summands can be converted into a partition of $n-k$ by subtracting 1 from each summand and removing any summands that become 0. Since the original partitions were distinct, the resulting partitions will also be distinct. On the other hand, any partition of $n-k$ with k or fewer summands can be converted to a partition of n having exactly k summands by adding 1 to each existing summand and then adding 1s as new summands until there are k summands. Distinct original summands will result in distinct new summands.
The base condition $p(n, k) = 0$ if $k > n$ still holds. In addition, $p(n, 1)$ represents the number of partitions of n into exactly 1 summand. The partition $n = n$ is the only possibility.

7. **(c)** **OD CD ¬Ø** $\quad 3!S(4, 3) = 36$

	X	Y	Z		X	Y	Z
1	a b	c	d	19	a b	d	c
2	a c	b	d	20	a c	d	b
3	a d	c	b	21	a d	b	c
4	b c	a	d	22	b c	d	a
5	b d	a	c	23	b d	c	a
6	c d	a	b	24	c d	b	a
7	c	a b	d	25	d	a b	c
8	b	a c	d	26	d	a c	b
9	c	a d	b	27	b	a d	c
10	a	b c	d	28	d	b c	a
11	a	b d	c	29	c	b d	a
12	a	c d	b	30	b	c d	a
13	c	d	a b	31	d	c	a b
14	b	d	a c	32	d	b	a c
15	c	b	a d	33	b	c	a d
16	a	d	b c	34	d	a	b c
17	a	c	b d	35	c	a	b d
18	a	b	c d	36	b	a	c d

9. **OI CD Ø** $n = 12, k = 6 \quad \binom{6+12-1}{12} = 6188$ ways

16. **OI CI Ø** $n = 8, k = 4$
$p(8, 1) + p(8, 2) + p(8, 3) + p(8, 4) = 1 + 4 + 5 + 5 = 15$

24.
$$\begin{aligned} S(9, 4) &= S(8, 3) + 4S(8, 4) \\ &= (S(7, 2) + 3S(7, 3)) + 4\,(S(7, 3) + 4S(7, 4)) \\ &= S(7, 2) + 7S(7, 3) + 16S(7, 4) \\ &= (S(6, 1) + 2S(6, 2)) + 7\,(S(6, 2) + 3S(6, 3)) \\ &\quad + 16\,(S(6, 3) + 4S(6, 4)) \\ &= S(6, 1) + 9S(6, 2) + 37S(6, 3) + 64S(6, 4) \end{aligned}$$

30. $s(6, 3) = -225$

Exercises 8.2.4

3. **(b)** Orthogonal.

(1, 3)	(2, 1)	(3, 2)
(2, 2)	(3, 3)	(1, 1)
(3, 1)	(1, 2)	(2, 3)

5. **(b)** Not self-orthogonal.

1	2	3	1	2	3	(1, 1)	(2, 2)	(3, 3)
2	3	1	2	3	1	(2, 2)	(3, 3)	(1, 1)
3	1	2	3	1	2	(3, 3)	(1, 1)	(2, 2)

7. If two rows are swapped, the numbers in those rows will remain unchanged, so every row will still contain every number in $\{1, 2, \ldots, n\}$. The interchange does not move elements from one column to another, it just moves numbers to different positions within a column. Hence, every column will still contain every number in $\{1, 2, \ldots, n\}$. Thus, after the interchange of two rows, the matrix will still be a Latin square.
Similar comments show that interchanging two columns also preserves the Latin square property.

10. The dance instructor could use a Latin square of order 12. The 12 rows could represent the boys, the 12 columns could represent the girls, and the 12 numbers in the Latin square could represent the 12 dance lessons. A k in the ith row and j^{th} column means that boy i dances with girl j during lesson k.
Since $L(12) \geq 12! \cdot 11! \cdots 2! \cdot 1!$, there are certainly many choices for a Latin square of order 12.

15. **(d)** Definitely true. The maximum number of mutually orthogonal Latin squares of order n is at most $n - 1$. The number of distinct Latin squares of order n is at least $n! \cdot (n-1)! \cdots 2! \cdot 1!$, which is larger than $n - 1$ for $n > 1$ (it quickly becomes *much* larger).

18. Since L does not contain p, the lines, $L_1, L_2, \ldots, L_n, L_{n+1}$, that contain p must each contain a common point with L (by FPP2). Suppose L_i and L_j both share the point, q, with L. This would violate FPP2 since then L_i and L_j would both contain the distinct points p and q. There must therefore be at least $n + 1$ distinct points on L. Denote the point that is on both L and L_i as p_i, for $i = 1, 2, \ldots, n + 1$.
Suppose there is another point, x, on L. Since $x \neq p$, it must be on a common line, L_j, with p (by FPP1). The distinct points x and p_j both are on L and on L_j, violating FPP1 (and FPP2). This contradiction means that the only points that are on L are $p_1, p_2, \ldots, p_{n+1}$, completing the proof.

21. Suppose every point is on either L_1 or L_2. By FPP2, there is a unique point, p, that is on both L_1 and L_2. By FPP3, there must exist 4 points, no three of which are on a common line. This means that two of those points, p_1 and p_1', must be on L_1, and two of the points, p_2 and p_2', must be on L_2. Also, $p \notin \{p_1, p_2, p_1', p_2'\}$ (or there would be three points on a common line).

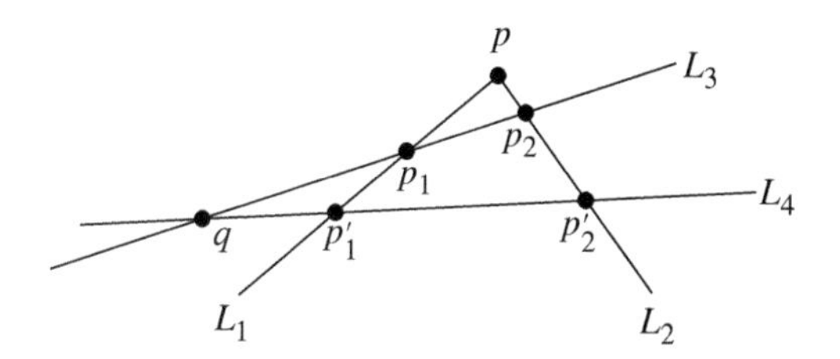

By FPP1, there must be a line, L_3, that contains p_1 and p_2 and another line, L_4, that contains p_1' and p_2'. The lines L_3 and L_4 must contain a common point, q, by FPP2. That point cannot be in $\{p_1, p_1', p_2, p_2'\}$ (for example, if the common point were p_1, then both L_1 and L_4 would contain p_1 and p_1', contradicting FPP2). Since q is on a common line, L_3, with p_1, it cannot be on L_1. Since q is on the common line, L_3, with p_2, it cannot be on L_2. Thus, q is the desired point.

26. (d) The duality principle does not make this claim. It makes an assertion about the statements of theorems. However, it is true that interchanging points and lines in an actual finite projective plane results in another finite projective plane. The reason is that the axioms (together with FPP3′) make points and lines entirely symmetrical objects. It is merely a matter of convention that we use dots and lines to represent the plane.

(e) Theorem 8.26 is *not* an existence theorem. It merely states that *if* a finite projective plane with one condition exists, then the other conditions are also true. In fact, the next section will show that there is no finite projective plane with 43 points and 43 lines, that is, there does not exist a finite projective plane of order 6.

Exercises 8.3.2

4. (a) (10, 15, 6, 4, 2) Potential.

(b) (15, 12, 5, 3, 2) Not possible. $bk \neq vr$, $r(k-1) \neq \lambda(v-1)$, Fisher's inequality fails.

7. (b) (22,7,2) Not possible. These parameters pass the tests in Theorem 8.35. Since v is even and $k - \lambda = 7 - 2 = 5$ is not a square, it fails the test in the Bruck–Ryser–Chowla theorem.

(c) (15,7,3) Potential. The relationship $r(k-1) = \lambda(v-1)$ is $7 \cdot 6 = 3 \cdot 14$, which is true. The Bruck–Ryser–Chowla theorem requires the equation $z^2 = 4x^2 - 3y^2$ to have an integer solution. Among many other solutions, $x = y = z = 1$ is easy to spot.

10. Since $v = 43$ is odd, the design will not exist unless the equation $z^2 = 6x^2 - y^2$ has a solution in integers, x, y, and z, not all 0. The equation can be written as $6x^2 = y^2 + z^2$. It is beyond the scope of this text to prove that this Diophantine equation has no non-trivial integer solutions.

15. (a) The residual design has parameters (10,15,6,4,2). It is as follows:

B_1^*	B_2^*	B_3^*	B_4^*	B_5^*	B_6^*	B_7^*	B_8^*	B_9^*
g	i	i	h	g	h	h	g	h
h	k	l	i	j	k	l	i	j
i	m	o	m	m	n	m	k	k
j	n	p	o	o	o	p	l	l

B_{10}^*	B_{11}^*	B_{12}^*	B_{13}^*	B_{14}^*	B_{15}^*
g	g	i	g	j	j
l	k	j	h	k	l
m	o	n	n	m	n
n	p	p	p	p	o

Exercises 8.4.1

1. (a) The algorithm produces the following table:

	T				K		
w	X	Y	Z	$B(w)$	X	Y	Z
0	0	0	0	0	**1**	**1**	**1**
1	0	0	0	0	0	0	0
2	0	0	$1+B(0)=1$	1	0	0	1
3	$4+B(0)=4$	0	$1+B(1)=1$	4	1	0	0
4	$4+B(1)=4$	$5+B(0)=5$	$1+B(2)=2$	5	0	1	0
5	$4+B(2)=5$	$5+B(1)=5$	$1+B(3)=5$	5	1	0	1
6	$4+B(3)=8$	$5+B(2)=6$	$1+B(4)=6$	8	2	0	0
7	$4+B(4)=9$	$5+B(3)=9$	$1+B(5)=6$	9	1	1	0
8	$4+B(5)=9$	$5+B(4)=10$	$1+B(6)=9$	10	0	2	0
9	$4+B(6)=12$	$5+B(5)=10$	$1+B(7)=10$	12	3	0	0
10	$4+B(7)=13$	$5+B(6)=13$	$1+B(8)=11$	13	2	1	0

An optimal packing is 2 copies of item X and 1 copy of item Y, for a total benefit of 13.

3. greedy Item X has maximum benefit, so it is added first. No other items will fit, so the total benefit is 4.

sophisticated greedy The expanded data table is shown below.

item	X	Y	Z
benefit	4	3	2
volume	5	4	3
quantity	1	1	1
b/v ratio	0.8	0.75	0.667

This algorithm will choose item X, for a total benefit of 4.

Knapsack The algorithm produces the following tables.

	T				K		
w	X	Y	Z	$B(w)$	X	Y	Z
0	0	0	0	0	**1**	**1**	**1**
1	0	0	0	0	0	0	0
2	0	0	0	0	0	0	0
3	0	0	$2+B(0)=2$	2	0	0	1
4	0	$3+B(0)=3$	$2+B(1)=2$	3	0	1	0
5	$4+B(0)=4$	$3+B(1)=3$	$2+B(2)=2$	4	1	0	0
6	$4+B(1)=4$	$3+B(2)=3$	$2+B_Z(3)=2$	4	1	0	0
7	$4+B(2)=4$	$3+B(3)=5$	$2+B(4)=5$	5	0	1	1

$B_Z(3) = 0$ because neither X nor Y will fit in a knapsack with capacity 3. The optimal packing is item Y and item Z, for a total benefit of 5.

6. **(a)** The optimal packing is 2 item X and 1 item Y, for a total benefit of 13.

	T				K		
w	X	Y	Z	$B(w)$	X	Y	Z
0	0	0	0	0	**2**	**2**	**2**
1	0	0	0	0	0	0	0
2	0	0	$1+B(0)=1$	1	0	0	1
3	$4+B(0)=4$	0	$1+B(1)=1$	4	1	0	0
4	$4+B(1)=4$	$5+B(0)=5$	$1+B(2)=2$	5	0	1	0
5	$4+B(2)=5$	$5+B(1)=5$	$1+B(3)=5$	5	1	0	1
6	$4+B(3)=8$	$5+B(2)=6$	$1+B(4)=6$	8	2	0	0
7	$4+B(4)=9$	$5+B(3)=9$	$1+B(5)=6$	9	1	1	0
8	$4+B(5)=9$	$5+B(4)=10$	$1+B(6)=9$	10	0	2	0
9	$4+B_X(6)=10$	$5+B(5)=10$	$1+B(7)=10$	10	1	1	1
10	$4+B(7)=13$	$5+B(6)=13$	$1+B(8)=11$	13	2	1	0

The recursive table for B_X must be produced:

	T				K		
w	X	Y	Z	$B_X(w)$	X	Y	Z
0	0	0	0	0	**1**	**2**	**2**
1	0	0	0	0	0	0	0
2	0	0	$1+B_X(0)=1$	1	0	0	1
3	$4+B_X(0)=4$	0	$1+B_X(1)=1$	4	1	0	0
4	$4+B_X(1)=4$	$5+B_X(0)=5$	$1+B_X(2)=2$	5	0	1	0
5	$4+B_X(2)=5$	$5+B_X(1)=5$	$1+B_X(3)=5$	5	1	0	1
6	$4+B_{XX}(3)=5$	$5+B_X(2)=6$	$1+B_X(4)=6$	6	0	1	1

Here is the table for B_{XX}.

	T		K	
w	Y	Z	$B_{XX}(w)$	Y	Z
0	0	0	0	**2**	**2**
1	0	0	0	0	0
2	0	$1+B_{XX}(0)=1$	1	0	1
3	0	$1+B_{XX}(1)=1$	1	0	1

10. Notice that a typical mall will have (for all practical purposes) unlimited quantities of each item. The knapsack problem can be represented with a knapsack having capacity 100 and item benefits, volumes, and quantities as shown.

item	watch	jean skirt	mug
benefit	5	4	3
volume	30	20	10
quantity	∞	∞	∞

item	book	swim suit	ring
benefit	3	3	5
volume	14	21	55
quantity	∞	∞	∞

item	earrings	backpack	stationary
benefit	2	2	1
volume	12	28	5
quantity	∞	∞	∞

The optimal solution is to purchase 10 mugs, for a total benefit of 30, and $100 spent.

This indicates one weakness in the knapsack model that has been presented: The model does not allow for the benefit level to change after packing a few of a particular item. You might attempt to fix this by setting the quantity of mug to a smaller number, then adding a new item "extra mugs," with unlimited quantity but a lower benefit. Since the "extra mugs" have the same volume, they will not be packed until all the "mugs" have been packed.

If the quantity of "mugs" is set at 3 and "extra mugs" are given benefit 2 (but unlimited quantity), the optimal benefit will be 24 and the knapsack will contain 3 mugs and 5 books.

13. **(a)** This is false. Consider the knapsack with total capacity $v = 3$, with two potential items.

item #	X	Y
benefit	6	2
volume	3	1
quantity	1	3

The optimal benefit is 6, but it can be achieved by packing one item X or three item Y's. So the optimal packing is not unique.

15. The proposed heuristic produces the inequality $11\left\lfloor\frac{9}{8.5}\right\rfloor \geq 5\left\lfloor\frac{9}{4}\right\rfloor$, which simplifies to $11 \geq 10$. Since this is true, the proposed heuristic would suggest packing an item X. There is no room for any other items, so the total benefit is 11. However, it is easy to see that packing two item Y's and one item Z produces a total benefit of 11.1, so the proposed heuristic fails to generate the optimal packing.

Exercises 8.5.4

1. The messages are 4 bits long, the corresponding code words are 7 bits long.

0000	0000000	0001	0001111
0100	0100101	0101	0101010
1000	1000011	1001	1001100
1100	1100110	1101	1101001
0010	0010110	0011	0011001
0110	0110011	0111	0111100
1010	1010101	1011	1011010
1110	1110000	1111	1111111

2. (b) 1111 1110111 has $x_5^c = 0 \neq x_5^r$, $x_6^c = 0 \neq x_6^r$, $x_7^c = 0 \neq x_7^r$ so the decoding table indicates an error in bit 4.

6. (a) 3 (differ in positions 2, 3 and 4)

7. (b) 0111001

10. (b) 101 (error in position 14)

11. (a) False. M is the number of code words (or messages), not the number of message *bits*.

14. (a)

0000000	0001111	0010110	0011001	0100101	0101010	0110011	0111100
1000000	1001111	1010110	1011001	1100101	1101010	1110011	1111100
0100000	0101111	0110110	0111001	0000101	0001010	0010011	0011100
0010000	0011111	0000110	0001001	0110101	0111010	0100011	0101100
0001000	0000111	0011110	0010001	0101101	0100010	0111011	0110100
0000100	0001011	0010010	0011101	0100001	0101110	0110111	0111000
0000010	0001101	0010100	0011011	0100111	0101000	0110001	0111110
0000001	0001110	0010111	0011000	0100100	0101011	0110010	0111101

1000011	1001100	1010101	1011010	1100110	1101001	1110000	1111111
0000011	0001100	0010101	0011010	0100110	0101001	0110000	0111111
1100011	1101100	1110101	1111010	1000110	1001001	1010000	1011111
1010011	1011100	1000101	1001010	1110110	1111001	1100000	1101111
1001011	1000100	1011101	1010010	1101110	1100001	1111000	1110111
1000111	1001000	1010001	1011110	1100010	1101101	1110100	1111011
1000001	1001110	1010111	1011000	1100100	1101011	1110010	1111101
1000010	1001101	1010100	1011011	1100111	1101000	1110001	1111110

18. (a) $M \leq \left\lfloor \dfrac{2^{14}}{\sum_{i=0}^{3}\binom{14}{i}} \right\rfloor = \left\lfloor \dfrac{8192}{235} \right\rfloor = 34$

Thus, the code can have at most 34 code words.

20. (a)

i. There are seven positions from which to choose the single bit to change. There is a probability of p that a bit is changed, and a probability of $1 - p$ that a bit remains unchanged. Therefore, the probability of exactly 1 bit being changed is

$$\binom{7}{1}p(1-p)^6 = 7p(1-p)^6$$

ii. If $p = 0.1$, the probability is 0.372009, whereas when $p = 0.5$ the probability is 0.0546875. This may seem paradoxical – the probability of exactly 1 bit being changed is lower when p is larger! Look at your solution to parts (b) and (c) to resolve this puzzle.

Exercises 8.6.3

1. (a) $\forall k \in S, \forall \{i_1, i_2, \ldots, i_k\} \subseteq S$ with $1 \le i_1 < i_2 < \cdots < i_k \le n, M(k, i_1, i_2, \ldots, i_k)$

2. (a) No system. $|A_1 \cup A_2| < 2$

(b) The list, a, b, c, is the unique system of distinct representatives

5. (c) No, since it does not satisfy the marriage condition. (Five sets, four elements.)

7. (a) Case 2 (since $|A_2| \not\geq 2$). It is possible to choose $m = 1$ and then rename A_2 as A_1, or to choose $m = 2$ and use the (possibly renamed) sets A_1, A_2 or A_2, A_3 as A_1 and A_2. The solution using A_1 and A_2 (without renaming) will be completed.
Removing $S = \{b, c\}$ from A_3 yields $B_3 = \{a\}$. The inductive step (applied to A_1 and A_2) assigns $r_1 = c, r_2 = b$ and a second application of the inductive step (applied to B_3) assigns $r_3 = a$. The system of distinct representatives is c, b, a.

8. (a) Since $t = n = 2$, the lower bound is $\frac{2!}{(2-2)!} = 2$. There are actually two systems of distinct representatives: a, b and b, a.

(b) Since $t = 2 < n$, the lower bound is $2! = 2$. There are actually three systems of distinct representatives: a, b, c and b, c, a and c, b, a.

14. (a) False. The marriage condition requires this to be true for *every* subcollection of k sets.

(e) False. The ranges in the table indicate that the correct value is not known, but it lies somewhere in the indicated range. Ramsey's theorem states that there is some unique number, $R(5, 5)$, such that every set with at least that size satisfies the (5, 5) Ramsey condition.

16. One person noted that the theoretical studies done by mathematicians are no more irresponsible than spending time playing (or watching) football, composing (or listening to) music, making movies, etc. (Also, the mathematics generally costs much less.) If time spent on artistic endeavors is justifiable, then so is time spent proving theorems.

Ann hinted at another aspect of the debate: *The principle must be one that can be applied to solve important scientific problems.* Notice that the scoreboards, video cameras, satellite links, and other equipment needed to broadcast a football game would not be possible without mathematicians having developed the mathematics needed for this technology. The stadium is safe because it was built using mathematically based principles from civil engineering.

Ann asked that the principle be explained *in language a lay person can understand.* This is exactly why the result was presented to the press as a problem about guests at a party, thus causing the objection by the letter writer.

H.9 Formal Models in Computer Science

Exercises 9.1.4

3. If a sample space of messages has many low probability messages and just a few high probability messages, there will be very little surprise when a typical message is sent (it will be one of the few in most cases). The definition assigns a relatively small information value as the average information for such a sample space. However, if every message is equally likely, then there will be the maximum variation in which message is sent. The surprise factor will be higher (because there is less predictability). The information value is highest in this case (Theorem 9.2).

4. The answer depends on whether or not n is a power of 2.

(a) $n = 14$ is not a power of 2. The number of bits is

$$\lceil \log_2(14) \rceil = \lceil 3.807351 \rceil = 4$$

In fact, $14_{10} = 1110_2$.

8. Write $\log_2(x) = \frac{1}{\ln(2)} \ln(x)$. Then use the standard result:

$$\begin{aligned} \lim_{x \to 0^+} x \ln(x) &= \lim_{x \to 0^+} \frac{\ln(x)}{\frac{1}{x}} \\ &= \lim_{x \to 0^+} \frac{\frac{1}{x}}{\frac{-1}{x^2}} \\ &= \lim_{x \to 0^+} (-x) = 0 \end{aligned}$$

$$\begin{aligned} &\lim_{x \to 0^+} \left[-x \log_2(x) - (1-x) \log_2(1-x)\right] \\ &= \lim_{x \to 0^+} \left[-\frac{x}{\ln(2)} \ln(x) - \frac{(1-x)}{\ln(2)} \ln(1-x)\right] \\ &= \frac{1}{\ln(2)} \lim_{x \to 0^+} [-x \ln(x)] - \frac{1}{\ln(2)} \lim_{x \to 0^+} [(1-x) \ln(1-x)] \\ &= \frac{1}{\ln(2)} \cdot 0 - \frac{1}{\ln(2)} \cdot (1 \cdot 0) \\ &= 0 \end{aligned}$$

10. Let $S = \{L, C, B, W\}$ with $p_L = \frac{3}{16}, p_C = p_B = \frac{1}{4}$ and $p_W = \frac{5}{16}$. Then

$$\begin{aligned} I(S) = &-\frac{3}{16} \log_2\left(\frac{3}{16}\right) - \frac{1}{4} \log_2\left(\frac{1}{4}\right) - \frac{1}{4} \log_2\left(\frac{1}{4}\right) \\ &- \frac{5}{16} \log_2\left(\frac{5}{16}\right) \\ &\simeq 1.97722 \end{aligned}$$

14. (d) False. Shannon's definition of information is *not* about semantics. The "surprise factor" in the formal definition is a purely probabilistic notion, which can be informally motivated by intuition about semantic content.

Exercises 9.2.3

1. (a) start in Zero, then: One, Zero, Zero, One, Zero, One, Two, Two. The string *is* recognized.

5. (c) This can be done with 4 states: *Empty, Zero, One, Both.* The start state is *Empty*, the only final state is *Both.*

State	Input 0	Input 1
Empty	Zero	One
Zero	Zero	Both
One	Both	One
Both	Both	Both

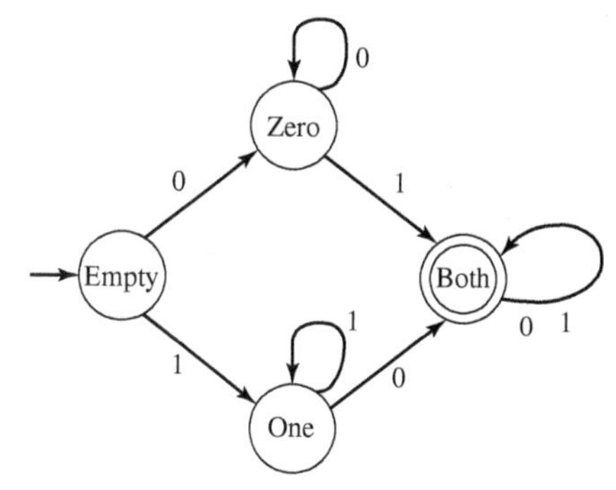

10. Here is a solution that uses a single final state. There are no transitions from the Lose and Win states because the input consists of only two numbers.

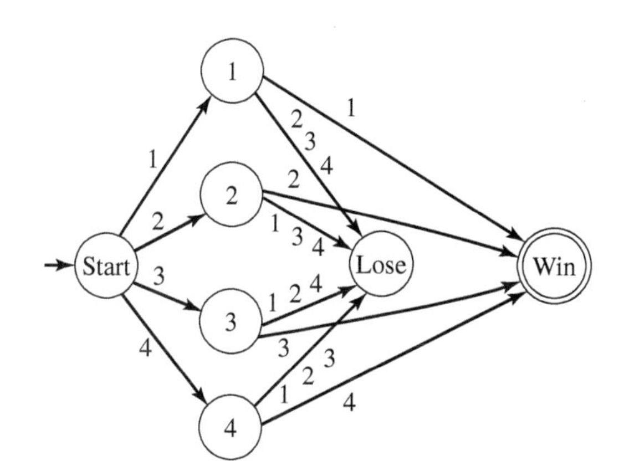

12. (a)

State	Input	New State	Output
c_0	11	c_1	0
c_1	01	c_1	0
c_1	10	c_1	0
c_1	10	c_1	0
c_1	00	c_0	1

The output string is therefore 10000 (13 + 3 = 16).

17. (b) Three state are sufficient. The start state is *No*, indicating that the pair *ab* is not in process. The state *a* indicates that an *a* has just been received. The state *ab* indicates that the pair of characters has already appeared.

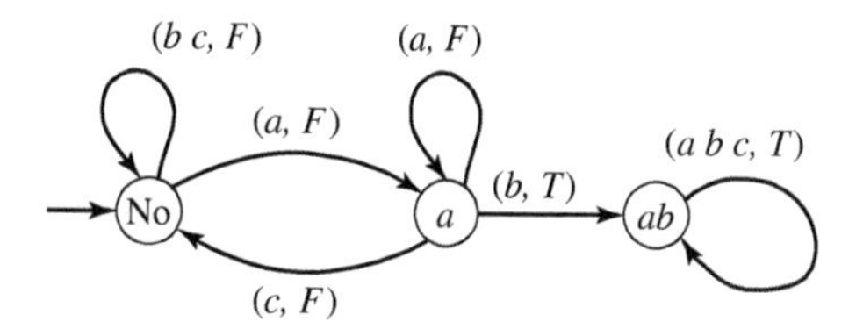

19. The set of states, $\mathcal{S}$, contains two elements: $\mathcal{S}$ = {initial, flip bits}. The start state is initial. The input and output sets are both {0,1}. The input and output functions are implicitly defined by the state diagram.

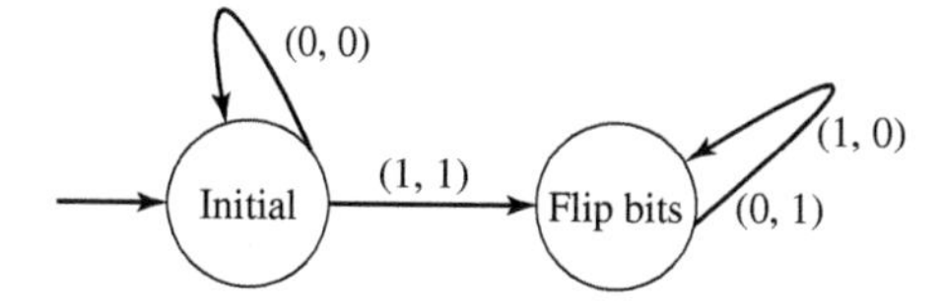

Exercises 9.3.2

1. (b) This is not a legal set of productions for a regular grammar. The production Y → Y has been ruled out in this textbook.

3. Notice that $\{uvw\} \cup \{vw\} = \{uvw, vw\}$. Also, $\{u, \lambda\}\{vw\} = \{uvw, \lambda vw\} = \{uvw, vw\}$, so $\{uvw\} \cup \{vw\} = \{uvw, vw\} = \{u, \lambda\}\{vw\}$.

5. (c) $S \Rightarrow 1Y \Rightarrow 100Y \Rightarrow 10000Y \Rightarrow 100001$

7. (a) $\{a^n c \mid n \geq 1\} \cup \{b^k \mid k \geq 2\}$

9. More than one solution is possible. There may be different choices for nonterminals, and even with the same non-terminals there may be multiple sets of productions that work. Notice the need to avoid using λ to finish (it is not in Σ).

(a) Δ = {S, T} with S as the start symbol.
$\Pi = \{S \rightarrow aS \mid bT \mid a \mid b, \quad T \rightarrow aS \mid a\}$

11. (c) This is false in general. The presence or absence of λ depends on the set of productions.

13. Note that $A^+ \subseteq A^*$, for any set, A. Consequently, if $x \in A^+$, then $x \in A^*$.

Suppose that $s \in (\Sigma^+)^*$. Then either $s = \lambda$ or there is a positive integer, n, with $s \in (\Sigma^+)^n$. If $s = \lambda$, then $s \in (\Sigma^*)^*$, since $(\Sigma^*)^* = \{\lambda\} \cup \bigcup_{i=1}^{\infty} (\Sigma^*)^i$. If $s \neq \lambda$, then s is the concatenation if n strings in Σ^+. But every string in Σ^+ is also a string in Σ^* (since $\Sigma^+ \subseteq \Sigma^*$). Thus, $s \in (\Sigma^*)^n \subseteq (\Sigma^*)^*$. This implies that $(\Sigma^+)^* \subseteq (\Sigma^*)^*$.

Now suppose that $s \in (\Sigma^*)^*$. Then either $s = \lambda$ or there is a positive integer, n, with $s \in (\Sigma^*)^n$. If $s = \lambda$, then $s \in (\Sigma^+)^*$, since $(\Sigma^+)^* = \{\lambda\} \cup \bigcup_{i=1}^{\infty} (\Sigma^+)^i$. If $s \neq \lambda$, then s is the concatenation of n strings in Σ^* (some of which might be λ). Let the substrings be $s_1, s_2, \ldots, s_n$, where $s_i \in \Sigma^{k_i}$. Since $s \neq \lambda$, at least one of the n substrings must be non-null. Thus, $k = k_1 + k_2 + \cdots + k_n > 0$. Since $s \in \Sigma^k$, and $k > 0$, $s \in \Sigma^+ \subseteq (\Sigma^+)^*$. This, together with the discussion for the case $s = \lambda$, implies that $(\Sigma^*)^* \subseteq (\Sigma^+)^*$.

The two set inclusions imply that $(\Sigma^*)^* = (\Sigma^+)^*$.

Exercises 9.4.3

3. `(^| )[aeiou][a-z]*[aeiou]($| )` or
`(^| )(a|e|i|o|u)[a-z]*(a|e|i|o|u)($| )`

5. (b) `[a-zA-Z]*(aa|ee|ii|oo|uu)[a-zA-Z]*`

14. It is tempting to use a regular expression that starts with

```
[0-9][0-9][0-9][0-9]( |-)?[0-9][0-9][0-9][0-9]( |-)?
```

but that would allow numbers such as 1111-2222 33334444 02/05 to be matched. The regular expression will need to have a form like

```
(A|B|C)  (0[1-9]|1[0-2])/[0-9][0-9]
```

where A is four copies of "`[0-9][0-9][0-9][0-9]`", separated by single spaces, B is four copies of "`[0-9][0-9][0-9][0-9]`", separated by single hyphens, and C is four copies of "`[0-9][0-9][0-9][0-9]`", all run together.

18. This cannot be done using regular expressions. It is tempting to try something like

```
[^P][^e][^r][^c][^i][^v][^a][^l]
```

but that is looking for a string with eight characters, each of which is any character except one position-specific letter.

Regular expressions are not powerful enough to specify a string that cannot be present; they can only specify a group of characters that cannot be present in some position.

Exercises 9.5.2

1. (a) Let $\Sigma = \{0, 1\}$ and $\Delta = \{Even, Odd\}$. The start symbol is $Even$. The productions will be those listed in the following table.

Transition	First Production	Second Production
$t(Even, 0) = Even$	$Even \to 0\ Even$	
$t(Even, 1) = Odd$	$Even \to 1\ Odd$	$Even \to 1$
$t(Odd, 0) = Odd$	$Odd \to 0\ Odd$	$Odd \to 0$
$t(Odd, 1) = Even$	$Odd \to 1\ Even$	

3. (a) $\Sigma = \{a, b, c\}$ and the start state is S. The set of states is initially $\mathcal{S} = \{S, B\}$. (I have modified the notational suggestions of the proof so that lowercase b does not have two meanings.) We need to add the state F to $\mathcal{S}$ because of the production $B \to c$. $\mathcal{F} = \{F\}$. No states other than BH are added when defining t, so $\mathcal{S} = \{S, B, F, BH\}$.
In tabular form ,the transition function looks like the next table.

State	Input a	b	c
S	B	BH	BH
B	BH	B	F
F	BH	BH	BH
BH	BH	BH	BH

The state diagram is shown next.

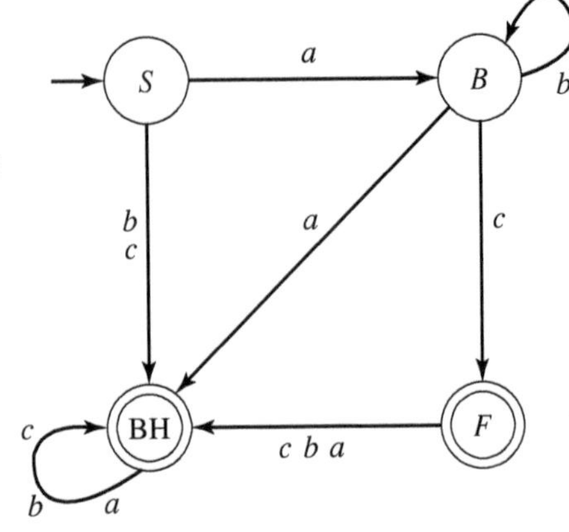

5. (b) Start by creating finite automata for the five regular expressions a, b, c, d, and e.

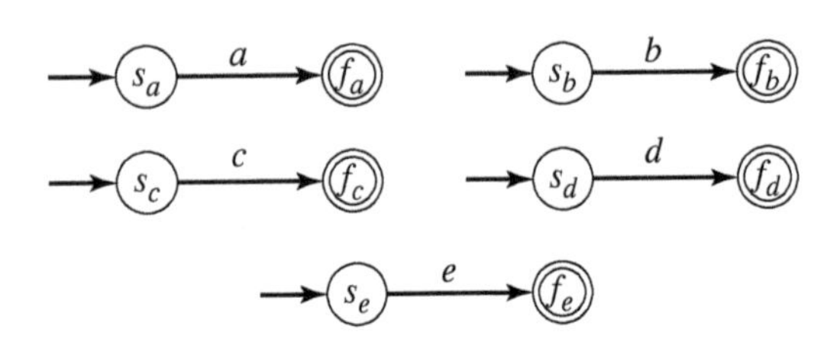

Next, the finite automata corresponding to ab and to cd can be formed using the concatenation rule.

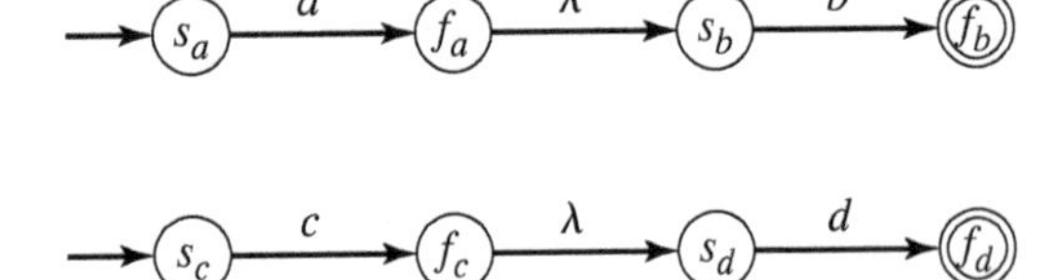

The alternative rule can now be used to construct a nondeterministic finite automaton for the regular expression, $(ab \mid cd)$.

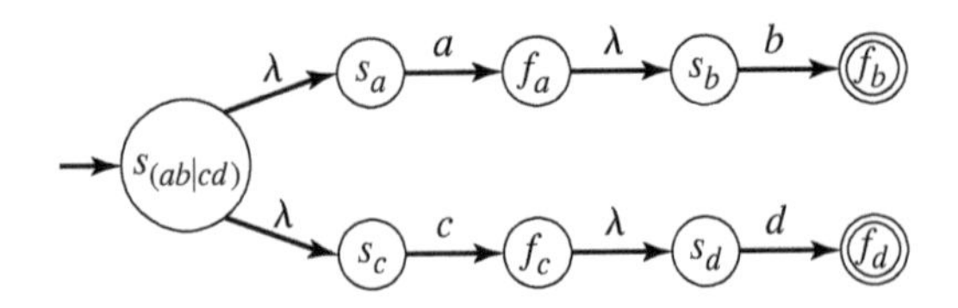

Finally, the concatenation rule can be used again to construct the final nondeterministic finite automaton for the regular expression, $(ab \mid cd)e$.

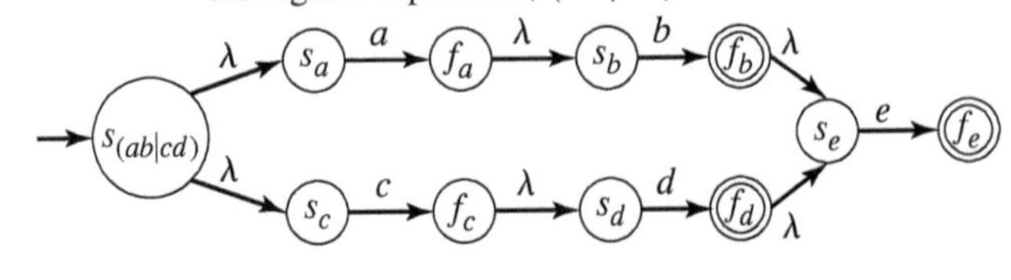

6. (b) A state diagram corresponding to the state table is shown.

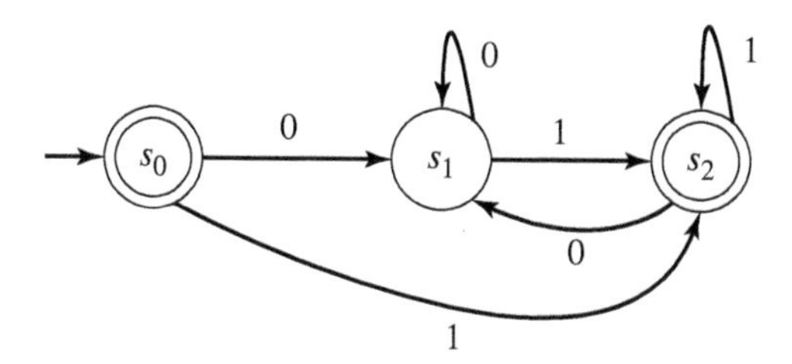

The modified state diagram with new initial state, s, and new final state, f_0, is shown next. Note the two λ-transitions into the new final state.

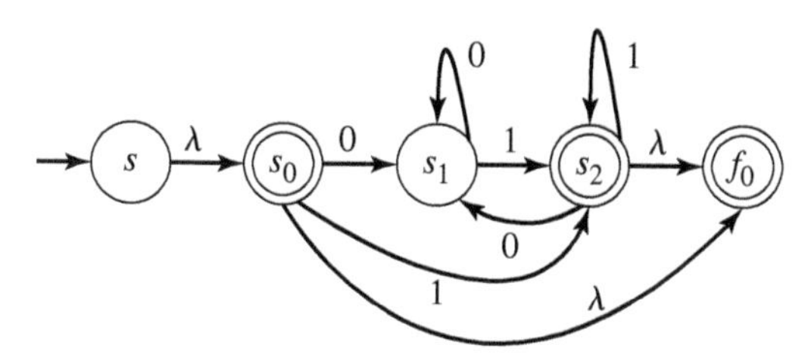

The reduction loop might be easiest to start with $y = s_0$. The choices for x and z are $x = s, z = s_1, x = s$, $z = s_2$, and $x = s, z = f_0$.

For $x = s, z = s_1$, $Lo(x, z) = \emptyset$, $Lo(x, y) = \lambda$, $Lo(y, y) = \emptyset$, and $Lo(y, z) = 0$. Thus, $Ln(x, z) = \emptyset \mid \lambda\emptyset^*0$, which simplifies to 0.

For $x = s, z = s_2$, $Lo(x, z) = \emptyset$, $Lo(x, y) = \lambda$, $Lo(y, y) = \emptyset$, and $Lo(y, z) = 1$. Thus, $Ln(x, z) = \emptyset \mid \lambda\emptyset^*1$, which simplifies to 1.

For $x = s, z = f_0$, $Lo(x, z) = \emptyset$, $Lo(x, y) = \lambda$, $Lo(y, y) = \emptyset$, and $Lo(y, z) = \lambda$. Thus, $Ln(x, z) = \emptyset \mid \lambda\emptyset^*\lambda$, which simplifies to λ.

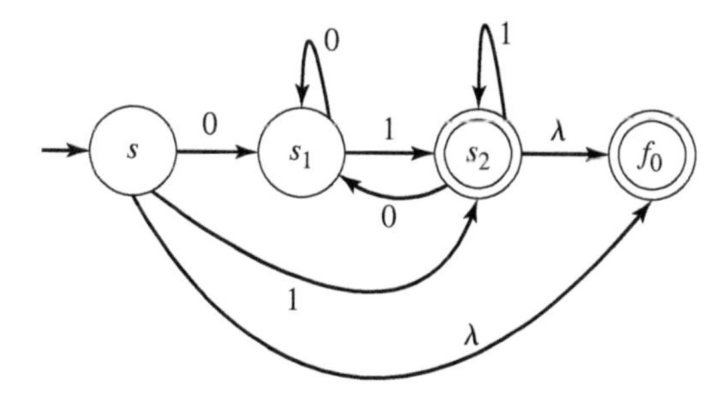

Now let $y = s_2$ (another final state). The possible transitions that pass through y are discussed next.

If $x = s$ and $z = f_0$, then $Lo(x, z) = \lambda$, $Lo(x, y) = 1$, $Lo(y, y) = 1$, and $Lo(y, z) = \lambda$. Thus, $Ln(x, z) = \lambda \mid 11^*\lambda$, which simplifies to $\lambda|1^+$.

If $x = s$ and $z = s_1$, then $Lo(x, z) = 0$, $Lo(x, y) = 1$, $Lo(y, y) = 1$, and $Lo(y, z) = 0$. Thus, $Ln(x, z) = 0 \mid 11^*0$, which simplifies to $0 \mid 1^+0$.

If $x = s_1$ and $z = f_0$, then $Lo(x, z) = \emptyset$, $Lo(x, y) = 1$, $Lo(y, y) = 1$, and $Lo(y, z) = \lambda$. Thus, $Ln(x, z) = \emptyset \mid 11^*\lambda$, which simplifies to 1^+.

If $x = s_1$ and $z = s_1$, then $Lo(x, z) = 0$, $Lo(x, y) = 1$, $Lo(y, y) = 1$, and $Lo(y, z) = 0$. Thus, $Ln(x, z) = 0 \mid 11^*0$, which simplifies to $0 \mid 1^+0$.

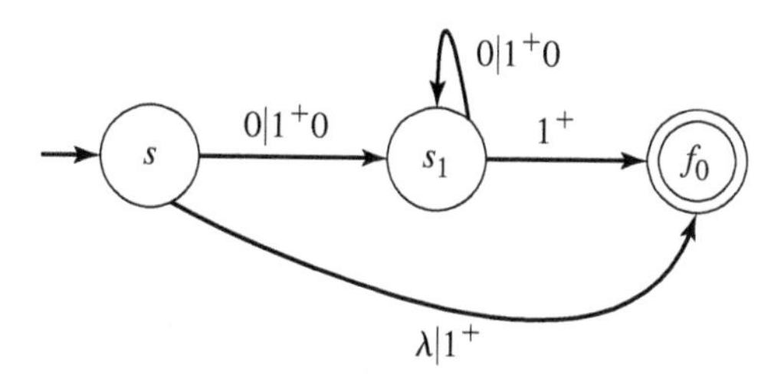

The final iteration uses $y = s_1$. There is only one choice for x and z: $x = s$ and $z = f_0$. In this instance, $Lo(x, z) = \lambda|1^+$, $Lo(x, y) = 0 \mid 1^+0$, $Lo(y, y) = 0 \mid 1^+0$, and $Lo(y, z) = 1^+$. Thus, $Ln(x, z) = (\lambda|1^+) \mid \left((0|1^+0)(0|1^+0)^*1^+\right)$. This simplifies to $\lambda \mid 1^+ \mid (0|1^+0)^+1^+$.

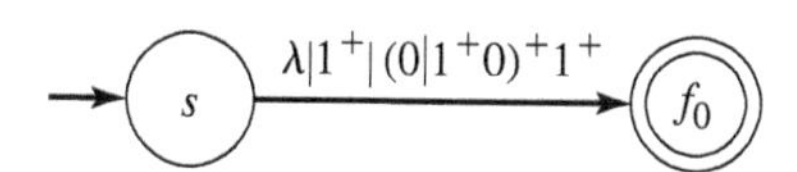

7. (d) False. Although a nondeterministic finite automaton does not *need* to have transitions associated with every input symbol, it is still permitted.

Exercises 9.6.3

1. $\Delta = \{S, X\}$, $\Pi = \{S \to aXcd,\ X \to aXc \mid b\}$

5. (a) These productions can be part of a context-sensitive grammar, but not a context-free grammar (since there are terminal symbols on the left). Notice that α_1 and/or α_2, in the description of context-sensitive productions, can be λ.

6. (a) These productions cannot be part of any of the grammars in the table, since it is not valid for α to be λ in any production of the form $\alpha \to \beta$.

11. (a) False. The language $\{a^n b^n c^n | n \geq 0\}$ is a counterexample.

14. (a) Halts in state s_1 (not recognized). Tape $= \cdots \lambda AAABB(C)C\lambda \cdots$

H.10 Graphs

Exercises 10.1.3

2. (a) When either $n \geq 2$ or $m \geq 2$ the graphs are all bipartite: Let V_α be the vertices whose row and column sum is even and V_β be the vertices whose row and column sum is odd.

4. (b)

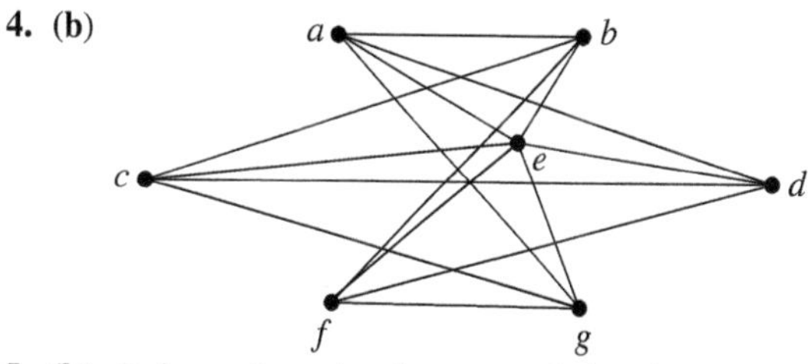

5. (b) S_n is regular only when n equals 1 or 2.

6. **(b)** C_5 is represented by open vertices and dashed edges, its line graph by solid vertices and solid edges. The line graph looks like another copy of C_5.

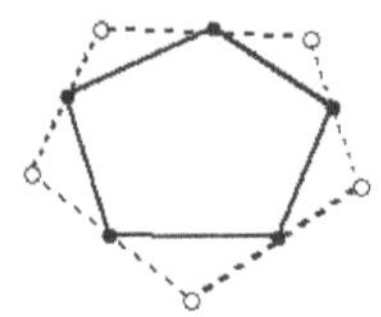

7. Since K_n is complete, it will have $n-1$ edges incident with the first vertex, $n-2$ additional edges incident with the second vertex, $n-3$ additional edges incident with the third vertex, and so on. The edges run out when the penultimate vertex is reached. The number of edges is therefore

$$1+2+\cdots+n-2+n-1=\frac{n(n-1)}{2}$$

An alternate proof: We want all possible unordered pairs of vertices from a set of size n. Each unordered pair corresponds to an edge of K_n. There are therefore

$$\binom{n}{2}=\frac{n(n-1)}{2} \text{ edges}$$

15. The union, $G \cup \overline{G}$, is K_n, and hence has $\binom{n}{2}$ edges. Since G has ϵ edges, its complement must have $\binom{n}{2}-\epsilon$ edges.

19. **(b)** False. Loops and/or multiple edges are permitted in multigraphs.

25. **(a)** Maximal clique: Any two adjacent vertices
Maximal independent set: $\{a, b, c\}$.

Exercises 10.2.3

3. **(a)** It is a trail and a path.

(b) It is a trail, a closed walk, a circuit, and a cycle.

5. **(a)** The Königsberg graph has connectivity 2 (b and d) and edge connectivity 3 (ad, bd, and cd).

7. **(a)**

$$\begin{array}{c} \\ a \\ b \\ c \\ d \end{array}\begin{array}{c} \begin{array}{cccc} a & b & c & d \end{array} \\ \begin{pmatrix} 0 & 2 & 0 & 1 \\ 2 & 0 & 2 & 1 \\ 0 & 2 & 0 & 1 \\ 1 & 1 & 1 & 0 \end{pmatrix}\end{array}$$

9. **(a)**

$$A^3:\quad \begin{array}{c} \\ a \\ b \\ c \\ d \end{array}\begin{array}{c} \begin{array}{cccc} a & b & c & d \end{array} \\ \begin{pmatrix} 4 & 22 & 4 & 11 \\ 22 & 8 & 22 & 11 \\ 4 & 22 & 4 & 11 \\ 11 & 11 & 11 & 8 \end{pmatrix}\end{array}$$

(b) The entry in row 1, column 4 of A^3 is the answer. There are 11 walks. They are $ae_1be_1ae_5d$, $ae_1be_2ae_5d$, $ae_2be_2ae_5d$, $ae_2be_1ae_5d$, $ae_1be_3ce_7d$, $ae_1be_4ce_7d$, $ae_2be_3ce_7d$, $ae_2be_4ce_7d$, $ae_5de_5ae_5d$, $ae_5de_6be_6d$, and $ae_5de_7ce_7d$.

15. **(a)** False. Consider the Königsberg graph in Exercise 3. It is clear that ae_1be_2a is a trail. However, this is not a path because vertex a is repeated.

Exercises 10.3.3

3. With two vertices, the complete graph K_2 has only one edge. There is a Hamilton path but no edge to use to return to the initial vertex (recall that a cycle is a circuit, which is a trail and consequently has no repeated edges). When $n \geq 3$, a cycle will never use an edge twice (or it would use vertices twice).

6. H_3

(a) There is neither an Euler circuit nor an Euler trail since there are eight vertices with odd degree.

(b) There are Hamilton cycles. One is shown.

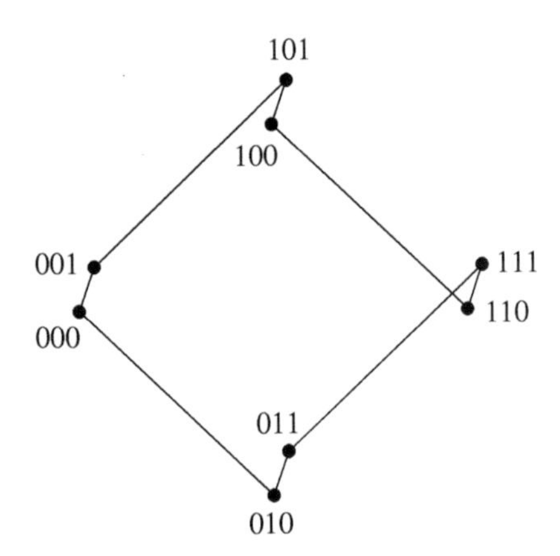

10. The path needs to connect three regions, but it must do so by passing through vertex 6 each time it changes region. Since two changes are required, vertex 6 must be crossed two times. This rules out a Hamilton path.

12. **(d)** False. To make this statement, you must first verify that G is a *connected* multigraph.

17. It is possible, as the following graph demonstrates.

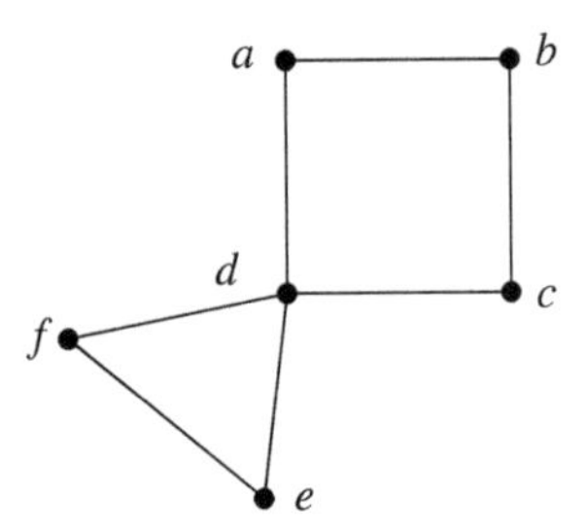

Exercises 10.4.3

1. **(b)**

$$\begin{array}{c} \\ a \\ b \\ c \\ d \\ e \end{array}\begin{array}{c} \begin{array}{ccccc} a & b & c & d & e \end{array} \\ \begin{pmatrix} 0 & 0 & 0 & 1 & 1 \\ 0 & 0 & 0 & 1 & 1 \\ 0 & 0 & 0 & 0 & 1 \\ 1 & 1 & 0 & 0 & 0 \\ 1 & 1 & 1 & 0 & 0 \end{pmatrix}\end{array}$$

$$\begin{array}{c} \\ a \\ b \\ c \\ d \\ e \end{array}\begin{array}{c} \begin{array}{ccccc} e_1 & e_2 & e_3 & e_4 & e_5 \end{array} \\ \begin{pmatrix} 1 & 0 & 0 & 1 & 0 \\ 0 & 1 & 1 & 0 & 0 \\ 0 & 0 & 0 & 0 & 1 \\ 1 & 1 & 0 & 0 & 0 \\ 0 & 0 & 1 & 1 & 1 \end{pmatrix}\end{array}$$

2. (b)
$$\begin{pmatrix} 1 & 0 & 0 & 1 & 0 \\ 0 & 1 & 0 & 1 & 0 \\ 0 & 1 & 0 & 0 & 1 \\ 1 & 0 & 0 & 0 & 1 \\ 0 & 0 & 1 & 0 & 1 \end{pmatrix} \begin{pmatrix} 1 & 0 & 0 & 1 & 0 \\ 0 & 1 & 1 & 0 & 0 \\ 0 & 0 & 0 & 0 & 1 \\ 1 & 1 & 0 & 0 & 0 \\ 0 & 0 & 1 & 1 & 1 \end{pmatrix} - \begin{pmatrix} 2 & 0 & 0 & 0 & 0 \\ 0 & 2 & 0 & 0 & 0 \\ 0 & 0 & 2 & 0 & 0 \\ 0 & 0 & 0 & 2 & 0 \\ 0 & 0 & 0 & 0 & 2 \end{pmatrix} = \begin{pmatrix} 0 & 1 & 0 & 1 & 0 \\ 1 & 0 & 1 & 0 & 0 \\ 0 & 1 & 0 & 1 & 1 \\ 1 & 0 & 1 & 0 & 1 \\ 0 & 0 & 1 & 1 & 0 \end{pmatrix}$$

4. (a) This cannot be an adjacency matrix, since vertex b has two neighbors according to row 2, but it has three neighbors according to column 2 (that is, this matrix is not symmetric).

7. (b) 3, 2, 2, 2, 1

10. (a) The only isomorphism is shown.

V	$\zeta(V)$
a	u
b	z
c	x
d	w
e	v
f	y

11. (a) The degree sequences are identical (3, 3, 3, 3, 3, 3, 3, 3) but they are not isomorphic. Notice that the second graph contains a triangle (a circuit of length 3). The smallest circuit in the first graph has length 4. Since isomorphism preserves circuits, these cannot be isomorphic.

12. (b) True. This implication is true since the hypothesis is always false. There cannot exist an isomorphism between a simple graph and a multigraph because the edge sets will either have a different number of elements, or else the adjacencies will not map properly.

(f) True. This implication is true since the conclusion is always true. Vertex degree is isomorphism invariant.

22. (a)

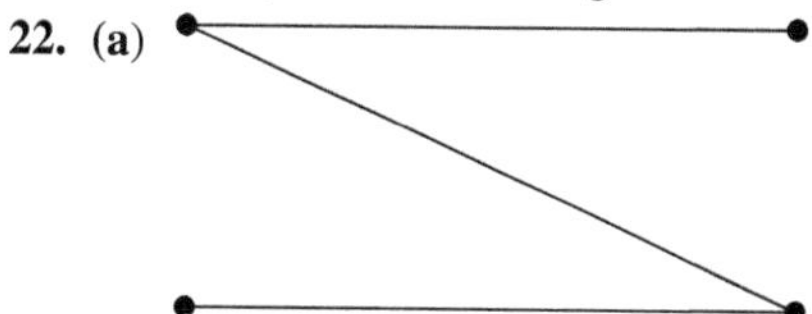

Exercises 10.5.4

5. (a) Notice that removing any vertex will also remove four edges. If it can be shown that any subgraph derived by removing just one edge is planar, then any subgraph with fewer vertices must also be planar. The symmetry of the graph means that removing any edge is equivalent to removing any other. However, the planar embedding we are used to seeing shows two apparently different kinds of edges (inside and outside). The following diagram shows the result of removing either of these apparently different kinds of edges.

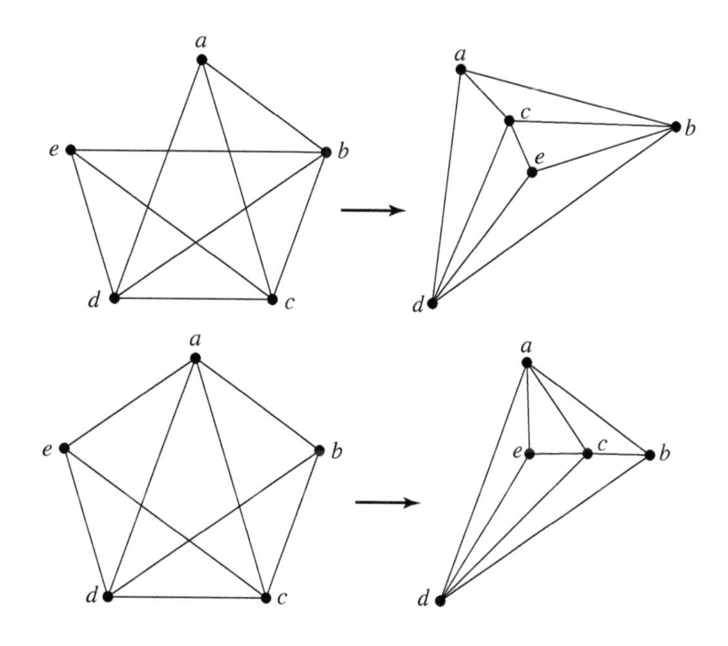

9. The graph is nonplanar. (Remove the central vertex and its incident edges. The resulting subgraph is homeomorphic to K_5.)

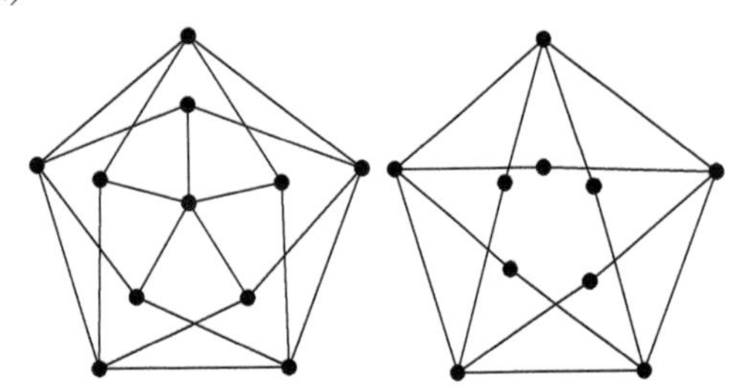

10. (a) The icosahdedron graph has chromatic number 4. It cannot be a larger number since it is planar (by the Four Color theorem). Notice the triangle around the outside. It will require three colors for the three vertices on the outside. There is only one way to color the next three vertices on the following diagram if we try to limit the coloring to three colors. There are now three vertices that require a fourth color.

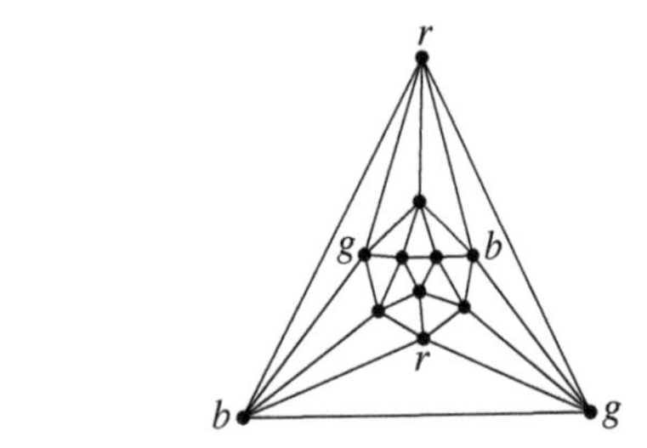

13. (a)

Isomorphic to

18. (c) False. The chromatic number is the *minimal* number of colors in any proper coloring of the graph.

Exercises 10.6.3

1. (a) This is weakly connected, but not strongly connected (vertex b has no outward arcs, so no directed walks exist from b to any other vertex).

2. (a)
$$\begin{pmatrix} 0 & 1 & 0 & 1 & 0 & 0 \\ 0 & 0 & 0 & 0 & 0 & 0 \\ 0 & 1 & 0 & 0 & 0 & 1 \\ 0 & 0 & 1 & 0 & 1 & 0 \\ 1 & 0 & 0 & 0 & 0 & 1 \\ 0 & 1 & 0 & 0 & 0 & 0 \end{pmatrix}$$

5. All three questions require A^3, which is shown.

$$\begin{pmatrix} 0 & 0 & 0 & 0 & 2 & 0 \\ 0 & 0 & 0 & 2 & 0 & 2 \\ 0 & 0 & 0 & 0 & 2 & 0 \\ 1 & 0 & 1 & 0 & 0 & 0 \\ 0 & 4 & 0 & 0 & 0 & 0 \\ 1 & 0 & 1 & 0 & 0 & 0 \end{pmatrix}$$

(a) There are two directed walks of length 3 from b to d.

10. If $\sum_{i=0}^{n-1} A^i$ has a zero in the ijth entry, then there is no directed walk (or path) between v_i and v_j that has length less than or equal to $n-1$. By the previous exercise (contrapositive), there is no directed walk between v_i and v_j, so D is not strongly connected.

If $\sum_{i=0}^{n-1} A^i$ has a nonzero value in the ijth position, then for at least one value of i, there is a directed walk of length i between v_i and v_j. If this is true for all choices of i and j, D must be strongly connected.

11. (b) False. To be multiple arcs, both must start and end at the same vertices: $\phi(e_1) = (v_i, v_f)$ and $\phi(e_2) = (v_i, v_f)$.

12. (a) The shortest path has length 7 and is *acdef.* The following table shows the details of Dijkstra's algorithm:

				d						*p*					
n	*B*	*r*	*A*	*a*	*b*	*c*	*d*	*e*	*f*						
0	∅			[0]	∞	∞	∞	∞	∞						
1	$\{a\}$	a	$\{b, c\}$		3	[2]					a	a			
2	$\{a, c\}$	c	$\{d, f\}$		[3]		4		8				c		c
3	$\{a, b, c\}$	b	$\{f\}$				[4]								
4	$\{a, b, c, d\}$	d	$\{e\}$					[6]						d	
5	$\{a, b, c, d, e\}$	e	$\{f\}$						[7]						e

19. (a) This is not a tournament graph (and hence not a transitive tournament graph either). One of the many reasons for this is that there is not an arc between vertices a and c.

22. The tournament graph is shown below.

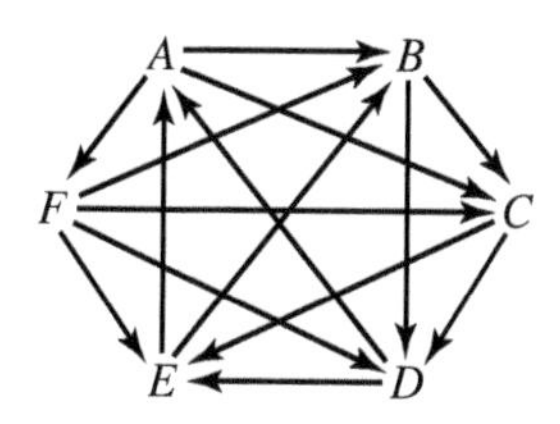

(a) Four of the many Hamilton paths are: FEABCD, EBCDAF, DEAFBC, and EAFBCD.

(b) The following table shows the numbers of wins.

A	B	C	D	E	F
3	2	2	2	2	4

This table favors a path starting at Faramir, but such a path cannot have Assumpta ranked second.
This example serves as motivation for the desirability of transitive tournament graphs.

H.11 Trees

Exercises 11.1.1

3. (a) Since $n = im + 1$, solve for i to show $i = \frac{n-1}{m}$. Since $n = i + l, l = n - i = n - \frac{n-1}{m} = \frac{n(m-1)+1}{m}$

6. (a) Theorem 11.14 implies that

$$10250 \leq \frac{m^6 - 1}{m - 1}$$

Since $\frac{6^6-1}{6-1} = 9,331$ and $\frac{7^6-1}{7-1} = 19,608$, $m = 7$ is the smallest value of m which makes this possible.

(b) Theorem 11.14 implies that

$$10250 \leq m^5$$

Since $6^5 = 7,776$ and $7^5 = 16,807$, the answer is still $m = 7$. This is true because most of the nodes in a maximal complete m-ary tree are leaf nodes. (When $m = 7$, a height five m-ary tree has 2,801 interior nodes and 16,807 leaves.)

10. The key ideas are follows: First check if you are at a leaf. If so, return $h = 0$, otherwise, return one more than the height of x's tallest subtree.

```
integer height(x)
  if x is a leaf
     return 0
  else
     max = 0
     for each child c of x
        ch = height(c)
        if ch > max
           max = ch
     return max + 1
end height
```

11. The extreme cases will be a complete binary tree and a chain.

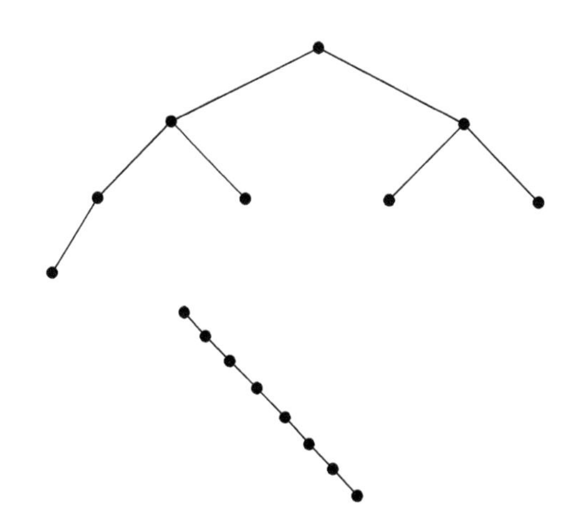

A chain with n vertices will have height $n - 1$. Hence, $h \leq n - 1$.

A maximal complete binary tree with height h has $n = \frac{2^{h+1}-1}{2-1} = 2^{h+1} - 1$ nodes. Solve for h: $h = \log_2(n + 1) - 1$. If the tree is complete but not maximal, $\log_2(n + 1)$ will not be an integer. Since h must be an integer, the number $\log_2(n + 1) - 1$ must be rounded up to provide a valid height.

Combining the results leads to

$$\lceil \log_2(n + 1) - 1 \rceil \leq h \leq n - 1$$

16. We need to show that there are no cycles in the graph. A proof by mathematical induction is in order.

Base Step A simple connected graph with 1 vertex has no edges (0 = 1-1) and is a tree since there are no cycles.

Inductive Step Assume that every simple, connected graph with n vertices and $n - 1$ edges is a tree.

Let G be a simple, connected graph with $n + 1$ vertices and n edges. Suppose that there is a cycle in G.

Exercise 15 implies that there must be at least one vertex with degree 1 and that vertex cannot be part of a cycle. Consequently, the cycle cannot contain every vertex of G.

Therefore the cycle must lie entirely within a proper subgraph of G. Create the subgraph G_1 by removing a vertex of degree 1 and the edge it is incident with. This is always possible by Exercise 15. The subgraph G_1 will still be simple and connected and will have n vertices and $n - 1$ edges. It will also still contain the cycle. The inductive hypothesis implies that the subgraph G_1 is a tree, contradicting the existence of a cycle in G_1. The erroneous assumption must have been the assumption that G contains a cycle. If G has no cycles, it must be a tree, completing the proof.

18. **(a)** False. The height of the subtree is related to a path going from the node toward nodes that are even further away from the root. The level is the length of the path from the node back to the root.

21. **(a)**

node	a	b	c	d	e	f
eccentricity	3	3	2	2	3	3

centers: c, d

Exercises 11.2.4

1. **(a)** **preorder** $a\ b\ c\ d\ e\ f$
inorder $b\ a\ e\ d\ f\ c$
postorder $b\ e\ f\ d\ c\ a$

3. **(a)**

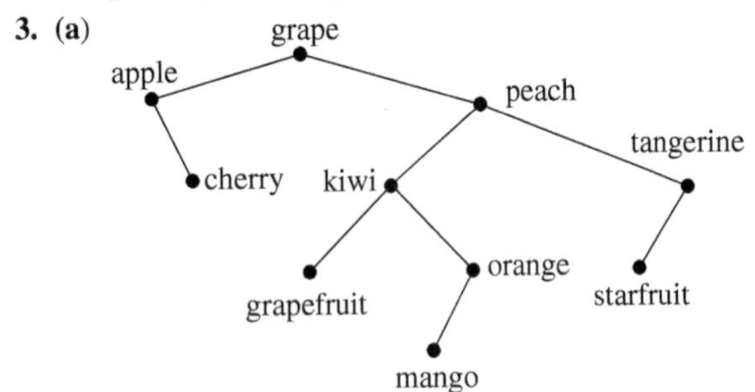

9. **(a)** The postfix expression is $x\ \ y\ \ z\ \ \hat{}\ \ \hat{}\ \ x\ \ y\ \ \hat{}\ \ z\ \ \hat{}\ \ +$.

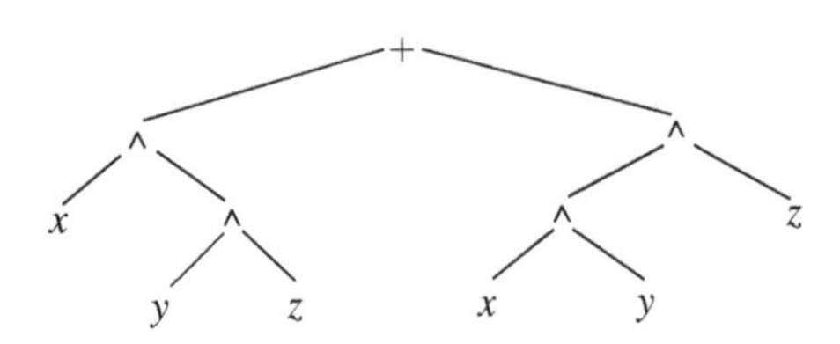

12. The solution strategy is to build a binary search tree from the list of letters, and then do an inorder traversal of the tree, writing the letter attached to each node as the traversal progresses.

(a) b c e f q r s t w

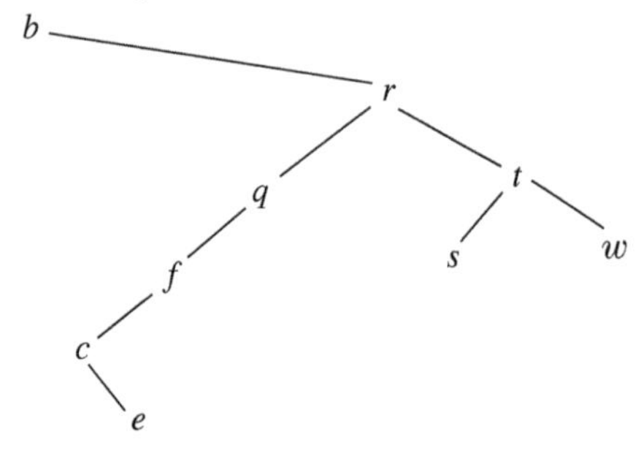

14. **(a)** This is a heap because every parent is smaller than all of its children.

16. Heap sort

(a) The (value at the) root node is always a smallest element in the tree.

(b) The algorithm is often called **percolateDown**. A pseudocode description is presented next. The strategy is to start at the (probably out-of-place) root and compare it to its two children. If either of the children is smaller, swap the current root element with the smaller of the two children. Now carry out the same procedure with the possibly out-of-place element at level 1. Keep swapping until the element that started at the root is smaller than (or equal to) both its current children. The test `minChild exists` is used to determine when the out-of-place node reaches a leaf position.
Since **percolateDown** only looks at nodes in one path from the root to a leaf, and the tree is a complete tree, it will examine at most $\log_2(n)$ nodes, so the algorithm is in $\Theta(\log_2(n))$.

```
void percolateDown(node root)
   currentNode = root
   minChild = min(currentNode's
        leftChild, currentNode's
        rightChild)
   while (minChild exists) and
         (minChild < currentNode)
      swap currentNode with minChild
      minChild = min(currentNode's
      new leftChild,
      currentNode's new rightChild)
end percolateDown
```

(c) Build a complete binary heap. Then start a loop that removes the root node, replacing it with the right-most element in the final level, and then invokes **percolateDown** to restore the heap property. Continue the loop until the heap is empty.
The elements come off the heap in increasing order, starting with the smallest [by part (a)]. The loop is processed n times (once per item in the heap). Percolate down will take at most $\log_2(k)$ swaps when the heap has k elements. Thus, the complexity of the loop is

$$\sum_{k=1}^{n} \log_2(k) \leq n \log_2(n) \in \mathcal{O}(n \log_2(n))$$

The first phase has big-$\mathcal{O}$ complexity $\mathcal{O}(n)$, so Theorem 4.14 implies that the complexity of the entire algorithm is in $\mathcal{O}(n \log_2(n))$.

Exercises 11.3.4

1. **(a)**

3. This is not consistent. The subtree below would need a production of the form X → bY, which is not present in the grammar.

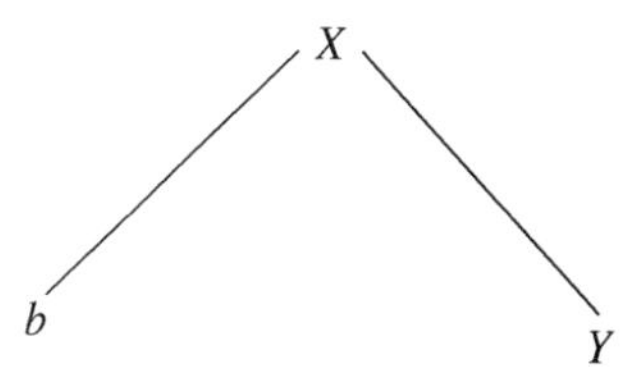

7. The frequency table is

d	e	i	p	r	␣
1	2	2	3	1	1

so the initial queue looks like the diagram below.

Front					Back
1 d	1 r	1 ␣	2 e	2 i	3 p

The final labeled tree is shown below. (The first subtree will go into the queue after the 2-i node.)

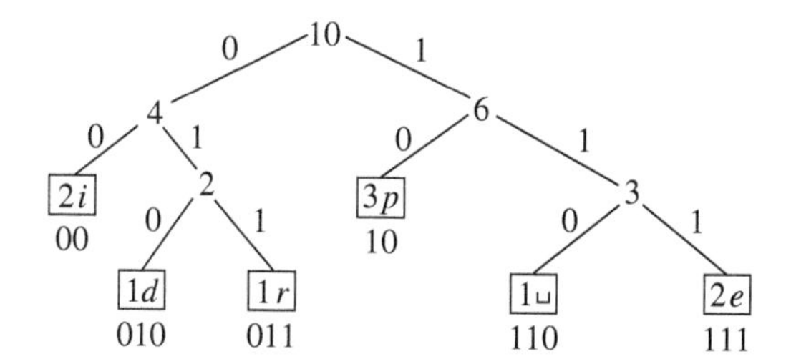

The encoding table is

d	e	i	p	r	␣
010	111	00	10	011	110

10. (a) The frequency table is

A	C	G	T
4	3	2	1

so the initial queue looks like the following diagram.

Front			Back
1 T	2 G	3 C	4 A

The final labeled tree is shown. (The first subtree will go into the queue after the 3C node.)

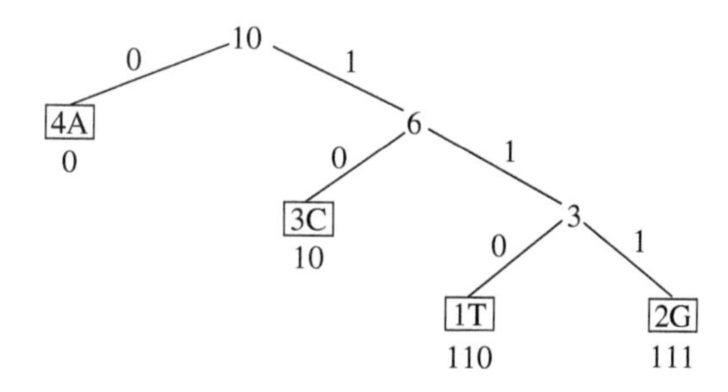

The encoding table is shown.

A	C	G	T
0	10	111	110

The 2-bit-per-character encoding will require $10 \cdot 2 = 20$ bits. The Huffman encoding needs 19 bits.

14.

```
<!ELEMENT letter (greeting, body,
    closing, signature)>
<!ELEMENT greeting (#PCDATA)>
<!ELEMENT body (paragraph)+>
<!ELEMENT paragraph
    (#PCDATA | emphasis)+>
<!ELEMENT emphasis (#PCDATA)>
<!ELEMENT closing (#PCDATA)>
<!ELEMENT signature (#PCDATA)>
```

17. (a)

```
<!ELEMENT email (to+, cc*,
    replyto, subject, body)>
<!ELEMENT to (#PCDATA)>
<!ELEMENT cc (#PCDATA)>
<!ELEMENT replyto (#PCDATA)>
<!ELEMENT subject (#PCDATA)>
<!ELEMENT body (paragraph)*>
<!ELEMENT paragraph (#PCDATA)>
```

(b)

```
<?xml version = "1.0" ?>
<!DOCTYPE email SYSTEM "email.dtd">
<email>
  <to>gosrac@bethel.edu</to>
  <to>gosnat@bethel.edu</to>
  <replyto>gossett@bethel.edu
  </replyto>
  <subject>Testing a DTD</subject>
  <body>
      <paragraph>This paragraph is
      for testing purposes. Should
      this have been a real email,
      something  more interesting
      would have been written.
      </paragraph>
      <paragraph>This is a second
      scintillating paragraph.
      </paragraph>
  </body>
</email>
```

Exercises 11.4.3

1. Answers will vary widely. The common factor is that each spanning tree will always have one less edge than there are vertices in the original graph. Here is one set of solutions.

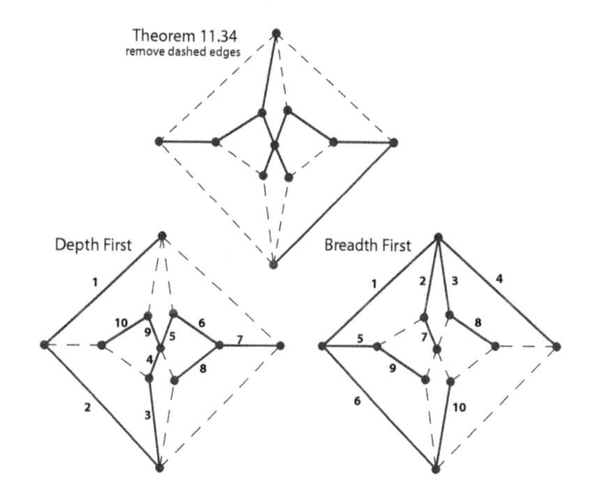

12. There are many correct solutions. The primary common feature is the sum of the weights on the edges of the minimal spanning tree. For Prim's algorithm, it is important to check that new edges are always connected to the previous partial spanning tree. The total weights are listed in the next table.

graph	a	b	c	d
total minimal edge weight	50	37	27	55

For part (c), both algorithms produce this solution.

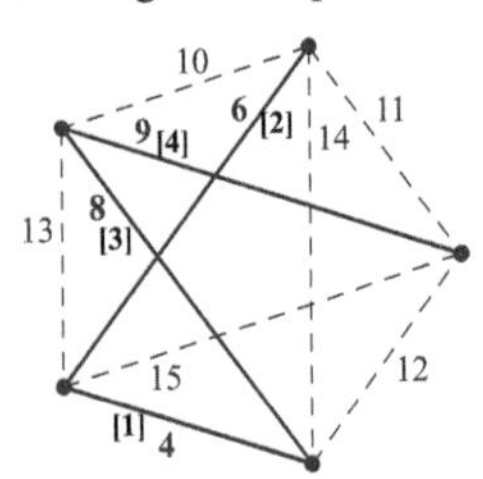

14. The total weights are listed in the next table.

graph	a	b	c	d
total minimal edge weight in modified graph	43	63	6	30
total maximal edge weight of original graph	97	137	54	132

16. (b) There are 3 spanning trees. The matrix K is

$$K = \begin{pmatrix} 2 & -1 & -1 & 0 \\ -1 & 2 & -1 & 0 \\ -1 & -1 & 3 & -1 \\ 0 & 0 & -1 & 1 \end{pmatrix}$$

with cofactor determinant 3.
The distinct spanning trees are shown next.

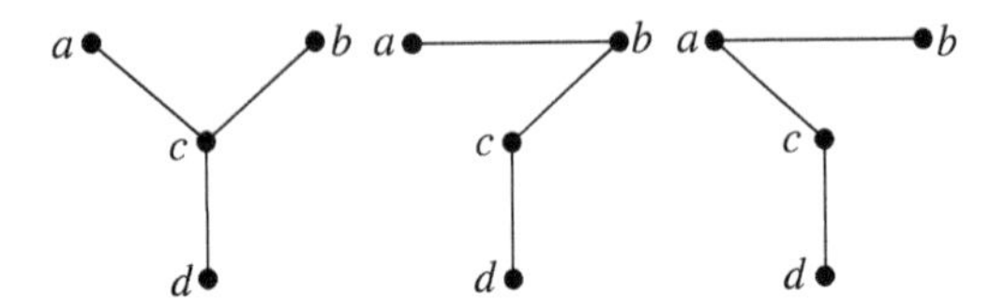

17. (a) The adjacency matrix for C_5 is shown.

$$\begin{pmatrix} 0 & 1 & 0 & 0 & 1 \\ 1 & 0 & 1 & 0 & 0 \\ 0 & 1 & 0 & 1 & 0 \\ 0 & 0 & 1 & 0 & 1 \\ 1 & 0 & 0 & 1 & 0 \end{pmatrix}$$

The matrix, K, is

$$\begin{pmatrix} 2 & -1 & 0 & 0 & -1 \\ -1 & 2 & -1 & 0 & 0 \\ 0 & -1 & 2 & -1 & 0 \\ 0 & 0 & -1 & 2 & -1 \\ -1 & 0 & 0 & -1 & 2 \end{pmatrix}$$

There are $C_{11} = (-1)^2 \det A_{11} = 5$ distinct spanning trees.

H.12 Functions, Relations, Databases, and Circuits

Exercises 12.1.3

1. (a) Not a function: (3, 3) and (3, 4) are both in the relation

2. (a) Not onto: 2 is not a second coordinate

3. (a) Not one-to-one: (2, 4) and (3, 4) are both in the relation

4. (a) $\{(1,1),(4,2),(3,3),(4,3),(5,4),(5,5)\}$

9. (c) This composition is not defined, since some elements (such as 2) in the image of $\mathcal{R}$ are not in the domain of $\mathcal{S}$.

13. (a) $\mathcal{T} \circ \mathcal{S} \circ \mathcal{R} =$ $\{(GA, IL), (GA, KA), (MI, IL), (MI, KA)\}$

15. (d) False. $\{1, 2, 3\}$ is the *image* of $\mathcal{F}$ but not necessarily the range. In other words, not every function is onto.

Exercises 12.2.3

1.

Relation	(a) Properties	(b) Equivalence Relation?	(c) Partial Order?
$\mathcal{R}_1$	symmetric, transitive, antireflexive	no	no

In blended families, the properties are different. Transitivity holds for full siblings, but not for half-siblings.

4. (a) The relation $\{(1,2),(2,2)\}$ on $\{1,2\}$ is neither reflexive nor antireflexive. So "not reflexive" does not imply "antireflexive". Also, the relation $\emptyset$ on $\emptyset$ is antireflexive and reflexive, so "antireflexive" does not imply "not reflexive" (the definitions are vacuously satisfied).

7. (a) $\mathcal{S}_1 = \{(a,b),(a,c),(b,c),(c,c)\}$

reflexive closure
$\{(\mathbf{a},\mathbf{a}),(a,b),(a,c),(\mathbf{b},\mathbf{b}),(b,d),(c,c),(\mathbf{d},\mathbf{d})\}$
Note: (d,d) is in the reflexive closure since $\mathcal{A}$ has not changed.

symmetric closure
$\{(a,b),(a,c),(\mathbf{b},\mathbf{a}),(b,c),(\mathbf{c},\mathbf{a}),(\mathbf{c},\mathbf{b}),(c,c)\}$

transitive closure $\{(a,b),(a,c),(b,d),(c,c)\}$ (nothing added)

10. (a) Every integer in the set $\mathcal{F}$ (and any integer in general) is equal to itself, so the relation is reflexive. The relation is trivially antisymmetric because two integers will not be in the relation unless they are equal. Suppose $(x,y) \in \mathcal{R}_1$ and $(y,z) \in \mathcal{R}_1$. Then $x = y$ and $y = z$. Thus, $x = z$ and so $(x,z) \in \mathcal{R}_1$. This implies that the relation is transitive.
The Hasse diagram is very simple: It consists of a row containing the numbers -10 to 10, with no edges.

12. Let $x \bmod 5 = y \bmod 5 = r$. Then there are integers q_x and q_y such that $x = 5q_x + r$ and $y = 5q_y + r$. But then $x - 5q_x = r = y - 5q_y$, so $x - y = 5(q_x - q_y) = 5n$.

15. **(a)** **Reflexive** Let $(a, a) \in \mathbb{Z} \times \mathbb{Z}$. Since $a + a = 2a$, which is even, $(a, a) \in \mathcal{S}$.

Symmetric Let $(a, b) \in \mathcal{S}$. Then $a + b = 2k$ for some $k \in \mathbb{Z}$. It is then trivially true that $b + a = 2k$, so $(b, a) \in \mathcal{S}$.

Transitive Let $(a, b) \in \mathcal{S}$ and $(b, c) \in \mathcal{S}$. Then $a + b = 2m$ and $b + c = 2n$ for some $m, n \in \mathbb{Z}$. Thus, $a + c = (2m - b) + (2n - b) = 2(m + n - b) = 2k$ for some $k \in \mathbb{Z}$. This implies that $(a, c) \in \mathcal{S}$.

(b) Integers of the same parity form the equivalence classes of the relation. That is, the even integers are in one equivalence class and the odd integers are in another.

20. Let $[x] = [x']$ and $[y] = [y']$. Then $x - \lfloor x \rfloor = x' - \lfloor x' \rfloor$ and $y - \lfloor y \rfloor = y' - \lfloor y' \rfloor$. Thus, $x' = x - \lfloor x \rfloor + \lfloor x' \rfloor$ and $y' = y - \lfloor y \rfloor + \lfloor y' \rfloor$. Consequently,

$$\begin{aligned} x' + y' &= (x - \lfloor x \rfloor + \lfloor x' \rfloor) + (y - \lfloor y \rfloor + \lfloor y' \rfloor) \\ &= (x + y) + (\lfloor x' \rfloor + \lfloor y' \rfloor - \lfloor x \rfloor - \lfloor y \rfloor). \end{aligned}$$

Therefore, $x' + y'$ and $x + y$ differ by an integer (more specifically, the integer $\lfloor x' \rfloor + \lfloor y' \rfloor - \lfloor x \rfloor - \lfloor y \rfloor$). Thus, $x' + y' = x + y + n$ and so $x'+y'-\lfloor x'+y' \rfloor = x+y+n-\lfloor x+y+n \rfloor = x+y-\lfloor x+y \rfloor$. But this means that $[x' + y'] = [x + y]$, and so $\oslash$ is well defined.

Exercises 12.3.6

1. **(a)** This cannot represent a relation in a relational database because the Understudy column is not single valued. In particular, the Hamlet row has two values in the Understudy column.

4. **(Teaching Assignments)[Course, Instructor]**

Course	Instructor
MAT241	Gossett
MAT222	Kinney
MAT124M	Conrath
MAT124M	Kinney

6. **(a)** The join is on Student.

Yearbook

Student	Task	Homeroom	Feature	Cohort
Kim	Teacher Photos	H4	Photos	Senior
Rosa	Student Photos	D5	Editorials	Junior

11. **(c)** The functional dependencies are listed below.

Part ID → Part Name
Part ID → Part Location
Part Name → Part Location

Part ID is the primary key. The functional dependency Part Name → Part Location prevents this relation from being in third normal form. It *is* in second normal form, however, since no attribute is functionally dependent on a proper subset of the primary key.

12. **(c)** The projections are shown.

(Widgets and Flanges)[Part ID]

Part ID
W1256
W1257
W2276
F4
F6
⋮

(Widgets and Flanges)[Part Name, Part Location]

Part Name	Part Location
Widget (metric)	Warehouse W
Widget (English)	Warehouse W
Flange (4 inch)	Warehouse F
Flange (6 inch)	Warehouse F
⋮	⋮

The third assertion in Proposition 12.37 implies that the join will be the Cartesian product {Part ID} × {Part Name, Part Location}. This means the join will contain many tuples that are not in the original relation. One such tuple is (W1256, Flange (4 inch), Warehouse F). This is *not* a lossless decomposition.

14. **(c)** The functional dependencies are listed below.

Part ID → Part Name
Part ID → Part Location
Part Name → Part Location

Part ID is the primary key. The functional dependency Part Name → Part Location prevents this relation from being in third normal form. The algorithm indicates that Widgets and Flanges should be replaced by {(Widgets and Flanges)[Part Name, Part Location], Widgets and Flanges)[Part ID, Part Name]}. Both new relations are in third normal form. Part Name is the primary key in the first relation and a foreign key in the second.

(Widgets and Flanges)[Part Name, Part Location]

Part Name	Part Location
Widget (metric)	Warehouse W
Widget (English)	Warehouse W
Flange (4 inch)	Warehouse F
Flange (6 inch)	Warehouse F
⋮	⋮

(Widgets and Flanges)[Part ID, Part Name]

Part ID	Part Name
W1256	Widget (metric)
W1257	Widget (metric)
W2276	Widget (English)
F4	Flange (4 inch)
F6	Flange (6 inch)
⋮	⋮

19. **(b)** This is true. The join of the sub-relations in the decomposition produces $\mathcal{T}$ again, hence the name *lossless*.

(d) This is false. If the two relations have no common attributes, then $\mathcal{R} * \mathcal{T} = \mathcal{R} \times \mathcal{T}$, which is as large as the join can get (see Proposition 12.37 on page 771).

Exercises 12.4.4

3. There are $2^{2^3} = 2^8 = 256$ (Corollary 12.43).

5. **(b)** This is not a minterm because, among other reasons, it does not contain either x or $\overline{x}$.

7. **(b)** This is not in disjunctive normal form because it is not 0 and it does not contain any minterms.

9. The binary function $f(x, y) = x \cdot \overline{y} + \overline{x} \cdot y$ evaluates to 1 at $(1, 0)$ and $(0, 1)$. It evaluates to 0 at $(0, 0)$ and $(1, 1)$.

11. **(a)**

$$\begin{aligned} &\overline{x_1 + \overline{x_2}} \cdot (x_1 + \overline{x_3}) \\ &= \left(\overline{x_1} \cdot \overline{\overline{x_2}}\right) \cdot (x_1 + \overline{x_3}) && \text{De Morgan} \\ &= (\overline{x_1} \cdot x_2) \cdot (x_1 + \overline{x_3}) && \text{involution} \\ &= \overline{x_1} \cdot x_2 \cdot (x_1 + \overline{x_3}) && \text{associativity (this step is optional)} \end{aligned}$$

12. **(a)**

$$\begin{aligned} &(\overline{x_1} \cdot 1 + x_2 \cdot x_3) \cdot x_1 \\ &= ((\overline{x_1} \cdot 1) \cdot x_1 + (x_2 \cdot x_3) \cdot x_1) && \text{distributivity} \\ &= \overline{x_1} \cdot 1 \cdot x_1 + x_2 \cdot x_3 \cdot x_1 && \text{associativity (three times)} \end{aligned}$$

13. **(a)** A short solution.

$$\begin{aligned} x_1 \cdot x_3 \cdot \overline{x_1} &= x_1 \cdot \overline{x_1} \cdot x_3 && \text{commutativity} \\ &= (x_1 \cdot \overline{x_1}) \cdot x_3 && \text{associativity} \\ &= 0 \cdot x_3 && \text{complement axiom} \\ &= x_3 \cdot 0 && \text{commutativity} \\ &= 0 && \text{domination} \end{aligned}$$

A long solution.

$$\begin{aligned} x_1 \cdot x_3 \cdot \overline{x_1} &= x_1 \cdot x_3 \cdot \overline{x_1} \cdot x_2 + x_1 \cdot x_3 \cdot \overline{x_1} \cdot \overline{x_2} && \text{replacement algorithm } (i = 2) \\ &= x_1 \cdot \overline{x_1} \cdot x_2 \cdot x_3 + x_1 \cdot \overline{x_1} \cdot \overline{x_2} \cdot x_3 && \text{sort within terms} \\ &= (x_1 \cdot \overline{x_1}) \cdot x_2 \cdot x_3 + (x_1 \cdot \overline{x_1}) \cdot \overline{x_2} \cdot x_3 && \text{associativity (twice)} \\ &= 0 \cdot x_2 \cdot x_3 + 0 \cdot \overline{x_2} \cdot x_3 && \text{complement axiom (twice)} \\ &= 0 \cdot (x_2 \cdot x_3) + 0 \cdot (\overline{x_2} \cdot x_3) && \text{associativity (twice)} \\ &= (x_2 \cdot x_3) \cdot 0 + (\overline{x_2} \cdot x_3) \cdot 0 && \text{commutativity (twice)} \\ &= 0 + 0 && \text{domination (twice)} \\ &= 0 && \text{identity axiom} \end{aligned}$$

14. **(b)** **Move all complements onto single variables.** Already in this form.
Transform to a sum of products. Already in this form.
Transform into disjunctive normal form.

$$\begin{aligned} 1 &= 1 \cdot x_1 + 1 \cdot \overline{x_1} && \text{replacement algorithm } (i = 1) \\ &= x_1 \cdot 1 + \overline{x_1} \cdot 1 && \text{commutativity (twice)} \\ &= x_1 + \overline{x_1} && \text{identity (twice)} \\ &= x_1 \cdot x_2 + x_1 \cdot \overline{x_2} + \overline{x_1} && \text{replacement algorithm } (i = 2) \\ &= x_1 \cdot x_2 + x_1 \cdot \overline{x_2} + \overline{x_1} \cdot x_2 + \overline{x_1} \cdot \overline{x_2} && \text{replacement algorithm } (i = 2) \\ &= x_1 \cdot x_2 \cdot x_3 + x_1 \cdot x_2 \cdot \overline{x_3} + x_1 \cdot \overline{x_2} + \overline{x_1} \cdot x_2 + \overline{x_1} \cdot \overline{x_2} && \text{replacement algorithm } (i = 3) \\ &= x_1 \cdot x_2 \cdot x_3 + x_1 \cdot x_2 \cdot \overline{x_3} + x_1 \cdot \overline{x_2} \cdot x_3 + x_1 \cdot \overline{x_2} \cdot \overline{x_3} + \overline{x_1} \cdot x_2 + \overline{x_1} \cdot \overline{x_2} && \text{replacement algorithm } (i = 3) \\ &= x_1 \cdot x_2 \cdot x_3 + x_1 \cdot x_2 \cdot \overline{x_3} + x_1 \cdot \overline{x_2} \cdot x_3 + x_1 \cdot \overline{x_2} \cdot \overline{x_3} \\ &\quad + \overline{x_1} \cdot x_2 \cdot x_3 \overline{x_1} \cdot x_2 \cdot \overline{x_3} + \overline{x_1} \cdot \overline{x_2} && \text{replacement algorithm } (i = 3) \\ &= x_1 \cdot x_2 \cdot x_3 + x_1 \cdot x_2 \cdot \overline{x_3} + x_1 \cdot \overline{x_2} \cdot x_3 + x_1 \cdot \overline{x_2} \cdot \overline{x_3} \\ &\quad + \overline{x_1} \cdot x_2 \cdot x_3 + \overline{x_1} \cdot x_2 \cdot \overline{x_3} + \overline{x_1} \cdot \overline{x_2} \cdot x_3 + \overline{x_1} \cdot \overline{x_2} \cdot \overline{x_3} && \text{replacement algorithm } (i = 3) \end{aligned}$$

The terms are already minterms.

Sort and reduce into a unique disjunctive normal form.
The minterms are already sorted. There are no duplicate minterms. The final expression is thus:

$$\begin{aligned} &x_1 \cdot x_2 \cdot x_3 + x_1 \cdot x_2 \cdot \overline{x_3} + x_1 \cdot \overline{x_2} \cdot x_3 \\ &+ x_1 \cdot \overline{x_2} \cdot \overline{x_3} + \overline{x_1} \cdot x_2 \cdot x_3 + \overline{x_1} \cdot x_2 \cdot \overline{x_3} \\ &+ \overline{x_1} \cdot \overline{x_2} \cdot x_3 + \overline{x_1} \cdot \overline{x_2} \cdot \overline{x_3}. \end{aligned}$$

This makes sense: The function should evaluate to 1 at all 8 possible ordered triples.

17. (b) Move all complements onto single variables.

$$\begin{aligned} &\overline{x_1 \cdot x_4 + x_1 \cdot \overline{x_2}} \cdot x_2 \\ &= \overline{(x_1 \cdot x_4) + (x_1 \cdot \overline{x_2})} \cdot x_2 && \text{associativity (twice)} \\ &= \left(\overline{(x_1 \cdot x_4)} \cdot \overline{(x_1 \cdot \overline{x_2})}\right) \cdot x_2 && \text{De Morgan} \\ &= \left((\overline{x_1} + \overline{x_4}) \cdot (\overline{x_1} + \overline{\overline{x_2}})\right) \cdot x_2 && \text{De Morgan (twice)} \\ &= ((\overline{x_1} + \overline{x_4}) \cdot (\overline{x_1} + x_2)) \cdot x_2 && \text{involution} \\ &= (\overline{x_1} + \overline{x_4}) \cdot (\overline{x_1} + x_2) \cdot x_2 && \text{associativity} \end{aligned}$$

Transform to a sum of products.

$$\begin{aligned} &(\overline{x_1} + \overline{x_4}) \cdot (\overline{x_1} + x_2) \cdot x_2 \\ &= (\overline{x_1} + \overline{x_4} \cdot x_2) \cdot x_2 && \text{distributivity (+ over −)} \\ &= \overline{x_1} \cdot x_2 + (\overline{x_4} \cdot x_2) \cdot x_2 && \text{distributivity} \\ &= \overline{x_1} \cdot x_2 + \overline{x_4} \cdot x_2 \cdot x_2 && \text{associativity} \end{aligned}$$

Transform into disjunctive normal form.

$$\begin{aligned} &\overline{x_1} \cdot x_2 + \overline{x_4} \cdot x_2 \cdot x_2 \\ &= \overline{x_1} \cdot x_2 + \overline{x_4} \cdot x_2 && \text{associativity, idempotence} \\ &= \overline{x_1} \cdot x_2 + \overline{x_4} \cdot x_2 \cdot x_1 + \overline{x_4} \cdot x_2 \cdot \overline{x_1} && \text{replacement algorithm } (i = 1) \\ &= \overline{x_1} \cdot x_2 \cdot x_3 + \overline{x_1} \cdot x_2 \cdot \overline{x_3} \\ &\quad + \overline{x_4} \cdot x_2 \cdot x_1 + \overline{x_4} \cdot x_2 \cdot \overline{x_1} && \text{replacement algorithm } (i = 3) \\ &= \overline{x_1} \cdot x_2 \cdot x_3 + \overline{x_1} \cdot x_2 \cdot \overline{x_3} + \overline{x_4} \cdot x_2 \cdot x_1 \cdot x_3 \\ &\quad + \overline{x_4} \cdot x_2 \cdot x_1 \cdot \overline{x_3} + \overline{x_4} \cdot x_2 \cdot \overline{x_1} && \text{replacement algorithm } (i = 3) \\ &= \overline{x_1} \cdot x_2 \cdot x_3 + \overline{x_1} \cdot x_2 \cdot \overline{x_3} + \overline{x_4} \cdot x_2 \cdot x_1 \cdot x_3 \\ &\quad + \overline{x_4} \cdot x_2 \cdot x_1 \cdot \overline{x_3} + \overline{x_4} \cdot x_2 \cdot \overline{x_1} \cdot x_3 + \overline{x_4} \cdot x_2 \cdot \overline{x_1} \cdot \overline{x_3} && \text{replacement algorithm } (i = 3) \\ &= \overline{x_1} \cdot x_2 \cdot x_3 \cdot x_4 + \overline{x_1} \cdot x_2 \cdot x_3 \cdot \overline{x_4} + \overline{x_1} \cdot x_2 \cdot \overline{x_3} + \overline{x_4} \cdot x_2 \cdot x_1 \cdot x_3 \\ &\quad + \overline{x_4} \cdot x_2 \cdot x_1 \cdot \overline{x_3} + \overline{x_4} \cdot x_2 \cdot \overline{x_1} \cdot x_3 + \overline{x_4} \cdot x_2 \cdot \overline{x_1} \cdot \overline{x_3} && \text{replacement algorithm } (i = 4) \\ &= \overline{x_1} \cdot x_2 \cdot x_3 \cdot x_4 + \overline{x_1} \cdot x_2 \cdot x_3 \cdot \overline{x_4} + \overline{x_1} \cdot x_2 \cdot \overline{x_3} \cdot x_4 + \overline{x_1} \cdot x_2 \cdot \overline{x_3} \cdot \overline{x_4} \\ &\quad + \overline{x_4} \cdot x_2 \cdot x_1 \cdot x_3 + \overline{x_4} \cdot x_2 \cdot x_1 \cdot \overline{x_3} + \overline{x_4} \cdot x_2 \cdot \overline{x_1} \cdot x_3 + \overline{x_4} \cdot x_2 \cdot \overline{x_1} \cdot \overline{x_3} && \text{replacement algorithm } (i = 4) \\ &= \overline{x_1} \cdot x_2 \cdot x_3 \cdot x_4 + \overline{x_1} \cdot x_2 \cdot x_3 \cdot \overline{x_4} + \overline{x_1} \cdot x_2 \cdot \overline{x_3} \cdot x_4 + \overline{x_1} \cdot x_2 \cdot \overline{x_3} \cdot \overline{x_4} \\ &\quad + x_1 \cdot x_2 \cdot x_3 \cdot \overline{x_4} + x_1 \cdot x_2 \cdot \overline{x_3} \cdot \overline{x_4} + \overline{x_1} \cdot x_2 \cdot x_3 \cdot \overline{x_4} + \overline{x_1} \cdot x_2 \cdot \overline{x_3} \cdot \overline{x_4} && \text{sort within terms} \end{aligned}$$

Sort and reduce into a unique disjunctive normal form.
Sort the minterms.

$$\begin{aligned} &x_1 \cdot x_2 \cdot x_3 \cdot \overline{x_4} + x_1 \cdot x_2 \cdot \overline{x_3} \cdot \overline{x_4} + \overline{x_1} \cdot x_2 \cdot x_3 \cdot x_4 \\ &+ \overline{x_1} \cdot x_2 \cdot x_3 \cdot \overline{x_4} + \overline{x_1} \cdot x_2 \cdot \overline{x_3} \cdot x_4 + \overline{x_1} \cdot x_2 \cdot \overline{x_3} \cdot \overline{x_4} \end{aligned}$$

Remove duplicate minterms.

$$x_1 \cdot x_2 \cdot x_3 \cdot \overline{x_4} + x_1 \cdot x_2 \cdot \overline{x_3} \cdot \overline{x_4} + \overline{x_1} \cdot x_2 \cdot x_3 \cdot x_4 + \overline{x_1} \cdot x_2 \cdot x_3 \cdot \overline{x_4} + \overline{x_1} \cdot x_2 \cdot \overline{x_3} \cdot x_4 + \overline{x_1} \cdot x_2 \cdot \overline{x_3} \cdot \overline{x_4}$$

18. Maxterms and Conjunctive Normal Form

(a)

DEFINITION H.1 ***Maxterm***

Let $x_1, x_2, \ldots, x_n$ be n binary variables. A *maxterm* is a binary expression in the form

$$\hat{x_1} + \hat{x_2} + \cdots + \hat{x_n}$$

where $\hat{x_i}$ is either x_i or $\overline{x_i}$, for $i = 1, 2, \ldots, n$.

(b)

DEFINITION H.2 ***Conjunctive Normal Form***

A binary expression is in *conjunctive normal form* if it is either a product of distinct maxterms or it is the expression, 1.

(c)

PROPOSITION H.3 ***Evaluating Maxterms***

Let $\hat{x_1} + \hat{x_2} + \cdots + \hat{x_n}$ be a maxterm in the binary variables, $x_1, x_2, \ldots, x_n$. Define an n-variable binary function, f, as

$$f(x_1, x_2, \ldots, x_n) = \hat{x_1} + \hat{x_2} + \cdots + \hat{x_n}$$

Then f has the value 0 at only one element in its domain; it has the value 1 at all other elements of its domain. The n-tuple at which f has the value 0 is determined by setting

$$x_i = \begin{cases} 0 & \text{if } \hat{x_i} = x_i \\ 1 & \text{if } \hat{x_i} = \overline{x_i} \end{cases}$$

for $i = 1, 2, \ldots, n$.

Proof: The function, f, is defined in the n terms, $\hat{x_1}, \hat{x_2}, \ldots, \hat{x_n}$. If any one of those terms evaluates to 1, then f will also evaluate to 1. The only way to make each of the factors evaluate to 0 is to make the assignments specified in the statement of the proposition. □

(d)

THEOREM H.4 ***Every Binary Function Can be Expressed in Conjunctive Normal Form***

Let f be a binary function in the binary variables, $x_1, x_2, \ldots, x_n$. Then there is a Boolean expression in conjunctive normal form that is equal to f when viewed as a function.

(e)

THEOREM H.5 ***Every Binary Expression is Equivalent to a Unique Expression in Conjunctive Normal Form***

Every binary expression is equivalent to a unique expression in conjunctive normal form. The uniqueness requires a preestablished lexicographical ordering of the variables.

Exercises 12.5.4

1. (b) $x \cdot y \cdot z + x \cdot y \cdot \overline{z} + x \cdot \overline{y} \cdot z + x \cdot \overline{y} \cdot \overline{z} + \overline{x} \cdot y \cdot z$

Phase 1:

1	111	√	1, 2	1 1 −	√	1, 2, 3, 5	1 − −
2	110	√	1, 3	1 − 1	√		
3	101	√	1, 4	− 1 1			
4	011	√	2, 5	1 − 0	√		
5	100	√	3, 5	1 0 −	√		

Phase 2:

	1 $x \cdot y \cdot z$	2 $x \cdot y \cdot \overline{z}$	3 $x \cdot \overline{y} \cdot z$	4 $\overline{x} \cdot y \cdot z$	5 $x \cdot \overline{y} \cdot \overline{z}$
-11	X			X	
1--	X	X	X		X

The simplified function is

$$f(x, y, z) = x + y \cdot z$$

4. (a) Short version:

$$\begin{aligned} & w \cdot \left(\overline{x \cdot y + x}\right) & \\ = \; & w \cdot \left(\overline{x + x \cdot y}\right) & \text{commutativity} \\ = \; & w \cdot (\overline{x}) & \text{absorption} \\ = \; & w \cdot \overline{x} & \text{associativity} \end{aligned}$$

Long version:

$$\begin{aligned} & w \cdot \left(\overline{x \cdot y + x}\right) & \\ = \; & w \cdot (\overline{x \cdot y} \cdot \overline{x}) & \text{De Morgan} \\ = \; & w \cdot ((\overline{x} + \overline{y}) \cdot \overline{x}) & \text{De Morgan} \\ = \; & w \cdot (\overline{x} \cdot (\overline{x} + \overline{y})) & \text{commutativity} \\ = \; & (w \cdot \overline{x}) \cdot (\overline{x} + \overline{y}) & \text{associativity} \\ = \; & (w \cdot \overline{x}) \cdot \overline{x} + (w \cdot \overline{x}) \cdot \overline{y} & \text{distributivity} \\ = \; & w \cdot (\overline{x} \cdot \overline{x}) + (w \cdot \overline{x}) \cdot \overline{y} & \text{associativity} \\ = \; & w \cdot \overline{x + x} + (w \cdot \overline{x}) \cdot \overline{y} & \text{De Morgan} \\ = \; & w \cdot \overline{x} + (w \cdot \overline{x}) \cdot \overline{y} & \text{idempotence} \\ = \; & w \cdot \overline{x} + w \cdot \overline{x} \cdot \overline{y} & \text{associativity} \\ = \; & w \cdot \overline{x} \cdot 1 + w \cdot \overline{x} \cdot \overline{y} & \text{identity} \\ = \; & (w \cdot \overline{x}) \cdot 1 + (w \cdot \overline{x}) \cdot \overline{y} & \text{associativity (twice)} \\ = \; & (w \cdot \overline{x}) \cdot (1 + \overline{y}) & \text{distributivity} \\ = \; & (w \cdot \overline{x}) \cdot (\overline{y} + 1) & \text{commutativity} \\ = \; & (w \cdot \overline{x}) \cdot 1 & \text{domination} \\ = \; & (w \cdot \overline{x}) & \text{identity} \\ = \; & w \cdot \overline{x} & \text{associativity} \end{aligned}$$

6. (b) $(\overline{x} + \overline{y}) \cdot (\overline{x + y})$

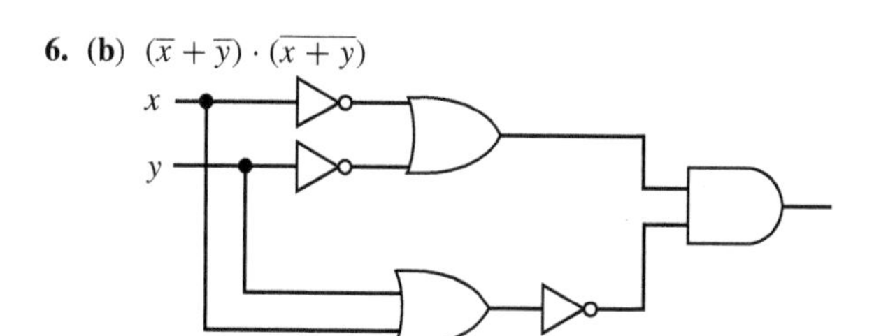

8. (b) $x + y \cdot z$

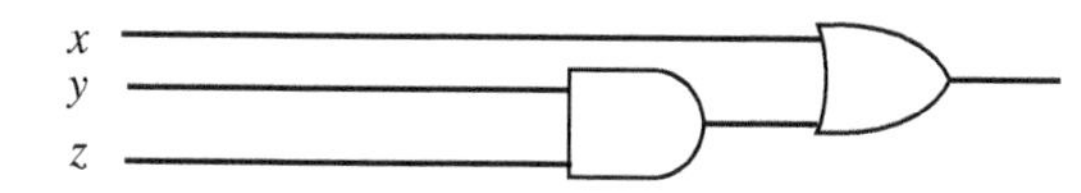

13. The general process that was outlined in the textbook will be used.

Step 1. There are three input variables and one output variable. Let the inputs be A, B, and M, representing the different family members. There will be one output variable: E. An input variable will have the value 1 if the person wants to go out to eat on Saturday and the value 0 if person does not want to go out to eat on Saturday. The output variable will have the value 1 if the family eats out on Saturday and the value 0 if the family eats at home on Saturday.

Step 2. The binary function is shown below.

A	*B*	*M*	*E*
0	0	0	0
0	0	1	0
0	1	0	0
0	1	1	1
1	0	0	0
1	0	1	1
1	1	0	0
1	1	1	1

Step 3. The binary expression in disjunctive normal form is

$$E(A, B, M) = \overline{A} \cdot B \cdot M + A \cdot \overline{B} \cdot M + A \cdot B \cdot M$$

Step 4. Quine–McCluskey
Phase 1:

1	111	✓	1, 2	1−1
2	101	✓	1, 3	−11
3	011	✓		

Phase 2

	1 $A \cdot B \cdot M$	2 $A \cdot \overline{B} \cdot M$	3 $\overline{A} \cdot B \cdot M$
1-1	X	X	
-11		X	X

The simplified function is:

$$E(A, B, M) = A \cdot M + B \cdot M$$

Step 5. The following diagram was created using Logisim.

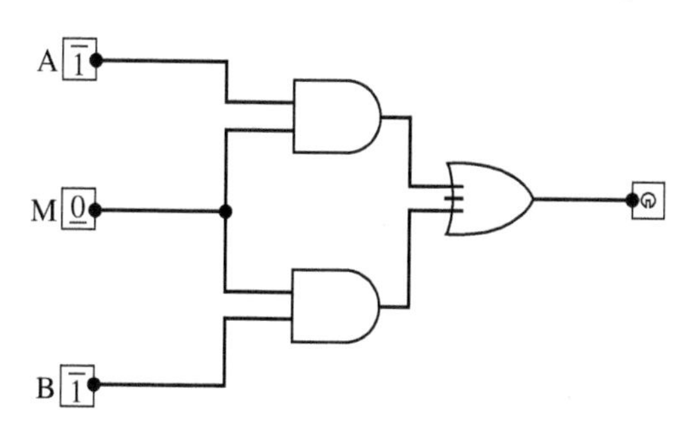

20. (b)

P	*Q*	$P \wedge Q$	$P \uparrow Q$	$(P \uparrow Q) \uparrow (P \uparrow Q)$
T	T	T	F	T
T	F	F	T	F
F	T	F	T	F
F	F	F	T	F

21. (b)

$$\begin{aligned} P \vee Q \vee R &\leftrightarrow P \vee (Q \vee R) \\ &\leftrightarrow P \vee ((Q \uparrow Q) \uparrow (R \uparrow R)) \\ &\leftrightarrow (P \uparrow P) \uparrow (((Q \uparrow Q) \uparrow (R \uparrow R)) \uparrow \\ &\quad ((Q \uparrow Q) \uparrow (R \uparrow R))) \end{aligned}$$

or

$$\begin{aligned} P \vee Q \vee R &\leftrightarrow (P \vee Q) \vee R \\ &\leftrightarrow ((P \uparrow P) \uparrow (Q \uparrow Q)) \vee R \\ &\leftrightarrow (((P \uparrow P) \uparrow (Q \uparrow Q)) \uparrow \\ &\quad ((P \uparrow P) \uparrow (Q \uparrow Q))) \uparrow (R \uparrow R) \end{aligned}$$

H.13 Appendices

Exercises C.1.3

1. It is tempting to assume that Ebenezum made the first statement and Jedediah the second. However, the exercise specifically states that I do not immediately know which statement was uttered by Ebenezum. The only implied information is that each brother made (a different) one of the statements.

Suppose that Ebenezum did make the first statement. That statement is consistent with what we know about his character. Then Jedediah must have made the second statement. Since Ebenezum never lies, Jedediah must be lying.

Suppose now that Ebenezum made the second statement. Then we must assume that Jedediah is lying, since Ebenezum always tells the truth. Jedediah has thus (falsely) claimed to always tell the truth. But we know that he *doesn't* always tell the truth, so his lie is consistent with the information from Ebenezum.

In either case, this must be one of Jedediah's lying days.

7. One possibility is to cause her to mentally step back and analyze the answer she would make to a particular question. Don't ask her to answer the question under consideration. Rather, have her tell how she would answer *if you ever did ask it*. One such question is,

> If I were to ask which road leads to the capital city, which road would you tell me is the one?

Suppose she is a knight. Then the road she points to is the same road she would point to if you were to directly ask *Which road leads to the capital?*

On the other hand, if she is a knave, and you asked directly *Which road leads to the capital?*, she would lie and point to the other road. However, you didn't ask her to answer the question. You asked her to predict how she *would* answer it. She, of course. will lie about how she would answer. The net result is she will point to the road that leads to the capital.

In either case, you can take the road that is indicated will lead to the capital city.

Exercises C.2.3

4. (a)

Symbolic Form	Name	English Translation	T/F
$P \to Q$	implication	If I am pregnant, then I am a female.	T
$\neg Q \to \neg P$	contrapositive	If I am not a female, then I am not pregnant.	T
$Q \to P$	converse	If I am a female, then I am pregnant.	F
$\neg P \to \neg Q$	inverse	If I am not pregnant, then I am not a female.	F

5. (a) If Socrates was a man, then he was mortal.
Socrates was mortal.
———————————————
Therefore Socrates was a man. **INVALID—fallacy of affirming the consequent**

6. (b) Invalid: we have not been told that all students live at home (first diagram). Zelda may be one who does not (second diagram).

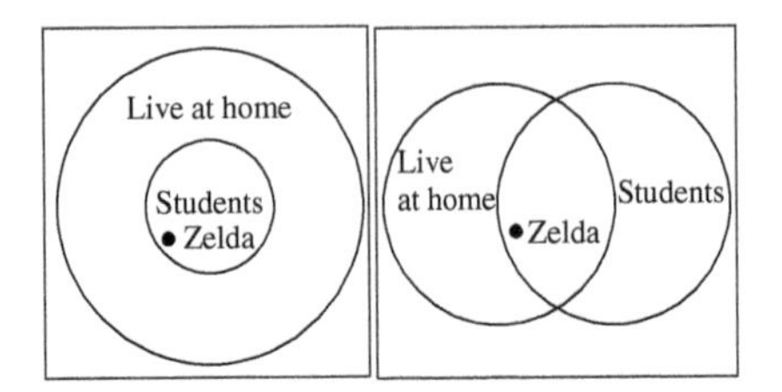

9. We need to compare the truth table for implication with the nature of the speaker. If he is a scholar, then the hypothesis is false, so the implication is true, independent of the truth value of the conclusion. A scholar could make a single true statement, so no inconsistencies arise. On the other hand, if the speaker is a dunce, the hypothesis is true and the conclusion is false, making the implication false. A dunce could make a false statement, so no inconsistencies arise in this case. We cannot determine the nature of the speaker.

'If I am a dunce, then I am a scholar'

Nature	Hypothesis	Conclusion	Implication	Consistency
Scholar	F	T	T	*Consistent*
Dunce	T	F	F	*Consistent*

Bibliography

[1] Robert W. Allen and Lorne Greene. The Propaganda Game. WFF 'N Proof, New Haven, Conn., 1970.

[2] Anonymous. Chinese remainder theorem. http://www.chinapage.com/math/crt.html.

[3] Anonymous. On the 2rot13 encryption algorithm. http://www.pruefziffernberechnung.de/Originaldokumente/2rot13.pdf.

[4] Ron Artstein. Hypothesis testing. http://cswww.essex.ac.uk/LAC/LAC05-06/hypothesis-testing.html, October 2005.

[5] Leigh Atkinson. Where do functions come from? *The College Mathematics Journal*, 33(2):107–112, March 2002.

[6] Norman Balabanian and Bradley S. Carlson. *Digital Logic Design Principles*. Wiley, 2001.

[7] W. W. Rouse Ball and H. S. M. Coxeter. *Mathematical Recreations & Essays*. MacMillan, 11 edition, 1960. First issued in 1892.

[8] Arthur T. Benjamin and Jennifer J. Quinn. *Proofs that really count: the art of combinatorial proof*. Number 27 in The Dolciani mathematical expositions. The Mathematical Association of America, 2003.

[9] William Berlinghoff. *Mathematics the Art of Reason*. Heath, 1968.

[10] Norman Biggs, E. Keith Lloyd, and Robin J. Wilson. *Graph Theory: 1736 – 1936*. Clarendon Press, 1976.

[11] C. Böhm and G. Jacopini. Flow diagrams, turing machines and languages with only two formation rules. *CACM*, May 1966.

[12] J.A. Bondy and U. S. R. Murty. *Graph Theory with Applications*. American Elsevier, 1976.

[13] Carl Boyer. *A History of Mathematics*. Wiley, 1968.

[14] John D. Bransford and Barry S. Stein. *The IDEAL Problem Solver*. W.H. Freeman and Company, 1984.

[15] Ricard A. Brualdi. *Introductory Combinatorics*. North-Holland, 1977.

[16] Anne M. Burns. Persian recursion. *Mathematics Magazine*, 70(3):196–199, June 1997.

[17] Peter J. Cameron. Mutually orthogonal latin squares. http://designtheory.org/library/encyc/mols/m/, December 2006.

[18] Daniel A. Cohen. *Introduction to Computer Theory*. John Wiley adn Sons, 1986.

[19] Charles J. Colbourn and Jeffrey H. Dinitz, editors. *The CRC Handbook of Combinatorial Designs*. CRC Press, Inc., 1996.

[20] David Crowdis and Brandon Wheeler. *Introduction to Mathematical Ideas*. McGraw-Hill, 1969.

[21] Antonella Cupillari. *The Nuts and Bolts of Proof*. Wadsworth, 1989.

[22] Deitel, Deitel, Nieto, Lin, and Sadhu. *XML: How to Program*. Prentice Hall, 2001.

[23] Rene Descartes. *Discourse on Method* and *Meditations*. The Library of Liberal Arts. Bobbs-Merrill, 1960. Laurence J. Lafleur (translator).

[24] Reinhard Diestel. *Graph Theory*. Number 173 in Graduate texts in mathematics. Springer-Verlag, 1997.

[25] Underwood Dudley. *A Budget of Trisectors*. Springer-Verlag, 1987.

[26] William Dunham. *Journey Through Genius: the great theorems of mathematics*. John Wiley and Sons, 1990.

[27] William Dunham. *Euler: the master of us all*, volume 22 of *Dolciani Mathematica Expositions*. The Mathematical Association of America, 1999.

[28] Margery G. Dunn, editor. *Exploring Your World, the adventure of geography*. The National Geographic Society, 1989.

[29] Susanna S. Epp. *Discrete Mathematics with Applications*. Brooks/Cole, 3 edition, 2004.

[30] P. Erdös and A. Szekeres. A combinatorial problem in geometry. *Composito Mathematica*, 2:463–470, 1935.

[31] Martin J. Erickson. *Introduction to Combinatorics*. Wiley-Interscience series in discrete mathematics and optimization. John Wiley and Sons, 1996.

[32] Leonhard Euler. *Introduction to Analysis of the Infinite*, volume 1. Springer-Verlag, 1988. translated by John Blanton.

[33] Leonhard Euler. *Letters to a German Princess*. Thoemmes Press, 1998 (originally translated into English circa 1800).

[34] Howard Eves. *An Introduction to the History of Mathematics*. Saunders, 6 edition, 1990.

[35] Jill D. Foley. *Unisured in the United States: the nonelderly population without health insurance*. Employee Benefit Research Institute, April 1991.

[36] National Center for Health Statistics. Health, United States, 2007: with chartbook on trends in the health of Americans. http://www.cdc.gov/nchs/data/hus/hus07.pdf Table 31, Hyattsville, MD: 2007.

[37] Jr. Frederick P. Brooks. *The Mythical Man-Month*. Addison-Wesley, 20th anniversary edition edition, 1995.

[38] Quentin F.Stout and Patricia A. Woodworth. Relational databases. *American Mathematical Monthly*, 90:101–118, February 1983.

[39] D. Gale and L. S. Shapley. College admissions and the stability of marriage. *American Mathematics Monthly*, 69, January 1962.

[40] J. A. Gallian. Assigning driver's license numbers. *Mathematics Magazine*, 64(1):13–22, February 1991.

[41] Martin Gardner. *Penrose Tiles to Trapdoor Ciphers ... and the Return of Dr. Matrix*. The Mathematical Association of America, revised edition, 1997.

[42] Joseph Gastwirth. The statistical precision of medical screening procedures: Application to polygraph and AIDS antibodies test data. *Statistical Science*, 2(3), 1987.

[43] Ronald L. Graham, Donald E. Knuth, and Oren Patashnik. *Concrete Mathematics: a foundation for computer science*. Addison-Wesley, 1989.

[44] Ralph P. Grimaldi. *Discrete and Combinatorial Mathematics*. Addison-Wesley Longman, Inc., fourth edition, 1999.

[45] Jan Haliday and Peter Fuller. *The Psychology of Gambling*. Penguin, 1974.

[46] Marshall Hall, Jr. *Combinatorial Theory*. Wiley-Interscience. John Wiley and Sons, second edition, 1986.

[47] Arthur Hallerberg. *Mathematical proof.* Hafner Press, 1974 (out of print).

[48] Frank Harary. *Graph Theory.* Addison-Wesley, 1972.

[49] James L. Hein. *Discrete Structures, Logic, and Computability.* Jones and Bartlett, 1995.

[50] Mark S. Hoffman, editor. *The World Almanac and Book of Facts, 1992 edition.* Pharos Books, 1991.

[51] Ross Honsberger. *More Mathematical Morsels.* Number 10 in Dolciani Mathematical Expositions. Mathematical Association of America, 1991.

[52] Preston Hunter. Major religions of the world ranked by number of adherents. http://www.adherents.com/Religions_By_Adherents.html, August 2007.

[53] E. V. Huntington. Postulates for the algebra of logic. *Transactions of the american mathematical society*, 5:288–309, 1904.

[54] J. B. Saxe J. L. Bently, D. Haken. A general method for solving divide-and-conquer recurrences. *SIGACT News*, 12(3):6–44, 1980.

[55] Otto Johnson, editor. *1988 Information Please Almanac.* Information Please Almanac Atlas and Yearbook. Houghton Mifflin Company, 41 edition, 1987.

[56] Flavious Josephus. *Josephus The Jewish War.* Zondervan, 1982. Gaalya Cornfeld General Editor.

[57] Jszigetvari. A three rotor enigma cypher machine used by luftwaffe in 1944. http://commons.wikimedia.org/wiki/Image:Enigma_-_Military.jpg. Permission is granted to copy, distribute and/or modify this document under the terms of the GNU Free Documentation license, Version 1.2 or any later version published by the Free Software Foundation; with no Invariant Sections, no Front-Cover Texts, and no Back-Cover Texts. A copy of the license is included in the section entitled "GNU Free Documentation license" (http://commons.wikimedia.org/wiki/Commons:GNU_Free_Documentation_License).

[58] Ryan Kaldari. The Parthenon in Nashville, Tennessee, USA is a full-scale replica of the original Greek Parthenon. http://en.wikipedia.org/wiki/Image:Parthenon.at.Nashville.Tenenssee.01.jpg. I, the copyright holder of this work, hereby release it into the public domain. This applies worldwide. In case this is not legally possible: I grant anyone the right to use this work for any purpose, without any conditions, unless such conditions are required by law.

[59] Dean Kelley. *Automata and Formal Languages: an introduction.* Prentice Hall, 1995.

[60] Walt Kelly. *Ten Ever-lovin' Blue-Eyed Years with Pogo.* Fireside Books. Simon and Schuster, 1959.

[61] Michael Kifer, Arthur Bernstein, and Philip M. Lewis. *Database Systems: an application-oriented approach.* Addison-Wesley, second edition, 2005. complete version.

[62] Morris Kline. *Mathematical Thought from Ancient to Modern Time.* Oxford University Press, 1972.

[63] Donald E. Knuth. *The Art of Computer Programming*, volume 2. Addison-Wesley, 3 edition, 1998.

[64] Ramanujachry Kumanduri and Cristina Romero. *Number Theory with Computer Applications.* Prentice Hall, 1998.

[65] Alison Landes, Carol D. Foster, and Betsie B. Caldwell, editors. *Homeless In America.* The Information Series on Current Topics. Information Plus, 1991 edition, 1991.

[66] Henry B. Laufer. *Discrete Mathematics and Applied Modern Algebra.* Prindle, Weber & Schmidt, 1984.

[67] F. J. MacWilliams and N. J. A. Sloane. *The Theory of Error-Correcting Codes.* North-Holland, 1977.

[68] Mark Mandelkern. Constructive mathematics. *Mathematics Magazine*, 58(5), November 1985.

[69] George Markowsky. Misconceptions about the golden ratio. *The College Mathematics Journal*, 23(1):2–19, January 1992.

[70] Philip J. Pratt and Joseph J Adamski. *Database Systems Management and Design.* Boyd aznd Fraser, third edition, 1994.

[71] Gordon Raisbeck. *Information Theory: an introduction for for scientists and engineers.* M.I.T. Press, 1963.

[72] Herbert John Reyser. *Combinatorial Mathematics*, volume 14 of *The Carus Mathematical Monographs.* The Mathematical Association of America, 1963.

[73] Peter Rob and Carlos Coronel. *Database Systems: design, implmentation, and management.* Course Technology, third edition, 1997.

[74] Fred S. Roberts. *Applied Combinatorics.* Prentice Hall, 1984.

[75] Steven Roman. *Coding and Information Theory.* Springer, 1992.

[76] Kenneth H. Rosen. *Discrete Mathematics and its Applications.* McGraw-Hill, 4 edition, 1999.

[77] Kenneth H. Rosen, John G. Michaels, Jonathan L. Gross, Jerrold W. Grossman, and Douglas R. Shier, editors. *Handbook of Discrete and Combinatorial Mathematics.* CRC Press, Inc., 2000.

[78] James T. Rosenbaum and Richard Wernick. The utility of routine screening of patients with uveitis for systemic lupus erythematosus or tuberculosis. *Arch Ophthalmol*, 108(9), September 1990.

[79] Hans Sagan. *Space-Filling Curve.* Springer-Verlag, 1994.

[80] Dr. Seuss. *On Beyond Zebra.* Random House, 1955.

[81] Claude E. Shannon and Warren Weaver. *The Mathematical Theory of Communication.* University of Illinois Press, 1949.

[82] Lloyd Shaw. *Cowboy Dances.* The Caxton Printers, Ltd., 1952.

[83] Joseph H. Silverman. *A Friendly Introduction to Number Theory.* Prentice Hall, 1997.

[84] Steven Skiena. *Implementing Discrete Mathematics: Combinatorics and graph theory with Mathematica.* Addison-Wesley, 1990.

[85] David A. Smith. Chinese remainder theorem. http://www.libraryofmath.com/chinese-remainder-theorem.html. Published by Library of Math – Online math organized by subject into topics.

[86] David A. Smith. Fermat's little theorem. http://www.libraryofmath.com/fermat-little-theorem.html. Published by Library of Math – Online math organized by subject into topics.

[87] Raymond Smullyan. *What is the Name of this Book? The Riddle of Dracula and Other Logical Puzzles.* Prentice Hall, 1978.

[88] Raymond Smullyan. *Alice in Puzzleland.* Penguin Books, 1984.

[89] Daniel Solow. *How to Read and Do Proofs.* Wiley, 1982.

[90] Elizabeth Stapel. Functions versus Relations. http://www.purplemath.com/modules/fcns.htm, 2006.

[91] Daniel F. Stubbs and Neil W. Webre. *Data Structures with Abstract Data Types and Pascal*. Brooks/Cole, 1984.

[92] *People vs. Collins* 68 Cal. 2d 319,335, 438 P.2d 33,45, 66 Cal. Rptr 497,507 (1968).

[93] Frank Swetz. *The Nine Chapters of the Mathematical Art: an amazing book*. Historical Notes: Mathematics Through the Ages. COMAP, 1992.

[94] Richard J. Trudeau. *Introduction to Graph Theory*. Dover Publications, 1993.

[95] Vital Statistics of the United States, 1987, volume 1 - Natality. U.S. Dept. of Health and Human Services, Public Health Service, Centers for Disease Control, National Center for Health Statistics, 1989. Manning Feinleib, Director.

[96] Karen Doyle Walton. *Imagine That! A History of Imaginary Numbers*. Historical Notes: Mathematics Through the Ages. COMAP, 1992.

[97] Sherwood Washburn, Thomas Marlowe, and Charles T. Ryan. *Discrete Mathematics*. Addison-Wesley, 2000.

[98] William Waterhouse. Why square roots are irrational. *The American Mathematical Monthly*, 93(3), March 1986.

[99] Mark Allen Weiss. *Data Structures and Problem Solving Using Java*. Addison-Wesley, 1998.

[100] Douglas B. West. *Introduction to Graph Theory*. Prentice Hall, 1996.

[101] Wikipedia. Caesar cipher. http://en.wikipedia.org/wiki/Caesar_code.

[102] Wikipedia. Enigma machine. http://en.wikipedia.org/wiki/Enigma_machine.

[103] Wikipedia. Euclid's lemma. http://en.wikipedia.org/wiki/Euclid's_lemma.

[104] Wikipedia. Fermat's last theorem. http://en.wikipedia.org/wiki/Fermat's_last_theorem.

[105] Wikipedia. Padding schemes. http://en.wikipedia.org/wiki/Rsa_encryption#Padding_schemes.

[106] Wikipedia. Zero to the zero power. http://en.wikipedia.org/wiki/0%5E0#Zero_to_the_zero_power.

[107] Wayne L. Winston. *Operations Research: applications and algorithms*. Duxbuery Press; International Thompson Publishing, 3 edition, 1994.

[108] Ernst F. Winter. *Erasmus–Luther: Discouse on Free Will*. Frederick Unger Publishing Co., Inc., 1967.

[109] Niklaus Wirth. *Data Structures + Algorithms = Programs*. Prentice Hall, 1976.

Index

Printed in the USA/Agawam, MA
June 25, 2019

705870.008